Heidrun E. K. Hartmann (Ed.)

Illustrated Handbook of Succulent Plants: Aizoaceae F–Z

With 85 Figures and 384 Colour Photos, Printed in 48 Colour Plates

Springer

Dr. Heidrun E. K. Hartmann

Institut für Allgemeine Botanik
Ohnhorststraße 18
22609 Hamburg
Germany

e-mail: hartmann@botanik.uni-hamburg.de

With contributions by:
P. Chesselet, D. T. Cole, N. A. Cole, M. Gerbaulet, L. B. Groen, S. A. Hammer, H. E. K. Hartmann, H.-D. Ihlenfeldt, C. Klak, I. M. Niesler, S. M. Pierce, N. Zimmermann

ISBN 978-3-642-62595-4

Library of Congress Cataloging-in-Publication Data

Illustrated handbook of succulent plants. Aizoaceae / Heidrun E.K. Hartmann (ed.).
p. cm.
Includes bibliographical references.
Contents: [1] A-E -- [2] F-Z.
ISBN 978-3-642-62595-4 ISBN 978-3-642-56265-5 (eBook)
DOI 10.1007/978-3-642-56265-5
1. Aizoaceae--Classification. I. Hartmann, Heidrun E. K.

QK495.A32 I44 2001
583'.53--dc21

2001040053

Originally published by Springer-Verlag Berlin Heidelberg New York in 2002
Softcover reprint of the hardcover 1st edition 2002

Cover design: *design & production* GmbH, 69121 Heidelberg, Germany
Typesetting: Data conversion by Büro Stasch, Bayreuth, Germany (*stasch@stasch.com*)
SPIN: 11332442 31/3111 - 5 4 3 2 1 – Printed on acid-free paper

Illustrated Handbook of Succulent Plants

Series Editors: U. Eggli, H. E. K. Hartmann

Springer-Verlag Berlin Heidelberg GmbH

Illustrated Handbook of Succulent Plants
(Eggli/Hartmann Eds.)

Illustrated Handbook of Succulent Plants: Monocotyledons
Ed. by Urs Eggli (2001)

Illustrated Handbook of Succulent Plants: Dicotyledons
Ed. by Urs Eggli (2002)

Illustrated Handbook of Succulent Plants: Aizoaceae A–E
Ed. by H. E. K. Hartmann (2001)

Illustrated Handbook of Succulent Plants: Aizoaceae F–Z
Ed. by H. E. K. Hartmann (2001)

Illustrated Handbook of Succulent Plants: Asclepiadaceae
Ed. by F. Albers and U. Meve (2002)

Illustrated Handbook of Succulent Plants: Crassulaceae
Ed. by Urs Eggli (2003)

Preface

Members of the iceplant family, Aizoaceae, are appreciated by many people and disliked by some. Farmers value the plants for their browsing stock, hobbyists like to grow and flower mainly the "stone plants" of the family, and tourists enjoy the mass flowering of millions of annuals after good rains in the south western part of South Africa. Only taxonomists show some reservation against members of the Aizoaceae because of the size of the family and the confused and entangled systematics of many groups.

For all groups of users, this book will offer some improvements. A worldwide web has been knitted among hundreds of interested people resulting in the production of this lexicon. As a consequence of the expanded collaboration over five continents, innumerable new results can be presented here, including many new delimitations of genera as well as new combinations and names of species. All conclusions are based on thousands of days spent on field work, thousands of original literature references inspected, thousands of herbarium sheets studied, thousands of scanning electron microscopic pictures compared, thousands of photographs taken, and thousands of taxonomic decisions made. As a result of all united efforts, an up-to-date survey of the state of the art can be presented, forming the basis for further research in the family and providing a challenge to every reader to add his special knowledge towards an improved second edition.

This book continues in part the tradition of Jacobsen's "Handbook of Succulent Plants" vol. 3 (1960), and of the "Lexicon of Succulent Plants", published from 1955 onwards in different editions in German and English, in which Jacobsen treated up to 122 genera of Mesembryanthema under the heading "Mesembryanthemaceae".

In these two volumes dedicated to the family Aizoaceae, the Mesembryanthema are included. 234 genera have been named in the family Aizoaceae in the present circumscription, all of which are described here, but only 136 are accepted as valid. 29 new combinations, one new name, and two new species are published here; 189 taxa are lectotypified, 26 are neotypified.

After a long trip in troubled waters, the manuscript was revised and updated in the year 2000, the closing date for inclusion of known publications being 31.12.2000. I am most grateful to all authors and colleagues who have helped to keep up with the numerous new descriptions, including contributions of taxa in the process of description now, and unpublished information.

I should like to thank all collaborators and contributors of text or artwork. My special thanks go to Dr. Czeschlik and Dr. Lindenborn, Springer-Verlag, who made the publication possible at last.

H. E. K. Hartmann

Hamburg, September 2001

Acknowledgements

Maike Gerbaulet gratefully acknowledges travel grants from the Cactus and Succulent Society of America and the Mesem Study Group, England, constant support by the staff members of the Bolus Herbarium, professional maintenance of her living plant collection by Ernst van Jaarsveld, and special support by Chikako Ueda.

H. E. K. Hartmann is most grateful to about 500 officials, colleagues, friends, and students for all kindness and friendly support without which a work of this dimension would never have been possible.

Over the 32 years of studies in Aizoaceae including field work in Argentina, Botswana, Ghana, India, Kenya, Lesotho, Mexico, Namibia, South Africa, Sudan, Swaziland, Tanzania, USA, and Zimbabwe numerous organizations and people have helped her work in many ways: giving permission to work in regions with restricted access like Diamond or other Mining Areas, Nature Reserves, Botanical Gardens, or Military grounds, getting permits from Nature Conservation Bodies for collecting protected plants in all countries visited, organizing sometimes extensive field trips, allowing to stay over night and feeding her well, searching for plants in sometimes rather inaccessable places and under difficult conditions, collecting data on distribution and taking her to places, keeping her company and helping during field work, driving cars and repairing them miraculously, finding very rare and difficult to trace literature, checking herbarium specimens, drying and shipping the collected herbarium material, doing scanning electron microscopic work, caring for living plants and growing them, measuring and drawing leaves, fruits, and flowers, turning drawings into ink versions for publication, discussing problems in finding roads or drawing the correct taxonomic conclusions, and, above all, being willing to share their knowledge with her.

The directors of the herbaria B, BM, BOL, BOLO, HEL, HBG, K, NBG, PRE, UPS are thanked for most generous permission to study or get on loan valuable type material or photographs which made progress possible. B, BOL, and K are especially acknowledged for very extensive loans over a much extended period.

Financial support from the German Science Foundation (DFG) since 1969, from the Council of Scientific and Industrial Research in South Africa, the award of the Kistenbosch Jubilee Fund 1977, from the National Botanic Gardens via the Karoo Botanic Garden, the Universities of Cape Town and Pretoria, the Hansische Universitätsstiftung of the University of Hamburg, the University of Hamburg, and the Cactus and Succulent Society of America is most gratefully acknowledged.

H.-D. Ihlenfeldt should like to thank the German Science Foundation for financial support over many years and the director and staff of BOL for the extensive loan of valuable herbarium material.

I. Niesler is grateful to S. Hammer for helpful information and material, to H. E. K. Hartmann for many discussions and support of her work. She should also like to thank the directors of BOL and K for extended loans of important material.

N. Zimmermann gratefully acknowledges a travel grant from the Cactus and Succulent Society of America and the loan of material of Schwantesia from BOL.

Contents

List of Abbreviations and Symbols

∅ .... diameter
± .... more or less
> .... bigger than
≥ .... bigger than or equal to
< .... smaller than
≤ .... smaller than or equal to
× .... times; hybrid sign in a name
! .... the cited herbarium sheet has been seen
= .... equals; between names: heterotypic synonym, based on a different type than the accepted name of the taxon
≡ .... homotypic synonym, based on the same type as the accepted and valid name
§ .... section

App. .... appendix
Art. .... article

b .... broad

c. .... circa, about
Chr .... chromosomes
comb. nov. .... combinatio nova, new combination

Distr .... Distribution

Ecol .... Ecology
Etym .... Etymology

f .... figure
Fl .... flower(s)
Fr .... fruit(s)

Gk .... Greek

h .... high
hort. .... hortus, garden

I .... internode(s)
ICBN .... International Code of Botanical Nomenclature
in litt. .... in litteris, in a letter
Inf .... inflorescence(s)
ISO .... isotype(s)

K .... calyx

l .... long
L .... leaf, leaves
La .... Latin
LT .... lectotype

nomen illeg. . nomen illegitimum, illegitimate name, not permitted under the ICBN
nomen invalid. nomen invalidum, invalid name, not fulfilling the requirements for valid publication as given by ICBN
nov. .... nova, novum, new
NT .... neotype

p.a. .... per annum, in a year
pers. comm. .. personal communication
p.p. .... pro parte, in part

S .... seed(s)
s.lat. .... sensu latiore, in the wider sense
s.n. .... sine numero, without a number
ssp. .... subspecies
s.str. .... sensu stricto, in the narrow sense
subg. .... subgenus

t. .... tabula, plate
t .... thick
T .... holotype

var. .... varietas, variety

$2n$ .... double chromosome set

Abbreviations of names of authors of plant names follows Brummitt R.K. & Powell C.E. (1992) Authors of Plant Names. Royal Botanic Gardens. Kew in many cases, e.g. Haw. for Haworth, L. for Linnaeus, or N.E.Br. for N.E. Brown. Deviations have been found useful in cases in which the authors are rare within the family.

Abbreviations of names of herbaria follows Holmgren et al. (1990) Index Herbariorum, Ed. 8

Names of genera are abbreviated by their first, capital, letter within the text covering that genus. An exception has been made for M, in which case M. always stand for the genus Mesembryanthemum; all other genera starting with an M are abbreviated by their first two letters, e.g. Ma. for Malephora or Machairophyllum, respectively.

Abbreviations of Geographical Terms

Ab .... Aberdeen
A.B. .... Alexander Bay
Ag .... Aggenys
Au .... Aus

B.E. .... Barkly East
Bi .... Bitterfontein
Bl .... Bloemfontein
Br .... Bredasdorp
Bu .... Burgersdorp
B.W. .... Beaufort West

Ca .... Carnarvon
Cl .... Calvinia
Co .... Carolina
Cr .... Cradock
C.T. .... Cape Town

E .... east, eastern
EC .... Eastern Cape Province
E.L. .... East London

Fr .... Fraserburg
FS .... Free State, Province

Ga .... Garies
GAU .... Gauteng, Province
Gb .... Gaborone
Ge .... George
Gh .... Grahamstown
Gr .... Griquatown
G.R. .... Graaff Reinet

Ha .... Harrismith
Ho .... Hondeklipbaai

Ja .... Jagersfontein
Jo .... Johannesburg

Ke .... Kenhardt
Ki .... Kimberley
K.K. .... Kap Kruis
KN .... KwazuluNatal, Province
Ku .... Kuruman

La .... Laingsburg
L.B. .... Lamberts Bay
Lo .... Loeriesfontein
L.T. .... Louis Trichardt
Lü .... Lüderitzbucht

Ma .... Maseru
Mi .... Middelburg
Mo .... Montagu
MM .... Mmabatho
MP .... Mpumalanga, Province
Mu .... Murraysburg

N .... north, northern
NC .... Northern Cape Province
No .... Noorduwer
NP .... Northern Province
NW .... North Western Province

OR .... Orange River
Ou .... Oudtshoorn

P.A. .... Prince Albert
Pc .... Potchefstroom
P.E. .... Port Elizabeth
Pi .... Pietersburg
Pm .... Posmasburg
P.N. .... Port Nolloth
Po .... Pofadder
Pr .... Pretoria

Rc .... Richmond
Ri .... Riversdale
Ru .... Rustenburg

S .... south, southern
S Africa .... Republic of South Africa
Sp .... Springbok
St .... Steytlerville
Su .... Sutherland
Sw .... Swakopmund

To .... Touwsrivier

Ui .... Uitenhage
Um .... Umtata
Un .... Uniondale
Up .... Upington

Va .... Vanrhynsdorp
Vr .... Vryburg
VR .... Vaal River

W .... west, western
Wa .... Warmbad
WC .... Western Cape Province
Wh .... Windhoek
Wi .... Willowmore
Wl .... Williston
Wo .... Worcester

Contributors[1]

C. Bruckmann
Hamburger Berg 25
20359 Hamburg
Germany

P. Chesselet
Compton Herbarium
Kirstenbosch Research Centre
Private Bag X7
Claremont 7735
South Africa (RSA)

Prof. D. T. & N. A. Cole
P.O. Box 447
Lonehill 2062
South Africa (RSA)

Dr. M. Gerbaulet
Phyllobolic Institute
Friedrich-Soltau-Straße 13
29456 Hitzacker
Germany

L. B. Groen
Paleis Het des
National Museum
Koninklijk Park 1
7315 JA Apeldoorn
The Netherlands

S. A. Hammer
845 Mason Road
Vista California 92084
U.S.A.

B. Hansen
Lokstedter Steindamm 74
22529 Hamburg
Germany

PD Dr. H. E. K. Hartmann
Institut für Allgemeine Botanik
Ohnhorststraße 18
22609 Hamburg
Germany

Prof. Dr. H.-D. Ihlenfeldt
Institut für Allgemeine Botanik
Ohnhorststraße 18
22609 Hamburg
Germany

C. Klak
Bolus Herbarium
University of Cape Town
Rondebosch 7700
South Africa (RSA)

I. Niesler
Institut für Allgemeine Botanik
Ohnhorststraße 18
22609 Hamburg
Germany

Dr. S. M. Pierce
Bolus Herbarium
University of Cape Town
Rondebosch 7700
South Africa (RSA)

Dr. M. Struck-Gerbaulet
Kochstraße 4
29451 Dannenberg
Germany

Dr. N. Zimmermann
Friedrich-Engels-Straße 45c
07749 Jena
Germany

M. B. Bayer, Paarl, South Africa; V. Bittrich, Campinas, Brazil; C. Bruckmann, Hamburg, Germany; R. Chinnock, Adelaide, Australia; D. Cole, Lonehill, South Africa; M. Dillon, Chicago, USA; M. Gerbaulet, Hitzacker, Germany; B. Groen, Wageningen, Netherlands; H. E. K. Hartmann, Hamburg, Germany; H.-D. Ihlenfeldt, Hamburg, Germany; E. van Jaarsverld, Cape Town, South Africa; R. Kabutz, Pretoria, South Africa; C. Klak, Cape Town, South Africa; S. Liede, Bayreuth, Germany; U. Meve, Bayreuth, Germany; I. Niesler, Hamburg, Germany; S. Pierce, Cape Town, South Africa; U. Schmiedel, Hamburg, Germany; M. Struck, Dannenberg, Germany; H. Toelken, Adelaide, Australia; C. Taylor, Missouri, USA; J. Traeger, San Francisco, USA; R. Uijs, Pretoria, South Africa; N. Zimmermann, Jena, Germany.

B. Braeutigam, Hamburg, Germany, has prepared the ink versions of all drawings.

[1] ... of both volumes (A–E and F–Z).

Introduction

H. E. K. Hartmann

The Family **Aizoaceae**

Circumscription and Delimitation of the Family **Aizoaceae**

The plant family **Aizoaceae** Martynov nomen conservandum is understood here in the sense of Bittrich & Hartmann (1988: 239–254): **Aizoaceae** s.str., as described in the formal treatment.

Within the order Centrospermae, the **Aizoaceae** can best be recognized by a number of probable synapomorphies, i.e. features ± exclusively present in the primitive members of the family and absent in other families of the order (Bittrich & Hartmann 1988, Bittrich 1990: 493–494). These are

- possession of epidermal bladder cells
- perigynous flowers
- very long funicles
- absence of wood rays in the normal and abnormal secondary xylem
- possession of expanding tissue acting on a hygrochastic base

As discussed by Bittrich & Hartmann (1988) and Bittrich (1990), only the possession of epidermal bladder cells is completely unique within the Centrospermae, and consequently only this one feature is a true synapomorphy. For all other probable synapomorphies listed above, an occurrence in other families can also be shown. This does not render the list above invalid, but more investigations in other families are necessary in order to determine the evolutionary direction of the character states.

Subdivision of the **Aizoaceae**

Based on the division given by Bittrich & Hartmann (1988), five subfamilies are recognized here:

- **Aizooideae**
- **Mesembryanthemoideae**
- **Ruschioideae**
- **Sesuvioideae**
- **Tetragonioideae**

Several of these subfamilies have been considered to represent families of their own: **Aizoaceae** in a stricter sense than used here, *Mesembryanthemaceae* for the two subfamilies **Mesembryanthemoideae** and **Ruschioideae** together, *Sesuviaceae*, and *Tetragoniaceae* (tables in Bittrich & Hartmann 1988: 246, 247). The contents of the families varied greatly, hence older placements of genera must be checked and adapted according to recent treatments.

The five subfamilies can be arranged in two groups. One group, with petaloid staminodes or petals, comprises the subfamilies **Mesembryanthemoideae** and **Ruschioideae** and is, as mentioned above, sometimes ranked as a family by the conserved name *Mesembryanthemaceae* Fenzl (Bittrich & Hartmann 1988, Bittrich 1990). The group is certainly monophyletic, but it has not yet been possible to establish a synapomorphy.

The remaining three subfamilies may be grouped together, too, if the basic chromosome number of $n = 8$ (or 7), the petaloid colouring of the inner side of the calyx lobes, or the possession of calcium oxalate druses in the cell lumina (in contrast to raphide bundles) can be shown to be synapomorphic for this group (for discussion, see Bittrich 1990).

Bittrich (1990) suggests also an alternative scenario with the **Sesuvioideae** as a sister-group to the remaining four subfamilies. These latter may fall into one branch consisting of the **Aizooideae** and **Tetragonioideae** on one side, and the **Mesembryanthemoideae** and **Ruschioideae** on the other side. For the branch of four subfamilies, the possession of hygrochastic expanding tissue appears the most probable synapomorphy (for more possible synapomorphies, see Bittrich 1990: 497).

Until a well-based study is presented to allow a well-founded decision, and for practical reasons, the five subfamilies are kept up here. The **Mesembryanthemoideae** and **Ruschioideae** together are named Mesembryanthema (Hartmann 1991), a group without a taxonomic rank.

The subfamilies as represented here certainly do not meet equal standards as regards the knowledge of their character states, nor do they present groups of equal rank in a cladistic sense. In particular the relations between the **Tetragonioideae** and the **Aizooideae**, as discussed by Bittrich (1990), need more intensive study, and the problem of the position of the **Sesuvioideae** can only be solved when the weight given to such deviating features as the complete aril and the circumscissile capsule can be determined.

Diversity and Ecology of the Family

The family **Aizoaceae** shows a remarkable diversity in many respects, particularly in ecology, distribution, growth forms, and epidermal anatomy. In addition to the details given, e.g., in Hartmann (1991), some notes on ecology and distribution illustrate the wide array of features met with.

The extremes of the ecological niches inhabited by members of the family are dry subtropical deserts, wet tropical coasts, and snow-covered subtropical mountains.

The deserts receive less than 100 mm rainfall per year; some of them, however, are exposed to regular fogs from the sea, increasing the overall available humidity. Rain falls either in winter, e.g. in the southern Namib and adjacent areas, where e.g. **Fenestraria** and **Psammophora** grow, or in summer, e.g. along the edges of the Sahara and into East Africa, where, e.g., species of **Zaleya** or **Aizoon** are found.

Wet tropical coasts are inhabited by some members of the family all around the world on all continents. On rocks by the sea, exposed to salty spray, e.g., **Delosperma napiforme** and **Sesuvium edmonstonei** occur. In brackish marshy areas, often near mangrove vegetation, and exposed to tidal floods, **Sesuvium portulacastrum** can form huge stands.

Snow covers the high peaks of the Drakensberg Mountains in Lesotho for part of the year; some species of **Delosperma** occur in these conditions, some of them as endemics, e.g. **Delosperma nubigenum**, which can consequently also be grown outdoors in moderate climates if protected from too much humidity.

The climate in which the highest number of genera and species is naturally found can be described as semi-arid with 100–400 mm precipitation per year, the rain falling mainly in winter. The annual amounts of rain can vary considerably, in one locality e.g. between 110 and 330 mm on a single farm near Fraserburg, and the intermediate drying of the soil represents an important factor in the life cycle. Under these conditions, members of the family form a major portion of the vegetation cover, in the number of species and in biomass. **Aizoaceae** are therefore important elements in farming management in extensive winter rainfall areas and adjacent regions.

Distribution

Determined by the ecological conditions described, **Aizoaceae** inhabit the tropics and subtropics worldwide.

In the New World, the distribution area stretches from the southern States of USA over Central America and the Carribean Islands along the coasts of South America to Chile > 40° S in the west and to Buenos Aires in the east.

In Asia, the southern countries are inhabited by members of the **Aizoaceae** from Turkey to Malaysia and as far north as Afghanistan and China, albeit mainly as weeds.

In Australia, New Zealand, and the adjacent islands members of the family are found in various habitats, and best known is probably the New Zealand spinach, **Tetragonia tetragonioides**, the only species of the family grown as a vegetable.

In Africa, including the Cap Verdes, the Canary and Madeira islands, only the low lying parts of the tropics around the Congo Basin are devoid of **Aizoaceae**.

Only few species occur in Europe, e.g. in southern Spain and Italy, in perhaps natural stands. There is little doubt that various species and genera found in localities further north are secondary immigrants, e.g. in southern England and on the Canary Islands. Yet, it remains uncertain which taxa have immigrated when.

Later immigration is also suspected for some of the species found in California and perhaps also in South America, exemplified by the genus **Carpobrotus**. It must be remembered, though, that several species of the genus **Tetragonia** occur naturally along the western coast, i.e. the Pacific ocean. In spite of the very large distribution area, the centres of diversity lie close together in the south western part of Africa (compare, e.g., the latest map in Hartmann 1994: 1166 for Mesembryanthema). Latest counts on the basis of the genera accepted here show, e.g. for a square of 30 × 30' (i.e. a quarter of a degree square), 36 different genera in the Gariep centre, i.e. in the lower part of the Orange or Gariep River.

Importance of Characters for the Delimitation of Taxa Within the Aizoaceae

Haworth (1795–1831) used mainly vegetative characters to delimitate species, following Linnaeus (1753, 1762), who referred to Dillenius' drawings (1732) in many cases. Brown (1925: 212) was the first to recognize that similar or apparently identical growth forms and leaf shapes were correlated with different types of flowers and fruits. Since then, capsule morphology has played an increasingly important role in delimiting genera, based on a first general typification of Schwantes (1952). The number of genera with a uniform or rather uniform type of fruit has increased over the decades (e.g. **Argyroderma**, Hartmann 1978; **Antimima**, Dehn 1989), albeit also genera have been found recently in which certain features of the capsule morphology vary to a certain degree, e.g. in **Khadia** (Chesselet & Hartmann 1995). The species of **Khadia** differ in the anatomy and morphology of the closing bodies, which, nevertheless, appear to result in similar function.

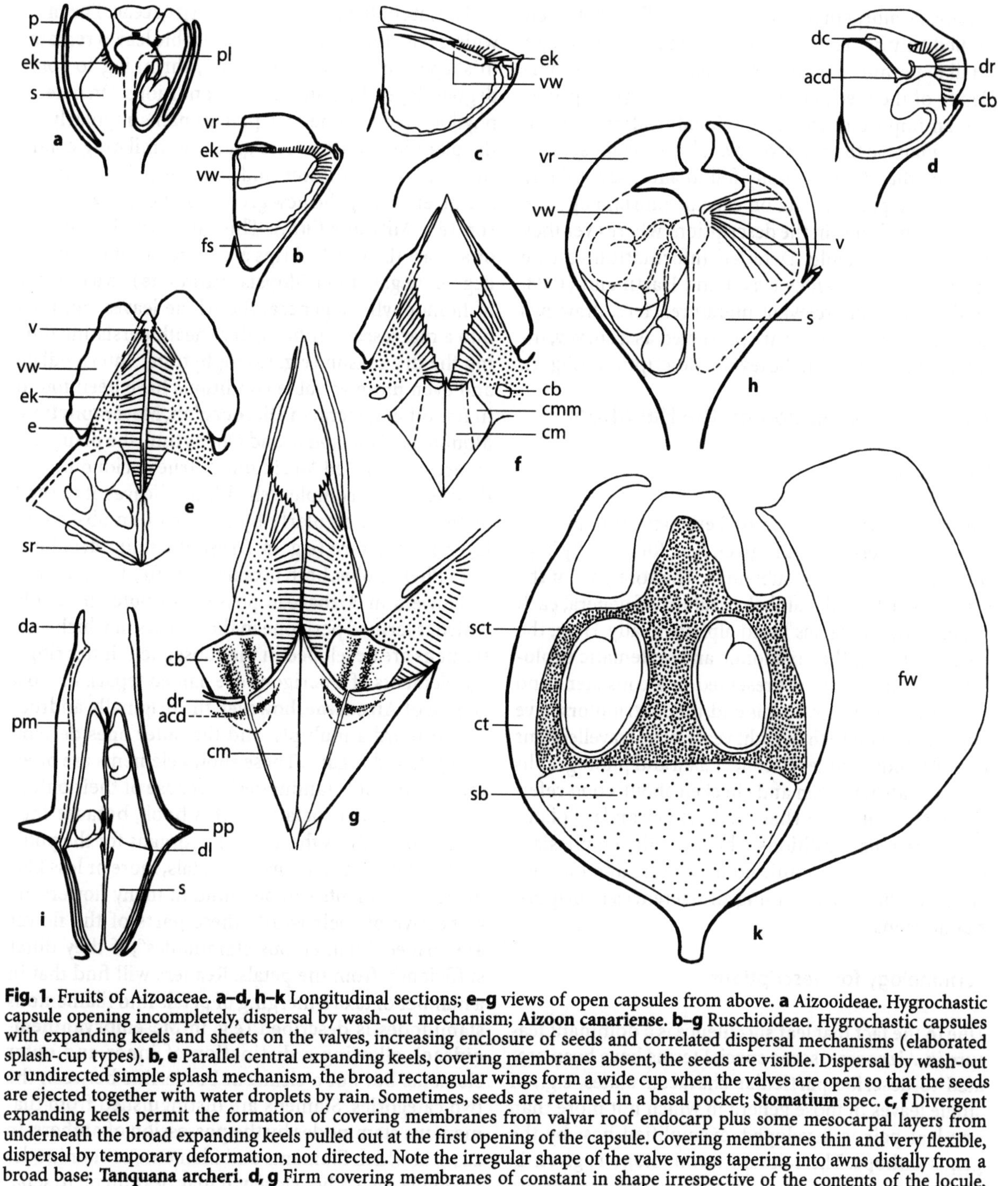

Fig. 1. Fruits of Aizoaceae. **a–d, h–k** Longitudinal sections; **e–g** views of open capsules from above. **a** Aizooideae. Hygrochastic capsule opening incompletely, dispersal by wash-out mechanism; **Aizoon canariense. b–g** Ruschioideae. Hygrochastic capsules with expanding keels and sheets on the valves, increasing enclosure of seeds and correlated dispersal mechanisms (elaborated splash-cup types). **b, e** Parallel central expanding keels, covering membranes absent, the seeds are visible. Dispersal by wash-out or undirected simple splash mechanism, the broad rectangular wings form a wide cup when the valves are open so that the seeds are ejected together with water droplets by rain. Sometimes, seeds are retained in a basal pocket; **Stomatium** spec. **c, f** Divergent expanding keels permit the formation of covering membranes from valvar roof endocarp plus some mesocarpal layers from underneath the broad expanding keels pulled out at the first opening of the capsule. Covering membranes thin and very flexible, dispersal by temporary deformation, not directed. Note the irregular shape of the valve wings tapering into awns distally from a broad base; **Tanquana archeri. d, g** Firm covering membranes of constant in shape irrespective of the contents of the locule, undulate in radial view with a ± pronounced recurving at the distal end touching the mostly large closing bodies and resulting in a complete blocking of the distal opening of the locule. Additional closing devices at the base of the recurving as rodlets, ledges, or bulges. Dorsal appendages on the exposed surface of the covering membranes stabilize the construction aiding directed dispersal. Dispersal by ejection of seeds through slits in ± proximal elevated parts of the covering membranes, optimal ballistic curves can be reached by the seeds. **d Antimima** spec., *acd:* a closing rodlet here. **g Odontophorus angustifolius,** *acd:* a closing ledge here. **h** Mesembryanthemoideae. Hygrochastic capsule opening hygroscopically: the upper half of the septum expands when wet; false septa form broad valve wings; dispersal by rain as simple splash-cup or wash-out mechanisms; **Phyllobolus trichotomus. i** Sesuvioideae. Capsule opening by removal of the top half of the capsule in dry conditions; **Sesuvium sesuvioides. k** Tetragonioideae. Winged nut, dispersal anemochoric or perhaps hydrochoric; **Tetragonia decumbens.** *acd:* Additional closing device at base of distal recurving of covering membrane; *cb:* closing body; *cm:* covering membrane; *cmm:* margin of covering membrane of mesocarpal origin; *ct:* corky, spongy tissue; *da:* dorsal appendage at apex of perigone-lobe; *dc:* subcentral appendage of covering membrane; *dl:* dehiscence line; *dr:* distal recurving of covering membrane; *ek:* expanding keel; *e:* expanding sheet; *fs:* basal false septum; *fw:* fruit wing; *p:* persistent dry perigone-lobe; *pl:* placenta; *pm:* translucent margin of perigone-lobe, *pp:* protrusion of perigone; *s:* septum; *sb:* spongy body of white, slightly cutinised tissue, separated distinctly from *ct*; *sct:* sclerotic tissue of nut; *sr:* septal rim; *v:* valve; *vr:* valve rim; *vw:* valve wing

Capsule morphology has, in spite of all insight, been studied only in a restricted number of cases in enough detail to permit conclusions, perhaps due to the complexity of the mechanism and morphology. Capsules play an important role in the identification of taxa: they remain on the plant for several years in most cases and can, therefore, be easily found and used reliably over a long period. Data on capsule morphology are therefore included in the descriptions where ever they have been available. Detailed information can be drawn from Schwantes (1952) and Hartmann (1988, 1991), both citing relevant literature. Here, drawings of several typical forms are presented as a survey, including names used in the text for descriptions (Fig. 1).

Technical Information on the Handbook

General Data

An introduction to the family, keys to the subfamilies, and their descriptions are given first, followed by keys to the genera for each subfamily. The core part of the book consists of the alphabetical list of genera, each genus with its species listed alphabetically. Here, the current state of the taxonomic and systematic exploration of the family is presented in a consistent and informative manner. To this end, the contributors have worked over more than eight years, have travelled tens of thousands of miles to collect data and photographs in the field, have seen and checked about 5 000 references in literature and have grown thousands of plants under various conditions. Above all, the emphasis is on the correct taxonomic nomenclature and identification of the taxa, including all references to original publications.

Terminology for Descriptions

Due to the fact that leaf succulence is a dominant feature in **Aizoaceae**, some special terms have been created to describe particular character states, e.g., of single leaves in those genera in which not more than one leaf-pair is developed, exemplified best by the genus **Lithops**. The relevant terminology is given in the appropriate places; only some general terms are discussed here.

"Spines" in **Aizoaceae** are always organs derived from stems, mostly from pedicels or peduncles (e.g. in **Ruschia** subg. **Spinosae**); "prickles" are remnants of organs of various origin and generally blunter than spines (e.g. in **Drosanthemum**); "spinules" are sharp, pointed, tiny organs, mostly at nodes and derived from the central bundle of a withered leaf or the triangle between leaf bases (e.g. in **Drosanthemum**). In some rare cases, leaves can turn into spinulate organs with age, too.

The fate of the leaf can vary in **Aizoaceae**. The most widespread type of hard, xeromorphic leaves remains in shape during the dry season, shrinking to some degree, depending on drought conditions. On heterophyllous plants, one leaf pair remains regularly on the plant in a more or less green, assimilating condition, the other – lower and outer – pair forming a sheath enclosing the new growth of the following season (e.g. **Mitrophyllum**). This latter whole leaf pair or at least the free leaf tips are marcescent to various degrees (e.g. in **Cheiridopsis, Antimima**). In a number of homophyllous genera, the whole leaves can turn into a marcescent or hard, dry sheath persisting during the dry season (e.g. **Conophytum**). Intermediate and even more variable conditions have been found in few other genera; in **Dicrocaulon, Meyerophytum, Monilaria, Diplosoma**, and **Oophytum**, the plants are sometimes termed "deciduous". True deciduous conditions, i.e. the complete shedding of leaves at the end of the season, regularly by means of an abscission layer, is very rare (e.g. in **Drosanthemum deciduus**, or in **Phyllobolus**, both taxa named after this feature).

Flowers in **Aizoaceae** possess an outer perianth, which is termed perigone in those cases in which only stamens are developed in the subsequent inner rings; the elements of a perigone are named tepals. In most genera of **Aizoaceae**, however, the originally androecial primordia multiply, and the outer ones develop into petaloid organs. These showy elements are often called "petaloid staminodes", because of their ontogeny, and named "petals" in this book, because they act ecologically as the attracting organs for the pollinators. Between stamens and petals, more or less filamentous elements can be found in many flowers. Irrespective of their width, these parts of the flower are termed "filamentous staminodes", if they differ sufficiently from the petals. Readers will find that in a number of taxa a merging of petals into filamentous staminodes is described (e.g. in **Mesembryanthemoideae**), allowing a distinction of inner and outer elements but not a clear-cut borderline between the two categories. Stamens are distinguished from filamentous staminodes by the possession of anthers. In cases in which filaments and filamentous staminodes resemble each other closely, it may be difficult to separate members of the two categories because anthers fall off in aging flowers. At the base of the stamens, nectaries of different shapes can be found. Two types prevail: coilomorphic nectaries are shell-shaped and characterize members of the **Mesembryanthemoideae**, lophomorphic nectaries are crest-shaped, dominate in the **Ruschioideae**, and can occur as a ring or as single nectaries. Of the gynoecium, the stigmas are the parts most easily visible, representing the pollen-receiving free portions; styles, the connate lower portion of stigmas, occur very rarely in **Aizoaceae** (e.g. in **Argyroderma**). During anthesis, the flowers open and close by basal growth of all elements, resulting in an increase in diameter during the flowering period, a feature to be observed when comparing given values. Also important is the fact that the male phase of the flower precedes the female phase in all flowers seen until now, the stamens appearing very prominent at first, hiding the gynoecium more or less completely. In many flowers, the stigmas spread during the female phase, often elongating and spreading widely over the withered stamens. These developmental changes should be considered when floral characters are compared.

Fruits of **Aizoaceae** are mostly capsules, very rarely berries (e.g. **Carpobrotus**), nuts (e.g. **Tetragonia**), or drupes (e.g. **Gunniopsis**). For the majority of the hygrochastic capsules of the family with many very special morphological components, an elaborate terminology has been developed. Terms used in this book are explained in Fig. 1 and, in addition, specified in cases of necessity in some genera (e.g. **Caryotophora**). Fruit types follow those described by Hartmann (1988), descriptions are repeated here with those genera after which the type is named.

Taxonomic Principles

Rank and Limits of Genera

During the intensified investigations in the family in the past decades, the question of generic rank has been discussed several times (e.g. by Hartmann 1983: 46–48, Bittrich 1987: 65–66, Hartmann BJS 1992: 43–46) suggesting criteria for the establishment of a taxon at genus rank.

Many genera have been described in the early period of splitting up the old genus **Mesembryanthemum** based on one or few extraordinary character expressions, e.g. **Fenestraria** because of the club-shaped leaves with a terminal window, **Amphibolia** because of the combination of a closing body and broad valve wings in the fruit, or **Esterhuysenia** without valve wings and closing bodies in the fruit. In the absence of detailed investigations, several of these often monotypic genera have been reduced to synonyms of various big genera (e.g. Rowley 1978: 6–9, 62), enlarging these latter ones past recognition and definable limits.

In the course of further studies, an overall genus concept will probably be developed, but until then a pragmatic view has to be taken. For the present treatment, the guide line has been to delimit genera as such when some to few clear distinctive character states can be found which, on one hand, permit a clear circumscription of the genus, and on the other hand allow a distinct separation from other genera. In most cases, this leads to traditional use, but in some cases remarkable new arrangements had to be made. Thus, e.g., the genera **Eberlanzia** and **Amphibolia** can clearly be separated based on fruit and leaf character states, both genera can neither be united nor can they be placed in some other genus without unjustified distortions (Hartmann Bradleya 1998).

On the same line of arguments, namely that taxa are recognized here as genera based on clear differences, the genera **Peersia** and **Deilanthe** have been re-established. Based on results of Dehn (1989: 189–216), the genus **Antimima** has been enlarged to about 100 species (Hartmann Bothalia 1998) that can probably be grouped into several subunits. For all these genera, a positive placement of the species included was possible, i.e., for all units below genus rank, the features found agree with the critical character expressions of the genus.

In contrast to these re-established genera and their species, a number of genera remains containing several to many aberrant species. These species can, at this date, only be pointed out as being not in agreement with the established generic character states; at the same time, they cannot be placed elsewhere due to lack of data. The situation will be found particularly disturbing and dissatisfying in the large genera **Delosperma**, **Drosanthemum**, **Lampranthus**, and **Ruschia**, which need a revision urgently. In these, and in other cases where no recent results are available, a very conservative view has been taken, keeping up most species in the genus into which they had originally been placed.

Rank and Limits of Species

Of the numerous definitions of species, a simple one will serve the purpose of this handbook: species are biological units comprising all members of a group that can cross and produce viable offsprings in habitat. As an approximation to this standard, which is not readily testable in perennial plants, a certain degree of similarity in all possible character expressions is considered sufficient to circumscribe a species. Yet, even this level of treatment has seldom been reached. Instead, most species in **Aizoaceae** have been described based on a single specimen, available in the dry state as a herbarium specimen only, or as a growing plant. These latter have often been in cultivation for some period under conditions differing markedly from those of the natural habitat. In extreme cases, dry and cultivated plants have been described as different species. Early studies in habitat including the variation in character states have been extremely rare (e.g. Dinter 1923, 1928, Marloth 1910, 1914). This is particularly true for the period between 1920 and 1940, when most species were described by Bolus (Cape Town), Brown (London), and Schwantes (Kiel). Since for these descriptions no information regarding the variability was known, boundaries of species had necessarily to be uncertain, resulting in severe discrepancies about limits, affinities, and identities of taxa, and causing numerous discussions and hot debates among the main authors. The situation turned only when a revision of the genus **Argyroderma** was presented (Hartmann 1973, published 1978), which was based on extensive field work including ecology and population biology, considering a wide array of characters, using statistical methods, and offering a species concept for succulent plants for the first time. It was not surprising that the number of species in **Argyroderma** was consequently reduced from 62 to eleven, including one hybrid species. Later revisions of several bigger genera conducted in the same way, e.g. **Cheiridopsis** or **Cephalophyllum**, have reduced the number of species by about a third. Whether a similar effect will be found once the large genera, e.g. **Delosperma, Lampranthus**, and **Ruschia** have been studied, remains to be seen. The small genera with few species are mostly too well known to house much material for a reduction of the species number.

Ranks Below Species

In two ways, subspecific taxa have been created. Older authors, e.g. Linnaeus 1762, described forms deviating from the typical or α-form by adding β or more forms indicating minor differences not demanding recognition at species level; these taxa are considered to represent varieties (ICBN 2000: Article 35.4). Later authors used subspecific ranks to describe various types of differences, in most cases not applying a uniform standard. The ranks of subspecies, variety, and forma have been used extensively in those genera that are most widely grown in cultivation in Europe and America: **Conophytum** and **Lithops**.

Based on definitions by Du Rietz (1930) and Stace (1985), the following guidelines are used here mainly:

- The rank of subspecies is used for taxa differing only a little in their character states, growing in different geographical areas, and occurring in different ecological situations.
- Varieties of a species show a lesser degree of difference than subspecies, grow rarely sympatrically but mostly in a very restricted area only, and are not necessarily ecologically differentiated.
- Formae are in general sports or single aberrant plants within an otherwise rather uniform population, the difference being only expressed in a single feature, e.g. green instead of brownish leaves.

In this treatment, the lowest rank considered is that of subspecies, published varieties are listed under the relevant species or subspecies.

General Scheme for the Treatment of Taxa

All genera are arranged alphabetically, the name being followed by the name of the subfamily into which the genus is placed. All valid and accepted names are printed in **bold**, synonyms in *italics*. For each treatment, the contributor(s) is (are) given.

The basic scheme for the treatment of an accepted taxon is as follows:

- nomenclatural part including the name of the taxon with author and bibliographical reference(s), its synonym(s) and types with full references, etymology (only for genera)
- detailed description (see below)
- notes (see below)

For genera, the name of the subfamily into which they are placed, is stated in the headline.

Each synonym appears at least twice: once together with the accepted name under which nomenclatural and bibliographical details are to be found, and secondly in the alphabetical order either as genus or as species under the genus in which it had been placed. For subfamilial and generic synonyms a brief explanation is given for the reason(s) which render them invalid or illegitimate. The specific synonyms are principally listed together with the accepted names under the genus in alphabetical order. An exception from this rule is **Mesembryanthemum** for which synonyms are presented in two parts:

- All names synonymous to names of taxa currently included in the genus **Mesembryanthemum** are given in the account itself.
- All synonyms of names of taxa now placed in other genera, and those names under **Mesembryanthemum** that have been validly published but can at present not be placed with confidence in any of the existing other genera, are separately listed immediately after the account on the genus **Mesembryanthemum**.

Nomenclature and Typification

Homotypic synonyms, that is names of taxa based on the same type, are separated by a triple hyphen (≡); all names are followed by the citation of their common type. A possible source of homotypic synonyms are species which have been transferred into different genera, e.g.,

Aspazoma amplectens (L.Bolus) N.E.Br. in Phillips 1926: 244 ≡ *Mesembryanthemum amplectens* L.Bolus ASAM 1913: 153 **Lectotype,** here designated: Pearson 6064 (BOL!).

Another source are taxa that were described more than once based on the same type, as was done in **Tetragonia tetragonioides** and *T. expansa*.

In contrast, heterotypic synonyms are names of taxa based on different types which are presently incorporated in a single taxon. These synonyms are separated by a double hyphen (=), and the types are cited right behind each name, e.g.,

Cephalophyllum parvibracteatum (L.Bolus) H.E.K.Hartmann T Frames 18967 (BOL!) = *C. kliprandense* L. Bolus T Stayner 1223/62 (BOL!).

All valid taxa have to be typified (ICBN 2000: Art. 7.1), be they legitimate or illegitimate. **T** here denotes a holotype, **LT** a lectotype, and **NT** a neotype. In a number of cases a lectotype or neotype is chosen here for the first time, which then is indicated as required (ICBN 2000: Art. 7.11) by adding "here designated".

Type specimens seen are marked with an exclamation mark printed after the abbreviation of the herbarium housing the sheet, which is cited in parentheses, e.g. (BOL!).

Standard of Descriptions

Descriptions are given for the family, the subfamilies, the genera, and all species and subspecies according to the state of knowledge at the date of completion of the manuscript. This implies necessarily differences in standard because for a number of genera revisions and plenty of fresh material have been available for this book, but numerous genera have never been revised after their establishment, though much material has been available in various herbaria.

For all taxa, the original descriptions and all further combinations published have been seen (a total of about 5 000 references, the few exceptions not seen are marked in the References), and results have been compared with these. In cases where revisions and plants were available, well-balanced descriptions can be offered, mostly also based on extensive field studies, which have partly been conducted for the purpose of writing the texts for this handbook.

In contrast to this favourable state of knowledge, there are numerous taxa where no revisions were available and no studies were possible, though much material was at hand, but time was lacking. In these cases, the descriptions have been taken back to the original texts, their possibly accompanying figures, and to the types, when these have been preserved. The descriptions thus derived will disagree with many traditional usages of names, e.g. in **Lampranthus**, but the chosen procedure is the only basis on which to promote an unbiassed study of the group. All later published descriptions are to be seen as interpretations, which may or may not be correct, as has been demonstrated in a number of cases, e.g. **Lampranthus scaber** (Liede & Meve 1990).

In the case of complete data, the descriptions follow a scheme from growth form over leaves (**L**) to flowers (**Fl**), containing often information on the calyx (**K**), to fruits (**Fr**), seeds (**S**), chromosomes (**Chr**), ecology (**Ecol**), and distribution (**Distr**). Each item is introduced by the bold letters given here in brackets. In addition, internodes (**I**) and inflorescences (**Inf**) can occur. In cases of incomplete data sets, the key words for which data are missing are left out, excepting **Ecol** and **Distr**, in which cases "unknown" or "uncertain", or other available information can be inserted, e.g. the continent of origin.

Additional information on the taxonomic history, similarities, cultivation, and usage is given under "*Note*".

Orthography

Orthography is in British English; some expressions used for South African types of vegetation, e.g. "veld", are included as well. The word "Karoo" is spelt in different ways in English. In general, the spelling "Karroo" is used in the text here; in various names, however, the original spelling like "karooides" or "karroid" is retained.

Keys

Keys to the subfamilies, to their main groups and down to the genera are given at the beginning, together with the relevant descriptions. Keys to sub-genera, sections, species, and subspecies are found under the genera in those cases in which enough material has been available to construct a key. All keys are based on features that are visible through most of the year, which excludes flowers in many cases because they are short-lived and only distinctive in some taxa or when much material can be compared, which is very rarely the case.

Distribution

A map is provided for each genus delineating the area in which material of the genus has been found, based on literature, herbarium sheets, and on well over 15 000 collections by the contributors and their collaborators. Place names are abbreviated on these maps; the complete place names referred to in distribution maps can be found on a map covering most of southern Africa (Fig. 2), which also shows the old districts of Namibia and the boundaries of the new provinces of the Republic of South Africa.

In addition, for each species and subspecies the distribution is cited in the sequence of district, province, country for most parts of the world. If more than five districts in a province are included, a general descrip-

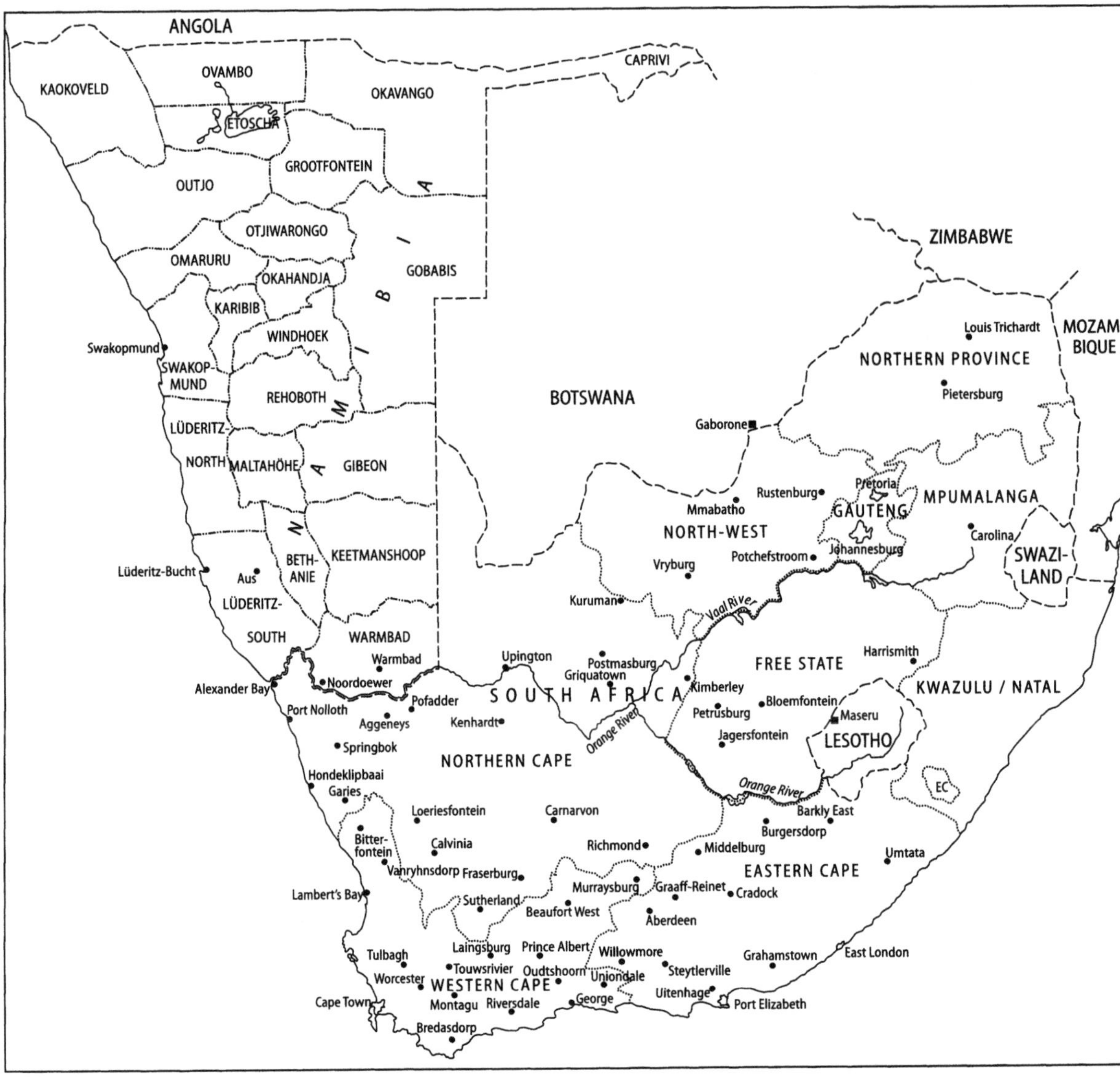

Fig. 2. Map of southern Africa with place names referred to in distribution maps

tion of the area is given. During the production of this work, the names of numerous districts were changed, sometimes also the boundaries. Reference is here made to the old districts by name and in their boundaries because these old names have been used in most cases in the original descriptions.

Illustrations

768 coloured figures illustrate a selection of species, 384 per volume. The majority of the photographs has been taken in habitat, and it has been ascertained that the plants have been reliably identified by the authors. This procedure has the consequence that only revised or well known taxa are figured. Since the aim was also to present plants in the field, it means that photographs taken in the open, often under difficult or adverse conditions, may not be as brilliant as those taken in a collection or garden. Species known only from literature or from herbarium material and not reconfirmed on fresh material are only described. In this way it is hoped to break the sequence of misidentifications in several cases and at the same time to stimulate new studies in little known yet sometimes widely occurring genera like **Lampranthus, Delosperma,** and **Ruschia.**

Aizoaceae

H. E. K. Hartmann

Aizoaceae Martynov 1820: 15 nomen conservandum (ICBN 2000: App. IIB) **Typus Aizoon** L.; Bittrich & Hartmann (1988); Hartmann (1991); Hartmann (1993) = *Ficoidaceae* Jussieu 1789: 315 nomen rejiciendum **Typus Mesembryanthemum** L. (as *Ficoides* Miller 1754: without page) ≡ *Mesembryanthemaceae* Fenzl 1836: 349 nomen conservandum (ICBN 2000: App. IIB) **Typus Mesembryanthemum** L. nomen conservandum = *Sesuviaceae* Horaninow 1834: 83 **Typus Sesuvium** L. = *Tetragoniaceae* Link 1831: 17 nomen conservandum (ICBN 2000: App. IIB) **Typus Tetragonia L.**

Predominantly perennial shrubs, rarely annual to biennial herbs, rarely trees, and rarely plants reduced to a single leaf-pair, growing sometimes sunken in the ground; **L** mostly opposite, simple and entire, rarely lyrate, epidermis with bladder cells of various shapes and sometimes reduced, or uniform and xeromorphic, often papillate to pubescent, base of petioles rarely with stipuliform appendages or leaves sessile, sometimes with a connate leaf sheath; **Inf** of principally dichasial pattern, complete or in various derived forms, mostly terminal, often seemingly axillary, often reduced to a single flower; **Fl** actinomorphic, mostly bisexual, perigynous to hypogynous or epigynous, perianth elements (3–)5(–8), free portions often unequal and with dorsal subapical appendages, basally connate and adnate to the filaments, thus forming a perianth-stamen tube, inner surface of upper portion petaloid or green, named calyx when petaloid elements are present; androecial elements 4–many, if many, the outer primordia developing often into petaloid organs (= petals), filaments rarely connate, rarely connate with the petals forming a tube, anthers dehiscing by longitudinal slits; ovary syncarpous, (1–)5(–∞)-carpellate, placentation axile, basal or parietal, ovules (1–)∞ per carpel, anacampylotropous or campylotropous, bitegmic, crassinucellate, rarely pendulous; **Fr** mostly a hygrochastic loculicidal, rarely septicidal or xeromorphic capsule, rarely schizocarpous, sometimes a hard and indehiscent 1-seeded nut, rarely a drupe, occasionally in aggregates, or a circumscissile capsule; **S** mostly ± ovoid, rarely arillate, with a curved peripheral embryo, and voluminous starchy perisperm, endosperm reduced to a layer enveloping the radicle; **Ecol** mainly in subtropical and mediterranean climates with moderate to low rainfall regimes mainly as winter rains; **Distr** worldwide in the subtropics, the highest number of genera and the centre of diversity at species level in southwestern Africa.

Note: The latest Code (ICBN 2000: App. IIB) lists the family name as **Aizoaceae** Martynov, as given above. For several decades, the name **Aizoaceae** had been used as *Aizoaceae* Rudolphi 1830: 53 nomen conservandum, until Reveal (1998) published a list of older valid family names. The publication of Martynov is a dictionary, in the Latin part of which Martynov gave an entry "AIZOONIDES" on page 15, citing Aizoon as the name-giving genus and referring to Augier (as Г. Ожъера), whose list of genera is repeated and a short description given. Since the name in Martynov 1820, although identical in spelling with that of Augier, is in Latin, the family name is validly published with only the ending to be corrected (ICBN 2000: article 18.4). The basionym is, therefore, Augier (1801: 129, 137–138), who gave with the French name "Aizoonides", explicitly called "family", a description in French and a list of the genera Aizoon, Tounatea, Smyda, Calligonum, Sloanea, and Sesuvium, all without authors, but with short descriptions.

This treatment follows the delimitation of the family by Bittrich & Hartmann (1988), who excluded members of the family **Molluginaceae** Rafinesque and established a circumscription of the **Aizoaceae** s.str. including five subfamilies. Some of these latter have at times been considered families of their own, e.g. *Sesuviaceae* or *Tetragoniaceae*. Applications of all family names listed below have been variable, as shown explicitly by Bittrich & Hartmann 1988. The older name *Ficoidaceae* is not available since **Aizoaceae** became a nomen conservandum.

As the main synapomorphy, the possession of bladder cells has to be named, because other features occurring in the family cannot definitely be shown to be synapomorphic since they occur also in other families of the order Centrospermae (Bittrich 1990: 491–507).

221 genera have been described as members of this family, of which 132 are accepted here.

Shrubs of many genera form the dominant part of the vegetation of the winter rainfall area of southern Africa (e.g. **Amphibolia, Aridaria, Antimima, Eber-**

lanzia, Lampranthus, Leipoldtia, Ruschia, Vanzijlia) some genera are also important elements of the Karroo vegetation outside this region (e.g. **Ruschia, Trichodiadema**), thus contributing considerably to the carrying capacity of the natural vegetation. Few species possess chemical compounds which are used in traditional healing (e.g. **Sceletium, Zaleya**). A selection of small-growing species (e.g. **Conophytum, Lithops**) is kept in horticulture, mainly indoors or in greenhouses worldwide in moderate climates. Some genera are found as garden elements or sand-binders in suitable climates (e.g. **Carpobrotus, Malephora**). The sour-tasting fruits of some species of **Carpobrotus** are sold semi-dried in the market to be used for jellies or preserves. Only one species, **Tetragonia tetragonioides**, is grown as a vegetable in appropriate conditions.

Ficoidaceae = **Aizoaceae**

H. E. K. Hartmann

De Jussieu (1789: 315) was the first to describe the family *Ficoideae* at that rank, including the genera "**Reaumuria** Hasselq., **Nitraria** Schob., **Sesuvium** L., **Aizoon** L. *Ficoidea* Niss., **Glinus** L., **Orygia** Forsk., **Mesembryanthemum** L. *Ficoides* T., **Tetragonia** L.". It must be noted that *Ficoidea* Nissole (1730: 319) is an obligate synonym of **Aizoon** L., whereas *Ficoides* Tournefort (1730: 238–241) is an obligate synonym of **Mesembryanthemum** L.; both names were published before Linnaeus 1753, only one was validated late: *Ficoides* Miller 1754. Only this name can be considered when a type species of *Ficoidaceae* has to be found.

The question of the family name has been discussed extensively (e.g. Sprague 1922: 71, Rowley 1951: 27–28, Friedrich 1955: 56–66, Schwantes 1957: 157–158, 167–169). It was only settled when **Aizoaceae** Rudolphi became the nomen conservandum for the relevant genera (ICBN 1961), including **Aizoon.** The new nomen conservandum **Aizoaceae** Martynov does not interfere with the rejection of *Ficoidaceae* de Jussieu. If *Mesembryanthemaceae* are considered as a separate family, the name *Ficoidaceae* becomes an obligate synonym, since both names are based on the same genus.

Mesembryanthemaceae = **Aizoaceae**

H. E. K. Hartmann

The family *Mesembryanthemaceae* was first described as such by Fenzl (1836: 349), albeit with a wrong ending: *Mesembryanthemeae.* This error was to be corrected, since the rank was clearly indicated in a summary preceding the description (Fenzl 1836: 347).

The family name has been widely used for **Mesembryanthemum** and all genera segregated from it (e.g. Herre 1971). When establishing the circumscription of the **Aizoaceae** s.str. excluding the **Molluginaceae**, Bittrich & Hartmann (1988: 239–254) included all members of *Mesembryanthemaceae* in **Aizoaceae**, mainly based on the possession of bladder cells, the synapomorphy of the family **Aizoaceae** within the order Centrospermae (Caryophyllales). Two subfamilies are recognized within this complex: **Mesembryanthemoideae** and **Ruschioideae**. Both share a number of characters and together certainly represent a monophylum, but for many character states it cannot be decided whether they are primitive or advanced (e.g. Bittrich 1990, Hartmann 1991). It is therefore at present not possible to demonstrate monophyly for the remaining members of the **Aizoaceae** after the *Mesembryanthemaceae* have been separated. Notwithstanding this taxonomic pitfall, both families are often kept apart in herbaria for practical reasons. In this unsatisfactory situation, Hartmann (1991: 75–157) decided to use the unranked term "Mesembryanthema" for the united subfamilies **Mesembryanthemoideae** and **Ruschioideae**, thus stressing the evolutionary proximity of both taxa. It seems, nevertheless, possible that the **Aizoaceae** might be divided into two families but this can only be demonstrated once the knowledge of the subfamilies **Aizooideae, Sesuvioideae,** and **Tetragonioideae** has been expanded considerably.

Sesuviaceae = **Aizoaceae**

H. E. K. Hartmann

Horaninow (1834: 83) was the first to use the family name "*Sesuviaceae* (Ficoideae)", including the genera "**Mesembryanthemum, Tetragonia, Aizoon, Sesuvium, Trianthema, Glinus; – Reaumuria? Nitraria?**". A little later (Horaninow 1843: 29), four tribes were listed: *Paronychieae, Portulaceae, Ficoideae,* and "*Neuradeae et Surianeae?*" containing 18 subgroups plus two with a question mark, but no genus names. Retaining the subdivision of the family as well as its circumscription more or less, Horaninow used the family name **Portulacaceae** for the broadened assembly of genera some years later (1847: 138–140) without giving an author or a reason for his change.

The family name has very rarely been used, probably due to the fact that the names *Ficoideae* and **Aizoaceae** were available at the rank of family for most of the assembly of species listed by Horaninow at that time as well. Nevertheless, various authors refer to the "**Sesuvium**-group", comprising the genera **Cypselea**, **Sesuvium**, **Trianthema**, and **Zaleya**, when investigating **Aizoaceae** s.lat. (e.g. Hofmann 1973), i.e. when the **Molluginaceae** are included.

Bittrich & Hartmann (1988: 240–245) argued that the possession of bladder cells in the epidermis and the perianth-stamen tube in the **Sesuvium**-group sensu Hofmann (1973) justify its inclusion in the family **Aizoaceae** as a separate subfamily **Sesuvioideae**.

Tetragoniaceae = Aizoaceae

H. E. K. Hartmann

Tetragoniaceae Link 1831: 17 nomen conservandum has been established lately as the correct name of the family again (ICBN 2000: App. IIB), including the genera **Tetragonia** and **Aizoon**. For some time, Nakai (1942: 103) had been used as the author of the family name as a nomen conservandum. Nakai had ascribed the family name to Reichenbach in Mössler (1927: 52). Reichenbach had indeed placed the "*Tetragoniaceae*" as a subgroup "d" under "Family 97: *Aizoideae*". It is obvious that the ending "-aceae", denoting family rank today, was not applied in that sense by Reichenbach, who incidentally used various endings for the subdivisions of the families. The family name can often be found in floras. Bittrich & Hartmann (1988: 240–245) included the group in **Aizoaceae** based on the occurrence of bladder cells in the epidermis, an inferior ovary, and further similarities with the **Aizoon**-group or subfamily **Aizooideae** of the **Aizoaceae**. Bittrich (1990: 491–507) stressed the proximity of the subfamily **Tetragonioideae** to the subfamily **Aizooideae**, based on the possession of bladder hairs in both subfamilies, resulting in a monophylum including both subfamilies.

Key to the Subfamilies of the Aizoaceae

H. E. K. Hartmann

1. Petals of staminodial origin present 2
– Petals absent; perigone-lobes petaloid inside, sepaloid outside 3
2. Placentation central; nectaries shell-shaped to tubular (koilomorphic); petals and stamens mostly united into a tube ..... **Mesembryanthemoideae**
– Placentation basal or parietal, extremely rarely central; nectaries crest-shaped (lophomorphic), petals mostly free ..... **Ruschioideae**
3. Fruit indehiscent, a winged or horned nut, rarely a drupe; plants often andro-monoecious ..... **Tetragonioideae**
– Fruit a capsule; inflorescences of hermaphrodite flowers only 4
4. Capsule circumscissile; aril present, covering the seed completely ..... **Sesuvioideae**
– Capsule loculicidal or septicidal, often hygrochastic; seeds never arillate ..... **Aizooideae**

Aizooideae

H. E. K. Hartmann

Aizoaceae Rudolphi subfamily **Aizooideae**; Bittrich & Hartmann 1988: 239–254.

Annual to perennial, prostrate to erect herbs, subshrubs, or shrubs; **L** opposite to alternate, flat to cylindrical, mostly free, bladder hairs present; **Inf** mostly frondose often representing the entire plant, **Fl** seemingly solitary, perigone 4–5-lobed, basally mostly connate, sepaloid outside, petaloid inside, nectary coating the perianth-stamen tube, stamens mostly < 30; ovary 2–5-locular; **Fr** loculicidal or septicidal mostly hygrochastic capsules, rarely xerochastic; **S** 1–∞ per locule; **Ecol** mostly in open, dry areas in gravelly soils, often in disturbed places; **Distr** mostly in mediterranean and subtropical climates on all continents.

Note: Six genera are distinguished, occurring mainly in S Africa and Australia, in E Africa and around the Mediterranean Sea, introduced worldwide into suitable climates, mainly as weeds in arid country.

Key to the Genera of Aizoaceae Subfamily Aizooideae

1. Fruit a purely septicidal or septicidal and loculicidal hygrochastic capsule .... **Gunniopsis**
- Fruit a purely loculicidal hygrochastic or xerochastic capsule, often opening tardily 2
2. Septum incomplete, 2 locules with a single basal ovule each .... **Acrosanthes**
- Septa complete, locules more than 2, if 2, ovules pendulous 3
3. Capsule 5–10-locular, with 2–many ovules in each locule; distinct expanding keels present 4
- Capsule 2–5 locular with a single pendulous ovule in each locule; expanding tissue rudimentary to absent . 5
4. Capsules depressed at apex, if not, valves with prominent rims .... **Aizoon**
- Capsules convex at apex, ± rounded .... **Aizoanthemum**
5. Stamens as many as the perigone-lobes .... **Plinthus**
- Stamens twice the number of the perigone-lobes, in pairs alternating with perigone-lobes .... **Galenia**

Anettarioideae = Ruschioideae

H. E. K. Hartmann

Over several decades, Rappa developed a system of Mesembryanthema, at times in collaboration with Camarrone. A first formal subdivision of the family *Mesembryanthemaceae* was published in 1953 (1–39), based on features of the nectary. The circumscriptions of the taxa at genus rank, e.g. *Tetracoilanthus*, were solely related to character states of the nectary, and not for all distinguished taxa were species names available at that date. Albeit the ranks were cited (1953: 31), no Latin diagnoses were supplied, this rendering the names invalidly published.

For the subfamily *Anettarioideae* Rappa & Camarrone 1953: 31, as *Anettari*, five species of the genus **Glottiphyllum** were listed. The subfamilial name is therefore a synonym of A. subfamily **Ruschioideae.**

Aptenioideae = Mesembryanthemoideae

M. Gerbaulet & M. Struck

The subfamily name *Aptenioideae* Schwantes (1947) was invalidly published, since it was not accompanied by a Latin diagnosis. When revising the suprageneric classification of the **Aizoaceae** s.str., Bittrich & Hartmann (1988) validated the name *Aptenioideae* Schwantes ex Bittrich & H.E.K. Hartmann. However, already in 1962 (53) Ihlenfeldt, Schwantes & Straka had published the name **Mesembryanthemoideae** Ihlenf., Schwantes & Straka which is the earliest valid name under the family name **Aizoaceae** for this subfamily. Hence, *Aptenioideae* Schwantes ex Bittrich & H.E.K. Hartmann is an illegitimate name. This was rectified by Hartmann & Bittrich (1991).

Caryotophoroideae = Ruschioideae

H. E. K. Hartmann

The name for the subfamily *Caryotophoroideae* was published invalidly at first (Ihlenfeldt et al. in Schwantes in Jacobsen 1960: 962) and validated two years later (Ihlenfeldt et al. in Ihlenfeldt, Schwantes & Straka 1962: 54) with a Latin description. Only one genus was included, **Caryotophora**; the rank of subfamily was chosen because the oligomerous gynoecium develops into a hard schizocarpous fruit, a unique situation within the Mesembryanthema. Bittrich (1987: 50–53) investigated the problem and decided that the *Caryotophoroideae* should better be placed in **Ruschioideae**, based on the fruit ontogeny.

Coilomorphioideae = Mesembryanthemoideae

M. Gerbaulet & M. Struck

In an attempt of a suprageneric classification of the *Mesembryanthemaceae* based on nectary morphology as the principal taxonomic character Rappa & Camarrone (1953, 1960) invalidly published the subfamily name *Coilomorphioideae* as *Coilomorfi* (Rappa & Camarrone 1953: 30), the circumscription of which, incidentally, is identical with that of **Mesembryanthemoideae.** The name *Coilomorphioideae* alludes to the coilomorphic, that is shell-shaped to tubular, nectaries which constitute one of the synapomorphies of this subfamily.

Hymenogynoideae = Ruschioideae

H. E. K. Hartmann

The name *Hymenogynoideae* for a subfamily was at first published by Schwantes (KuaS 1957: 168) as an invalid name and validated in 1962: *Hymenogynoideae* Schwantes in Ihlenfeldt et al.(1962: 54) with a Latin description. The only genus, **Hymenogyne**, had already been separated from **Mesembryanthemum** by Haworth. The rank of subfamily was mainly chosen because of the peculiarities of the fruit which develops false septa and retains the seeds in clausae. Bittrich (1987: 50–52) after studying the ontogeny of the fruit concluded that the *Hymenogynoideae* should better be placed under **Ruschioideae.**

Lophomorfioideae = Ruschioideae

H. E. K. Hartmann

As explained under *Anettarioideae*, the subdivision of the *Mesembryanthemaceae* by Rappa and Camarrone (1953: 1–39) was mainly a mechanical distinction of different types of nectaries, to which secondarily names at given ranks were added and genera listed, if these could be found.

It can be remarked that no species was found to possess hexagonal holonectaries, a feature used to establish the so-called subtribe *Esaolonettari* (Rappa & Camarrone 1953: 31), which was consequently left without any species.

All other genera cited under *Lofomorfioideae* Rappa & Camarrone 1953 are now placed in the subfamily **Ruschioideae,** to which the latter name is therefore a synonym, published invalidly because a Latin diagnosis is lacking.

Mesembryanthemoideae

M. Gerbaulet & M. Struck

Mesembryanthemoideae Ihlenf., Schwantes & Straka 1962: 53; Bittrich 1987: 5–116; Hartmann 1991: 75–157 **Typus Mesembryanthemum** L. typus conservandum = *Aptenioideae* Schwantes SKK 1947: 39 nomen invalid. ≡ *Aptenioideae* Schwantes ex Bittrich & H.E.K.Hartmann 1988: 250 nomen illeg. **Typus Aptenia** N.E.Br. = *Coilomorphioideae* Rappa & Camarrone 1953: 30 (as "Coilomorfi") nomen invalid. **Typus** not given.

Annual or biennial herbs, small to large shrubs or geophytes, prostrate, decumbent or erect, sometimes tufted, rarely creeping, roots sometimes thickened, **I** herbaceous, woody or succulent and either green and assimilating or with a conspicuous cork layer, terete, angled or winged, cortex mostly with additional vascular bundles; **L** flat to (almost) cylindrical or slightly channelled, decussate, but often becoming alternate in the **Inf**, or alternate throughout, decussate leaves mostly shortly connate at the base, persistent, marcescent, or deciduous, cylindrical leaves often with enlarged central water-storing cells, epidermal bladder cells large to much flattened and seemingly absent, usually mesomorphic but sometimes xeromorphic; **Fl** solitary or in cymes, ∅ 1–6 cm, **K** 4–5, petals pink, salmon, orange, yellow, cream, sand-coloured, greenish, or white, **K** and petals mostly connate into a short or long tube, rarely free to the base, petals filiform or narrowly lanceolate, rarely much reduced, filamentous staminodes present or absent, stamens smooth and without papillae, petals often merging into filamentous staminodes and stamens, sometimes reproductive parts concealed by the filamentous staminodes or by the inner petals, nectaries 4 or 5, shell-shaped or tubular, separate, rarely reduced, ovary semi-inferior to inferior, septate, with axile placentation; **Fr** a loculicidal hygrochastic capsule or rarely nut-like, capsule opening by means of expanding keels, usually with valve wings which may be inflexed over the valves or reflexed and fused in pairs, rarely without valve wings, usually 4 or 5 locules, rarely only 3; **S** broadly or narrowly ovate, triangular, D-, kidney-, horseshoe- or pear-shaped, black, brown, yellow or whitish, testa rough or smooth, with or without a distinct crest; **Chr** $x = 9$; **Ecol** many species in disturbed places, mainly within the winter rainfall region, but some also in the summer rainfall region; **Distr** SW Angola; SW Australia; Azores; Canary Islands; Cape Verdes; Iraq; Iran; Mauritania; Mediterranean basin; Namibia; N Africa; Palestine; EC, NC, WC, S Africa; St. Helena; Turkey; California, USA.

Note: This subfamily was first recognized by Schwantes (1947: 39) under the name *Aptenioideae*, a name invalidly published, since it was not accompanied by a Latin diagnosis. Some 15 years later Ihlenfeldt et al. (1962: 53) in their taxonomic account on the family *Mesembryanthemaceae* correctly applied the autonym rule by citing Schwantes' nomenclatural proposal as a synonym of the name **Mesembryanthemoideae.** Although not required for autonyms, they also provided a Latin diagnosis for this subfamily. When revising the suprageneric classification of the **Aizoaceae** s.str., Bittrich & Hartmann (1988: 250–251) erroneously rejected the name **Mesembryanthemoideae** Ihlenfeldt, Schwantes & Straka as invalid, but took up and validated the older name *Aptenioideae* Schwantes ex Bittrich & H.E.K. Hartmann. Nevertheless, as discussed in detail by Hartmann & Bittrich (1991), the name **Mesembryanthemoideae** Ihlenfeldt, Schwantes & Straka is in fact valid, both, as an autonym under the family name *Mesembryanthemaceae*, and as the earliest valid name for the subfamily under the family name **Aizoaceae.** By providing a Latin diagnosis and even author's names Ihlenfeldt et al. (1962) validated "their" subfamiliar name for a future non-autonymous state. Another attempt of a suprageneric classification of the *Mesembryanthemaceae* by Rappa & Camarrone (1953, 1960), based on nectary morphology as the principal taxonomic character, proved to be less satisfactory.

Key to the Genera of the Mesembryanthemoideae

1. Leaves flat, roundish or oval and decussate, connate into a disc or a cone, upper ones much smaller and free, epidermis smooth or with small mesomorphic bladder cells at the margins .... **Synaptophyllum**
- Leaves not connate into a disc or cone, epidermal bladder cells present all over, although they may be much reduced 2
2. Plants perennial with at least some internodes with persisting green succulent cortex 3
- Plants annual, biennial, or perennial without persisting green succulent cortex on the stems or geophytes . 6
3. Epidermal bladder cells alike on stems and leaves, much reduced and epidermis smooth, warty, or rarely hair-like .... **Psilocaulon**
- Epidermis of the stems with tall, cylindrical, closely packed, xeromorphic bladder cells, epidermis of the leaves with mesomorphic, loosely arranged and somewhat flattened bladder cells 4
4. Plants prostrate to decumbent ground-coverers with flat leaves or ± erect climbers with almost cylindrical or channelled leaves .... **Aptenia**
- Plants decumbent to erect small to large shrubs with almost cylindrical leaves 5
5. Leaves basally free, each leaf with a tubular, stem-clasping sheath; flower ∅ to 45 mm, petals basally connate into a short tube, filamentous staminodes present; seeds per locule 2, large, to 2.2 mm long ... **Aspazoma**
- Leaves free or connate but never with a tubular, stem clasping sheath; flower ∅ 5–25 mm, petals free, filamentous staminodes absent; seeds per locule numerous, rarely only 1, small, 1–1.5 mm long .... **Brownanthus**
6. Plants annual or biennial, the latter in some cases with small but always distinct epidermal bladder cells .... **Mesembryanthemum**
- Plants perennial but sometimes short-lived, the latter always with much flattened epidermal bladder cells . 7
7. Geophytes with much flattened epidermal bladder cells, leaves decussate and free, stigmas spreading when fertile, locules of the fruits to 10 mm deep .... **Caulipsolon**
- Plants perennial, if geophytes, then with conspicuous epidermal bladder cells, decussate leaves shortly connate at their bases, stigmas erect, locules of the fruits less than 10 mm deep 8
8. Epidermis with a thick wax layer which can easily be wiped off, bladder cells flattened 9
- Wax layer of the epidermis less conspicuous, epidermal bladder cells rarely flattened 10
9. Stems herbaceous or only weakly lignified, fruits very robust, lower part of the fruits conspicuously longer than upper part .... **Prenia**
- Stems woody, leaves strictly decussate, fruits rather brittle, lower part of the fruits about as long as upper part .... **Aridaria**
10. Old dry leaves "skeletonized", retaining their shape ± completely by means of lignified leaf veins .. **Sceletium**
- Old dry leaves sometimes becoming spiny but never "skeletonized" .... **Phyllobolus**

Ruschioideae

H. E. K. Hartmann

Ruschioideae Schwantes in Ihlenfeldt, Schwantes & Straka 1962: 54 ≡ *Ruschioideae* Schwantes 1947: 6 nomen invalid. **Typus Ruschia** Schwantes = *Anettarioideae* Rappa & Camarrone 1953: 30 (as "Anettari") nomen invalid. **Lectotypus**, here designated: **Glottiphyllum** N.E.Br. = *Caryotophoroideae* Ihlenfeldt, Schwantes & Straka in Schwantes in Jacobsen 1960: 962 nomen invalid. ≡ *Caryotophoroideae* Ihlenfeldt, Schwantes & Straka 1962: 54 **Typus Caryotophora** Leistner = *Hymenogynoideae* Schwantes 1957: 168 nomen invalid. ≡ *Hymenogynoideae* Schwantes in Ihlenfeldt, Schwantes & Straka 1962: 54 **Typus Hymenogyne** Haw. = *Lofomorfoideae* Rappa & Camarrone 1953: 30 (as "Lofomorfi") nomen invalid. **Typus** not chosen.

Annuals or perennials, shrubs to trees, climbing, creeping, prostrate, compact, or reduced to a single leaf-pair; **L** mostly trigonous to triquetrous, often with a marked hood or hump, rarely flat, basally mostly connate, in several genera almost free, with bladder cells or with a firm, xeromorphic epidermis; **Fl** in distinct **Inf** separated from the basal vegetative part of the plant, almost the entire plant representing an **Inf**, or solitary, sometimes only appearing so but in reality a perennial **Inf**, **K** 4–8, mostly 5, the lobes mostly free, petals mainly purple or yellow, both colours in all shades, also white, red, orange, or copper, filamentous staminodes absent or present, rarely concealing the stamens mostly equipped with long papillae near their bases, nectaries lophomorphic or crest-like, either separated or connate into a ring, rarely apparently absent, placenta parietal to basal, very rarely central; **Fr** mostly a hygrochastic capsule with expanding keels mostly extending into the locule and expanding sheets on the valves which can possess valve wings, covering membranes absent or present, often with additional closing devices; **S** smooth to echinate, whitish to dark brown; **Chr** $x = 9$; **Ecol** many genera in winter rainfall areas with < 300 mm precipitation p.a., a number in regions with bimodal rainfall and some in summer rainfall areas; **Distr** S parts of Australia; Botswana; Chile; Lesotho; Namibia; New Zealand; all provinces of S Africa; Swaziland; secondarily also on the Channel Islands; in S England; S Europe; S coast of the Mediterranean Sea; California, USA and probably other areas.

Note: With 112 genera this subfamily contains most of the genera of the **Aizoaceae**. The distinguishing features are the parietal to basal placentation, the crest-shaped nectaries, and probably also the free calyx lobes and petals. This subfamily inhabits the widest range of ecological niches of all subfamilies of the **Aizoaceae** and has developed the most diverse growth forms and leaf surfaces. Only in the **Ruschioideae** a firm, xeromorphic epidermis is found, probably developed in several independent lines (Ihlenfeldt & Hartmann 1982, Jürgens 1990).

Key to the Genera of Ruschioideae

1. Fruit a fleshy berry .... **Carpobrotus**
– Fruit dry **2**
2. Fruit a nut, finally disintegrating into nutlets, or breaking into mericarps **3**
– Fruit a loculicidal capsule **5**
3. Fruit woody, dispersal units nut, or nutlets **4**
– Fruit leathery, breaking into one-seeded mericarps as dispersal units .... **Hymenogyne**
4. Fruit > 10 mm in ∅, finally breaking completely into mericarps; leaves flat .... **Caryotophora**
– Fruit < 7 mm in ∅; the sclerotinized vascular bundles of the fruit remaining on the plant; leaves ± club-shaped to falcate .... **Ruschianthemum**
5. Capsule xerochastic, expanding tissue absent **6**
– Capsule hygrochastic, expanding tissue present **8**
6. Locules 5–7 .... **7**
– Locules > 8 .... **Conicosia**
7. Flat base of bottom of fruit widened, containing seeds, formed like a ring-shaped pocket, septa remaining connate with valves .... **Saphesia**
– Fruits with separate seed-pockets at half the height of the capsule, septa free from valves **Skiatophytum**
8. Expanding tissue reduced, consisting either of expanding sheets or expanding keels **9**
– Expanding tissue consisting of both, expanding sheets and keels **12**
9. Expanding keels only **10**
– Expanding sheets only **11**
10. Locules open, seeds visible, seed pockets at outer base of locules, pedicels glabrous, leaf margins with bladder cells only .... **Apatesia**
– Locules covered by septa arching over them, seeds hidden, placenta raised, but without seed pockets, pedicel and calyx with long hairs, leaf margins with long hairs .... **Carpanthea**
11. Leaves lyrate .... **Aethephyllum**
– Leaves entire .... **Dorotheanthus**
12. Capsule opening but once, rarely again, but never repeatedly, expanding keels breaking after first opening **13**
– Capsule opening repeatedly, expanding keels remaining functional **14**

13. Locules 5–6, valves free to base, valve wings broad, hard, erect ... **Stoeberia**
- Locules (4–6)–12, valves united laterally at base, valve wings absent ... **Stayneria**
14. Covering membranes present ... 15
- Covering membranes absent, sometimes narrow rims present ... 16
15. Covering membranes of constant shape, even after the seeds have been removed ... 44
- Covering membranes declining into empty locules ... 45
16. Expanding keels distinctly separated from expanding sheets, keels parallel, high, ± sharply pleated ... 19
- Expanding keels merging into expanding sheets ... 17
17. Heterophyllous shrubs ... **Mitrophyllum**
- Homophyllous plants ... 18
18. Annual herbs ... **Cleretum**
- Perennial plants ... 42
19. Calyx and leaves sticky, sand adhering to it ... **Psammophora**
- Calyx and leaves not sticky ... 20
20. Valve wings absent or as awns ... 21
- Valve wings present, ± rectangular ... 24
21. Capsule opening incompletely (valves at most erect), breaking into mericarps early, leaves ± club-shaped with apical windows ... **Frithia**
- Capsule opening completely (valves spreading) and persistent, leaves without apical windows ... 22
22. Pedicel very long (5–12 cm), flowers mostly ternate, leaves keeled, toothed, plants compact **Carruanthus**
- Pedicel shorter (–2.5 cm), flowers solitary, leaves not keeled, not toothed, plants shrubby, caespitose to decumbent or creeping ... 23
23. Shrubs, flowers opening midday ... **Jensenobotrya**
- Plants creeping or caespitose, flowers opening at night ... **Neohenricia**
24. Valve wings at most as broad as expanding keels ... 25
- Valve wings broader than expanding keels ... 26
25. Plants creeping ... **Mossia**
- Plants compact to caespitose ... 43
26. Bracteoles present (different from foliage leaves) ... 38
- Bracteoles absent (indistinguishable from foliage leaves) ... 27
27. Base of capsule with very long soft hairs, leaves hairy ... **Muiria**
- Base of capsule and leaves not velvety but glabrous, with bladder cells, or papillate ... 28
28. Flowers with a short hypanthium ... 29
- Flowers with free elements from above the ovary ... 31
29. Leaves smooth, glabrous ... **Peersia**
- Leaves finely granulate and rough ... 30
30. Plants mat-forming, if more compact, leaves trigonous in section ... **Rhinephyllum**
- Plants compact, leaves mostly dentate, if not, leaves apically broadly widened ... **Stomatium**
31. Base of capsule and leaves with bladder cells ... 32
- Base of capsule and leaves xeromorphic, glabrous ... 33
32. Plants heterophyllous, always compact ... **Oophytum**
- Plants homophyllous, caespitose to shrubby ... **Delosperma**
33. Leaves obliquely triquetrous ... **Schwantesia**
- Leaves isometric ... 34
34. Plants caespitose ... **Corpuscularia**
- Plants compact ... 35
35. Plants forming clumps above the ground ... **Dinteranthus**
- Plants sunken in the ground ... 36
36. Predominant number of locules per fruit in a population 5 or 6 ... **Lithops**
- Predominant number of locules per fruit 7 or more ... 37
37. Seeds tuberculate, tubercles distant ... **Dinteranthus**
- Seeds smooth to rugose ... **Lithops**
38. Flowers in inflorescences raised well above the plant ... 39
- Flowers solitary ... 40

39. Flowers remaining open during and after anthesis, petals stiff ..... **Nelia**
- Flowers closing over night, petals ± lax ..... **Hartmanthus**
40. Locules ≤ 7 ..... **Conophytum**
- Locules ≥ 10 41
41. Plants above ground, leaves almost semi-globose, ± appressed ..... **Namibia**
- Plants at least partly sunken in the ground, leaves triangular and pointed, spreading ..... **Nananthus**
42. Locules mostly 5, stamens > 10 ..... **Delosperma**
- Locules 4, stamens < 10 ..... **Ectotropis**
43. Capsules dark brown, top rounded ..... **Antegibbaeum**
- Capsules light brown, top flat ..... **Ruschianthus**
44. Covering membranes without distal recurving and always without any additional closing devices like rodlets, ledges or bulges at their distal ends 45
- Covering membranes with a ± distinct distal recurving, with or without additional closing devices like rodlets, ledges or bulges, if straight, always with some additional closing device 91
45. Covering membranes declining into the empty locule along a preformed edge 78
- Covering membranes straight or only slightly and gradually declining into empty locules 46
46. Plants with distinct prominent bladder-cells, glistening in the sun, often appearing scaly on herbarium material 47
- Plants rough, glabrous or hairy 63
47. Leaves flat, petiolate; plants annual 48
- Leaves terete to trigonous or semi-globose 51
48. Leaves lyrate 49
- Leaves entire 50
49. Top of capsule flat, expanding sheets only ..... **Aethephyllum**
- Top of capsule with ribs, expanding sheets merging into expanding keels ..... **Cleretum**
50. Valve wings ± broad, rectangular ..... **Cleretum**
- Valve wings tapering into awns ..... **Dorotheanthus**
51. Plants heterophyllous 52
- Plants homophyllous 57
52. Expanding keels divergent, emerging gradually from expanding sheets 53
- Expanding keels parallel, sheets separate from expanding keels ..... **Oophytum**
53. Plants compact and low (height < 4 cm) ..... **Diplosoma**
- Plants shrubby, if compact, height > 5 cm 54
54. Basal parts of old leaf sheaths sclerotic, completely enveloping the stem ..... **Monilaria**
- Old leaf-sheaths papery-pergamentaceous 55
55. Flowers basally enclosed by connate hypsophylls forming a cup ..... **Dicrocaulon**
- Flowers exserted above leaves, no cups 56
56. Both leaf-pairs of a season connate for > 30% ..... **Meyerophytum**
- One leaf-pair of a season connate for < 25%, spreading, the other for > 50%, erect ..... **Mitrophyllum**
57. Leaves with an apical diadem ..... **Trichodiadema**
- Leaves without an apical diadem 58
58. Bladder cells on leaves thin-walled, collapsing when dry, surface glittering 59
- Bladder cells thick-walled, constant in shape, surface dull 62
59. Surfaces of primary internodes with prominent bladder cells ..... **Drosanthemum**
- Surfaces of primary internodes glabrous 60
60. Leaves fat-digitiform, length > 4 cm ..... **Jacobsenia**
- Leaves subterete to trigonous, length < 3 cm 61
61. Locules 5 ..... **Mestoklema**
- Locules 6–11 ..... **Malephora**
62. Locules ≥ 8 ..... **Jacobsenia**
- Locules 6 ..... **Drosanthemum**
63. Flowers white over pink to purple 64
- Flowers yellow 72
64. Fruits brown, hard 65
- Fruits whitish to light brown, ± soft 69

65. Fruits in raised inflorescences .... **Nelia**
– Fruits solitary **66**
66. Plants shrubby .... **Polymita**
– Plants compact to caespitose **67**
67. Leaves digitiform, more than 3 × as long as broad, valve wings broad .... **Antegibbaeum**
– Leaves ± trigonous, with convex sides, plants compact, valve wings narrow to absent **68**
68. Plants with thick tap-roots .... **Ebracteola**
– Plants with thinner roots **90**
69. Erect shrubs **70**
– Compact, creeping, or caespitose plants **71**
70. Calyx lobes 4 .... **Juttadinteria**
– Calyx lobes 5 .... **Hartmanthus**
71. Flowers bracteolate .... **Hammeria**
– Flowers ebracteolate **77**
72. Closing body bipartite .... **Rhombophyllum**
– Closing body absent or present as a single body **73**
73. Closing body a broad sill .... **Bergeranthus**
– Closing body absent or present as a round body **74**
74. Fruits on short stalks, closing body big and round .... **Bijlia**
– Fruits on long stalks, closing body absent or small **75**
75. At least some leaves with teeth along keel and/or margins .... **Chasmatophyllum**
– All leaves without teeth **76**
76. Leaves soft, fruits whitish, locules always with false septa .... **Malephora**
– Leaves firm, fruits brown, locules without false septa .... **Hereroa**
77. Nectary a crenulate ring .... **Disphyma**
– Nectary as 6 glands .... **Gibbaeum**
78. Valve wings absent, rims of valves high .... **Ebracteola**
– Valve wings present, at least basally very broad **79**
79. Capsules brown **80**
– Capsules whitish-grey **83**
80. Fruits persisting on ± long stalks .... **Prepodesma**
– Fruits breaking off their stalks, dispersed as tumble fruits **81**
81. Leaves velvety .... **Deilanthe**
– Leaves smooth to rough from warts **82**
82. Leaves sharply pointed and triangular in section, surface finely warty .... **Rabiea**
– Leaves flattened or rounded at tips and margins, surface with prominent warts .... **Aloinopsis**
83. Closing body present **84**
– Closing body absent **85**
84. Base of capsule papillate, leaves dotted .... **Ihlenfeldtia**
– Base of capsule smooth, leaves smooth .... **Tanquana**
85. Base of capsule papillate; leaves papillate, keeled, toothed .... **Vanheerdea**
– Base of capsule smooth, leaves smooth or warty, never toothed **86**
86. Locules 5–7, rarely in single fruits 8 .... **87**
– Locules ≥ 8 in all fruits **89**
87. Locules very shallow, covering membranes therefore held up by the funicles; flowers appearing after the leaf-pairs, thus seemingly lateral .... **Didymaotus**
– Locules deeper, covering membranes bending into empty locule; flowers distinctly terminal **88**
88. Leaves warty, plants mostly sunken in the ground .... **Titanopsis**
– Leaves deltoid, keeled, gibbose, plants above ground .... **Lapidaria**
89. Shrubs .... **Zeuktophyllum**
– Compact plants .... **Dracophilus**
90. Top of capsule convex from high valve rims .... **Cerochlamys**
– Top of capsule flat .... **Vlokia**
91. Closing body absent **92**
– Closing body present, sometimes as a broad sill **109**

92. Valve wings absent or narrow, i.e ≤ width of expanding keels 99
- Valve wings, at least at base, much broader than expanding keels 93
93. Valve wings rectangular 95
- Valve wings tapering into awns towards the tip of the valve 94
94. Capsules multilocular, plants compact .... **Pleiospilos**
- Capsules 5-locular, if multilocular, plants shrubby .... **Erepsia**
95. Valve wings in open capsules stiff, stretching over locules and not moving outwards with valves .. **Faucaria**
- Valve wings in open capsules flexible, moving outwards with the valve 96
96. Seeds echinate .... **Braunsia**
- Seeds with low papillae 97
97. Plants compact .... **Orthopterum**
- Plants shrubby, rarely caespitose or creeping 98
98. Filaments broadened and connate at base .... **Circandra**
- Filaments filiform, free .... **Lampranthus**
99. Leaves xeromorphic 102
- Leaves mesomorphic 100
100. Leaves flat, petiolate, plants annual .... **Carpanthea**
- Leaves terete to trigonous, plants perennial 101
101. Flowers yellow .... **Scopelogena**
- Flowers white to purple .... **Oscularia**
102. Seeds echinate .... **Namaquanthus**
- Seeds smooth to moderately papillate 103
103. Surface of ovary concave, forming a hypanthium in the flower .... **Erepsia**
- Surface of ovary flat to raised 104
104. Leaves connate into a sheath which is continuous with the cortex of the preceding internode 105
- Leaves ± free or only somewhat fused 107
105. Free parts of leaves < 10 mm long .... **Smicrostigma**
- Free parts of leaves 20–30 mm long 106
106. Fruits reddish-brown, not closing again completely after the first opening, shrubs to 2 m high ... **Stayneria**
- Fruits with a white cover on top, closing completely again after the first opening, shrubs to c. 40 cm high .... **Zeuktophyllum**
107. Flowers white to pink, leaves grey from an extremely thick wax cover .... **Wooleya**
- Flowers purple or magenta, leaves bright green 108
108. Leaves club-shaped, apically rounded, flowers ebracteolate, 4 calyx lobes .... **Enarganthe**
- Leaves with a marked tooth or mucro at the tip, flowers with bracteoles, 5 calyx lobes ... **Esterhuysenia**
109. Closing body bipartite 110
- Closing body not divided 111
110. Closing body U-shaped, leaves mesomorphic, plants creeping .... **Disphyma**
- Closing body of two ± flat plates, leaves xeromorphic, plants compact to shrubby ... **Rhombophyllum**
111. Leaves soft, mesomorphic with bladder cells 112
- Leaves hard, xeromorphic without bladder cells 114
112. Plants annual, leaves flat .... **Dorotheanthus**
- Plants perennial 113
113. Covering membranes ± straight, plants heterophyllous .... **Mitrophyllum**
- Covering membranes undulate, leaves anisophyllous in a pair .... **Glottiphyllum**
114. Expanding keel clearly separate from expanding sheet 115
- Expanding keel merging into expanding sheet .... **Bijlia**
115. Valve wings broad (at least at base) 116
- Valve wings absent or narrow (≤ width of expanding keels) 127
116. Valve wings tapering from a broad base 125
- Valve wings ± rectangular 117
117. Closing bodies not larger than seeds, leaving space at the sides 118
- Closing bodies larger than seeds, ± blocking the distal opening of the locule 121
118. Capsule pentamerous 120
- Capsule multilocular 119

119. Capsules dispersed as tumble-fruits, leaves with apical window, plants under ground .... **Fenestraria**
- Capsules persistent on pedicels, leaves ± fusiform without a window, plants above ground ..... **Jordaaniella**
120. Top of capsule with remains of hard bases of stigmas in the centre .... **Arenifera**
- Top of capsule flat in the centre .... **Amphibolia**
121. Base of capsule velvety or rough from papillae .... **Cheiridopsis**
- Base of capsule glabrous **122**
122. First leaves of side branches connate > 50% .... **Vanzijlia**
- First leaves of side branches hardly connate like all others (seedlings occasionally with connate leaf pairs, though) **123**
123. Capsules with closing bulges .... **Cephalophyllum**
- Capsules with closing rodlets **124**
124. Top of capsule turret-shaped; procumbent branches from compact centre not rooting .... **Hallianthus**
- Top of capsule rounded, but not turret-shaped, erect to ascending, or saltatory and rooting ...... **Leipoldtia**
125. 5 locules **128**
- Many locules **126**
126. Leaves sharply triquetrous, grey .... **Machairophyllum**
- Leaves trigonous to semi-ovate, gibbose, dark green .... **Pleiospilos**
127. Closing body large, completely locking the locule **129**
- Closing body smaller, leaving some space at the sides **135**
128. Leaves almost free, appearing inflated .... **Eberlanzia**
- Leaves basally connate, triangular .... **Polymita**
129. Capsule pentamerous **130**
- Capsule multilocular **132**
130. Flowers yellow .... **Bergeranthus**
- Flowers pink to purple **131**
131. Leaves awl-shaped, the keel visible as a line in turgid conditions .... **Marlothistella**
- Leaves never awl-shaped, a keel always prominently visible .... **Antimima**
132. Leaves glabrous **133**
- Leaves velvety, hairy, or rough **134**
133. Flowers with a hypanthium; capsule 12–24 locular .... **Argyroderma**
- Flowers without a hypanthium; capsule 8-locular .... **Octopoma**
134. Leaves and base of capsule velvety .... **Odontophorus**
- Leaves and base of capsule rough .... **Cheiridopsis**
135. Flowers with ± 2 000 elements .... **Cylindrophyllum**
- Flowers with up to 300 elements **136**
136. Flowers yellow **137**
- Flowers white to purple **138**
137. Leaves sharply triquetrous, grey .... **Machairophyllum**
- Leaves ± scimitar-shaped with a rounded keel .... **Hereroa**
138. Plants compact, caespitose or creeping **139**
- Shrubs **140**
139. Petals white or purple, never striped .... **Khadia**
- Petals with a central, longitudinal deep purple stripe, i.e. vittate .... **Acrodon**
140. Seeds echinate or markedly papillate .... **Astridia**
- Seeds smooth to rough **141**
141. Flowers in annually enlarged, persisting inflorescences .... **Ottosonderia**
- Flowers solitary or in annually formed and ripening inflorescences, new ones formed every year **142**
142. Fruits mostly 5, rarely 6 locules **143**
- Fruits with mostly 8, rarely up to 10 locules **145**
143. Valves without wings .... **Ruschia**
- Valves with narrow wings broadest in their middle **144**
144. Stems reddish-brown .... **Ruschia**
- Stems whitish-grey .... **Eberlanzia**
145. Flowers white .... **Octopoma**
- Flowers pink to purple .... **Schlechteranthus**

Sesuvioideae

H. E. K. Hartmann

Sesuvioideae Lindley 1853: 527; Adamson 1962: 243–253; Bittrich & Hartmann 1988: 239–254 ≡ *Sesuviaceae* Horaninow 1834: 83 **Typus Sesuvium** L.

Prostrate to erect perennials or annuals; **L** flat to terete, ± unequal, with or without stipules, often with membranous sheaths at the base of the petiole, leaf surface with ± prominent bladder cells, hairy or smooth; **Inf** terminal but apparently axillary through sympodial branching, **Fl** solitary or in clusters, rarely in ± loose dichasia, perianth with subapical appendages with unifacial tips, rarely without, perigynous, 1–5 styles, ovules 2–many; **Fr** a circumscissile capsule; **S** completely enveloped by an aril; **Chr** $x = 8$; **Ecol** coastal to inland in mostly moister climates than the other subfamilies; **Distr** worldwide in the tropics and subtropics.

Note: Members of the subfamily are probably best known from tropical coasts in brackish soil, sometimes on rocks, often near mangrove, where the plants can form conspicuous ground covers, red to purple in many cases. Other members are found in disturbed areas near roads, being typical weeds.

Key to the Genera of the Subfamily Sesuvioideae

1. Style 1 ..... **Trianthema**
- Styles 2–5 2
2. Septa complete in ripe fruits; seeds brown to blackish 3
- Septa incomplete or absent in ripe fruits; seeds light coloured ..... **Cypselea**
3. Operculum of capsule breaking into 2 parts, 4 seeds per fruit ..... **Zaleya**
- Operculum of capsule in one piece; numerous seeds per fruit ..... **Sesuvium**

Tetragonioideae

H. E. K. Hartmann

Tetragonioideae Lindley 1853: 527; Adamson 1955: 110–154; Bittrich & Hartmann 1988: 239–254 ≡ *Tetragoniaceae* Link 1831: 17 **Typus Tetragonia** L.

Herbaceous to woody shrubs, rarely annual herbs; **L** mostly alternate, petiolate, entire, flat, bladder hairs present; **Fl** solitary or in groups, sometimes umbellate or in cymose panicles, sessile to shortly pedicellate, **Inf** often andromonoecious, perianth segments (3–)4(–5), inferior ovary with (1–)3–9 locules with one pendulous ovule on a short funicle each; **Fr** woody, indehiscent, winged, ridged, spiny, or horned, rarely drupaceous; **Chr** $x = 8$; **Ecol** worldwide in tropical and subtropical climates with a higher number of species in winter rainfall; **Distr** in the tropics of the Old and New World and Australia.

Note: Bittrich (1990) places the subfamily very close to **Aizoaceae** subf. **Aizooideae**, based on the possession of bladder hairs of both taxa, which is a synapomorphy for the group within the family **Aizoaceae**.

This subfamily contains the only species of **Aizoaceae** that is grown as a vegetable to some extent in suitable climates, sometimes even in moderate climates: **Tetragonia tetragonioides**. Several species are good fodder in South Africa, their frequency in the wild being understood as a measurement of quality of farming.

Key to the Genera of the Subfamily Tetragonioideae

1. Fruit derived from a single ovary, winged, ridged, or horned, perianth with free or only basally connate lobes ..... **Tetragonia**
- Fruit compound, derived from a complete inflorescence including the bracts and bracteoles; spiny; perianth connate for ± half its length ..... **Tribulocarpus**

Ranks Below Subfamily

H. E. K. Hartmann

Below the rank of subfamily, five tribes have been validly published in the **Ruschioideae** (Ihlenfeldt, Schwantes & Straka 1962: 54–55). Characters used for the delimitations are very restricted, as the descriptions below, based only on the original texts, show, and a thorough study should be undertaken before a soundly based taxonomic subdivision of the subfamilies can be reached.

Tribus **Ruschieae** Schwantes ex Ihlenfeldt, Schwantes & Straka 1962: 54 **Typus Ruschia.**

Fl isomerous or pleiomerous; **Fr** a hygrochastic loculicidal capsule; **S** numerous in all locules.

Note: Most genera of the subfamily **Ruschioideae** answer this very general description.

Tribus **Apatesieae** Schwantes ex Ihlenfeldt, Schwantes & Straka 1962: 55 **Typus Apatesia.**

Fl pleiomerous gynoecium; **Fr** abaxial seed pockets formed containing 1–2 seeds each, hygrochastic loculicidal capsules to schizocarpous or xerochastic fruits.

Note: The genera **Apatesia, Conicosia,** and *Herrea* have been included in this tribe.

Tribus **Skiatophyteae** (Straka) Ihlenfeldt in Ihlenfeldt, Schwantes & Straka 1962: 55 ≡ *Skiatophytinae* Straka 1955: 161 nomen invalid. **Typus Skiatophytum.**

Annual to biennial herbs; **L** flat; **Fl** white, gynoecium isomerous, false septa present; **Fr** pockets developed from false septa, each containing a single seed, expanding keels reduced in function opening once only.

Note: The description of the subtribe *Skiatophytinae* by Straka was invalid because it was in German only.

Tribus **Saphesieae** (Schwantes) Ihlenfeldt in Ihlenfeldt, Schwantes & Straka 1962: 55 ≡ *Saphesieae* Ihlenfeldt in Schwantes 1960: 954 nomen invalid. ≡ *Saphesiinae* Schwantes 1960: 954 nomen nudum **T Saphesia.**

Fl gynoecium isomerous; **Fr** few locules with incomplete seedpockets, xerochastic; **S** many per locule.

Note: The tribal name **Saphesieae** seems to appear first in Schwantes in Jacobsen 1960: 954 in the synonymy of the name *Saphesieae* (Schwantes) Ihlenfeldt, which itself is an invalid name because it has an English description only. The validation was undertaken in 1962, as given above.

Tribus **Carpobroteae** Schwantes in Ihlenfeldt, Schwantes & Straka 1962: 55 **Typus Carpobrotus.**

Fr a pleiomerous, juicy berry, edible; **S** many in all locules.

Note: At the time of description, only the genus **Carpobrotus** was included in this tribe.

Below the rank of tribes, Schwantes described numerous subtribes (1947: 34–40), 22 of them in the tribe *Ruschieae*. All new names in the cited publication lack Latin descriptions and are therefore invalidly published. In some cases, when revisions were presented, Latin descriptions were provided, e.g. **Dorotheanthinae** Schwantes ex Ihlenfeldt & Struck (1987: 433) with the type genus **Dorotheanthus; Leipoldtiinae** H.E.K.Hartmann BB 1983: 50, type genus **Leipoldtia; Ruschiinae** Dehn 1993: 185, type genus **Ruschia.** In view of the uncertainties considering the circumscriptions of monophyletic groups at and below subfamilial level, Hartmann (1991: 130–132) decided to present groupings of genera without any ranking in the Mesembryanthema. Albeit in the meantime several changes to these groups are necessary, no new lists are presented here, due to lack of data comparable over all genera, and readers are referred to Hartmann (1993: 48–49) for the latest update of the groupings.

Faucaria Ruschioideae

L. E. Groen & H.E.K. Hartmann

Faucaria Schwantes ZSK 1926: 176; Groen & van der Maesen 1999: 35–58 **Typus F. tigrina** (Haw.) Schwantes **Etym** La *faux, fauces,* throat or the entrance thereof, here describing the toothed leaves formed like the open jaws of an animal.

Compact, in some species developing short, thick rhizomes with age, others with a thickish tap-root; **L** trigonous, keel marked by a white, cartilaginous fringe, margins mostly with teeth, rarely smooth, the lower leaf surface forming a chin to different extension, epidermis smooth, rarely with elevations, with layers of crystal sand in certain parts of the leaves, mostly over the subhypodermal tannin idioblasts, but in many cases also extending further, either as white dots or flecks, more rarely as lines connecting the dots, often with cuticular folds, but also with wax as a ± homogeneous cover; **Fl** solitary, ebracteolate, sessile between the leaves, **K** 5, petals yellow to golden, filamentous staminodes absent, filaments yellow, often basally white, stigmas as long as the stamens, nectary glands 5; **Fr** pedicels up to 3.5 mm *l*, laterally compressed, base broadly bell-shaped, often apically with ring-shaped impression above which the uppermost part of the connate calyx and the calyx lobes widen again, top either nearly semi-globose and with a central dip, or nearly conical with rising valves forming a central peak, covering membranes reduced along the upper edges of the septa, valves with very broad valve wings opening but once into an erect position because the hard, xeric valve wings standing out radially into the centre prevent a proper closing of the valves when drying again, 5(–6) locules; **S** pear-shaped to tetrahedral, dark brown, papillae knobby with thicker heads, 1.4–1.7 mm *l*, 0.9–1.4 mm *b*; **Ecol** in stony ground, rarely in loam, often under shrubs, > 200 mm rainfall in March and November or at all seasons; **Distr** EC, W districts of WC, S Africa.

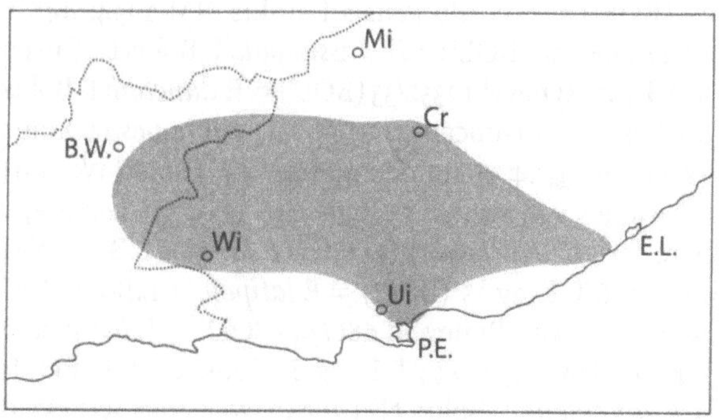

Distribution of **Faucaria** Schwantes

Note: The peculiar fruits in combination with the firm, smooth leaves characterize the genus beyond doubt. The genus **Orthopterum,** similar in some respect, differs by the possession of widely opening fruits closing completely again when drying. In contrast to Groen & van der Maesen (1999), **F. britteniae, F. felina,** and **F. tuberculata** are understood to represent separate species, based on the results of the cladogram in Groen & van der Maesen (1999: 55).

Key to the Species of Faucaria

1. Margins and keel with conspicuous white cartilaginous edges, teeth also cartilaginous, if present .... **F. bosscheana**
– Margins at most with cartilagineous teeth and adjacent sections, keel regularly cartilaginous, teeth never cartilaginous but composed of thick protrusions with papillae **2**
2. Top of capsule with a raised centre in most fruits of a population **3**
– Top of capsule flat or with a central dip in most fruits of a population **5**
3. Leaves with several to many long teeth **4**
– Leaves with few, distant, short teeth (< 1 mm long) or none at all .... **F. subintegra**
4. Keel at right angle to the upper leaf surface .... **F. britteniae**
– Keel at an angle < 90° to the upper leaf surface **7**
5. Leaves spreading, teeth tiny (< 1 mm long) .... **F. nemorosa**
– Leaves ± erect, teeth long (> 3 mm long) **6**
6. Each margin with 8–10 teeth .... **F. tigrina**
– Each margin with up to 5 teeth .... **F. gratiae**
7. Leaves with tubercles on upper surface .... **F. tuberculosa**
– Leaves without tubercles on upper leaf surface .... **F. felina**

F. acutipetala = **F. felina**
F. albidens = **F. bosscheana**

F. bosscheana (A.Berger) Schwantes ZSK 1926: 177 ≡ *M. bosscheanum* A.Berger 1908: 269 **Lectotype**, here designated (not neotype, as in Groen & van der Maesen 1999): Wildeman 1908: t. 230 = *F. albidens* N.E.Br. JB 1927: 349 **LT** Frith s.n. (K!) = *F. haagei* Tischer MDK 1929: 20 ≡ *F. bosscheana* (A.Berger) Schwantes var. *haagei* (Tischer) H.Jacobsen in Jacobsen & Rowley 1955: 80 **LT** Tischer MDK 1929: f on p 21 = *F. kendrewensis* L.Bolus NM 1934: 469 **T** van der Berg 1619/34 (BOL!) = *F. paucidens* N.E.Br. in Brown et al. 1931: 210 **Lectotype**, here designated: Brown et al. 1931: f 100 = *F. peersii* L.Bolus NM 1937: 108 **T** Peers 21979 (BOL!).

Compact; **L** light greyish-green, the keel and the margins conspicuously white from cartilaginous tissue, the up to 4 teeth per margin formed by the same tissue, but most populations with many edentate leaves, margins and keel with almost semi-globose cells, faces with ± flat cells, in all epidermal cells the outer walls with a continuous layer of crystal sand, wax cover rugose; **Fl** ∅ 30–35 mm, 85–100 petals, 175–190 stamens, glands green; **Fr** base with a subapical constriction, top rather low, the valves bending centrally forming a dip there, 7–11 mm *l*, 8–10 mm ∅; **S** 1.45–1.65 mm *l*, 1.2–1.35 mm *b*; **Ecol** in crevices of rocks or between stones on slopes, often on sandstone, in open vegetation types, 200–300 mm rainfall p.a. mainly in March and November; **Distr** Aberdeen, Jansenville, Graaff Reinet, Steytlerville, Willowmore, EC, Beaufort West, Prince Albert, WC, S Africa.

Note: In teeth number, the species varies considerably, but the teeth are always formed by sharp cartilaginous protrusions, thus permitting, together with the white margins and keel, an instant identification of the species.

F. bosscheana var. *haagei* = **F. bosscheana**

F. britteniae L.Bolus SAG 1933: 22 ≡ *F. felina* (Weston) Schwantes ssp. *britteniae* (L.Bolus) L.E.Groen in Groen & van der Maesen 1999: 41 **T** L.Britten 5395 (BOL!) = *F. coronata* L.Bolus NM 1934: 446 **T** G.V.Britten 34 (BOL!) = *F. grandis* L.Bolus NM 1938: 125 **T** G.V.Britten 21986 (BOL!) = *F. smithii* L.Bolus NM 1936: 32 **T** Smith 90 (BOL!) = *F. speciosa* L.Bolus NM 1934: 445 (BOL!) **T** L.Britten & Harcourt-Wood 691/34 (BOL!).

Compact, almost semi-globose with age; **L** keel sharp, white, forming a pronounced chin in almost rectangular arrangement as seen from the side, margins with 6–13 teeth 2.5–5 mm *l*, consisting of a broad, white base and a long, sharp awn-shaped end with elevations and distant papillae, epidermis with well developed cuticular folds covered by a ± continuous wax layer, partially greyish-green, **L** 30–55 mm *l*, 12–33 mm *b*; **Fl** 110–160 petals, 260–350 stamens, glands green to brown; **Fr** base very broadly bell-shaped, top with a high centre, 9–12.5 mm *l*, 9.5–13 mm ∅; **S** 1.4–1.55 mm *l*, 0.9–1.2 mm *b*; **Ecol** in Valley Bushveld and drier grass veld, in loamy flats or on slopes among stones, 350–700 mm rainfall p.a.; **Distr** Albany, Fort Beaufort, EC, S Africa.

Note: **F. britteniae** differs from **F. felina** mainly in the basally broader and somewhat stouter leaves, in the keel meeting the upper leaf surface nearly at a right angle at the apex, and in the very numerous small dots, absent in **F. felina**.

F. candida = **F. felina**
F. coronata = **F. britteniae**
F. cradockensis = **F. felina**
F. crassisepala = **F. felina**
F. duncanii = **F. felina**

F. felina (Weston) Schwantes ZSK 1926: 177 (see **photo**) ≡ *M. felinum* Weston 1770: 172 ≡ *M. ringens* L. var. β *felinum* L. 1753: 487 **LT** Dillenius 1732: t. 187 f 230 = *F. acutipetala* L.Bolus NM 1934: 443 **T** van Ryneveld 684/34 (BOL!) = *F. candida* L.Bolus NM 1937: 105 **T** Mclane 651/37 (BOL!) = *F. cradockensis* L.Bolus NM 1934: 448 **T** Rogers 687/34 (BOL!) = *F. crassisepala* L.Bolus NM 1934: 442 **T** van Ryneveld 2352/33 (BOL!) = *F. duncanii* L.Bolus SAG 1933: 22 **T** Duncan 19645 (BOL!) = *F. jamesii* L.Bolus ex Tischer 1934: 99 nomen nudum ≡ *F. felina* (Weston) Schwantes var. *jamesii* L.Bolus NM 1933: 395 Lectotype here designated: James 7 (BOL!) = *F. kingiae* L.Bolus NM 1937: 106 **T** King 15 (BOL!) = *F. latipetala* L.Bolus NM 1934: 443 **T** van Ryneveld 686/34 (BOL!) = *F. laxipetala* L.Bolus NM 1934: 444 **T** Long & Cook 286/33 (BOL!) = *F. longidens* L.Bolus NM 1934: 470 **T** van Ryneveld 2355/33 (BOL!) = *F. longifolia* L.Bolus NM 1934: 471 **T** van Ryneveld 690/34 (BOL!) = *F. lupina* (Haw.) Schwantes ZSK 1926: 177 ≡ *M. lupinum* Haw. 1824: 111 **LT** drawing of a plant sent by Bowie, 1823 (K!) = *F. militaris* Tischer MDK 1932: 280 **LT** Tischer MDK 1932: f on p. 280 = *F. montana* L.Bolus NM 1934: 448 **T** James 685/34 (BOL!) = *F. multidens* L.Bolus NM 1934: 450 **T** van Ryneveld 1412/33 (BOL!) = *F. multidens* L.Bolus var. *paardepoortensis* L.Bolus NM 1937: 107 **T** King 4 (BOL!) = *F. plana* L.Bolus NM 1937: 107 **T** without collector 21978 (BOL!) = *F. ryneveldiae* L.Bolus NM 1934: 446 **T** van Ryneveld 688/34 (BOL!) = *F. uniondalensis* L.Bolus NM 1937: 106 **T** Markoetter 10662 (BOL!).

Compact, developing very short, thick rhizomes with age; **L** slender, spreading, margins with 3–8 teeth in the upper half, rarely more or less, teeth from a very broad base rising from a high area on the leaf, between these raised tooth-bases often distinct furrows visible, teeth ending in papillate awns of 0.5 to 3 mm length, faces of leaves smooth to warty from elevations above tannin idioblasts, dark green, keel in an oblique angle with the tip of the leaf, white, several rows of elevations marking the white band, rows of papillae at the actual edge of the keel, epidermis with cuticular folds, wax cover thin, crystals in the outer walls in certain areas resulting in white dots, up to 70 mm *l*, 5–20 mm *b*; **Fl** 100–135 yellow, rarely white, petals, 210–320 stamens, glands green to brown; **Fr** base broadly bell-shaped, top with the valves raised in the centre in most fruits of one population, 10–16 mm *l*, 9–11 mm ∅; **S** 1.4–1.55 mm *l*, 1–1.3 mm *b*; **Ecol** mostly in fine grained soils in grassy or shrubby vegetation types, rainfall 200–700 mm p.a., mainly in March and November; **Distr** W districts of EC c. to 26° E, E districts of WC, S Africa.

Note: The lower number of teeth separates the species from **F. tigrina**, which also has more erect leaves. **F. felina** is most similar to **F. britteniae**, which grows further east. Although the variation in **F. felina** is great, as is reflected by the numerous synonyms, the species can be identified by the slender, spreading leaves with an oblique keel, never at a right angle to the upper leaf surface at the apex, the teeth only in the upper part of the leaf, and the high centre of the top of the capsule, if these features are taken in combination. This species and **F. bosscheana** occur in a wide area together, but they never grow sympatrically. The name *F. jamesii* occurs in lists of seeds sold by Winter, Frankfurt, and Bloßfeld, Potsdam, cited by Tischer (1934: 99); in the lists, the name is given without an author in the first place, with an author in the second, and Tischer does not refer to *F. felina* var. *jamesii*. The synonymy is established here following practice in cultivation, where often only the last epithet is used.

F. felina ssp. *britteniae* ≡ **F. britteniae**
F. felina var. *jamesii* = **F. felina**
F. felina ssp. *tuberculosa* ≡ **F. tuberculosa**
F. grandis = **F. britteniae**

F. gratiae L.Bolus NM 1933: 394 (see **photo**) **T** G.V.Britten 35 (BOL!) = *F. hooleae* L.Bolus NM 1934: 447 **T** Hoole & G.V.Britten 684/34 (BOL!).

Compact; **L** ± erect, spreading only when pressed by the subsequent leaf, all leaf faces with rough white dots, keel white, margins white throughout or with white bases of 2–5 teeth, which develop into awns up to 4 mm *l*, surfaces often dark green or purplish, 20–25 mm *l*, 10–20 mm *b*; **Fl** 60–85 petals, 110–140 stamens, glands dark brown to orange; **Fr** base bell-shaped, top rather low, 7–9 mm *l*, 6–9 mm ∅; **S** 1.2–1.3 mm *l*, 0.85–1 mm *b*; **Ecol** in shallow pans on sandstone plates; **Distr** Albany, EC, S Africa.

Note: The species resembles **F. tigrina** when found in habitat because both species keep the leaves of a pair rather closed and erect. **F. gratiae** differs from the former in the much smaller seeds and fruits as well as in fewer teeth restricted to the upper half of the leaf, the absence of confluent flecks, and the distinctly lower number of elements in the androecium. In the white keel and margins, **F. gratiae** resembles **F. bosscheana**.

F. haagei = **F. bosscheana**
F. hooleae = **F. gratiae**
F. jamesii = **F. felina**
F. kendrewensis = **F. bosscheana**
F. kingiae = **F. felina**
F. latipetala = **F. felina**
F. laxipetala = **F. felina**
F. longidens = **F. felina**
F. longifolia = **F. felina**
F. lupina = **F. felina**
F. militaris = **F. felina**
F. montana = **F. felina**
F. multidens = **F. felina**
F. multidens var. *paardepoortensis* = **F. felina**

F. nemorosa L.Bolus ex L.E.Groen 1994: 41 (see **photo**) ≡ *F. nemorosa* L.Bolus nomen nudum in schedis **T** Archibald 960 (BOL!).

Compact to caespitose, a thickish hypocotyl present, branches to 15 cm *l*, 7–13 mm ∅; **L** very slender, keel asymmetrically placed, margins with 2–4 teeth mostly, which are < 1 mm *l*, often toothless, epidermis greyish-green, **L** 25–45 mm *l*, 9–14 mm *b*; **Fl** petals 70–90, 20–25 mm *l*, 2–2.3 mm *b*, c. 200 stamens, nectary glands olive-green; **Fr** 4–9.5 mm ∅; **S** 1.3–1.5 mm *l*, 0.9–1 mm *b*; **Ecol** in high shrubby vegetation, always under shrubs; **Distr** Albany, EC, S Africa.

Note: The species has been described after three herbarium sheets at BOL and living material at Kirstenbosch; it grows sympatrically with **F. felina**.

F. paucidens = **F. bosscheana**
F. peersii = **F. bosscheana**
F. plana = **F. felina**
F. ryneveldiae = **F. felina**
F. smithii = **F. britteniae**
F. speciosa = **F. britteniae**
F. subindurata = **F. subintegra**

F. subintegra L.Bolus NM 1934: 449 (see **photo**) **T** Brownlee 1046/33 (BOL!) = *F. subindurata* L.Bolus NM 1934: 444 **T** Smith 102 (BOL!).

Compact to caespitose plants; **L** spreading, smooth, margins mostly with 1–3 fine teeth on protrusions, green to yellowish, white dots hardly visible, 20–35 mm *l*, 10–15 mm *b*; **Fl** petals 70–83, stamens 190–250, nectary glands dark green to brown; **Fr** top broadly conical, the highest point in the centre, 7–10 mm *l*, 6–12 mm ∅; **S** 1.1–1.4 mm *l*, 1–1.1 mm *b*; **Ecol** in shallow pockets on sandstone, open patches in otherwise dense, high Bushveld, 500–900 mm rainfall p.a.; **Distr** East London, Transkei, EC, S Africa.

Note: Like **F. bosscheana**, this species has few or no teeth, but it lacks the conspicuous white keel and margins of **F. bosscheana**. The leaves are very similar to those of **F. nemorosa** and **Orthopterum coeganum**. The names **F. subintegra** and *F. subindurata* were published on the same date, **F. subintegra** is considered more apt for the plants in question.

F. tigrina (Haw.) Schwantes ZSK 1926: 177 (see **photo**) ≡ *M. tigrinum* Haw. 1795: 164 **NT** Edwards 1818 (not 1817): t. 260 = *F. tigrina* (Haw.) Schwantes forma *splendens* H.Jacobsen in Jacobsen & Rowley: 81 ≡ *F. tigrina* (Haw.) Schwantes var. *splendens* (H.Jacobsen) H.Jacobsen 1981: 456 nomen invalid. **T?** = *F. tigrina superba* C.Weber 1968: 12 nomen illeg.

Compact to semi-globose with age; **L** ± erect, spreading only when pressed by new leaves in the centre, margins with mostly 9–10 teeth with lateral papillae along the entire side and ending in papillate awns, papillae pointing downwards, faces of all sides roughish, densely covered with white dots or flecks, dark green and often purplish suffused, epidermis with cuticular folds, **L** 30–40 mm *l*, 15–20 mm *b*, appearing short and stout; **Fl** 100–120 petals yellow, rarely red, c. 270 stamens, glands dark brown; **Fr** top flattened to slightly raised, ∅ 9–16 mm, easily detached; **S** 1.1–1.5 mm *l*, 1–1.25 mm *b*; **Ecol** on shallow sandstone at the edge of scrub, c. 700 mm rainfall p.a.; **Distr** Albany, EC, S Africa.

Note: The white dots join up, forming lines running towards the margins and ending in the white bases of the teeth often, a remarkable sight. The white keel remains inconspicuous.

F. tigrina forma *splendens* = **F. tigrina**
F. tigrina var. *splendens* = **F. tigrina**
F. tigrina superba = **F. tigrina**

F. tuberculosa (Rolfe) Schwantes ZSK 1926: 177 ≡ *M. tuberculosum* Rolfe CBM 1916: t. 8674 B ≡ *F. felina* (L.) Schwantes ssp. *tuberculosa* (Rolfe) L.E.Groen in Groen & van der Haesen 1999: 42 **T** Rolfe CBM 1916: t. 8674 B.

Compact, branched or with a single leaf-pair only; **L** 4–8 per branch, keel at an oblique angle with the tip of the leaf, 3–6 short, conical whitish teeth in the upper halves, awns stout and to 2 mm *l* or absent, tooth-like tubercles usually present on the upper surface, epidermis with cuticular folds, **L** to 50 mm *l*, 15–20 mm *b*; **Fl** 100–135 golden yellow petals, 200–220 stamens, nectary glands large, yellow-brown; **Fr** as in **F. felina**; **Ecol** in grassland, c. 500 mm rainfall p.a., mainly in November to April; **Distr** W districts of EC, S Africa.

Note: The differences in leaf characters deserve the recognition of a separation of the taxon from the somewhat similar species **F. felina**. Most tuberculate plants in cultivation are hybrids, the true species is very rare in natural habitats.

F. uniondalensis = **F. felina**

Fenestraria Ruschioideae

H. E. K. Hartmann

Fenestraria N.E.Br. GC 1925: 433; Hartmann 1982: 145–183. **Typus F. rhopalophylla** (Schlechter & Diels) N.E.Br. **Etym** La *fenestra*, window, referring to the windows on the round leaf tips.

Compact to caespitose, growing in the ground; **I** short, thickened, fleshy, with numerous thick, fleshy adventitious roots tapering into thread-like ends, often plagiotropous, to 1 m *l*; **L** clavate, dark green, only the apical window visible in natural habitat, crystal sand layers in outer epidermal wall, wax cover continuous-rugose; **Fl** 1(–2), erect, bracteolate, nectary as a lophomorphic ring, green; **Fr** decumbent on a long stalk, capsule breaking off, dispersed as a tumble-fruit, valve wings broad, rectangular, closing body small, brown, covering membranes complete with distal closing ledges; **S** light brown, smooth except for hilum and micropyle; **Chr** $2n = 18$; **Ecol** on coastal aeolian sands or gravel, rarely in pockets above dolomite; < 100 mm rainfall p.a. in winter, fog; **Distr** Lüderitz-South, Namibia; Namaqualand, NC, S Africa.

Distribution of **Fenestraria** N.E.Br.

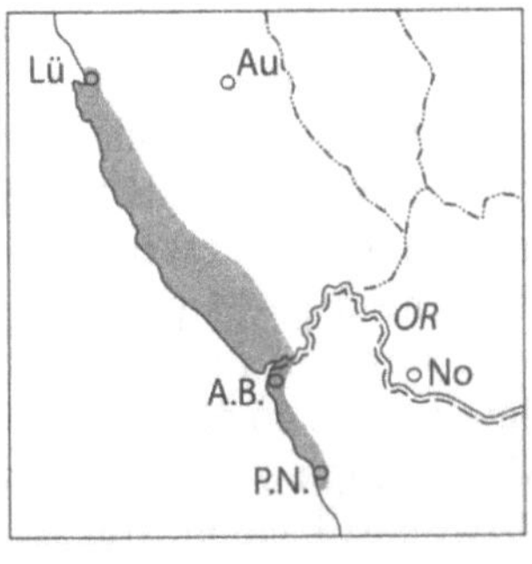

Note: Fruits of the genus are most similar to those of **Jordaaniella**, both differing in growth forms and leaves. With age, the horizontal stems rot away, leaving separate plants with their own storage roots which can easily be taken apart without damage. Care has to be taken not to overwater because the roots are soft, moulding easily.

F. aurantiaca = **F. rhopalophylla**
F. aurantiaca forma *rhopalophylla* ≡ **F. rhopalophylla**

F. rhopalophylla (Schlechter & Diels) N.E.Br. GC 1925: 433 ≡ *M. rhopalophyllum* Schlechter & Diels 1907: 692 T Schultze 11 (B!).

L as for genus, even the stomata with crystals in the outer epidermal wall; **Fl** ∅ 25–70 mm, petals yellow or white; **Fr** stalk 30–40 mm *l*, top somewhat turret-shaped through elevated rims, base rounded, locules (8–)10(–16); **S** 0.58–0.76 mm *l*, 0.45–0.57 mm *b*; **Ecol** as for genus; **Distr** as for genus.

Note: Two subspecies are distinguished; for reasons of priority, **F. rhopalophylla** is the valid species name.

Key to the Subspecies of F. rhopalophylla

1. Flowers white, ∅ mostly < 50 mm .... **F. rhopalophylla** ssp. **rhopalophylla**
– Flowers yellow to coppery, ∅ –70 mm .... **F. rhopalophylla** ssp. **aurantiaca**

F. rhopalophylla (Schlechter & Diels) N.E.Br. ssp. **aurantiaca** (N.E.Br.) H.E.K.Hartmann 1982: 180 (see **photo**) ≡ *F. aurantiaca* N.E.Br. GC 1927: 263 ≡ *F. aurantiaca* N.E.Br. forma *aurantiaca* G.D.Rowley 1973: 6 nomen illeg. **Lectotype,** here designated: Buckerfield 860/18 (BOL!).

Fl ∅ 40–72 mm, petals yellow to orange, externally purplish suffused, often pink at tips, lighter to white at base, filaments epapillate; **Ecol** on deep aeolian sand, rarely in pockets above dolomite outcrops; **Distr** S part of the area of the species, Lüderitz-South, Namibia; N coastal Namaqualand, NC, S Africa.

Note: A rich variation in colour of petals is known, also in nature, most remarkable is perhaps the occasional extension of pink further down from the tip towards the base. If this overlaps with a rich orange from the base, very bright reddish colours can be found. No type has been designated by the author, but in Brown's notes, a drawing agreeing with one at BOL representing the plant of the collection Buckerfield 860/18 exists. This drawing at K is therefore chosen as the lectotype (not type, as in Hartmann 1982).

F. rhopalophylla (Schlechter & Diels) N.E.Br. ssp. **rhopalophylla** ≡ *F. aurantiaca* N.E.Br. forma *rhopalophylla* (Schlechter & Diels) G.D.Rowley 1973: 6 nomen illeg.

Fl ∅ 25–52 mm, petals white, filaments with papillae; **Ecol** in gravel or crevices of sandy loam above limestone; **Distr** N part of area of the species, Lüderitz-South, Namibia.

Note: Spectacular through the porcelain white of the petals.

Frithia Ruschioideae

H. E. K. Hartmann

Frithia N.E.Br. GC 1925: 433; Hammer 1989: 199–201; Zimmermann 1996: 146–152; Burgoyne, Smith & du Plessis 2000: 1–7 **Typus F. pulchra** N.E.Br. **Etym** *F. Frith*, 1872–1954, a railway service gardener at Johannesburg, who took the plant alive to Mr. Brown.

Compact, growing sunken in the ground, rosettes up to 30 mm in ∅; **L** alternate, club-shaped, but with an almost flat top consisting of a translucent window with irregular edges, sides of leaves paved by raised big cells, the stomata hidden in the narrow valleys in between, wax cover almost complete, surface rugulose; **Fl** solitary, ∅ to 35 mm, bases of of all androecial elements forming a tube, petals decreasing in length towards the centre, but keeping almost the same width, thus not forming proper filamentous staminodes; **Fr** yellow to ochre, globose to oval in shape, the top forming only a tiny lid, the valves lifted only a little by short, narrow expanding keels, valve wings and covering membranes absent, the capsules soon breaking into double clausae, each of these consisting of two halves of adjacent locules with the septum in between, double clausae dispersed as tumble units by wind, 5 locules, ∅ up to 5 mm, to 7 mm *l*; **S** brown, round with an elongated micropyle protruding like a handle, testa cells raised; **Ecol** in gravelly flats of withered quartzite or sandstone, > 500 mm rainfall p.a. in summer; **Distr** E NW, GAU, S Africa.

Note: Superficially, plants of the genus resemble those of **Fenestraria** because of the windows on top of the truncate leaves; they differ from the latter by their alternate arrangement of the leaves and the less glossy windows. Two species are recognized at present.

Distribution of **Frithia** N.E.Br.

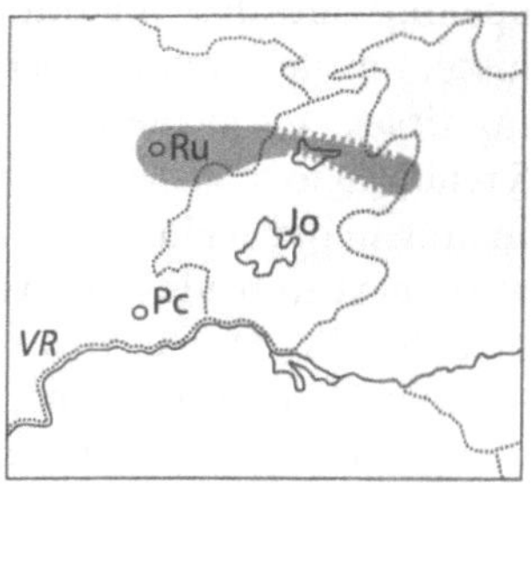

Key to the Species of Frithia

1. Petals deep reddish purple, basally white, inner orange, leaves 15–25 mm long .... **F. pulchra**
– Petals white, sometimes tinged pink or apically purple, inner yellow, leaves 10–15 mm long ... **F. humilis**

F. humilis Burgoyne 2000: 1 (see **photo**) **T** Burgoyne 6693 (PRE) = *F. pulchra* N.E.Br. var. *minor* De Boer Succulenta 47, 1968: 147 nomen invalid. **Lectotype**, here designated: de Boer 1968: photograph on p 148.

Plants < 3 cm in ∅; **L** 10–15 mm *l*, c. 3 mm in ∅ at apex, the area of the lateral grey rows of cells reaching with long, thin protrusions into the apical window, epidermal cells of the top almost brick-shaped with sharp edges; **Fl** ∅ 11–20 mm, 20–30 outer petals white to pinkish, rarely apically purple, 12–20 inner ones yellow, anthesis XII–III in habitat; **Ecol** in coarse sand along exposed sandstone plates; **Distr** Bronkhorstspruit, GAU, S Africa.

Note: The flowering period appears to be induced by rainfall because plants have been found in flower from mid-December until the end of March.

F. pulchra N.E.Br. 1926: 162 (see **photo**) ≡ *M. pulchrum* (N.E.Br.) hort. Kew ex Hill 1931: 129 **T** Frith s.n. (K).

3 cm in ∅; **L** 12–25 mm *l*, apical ∅ 4–5 mm, the edge of the window undulate, not with fingers from lateral tissue, epidermal cells milestone or fingertip-shaped; **Fl** ∅ 25–35 mm, petals deep reddish purple, bases lighter to white, inner ones orange, anthesis in II–III in habitat; **Ecol** in flats with quartzite pebbles; **Distr** E part of the area of the genus.

Note: In habitat, the species can form impressive monotypic stands when in flower, although the sunken plants are not conspicuous individually.

Galenia Aizooideae

H. E. K. Hartmann

Galenia L. 1753: 359; Fenzl 1839: 288; Adamson 1956: 87–127; Friedrich 1970: 47–55 **Typus G. africana** L. **Etym** *C. Galen*, 129–199?, Greek medical doctor, philosopher, and natural scientist, later in Rome = *Kolleria* C.Presl 1831: 24 **Typus** *K. herniariaefolia* C.Presl ≡ **G. herniariaefolia** (C.Presl) Walpers = *Sialodes* Ecklon & Zeyher 1837: 329 **Typus** *Sialodes glauca* Ecklon & Zeyher ≡ **G. ecklonis** Walpers = *Tephras* E. Meyer ex Sonder in Harvey & Sonder 1862: 477 nomen nudum **Typus** *Tephras papulosa* E. Meyer ex Sonder ≡ **G. pruinosa** Sonder in Harvey & Sonder.

Shrubs or perennial herbs, often also the flowering branches dying back, **I** papillate to smooth, branches often secund; **L** opposite or alternate, lanceolate to needle-shaped, epidermis with ± low papillae, glistening in the fresh state, turning white when dry; **Fl** in ± rich cymes, these often well separated from the sometimes big vegetative bush, sessile, perigone small, (3–)4–5 lobes, 8–10 stamens, in groups of 2 alternate with the perigone lobes, the filaments elongating by 3 × their length when the anthers open, those of a pair opening one after the other; **Fr** dry, leathery, indehiscent or dehiscent, sometimes with some expanding tissue, 3–5 locules, each with 1 pendulous seed; **S** reniform to orbicular, mostly flattened; **Ecol** predominantly in dry regions in lowland, often in disturbed areas next to roads in gravelly soils; **Distr** S Angola; Namibia; EC, FS, NC, WC, S Africa.

Note: Plants of a few species are prominent settlers along roadsides, where their yellow colour catches the eye. They are also among the tallest members of the Aizoaceae, reaching heights of up to 2 m and diameters of up to 2 m, too.

This treatment is based on the taxonomic conclusions of Adamson (1956: 87–127) and Friedrich (1970: 47–55), including the division of the genus

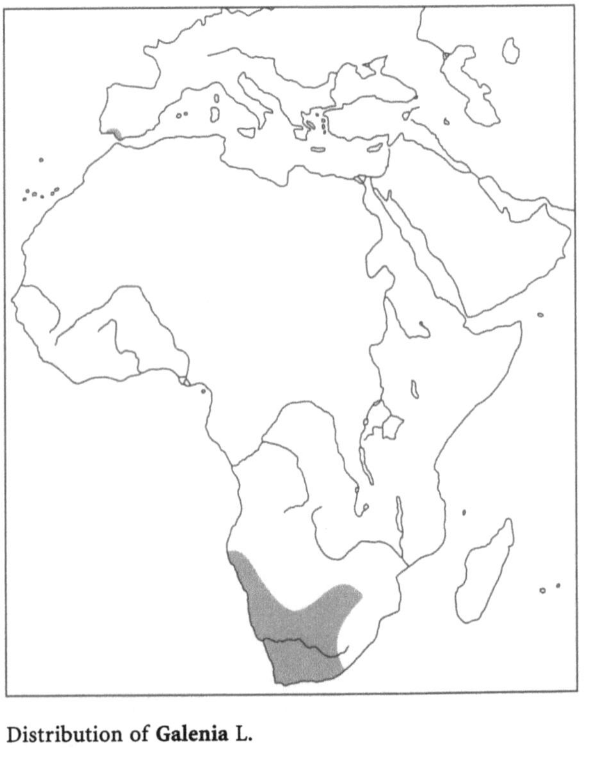

Distribution of **Galenia** L.

into two subgenera. As already pointed out by Bittrich (1990: 217–129), members of the subgenus **Kolleria** agree in several features with those of **Aizoon.** The plants are important for farming. Available studies have not considered the differences in the shapes of papillae and hairs, nor have seeds been investigated thoroughly, and very rarely have plants been grown to study the changes in leaf shapes and sizes along the stems and at the turn from vegetative to flowering phase.

Key to the Subgenera of Galenia

1. Flowers in terminal cymes branching to all sides, fruits not or tardily dehiscent .... G. subg. **Galenia**
– Flowers in secund arrangement in cymes, fruits dehiscent .... G. subg. **Kolleria**

G. subg. **Galenia** = *Sialodes* Ecklon & Zeyher 1837: 329 ≡ *G.* § *Galenia* Walpers 1843: 232 **Typus** *Sialodes glauca* Ecklon & Zeyher ≡ **G. ecklonis** Walpers 1843: 232.

Erect to suberect shrubs, shoots often dying back in the dry season, plants green or yellow; **L** opposite, linear or a little broader, usually glabrous or nearly so, sometimes with sticky glandular hairs; **Fl** in terminal cymes with branches pointing into all directions, often composed and appearing thyrse- or panicle-like, perigone lobes 4, the pairs dissimilar, 8 stamens; **Fr** not or tardily dehiscent, 2 locules; **Ecol** in gravel along roadsides or in rocky, stony places; **Distr** S Angola; Namibia; EC, FS, NC, EC, S Africa.

Note: Following Adamson (1957: 89), only the species answering the above description are included in this subgenus, leaving the other subgenus with a rather heterogeneous composition of character states and species. At present, the fruit dehiscence is the differentiating feature.

G. subg. **Kolleria** Fenzl 1839: 288 ≡ *Kolleria* C.Presl 1831: 24 ≡ *G.* § *Kolleria* (presl) Walpers 1843: 232 **Typus** *Kolleria herniariaefolia* C.Presl ≡ **G. herniariaefolia** (C.Presl) Fenzl = *Tephras* E. Meyer ex Sonder in Harvey & Sonder 1862: 477 nomen nudum **Typus** *Tephras papulosa* E. Meyer ex Sonder ≡ **G. pruinosa** Sonder in Harvey & Sonder 1862: 477.

Diffuse or prostrate shrubs; **L** usually secund, opposite or alternate, most often ovate to spatulate, grey to white; **Fl** in cymes with secund branches, perigone with 3–5 lobes, usually alike; **Fr** dehiscent, 2–5 locules; **Ecol** in drier areas; **Distr** as for genus.

Note: Within the subgenus, several groups can be recognized.

G. § *Galenia* ≡ **G.** subg. **Galenia**
G. § *Kolleria* ≡ **G.** subg. **Kolleria**

G. acutifolia Adamson 1956: 112 subg. **Kolleria** T Esterhuysen 813 (BOL).

Shrubs with many decumbent to ascending branches from a thick, woody rootstock, young I with distant, appressed hairs, older I glabrescent; **L** opposite, greyish-green, obovate or rhombic-ovate, acute, narrowing into a petiole 2–3 mm *l*, **L** blades 8–12 mm *l*, 4–5 mm *b*; **Fl** on short branches, sessile, bracts ± sessile, bracteoles slightly longer than the perigone, perigone lobes 5, pale cream inside, somewhat hairy outside, 2–2.5 mm *l*, anthers yellow; **Fr** 5 locules; **Ecol** unknown; **Distr** Barkly West, NC, S Africa.

Note: Reported to differ from the similar **G. prostrata** by a more woody habit, less prostrate branches, less hairy condition, more crowded flowers and the narrower, more acute leaves.

G. affinis Sonder in Harvey & Sonder 1862: 476 subg. **Kolleria** ≡ *Aizoon fruticosum* Ecklon & Zeyher 1837: 327 nomen illeg. non L.f. 1781: 261 T Ecklon & Zeyher 2139 (S) = *Tephras pubescens* E. Meyer nomen nudum in schedis T Drege s.n. (K).

Shrubs 30–60 cm *h*, branches prostrate to ascending, younger ones secund, spreading, I pale, with short appressed hairs; **L** alternate, obovate to spatulate, folded upwards, with acute, recurved points, with some or many closely appressed hairs, 5–12 mm *l*, 2–5 mm *b*; **Fl** on short, pinnate, secund branches, bracts longer than the perigone with 5 lobes, hairy outside, 1.5–2 mm *l*, anthers yellow; **Fr** 3 locules, rarely to 5; **Ecol** on rocky koppies on the Karoo; **Distr** S districts of WC, S Africa.

Note: Sonder gave the height as "several feet", which is not confirmed by Adamson, who also reported a wide and dispersed distribution. The species is said to be very similar to **G. fruticosa**, which has silky hairy leaves, 4 calyx lobes, and 2 locules.

G. africana L. 1753: 359 (see **photo**) subg. **Galenia** ≡ *G. tenuifolia* Salisbury 1796: 283 nomen illeg. T LINN 507.1 (LINN) = *G. africana* L. var. *halimifolia* Fenzl ex Sonder in Harvey & Sonder 1862: 478 nomen nudum ≡ *G. halimifolia* Fenzl in schedis nomen nudum T Drege s.n. (?) = *G. africana* L. var. *pentandra* Hiern 1898: 411 T Welwitsch 1093 (BM, ISO K!) = *G. africana* L. var. *secundata* Adamson 1956: 92 T Bryant 115 (K!) = *G. linearis* Thunberg 1800: 77 T Thunberg (UPS).

Shrubs to 1.5 m *h*, ∅ to 2 m, I pale ochre, glabrou except for the youngest ones, which can have som hairs; **L** opposite, yellowish-green, later turning bright yellow, the older falling off, the epidermis with papillae and hairs with glands, which give the plant a sticky touch, 15–50 mm *l*, 2–5 mm *b*; **Fl** in cymes of 30–100 mm length, borne above the plant, lighter in colour, perigone yellow-green, bracteoles not longer than the flower, perigone lobes 4, 1–1.5 mm *l*, the outer cucullate, the inner flat with a cylindrical mucronate point, anthers yellow or orange, rarely red; **Fr** 2 mm; **Ecol** in gravel or looser soils, often along roadsides; **Distr** S Angola; Namibia; W parts of EC, W part of NC, WC, S Africa.

Note: Occurring in a large area, the species appears to have expanded its range in the past decades considerably. **G. africana** is often reported to be poisonous for cattle. Other sources mention that some forms only are dangerous. Salisbury (1796: 283) replaced **G. africana** L. with *G. tenuifolia*, using the text of Linnaeus 1762: 515, which is identical with that of Linnaeus 1753: 359; both names refer therefore to the same type.

G. africana var. *halimifolia* = **G. africana**
G. africana var. *pentandra* = **G. africana**
G. africana var. *secundata* = **G. africana**
G. aizoides = **G. secunda**
G. asiatica Burman f. 1768: 88, t. 31
= **Salvadora persica** L. 1753: 122 (Salvadoraceae)
G. celesioides Sprengel 1818: 18
= **Microtea paniculata** (Salzmann) DC. 1849: 18 (Phytolaccaceae)

Note: This naming follows the Kew Index for *Microtea celesioides*. De Candolle 1849: 18 is cited as a synonym only under **Microtea paniculata** with the addition "Moq. olim", for which no further reference has been found.

G. collina (Ecklon & Zeyher) Walpers 1843: 232 subg. **Kolleria** ≡ *Kolleria collina* Ecklon & Zeyher 1837: 327 ≡ *G. humifusa* Fenzl ex Sonder in Harvey & Sonder 1862: 476 **T** Ecklon & Zeyher 2143 (S).

Shrubs with decumbent, woody branches, many subsecund, **I** grey; **L** arched, linear to oblong-spatulate, 7–11.5 mm *l*, 1.1–3.4 mm *b*; **Fl** sessile, solitary, perigone lobes 4, 2.5–3 mm *l*, white or yellowish inside; **Fr** 3–5?; **Ecol** on hills; **Distr** W districts of WC, S Africa.

Note: Adamson (1956: 102) included also plants from Namaqualand and Namibia, a decision not accepted by Friedrich (1970: 52) because the type of **G. collina** possesses 4 perigone lobes. Following Friedrich (1970: 52), **G. collina** is identical with **G. herniariaefolia**, also from the SW part of the Western Cape Province.

G. collina var. *meziana* ≡ **G. meziana**
G. contaminata = **G. secunda**

G. crystallina (Ecklon & Zeyher) Sonder in Harvey & Sonder 1862: 477 subg. **Kolleria** ≡ *Aizoon crystallinum* Ecklon & Zeyher 1837: 327 **T** Ecklon & Zeyher 2138 (S) = *G. namaquensis* Schellenberg & Schlechter 1912: 500 **T** Schlechter 11290 (B, ISO K!).

Shrubs with decumbent to upright, alternate branches, **I** covered densely with translucent papillae giving them, and the leaves, a crystalline appearance; **L** alternate, oblong-cuneate, obtuse; **Fl** in secund **Inf**, sessile, bracteoles c. as long as the perigone of 4–5 lobes; **Fr** 2 locules; **Ecol** in sandy places; **Distr** Namaqualand, NC, Clanwilliam, WC, S Africa.

Note: Whether the species does indeed include populations from further north remains doubtful, but the inclusion of *G. namaquensis* follows Adamson (1956: 119).

G. cymosa Adamson 1956: 122 subg. **Kolleria** **T** Adamson 5135 (BOL).

Diffusely branching shrubs to 50 cm *h*, with opposite, secund branches often divaricate, young **I** with few appressed scales, older **I** glabrous; **L** opposite, the upper alternate, darkish-green, spatulate or obovate, petiolate, finely papillate, 5–20 mm *l*, 4–6 mm *b*; **Fl** in secund **Inf** with widely forking branches, upper alternate, bracts not longer than the perigone of 5, less often 4, lobes, 2–3 mm *l*, papillate outside, anthers yellow; **Fr** 3–5 locules; **Ecol** in sheltered places among rocks; **Distr** Albany, Fort Beaufort, EC, E districts of WC, S Africa.

Note: Similar to **G. pruinosa**, but less woody and more slender, the internodes not white, the inflorescence secund and laxer.

G. dinteri = **G. meziana**

G. dregeana Fenzl ex Sonder in Harvey & Sonder 1862: 477 subg. **Kolleria** **T** Drege (B).

Erect, stout, woody shrubs to 45 cm *h*, **I** pale yellow, with translucent scales; **L** opposite to subalternate, petioles thick, 2 mm *l*, suborbicular, coarsely papillate, blades 7–15 mm *l*, 7–13 mm *b*; **Fl** in dense glomerate cymes on terminal panicles, perigone with 5 lobes, 2.25 mm *l*, densely covered with scales outside, anthers deep orange to pink; **Fr** 2 locules; **Ecol** on dry river banks; **Distr** Lüderitz-South, Namibia; Namaqualand, NC, S Africa.

Note: The species can easily be identified by its nearly orbicular leaves.

G. ecklonis Walpers 1843: 232 subg. **Galenia** ≡ *Sialodes glauca* Ecklon & Zeyher 1837: 329 ≡ *G. glauca* Sonder in Harvey & Sonder 1862: 478 non (Ecklon & Zeyher) Walpers 1843: 232 nomen illeg. **T** Ecklon & Zeyher 2149 (K!).

Procumbent branches from a woody rootstock, **I** glabrous or nearly so, branches 10–40 mm *l*; **L** linear-subulate, rolled upwards, 5–10 mm *l*, often falcate, the pair connate at the base; **Fl** in lax terminal cymes, perigone lobes 4, 3–4 mm *l*, white or red inside, mucronate at the tip, anthers pale yellow; **Fr** 2 locules; **Ecol** in sand; **Distr** SW districts of WC, S Africa.

Note: A low growing species best distinguished by the rather big flowers and the mucronate perigone tips.

G. elongata = **G. secunda**

G. exigua Adamson 1956: 101 subg. **Kolleria T** Acocks 1900 (PRE).

Shrubs with many procumbent stems, branches 10–20 cm *l*; **L** alternate, narrowing to the base, linear-oblanceolate, very finely papillate, 8–15 mm *l*, 1–3 mm *b*; **Fl** perigone much shorter than the leaf-like bracts, perigone lobes 5, 1.5–2 mm *l*, the outer cucullate, the inner incurved at the tip, anthers yellow or pink; **Fr** broadly pentagonal, more than twice as broad as *h*, 5 locules; **Ecol** open places in Karroo; **Distr** Herbert, Hopetown, Kimberley, NC, Jacobsdal, FS, S Africa.

Note: Similar to **G. filiformis**, distinguished mainly by the broad, rather flat fruit.

G. fallax = **G. fruticosa**

G. filiformis (Thunberg) N.E.Br. 1922: 155 subg. **Kolleria** ≡ *M. filiforme* Thunberg 1800: 89 **T** Thunberg 11981 (UPS!).

Low shrubs with prostrate branches 20–80 cm *l*, young **I** with some scales (from appressed papillae); **L** alternate, obovate, narrowing to the base into a very short petiole, covered by roundish papillae, white when dry, 5–10 mm *l*, 1–4 mm *b*; **Fl** on short shoots in secund arrangement, in monochasia, perigone lobes 5, c. 2 mm *l*, anthers red, rarely yellow; **Fr** 2 locules; **Ecol** in sand or gravel, often in river beds; **Distr** WC, S Africa.

Note: The type does not possess any hairs, but roundish bladder cells. Adamson (1956: 100) reported hairs on other populations.

G. fruticosa (L.f.) Sonder in Harvey & Sonder 1862: 477 subg. **Kolleria** ≡ *Aizoon fruticosum* L.f. 1781: 261 **T** Thunberg (UPS) = *G. fallax* Pax in Engler 1915: f 94 J-L nomen nudum ≡ *G. fallax* Pax ex Adamson 1956: 120 nomen invalid. T? = *G. fruticosa* (L.f.) Sonder var. *prostrata* Adamson 1956: 118 **T** Salter 7012 (ISO K!) = *G. salsoloides* Fenzl ex Sonder in Harvey & Sonder 1862: 478 nomen nudum = *G. sericea* Fenzl ex Sonder in Harvey & Sonder 1862: 478 nomen nudum.

Erect shrubs, rarely procumbent, to 50 cm *h*, often turning spinescent with age, young **I** silky-pubescent; **L** opposite, oblong-cuneate to -lanceolate, folded upwards, 5–15 mm *l*, 3–5 mm *b*, acute; **Fl** in rigid, terminal cymes shaped like panicles with short secund branches, sessile, bracts as long as or longer than the perigone, perigone mostly 4-lobes, 5 lobes also present, anthers red or yellow; **Fr** 2 locules; **Ecol** in open places on shales; **Distr** Keetmanshoop, Lüderitz-South, Namibia; from Namaqualand, NC, to Robertson, WC, S Africa.

Note: G. fallax is placed as a synonym here, following Friedrich (1970: 51–52) and considering that the species has, until now, not been validly described: the drawing by Pax lacks any description, and Adamson (1956: 120–121) gave an English text only.

G. fruticosa var. *prostrata* = **G. fruticosa**

G. glandulifera Bittrich Bothalia 1990: 217 subg. **Kolleria T** Gerbaulet & Struck 24236 (HBG!).

Low perennial herbs to 10 cm *h*, 10–60 cm ∅, main stem erect, lateral ones decumbent or ascending, ending in cymes as **Inf**, **I** and leaves with large translucent papillae (bladder cells) and oblong, ovoid bottle-shaped secretory glands on multiseriate stalks; **L** opposite on main stem, alternate on lateral ones, petioles 2–7 mm *l*, blade obovate, obtuse, the semiglobose bladder cells with distant platelets of wax, c. 20 mm *l* and 8 mm *b* on main stem, 8 mm *l*, 5 mm *b* on lateral branches; **Fl** in almost monopodial cymes up to 30 cm *l*, bracts recaulescent with their axillary shoot up to the terminal **Fl**, sessile, perigone lobes 5, magenta or purple inside, nectary a flat disc at the base of the filaments; **Fr** depressed centrally, with expanding sheets and keels opening the valves slightly after wetting, pentangular, 2 mm ∅; **S** reniform, dark brown to black, with small tubercles, 0.75 mm *l*, 0.5 mm *b*; **Chr** $2n = 16$; **Ecol** unknown; **Distr** W districts of EC, S districts of NC, E districts of WC, S Africa.

Note: Bittrich (Bothalia 1990: 218) described the secretion produced by the glands as smelling of coumarin, yellow at first, turning brown later. He also pointed out that a similar sticky substance occurs in **G. africana**, where probably T-shaped cells are responsible for the production of the secretion. He further suggested that these glands may indicate a closer relationship between the two species, now placed in different subgenera.

G. glauca ≡ **G. ecklonis**; = **G. pallens**
G. halimifolia = **G. africana**

G. hemisphaerica Adamson 1956: 122 subg. **Kolleria T** Dinter 6440 (BOL).

Hemisphaerical shrubs, 20–25 cm *h*, 30–50 cm ∅, stems basally woody, much branched, **I** pale brown, young **I** papillate, older smooth, glabrous; **L** opposite, deciduous, flat, wedge-shaped, obtuse, papillate, 8–15 mm *l*, 2–4 mm *b*; **Fl** in terminal, lax dichasial **Inf**, bracteoles shorter than the **Fl**, perigone lobes 5, 1.5–2 mm *l*, papillate outside; **Fr** 2 locules, rarely 3; **Ecol** in dry open places in sand; **Distr** Lüderitz-South, Namibia, Namaqualand, NC, S Africa.

Note: The species is mainly based on the hemisphaerical shape of the shrubs, which roll in the wind when dry. Friedrich (1970: 54) considered plants of this species as extreme forms of **G. pruinosa.**

G. herniariaefolia (C.Presl) Walpers 1843: 232 subg. **Kolleria** ≡ *Kolleria herniariaefolia* C.Presl 1831: 24 ≡ *Aizoon herniarium* Reichenbach ex Steudel 1841: 46 T Sieber 164 (K) = *Aizoon microphyllum* Bartling 1832: 541 T s.n. (?).

Low shrubs with prostrate, basally woody branches to 75 cm *l*, side branches alternate, young **I** with scattered hairs, older glabrous; **L** alternate, linear-lanceolate, obtuse, basally narrowed, 6–9 mm *l*, to 2.25 mm *b*; **Fl** sessile in secund lateral branches, perigone to 2.25 mm *l*, 4 lobes, hairy outside, white inside, anthers pink; **Fr** with leathery valves, glabrous, 1/2 as long as the perigone, mostly 4 locules; **S** 1 per locule, black, shiny; **Ecol** in sandy places; **Distr** from Cape Town to George, WC, SW Africa.

Note: Friedrich (1970: 52–53) considered **G. herniariaefolia** and **G. collina** as conspecific. It should be examined whether Barthling based *Aizoon microphyllum* possibly on the same type as Presl.

G. heterophylla = **G. pubescens**

G. hispidissima Fenzl ex Sonder in Sonder & Harvey 1862: 478 subg. **Galenia T** Drege 7055 (W).

Low-growing, many-stemmed plants up to 15 cm *h*, woody at base, younger **I** covered with stiff, spreading hairs 1–2 mm *l*; **L** linear, flat, green, acute, erect hairs along margins and on lower leaf surface, 7–25 mm *l*; **Fl** in dense, cymose panicle-like **Inf**, perigone lobes 2.2 mm *l*, outside with stiff hairs, anthers orange or red; **Fr** 2 locules; **Ecol** in sandy soils among bushes on lower mountain slopes; **Distr** Ceres, Clanwilliam, Piketberg, WC, S Africa.

Note: Low growth and stiff hairs characterize this species with a restricted distribution area.

G. humifusa ≡ **G. collina**
G. leucoclada = **G. pruinosa**
G. leucodermis = **G. pruinosa**
G. linearis = **G. africana**

G. meziana K. Mueller 1908: 68 subg. **Kolleria** ≡ *G. collina* (Ecklon & Zeyher) Walpers var. *meziana* (K.Mueller) Adamson T Schlechter s.n., lost? (B) **NT** Schlechter 34 (K!) = *G. dinteri* Schellenberg 1912: 501 T Dinter 1132 (B) = *G. steingroeveri* Engler 1915: 147 nomen nudum.

Low shrubs with prostrate branches, young **I** finely papillate; **L** opposite, alternate in the **Inf**, soon falling off, spatulate, scaly-papillate on both sides, to 9 mm *l*, c. 1.5 mm *b*; **Fl** in axillary groups, perigone 1.5–2 mm *l*, 5 lobes, slightly hairy outside, stamens 5 or 10; **Fr** 3–4 locules; **Ecol** in sandy or gravelly places; **Distr** Lüderitz-South, Namibia; Namaqualand, NC, S Africa.

Note: As understood here, the species is restricted to Namibia and the adjacent part of South Africa, following Friedrich (1970: 52–53) and not Adamson (1956: 103). The relation of this taxon to **G. collina** and **G. herniariaefolia**, both from further south and near the Cape of Good Hope, needs more attention. The species was originally based on the possession of only five stamens, but this feature seems to vary more widely than anticipated by Mueller (1908: 68).

G. namaensis Schinz 1895: 76 subg. **Galenia T** Fenchel 171 (Z, ISO K!).

Shrubs up to 60 cm *h*, the young parts covered by fine papillae, **I** later white or light yellow and smooth; **L** linear, folded upwards, mostly falcate with a hooked tip, 3–20 mm *l*, 2 mm *b*; **Fl** in rich cymes raised above the plant, bracteoles c. as long as the flowers, perigone c. 2 mm *l*, white or greenish inside, the outer cucullate, anthers yellow, orange or rarely pink; **Fr** apically not indented; **S** black, reniform; **Ecol** rocky places, especially in dry river beds; **Distr** Namibia; FS, NC, S Africa.

Note: Friedrich (1970: 50–51) pointed out that the species appears to be rather rare but has often been confused with **G. africana,** to which also some of the material cited by Adamson (1956: 95) belongs. **G. africana** can be recognized by the stalked glands.

G. namaquensis = **G. crystallina**

G. pallens (Ecklon & Zeyher) Walpers 1843: 232 subg. **Kolleria** ≡ *Kolleria pallens* Ecklon & Zeyher 1837: 328 **T** Ecklon & Zeyher 2144 (W) = *G. pubescens* (Ecklon & Zeyher) Druce var. *pallens* Adamson **T** Zeyher 2636 (ISO K!) = *G. glauca* (Ecklon & Zeyher) Walpers 1843: 232 non Sonder in Harvey & Sonder 1862: 478 ≡ *Kolleria glauca* Ecklon & Zeyher 1837: 328 T Ecklon & Zeyher 2145 (W).

Low shrubs with procumbent, woody, opposite branches, **I** with a dense cover of dense, yellowish hairs; **L** alternate, obovate to subspatulate, young ones opposite, also with dense, yellowish hairs; **Fl** in groups of 3 mostly, the lateral ones with very short pedicels, perigone lobes 4; **Fr** 4 locules; **Ecol** among shrubs in fields; **Distr** Uitenhage, EC, S Africa.

Note: The inclusion of *G. glauca* as a synonym follows all older literature (Adamson 1956: 110).

G. papulosa (Ecklon & Zeyher) Sonder in Harvey & Sonder 1862: 475 (see **photo**) subg. **Kolleria** ≡ *Aizoon papulosum* Ecklon & Zeyher 1837: 326 T Ecklon & Zeyher 2137 (B) = *Galenia papulosa* (Ecklon & Zeyher) Sonder var. *microphylla* Adamson 1956: 114 T Adamson 5249A (BOL) = *Galenia papulosa* (Ecklon & Zeyher) Sonder var. *tristyla* O.Kuntze 1886: 264 T?

Shrubs with ascending to decumbent branches, 10–40 cm *h*, young **I** densely covered with round, glistening papillae; **L** alternate, rarely opposite, obovate to cuneate, covered with big, coarse, round papillae, 10–30 mm *l*, 8–17 mm *b*; **Fl** in ± lax, secund cymes, bracts longer than the flowers, bracteoles c. the same length, perigone lobes 5, 1.5–2 mm *l*, anthers yellow; **Fr** mostly 5 locules; **Ecol** in gravel or sand; **Distr** from Swakopmund, Namibia, to Swellendam, WC, S Africa.

Note: As understood in literature (Adamson 1956: 113–114, Friedrich 1970: 53–54), the species occurs in a wide area, distinguished from all others by the obtuse, transparent papillae. Whether these merge indeed into forms found in **G. sarcophylla** and **G. squamulosa**, as reported by Friedrich (1970: 113–114), for which mostly scales are reported to occur, remains unclear.

G. papulosa var. *microphylla* = **G. papulosa**
G. papulosa var. *tristyla* = **G. papulosa**

G. portulacacea Fenzl ex Sonder in Harvey & Sonder 1862: 475 subg. **Kolleria** T Drege s.n. (B).

Prostrate to suberect shrubs with ascending shoots, woody at the base only, **I** with white, hair-like appressed scales; **L** opposite, alternate in the flowering region, oblong to spatulate, flat, obtuse, papillate, upper ones plicate, covered by white scales when young, 10–30 mm *l*, 3–6 mm *b*, narrowing towards the base; **Fl** in groups of 5–7 in lax terminal cymes, perigone lobes 5, c. 3 mm *l*, anthers yellow; **Fr** 3 or 5 locules; **Ecol** on sand; **Distr** from Namaqualand, NC, to Mosselbay, WC, S Africa.

Note: Similar to **G. sarcophylla**, but less succulent and with bigger leaves.

G. procumbens L.f. 1781: 227 subg. **Galenia** T "Linn. f. 91" as LINN 570.3 (LINN).

Erect to procumbent, much-branched shrubs, 30–100 cm *h*, youngest **I** without or with few appressed hairs, branches ascending or divaricate, sometimes partly secund; **L** linear, falcate, usually deeply concave above, often crowded on lateral branches, old **L** falling off, 30–60 mm *l*; **Fl** in small, lax cymes, not overtopping the neighbouring branches, cymes terminally monochasial, perigone lobes 4, 1.5–2 mm *l*, white or tinged with pink inside, the outer cucullate, anthers orange, rarely red; **Fr** flattened, indented at the top and with rounded lateral bulges, 2 locules; **Ecol** in rocky and stony places; **Distr** EC, NC, WC, S Africa.

Note: The growth form of this species appears most variable, procumbent forms being more common in overgrazed areas. Some doubts remain as to the identity of the type given here following Adamson (1956: 96) because this may differ from the sheet cited by Linnaeus fil. ("Thunberg"), probably referring to material collected by Thunberg and kept at UPS.

G. propinqua = **G. secunda**

G. prostrata Schellenberg 1913: 501 subg. **Kolleria** T Bolus 6815 (B).

Slender, prostrate, many-stemmed shrubs, young **I** with white appressed hairs, ∅ 50–80 cm; **L** opposite, rhomboid, acute, thinly covered with appressed hairs, 8 mm *l*, 4 mm *b*; **Fl** solitary and sessile in lower bracts, in groups in upper bracts, perigone lobes 5, the outer with hairs, extending into a long, terete protrusion beyond the actual leaf-tip, pink inside, anthers mostly pink; **Fr** 5 locules; **Ecol** unknown; **Distr** Barkly West, Kimberley, NC, S Africa.

Note: The long, acute tips of the perigone lobes characterize the species without doubt, similar features occur only in **G. pallens**, where four perigone lobes are found.

G. pruinosa Sonder in Harvey & Sonder 1862: 477 subg. **Kolleria** ≡ *Tephras papulosa* E.Meyer ex Sonder in Harvey & Sonder 1862: 477 T Drege (W) = *G. leucoclada* Schellenberg & Schlechter 1912: 500 **LT** Schlechter 11279 (B) = *G. leucodermis* Dinter nomen nudum in schedis (Dinter 6164 K!).

Shrubs to 50 cm *h*, **I** prominently white, with appressed scales when young, branches opposite, divaricate; **L** opposite, obovate-cuneate, obtuse, 5–15 mm *l*, 3–6 mm *b*; **Fl** in terminal, pedunculate cymes appearing as a panicle, **Fl** in separated groups, perigone lobes mostly 5, c. 2 mm *l*, papillate and with appressed scales outside, anthers yellow; **Fr** 2, rarely 3 locules; **Ecol** in plains; **Distr** Lüderitz-South, Warmbad, Namibia; Namaqualand, NC, Clanwilliam, Piquetberg, WC, S Africa.

Note: The species can be easily identified by its white stems and branches.

G. pubescens (Ecklon & Zeyher) Druce 1917: 624 subg. **Kolleria** ≡ *Aizoon pubescens* Ecklon & Zeyher 1837: 326 T Ecklon & Zeyher 2133 (?) **LT** Zeyher 2638 (S) ≡ *G. spathulata* Fenzl ex Sonder in Harvey & Sonder 1862: 475 **Lectotype**, here designated: Ecklon & Zeyher 2133 (S) = *G. heterophylla* Fenzl ex Sonder in Harvey & Sonder 1862: 475 nomen nudum T Zeyher 2635 (?) = *G. pubescens* (Ecklon & Zeyher) Druce var. *cerosa* Adamson T Esterhuysen 811 (BOL) = *G. pubescens* (Ecklon & Zeyher) Druce var. *fourcadei* Adamson T Fourcade 4660a (BOL) = *G. pubescens* (Ecklon &

Zeyher) Druce var. *lignosa* Adamson T Mogg 8109 (PRE, ISO K!).

Low shrubs with prostrate branches, Ø 100 cm or more, young I with short, appressed hairs, older I glabrescent; L alternate, with a short petiole, obovate-spatulate, papillate and hairy, 6–22 mm *l*, 3–17 mm *b*; **Fl** in alternate secund lateral **Inf**, perigone lobes 5, 2–2.5 mm *l*, appressed hairs outside, pinkish inside, anthers pink; **Fr** 5, rarely 3 locules; **Ecol** in fields; **Distr** EC, FS, NC, NW, WC, S Africa.

Note: The lectotypification of Jacobs (1988: 141–142) is followed here, although it lacks a discussion why the type originally cited with the description of the basionym *Aizoon pubescens*, Ecklon & Zeyher 2133, has not been included in his study of the problem. This collection is also cited for *G. spathulata* and is chosen as the lectotype here. The species is also reported for Australia as an immigrant near settled areas (Prescott & Venning 1984: 50).

G. pubescens var. *cerosa* = **G. pubescens**
G. pubescens var. *fourcadei* = **G. pubescens**
G. pubescens var. *lignosa* = **G. pubescens**
G. pubescens var. *pallens* = **G. pallens**

G. rigida Adamson 1956: 94 subg. **Galenia T** Adamson 5240 (BOL).

Erect shrubs with rigid branches, up to 50 cm *h*, I pale, covered with wax, sometimes sticky; **L** rigid, linear-oblong, flat, concave above, glabrous and often sticky, 10–12.5 mm *l*, 2–3 mm *b*; **Fl** in an umbellate cyme, flat above, rather dense, on an erect, unbranched shoot 5–10 cm *l*, bracteoles very small, perigone white inside, cucullate at tips, 1.5–2 mm *l*; **Fr** unknown; **Ecol** in rocky places on dolerite; **Distr** Sutherland, NC, Clanwilliam, WC, S Africa.

Note: Remarkable for the species is the elevated position of the inflorescence on top of an unbranched stem.

G. sarcophylla Fenzl ex Sonder in Harvey & Sonder 1862: 475 subg. **Kolleria T** Zeyher 719 (W).

Low shrubs with prostrate to suberect branches, to 30 cm *h*, to 50 cm Ø, young I glistening from coarse papillae and hairy, greyish-green, older ones red or yellow; **L** lower ones opposite, upper ones alternate, obovate or spatulate, narrowing into a petiole at the base, upper ones sessile, 6–16 mm *l*, 3–8 mm *b*; **Fl** in secund cymes, the upper approximate, perigone lobes 5, 2 mm *l*, covered by scales outside, anthers yellow, rarely pink; **Fr** 3–4 locules, rarely 5; **Ecol** in sand or gravel; **Distr** Namibia; from Namaqualand, NC, to Bloemfontein, FS, and Graaff Reinet, EC, S Africa.

Note: The present circumscription follows Adamson (1956: 114–115). Friedrich (1970: 53) included the name as a synonym under **G. papulosa** since he had found a wide variation in the papillae. **G. sarcophylla** possesses bladder-shaped, hair-like and scaly papillae (Friedrich 1970: 53).

G. salsoloides = **G. fruticosa**

G. secunda (L.f.) Sonder in Harvey & Sonder 1862: 474 subg. **Kolleria** ≡ *Aizoon secundum* L.f. 1781: 261 **LT** Thunberg 12059 (UPS!) = *G. secunda* (L.f.) Sonder in Harvey & Sonder var. *strigulosa* Sonder in Harvey & Sonder 1862: 474 ≡ *G. aizoides* Fenzl ex Sonder in Harvey & Sonder 1862: 474 nomen nudum T Drege s.n. (?) = *Aizoon contaminatum* Ecklon & Zeyher 1837: 326 T Ecklon & Zeyher 2135 (S) = *Aizoon elongatum* Ecklon & Zeyher 1837: 326 T Ecklon & Zeyher 2132 (S) = *Aizoon propinquum* Ecklon & Zeyher 1837: 326 T Ecklon & Zeyher 2134 (S).

Low shrubs with branches 10–50 cm *l*, young I covered densely with white, appressed, branched hairs, branches secund; **L** alternate, obovate, ± acute, densely hairy like the stem, 7–20 mm *l*, 4–12 mm *b*; **Fl** in lateral secund cymes resembling fronds in shape, perigone lobes 5, 2–3 mm *l*, densely hairy outside, yellow or white inside; **Fr** rounded, depressed on top, 5, rarely 3 locules; **Ecol** in open places in sand or gravel; **Distr** Namibia; from Kenhardt to Kimberley, NC, W districts of EC, WC, S Africa.

Note: Not only has the species a wide distribution area in Southern Africa, it has also been reported as an immigrant from Spain (Arvidsson & Liden 1977: 130) and Australia (Prescott & Venning 1984: 52).

G. secunda var. *strigulosa* = **G. secunda**
G. sericea = **G. fruticosa**
G. spathulata ≡ **G. pubescens**

G. squamulosa (Ecklon & Zeyher) Fenzl ex Sonder in Harvey & Sonder 1862: 477 subg. **Kolleria** ≡ *Aizoon squamulosum* Ecklon & Zeyher 1837: 326 T Ecklon & Zeyher 2136 (S).

Low shrubs with decumbent or prostrate branches 20–50 cm *l*, woody at base, I covered with flat, spreading scales; **L** opposite, alternate in the flowering parts, oblong to obovate, obtuse, flat, with spreading scales, 10–20 mm *l*, 3–7 mm *b*; **Fl** in terminal cymes, perigone lobes 5, 2 mm *l*, densely covered with scales outside, yellow inside, anthers yellow; **Fr** 2 locules; **Ecol** in sandy places; **Distr** Namibia; Namaqualand, NC.

Note: Adamson (1956: 117) gave also Gordonia and Victoria West, NC, for the distribution. This needs further study, also in view of Friedrich's inclusion of the name as a synonym under **G. papulosa** (1970: 53). From that latter species **G. squamulosa** is said to differ mainly by the possession of broad scales on leaves and internodes.

G. steingroeveri = **G. meziana**

G. subcarnosa Adamson 1956: 103 subg. **Kolleria T** Acocks 14358 (PRE).

Low shrubs with prostrate branches forming mats of 10–25 cm ∅, young **I** finely pruinose or finely papillate with few appressed scales; **L** acute or subacute, concave above, 3–7 mm *l*, 1–3 mm *b*; **Fl** on short, alternate secund lateral branches, perigone lobes 5, 1.5 mm *l*, finely papillate; **Fr** pentangular, broader than high, 5 locules; **Ecol** in open rocky places; **Distr** from Kimberley to Sutherland, NC, Middelburg, EC, S Africa; Matopos, Zimbabwe.

Note: This is the only species reported naturally from Zimbabwe.

G. tenuifolia ≡ **G. africana**

Gasoul ≡ Mesembryanthemum

M. Gerbaulet

This genus (Adamson 1763: 243) is based on the conserved type of **Mesembryanthemum, M. nodiflorum.** The name *Gasoul* became illegitimate after the conservation of the former name.

G. aitonis ≡ **Mesembryanthemum aitonis**
G. crystallinum
≡ **Mesembryanthemum crystallinum**
G. nodiflorum ≡ **Mesembryanthemum nodiflorum**

Gibbaeum Ruschioideae

H. E. K. Hartmann

Gibbaeum N.E.Br. GC 1921: 273; Nel 1953; Glen 1974 **Typus G. pubescens** (Haw.) N.E.Br. **Etym** Gk *gibba*, hump, referring to the pronounced, prolonged hump or chin of the lower leaf surface of the longer of the two leaves = *Argeta* N.E.Br. GC 1927: 113 **Typus** *Argeta petrensis* N.E.Br. ≡ **G. petrense** (N.E.Br.) Tischer = *Imitaria* N.E.Br. JB 1927: 348 **Typus** *Imitaria muirii* N.E.Br. ≡ **G. nebrownii** Tischer = *Mentocalyx* N.E.Br. GC 1927: 251 **Typus** *Mentocalyx muirii* N.E.Br. ≡ **G. schwantesii** Tischer = *Rimaria* N.E.Br. GC 1925: 413 non L.Bolus NM 1937: 69 **Typus** *Rimaria heathii* N.E.Br. ≡ **G. heathii** (N.E.Br.) L.Bolus.

Compact, caespitose or mat-forming plants, some with a thickish rhizome or rootstock with age, most species anisophyllous, few isophyllous; **L** always developing a chin or hump, or a finger-shaped protrusion of the lower leaf surface, at least on the longer leaf of an anisophyllous pair, basally connate forming a sheath, surface smooth to hairy, but always with some ornamentation on the surface of the epidermal cells, epidermis very variable in the formation of the outer wall; **Fl** apparently ebracteolate, **K** mostly 6, petals white to purple in different shades, filamentous staminodes absent or present, nectary glands mostly 6; **Fr** base elongate funnel-shaped, the stalk often winged, top ± flat, covering membranes ± complete, expanding keels diverging a little, valve wings broader than the expanding keels, rectangular, mostly 6 locules; **Ecol** mostly in flats or on gentle slopes with quartz pebbles, 125–250 mm rainfall p.a. mainly in March and November; **Distr** from Worcester to Calitzdorp, WC, S Africa.

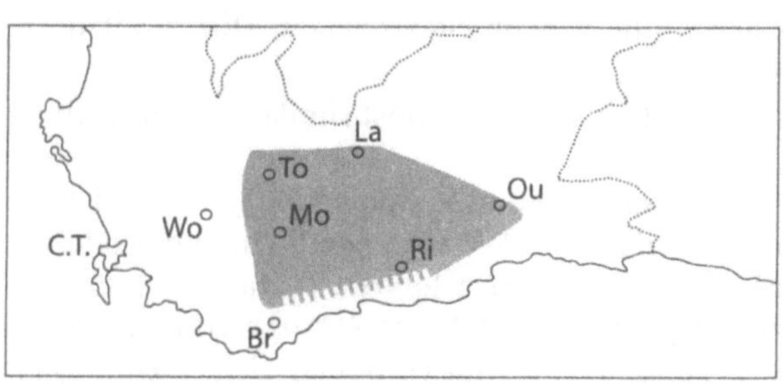

Distribution of **Gibbaeum** N.E.Br.

Note: Plants of the genus are conspicuous on the quartz fields of the Little Karroo, sympatry of two to three species is common. The plants are also popular in cultivation, but an accepted published revision is lacking. The circumscription of the genus follows convention, including a wide variation, particularly in leaf character states. *Argeta* appears to have been seen as a separate genus because of the isophylly and the firm, smooth leaves; *Imitaria* was based on the tube developed by the bases of the petals and the stamens, resembling **Conophytum;** *Mentocalyx* was described independently probably because it possesses soft, widely spreading leaves, and *Rimaria* because of the higher number of locules, on which ground it was at that time combined with species now placed in **Vanheerdea.** As features uniting all species of **Gibbaeum** can be seen the tendency to anisophylly, the ornaments or hairs on the epidermal cells of the leaves, the possession of separate nectary glands, and the development of rather soft fruits of the Drosanthemum-type. Haworth (1821: 104) already suggested that his section *Gibbosa* in **Mesembryanthemum** was a genus of its own. Since in the introduction of the arrangement (1821: 80) he explicitly mentions *M. glabrum* as the only definitely separated genus from **Mesembryanthemum, Gibbaeum** is not validly published there but must be cited as **Gibbaeum** N.E.Br. Wulff

(1944: 149–189) developed a subdivision of the genus into sections and subsections, based on chromosome numbers and leaf epidermis mainly. This is listed here, although later unpublished data (Glen 1974) suggest rearrangements. A concise revision will also have to include the illegitimate nomenclatural changes suggested by Heath (1993: 27–28), based on the erroneous assumption that Haworth was the author of the genus. Furthermore, a subgenus has been erected recently, which must also be considered when the taxonomy of the genus is revised.

List of validly published infrageneric taxa in **Gibbaeum** above the rank of species:

G. subg. **Imitaria** (N.E.Br.) van Jaarsveld & S.A.Hammer Bradleya 1996: 16 ≡ *Imitaria* N.E.Br. JB 1927: 348 **Typus** *Imitaria muirii* ≡ **G. nebrownii** Tischer

G. § **Archaeogibbaeum** Wulff 1944: 156
- sub§ **Protogibbaeum** Wulff 1944: 156, containing **G. perviride** and **G. luteoviride.**
- sub§ **Muiriopsis** Wulff 1944: 159, containing **G. gibbosum, G. molle,** and **G. pilosulum.**

G. § **Gibbaeum** (as *Eugibbaeum* Wulff 1944: 161)
- sub§ **Gibbaeotypus** Wulff 1944: 161, containing **G. pubescens, G. shandii,** and **G. geminum.**
- sub§ **Imitariopsis** Wulff 1944: 165, containing **G. angulipes, G. dispar,** and **G. album.**
- sub§ **Mentocalyx** (N.E.Br.) Wulff 1944: 167 ≡ *Mentocalyx* N.E.Br. GC 1927: 251, containing **G. velutinum, G. schwantesii,** and **G. pachypodium.**

G. § **Neogibbaeum** Wulff 1944: 169
- sub§ **Argeta** (N.E.Br.) Wulff 1944: 169 ≡ *Argeta* N.E.Br. GC 1927: 113, containing **G. petrense.**
- sub§ **Rimaria** (N.E.Br.) Wulff 1944: 172 ≡ *Rimaria* N.E.Br. GC 1925: 413, containing **G. heathii, G. comptonii,** and **G. luckhoffii.**

G. album N.E.Br. GC 1926: 215 (see **photo**) **T** syntypes Muir 3093, 3623 (K) = *G. album* N.E.Br. var. *roseum* N.E.Br. GC 1927: 430 ≡ *G. album* N.E.Br. forma *roseum* (N.E.Br.) G.D.Rowley NCSJ 1958: 77 **T** Muir 3975 (K).

Compact, always above the ground, 4–15 leaf-pairs, developed mainly towards the outer edge hence pressed laterally to the soil, the interior of old plants occupied by persisting old capsules and remains of old leaves, isophyllous; **L** lower leaf surface developed into a prominent chin, to almost semi-globose, yet the humps a little triangular, a slightly oblique keel running over the visible surface as if the leaf was folded by fingers there, the upper leaf surfaces appressed tightly, light greyish-white with a metallic colour, but not shining, because the outer surface is formed by short, highly branched, interlocking hairs consti-tuting a second layer above the epidermal cells; **Fl** Ø c. 30 mm, pedicels to 15 mm *l*, petals transgrading into filamentous staminodes, white to pink; **Fr** base with 2 wings transgrading into the winged pedicel, thus obscuring the shape of the capsule base, top a little raised, rims low, Ø 5–6 mm; **Chr** $2n = 18$; **Ecol** in fields of quartzite pebbles; **Distr** Riversdale, WC, S Africa.

Note: The leaves of this species resemble those of **Argyroderma congregatum** and **Antimima dualis** from the Knersvlakte.

G. album var. *roseum* = **G. album**
G. album forma *roseum* = **G. album**

G. angulipes (L.Bolus) N.E.Br. GC 1927: 430 (see **photo**) ≡ *M. angulipes* L.Bolus ABH 1925: 2 **T** Muir 17695 (BOL).

Creepers forming mats of up to 2 m Ø anisophyllous; **L** erect, apically a little recurved, terete, apically moderately acuminate, one leaf with a more pronounced hump overtopping the other, which was appressed when younger and has a lower or no hump at all, epidermis with branched papillae in the upper part, straight and brushed downwards in the lower part of the leaf, with occasional semi-globose tiny bulges; **Fl** Ø c. 25 mm, pedicels 20–24 mm *l*, petals pinkish-purple; **Fr** pedicels terete, base short funnel-shaped, top semi-globose, rims low; **Chr** $2n = 18$; **Ecol** in sand mixed with some pebbles of quartzite, settlers on disturbed ground, when not interferred with; **Distr** Riversdale, WC, S Africa.

Note: **G. angulipes** forms huge mats, developing flowers on all short side branches all over the plants.

G. argenteum = **G. pubescens**
G. austricolum ≡ **G. haaglenii**

G. blackburniae L.Bolus NM 1937: 65 **T** Blackburn 21819 (BOL).

Compact, above the ground, 15–20 heads somewhat anisophyllous; **L** a pair forming a subglobose body, the fissure between the upper leaf surfaces to 6 mm wide, one leaf a little smaller, each almost semiglobose in shape, but with a low keel in oblique position; **Fl** pedicels 20–25 mm *l*, **K** 6–7, petals pink, inconspicuously vittate, to 15 mm *l*, to 1.5 mm *b*, filamentous staminodes present; **Fr** Ø 5–6 mm; **Chr** $2n = 18$; **Ecol** on clayey ridges; **Distr** Calitzdorp, WC, S Africa.

Note: This species is most similar to **G. heathii.**

G. comptonii (L.Bolus) L.Bolus NM 1937: 65 ≡ *Rimaria comptonii* L.Bolus NM 1932: 369 **T** syntypes Compton 2324/31, L.Bolus 728/32, Joubert s.n. (all BOL) = *Rimaria comptonii* forma L.Bolus NM 1933: 411 nomen illeg. ≡ *G. comptonii* forma (L.Bolus) H.Jacobsen 1949: 58 nomen illeg.

Plants with 8–10 cm Ø, furnished in the centre with the remains of old stems, surrounded by growing ones around the outer edge, the thick root to 8 cm *l*, apically 25 mm Ø, slightly anisophyllous; **L** smooth to the touch, each leaf almost semi-globose with an obscure keel; **Fl** Ø c. 29 mm, pedicels to 20 mm *l* in fruit, petals pink, inconspicuously vittate, to 14 mm *l*, to 1.5 mm *b*, filamentous staminodes none; **Fr** 7 mm *l*, elevated for 2 mm above, Ø 10 mm, expanded to 17 mm, 6 locules; **Chr** $2n = 18$; **Ecol** not given; **Distr** Ladismith, WC, S Africa.

Note: The species appears most similar to **G. heathii**. The "forma" is not a name because it has no epithet since its first publication.

G. cryptopodium = **G. nuciforme**
G. digitiforme = **Phyllobolus digitatus** ssp. **digitatus**

G. dispar N.E.Br. GC 1926: 215 (see **photo**) T Muir 3797 (K).

Compact, growing above the ground, the new leaves developed around the outer edge, the interior of the plant filled with remains of old leaves and erect, persisting fruits; **L** almost equal in size, the upper leaf surfaces appressed closely and invisible, the lower leaf surfaces produced into a well developed chin, the pair resembling a lozenge with extremely rounded corners as seen from above, or a rather cornered egg, a keel only sometimes visible as an oblique ridge over the tops, epidermis cells of the upper leaf surface with short branched, not interlocking papillae, in the upper part of the leaf, longer papillae with distal branches in the middle, straight papillae in the lower part of the leaf, lower leaf surface with longer branched papillae; **Fl** Ø to 30 mm, petals pink, filamentous staminodes none; **Fr** stalks 5 mm *l*, terete, base funnel-shaped, top flat to conical, Ø 4–5 mm Ø; **Chr** $2n = 18$; **Ecol** on shaly bands; **Distr** Ladismith, WC, S Africa.

Note: In growth and leaf shape, **G. dispar** is most similar to **G. album**, which lacks the hairs typical of the former species.

G. dubium = **G. heathii**

G. esterhuyseniae L.Bolus NM 1958: 327 T Esterhuysen 5655 (BOL).

Compact, a little anisophyllous; **L** differing in size, but hardly in shape, both leaves of a pair ± trigonous above, deeper than wide, spreading with age, epidermis smooth; **Fl** Ø 35–50 mm, pedicels 9–14 mm *l*, petals pink, to 22 mm *l*, 1–1.5 mm *b*; **Fr** unknown; **Ecol** unknown; **Distr** Robertson, WC, S Africa.

Note: The species differs from **G. haaglenii**, occurring also in the southern part of the area of the genus, in its glabrous leaf surfaces and the less pronounced dorsal humps.

G. fissoides ≡ **Antegibbaeum fissoides**

G. geminum N.E.Br. GC 1922: 129 (see **photo**); GC 1926: 215 T Pole Evans 6925 (K).

Creeping plants forming extensive, loose mats, the branches often 1 m *l* or longer; **L** erect, those of a pair different: a longer one finger-shaped with a lateral excavated impression basally where the young shorter leaf once was placed, the shorter one semi-finger-shaped because its upper leaf surface was completely appressed to the longer leaf during the earlier states of development, epidermis cells turret-shaped with branched short protrusions, interlocking on the upper leaf surface; **Fl** Ø 13 mm, pedicels to 15 mm *l*, petals purple or a little paler, filamentous staminodes pink, apically white; **Fr** stalks to 20 mm *l*, terete with very thin, narrow wings, base funnel-shaped, top almost semi-globose, rims low; **Ecol** in loamy soils above shaly ground; **Chr** $2n = 36, 54$; **Distr** Worcester to Riversdale, WC, S Africa.

Note: In its creeping growth, **G. geminum** agrees with **G. angulipes**, differing in the less angular leaves and the more pronounced anisophylly from the latter. **G. geminum** appears to settle less readily on disturbed ground.

G. gibbosum (Haw.) N.E.Br. GC 1922: 151 (see **photo**) ≡ *M. gibbosum* Haw. 1795: 137 T?

Compact, growing flat in the ground or as a semi-globose body above the level of the soil surface, leaves and flowers, respectively fruits, distributed all over the plants, the old leaves forming white sheaths around the bases of the leaf-pairs, anisophyllous; **L** the longer leaf trigonous above, the lower leaf surface well produced, almost scimitar-shaped, the upper leaf surface pressed closely against the smaller leaf and thus invisible, the shorter leaf with a less pronounced hump on the lower leaf surface, the upper leaf surface completely invisible during the dry season, the leaves of a pair spreading in wet conditions, otherwise forming a rounded pyramid during the dry season, epidermis cells cobblestone-shaped, densely arranged; **Fl** Ø 25–40 mm, pedicels 15–17 mm *l*, **K** 6(–7), petals 35–43, purple to pink, filamentous staminodes 30–40; **Fr** base funnel-shaped, the 2 wings merging into the wings of the stalk, top raised and rounded a little, 6 locules; **Chr** $2n = 18$; **Ecol** in gravel of mixed composition, often next to shrubs; **Distr** Ceres, Montagu, Laingsburg, Worcester, WC, S Africa.

Note: The species is one of the more prominent during the dry season because the yellowish- to bluish-green leaves form distinct patches.

G. haagei ≡ **G. haaglenii**; = **G. petrense**
G. haagei var. *parviflorum* = **G. haaglenii**

G. haaglenii H.E.K.Hartmann **nom. nov.** ≡ *G. haagei* Schwantes Beiträge zur Sukkulentenkunde und -pflege 1938: 89 nomen illeg. (as replaced synonym) ≡ *G. austricola* Glen in litt. nomen nudum **Lectotype**, here designated: Schwantes BSKP 1938: f on p 89 = *G. haagei* Schwantes var. *parviflorum* L.Bolus 1966: 128 **T** van Breda 1754/62 (BOL).

Compact, the assimilating leaves separated with age by persisting old leaves remaining on the old stems, pressing the growing branches outwards, anisophyllous; **L** both leaves with rounded keels and margins, the longer one with a gibbose keel, the shorter almost semi-lunate in section, epidermis in the upper parts of the leaves with stellate protrusions resembling the hairy cover of **G. album**, the arrangement becoming looser towards the middle, the hairs longer, the protrusions placed apically, until near the leaf base the hairs are almost straight and decumbent; **Fl** pedicel c. 10 mm *l*, ∅ 25–30 mm, petals c. 40, purple, filamentous staminodes up to 50; **Fr** base funnel-shaped, 2 lateral wings merging into broad wings along the stalk, thus obscuring the shape of the capsule base, top raised from high rims which recurve with age, 6 locules, ∅ 6–7 mm; **Ecol** in flats with quartzite; **Distr** Bredasdorp, WC, S Africa.

Note: Schwantes used the name *G. haagei* twice, once invalidly for a species now placed in the synonymy of **G. petrense** (Schwantes in Jacobsen 1935: 179), and the second time for the species described here (Schwantes BSKP 1938: 89). Glen, in his unpublished thesis of 1974, recognized that this later name was illegitimate because it was taking up an epithet used for a different species already, albeit invalidly. He suggested the name *G. austricola*, which has never been published. The new name **G. haaglenii** is chosen here combining the old name with Glen's, who noticed the mistake. Since this name is an avowed substitute for the illegitimate *G. haagei* Schwantes 1938: 89, another Latin description is unnecessary.

G. heathii (N.E.Br.) L.Bolus NM 1937: 65 (see **photo**) ≡ *M. heathii* N.E.Br. 1920: 67 ≡ *Rimaria heathii* N.E.Br. GC 1926: 135 **T** Pillans 890 (K) = *Rimaria heathii* N.E.Br. var. *major* L.Bolus SAG 1932: 331 ≡ *G. heathii* N.E.Br. var. *majus* (L.Bolus) L.Bolus ex H.Jacobsen 1960: 1154 nomen invalid. **T** Luckhoff 20811 (BOL) = *Rimaria heathii* N.E.Br. var. *elevata* L.Bolus SAG 1932: 331 ≡ *G. heathii* N.E.Br. var. *elevatum* (L.Bolus) L.Bolus ex H.Jacobsen 1960: 1154 nomen invalid. **T** L.Bolus 771/32 (BOL) = *M. fissum* N.E.Br. 1920: 67 nomen illeg. non Haw. 1795: 134 ≡ *Rimaria dubia* N.E.Br. GC 1926: 155 ≡ *G. dubium* (N.E.Br.) H.Jacobsen 1949: 58 **T** collector unknown s.n. (K).

Compact plants, mostly growing above the ground, rarely sunken partially, leaves formed all over the plant, but with age, the old stems densely covered in remains of dead leaves well visible in the centre, only a little anisophyllous; **L** the pair almost globose, but each leaf ± semi-globose, epidermis in a cobblestone-pattern, each cell with several central bulges; **Fl** ∅ 20–32 mm, pedicels enclosed between the leaves, later the subsequent leaf-pairs, **K** mostly 7–8, rarely 6 or 9, petals white to pale pink, filamentous staminodes none, c. 140 stamens, filaments white, anthers and pollen pale yellow; **Fr** base funnel-shaped, the 2 wings merging into the very broad wings of the stalk, thus obscuring the shape of the capsule base, top a little conically raised, often skewed because of the wedged position between the leaves, ∅ c. 10 mm, 7–8, rarely 6 or 9, locules; **Chr** $2n = 18$; **Ecol** in loamy soil with variable proportions of quartzite pebbles; **Chr** $2n = 18$; **Distr** Ladismith, Montagu, Swellendam, WC, S Africa.

Note: The inclusion of *G. dubium* under **G. heathii** is based on considerations of Brown (GC 1926: 155). The varieties described by Bolus under *Rimaria heathii* have been transferred without intention by Jacobsen 1960: 1154, assuming that Bolus had made transfers before.

G. heathii var. *elevatum* = **G. heathii**
G. heathii var. *majus* = **G. heathii**

G. helmiae L.Bolus NM 1933: 402 **T** Helm 1717/32 (BOL).

Compact, to 7 branches, old leaves enclosing the fresh leaf-pairs basally; **L** of a pair forming a globose to ovate body, the short fissure shifted to one side a little, thus separating a longer and a shorter leaf, tips of both always rounded, epidermis with erect papillae; **Fl** ∅ 20–28 mm, pedicels enclosed in the body, petals bright pink, filamentous staminodes white; **Fr** ∅ 5.5–6.5 mm, 6 locules, rarely 7; **Chr** $2n = 18$; **Ecol** in flats with white quartz pebbles; **Distr** Oudtshoorn, WC, S Africa.

Note: The species is very similar to **G. pilosulum**.

G. johnsonii ≡ **G. johnstonii**

G. johnstonii van Jaarsveld & S.A.Hammer Bradleya 1996: 14 ≡ *G. johnsonii* van Jaarsveld & S.A.Hammer in Pavelka 1997: 39, typographical error ≡ *G. nebrownii* Tischer var. *johnsonii* (van Jaarsveld & S.A.Hammer) Pavelka 1997: 39, nomen invalid. **T** van Jaarsveld & Beran 14205 (NBG).

Compact plants sunken halfway into the ground, 20–28 bodies, the truncate leaf-pair enclosed at least basally by papery but tough, white sheaths developed from the old leaves of the previous seasons; **L** of a pair different, the apex markedly open or windowed, bright to purplish green, to 28 mm *l*, epidermis with rather long hairs from a broad base, hairs branching apically; **Fl** appearing in July in habitat together with the new leaf-pair emerging from the sheaths, morphology and numbers as for **G. nebrownii**; **Fr** 6-locular; **Ecol** in shale bands; **Distr** Montagu, WC, S Africa.

Note: In most features and in the measurements given, this species agrees extremely well with **G. nebrownii**; this applies even to the branched hairs (Hammer, in litt., in contrast to the description). The differences lie in the higher branching of the plants of **G. johnstonii** and in its early flowering from the developing fresh leaves bursting the sheaths. Both species flower from the latest leaf-pair during the season. The transfer of the species to a variety of **G. nebrownii** by Pavelka is invalid because no type it cited; in addition, the wrong spelling "johnsonii" is used.

G. luckhoffii L.Bolus NM 1931: 232 ≡ *Rimaria luckhoffii* (L.Bolus) L.Bolus NM 1932: 368 **T** Luckhoff 19393 (BOL).

Compact, 5–7 cm *h*; **L** forming bodies to 18 mm *h*, ∅ to 15 mm; **Fl** ∅ to 20 mm, pedicels emerging from the leaves for c. 3 mm, **K** 6–7, petals pink, filamentous staminodes none; **Fr** ∅ 9 mm, expanded 14 mm, 7 locules; **Chr** $2n = 18$; **Ecol** in open barren area; **Distr** Ladismith, WC, S Africa.

Note: Similar to **G. heathii**, smaller in most measurements given.

G. luteoviride (Haw.) N.E.Br. GC 1926: 215 ≡ *M. luteoviride* Haw. 1812: 226 ≡ *M. perviride* Haw. var. β Haw. 1803: 37 ≡ *G. perviride* (Haw.) N.E.Br. var. *luteoviride* (Haw.) N.E.Br. GC 1922: 151 **T?**

With stems 5.4 to 7 cm *l*, anisophyllous; **L** triquetrous above, semi-cylindrical below, one with a longer hump at the tip, overtopping the other leaf, glabrous; **Fl** petals pale reddish, darker vittate, filaments white, anthers sulphur coloured; **Fr** 6 locules; **Chr** $2n = 18$; **Ecol** unknown; **Distr** unknown.

Note: Separated from *M. perviride* later, and perhaps only a yellowish-green coloured version of **G. gibbosum**.

G. marlothii N.E.Br. GC 1928: 492 **T** Marloth 13157 (K).

Compact; **L** less unequal than those of **G. gibbosum**, but forming an almost ovate, erect body like that species, pressed together during the dry season, spreading when water is received, epidermis glabrous; **Fl** ∅ 20–25 mm, pedicels 2-edged, petals purple, filamentous staminodes present; **Fr** 6 locules; **Ecol** unknown; **Distr** Ceres, WC, S Africa.

Note: Differing from **G. gibbosum** in the smaller size and the less unequal leaves.

G. molle N.E.Br. GC 1926: 216 **T** syntypes Muir 3210, 3796 (K).

Compact with a deeply descending rootstock, many growths above; **L** of a pair connate into a body resembling a shark's head, but often the leaves of a pair equal and with the short fissure on top, epidermis smooth; **Fl** pedicels enclosed completely by the leaves, petals rose-pink, filamentous staminodes not mentioned; **Fr** 7 locules; **Ecol** on Bokkeveld shales; **Chr** $2n = 18$; **Distr** Ladismith, Montagu, WC, S Africa.

Note: The species has been considered as synonym of **G. pilosulum**, but the description lacks any information on hairs, perhaps due to old age of the material.

G. muirii N.E.Br. GC 1926: 216 non Schwantes in Jacobsen 1935: 180 **T** Muir 3704 (K).

Plants with longer, prostrate branches and erect secondary growths, stems sheathed by remains of old leaves; **L** pair almost ovate, erect, one leaf longer and the tip more pronounced and rounded than in the other, epidermis smooth; **Fl** pedicels exceeding the leaves a little, petals rose or rose-purple, 14–18 mm *l*, c. 1.1 mm *b*, filamentous staminodes white; **Fr** 6 locules; **Ecol** unknown; **Distr** Montagu, WC, S Africa.

Note: Following the data in the description, the species appears most similar to **G. gibbosum**.

G. nebrownii Tischer 1937: 151 (see **photo**) ≡ *Imitaria muirii* N.E.Br. JB 1927: 348 **T** Muir 4021 (BOL).

Compact plants, for most of the year sunken in the ground, the ovate leaf-pair enclosed at least basally by papery but tough, white sheaths developed from the old leaves of the previous seasons; **L** of a pair different, the apex open or windowed to different degrees, 13.5–25 mm *l*, epidermis with rather long hairs from a broad base, branching apically 1–3 times, wax cover breaking into plates, fine sculptures amorphous; **Fl** petals 35–55, basally connate forming a tube of c. 6 mm length, free parts spreading, pale pink, whitish towards the entrance of the tube, filamentous staminodes 25–35, connate with the petal tube for c. 4 mm, white, erect, stamens 14–20, shorter ones c. at the entrance to the tube, longer ones a little higher, the 6 nectary glands at the base of the tube; **Fr** stalks winged, base a little rounded, top raised, but with low rims, 6 locules; **Ecol** in crevices on exposed, vertical shales; **Chr** $2n = 18$; **Distr** Montagu, WC, S Africa.

Note: In the flower developing a tube of petals and filamentous staminodes, the species differs from most others in the genus, but it shares its anisophylly and the development of medium long, apically branching hairs with several other species. The transfer of the species to **Gibbaeum** by Tischer is therefore followed here. **G. johnstonii** is most similar to this species.

G. nebrownii var. *johnsonii* ≡ **G. johnstonii**
G. nelii = **Antegibbaeum fissoides**

G. nuciforme (Haw.) Glen & H.E.K.Hartmann **comb. nov.** (see **photo**) ≡ *M. nuciforme* Haw. Observations on Mesembryanthemum ... 1795: 129 (as basionym) ≡ *Conophytum nuciforme* (Haw.) N.E.Br. GC 1922: 84 ≡ *Derenbergia nuciformis* (Haw.) Schwantes ZSK 1925: 138 **Neotype**, here designated: Glen 672 (BOL) = *M. cryptopodium* Kensit 1909: 150 ≡ *G. cryptopodium* (Kensit) L.Bolus NM 1933: 402 ≡ *Derenbergia cryptopodia* (Kensit) Schwantes ZSK 1925: 138 **T** Pillans 892 (BOL).

Compact in the ground, only the upper parts of the leaves raised during the rainy season, sheathed completely by the white remains of the old leaf-pair during the dry season, withdrawn into the ground and then level with the soil surface; **L** of a pair forming an ovate or subglobose body, connate for most of their length, the upper leaf surfaces therefore invisible, the short fissure only apically developed, yellowish-green during the rainy season, epidermis cells with short, fine hairs, at first erect, later decumbent and finally rubbed off, hence the leaves when withdrawing at the end of the season often glabrous, old leaves turning parchment-like, yellowish to ochre, sheathing the fresh leaves at first, but remaining around the bases of the branches for several years in habitat, contributing to the appearance of an untidy small mat; **Fl** pedicels enclosed in the body, petals pink, vittate; **Fr** stalks 10–20 mm *l*, with very narrow wings, base funnel-shaped, top conical, rims very low, 6 locules; **Chr** $2n = 18$; **Ecol** in loam covered with pebbles of quartzite in varying density; **Distr** Laingsburg, Montagu, Worcester, WC, S Africa.

Note: Following a note of Brown at K, he had recognized the identity of *M. nuciforme* and *M. cryptopodium*, but he never published the necessary new combination. Glen, in his unpublished revision (1974: 98), prepared the combination, but there is no valid publication of this, although Hammer (1993: 253) cited *G. nuciforme* (Haw.) Glen. This is, in the absence of an exact citation of the basionym, not a valid publication either. The authorship given here indicates the rather winding history of the species, the neotype is that chosen by Glen (1974), which agrees also with the figure of *M. cryptopodium* (Kensit 1909: t. 21 f C).

G. pachypodium (Kensit) L.Bolus NM 1928: 152 (see **photo**) ≡ *M. pachypodium* Kensit 1909: 152 **T** Pillans 898 (BOL).

Forming dense mats; **L** of a pair almost equal in size, erect to spreading, with soft hairs; **Fl** pedicels 55–100 mm *l*, petals pink, to 20 mm *l*, c. 1.5 mm *b*, filamentous staminodes numerous; **Fr** unknown; **Ecol** in stony flats; **Chr** $2n = 18, 36$; **Distr** Ladismith, WC, S Africa.

Note: Although the plants are the tallest in the genus, the species seems not to have been collected often.

G. perviride (Haw.) N.E.Br. GC 1922: 151 ≡ *M. perviride* Haw. 1795: 136 **T?**

Compact, perhaps developing a stem with age, anisophyllous; **L** linear-oblong, keel inconspicuous near the tip, glabrous, deep glossy green; **Fl** unknown; **Fr** unknown; **Chr** $2n = 18$; **Ecol** unknown; **Distr** unknown.

Note: Described by Haworth (1795: 136) together with *M. gibbosum*, a variety β later separated as the more yellowish-green *M. luteoviride*.

G. perviride var. *luteoviride* ≡ **G. luteoiride**

G. petrense (N.E.Br.) Tischer 1937: 151 (see **photo**) ≡ *Argeta petrensis* N.E.Br. GC 1927: 114 **T** Muir 3622 (K) = *G. haagei* Schwantes in Jacobsen 1935: 179 nomen invalid. ≡ *G. tischleri* Wulff 1949: 55 **Lectotype**, here designated: Wulff 1949: figure on p 55.

Compact above the ground, flowers all over the plant; **L** almost equal in size, spreading a little, trigonous with convex sides, keel and margins rather sharp, old leaves turning yellow and later white, parchment-like, epidermis with elevated cells shaped like cobblestones, but the upper edges adorned with round bulges, often 5 per cell; **Fl** pedicels enclosed by the base of the leaf-pair, petals light purple or dark pink, filamentous staminodes white; **Fr** often skewed and obliquely compressed, base obscured by 2 broad wings continuing along the stalk, top conical, rims low, 6 locules; **Chr** $2n = 18, 36$; **Ecol** in flats or on slopes rich in quartzite pebbles; **Distr** Riversdale, WC, S Africa.

Note: The leaves appear hard to the touch, similar to those of **G. album**, but they lack the distinct protrusions on the epidermal cells of the latter species, developing only protrusions which remind to some degree of star-shaped biscuits, with their rounded corners, placed densely in the manner of cobblestones. *G. haagei* is invalid because it lacks a Latin description; the name has been used again later for a different species, which caused a great deal of confusion. *G. tischleri*, the avowed substitute for *G. haagei*, is the valid name if the tetraploid plants are considered to represent a species of their own.

G. pilosulum (N.E.Br.) N.E.Br. GC 1926: 234 ≡ *M. pilosulum* N.E.Br. 1920: 98 ≡ *Conophytum pilosulum* (N.E.Br.) N.E.Br. GC 1922: 214 **T** Pole Evans 6927 (K).

Compact, the assimilating leaves enclosed by sheaths developed from the old leaf-pairs, yellow at first, later whitish, parchment-like, anisophyllous; **L** the apical fissure short, 3–5 mm *l*, epidermis with very long, erect, straight hairs; **Fl** ∅ 15 mm, petals bright purple; **Fr** unknown; **Chr** $2n = 18$; **Ecol** unknown; **Distr** Ladismith, WC, S Africa.

Note: The long, erect, fine hairs distinguish the species from all others; in shape of the body and colouring, the species is very similar to **G. nuciforme**.

G. pubescens (Haw.) N.E.Br. GC 1922: 129 (see **photo**) ≡ *M. pubescens* Haw. 1795: 138 T? = *G. argenteum* N.E.Br. GC 1921: 273 **T** Pole Evans 6932 (K).

Caespitose to almost semi-orbicular, the remains of the old leaves forming persisting sheaths, the free parts in four distinct rows, supporting the assimilating leaf-pair, like a column anisophyllous; **L** pair resembling a shark's head, one leaf longer and more prominently keeled, the other c. half as long, epidermis with long hairs adorned with tiny, globose protrusions along the entire length, bent downwards, resembling fur that has been combed; **Fl** pedicels narrowly winged, c. 20 mm *l*, ∅ 20–35 mm, petals 35–55, purple, filamentous staminodes 40–55, purple, stamens 85–135, filaments white; **Fr** pedicels with narrow lateral wings, later terete, base funnel-shaped, top almost semi-globose, rims low, 6 locules; **Chr** $2n = 18$; **Ecol** in flats with a dense cover of quartzite pebbles; **Distr** Ladismith, Swellendam, WC, S Africa.

Note: In habitat, the raised clumps of **G. pubescens** are very prominent and conspicuous in the low, shrubby vegetation because of the whitish colour of the leaves. Brown (1922: 129) placed *G. argenteum* in synonymy here, this view is followed based on the fact that the latter species is also described to possess long hairs "all pointing downwards", a feature apparently typical of **G. pubescens**, and in contrast to the epidermis of **G. shandii**.

G. schwantesii Tischer 1937: 152 ≡ *Mentocalyx muirii* N.E.Br. GC 1927: 252 ≡ *G. muirii* Schwantes in Jacobsen 1935: 180 nomen illeg. non N.E.Br. 1926: 216 **T** Muir 3892 (K).

Compact, anisophyllous; **L** spreading widely, so that the upper leaf surface is visible, the longer leaf mostly hook-shaped apically, the shorter originally appressed to it, hence a clear-cut margin visible on the longer leaf, epidermis with branching ends of elevated, basally almost conical cells; **Fl** pedicels 40–50 mm *l*, ∅ 33–54 mm, petals white, basally faintly pink, filamentous staminodes white, basally faintly pink, filaments white; **Fr** 6 locules; **Chr** $2n = 36$; **Ecol** in flats with quartzite pebbles; **Distr** Riversdale, WC, S Africa.

Note: Following Tischer (1937: 151–2), the species is kept up here, distinguished from the similar **G. velutinum** by more slender leaves, the longer more strongly incurved than the shorter, by the distinctly branching hairs on the epidermis, and by the white to pale pinkish petals.

G. shandii N.E.Br. GC 1922: 129 (see **photo**) ≡ *M. shandii* N.E.Br. GC 1921: 151 nomen nudum **T** Pole Evans 6921 (K).

Caespitose, anisophyllous; **L** body resembling a shark's head, the longer leaf almost finger-shaped, the shorter appressed, semi-lunate in section, epidermis with short, richly branching hairs, the short protrusions resembling those of **G. album**, but less interwoven here; **Fl** petals purple, filamentous staminodes white; **Fr** stalks terete, base short funnel-shaped, top conical, 6 locules; **Chr** $2n = 36$; **Ecol** on slopes with quartzite; **Distr** Swellendam, WC, S Africa.

Note: The species is often considered a synonym of **G. pubescens**, from which it differs mainly in two features: the nearly stellate hairs, as already noted by Brown (1922: 129), and the conical tops of the capsules.

G. tischleri = **G. petrense**

G. velutinum (L.Bolus) Schwantes ZSK 1927: 106 (see **photo**) ≡ *M. velutinum* L.Bolus ABH 1922: 124 non Dinter 1923: 149 nomen illeg. ≡ *Mentocalyx velutinum* (L.Bolus) Schwantes MDK 1929: 17 **T** collector unknown, 15195 (BOL).

Compact, flowers all over the plant, anisophyllous; **L** spreading widely exhibiting the upper leaf surface, longer leaf with a pronounced, triquetrous protrusion of the lower leaf surface, shorter leaf with only an indication of a keel, half as thick as the longer one, both with short hairs developing some protrusions apically and laterally, but not branching in a stellate manner at the end; **Fl** pedicels 20–25 mm *l*, ∅ 35–40 mm, petals 50–55, purple with a darker central stripe, filamentous staminodes 55–60, stamens c. 250; **Fr** base merging gradually into the stalk, thus appearing very long funnel-shaped, stalk only basally with very narrow wings, top with very high rims of c. 3 mm, 6 locules, ∅ 12–15 mm, length of fruit 30–40 mm; **Chr** $2n = 36$; **Ecol** in crevices of oblique steps of exposed shale in Rhenoster-bosveld; **Distr** Swellendam, WC, S Africa.

Note: Under a lens, the short hairs appear unbranched because the protrusions can only be guessed as roughenings of the surface, but the difference to the much longer, apically branching hairs of **G. schwantesii** is very obvious. Petals in this species appear to be always rather deeply purple and vittate, but the flowers possess filamentous staminodes, which are not mentioned in the original description.

Glottiphyllum RUSCHIOIDEAE

H. E. K. HARTMANN

Glottiphyllum N.E.Br. GC 1921: 311; Hartmann & Gölling 1994: 1–49 **Typus G. linguiforme** (L.) N.E.Br. **Etym** Gk *glottis*, tongue, *phyllon*, leaf, in allusion to the tongue-shaped leaves of some species.

Compact to mat-forming, few to many branches, mostly with short, fat storage rhizomes; tap-roots and adventitious roots thickened, stems ascending to decumbent, isophyllous in seedlings and in adults of some species, anisophyllous in adults of most species; **L** decussate when isophylly prevails, in ± two rows (distichous) on anisophyllous plants, shapes triangular or oval in section, thick or flat, then strap- or tongue-shaped, very soft in texture, epidermis with bladder cells, often elongated into hairs along the margins and reduced on sides, covered with wax; **Fl** seemingly solitary, but plants in reality in a monochasial anthoclade, pedicellate or sessile, **K** 4, petals deep yellow, rarely basally white, forming a broad, basally connate funnel of 30–70 elements, stamens 150–250, rarely to 500, staminodes absent, filaments lacking papillae, nectary apparently absent; **Fr** ± spongy, mostly with elongate bases and high tops, endocarpal closing bodies big, spongy, valve wings narrow, as awns, or absent, covering membranes undulate, resilient, the distal recurving touching the closing body, (5–)8–9(–14) locules; **S** papillate to echinate, rarely smooth, > 1 mm *l*; **Chr** $2n = 18$; **Ecol** in regions with bimodal rainfall, peaks in March and November, 125–500 mm rainfall p.a.; mostly in shale and sandstone, some species in clay, quartzitic pebbles or rocks; **Distr** EC, central and eastern WC, S Africa.

Note: Whereas the delimitation of the genus is clear, limits of species have been confused, partly due to the widespread formation of hybrids, in cultivation and in nature. Variability in leaf shapes is also common within species, but plants f a populations are very similar. The vegetative growth of the plants of **Glottiphyllum** is restricted to the seedling state until it has formed three leaf-pairs of equal sizes and shapes (isophylly). Anisophylly and the arrangement of the leaves in two rows, characteristic of most species, occurs in the adult, flowering state. All branches from the first flower onwards are thus anthoclades, i.e. flowering branches, and the visible leaves represent bracteoles of the terminal flower in each case.

Distribution of **Glottiphyllum** N.E.Br.

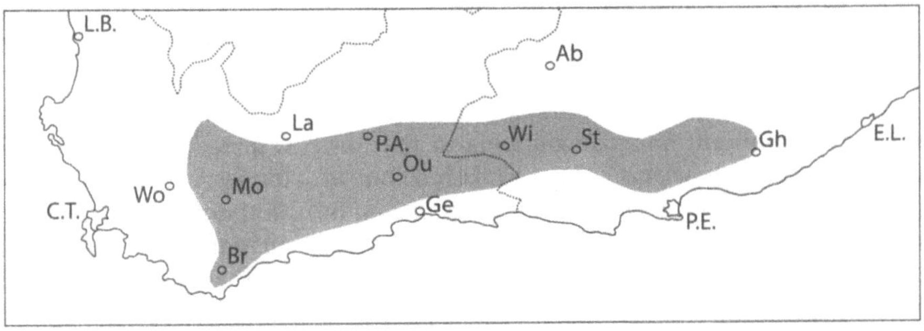

Key to the Species of Glottiphyllum

1. Adult leaves clearly decussate 2
– Adult leaves distichous or nearly so 8
2. Adult leaves markedly anisophyllous 3
– Adult leaves only inconspicuously anisophyllous 4
3. Longer leaf with a basal notch, shorter leaf extremely short .... **G. peersii**
– Both leaves with broad, fat marginal protrusions or teeth, shorter leaf reaching up to 60% of the longer leaf .... **G. difforme**
4. Leaf base broadened, capsules with very high tops and disintegrating stalks .... **G. cruciatum**
– Leaf base not broader than leaf, capsules with persistent stalks 5
5. Leaves bright green, no wax discernable, no valve wings or awns 6
– Leaves grey from a thick wax layer, valves with awns 7
6. Capsules somewhat woody, rounded, valves with low rims; leaves shorter than 60 mm .... **G. surrectum**
– Capsule base very spongy, elongate-conical, valves with high rims; leaves longer than 65 mm ...... **G. regium**
7. Leaves tapering regularly towards a pointed tip; capsules persisting in situ for several years .... **G. fergusoniae**

- Leaves scimitar-shaped, laterally compressed, with an impression from the opposite, shorter leaf; capsules tumble-fruits, leaving the mother plant soon .... **G. salmii**
8. Leaves covered by a wax layer that can be rubbed off **9**
- Leaves green, not or only very thinly covered by wax **13**
9. Capsules persisting on plants for several years .... **G. neilii**
- Capsules soon free as tumble fruits **10**
10. Margins of leaves rounded, many leaves per branch .... **G. oligocarpum**
- Margins of leaves distinct, few leaf-pairs per branch **11**
11. Leaves flat .... **G. linguiforme**
- Leaves thickened, apically with a chin (hunched) **12**
12. Seeds echinate, i.e. with very long papillae in dorsal region .... **G. carnosum**
- Seeds with long papillae of equal size .... **G. suave**
13. Capsules with high, spongy tops with thickened valves with low rims .... **G. depressum**
- Capsules with high or low rims, valves not thickened **14**
14. Leaves strap-shaped with sharp margins; seeds almost smooth on sides; locules 9 or more **15**
- Leaves with rounded margins marked by a row of dark spots from subhypodermal idioblasts; seeds papillate throughout; locules 8 or less .... **G. nelii**
15. Seeds 1.1–1.2 mm long, fruits stalked .... **G. longum**
- Seeds more than 1.6 mm long, tumble fruits .... **G. grandiflorum**

G. adscendens = **G. longum**
G. album L.Bolus ex Jacobsen 1955: 1388 nomen invalid. *Note:* The name has been used for different species with white flowers.
G. angustum = **G. cruciatum**
G. apiculatum = **G. cruciatum**
G. armoedense = **G. cruciatum**
G. arrectum = **G. surrectum**
G. barrydalense = **G. depressum**
G. buffelsvleyense = **G. depressum**

G. carnosum N.E.Br. GC 1927: 490 (see **photo**) **T** Muir 3894 (K!) = *G. jordaanianum* Schwantes 1954: 73 **LT** Schwantes 1954: 73.

Compact, 3–5 decumbent branches; **L** distichous, oval, gibbose with a humped lower side, ± erect, wax cover of dense platelets thick, sides of bladder cells moderately undulate, **L** of seedlings broadly acuminate; **Fr** tumble fruit, base somewhat woody, valves thick and high, rims low, valve wings as awns, 6–7(–8) locules; **S** papillate, longer in dorsal region, papillae densely covered with rodlets, 1.4–1.5 mm long, 1–1.2 mm *b*; **Ecol** on sandstone slopes, with **G. regium** and **G. cruciatum**; **Distr** Calitzdorp, Oudtshoorn, WC, S Africa.

Note: The species is similar to **G. linguiforme** and **G. suave** in oval leaf shape and wax cover, differing in capsule and seed characters mainly. The leaves are often smaller than those of **G. linguiforme** and, also in contrast to **G. suave**, often equipped with an undulate margin, but both distinctions are not constant throughout the populations.

G. cilliersiae = **G. linguiforme**
G. compressum = **G. regium**
G. concavum = **G. surrectum**

G. cruciatum (Haw.) N.E.Br. GC 1922: 9 (see **photo**) ≡ *M. cruciatum* Haw. 1795: 173 **NT** Salm-Dyck 1837: § 7 f 7 = *G. angustum* (Haw.) N.E.Br. GC 1921: 327 ≡ *M. angustum* Haw. 1795: 176 **T?** = *M. angustum* Haw. var. *pallidum* Haw. 1803: 101 **T?** = *G. apiculatum* N.E.Br. GC 1927: 449 **T** Marloth 12187 (K!) = *G. armoedense* Schwantes 1954: 75 **LT** Schwantes 1954: 75 = *M. linguaeforme* Haw. var. *subcruciatum* Haw. 1821: 98 **T?** = *G. longipes* N.E.Br. GC 1927: 429 **LT** Salm-Dyck 1837: § 7 f 7 = *G. praepingue* (Haw.) N.E.Br. GC 1922: 9 ≡ *M. praepingue* Haw. 1795: 179 non Salm-Dyck 1849: § 285 nomen illeg. **T?** = *G. rosaliae* L.Bolus NM 1958: 328 **LT** du Plessis 54556 (BOL!).

Mat-forming, stems decumbent; **L** decussate, ± erect, basally dilated and broader, acute, often curved or twisted, sides of epidermal bladder cells moderately undulate; **L** of seedlings acutely pointed; **Fl** long stalked; **Fr** stalks disintegrating, capsule a tumble fruit, top very high, valves very thick and spongy, base elongate-rounded, capsule thus resembling an acorn, valve wings broadened awns but no wings, mostly 8 locules; **S** echinate, i.e. long papillae in the dorsal region, the short rodlets on them reduced, 1.4–1.6 mm long, 1–1.2 mm *b*; **Ecol** mostly under shrubs in sandy-loamy soil, 300–400 mm rainfall p.a., often with **G. linguiforme** and **G. carnosum**; **Distr** Oudtshoorn, WC, S Africa.

Note: The acorn-shaped capsule distributed as a tumble fruit separates the species from all others.

G. cultratum = **G. longum**
G. davisii = **G. longum**

G. depressum (Haw.) N.E.Br. GC 1921: 327 (see **photo**) ≡ *M. depressum* Haw. 1803: 33 ≡ *M. depressum* Haw. var. *pallens* Haw. 1821: 99 **NT** Sims 1816: t. 1866 = *G. barrydalense* Schwantes 1954: 79 **LT** Schwantes 1954: 79 = *G. buffelsvleyense* Schwantes 1954: 76 **LT**

Schwantes 1954: 76 = *M. depressum* Haw. var. *lividum* Haw. 1821: 99 T? = *G. fragrans* (Salm-Dyck) Schwantes 1926: 242 ≡ *M. fragrans* Salm-Dyck 1820: 8 ≡ *M. linguiforme* L. var. *fragrans* (Salm-Dyck) A.Berger 1908: 242 **LT** Salm-Dyck 1842: § 8 f 2 = *G. framesii* L. Bolus NM 1929: 81 **T** Frames 824/29 (BOL!) = *G. haagei* Tischer 1928: 394 **LT** Tischer 1928: f on p 395 = *G. jacobsenianum* Schwantes 1954: 71 **LT** Schwantes 1954: 71 = *M. linguiforme* L. var. *angustum* (Haw.) A.Berger 1908: 241 nomen invalid. T? = *M. longum* Haw. var. *angustius* Haw. 1821: 96 **T?** = *M. longum* Haw. var. *declive* Haw. 1821: 96 T? = *M. longum* Haw. var. *flaccidum* Haw. 1821: 96 nomen illeg. non Haworth 1819: 222 T? = *M. longum* Haw. var. *purpurascens* Haw. 1821: 97 T? = *G. marlothii* Schwantes ZSK 1926: 242 **T** Marloth 10991 (HBG!) = *G. muirii* N.E.Br. GC 1927: 489 **T** Muir 3876 (K!) = *G. nysiae* Schwantes 1954: 74 **LT** Schwantes 1954: f on p 74 = *G. platycarpum* L.Bolus NM 1929: 81 **T** Ferguson 553/29 (BOL!) = *G. proclive* N.E.Br. GC 1927: 470 ≡ *M. angustum* Haw. var. *heterophyllum* Salm-Dyck 1849: § 7 f 6b nomen illeg. non Haw. 1821: 102 ≡ *M. linguiforme* L. var. *angustum* (Haw.) A.Berger p.p. (subvar. *heterophyllum* (Haw.) A.Berger 1908: 242) **T** Salm-Dyck 1849: § 7 f 6b = *G. rufescens* (Haw.) Tischer 1929: 106 ≡ *M. rufescens* Haw. 1819: 221 ≡ *M. linguaeforme* Haw. var. *rufescens* Haw. 1821: 98 nomen invalid. T? = *G. starkeae* L. Bolus NM 1931: 404 **LT** Starke 102/27 (BOL!) = *G. taurinum* (Haw.) N.E.Br. GC 1922; 9 ≡ *M. taurinum* Haw. 1819: 224 T? = *G. uniondalense* L. Bolus NM 1932: 362 **T** Leipoldt 19655 (BOL!).

Compact, moderately branched; **L** decumbent to ascending, several pairs to a branch, tips of leaves with a ± distinct hook from the marked impression of the opposite leaf, bladder cells in horizontal arrangement, anticlinal walls strongly undulate, **L** of seedlings acutely pointed; **Fr** spongy, tumble fruit, top very high, mostly higher than base, valves very thick, rims low, valve wings broad or reduced to awns; **S** papillate, dimorphic rodlets on the papillae, 1.1–1.35 mm long, 0.85–1.1 mm *b*; **Ecol** under shrubs on shaly sandstone or loamy soil, rarely in the open, rainfall p.a. < 400 mm, often the only species of the genus in the area, in particular in shrubby vegetation types; **Distr** from Ceres to Uniondale, WC, S Africa.

Note: Within the very wide geographical area, variation is considerable; yet, no arguments for a subdivision can be found (Hartmann & Gölling 1994).

G. difforme (L.) N.E.Br. GC 1922: 22 (see **photo**) ≡ *M. difforme* L. 1753: 487 non Thunberg 1800: 90 nomen illeg. **LT** Dillenius 1732: t. 194 f 241, 242 = *M. bidentatum* Haw. 1819: 89 **NT** Salm-Dyck 1837: § 7 f 1 = *M. bigibberatum* Haw. 1826: 328 **NT** Salm-Dyck 1854: § 7 f 4 = *M. heterophyllum* Haw. 1795: 420 T? = *G. semicylindricum* (Haw.) N.E.Br. GC 1922: 22 ≡ *M. semicylindricum* Haw. 1795: 238 **LT** Dillenius 1732: t. 194 f 241 = *G. subditum* N.E.Br. GC 1927: 409 ≡ *M. praepingue* Salm-Dyck 1849: § 7 f 5 nomen illeg. non Haworth 1795: 179 **LT** Salm-Dyck 1849: § 7 f 5.

Mat-forming, markedly anisophyllous; **L** decussate, erect with white, broadly dentate margins, bladder cells inconspicuous, moderately undulate, elongate along green parts of margins, **L** of seedlings equal and pointed; **Fr** stalks persistent, base of capsule spongy, valve rims visibly high, valve wings completely reduced; **S** echinate with long papillae in dorsal region, dimorphic rodlets on papillae dense, rarely reduced, **S** 1.1–1.45 mm long, 0.8–1.15 mm *b*; **Ecol** in open spaces between shrubs on shaly sandstone or in silty flats, 100–200 mm rainfall p.a., dry season in winter; **Distr** Aberdeen, Jansenville, Steytlerville, Willowmore, EC, Victoria West, NC, S Africa.

Note: The species often forms prominent monotypic stands. *M. heterophyllum* is placed here because the longer leaf is described as being similar to that of *M. difforme*, which is unique in the genus with its broad teeth on the margins.

G. erectum = **G. longum**

G. fergusoniae L.Bolus NM 1929: 58 (see **photo**) **T** Ferguson 18904 (BOL!).

Compact; **L** erect, decussate, pointed, base broad and pustulate, bladder cells inconspicuous with straight anticlinal walls, not larger along margins, wax in horizontal, rugose plates, often brownish; **L** of seedlings like adult leaves; **Fr** persisting on plant, held in the centre by remains of old leaves, base rounded or elongate, top moderately raised with low rims, valves with awns; **S** with long papillae with dimorphic rodlets, papillae longer in dorsal region, **S** 1.2–1.4 mm long, 0.95–1.1 mm *b*; **Ecol** in crevices in shale and sandstone, rarely on Witteberg Quartzite, always exposed, 200–300 mm rainfall p.a. mainly in March and November; **Distr** Montagu, Swellendam, WC, S Africa.

Note: Distinguished from all other species by its pointed, erect leaves which often appear brownish in habitat.

G. fragrans = **G. depressum**
G. framesii = **G. depressum**

G. grandiflorum (Haw.) N.E.Br. GC 1921: 327 (see **photo**) ≡ *M. grandiflorum* Haw. 1826: 327 ≡ *M. linguiforme* L. var. *grandiflorum* (Haw.) A.Berger 1908: 242 **NT** Salm-Dyck 1842: § 8 f 3 = *G. lombergii* N.E.Br. nomen nudum in schedis (K!).

Compact, with > 4 leaves to a branch; **L** distichous, strap-shaped with rounded margins, bladder cells reduced, in strictly horizontal arrangement along the margins; **L** of seedlings decussate, spreading, acute;

Fr tumble fruit, top and rims very high (3–4 mm), valves with awns, mostly 9 locules; **S** laterally compressed, testa cells low and elongate, almost smooth on sides, very few short rodlets, 1.6 mm *l*, 1.15 mm *b*; **Ecol** mostly in finely grained loamy or silty soils, < 400 mm rainfall p.a. in all seasons; **Distr** Albany, Alexandria, Kirkwood, Somerset East, EC, S Africa.

Note: In vegetative growth, the species is very similar to **G. longum**; distinctive characters of **G. grandiflorum** are very big seeds and a sessile capsule.

G. haagei = **G. depressum**
G. herrei = **G. suave**
G. jacobsenianum = **G. depressum**
G. jordaanianum = **G. carnosum**
G. latifolium = **G. linguiforme**
G. latum = **G. longum**
G. latum var. *cultratum* = **G. longum**
G. linguaeforme = **G. longum**

G. linguiforme (L.) N.E.Br. GC 1921: 327 (see **photo**) ≡ *M. linguiforme* L. 1753: 488 non Haworth 1795: 182 nomen illeg. **LT** Dillenius 1732: t. 183 f 224 = *G. cilliersiae* Schwantes 1938: 91 **T** lost = *G. latifolium* N.E.Br. 1931: 228 **T** Muir 4197 (K!) = *G. ryderae* Schwantes 1954: 77 **T** lost = *M. scalpratum* Haw. 1795: 187 nomen illeg. ≡ *M. linguiforme* L. var. *scalpratum* (Haw.) A.Berger 1908: 242 **Lectotype,** here designated: Dillenius 1732: t. 183 f 224.

Compact, mostly 3 decumbent branches; **L** distichous, ascending to erect, rather flat, oval, narrowing to the pustulate base, bladder cells moderately undulate, wax cover of small platelets dense and thick, greyish, often pinkish suffused, **L** of seedlings decussate, acutely pointed; **Fr** tumble fruit, top high, valves thick, rims low, base woody rather than spongy, locules 8(–9); **S** papillate, papillae with dimorphic rodlets, **S** 1.4–1.5 mm long, 1.05–1.2 mm *b*; **Ecol** on clayey to sandy soil with some quartzite pebbles, under bushes, 100–200 mm rainfall p.a. mainly in March and November; **Distr** Calitzdorp, Oudtshoorn, WC, S Africa.

Note: Similar to **G. carnosum** and **G. suave** in the oval leaf shapes and the thick wax cover of tiny platelets, but distinguished from both by mostly larger, flat leaves with smooth margins. Differing from **G. suave** in its bigger seeds, and from **G. carnosum** by its higher locule number. Haworth (1795: 187–188) published the new name *M. scalpratum* to replace **M. linguiforme** L., hence the former name is illegitimate (ICBN 2000: Art. 52.1).

G. lombergii = **G. grandiflorum**
G. longipes = **G. cruciatum**

G. longum (Haw.) N.E.Br. GC 1922: 9 (see **photo**) ≡ *M. longum* Haw. 1795: 177 ≡ *M. linguiforme* L. var. *longum* (Haw.) A.Berger 1908: 240 **LT** Dillenius 1732: t. 185 f 227 = *M. adscendens* Haw. 1812: 220 ≡ *M. linguiforme* L. var. *adscendens* (Haw.) A.Berger 1908: 241 **NT** Salm-Dyck 1849: § 8 f 4 = *G. cultratum* (Salm-Dyck) N.E.Br. GC 1921: 336 ≡ *M. cultratum* Salm-Dyck 1820: 7 ≡ *G. latum* N.E.Br. var. *cultratum* (Salm-Dyck) N.E.Br. GC 1927: 509 ≡ *M. linguiforme* L. var. *cultratum* (Salm-Dyck) A.Berger 1908: 240 **LT** Salm-Dyck 1849: § 8 f 5 = *G. davisii* L. Bolus NM 1933: 403 **T** Davis 2359/32 (BOL!) = *G. erectum* N.E.Br. GC 1927: 291 **T** Leslie s.n. (K!) = *G. latum* N.E.Br. GC 1921: 327 ≡ *M. latum* Haw. 1795: 186 non L.Bolus NM 1929: 141 nomen illeg. **LT** Dillenius 1732: t. 184 f 225 = *M. latum* Haw. var. *breve* Haw. 1821: 99 **T?** = *M. linguaeforme* Haw. 1795: 182 **T?** = *M. linguaeforme* Haw. var. *assurgens* Haw. 1821: 98 **T** = *M. linguaeforme* Haw. var. *prostratum* Haw. 1821: 98 **T?** = *M. linguiforme* L. var. *depressum* (Haw.) A.Berger 1908: 240 nomen invalid. **LT** Salm-Dyck 1849: § 8 f 7 = *M. longum* Haw. var. *attollens* Haw. 1821: 97 **T?** = *M. longum* Haw. var. *flaccidum* Haw. 1819: 222 non Haworth 1821: 96 nomen illeg. **T?** = *G. longum* (Haw.) N.E.Br. var. *hamatum* N.E.Br. GC 1927: 471 **T** Jackson in Andrews 1808: t. 540 = *G. longum* (Haw.) N.E.Br. var. *heterophyllum* (Haw.) G.D.Rowley 1973: 1 ≡ *M. angustum* Haw. var. *heterophyllum* Haw. 1821: 102 non Salm-Dyck 1849: § 7 f 6b nomen illeg. **T** Jackson in Andrews 1808: t. 540 = *M. lucidum* Haw. 1819: 89 **T?** = *M. medium* Haw. 1819: 88 **T?** = *G. obliquum* (Willdenow) N.E.Br. GC 1921: 336 ≡ *M. obliquum* Willdenow 1799: 1027 non Haw. 1821: 183 nomen illeg. ≡ *M. linguiforme* L. var. *obliquum* (Willdenow) A.Berger 1908: 241 **LT** Dillenius 1732: t. 185 f 226 ≡ *M. linguiforme* Haw. 1795: 182 nomen illeg. non L. 1753: 488 **Lectotype,** here designated: Dillenius 1732: t. 184 f 226 = *G. propinquum* N.E.Br. KB 1929: 57 **T** van der Bijl 32 (K!) = *G. pustulatum* (Haw.) N.E.Br. GC 1921: 327 ≡ *M. pustulatum* Haw. 1819: 88 ≡ *M. linguiforme* L. var. *pustulatum* (Haw.) A.Berger 1908: 240 **LT** Salm-Dyck 1849: § 8 f 10 = *M. pustulatum* Haw. var. *lividum* Salm-Dyck 1849: § 8 f 10b **T** Salm-Dyck 1849: § 8 f 10b = *G. uncatum* (Salm-Dyck) N.E.Br. GC 1921: 336 ≡ *M. uncatum* Salm-Dyck 1849: § 8 f 6 ≡ *M. linguiforme* L. var. *uncatum* (Haw.) A.Berger 1908: 241 ≡ *M. longum* Haw. var. *uncatum* Haw. 1821: 97 **T** Salm-Dyck 1849: § 8 f 6.

Compact to caespitose, three to five, later many branches; **L** pairs several to many per branch, ascending to decumbent, flat, strap-shaped with rounded, basally ciliate margins, bladder cells low, sides moderately undulate, wax cover thin, in dispersed platelets, **L** of seedlings decussate, pointed; **Fr** stalks persistent, 10–30 mm long, tops flat, rims low, base spongy, valve wings as awns, 9–11 locules; **S** flat with flat testa cells laterally, papillae low but elongate-oval in shape, short rodlets on testa mostly absent, **S** 1.05–1.3 mm *l*, 0.75–0.9 mm *b*, the most slender and smallest in the genus; **Ecol** in sandy-loamy soils under shrubs, 300–600 mm rainfall p.a.; **Distr** S districts of WC and EC, S Africa.

Note: The species occurs over the entire southern range of the genus, but so far little variation has been found. When in leaf, **G. longum** is so similar to **G. grandiflorum** that plants cannot be distinguished.

G. longum var. *hamatum* = **G. longum**
G. longum var. *heterophyllum* = **G. longum**
G. marlothii = **G. depressum**
G. muirii = **G. depressum**

G. neilii N.E.Br. GC 1928: 49 (see **photo**) ≡ *M. neilii* (N.E.Br.) hort. Kew ex Hill 1931: 129 **T** van der Bijl s.n. 10.5.1926 (K!).

Compact with two to three decumbent branches; **L** mostly distichous, old ones persisting for several years, mostly two fresh leaf-pairs, with a thick chin, margins sharp, bladder cells with undulate anticlinal walls, wax cover of platelets very thick, surface therefore grey, **L** of seedlings decussate and acuminate; **Fr** always persisting in the centre of the plant, held tightly by the old leaves, top flattish, base spongy, elongate funnel-shaped, sometimes tapering into a stalk, rims pronounced, locules 8(–10); **S** papillate, with dimorphic rodlets, 1.2–1.6 mm *l*, 1.05–1.25 mm *b*; **Ecol** in quartz patches in open vegetation, < 200 mm rainfall p.a.; **Distr** Prince Albert, WC, S Africa.

Note: **G. neilii** shares some features of the capsule with **G. nelii**: an elongated, funnel-shaped base that can taper into a stalk and the persistence of the fruit on the plant over many years.

G. nelii Schwantes MDG 1928: 92 (see **photo**) **NT** Nel 9827 (HBG) = *G. pallens* L.Bolus NM 1933: 389 **T** Peers 20423 (BOL!) = *G. pygmaeum* L.Bolus NM 1933: 390 **T** Peers 20422 (BOL!).

Compact, up to five branches, more with age; **L** several to many on a branch, distichous with dark, rounded margins dotted from subhypodermal idioblasts, mostly rounded, rarely sharp, bright green, wax scarce, bladder cells with moderately undulate anticlinal walls, **L** of seedlings short, acuminate, decussate; **Fr** persisting in the centre of the plant, held tightly by old leaves, tops high with very well developed rims to 4 mm *h*, base spongy, conical, often with edges, valves with narrow awns or without, 7–8 locules; **S** with shorter papillae on testa than **G. neilii**, covered densely with dimorphic rodlets, 1.3–1.6 mm *l*, 1–1.35 mm *b*; **Ecol** mostly on quartz under bushes or in the open, 100–200 mm rainfall p.a.; **Distr** Prince Albert, WC, Jansenville, Willowmore, EC, S Africa.

Note: The eastern populations possess leaves in mostly decumbent position with rounder margins marked by conspicuous dark dots. The western populations tend to exhibit sharper edges on slightly more ascending leaves.

G. nysiae = **G. depressum**
G. obliquum = **G. longum**
G. ochraceum ≡ **Malephora ochracea**

G. oligocarpum L.Bolus NM 1934: 450 (see **photo**) **T** Nel 9826 (BOL!).

Compact, 3–5 branches; **L** distichous, several pairs to one branch, rounded apically and along margins, bladder cells moderately undulate, wax cover of fine platelets very thick, surface therefore grey, **L** of seedlings rounded, decussate; **Fr** without stalks, tops raised, rims conspicuously high, bases spongy, conical, valves without wings or awns, locules 6; **S** echinate, rodlet dimorphism on testa inconspicuous, short rodlets rare, long rodlets < 5 µm, **S** 1.5–1.6 mm *l*, 1.15–1.25 mm *b*; **Ecol** in locally dry, open quartz patches, often in rough pebbles, 200–300 mm rainfall p.a.; **Distr** Jansenville, Steytlerville, Willowmore, EC, S Africa.

Note: This species is unique in two features within the genus: the seedlings develop rounded leaves, and the fruit possesses about six locules.

G. pallens = **G. nelii**
G. parvifolium = **G. surrectum**

G. peersii L.Bolus NM 1932: 362 (see **photo**) **LT** Peers 19654 (BOL!).

Mat-forming with ascending, later decumbent stems, **I** visible, but short, anisophyllous; **L** erect, decussate, long leaf of a pair 40–65 mm, short one 8–14 mm *l*, margins white from corky tissue, long **L** basally with a constriction, apically keeled, bladder cells with almost straight anticlinal walls, larger ones on margins, wax cover inconspicuous, surface therefore bright green, **L** of seedlings equal; **Fr** stalks persistent, tops raised, rims distinct, base spongy and conical, valves without wings or awns, locules 8; **S** with flatter elongate-oval cells on sides, not dorsally, rodlets dispersed, few short rodlets, **S** 1.2–1.4 mm long, 0.9–1.1 mm *b*; **Ecol** on quartz fields, often dominating, 100–200 mm rainfall p.a.; **Distr** Prince Albert, WC, Willowmore, EC, S Africa.

Note: The species shares the expressed anisophylly and the white corky margins as well as the mat-forming habit and the erect leaves with **G. difforme**, differing from the latter mainly in leaf shapes and seed characters.

G. platycarpum = **G. depressum**
G. praepingue = **G. cruciatum**
G. proclive = **G. depressum**
G. propinquum = **G. longum**
G. pustulatum = **G. longum**
G. pygmaeum = **G. nelii**

G. regium N.E.Br. GC 1928: 123 (see **photo**) T Muir 4192 (K!) = *G. compressum* L.Bolus NM 1931: 249 **LT** Nel 8979 (BOL!).

Compact mats with ascending, later decumbent stems, **I** longish, but coated by remains of old leaves; **L** decussate, very big (to 130 mm long), almost equal, trigonous to terete, bladder cells reduced, their anticlinal walls straight, horizontally elongated along margins, wax cover inconspicuous, bright green; **Fr** stalks long (> 30 mm), tops raised, rims very high, base an elongate cone tapering into the stalk, spongy, valve wings or awns absent, locules 8(–10); **S** papillate, dimorphic rodlets, but short ones rare, **S** 1.5 mm *l*, 1.1 mm *b*; **Ecol** on sandstone slopes or in sandy loam, often in dominating monotyic stands, 100–200 mm rainfall p.a.; **Distr** Calitzdorp, Oudtshoorn, WC, S Africa.

Note: The biggest leaves of the genus are found in this species.

G. rosaliae = **G. cruciatum**
G. rubrostigma = **G. surrectum**
G. rufescens = **G. depressum**
G. ryderae = **G. linguiforme**

G. salmii (Haw.) N.E.Br. GC 1922: 9 (see **photo**) ≡ *M. salmii* Haw. 1819: 89 **NT** Link & Otto 1823: t. 44 = *M. salmii* Haw. var. *angustifolium* Haw. 1821: 101 T? = *M. salmii* Haw. var. *decussatum* Haw. 1821: 100 T? = *M. salmii* Haw. var. *semicruciatum* Salm-Dyck 1820: 12 non (Haw.) A.Berger 1908: 236 T?

Compact with three to five leaf-pairs on each branch, only a little anisophyllous; **L** decussate, erect to ascending, scimitar-shaped with a well developed apical keel, humped, often mucronate, bladder cells with moderately undulate anticlinal walls, ciliate along the margins, wax cover thick, rugose; **Fr** stalks short, disintegrating, tumble fruits, tops high, rims very low, base moderately spongy, roundish, valves with awns, locules 8; **S** roundish with an inconspicuously protruding micropyle, papillae of equal length, moderately long and covered densely with dimorphic rodlets, 1.4–1.5 mm *l*, 1.1–1.2 mm *b*; **Ecol** in open places on shaly sandstone slopes, 100–200 mm rainfall p.a.; **Distr** Uniondale, WC, Willowmore, EC, S Africa.

Note: In vegetative growth, the species resembles a well developed seedling having branched more than usual.

G. semicylindricum = **G. difforme**
G. starkeae = **G. depressum**

G. suave N.E.Br. GC 83, 1928: 13 (see **photo**) T Muir 3989 (K!) = *G. herrei* L. Bolus NM 1931: 249 **T** Herre 10095 (BOL!).

Compact with three, rarely more decumbent branches, mostly two leaf-pairs to each; **L** oval above, margins sharp, thick below, but not humped, distichous, bladder cells with strongly undulate anticlinal walls, wax in a dense layer of small platelets, **L** of seedlings decussate and acute; **Fr** tumble fruits, tops moderately high (2.5 mm) with low rims, bases moderately spongy, conical, valve wings narrow, locules 8 or 9; **S** testa papillate, densely covered by dimorphic rodlets, short rodlets fewer, **S** 1.1–1.35 mm *l*, 0.95–1.1 mm *b*; **Ecol** on quartz or shaly sandstone, 100–200 mm rainfall p.a.; **Distr** Ladismith, Laingsburg, WC, S Africa.

Note: **G. suave**, **G. linguiforme**, and **G. carnosum** form a complex of species with greyish leaves from a thick cover of wax in platelets. **G. suave** possesses thicker leaves and smaller seeds than the other two species.

G. subditum = **G. difforme**

G. surrectum (Haw.) L. Bolus NM 1950: 236 (see **photo**) ≡ *M. surrectum* Haw. 1821: 101 **LT** drawing at Kew "216" (K!) = *G. arrectum* N.E.Br. GC 1922: 9 ≡ *M. arrectum* (N.E.Br.) hort. Kew ex Hill 1931: 126 T Pole Evans 6922 (K!) = *G. concavum* N.E.Br. GC 1927: 450 **T** Page 17361 (K!) = *G. parvifolium* L. Bolus NM 1931: 250 **T** Nel 9469 (BOL!) = *G. rubrostigma* L. Bolus NM 1929: 72 **LT** Rennie 1419/28 (BOL!).

Mat-forming, **I** separating leaves, but enclosed by the old leaf sheaths; **L** decussate, almost terete, basally pustulate, bladder cells horizontal, moderately raised, with almost straight anticlinal walls, margins sparsely ciliate, wax cover thin, breaking above the anticlinal walls into inconspicuous plates; **Fr** stalks short but persistent, rounded to globular, woody, the rims inconspicuous, 8 or fewer locules, valve wings or awns absent; **S** with papillae of equal length on the entire testa, rodlets dense, but probably not dimorphic, no small rodlets, **S** 1.3–1.4 mm *l*, c. 1.1 mm *b*; **Ecol** on decomposed shaly sandstone, 100–200 mm rainfall p.a.; **Distr** Ladismith, Laingsburg, Prince Albert, WC, S Africa.

Note: Slender leaves not widened at the base and rather round, woody capsules distinguish this species from **G. cruciatum**, which is similar in growth form and general appearence. Similar in leaf shape to **G. regium**, **G. surrectum** can always be recognized by its smaller leaves.

G. taurinum = **G. depressum**
G. uncatum = **G. longum**
G. uniondalense = **G. depressum**

Gunnia = Gunniopsis

H. E. K. Hartmann

Mueller (1859: 9) described the genus *Gunnia* F. Mueller to comprise a single species which is now placed in **Gunniopsis.** The name **Gunnia,** however, had been used before by Lindley (1834: t. 1699, in an appendix to the plate of **Oncidium ampliatum** Lindley), making the later homonym of Mueller an illegitimate name. *Gunnia* F.Mueller was later renamed *Neogunnia* by Pax & Hoffmann (in Engler & Prantl 1934: 225).

G. drummondii = **Gunniopsis septifraga**
G. septifraga ≡ **Gunniopsis septifraga**

Gunniopsis Aizooideae

H. E. K. Hartmann

Gunniopsis Pax in Engler & Prantl 1889: 44; Chinnock 1983: 133–179 ≡ *Aizoon* subg. *Gunniopsis* Pax & Hoffmann ex Adamson 1959: 30 **Typus G. quadrifida** (F.Mueller) Pax ≡ *Aizoon quadrifidum* (F.Mueller) F.Mueller **Etym** *R.C. Gunn*, 1808–1881, police magistrate, coroner, and naturalist, mainly on Tasmania = *Gunnia* F.Mueller 1859: 9 nomen illeg. non Lindley 1834: t. 1699 ≡ *Neogunnia* Pax & Hoffmann in Engler & Prantl 1934: 225 **Typus** *Gunnia septifraga* F.Mueller ≡ *Neogunnia septifraga* (F.Mueller) Pax & Hoffmann ≡ **G. septifraga** (F.Mueller) Chinnock.

Annual to perennial herbs, some with a woody base, or woody shrubs; **L** opposite, basally free or connate, bladder cells ± prominent; **Fl** appearing solitary often because the bracts are leaf-shaped, but **Inf** really in sometimes very big cymes, perigone with 4 lobes, basally connate to free, stamens 4–30, in 4 groups or in several whorls, pollen punctate and with low, dispersed spinules, gynoecium semi-inferior to superior; **Fr** with expanding keels lateral on the valves which often split at the first opening, forming 8 parts, but only 4 valves present; **S** ovoid to almost circular, smooth to ornamented; **Ecol** in dry areas; **Distr** S part of Australia.

Note: This treatment is based entirely on the revision of the genus by Chinnock (1983: 133–179), including the literature references given there. As already indicated by Adamson (1959: 30) and confirmed by Chinnock (1983), the genus can be clearly separated from **Aizoon,** with which it had been united for a long time. A unique feature in **Gunniopsis** are the septicidal valves, not found in any other taxon of the family and therefore considered a synapomorphy of the genus.

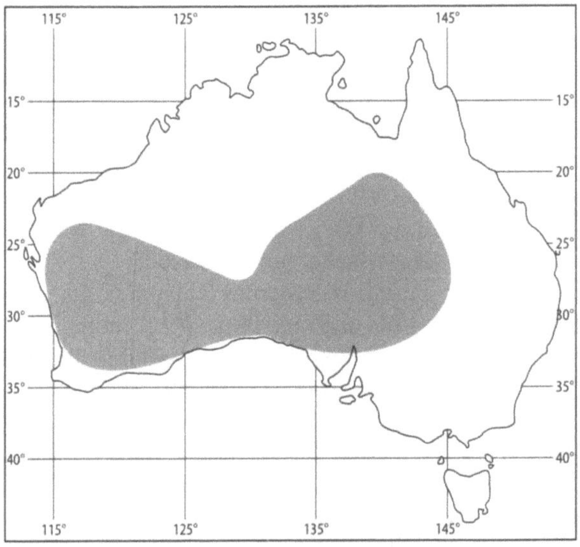

Distribution of **Gunniopsis** Pax in Engler & Prantl

Key to the Species of Gunniopsis

1. Stamens 4–12 2
– Stamens 30 or more 4
2. Leaves oblong to oblanceolate .... **G. septifraga**
– Leaves linear 3
3. Valves of capsule splitting .... **G. propinqua**
– Valves of capsule entire .... **G. rubra**
4. Leaves linear, terete 5
– Leaves flattened, ovate 6
5. Leaves and stems with white to yellow, erect hairs, perigone white inside .... **G. quadrifida**
– Leaves and stems glabrous, leaf bases ciliate, perigone yellow-green inside .... **G. tenuifolia**
6. Branches glabrous 7
– Branches with hairs or scales 13

7. Branches wiry, ending in a solitary flower, perennials with prominent tap-roots .... **G. glabra**
- Branches succulent, flowers in cymes, annuals **8**
8. Perigone bright yellow inside **9**
- Perigone white or pink inside **10**
9. Branches and leaves shiny .... **G. intermedia**
- Branches and leaves dull .... **G. papillata**
10. Perigone pink inside .... **G. kochii**
- Perigone white inside **11**
11. Stamens in four bundles .... **G. rodwayi**
- Stamens in whorls **12**
12. Valves of capsules entire .... **G. divisa**
- Valves of capsules deeply divided .... **G. calva**
13. Leaves and branches with flat, appressed ± circular scales .... **G. calcarea**
- Leaves and branches hairy **14**
14. Stems woody at base at least .... **G. zygophylloides**
- Stems herbaceous **15**
15. Perigone pink inside .... **G. kochii**
- Perigone white or yellow inside .... **G. papillata**

G. calcarea Chinnock 1983: 156 (see **photo**) **T** Chinnock 5086 (AD).

Compact shrubs 10–20 cm *h*, 10–80 cm ∅, **I** with flat star-shaped scales when dry; **L** linear-lanceolate, covered densely by star-shaped papillae when young, later only sparsely scaly, 8–28 mm *l*, 1.7–4.5 mm *b*; **Fl** in loose cymes or solitary, perigone fused basally, green outside, turning red after fertilization, scaly, inside yellow, finely papillate, stamens in 2–3 whorls, filaments basally flat; **Fr** valves with acute distal elevations, valves cucullate when open, emarginate, 4–7 mm *l*, 3–6.7 mm *b*; **S** almost kidney-shaped, testa with prominently ovate cells in rows, dark brown to blackish-brown, 1–1.1 mm *l*, 0.8–1 mm *b*; **Ecol** on calcareous brown soils; **Distr** SW corner of Western Australia, SE corner of South Australia, Australia.

Note: The species is distinct in its preference for limestone and the possession of scaly, star-shaped papillae.

G. calva Chinnock 1983: 164 (see **photo**) **T** Eichler 18818 (AD).

Erect to prostrate herbs, **I** glabrous, often pustulate from salt crystals protruding from epidermal and sub-epidermal layers, striate when dry; **L** obovate to spatulate, obtuse, 10–35 mm *l*, 3–12 mm *b*, basal, vegetative ones somewhat bigger than the bracts and bracteoles higher up; **Fl** in cymes of several storeys, perigone fused basally, green and glabrous outside, white and finely papillate inside, stamens in 2–3 whorls, filaments flat below; **Fr** valves with distal protrusions when closed, bifid when open, appearing 8-locular when closed; **S** pear-shaped, testa cells raised, ovate, in rows, pale brown, shiny, 0.5–0.7 mm *l*, 0.4–0.6 mm *b*; **Ecol** around the margins of salt lakes; **Distr** SC part of South Australia, Australia.

Note: A species with a restricted distribution, differing at first sight from the also annual and glabrous **G. glabra** and **G. divisa** by the elaborate cymes.

G. divisa Chinnock 1983: 164 **T** Tyson 3 (MEL).

Erect to prostrate herbs, **I** glabrous; **L** linear, flat, 1–2 pairs clustered close to the flowers, 5–15 mm *l*, 1.3–2 mm *b*; **Fl** solitary at the end of branches or in small cymes, 1–2 pairs of bracteoles below the flower, perigone lobes free, green and glabrous outside, white and papillate inside, stamens in 2–3 whorls, filaments terete; **Fr** almost globose, valves acute, entire, expanding keels prominent along the margins of the valves, 4–7 mm *l*, 4–5.5 mm ∅; **S** pear-shaped, dorsal part broadened laterally, testa cells elongate, elevated, in distinct rows, dark brown, 0.7–0.8 mm *l*, 0.6–0.7 mm *b*; **Ecol** on a conglomerate rise; **Distr** W part of Western Australia, Australia.

Note: The species is based on material cited also as a syntype for **G. glabra** from which it differs by being annual and having entire valves.

G. glabra (Luehmann ex Ewart) Gardner 1930: 42 (see **photo**) ≡ *Aizoon glabrum* Luehmann ex Ewart 1908: 128 **LT** Heal s.n. (MEL).

Perennial plants with several short vegetative shoots forming a dense cushion and some elongate flowering branches bearing one flower each, both types of shoots probably annual only; **L** basal ones subterete, on flowering branches lanceolate, conduplicate, sheath-like and turning papery later, glaucous green, 15–45 mm *l*, 2–7 mm *b* in basal position, 4.6–26 mm *l*, 1.8–3 mm *b* on flowering branches; **Fl** solitary, terminating the fertile long shoots, perigone basally fused, glaucous to greyish-green outside, white and papillate inside, stamens in up to 9 whorls, filaments terete; **Fr** almost globose, valve splitting when open, capsule appearing 8-valved when closed,

4–7 mm *l*, 4–8 mm ∅; **S** ovoid, the dorsal region broadened, testa with big, distant papillae shaped as if dough had been pulled up with two fingers, but the tip blunt, black, 1–1.5 mm *l*, 0.7–0.8 mm *b*; **Ecol** on sandy soils, rarely on loam, among the margins of salt lakes; **Distr** SW part of Western Australia, Australia.

Note: The species is distinct in the genus because of its tap-root and the differentiation into a vegetative, rather compact basal part and emerging elongate flowering branches. These develop during spring, but the flowers only open in summer.

G. intermedia Diels in Diels & Pritzel 1904: 197 (see **photo**) **T** Diels 5462 (B).

Erect to prostrate herbs, 3–50 cm *h*, **I** green to reddish-brown when young, later straw-coloured and striate; **L** oblanceolate to spatulate, obtuse, channelled above, shiny, 21–72 mm *l*, 3.2–20 mm *b*; **Fl** in cymes, perigone fused basally, outside green to yellowish-green, inside bright yellow, fading to white, minutely papillate, stamens in 2 whorls, filaments basally flattened; **Fr** prominently 4-winged, valves cucullate when open, emarginate, 3.5–5.5 mm *l*, 3.3–4.8 mm *b*; **S** ovoid, testa cells elevated, basal part with 6–8 bulges, tip drawn out into a blunt end, fine sculpture dense, apparently amorphous, dark blackish-brown, 0.8–1 mm *l*, 0.6–0.8 mm *b*; **Ecol** on flats or sandy rises near salt lakes, often in shrubland; **Distr** SW part of Western Australia, Australia.

Note: Seeds of this species appear unique in the genus with their oddly shaped papillae and the remarkable fine sculpture of the testa.

G. kochii (R.Wagner) Chinnock 1983: 162 (see **photo**) ≡ *Aizoon kochii* R.Wagner 1904: 80 **T** Koch 354 (W).

Erect to prostrate herbs, **I** with densely placed, dome-shaped papillae; **L** oblanceolate, mostly papillate, 10–25 mm *l*, 3–9 mm *b*; **Fl** the plant consisting of several rich cymes only, perigone fused basally, outside green, papillate, inside pink, minutely papillate, stamens in 2–3 whorls, the outer filaments sometimes flat at their bases; **Fr** valves with prominent distal protrusions when closed, each bifid when open; **S** pear-shaped, dorsally very broad, testa cells ovate to round, a little elevated as if on a short stalk, anticlinal walls undulate, fine sculpture present, rodlets?, 0.5–0.7 mm *l*, 0.4–0.6 mm *b*; **Ecol** mostly on stony soils; **Distr** C part of South Australia, Australia.

Note: A species with a restricted distribution.

G. papillata Chinnock 1983: 160 (see **photo**) **T** Weber 5747 (AD).

Erect to prostrate herbs, **I** with dispersed, short, finger-shaped papillae when young; **L** oblanceolate, obtuse, papillate to various degrees, 18–36 mm *l*, 2.5–12 mm *b*; **Fl** the entire plant representing a richly branched cyme, perigone fused basally, outside green, papillate, inside white or yellow, stamens in 2–3 whorls, filaments terete or basally flat; **Fr** valves with distal protrusions when closed, bifid when open, 4.7–7 mm *l*, 4.5–5 mm ∅; **S** almost round, testa with distant, ovate to round, almost stalked papillae, the anticlinal walls undulate, dark brown, 0.9–1.1 mm *l*, 0.7–0.9 mm *b*; **Ecol** often on ephemeral river beds on saline loams or clays; **Distr** NE corner of New South Wales, S edge of Northern Territory, SW and W part of Queensland, South Australia, Australia.

Note: Forms with pink perigone lobes inside are considered erroneous identifications (Chinnock 1983: 161–162).

G. propinqua Chinnock 1983: 175 (see **photo**) **T** Wilson 7436 (PERTH).

Prostrate herbs, 3–10 cm *h*, 9–12 mm ∅, **I** smooth, often with protruding salt pustules, green, turning straw-coloured with age; **L** linear-lanceolate, channelled above, basally connate into a short sheath, papillate, 5–9.5 mm *l*, 2.3–5 mm *b*; **Fl** in rich cymes, perigone fused basally, outside green, inside pink or white, stamens 4–12 in groups of 1–3 alternating with the perigone lobes, filaments slightly flattened; **Fr** ovate to almost globose, valves deeply divided when open, 3 mm *l*, 3–3.6 mm ∅; **S** shell-shaped to triangular as seen from the back, testa cells ovate to globose, in distinct rows, 0.4–0.7 mm *l*, 0.35–0.5 mm *b*; **Ecol** on sandy stony loams or on lateritic outcrops; **Distr** CS part of Western Australia, Australia.

Note: The species looks similar to **G. septifraga** in growth, but differs in the ornamented, dorsally broadened seeds and the more deeply divided valves.

G. quadrifida (F.Mueller) Pax in Engler & Prantl 1894: 44 (see **photo**) ≡ *Sesuvium quadrifidum* F.Mueller 1859: 9 ≡ *Aizoon quadrifidum* (F.Mueller) F.Mueller 1861: 148 **T** Hergolt s.n. (?) **NT** Weber 5790 (AD).

Roundish shrubs 30–100 cm *h* and ∅, fresh **I** with white or yellow erect hairs like the leaves; **L** linear, semi-terete, obtuse, 10–50 mm *l*, 1–3 mm *b*; **Fl** solitary or in cymes, pedicellate, perigone basally fused, hairy outside, white and papillate inside, stamens numerous, in 4 bundles alternating with the perigone lobes, anthers white or green, rarely pink; **Fr** the 4 valves deeply bifid, fruit appearing therefore 8-locular when closed, 4.5–8.5 mm *l*, 5–9.5 mm *b*; **S** obliquely pear-shaped, testa with distant papillae, anticlinal walls undulate, black, 1.3–1.8 mm *l*, 1.1–1.5 mm *b*; **Ecol** on outwash plains, along drainage channels and on rises adjacent to salty or freshwater lakes, on sandy or clayey loam soils; **Distr** NE part of New South Wales, S part of Northern Territory, SW part of Queensland, South Australia, S part of Western Australia, Australia.

Note: The species can dominate suitable habitats, it can also occur together with members of the family Chenopodiaceae, especially in salty places. From west to east a general decrease in leaf size and robustness of plants has been observed.

G. rodwayi (Ewart) Gardner 1930: 42 (see **photo**) ≡ *Aizoon rodwayi* Ewart 1908: 128 **LT** Rodway s.n. (MEL).

Erect to prostrate herbs, **I** glaucous greyish-green, later with protruding salt pustules, turning straw-coloured when dry; **L** ovate to lanceolate, channelled, basally connate forming a clasping sheath, glaucous, greyish-green, 5.5–40 mm *l*; **Fl** in terminal cymes on branches in c. 3 storeys and additional lateral side branches, perigone fused basally, stamens in 4 bundles or almost forming a ring of 2–3 whorls, filaments terete; **Fr** almost globose with high ribs on the valves, these splitting when open, the capsule appearing 8-locular when closed; **S** pear-shaped to handle-shaped, the dorsal part broadened rather abruptly against the much thinner base, testa cells raised like dough taken with two fingers, tip blunt, sides with marked ribs as if subtending the papilla, black, 1.2 mm *l*, 0.7–0.8 mm *b*; **Ecol** in sand around salt lakes; **Distr** SW part of Western Australia, Australia.

Note: Occurring often together with **G. quadrifida** and at the upper edges of low shrublands. The growth depends largely on moisture availability, diameters of 50 cm have been found in favourable conditions (Chinnock 1983: 170).

G. rubra Chinnock 1983: 174 (see **photo**) **T** Chinnock 5268 (AD).

Prostrate herbs, 1 cm *h*, 2–12 cm ∅, **I** glabrous when fresh, striate when dry, green to red; **L** subterete, obtuse, flattened above, 3.5–10 mm *l*, 0.7–1 mm *b*; **Fl** in cymes, the central flower sessile, the lateral ones shortly pedicellate, perigone lobes free, outside green to red, rugose, inside green, smooth, stamens 4, alternating with the perigone lobes, filaments terete; **Fr** globose, valves entire, 2–2.6 mm *l*, 1.5–2.5 mm ∅; **S** pear-shaped, testa cells ± flat, light brown, shiny, 0.4 mm *l*, 0.25–0.3 mm *b*; **Ecol** on light brown sandy loam in open woodland of **Eucalyptus**; **Distr** SW part of Western Australia, Australia.

Note: Although **G. rubra** occurs in the same area as **G. septifraga**, it is found on "normal" soils and not in saline situations.

G. septifraga (F.Mueller) Chinnock 1983: 172 (see **photo**) ≡ *Gunnia septifraga* F.Mueller 1859: 9 ≡ *Neogunnia septifraga* (F.Mueller) Pax & Hoffmann in Engler & Prantl 1934: 225 **T** Hergolt s.n. (MEL) = *Gunnia drummondii* Bentham 1867: 327 ≡ *Neogunnia drummondii* (Bentham) Pax & Hoffmann in Engler & Prantl 1934: 225 **T** Drummond 241 (K).

Prostrate, caespitose herbs, 0.5–5 cm *h*, 2–15 cm ∅, **I** sparsely papillate, later striate with age; **L** subterete, obtuse, yellowish-green, turning straw-coloured with age, glabrous, 5–15 mm *l*, 1–4.8 mm *b*; **Fl** in numerous cymes forming the plant, perigone fused basally, stamens 4, rarely 8, alternating with the perigone lobes, filaments a little flattened towards the base; **Fr** terete and tapering into a blunt tip, valves splitting apically when open to various degrees, 2.5–4.5 mm *l*, 2.3–3.7 mm ∅; **S** pear-shaped, testa cells only a little elevated, rugose, whitish to hyaline, 0.5–0.7 mm *l*, 0.3–0.4 mm *b*; **Ecol** in salt pans or lakes or at their margins; **Distr** SE corner of New South Wales, S part of Northern Territory, South Australia, S part of Western Australia, Australia.

Note: This species grows in the most saline habitats of all, often forming dominating stands.

G. tenuifolia Chinnock 1983: 152 (see **photo**) **T** Chinnock 5492 (AD).

Flattish shrubs 50–90 cm *h*, 50–170 cm ∅, **I** minutely papillate when young, later ochre to straw-coloured, old **I** with a deeply fissured bark; **L** terete, acute, 28–65 mm *l*, 0.7–2 mm *b*, glabrous except for margins at base; **Fl** in terminal, aggregated cymes, bracts and bracteoles distinctly smaller than leaves, 6–16 mm *l*, hence the **Inf** well separated from the vegetative part of the plant, perigone basally fused, outside green to reddish, inside yellowish-green, finely papillate, stamens in rich bundles alternating with the perigone lobes; **Fr** valves emarginate when open, the parts rounded apically, 3.5–5.5 mm *l*, 3.5–6.7 mm *b*; **S** asymmetrically kidney-shaped, testa with low papillae, anticlinal walls undulate, blackish-brown, 1.4–1.8 mm *l*, 1.2–1.5 mm *b*; **Ecol** on stony slopes or in faults or depressions; **Distr** C part of South Australia, Australia.

Note: The species is similar to **G. quadrifida**, differing from the latter in the absence of hairs and the finer sculptured seeds as well as in the capsule valves, which are rounded in **G. tenuifolia**.

G. zygophylloides (F.Mueller) Diels in Diels & Pritzel 1904: 197 (see **photo**) ≡ *Aizoon zygophylloides* F.Mueller 1871: 129 **T** Herschel Babbage 99635 (MEL).

Shrubs 15–40 cm *h*, 15–80 cm ∅; **I** with dispersed, short hairs, later straw-coloured; **L** ovate to obovate, acuminate, base auriculate when dry, papillate and hairy to some extent, 16–60 mm *l*, 5–31 mm *b*; **Fl** in small, apparently lateral cymes, perigone basally connate, green and sparsely hairy outside, yellow and finely papillate inside; **Fr** valves cucullate when open, emarginate, 5–8 mm *l*, 4.5–8.5 mm ∅; **S** ovoid, testa with distinct dorsal rows of papillate cells, black, 1–1.2 mm *l*, 0.8–1 mm *b*; **Ecol** on shaly or quartzitic soils or in gravel along river courses; **Distr** S part of Northern Territory, N part of South Australia, Australia.

Note: The leaves possess moderately elevated bladder cells, which appear as low papillae, and short hairs bearing a broader distal head; the stem seems to possess finger-shaped hairs only.

Gymnopoma = Skiatophytum

H. E. K. Hartmann

In March 1928, N. E. Brown (GC 1928: 194) described *Gymnopoma* as a new genus, based on *M. tripolium*, which had already been named **Skiatophytum tripolium** by L. Bolus in December 1927 (SAG 1927: 435). The generic name **Skiatophytum** is, therefore, three months older and has priority over *Gymnopoma*, although the description of the former is only given in the form of a legend to the accompanying figure. In May 1928 (NM 1928: 124, 134), Bolus confirmed her nomenclatural conclusions formally.

G. tripolium ≡ **Skiatophytum tripolium**

Gynicidia = Mesembryanthemum

M. Gerbaulet

In 1960 the Nomenclatural Congress declared *Gynicidia* Necker (1790: 81), amongst other genera proposed by this author, invalidly published because of the equivocal nature of the protologue. Rothmaler (1941: 411, 413) recognized *Gynicidia* as a younger synonym of *Gasoul*, which in turn became illegitimate after the conservation of the name **Mesembryanthemum.** No species have been assigned to *Gynicidia*.

Halenbergia = Mesembryanthemum

M. Gerbaulet

This monotypic genus (Dinter KuaS 1937: 158) was based on a specimen of **M. hypertrophicum** from Halenberg, Namibia, which appears to have been exceedingly fat. The species is currently considered as part of **Mesembryanthemum** subg. **Opophytum** (Bittrich 1987: 73).

H. hypertrophica
≡ **Mesembryanthemum hypertrophicum**

Halimum ≡ *Halimus* = Sesuvium

H. E. K. Hartmann

The names *Halimum* and *Halimus* have been considered the same (e.g. Hiern 1898: 411–414). Kuntze (1891: 263) ascribed the name *Halimus* to "Rumpf" 1750: 165, which was published before 1753, the starting date for priority, and also obviously a mistake because Rumph gave **Crithmus indicus** as the name of the plant described and figured (Rumph 1750: 165, t. 72 f 1).

Linnaeus (1759: 1058) created the name **Sesuvium** for *Halimum* Loefling (1758: 191) because Linnaeus had used the name *Halimus* (1735: Y. Polygama) for a plant he later included in **Atriplex.** Furthermore, **Halimus** P.Browne 1756: 206 had been published before Loefling's application of the name. It remains uncertain whether Linnaeus had been aware of this latter legitimate name. *Halimus* has been used again by Wallroth (1822: 117), in the sense of *Halimus* L. 1735 and 1737, a name replaced by Aellen (1938: 121) with *Halimione* Aellen, which is considered to belong to **Atriplex** (Chenopodiaceae).

Only those specific names are of interest here that belong to **Sesuvium.** The genus name *Halimum*/*Halimus* has been used in that sense by two authors: Kuntze 1891: 263 and Hiern 1898: 411–414. From these, the synonyms listed here are derived.

H. congensis ≡ **Sesuvium congense**
H. crystallinus
= **Sesuvium mesembryanthemoides**
H. edmonstonei ≡ **Sesuvium edmonstonei**
H. maritimus ≡ **Sesuvium maritimum**
H. mesembrianthemoides
≡ **Sesuvium mesembrianthemoides**
H. portulacastrus
≡ **Sesuvium portulacastrum**
H. portulacastrus var. *crithmoides*
≡ **Sesuvium crithmoides**
H. sesuvioides ≡ **Sesuvium sesuvioides**
H. sesuvioides var. *angustifolius*
≡ **Sesuvium sesuvioides**
H. sesuvioides var. *reduplicatus*
≡ **Sesuvium sesuvioides**
H. sesuvioides var. *welwitschii*
≡ **Sesuvium sesuvioides**

Hallianthus RUSCHIOIDEAE

H. E. K. HARTMANN

Hallianthus H.E.K.Hartmann BJS 1983: 167; Hartmann BJS 1983: 143–169 **Typus H. planus** (L.Bolus) H.E.K.Hartmann **Etym** *H. Hall*, 1906–1986, gardener at Kirstenbosch, field botanist and conservationist, Gk *anthos*, flower.

Procumbent with a central head from which long, trailing branches emerge, bearing flowers distally, no adventitious roots, **I** dark brown to red, cortex shrivelling with age, a broad xylem developing in the interior; **L** triangular, trigonous to triquetrous to crescent-shaped, almost free, dark green, epidermis almost flat, wax cover in flat plates, one per cell; **Fl** in cymes of 3–5, rarely solitary, petals white or purple, rarely apically purple and basally white, many filamentous staminodes in a central cone surrounding the stamens, apically purple or yellow, elongating with age, pollen breviaxial with foveolate exine and big spinules; **Fr** stalks decumbent, the capsule bent to face upwards, Leipoldtia-type with a very high upper part, the covering membranes developing high subcentral protrusions, a closing rodlet at the distal opening and a rim above it, ± 10 locules; **S** light brown, pear-shaped, testa smooth to rough from slightly elevated cells; **Chr** $2n = 18$; **Ecol** on rocky outcrops or steps of granite, sandstone or quartz, in crevices or shallow pockets, winter rainfall < 200 mm p.a.; **Distr** Namaqualand, NC, Vanrhynsdorp, WC, S Africa.

Note: The genus has been separated from **Leipoldtia**, based on the flowers with their filamentous staminodes growing visibly during anthesis, similar to those of **Vanzijlia**, and on the inner capsule morphology resembling that of some species of **Cheiridopsis**; the pollen is a rare type within the subtribe Leipoldtiinae.

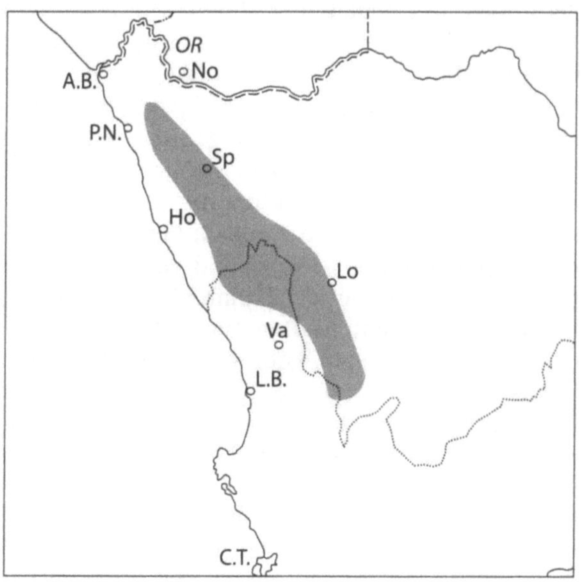

Distribution of **Hallianthus** H.E.K.Hartmann

H. planus (L.Bolus) H.E.K.Hartmann BJS 1983: 167 (see **photo**) ≡ *M. planum* L.Bolus ABH 1925: 12 ≡ *Leipoldtia plana* (L.Bolus) L.Bolus FPSA 1927: t. 260 **T** Watermeyer 1220/23 (BOL!) = *Leipoldtia compressa* L.Bolus NM 1930: 188 **T** Bolus 19116 (BOL!) = *Leipoldtia compressa* var. *lekkersingensis* L.Bolus 1967: 67 **T** Littlewood 939/62 (BOL!).

Growth as for genus, branches up to 20 years old, most leaves terminal in groups; **L** shapes as for genus, calcium oxalate crystals absent in stomata and subsidiary cells, which appear less raised than in **Leipoldtia**; **Fl** ∅ 26–35 mm, white petals often more slender than purple ones resulting in a more star-shaped appearance in white flowers whereas purple ones show a more continuous coloured area, 40–80 petals, 50–120 filamentous staminodes and 84–150 stamens, the lower numbers mostly in purple flowers; **Fr** top higher than basal part, 7–12 locules with a mean at 10, ∅ 6–8 mm; **S** 0,8–1,05 mm *l*, 0.4–0,6 mm *b*; **Ecol** in crevices or small pans always in exposed habitats with a good moisture supply; **Distr** as for the genus.

Note: Of the numerous distant populations, a great number can easily be recognized individually, e.g. by leaf shape or sizes. Obviously, gene exchange among the distant populations is rare so that a particular set of character states can be stabilized in one locality. The differences in flower colour and the number of elements appear insufficient for the distinction of two taxa, in particular in view of the fact that white and purple can also occur in one flower.

Hammeria RUSCHIOIDEAE

H. E. K. HARTMANN

Hammeria Burgoyne 1998: 204; Klak 2000: 38–39 **Typus** *H. salteri* Burgoyne = **H. meleagris** (L.Bolus) Klak **Etym** S.A. Hammer, * 1951, American specialist on Mesembryanthema, especially on the genus **Conophytum**.

Compact to creeping or erect and to 20 cm *h* when young, in this case spreading later, somewhat heterophyllous; **L** triquetrous, often rough from raised tannin idioblasts below the hypodermis,

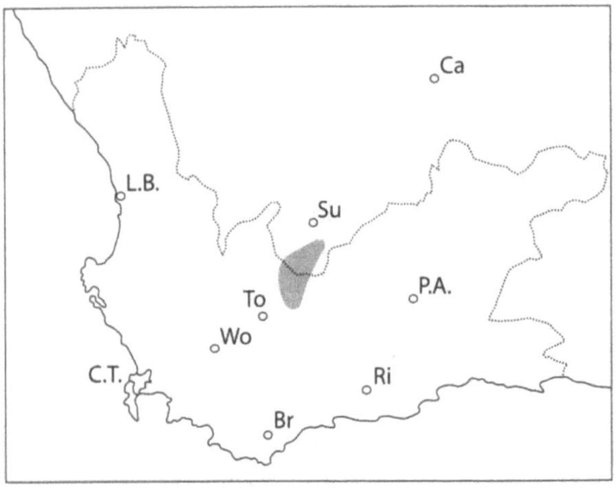

Distribution of **Hammeria** Burgoyne

dentate along the keel; **Fl** mostly solitary, petals pink to purple, vittate, filamentous staminodes pink, basally paler to white, surrounding the central cone of stamens, nectary in a ring; **Fr** similar to the Titanopsis-type with flexible covering membranes and broad valve wings, expanding keels almost parallel, yet widely separate, or diverging from the base, endocarpal closing body absent; **Ecol** < 100 mm rainfall p.a. in March and November; **Distr** Calvinia, NC, Ceres, Laingsburg, WC, S Africa.

Note: The genus was established to hold mainly the species *Ruschia salteri*, which had to be excluded from **Ruschia** because it lacks closing bodies and possesses broad valve wings. Klak (2000: 38–39) was recently able to transfer some species of **Lampranthus** to **Hammeria**, an action resulting in a new name for the type species and new synonyms for it. The description of a second species reported with closing bodies seems to cast doubt on the diagnostic character set of the genus.

Key to the Species of Hammeria

1. Plants caespitose to creeping, capsules without closing bodies .... **H. meleagris**
– Plants erect to sprawling, capsules with closing bodies .... **H. gracilis**

H. gracilis Burgoyne 1998: 207 T Hammer & Vlok 1406 (PRE!).

Young plants with erect branches, becoming straggly later, somewhat heterophyllous; **L** of one pair with distinct, yet small teeth along the apical part of the keel, wax cover continuous, breaking into horizontal plates later; **Fl** 1–3, pedicels to 35 mm *l*, **K** 5, petals pale magenta, vittate, filamentous staminodes?; **Fr** expanding keels diverging, placental closing bodies small, 5 locules; **S** smooth, 0.7 mm *l*, 0.5 mm *t*; **Ecol** in stony soils; **Distr** Sutherland, WC, S Africa.

Note: Examination of the type showed that the placenta ends in a knobby tip, reported as "small closing body", and that the covering membranes retain their shape. Types of species of **Ruschia** without fruits should be studied in more detail, perhaps revealing more members of **Hammeria**.

H. meleagris (L.Bolus) Klak 2000: 39 (see **photo**) ≡ *Lampranthus meleagris* (L.Bolus) L.Bolus NM 1939: 169 ≡ *M. meleagre* L.Bolus NM 1928: 17 T Pillans 6130 (BOL) = *Lampranthus longisepalus* (L.Bolus) L.Bolus NM 1939: 169 ≡ *M. longisepalum* L.Bolus NM 1930: 198 T Leipoldt 19135 (BOL) ≡ *Ruschia salteri* L.Bolus NM 1932: 371 ≡ *Lampranthus tanquanus* (L.Bolus) H.E.K.Hartmann Bradleya 1998: 70 **LT** Salter 2668 (BOL!) = *Lampranthus stoloniferus* L.Bolus 1965: 307 T Stayner 214/65 (BOL).

Compact when young, later developing some decumbent to creeping long shoots with few-leaved short shoots on them, **I** 20–35 mm, somewhat heterophyllous; **L** trigonous, lunate in side view, but the keel rounded, on some leaf-pairs the epidermis with elevations elongating into teeth along keel and margins, upper leaf surfaces appressed in the resting state, spreading in wet conditions, to 15 mm *l*, to 5 mm *b*, to 8 mm ∅; **Fl** solitary, pedicels erect, becoming S-shaped and decumbent with the ripening of the fruit, petals pink, vittate, 7–13 mm *l*, to 2 mm *b*, filamentous staminodes few, pale, filaments apically purple, anthers and pollen white; **Fr** stalk bent like a lying S, top almost flat, base a very short funnel, covering membranes almost complete, but thin and flexible and falling into the locules as in fruits of the Titanopsis-type once the capsules are empty, no distal closing device, closing body absent, the placenta forming a sausage-shaped bulge at its upper end, valve wings broad and almost rectangular, ∅ 6.5–7 mm, 5 locules; **S** c. 1 mm *l*; **Ecol** in finely grained Ecca shales or amongst rougher shales, < 100 mm rainfall p.a. **Distr** Calvinia, NC, Ceres, Sutherland, WC, S Africa.

Note: In the course of her ongoing study in **Lampranthus**, Klak (2000: 38–39) was able to identify the types of three species as belonging with *H. salteri*, one of them older and, therefore, the base of the valid name: **H. meleagris**.

Hartmanthus Ruschioideae

H. E. K. Hartmann

Hartmanthus S.A.Hammer Haseltonia 1995: 79 **Typus H. pergamentaceus** (L.Bolus) S.A.Hammer **Etym** *H.E.K. Hartmann,* * 1942, botanist at the University of Hamburg, Germany, *anthos,* flower.

Densely leaved, compact shrubs, with some elongate ± erect stems representing the dichasial **Inf, I** orange-coloured, flattened in the upper parts; **L** sharply triquetrous, pale glaucous, very smooth surface firm to the touch, withering into parchment-like dry leaves persisting for many years on the plant, epidermis with low papillae, higher papillae surrounding the sunken stomata in a cone, wax cover with dispersed, low platelets emerging from the continuous basal layer, to 70 mm *l*, c. 15 mm *b*, to 20 mm *t*; **Fl** in perennial cymes, the bracteoles persisting over several years, similar to those of **Ottosonderia,** ∅ 45–55 mm, **K** 5, 75–117 white to pink petals, 0–12 mostly white filamentous staminodes, stamens in a central cone, nectaries lophomorphic, 5 or an irregular ring; **Fr** top flat to forming a low cone, base elongate funnel-shaped, covering membranes thin and flexible, developed from tissue between the valve proper and the endocarpal expanding keels, therefore varying in extension from about half to completely covering the locule, neither closing devices nor a closing body present, expanding keels parallel for most of their length, valve wings very broad, 10–20 mm *l*, ∅ 5–10 mm, 5 or 8–10 locules; **S** elongate-ovate, reddish-brown, 0.9–1 mm *l*, c. 0.5 mm *b*, testa almost smooth; **Ecol** on quartzitic slopes, < 100 mm rainfall p.a. in winter with occasional thunderstorms in and around March to April; **Distr** Lüderitz-South, Namibia; NW part of Namaqualand, NC, S Africa.

Note: The genus was separated from **Delosperma** because of its thick, persisting, xeromorphic leaves; it also differs in its fruit from that genus: the covering membranes are much more developed than in any typical member of **Delosperma** and are very similar to those of **Juttadinteria** and related genera.

Distribution of **Hartmanthus** S.A.Hammer

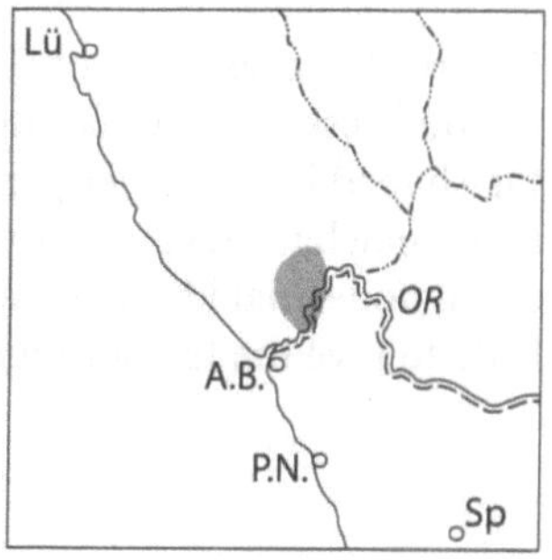

Key to the Species of Hartmanthus

1. Strictly 5 locules .... **H. pergamentaceus**
– Most fruits in a population between 8 and 10 locules .... **H. hallii**

H. hallii (L.Bolus) S.A.Hammer Haseltonia 1995: 80 (see **photo**) ≡ *Delosperma hallii* L.Bolus JSAB 1960: 49 T Hall 1923 (BOL).

Shrubs to 20 cm *h*, ∅ to 25 cm, forming big cushions with only short protruding cymes; **L** as for genus, epidermis rough like sandpaper, the cells around the stomata connate and forming a ring; **Fl** 200–250 stamens, nectary in an irregular ring of tiny glands; **Fr** covering membranes extensive to complete, 5–10 locules, most fruits in a population in the classes 8,9, or 10 locules; **S** c. 0.9 mm *l*, 0.45–0.5 mm *b*; **Ecol** as for genus, **Distr** Lüderitz-South, Namibia.

Note: Differing from **H. pergamentaceus** mainly by its nectary ring and the high number of locules, both features may be directly related. Similar correlations between the number of nectaries and locules are known from **Malephora,** certainly a parallel development. Shrubs of **H. hallii** appear mostly more compact, often becoming broader than high, than those of **H. pergamentaceus.** The roughness of the leaves is reduced in cultivation (Hammer Haseltonia 1995: 77–82).

H. pergamentaceus (L.Bolus) S.A.Hammer Haseltonia 1995: 80 (see **photo**) ≡ *Delosperma pergamentaceum* L.Bolus FPSA 1927: 278 **T** Pillans 5693 (BOL) = *Delosperma pergamentaceum* L.Bolus var. *roseum* Lavis 1967: 313 **T** Roux 519 (BOL).

Shrubs developing erect, high branches from a compact centre; **L** epidermis smooth, the cells surrounding the stomata mostly separated for some length; **Fl** 240–380 stamens, 5 koilomorphic, separated nectaries; **Fr** covering membranes extending over c. half the locule, strictly 5 locules; **S** 0.95–1.05 mm *l*, 0.55–0.6 mm *b*; **Ecol** as for genus; **Distr** as for genus.

Note: **H. pergamentaceus** possesses always five nectaries and five locules. In habitat, young plants tend to develop more prominently erect branches than those of **H. hallii,** but this feature is lost with age. Yet, plants of **H. pergamentaceus** most often become higher than broad.

Henricia ≡ Neohenricia

H. E. K. HARTMANN

When Bolus described the new genus *Henricia* (NM 1936: 39), she was unaware that the name had been used earlier by Cassini (1817: 11) for a genus in the family Asteraceae. As a consequence, *Henricia* L.Bolus is an illegitimate name. The error was corrected by naming the younger genus **Neohenricia** (L.Bolus JSAB 1938: 51).

H. sibbettii ≡ **Neohenricia sibbettii**

Hereroa RUSCHIOIDEAE

H. E. K. HARTMANN

Hereroa (Schwantes) Dinter & Schwantes ZSK 1927: 23 ≡ *Bergeranthus* Schwantes subg. *Hereroa* Schwantes ZSK 1926: 180 **Typus** *Bergeranthus puttkamerianus* (Dinter & Berger) Schwantes ≡ **H. puttkameriana** (Dinter & A.Berger) Dinter & Schwantes **Etym** *Herero*, tribe in Namibia.

Compact to caespitose or shrubby plants; **L** subterete or compressed, either falcate to acinaciform or scimitar-shaped in the upper part, epidermis cells smooth to somewhat elevated, often elevations above the subhypodermal idioblasts appearing as dots raised a little to give the leaf a rough appearance; **Fl** in moderate cymes, rarely solitary, opening in the afternoon or later, pedicels often longer than the leaves, in particular in wetter sites, **K** 5, petals mostly golden, filamentous staminodes absent, stamens in a central ring, erect, nectary as 5 separate, mostly lunate glands; **Fr** base short, funnel-shaped most often, top with rims of different heights, covering membranes complete or a little reduced by a shortening of the radial extension, additional closing devices at most as ledges, often absent, closing bodies absent, as knobs formed by the end of the placenta, or as small endocarpal protrusions often covered by a layer similar to that of the expanding sheet, expanding keels mostly distant from one

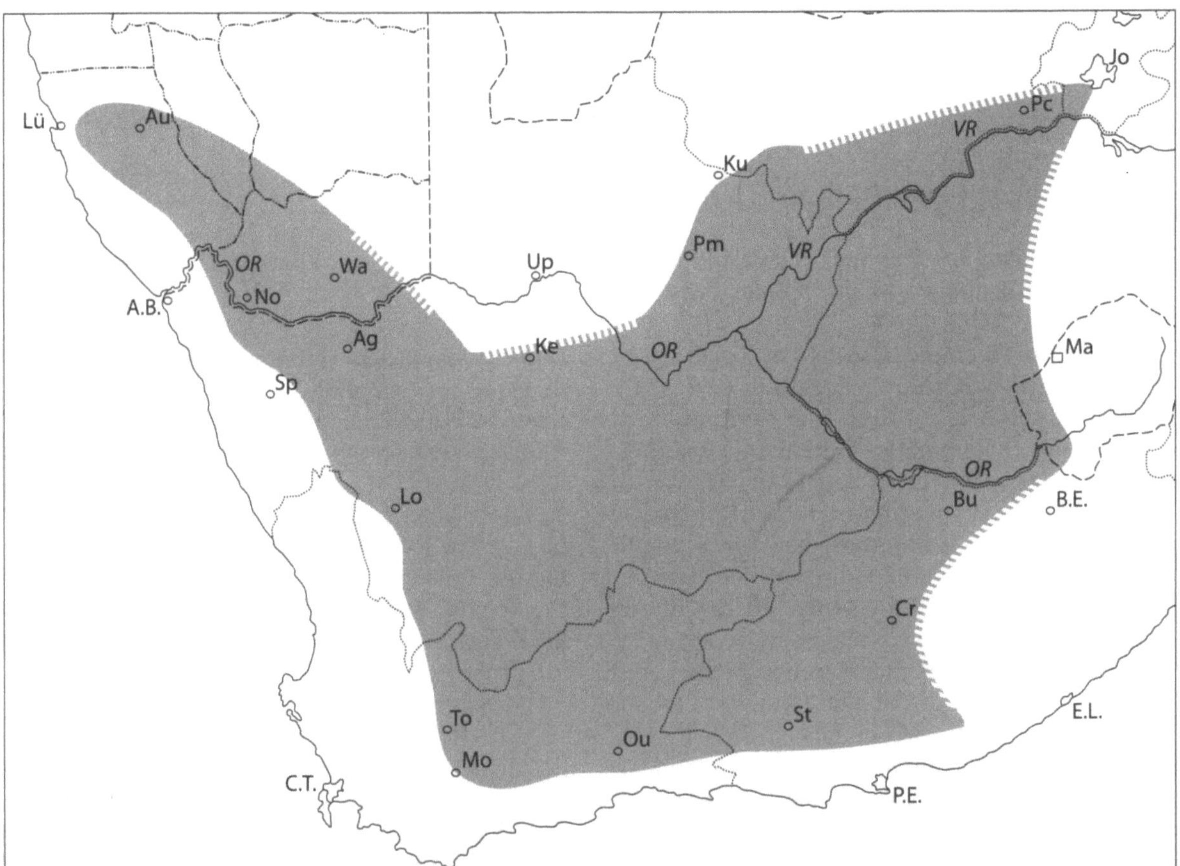

Distribution of **Hereroa** (Schwantes) Dinter & Schwantes

another, valve wings irregularly shaped: the tips and bases always narrower than the middle, 5, very rarely 6 locules; **S** light brown; **Ecol** mainly in the area receiving rainfall predominantly in March and November; **Distr** Lesotho; Keetmanshoop, Lüderitz-South, Warmbad, Namibia; EC, FS, GAU, NC, NW, WC, S Africa.

Note: At first described as a subgenus, the taxon was soon raised to genus rank, after two species had been excluded and transferred to **Ebracteola.** As understood here, the genus comprises all forms without or with small closing bodies and narrow valve wings with tapering tips and mostly narrow basal ends. The leaves elongate considerably reaching double their length in cultivation, but not gaining in width or thickness, as can be seen on various types and when comparing the descriptions and figures of **H. hesperantha** and its synonyms. Furthermore, leaves of a pair are often different in habitat, mainly in the formation and extension of the dorsal bulge or hump.

H. acuminata L.Bolus NM 1933: 390 **T** Jordaan 19657 (BOL!).

Compact, anisophyllous; **L** trigonous, a little incurved, one leaf almost straight, the other with a subapical hunch or hump, thus the apical part almost dolabriform, epidermis cells smooth, elevations inconspicuous, 50 mm *l*, 8 mm *b*, 8 mm *t*; **Fl** apparently solitary, but perhaps only young cymes, bracteoles at the base of the pedicel, with a long sheath and short free parts, petals yellow, apically reddish, to 16 mm *l*, 0.5–1 mm *b*, anthers and pollen golden, glands longitudinally rugose; **Fr** unknown; **Ecol** unknown; **Distr** Ladismith, WC, S Africa.

Note: Compact growth and anisophylly are typical of this species.

H. albanensis ≡ **Rhombophyllum albanense**
H. angustifolia = **H. puttkameriana**

H. aspera L.Bolus NM 1933: 391 (see **photo**) **Lectotype,** here designated: Davis 2448/32 (BOL!).

Shrubs to 8 cm *h*; **L** terete, obtuse, greyish-white, epidermis cells smooth, elevations on dry leaves present, to 25 mm *l*, c. 3 mm *b*, 3.5 mm *t*; **Fl** mostly solitary, bracteoles at the middle of the pedicels, ∅ 36 mm, petals golden, to 16 mm *l*, to 1.5 mm *b*, filaments golden, basally pale, glands distant; **Fr** base short funnel-shaped, top with rims to 1 mm *h*, covering membranes straight and distally a little recurved, closing body a small, endocarpal protrusion, expanding keels well separated basally, valve wings narrow, tapering to tips and base, 5 locules; **Ecol** in gravelly flats; **Distr** Swellendam, WC, S Africa.

Note: Rough, small, terete leaves characterize this shrubby species, similar in leaf shape to **H. odorata** and **H. tenuifolia.**

H. brevifolia L.Bolus NM 1929: 116 **T** Tugwell 18954 (BOL!).

Shrubs forming dense cushions up to 7 cm *h*, ∅ to 20 cm, **I** basally ∅ 25 mm; **L** of a pair differing a little in size, apically subtruncate to subacute, rarely submarginate, epidermis granular, 20–30 mm *l*, 5–9 mm *b*, 3–7 ∅, the subapical thickening 6–7 mm *t*; **Fl** in cymes, opening around noon, ∅ 29 mm, pedicels 60 mm *l*, bracteoles at their base, petals 9–13 mm *l*, yellow, apically deeper in colour, the outside paler, filaments white, anthers and pollen yellow, 5 glands conspicuously crenulate; **Fr** expanded 12 mm ∅, closing body small, expanding keel diverging widely, valves ending in awns; **Ecol** unknown; **Distr** Prince Albert, WC, S Africa.

Note: The type consists of pieces of a flower only.

H. calycina L.Bolus JB 1928: 200 **T** Grant & Blenkiron 165/27 (BOL!).

Compact, forming short, thick rhizomes with age, probably a little anisophyllous; **L** subfalcate to incurved, some leaves trigonous, others with a low subapical hump, epidermis rough, some cells along keel and margins protruding and almost denticulate, to 40 mm *l*, 8 mm *b* and *t*; **Fl** in cymes, ∅ 28 mm, petals yellow, outside reddish, 12 mm *l*, to 0.5 mm *b*, filaments basally paler, anthers yellow, glands yellowish-green; **Fr** unknown; **Ecol** unknown; **Distr** Queenstown, EC, S Africa.

Note: Material from the area with similar rough leaves differs in having very long pedicels on the fruits, which may have developed after the anthesis.

H. cana ≡ **Bijlia dilatata**

H. carinans (Haw.) Dinter & Schwantes ex H.Jacobsen 1933: 138, L. Bolus NM 1950: 233 ≡ *M. carinans* Haw. 1821: 90 ≡ *Bergeranthus carinans* (Haw.) Schwantes ZSK 1926: 180 **T?**

Compact; **L** erect to incurved, acinaciform, glaucous, with dots and rough; **Fl** petals yellow? **Fr** unknown; **Ecol** unknown; **Distr** unknown.

Note: Haworth coined the name for a plant sent to him by Salm-Dyck as *M. canum*, placed by the latter near *M. dolabriforme*. Haworth (1821: 90) suggested *M. albidum*, now in **Machairophyllum**, and *M. magnipunctatum*, now in **Pleiospilos**, as possible relatives.

H. concava L.Bolus JB 1928: 200 **T** L.Bolus 17514 (BOL).

Plants to 7 cm *h*, a little anisophyllous; **L** one leaf trigonous and acute, the other with a low subapical thickening, 30–40 mm *l*, 4–5 mm *b*, 2–3 mm *t*; **Fl** ∅ 30 mm, petals golden, apically red, just shorter than the calyx lobes, 8–13 mm *l*, 1–1.5 mm *b*, filaments yellow, basally white, anthers yellow, glands conspicuous, crenulate; **Fr** unknown; **Ecol** on stony hills; **Distr** Beaufort West, WC, S Africa.

Note: The species is not known well.

H. crassa L.Bolus NM 1932: 363 **Lectotype,** here designated: Archer 443 (chosen from 2 syntypes, BOL!).

Dense shrubs to 12 cm *h*, ∅ to 30 cm, **I** < **L**, reddish to grey, ∅ to 7 mm; **L** thick, falcate, keel rounded, epidermis a little rough, cells elevated a little only, 25–32 mm *l*, to 7 mm *b*, to 11 mm *t*; **Fl** solitary or in elongate monochasia, bracteoles enclosing the base of the flower, connate for most of their length, margins undulate and overlapping, petals golden, red outside, 10–16 mm *l*, up to 0.75 mm *b*; **Fr** base short, funnel-shaped, top with rims to 2 mm *h*, covering membranes almost straight, no closing devices at their ends, closing body absent, the end of the placenta flat, expanding keels well separated from each other basally, valve wings narrow and broadest in their middle, 5 locules; **S** 1.15 mm *l*, 0.82 mm *b*; **Ecol** on slopes with quartz rocks; **Distr** Beaufort West, Laingsburg, Prince Albert, WC, S Africa.

Note: The species is most similar in fruit morphology to **H. fimbriata** and **H. joubertii**, all sharing the absence of closing bodies and a reduction in the extension of the covering membranes. **H. crassa** can best be recognized in the complex by the almost connate bracteoles forming a cup below the flower. The pedicel elongates during the ripening of the fruit, but it remains short, being about as long as the diameter of the fruit.

H. dolabriformis ≡ **Rhombophyllum dolabriforme**
H. dyeri ≡ **Rhombophyllum dyeri**

H. fimbriata L.Bolus JB 1928: 198 **T** van der Bijl 18799 (BOL!).

Shrubs to 35 cm *h*, erect branches with leaves up to the **Inf**, **I** c. 20 mm *l*, ochre to brown, ∅ to 8 mm; **L** falcate, keel rounded, up to 50 mm *l* in the basal part of the long shoots, only c. 20 mm *l* in the upper parts of the shoots, 6–8 mm *b*, to 12 mm *t*; **Fl** mostly solitary, but several appearing in one season, separated by 2–4 leaf-pairs, pedicels to 20 mm *l*, raised well above the bracteoles forming a small cup, ∅ to 44 mm, petals golden, reddish outside, to 19 mm *l*, the glands almost connate; **Fr** base short, funnel-shaped, top with rims to 4 mm *h*, covering membranes reduced, without distal closing devices, closing body absent, placenta flat on the outer wall, expanding keels separated from one another basally, valve wings narrow, broadest in their middle, 5 locules, ∅ 10–12 mm; **S** c. 1 mm *l*, c. 0.7 mm *b*; **Ecol** in shaly rocky places; **Distr** Prince Albert, WC, S Africa.

Note: The big, densely leaved shrubs are conspicuous in habitat, also in their grey leaf colour; within the complex with **H. crassa** and **H. joubertii**, sharing the same fruit type, **H. fimbriata** can be recognized by long-stalked, big fruits and the possession of leaves on the entirely erect shoots.

H. glenensis (N.E.Br.) L.Bolus NM 1950: 233 (see **photo**) ≡ *Bergeranthus glenensis* N.E.Br. in Brown, Tischer & Karsten 1931: 120 **Lectotype,** here designated: Brown, Tischer, Karsten: f 23.

Compact with short, thick rhizomes; **L** dark green with visible darker dots, terete in lower half, a little trigonous towards the tips, narrowing gradually in breadth and thickness from the base, stomata sunken only a little, **L** 25–40 mm *l*, 5–8 mm *b*, 4–6 mm *t*; **Fl** 1–3, peduncle shorter than pedicels, ∅ to 30 mm, petals bright yellow, stamens entirely yellow, stigmas filiform; **Fr** up to 50 mm very long, top with high rims, base short, funnel-shaped, covering membranes almost straight, firm, with a low rim at the distal end, closing ledges inconspicuous, closing body very tiny, expanding keels ending in short awns, valve wings narrow, broadest in their middle, 5 locules; **S** reddish-brown, with low papillae, 0.9–1.1 mm *l*, 0.7–0.8 mm *b*; **Ecol** mostly in crevices of rocks or in pans, in grassland, often with shrubs, > 300 mm rainfall p.a. mainly in summer; **Distr** Bloemfontein, Fauresmith, Harrismith, Virginia, FS, Johannesburg, GAU, Kimberley, NC, S Africa.

Note: The species can best be recognized by its terete leaves tapering in width and thickness to the tip, lacking a humped keel typical of many members of the genus; this feature had probably lead to the original placement of the species in the genus **Bergeranthus**, which differs from **Hereroa** by broad closing bodies.

H. gracilis L.Bolus NM 1931: 268 **T** Nel 8988 (BOL!) = *H. gracilis* L.Bolus var. *compressa* L.Bolus NM 1931: 268 **T** Mathews 135/33 given as SUG 9472 in the description (BOL!).

Compact, later developing a thick rhizome, **I** invisible, enclosed; **L** with a subterminal thickening on one, often without it on the other leaf of a pair, rough in dry state, 35–50 mm *l*, 3–4 mm *b*, 3–6 mm *t*; **Fl** in cymes raised well above the plant, ∅ 24 mm, bracteoles small, pointed, petals bright yellow, outside reddish, filaments white, apically bright yellow, glands undulate, producing much nectar; **Fr** base short, funnel-shaped, top with rims to 1 mm *h*, covering membranes straight, without any additional closing devices, closing body a tiny endocarpal protrusion, expanding keels almost touching each other basally, valve wings very narrow, broadest in the middle, 5 locules, ∅ c. 10 mm; **Ecol** unknown; **Distr** Uniondale, WC, S Africa.

Note: The species appears most similar to **H. rehneltiana**, differing mainly in the more slender leaves and the narrower valve wings.

H. gracilis var. *compressa* = **H. gracilis**

H. granulata (N.E.Br.) Dinter & Schwantes ZSK 1927: 24 ≡ *M. granulatum* N.E.Br. 1920: 66 ≡ *Bergeranthus granulatus* (N.E.Br.) Schwantes ZSK 1926: 180 T collector unknown (K).

Caespitose, 3–4.5 cm *h*, 7–8 cm *h* when flowering; L thickening subapically, scimitar-shaped in the upper part, semi-terete in the lower, dot-like tubercles and flat dots, dull green with a glaucous tinge, 20–50.5 mm *l*, 5–7 mm *b*, at most 5–5.5 mm *t*; **Fl** 1–3, pedicels 40–50 mm *l*, Ø 25 mm, bracteoles acute, petals light yellow; **Fr** unknown; **Ecol** unknown; **Distr** unknown.

Note: Brown (1920: 66) based his species on material of unknown origin, citing also the description of *M. carinans* by Berger (1908: 245) as a reference and stating that *M. carinans* is two to three times larger than this species.

H. herrei Schwantes ZSK 1928: 302 T without collector 1432 (?).

In habit and **L** as **H. granulata**, perhaps less thickened subapically; **Fr** covering membranes not touching each other along their contact line, expanding keels parallel and close at their bases, dark brown, valve wings very narrow (< 0.75 mm), Ø 7 mm, expanded 12 mm Ø, 5 locules; **S** brownish-yellow, ovate to globose, testa with low papillae, 0.75 mm *l*; **Ecol** unknown; **Distr** Graaff Reinet, EC, S Africa.

Note: Data taken from the description only, because no material is available in the collection of Schwantes at HGB.

H. hesperantha (Dinter & A.Berger) Dinter & Schwantes ZSK 1927: 24 (see **photo**) ≡ *M. hesperanthum* Dinter & A.Berger BJS 1914: 589 T Dinter 1099 (B) = *M. karasbergense* L.Bolus ABH 1914: 17 T Pearson 8270 (BOL!) = *M. bergerianum* Dinter 1914: 37 T Dinter 3249 (B).

Rather dense shrubs 15–30 cm *h*, the flowering branches overtopping the plant; **L** trigonous with convex sides and rounded edges, thickest in the middle, epidermal cells flat, but elevations prominent, covered by a dense layer of wax, 10–30 mm *l*, c. 5 mm *b*, to 7 mm *t*; **Fl** 1–3, bracteoles 5–10 mm *l*, like miniature foliage leaves, golden petals > **K**; **Fr** with high valve rims, base short funnel-shaped, covering membranes ± complete with or without a low distal closing ledge, closing body very small, round, placental of origin, but placed on a protrusion of endocarpal origin, expanding keels not touching each other basally, valve wings narrow, broadest in their middle, 5 locules, very rarely 6, 15–23 mm *l*, 8.5–9.5 mm Ø; **S** c. 0.8 mm *l*, 0.6–0.65 mm *b*; **Ecol** in gravelly plains rich in decomposed gneiss and/or quartzite, < 100 mm rainfall p.a.; **Distr** Keetmanshoop, Lüderitz-South, Warmbad, Namibia.

Note: Whereas **H. puttkameriana** inhabits crevices in rocky outcrops, **H. hesperantha** settles in the surrounding plains.

H. incurva L.Bolus NM 1929: 83 T Pillans 415/16 (BOL).

Low plants; **L** incurved, acute to acuminate, glaucous green, apically reddish suffused, finely and densely punctate, to 35 mm *l*, to 7 mm *b*, 6 mm *t*; **Fl** solitary, Ø 36 mm, pedicels well above the leaves, to 50 mm *l*, bracteoles at their bases, petals yellow or golden, to 17 mm *l*, to 1 mm *b*, filaments pale yellow, anthers and pollen white; **Fr** unknown; **Ecol** unknown, **Distr** unknown, S Africa.

Note: Similar to **H. willowmorensis.**

H. joubertii L.Bolus NM 1935: 484 (see **photo**) **T** Joubert 122 (BOL!).

Low, densely leaved shrubs to 30 cm Ø, **I** ochre to grey; **L** subfalcate, those of a pair equal, epidermis almost smooth, elevations inconspicuous, 30–45 mm *l*, to 9 mm *b*, to 9 mm *t*; **Fl** solitary or in cymes, bracteoles spreading and only a little connate, pedicels elongate, petals 12–18 mm *l*, filaments white, apically yellow, glands almost connate; **Fr** base short, funnel-shaped, top with rims to 2 mm *h*, covering membranes reduced and without any distal closing devices, closing body absent, expanding keels almost touching each other basally, valve wings very narrow, broadest in their middle, 5 locules, Ø 8–10 mm; **S** 0.7–0.8 mm *l*, 0.55–0.62 mm *b*; **Ecol** in flats with broken shale; **Distr** Laingsburg, Prince Albert, WC, S Africa.

Note: In the complex of species with similar fruits and shrubby growth, **H. joubertii** can best be recognized by the short internodes, the long, fat leaves, and the spreading bracteoles.

H. latipetala L.Bolus JB 1928: 199 T van der Bijl 18800 (BOL!).

Erect shrubs to 9 cm *h*, Ø to 13 cm, **I** 5–9 mm *l*, Ø to 4 mm; **L** falcate, epidermis wrinkled in the dry state, but not rough, to 35 mm *l*, below the apex at the broadest point to 11 mm *t*; **Fl** in cymes of up to 7, bracteoles up to 5 mm *l*, pointed, the **Inf** thus clearly separated from the vegetative part of the plant, Ø to 32 mm, petals golden, red outside, to 13 mm *l*, 1 mm *b*, filaments golden, basally white, glands greenish-brown, distant; **Fr** base short funnel-shaped, top with rims to 1.5 mm *h*, covering membranes straight and distally a little recurved, but no closing devices present, closing body a prominent, protruding nose with a rim around it, both covered by tissue similar to that of the expanding sheet, expanding keels almost touching each other basally, valve wings narrow, tapering towards the tip and the base, 5 locules; **Ecol** unknown; **Distr** Prince Albert, WC, S Africa.

Note: The species differs from the other shrubby ones in its long, falcate leaves and the big closing bodies.

H. muirii L.Bolus SAG 1927: 365; NM 1928: 132 **T** Muir 3866 (BOL!).

Compact to "forming flat tufts" or cushions, **I** enclosed and invisible; **L** semi-terete, obtuse, a little rough, to 50 mm *l*, 8 mm *b*; **Fl** in cymes, bracteoles acuminate, petals golden, outside reddish; **Fr** base short, funnel-shaped, top with rims to 1 mm *h*, covering membranes complete, a little distally recurved, but no additional closing devices, closing body rather big and knobby, covered by a light ochre to translucent layer of tissue like that of the expanding sheets, expanding keels almost touching each other basally, valve wings very narrow, broadest in the middle, 5 locules, ∅ 8–9 mm; **S** 0.55–0.6 mm *l*, 0.42–0.45 mm *b*; **Ecol** among pebbles of quartzite with some limestone contents; **Distr** Swellendam, WC, S Africa.

Note: The almost terete leaves and the closing body, rather big for the genus, help to identify the species readily.

H. nelii Schwantes MDK 1929: 14 **T** Nel 8178 (collection Schwantes, HBG!).

Shrubs with a densely leaved centre, flowering branches elongated, raised above the cushion-shaped vegetative part, **I** here visible, to 5 mm *l*; **L** falcate, compressed, epidermis a little rough, 25–30 mm *l*, 5 mm *b*, to 10 mm *t*; **Fl** 1–3 in terminal cymes, bracteoles spreading, petals yellow; **Fr** base short, funnel-shaped, top with rims to 4 mm *h*, covering membranes straight, but without additional closing devices at their ends, closing body absent, but the end of the placenta forming a narrow, crenulate bulge, expanding keels almost touching one another at their bases, valve wings narrow, even more so in the distal part, not tapering towards the base, 9–13 mm ∅; **S** 0.7–0.86 mm *l*, 0.55–0.65 mm *b*; **Ecol** in flats on broken shales; **Distr** Ceres, WC, S Africa.

Note: The species has light grey leaves like **H. crassa** and similar taxa, but differs in the bigger covering membranes of the fruit and a distinct crenulate bulge at the end of the placenta, indicating no close relationship. Seeds of fresh material are bigger than the measurements given by Schwantes in the description.

H. odorata (L.Bolus) L.Bolus NM 1950: 222 ≡ *M. odoratum* L.Bolus ABH 1927: 100 nomen illeg. non N.E.Br. KB 1913: 119 ≡ *Aridaria odorata* (L.Bolus) Schwantes in Jacobsen 1949: 58 **T** Muir 3927 (BOL!).

Shrubs with decumbent branches, erect ones flowering, with few leaves only, short shoots forming a densely leaved centre, **I** dark brown; **L** almost terete, obtuse, epidermis almost smooth, to 30 mm *l*, c. 6 mm *b* and *t*; **Fl** solitary or in terminal cymes, stalks elongate, bracteoles spreading, petals yellow, apically reddish, to 11 mm *l*, to 0.5 mm *b*; **Fr** base short, funnel-shaped, top with rims up to 2 mm *h*, covering membranes reduced and without any distal closing devices, closing body absent, expanding keels almost touching each other basally, valve wings very narrow, broadest in the middle, 5 locules; ∅ c. 9 mm; **S** c. 0.7 mm *l*, 0.55–0.6 mm *l*; **Ecol** in plains with shaly soil; **Distr** Swellendam, WC, S Africa.

Note: The apically rounded leaves are very similar to those of **H. aspera**, which differs in the rough leaf surface; leaves of **H. odorata** are also similar to those of **H. tenuifolia**, which are longer growing on more erect stems.

H. pallens L.Bolus NM 1939: 156 (see **photo**) **T** Smith 1487 (BOL!).

Compact, branched moderately, anisophyllous; **L** trigonous to triquetrous, the keel thicker above the middle, erect to a little incurved, one leaf-pair straight, roundish, c. 30 mm *l*, to 4 mm *t*, the other pair with a subterminal thickening, to 40 mm *l*, to 10 mm *t*, grey, epidermis smooth, rough only when dry, wax cover a thin layer; **Fl** mostly solitary, bracteoles in the lower half of the stalk or at the base, free parts spreading, petals golden yellow, 8–11 mm *l*, 1.25–1.75 mm *b*; **Fr** robust, top with low rims, base forming a broad funnel, covering membranes incomplete, but persistent in position after the seeds are dispersed, closing body tiny, expanding keels just not touching each other at their bases, valve wings narrow, broadest at their base, 5 locules; **S** light brown, almost smooth, 0.8–0.9 mm *l*, ± 0.6 mm *b*; **Ecol** in open flat areas covered by quartzitic gravel, < 100 mm rainfall p.a. mainly in March; **Distr** Kenhardt, Namaqualand, NC, S Africa.

Note: The plants look very similar to **H. puttkameriana**, a species occurring further north in Namibia. The leaves of **H. pallens** are shorter and narrower, but it should be remembered that material of **H. puttkameriana** had been in cultivation when described.

H. puttkameriana (Dinter & A.Berger) Dinter & Schwantes ZSK 1927: 24 (see **photo**) ≡ *M. puttkamerianum* Dinter & A.Berger BJS 1914: 586 ≡ *Bergeranthus puttkamerianus* (Dinter & A.Berger) Schwantes ZSK 1926: 180 **T** Dinter 1100 (B) = *H. puttkameriana* (Dinter & A.Berger) Dinter & Schwantes var. *glabrior* Dinter ex Graessner 1937: 17 nomen nudum = *H. angustifolia* L.Bolus NM 1930: 197 **T** Pillans 5939 (BOL!).

Compact, branched highly, but not spreading; **L** terete in lower half, inconspicuously keeled in upper half, almost triangular in ∅, glaucous green, a little rough from raised dots, keel finely rough, epidermis almost even, stomata at the bases of the subsidiary cells, fine crystals in a thin layer 60–70 mm *l*, 6–7 mm *b* and *t*; **Fl** 1–3, ∅ 24–32 mm, bracteoles near base, to 18 mm *l*, petals linear, 71–97, rich yellow, 209–287 yellow, basally white stamens, 5 semi-circular glands; **Fr** top with high valve rims, base short, funnel-shaped, covering membranes convex, not always complete and with a distal low rim with a closing ledge underneath, endocarpal closing body knobby above, hollow, in some populations very small, expanding keels broad, flattening outwards, valve wings narrow, broadest in their middle, 5 locules, oldest fruit of a cyme 18–30 mm *l*, 8–9 mm ∅; **S** brown, testa with dome-shaped papillae, **S** 0.75–0.85 mm *l*, 0.55–0.7 mm *b*; **Ecol** in crevices of gneissic rocks, < 100 mm rainfall p.a. in March mainly; **Distr** Lüderitz-South, Namibia.

Note: The two species found in the area can best be distinguished by their growth, **H. puttkameriana** always growing compact with ± erect leaves. It may, at certain stages, be confused with **Ebracteola derenbergiana** with distinctly shorter inflorescences, pink to purple flowers opening during the day, and a central group of old fruits remaining on the plant over many years, while the plants grow further marginally, a feature not observed in **H. puttkameriana.**

H. puttkameriana var. *glabrior* = **H. puttkameriana**

H. rehneltiana (A.Berger) Dinter & Schwantes ZSK 1927: 24 ≡ *M. rehneltianum* A.Berger 1908: 245 ≡ *Bergeranthus rehneltianus* (A.Berger) Schwantes ZSK 1926: 180 **Lectotype,** here designated: Berger 1908: f 52.

Compact, later developing a thick rhizome and becoming caespitose; **L** with a subapical thickening, almost scimitar-shaped in the upper half, a little rough, 60–100 mm *l*, to 10 mm *b*, c. 6 mm *t*; **Fl** in cymes raised well above the plant, pedicels 30–80 mm *l*, bracteoles acuminate, petals yellow, filaments with white papillae near their bases; **Fr** base short, funnel-shaped, top with rims to 1 mm *h*, covering membranes straight, without any additional closing devices, closing body laterally broadened, consisting of tissue similar to that of the expanding sheets, forming a distinct, dark, crenulate wall over a white protrusion from the endocarp (visible only in section), expanding keels nearly touching each other basally, valve wings narrow, rectangular in the proximal 2/3, abruptly narrowed in the distal 1/3, 5 locules, c. 10 mm ∅; **Ecol** in grassy vegetation together with Restionaceae and bushes; **Distr** Uniondale (?), WC, S Africa.

Note: The very big leaves as described are probably results of conditions in cultivation; the distribution of the species is incompletely known.

H. stanfordiae L.Bolus NM 1935: 485 (see **photo**) **T** Stanford 21431 (cited) 21466 (annotated as type at BOL!).

Compact to loosely caespitose with age; **L** laterally flattened for most of its length, crescent-shaped at least in one leaf of a pair, the other often slender and less incurved, 25–35 mm *l*, c. 5 mm *b*, to 8 mm *t* at the broadest point above the middle; **Fl** often solitary, bracteoles in the lower part of the stalk, petals yellow, to 14 mm *l*, 0.5–1 mm *b*; **Fr** raised well above the plant, ∅ 7.5–8.5 mm, covering membranes slightly convex, persistent in shape, distal rims low, closing body very small, expanding keels basally not touching each other, valve wings narrow, broadest in their middle, 5 locules; **S** brown, papillate, 0.85–0.95 mm *l*, 0.6–0.7 mm *b*; **Ecol** in crevices of shaly rocks, 100–300 mm rainfall p.a. mainly in March and November; **Distr** Carnarvon, Victoria West, NC, S Africa.

Note: Among the species with compact growth, **H. stanfordiae** can be easiest recognized by its extremely compressed or laterally flattened, medium long leaves; also typical are the papillate seeds of rather big dimensions.

H. stanleyi ≡ **Chasmatophyllum stanleyi**

H. stenophylla L.Bolus NM 1950: 232 **T** collector unknown (BOL).

L almost equal in a pair, gradually narrowing to the pointed tips, rounded at the back, rich green, 25–40 mm *l*, 3–4 mm *b* and *t*; **Fl** in cymes, pedicels to 12 mm *l*, petals golden, 8–13 mm *l*, almost to 1 mm *b*, filaments with yellow papillae, anthers and pollen paler, semi-circular glands contiguous; **Fr** unknown; **Ecol** unknown; **Distr** unknown, S Africa.

Note: The plants had been grown in Bolus' garden, hence no origin has been given.

H. tenuifolia L.Bolus NM 1931: 251 **T** Nel (corrected on the sheet from Herre) 5860 (BOL!).

Shrubs to 25 cm *h*, **I** grey, to 30 mm *l*, 2.5 mm ∅; **L** semi-terete, obtuse, a little thicker apically than basally, glaucous green, a little rough, 45 mm *l*, 5 mm *b* and *t*; **Fl** in cymes, ∅ 29 mm, scented, pedicels 25–30 mm *l*, bracteoles above the middle, bright yellow, reddish outside, to 13 mm *l*, to 5 mm *b*, filaments glabrous, white, upper part and anthers yellow, nectary glands producing much nectary; **Fr** unknown; **Ecol** unknown; **Distr** uncertain in the districts Montagu, Robertson, Swellendam, WC, S Africa.

Note: The rounded, terete leaves resemble those of **H. aspera** and **H. odorata.**

H. teretifolia L.Bolus JB 1928: 229, L.Bolus SAG 1928: 126 nomen nudum **T** Watermeyer 664/23 (BOL).

Stems with up to 10 mm ∅ basally, upper **I** protruding, to 15 mm *l*; **L** terete in the middle, a little compressed above, obtuse, green, to 50 mm *l*, to 8 mm *t*; **Fl** ∅ 37 mm, scented, pedicels c. as long as the uppermost leaves, petals yellow, paler towards the base, reddish outside, to 15 mm *l*, 0.5–0.75 mm *b*, numerous filamentous staminodes surrounding the central column of stamens, filaments apically pale yellow, anthers yellow, glands distant; **Fr** unknown; **Ecol** unknown; **Distr** probably in Calvinia or Ceres, WC, S Africa.

Note: The numerous filamentous staminodes cast doubt on the placement of the species in **Hereroa.**

H. triebneri Graessner Samenliste 1937: 17, nomen nudum, lacking any description or reference; hence, the name cannot be connected with any of the species.
H. tugwelliae ≡ **Bijlia tugwelliae**
H. uncipetala = **H. wilmaniae**
H. vallis-pacis = **Ebracteola montis-moltkei**

H. willowmorensis L.Bolus NM 1932: 323 **T** Brauns 8278 (BOL!).

Compact; **L** with convex sides, keel only as a line, the thickest part near the base, tapering from there to a rather acuminate, somewhat incurved tip, 20–25 mm *l*, much longer in cultivation, c. 6 mm *b* and *t*; **Fl** 1(–3), bracteoles at base, connate for most of their part, free, rounded parts only c. 1 mm *l*, forming a short funnel around the stalk, petals golden, 10–13 mm *l*, c. 1.5 mm *b*, filaments apically golden; **Fr** top almost flat, base light grey, funnel-shaped, connate bracteoles at bases of stalks, covering membranes straight, bending into the empty locule, closing body an inconspicuous, crenulate short band, 5 locules; **S** light brown, almost smooth, 0.7–0.75 mm *l*, 0.55–0.6 mm *b*; **Ecol** in open shaly or gravelly places, 100–300 mm rainfall p.a. mainly in March and November; **Distr** Aberdeen, Willowmore, EC, Beaufort West, WC, S Africa.

Note: The small, incurved leaves and the very short bracteoles forming a funnel are typical of the species. The most remarkable elongation of the leaves in cultivation can well be seen on the type plant; to some degree, it occurs also in habitat in shady places.

H. wilmaniae L.Bolus NM 1929: 82 (see **photo**) **Lectotype**, here designated: Wilman 17264 (chosen from two syntypes, BOL!) = *Prepodesma uncipetala* N.E.Br. GC 1931: 389 ≡ *H. uncipetala* (N.E.Br.) L.Bolus NM 1938: 135 **T** Pole Evans 6891 (K!) = *H. wilmaniae* L.Bolus var. *langebergensis* L.Bolus NM 1929: 82 **T** Wilman 17352 (BOL!).

Compact to caespitose with a thick, fleshy rhizome; **L** nearly scimitar-shaped, with a subterminal thickening, epidermis smooth and even, **L** 40–60 mm *l*, 6.7–8 mm *b*, 9–10.5 mm *t*; **Fl** solitary or ternate, pedicels to 30 mm *l*, ∅ 31–50 mm, 13–16 mm *l*, basally 5.6–9 mm *b*, petals 7–25 mm *l*, 1–1.5 mm *b*, golden, apically reddish suffused, inner filaments with long papillae, yellow to basally white; **Fr** base short funnel-shaped, top flat with rims to 1.5 mm *h*, covering membranes straight, falling into the locules when dry, closing body of two white bulges formed by placental tissue entirely, expanding keels not touching each other basally, valve wings rather broad; **Ecol** in grassland in rocky places; **Distr** Herbert, Postmasburg NC, S Africa.

Note: The type of *Prepodesma uncipetala* is a young vegetative plant with a moderate tap-root only, whereas the other type sheets show plants with well developed, divided thick rootstocks not forming horizontal thick rhizomes as in other species of **Hereroa.**

H. wilmaniae var. *langebergensis* = **H. wilmaniae**

Herrea = Conicosia

H. E. K. Hartmann

The genera *Herrea* and **Conicosia** differ mainly in the cross sections of their leaves, the presence or absence of a preformed breaking-point of the stalk, the root, and life strategies (Ihlenfeldt & Gerbaulet 1990: 483). They share the same fruit morphology, flower types, seed morphology, epidermis, and shapes of leaves. Ihlenfeldt & Gerbaulet therefore conclude that both taxa belong to one genus representing two subgenera, reducing *Herrea* to **Conicosia** subg. **Herrea.**

H. acocksii = **Conicosia elongata**
H. affinis = **Conicosia elongata**
H. alba = **Conicosia elongata**
H. albolutea = **Conicosia elongata**
H. blanda = **Conicosia elongata**
H. brevisepala = **Conicosia elongata**
H. caledonica = **Conicosia pugioniformis** ssp. **pugioniformis**
H. conicosioides = **Conicosia pugioniformis** ssp. **pugioniformis**
H. elongata ≡ **Conicosia elongata**
H. elongata var. *minor* = **Conicosia elongata**
H. excavata = **Conicosia elongata**
H. fusiformis = **Conicosia elongata**
H. gydouwensis = **Conicosia elongata**
H. inaequalis = **Conicosia elongata**
H. klaverensis = **Conicosia elongata**
H. laticalyx = **Conicosia elongata**
H. macrocalyx = **Conicosia elongata**
H. nelii = **Conicosia elongata**
H. obtusa = **Conicosia elongata**
H. grandis = **Conicosia elongata**
H. plana = **Conicosia elongata**
H. porcina = **Conicosia elongata**
H. robusta = **Conicosia elongata**
H. ronaldii = **Conicosia elongata**
H. roodiae = **Conicosia elongata**
H. stipitata = **Conicosia elongata**

Herreanthus = Conophytum

S. A. Hammer

Schwantes based this monotypic genus (GW 1928: 514–515) on a single species, *H. meyeri* Schwantes, which is essentially a giant **Conophytum**, now known as **C. herreanthus** S.A.Hammer. The species agrees with some members of **Conophytum**, in particular with **C. blandum**, in the xeromorphic epidermis, the firm apiculate leaves with a low degree of basal fusion, and the large spongy-based whitish fruits. The recently described **C. herreanthus** ssp. **rex** S.A.Hammer lacks the deeply sunken stomata of the leaves of **C. herreanthus** ssp. **herreanthus**, resembling in this feature many other species of **Conophytum.**

H. meyeri
≡ **Conophytum herreanthus** ssp. **herreanthus**

Hydrodea = Mesembryanthemum

M. Gerbaulet

This genus comprises a number of species with much reduced flowers (N.E.Br. GC 1925: 412), which are now considered as members of a single species of **Mesembryanthemum** subg. **Opophytum** (Bittrich 1987: 73).

H. bossiana = **Mesembryanthemum cryptanthum**
H. cryptantha ≡ **Mesembryanthemum cryptanthum**
H. hampdenii = **Mesembryanthemum cryptanthum**
H. sarcocalycantha
= **Mesembryanthemum cryptanthum**

Hymenocyclus = Malephora

H. E. K. Hartmann

Schwantes described the genus to hold the species of Haworth's group Crocea and a new species, *Hymenocyclus herrei* (MDG 1927: 27). This publication came out 10 days later than the description of **Malephora** by Brown (GC 1.1.1927). Schwantes (ZSK 1927: 106) noticed the congruence later in the year, but doubted even later that the type species of Brown was correctly identified (Jacobsen FR 1938: 231). Bolus and Schwantes used the name *Hymenocyclus* for many years.

H. crassus ≡ **Malephora crassa**
H. croceus ≡ **Malephora crocea**
H. englerianus ≡ **Malephora engleriana**
H. framesii ≡ **Malephora framesii**
H. herrei ≡ **Malephora herrei**
H. latipetalus ≡ **Malephora latipetala**
H. luteus ≡ **Malephora lutea**
H. luteolus ≡ **Malephora luteola**
H. monticola ≡ **Ottosonderia monticola**
H. purpureo-croceus
≡ **Malephora purpureo-crocea**
H. smithii ≡ **Malephora smithii**
H. thunbergii ≡ **Malephora thunbergii**
H. uitenhagensis ≡ **Malephora uitenhagensis**

Hymenogyne Ruschioideae

H. E. K. Hartmann

Hymenogyne Haw. 1821: 192 Ihlenfeldt & Gerbaulet 1990: 457–498; Gerbaulet & Ihlenfeldt 1990: 511–512 **Typus H. glabra** (Aiton) Haw. **Etym** Gk *hymen*, skin, *gyne*, female, probably referring to the seemingly winged seeds = *Thyrasperma* N.E.Br. GC 1925: 412 **Lectotypus** *Thyrasperma helianthoides* N.E.Br. ≡ **H. conica** L.Bolus.

Annuals, the entire plant consisting of a rich dichasium; **L** flat, those of a pair forming basally an open sheath, moderately succulent, with bladder cells along the margins, epidermis homogeneous, but often with some tannin cells, with wax platelets; **Fl** petals

Distribution of **Hymenogyne** Haworth

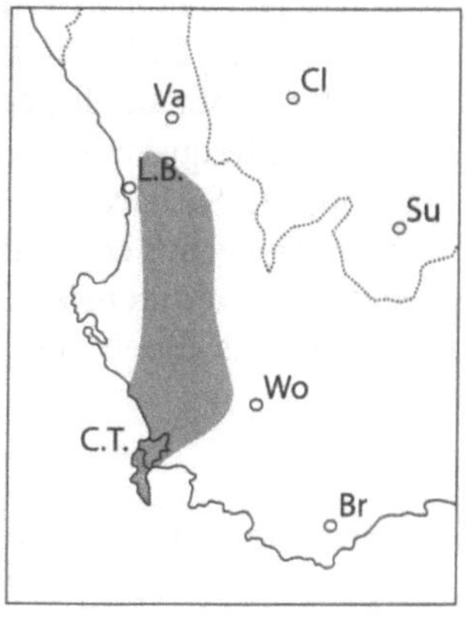

smooth along their margins, filaments basally flattened and ciliate, stigmas basally connate forming a style; **Fr** stalk decumbent, locules with two seeds each, separated by a complete false septum breaking into one-seeded schizocarps when ripe, the seed enclosed during dispersal by the septum on one side and the false septum on the other side, both forming a cartilaginous wing around the seed, 8–12 locules; **S** very big, dark brown, kidney-shaped, flattened laterally, testa papillate; **Chr** $2n = 18$; **Ecol** in sandy places, winter rainfall > 400 mm p.a.; **Distr** western districts in WC, S Africa.

Note: The genus is unique in its fruit structures, which led to its early recognition by Haworth and its placement in a separate tribe (invalidly published by Schwantes 1947: 39) raised to the rank of subfamily later (invalidly published by Schwantes 1957: 168, validated by Ihlenfeldt, Schwantes & Straka 1962: 54). Bittrich (1987: 50–52) showed that the set of character states agrees better with the Ruschioideae, in particular the Apatesia-group (e.g. Hartmann 1991). Also unique is the presence of tannin idioblasts in the epidermis – these cells are found in most Ruschioideae, where they occur inside the leaves of many genera.

Key to the Species of Hymenogyne

1. Style forming a narrow cone, petals light yellow .... **H. conica**
– Style forming a turret ending in a funnel, petals dark yellow .... **H. glabra**

H. conica L.Bolus NM 1950: 231 ≡ *Thyrasperma helianthoides* N.E.Br. in Phillips 1926: 246 nomen invalid. **LT** Stephens 7169 (K).

L with marginal bladder cells, but only rarely with tannin idioblasts; **Fl** petals light yellow, style forming a long and narrow cone; **Fr** as for genus; **S** as for genus; **Ecol** as for genus; **Distr** W districts in WC, S Africa.

Note: Some populations lack tannin idioblasts in the epidermis completely.

H. glabra (Aiton) Haw. 1821: 192 ≡ *M. glabrum* Aiton 1789: 193 **NT** H.Bolus 1158 (K) = *H. stephensiae* N.E.Br. GC 1928: 253 nomen invalid.

L with marginal bladder cells and tannin idioblasts; **Fl** petals dark yellow, style connate, forming a funnel distally; **Fr** as for genus; **S** as for genus; **Ecol** as for genus; **Distr** Clanwilliam, WC, S Africa.

Note: A population with cleistogamous flowers has been reported.

H. stephensiae = **H. glabra**

Ihlenfeldtia Ruschioideae

H. E. K. Hartmann

Ihlenfeldtia H.E.K.Hartmann BJS 1992: 47 **Typus I. excavata** (L. Bolus) H.E.K.Hartmann **Etym** *H.-D. Ihlenfeldt* * 1932, professor of botany at the University of Hamburg, working on Mesembryanthema and Pedaliaceae.

Compact; **L** acutely keeled, papillate; **Fl** solitary, bracteolate, ∅ > 70 mm, petals in all shades of yellow, often purplish suffused, rarely creamy white, red with age; **Fr** Titanopsis-type, valve wings narrowing distally, closing bodies small, with a translucent layer on top; **S** smooth, small, pear-shaped; **Chr** $2n = 18, 36$; **Ecol** on white quartz slopes; **Distr** Kenhardt, Namaqualand, NC, S Africa.

Note: The genus comprises some species formerly included in **Cheiridopsis** (Hartmann BJS 1992: 29–50). **Ihlenfeldtia** is closely related to **Tanquana** and **Vanheerdea**. The two species occur in different areas: allopatric distribution.

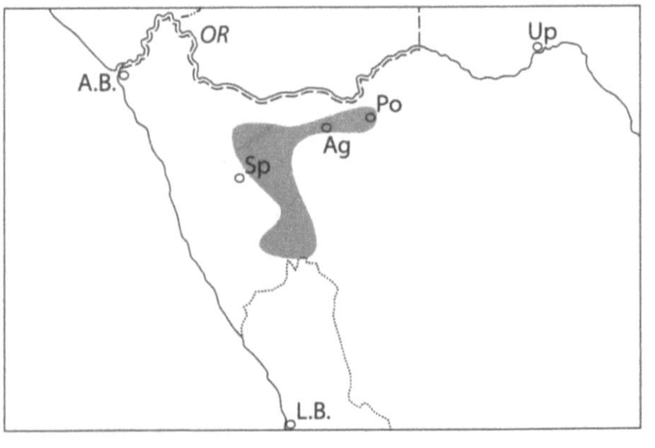

Distribution of **Ihlenfeldtia** H.E.K.Hartmann

Key to the Species of Ihlenfeldtia

1. Bracteoles at base of stalk, capsule base funnel-shaped .... **I. excavata**
– Bracteoles reaching up to the base of the fruit, capsule base broadened, spongy .... **I. vanzylii**

I. excavata (L.Bolus) H.E.K.Hartmann BJS 1992: 48 (see **photo**) ≡ *Cheiridopsis excavata* L.Bolus NM 1937: 48 **T** van Heerde 21836 (BOL!) = *Cheiridopsis dilatata* L.Bolus 1967: 65 **T** van Heerde 27828 (BOL!) = *Cheiridopsis albirosea* L.Bolus 1967: 65 **T** van Heerde 27829 (BOL!) = *Cheiridopsis vanbredai* L.Bolus 1963: 11 **T** van Breda 1768/61 (BOL!).

Compact, up to 110 branches, rarely growing sunken in the ground; **L** mucronate, often prickly from stout teeth on the excavated keel; **Fl** petals in various shades and colours, as for genus; **Fr** with a funnel-shaped base, 10–15 locules, bracteoles at base of stalk; **S** elongate pear-shaped, 0.63–0.85 mm *l*; **Chr** $2n = 18$; **Ecol** at the border of the winter rainfall region, c. 125 mm p.a.; **Distr** Namaqualand, NC, S Africa.

Note: Populations differ in flower colours and expression of teeth.

I. vanzylii (L.Bolus) H.E.K.Hartmann BJS 1992: 49 (see **photo**) ≡ *Cheiridopsis vanzylii* L.Bolus NM 1929: 116 **T** Fuller 4 (BOL!) = *Cheiridopsis richardiana* L.Bolus NM 1929: 135 **T** Fuller (van Zyl) 5 (BOL!) = *Cheiridopsis papillata* L.Bolus NM 1958: 299 **T** de Klerk 11993 (BOL!).

Moderately branched (to 10 branches); **L** without a mucro or teeth, keel curving down, sides convex; **Fl** chrome yellow, often with a white base; **Fr** with a broadened base of corky tissue, bracteoles reaching up to the capsule itself; **S** roundish, 0.6–0.65 mm *l*; **Chr** $2n = 36$; **Ecol** on quartzitic slopes, < 125 mm rain p.a.; **Distr** Kenhardt, NC, S Africa.

Note: The species can be understood as a neotenic form of **I. excavata**, i.e. the juvenile shape has been retained while flowering was reached.

Imitaria = Gibbaeum

H. E. K. Hartmann

The genus was described by Brown (JB 1927: 348) as being intermediate between **Conophytum** and **Gibbaeum**, perhaps of hybrid origin. Since the plants possess highly branched epidermal papillae, similar to those of other species in **Gibbaeum**, the genus cannot be upheld (Tischer 1937: 150–152). The connate bases of petals and stamens can be understood as an adaptation to the enclosed position of the flowers, representing a parallel development to that found in **Conophytum** and some other genera.

I. muirii ≡ **Gibbaeum nebrownii**

Jacobsenia Ruschioideae

H.-D. Ihlenfeldt

Jacobsenia L.Bolus & Schwantes in Schwantes KuaS 1954: 69; Ihlenfeldt 1997: 109–126 **Typus J. kolbei** (L.Bolus) L.Bolus & Schwantes in Schwantes **Etym** *H. Jacobsen*, 1898–1978, curator of the Botanic Garden of the University of Kiel, Germany = *Anisocalyx* L.Bolus NM 1958: 385 nomen illeg. non Hance ex Walpers 1852: 195 ≡ *Drosanthemopsis* Rauschert 1982: 555 **Typus** *Anisocalyx salarius* L.Bolus ≡ *Drosanthemopsis salaria* (L.Bolus) Rauschert = **J. vaginata** (L.Bolus) Ihlenf.

Shrubs to 50 cm *h* or almost compact to 15 cm *h*, **I** to 10 mm ∅, juvenile plants caespitose, adult plants with erect shoots with elongated **I** and distinct lateral short shoots or adult plants of ± compact habit and without a clear distinction between long and short shoots; **L** with large bladder cells with thickened outer walls, elliptic to ± circular, often with a spike-like central or with numerous antler-like protrusions, leaves of a pair sometimes slightly anisophyllous, each stem with 2(–4) leaf-pairs per season, in cultivation up to 8 pairs, leaf-pairs connate for c. 1/5 of their length, erect or spreading, ± digitiform, often ± curved, upper surface sometimes flattened, lower surface sometimes somewhat keeled, at least the last leaf-pair of the season persisting on the stem until the next season, the very base of the connate portion of a leaf-pair developing into a narrow ring-shaped persisting sclerotic sheath, usually brown-red, **L** to 70 mm *l*, ∅ to 10 mm; **Fl** solitary, terminal, emerging towards the end of the season from the connate portion of the 2nd leaf-pair of the season, sometimes from the 1st pair, ∅ to 80 mm, pedicels 10–70 mm *l*, **K** 5–6, petals in c. 4–5 rows, white, light yellow or very rarely red, inner petals shorter, filamentous staminodes mostly absent or very few, filaments white or yellow, subbasally papillate, holonectary, VII–IX; **Fr** expanding keels merging into expanding sheets, the latter sometimes rather indistinct, covering membranes over 1/2 to 3/4 of the locule, closing bodies absent, opening between the covering membranes and the outer rim of the capsule narrowed by a cluster of sterile funicles, locules 5–16, number sometimes very variable, ∅ 5–17 mm; **Chr** $2n = 18$; **S** testa brown to light brown, minutely papillate, c. 0.7 mm *l*, 0.5 mm *b*; **Ecol** in shallow rocky soils, sometimes in reefs of schist, or in loamy strongly saline sand, 100–200 mm rainfall p.a. mainly in winter; **Distr** SW Namaqualand, NC, Vanrhynsdorp, Vredendal, WC, S Africa.

Note: The genus **Jacobsenia** has been established twice with the same type species: Schwantes (KuaS 1954: 69) and two months later Bolus (NM 1954: 255). Both descriptions cited "L.Bolus & Schwantes" as the authors of the new genus. The correct citation of the generic name, therefore, should read as "**Jacobsenia** L.Bolus & Schwantes in Schwantes KuaS 1954: 69". The same applies to the correct citation of the type species of the genus. Cultivation of all species of this genus is easy, but the plants do not flower freely in cultivation.

Distribution of **Jacobsenia** L.Bolus & Schwantes

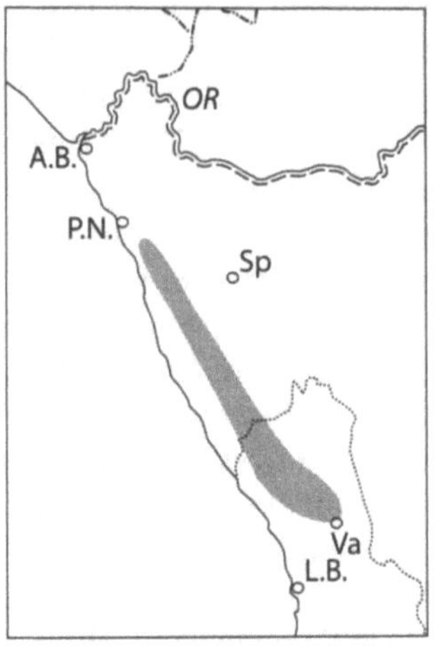

Key to the Species of Jacobsenia

1. Adult plants with erect long shoots bearing lateral short shoots, to 50 cm *h*; sepals 5, locules 5 ... **J. kolbei**
- Adult plants compact shrubs without distinct erect long shoots, to 15 cm *h*, stems in old plants procumbent, though sometimes ascending; sepals usually 6, locules 6–16 ... **2**
2. Bladder cells with a single central protrusion; shape of capsule seen from above 6-angled, locules mostly (5–)6(–8), Ø 8–17 mm ... **J. hallii**
- Bladder cells with many antler-like protrusions; shape of capsule seen from above ± circular, locules 6–16, Ø 5–9 mm ... **J. vaginata**

J. hallii L.Bolus NM 1958: 329 (see **photo**) ≡ *Drosanthemum jacobsenii* G.D.Rowley 1978: 8 **T** Hall 405/54 (BOL!).

Compact, to 15 cm *h*, **I** to 10 mm Ø; **L** with large ± circular bladder cells with a single long central spike-like protrusion, leaf-pairs sometimes slightly anisophyllous, leaves often slightly curved, moderately spreading, ± digitiform, upper surface flattened, lower surface slightly keeled, to 70 mm *l*, Ø to 10 mm; **Fl** emerging from the 2nd leaf-pair of the season, Ø to 80 mm, pedicels 10–20 mm *l*, **K** 6, petals white or light yellow, filaments dark yellow, VIII–IX; **Fr** capsule as seen from above 6-angled, locules mostly (5–)6(–8), Ø 8–17 mm; **Ecol** in shallow rocky soils with large quartz pebbles; **Distr** S Namaqualand, NC, Vredendal, WC, S Africa.

Note: In the known localities, the populations of **J. hallii** are in contact with populations of **J. kolbei**. At these sites, intermediate specimens of probably hybrid origin have been observed, though, in principal, the species are separated by different flowering periods (the peak in flowering in **J. hallii** is about four weeks later than in **J. kolbei**), which, however, show some overlapping.

J. kolbei (L.Bolus) L.Bolus & Schwantes in Schwantes KuaS 1954: 69 (see **photo**) ≡ *M. kolbei* L.Bolus ABH 1922: 131 ≡ *Drosanthemum kolbei* (L.Bolus) G.D.Rowley 1978: 8 **T** Kolbe 2388/17 (BOL!).

To 50 cm *h*, **I** to 10 mm Ø, adult plants with erect long shoots and lateral short shoots; **L** with large, usually smooth elliptic bladder cells, leaf-pairs often slightly anisophyllous, short shoots usually with 2 leaf-pairs per season, long shoots often with more leaf-pairs, up to 8 in cultivation, **L** often distinctly curved, erect, or especially in short shoots spreading, usually terete, to 60 mm *l*, Ø to 10 mm, on short shoots often much smaller; **Fl** emerging from the 2nd leaf-pair of the season, Ø to 60 mm, pedicels 25–70 mm *l*, **K** (4–)5, petals white, very rarely red, filaments white, VII–VIII; **Fr** capsule seen from above 5-angled, locules 5(–6), Ø 7–11 mm; **Ecol** in shallow rocky soils, in the S part of the distribution area frequently in reefs of shist; **Distr** Vanrhynsdorp, Vredendal, WC, S Africa.

Note: This species is widespread and fairly frequent in the southern part of the distribution area of the genus.

J. vaginata (L.Bolus) Ihlenf. 1997: 124 (see **photo**) ≡ *Drosanthemum vaginatum* L.Bolus NM 1929: 138 ≡ *Anisocalyx vaginatus* (L.Bolus) L.Bolus 1962: 303 nomen illeg. ≡ *Drosanthemopsis vaginata* (L.Bolus) Rauschert 1982: 555 **T** Bolus 1759/29 (BOL!) = *Anisocalyx salarius* L.Bolus NM 1958: 385 nomen illeg. ≡ *Drosanthemopsis salaria* (L.Bolus) Rauschert 1982: 555 **T** Hall 634/57 (BOL!).

Compact, to 8 cm, rarely to 20 cm *h*, **I** in old plants procumbent, rarely ascending, to 10 mm Ø; **L** with

± circular bladder cells, top flat, with many antler-like protrusions, especially along the margin of the top, each stem usually with 2 leaf-pairs per season, **L** usually ascending at c. 45° from the stem, upper surface flattened, lower surface towards the apex distinctly keeled, to 35 mm *l*, ∅ to 11 mm; **Fl** usually emerging from the 1st leaf-pair of the season, ∅ to 40 mm, pedicels to 10 mm *l*, **K** mostly 6, petals white, filaments white, VIII–IX; **Fr** capsule seen from above ± circular, expanding sheets rather indistinct, locules (6–)9(–16), ∅ 5–9 mm; **Ecol** in loamy, strongly saline sand with quartz pebbles, < 100 mm rainfall p.a.; **Distr** SW Namaqualand, NC, S Africa.

Note: The type locality of **J. vaginata** recorded by Bolus in the original description ("between Vanrhynsdorp and Bitterfontein", i.e. in the Knersvlakte) is erroneous. The species has never been recollected in this very well surveyed area; according to present knowledge, the species only occurs in SW Namaqualand. The synonymy of **J. vaginata** and *Anisocalyx salarius* was already realized by Bolus (1962: 303); the combination *Drosanthemopsis salaria* (L.Bolus) Rauschert, therefore, was superfluous. There are populations which nearly exclusively have 6 locules, whereas populations from nearby may exhibit a range of 6–16 locules with a maximum value of 9 locules.

Jensenobotrya Ruschioideae

H. E. K. Hartmann

Jensenobotrya Herre 1951: 79 **Typus J. lossowiana** Herre **Etym** *E. Jensen*, amateur botanist at Lüderitzbucht, Gk *botrys*, bunch of grapes; the name combines the collector's name with his description of the plants hanging down like bunches of grapes.

Densely leaved, decumbent shrubs hanging down the cliffs like bunches of grapes, **I** remaining soft, spongy; **L** shortly clavate to almost globose, basal ∅ 5–7 mm, apical ∅ 12–15 mm, greyish-pink, epidermis cells almost flat, stomata not sunken, outer epidermal wall with a central layer of fine crystal sand, wax in a thick layer breaking into horizontal plates covered by ± amorphous small particles; **Fl** ebracteolate, pedicel ca. 10 mm *l*, ∅ 20–35 mm, **K** 5, pink petals, often basally white, filamentous staminodes absent, white filaments with yellow anthers, nectary as a dark green lophomorphic ring; **Fr** withering quickly, ochre, base as long as broad in its upper ∅, thus funnel-shaped with slightly convex sides, inner morphology of the Delosperma-type, with parallel expanding keels, placenta ending in a small knob, valve wings broad, fixed only distally to the valve, from where it spreads like a ribbon, 5 locules; **S** almost globose, yellow with dark brown papillae placed distantly giving the seed a speckled appearance; **Ecol** in crevices of gneiss at the sea; **Distr** Lüderitz-South, Namibia.

Note: The genus does not only grow in a geographically isolated spot, it is also difficult to place within the Mesembryanthema. Herre (1951: 80) noted already that ribbon-shaped false septa had been pulled out of the fruit, a feature perhaps caused by preformed tearing when the fruit opens for the first time.

Distribution of **Jensenobotrya** Herre

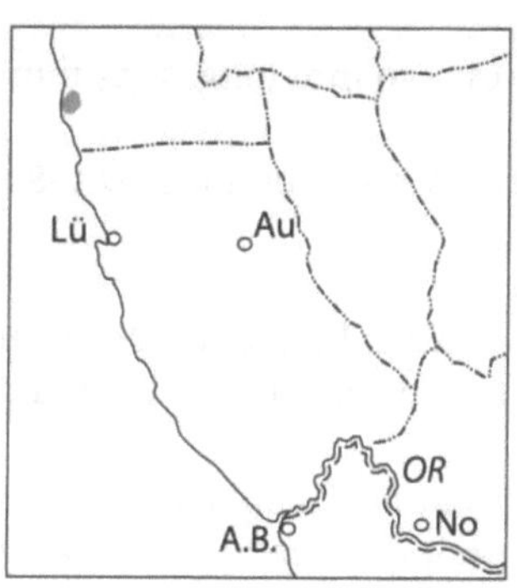

J. lossowiana Herre 1951: 79 (see **photo**) **T** Jensen s.n. as liquid preserve under SUG 12618 (STE?).

Plants very dense in habitat, growing more loosely in cultivation; **L** elongating into clavate shape in cultivation; **Fl** 40–60 petals, 170–250 stamens; **Fr** c. 10 mm ∅; **S** c. 1 mm *l*; **Ecol** as for genus; **Distr** as for genus.

Note: Since the plants are easy to keep and to flower in cultivation, this is one of the examples of a taxon rare in habitat but very common in numbers out of its area.

J. vanheerdei = **Stoeberia carpii**

Jordaaniella Ruschioideae

H. E. K. Hartmann

Jordaaniella H.E.K.Hartmann BB 1983: 57; Hartmann BJS 1984: 321–360 **Typus J. clavifolia** (L.Bolus) H.E.K. Hartmann **Etym** *P.G. Jordaan*, 1913–1987, professor of Botany at the University of Stellenbosch.

Creeping to ascending, sometimes climbing in other shrubs, **I** always long, visible and with a smooth, shining surface, spongy inside from a thick cork layer, up to 15 **I** formed per season, adventitious roots to 50 cm *l*;

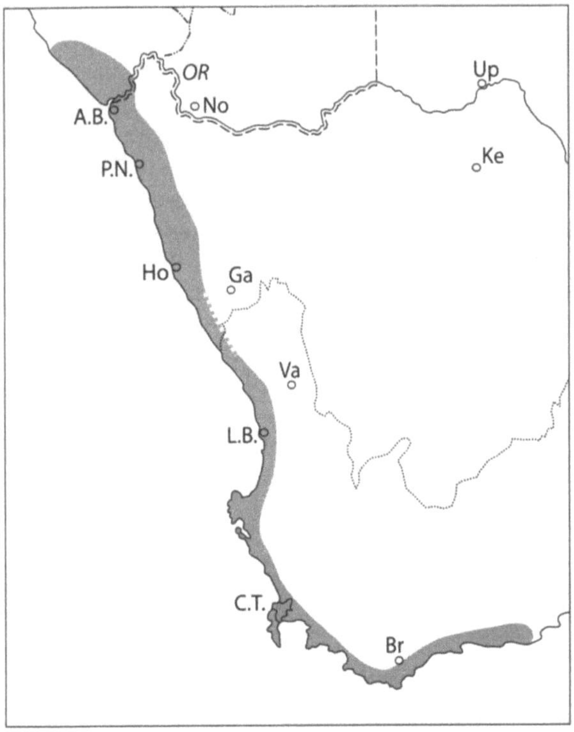

Distribution of **Jordaaniella** H.E.K.Hartmann

L terete or nearly so, epidermis cells a little elevated, covered by wax plates and sometimes with additional platelets or rodlets, calcium crystals c. 2 µm in Ø, also in the guard cells (but absent there in cultivation); **Fl** mostly solitary on short shoots, bracteolate, (4–)5(–6) calyx lobes, petals yellow, salmon, orange, copper, purple, or white, filaments yellow or white; **Fr** stalks long, persistent, top highly elevated, the high rims of the valves forming a ridge around or near the outer border, descending into a central crater, base flat to short funnel-shaped with a broad central columella, closing body small or absent, covering membranes almost straight, distally with low rims, valve wings very broad; **S** shortly pear-shaped to roundish, light brown and smooth with darker, rougher zones at hilum and micropyle; **Ecol** on aeolic sands near the sea, winter rains and sea fogs; **Distr** Lüderitz-South, Namibia; coast of Namaqualand, NC, and WC, S Africa.

Note: **J. spongiosa** is added to the genus here, agreeing with the other species in fruit morphology, stem anatomy, leaf characters, flowers, and seed morphology. All species flower reluctantly only after considerable vegetative growth, requiring much space in cultivation. In suitable climates the plants form huge mats as ground covers, beautiful also in the colouring of the leaves in autumn.

Key to the Species of Jordaaniella

1. Capsules on erect stalks, leaves ± finger-shaped, very fat and big, > 10 cm long .... **J. spongiosa**
- Capsules on decumbent stalks, leaves fusiform, < 10 cm long 2
2. Internodes of long shoots > 20 mm long, Ø < 5 mm, l:Ø ≥ 5, leaves slender, without a pustule 3
- Internodes of long shoots < 20 mm long, Ø 2.5–13 mm, l:Ø ≤ 4 4
3. Leaves slender, < 50 mm long, < 5 mm broad, often purplish suffused, capsules with a rounded to semi-globose base .... **J. dubia**
- Leaves more robust, > 60 mm long, 5–7 mm broad, green and erect, also during the resting period, capsules with a funnel-shaped base .... **J. uniflora**
4. Internodes of long shoots fat, Ø ≥ 5 mm, l:Ø ≤ 2, leaves > 50 mm long, > 7.5 mm broad, locules usually > 15 .... **J. cuprea**
- Internodes of long shoots thin, Ø < 5 mm, l:Ø 3–4, leaves < 50 mm long, Ø 5–8 mm, locules 14–15 .... **J. clavifolia**

J. clavifolia (L.Bolus) H.E.K.Hartmann BB 1983: 57 (see **photo**) ≡ *M. clavifolium* L.Bolus ABH 1927: 95 ≡ *Cephalophyllum clavifolium* (L.Bolus) L.Bolus ex N.E.Br. JB 1928: 326 **T** Pillans 17774 (BOL!).

Creeping, **I** 10–20 mm *l*, 2.5–5 mm Ø, l:Ø 3–4; **L** 18–50 mm *l*, 4–8 mm *b*, broadest usually above the middle, hence club-shaped, in habitat mostly lying on the soil during the resting period, then yellowish green to orange, wax cover in almost smooth plates; **Fl** yellow, Ø 45–75 mm, 65–75 petals, 122–280 stamens, filaments papillate near base; **Fr** with a rounded base, closing body small, mostly 14–15 locules, 20–26 mm *l*, Ø 8.5–9.5 mm; **S** 0.7–0.8 mm *l*, 0.55 mm *b*; **Chr** $2n = 36$; **Ecol** Strandveld, < 100 mm rainfall p.a.; **Distr** coast of Namaqualand, NC, S Africa.

Note: In habitat, the decumbent leaves distinguish the species from **J. cuprea** during the resting period.

J. cuprea (L.Bolus) H.E.K.Hartmann BJS 1984: 355 (see **photo**) ≡ *Cephalophyllum cupreum* L.Bolus NM 1929: 114 **T** Pillans 17753 (BOL!) = *Cephalophyllum luxurians* Dinter FR 1931: 169 nomen nudum ≡ *M. luxurians* Dinter ex Range 1934: 13 nomen nudum, sheets: Dinter 6483 (M! K!).

Creeping, **I** < 15 mm *l*, Ø > 5 mm, l:Ø < 2, short shoots with short **I** common, plants therefore often nearly compact; **L** 50–100 mm *l*, 8–14 mm *b*, in habitat erect during the resting period, green to dark purple, with a pustule, wax cover continuous with platelets and thread-like rodlets; **Fl** 5(–6) calyx lobes, Ø 30–100 mm, petals 66–115, light to dark yellow, apically orange to pink, often salmon coloured, 180–420 stamens; **Fr** with a rounded base, closing body small, 14–20 locules, numbers very variable, 30–50 mm *l*, ∅ 11–16 mm; **S** 0.7–0.83 mm *l*, 0.5–0.7 mm *b*; **Chr** 2*n* = 18; **Ecol** strandveld, < 100 mm rainfall p.a.; **Distr** coast of Lüderitz-South, Namibia; coast of Namaqualand, NC, S Africa.

Note: Probably the most suitable species for cultivation because several populations remain rather compact, flowering a little more readily than the other species. In the sandy coastal habitats, the species is very important for the stabilisation of soil due to its capacity to regrow after having been covered by sand. Thick sand layers cannot be penetrated, however, and the plants die.

J. dubia (Haw.) H.E.K.Hartmann BJS 1984: 356 (see **photo**) ≡ *M. dubium* Haw. 1803: 39 nomen illeg. non Haw. 1795: 470 ≡ *Cephalophyllum dubium* (Haw.) N.E.Br. ex Schwantes GF 1928: 69 **NT** Wisura 2200 (NBG!) = *M. albertiniense* L.Bolus ABH 1925: 3 ≡ *Cephalophyllum albertiniense* (L.Bolus) Schwantes GF 1928: 69 **T** Muir 919 (BOL!) = *M. anemoniflorum* L.Bolus ABH 1923: 159 ≡ *Cephalophyllum anemoniflorum* (L.Bolus) N.E.Br. JB 1928: 326 **T** Strangman 2597/16 (BOL!) = *M. maritimum* L.Bolus ABH 1925: 2 non L.Bolus NM 1937: 55 nomen illeg. ≡ *Cephalophyllum maritimum* (L.Bolus) Schwantes GF 1928: 69 **T** Muir 3095 (BOL!) = *Cephalophyllum pittenii* L.Bolus SAG 1928: 156 ≡ *Cephalophyllum vanputtenii* L.Bolus JSAB 1967: 308 **T** van Putten 1041/25 (BOL!) = *M. procumbens* Haw. 1821: 111 ≡ *Cephalophyllum procumbens* (Haw.) Schwantes GF 1928: 69 **T?** = *Cephalophyllum watermeyeri* L.Bolus NM 1932: 338 **T** Watermeyer 1305/29 (BOL!).

Creeping, **I** > 20 mm *l*, ∅ < 5 mm, l:∅ ≥ 6, purple when young; **L** < 50 mm *l*, < 5 mm *b*, without a pustule, often purplish suffused, wax cover continuous, sometimes broken into horizontal plates, always with platelets and thread-shaped rodlets; **Fl** ∅ 35–75 mm, petals 30–70, mostly yellow, some populations salmon, white, or purple, filaments papillate near base; **Fr** with a rounded to bell-shaped base, closing body small, 10–15 locules, 20–50 mm *l*, ∅ 8–12.5 mm; **S** 0.67–0.86 mm *l*, 0.5–0.67 mm *b*; **Chr** 2*n* = 18; **Ecol** in open places in shrubby associations of Fynbos or Coastal Rhenosterbosveld, > 200 mm rainfall p.a. in winter or all seasons; **Distr** SW and S coasts of WC, S Africa.

Note: The more isolated populations in the south vary considerably in flower colour, but in no other character states. It is assumed that gene exchange is reduced between the populations, perhaps due to great distances between them, resulting in the establishment of deviating flower colours. This view is supported by the observation that all plants of *M. albertiniense*, flowering purple, found until now are sterile. Vegetative propagation also plays an important role: pieces of plants can establish new populations. Their fast growth can contribute considerably to the stabilization of sand, but they cannot stand much shading and are vulnerable to heavy walking. The name *M. dubium* had been used by Haworth in 1795 for a plant figured by Petiver (1709: t. 39 f 3), which can be identified as **Aizoon canariense**. When using the name again in 1803, Haworth did not refer to the older usage at all. The epithet, albeit illegitimate under **Mesembryanthemum**, is available for the species under **Jordaaniella**.

J. spongiosa (L.Bolus) H.E.K.Hartmann **comb. nov.** (see **photo**) ≡ *M. spongiosum* L.Bolus Annals of the Bolus Herbarium 1927: 74 (as basionym) ≡ *Cephalophyllum spongiosum* (L.Bolus) L.Bolus SAG 1928: 156 **Lectotype**, here designated: Pillans 17752 III (chosen from three sheets, BOL!).

Decumbent to ascending, often straggling in other shrubs, **I** up to 10 mm ∅; **L** thick, finger-shaped, c. 10 cm *l*, ∅ 10–15 mm, wax cover of horizontal plates with some platelets and rodlets; **Fl** 70–100 mm ∅, petals 230–270, red with an orange base, stamens 700–1050, orange filaments smooth, nectary ring inconspicuous; **Fr** with a short, funnel-shaped base, closing body tiny to absent, 18–28 locules, 18–22 mm ∅, 44–52 mm *l*; **S** 0.6–0.7 mm *l*, 0.43–0.48 mm *b*, less bulgy than the other species; **Chr** probably 2*n* = 36; **Ecol** in strandveld with other shrubs, rainfall < 100 mm p.a.; **Distr** coast of Namaqualand, NC, S Africa.

Note: The species stands out in the family by its strikingly large, brightly coloured flowers with unusually numerous elements, exceeded in size only by the whitish flowers of **Cylindrophyllum**. Easy to grow, it nevertheless flowers reluctantly in cultivation in Europe in the greenhouse around Christmas. Within the genus, **J. spongiosa** can be recognized by its large size of leaves, stems, fruits, and flowers, its ascending growth, and smaller seeds.

J. uniflora (L.Bolus) H.E.K.Hartmann BJS 1984: 357 (see **photo**) ≡ *Cephalophyllum uniflorum* L.Bolus NM 1929: 88 **T** Pillans 5768 (BOL!).

Creeping, **I** 20–30 mm *l*, ∅ 3.5–4.5 mm, *l* : ∅ 5–7; **L** green and erect, also during the resting period, without a pustule, > 60 mm *l*, 5–7 mm *b*, wax in horizontal plates, rarely with few amorphous small particles; **Fl** ∅ 40–45 mm, petals c. 100, yellow, stamens c. 320, filaments papillate near base; **Fr** base funnel shaped, closing bodies small, 15–16 locules, 35–40 mm *l*, ∅ 11–12.5 mm; **S** 0.64–0.7 mm *l*, 0.48–0.59 mm *b*; **Chr** 2*n* = 18; **Ecol** N part of strandveld in open patches amongst shrubs, rainfall < 100 mm p.a.; **Distr** coast of Namaqualand, NC, S Africa.

Note: In habitat, the species forms conspicuous big carpets of robust, erect dark green leaves, always distinctly longer than the internodes.

Juttadinteria Ruschioideae

H. E. K. Hartmann

Juttadinteria Schwantes ZSK 1926: 182 Walgate 1939: 171–188; Friedrich 1970: 58–64 **Typus J. simpsonii** (Dinter) Schwantes **Etym** *Jutta Dinter*, wife of Prof. K. Dinter.

Shrubs with 3–5 branches, rarely more compact, to 40 cm high; **I** short, often enclosed; **L** 6–8 to a branch, trigonous, broadly boat-shaped to nearly linear, always with a triangular apical part, keel and margins smooth to dentate, epidermis even, with dome-shaped cells of different sizes, waxes smooth on top of cells, in platelets on sides, stomata sunken, surrounded by smaller epidermal cells; **Fl** ebracteolate, four calyx lobes, petals white, stamens in a pronounced central cone when young, later forming a cylinder, papillate near base, nectary ring dark green; **Fr** flat on top, with age elevated in the centre, base funnel-shaped, covering membranes as narrow ledges or short, never covering the locules completely, closing body absent, valve wings broad, rectangular, 8(–12) locules; **S** pear-shaped, micropyle elongate and protruding, testa papillate, brown, 0.7–1 mm *l*, 0.5–0.75 mm *b*; **Ecol** in calcareous sands and loams, often of coarser structure, in flats or on slopes, < 100 mm rainfall p.a. in winter, fog from the nearby sea; **Distr** Lüderitz-South, Namibia; N Namaqualand, NC, S Africa.

Distribution of **Juttadinteria** Schwantes

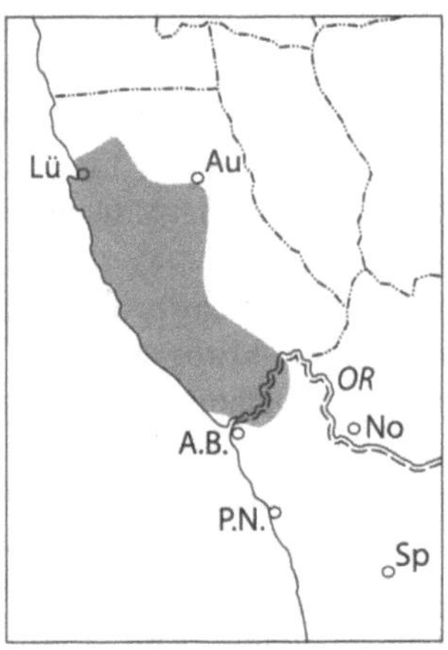

Note: All species flower freely during most of the year, they are not dependent on a strict dry season. Up to now, only one species per habitat has been found. It seems, therefore, that allopatric speciation has taken place: in different areas a separation of several populations from an ancestral stock has occurred. The most distinct forms are found near the outer edges, sometimes as clearly separated species, as in the south, sometimes merging towards the centre, as in **J. ausensis**. Along contact lines, differentiation is incomplete, many characters have to be taken into account to identify material.

The genus is understood here in its restricted sense, **Juttadinteria** s.str., excluding the genera **Namibia** and **Dracophilus**. These were originally included in **Juttadinteria** s.lat. (Schwantes 1926), a view taken also by Friedrich (1970). Schwantes (1927) and Walgate (1939) recognized the three genera distinguished here, too. In order to permit identification of material for which the genus is not known, a key to the three genera of **Juttadinteria** s.lat. is given:

Key to the Genera of Juttadinteria s.lat.

1. Locules open, without any covering membranes at all, leaves very fat, surface velvety to the touch, papillae pointed at their tips, epidermis cells of equal sizes .... **Namibia**
- Locules with covering membranes of different extension, sometimes only as rims to the septa, papillae on leaves rounded and dome-shaped, epidermis cells smaller around the stomata, hard to the touch 2
2. Calyx of four lobes, locule numbers mostly 8, rarely up to 10, low shrubs .... **Juttadinteria** s.str.
- Calyx lobes five, locule numbers mostly ten, rarely less, compact to cushion-shaped .... **Dracophilus**

Key to the Species of Juttadinteria s.str.

1. Leaf margins parallel for at least 2/3 of the leaf length, > 250 stamens per flower 2
- Leaf margins tapering from a broader base **or** widening to form a rhomboidal upper leaf surface, < 250 stamens per flower 3
2. Leaves 5–8 cm long, fat, margins very rarely dentate, capsule ∅ above 9,5 mm, 8–9 locules ...... **J. albata**
- Leaves 2–4 cm long, slender, margins heavily to sparsely dentate, capsule ∅ 8–9,5 mm, rarely higher, 8(–12) locules .... **J. ausensis**
3. Leaves acute, slender, length > 3 × width .... **J. attentuata**
- Leaves boat-shaped, rounded to trigonous, length < 3 × width 4
4. Leaves heavily dentate, short, shortly boat-shaped .... **J. simpsonii**
- Leaves with smooth edges or only minutely dentate, slender, elongate-trigonous .... **J. deserticola**

J. albata (L.Bolus) L.Bolus SAG 1928: 249 (see **photo**) ≡ *M. albatum* L.Bolus ABH 1927: 97 T Pillans 5039 (BOL!).

Ascending to erect branches, **I** enclosed, also in cultivation; **L** 50–80 mm *l*, > 20 mm thick; **Fl** 55–70 petals, ∅ 33–55 mm; **Fr** > 19 mm *l*, ∅ 9–11 mm, 8–9 locules, covering membranes as ledges of c. 2.5 mm; **S** 0.8–0.9 mm *l*, 0.55–0.6 mm *b*; **Ecol** on soils with high calcareous contents, on slopes; **Distr** S part of Lüderitz-South, Namibia; NW Namaqualand, NC, S Africa.

Note: The combination of fat leaves and rather small seeds distinguishes **J. albata** from all other species; some populations show very large flowers.

J. attenuata Walgate NM 1939: 178 (see **photo**) T Holloway 3 (BOL!).

Ascending to erect branches, **I** often long and visible in cultivation; **L** slender, tapering to a sharp tip from a broader base, length > 3 × width, to 40 mm *l*, to 10 mm *t*; **Fl** c. 30 mm ∅, 150–250 stamens, 45–55 petals; **Fr** 16–19 mm *l*, covering membranes as narrow ledges to 1.5 mm *b*, 8 locules; **Ecol** in calcareous gravel, mostly in flats; **Distr** Lüderitz-South, Namibia; NW Namaqualand, NC, S Africa.

Note: Slender, pointed leaves characterize this species within the genus.

J. ausensis (L.Bolus) Schwantes ZSK 1926: 183 (see **photo**) ≡ *M. ausense* L.Bolus ABH 1922: 123 **Lectotype**, here designated: Lewis 543/18, chosen from two syntypes (BOL!) = *J. suavissima* (Dinter) Schwantes ZSK 1926: 183 ≡ *M. suavissimum* Dinter 1923: 138 non L.Bolus 1930: 220 nomen illeg. **T** Dinter 3739(B, lost) **Lectotype**, here designated: Range 1885 (B!)[1].

Erect to ascending, rarely decumbent branches, **I** mostly visible and long, more so in cultivation; **L** slender, with a long basal part with parallel margins and a short apical triangle of equal sides, heavily dentate in the northern populations, sparsely dentate with tiny teeth in the southern ones; **Fl** ∅ mostly 40–55 mm, > 250 stamens, 45–65 petals; **Fr** 15–25 mm *l*, ∅ 8–10 mm, covering membranes to 2.5 mm *b*, 8–12 locules; **Ecol** in sandy flats with pebbles of dolomite, quartz, and limestone; **Distr** Lüderitz-South, Namibia.

Note: The variation in many obvious characters, e.g. teeth, appears contradictory to original descriptions. Neither Bolus nor Dinter were aware of the constant and therefore diagnostic characters: the slender leaves with a high portion of parallel margins and the high number of stamens. As in **J. albata**, the diameter of the flowers can be very big, and the number of petals can be very high, resulting in a beautiful show. Erect forms are more common in the northern and north-eastern part of the area of the genus, decumbent ones in the south-western and southern parts.

J. buchubergensis = **J. deserticola**
J. cinerea ≡ **Namibia cinerea**
J. dealbata ≡ **Dracophilus dealbatus**
J. decumbens = **J. deserticola**
J. delaetiana ≡ **Dracophilus delaetianus**

J. deserticola (Marloth) Schwantes ZSK 1926: 183 (see **photo**) ≡ *M. deserticola* Marloth 1910: 34 T Marloth 4688 (BOL!) = *J. buchubergense* Dinter FR 1931: 169 nomen nudum ≡ *M. buchubergense* Dinter ex Range 1934: 15 nomen nudum non Dinter ex Range 1934: 13 nomen nudum nec Dinter ex Range 1934: 18 = *J. decumbens* Tischer 1933: 48 **Lectotype**, here designated: Tischer 1933: f on p 48 = *J. elizae* (Dinter & A.Berger) L.Bolus SAG 1928: 249 ≡ *M. elizae* Dinter & A.Berger 1914: 587 **T** Dinter 1010 lost = *J. insolita* (L.Bolus) L.Bolus SAG 1928: 249 ≡ *M. insolitum* L.Bolus ABH 1927: 75 T Pillans 5692 (BOL!) = *J. tetrasepala* L.Bolus NM 1932: 343 T Herre 9046 (BOL!).

Decumbent, rarely ascending or erect, **I** rarely visible, also in cultivation; **L** boat-shaped to pointed, length < 30 mm, width < 20 mm, margins only in some northern populations with minute teeth; **Fl** ∅ 20–30 mm, 150–250 stamens, 40–60 petals; **Fr** 11–19 mm *l*, covering membranes absent or to 1 mm *b*, ∅ 6.5–9.5 mm, locules c. 8; **Ecol** in stony flats covered by gravel from quartz, dolomite, and limestone, rarely on phonolite (black) or sedimentary schists, sea fog; **Distr** Lüderitz-South, Namibia; W Namaqualand, S Africa.

Note: The very wide continuous distribution from Lüderitzbucht to south of Alexanderbay can now be demonstrated by numerous collections. Northern and southern populations differ in their leaf shapes, the northern being more rounded apically, appearing broader, the southern more pointed, appearing slender. A subgroup can be distinguished west of the Klinghardt Mountains: capsules of these populations are broader (8–9.5 mm) than coastal ones (6.5–8.2 mm) and their locule number is 8 or higher, whereas the coastal populations possess 8 or lower locule numbers. Only in these Klinghardt populations is an erect habit common, but not exclusive, and leaves are pointed throughout, thus resembling the southern populations nearer to the coast.

J. elizae = **J. deserticola**
J. insolita = **J. deserticola**
J. kovisimontana = **J. simpsonii**

[1] Since Dinter saw Range 1885 before the description of *J. suavissima* was published, the sheet can be chosen here as the lectotype, although Dinter noted "M. deserticola or M. suavissima". The slender leaf shape and tiny teeth support the placement in **J. ausensis**.

J. longipetala = **Namibia cinerea**
J. montis-draconis = **Dracophilus dealbatus**
J. pomonae ≡ **Namibia pomonae**
J. proxima = **Dracophilus dealbatus**
J. rheolens = **Dracophilus dealbatus**

J. simpsonii (Dinter) Schwantes ZSK 1926: 183 (see **photo**) ≡ *M. simpsonii* Dinter 1923: 134 **T** Dinter 3742 lost **Lectotype**, here designated: Dinter 54 (HBG!)[2] = *J. kovisimontana* (Dinter) Schwantes ZSK 1926: 183 ≡ *M. kovisimontanum* Dinter 1923: 135 **T** Dinter 3881 lost.

Up to 15 ascending branches forming a shrub up to 25 cm *h*; **I** short, visible, longer in cultivation; **L** broadly boat-shaped, margins and keel, often also the faces, with teeth, mostly fiercely dentate, sometimes smoother; **Fl** 180–305 stamens, 40–60 petals, ∅ 20–35 mm; **Fr** 13–18 mm *l*, covering membranes as ledges to 3 mm, sometimes overlapping in the centre for c. 1/2 their length, 8(–12) locules; **Ecol** in sandy pockets amongst dolomite rocks; **Dist** Lüderitz-South, Namibia.

Note: The peculiar roughness in *J. kovisimontana* is due to the occurrence of incompletely developed teeth forming wart-like protrusions, a feature occurring irregularly in several populations. Both names appear in the same publication, **J. simpsonii** is chosen here as the valid name because its character set is more typical of the species and a lectotype can be chosen.

J. suavissima = **J. ausensis**
J. tetrasepala = **J. deserticola**
J. tugwelliae = **Bijlia tugwelliae**

Kensitia = Erepsia

H. E. K. Hartmann

The genus name *Kensitia* was used by Fedde (1940: 11) to replace the name *Piquetia* N.E.Br. (GC 1925: 433), which was illegitimate because of the older name **Piquetia** Hallier (1921: 162), a genus in the family Theaceae. Liede (1990a: 443–473) included the only species placed in *Kensitia* in **Erepsia** considering the similarities in leaf and fruit structures more important than the distinctive petal shapes of *Kensitia.*

K. pillansii ≡ **Erepsia pillansii**

Khadia Ruschioideae

H. E. K. Hartmann & P. Chesselet

Khadia N.E.Br. GC 1930: 279; Chesselet & Hartmann 1995: 46–49; Chesselet et al. 1998: 25–33 **Typus K. acutipetala** (N.E.Br.) N.E.Br. **Etym** *Khadi*, a drink said to be made from the roots of **Khadia**, but probably also from other fleshy roots.

Compact, when young, the thick hypocotyl and rootstock branching later, developing a thick, short rhizome as in some species of the genus **Iris**, to 15 mm ∅ or forming long shoots with **I** to 40 mm *l* lying on rocks; **L** almost free, glaucous to dark green, epidermis cells only a little elevated, stomata a little sunken, outer walls xeromorphic, central layer of calcium crystals to 6 µm *t*, the continuous wax layer breaking into horizontal plates, mostly one per cell; **Fl** solitary, no bracteoles discernable in size or shape, a new branch developing from the axil of one of the leaves of the distal pair, petals white to pink to purple, these latter two also possibly white towards their bases, filamentous staminodes and filaments white to pink, lophomorphic holonectary as a dark green ring; **Fr** woody, dark brown, long persistent, base funnel-shaped to somewhat bell-shaped, top flat to conically raised, not always closing completely again, covering membranes complete with a distal rim and closing

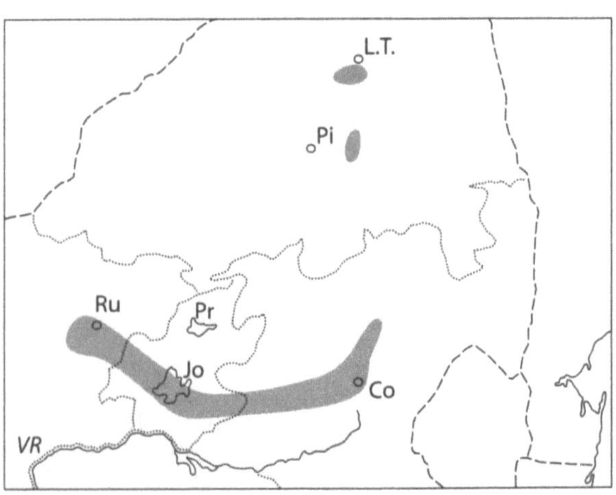

Distribution of **Khadia** N.E.Br.

[2] The type appears to be lost. The sheet chosen as lectotype bears the annotation "54 M. simpsonii 18.8.22 sp.n." in Dinter's handwriting. As in the case of **Dracophilus delaetianus**, it stems probably from cultivated material Dinter kept at Lichtenstein for further observations.

ledges below, at least partially thickened by spongy tissue, closing bodies most often composed of a central endocarpal layer covered by an extension of the expanding sheet with the placenta placed below, rarely completely placental, valve wings absent, 4–11 locules; S pear-shaped, dark brown, testa cells with papillae with almost straight walls leaving a small area around the elevated part, the anticlinal walls always undulate, 0.85–1.2 mm *l*, 0.7–1 mm *b*; **Ecol** in grassland next to stony places, often in pockets of soil on quartzite of the Soutpansberg Group, the Steenkampsberg Formation, the Magaliesberg Formation, and the Witwatersrand Supergroup, 1500–2300 m altitude, summer rainfall of c. 700 mm p.a.; **Distr** GAU, KN, MP, NP, NW, S Africa.

Note: The genus occurs in an area remote from the main centres of diversity of the subfamily. It is also difficult to place satisfactorily within the system of the subfamily because the closing bodies are unusually constructed, resembling most closely those of the genus **Ebracteola**. Spongy covering membranes are also found in other genera, e.g. **Namaquanthus** or **Schlechteranthus**. The former placement of **Khadia** in the **Ruschia**-group cannot be upheld because of differences in critical features. Six species can be distinguished, one of them newly discovered. Populations often inhabit almost peaty pockets along horizontally exposed rocks. In cultivation, they tend to develop very narrow and small leaves if not kept adequately.

Key to the Species of Khadia

1. Plants at least with some long, trailing branches, internodes to 40 cm long ..... **K. borealis**
– Plants with internodes always shorter than the leaves, often enclosed and invisible 2
2. Most capsules of a population 8–9 locules (30–55% of all capsules), petals magenta or pink over their entire length ..... **K. acutipetala**
– Most capsules of a population 5–6 locules (> 65% of a population) petals white or pink, in the latter case paler at the base 3
3. Locules and calyx lobes predominantly 5 ..... **K. media**
– Locules and calyx lobes predominantly 6 4
4. Flowers white, opening in the afternoon ..... **K. carolinensis**
– Flowers pink to purple, base often paler, opening over midday 5
5. Leaves 25–40 mm long, glaucous ..... **K. beswickii**
– Leaves 11–14 mm long, dark green ..... **K. alticola**

K. acutipetala (N.E.Br.) N.E.Br. GC 1931: 279 (see **photo**) ≡ *M. acutipetalum* N.E.Br. 1908: 407 **T** Turner s.n. (K) = *K. nationiae* (N.E.Br.) N.E.Br. GC 1931: 279 ≡ *M. nationiae* N.E.Br. in Burtt Davy 1926: 161 **T** Nation s.n. (K).

Growth rather compact, the rootstock and hypocotyl branching in almost vertical direction, no horizontal rhizome present, even in old plants; **L** triquetrous above, base almost semi-terete, dark green to glaucous, 15–35 mm *l*, 4–6 mm *b*, 3.5–5 mm *t*; **Fl** Ø 25–35 mm, petals deep purple to pink, filamentous staminodes and filaments often purple; **Fr** breaking off the stalks when ripe, lying nearby in habitat, base in those cases flattish, otherwise broadly funnel- to bell-shaped, top conically raised to different degrees, rims bent back to some extent, often dispersed as tumble-fruits, covering membranes spongy over their entire length thickening towards the distal end where almost a closing bulge is formed, tiny closing body consisting only of the placenta and tissue extending to there from the expanding keels, 6–11, mostly 8 or 9 locules; **S** papillae densely placed all over the testa, fine sculptures on testa cells amorphous, **S** 0.85–1 mm *l*; **Ecol** on Magaliesberg quartzite, in pockets next to outcrops; **Distr** GAU, Rustenburg, NW, S Africa.

Note: Typical of the species is the only moderately expressed maximum in the distribution of locule numbers in a given population, either at 8 or 9 locules, or in both classes. Morphological differences (particularly of the capsule, e.g. in the formation of the conical top) between populations appear very prominent in some cases, but they occur within populations as well, and intermediate states and combinations can be found, too.

K. alticola Chesselet & H.E.K.Hartmann 1995: 48 (see **photo**) **T** Hartmann & Hartzer 32729 (PRE!).

Mat-forming with ± horizontal thick rhizomes; **L** triquetrous, dark to bright green with translucent dots, 11–15 mm *l*, 5–7 mm *b*; **F** petals, filamentous staminodes and filaments pink; **Fr** top semi-orbicular from mostly half-open valves not closing again after their first opening, base funnel-shaped with straight sides, valve rims bent back very conspicuously, covering membranes with spongy thickening in their distal third ending in a thick closing ledge or a thin closing bulge below, closing body composed of placenta, endocarpal layer in the middle, and tissue extending from the expanding keel above, 6(–7) locules; **S** papillae less elevated than in the other species, resembling in their undulate margins the bladder cells on the leaves of some species of **Glottiphyllum**, no fine sculptures discernible, 1.12 mm *l*, 0.82 mm *b*; **Ecol** along the edges of horizontally exposed rocks of Steenkampsberg Quartzite in grassland; **Distr** Belfast, ET, Utrecht, KN, S Africa.

Note: In the past, plants of this species have been confused with **K. beswickii**, also with pinkish flowers, but mostly with paler bases and often almost whitish, and with longer leaves, as well as with **K. carolinensis**, with white petals.

K. beswickii (L.Bolus) N.E.Br. GC 1931: 279 (see **photo**) ≡ *M. beswickii* L.Bolus ABH 1922: 124 **T** Beswick 218/20 (BOL!) = *K. nelsonii* N.E.Br. GC 1931: 279 **T** Nelson s.n. (K).

Mats rather dense, rhizomes very short; **L** glaucous, triquetrous, roundish below; **Fl** petals pink, basally white, often very pale; **Fr** base ± bell-shaped, top rounded, low, rims of valves turned backwards a little, covering membranes completely spongy, the distal closing ledge very inconspicuous to absent, closing body consisting only of the placenta and a layer of expanding sheet tissue with a knobby surface, mostly 6 locules; **S** papillae moderately high, fine sculptures on their surfaces present, **S** 1–1.25 mm *l*, 0.95–1 mm *b*; **Ecol** in grassland, often disturbed; **Distr** Nigel, GAU, S Africa.

Note: Plants identified as **K. beswickii** in the past do mostly belong to **K. alticola**, although specimens of **K. acutipetala** with 6 sepals have also been named **K. beswickii.** The confusion is probably based on the pink petals occurring in all three species, and on the fact that **K. beswickii** was reported to have come from "near Johannesburg". Both populations found to agree with **K. beswickii** from the area in question appear to be in danger of extinction due to recent expansion of settlement.

K. borealis L.Bolus NM 1936: 6 (see **photo**) **LT** Crundall 21638 (BOL!).

Mat-forming, **I** to 40 mm *l* when covering rocks; **L** triquetrous, dark green, 20–30 mm *l*, to 4 mm *b* and *t*; **Fl** ∅ to 20 mm, petals, filamentous staminodes and filaments white, rarely pink; **Fr** top almost flat, rims very low, hardly bent backwards, base funnel-shaped with concave sides, covering membranes convex, no spongy tissue, closing ledges very conspicuous, closing body comprising the uppermost tip of the placenta and a tiny section of tissue from the expanding sheet, 4–8, but mostly 5–6 locules; **S** papillae of testa low, a little distant, covered densely by rodlet-shaped sculptures, often arranged in groups standing together like sheaves in a field or single; **Ecol** in open grassland, near and on rock plates in sand derived from quartzite, shallow well-drained soil, in full sun on top of the mountain; **Distr** Soutpansberg, NT, S Africa.

Note: The expansively creeping habit and the small flowers distinguish this species from all others.

K. carolinensis (L.Bolus) L.Bolus NM 1936: 6 (see **photo**) ≡ *M. carolinense* L.Bolus ABH 1922: 128 ≡ *Rabiea carolinensis* (L.Bolus) N.E.Br. GC 1931: 53 **LT** Rogers 1833/17 (BOL!).

Dense mats consisting of highly branched, short, thick, horizontal rhizomes rooting a little only, the central tap-root dominating, **I** possibly elongating in cultivation; **L** trigonous with rounded margins, thus almost semi-terete, ± bright green, to 25 mm *l*, to 6 mm *b* and *t*; **Fl** opening in the afternoon, ∅ 30–40 mm, petals, filamentous staminodes and filaments white; **Fr** top roundish from high, gapping valve rims, base elongate funnel-shaped, merging into the robust, persisting stalk, covering membranes not spongy, the distal closing device a broad ledge or a longish bulge, closing body either purely placental or composed of placenta, endocarpal centre, and a top layer of tissue from the expanding sheet, mostly 6 locules; **S** papillae very low, their margins grossly undulate, resembling the bladder cells on leaves of some species of **Glottiphyllum**, 1.15 mm *l*, 0.9 mm *b*; **Ecol** in pockets with finely grained soil in grassland next to vertically exposed bands of quartzitic rocks of the Transvaal Series; **Distr** Belfast, Carolina, ET, S Africa.

Note: The combination of short, rounded leaves and white flowers opening in the afternoon distinguishes **K. carolinensis** from the other species of the genus.

K. media Winter & Hahn 1999: 16 **T** Hahn & Winter 1494 (PRE).

Compact with a very thick tap-root branching at its top; **L** of a pair only occasionally asymmetrical, dotted along the margins, tapering into a hyaline tip; **Fl** ∅ 15 mm, **K** 5-lobed, petals white, rarely light pink, filamentous staminodes flat and not filiform, inner ones appressed to the central cone of stamens; **Fr** mostly 5 locules, base concave, attenuating into the pedicel, top with recurved valve rims, covering membranes with recurved distal rims above closing ledges; **S** orange; **Ecol** in shallow soils on large rock outcrops of granite, gneiss, gabbro, and diabase, 1500–1870 m; **Distr** Pietersburg, NP, S Africa.

Note: The dominance of five sepals and five locules singles the species out in the group with compact growth in **Khadia**.

K. nationiae = **K. acutipetala**
K. nelsonii = **K. beswickii**

Kolleria = Galenia

H. E. K. Hartmann

Presl (1831: 24 and *t.* 14) established the genus *Kolleria* C.Presl based on the species *K. herniariaefolia* only. Whereas Ecklon & Zeyher (1837: 327–328) added three more species to the genus, all later authors placed the genus as a synonym into **Galenia**, albeit as a subgenus or section. Attention is drawn, however, to the fact that different authors used varying circumscriptions of **Galenia** subg. or § **Kolleria**, and that no comprehensive study is available for the entire genus **Galenia** considering all characters and using modern methods.

K. collina ≡ **G. collina**
K. glauca = **G. pallens**
K. herniariaefolia ≡ **G. herniariaefolia**
K. pallens ≡ **G. pallens**

Lampranthus Ruschioideae

H. E. K. Hartmann

Lampranthus N.E.Br. GC 1930: 71; Hartmann 1998: 66–71, 74–79 **Typus L. multiradiatus** (Jacquin) N.E.Br. **Etym** Gk *lampros*, shining, *anthos*, flower, referring to the bright shiny flowers of most species of the genus = *Aristanthus* Schwantes ZSK 1927: 28 nomen nudum = *Mesembryanthus* Necker ex Rothmaler 1940: 413 **Typus** *Mesembryanthus tenuifolius* (L.) Rothmaler ≡ **L. tenuifolius** (L.) N.E.Br.

Erect to creeping shrubs, often with markedly red stems; **L** almost free, only moderately xeromorphic; **Fl** solitary or in bigger dichasia, ∅ up to 60 mm, petals often shiny and bright in colour, filamentous staminodes mostly present, nectary in a ring, mostly inconspicuous and crenulate; **Fr** dark brown, long persistent, base long funnel- to bell-shaped, top raised, covering membranes complete, with a distal rim and mostly with closing ledges or rodlets below, closing bodies absent, but the endocarp often bulging at the top of the placenta, pushing it forward so that its tip resembles a small closing body, valves with broad, mostly rectangular valve wings, 5 locules; **S** pear-shaped, ± smooth to somewhat papillate; **Ecol** most species in the wetter winter rainfall areas near the Cape of Good Hope; **Distr** Lüderitz-South, Namibia; EC, KN, Namaqualand, NC, WC, S Africa.

Note: **L. multiradiatus** has been named by Brown (1930) as the type species, later lectotypifications (e.g. Herre 1971) are therefore superfluous and illegitimate. Schwantes (1927) suggested the name *Aristanthus* for a group of species typified by *M. tenuifolium*, but he did not describe the genus nor make any combina-

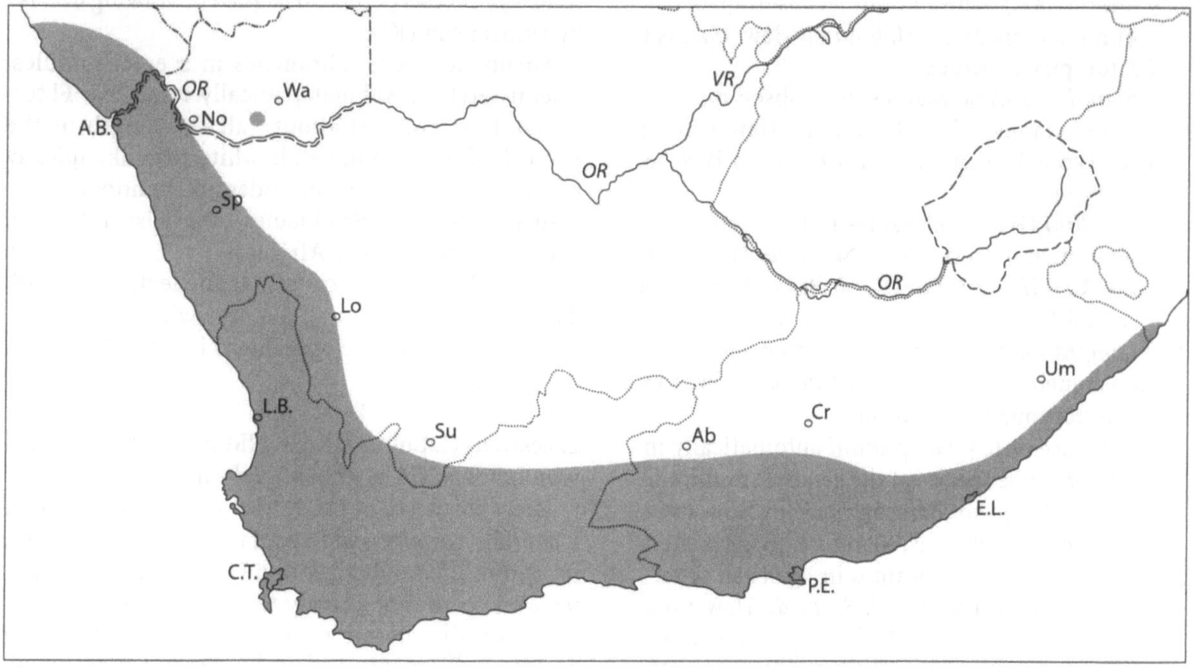

Distribution of **Lampranthus** N.E.Br.

tions because he assumed that this group was, following Brown's circumscription, representing the old genus **Mesembryanthemum.** Although several species of **Lampranthus** are used as ornamental plants worldwide in gardens, even in moderate climates, the genus has not been studied sufficiently up to now to permit a clear delimitation of all species. The unpublished study of Glen (1978) offers some possible arrangements including a suggested division into five sections, considering the genus **Oscularia** as one of them. This view is not followed here, **Oscularia** being understood as a separate genus not closely related to **Lampranthus.** Those species of **Lampranthus** with very light grey leaves thicker than broad, with or without teeth, with five separate nectary glands in the flower, even if touching each other, and fruits of a more fragile construction than that of **Lampranthus** are therefore transferred to **Oscularia.** A subdivision of **Lampranthus** into seven sections was published by Schwantes in Jacobsen (1960: 1190–1191); all these are invalid because they lack full and direct references to the author and place of publication of the basionym; furthermore, the name of one section are illegitimate in the given circumscription represented by the names of included species.

L. § *Adunci* (Salm-Dyck) Schwantes in Jacobsen 1960: 1190 nomen invalid. ≡ *M.* § *Adunca* Salm-Dyck 1820: 25 **Lectotypus L. aduncus**

L. § *Amoeni* (Salm-Dyck) Schwantes in Jacobsen 1960: 1190 nomen invalid. ≡ *M.* § *Amoena* Salm-Dyck ex A.Berger 1908: 159 ≡ *Eximia* Haw. 1821: 145 ≡ *Conferta* Salm-Dyck 1820: 29 **Lectotypus L. amoenus**

L. § *Aurei* (Haw.) Schwantes in Jacobsen 1960: 1191 nomen invalid. ≡ *M.* § *Aurea* Haw. 1821: 154 **Lectotypus L. aureus**

L. § *Blandi* (Haw.) Schwantes in Jacobsen 1960: 1191 nomen invalid. ≡ *M.* § *Blanda* Haw. 1821: 146 **Lectotypus L. blandus** = *M.* § *Conferta* Haw. 1812: 240 **Typus?**

L. § *Corallini* (Haw.) Schwantes in Jacobsen 1960: 1190 nomen invalid. ≡ *M.* § *Corallina* Haw. 1821: 154 ≡ *M.* § *Hawortiana* DC. 1828: 439 **Lectotypus L. coralliflorus**

L. § *Lampranthi* Schwantes in Jacobsen 1960: 1190 nomen illeg. ≡ *M.* § *Tenuifolia* Salm-Dyck 1820: 24 **Lectotypus L. tenuifolius.**

The use of *L.* § *Lampranthi* automatically includes the type species of the genus, **L. multiradiatus,** which was not done explicitly by Schwantes; the lectotypification is, therefore, illegitimate.

L. § *Scabridi* (Haw.) Schwantes in Jacobsen 1960: 1190 nomen invalid. ≡ *M.* § *Scabrida* Haw. 1812: 267 ≡ *M.* § *Scabra* Salm-Dyck 1820: 27 nomen illeg. ≡ *M.* § *Asperifolia* Salm-Dyck ex Haw. 1821: 138 **Lectotypus L. scaber**

As typical features of the genus winged valves and the absence of closing bodies are understood here. In agreement with this concept, a number of species has been transferred to **Lampranthus** from different genera.

L. acrosepalus (L.Bolus) L.Bolus NM 1939: 170 ≡ *M. acrosepalum* L.Bolus NM 1936: 11 **T** Pillans 7752 (BOL).

Erect shrubs, 18–25 cm *h*, **I** 20–40 mm *l*, ∅ 3 mm; **L** triquetrous, acuminate, 17–30 mm *l*, 1–2 mm *b*, to 2 mm *t*; **Fl** solitary, pedicels 10–20 mm *l*, petals golden, 9–30 mm *l*, 0.5–2 mm *b*, filamentous staminodes not mentioned, filaments yellow, anthers and pollen pale yellow, nectary ring inconspicuous, stigmas deep red; **Fr** unknown; **Ecol** on plateau; **Distr** Piquetberg, WC, S Africa.

Note: Dark red or purple stigmas are rare in the genus.

L. acutifolius (L.Bolus) N.E.Br. GC 1930: 211 ≡ *M. acutifolium* L.Bolus NM 1929: 59 **T** Pillans 18871 (BOL).

Shrubs to 12 cm *h*, **I** 1–20 mm *l*; **L** subfalcate, apically incurved, carinate, to 10 mm *l*, 2 mm *b* and *t*; **Fl** 1–3, ∅ 30 mm, pedicels to 20 mm *l*, bracteoles at their middle, **K** basally scabrous, petals pinkish-purple, 14 mm *l*, to 1.5 mm *b*, filamentous staminodes absent, anthers and pollen yellow, nectary ring inconspicuous; **Fr** unknown; **Ecol** on a mountain; **Distr** Malmesbury.

Note: The species is placed in the group Scabrida, based on the rough leaves, by Bolus.

L. aduncus (Haw.) N.E.Br. GC 1930: 211 ≡ *M. aduncum* Haw. 1803: 87 non Jacquin 1804: t. 51, f 2 nec Willdenow 1809: 534 **Lectotype,** here designated: drawing 920/157 by Duncanson (K!).

Shrubs to 15 cm *h*, branches in ± erect bundles; **L** semi-terete to subulate, apically recurved; **Fl** terminal, bracteoles at about half the length of the pedicels, ∅ c. 20 mm, petals white to pink, apically reddish, filamentous staminodes not mentioned, filaments apically red; **Fr** 5 locules, otherwise unknown; **Ecol** unknown; **Distr** S Africa.

Note: The drawing chosen as the lectotype is not dated, but it agrees in all features with the description. The central cone of stamens is a rare feature in the genus.

L. aestivus (L.Bolus) L.Bolus NM 1939: 168 ≡ *M. aestivum* L.Bolus NM 1930: 188 **T** Edwards 19098 (BOL).

Erect shrubs to 25 cm *h*, **I** 20–35 mm *l*, ∅ to 3 mm; **L** laterally compressed to subfalcate, glaucous green, 10–15 mm *l*, 2 mm *b*, 3.5 mm *t*; **Fl** in cymes, ∅ 28 mm, pedicels to 20 mm *l*, rarely to 40 mm *l*, bracteoles at their middle, petals pinkish-purple, 12 mm *l*, to 2.5 mm *b*, filamentous staminodes not mentioned, filaments straw-coloured, anthers yellow, nectary ring inconspicuous; **Fr** top 2 mm *h*, 6 mm *l*, ∅ 8 mm; **Ecol** unknown; **Distr** Caledon, WC, S Africa.

Note: Big capsules characterize many species of the genus.

L. affinis L.Bolus 1962: 13 **T** Stayner 794/60 (BOL).

Shrubs to 20 cm *h*, **I** 15–40 mm *l*, 1.5–2 mm ∅; **L** semi-terete, acute, subglaucous, green, to 25 mm *l*, 1.5–2 mm *t*; **Fl** solitary or 2, pedicels 30–40 mm *l*, bracteoles at their middle, pale bright pink, shining, paler at base, 18–23 mm *l*, merging into filamentous staminodes 8–14 mm *l*, 0.5–1 mm *b*, outer erect, inner pressed against the filamentous staminodes surrounding the central cone of stamens closely, filaments white, anthers and pollen white; **Fr** unknown; **Ecol** unknown; **Distr** Prince Albert, WC, S Africa.

Note: The description is followed by a short diagnosis of a forma without a name. **L. affinis** is reported to be similar to **L. coralliflorus**, differing mainly from other species in the genus by the erect inner petals.

L. albus ≡ **Oscularia alba**

L. algoensis L.Bolus JSAB 1960: 159 **T** Feinauer 691/59 (BOL).

Erect shrubs to 35 cm *h*, **I** 10–20 mm *l*, ∅ to 2 mm; **L** trigonous, apically purplish, apiculate, recurved, 10–14 mm *l*, 1–2 mm *b* and *t*; **Fl** solitary, pedicels to 15 mm *l*, bracteoles at their middle, petals red, basal 1/3 white, filamentous staminodes present, nectary ring obscure; **Fr** ∅ expanded 14 mm, 5 locules; **Ecol** unknown; **Distr** Port Elizabeth, EC, S Africa.

Note: The petals probably merge into the filamentous staminodes.

L. alpinus ≡ **Esterhuysenia alpina**

L. altistylus N.E.Br. GC 1930: 211 ≡ *M. longistylum* L.Bolus NM 1928: 2 nomen illeg. non DC. 1804: t. 147 **T** L.Bolus 18823 (BOL).

Creeping shrubs, **I** blackish-brown, to 10 mm *l*; **L** trigonous, scabrous, finely papillate, 4 mm *l*, 1.5–2 mm *t*; **Fl** solitary, ∅ 19 mm, pedicels 10–20 mm *l*, bracteoles at their middle, petals 7 mm *l*, 1.5 mm *b*, filamentous staminodes few, filaments purplish-pink, basally paler, anthers pale, pollen golden, nectary ring denticulate; **Fr** unknown; **Ecol** in flats; **Distr** Malmesbury, WC, S Africa.

Note: Characteristic of the species are the scabrid leaves, an apically constricted calyx base, and the prominently denticulate nectary.

L. amabilis L.Bolus NM 1958: 330 **T** Esterhuysen 23153 (BOL).

Erect shrubs to 30 cm *h*, **I** 7–30 mm *l*, ∅ 2 mm; **L** trigonous, 15–23 mm *l*, 3 mm *b* and *t*; **Fl** solitary, pedicels 15–30 mm *l*, bracteoles near their middle, petals pale to bright pink, 3–25 mm *l*, 3–4 mm *b*, filamentous staminodes few, filaments golden; **Fr** pentangular, 9 mm *l*, ∅ 8 mm, top elevated for 3 mm, 5 locules; **Ecol** on stony slopes amongst low bush; **Distr** Bredasdorp, WC, S Africa.

Note: Placed in the group Tenuifolii by Bolus.

L. amoenus (Salm-Dyck ex DC.) N.E.Br. GC 1930: 211 ≡ *M. amoenum* Salm-Dyck ex DC. 1828: 436 **Lectotype**, here designated: Salm-Dyck 1837: § 27 f 2.

Shrubs with erect branches to 15 cm *h*; **L** grouped together on short shoots or at the base of flowering shoots, triquetrous, mucronate, punctate, deep green, sometimes reddish maculate, 27–40 mm *l*, 4.5 mm *b*; **Fl** in terminal ternate cymes, petals purple, stamens white; **Fr** unknown; **Ecol** unknown; **Distr** unknown.

Note: De Candolle based the description taken as the base of the present text on the name *M. amoenum* received from Salm-Dyck "in litteris". Salm-Dyck (1827: § 27 f 2) referred to the species as "*Mesembryanthemum amoenum* Nob." indicating clearly that the plant figured is the same as the one referred to by De Candolle.

L. amphibolius (G.D.Rowley) H.E.K.Hartmann Bradleya 1998: 67 non **L. hallii** L.Bolus ≡ *Ruschia amphibolia* G.D.Rowley NCSJ 1978: 7 non **Ruschia hallii** L.Bolus ≡ *Stoeberia hallii* L.Bolus JSAB 1960: 161 (as replaced synonym) ≡ *Amphibolia hallii* (L.Bolus) L.Bolus ex Toelken & Jessop 1976: 64; L.Bolus 1965: 169 nomen invalid. **T** Hall 1741 (BOL!).

Creeping, **L** > **I**; **L** almost terete, those of a pair connate for 3–4 mm, free parts 20–35 mm *l*, 4–6 mm *b* and *t*; **Fl** solitary, bracteoles below the middle, petals pink, filamentous staminodes?, filaments purplish-pink, basally white; **Fr** top ± conically raised, base bell-shaped, covering membranes convex, with a small closing rodlet at the distal end of each, closing body absent, valve wings broader than the expanding keels, 5 locules; **Ecol** < 100 mm rainfall p.a. in winter; **Distr** Ceres, WC, S Africa.

Note: As in **L. gydouwensis**, the lifted free end of the placenta has probably been mistaken as an endocarpal closing body when the species was placed in **Stoeberia** and **Amphibolia**. The presence of broad valve wings and deep locules, and the absence of an endocarpal closing body indicate that the species belongs to the genus **Lampranthus**. Since the name **L. hallii** had been used for a different species in this genus already, the younger epithet of Rowley has to be taken up at species level.

L. antemeridianus (L.Bolus) L.Bolus NM 1939: 168 ≡ *M. antemeridianum* L.Bolus NM 1928: 135 **T** Muir 1505/21 (BOL) = *M. antemeridianum* L.Bolus "forma flore pleno" NM 1933: 404 nomen invalidum **T** Ferguson 20821 (BOL).

Shrubs with ascending to decumbent branches 16 to 20 cm *l*, **I** ∅ to 3 mm; **L** trigonous, 20–25 mm *l*, 2 mm *b* and *t*; **Fl** solitary, ∅ 45 mm, pedicels 3 035 mm *l*, petals golden, salmon-coloured below or pale pink, to 21 mm *l*, 2.5 mm *b*, filamentous staminodes not mentioned, filaments epapillate, yellow, anthers and pollen golden, nectary ring inconspicuous; **Fr** ∅ 18 mm; **Ecol** in sand dunes; **Distr** Riversdale, WC, S Africa.

Note: Bolus (1939: 168) cited as basionym her forma, an invalid name because it lacks an epithet proper. Furthermore, the details given agree in all features with those of the species, a separation is therefore unfounded.

L. antonii L.Bolus 1966: 232 **T** Hall 1053 (BOL).

Creeping, **I** 25–35 mm *l*; **L** trigonous to semi-terete, 20–50 mm *l*, 2–6 mm *b*, 2–5 mm *t*; **Fl** solitary, ∅ 50–66 mm, pedicels 7–13 mm *l*, petals white, palest pink with age, 15–27 mm *l*, 1.5–2 mm *b*, filamentous staminodes absent, filaments epapillate, white; **Fr** base globose to obconical, top raised for 3 mm, 10-sided, to 11 mm *l*, to 9 mm ∅, to 14 mm ∅ when expanded, 5 locules; **Ecol** "common on moist rocks"; **Distr** Ceres, WC, S Africa.

Note: The plants are reported to mimic those of **Cephalophyllum** by Bolus.

L. arbuthnotiae (L.Bolus) L.Bolus NM 1939: 170 ≡ *M. arbuthnotiae* L.Bolus ABH 1925: 8 **T** Arbuthnot 17397 (BOL).

Creeping with erect flowering branches; **L** triquetrous, surface with pronounced dots, glaucous green, 8 mm *l*, 1 mm *b* and *t*; **Fl** 1, ∅ 14–16 mm, pedicels 6–12 mm *l*, bracteoles near the base, petals pinkish-purple, basally white, filamentous staminodes present, filaments densely barbate, purple, basally white, anthers and pollen golden, nectary ring dentate; **Fr** unknown; **Ecol** unknown; **Distr** "Cape Province, South-Western region", WC, S Africa.

Note: This is one of the species with rough leaves.

L. arenarius H.E.K.Hartmann Bradleya 1998: 67 ≡ *Ruschia arenosa* L.Bolus NM 1933: 411 **LT** L.Bolus 19304 II.

Erect shrubs, 20–30 cm *h*, **I** 10–40 mm *l*, 2–3 mm ∅; **L** erect, semi-terete, acute, glaucous green, 23 mm *l*, 3 mm *b*, 4 mm *t*; **Fl** 1–9, ∅ to 15 mm, pedicel 25 mm *l*, bracteoles at c. half their length, petals pink, 6 mm *l*, 2.5 mm *b*, filamentous staminodes pale, apically pink, anthers and pollen golden; **Fr** base longish, funnel-shaped, top raised, but rims low, covering membranes convex, distal closing ledges long, end of placenta semi-globose, but no closing body present, valves with broad rectangular wings, expanding keels diverging, not touching each other basally, 5 locules; **Ecol** in sandy places; **Distr** Calvinia, NC, S Africa.

Note: Since a species named **L. arenosus** existed already in **Lampranthus**, a new name had to be made for this species, which has been transferred from **Ruschia** based on the fruit morphology.

L. arenicola L.Bolus 1964: 38 **T** Esterhuysen 30524 (BOL).

Erect shrubs to 34 cm *h*, ∅ to 39 cm, **I** 5–13 mm *l*, 1 mm ∅, stems basally to 23 mm ∅; **L** semi-terete to trigonous, apically recurved, 7–10 mm *l*, 1 mm *b*, 1.5 mm *t*; **Fl** 3 to many in a cyme, ∅ to 14 mm, pedicels 10–20 mm *l*, bracteoles near their middle, similar to foliage leaves, petals purplish-pink, 4–6 mm *l*, 1 mm *b*, filamentous staminodes present, filaments papillate, pale, apically pink, nectary ring inconspicuous; **Fr** base cyathiform, top elevated for 1.5 mm, 5 mm *l*, ∅ 6 mm, expanded 10 mm ∅, closing body small or absent, closing rodlet present; **Ecol** in sand; **Distr** Laingsburg, WC, S Africa.

Note: Esterhuysen reports "Few plants only seen".

L. arenosus (L.Bolus) L.Bolus NM 1939: 168 ≡ *M. arenosum* L.Bolus NM 1932: 367 nomen illeg. non Schinz 1897: 79 **T** L.Bolus 20202 (BOL).

Shrubs to 25 cm *h*, diffusely branched and suberect, twining in other shrubs, **I** brown, 20–30(–60) mm *l*, ∅ 3 mm; **L** trigonous, pale glaucous green or yellowish-green, 15–20 mm *l*, 1.5–2.5 mm *b*, 2–3 mm *t*; **Fl** 1–5 in cymes, ∅ 45 mm, pedicels 15–35 mm *l*, bracteoles at their middle, petals pink, 7–12 mm *l*, 2.5 mm *b*, filamentous staminodes absent, filaments pale pink, inner basally densely papillate, anthers golden; **Fr** top elevated for 3 mm, 5 locules; **Ecol** in sand; **Distr** Vredenburg, WC, S Africa.

Note: High rising valve wings are rare in **Lampranthus.**

L. argenteus (L.Bolus) L.Bolus 1967: 306 ≡ *M. argenteum* L.Bolus ABH 1925: 8 **T** Compton 962/22 (BOL).

Richly branched shrubs to 23 cm *h*, branches spreading, to 30 cm *l*, **I** 20–40 mm *l*, 3 mm ∅; **L** trigonous, glaucous green, 10–13 mm *l*, 3–5 mm *b*, 2–3 mm *t*; **Fl** in cymes, ∅ 12 mm, pedicels 7–11 mm *l*, petals pale pink, margins deeper in colour, filamentous staminodes not mentioned, filaments pale, apically pink, anthers and pollen yellow, nectary as a crenulate ring; **Fr** unknown; **Ecol** unknown; **Distr** Clanwilliam, WC, S Africa.

Note: Leaves of this species are very smooth and light grey.

L. argillosus L.Bolus NM 1950: 227 ≡ *M. tulbaghense* L.Bolus NM 1932: 364 nomen illeg. non A.Berger 1922: 637 **T** Pillans, L.Bolus 18470 (BOL).

Erect shrubs to 20 cm *h*, **I** 5–25 mm *l*, ∅ to 6 mm basally; **L** trigonous, obtuse, dirty green, 8–13 mm *l*,

2 mm *b*, to 2.5 mm *t*; **Fl** in cymes, pedicels 7–13 mm *l*, bracteoles at their middle, petals purplish-pink, to 9 mm *l*, 1.5 mm *b*, filamentous staminodes reddish-purple, apically pale, pollen straw-coloured, nectary as a crenulate ring; **Fr** turbinate, top elevated for 2 mm, 7 mm *l*, ∅ 6 mm,; **Ecol** unknown; **Distr** Tulbagh, WC, S Africa.

Note: The name *M. tulbaghense* was illegitimate when published because it had been used before by Berger (1922: 637). It could have been used as a basionym in **Lampranthus**, had not Berger's species been transferred to there first by Brown; a new name was therefore required.

L. aurantiacus ≡ **L. glaucoides**

L. aureus (L.) N.E.Br. GC 1930: 211 ≡ *M. aureum* L. 1759: 1060 non Thunberg 1791: 16 **Neotype**, here designated: Curtis 1794: t. 262.

Shrubs, **I** cylindrical; **L** trigonous, those of a pair basally broader and a little connate, dotted; **Fl** 3 per terminal cyme, petals golden, stigmas blackish-purple; **Fr** 5 locules; **Ecol** unknown; **Distr** S Africa.

Note: Golden petals and blackish-purple stigmas have traditionally been understood to characterize this species, of which two older coloured figures exist: Curtis 1794: t. 262 and De Candolle 1804: t. 11. The former is chosen as the neotype here because Curtis referred to Aiton (1789: 190), who reported that the plants had been in cultivation with Miller in 1750, indicating a very long period of acquaintance with this species, which is said to grow easily from cuttings.

L. austricola (L.Bolus) L.Bolus NM 1939: 168 ≡ *M. austricola* L.Bolus NM 1934: 453 **T** Salter 4264 (BOL).

Erect shrubs, 15–30 cm *h*, **I** 5–13 mm *l*; **L** falcate, glaucous green, scabrous, 10–12 mm *l*, 2 mm *b*, 2.5 mm *t*; **Fl** ∅ 28 mm, pedicels 7–13 mm *l*, bracteoles near their bases, similar to the foliage leaves, petals pale pink, to 13 mm *l*, 1.5 mm *b*, inner bending over stamens and merging into them, filamentous staminodes pale, filaments white, anthers and pollen yellow, nectary ring pentagonal; **Fr** unknown; **Ecol** unknown; **Distr** uncertain in WC, S Africa.

Note: The filamentous staminodes bending into the interior of the flower resemble conditions found in flowers of **Erepsia**.

L. baylissii L.Bolus 1965: 240 **Syntypes** Bayliss 1083, Middlemost 569/64 (BOL).

Erect, richly branched shrubs to 15 cm *h*, **I** 2–10 mm *l*, ∅ basally to 5 mm; **L** subfalcate, to 20 mm *l*, 1.5 mm *b*, 2.5 mm *t*; **Fl** solitary, pedicels 8–10 mm *l*, petals bright yellow, 5–11 mm *l*, 1–2 mm *b*, filamentous staminodes absent, filaments epapillate, pale or golden, anthers and pollen golden; **Fr** base obconical, 5-sided, valve wings present; **Ecol** on flat rocks; **Distr** Knysna, WC, S Africa.

Note: In cultivation, the plant grew to 30 cm high and reached a diameter of c. 18 cm, as reported by Hall (Bolus 1965: 240).

L. berghiae (L.Bolus) L.Bolus NM 1939: 170 ≡ *M. berghiae* L.Bolus NM 1938: 127 **T** Bergh 21842 (BOL).

Erect shrubs to 50 cm *h*, **I** 10–40 mm *l*, ∅ 2–4 mm; **L** trigonous, obtuse, glaucous-green, 20–30 mm *l*, 2–2.5 mm *b*, 3–4 mm *t*; **Fl** in cymes, pedicels 20–50 mm *l*, bracteoles at their middle, petals pink or pale pink, 9–21 mm *l*, 1.5 mm *b*, filamentous staminodes absent, filaments pale pink or white, anthers and pollen yellow, nectary ring broad; **Fr** top to 4 mm *h*, ∅ 13 mm, 5 locules; **Ecol** in sand among Restionaceae; **Distr** Clanwilliam, WC, S Africa.

Note: One of the species with big capsules.

L. bicolor (L.) N.E.Br. GC 1930: 211 ≡ *M. bicolor* L. 1753: 485 ≡ *L. bicolor* (L.) H.Jacobsen 1949: 58 nomen illeg. ≡ *Mesembryanthus bicolor* (L.) Rothmaler 1941: 414 nomen illeg. **Lectotype**, here designated: Dillenius 1732: 267, t. 202 f 258.

Erect shrubs to 60 cm *h* and higher; **L** trigonous, basally separated by a line, green, a little rough; **Fl** petals golden inside, red outside, filamentous staminodes not mentioned; **Fr** top raised a little, base elongate funnel-shaped; **Ecol** unknown; **Distr** unknown.

Note: Jacobsen (1949: 58) seems to have overlooked the combination **L. bicolor** by Brown (1930: 211), hence the younger homonym is illegitimate.

L. bicolor var. *inaequalis* ≡ **L. inaequalis**

L. blandus (Haw.) Schwantes in Jacobsen FR 1938: 229 ≡ *M. blandum* Haw. 1819: 95 **Lectotype**, here designated: Ker & Gawler 1821: t. 582.

Erect, branched shrubs to 60 cm *h*, terminal **I** laterally compressed ; **L** triquetrous, leaving a triangle of stem tissue free between their bases, < 40 mm *l*, glaucous-green; **Fl** in terminal cymes of 3, petals bright white, light purple with age; **Ecol** unknown; **Distr** S Africa.

Note: Figures of this species have been published in 1821 twice, both from material supplied by Haworth, originally collected by Burchell: Loddiges (1821: t. 599) and Ker & Gawler (1821: t. 582). The latter figure is accompanied by descriptions of Haworth and is therefore chosen as the lectotype. The species has been studied by Haworth in the collections of Burchell, Hooker, and Ross, and is considered by the author to be most similar to *M. conspicuum* and *M. spectabile*.

L. borealis L.Bolus JSAB 1960: 160 **T** Herre 14513 (BOL!) = *L. brachyandrus* forma L.Bolus 1960: 58 nomen invalid. T Holloway 66 (BOL!) = *M. comptonii* L.Bolus var. *angustifolium* L.Bolus NM 1933: 409 ≡ *L. comptonii* (L.Bolus) N.E.Br. var. *angustifolius* (L.Bolus) L.Bolus ex H. Jacobsen FR 1938: 96 **T** Herre 8997 (BOL).

Rather low shrubs, branches 9–18 cm *l*, **I** 5–25 mm *l*, ∅ 1.5–3 mm, herbaceous parts glaucous; **L** trigonous to terete, slender, apically rounded, laterally compressed, basally almost separated by a triangle of stem tissue, to 24 mm *l*, 3 mm *b*, to 5 mm *t*; **Fl** solitary, but often arranged in 3, ∅ 40–50 mm, pedicels 20–50 mm *l*, bracteoles at their middle, petals palest pink, 0.5–5 mm *b*, filamentous staminodes absent, filaments white, anthers and pollen yellow, nectary ring crenulate; **Fr** stalks bracteolate, base bell-shaped, top high, rims medium high forming almost a crater in side view, covering membranes straight, with low closing ledges distally, expanding keels diverging from their bases where they meet, valve wings very broad, rectangular, 5 locules; **S** brown, papillate, 1.1 mm *l*, c. 1 mm *b*; **Ecol** on rocky slopes; **Distr** Lüderitz-South, Namibia.

Note: Some confusion exists as to the boundaries of this species (Friedrich 1960: 566, 1970: 64–66). Plants of **L. borealis** can clearly be separated from **L. brachyandrus** with erect, straight branches and from **L. hoerleinianus** with thicker leaves. Differing from both are also the flowers of **L. borealis** in lacking filamentous staminodes. In this circumscription the species has, until now, not been found south of Namibia.

L. brachyandrus (L.Bolus) N.E.Br. GC 1930: 211 ≡ *M. brachyandrum* L.Bolus NM 1928: 3 **T** Pillans 5803 (BOL).

Erect shrubs with straight branches and only short side branches, to 40 cm *h*, **I** glaucous to pink, 20–20 mm *l*, ∅ to 6 mm; **L** trigonous, slender, obtuse, 25–32 mm *l*, 6 mm *b*, 7 mm *t*; **Fl** in cymes, ∅ to 44 mm, smelling of **Heliotropium**, pedicels 10–20 mm *l*, petals purplish-pink, to 18 mm *l*, to 2 mm *b*, filamentous staminodes not mentioned, filaments purple, anthers and pollen golden, nectary ring blackish-green; **Fr** 5 locules; **Ecol** near the coast; **Distr** Namaqualand, NC, S Africa.

Note: Data taken from description and type only. The slender leaves agree with those of **L. borealis**, from which **L. brachyandrus** differs in the erect, straight branches.

L. brevistaminus (L.Bolus) L.Bolus NM 1939: 168 ≡ *M. brevistaminum* L.Bolus NM 1929: 155 **T** Stokoe 855/25 (BOL).

Erect shrubs to 14 cm *h*, richly branched, round in outline, branches twisted, **I** to 10 mm *l*, 1–3 mm ∅; **L** clavate to falcate, glaucous green, dotted, to 7 mm *l*, 2 mm *b* and *t*; **Fl** solitary, ∅ 34 mm, pedicels with bracteoles at their middle, petals purplish-pink, 4–15 mm *l*, to 2 mm *b*, filamentous staminodes absent, filaments straw-coloured, anthers golden, nectary ring inconspicuous; **Fr** 5 locules; **Ecol** unknown; **Distr** Worcester, WC, S Africa.

Note: Short stamens and the lack of filamentous staminodes occur in a number of species.

L. brownii (Hooker f.) N.E.Br. GC 1930: 211 ≡ *M. brownii* Hooker f. 1888: t. 6985 **T** s.n. from which Hooker 1888: t. 6985 is drawn (K!).

Erect shrubs to 30 cm *h*, **I** brown to maroon, shining; **L** terete, almost awl-shaped, pale glaucous green, the bases leaving a narrow, high triangle between them, smooth, 13.5 mm *l*; **Fl** 1–3, but several of these in bundles imitating big dichasia, ∅ 27 mm, bracteoles at the middle of the pedicels, petals purple, turning ochre later, filamentous staminodes not mentioned; **Fr** expanding keels basally in contact, diverging immediately above, covering membranes with knobby closing devices between closing rodlets and closing bulges, 5 locules; **Ecol** unknown; **Distr** S Africa.

Note: Data concerning the fruit are taken from a later addition to the type sheet marked "Mr Bates Sept. 1924". The fruiting stalk of this part agrees in leaf and stem features exactly with that of the type material placed next to it and is therefore considered identical. It is well known that Bates had a collection of material derived from types, some of of which was returned to Kew, where it is still kept today.

L. caespitosus (L.Bolus) N.E.Br. GC 1930: 211 ≡ *M. caespitosum* L.Bolus ABH 1920: 4 **T** Rohland 1712/14 (BOL) = *M. caespitosum* L.Bolus forma *luxurians* L.Bolus NM 1935: 487 ≡ *L. caespitosus* (L.Bolus) N.E.Br. forma *luxurians* (L.Bolus) H.Jacobsen 1955: 1427 nomen invalid. ≡ *L. caespitosus* (L.Bolus) N.E.Br. var. *luxurians* (L.Bolus) H.Jacobsen in H.Jacobsen & Volk 1950: 100 nomen invalid. T van Breda 197 (BOL).

Mat-forming shrubs with elongate, prostrate branches, flowering branches erect to ascending; **L** subfalcate, glaucous, 10 mm *l*, 2 mm *b* and *t*; **Fl** solitary, ∅ to 22 mm, petals pink, 9 mm *l*, 3 mm *b*, filamentous staminodes few, filaments white, apically pink, anthers yellow, nectary ring crenulate, deep green; **Fr** unknown; **Ecol** unknown; **Distr** Paarl, Worcester, WC, S Africa.

Note: The plants were in cultivation on a rather shaded bank at Kirstenbosch.

L. caespitosus forma *luxurians* = **L. caespitosus**
L. caespitosus var. *luxurians* = **L. caespitosus**

L. calcaratus (Wolley-Dod) N.E.Br. GC 1930: 211 ≡ *M. calcaratum* Wolley-Dod 1900: 170 **Syntypes** Wolley-Dod 2857, 2860 (K).

Suberect shrubs with many ascending branches; L terete, acutely pointed, basally with a straight spur, 9–18 mm *l*; **Fl** solitary, pedicels 15–30 mm *l*, petals bright rose; **Fr** unknown; **Ecol** in flats; **Distr** Wynberg, WC, S Africa.

Note: The basally spurred leaves appear unique in the genus.

L. candidus L.Bolus NM 1939: 156 **T** Pillans 8852 (BOL).

Erect shrubs with many branches 25–30 cm *h*, **I** 15–20 mm *l*, 1–3 mm ∅; **L** subfalcate, scabrous, glaucous, to 6 mm *l*, 1.5 mm *b*, 2 mm *t*; **Fl** 1–3, pedicels 10–17 mm *l*, bracteoles at their middle, petals white, 3–10 mm *l*, 0.25–1.5 mm *b*, filamentous staminodes absent, filaments white, anthers and pollen yellow, nectary ring inconspicuous, stigmas blackish-purple; **Fr** turbinate, 4–5 mm *l*, ∅ 4 mm, 5 locules; **Ecol** unknown; **Distr** Piquetberg, WC, S Africa.

Note: The species is named after the white flowers.

L. capillaceus (L.Bolus) N.E.Br. GC 1930: 212 ≡ *M. capillaceum* L.Bolus NM 1928: 137 **T** Stokoe 18781 (BOL).

Shrubs to 14 cm *h*, branches and pedicels hairy, **I** 10–25 mm *l*; **L** subfalcate, subglaucous, 5 mm *l*, 1.5 mm *b*, 2 mm *t*; **Fl** in cymes, ∅ to 20 mm, pedicels 30–40 mm *l*, bracteoles above their middle, petals purplish-pink, dorsally paler, to 8 mm *l*, 1 mm *b*, filamentous staminodes not mentioned, filaments pale, inner basally with blackish-purple papillae, anthers and pollen yellow, nectary ring inconspicuous; **Fr** 5 locules; **Ecol** unknown; **Distr** Tulbagh, WC, S Africa.

Note: Short stamens possess dark papillae near their bases in this species.

L. capornii ≡ **Ruschia capornii**

L. caudatus L.Bolus NM 1950: 226 **T** Leighton 1631 (BOL).

Erect shrubs to 30 cm *h*, basal ∅ of stem to 8 mm; **L** recurved, scabrous, glaucous green, 9–14 mm *l*, 1.5–2 mm *b* and *t*; **Fl** 1–3, pedicels to 20 mm *l*, **K** base scabrous, petals pale pink, to 10 mm *l*, to 1.5 mm *b*, filamentous staminodes few, anthers and pollen yellow, nectary ring inconspicuous; **Fr** 5 locules; **Ecol** in salty places; **Distr** Bredasdorp, WC, S Africa.

Note: Plants in especially salty places develop thicker leaves.

L. cedarbergensis ≡ **Oscularia cedarbergensis**

L. ceriseus (L.Bolus) L.Bolus NM 1939: 168 ≡ *M. ceriseum* L.Bolus NM 1932: 327 **T** L.Bolus 19926 (BOL).

Erect shrubs to 11 cm *h*, **I** 10–15 mm *l*, ∅ to 2 mm; **L** triquetrous, acute, glaucous green, dotted, to 15 mm *l*, to 2 mm *b*; **Fl** 1–5 in cymes, ∅ 35 mm, pedicels 20–30 mm *l*, bracteoles at their middle, petals cerise or brightest pink, filamentous staminodes absent, filaments reddish-purple, anthers and pollen golden, nectary ring inconspicuous; **Fr** 5 locules; **Ecol** unknown; **Distr** Riversdale, WC, S Africa.

Note: A species with moderately scabrid leaves.

L. citrinus (L.Bolus) L.Bolus NM 1939: 168 ≡ *M. citrinum* L.Bolus NM 1932: 326 **T** Leipoldt 19925 (BOL).

Erect shrubs to 12 cm *h*, **I** 10–15 mm *l*, ∅ 2 mm; **L** trigonous, green dots slightly raised, 10–15 mm *l*, 1.5 mm *b*, 2 mm *t*; **Fl** solitary, ∅ 48 mm, pedicels 10–20 mm *l*, petals lemon-yellow, outside apically red, to 21 mm *l*, to 3 mm *b*, filamentous staminodes absent, filaments pale yellow, glabrous, nectary ring inconspicuous; **Fr** top slightly raised, 7 mm *l*, ∅ 8 mm, expanded 12 mm ∅; **Ecol** unknown: **Distr** Malmesbury, WC, S Africa.

Note: One of the species with solitary, big capsules.

L. coccineus (Haw.) N.E.Br. GC 1930: 212 ≡ *M. coccineum* Haw. 1795: 247 **Lectotype**, here designated: DC. 1801: 83.

Shrubs with two-angled young branches, **I** later cylindrical, dark; **L** trigonous, compressed, obtuse, glaucous, dotted; **Fl** solitary or in loose cymes, pedicels with bracteoles at their middle, petals sanguineous-orange, fading later; **Fr** ficiform, 5 locules; **Ecol** unknown; **Distr** S Africa.

Note: The lectotypification is based on the remark of Haworth 1803: 85 "M. coccineum, Plantes Grasses, page 83. cum icone bono".

L. compressus ≡ **Oscularia compressa**
L. comptonii ≡ **Oscularia comptonii**
L. comptonii var. *angustifolius* = **L. borealis**
L. comptonii forma *roseus* ≡ **Oscularia comptonii**
L. comptonii var. *roseus* ≡ **Oscularia comptonii**

L. conspicuus (Haw.) N.E.Br. GC 1930: 212 ≡ *M. conspicuum* Haw. 1812: 240 non Salm-Dyck 1840: § 27 f 1 nomen illeg. **T** drawing 938/30 by Duncanson (K!).

Branched shrubs, **I** on flowering branches compressed, deep maroon to brown; **L** trigonous, from a broader base, crowded on short shoots in the axils of longer leaves, the bases of the leaf-pairs leaving a triangle free between them, green; **Fl** in many-flowered cymes, petals deep and bright purple, filaments pink; **Fr** turbinate, 5 locular; **Ecol** unknown; **Distr** S Africa.

Note: The drawing chosen as the lectotype has been annotated by Brown as being "Doubtless from a type plant".

L. convexus (L.Bolus) L.Bolus NM 1939: 168 ≡ *M. convexum* L.Bolus NM 1928: 18 **T** Hutchinson 406 (BOL).

Erect shrubs to 23 cm *h*, basal ∅ of stem to 10 mm, I 12 mm *l*, ∅ 4 mm, blackish-grey; **L** subterete to trigonous, pale green, 15–27 mm *l*, 2 mm *b*, 3 mm *t*; **Fl** in cymes, ∅ 18 mm, pedicels with bracteoles at their middle, petals pink, 8 mm *l*, 1.5 mm *b*, filaments pink, anthers and pollen golden, nectary ring crenulate; **Fr** ∅ expanded 10 mm, 5 locules; **Ecol** unknown; **Distr** Tulbagh, WC, S Africa.

Note: The stamens are collected in a cone, no filamentous staminodes are reported.

L. copiosus ≡ **Oscularia copiosa**

L. coralliflorus (Salm-Dyck) N.E.Br. GC 1930: 212 ≡ *M. coralliflorum* Salm-Dyck 1842: § 40 f 2 ≡ *M. corallinum* Haw. 1821: 154 nomen illeg. non Thunberg 1791: 12 ≡ *Erepsia coralliflora* (Salm-Dyck) Schwantes GF 1928: 68 **Lectotype**, here designated: Salm-Dyck 1842: § 40 f 2.

Shrubs with erect branches, to 90 cm *h*, richly branched, **I** light brown to red, later grey to red; **L** subterete to subclavate, obtuse, minutely mucronate, green, 54 mm *l*, 3.75 mm *b*; **Fl** solitary, ∅ 54 mm, pedicels to 17.5 mm *l*, bracteoles at their middle, **K** 5, rarely 4, petals bright purple, filamentous staminodes pale pink, filaments pinkish-white, anthers light yellow; **Fr** (4–)5 locules, otherwise unknown; **Ecol** unknown; **Distr** S Africa.

Note: A little known species although its early introduction into cultivation suggests that it stems from near the Cape of Good Hope. The present placement is not a confirmative one but based on the definite exclusion of the species from **Erepsia** (Liede 1990: 472).

L. creber L.Bolus NM 1958: 332 **T** du Plessis 15 (BOL).

Shrubs to 35 cm *h*, stems basally to 10 mm ∅, **I** brown, shining, 20–70 mm *l*, ∅ 1–1.5 mm ; **L** trigonous, rough, keel minutely apiculate, 6–9 mm *l*, 1–1.5 mm *b* and *t*; **Fl** pedicels 3–6 mm *l*, in fruit to 12 mm *l*, petals pink, filamentous staminodes pink, anthers and pollen yellow, nectary ring crenulate; **Fr** pentagonal, top 3–4 mm *h*, base obconical, 8–9 mm *l*; **Ecol** unknown; **Distr** Riversdale, WC, S Africa.

Note: Named after the rough leaves.

L. curviflorus (Haw.) H.E.K.Hartmann Bradleya 1998: 67 ≡ *L. curviflorus* (Haw.) N.E.Br. ex H.Jacobsen 1955: 1429 nomen invalid. ≡ *M. curviflorum* Haw. 1821: 147 ≡ *M. blandum* Haw. var. *curviflorum* (Haw.) A.Berger 1908: 163 **T** Ross (K).

Erect shrubs, **I** terete, reddish to reddish-brown, later grey; **L** triquetrous-compressed, glaucous; **Fl** pedicels as long as the bracteoles, petals incurved (hence the name), white; **Fr** unknown; **Ecol** unknown; **Distr** S Africa (?).

Note: The species was contrasted against *M. blandum*, now **L. blandus**, and separated mainly because the petals were incurved and the plants were twice as big. Jacobsen (1955: 1429) ascribed the name to Brown, who seems not to have published the combination. Since Jacobsen did not give a clear reference to a valid basionym, that combination was invalid.

L. curvifolius (Haw.) N.E.Br. GC 1930: 212 ≡ *M. curvifolium* Haw. 1803: 88 ≡ *L. curvifolius* (Haw.) Schwantes in Jacobsen 1949: 58 **T** unpublished drawing at K (Glen 1978, unpublished) = *M. ceratophyllum* Willdenow 1814: 36 **T**?

Shrubs > 15 cm *h*, younger branches rough, purple, older reddish-black; **L** younger densely placed, older with longer **I** between them, subulate, semicylindrical, bent like the horns of oxen, c. 25 mm *l*; **Fl** solitary, bracteoles above the middle of the pedicel, petals bright red, filamentous staminodes basally and apically white, the middle reddish, anthers yellow; **Fr** 5 locules; **Ecol** unknown; **Distr** unknown.

Note: M. ceratophyllum is included here, following Salm-Dyck (1836), but not the names referring to *M. curvifolius* var. *minus*, which are understood as the species **L. flexifolius**.

L. curvifolius var. *minor* ≡ **L. flexifolius**

L. cyathiformis (L.Bolus) N.E.Br. GC 130: 212 ≡ *M. cyathiforme* L.Bolus ABH 1923: 165 ≡ *Erepsia cyathiformis* (L.Bolus) Schwantes GF 1928: 68 **T** Arbuthnot 17255 (BOL).

Shrubs with erect branches, to 22 cm *h*, **I** 25–32 mm *l*; **L** trigonous, spreading, subflaccid, i.e. not able to hold its own weight, 25–32 mm *l*, 4 mm *b* and *t*; **Fl** solitary, pedicels 15–50 mm *l*, bracteoles at their middle, similar to the foliage leaves, petals pinkish-purple, basally white, to 20 mm *l*, 3–4 mm *b*, inner ones 8 mm *l*, filamentous staminodes pale, filaments white, anthers and pollen pale yellow; **Fr** ∅ 13 mm, 5 locules, otherwise unknown; **Ecol** unknown; **Distr** "Cape Province", S Africa.

Note: The present placement of the taxon here is based on the definite exclusion of the species from **Erepsia** (Liede 1990: 472).

L. debilis (Haw.) N.E.Br. GC 1930: 212 ≡ *M. debile* Haw. 1826: 330 **Lectotype**, here designated: drawing 401 by Duncanson of plants grown in 1823 from seed collected by Bowie (K!).

Caespitose, creeping with filiform branches rooting at their nodes; **L** aggregate at the nodes, clavate to acinaciform to triquetrous, obtuse, glaucous, smooth; **Fl** few, terminal, very pale reddish to whitish; **Fr** unknown; **Ecol** unknown; **Distr** S Africa.

Note: The species is reported to be similar to **L. reptans**, which is rougher. The choice of the drawing as a lectotype is based on a remark of Haworth (1831: 422, cited also by Brown on the drawing), who reported to have seen material in 1830 in flower at Kew.

L. deflexus (Aiton) N.E.Br. GC 1930: 212 ≡ *M. deflexum* Aiton 1789: 187 T Masson? (?).

Bending; **L** triquetrous, acute, glaucous, with inconspicuous dots; **Fl** margins of inner **K** lobes membranous; **Fr** unknown; **Ecol** unknown; **Distr** S Africa.

Note: Introduced by Masson 1774.

L. deltoides ≡ **Oscularia deltoides**

L. densifolius (L.Bolus) L.Bolus NM 1950: 227 ≡ *M. densifolium* L.Bolus NM 1928: 136 T Mathews 1705/24 (BOL).

Erect, richly branched shrubs, **I** ∅ to 4 mm; **L** trigonous, to 17 mm *l*, 2–3 mm *b* and *t*; **Fl** 1–3, showy, ∅ 50–60 mm, petals pink to purplish, to 25 mm *l*, to 2 mm *b*, filamentous staminodes very few, filaments white, anthers yellow; **Fr** pentagonal, ∅ expanded 18 mm, 5 locules; **Ecol** unknown; **Distr** Malmesbury, WC, S Africa.

Note: One of the species with solitary, big capsules.

L. densipetalus (L.Bolus) L.Bolus NM 1939: 168 ≡ *M. densipetalum* L.Bolus NM 1930: 218 T L.Bolus 19249 (BOL).

Erect, diffusely branched shrubs to 26 cm *h*, ascending to decumbent branches to 30 cm *l*, **I** maroon, 10–20 mm *l*, ∅ 3 mm; **L** slightly compressed, bases of a pair separated by a triangle, to 20 mm *l*, 3 mm *t*; **Fl** solitary, ∅ 35 mm, pedicels 14 mm *l*, bracteoles in their middle, petals white, filamentous staminodes absent, anthers, pollen yellow, nectary ring crenulate; **Fr** 7 mm *l*, ∅ 9 mm, 5 locules; **Ecol** unknown; **Distr** Namaqualand, NC, S Africa.

Note: Like the preceding, this is one of the species with solitary, big capsules.

L. dependens (L.Bolus) L.Bolus NM 1939: 170 ≡ *M. dependens* L.Bolus ABH 1922: 134 T Page 17156 (BOL).

Decumbent branches elongating, young **I** tetragonal, 10–55 mm *l*, 3 mm ∅; **L** recurved, basally gibbous, terete, minutely papillate, shining, inconspicuous white dots, 35–55 mm *l*, 3–4 mm *t*; **Fl** in cymes, ∅ to 60 mm, pedicels 30–45 mm *l*, bracteoles 10–24 mm *l*, petals lilac, to 25 mm *l*, to 2 mm *b*, inner shorter and more narrow, filamentous staminodes absent, filaments white, anthers white, pollen yellow, nectary ring crenulate, blackish-green; **Fr** 5 locules; **Ecol** hanging from cliffs; **Distr** Montagu, WC, S Africa.

Note: The lilac or mauve petals and a dark nectary are unusual features in members of **Lampranthus**.

L. diffusus (L.Bolus) N.E.Br. GC 1930: 212 ≡ *M. diffusum* L.Bolus NM 1928: 137 T Way 18759 (BOL).

Diffuse branches weak, to 16–20 cm *l*, **I** 10–40 mm *l*; **L** semi-terete, subglaucous, a little rough, 15–20 mm *l*, to 2 mm *b* and *t*; **Fl** 1–3, ∅ 27 mm, pedicels with bracteoles at their middle, petals to 13 mm *l*, to 2 mm *b*, filamentous staminodes not mentioned, filaments pale, anthers, pollen yellow, nectary ring crenulate; **Fr** ∅ 9 mm, 5 locules; **Ecol** unknown; **Distr** Ceres, WC, S Africa.

Note: One of the species with mostly solitary, big capsules.

L. dilutus (L.Bolus) N.E.Br. GC 1930: 212 ≡ *M. pallidum* L.Bolus NM 1929: 119 nomen illeg. non N.E.Br. 1920: 96 T Mathews 1196/26 (BOL).

Diffuse branches to 25 cm *l*, flowering branches 10–15 cm *l*, **I** 10–15 mm *l*, 2 mm ∅; **L** ± subfalcate to subglobose, glaucous-green, finely rough, 5–7 mm *l*, 2 mm *b*, 1.5–3 mm *t*; **Fl** 1–3, smelling of almonds, ∅ 23 mm, pedicels 8 mm *l*, minutely tuberculate, petals palest pink to white, filamentous staminodes few, filaments white, nectary ring inconspicuously dentate; **Fr** top elevated for 2 mm, 5 locules; **Ecol** unknown; **Distr** Piquetberg, WC, S Africa.

Note: The odour of almonds and the opening of the flower until the later afternoon are unusual features in this genus.

L. disgregus N.E.Br. GC 1930: 212 ≡ *M. disgregum* N.E.Br. KB 1929: 59 ≡ *M. stipulaceum* L.Bolus FPSA 1927: t. 253 non L. 153: 484 **Lectotype**, here designated: Bolus FPSA 1927: t. 253.

Shrubs to 60 cm *h*, branches erect to decumbent; **L** semi-terete to terete, smooth, 30–40 mm *l*, 2–5 mm *t*; **Fl** solitary, ∅ to 65 mm, bracteoles below the middle of the pedicels, petals rose-purple to almost red, 7–10 mm *l*, up to 3 mm *b*, filaments pale purple, anthers and pollen bright yellow, nectary ring inconspicuous, crenulate; **Fr** covering membranes well developed, valve wings very broad, expanded ∅ 15 mm, 5 locules; **Ecol** unknown; **Distr** Port Elizabeth, Uitenhage, EC, S Africa.

Note: Described by Bolus (FPSA 1927: t. 253) as *M. stipulaceum*, the taxon was considered to represent a different species by Brown (KB 1929: 59); Bolus (NM 1932: 192) stressed the height of the plant in order to support her original identification of the material. Similar to **L. zeyheri** Bolus (1927).

L. dissimilis (G.D.Rowley) H.E.K.Hartmann Bradleya 1998: 67 ≡ *Ruschia dissimilis* G.D.Rowley NCSJ 1978: 62 non **Ruschia stayneri** L.Bolus ≡ *Amphibolia stayneri* L.Bolus ex Toelken & Jessop 1976: 64 ≡ *Amphibolia stayneri* L.Bolus 1966: 126 nomen illeg. (as replaced synonym) T Stayner 258/65 (BOL!).

Shrubs ± erect branches to 12 cm *l*, root thickened and turnip-shaped, the stems probably annual or short-lived, **I** maroon; **L** subulate, tip recurved, surface scabrous from elevated regions above tanning idioblasts, 15–17 mm *l*, 1.5–2 mm *t*; **Fl** solitary, bracteoles at base of stalk, petals purplish-pink, vittate on both sides, filamentous staminodes apically purplish-pink, basally pale; **Fr** top raised, rims rather low (from age?), base funnel-shaped, covering membranes convex, distal rim low, closing ledge below it low, closing body absent, but the end of the placenta broad and hood-shaped, giving the impression of a closing body, expanding keels almost touching each other basally, valve wings broad, rectangular, 5 locules; **Ecol** c. 300 mm rainfall p.a., probably in March and November; **Distr** Ceres, WC, S Africa.

Note: As in some other species placed originally in **Amphibolia**, **L. dissimilis** cannot be kept there because it lacks the closing bodies and shows an enlarged placenta-end instead, lifted away from the wall, in this way simulating a closing body at first sight. The broad valve wings contradict a transfer to **Ruschia**, undertaken by Rowley (1978), but prove that the species belongs to **Lampranthus**. Since the name **L. stayneri** has been used in this latter genus before, the second-oldest epithet has to be chosen as the basionym, namely *Ruschia dissimilis* G.D.Rowley.

L. diutinus (L.Bolus) N.E.Br. GC 1930: 212 ≡ *M. diutinum* L.Bolus ABH 1923: 167 **T** Muir 2670 (BOL).

Creeping, flowering branches ascending to erect, **I** compressed and winged, 10–25 mm *l*; **L** subfalcate, triquetrous, apiculate, green, smooth, to 20 mm *l*, 2.5 mm *b*, 3.5 mm *t*; **Fl** 1–2, open later in the day than the other species, ⌀ 40–48 mm, pedicels with their bracteoles at the middle, 30–60 mm *l*, petals pinkish-purple, 18–20 mm *l*, to 3 mm *b*, inner 7–13 mm *l*, inner ones linear, filaments white, epapillate, anthers, pollen yellow, nectary ring conspicuously crenulate; **Fr** unknown; **Ecol** unknown; **Distr** Riversdale, WC, S Africa.

Note: Flowers remain open until later in the day than in many other species.

L. dregeanus (Sonder) N.E.Br. GC 1930: 212 ≡ *M. dregeanum* Sonder in Harvey & Sonder 1862: 414 ≡ *Erepsia dregeana* (Sonder) Schwantes GF 1928: 68 **LT** Ecklon & Zeyher 2011 (SAM).

Shrubs with stems of 60 cm or longer, **I** purplish or red, subpruinose, young ones 2-edged, reddish; **L** much compressed, 6.75–9 mm *l*, 3.75–4.5 mm *b*; **Fl** solitary, often terminating short branches, pedicels 25–30 mm *l*, petals dry pale yellowish, 9–11.25 mm *l*; **Fr** 9 mm *l*, 5 locules; **Ecol** sandy and stony places; **Distr** Tulbagh, Worcester, WC, S Africa.

Note: A little known species; placed here after it has been excluded definitely from **Erepsia** (Liede 1990: 472). The acinaciform, grey leaves and the reddish stems resemble those of **Oscularia**.

L. drepanophyllus ≡ **Esterhuysenia drepanophylla**

L. dubitans (L.Bolus) L.Bolus NM 1939: 168 ≡ *M. dubitans* L.Bolus NM 1929: 119 ≡ *Ruschia dubitans* (L.Bolus) L.Bolus NM 1950: 220 **Lectotype**, here designated: Giffen II 450/24 ≡ Cook 1459/24 (chosen from syntypes BOL!).

Shrubs to 16 cm *h*, **I** 2–15 mm; **L** trigonous to subterete, apically recurved, epidermis rough from raised dots, 10–25 mm *l*, 3 mm *b* and *t*; **Fl** solitary, ⌀ 12 mm, bracteoles at c. half the length of the pedicel, petals pink, margins deeper in colour, 6 mm *l*, 1 mm *b*, filamentous staminodes few, anthers and pollen white; **Fr** almost pear-shaped, base elongate, top almost conical, valve wings very broad, 5 locules; **Ecol** in crevices and in plains; **Distr** Worcester (?), WC, S Africa.

Note: The species is transferred to here from **Ruschia** based on the fruit characters: broad valve wings and the lack of closing bodies.

L. dulcis (L.Bolus) L.Bolus NM 1939: 168 ≡ *M. dulce* L.Bolus NM 1932: 328 **T** Leipoldt 19332 (BOL).

Erect, laxly branched shrubs to 45 cm *h*, **I** shining, pale brown, 20–45 mm *l*, ⌀ to 4 mm; **L** falcate to incurved, 20–35 mm *l*, to 2.5 mm *b*, to 4 mm *t*; **Fl** 1–3, pedicels to 55 mm *l*, bracteoles at their middle, petals pale pink, 5–20 mm *l*, 1.5 mm *b*, inner ones more narrow, filaments white, anthers and pollen lemon yellow, nectary ring crenulate; **Fr** unknown; **Ecol** unknown; **Distr** Clanwilliam, WC, S Africa.

Note: One of the glaucous, smooth-leaved species.

L. dunensis ≡ **Erepsia dunensis**

L. dyckii (A.Berger) N.E.Br. GC 1930: 212 ≡ *M. dyckii* A.Berger 1922: 635 ≡ *M. conspicuum* Salm-Dyck 1840: § 27 f 1 nomen illeg. non Haw. 1812: 240 ≡ *Ruschia dyckii* (A.Berger) H.Jacobsen 1974: 628 nomen invalid. **Lectotype**, here designated: Salm-Dyck 1840: 27 f 1.

Shrubs with erect branches, to 45 cm *h*, branches twisted, recurved, **I** greyish-black, basally with annular remains of old leaves, to 20 mm *l*; **L** in heads, semicylindrical to trigonous, punctate, smooth, green, reddish maculate, to 54 mm *l*, to 4.5 mm *t*; **Fl** 1(–3), pedicels to 80 mm *l*, bracteoles in their middle, petals showy, purple, filamentous staminodes not mentioned, filaments purplish-pink, anthers pale; **Fr** obconical, 5 locules; **Ecol** unknown; **Distr** S Africa.

Note: The drawing of Salm-Dyck shows clearly the annular remains of old leaves below several crowded assimilating leaf-pairs, a feature rare in other species of **Lampranthus**. Jacobsen (1974: 628) transferred the name to **Ruschia**, in his list of synonyms of **Lampranthus**, the combination is therefore, and for the lack of a direct and full citation of the author and original publication, invalidly published. It appears remarkable that Jacobsen later (e.g. 1981: 609) gave "invalid" behind **L. dyckii** not placing it in either genus, **Ruschia** or **Lampranthus**.

L. ebracteatus L.Bolus 1962: 294 **T** Hall 613/57 (BOL).

Erect shrubs to 25 cm *h*, **I** 5–20 mm *l*, ∅ to 5 mm; **L** sublunate, 10–20 mm *l*, 3–6 mm *b*, 5–7 mm *t*; **Fl** 3–14 in aggregate cymes, pedicels ebracteolate, 5–10 mm *l*, petals pale pink, 7–10 mm *l*, to 2 mm *b*, filamentous staminodes white, apically pink, anthers and pollen pale yellow, glands of the nectary approaching each other or 2 connate, **Fr** 5 locules; **Ecol** in flats at the river mouth; **Distr** Vredendal, WC, S Africa.

Note: The species combines a number of features rare in the genus and may not belong here.

L. edwardsiae (L.Bolus) L.Bolus NM 1939: 170 ≡ *M. edwardsiae* L.Bolus ABH 1923: 163 **T** Edwards 2600/14 (BOL).

Densely branched shrubs to 25 cm *h*; **L** triquetrous, often falcate, apiculate, glaucous, smooth, to 25 mm *l*, to 2 mm *b*, to 7 mm *t*; **Fl** in cymes, open over midday, ∅ 12 mm, pedicels with bracteoles at their middle, petals pink, margins deeper in colour, 4 mm *l*, 1 mm *b*, filamentous staminodes with long papillae, anthers yellow, nectary ring irregularly divided; **Fr** unknown; **Ecol** unknown; **Distr** Clanwilliam, WC, S Africa.

Note: Reported to be similar to **Oscularia lunata** (as *L. lunatus*), from which it differs, e.g., by the conspicuously crenulate nectary ring.

L. egregius (L.Bolus) L.Bolus NM 1939: 170 ≡ *M. egregium* L.Bolus NM 1936: 12 **T** Esterhuysen 21640 (BOL).

Erect shrubs to 40 cm *h*, **I** reddish-brown, 15–55 mm *l*, ∅ up to 4 mm; **L** subcompressed with convex sides, dirty green, 20–25 mm *l*, 3–4 mm *b*, 3–5 mm *t*; **Fl** in cymes, rarely solitary, pedicels 20–65 mm *l*, bracteoles in their middle, petals 5–30 mm *l*, to 2.5 mm *b*, pinkish-purple, basally paler, filamentous staminodes pale pink, filaments palest pink or white, anthers and pollen palest yellow, nectary ring broad, subtetragonal, divisions deep, dense and obtuse; **Fr** turbinate, pentagonal, top raised for 3 mm, 11 mm *l*, ∅ 11 mm, expanded 16 mm; **Ecol** unknown; **Distr** Montagu, WC, S Africa.

Note: Big fruits occur in this species.

L. elegans (Jacquin) Schwantes in Jacobsen 1938: 229 ≡ *M. elegans* Jacquin 1804: t. 436 **Lectotype**, here designated: Jacquin 1804: t. 436.

Diffusely branched shrubs, branches ascending to decumbent, to 90 cm *l*, **I** terete, young ones yellow, rough or papillate, older ones grey, to 5 mm ∅; **L** semiterete to subangular, rough to the touch, glaucous, mucronate, 22.5 mm *l*; **Fl** solitary to ternate, petals purple when young, later with paler bases, filamentous staminodes purple, anthers pale yellow; **Fr** 5 locules; **Ecol** unknown; **Distr** Simonstown (?), WC, S Africa.

Note: Brown noted on a print of the lectotype (K!) "= M. emarginatoides, Haw".

L. emarginatoides (Haw.) N.E.Br. GC 1930: 212 ≡ *M. emarginatoides* Haw. 1795: 343 ≡ *M. retroflexum* Haw. 1803: 71 nomen illeg. **Neotype**, here designated: drawing 1000/399 by Duncanson of 1820: (K!).

Shrubs with erect branches, **I** purple when young, later maroon; **L** aggregate and proleptic, i.e. leaves on lateral short shoots developing while the bract is still green, distinctly triquetrous, acuminate, scabrous-punctate and very rough, whitish-green; **Fl** solitary, petals apically and basally deeper pink than in the middle, filamentous staminodes deep purple; **Fr** angular; **Ecol** unknown; **Distr** S Africa.

Note: The original description is based on material seen at Mr. Lee's, and the drawing chosen als neotype here has only been made in 1820 at Kew. However, the very rough, whitish leaves suggest that the same species was involved indeed. The "very white" leaves rough to the touch cast some doubt on the placement of the species in **Lampranthus**. The species is similar in leaf to **L. emarginatus**, but the flowers of the latter lack filamentous staminodes and possess widely spreading filaments instead.

L. emarginatus (L.) N.E.Br. GC 1930: 212; Liede & Meve 1990: 39 ≡ *M. emarginatum* L. 1762: 692 **LT** Dillenius 1732: 260, t. 197 f 250 = *M. polyphyllum* Haw. 1821: 141 **T?** = *M. puniceum* Jacquin 1804: t. 442 ≡ *M. emarginatum* L. var. *puniceum* (Jacquin) A.Berger 1908: 147 ≡ *L. emarginatus* (L.) N.E.Br. var. *puniceus* (Jacquin) Schwantes in Jacobsen 1938: 97 **T** Jacquin 1804: t. 442 = *M. versicolor* Haw. 1803: 70 **T?**

Twisting, bending branches, **I** terete, lower ones quadrangular; **L** aggregate, subulate, scabrous; **Fl** solitary, with spinose **K**, petals emarginate, purple; **Fr** unknown; **Ecol** unknown; **Distr** unknown.

Note: The "spinose calyx" probably refers to the fact that the calyx lobes spread while the fruit is ripening, as shown in the figure of Dillenius. The assessment of *M. polyphyllum* as a synonym of **L. emarginatus** follows Jacobsen (1974: 639, 1981: 620). The placement of *M. puniceum* as a variety of **L. emarginatus** was first undertaken by Berger (1908: 147); Sonder (1862: 439) placed the species as a synonym of *M. violaceum*, a view not accepted by Brown, who noted also on the print of the type (K!) that he likened the plant to *M. brownii* and perhaps to *M. coccineum*. The position of *M. versicolor* here is based on comparisons of Haworth (1803: 70), not taken up later. *M. versicolor* was distinguished based on the changing of the flower colour during the day from red in the morning to white at midday and red again in the evening. This remarkable feature has not been observed again on any species.

L. emarginatus var. *puniceus* = **L. emarginatus**

L. ernestii (L.Bolus) L.Bolus NM 1939: 168 ≡ *M. ernestii* L.Bolus NM 1932: 327 T Galpin 2663/30 (BOL).

Low growing shrubs, elongated branches creeping, **I** to 20 mm *l*; **L** incurved, trigonous, glaucous, scabrous, to 14 mm *l*, to 2.5 mm *b*, to 3 mm *t*; **Fl** pedicels 5–10 mm *l*, petals purplish-pink, 8–10 mm *l*, filamentous staminodes appressed to the stamens, filaments white, anthers and pollen yellow, nectary ring crenulate; **Fr** pentangular, top elevated for 1.5 mm, ∅ 6 mm, expanded 10 mm, 5 locules; **Ecol** unknown; **Distr** Bredasdorp, WC, S Africa.

Note: No information on the inflorescences is available for this rather small-fruited species.

L. erratus N.E.Br. GC 1930: 212 ≡ *M. virens* Salm-Dyck 1840: § 24 f 1 nomen illeg. non Haw. 1821: 121 **Lectotype**, here designated: Salm-Dyck 1840: § 24 f 1.

Shrubs with suberect branches, **I** compressed when young, brown, later grey with age; **L** compressed-triquetrous, punctate, smooth, 27–34 mm *l*, 6.75 mm *b*; **Fl** in terminal, elongate cymes, pedicels to 27 mm *l*, bracteoles in their middle, **K** 6, petals pale purple, vittate, filamentous staminodes white, apically purplish, anthers pale yellow; **Fr** 6 locules; **Ecol** unknown; **Distr** S Africa.

Note: Salm-Dyck reported that the species had been grown at Kew in 1821 from seed from the Cape of Good Hope; six locules do not occur in species placed definitely in **Lampranthus**.

L. esterhuyseniae L.Bolus NM 1958: 331 T Esterhuysen 23638 (BOL).

Erect shrubs to 24 cm *h*, to 30 cm ∅, richly branched, **I** 10–30 mm *l*, 1–1.5 mm ∅, basally 3–5 mm ∅; **L** trigonous, scabrous, glaucous green, 10–20 mm *l*, to 2 mm *b* and *t*; **Fl** 3–9 per cyme, pedicels 25–40 mm *l*, bracteoles at their middle, petals purplish-pink, 8–15 mm *l*, 1–2 mm *b*, inner ones 4–7 mm *l*, 0.5 mm *b*, basally papillate, perhaps representing the filamentous staminodes, filaments pink, only the inner papillate, nectary ring crenulate; **Fr** 5 locules; **Ecol** unknown; **Distr** Namaqualand, NC, S Africa.

Note: Lack of staminodes characterizes this species.

L. excedens ≡ **Oscularia excedens**

L. eximius L.Bolus NM 1954: 273 T Hall 1364/50 (BOL).

Branches decumbent to creeping, **I** to 45 mm *l*, ∅ to 3 mm, flowering branches erect, 15–29 cm *h*, **I** there 15–35 mm *l*, 1–2 mm ∅; **L** trigonous to triquetrous, acute, smooth, 15–22 mm *l*, 2 mm *b* and *t*; **Fl** in produced cymes, pedicels 30–70 mm *l*, bracteoles at their middle, **K** lobes with brown margins, petals pinkish-purple or, at first, red, 5–23 mm *l*, 0.25–2.5 mm *b*, filamentous staminodes few, pale, filaments white, anthers and pollen yellow, nectary ring flattened; **Fr** 5 locules; **Ecol** in sand near the coast; **Distr** Hopetown, Malmesbury, Vredenburg, WC, S Africa.

Note: More or less conspicuous brown margins on the calyx lobes characterize this species.

L. explanatus (L.Bolus) N.E.Br. GC 1930: 212 ≡ *M. explanatum* L.Bolus ABH 1923: 166 **Syntypes** Page 15200, Page 15201 (BOL).

Creeping branches to 35 cm *l*, **I** 15–35 mm *l*, ∅ to 2 mm; **L** almost semi-terete, young one trigonous, acute, glaucous green, to 23 mm *l*, 2 mm *b* and *t*; **Fl** solitary or in cymes, ∅ to 44 mm, pedicels 15–35 mm *l*, with bracteoles at their middle, petals bright yellow to 21 mm *l*, to 2 mm *b*, stamens spreading, filaments straw-coloured, epapillate, anthers yellow, nectary ring subannular, inconspicuous; **Fr** unknown; **Ecol** on slopes; **Distr** Cape Town, Strand, WC, S Africa.

Note: This species lacks filamentous staminodes.

L. falcatus (L.) N.E.Br. GC 1930: 212 ≡ *M. falcatum* L. 1753: 484 non Thunberg 1800: 90 **Lectotype**, here designated: Dillenius 1732: t. 213 f 276 = *M. falcatum* L. var. *galpinii* L.Bolus NM 1931: 234 ≡ *L. falcatus* (L.) N.E.Br. var. *galpinii* (L.Bolus) L.Bolus in Jacobsen FR 1938: 98 T Galpin 2669/30 (BOL).

Erect shrubs to 15 cm *h*; **L** acinaciform, incurved, glabrous, glaucous; **Fl** mostly solitary, petals purple, filamentous staminodes present; **Fr** unknown; **Ecol** unknown; **Distr** unknown.

Note: Bradley (1727: 9, t. 42), cited by both, Dillenius (1732: 288) and Linnaeus, described the leaves as grain-sized, to which Dillenius remarks that younger plants possess longer and more distantly spaced leaves (his figure 276) than older ones (his figure 275).

L. falcatus var. *galpinii* = **L. falcatus**
L. falciformis ≡ **Oscularia falciformis**
L. falciformis var. *maritimus* = **Oscularia falciformis**

L. fergusoniae (L.Bolus) L.Bolus NM 1939: 168 ≡ *M. fergusoniae* L.Bolus NM 1934: 427 nomen illeg. non N.E.Br. 1930: 32 T Ferguson 21002 (BOL) = *L. fergusoniae* (L.Bolus) L.Bolus var. *crassistigma* L.Bolus JSAB 1960: 159 T Hall 35655 (BOL).

Branches decumbent or creeping, **I** 5–15 mm *l*; **L** compressed, acute, a little rough, 17 mm *l*, 15 mm *t*;

Fl solitary, pedicels 3–10 mm *l*, petals 5–16 mm *l*, to 2.25 mm *b*, filamentous staminodes absent, filaments golden, epapillate, anthers and pollen golden, nectary composed of 5 crenulate glands almost touching each other; **Fr** pentangular, 7 mm *l*, top elevated for 2 mm, Ø expanded 10 mm, 5 locules; **Ecol** in calcareous dunes; **Distr** Riversdale, WC, S Africa.

Note: The possession of five separate nectary glands resembles conditions in **Oscularia**.

L. fergusoniae var. *crassistigma* = **L. fergusoniae**

L. filicaulis (Haw.) N.E.Br. GC 1930: 212 ≡ *M. filicaule* Haw. 1803: 87 **T?**

Prostrate branches filiform, rooting at each node; **L** incurved, semi-terete, subulate, green, awn-shaped tips a little recurved; **Fl** subterminal, pedicels blackish, bracteoles at their middle, petals pale reddish inside, deeper coloured on the outside, filaments white, apically red; **Fr** 5 locules; **Ecol** unknown; **Distr** S Africa.

Note: The species appears to be similar to **L. debilis** in growth form.

L. flexifolius (Haw.) N.E.Br. GC 1930: 212 (see **photo**) ≡ *M. flexifolium* Haw. 1819: 96 ≡ *M. curvifolium* Haw. var. *minus* Salm-Dyck 1836: § 47 f 2 ≡ *L. curvifolius* (Haw.) N.E.Br. var. *minor* (Salm-Dyck) G.D.Rowley in Jacobsen & G.D.Rowley 1958: 77 **Lectotype**, here designated: drawing s.n. by Duncanson of material from Bowie, collected 1817 (K!).

Low shrubs with thin, decumbent branches, not rooting, **I** purple to maroon later; **L** triquetrous-subulate, basally broadest, apically somewhat recurved, dark green; **Fl** petals white, to apically pale pink, filamentous staminodes in a central column, white, filaments apically reddish; **Fr** unknown; **Ecol** unknown; **Distr** S Africa.

Note: The plant figured by Duncanson shows a remarkable difference in leaf sizes and shapes between the more basal leaves of a branch and those further up, the latter being smaller and more erect. An inclusion under **L. curvifolius** seems unjustified considering the rather crowded leaf arrangement and the much more narrow leaves of the latter.

L. flexilis (Haw.) N.E.Br. GC 1930: 212 ≡ *M. flexile* Haw. 1821: 140 **T?**

Erect branches spreading widely; **L** small and shorter than **I**, inflexed like horns, glaucous, subtriquetrous, obtuse, basally narrowed; **Fl** as in **L. polyanthon**; **Fr** unknown; **Ecol** unknown; **Distr** unknown.

Note: The original description was based on material given to Haworth by Parmentier under the name of *M. laxum*.

L. foliolosus ≡ **L. foliosus**

L. foliosus L.Bolus NM 1950: 226 ≡ *L. foliolosus* L.Bolus ex H. Jacobsen 1960: 1190 orthographic variant **T** Stokoe 22742 (BOL).

Creeping, **I** Ø 2 mm; **L** trigonous with convex sides or subfalcate, 15–20, in cultivation to 26 mm *l*, 1–3 mm *b* and Ø, **Fl** solitary, pedicels 10–20 mm *l*, petals pinkish-purple, to 22 mm *l*, to 3 mm *b*, filamentous staminodes absent, filaments white, apically yellow, anthers and pollen golden, nectary as a pentagonal, crenulate ring; **Fr** top elevated for 2 mm, pentagonous, 8 mm *l*, Ø 7–8 mm, expanded 10–12 mm Ø, 5 locules; **Ecol** unknown; **Distr** Caledon, WC, S Africa.

Note: Rather big, solitary capsules are found in this species.

L. formosus (Haw.) N.E.Br. GC 1930: 212 ≡ *M. formosum* Haw. 1821: 145 **Lectotype**, here designated: drawing 951 without date or artist (K!).

Low shrubs with decumbent branches; **L** acute, 27–70 mm *l*, 4.5–6.7 mm *b*; **Fl** subternate, Ø < 54 mm, petals bright red, filamentous staminodes very few, filaments white; **Fr** 5–6 locules; **Ecol** unknown; **Distr** S Africa.

Note: The drawing bears a note "Bowie 1817": it is, therefore, chosen as the lectotype here.

L. framesii (L.Bolus) N.E.Br. GC 1930: 212 ≡ *M. framesii* L.Bolus NM 1928: 137 non (L.Bolus) N.E.Br. 1930: 32 **T** Frames 318/27 (BOL).

Erect shrubs to 20 cm *h*, basal Ø to 20 mm, branches diffusely arranged, **I** 10–15 mm *l*, Ø 1 mm; **L** acute, finely rough, to 10 mm *l*, 1.5 mm *b*, 2 mm *t*; **Fl** in cymes, Ø 26 mm, pedicels 20–30 mm *l*, bracteoles at their middle, petals pink to white, to 12 mm *l*, to 2 mm *b*, filaments pale with reddish-purple papillae, anthers yellow, nectary ring inconspicuously crenulate; **Fr** 5 locules; **Ecol** unknown; **Distr** Caledon, WC, S Africa.

Note: The species is reported to be similar to **L. furvus**.

L. francesiae H.E.K.Hartmann Bradleya 1998: 68 ≡ *Ruschia leightoniae* L.Bolus NM 1935: 499 **T** Leighton 21422 (BOL!).

Shrublets to 12 cm *h*, **I** reddish-brown, erect; **L** triquetrous, epidermis cells raised and on elevations, 10–25 mm *l*, 3–3.5 mm *b*, c. 4 mm *t*; **Fl** solitary, bracteolate at the middle of the long stalk, petals pink with deeper coloured margins, filamentous staminodes purplish, basally paler, filaments white, apically pink; **Fr** top raised from centrally higher rims, base elongate, funnel-shaped, covering membranes complete with well developed closing rodlets at their distal ends, no closing body developed, valve wings very broad, ± rectangular, 5 locules; **Ecol** unknown; **Distr** Clanwilliam, WC, S Africa.

Note: Found together with **L. leightoniae**, this species had been described as *Ruschia leightoniae*; the transfer to **Lampranthus** required a new name, for which the first name of Mrs Leighton was chosen.

L. franciscii L.Bolus 1961: 113 **T** Stayner 240/60 (BOL).

Erect shrubs to 60 cm *h*, 90 cm ∅, basal stem 30 mm ∅, **I** 25–60 mm *l*; **L** trigonous to subfalcate, somewhat velvety to the touch, pale green, 25–45 mm *l*, sheath 4–5 mm *l*, 3–6 mm *b*, 4–6, rarely to 7 mm *t*; **Fl** in rich cymes of 11–25 flowers, pedicels 40–50 mm *l*, bracteoles in their middle, fragrant, petals bright pink, inconspicuously vittate, 13–18 mm *l*, 0.75–1.25 mm *b*, filamentous staminodes absent, filaments white, anthers and pollen straw-coloured, nectary ring inconspicuously crenulate; **Fr** reddish-brown, base obconical, to 5 mm *l*, top elevated for 2 mm, ∅ 7–9 mm, expanded 13 mm ∅, 5 locules; **Ecol** unknown; **Distr** Ceres, WC, S Africa.

Note: Unusual for **Lampranthus** is the possession of long leaf sheaths.

L. fugitans L.Bolus NM 1958: 330 (see **photo**) **T** Hall 496/56 (BOL).

Low shrubs forming big cushions, ∅ of plants to 200 cm, to 25 cm *h*, **I** reddish-brown; **L** triquetrous, apically pointed and purplish suffused, 20–35 mm *l*, 3–5 mm *b*, 4–5 mm *t*; **Fl** solitary, ∅ 40–50 mm, petals deep purple to paler, to 20 mm *l*, to 2 mm *b*, numerous purple filamentous staminodes surrounding the stamens in a column, filaments purple, anthers and pollen golden; **Ecol** in crevices of rocks, also in the spray of the sea, > 600 mm rainfall p.a. in summer; **Distr** Port Shepstone, KN, Transkei, EC, S Africa.

Note: A conspicuous species on the rocks in Pondoland, probably most suited for mild climates in similar situations. In contrast to the information in the description, the flowers are found not to whither very quickly, nor are the filamentous staminodes absent.

L. furvus (L.Bolus) N.E.Br. GC 1930: 212 ≡ *M. furvum* L.Bolus SAG 1928: 16; NM 1928: 134 **T** Lavis 18760 (BOL).

Laxly branched shrubs to 12 cm *h*, branches ascending, **I** 15–20 mm *l*; **L** subcompressed to subfalcate, to 10 mm *l*, to 2 mm *b*; **Fl** solitary, composing a secondary **Inf**, pedicels with bracteoles at their middle, petals purplish-pink, filaments pale, nectary ring crenulate, stigmas dark purple; **Fr** 5 locules; **Ecol** near the sea; **Distr** Caledon, WC, S Africa.

Note: The Latin description was published in May 1928, whereas the English text appeared earlier (SAG 1928: 16), the name being valid at that date.

L. galpiniae (L.Bolus) L.Bolus NM 1939: 169 ≡ *M. galpiniae* L.Bolus NM 1932: 328 **T** Galpin 2/31 (BOL).

Erect shrubs 10–12 cm *h*, **I** 15–20 mm *l*; **L** subfalcate to subterete, smooth, green, reddish suffused, 15–20 mm *l*, to 2 mm *t*; **Fl** solitary, pedicels bracteolate, petals white, basally pink, inner very short ones pink, to 29 mm *l*, to 3.5 mm *b*, filaments white, nectary ring crenulate; **Fr** top elevated for 3 mm, expanded 13 mm ∅; **Ecol** near the sea; **Distr** Bredasdorp, WC, S Africa.

Note: A solitary, big fruit is developed in this species.

L. glaucoides (Haw.) N.E.Br. GC 1930: 212 ≡ *M. glaucoides* Haw. 1795: 330 ≡ *M. aurantium* Haw. 1803: 84 ≡ *M. aurantiacum* DC. 1828: 437 ≡ *L. aurantiacus* (DC.) Schwantes in Jacobsen FR 1938: 229 **T?**

Erect shrubs, upper branches zigzagging; **L** triquetrous, glaucous, dotted, keel roundish; **Fl** bracteoles at or above the middle of the pedicel, reaching the base of the flower, petals purple, turning orange when drying; **Fr** unknown; **Ecol** unknown; **Distr** Mitchells Plain, WC, S Africa.

Note: Haworth (1803) changed the validly published name of the species to *M. aurantium*, which was altered to *M. aurantiacum* by De Candolle (1828), but following the literature citations and descriptions, different features were not involved.

L. glaucus (L.) N.E.Br. GC 1930: 212 ≡ *M. glaucum* L. 1753: 486 ≡ *Mesembryanthus glaucus* (L.) Rothmaler 1941: 413 nomen illeg. **Lectotype**, here designated: Dillenius 1732: t. 196 f 248 = *M. glaucum* L. var. *tortuosum* Salm-Dyck § 25 f 1 ≡ *L. glaucus* (L.) N.E.Br. var. *tortuosus* Schwantes ex H.Jacobsen 1955: 1434 nomen invalid. **T?** = *M. brachyphyllum* Welwitsch 1873: 290 non Pax in schedis **T** Welwitsch 307 (BM, K).

Shrubs with erect to winding or decumbent branches, **I** yellowish to grey; **L** triquetrous to sword-shaped, each with a gibbous base, inconspicuously dotted and scabrous; **Fl** 1–3, petals sulphur-coloured; **Fr** 5 locules; **Ecol** flowering in winter; **Distr** S Africa.

Note: This description is drawn from Dillenius 1732: 256–257, t. 196 f 248 and Bradley 1727: 15 and t. 37 (not 32, as cited by Linnaeus). Both figures show rather distinctly boat-shaped leaves. The species is also reported from West Australia, South Australia and Tasmania, Australia (Venning in Prescott & Venning 1984: 33) as well as from Algarvia, Portugal (as *M. brachyphyllum* Welwitsch 1873: t. 136).

L. glaucus var. *tortuosus* = **L. glaucus**

L. globosus (L.Bolus) L.Bolus NM 1939: 169 ≡ *M. globosum* L.Bolus NM 1931: 234 nomen illeg. non N.E.Br. KB 1913: 119 **T** Mathews 19384 (BOL) = *L. globosus* (L.Bolus) L.Bolus forma L.Bolus 1960: 59 nomen invalidum **T** Stayner 21659 (BOL).

Erect shrubs to 20 cm *h*, richly branched, **I** 10–15 mm *l*, ∅ 1–3 mm; **L** falcate to subterete, apiculate, pale glaucous green, to 12 mm *l*, 1.5–2 mm *t*; **Fl** 1–3, pedicels 20–30 mm *l*, all bracteolate at their middle, petals purple, to 18 mm *l*, to 1.5 mm *b*, filamentous staminodes absent, anthers yellow, nectary ring crenulate; **Fr** top raised for 1.5 mm, 6 mm *l*, ∅ 5 mm, expanded 9 mm ∅, 5 locules; **Ecol** unknown; **Distr** Vanrhynsdorp, WC, S Africa.

Note: The forma bears no name.

L. glomeratus (L.) N.E.Br. GC 1930: 212 ≡ *M. glomeratum* L. 1759: 1060 **Lectotype**, here designated: Dillenius t. 213 f 274 = *M. inflexum* Haw. 1819: 98 **T**?

Shrubs to 60 cm *h*; **L** triquetrous to falcate, dotted; **Fl** solitary, but several arranged in a group; **Fr** unknown; **Ecol** unknown; **Distr** Cape Town (?), WC, S Africa.

Note: Linnaeus based his description on Dillenius (1732: 287), who in turn cited Bradley 1727: 14 t. 36. Bradley mentioned that this species may be one of the oldest known in England from the Cape of Good Hope, hence it is suggested that **L. glomeratus** grows near Cape Town. The fact that Linnaeus called the leaves almost terete whereas Dillenius described them explicitly as triquetrous to falcate is a point of concern regarding the identity of the material seen by both. The assessment of *M. inflexum* as a synonym of **L. glomeratus** follows Jacobsen (1974: 636, 1981: 617).

L. godmaniae (L.Bolus) L.Bolus NM 1939: 169 ≡ *M. godmaniae* L.Bolus NM 1929: 141 **Syntypes** L.Bolus 18992, Godman 1806/29 (BOL) = *M. godmaniae* L.Bolus var. *grandiflorum* L.Bolus NM 1932: 326 ≡ *L. godmaniae* L.Bolus var. *grandiflorus* (L.Bolus) L.Bolus ex H.Jacobsen 1955: 1435 nomen invalid. **T** Compton 3719 (BOL).

Shrubs with erect branches to 35 cm *h*, **I** maroon, later light to dark grey, 20–40 mm *l*; **L** trigonous, glaucous, apically purple, 20–30(–40) mm *l*, to 5 mm *b* and *t*; **Fl** in cymes, ∅ to 42 mm, pedicels 20–70 mm *l*, bracteoles near the top, petals pink, to 19 mm *l*, to 2 mm *b*, filamentous staminodes numerous, pink, filaments pink, anthers and pollen yellow, nectary ring broad, flat and inconspicuous; **Fr** expanded 13 mm ∅; **Ecol** unknown; **Distr** Namaqualand, NC, S Africa.

Note: In this species, rather big capsules are arranged in big cymes.

L. godmaniae var. *grandiflorus* = **L. godmaniae**

L. gracilipes (L.Bolus) N.E.Br. GC 1930: 212 L.Bolus NM 1939: 169 nomen illeg. ≡ *M. gracilipes* L.Bolus NM 1929: 73 **T** without collector 17883 (BOL) = *L. gracilipes* (L.Bolus) N.E.Br. forma *luxurians* L.Bolus 1966: 232 **T** Hall 3039 (BOL).

Shrubs to 20 cm *h*, **I** quadrangular, 10–15 mm *l*, ∅ 2–3 mm; **L** subfalcate, but dorsally rounded, smooth, 7 mm *l*, to 3 mm *b*, to 4 mm *t*; **Fl** in cymes, ∅ 32 mm, pedicels 15–20 mm *l*, bracteoles at their middle, petals pink, to 15 mm *l*, to 1.5 mm *b*, filaments purple, anthers yellow, nectary ring obscurely pentagonal, crenulate; **Fr** 5 locules; **Ecol** probably in shady crevices; **Distr** Piquetberg, WC, S Africa.

Note: Bolus (NM 1939: 169) was obviously not aware that the species had been transferred to **Lampranthus** by Brown (1930: 212) before.

L. gracilipes forma *luxurians* = **L. gracilipes**
L. guthriae ≡ **Oscularia guthriae**

L. gydouwensis L.Bolus 1963: 13 ≡ *Amphibolia gydouwensis* (L.Bolus) L.Bolus ex Toelken & Jessop 1976: 64; L.Bolus 1967: 306 nomen invalid. ≡ *Ruschia gydouwensis* (L.Bolus ex Toelken & Jessop) G.D.Rowley NCSJ 1978: 7 **Lectotype**, here designated: Leipoldt 4801 II (chosen from two sheets, BOL!).

Shrubs to 40 cm *h* with ± erect branches, **I** dark; **L** of a pair basally connate for 3–5 mm, a line separating them, free parts almost terete, 30–50 mm *l*, 3 mm *b*, 4 mm *t*; **Fl** mostly in ternate cymes, sometimes several close together making up a larger **Inf**, bracteoles below the middle, petals purplish-pink, filamentous staminodes and stamens pink, basal halves white; **Fr** top almost conical, base long, funnel-shaped, covering membranes convex, a rodlet at each distal end, endocarp above the end of the placenta a little thickened, but no closing body present; **S** dark brown, c. 0.75 mm *l*; **Ecol** > 200 mm rainfall p.a. in winter; **Distr** Ceres, WC, S Africa.

Note: When the species was placed in **Amphibolia**, the protruding end of the placenta, lifted from the wall, was probably mistaken for an endocarpal closing body. There is no doubt that the species was correctly placed in **Lampranthus**.

L. hallii L.Bolus NM 1954: 272 **T** Hall 521/52 (BOL).

Shrubs to 36 cm *h*, **I** 20–60 mm *l*, 2–5 mm ∅; **L** laterally compressed, acute, bright green, smooth, 20–78 mm *l*, 2–4 mm *b*, to 5 mm *t*; **Fl** solitary or ternate, petals bright pale pink, 4–15 mm *l*, 0.25–1.25 mm *b*, filamentous staminodes absent, filaments white, anthers and pollen pale lemon-yellow, nectary ring crenulate; **Fr** top elevated for 3 mm, to 10 mm *l*, ∅ to 10 mm; **Ecol** unknown; **Distr** Riversdale, WC, S Africa.

Note: The very narrow inner petals may perhaps be filamentous staminodes.

L. haworthii (Haw.) N.E.Br. GC 1930: 212 ≡ *M. haworthii* Donn ex Haw. 1803: 65 ≡ *M. haworthii* Donn 1793: 66 nomen nudum ≡ *Erepsia haworthii* (Haw.) Schwantes GF 1928: 68 **T** drawing 959/132 by Duncanson (K!).

Dense shrubs with many erect branches, 60 cm *h*, I terete, green to reddish, old ones with reddish-brown remains of old leaf sheaths; L cylindrical to spindle-shaped, the sharp apices purplish suffused, c. 40 mm *l*; Fl 1–3, pedicels with bracteoles at their middle, Ø c. 80 mm, petals pale purple, basally white, filamentous staminodes white, filaments white-purple, anthers white to light yellow; **Fr** pyriform, 5 locules; **Ecol** unknown; **Distr** S Africa.

Note: Donn (1793: 66) listed the species in his catalogue of plants cultivated in "the Walkerian Garden, Cambridge" supplying the following information: flowering June–September, perennial. This is insufficient for a description. Instead, Haworth's text (1803: 65) seems to represent the oldest description; the authors are, therefore, corrected accordingly. Bolus (1949: t. 1064) gave a distribution from Worcester to Beaufort West and Graaff Reinet.

L. henricii (L.Bolus) N.E.Br. GC 1930: 212 ≡ *M. henricii* L.Bolus ABH 1923: 168 **Syntypes** H. Bolus 9006, Compton 762/23 (BOL).

Ascending branches c. 25 cm *l*, **I** to 4 mm Ø; L subfalcate to subclavate, smooth, glaucous green, to 9 mm *l*, 2.5 mm *b*, 3 mm *t*; **Fl** solitary to ternate, pedicels with bracteoles at their middle, 8 mm *l*; petals pinkish-purple, basally paler, stamens (?) apically purplish, anthers pale, nectary ring prominent, divided, yellowish-green; **Fr** unknown; **Ecol** unknown; **Distr** Clanwilliam, WC, S Africa.

Note: Bolus noted that the stamens were not well visible.

L. herrei ≡ **Ruschia sandbergensis**

L. hiemalis (L.Bolus) L.Bolus NM 1939: 169 ≡ *M. hiemale* L.Bolus NM 1930: 189 **T** Leipoldt 19117 (BOL).

Erect shrubs to 40 cm *h*, **I** blackish-brown, 20–40 mm *l*, basally to 6 mm Ø; **L** terete, keel and margins as lines, obtuse, to 35 mm *l*, 5 mm *b*, 6 mm *t*; **Fl** in small cymes, Ø to 15 mm, pedicels to 7 mm *l*, bracteoles placed above the middle, petals pink, margins deeper in colour, to 6 mm *l*, to 1 mm *b*, filamentous staminodes white, basally with dense, interwoven papillae, filaments white, apically purple, anthers and pollen golden, nectary ring denticulate; **Fr** unknown; **Ecol** unknown; **Distr** Clanwilliam, WC, S Africa.

Note: The species is reported to be similar to **L. argenteus**.

L. hoerleinianus (Dinter) Friedrich 1960: 566; Friedrich 1968: 218 nomen invalid. ≡ *M. hoerleinianum* Dinter FR 1923: 153 **T** Dinter 3788 I (B!).

Shrubs to 70 cm *h* and Ø, peripheral branches rooting, **I** light greyish-brown, smooth, 25–35 mm *l*, terminal ones 3–4 mm Ø; **L** trigonous, smooth, spreading, pale bluish-green, 25–30 mm *l*, 9 mm *t*; **Fl** pedicels 6–8 mm *l*, **K** papillate, petals 17 mm *l*, purple, lower side purplish-pink, c. 35, stamens purplish-pink, basally pale; **Fr** 5 locules; **Ecol** in sand above rocks or in pockets; **Distr** Lüderitz-South, Namibia.

Note: Friedrich (1960, 1968) published the combination twice and placed **L. brachyandrus** as a synonym under **L. hoerleinianus** in 1968.

L. holensis L.Bolus 1962: 294 **T** Hall 2568 (BOL) = *L. holensis* L.Bolus forma 1962: 294 nomen invalid. **T** Hall 2506 (BOL).

Branches to 20 cm *l*, **I** 10–30 mm *l*, Ø 5–6 mm; **L** trigonous, acute, very smooth, pale glaucous green, 45–55 mm *l*, 4–5 mm *b* and *t*; **Fl** solitary, Ø to 60 mm, **K** 5(–6), petals white, to 26 mm *l*, to 2 mm *b*, inner (probably filamentous staminodes) 0.5 mm *b*, filaments reddish-purple, anthers and pollen golden; **Fr** blackish, base clavate, top elevated for 3 mm, to 9 mm *l*, Ø 10 mm, expanded to 20 mm Ø, 5 locules; **Ecol** unknown; **Distr** Vredendal, WC, S Africa.

Note: The "forma" mentioned below the original description differs in having more narrow valve wings and petals as well as white filaments.

L. hollandii (L.Bolus) L.Bolus NM 1939: 169 ≡ *M. hollandii* L.Bolus NM 1931: 274 **T** Galpin 19640 (BOL).

Erect shrubs to 20 cm *h*, **I** 20–30 mm *l*, to 3 mm Ø; **L** trigonous, glabrous, 20–27 mm *l*, 2 mm *b*, 2–3 mm *t*; **Fl** in irregular cymes, pedicels 5–25 mm *l*, petals bright pink or pinkish-purple, to 21 mm *l*, to 2 mm *b*, filamentous staminodes absent, filaments white, anthers and pollen yellow, nectary ring crenulate; **Fr** unknown; **Ecol** near the sea; **Distr** Alexandria, EC, S Africa.

Note: The species is reported to be common along the sea.

L. hurlingii (L.Bolus) L.Bolus NM 1939: 169 ≡ *M. hurlingii* L.Bolus NM 1934: 427 **T** Hurling & Neil 21015 (BOL).

Laxly branched shrubs 20–32 cm *h*, **I** 15–25 mm *l*, to 3 mm Ø; **L** trigonous, acute, glaucous green, rough from dots, 20–25 mm *l*, to 2.5 mm *b* and *t*; **Fl** solitary or in cymes, pedicels 25–40 mm *l*, bracteoles at their middle, petals golden, 7–12 mm *l*, to 2 mm *b*, filamentous staminodes absent, filaments epapillate, anthers and pollen golden, nectary ring inconspicuous, dirty yellow; **Fr** top elevated for 1 mm, 7 mm *l*, 9 mm Ø, expanded 14 mm Ø, 5 locules; **Ecol** unknown; **Distr** Robertson, WC, S Africa.

Note: Big capsules are reported for this species.

L. imbricans (Haw.) N.E.Br. GC 1930: 212 ≡ *M. imbricans* Haw. 1819: 94 **Lectotype**, here designated: drawing 181 by Duncanson of 23.6.1826 (K!).

Erect shrubs to 30 cm *h*; **L** linear, triquetrous, glabrous, whitish-glaucous, placed densely and imbricately near the tips of side branches; **Fl** petals light purple; **Fr** unknown; **Ecol** unknown; **Distr** S Africa?

Note: Haworth compared the species with *M. versicolor*, *M. retroflexum* and **L. marginatum** in 1819. In 1821 (422), he reported that he had originally seen the plant in Chelsea garden, where it "perished", and that only recently had he seen young plants in Chelsea garden again, "in the young state", and these were in flower. It appears, therefore, not unlikely that the figure drawn by Duncanson in 1826 represents flowering material from that same source. Brown remarks on the drawing: "Not M. imbricans, Haw. as the stamens are not collected, see Rev. p.139, & Suppl p. 74". Since in neither place a reference to the stamens is given, Brown's exclusion is not accepted here.

L. immelmaniae (L.Bolus) N.E.Br. GC 1930: 212 ≡ *M. immelmaniae* L.Bolus NM 1928: 136 **T** Immelman 18715 (BOL).

Erect shrubs to 20 cm *h*, **I** 10–20 mm *l*, ∅ 2 mm; **L** almost semi-terete, acuminate, glaucous green, to 22 mm *l*, 1–2 mm *b* and *t*; **Fl** in small cymes, ∅ to 40 mm, pedicels 20–45 mm *l*, petals bright pink, 19 mm *l*, 1 mm *b*, filamentous staminodes absent, filaments golden, anthers and pollen golden, nectary ring obscure; **Fr** 5 locules; **Ecol** unknown; **Distr** Piquetberg (?), WC, S Africa.

Note: Lack of filamentous staminodes characterizes this species.

L. inaequalis (Haw.) N.E.Br. GC 1930: 212 ≡ *M. inaequale* Haw. 1819: 266 ≡ *M. bicolor* L. var. *inaequale* (Haw.) A.Berger 1908: 152 ≡ *L. bicolor* (L.) N.E.Br. var. *inaequale* (Haw.) Schwantes in Jacobsen FR 1938: 96 **Lectotype,** here designated: drawing 225 by Duncanson 10.8.1826 (K!).

Shrubs to 30 cm *h*, branches decumbent and diffusely arranged; **L** subtriquetrous, subulate; **Fl** solitary (only?), **K** lobes markedly unequal, the membranes longer than the lobes, petals orange in the bud, yellow with orange tips when open, paler outside; **Fr** unknown; **Ecol** flowering in autumn; **Distr** S Africa.

Note: Haworth cited also a drawing of Bradley (1716: 8, f 7), which differs by having equal calyx lobes, as opposed to unequal ones (name-giving) in **L. inaequalis.** Furthermore, Haworth compared this species with **L. bicolor** with equal calyx lobes and **L. coccineus** with membranes along the calyx lobes that are shorter than the lobes themselves.

L. inconspicuus (Haw.) Schwantes in Jacobsen FR 1938: 229 ≡ *M. inconspicuum* Haw. 1826: 12 **Lectotype,** here designated: drawing 1032/993 by Duncanson 21.9.1825, based on Bowie 662, collected in 1823 (K!).

Densely branched shrubs to 15 cm *h*, **I** firm, with blackish papillae (seen under a lens); **L** incurved, subpapillate to shiny, green, glaucous, c. 14 mm *l*; **Fl** on short ebracteolate pedicels, ∅ < 6.75 mm, **K** 4, petals deep red, filaments erect, apically deeper in colour, anthers white; **Fr** unknown; **Ecol** unknown; **Distr** S Africa.

Note: The plants are reported to have flowered in 1825 in the autumn, but again in June in 1826 at Kew, suggesting that the first anthesis took place after the import of the living plants. The small flowers and the apically hooked leaves cast doubt on the placement of the species in **Lampranthus**, the figure resembling species of **Ruschia.**

L. incurvus (Haw.) Schwantes in Jacobsen FR 1938: 230 ≡ *M. incurvum* Haw. 1803: 69 ≡ *M. roseum* Willdenow var. *confertum* Salm-Dyck 1849: § 29 f 4 **T**? = *M. confertum* Haw. 1812: 301 **T**?

Shrubs to 30 cm *h*, branches diffuse, decumbent, **I** terete; **L** triquetrous, compressed, acinaciform, glabrous, dotted, glaucous; **Fl** unknown; **Fr** unknown; **Ecol** unknown; **Distr** S Africa.

Note: Haworth's description was based on material sent to him by Mr. Howard, who raised the plant from seed in 1802. Salm-Dyck (1849: § 29 f 4) placed the species as a variety in *M. roseum* giving it a new epithet. *M. confertum* is a synonym based on Haworth 1821: 137; this later name, however, was based on material from Howard grown in 1805 (Haworth 1812: 301).

L. intervallaris L.Bolus JSAB 1960: 160 **T** Hall 1785 (BOL).

Laxly branched shrubs, **I** 15–25 mm *l*, in shade to 45 mm *l*; **L** trigonous, acute, very soft, green, 15–20 mm *l*, 1.5–2 mm *b*, 2–3 mm *t*; **Fl** in small cymes, 22–28 mm ∅, pedicels 10–20 mm *l*, in shade to 35 mm *l*, bracteoles at their middle, petals pale pink, deeper coloured towards the base, 10–13 mm *l*, to 1.5 mm *b*, filamentous staminodes and filaments pale pink, apically paler, anthers and pollen straw-coloured, nectary ring deeply crenulate; **Fr** with raised rims, 5 locules; **Ecol** unknown; **Distr** Clanwilliam, WC, S Africa.

Note: The name is based on the unusually long internodes.

L. laetus (L.Bolus) L.Bolus NM 1939: 169 ≡ *M. laetum* L.Bolus NM 1930: 189 **T** Page 17157 (BOL).

Erect shrubs to 20 cm *h*, **I** 4–5 mm *l*, on flowering branches 10–20 mm *l*, basally to 5 mm ∅; **L** trigonous, acute, dirty green, a little rough, 7–13 mm *l*, to 1.5 mm *b*, 1 mm *t*; **Fl** in rich cymes, 20–23 mm ∅, pedicels to 30 mm *l*, bracteoles at their middle, petals pinkish-purple, 5–10 mm *l*, to 1.25 mm *b*, filamentous staminodes pinkish-purple, basally papillate, filaments pink, anthers pale, nectary ring crenulate; **Fr** unknown; **Ecol** on rocky cliffs in a ravine; **Distr** Montagu, WC, S Africa.

Note: Blackish-brown internodes are reported for this species.

L. lavisii (L.Bolus) L.Bolus NM 1939: 169 ≡ *M. lavisii* L.Bolus NM 1929: 156 **T** Lavis 18762 (BOL) = *L. lavisii* (L.Bolus) L.Bolus var. *concinnus* L.Bolus JSAB 1960: 158 **T** Mackenzie 595/58 (BOL).

Erect shrubs 25–35 cm *h*, **I** brown to blackish-brown, 10–20 mm *l*, ∅ to 3 mm; **L** trigonous, acute, smooth, green, 25–35 mm *l*, to 2 mm *b*, 2–2.5 mm *t*; **Fl** in cymes, ∅ 40 mm, pedicels 15–30 mm *l*, bracteoles at their middle, petals bright pink, to 17 mm *l*, to 2 mm *b*, filamentous staminodes absent, filaments white, anthers and pollen straw-coloured, nectary ring inconspicuous and crenulate; **Fr** turbinate, pentangular, 3 mm *h*, expanded 15 mm ∅; **Ecol** unknown; **Distr** Humansdorp, EC, S Africa.

Note: The species was later (Bolus JSAB 1960: 158) united with the younger **L. hollandii**.

L. lavisii var. *concinnus* = **L. lavisii**

L. laxifolius (L.Bolus) N.E.Br. GC 1930: 212 ≡ *M. laxifolium* L.Bolus ABH 1927: 86 **T** Muir 2173/23 (BOL).

Erect shrubs 15–25 cm *h*, **I** ∅ 2 mm; **L** falcate to incurved, triquetrous, smooth, yellowish-green, 18–26 mm *l*, to 3 mm *b*, 2–3 mm *t*; **Fl** mostly solitary, rarely ternate, 30–35 mm ∅, petals pinkish-purple, basally white, 5–16 mm *l*, to 2 mm *b*, filaments white, anthers and pollen yellow, nectary ring inconspicuous, crenulate; **Fr** 5 locules; **Ecol** unknown; **Distr** Riversdale, WC, S Africa.

Note: Yellowish-green leaves are unusual in the genus.

L. leightoniae (L.Bolus) L.Bolus NM 1939: 169 ≡ *M. leightoniae* L.Bolus NM 1935: 488 **T** Leighton 21424 (BOL).

Erect shrubs to 15 cm *h*, **I** basally to 5 mm ∅, apically to 10 mm *l*; **L** trigonous to subterete, abruptly acute, dirty green, to 13 mm *l*, 2.5 mm *b*, 3 mm *t*; **Fl** solitary, rarely binate, pedicels 10–15 mm *l*, bracteoles at their middle, petals 3–12 mm *l*, to 1.5 mm *b*, filamentous staminodes merging into the petals, pale, filaments white, pollen lemon-yellow, nectary ring flat, inconspicuously crenulate; **Fr** 5 locules; **Ecol** unknown; **Distr** Clanwilliam, WC, S Africa.

Note: The filamentous staminodes are shorter than the stamens, an unusual feature in **Lampranthus**.

L. leipoldtii (L.Bolus) L.Bolus NM 1939: 169 ≡ *M. leipoldtii* L.Bolus NM 1929: 156 non (N.E.Br.) hort. Kew ex Hill 1931: 128 **T** Leipoldt 18997 (BOL).

Erect shrubs to 20 cm *h*, **I** deep brown, 10–30 mm *l*, ∅ to 2 mm; **L** triquetrous, dirty green, with raised dots, 10–15 mm *l*, 1 mm *b*, to 1.5 mm *t*; **Fl** solitary or many, ∅ 34 mm, pedicels 35–55 mm *l*, bracteoles at their middle, petals pink or pale pink, to 15 mm *l*, to 2 mm *b*, inner 5–9 mm *l*, 0.75 mm *b*, filamentous staminodes none, filaments whitish, nectary ring inconspicuously undulate; **Fr** expanded 11 mm ∅, valve wings narrow, 5 locules; **Ecol** unknown; **Distr** Worcester, WC, S Africa.

Note: Lack of filamentous staminodes characterizes this species.

L. leptaleon (Haw.) N.E.Br. GC 1930: 212 ≡ *M. leptaleon* Haw. 1821: 140 **T**?

Shrubs with deflexed branches, less branched than **L. deflexus**, **I** filiform; **L** subtriquetrous, smooth, glaucous, keel a little rough; **Fl** with 2 filiform bracteoles, **K** lobes almost spinose, petals red and paler, filamentous staminodes pink; **Fr** unknown; **Ecol** unknown; **Distr** unknown.

Note: Reported by Haworth (1821: 140) to differ from *M. deflexum* in its fewer and more slender branches.

L. leptosepalus (L.Bolus) L.Bolus NM 1939: 169 ≡ *M. leptosepalum* L.Bolus NM 1935: 489 **T** Compton 21430 (BOL).

Erect shrubs to 14 cm *h*, **I** 2-angled, basally terete, 10–20 mm *l*, 1–1.5 mm ∅; **L** trigonous, acute, dirty green, dotted, 10–15 mm *l*, 1.5–2 mm *b* and *t*; **Fl** mostly solitary, pedicels 30–50 mm *l*, bracteoles at their middle, petals pinkish-purple, 3–16 mm *l*, to 2 mm *b*, filamentous staminodes none, filaments white, basally with blackish-purple papillae, pollen yellow, nectary ring inconspicuous, obtusely pentagonal, obscurely crenulate; **Fr** 5 locules; **Ecol** unknown; **Distr** Clanwilliam, WC, S Africa.

Note: The reported blackish-purple papillae on the filaments are an unusual feature in the genus.

L. lerouxiae ≡ **Ruschia lerouxiae**

L. lewisiae (L.Bolus) L.Bolus NM 1939: 170 ≡ *M. lewisiae* L.Bolus NM 1937: 54 **T** Lewis 21820 (BOL).

Erect shrubs to 7.5 cm *h*, **I** 4–8 mm *l*, 1.5 mm ∅; **L** subterete, apiculate, pale green, obscurely scabrous, 12–14 mm *l*, 2 mm *b* and *t*; **Fl** solitary, pedicels raised for 15–18 mm above the leaves, petals pink, 15 mm *l*, filamentous staminodes and filaments pink, apically paler, anthers and pollen golden, nectary ring crenulate; **Fr** 5 locules; **Ecol** on mountain top; **Distr** Ceres, WC, S Africa.

Note: The filamentous staminodes are reported to be at most as long as the stamens in this species.

L. liberalis (L.Bolus) L.Bolus NM 1939: 170 ≡ *M. liberale* L.Bolus NM 1938: 128 **T** without collector 21843 (BOL).

Erect, richly branched shrubs 15–25 cm *h*, **I** 10–15 mm *l*, 1–2 mm ∅, basally to 3 mm ∅; **L** trigonous, often apically recurved, acute, finely rough, dirty glaucous green, 6–9 mm *l*, to 1.5 mm *b* and *t*;

Fl solitary, rarely binate, pedicels to 25 mm *l*, rarely to 45 mm *l*, bracteoles at their middle, petals pink, inner merging into filamentous staminodes, to 14 mm *l*, to 1.25 mm *b*, filamentous staminodes numerous, pale, apically purplish-pink, anthers and pollen yellow, nectary ring crenulate; **Fr** unknown; **Ecol** unknown; **Distr** S Africa.

Note: Exceptionally numerous petals and filamentous staminodes characterize this species.

L. littlewoodii L.Bolus 1961: 114 **T** Littlewood 581/59 (BOL).

Straggling shrubs to 23 cm *h*, ∅ to 35 cm, stem basally to 8 mm ∅, **I** 40–60 mm *l*, to 3 mm ∅, 12–14 leaves set closely near the tips of the branches; **L** trigonous to falcate, obtuse, pale glaucous green, 10–25 mm *l*, 1.5–2 mm *b*, 2–3 mm *t*; **Fl** in cymes, 23–30 mm ∅, pedicels 10–15 mm *l*, to 30 mm *l*, when fruiting, bracteoles at their middle, petals bright pink, 10–14 mm *l*, 1–2 mm *b*, inner 4–8 mm *l*, 0.25–0.5 mm *b*, filamentous staminodes very few, filaments pale pink, anthers and pollen pale straw-coloured, nectary ring crenulate; **Fr** base obconical, top raised for 2 mm, 5–6 mm *l*, ∅ 6–7 mm, expanded 9–10 mm ∅; **Ecol** unknown; **Distr** Namaqualand, NC, S Africa.

Note: The petals merge gradually into more narrow, shorter, erect elements which can be seen as filamentous staminodes.

L. longisepalus = **Hammeria meleagris**

L. longistamineus (L.Bolus) N.E.Br. GC 1930: 212 ≡ *M. longistamineum* L.Bolus NM 1928: 148 **T** Leipoldt 18810 (BOL).

Erect shrubs to 13 cm *h*, branches ± twisted, leaves crowded near the tips, **I** 4 mm *l*, basally to 4 mm ∅; **L** trigonous, obtuse, densely dotted, to 25 mm *l*, 6 mm *b*, 7 mm *t*; **Fl** 1–3-nate, ∅ to 30 mm, pedicels to 10 mm *l*, petals golden, to 14 mm *l*, to 2 mm *b*, filamentous staminodes none, filaments glabrous, anthers and pollen golden, nectary ring inconspicuous, crenulate; **Fr** expanded 11 mm ∅, 5 locules; **Ecol** unknown; **Distr** Clanwilliam, WC, S Africa.

Note: Lack of filamentous staminodes characterizes this species.

L. lunatus ≡ **Oscularia lunata**

L. lunulatus (A.Berger) L.Bolus NM 1958: 333 ≡ *M. lunulatum* A.Berger 1922: 638 **T** Diels 344 (B!).

Richly branched shrubs with erect branches, **I** maroon, later brown, a white triangular scar left at the node after the leaves are fallen off, 15–20 mm *l*; **L** triquetrous, incurved, 8 mm *l*, 1 mm *b*, dotted, smooth; **Fl** in ternate cymes, pedicels with bracteoles at base and above, petals only a little longer than the K lobes, white, inner ones reddish; **Fr** unknown; **Ecol** on stony slopes in open scrub, 75 m altitude; **Distr** Clanwilliam, WC, S Africa.

Note: In a note on the type sheet, Brown gives also the name "M. fredericii L.Bolus (1916)".

L. macrocarpus (A.Berger) N.E.Br. GC 1930: 212 ≡ *M. macrocarpum* A.Berger 1922: 634 **Syntypes** Murray s.n., Paterson 207(GRA).

Shrubs with spreading, twisting branches, **I** brown, 10–25 mm *l*; **L** subtriquetrous, finely scabrous, 10–25 mm *l*, 2–4 mm *b*; **Fl** mostly ternate, rarely less, ∅ to 50 mm, pedicels with bracteoles at their middle, petals longer than the **K** lobes, acute, stamens numerous, anthers pale yellow; **Fr** spongy, turbinate, ∅ 18 mm, 5 locular; **S** blackish-brown, verrucose; **Ecol** unknown; **Distr** Albany, Cradock, EC, S Africa.

Note: Spongy capsules are very unusual in the genus, most fruits being woody and hard.

L. macrosepalus (L.Bolus) L.Bolus NM 1939: 169 ≡ *M. macrosepalum* L.Bolus NM 1930: 219 **T** Lavis 18761 (BOL).

Low shrubs with branches to 20 cm *l*, **I** basally to 6 mm ∅; **L** triquetrous, obtuse, dirty green, 30–70 mm *l*, to 4 mm *b*, 5–7 mm *t*; **Fl** in small cymes, ∅ to 60 mm, pedicels 60–80 mm *l*, bracteoles at their middle, petals bright pink, to 25 mm *l*, 2.5 mm *b*, filamentous staminodes none, filaments white, anthers and pollen straw-coloured, nectary ring densely crenulate; **Fr** 5 locules; **Ecol** unknown; **Distr** Humansdorp, EC, S Africa.

Note: Lack of filamentous staminodes characterizes this species.

L. macrostigma L.Bolus NM 1954: 274 **T** Esterhuysen 18608 (BOL).

Erect shrubs to 22 cm *h*, **I** brown, 5–30 mm *l*, 1.5–2 mm ∅; **L** trigonous, hardly scabrous, glaucous green, 7–14 mm *l*, often 1.5 mm *b* and *t*; **Fl** mostly solitary, **K** base scabrous, petals white, to 12 mm *l*, to 1.75 mm *b*, merging into the filamentous staminodes, filaments pale green or white, anthers and pollen yellow, nectary composed of 5 crenulate, almost connate glands; **Fr** top elevated for 1.5 mm, 6 mm *l*, 7 mm ∅, broad valve wings; **Ecol** in sandy flats; **Distr** Clanwilliam, WC, S Africa.

Note: The five distinct glands remind of **Oscularia.**

L. magnificus (L.Bolus) N.E.Br. GC 1930: 212 ≡ *M. magnificum* L.Bolus ABH 1927: 93 **T** without collector 18532 (BOL).

Shrubs with prostrate to decumbent branches, flowering branches ascending to erect, 9–11 cm *h*, **I** shorter than **L**, ∅ to 2 mm; **L** semi-terete, acute, smooth, green, 25–45 mm *l*, 2–2.5 mm *t*; **Fl** mostly solitary, pedicels 30–55 mm *l*, bracteoles a little below their middle, petals coppery red, to 23 mm *l*, 2 mm *b*, filaments purple, anthers and pollen brownish-purple, nectary ring inconspicuous, crenulate; **Fr** 5 locules; **Ecol** unknown; **Distr** Malmesbury, WC, S Africa.

Note: Bolus (ABH 1927: 93) pointed out that "the purplish-brown colour of the anthers and pollen is very unusual in the genus".

L. marcidulus N.E.Br. GC 1930: 212 ≡ *M. flaccidum* L.Bolus ABH 1923: 164 nomen illeg. non Jacquin 1804: 38 t. 475 **T** Page 17258 (BOL).

Shrubs to 20 cm *h*, branches not able to hold their own weight (flaccid), **I** to 40 mm *l*; **L** triquetrous, acute, bright green, 20–40 mm *l*, 4–5 mm *b* and *t*; **Fl** solitary, ∅ 35 mm, pedicels 50–60 mm *l*, bracteoles at their middle, petals golden, inner with a deep orange line, 6–16 mm *l*, 0.25 mm *b*, filaments golden, without papillae, anthers yellow, nectary ring inconspicuous; **Fr** unknown; **Ecol** in open veld, scrambling in bushes; **Distr** Montagu, WC, S Africa.

Note: The species seems to lack filamentous staminodes.

L. marginatus (L.Bolus) H.E.K.Hartmann Bradleya 1998: 68 ≡ *Ruschia marginata* L.Bolus NM 1935: 500 **T** L.Bolus 1505/33 (BOL!).

Shrubs to 14 cm *h*, branches spreading to decumbent, to 30 cm *l*, **I** 5–20 mm *l*; **L** trigonous to semiterete, glaucous, a high triangle separating the bases of a leaf-pair, epidermis almost smooth, 20–25 mm *l*, 3 mm *b* and *t*; **Fl** in cymes, pedicels 25–30 mm *l*, bracteoles at their middle or below, petals bright pink, margins deeper in colour, 7–5–9 mm *l*, 1.5–1.75 mm *b*, filamentous staminodes few, filaments pale pink, apically purplish-pink, anthers and pollen straw-coloured; **Fr** base between funnel-and bell-shaped, with convex sides, top almost semi-globose, covering membranes with protruding closing rodlets at their distal ends, placenta ending in a roundish knob, no closing body, expanding keels almost touching each other, valve wings broad, rectangular; **Ecol** unknown; **Distr** Clanwilliam, Piquetberg, WC, S Africa.

Note: Absence of a closing body and presence of broad valve wings clearly place this species in **Lampranthus**, not in **Ruschia**, as originally published.

L. mariae (L.Bolus) L.Bolus NM 1939: 169 ≡ *M. mariae* L.Bolus NM 1932: 329 **T** Lavis 1764/31 (BOL).

Erect shrubs to 20 cm *h*, **I** 10–15 mm *l*, 1–2.5 mm ∅; **L** trigonous, falcate, glaucous green, to 7 mm *l*, 1.5 mm *t*; **Fl** solitary, ∅ 14 mm, pedicels 10 mm *l*, bracteoles above the middle, petals pinkish-purple, 5–7 mm *l*, to 2 mm *b*, filamentous staminodes very few, filaments pink, anthers white, nectary ring often deeply divided; **Fr** 5 locules; **Ecol** unknown; **Distr** Ceres, WC, S Africa.

Note: The few filamentous staminodes are c. as long as the stamens.

L. martleyi (L.Bolus) L.Bolus NM 1939: 169 ≡ *M. martleyi* L.Bolus NM 1932: 366 **T** Martley 20205 (BOL).

Erect shrubs to 30 cm *h*, **I** shining, brown, angular, 20–35 mm *l*, to 5 mm ∅; **L** triquetrous, glaucous green, obscurely scabrous, 15–20 mm *l*, 1.5 mm *b*, 2 mm *b*; **Fl** in small cymes, ∅ 40 mm, pedicels 15–60 mm *l*, bracteoles at their middle, petals shell-pink, 10–17 mm *l*, 1.5 mm *b*, filaments white, pollen lemon-yellow, nectary ring flat, obscurely crenulate; **Fr** 5 locules; **Ecol** unknown; **Distr** Piquetberg, WC, S Africa.

Note: Lack of filamentous staminodes characterizes this species.

L. maturus N.E.Br. GC 1930: 212 ≡ *M. praecox* L.Bolus NM 1928: 3 nomen illeg. non F. Mueller 1852: 384 **T** L.Bolus 1045/24 (BOL).

Branches 14–20 cm *l*, **I** brown, 5–10 mm *l*, ∅ to 3 mm; **L** trigonous, falcate, subtruncate, glaucous green, scabrous under the lens, to 7 mm *l*, to 2 mm *t*; **Fl** solitary, ∅ 20 mm, pedicels 6–12 mm *l*, petals pink, 9 mm *l*, 1.5 mm *b*, filamentous staminodes few, filaments purple, basally white, anthers and pollen pale yellow, nectary ring crenulate; **Fr** 5 locules; **Ecol** in clayey rocks; **Distr** Paarl, WC, S Africa.

Note: This species is reported to flower earlier than others in the genus.

L. matutinus (L.Bolus) N.E.Br. GC 1930: 212 ≡ *M. matutinum* L.Bolus NM 1928: 135 ≡ *L. matutinus* (L.Bolus) Schwantes in Jacobsen NCSJ 1949: 58 nomen illeg. **T** Immelman 18713 (BOL).

Creeping, **I** 10–25 mm *l*, 3 mm ∅, flowering branches 2.5–6 cm *l*; **L** semi-terete, yellow-green, sparsely dotted, 15–20 mm *l*, to 2 mm *b* and *t*; **Fl** solitary, ∅ 38 mm, open 9 a.m. to 1 p.m., pedicels 5–9 mm *l*, petals golden, apically reddish, to 18 mm *l*, 2 mm *b*, filamentous staminodes none, filaments glabrous, golden, anthers and pollen golden, nectary ring inconspicuous, crenulate; **Fr** 5 locules; **Ecol** unknown; **Distr** Piquetberg (?), WC, S Africa.

Note: Lack of filamentous staminodes characterizes this species.

L. maximiliani ≡ **Braunsia maximiliani**
L. meleagris ≡ **Hammeria meleagris**

L. microsepalus L.Bolus 1961: 261 **T** Duthie 1410 (BOL).

Branches 9–18 cm *l*, **I** 10–20 mm *l*, ∅ to 3 mm ; **L** trigonous, truncate, 8–10 mm *l*, 1–1.5 mm *t*; **Fl** solitary, ∅ to 22 mm, pedicels 10–15 mm *l*, petals pink with deeper coloured margins, 8–10 mm *l*, filamentous staminodes with abortive anthers present, filaments pink, anthers and pollen pale yellow, nectary ring inconspicuous, obscurely crenulate; **Fr** 5 locules; **Ecol** unknown; **Distr** Tulbagh, Worcester, WC, S Africa.

Note: Filamentous staminodes bearing abortive anthers are a rare feature in the subfamily Ruschioideae.

L. microstigma (L.Bolus) N.E.Br. GC 1930: 212 ≡ *M. microstigma* L.Bolus NM 1929: 142 **T** collector unknown 18971 (BOL).

Erect shrubs to 40 cm *h*, **I** 20–30 mm *l*, Ø 2 mm; **L** semi-terete, obscurely scabrous, 15–20 mm *l*, to 1.5 mm *b* and *t*; **Fl** in small cymes, 35–40 mm Ø, pedicels with bracteoles at their middle, petals pinkish-purple, to 16 mm *l*, to 1.5 mm *b*, filamentous staminodes white, anthers and pollen yellow, the inconspicuous nectary ring crenulate; **Fr** subpyriform, 7 mm *l*, acutely angled, expanded 12 mm Ø, covering membranes almost plane, 5 locules; **Ecol** unknown; **Distr** Piquetberg (?), WC, S Africa.

Note: The filamentous staminodes are reported to bend over the stamens.

L. middlemostii (L.Bolus) L.Bolus NM 1939: 169 ≡ *M. middlemostii* L.Bolus NM 1932: 329 **T** Middlemost 2071/28 (BOL).

Laxly branched shrubs probably straggling in other bushes, branches to 30 cm *l*, **I** 20–35 mm *l*, Ø to 3 mm; **L** triquetrous, falcate, obscurely scabrous, to 8 mm *l*, 1.5 mm *b*, to 2.5 mm *t*; **Fl** mostly solitary, pedicels 5 mm *l*, petals pale pink or white, basally pink, deeper coloured with age, to 9 mm *l*, to 2 mm *b*, nectary ring crenulate; **Fr** 5 locules; **Ecol** unknown; **Distr** Caledon, WC, S Africa.

Note: The species is reported to be similar to **L. falcatus.**

L. montaguensis (L.Bolus) L.Bolus 1967: 306 ≡ *M. montaguense* L.Bolus ABH 1925: 9 **T** Page 17692 (BOL).

Diffusely branched shrubs, branches to 30 cm *l*, **I** Ø to 5 mm; **L** trigonous, subfalcate, smooth, glaucous green, minutely papillate, 17–10 mm *l*, 2–3 mm *b*, 3 mm *t*; **Fl** in reduced cymes or solitary, Ø 19–23 mm, pedicels 5–15 mm *l*, **K** 6, pinkish-purple, obscurely vittate, 7–10 mm *l*, 1–2 mm *b*, filamentous staminodes white, apically purple, filaments basally white, anthers white, pollen light yellow, nectary ring denticulate; **Fr** unknown; **Ecol** unknown; **Distr** Montagu, WC, S Africa.

Note: The flowers are reported to be pentamerous, but 6 calyx-lobes are mentioned in the description.

L. monticola (L.Bolus) L.Bolus NM 1939: 169 ≡ *M. monticola* L.Bolus NM 1930: 171 nomen illeg. non Sonder 1862: 428 **T** Cook 1378/28 (BOL).

Caespitose, to 6 cm *h*, prostrate branches to 20 cm *l*, **I** 5–15 mm *l*, to 2 mm Ø; **L** trigonous, subfalcate, apiculate, glaucous, apically purple, 5–10 mm *l*, to 2 mm *b* and *t*; **Fl** in small cymes, Ø to 20 mm, pedicels 4–10 mm *l*, petals pink to pale pink, to 9 mm *l*, 1–1.5 mm *b*, filamentous staminodes longer than stamens, filaments pale, anthers and pollen yellow, nectary ring conspicuously crenulate; **Fr** unknown; **Ecol** in fissures of rocks; **Distr** Cape Town, WC, S Africa.

Note: In a later publication, Bolus (NM 1958: 388) stated that the species is synonymous with **L. stenus** and that the type locality given with the original description may be wrong, the species coming from sandy flats extending to the coast.

L. mucronatus L.Bolus 1966: 201 **T** Stayner 998/61 (BOL).

Rather compact plants to 15 cm *h*, Ø to 22 cm, **I** to 10 mm *l*, Ø 4–6 mm; **L** triquetrous, falcate, apically with a sharp mucro turning into an awn becoming almost spinescent with age, 20–35 mm *l*, to 5 mm *b*, 4–6 mm *t*; **Fl** solitary, pedicels 5 mm *l*, elongating to 10–25 mm in fruits, petals white, 10–15 mm *l*, to 2.5 mm *b*, filaments glabrous, white, anthers and pollen yellow, nectary well visible, wavy, as if composed of 5 connate glands; **Fr** base obconical, top flat, valve wings narrow, 5 mm *l*, Ø to 10 mm, expanded to 16 mm Ø; **Ecol** on a ridge; **Distr** Ceres, WC, S Africa.

Note: Fruiting material was collected by Stayner four years later in the same spot, where the plants grew in profusion, but not elsewhere.

L. multiradiatus (Jacquin) N.E.Br. GC 1930: 71 ≡ *M. multiradiatum* Jacquin 1800–1809: 44 ≡ *M. incurvum* Haw. var. *multiradiatum* (Jacquin) DC. 1828: 434 **Lectotype**, here designated: Jacquin 1800–1809: t. 53.

Erect shrubs to 30 cm *h*, **I** green to purplish, later grey, terete; **L** triquetrous, glaucous, 27 mm *l*; **Fl** solitary to ternate, pedicels bracteolate at the middle, petals c. 27 mm *l*, flesh-coloured, filaments white, anthers pale yellow; **Fr** base turbinate, 5 locules; **Ecol** unknown; **Distr** S Africa.

Note: The species is also reported from Australia (Venning in Prescott & Venning 1984: 34).

L. multiseriatus (L.Bolus) N.E.Br. GC 1930: 212 ≡ *M. multiseriatum* L.Bolus ABH 1928: 128 **Syntypes** Muir 2416, La Grange 4584/14 (BOL).

Creeping, **I** 25–35 mm *l*, erect branches 5–6 cm *h*; **L** trigonous, smooth, glaucous green, 15–18 mm *l*, 1.5 mm *b*, to 2.5 mm *t*; **Fl** solitary, Ø 44 mm, pedicels 13–23 mm *l*, bracteoles at their middle, **K** base sparsely tuberculate, petals pink, 5–20 mm *l*, 0.5–2 mm *b*, filamentous staminodes absent, filaments pale yellow, anthers and pollen golden, nectary ring an inconspicuous, crenulate ring; **Fr** 5 locules; **Ecol** unknown; **Distr** Riversdale, WC, S Africa.

Note: The species differs from the similar **L. sororium** "by the petals in many series" (Bolus ABH 1928: 128).

L. mutans (L.Bolus) N.E.Br. GC 1930: 212 ≡ *M. mutans* L.Bolus NM 1928: 18 **T** Fourcade 3336 (BOL).

Erect shrubs to 30 cm *h*, **I** 10–20 mm *l*, 1–3 mm ∅; **L** subterete, subuncinate, apiculate, green, smooth, to 15 mm *l*, ∅ to 2 mm; **Fl** often ternate, ∅ 36 mm, pedicels to 30 mm *l*, bracteoles at their middle, petals changing colour: salmon, later pink, basally yellow, to 16 mm *l*, to 2 mm *b*, filamentous staminodes flat, basally papillate, filaments white, anthers and pollen yellow, nectary ring crenulate; **Fr** unknown; **Ecol** unknown; **Distr** Humansdorp, EC, S Africa.

Note: The name refers to the change in flower colour during anthesis.

L. mutatus (G.D.Rowley) H.E.K.Hartmann Bradleya 1998: 68 ≡ *Ruschia mutata* G.D.Rowley NCSJ 1978: 7 non **L. littlewoodii** L.Bolus ≡ *Stoeberia littlewoodii* L.Bolus JSAB 1960: 162 ≡ *Amphibolia littlewoodii* (L.Bolus) L.Bolus ex Toelken & Jessop 1976: 64; L.Bolus 1965: 170 nomen invalid. **T** Littlewood 522/59 (BOL!).

Erect, straggly shrubs to 30 cm *h*, **I** reddish; **L** trigonous, connate for 2–3 mm, a distinct line separating the two leaves of the pair, free parts 10–22 mm *l*, 3–4 mm ∅ and *b*, epidermis glabrous; **Fl** in ternate cymes, often with additional flowers below, bracteoles above the middle of the stalk, petals pinkish-purple, stamens purple, bases lighter coloured; **Fr** bracteoles below the middle of the stalk, top raised from high valve rims, base long, funnel-shaped, covering membranes convex, distally turned upwards a little, closing rodlets below protruding into the opening, closing body a round outgrowth from the end of the placenta, translucent, small, valve wings c. as broad as the short expanding keels; **S** brown, almost smooth; **Ecol** < 100 mm rainfall p.a. in winter; **Distr** Ceres, WC, S Africa.

Note: The species is placed in **Lampranthus** with some hesitation because of its placental closing body. The name **L. littlewoodii** had been used earlier by Bolus for a different species in this genus, the younger epithet has therefore to be taken up, and the species becomes **L. mutatus**.

L. nardouwensis (L.Bolus) L.Bolus NM 1939: 169 ≡ *M. nardouwense* L.Bolus NM 1933: 407 **T** Salter 3531 (BOL).

Laxly branched shrubs, branches to 23 cm *l*, **I** 10–20 mm *l*, ∅ 1–3 mm; **L** trigonous, 17 mm *l*, 3 mm *b*, 4 mm *t*; **Fl** in small cymes, pedicels 10–18 mm *l*, bracteoles at their middle, petals pink, 4–7 mm *l*, to 1.5 mm *b*, filamentous staminodes basally white, middle pink, anthers and pollen golden, nectary ring deeply divided; **Fr** 5 locules; **Ecol** unknown; **Distr** Clanwilliam, WC, S Africa.

Note: Remarkable is the description of the filaments bending over the stigmas and standing erect in their upper parts, resembling conditions found in the genus **Amphibolia**.

L. nelii L.Bolus 1965: 308 **T** Nel 27732 (BOL).

Shrubs to 14 cm *h*, **I** 6–20 mm *l*, ∅ 1–1.5 mm; **L** trigonous, aristate, 10–17 mm *l*, 1–1.5 mm *t*; **Fl** pedicels 10–13 mm *l*, petals pink, basally paler, 9–10 mm *l*, often to 1 mm *b*, filamentous staminodes basally and apically white, middle pink, filaments white, anthers and pollen ochre, nectary ring conspicuously crenulate; **Fr** 5 locules; **Ecol** unknown; **Distr** Ladismith, WC, S Africa.

Note: The stigmas exceed the stamens in this species in length.

L. neostayneri L.Bolus JSAB 1960: 58 **T** L.Bolus 19950 (BOL).

Erect shrubs 15–20 cm *h*, ∅ to 30 cm, **I** reddish-brown, 10–15 mm *l*, basal ones 5–10 mm ∅; **L** acuminate, green, scabrous, 10–15 mm *l*, 1–1.5 mm *b* and *t*; **Fl** in cymes, ∅ 25–40 mm, pedicels terete, 12–30 mm *l*, bracteoles below their middle, petals purplish-pink, 8–18 mm *l*, 1.5–2 mm *b*, filamentous staminodes few, filaments pink, nectary ring crenulate; **Fr** obtusely pentangular, top 3 mm *h*, base obconical, 7 mm *h*, ∅ 10 mm, expanded 15 mm ∅, 5 locules; **Ecol** unknown; **Distr** Piquetberg, WC, S Africa.

Note: Big capsules arranged in cymes characterize this species.

L. obconicus (L.Bolus) L.Bolus NM 1939: 169 ≡ *M. obconicum* L.Bolus NM 1930: 220 **T** L.Bolus 19230 (BOL).

Erect shrubs 20–30 cm *h*, **I** brown, shining, 30–50 mm *l*, on flowering branches 10–20 mm *l*; **L** trigonous, obscurely scabrous, dirty green, to 10 mm *l*, to 2 mm *b* and ∅, keel decurrent on stem; **Fl** in small cymes, ∅ to 27 mm, petals deep pink, to 11 mm *l*, to 1.5 mm *b*, filamentous staminodes pink, apically recurved and paler, anthers and pollen yellow; **Fr** top raised for 2 mm, expanded to 10 mm ∅, 5 locules; **Ecol** in sand; **Distr** Calvinia, NC, S Africa.

Note: Small cymes and small capsules appear to be typical of this species.

L. occultans L.Bolus NM 1950: 225 **T** Esterhuysen 10942 (BOL).

Erect shrubs, 30–40 cm *h*, **I** 0.5–3 mm ∅; **L** very slender, glaucous green, 10–17 mm *l*, to 1 mm *b*, to 1.5 mm *t*; **Fl** 1–3, pedicels 25–40 mm *l*, bracteoles at their middle, petals pink, to 10 mm *l*, 0.5–1 mm *b*, inner simulating filamentous staminodes and hanging over the stamens hiding them, with conspicuous red glands in the dry state, filaments white, anthers and pollen pale yellow, nectary as a pentagonal, crenulate ring; **Fr** top raised for 1.5 mm, to 6 mm *l*, expanded to 10 mm ∅; **Ecol** unknown; **Distr** Worcester, WC, S Africa.

Note: Reported to be similar to **L. rabiesbergensis**; unusual for the genus are the hidden stamens.

L. ornatus ≡ **Oscularia ornata**

L. otzenianus (Dinter) Friedrich 1960: 565 (see **photo**) ≡ *M. otzenianum* Dinter FR 1923: 152 ≡ *Psilocaulon otzenianum* (Dinter) L.Bolus NM 1928: 31 ≡ *Drosanthemum otzenianum* (Dinter) Friedrich 1968: 217 **T** Dinter 3789 (B) = *M. spathulatum* L.Bolus NM 1928: 136 nomen illeg. non Thunberg 1791: 5 nec Willdenow 1799: 1025 nomen illeg. ≡ *L. uniflorus* (L.Bolus) L.Bolus var. *spathulatus* (L.Bolus) L.Bolus NM 1958: 388 **T** Pillans 5719 (BOL!).

Bluish- to yellowish-green shrubs, to 30 cm *h*, 40–80 cm ∅, **I** whitish-grey, 20–40 mm *l*; **L** terete, obtuse, soft, epidermis smooth, 15–25 mm *l*, 7–9 mm *t*; **Fl** solitary, pedicels 12–15 mm *l*, **K** almost equal, ∅ 25–30 mm, petals 22–25, purplish-pink, filaments pink, anthers cream-coloured to pinkish; **Fr** base funnel- to narrowly bell-shaped, top rather flat, covering membranes laterally reduced, placenta parietal, not always reaching into the base of the locule, expanding keels broad, diverging somewhat, valve wings very broad, almost rectangular, narrowing basally, 5 locules; **S** brown, smooth, shiny, 0.9–1.1 mm *l*, 0.7–1 mm *b*; **Ecol** mostly in flats in loamy soil; **Distr** Lüderitz-South, Namibia; Namaqualand, NC, S Africa.

Note: The species is most similar to **L. uniflorus**, and both species may form a separate genus. The lack of papillae on stem, pedicels, and the base of the flower as well as the possession of a nectary ring exclude the species from **Drosanthemum. L. otzenianus** does not show the false septa present in populations of **L. uniflorus**.

L. paardebergensis ≡ **Oscularia paardebergensis**

L. paarlensis L.Bolus 1961: 261 **T** Middlemost 1360/37 (BOL).

Diffusely branched shrubs to 46 cm *h*, **I** brown, shining, to 20 mm *l*, to 2 mm ∅; **L** triquetrous, very slender, green, inconspicuously scabrous, 18–22 mm *l*, to 1.5 mm *b* and *t*; **Fl** solitary, ∅ to 30 mm, pedicels 15–28 mm *l*, petals bright pink or almost white, 10–13 mm *l*, to 1.5 mm *b*, filamentous staminodes and filaments white, apical half red, anthers and pollen yellow; **Fr** 5 locules; **Ecol** unknown; **Distr** Paarl, WC, S Africa.

Note: Solitary, big flowers suggest that big capsules may be developed in this species.

L. pakhuisensis (L.Bolus) L.Bolus NM 1939: 169 ≡ *M. pakhuisense* L.Bolus NM 1932: 364 **T** Salter 2667 (BOL).

Erect shrubs to 25 cm *h*, richly branched, **I** often to 15 mm *l*, rarely 20 mm *l*; **L** trigonous, subcompressed, to 15 mm *l*, 2 mm *b* and *t*; **Fl** solitary, pedicels 7–10 mm *l* above the bracteoles, 3 of the **K** lobes with broad, brown margins, petals pink, 3–12 mm *l*, to 2 mm *b*, filamentous staminodes pink, filaments pale, anthers straw-coloured, nectary ring crenulate; **Fr** 5 locules; **Ecol** unknown; **Distr** Clanwilliam, WC, S Africa.

Note: The possession of three calyx lobes with brown margins appears remarkable in the genus.

L. pakpassensis H.E.K.Hartmann Bradleya 1998: 70 ≡ *Ruschia pakhuisensis* L.Bolus NM 1933: 414 **T** L.Bolus 20479 (BOL!).

Shrubs to 30 cm *h*, **I** black with age, ∅ 4 mm; **L** semi-terete and obtuse to subfalcate, 8–10 mm *l*, 3 mm *b*, 4 mm *t*; **Fl** bracteoles above the middle of the pedicel, petals pink with deeper coloured margins, to 6 mm *l*, to 1.25 mm *b*, filamentous staminodes white, apically pink, anthers and pollen golden; **Fr** top rounded, base funnel-shaped, covering membranes with small closing rodlets, the top of the placenta inconspicuous, expanding keels meeting each other at their bases, valve wings very broad and rectangular, 5 locules; **Ecol** unknown; **Distr** Clanwilliam, WC, S Africa.

Note: The species has been transferred to **Lampranthus** because the fruit on the type shows wide valve wings and no closing body.

L. palustris (L.Bolus) L.Bolus NM 1939: 169 ≡ *M. palustre* L.Bolus NM 1932: 325 **T** L.Bolus 19927 (BOL).

Shrubs 12–24 cm *h*, **I** 10–15 mm *l*, 1–2 mm ∅; **L** triquetrous, glaucous green, keel and margins with prominent dots, 6–8 mm *l*, to 21 mm *b* and *t*; **Fl** solitary, pedicels 15–20 mm *l*, bracteoles a little below the middle, petals golden, to 19 mm *l*, to 2 mm *b*, filaments erect, golden, anthers and pollen golden, nectary ring inconspicuous; **Fr** 5 locules; **Ecol** in swampy places; **Distr** Stellenbosch, WC, S Africa.

Note: Similar in growth form to **L. glaucus**.

L. parcus N.E.Br. GC 1930: 212 ≡ *M. tenue* L.Bolus NM 1928: 1 nomen illeg. non Haw. 1821: 175 **T** Pillans 18844 (BOL).

Shrubs to 20 cm *h*, loosely branched, **I** 4–15 mm *l*; **L** trigonous, scabrous, 4–9 mm *l*, to 1.5 mm *b* and *t*; **Fl** ∅ 21 mm, pedicels to 10 mm *l*, terete, petals pink, 8 mm *l*, to 1 mm *b*, filamentous staminodes thin, white, apically pink, anthers straw-coloured, nectary ring conspicuous, blackish-green, **Fr** unknown; **Ecol** unknown; **Distr** Caledon, WC, S Africa.

Note: Reported to be similar to **Erepsia gracilis**.

L. pauciflorus (L.Bolus) N.E.Br. GC 1930: 212 ≡ *M. pauciflorum* L.Bolus ABH 1927: 90 **T** Phillips 18552 (BOL).

Loosely branching shrubs to 30 cm *h*, **I** 10–30 mm *l*; **L** triquetrous, subfalcate, smooth, glaucous green, to 18 mm *l*, to 2 mm *b*, to 3 mm *t*; **Fl** solitary, ∅ to 40 mm, pedicels 20–40 mm *l*, bracteoles at their middle, petals deep pink, inner very narrow, to 18 mm *l*, 1 mm *b*, few filamentous staminodes, anthers and pollen yellow, nectary ring inconspicuous, crenulate; **Fr** unknown; **Ecol** on rocks edges and stony screes of cliffs; **Distr** Knysna, WC, S Africa.

Note: "In the paucity of its flowers this species is remarkable in the genus" (Bolus ABH 1927: 90).

L. paucifolius (L.Bolus) N.E.Br. GC 1930: 212 ≡ *M. paucifolium* L.Bolus NM 1928: 138 **T** Compton 1509/26 (BOL).

Erect shrubs to 25 mm *h*, **I** 10–40 mm *l*, ∅ to 2 mm; **L** trigonous, acute, 17 mm *l*, 1–2 mm *b* and *t*; **Fl** in small cymes, ∅ 26 mm, petals purplish-pink, to 10 mm *l*, almost to 2 mm *b*, filamentous staminodes few, purple, filaments pink, basally paler, anthers yellow, nectary as a ring; **Fr** 5 locules; **Ecol** unknown; **Distr** Calvinia, NC, S Africa.

Note: Purple filamentous staminodes are rare in **Lampranthus.**

L. peacockiae (L.Bolus) L.Bolus NM 1939: 170 ≡ *M. peacockiae* L.Bolus ABH 1921: 71 **Syntypes** Peacock 1659/17, L.Bolus 15192, Guthrie s.n. (BOL).

Creeping with flexuose branches; **L** semi-terete, glaucous, greenish in cultivation, punctate, 6–9 mm *l*, 2 mm *b*; **Fl** solitary, pedicels with bracteoles at their middle, ∅ 25–30 mm, petals pale pink, tip and margins purplish, 10–20 mm *l*, to 2 mm *b*, filaments white, purplish in their upper half, nectary ring crenulate; **Fr** unknown; **Ecol** unknown; **Distr** Cape Town, Malmesbury, WC, S Africa.

Note: The species seems to lack filamentous staminodes.

L. peersii (L.Bolus) N.E.Br. GC 1930: 212 ≡ *M. peersii* L.Bolus ABH 1922: 131 non M. peersii L.Bolus ex hort. Kew ex Hill 1931: 131 nomen illeg. **T** Peers 3002/15 (BOL).

Erect shrubs to 30 cm *h*, **I** 20–40 mm *l*; **L** triquetrous, glaucous green, apically purplish, 20–40 mm *l*, 4 mm *b* and *t*; **Fl** in small cymes, ∅ 35–44 mm, pedicels 40–45 mm *l*, with bracteoles at their middle, petals golden, then reddish, at last pinkish-purple, 5–18 mm *l*, filamentous staminodes like threads, coloured like the petals, filaments pale, anthers golden, nectary glands semi-lunate, connate; **Fr** unknown; **Ecol** unknown; **Distr** Clanwilliam, WC, S Africa.

Note: The separate, semi-lunate nectary glands resembles those found in **Oscularia.**

L. perreptans L.Bolus NM 1939: 169 ≡ *M. humile* L.Bolus NM 1929: 73 nomen illeg. non Haw. 1795: 471 **T** van der Merwe 1195/28 (BOL).

Branches to 20 cm *l*, creeping, **I** 15–20 mm *l*, 2 mm ∅; **L** subfalcate, keel eccentric, dirty green, smooth, 15–20 mm *l*, to 3 mm *b* and *t*; **Fl** solitary, ∅ to 43 mm, pedicels to 5 mm *l*, petals pink, to 20 mm *l*, to 2.5 mm *b*, inner much more narrow, filamentous staminodes none, filaments straw-coloured, anthers and pollen golden, nectary ring crenulate; **Fr** 5 locules; **Ecol** unknown; **Distr** Swellendam (?), WC, S Africa.

Note: Lack of filamentous staminodes characterizes this species.

L. persistens (L.Bolus) L.Bolus NM 1939: 169 ≡ *M. persistens* L.Bolus NM 1931: 235 **T** Mathews 19383 (BOL).

Erect shrubs to 25 cm *h*, **I** very pale brown, 20–40 mm *l*, 3 mm ∅; **L** apically incurved, S-shaped, subterete, dirty green, apically reddish, smooth, 7–14 mm *l*, to 2 mm *b* and *t*, sheaths to 2.5 mm *l*, long persisting while the free parts disintegrate; **Fl** in rich cymes, pedicels 25–40 mm *l*, bracteoles at their middle, petals pinkish-purple, to 7 mm *l*, 1 mm *b*, filamentous staminodes pink, apically pale, anthers pale, nectary ring conspicuous, deeply divided; **Fr** top raised for 1.5 mm, 6 mm *l*, ∅ 6 mm, expanded 10 mm, valve wings very broad, covering membranes with 2 distal closing rodlets; **Ecol** unknown; **Distr** Calvinia, NC, S Africa.

Note: Remarkable is the well developed nectary, a rare feature in **Lampranthus.**

L. perspicuus ≡ **L. schlechteri**
L. piquetbergensis ≡ **Oscularia piquetbergensis**
L. pittenii ≡ **L. vanputtenii**

L. plautus N.E.Br. GC 1930: 212 ≡ *M. latum* L.Bolus NM 1929: 141 nomen illeg. non Haw. 1795: 186 **T** Pillans 6267 (BOL).

Diffusely branched shrubs, leaves crowded near the tips of the branches, **I** to 15 mm *l*; **L** trigonous, glaucous green, smooth, to 27 mm *l*, to 5 mm *t*, sheath to 2 mm *l*; **Fl** in small cymes, ∅ 34 mm, pedicels to 25 mm *l*, to 30 mm *l* when fruiting, bracteoles placed above the middle, petals purplish-pink, to 17 mm *l*, to 2 mm *b*, filamentous staminodes none, nectary ring inconspicuous; **Fr** 8 mm *l*, ∅ 11 mm, expanded 16 mm ∅, valve wings narrow; **Ecol** under a shrub; **Distr** Namaqualand, NC, S Africa.

Note: The big capsules are arranged in small cymes in this species.

L. pleniflorus L.Bolus 1962: 293 **T** Breda 1710/62 (BOL).

Branches to 35 cm *l*, **I** brown, shining, 15–45 mm *l*, ∅ 2–3 mm; **L** trigonous, 10–30 mm *l*, 2–3 mm *b*, 3–5 mm *t*; **Fl** in rich cymes, pedicels often reddish,

5–10 mm *l*, bracteoles at their middle, petals purplish-pink, margins deeper in colour, 4–5 mm *l*, 0.5–1 mm *b*, filamentous staminodes white, apically pink, filaments pale or white, apically pink, anthers and pollen reddish-golden, nectary ring deeply divided; **Fr** top elevated for 2 mm, base obconical, pentangular, ∅ 6 mm, expanded to 10 mm ∅, valve wings very broad; **S** subovate, brown, smooth, 1 mm *l*; **Ecol** unknown; **Distr** Worcester, WC, S Africa.

Note: The markedly divided nectary ring is remarkable for a member of **Lampranthus**.

L. plenus (L.Bolus) L.Bolus NM 1939: 169 ≡ *M. plenum* L.Bolus NM 1933: 407 **T** L.Bolus 20818 (BOL).

Erect to decumbent branches, **I** 10–50 mm *l*, ∅ 2–3 mm, 2-angled or narrowly winged; **L** trigonous, incurvate, green, smooth, 10–15 mm *l*, to 30 mm *l* in shady places, 2 mm *b*, 2–3 mm *t*; **Fl** in small cymes, pedicels 30–60 mm *l*, bracteolate above the middle, petals very numerous, pink, 6–19 mm *l*, to 2 mm *b*, filamentous staminodes bending over the stamens and hiding them, pink, filaments white, anthers and pollen straw-coloured, nectary ring depressed, inconspicuous, undulate; **Fr** top elevated for 3 mm, 9 mm *l*, 9 mm ∅, expanded 15 mm ∅, 5 locules; **Ecol** flowering in autumn; **Distr** Clanwilliam, WC, S Africa.

Note: The name refers to the flowers opening fully during the day, a feature common in the genus.

L. pocockiae (L.Bolus) N.E.Br. GC 1930: 212 ≡ *M. pocockiae* L.Bolus ABH 1927: 80 **T** Pocock 158 (BOL).

Erect shrubs to 9 cm *h*, **I** occasionally almost hairy, 40–80 mm *l*; **L** crowded near the tips of the branches, trigonous, moderately tuberculate, glaucous green to purplish, to 11 mm *l*, 2 mm *b* and *t*; **Fl** solitary, ∅ 20 mm, pedicels 7–9 mm *l*, bracteoles near their base, petals c. 50, 4–9 mm *l*, c. 1 mm *b*, filamentous staminodes very few, filaments pale, anthers and pollen yellow, nectary ring inconspicuously crenulate; **Fr** unknown; **Ecol** unknown; **Distr** Prince Albert, WC, S Africa.

Note: Bolus (ABH 1927: 80) stressed the possession of laxly disposed stamens, solitary flowers on short pedicels, and small flowers as typical features of this species.

L. polyanthon (Haw.) N.E.Br. GC 1930: 212 ≡ *M. polyanthon* Haw. 1812: 270 **T**?

Densely branched shrubs to 60 cm *h*, **I** reddish; **L** triquetrous, scabrous, glaucous; **Fl** many, arranged imbricately in a panicle, petals pale reddish; **Fr** unknown; **Ecol** flowering readily in cultivation in late summer; **Distr** unknown.

Note: Haworth (1812: 270) noted that "this brilliant species is often taken for *M. glomeratum*".

L. praecipitatus (L.Bolus) L.Bolus NM 1939: 169 ≡ *M. praecipitatum* L.Bolus NM 1932: 330 **T** G.Britten 25 (BOL).

Robust shrubs hanging down with branches to 40 cm *l*, **I** deep brown, to 6 mm ∅; **L** semi-terete, pale glaucous or pale green, very smooth, translucent, to 17 mm *l*, to 2 mm *b* and *t*; **Fl** in small cymes, pedicels terete, 10–25 mm *l*, bracteoles at their middle, petals white, to 23 mm *l*, to 2 mm *b*, filamentous staminodes white, merging into the petals, filaments white, anthers straw-coloured, nectary ring crenulate; **Fr** 5 locules; **Ecol** unknown; **Distr** Albany, EC, S Africa.

Note: This species had earlier been confused with **L. dependens** from Montagu (Bolus NM 1932: 330).

L. prasinus ≡ **Oscularia prasina**
L. primivernus ≡ **Oscularia primiverna**

L. productus (Haw.) N.E.Br. GC 1930: 212 (see **photo**) ≡ *M. productum* Haw. 1824: 425 **T** drawing 993/ 435 by Duncanson 4.10.1823 based on material raised from seed collected by Bowie (K!) = *M. lepidum* Haw. 1826: 130 ≡ *M. productum* Haw. var. *lepidum* (Haw.) Salm-Dyck 1837: § 40 f 4 ≡ *L. productus* (Haw.) N.E.Br. var. *lepidus* (Haw.) Schwantes in Jacobsen FR 1938: 100 **Lectotype**, here designated: drawing 225/24 by Duncanson of 28.7.1825 of a plant grown from seed collected by Bowie 687, 1824 (K!) = *M. productum* Haw. var. *purpureum* L.Bolus NM 1929: 43 ≡ *L. productus* (Haw.) N.E.Br. var. *purpureus* (L.Bolus) L.Bolus ex H.Jacobsen FR 1938: 100 **Syntypes** Dyer 1323, Fourcade 2554 (BOL).

Erect shrubs to 30 cm *h*, short shoots with 2–3 leaf-pairs developed during the flowering phase already, **I** smooth, maroon to dull brown; **L** recurved, sub-incurved, trigonous, glaucous, tips reddish, < 27 mm *l*; **Fl** in terminal small cymes, bracteoles at the middle of the pedicels, half the size of the leaves, ∅ 29–31 mm, **K** lobes very long, petals at most as long as the **K** lobes, very pale pink to white, numerous filamentous staminodes spreading during anthesis, filaments apically straw-coloured; **Fr** turbinate, 5 locules; **Ecol** in loamy to stony places in plains; **Distr** Albany, Kirkwood, Uitenhage, EC, S Africa.

Note: The drawing of the type plant of the species bears an inscription "Bowie … collected them on the stony plains of Uitenhage". Beside by the formation of densely leaved proleptic side branches, the species can also be recognized by its white flowers with many, later spreading filamentous staminodes and by the petals reaching at most the length of the very long calyx lobes. The drawing of the plant representing the variety shows a much elongated inflorescence, indicating shady cultivation.

L. productus var. *lepidus* = **L. productus**
L. productus var. *purpureus* = **L. productus**

L. profundus (L.Bolus) H.E.K.Hartmann Bradleya 1998: 70 ≡ *Ruschia profunda* L.Bolus NM 1936: 16 **T** Pillans 7711 (BOL!).

Shrubs to 36 cm *h*, **I** when young, compressed and winged, reddish, older ones round, 10–30 mm *l*, 4 mm ∅; **L** trigonous to subfalcate, keel serrulate, keel and margins cartilaginous, epidermis cells smooth, yellow-green, apically reddish-purple, 10–17 mm *l*, 3–4 mm *b*; **Fl** 1–3, ∅ 17 mm, pedicels 5–15 mm *l*, bracteoles at their middle, petals pink to palest pink, apically deeper in colour, anthers and pollen pale yellow, stigmas purple; **Fr** funnel-shaped, but sides convex, top roundish to semi-globose, covering membranes with small closing rodlets, placenta ending in a round knob, valves with very narrow wings in the available young fruit, expanding keels almost touching each other, 5 locules; **Ecol** unknown; **Distr** Piquetberg, WC, S Africa.

Note: The rounded base and top of the fruit and the absence of a closing body exclude the species from **Ruschia** and agree with features found in **Lampranthus**, but the narrow valve wings are a rare feature in the latter genus.

L. prominulus (L.Bolus) L.Bolus NM 1939: 169 ≡ *M. prominulum* L.Bolus NM 1933: 408 **T** without collector 20825 (BOL).

Erect shrubs to 35 cm *h*, **I** 5–20 mm *l*, 2–3 mm ∅; **L** trigonous, apically subrecurved, dirty glaucous green, surface with raised dots, 10–22 mm *l*, 1.5 mm *b*, to 2 mm *t*; **Fl** in small cymes, pedicels reddish, 20–40 mm *l*, bracteoles at their middle, **K** lobes brown to black, petals pink or purplish-pink, 4–13 mm *l*, 0.25–1.5 mm *b*, filamentous staminodes purple, apically translucent, anthers and pollen straw-coloured, nectary ring crenulate; **Fr** 5 locules; **Ecol** unknown; **Distr** George, WC, S Africa.

Note: Brown to black calyx lobes are a rare feature in the genus.

L. promontorii (L.Bolus) N.E.Br. GC 1930: 212 ≡ *M. promontorii* L.Bolus ABH 1925: 10 **T** Arbuthnot 17486 (BOL).

Erect shrubs to 15 cm *h*, rarely to 30 cm *h*, **I** to 2 mm ∅; **L** triquetrous, falcate, apiculate, smooth, 10–23 mm *l*, *t* 2 mm *b*, 3–5 mm *t*; **Fl** ternate, 14–23 mm ∅, pedicels 5–15 mm *l*, ebracteolate, petals bright yellow, apically reddish suffused, 2–11 mm *l*, to 1 mm *b*, filaments without papillae, anthers and pollen yellow, nectary ring inconspicuous, denticulate; **Fr** unknown; **Ecol** on slopes of hills; **Distr** Simonstown (?), WC, S Africa.

Note: Absence of filamentous staminodes and of papillae on the filaments characterize this species.

L. proximus L.Bolus 1966: 127 **T** Nel 27738 (BOL).

Shrubs like those of **L. prominulus**; **L** less prominently dotted; **Fl** petals to 10 mm *l*, to 2 mm *b*, filamentous staminodes fewer than in **L. prominulus**, anthers and pollen at first yellow; **Fr** base obconical, top raised for 2 mm, 5 mm *l*, ∅ to 6 mm, expanded to 11 mm ∅; **Ecol** flowering in early spring; **Distr** Paarl, WC, S Africa.

Note: Reported to be very similar to **L. prominulum**.

L. punctulatus ≡ **Ruschia punctulata**

L. purpureus L.Bolus NM 1939: 157 **T** Pillans 8670 (BOL).

Erect shrubs to 40 cm *h*, higher in shady places, **I** 20–40 mm *l*, 1–3 mm ∅; **L** trigonous, finely scabrous, glaucous green, 15–35 mm *l*, 1–1.5 mm *b*, to 2 mm *t*; **Fl** in small cymes, pedicels 35–45 mm *l*, bracteoles a little above the middle, petals pinkish-purple, 3–14 mm *l*, 0.25–1.5 mm *b*, filamentous staminodes none, filaments purple, anthers and pollen yellow, nectary ring crenulate; **Fr** pentangular, top raised for 2 mm, 9 mm *l*, 8 mm ∅; **Ecol** on hills; **Distr** Clanwilliam, WC, S Africa.

Note: Lack of filamentous staminodes characterizes this species.

L. rabiesbergensis (L.Bolus) L.Bolus NM 1939: 170 ≡ *M. rabiesbergense* L.Bolus NM 1936: 13 **T** Lewis 21642 (BOL).

Diffusely branched, **I** 20–40 mm *l*, ∅ to 4 mm; **L** almost terete, acute, glaucous green, scabrous, 15–20 mm *l*, 1–1.5 mm *t*; **Fl** solitary, pedicels to 60 mm *l*, bracteoles at their middle, **K** base scabrous, petals 3 to 15 mm *l*, to 2 mm *b*, purplish-pink, filamentous staminodes and filaments apically pink, anthers and pollen yellow, nectary ring densely and deeply divided; **Fr** 5 locules; **Ecol** on steep slopes; **Distr** Worcester, WC, S Africa.

Note: In contrast to **L. occultans**, with which the species is compared, the filamentous staminodes form a cone appressed to the stamens.

L. recurvus (L.Bolus) Schwantes in Jacobsen 1949: 58 ≡ *M. recurvum* L.Bolus NM 1932: 367 nomen illeg. non Moench 1802: 190 nec Haw. 1819: 90 ≡ *M. bolusiae* Schwantes in Jacobsen FR 1938: 228 **T** L.Bolus 20166 (BOL).

Diffusely branched with decumbent branches to 25 cm *l*, **I** 10–15 mm *l*, ∅ to 3 mm; **L** subfalcate, glaucous green, 8–11 mm *l*, 1.5 mm *b*, to 2.5 mm *t*; **Fl** in small cymes, pedicels 20–40 mm *l*, bracteoles at their middle, petals pink, 7–9 mm *l*, to 1.5 mm *b*, filamentous staminodes few, filaments white, apically purple, anthers whitish, pollen yellow to yellow-brown, nectary ring crenulate; **Fr** 5 locules; **Ecol** in damp places; **Distr** Tulbagh, WC, S Africa.

Note: The name refers to the recurved calyx lobes.

L. reptans (Aiton) N.E.Br. GC 1930: 212 ≡ *M. reptans* Aiton 1789: 185 **T** Masson (?).

Creeping; **L** triquetrous, acute, scabrous; **Fl** purple; **Fr** unknown; **Ecol** flowering in summer; **Distr** S Africa.

Note: The material seen by Aiton was introduced in 1774 to England by Masson.

L. roseus (Willdenow) Schwantes in Jacobsen FR 1938: 230 ≡ *M. roseum* Willdenow 1809: 535 ≡ *M. incurvum* Haw. var. *roseum* (Willdenow) DC. 1828: 433 **T** hort. Berlin (not found at B).

Shrubs; **L** triquetrous, obtuse, basally narrowed, glaucescent, with translucent dots; **Fl** solitary, petals flesh-coloured to palest pink; **Fr** unknown; **Ecol** unknown; **Distr** S Africa.

Note: The species is very insufficiently known.

L. ruber ≡ **Astridia rubra**

L. rubroluteus (L.Bolus) L.Bolus NM 1939: 170 ≡ *M. rubroluteum* L.Bolus NM 1936: 11 **T** Pillans 7749 (BOL).

Erect shrubs, 24–35 cm *l*, **I** 15–30 mm *l*, ∅ 4 mm; **L** subfalcate, 15–25, to 3 mm *b* and *t*; **Fl** solitary, pedicels 10–20 mm *l*, petals shining white, 7–19 mm *l*, 0.75–2.5 mm *b*, filamentous staminodes none, filaments deep purple, apically yellow, anthers and pollen golden, nectary ring crenulate; **Fr** base turbinate, conspicuously angulate, top elevated for 2–3 mm, 8–12 mm *l*, 12–14 mm ∅, expanded 16–18 mm ∅; **Ecol** among rocks, 1100 m altitude; **Distr** Piquetberg, WC, S Africa.

Note: Lack of filamentous staminodes characterizes this species.

L. rupestris (L.Bolus) N.E.Br. GC 1930: 212 ≡ *M. rupestre* L.Bolus SAG 1928: 15; L.Bolus NM 1928: 134 **T** Pillans 18780 (BOL).

Richly branched shrubs to 15 cm *h*, branches diffuse, often decumbent or creeping; **L** trigonous, acute, pale glaucous, with prominent dots, to 12 mm *l*, 2 mm *t*; **Fl** 1–3, ∅ 24 mm, pedicels filiform, bracteolate, petals white, lax, to 2 mm *b*, filaments glabrous, white, anthers straw-coloured, 5 connate glands; **Fr** unknown; **Ecol** among rocks; **Distr** Malmesbury, WC, S Africa.

Note: Small flowers on thin pedicels are typical of this species.

L. rustii (A.Berger) N.E.Br. GC 1930: 212 ≡ *M. rustii* A.Berger 1922: 636 **T** Rust 312 (B!).

Shrubs to 30 cm *h*, densely branched, **I** brown, 10–25 mm *l*; **L** triquetrous, uncinate, recurved, 14 mm *l*, to 1.5 mm *b*; **Fl** in rich cymes, pedicels 10–30 mm *l*, ∅ c. 20 mm, petals twice the length of the **K** lobes which possess brown margins; **Fr** unknown; **Ecol** unknown; **Distr** Riversdale, WC, S Africa.

Note: The brown margins of the calyx lobes are not very common in the genus.

L. salicola (L.Bolus) L.Bolus NM 1939: 170 ≡ *M. salicola* L.Bolus NM 1936: 12 **T** L.Bolus 21641 (BOL).

Erect shrubs, 20–30 cm *h*, **I** 10–20 mm *l*, 2–3 mm ∅; **L** semiterete, acute, smooth, glaucous green, to 15 mm *l*, to 2 mm *b*, to 3 mm *t*, in shady places to 25 mm *l* and a little thicker; **Fl** in small cymes, pedicels with bracteoles at their middle, 20–25 mm *l*, 60–80 mm *l* in shade, petals purplish-pink, to 20 mm *l*, to 2.5 mm *b*, filamentous staminodes none, filaments white, anthers and pollen yellow, nectary ring crenulate, deeply and obtusely divided; **Fr** 5 locules; **Ecol** in salty plains; **Distr** Piquetberg, WC, S Africa.

Note: Lack of filamentous staminodes and pink petals characterize this species.

L. salteri (L.Bolus) L.Bolus NM 1939: 169 ≡ *M. salteri* NM 1934: 428 **T** Salter 4099 (BOL).

Diffusely branched shrubs to 15 cm *h*, decumbent to spreading branches to 20 cm *l*, **I** 10–20 mm *l*, to 4 mm ∅; **L** trigonous, acute, green, sparsely dotted, 15–30 mm *l*, 1.5–3 mm *b*, 2–3.5 mm *t*; **Fl** solitary, pedicels 5–10 mm *l*, petals reddish-golden on the inner, purple on the outer side, 5–14 mm *l*, 0–5–1.75 mm *b*, filamentous staminodes none, filaments epapillate, golden, anthers and pollen golden, nectary ring almost divided into 5 glands; **Fr** unknown; **Ecol** unknown; **Distr** Bredasdorp, WC, S Africa.

Note: In this species, lack of filamentous staminodes is combined with golden filaments.

L. saturatus (L.Bolus) N.E.Br. GC 1930: 212 ≡ *M. saturatum* L.Bolus ABH 1923: 169 **T** van Zijl s.n. (BOL).

Shrubs to 25 cm *h*, ascending branches basally dense, laxer above, **I** 15–45 mm *l* (?, unclear); **L** almost terete, very smooth, somewhat yellowish, glaucous green, to 40 mm *l*, 4–5 mm *b*, to 6 mm *t*; **Fl** solitary, ∅ to 56 mm, pedicels 56 mm *l*, bracteolate at their middle, petals pinkish-purple, basally and below paler, 26 mm *l*, to 4 mm *b*, inner ones vittate, filaments deep red, anthers and pollen yellow, nectary ring inconspicuous; **Fr** 5 locules; **Ecol** unknown; **Distr** Vanrhynsdorp (?), WC, S Africa.

Note: The name refers to the deep (saturate) red filaments.

L. sauerae (L.Bolus) L.Bolus NM 1939: 169 ≡ *M. sauerae* L.Bolus NM 1933: 409 **T** Sauer 9114 (BOL).

Erect shrubs to 15 cm *h*, richly branched, **I** 5–15 mm *l*, 1–1.5 mm ∅; **L** subfalcate, acute, dots raised, dirty glaucous green, 5–10 mm *l*, 1 mm *b*, 1.5 mm *t*; **Fl** solitary, pedicels 20–30 mm *l*, petals ruby red, basally purple, copper red outside, 8–21 mm *l*, 0.25–1.25 mm *b*, filamentous staminodes few, basally white, apically purple, filaments white, anthers and pollen lemon-coloured, nectary ring flat, inconspicuously undulate; **Fr** 5 locules; **Ecol** unknown; **Distr** Malmesbury, WC, S Africa.

Note: Petals are very numerous in this species.

L. scaber (L.) N.E.Br. GC 1930: 212 (see **photo**); Liede & Meve 1990: 39–44 ≡ *M. scabrum* L. 1753: 483 **LT** Dillenius 1732: t. 197 f 251.

Subshrubs or geophytes with many erect dark brown branches from the base reaching up to 30 cm *h*, tuberous rootstock 10–15 cm *l*; **L** almost free, their tips bent back, deciduous, bright green during the season, epidermis with prominent white bladder cells, stomata not sunken; **Fl** solitary, **K** base scabrid, ∅ 28–30 mm, 40–45 lilac petals, 75–90 filamentous stamimodes with purplish tips forming an erect ring around the c. 120 stamens in the centre, pollen semi-tectate to reticulate; **Fr** pentagonal from above, base long, funnel-shaped, top raised moderately, ∅ 6–8 mm, 8–10 mm *h*, distal exit of locules blocked by sterile small seeds; **S** dark brown, D-shaped, 1.2–1.5 mm *l*, 0.6–0.8 mm *b*, epicuticular particles, no rodlets; **Chr** $2n = 18$; **Ecol** on granite hills in disturbed Rhenosterbosveld among grass, > 400 mm rainfall p.a. in winter; **Distr** Malmesbury, WC, S Africa.

Note: Leaves are regularly lost in this species, but branches can also disappear during the dry season, either partly (subshrub) or completely (geophyte). This is a very rare and strict adaptation to the seasonal rainfall in the area, it is also rare in the whole family. **L. scaber** can easily be recognized in habitat, but the application of the name had been confused since Linnaeus restricted it from the wider description in 1753 to a more narrow one in 1762 (Liede & Meve 1990).

L. schlechteri (Zahlbruckner) L.Bolus NM 1950: 227 ≡ *M. schlechteri* Zahlbruckner 1900: 35 non (Schwantes) N.E.Br. GC 1930: 32 nomen illeg. ≡ *M. sabulosum* Schlechter 1899: 130 nomen illeg. non Thunberg 1791: 17 ≡ *M. perspicuum* A.Berger 1922: 634 ≡ *L. perspicuus* (A.Berger) N.E.Br. GC 1930: 212 **T** Schlechter 9214 (B).

Shrubs with spreading branches, 15–20 cm *h*, **I** 2-edged, later terete, brownish-black, 20–40 mm *l*; **L** trigonous, acute, mucronate, erect to incurved, scabrous, punctate, 10–25 mm *l*, 1–2 mm *b*, to 4 mm *t*; **Fl** solitary or ternate, ∅ 40–50 mm, pedicels bracteolate, **K** base verruculose, petals c. twice the length of the **K** lobes, 35 mm *l*, 2–3 mm *b*, salmon-coloured, anthers yellow; **Fr** 5 locules; **Ecol** in sandy places, 160 m altitude; **Distr** Somerset West, WC, S Africa.

Note: On young shoots, the leaf bases of a pair are separated by a distinct triangle.

L. serpens (L.Bolus) L.Bolus NM 1939: 169 ≡ *M. serpens* L.Bolus NM 1930: 221 **T** Edwards 19247 (BOL).

Creeping with branches to 25 cm *l*, **I** subcompressed, 20–30 mm *l*, 5 mm ∅; **L** triquetrous, turned back, keel inconspicuously serrulate, smooth, green, apically purplish apiculate, 23 mm *l*, 5 mm *b*, to 6 mm *t*; **Fl** solitary, 45 mm ∅, pedicels 25–35 mm *l*, bracteoles basal, petals golden, towards the tips with yellow-golden margins, to 20 mm *l*, to 3 mm *b*, filamentous staminodes none, filaments epapillate, golden, anthers and pollen golden, 5 crenulate glands; **Fr** 5 locules; **Ecol** unknown; **Distr** Caledon, WC, S Africa.

Note: The five separate glands present a deviation from the typical inconspicuous, crenulate ring in **Lampranthus**.

L. simulans L.Bolus NM 1954: 273 **T** Esterhuysen 18960 (BOL).

Simulating **Erepsia** in growth form, erect, 14–16 cm *h*, herbaceous parts reddish punctate, **I** 15–30 mm *l*, 1–2 mm ∅; **L** 10–17 mm *l*, 4–6 mm *t*; **Fl** in small cymes, 20–32 mm ∅, pedicels 7–10 mm *l*, placed above the middle, petals silvery white, apically pink, 5–15 mm *l*, to 2 mm *b*, inner ones sparsely papillate, simulating filamentous staminodes, true filamentous staminodes none, stamens not incurved, filaments white, anthers yellow, nectary ring distantly crenulate; **Fr** base globose-obconical, subspongy, hard above, top elevated for 3 mm, ∅ 7 mm, expanded to 15 mm ∅, 7 locules; **Ecol** in sand on flats near the sea; **Distr** Bredasdorp, WC, S Africa.

Note: The high number of locules and a spongy fruit base cast some doubt on the placement of this species in **Lampranthus**.

L. sociorum (L.Bolus) N.E.Br. GC 1930: 212 ≡ *M. sociorum* L.Bolus ABH 1922: 132 **Syntypes** Peacock & L.Bolus 1666/17, Peacock & L.Bolus 1922/17, La Grange 4584/14 (BOL).

Creeping branches to 20 cm *l*, **I** 20–30 mm *l*, ∅ 2 mm, flowering branches erect, 2–4 cm *h*; **L** subfalcate, trigonous, glaucous green, 10–25 mm *l*, to 3 mm *b* and *t*; **Fl** solitary, 30–35 mm ∅, pedicels 6–20 mm *l*, petals pink, purplish towards the base, to 18 mm *l*, 2.5 mm *b*, filamentous staminodes apically pinkish-purple, filaments pale, apically yellow, anthers golden, nectary ring unequally and coarsely crenulate; **Fr** unknown; **Ecol** unknown; **Distr** Malmesbury, Riversdale, Tulbagh, WC, S Africa.

Note: Remarkable for this species is the crenulate nectary ring.

L. sparsiflorus L.Bolus NM 1958: 332 **T** Esterhuysen 24218 (BOL).

Shrubs to 38 cm *h*, basal **I** 1.5–6 mm ∅, on branches 5–35 mm *l*, 0.5–1.5 mm ∅; **L** trigonous, glaucous green, 13–20 mm *l*, 2.5 mm *b*, to 4 mm *t*; **Fl** solitary to rarely 2, pedicels 15–25 mm *l*, bracteoles mostly at their middle, petals purplish-pink, 11–15 mm *l*, 1.25 mm *b*, inner 3 series 6–9 mm *l*, 0.25–0.5 mm *b*, few inner ones paler and simulating filamentous staminodes, stamens collected and almost hidden as if placed in a pocket formed by the bases of the petals, nectary ring crenulate; **Fr** base obconical, pentangular, top raised for 1–1.5 mm, 6–8 mm *l*, 7 mm ∅, expanded 12 mm ∅, 5 locules; **Ecol** unknown; **Distr** Knysna, WC, S Africa.

Note: Note the incurved, hidden stamens resembling **Erepsia** in this respect; the species is reported to be similar to **L. hollandii.**

L. spectabilis (Haw.) N.E.Br. GC 1930: 212 ≡ *M. spectabile* Haw. 1795: 385 **Lectotype,** here designated: Curtis 1798: t. 396.

Shrubs with ascending branches > 30 cm *l*, **I** purple and angular when young, later brown and terete; **L** trigonous, glaucous, punctate, sheath forming a swelling at first, later the bases of the leaves leaving a triangle of stem tissue free, keel somewhat cartilagenous, 54–135 mm *l*, 7 mm *b*; **Fl** solitary, bright purple; **Fr** unknown; **Ecol** unknown; **Distr** S Africa.

Note: Haworth based his description on material "given by the Foreman of Mr. Curtis' Botanical Garden", hence the lectotypification. Curtis (1798: t. 396) stressed the ease with which plants of this species grow from cuttings and the production of the big, showy, purple flowers throughout the summer. He supposed that material of the species grown long since in Britain may have been collected by Masson.

L. spiniformis (Haw.) N.E.Br. GC 1930: 212 ≡ *M. spiniforme* Haw. 1795: 240 **T?**

Shrubs with erect, woody branches to 15 cm *h*; **L** terete, apically recurved and attenuate, hence subspiniform; **Fl** terminal, pedicels with bracteoles, petals short, purple, basally paler; **Fr** unknown; **Ecol** unknown; **Distr** S Africa.

Note: Additional remarks by Haworth (1803: 87–88) include more detailed information on growth, leaves and flowers.

L. staminodiosus (L.Bolus) Schwantes in Jacobsen 1949: 58 ≡ *M. staminodiosum* L.Bolus NM 1937: 76 nomen illeg. non N.E.Br. GC 1930: 32 **T** Lewis 21839 (BOL).

Erect shrubs to 13 cm *h*; **L** ± terete, obscurely scabrous, glaucous, apically purplish, to 8 mm *l*, to 1.5 mm *b* and *t*; **Fl** solitary, pedicels 10–18 mm *l*, petals reddish to blackish-purple, inner appressed to the filamentous staminodes, filaments with purple papillae, anthers and pollen bright yellow, nectary ring almost flat, with distant, obtuse lobes; **Fr** 5 locules; **Ecol** unknown; **Distr** Clanwilliam.

Note: The name refers to the very numerous filamentous staminodes.

L. stanfordiae L.Bolus NM 1939: 158 **T** Stanford 21851 (BOL).

Erect, 20–25 cm *h*, **I** 15–30 mm *l*, 1–2 mm ∅; **L** trigonous, obscurely scabrous, to 17 mm *l*, 1.5 mm *b*, 2.5 mm *t*; **Fl** in ternate cymes, pedicels to 30 mm *l*, bracteoles a little below their middle, **K** lobes broad and brown, petals golden, 10–17 mm *l*, 0.5–1.5 mm *b*, filamentous staminodes absent, anthers and pollen golden, nectary ring pentangular, inconspicuously crenulate; **Fr** unknown; **Ecol** unknown; **Distr** Malmesbury, WC, S Africa.

Note: The species can be characterized by the broad, brown margins of the calyx lobes.

L. stayneri (L.Bolus) N.E.Br. GC 1930: 212 ≡ *M. stayneri* L.Bolus ABH 1925: 6 **T** Stayner 81/21 (BOL).

Densely branched shrubs to 20 cm *h*, **I** 15–20 mm *l*; **L** triquetrous, basally semi-terete, glaucous green, apically reddish, smooth, 15–25 mm *l*, 3 mm *b* and *t*; **Fl** in ternate cymes or with aborted flowers, pedicels rarely to 20 mm *l*, bracteolate at their middle, petals pinkish-purple, 3–14 mm *l*, 1 mm *b*, filamentous staminodes pink, filaments pale, papillate, anthers and pollen deep yellow, nectary ring crenulate, **Fr** unknown; **Ecol** unknown; **Distr** Bedford, EC, S Africa.

Note: Reported to be similar to **L. productus.**

L. steenbergensis ≡ **Oscularia steenbergensis**

L. stenopetalus (L.Bolus) N.E.Br. GC 1930: 212 ≡ *M. stenopetalum* L.Bolus ABH 1925: 10 **T** Peers 17830 (BOL).

Erect shrubs, 16–23 cm *h*; **L** very acute, glaucous green, dots hardly prominent, 10–20 mm *l*, to 2 mm *b* and *t*; **Fl** in small cymes, pedicels 10–25 mm *l*, bracteolate, petals purple to diluted purple, or pink to diluted pink, margins recurving soon giving the flower a tassel-like appearance, 4–10 mm *l*, filamentous staminodes numerous, white, filaments white, anthers and pollen pale yellow, nectary ring crenulate, **Fr** unknown; **Ecol** unknown; **Distr** Clanwilliam, WC, S Africa.

Note: "Remarkable in the section [now genus] for the numerous petals and the variety of shades of colour exhibited by them. These tend to roll their margins back soon after the flower opens and so appear narrower than they really are" (Bolus ABH 1925: 10).

L. stenus (Haw.) N.E.Br. GC 1930: 212 ≡ *M. stenum* Haw. 1831: 421 **T?**

Diffusely branched shrubs, branches to 15 cm *l*, **I** basally whitish, upper parts reddish-brown to dull brown, tips purplish; **L** falcate, glaucous, with few dots, 7–20 mm *l*; **Fl** pedicels short, **K** subtuberculate, petals pale purple to reddish, filamentous staminodes white, anthers pale sulphur-coloured; **Fr** 5 locules; **Ecol** unknown; **Distr** S Africa.

Note: Bolus (NM 1958: 388) placed a sparsely flowering form here, suggesting also synonymy with **L. monticola.** Haworth (1831: 422) compared the species with *M. debile* and *M. reptans.*

L. stephanii (Schwantes) Schwantes in Jacobsen FR 1938: 230 ≡ *M. stephanii* Schwantes MDG 1927: 159 **T**?

Erect shrubs with **I** to 150 mm *l*; **L** semi-terete, 80–150 mm *l*, 3 mm *b*, 4 mm *t*; **Fl** terminal, pedicels long, petals white, 13 mm *l*, 1.5 mm *b*, filamentous staminodes white, inner ones papillate along their margins, nectary ring dark green, crenulate, inconspicous; **Fr** as for genus, 5 locules; **Ecol** unknown; **Distr** S Africa.

Note: The original description was based on cultivated material provided by Mr. Taylor; the species can best be recognized by long leaf-like protrusions of the calyx lobes (Schwantes MDG 1927: 159).

L. sternens L.Bolus NM 1939: 158 **T** Mathews 21852 (BOL).

Creeping branches forming a dense carpet on the ground, **I** 15–25 mm *l*, 1–1.5 mm ∅; **L** subfalcate, trigonous, deep green, with fine dots, 12–15 mm *l*, to 1.5 mm *b*, to 2 mm *t*; **Fl** solitary, pedicels 10–15 mm *l*, **K** lobes with broad, brown margins, petals golden yellow, 7–15 mm *l*, to 2 mm *b*, filaments apically golden, anthers and pollen yellow, nectary as if composed of 5 glands, hardly crenulate; **Fr** 5 locules; **Ecol** unknown; **Distr** Bredasdorp, WC, S Africa.

Note: Broad brown margins of the calyx lobes are typical of this species.

L. stipulaceus (L.) N.E.Br. GC 1930: 212 ≡ *M. stipulaceum* L. 1753: 484 ≡ *Erepsia stipulacea* (L.) Schwantes GF 1928: 68 **LT** Dillenius 1732: t. 209 f 267, 268.

Shrubs with firm, erect, robust branches, **I** 27–34 mm *l*, deep reddish-brown (date-coloured); **L** subterete, subfalcate, epidermis with low papillae, glaucous; **Fl** solitary, petals purple, inner smaller; **Fr** unknown; **Ecol** unknown; **Distr** S Africa.

Note: An insufficiently known species; the placement in **Lampranthus** is based on the positive exclusion of the species from **Erepsia** (Liede 1990: 473).

L. stoloniferus = **Hammeria meleagris**

L. suavissimus (L.Bolus) L.Bolus NM 1939: 169 ≡ *M. suavissimum* L.Bolus NM 1930: 220 nomen illeg. non Dinter 1923: 138 **T** Pillans 17929 (BOL) = *M. suavissimum* (L.Bolus) L.Bolus forma *ferum* L.Bolus NM 1930: 220 ≡ *L. suavissimus* (L.Bolus) L.Bolus forma *ferus* (L.Bolus) L.Bolus ex H.Jacobsen 1955: 1449 nomen invalid. **T** L.Bolus 19250 (BOL) = *M. suavissimum* L.Bolus var. *oculatum* L.Bolus NM 1930: 220 ≡ *L. suavissimus* (L.Bolus) L.Bolus var. *oculatus* (L.Bolus) L.Bolus ex H.Jacobsen 1955: 1449 nomen invalid. **T** Pillans 6060 (BOL).

Erect shrubs to 100 cm *h* in cultivation, **I** maroon, later black, to 6 mm ∅; **L** rather dense, obtuse, subfalcate, semi-terete, glaucous green, 25–35 mm *l*, to 4 mm *b* and *t*; **Fl** in cymes, ∅ 45–55 mm, pedicels bracteolate, petals most beautifully pink, to 25 mm *l*, to 2.5 mm *b*, filamentous staminodes none, filaments white, anthers and pollen yellow, nectary ring inconspicuous; **Fr** 5 locules; **Ecol** near the coast, in sand; **Distr** Namaqualand, NC, S Africa.

Note: The forma and the variety are included here since they seem to differ mainly in height (40–45 cm) from the typical form.

L. suavissimus forma *ferus* = **L. suavissimus**
L. suavissimus var. *oculatus* = **L. suavissimus**

L. subaequalis (L.Bolus) L.Bolus NM 1939: 170 ≡ *M. subaequale* L.Bolus NM 1937: 56 **T** Fourcade 1883/34 (BOL).

Erect shrubs to 28 cm *h*, **I** 10–20 mm *l*, 1–3 mm ∅; **L** trigonous, apiculate, 8–12 mm *l*, 1 mm *b*, to 1.5 mm *t*; **Fl** in cymes, pedicels bracteolate at their middle, 20–35 mm *l*, **K** lobes with brown margins, petals merging into filamentous staminodes, pale pink, 6–13 mm *l*, 0.5–1.5 mm *b*, filamentous staminodes almost covering the stamens, anthers and pollen lemon-coloured, nectary ring pentagonal, crenulate; **Fr** 5 locules; **Ecol** unknown; **Distr** Uniondale, WC, S Africa.

Note: The combination of numerous petals merging into filamentous staminodes and the brown margins of the calyx lobes appears distinctive for the species.

L. sublaxus L.Bolus NM 1939: 169 ≡ *M. laxum* L.Bolus NM 1933: 406 nomen illeg. non Willdenow 1809: 536 **T** L.Bolus 20165 (BOL).

Erect shrubs to 16 cm *h*, **I** 10–20 mm *l*, to 2 mm ∅; **L** trigonous, green, smooth, to 12 mm *l*, 1.5 mm *b*, to 2 mm *t*; **Fl** solitary or in small cymes, pedicels 25–30 mm *l*, 4 mm ∅, bracteoles at the middle, the inner 3 **K** lobes with brown margins, petals purplish-pink, to 10 mm *l*, to 1.5 mm *b*, filamentous staminodes bearing abortive anthers, deep pink, filaments pink, anthers and pollen straw-coloured to white, nectary ring crenulate; **Fr** 5 locules; **Ecol** unknown; **Distr** Calvinia (?), NC, S Africa.

Note: Brown margins of the inner calyx lobes are combined here with purplish-pink petals.

L. subrotundus L.Bolus NM 1939: 170 ≡ *M. subglobosum* L.Bolus NM 1937: 56 non Haw. 1803: 62, both nomina illegitima **T** collector unknown 1477a (BOL).

Erect shrubs to 20 cm *h*, **I** 10–40 mm *l*; **L** semi-terete, apiculate, scabrous, dirty glaucous green, 9–13 mm *l*, 1.25 mm *b*, 1.5 mm *t*; **Fl** in cymes arranged in corymbs, pedicels 15–35 mm *l*, bracteolate below the middle, **K** base semi-globose, petals pinkish-purple, 5–10 mm *l*, 0.25–1.75 mm *b*, filamentous staminodes none, filaments pale, anthers and pollen lemon-coloured, nectary of compressed, distant lobes obtusely tuberculate, elevated for 0.75 mm; **Fr** 5 locules; **Ecol** unknown; **Distr** S Africa.

Note: The nectary and the semi-globose base of the calyx are unusual in **Lampranthus**.

L. subtruncatus L.Bolus NM 1939: 159 T Pillans 8854 (BOL) = *L. subtruncatus* L.Bolus var. *wupperthalensis* L.Bolus 1961: 114 T Littlewood 380/59 (BOL).

Erect shrubs, 30–45 cm *h*, I grey, 10–25 mm *l*, 1–5 mm Ø; **L** trigonous to semi-terete, exapiculate, glaucous green, 15–20 mm *l*, to 1.5 mm *b* and *t*; **Fl** in small cymes, 3 inner **K** lobes with brown margins, petals pink, inner appressed to stamens, 4–12 mm *l*, 1–1.5 mm *b*, filamentous staminodes none, filaments white, inner incumbent, anthers and pollen lemon-coloured, nectary ring pentagonal, crenulate, **Fr** unknown; **Ecol** mountain slope; **Distr** Piquetberg, WC, S Africa.

Note: Lack of filamentous staminodes and brown margins of the inner calyx lobes appear to be typical of this species.

L. subtruncatus var. *wupperthalensis*
= **L. subtruncatus**
L. superans ≡ **Oscularia superans**

L. swartbergensis (L.Bolus) N.E.Br. GC 1930: 212 ≡ *M. swartbergense* L.Bolus NM 1928: 137 T Pocock 96 (BOL).

Erect shrubs to 30 cm *h*, I angular, 20–30 mm *l*; **L** trigonous, scabrous, glaucous green, 18 mm *l*, to 3 mm *b* and *t*; **Fl** solitary, Ø to 30 mm, pedicels 20–30 mm *l*, petals pinkish-purple, basally yellow, inner pale like filamentous staminodes, to 12 mm *l*, to 1.5 mm *b*, filaments white, anthers and pollen yellow, nectary as a ring; **Fr** 5 locules, Ø 12 mm; **Ecol** unknown; **Distr** Prince Albert, WC, S Africa.

Note: It remains uncertain whether filamentous staminodes proper can be distinguished in this species.

L. swartkopensis Strohschneider 1935: 292 T Brunnthaler s.n., 1909 (W).

Erect shrubs, 35–50 cm *h*, I green to purplish, later greyish-brown, **L** trigonous, apiculate, finely dotted, smooth, 20–40 mm *l*, 2–5 mm *t*; **Fl** 1–3, Ø c. 15 mm, **K** 4–5, petals reddish-lilac, filamentous staminodes basally pale, apically reddish, filaments whitish, anthers yellow; **Fr** base obconical, 5 locules; **Ecol** unknown; **Distr** "Swartkop", South Africa.

Note: The species is reported to be similar to **L. coccineus**, **L. spiniformis**, and **L. swartbergensis**.

L. tanquanus = **Hammeria meleagris**

L. tegens (F. Mueller) N.E.Br. GC 1930: 212 ≡ *M. tegens* F. Mueller 1866: 157 **Syntypes** Mueller 74972, 74973 (MEL, K).

Caespitose to sarmentose forming a carpet; **L** trigonous, smooth, glaucescent, 8–9 mm *l*, 1–3 mm *b*; **Fl** solitary, petals 20–30, pale pink, 15–20 mm Ø, 1/3 to 1/2 longer than the **K** lobes, filamentous staminodes 10–20, filaments 36–44, apically pink, anthers yellow; **Fr** 5 locules; **Ecol** on low lying, salty meadows; **Distr** NSW, Australia.

Note: Brown (in Ewart & White 1909: 544) discussed the identity of the species and found it "specifically the same" as a collection by Wooley Dod from Paarden Eiland in 1897. Brown added: "it is just possible that it has been introduced there from Australia". Venning in Prescott & Venning 1984: 33 suggested that **L. caespitosus** "is probably identical" with **L. tegens** and that South Africa is the place of origin of the taxon.

L. tenuifolius (L.) N.E.Br. GC 1930: 212 ≡ *M. tenuifolium* L. 1753: 484 ≡ *Mesembryanthus tenuifolius* (L.) Rothmaler 1941: 413 nomen illeg. **LT** HSC 220 (BM).

Shrubs with ascending branches; **L** longer than the yellowish **I**, semi-terete, glabrous; **Fl** solitary, petals gradually shorter to the interior, deep red, filaments apically yellow; **Fr** 5 locules; **Ecol** unknown; **Distr** S Africa.

Note: Wijnands (in litt.) considered his note (1983: 149) as the lectotypification of the species, a view followed here. It appears worthwhile, however, to recall that a sheet 649.4 is kept at LINN (!), which may represent the type, and that Linnaeus referred to Dillenius (1732: 264–266, t. 201) as well. Since this species has been used to typify the genus, the lectotypification should be reconsidered with particular care and on a broad base.

L. tenuis L.Bolus 1969: 144 T Esterhuysen 32044 (BOL).

Shrubs to 26 cm *h*, erect or falling over; **L** acute, 6–11 mm *l*, to 2 mm *b*; **Fl** solitary, Ø 16–20 mm, pedicels 5–6 mm *l*, petals pale purplish-pink, to 9 mm *l*, to 1.5 mm *b*, filamentous staminodes finely papillate, deep red, anthers yellow, nectary ring crenulate; **Fr** 5 locules; **Ecol** "amongst low bush on small hillocks in marshy area"; **Distr** Cape Town, WC, S Africa.

Note: The species is reported to propagate easily from shoots falling to the ground and rooting.

L. thermarum ≡ **Oscularia thermarum**

L. tulbaghensis (A.Berger) N.E.Br. GC 1930: 212 non L.Bolus NM 1950: 227 ≡ *M. tulbaghense* A.Berger 1922: 637 non L.Bolus NM 1932: 364 nomen illeg. **Syntypes** Schlechter 7477, Mac Owan 3052 (B).

Shrubs to 30 cm *h* with many branches, I brown to greyish, 5–20 mm *l*; **L** falcate, triquetrous, acute, roughly punctate, 5–10 mm *l*, to 1.5 mm *b*; **Fl** ± ternate,

∅ 20 mm, pedicels bracteolate, petals twice the length of the **K** lobes; **Fr** unknown; **Ecol** in rocky places, 650–800 m altitude, flowering in summer; **Distr** Tulbagh, WC, S Africa.

Note: Bolus (NM 1950: 227) was under the impression that the earlier combination **L. tulbaghensis** by Brown (GC 1930: 212) had been based on her illegitimate name *M. tulbaghense*, whereas the latter text clearly stated Berger as the author of the basionym. Furthermore, in 1930 the illegitimate name of Bolus 1932 had not been published yet and was not available to Brown then.

L. turbinatus (Jacquin) N.E.Br. GC 1930: 212 ≡ *M. turbinatum* Jacquin 1804: t. 476 **T** Jacquin 1804: t. 476 (K!).

Shrubs with low, diffusely spreading woody branches, c. 30 cm *h*; **L** triquetrous, recurving or spreading widely, c. 7 mm *l*; **Fl** solitary, pedicels c. 18 mm *l*, with small bracteoles at their middle, the **K** at its top and below the 5 free lobes with a distinct ring-shaped impression, **K** dark brown, petals reddish, filaments white, anthers pale yellow; **Fr** unknown; **Ecol** unknown; **Distr** S Africa.

Note: The species is reported to flower in June in the garden at Vienna.

L. uncus (L.Bolus) Schwantes in Jacobsen 1949: 58 ≡ *M. uncum* L.Bolus NM 1928: 136 ≡ *Ruschia unca* (L.Bolus) L.Bolus NM 1950: 220 **Lectotype**, here designated: Compton 2275/26 II (chosen from three sheets, BOL!) = *L. uncus* (L.Bolus) Schwantes var. *gydouwensis* (L.Bolus) Schwantes 1949: 58 ≡ *M. uncum* L.Bolus var. *gydouwense* L.Bolus NM 1937: 55 **T** Lewis 21821 (BOL!).

Shrubs to 25 cm *h*, **I** ochre to grey; **L** subulate, the tip recurved markedly, glabrous, greyish-green, 10–25 mm *l*, c. 3 mm *b* and *t*; **Fl** solitary, bracteoles at ± the middle of the pedicel, ∅ c. 18 mm, petals pale pink, with pink margins, c. 8 mm *l*, c. 1 mm *b*, filamentous staminodes white, apically purplish-pink, pollen dirty yellow; **Fr** top with low rims, base funnel-shaped, covering membranes probably complete with closing ledges at the outer end, closing body absent, expanding keels very short in type material, probably malformed, 5 locules; **Ecol** ca. 200 mm rainfall p.a. in winter; **Distr** Calvinia, NC, S Africa.

Note: Although the fruit enclosed with the type sheet is abnormal, it is ripe and can be assessed as possessing the features typical of **Lampranthus** (absence of a closing body, presence of broad, rectangular valve wings). Bolus' (1950) identification of material with fruits of the Ruschia-type with this species is an error to be seen in connection with the inclusion of *M. puncutulatum* in **L. uncus** as *Ruschia unca*. Undoubtedly, the recurved leaves of both species are similar, but the surfaces differ clearly, as do the capsules available on the type material.

L. uncus var. *gydouwensis* = **L. uncus**

L. uniflorus (L.Bolus) L.Bolus NM 1939: 170 (see **photo**) ≡ *M. uniflorum* L.Bolus ABH 1922: 135 ≡ *Drosanthemum uniflorum* Friedrich ex H.Jacobsen 1974: 418 nomen invalid. **T** Rogers 17066 (BOL!).

Shrubs to 30 cm *h*, ∅ to 50 cm, branching below the flower, **I** ochre, later grey to whitish, 30–45 mm *l*, ∅ to 4 mm; **L** clavate, somewhat incurved, bright light green, often yellowish-green, epidermis smooth, 20–28 mm *l*, 3–4 mm *t*; **Fl** ebracteate, pedicels 7–33 mm *l*, petals pinkish-purple, 7–11 mm *l*, to 2 mm *b*, filamentous staminodes present, nectary as a crenulate ring; **Fr** base long, funnel-shaped, top ± flat, covering membranes complete, thin, laterally not quite complete, placenta raised by a high false septum to c. half the depth of the locule, expanding keels almost parallel, but distant from each other, valve wings very broad, nearly rectangular, 5 locules; **S** brown, smooth, shiny; **Ecol** in sandy or coarser ground, but mostly in flats amongst other shrubs; **Distr** Calvinia, Namaqualand, NC, Beaufort West, Ceres, WC, S Africa.

Note: This species, together with **L. otzenianum**, differs from others in the genus by smooth leaf surfaces and the ebracteate, grey, rather soft fruit. The name **L. uniflorum** has been used for populations growing in the southern range, the name **L. otzenianum** for those further north. The high false septum is typical of the southern populations occurring as far north as 31° S, where sympatrical populations with both fruit types have been found. In this particular situation, leaves of **L. uniflorus** are shorter than those of **L. otzenianum**.

L. uniflorus var. *spathulatus* = **L. otzenianus**
L. utilis ≡ **Stoeberia utilis**

L. vallis-gratiae (Schlechter & A.Berger) N.E.Br. GC 1930: 212 ≡ *M. vallis-gratiae* Schlechter & A.Berger in Berger 1922: 634 **T** Schlechter 10303 (B!).

Shrubs, 15–20 cm *h*, **I** blackish-brown, later turning grey; **L** roughly punctate, glaucous, 4–6 mm *l*; **Fl** subternate, pedicels 10–20 mm *l*, bracteolate, petals longer than **K** lobes; **Fr** unknown; **Ecol** c. 700 m altitude; **Distr** Genadental, WC, S Africa.

Note: Scars are left at the nodes after the leaves have fallen off; furthermore, short shoots tend to die and break on older stems, the stem is, therefore, prickly.

L. vanheerdei L.Bolus 1964: 39 **T** van Heerde 27493 (BOL).

Shrubs with decumbent maroon branches to 35 cm *l*, **I** 15–30 mm *l*, ∅ 3 mm at base, 1–1.5 mm in flowering branches; **L** trigonous, glaucous green, 10–16 mm *l*, 1–1.25 mm *b*, to 2 mm *t*; **Fl** 3–7, pedicels 30–50 mm *l*, bracteolate above the middle, petals pinkish-purple, to 9 mm *l*, to 1.25 mm *b*, inner few appressed to the stamens, filaments pale pink, anthers and pollen whitish, nectary ring crenulate; **Fr** 5 locules, 5 mm *l*, Ø 6 mm, expanded 10 mm Ø; **Ecol** unknown; **Distr** Namaqualand, NC, S Africa.

Note: Well developed cymes are combined in this species with small capsules.

L. vanputtenii (L.Bolus) N.E.Br. GC 1930: 212 ≡ *M. pittenii* L.Bolus NM 1929: 98 ≡ *M. vanputtenii* L.Bolus 1967: 308 correction of name of collector **Syntypes** Compton 916/22, van Putten 1045/25 (BOL).

Erect shrubs, 20–30 cm *h*, **I** terete, grey, later blackish-brown, shining, 5–20 mm *l*, to 5 mm Ø; **L** trigonous, connate with the stem, smooth, glaucous green, 20–25 mm *l*, to 6 mm *b* and Ø, sheath 1.5 mm *l*; **Fl** in cymes, Ø to 70 mm, very beautiful, pedicels 10–20 mm *l*, bracteoles present, **K** 9–11 mm *l*, petals bright purplish-pink, basally white, 9–30 mm *l*, 0.5–3.5 mm *b*, filamentous staminodes none, filaments white, apically yellowish, nectary ring inconspicuous; **Fr** 5 locules; **Ecol** unknown; **Distr** Clanwilliam, WC, S Africa.

Note: Lack of filamentous staminodes characterizes this species.

L. vanzijliae (L.Bolus) N.E.Br. GC 1930: 212 ≡ *M. vanzijliae* L.Bolus ABH 1922: 126 T van Zijl 17141 (BOL).

Creeping with long branches, **I** 15–60 mm *l*; **L** triquetrous, apically incurved, green, inconspicuously punctate, 30–55 mm *l*, to 10 mm *b*; **Fl** solitary, Ø to 60 mm, pedicels 30–50 mm *l*, petals deep golden, 10–25 mm *l*, 1–3 mm *b*, filaments dark red, anthers yellow, nectary ring very inconspicuous; **Fr** globose turbinate, top convex, 21 mm *l* and Ø; **Ecol** unknown; **Distr** Worcester, WC, S Africa.

Note: "The deep orange-red of the petals has the glow of burnished copper, and the dark red filaments and yellow anthers enhance the brilliancy" (Bolus ABH 1922: 126).

L. variabilis (Haw.) N.E.Br. GC 1930: 212 ≡ *M. variabile* Haw. 1803: 85 T?

Shrubs with decumbent, diffuse branches, **I** reddish; **L** subtriquetrous, glaucous, scabrous; **Fl** in small cymes, pedicels terete, petals variable in colour: inside at first golden to safran-yellow, later yellow, soon paler or whitish dyed red, margins red, thereafter from whitish to flesh-coloured, red outside with deeper coloured margins, and finally absolutely red, but paler inside, except for the margins, filaments pale, basally red, anthers yellow; **Fr** turbinate, 5 locules; **Ecol** unknown; **Distr** unknown.

Note: The species is considered to be most similar to *M. violaceum* by Haworth (1803: 85).

L. verecundus (L.Bolus) N.E.Br. GC 1930: 212 ≡ *M. verecundum* NM 1928: 134 T Muir 3865 (BOL).

Laxly branched shrubs, **I** 10–20 mm *l*; **L** subfalcate, trigonous, 10–14 mm *l*, to 1.5 mm *b*, to 3 mm *t*; **Fl** solitary, Ø 24 mm, pedicels 10–20 mm *l*, **K** base almost verrucose, petals palest pink, to 10 mm *l*, to 2 mm *b*, filamentous staminodes white, nectary ring present; **Fr** 5 locules; **Ecol** unknown; **Distr** Riversdale, WC, S Africa.

Note: The reported nearly verrucose calyx base appears distinctive for this species.

L. vernalis (L.Bolus) L.Bolus NM 1939: 170 ≡ *M. vernale* L.Bolus ABH 1927: 93 T Arbuthnot 17305 (BOL).

Shrubs with lax branches 10–40 cm *l*, **I** 10–25 mm *l*, Ø to 4 mm; **L** semi-terete to trigonous, acute, subfalcate, finely glittering from papillae, bright green, very soft, 10–15 mm *l*, rarely to 20 mm *l*, 2.5 mm *t*; **Fl** in cymes, Ø to 44 mm, pedicels 25–50 mm *l*, bracteoles at their middle, petals bright pink, to 21 mm *l*, to 2,5 mm *b*, inner ones very narrow, filaments pink, anthers and pollen yellow, nectary ring inconspicuous; **Fr** 5 locules; **Ecol** unknown; **Distr** uncertain, WC, S Africa.

Note: The glittering leaves resemble those of **Drosanthemum**.

L. vernicolor ≡ **Oscularia vernicolor**
L. verruculatus ≡ **Scopelogena verruculata**

L. viatorum (L.Bolus) N.E.Br. GC 1930: 212 ≡ *M. viatorum* L.Bolus ABH 1922: 136 T Pearson 5575 (BOL).

Shrubs to 15 cm *h*, branches ascending or spreading, **I** maroon to light ochre-grey, 2 mm Ø; **L** triquetrous, aristate, recurved, 10–20 mm *l*, 4 mm Ø, epidermis with short papillae; **Fl** solitary, Ø 23 mm, pedicels 30–40 mm *l*, bracteolate at or above the middle, petals pink, 10 mm *l*, 1–1.5 mm *b*, nectary ring crenulate; **Fr** 5–6 locules; **Ecol** unknown; **Distr** Namaqualand, NC, S Africa.

Note: The papillate, recurved leaves suggest that the species may perhaps be better placed in **Ruschia** or **Antimima bina**, as suggested on the type sheet by O'Connor-Fenton.

L. villiersii (L.Bolus) L.Bolus NM 1939: 170 ≡ *M. villiersii* L.Bolus NM 1935: 489 T de Villiers 2547/32 (BOL).

Erect shrubs to 12 cm *h*; **L** semi-terete to subulate, scabrous, 15 mm *l*, 1.5 mm *b* and *t*; **Fl** in cymes, pedicels 15 mm *l*, bracteolate, petals purplish-pink, 5–12 mm *l*, 1–1.5 mm *b*, filamentous staminodes acuminate, filaments white, pollen lemon-coloured, nectary ring flat;

Fr top raised for 3 mm, 6 mm *l*, Ø 7 mm, 5 locules; **Ecol** unknown; **Distr** Caledon, WC, S Africa.

Note: Remarkable is the fact that the filamentous staminodes and the inner petals are described as being incumbent for a long time (diu incumbentia), resembling conditions found in the genus **Erepsia.**

L. violaceus (DC.) Schwantes in Jacobsen FR 1938: 230 ≡ *M. violaceum* DC. 1801: t. 84 **Lectotype,** here designated: de Candolle 1801: t. 84.

Shrubs up to 46 cm *h*, branches twisted, **I** blackish; **L** semi-cylindrical, acute, a little scabrous, glaucous; **Fl** solitary, pedicels very long, **K** lobes subpungent, petals purple, inner shorter, filaments white, anthers pale yellow; **Fr** 5 locules; **Ecol** unknown; **Distr** unknown.

Note: The print of De Candolle (1804: t. 84, K!) bears a note by Brown: "? = M. emarginatum, L.".

L. virgatus L.Bolus 1961: 262 **T** L.Bolus 27100 (BOL).

Shrubs to 45 cm *h*, branches erect, **I** shining, brown, 25–40 mm *l*, to 5 mm Ø at base; **L** triquetrous, green, inconspicuously scabrous, apiculate, 20–23 mm *l*, 1–1.5 mm *b* and *t*; **Fl** in cymes or solitary, Ø 34 mm, pedicels 25–55 mm *l*, petals few, bright pink, to 14 mm *l*, to 1.5 mm *b*, filamentous staminodes not very clear, nectary ring deeply and distantly crenulate; **Fr** 5 locules; **Ecol** unknown; **Distr** Clanwilliam, WC, S Africa.

Note: Remarkable is the deeply incised nectary ring.

L. vredenburgensis ≡ **Oscularia vredenburgensis**

L. walgateae L.Bolus NM 1939: 159 **T** Walgate 21853 (BOL).

Erect shrubs to 35 cm *h*, **I** mostly 15–25 mm *l*, 1–2 mm Ø; **L** semi-terete, dirty subglaucous green, 10–17 mm *l*, to 1.5 mm *b*, to 2.25 mm *t*; **Fl** mostly ternate, pedicels with bracteoles at their middle, petals purplish-pink, 6–15 mm *l*, 0.75–2.75 mm *b*, inner ones incurved over the stamens, filamentous staminodes none, filaments pale, basally with blackish-purple papillae, anthers and pollen yellow, nectary ring inconspicuous; **Fr** top elevated for 2 mm, 5 mm *l*, Ø to 5 mm, 5 locules; **Ecol** unknown; **Distr** Swellendam, WC, S Africa.

Note: Lack of filamentous staminodes and brown margins of the calyx lobes appear typical of this species.

L. watermeyeri (L.Bolus) N.E.Br. GC 1930: 212 ≡ *M. watermeyeri* L.Bolus ABH 1921: 72 **T** Watermeyer 2256/15 (BOL).

Erect shrubs to 25 cm *h*, **I** compressed when young, terete with age, Ø to 7 mm; **L** semi-terete to trigonous, incurved, 20–35 mm *l*, 5–6 mm *b* and *t*; **Fl** ternate, pedicels to 40 mm *l*, bracteoles at their middle, petals white, to 25 mm *l*, 1–2 mm *b*, filaments light yellow, anthers yellow, nectary ring finely crenulate; **Fr** unknown; **Ecol** unknown; **Distr** Calvinia, NC, S Africa.

Note: No filamentous staminodes are mentioned in the description.

L. woodburniae (L.Bolus) N.E.Br. GC 1930: 212 ≡ *M. woodburniae* L.Bolus NM 1928: 1 **T** Woodburn 595/25 (BOL).

Low plants with long, rooting branches, **I** often 30 mm *l*; **L** trigonous, smooth, glaucous, reddish apiculate, to 35 mm *l*, 5–6 mm *b*, 4–6 mm *t*; **Fl** solitary, Ø 40 mm, pedicels to 50 mm *l*, petals bright yellow, basally reddish-purple, to 18 mm *l*, 2–2.5 mm *b*, filamentous staminodes none, filaments glabrous, reddish-purple, anthers and pollen yellow, nectary ring hardly crenulate; **Fr** 5 locules; **Ecol** unknown; **Distr** Worcester, WC, S Africa.

Note: Reported to be very similar to **L. vanzijliae,** differing from the latter in smaller flowers with bright yellow, basally reddish-purple petals (Bolus NM 1928: 126).

L. wordsworthiae (L.Bolus) N.E.Br. GC 1930: 212 ≡ *M. wordsworthiae* L.Bolus ABH 1922: 126 **T** Wordsworth 1899/17 (BOL).

Branches prostrate to creeping, from a denser centre which is to 3 cm *h*, **I** 45 mm *l*; **L** semi-terete, acute, reddish apiculate, 15–28 mm *l*, 3–5 mm *b*; **Fl** solitary, Ø to 27 mm, pedicels very short, petals to 11 mm *l*, to 2 mm *b*, filamentous staminodes few, anthers yellow, nectary ring crenulate; **Fr** unknown; **Ecol** unknown; **Distr** Caledon, WC, S Africa.

Note: Reported to be "Allied to *M. curvifolium*, Haw." (Bolus ABH 1922: 126).

L. zeyheri (Salm-Dyck) N.E.Br. GC 1930: 212 ≡ *M. zeyheri* Salm-Dyck 1842: § 40 f 5 ≡ *M. haworthii* Ecklon & Zeyher 1837: 314 nomen illeg. non Haw. 1803: 65 **Lectotype,** here designated: Ecklon & Zeyher 2032 (S).

Shrubs with erect branches to 45 cm *h*, richly branched, **I** at first dull brown, later turning greyish-brown; **L** terete, mucronate, very bright green, very smooth, translucently punctate in light, 25–38 mm *l*, to 2.25 mm *b*; **Fl** solitary, Ø 37 mm, pedicels 50 mm *l*, bracteoles at their middle, petals in the brightest purple, filamentous staminodes pink, filaments white, anthers yellow; **Fr** 5 locules; **Ecol** flowering in late spring; **Distr** Uitenhage, EC, S Africa.

Note: The collections of Ecklon & Zeyher (1837: 308–329) of Mesembryanthema are arranged after the sections of Salm-Dyck; in this case the species is placed by all three authors in "§ Haworthiana". Ecklon & Zeyher also list *M. stipulaceum* and *M. laeve* in the section, both not taken up by Salm-Dyck.

L. zygophylloides ≡ **Drosanthemum zygophylloides**

Lapidaria RUSCHIOIDEAE

H. E. K. HARTMANN

Lapidaria (Dinter & Schwantes) Schwantes ex N.E.Br. GC 1928: 472 ≡ *Dinteranthus* Schwantes subg. *Lapidaria* Dinter & Schwantes MDG 1927: 223 **Typus** *Dinteranthus margaretae* (Schwantes) Schwantes ≡ **L. margaretae** (Schwantes) Dinter & Schwantes ex N.E.Br. **Etym** La *lapis, lapidis,* stone; plants resembling a group of stones.

Compact plants with 1(–3), rarely to several branches, 2–3 leaf-pairs to a growth per season, more in cultivation; **L** rounded-triangular, as thick as wide, keel and margins hard and prominent, greyish-white, often orange and purplish suffused or pink, epidermis cells elevated like cobblestones in a road, outer walls very thick and filled with crystals, the inner markedly bigger than the outer ones, smooth wax cover with groups of parallel, erect wax platelets on top of each cobblestone-shaped cell; **Fl** single, ebracteolate, pedicel laterally compressed, thus two-keeled, (6–)8 calyx lobes, petals golden; **Fr** Titanopsis-type, valve wings narrowing apically only a little, (6–)7(–9) locules; **S** small; **Chr** $2n = 18$; **Ecol** in quartz flats among loose stones, c. 250 mm rainfall p.a., mainly in March; **Distr** Warmbad, Namibia; Kenhardt, Namaqualand, NC, S Africa.

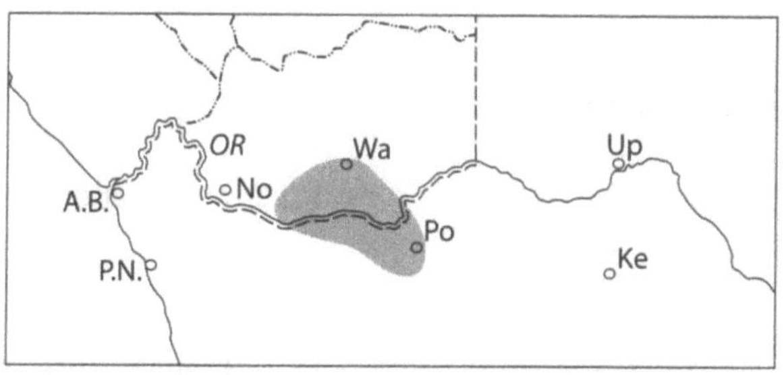

Distribution of **Lapidaria** (Dinter & Schwantes) Schwantes ex N.E.Br.

Note: The genus represents one of the elements characteristic of the Pofadder-centre; only one species. In leaf and fruit morphology, **Lapidaria** is very similar to **Didymaotus** from the southern Ceres Karroo. In wetter and darker cultivation, up to 12 leaf-pairs can persist as foliage leaves on the plant, i.e. they live longer than one season only, in contrast to the situation in habitat. In this way, the plant develops into a column, flowering terminally, the stem being continued by one of the originally lateral side branches.

L. margaretae (Schwantes) Dinter & Schwantes ex N.E.Br. GC 1928: 492 (see **photo**) ≡ *M. margaretae* Schwantes 1920: 55 ≡ *Argyroderma margaretae* (Schwantes) N.E.Br. GC 1926: 268 ≡ *Dinteranthus margaretae* (Schwantes) Schwantes ZSK 1926: 184 **LT** Dinter 5174 (B!) = *Argyroderma roseatum* N.E.Br. GC 1922: 105 **T** without collector (K!).

Compact; **L** thick, hard-edged; **Fl** with c. 100 golden petals, later white at base, 300–500 stamens, ± erect in the centre; **Fr** as for genus; **S** c. 0.6 mm *l*, c. 0.5 mm *b*; **Ecol** as for genus; **Distr** as for genus.

Note: Based on the hard, greyish leaves the species had been placed in the genus **Argyroderma** from the Knersvlakte, from which **Lapidaria** differs clearly in the erect ring of stamens. The erect wax platelets on top of the epidermis cells resemble those found in some species of **Lithops**.

Leipoldtia RUSCHIOIDEAE

H. E. K. HARTMANN

Leipoldtia L.Bolus FPSA 1927: t. 256; Hartmann & Rust 1995: 275–351; Hartmann & Stüber 1993: 1–75; Hartmann & Stüber 1995: 353–372 **Typus** *L. constricta* (L.Bolus) L.Bolus = **L. schultzei** (Schlechter & Diels) Friedrich **Etym** *C.L. Leipoldt*, 1880–1947, poet and plant collector in South Africa = *Rhopalocyclus* Schwantes GW 1928: 599 **Typus** *Rhopalocyclus nelii* Schwantes = **L. schultzei** (Schlechter & Diels) Friedrich.

Erect, ascending or saltatory shrubs with a well developed basal centre of regeneration, branches living for several years, replaced by young ones; **L** spreading, keeled, triangular in section, or sickle-shaped, epidermis smooth, convex, or papillate, subsidiary cells of stomata always more elevated than the normal cells; **Fl** in cymes or compound cymoids, often storeyed and thus enriched, **K** 5, bracteoles in terminal flowers absent or present, petals pink to purple with a central line, rarely white or yellow, staminodes 0–50; **Fr** of the Leipoldtia-type, with a flattish, funnel-shaped base, top rounded, rims on valves low, closing body big, valve wings almost rectangular, broad, covering membranes radially undulate, with closing rodlets at the distal recurving, c. 10 locules; **S** smooth, light brown, 0.55–1.15 mm *l*, 0.35–0.7 mm *b*; **Chr** $2n = 18, 36, 54$; **Ecol** mostly in flats or on gentle slopes, on gravelly soils to sandy loams; **Distr** Lüderitz-South, Namibia; E part of EC, W part of NC, inland parts of WC, S Africa.

Note: Although comprising only 10 species, the genus occurs in a very wide area forming an important element of the shrubby Karroo vegetation of the winter rainfall region.

Distribution of **Leipoldtia** L.Bolus

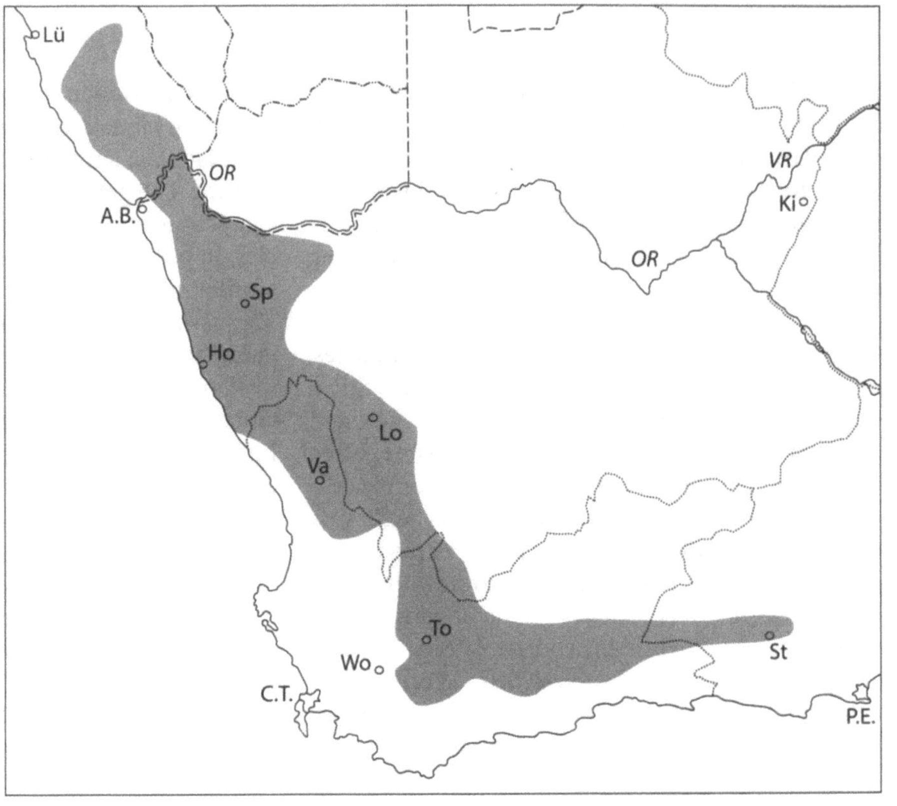

Key to the Subgenera of Leipoldtia

1. Inflorescences with lateral spines, fruits central .... L. subg. **Spinosae**
- Inflorescences without spines **2**
2. Leaves > 50 mm long **3**
- Leaves < 30 mm long .... L. subg. **Leipoldtia**
3. Flowers yellow, 40–50 mm ∅ .... L. subg. **Aureae**
- Flowers uniformly light purple, 35–40 mm ∅ .... L. subg. **Cephalophylloides**

L. subg. **Aureae** H.E.K.Hartmann BB 1983: 55 **Typus L. frutescens** (L.Bolus) H.E.K.Hartmann.

Erect to decumbent shrubs, **L** > **I**; **L** basally connate into a sheath, 8–10 times as long as broad, finger-shaped, remaining on the plant in the dry state; **Fl** 3–5 in a cyme, petals yellow, sometimes with a red line, filamentous staminodes absent, ∅ 35–50 mm; **Fr** brown, c. 12 locules (9–16); **S** big; **Chr** $2n = 36, 54$; **Ecol** on coastal plains or near-by mountains, < 100 mm rainfall p.a. in winter; **Distr** W Namaqualand, NC, S Africa.

Note: A single species is known, **L. frutescens.**

L. subg. **Cephalophylloides** H.E.K.Hartmann & Rust 1995: 348 **Typus L. rosea** L.Bolus.

Decumbent, rarely ascending shrubs, sometimes creeping, **L** > **I**; **L** c. 8–10 × as long as broad or thick, finger-shaped or rounded triangular, dry remains persisting on the stems, grey from a dense wax layer; **Fl** in forked, two-flowered reduced cymes, the long stalks connate for nearly half their lengths, petals light purple, filamentous staminodes absent, ∅ up to 40 mm; **Fr** 9–14 locules, grey; **Chr** $2n = 18$; **Ecol** in fine, shaly gravel with higher limestone contents, at the edge of the winter rainfall region, 100–200 mm p.a.; **Distr** Namaqualand, NC, Vanrhynsdorp, WC, S Africa.

Note: A single species is recognized, **L. rosea.**

L. subg. **Leipoldtia**

Erect, ascending, decumbent, or saltatory shrubs, **L** < **I**; **L** at most 5 × as long as thick or wide, basally hardly connate, old leaves falling off; **Fl** in simple or compound, often storeyed cymes with additional lateral dichasia, petals purple to white; **Fr** grey, c. 10 locules; **Chr** $2n = 18, 36$; **Ecol** and **Distr** as for genus.

Note: Six species are distinguished.

Key to the Species of Leipoldtia subg. Leipoldtia

1. Filamentous staminodes 0–5, spreading, fruit stalks erect, straight, internodes reddish 2
– Filamentous staminodes > 5, grouped together with the stamens into a cone, fruit stalks horizontal or bent, internodes ochre or yellowish, often with reddish dots 3
2. Petals purple, ∅ 6–21 mm, < 150 stamens .... **L. schultzei**
– Petals white, ∅ 24–36 mm, 150–190 stamens .... **L. calandra**
3. Petals deep purple, staminodes and filaments apically purple, growth low (20–25 cm), broadly funnel-shaped, > 10 long shoots per plant .... **L. uniflora**
– Petals pink to white, staminodes and filaments basally white or pink, growth mostly erect, < 10 long shoot per plant 4
4. Staminodes 20–50, base of capsule yellowish, funnel-shaped .... **L. laxa**
– Staminodes 2–20, base of capsule greyish, shallow 5
5. Leaves laterally strongly compressed, crescent-shaped, 2–5 long shoots to a plant, cymes elongating by few flowers per season, long shoots long-lived .... **L. lunata**
– Leaves triangular in section, straight, 5–10 long shoots per plant, cymes not enlarging but replaced by new branches from the base, long shoots living for some years only .... **L. weigangiana.**

L. subg. **Spinosae** H.E.K.Hartmann & Stüber 1995: 370 **Typus L. alborosea** (L. Bolus) H.E.K.Hartmann & Stüber.

Low flat shrubs 10–20 cm *h*, side branches ubiquitous, **I** whitish, shorter than leaves; **L** trigonous; **Inf** with central fruits and distal, brittle acute spines which loose their vestigial buds by breaking off; **Fl** light pink to white; **Fr** 6–9 locules; **S** elongate-slender; **Chr** $2n = 18$; **Ecol** winter rainfall < 100 mm p.a.; **Distr** Lüderitz-South, Namibia; W Namaqualand, NC, S Africa.

Note: The subgenus is restricted to the winter rainfall region, whereas the other spiniferous taxa, e.g. **Ruschia** subg. **Spinosae,** grow predominantly in the area of bimodal rainfall, some in the summer rainfall region. Two species are known.

Key to the Species of Leipoldtia subg. Spinosae

1. Leaves laterally compressed, therefore thicker than wide, often sickle-shaped, grey .... **L. alborosea**
– Leaves triangular in section, as thick as wide, slender, green .... **L. compacta**

L. alborosea (L.Bolus) H.E.K.Hartmann & Stüber 1993: 70 (see **photo**) subg. **Spinosae** ≡ *M. alboroseum* L.Bolus NM 1928: 148 **T** Pillans 5748 (BOL!) = *Eberlanzia vanheerdei* L.Bolus 1963: 14 **T** van Heerde 27275 (BOL!).

Flat open shrubs; **L** somewhat sickle-shaped, light grey and often with a pinkish tinge, 20–30 mm *l*, 5–6 mm *b*; **Inf** of 1, 3 or 7 flowers and 2, 4, 6, 8, or 12 spines; **Fr** 5.3–6.3 mm in ∅, 5–6.5 mm *h*; **S** 0.85–1 mm *l*, 0.45–0.57 mm *b*; **Ecol** on shaly slopes of diamictite, dolomite, or phyllite; **Distr** Lüderitz-South, Namibia; W Namaqualand, NC, S Africa.

Note: Although sparsely flowering, the species is attractive through leaves and spines throughout the year. In habitat, dominant in some areas.

L. amplexicaulis = **L. schultzei**
L. amplexicaulis forma *fera* = **L. schultzei**
L. aprica = **L. schultzei**
L. brevifolia = **L. schultzei**
L. britteniae = **L. schultzei**

L. calandra (L.Bolus) L.Bolus NM 1929: 11 subg. **Leipoldtia** ≡ *M. calandrum* L.Bolus ABH 1925: 7 **T** without collector BOL 17688 (BOL!).

Erect shrubs with > 10 long shoots per plant, **I** red; **L** trigonous, epidermis cells flat; **Fl** white, 52–56 spatulate petals, 130–210 stamens; **Fr** base funnel-shaped, greyish-brown, stalks straight and erect, terminal capsules with scars from bracteoles; **Chr** $2n = 36$; **Ecol** in gravelly flats, winter rainfall, 100–200 mm p.a.; **Distr** Namaqualand, NC, S Africa.

Note: The species is difficult to identify in the vegetative state, in particular since it occurs sympatrically with **L. schultzei,** a very widespread species.

L. ceresiana L.Bolus ex Graessner 1937: 17, nomen nudum.

Note: The name has been published without any further information and cannot be placed into synonymy with any of the existing species.

L. compacta L.Bolus ABH 1928: 129 subg. **Spinosae T** Pillans 5773 (BOL!) = *Ruschia gibbosa* L.Bolus NM 1931: 285 **T** M. Schlechter 8372 (BOL!).

Very flat cushion-shaped shrubs branching densely at their ends; **L** slender, c. 15 mm *l*, 3 mm ∅; **Inf** with a single fruit and two lateral spines, sometimes more highly branched; **Fr** 4.5–5 mm in ∅, 4 mm *h*;

S 0.7–0.8 mm *l*, c. 0.4 mm *b*; **Ecol** on alluvial, fine loamy soils; **Distr** W Namaqualand, NC, S Africa.

Note: Spines inconspicuous.

L. compressa = **Hallianthus planus**
L. compressa var. *lekkersingensis* = **Hallianthus planus**
L. constricta = **L. schultzei**
L. framesii = **L. laxa**

L. frutescens (L.Bolus) H.E.K.Hartmann BB1983: 55 (see **photo**) subg. **Aureae** ≡ *Cephalophyllum frutescens* L.Bolus NM 1954: 262 **T** Hall 595/52 (BOL!) = *Cephalophyllum frutescens* L.Bolus var. *decumbens* L.Bolus NM 1954: 263 **T** Hall 735/53 (BOL!).

Erect or decumbent shrubs; **L** connate for c. 10 mm, > 70 mm *l*, epidermal cells convex; **Inf** rich, **Fl** opening one after the other; **Fl** yellow, ∅ to 40 mm, stamens 230–450, spreading; **Fr** mostly 3, firm and persisting over several years, dark brown; **S** big, 0.9–1.15 mm *l*, 0.6–0.7 mm *b*; **Chr** $2n = 36, 54$; **Ecol** and **Distr** as for subgenus.

Note: Flowering only after a considerable size has been reached, branches mostly straggling in cultivation. Due to the subsequent anthesis of the flowers in the inflorescence, the shrub does not produce a showy aspect, but flowers over a long period of some months in Europe.

L. grandifolia = **L. weigangiana**
L. herrei = **L. schultzei**
L. jacobseniana = **L. schultzei**

L. klaverensis L.Bolus NM 1937: 53 (see **photo**) **T** Lewis 1891/32 (BOL!).

Erect shrubs with stiff branches to 20 cm *h*, **I** 15–30 mm *l*; **L** distant, erect, keeled, 10–13 mm *l*, 3–4 mm *b* and *t*; **Fl** solitary, pedicels 15–35 mm *l*, bracteoles at their middle, petals pink, inconspicuously vittate, 7–10 mm *l*, to 2 mm *b*, filamentous staminodes present or absent; **Fr** brown, top rounded, base funnel-shaped, covering membranes with tiny closing rodlets, valve wings extremely narrow or absent, 7–8 locules; **Ecol** in gravelly soil with some quartz; **Distr** from Namaqualand, NC, to Vanrhynsdorp, WC, S Africa.

Note: The species has been excluded from the genus **Leipoldtia** by Hartmann & Rust (1994: 349), based on the absence of rectangular valve wings and the harder brown fruits. This view is supported, but a new position for the species cannot be presented at present. The populations answering the critical features of **L. klaverensis** form two groups, a northern one with numerous filamentous staminodes and a southern one without filamentous staminodes in the flowers.

L. laxa L.Bolus NM 1928: 11 subg. **Leipoldtia T** Pillans 5776 (BOL!) = *L. framesii* L.Bolus NM 1929: 140 **T** Frames 18974 (BOL!).

Shrubs with mostly 5–10 erect long shoots, c. 35 cm *h*, **I** ochre, < 20 mm ∅; **L** triangular in section, epidermis cells a little convex; **Fl** with pink, lineal petals with a darker central line, more prominent towards the tip, ∅ 9–33 mm, 32–50 petals, 10–55 filamentous staminodes forming a central cone together with the stamens; **Fr** base funnel-shaped with straight or concave sides, stalks bent and horizontal, terminal capsules without scars; **S** 0.6–0.8 mm *l*, 0.35–0.5 mm *b*; **Chr** $2n = 18$; **Ecol** in gravelly flats, winter rainfall < 100 mm p.a.; **Distr** W Namaqualand, NC, S Africa.

Note: With its erect stems the species resembles the northern populations of **L. weigangiana.**

L. littlewoodii = **L. weigangiana** ssp. **littlewoodii**

L. lunata H.E.K.Hartmann & Rust 1995: 342 subg. **Leipoldtia T** Hartmann 8288 (HBG!).

Erect shrubs with only 2–5 long shoots, c. 35 cm *h*, **I** ochre, > 2 mm ∅; **L** crescent-shaped, thicker than wide, epidermal cells a little convex; **Fl** in reduced cymes, 3–4 additional ones forming per season, **Inf** therefore long-lived, petals pink, spatulate, central darker line only at base of petals, 0–15 white filamentous staminodes, forming a central cone together with the 110–230 white stamens; **Fr** with a shallow base, stalks bent and horizontal, mostly without scars on them; **S** 0.75–0.8 mm *l*, 0.45–0.5 mm *b*; **Chr** $2n = 18$; **Ecol** in crevices of granite, < 100 mm winter rainfall p.a.; **Distr** W Namaqualand, NC, S Africa.

Note: The species is very distinct in ecology, leaf shape, and growth form, remarkable is the annual elongation of the sparsely flowering cymes by a series of ≤ 10 sympodially developing flowers.

L. nelii = **L. schultzei**
L. pauciflora = **L. schultzei**
L. plana ≡ **Hallianthus planus**

L. rosea L.Bolus NM 1928: 13 (see **photo**) subg. **Cephalophylloides** ≡ *Cephalophyllum roseum* (L.Bolus) L.Bolus NM 1950: 222 **T** Frames 1178/26 (BOL!) = *Cephalophyllum calvinianum* L.Bolus 1961: 257 **T** Stayner 475/61 (BOL!).

Decumbent shrubs forming large mats with age, rooting with age; **L** whitish-grey, thin, < 70 mm *l*; **Fl** bright purple, stamens c. 290, spreading; **Fr** mostly with 10 locules, less persistent than those of **L. frutescens**; **Ecol** and **Distr** as for subgenus.

Note: One of the strikingly bright flowered species, but difficult to flower in cultivation.

L. schultzei (Schlechter & Diels) Friedrich 1960: 567 (see **photo**) subg. **Leipoldtia** ≡ *M. schultzei* Schlechter & Diels in Schultze 1907: 692 T Schultze 222[g] (B!) = *L. amplexicaulis* (L.Bolus) L.Bolus FPSA 1927: t. 256 ≡ *M. amplexicaule* L. Bolus ABH 1923: 164 T without collector 17045 (BOL!) = *L. amplexicaulis* (L.Bolus) L.Bolus forma *fera* L.Bolus NM 1954: 275 T Acocks 16942 (BOL) = *L. aprica* (A.Berger) L.Bolus NM 1950: 221 ≡ *M. apricum* A.Berger 1922: 637 T Schlechter 11355 (B!) = *L. brevifolia* L.Bolus NM 1929: 139 T L.Bolus 18975 (BOL!) = *L. britteniae* (L.Bolus) L.Bolus NM 1928: 11 ≡ *M. britteniae* L.Bolus ABH 1922: 132 T Britten 1802 (BOL!) = *L. constricta* (L.Bolus) L.Bolus FPSA 1927: t. 256 ≡ *M. constrictum* L.Bolus ABH 1925: 11 T without collector BOL 17677 (BOL!) = *L. herrei* (Schwantes) Schwantes in Jacobsen 1949: 58 ≡ *Rhopalocyclus herrei* Schwantes GW 1928: 599 T without collector SUG 1416 (HBG!) = *Rhopalocyclus nelii* Schwantes 1928: 599 ≡ *L. jacobseniana* Schwantes in Jacobsen 1949: 58 T without collector SUG 1455 (HBG!) = *L. nelii* L.Bolus NM 1937: 54 T Nel 9688 (BOL!) = *L. pauciflora* L.Bolus NM 1931: 302 T Fuller 56 (BOL!).

Decumbent, saltatory, or funnel-shaped shrubs, rarely with erect branches, often rooting, 3–many long shoots, 10–70 cm *h*, **I** red to maroon; **L** triangular in section, c. 15 mm *l*, epidermal cells convex; **Fl** in cymes, petals purple, somewhat spatulate, filamentous staminodes absent, 50–230 stamens spreading, filaments apically purple; **Fr** base funnel- to bell-shaped, stalks erect and straight, terminal capsules with scars from deciduous bracteoles; **S** 0.6–1 mm *l*, 0.35–0.6 mm *b*; **Chr** $2n = 18$ N, $2n = 36$ S populations; **Ecol** in flats with gravel, sand or loam, between rocks, rainfall mainly in winter or in March and November, 100–300 mm p.a.; **Distr** W part of EC, W parts of NC, inland parts of WC, S Africa.

Note: This species shows a particularly wide variation. The southern forms spread quickly on disturbed ground, e.g. on road banks, stabilizing the surface there rapidly.

L. uniflora L.Bolus JB 1928: 13 (see **photo**) subg. **Leipoldtia** T Pillans 5740 (BOL!).

Low, funnel-shaped shrubs with > 10 long shoots, 20–25 cm *h*, **I** light ochre; **L** triangular in section, 10–20 mm *l*, epidermis cells a little convex; **Fl** in cymes, ∅ 6–33 mm, 28–48 purple somewhat spatulate petals, 2–35 filamentous staminodes and 70–110 stamens forming a central cone in the centre, both apically purple; **Fr** with a shallow base, light grey, stalks bent and horizontal, terminal fruits without scars; **S** 0.55–0.75 mm *l*, 0.35–0.5 mm *b*; **Chr** $2n = 36$; **Ecol** in flats and on lower slopes, in gravelly soil with limestone, c. 100 mm rainfall p.a. in winter; **Distr** Namaqualand, NC, S Africa.

Note: Growing sympatrically with **L. laxa**, this species can easily be distinguished by its highly branched, low habit; on slopes it lives higher up than **L. laxa**.

L. weigangiana (Dinter) Dinter & Schwantes in Jacobsen 1937: 163 subg. **Leipoldtia** ≡ *M. weigangianum* Dinter 1923: 131 ≡ *Cephalophyllum weigangianum* (Dinter) Dinter & Schwantes 1928: 69 ≡ *Rhopalocyclus weigangianus* (Dinter) Dinter & Schwantes 1928: 599 ≡ *L. weigangiana* (Dinter) Dinter & Schwantes in Jacobsen 1937: 163 T Dinter 3638; other synonyms as for subspecies.

Shrubs with 5–10 mostly erect long shoots, to 35 cm *h*, **I** ochre or red under a white cover; **L** trigonous to laterally compressed, 10–15 mm *l*, epidermis cells somewhat convex; **Fl** in cymes, 6–33 mm ∅, 22–50 spatulate petals, 3–30 filamentous staminodes forming a central cone together with the 50–210 stamens, mostly white; **Fr** light grey, base a shallow bowl, stalks bent into a horizontal position, stalks of terminal fruits without scars; **Ecol** in flats or on mountains, always < 100 mm rainfall p.a. in winter; **Distr** Lüderitz-South, Namibia; Namaqualand, NC, S Africa.

Note: Three subspecies are distinguished.

Key to the Subspecies of L. weigangiana

1. Up to 5 long shoots per plant, rarely more .... **L. weigangiana** ssp. **grandifolia**
– Between 5 and 10 long shoots per plant **2**
2. Internodes red under a white cover of bark, ∅ of stems > 2 mm .... **L. weigangiana** ssp. **weigangiana**
– Internodes yellowish-grey, ∅ of stems < 2 mm .... **L. weigangiana** ssp. **littlewoodii**

L. weigangiana (Dinter) Dinter & Schwantes ssp. **grandifolia** (L.Bolus) H.E.K.Hartmann & Rust 1995: 346 (see **photo**) ≡ *L. grandifolia* L.Bolus NM 1938: 125 T Holloway 62 (BOL!).

Erect shrubs with up to 5 long shoots, **I** white; **L** laterally compressed; **Fl** pale pink, spatulate, 2–25 filamentous staminodes, 70–210 stamens; **S** 0.8–1 mm *l*, 0.45–0.6 mm *b*; **Chr** $2n = 36$; **Ecol** in flats or on slopes; **Distr** Lüderitz-South, Namibia; Namaqualand, NC, S Africa.

Note: With laterally compressed leaves, this subspecies resembles **L. lunata** to some extent and has probably been confused with it.

L. weigangiana (Dinter) Dinter & Schwantes ssp. **littlewoodii** (L.Bolus) H.E.K.Hartmann & Rust 1995: 347 (see **photo**) ≡ *L. littlewoodii* L.Bolus 1962: 301 T Littlewood 861/61 (BOL!).

Erect to flat funnel-shaped shrubs with 5–10 long shoots mostly, > 30 cm *h*, **I** yellowish-white; **L** triangular in section to laterally compressed; **Fl** purple, somewhat spatulate, 2–30 filamentous staminodes, 50–170 stamens; **S** 0.7–1 mm *l*, 0.44–0.6 mm *b*; **Chr** $2n = 36$; **Ecol** more often in flats than among rocks; **Distr** Lüderitz-South, Namibia; Namaqualand, NC, S Africa.

Note: This subspecies can in most places best be recognized by its oblique branches, growing more crowded than in the other subspecies.

L. weigangiana (Dinter) Dinter & Schwantes ssp. **weigangiana** (see **photo**).

Erect shrubs with 5–10 long shoots, to 35 cm *h*, **I** red under a white bark; **L** triangular in section to somewhat laterally compressed; **Fl** in rich cymes, petals pink, spatulate, 2–30 filamentous staminodes, 50–170 stamens; **S** 0.65–0.87 mm *l*, 0.35–0.55 mm *b*; **Chr** $2n = 18$; **Ecol** mainly in rocky places; **Distr** Lüderitz-South, Namibia.

Note: In growth form similar to **L. laxa.**

Lithops Ruschioideae

D. T. Cole & N. A. Cole[3]

Lithops (N.E.Br.) N.E.Br. GC 1922: 44, 1926: 80 Nel [1946]; Fearn 1981; Heine 1986; Cole 1988, 1997 **Typus L. lesliei** (N.E.Br.) N.E.Br. **Etym** Gk *lithos*, stone, *-ops*, face or appearance, referring to the resemblance of the faces or top surfaces of the plants to the stones among which they typically grow.

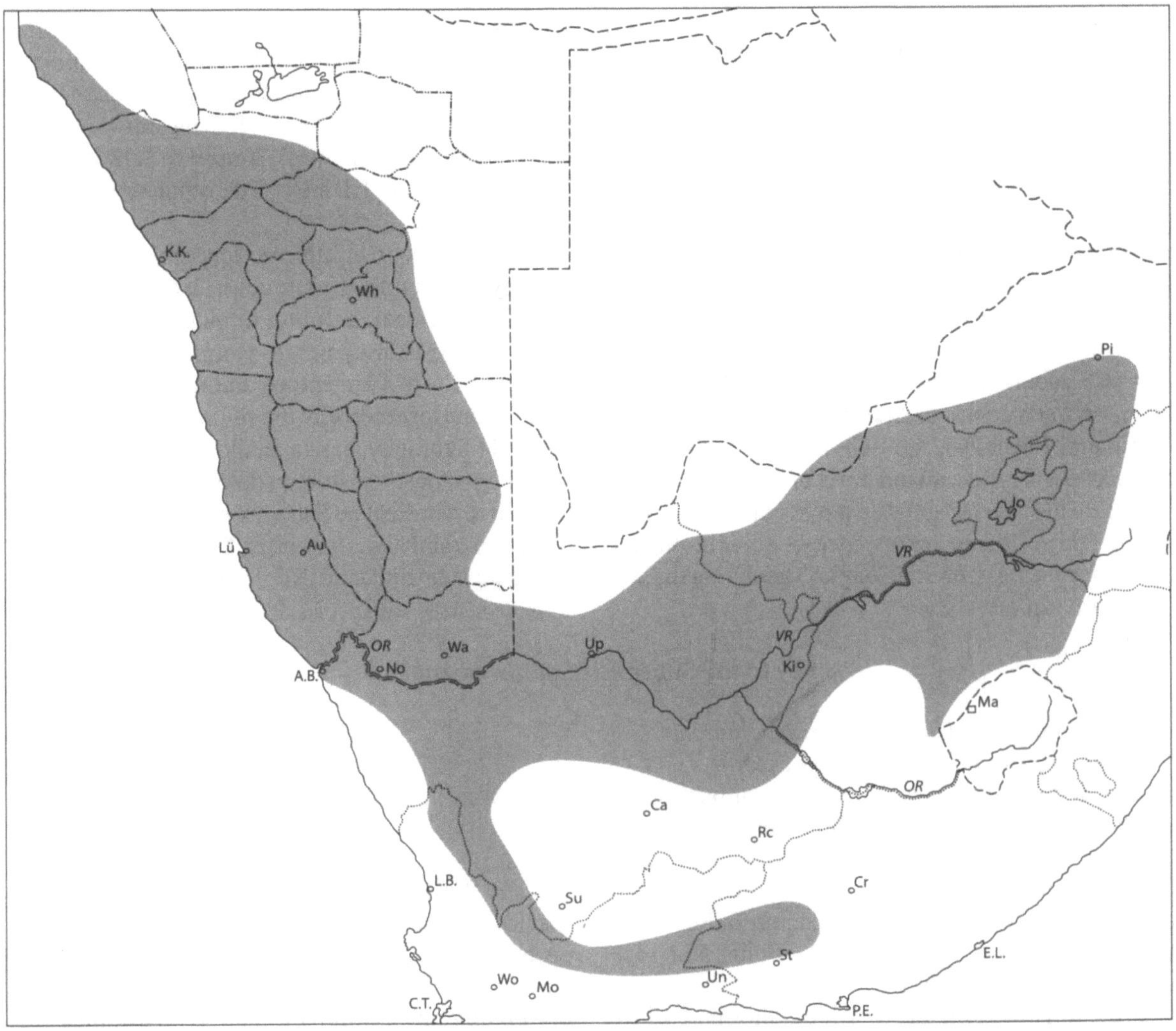

Distribution of **Lithops** N.E.Br.

[3] We are deeply indebted to Dr. H. E. K. Hartmann for devoting four very busy days with us, editing our work, working through the literature and setting up the lists of synonyms and references. She also wrote all the notes on distribution and prepared the map.

Compact to caespitose, sunken in the ground, often wedged between stones, usually with several to many heads, each consisting of one leaf-pair, but in a few taxa a single head is the norm, roots thickened, fleshy, tapering into threadlike ends, often plagiotropous, **I** absent or very short; **L** fused together in pairs, constituting a turbiniform head; **Fl** 1 per head, erect; **Fr** erect, seated in the cleft between the two leaf-lobes of the head; **Chr** $2n = 18$; **Ecol** nearly always in stony ground, surrounded by stones and larger rocks, rainfall 25 to 1 000 mm p.a.; **Distr** SE parts of Botswana; from Kaokoveld to Warmbad, Namibia; EC, FS, GAU, NC, NP, NW, WC, S Africa.

Note: Almost all species are extremely variable; nevertheless, the congruent features and patterns of resemblance make it possible for the plants in different colonies of a species to be grouped into subspecies and varieties (Cole 1988).

The following specific terms and abbreviations are used in the description of **Lithops: bicuneate** plant body double wedge-shaped, with fissure deep and usually widely gaping, leaves therefore divergent; **Chan** channels, the translucent or semi-translucent areas surrounding the islands on the faces of the plants; **cordate** plant body more or less heart-shaped, top and outer surfaces of the lobes strongly convex and rounded; **faces** the top surfaces of the plant bodies or their fruits; **fissure** (or cleft) the dorsal split between the two leaves of a plant body; the fissure may be shallow or deep, narrow or divergent; **head** a pair of leaves fused together to form a single plant-body; **Isl** islands, opaque areas, surrounded by channels, on the faces of the plants; **Mar** margins, the inner and outer opaque areas on the faces, which enclose the windows or channels; **reniform** the face of the head kidney-shaped; **Rub** rubrications, red dots, dashes, hooks or lines which appear in the windows or channels of many species; **truncate** plant body more or less flat-topped, the top surface usually slightly convex; **Win** windows, translucent or semi-translucent areas within the margins; when there are many or large islands, the windows are reduced to channels.

Key to the Species of Lithops (by N. A. Cole, see also Clark 1996: 1–9)

1. Flower colour very variable, almost white through yellow to bronze and pink; plants with small red warts on top surface .... **L. verruculosa**
- Flower colour yellow or white 2
2. Flower colour white, may have pink tinge 3
- Flower colour yellow, may have white centre 9
3. Profile cordate-truncate, flat to convex, or truncate-cordate, fissure fairly deep, margins distinct 4
- Profile truncate, flat to slightly convex, fissure shallow to medium, lobes conjunct 6
4. Window ± occluded or absent, fissure divergent, colour opaque pale grey, marbled, multiheaded ≥ 26, large .... **L. marmorata**
- Window open, fissure conjunct or slightly divergent, colours greenish-grey, multiheaded, small to medium 5
5. Profile usually distinctly convex, margin usually regular, ≥ 20 heads, surface smooth; petals pink-tinged .... **L. optica**
- Profile flat to somewhat convex, margin irregularly dentate, heads ≥ 8, surface smooth to slightly rugose .... **L. villetii**
6. Surface smooth, windows usually open, margins distinct and regular, heads ≥ 50, colour grey-green, rubrications usually absent .... **L. salicola**
- Surface smooth to rugose, windows absent or occluded, rubrications present, heads 6–15 7
7. Margins irregularly dentate, colours usually grey, beige, red-brown, rubrications scattered, heads up to 6, medium to very large .... **L. hallii**
- Margins obscure, rubrications network, heads up to 15, medium to large 8
8. Colours pale whitish grey tinged with blue or pink, fuscous lip-smear on inner margin or stitch-like markings in the outer margin, rubrications a network of lines or dots and dashes in the marginal indentations .... **L. julii**
- Colours brick red, beige, yellow, grey-green, brown, pink, grey-white, rubrications absent or a network of dots, hooks, triangles, short lines, no lip-smear .... **L. karasmontana**
9. Flower colour yellow with white centre 10
- Flower colour yellow 19
10. Profile truncate, flat to slightly convex, fissure deep, lobes conjunct to divergent, margins distinct, windows open, surface smooth, heads ≥ 30, colours greens and greys or salmon pink .... **L. olivacea**
- Profile bicuneate, bicuneate-cordate, cordate-bicuneate, cordate-truncate, truncate-cordate, cordate 11

11. Profile bicuneate, bicuneate-cordate, cordate-bicuneate, fissure deep, divergent 12
- Profile cordate-truncate, truncate-cordate, cordate 15
12. Margins not clear, windows cloudy, colours whitish grey tinged with greenish grey .... **L. meyeri**
- Margins distinct, regular to irregular, windows generally occluded, may be open, colours other than grey 13
13. Profile bicuneate, convex, pale grey-green like "frosted glass" or amethyst .... **L. divergens**
- Profile cordate, bicuneate-cordate 14
14. Profile essentially cordate, usually distinctly humped, margins distinct slender lines, irregularly dentate with small peninsulas, colours various shades of green and grey with windows having a glassy appearance .... **L. helmutii**
- Profile bicuneate-cordate, margins distinct, narrow, usually very irregular, with numerous peninsulas, colours greyish green, greenish, brownish, reddish grey, often with a reddish suffusion lining the margins .... **L. naureeniae**
15. Profile cordate, window open, fissure deep, somewhat divergent, uniform, greyish green .... **L. viridis**
- Profile cordate-truncate, truncate-cordate, window open or occluded, fissure fairly deep and conjunct to slightly divergent 16
16. Profile cordate-truncate, convex 17
- Profile truncate-cordate, margins distinct and finely dentate, colours pinkish, bluish-grey .... **L. comptonii**
17. Margins scalloped with rounded peninsulas .... **L. otzeniana**
- Margins distinct and regular or indistinct and irregular but not scalloped 18
18. Margins distinct and regular, fissure fairly deep, lobes conjunct, colours grey-green, milky grey, buff, beige, windows occluded, green, grey-green .... **L. geyeri**
- Margins usually indistinct and irregular, fissure deep, conjunct to slightly divergent, colours milky grey, windows usually ± occluded, ± grey, tinged, green tinged .... **L. herrei**
19. Profile cordate, ovate-cordate, cordate-truncate, windows usually absent or occluded 20
- Profile truncate, generally flat to slightly convex, fissure shallow to deep, usually conjunct 27
20. Profile cordate or ovate-cordate, margins and windows usually absent or partly occluded 21
- Profile cordate-truncate, truncate-cordate, convex, windows usually occluded or not distinguishable, fissure shallow to deep, usually conjunct, rarely slightly divergent at top 23
21. Profile cordate, face flush, smooth, covered with dusky dots, colours shades of brown (earth-coloured) and green .... **L. localis**
- Profile cordate or ovate-cordate, face elevated, colours greyish-white, buff, cream 22
22. Profile ovate-cordate, egg-shaped, with very shallow (1 mm) groove right across face, lips of fissure sometimes everted, surface smooth .... **L. steineckeana**
- Profile cordate, fissure deep, conjunct, surface smooth to slightly rugose, rubrications present, few to many, no dusky dots .... **L. ruschiorum**
23. Profile truncate-cordate, usually convex, margins distinct, sharply defined, irregularly dentate, fissure shallow, conjunct, rubrications numerous bold dark red lines, in windows and channels .... **L. dorotheae**
- Profile cordate-truncate, convex, surface smooth to rugose, fissure deep, conjunct, sometimes slightly divergent at top, windows occluded or not distinguishable 24
24. Margins not clearly distinguishable, dusky dots numerous 25
- Margins distinctly outlined, irregularly dentate, no rubrications, few to many dusky dots but only visible with a lens .... **L. gesineae**
25. Rubrications in the form of a network in the channels, numerous scattered dusky dots, small, channels broad to narrow, colours light grey or greenish-grey, channels greenish-grey or brown .... **L. werneri**
- Rubrications absent, dusky dots numerous, clearly visible, channels indistinct or narrow semi-translucent, colours opaque beige, grey-white tinged with pink, yellow, green 26
26. Channels indistinct, colours opaque beige, grey-white tinged with pink, yellow, green, plant-bodies medium size .... **L. francisci**
- Channels narrow, semi-translucent, colours pale creamy brown suffused with pink or green, plant-bodies small .... **L. coleorum**
27. Surface rugose, windows usually occluded or absent 28
- Surface smooth to slightly rugose 30
28. Margins usually distinct but irregularly dentate with peninsulas, windows occluded, rarely completely open, number of heads ≥ 10, rubrications forming bold network .... **L. bromfieldii**
- Margins absent or not clear, windows absent 29

29. Colours grey-white, beige, shades of brown, rubrications as dots and dashes or else a bold network, usually single-headed .... **L. gracilidelineata**
- Colours opaque brown, shades of orange brown, with channels and grooves greenish, reddish brown, rubrications absent or seldom visible without a lens, number of heads ≥ 10 .... **L. hookeri**
30. Surface slightly rugose, wrinkled or dotted with pin-pricks, colour near white, pale grey-white, tinged with yellow, green, beige or pink, uniform, no rubrications or dusky dots, number of heads ≥ 10 ..... **L. vallis-mariae**
- Other than above **31**
31. Rubrications present **32**
- Rubrications absent **35**
32. Rubrications bold red dots, dashes, few to many, dusky dots obscure or not visible without a lens .... **L. dinteri**
- Dusky dots numerous, clearly visible, rubrications bold **33**
33. Very obvious dusky dots scattered over entire face, dark grey-green to blue-green, rubrications few to many, plants fairly uniform brown and beige or opaque milky bluish or pinkish grey-white .... **L. fulviceps**
- Dusky dots few to many but not always clearly visible, rubrications usually branched lines or network **34**
34. Colours of face pale grey tinged with pink, blue, green, yellow, brownish-pink, bluish-white, milky grey-green, channels and rubrications densely and ramosely branched .... **L. pseudotruncatella**
- Colours of face greyish green, yellowish, pinkish, orange, brown-grey, purplish-brown, grey-brown, bluish, greenish-grey, bold rubrications, light coloured marginal bands .... **L. schwantesii**
35. Relatively uniform, windows and channels mainly green, grey-green with very irregular rust-brown islands forming a filigree pattern, numerous mini-windows or various shades of whitish-grey with dark slate-grey windows and channels with lines extending into the margins .... **L. lesliei**
- Variable, windows and channels various shades of brown or grey 36
36. Windows and channels various shades of brown, islands usually distinctly irregular, small to large, contrasting in colour with the windows, plant bodies large, 2 to many heads .... **L. aucampiae**
- Windows occasionally almost open, broad to narrow, irregularly shaped channels, translucent dark grey to greenish-grey, sometimes with reddish tint, islands pale grey. Plant bodies small, 1–2 heads ... **L. hermetica**

L. subg. *Leucolithops* Schwantes NCSJ 1951: 23 nomen invalid.
L. subg. *Xantholithops* Schwantes NCSJ 1951: 23 nomen illegit.

L. alpina = **L. pseudotruncatella** ssp. **pseudotruncatella**
L. annae = **L. gesineae**
L. archerae = **L. pseudotruncatella** ssp. **archerae**

L. aucampiae L.Bolus SAG 1932: 306 T Syntypes Aucamp, McCormick 616/31 (BOL).

2–5(–10) heads; L truncate in profile, 30–50 mm *l*, 20–35 mm *b*, fissure shallow, faces flush, elliptic-reniform, smooth to slightly rugose, **Win** usually ± occluded, occasionally ± completely open, **Rub** absent; **Fl** yellow, ∅ 20–50 mm; **Fr** profile boat-shaped, top ± flat, faces broadly elliptic; **Ecol** among stones or in sandy soil; **Distr** from Hopetown, NC, to Vryburg, NW, S Africa.

Note: A widespread species with different colour forms according with the soils in which they occur. Two subspecies are recognized.

Key to the Subspecies of L. aucampiae

1. Margins with numerous digitate laciniations projecting from the windows into the margins .... **L. aucampiae** ssp. **euniceae**
- Margins with irregularly incised but not digitate laciniations .... **L. aucampiae** ssp. **aucampiae**

L. aucampiae L.Bolus ssp. **aucampiae** (see **photo**) = *L. koelemanii* De Boer 1960: 28 ≡ *L. aucampiae* L.Bolus var. *koelemanii* (De Boer) D.T.Cole Excelsa 1973: 55 ≡ *L. aucampiae* L.Bolus ssp. *aucampiae* L.Bolus var. *koelemanii* (De Boer) D.T.Cole 1988: 102 ≡ *L. aucampiae* L.Bolus var. *koelemanii* (De Boer) De Boer 1964: 51 nomen invalid. = *L. loganiae* L.Bolus ex H.Jacobsen 1935: 200 nomen nudum.

Mar usually distinct but irregularly incised, **Isl** usually distinct, few to many, small to large, very irregular, opaque yellowish to reddish brown, **Chan** narrow to broad, irregular, obscurely translucent yellowish to reddish chocolate brown; **Fr** 6(–7) locules; **Ecol** various forms and colours of ironstone, sandstone, chert, quartzite, mainly brown, yellow-brown, black, with some red, grey, white; **Distr** Danielskuil, Griquatown, Niekerkshoop, Postmasburg, Olifantshoek, NC, Lykso, Severn, NW, S Africa.

Note: A widespread species, the subspecies differing in the soils they grow in.

L. aucampiae L.Bolus ssp. **euniceae** (De Boer) D.T.Cole 1988: 220 ≡ *L. aucampiae* L.Bolus var.

euniceae De Boer 1966: 54 T Cole 48 (PRE) = *L. aucampiae* L.Bolus var. *fluminalis* D.T.Cole NCSJ 1970: 8 ≡ *L. aucampiae* L.Bolus ssp. *euniceae* (De Boer) D.T.Cole var. *fluminalis* D.T.Cole 1988: 104[4] T Cole 54 (PRE).

Mar distinct with numerous digitate laciniations projecting from the windows, **Isl** usually distinct, mostly fairly small, yellowish, greenish or pinkish grey, **Chan** usually broad, occasionally reduced to narrow grooves, obscurely translucent dark red-brown; **Fr** mostly 6 locules; **Ecol** reddish sandy soil with fine-grained brown sandstone; **Distr** Hopetown, NC, S Africa.

Note: Two varieties were distinguished by Cole (1988: 103–105) in **L. aucampiae** ssp. **euniceae**: var. *euniceae*, var. *fluminalis*.

L. aucampiae var. *euniceae*
≡ **L. aucampiae** ssp. **euniceae**
L. aucampiae var. *fluminalis*
= **L. aucampiae** ssp. **euniceae**
L. aucampiae var. *koelemanii*
= **L. aucampiae** ssp. **aucampiae**
L. aurantiaca = **L. hookeri**
L. bella ≡ **L. karasmontana** ssp. **bella**
L. bella var. *eberlanzii*
≡ **L. karasmontana** ssp. **eberlanzii**
L. bella var. *lericheana*
= **L. karasmontana** ssp. **karasmontana**
L. brevis = **L. dinteri** ssp. **dinteri**

L. bromfieldii L.Bolus NM 1934: 452 T Bromfield 2286/33 (BOL) = *L. glaudinae* De Boer 1960: 129 ≡ *L. bromfieldii* L.Bolus var. *glaudinae* (De Boer) D.T.Cole Excelsa 1973: 50 T Cole 116 (PRE) = *L. insularis* L.Bolus NM 1937: 75 ≡ *L. bromfieldii* L.Bolus var. *insularis* (L.Bolus) Fearn 1970: 92 T Wilmot 1353/34 (BOL) = *L. mennellii* L.Bolus NM 1937: 75 ≡ *L. bromfieldii* L.Bolus var. *mennellii* (L.Bolus) Fearn 1970: 92 T Mennell 645/34 (BOL) = *L. rooi* Haage 1937: 16 nomen nudum = *L. rouxii* J. Oesterreich 1937: 100 nomen nudum = *L. upingtonensis* L.Bolus in litt. nomen nudum.

2–4(–10) heads; **L** truncate in profile, 20–30 mm *l*, 18–23 mm *b*, fissure shallow, faces flush, elliptic-reniform, usually distinctly rugose, **Mar** usually distinct but irregularly dentate or sinuate, **Isl** usually distinct, ± humped, pale pinkish yellow, cream, pale orange yellow or pinkish grey, **Win** usually ± occluded, **Chan** narrow to broad, translucent brownish or reddish green, greenish red or dull greyish green, **Rub** usually numerous, bold red lines forming an irregularly branched broken network; **Fl** yellow, ∅ 20–45 mm; **Fr** 5 locules, profile boat-shaped, top flat, faces broadly elliptic with hinge-rim; **Ecol** white and red-stained quartzite, greyish, brownish or reddish sheared quartzite; **Distr** Gordonia, Hay, Prieska, NC, S Africa.

Note: Four varieties were distinguished by Cole (1988: 105–109) in **L. bromfieldii**: var. *bromfieldii*, var. *glaudinae*, var. *insularis*, var. *mennellii*. The spelling *L. rooi* is apparently based on the pronounciation "ru" or "roo" of the originally French name *L. rouxii*.

L. bromfieldii var. *glaudinae* = **L. bromfieldii**
L. bromfieldii var. *insularis* = **L. bromfieldii**
L. bromfieldii var. *mennellii* = **L. bromfieldii**
L. christinae = **L. schwantesii** ssp. **schwantesii**
L. chrysocephala = **L. julii** ssp. **julii**

L. coleorum S.A.Hammer & Uijs Aloe 1994: 36–38 (see **photo**) T Hammer, Brink & Retief 1500 (PRU).

2–4(–8) heads; **L** cordate in profile, usually strongly convex, 10–15 mm *l*, 6–10 mm *b*, fissure deep, faces slightly elevated, elliptic-reniform, smooth, **inner Mar** usually distinct, **outer Mar** very irregular, **Isl** irregular, pale creamy brown, often suffused with pink or green, **Win** mostly absent, **Chan** usually narrow, up to 3 mm, semi-translucent brownish green to greenish brown, **Rub** absent, **dusky dots** numerous, distinct; **Fl** bright golden yellow, ∅ 15–25 mm; **Fr** 5 locules, profile rotund, top flat, faces broadly elliptic to round; **Ecol** sandstone rubble with some white or yellowish quartzite; **Distr** Ellisras, NP, S Africa.

Note: This is a newly discovered species not included in Cole (1988).

L. commoda = **L. karasmontana** ssp. **karasmontana**

L. comptonii L.Bolus SAG 1930: 212 L.Bolus NM 1937: 56 T Compton 1979/29 (BOL) = *L. weberi* Nel 1940: 16 ≡ *L. comptonii* L.Bolus var. *weberi* (Nel) D.T.Cole Excelsa 1973: 69 ≡ *L. otzeniana* Nel var. *weberi* (Nel) Fearn 1970: 92 **Lectotype**, here designated: Nel 1940: f on p 18.

2–4, also > 10 heads; **L** truncate-cordate in profile, 20–35 mm *l*, 15–23 mm *b*, fissure fairly deep, faces flush or slightly elevated, elliptic-reniform, smooth to slightly rugose, **Mar** distinct but usually finely dentate or sinuate, **Isl** few, mostly small and irregular, opaque pinkish beige, light grey or pale greenish grey, **Win** ± occluded or ± open, **Chan** broad to narrow, obscurely translucent slate grey, dark greenish grey or violet grey, **Rub** absent; **Fl** yellow, white centre, ∅ 20–30 mm; **Fr** 5 locules, profile shallow, boat-shaped, top ± flat, faces broadly elliptic to round; **Ecol** chert, ferruginous chert and calcrete, mostly yellow-brown, dark brown, grey; **Distr** Calvinia, NC, Ceres, Laingsburg, WC, S Africa.

Note: The species is restricted essentially to the Tanqua Karoo.

---

4 The transfer of a variety from one subspecies to another within the same species does not require an alteration in the authorship (ICBN 2000: Art. 49.1).

L. comptonii var. *divergens* ≡ **L. divergens**
L. comptonii var. *viridis* ≡ **L. viridis**
L. comptonii var. *weberi* = **L. comptonii**
L. dabneri = **L. hookeri**
L. damarana = **L. karasmontana** ssp. **karasmontana**
L. deboeri = **L. villetii** ssp. **deboeri**
L. dendritica = **L. pseudotruncatella** ssp. **dendritica**

L. dinteri Schwantes ZSK 1927: 97 **T** Rusch s.n. (?).

Fissure deep, faces flush, elliptic-reniform, ± smooth; **Fl** yellow, Ø 18–32 mm; **Fr** 5 locules, profile boat-shaped, top mostly ± flat, faces broadly elliptic to round, with hinge-rim; **Ecol** frequently in pegmatite; **Distr** S part of Namibia; Namaqualand, NC, S Africa.

Note: Three subspecies are recognized.

Key to the Subspecies of L. dinteri

1. Profile cordate-truncate, windows mostly ± occluded, rubrications in the form of numerous dots, short dashes and/or hooks, sometimes forming a broken network .... **L. dinteri ssp. frederici**
– Profile truncate **2**
2. Windows generally open, rubrications in the form of few to many dots and/or short dashes and/or hooks .... **L. dinteri ssp. dinteri**
– Windows generally absent, rubrications in the form of dashes, hooks and/or dots, sometimes forming a broken network .... **L. dinteri ssp. multipunctata**

L. dinteri Schwantes ssp. **dinteri** = *L. brevis* L.Bolus NM 1934: 344 ≡ *L. dinteri* Schwantes var. *brevis* (L.Bolus) Fearn 1970: 92, also as *L. dinteri* Schwantes ssp. *dinteri* var. *brevis* (L.Bolus) Fearn[5] in D.T.Cole 1988: 113 **T** Pillans 6608 (BOL).

1–3, also > 7 heads; **L** truncate in profile, 18–30 mm *l*, 13–20 mm *b*, **Mar** usually distinct and fairly regular, **Isl** usually absent or indistinct, opaque pale buff to yellowish, greenish or pinkish grey, **Win** usually completely open and translucent, **Chan** usually absent, various shades of translucent grey or grey-green, often with a reddish tint, **Rub** usually 5–15 or more blood-red dots and/or short dashes and/or hooks; **Ecol** coarse pegmatite, white, grey-white, pink, light brown; **Distr** Warmbad, Namibia.

Note: Two varieties were distinguished by Cole (1988: 112–114): **L. dinteri** var. *brevis*, var. *dinteri*.

L. dinteri Schwantes ssp. **frederici** (D.T.Cole) D.T.Cole 1988: 220 ≡ *L. dinteri* Schwantes var. *frederici* D.T.Cole Madoqua 1973: 33 **T** Cole 180 (PRE).

2–3, also > 6 heads; **L** cordate-truncate in profile, 12–20 mm *l*, 10–14 mm *b*, **Mar** usually absent or indistinct, **Isl** few and indistinct, various shades of opaque creamy buff or beige tinged with pink or yellow, **Win** ± densely occluded, clouded and indistinct, **Chan** indistinct, various shades of obscurely translucent greenish grey, **Rub** numerous dots and/or short dashes and/or hooks in the channels, dull dark red, sometimes bright blood red; **Ecol** fine pegmatite gravel, white, grey-white, light brown; **Distr** Namaqualand, NC, S Africa.

Note: This subspecies differs from **L. dinteri** ssp. **dinteri** in markings, size of the heads and profile of the leaf-pair.

[5] No change in authorship is required here, see Footnote 4.

L. dinteri var. *brevis* = **L. dinteri** ssp. **dinteri**
L. dinteri var. *frederici* ≡ **L. dinteri** ssp. **frederici**
L. dinteri var. *marthae*
= **L. schwantesii** ssp. **schwantesii**

L. dinteri Schwantes ssp. **multipunctata** (De Boer) D.T.Cole 1988: 227 ≡ *L. dinteri* Schwantes var. *multipunctata* De Boer 1966: 21 ≡ *L. pringlei* Roux in hort. nomen nudum **T** Pringle & Roux s.n. (?) **Neotype,** here designated: Cole 181 (PRE).

2–5(–13) heads; **L** truncate in profile, 18–33 mm *l*, 14–22 mm *b*, **Mar** indistinct and irregularly incised, **Isl** absent or only as indistinct opaque areas, opaque buff to pale brown, beige or pinkish grey, **Win** absent or not clearly distinguishable, **Chan** often narrow and indistinct, translucent dull grey-green, often with a reddish tint, **Rub** numerous bold blood-red lines, dashes, hooks and dots, sometimes forming a broken network; **Ecol** pegmatite, sheared fine-grained micaceous quartz, feldspar rock, white, pink, brown, reddish-brown; **Distr** Warmbad, Namibia.

Note: The distinctive feature of this subspecies is the much more extensive system of rubrications.

L. dinteri var. *multipunctata*
≡ **L. dinteri** ssp. **multipunctata**
L. diutina = **L. marmorata**

L. divergens L.Bolus NM 1934: 451 ≡ *L. comptonii* L.Bolus var. *divergens* (L.Bolus) Fearn 1970: 92 **T** Eksteen 2295/33 (BOL) = *L. divergens* L.Bolus var. *amethystina* De Boer in de Boer & Boom 1961: 40 ≡ *L. divergens* L.Bolus var. *divergens* forma *amethystina* (De Boer) Fearn 1970: 92 **Lectotype,** here designated: De Boer 1961: 40 f 1 (corrected 1961: 55).

2–4(–10) heads; **L** bicuneate in profile, 18–25 mm *l*, 10–14 mm *b*, fissure deep, lobes divergent, faces flush to slightly elevated, elongated elliptic, smooth, **Mar** usually regular but very slender lines, **Isl** very small and indistinct finely mottled flecks over the whole window, opaque pale greenish, bluish, pinkish or milky grey, **Win** usually extensively occluded by fine mottling of very small fleck-like islands (frosted glass effect), **Chan** very narrow and indistinct, various shades of translucent pale greyish green, **Rub** absent; **Fl** yellow, with white centre, ∅ 13–27 mm; **Fr** 5 locules, profile boat-shaped, top flat, faces round to broadly elliptic, with hinge-rim; **Ecol** white quartzite with some brown, red and black; **Distr** Vanrhynsdorp, WC, S Africa.

Note: Two varieties were distinguished by Cole (1988: 117–119): **L. divergens** var. *amethystina* and var. *divergens*.

L. divergens var. *amethystina* = **L. divergens**
L. divergens var. *divergens* forma *amethystina* = **L. divergens**

L. dorotheae Nel 1939: 77 ≡ *L. eksteeniae* L.Bolus NM 1939: 163 T Eksteen 1520/38 (BOL).

2–5(–7) heads; **L** truncate-cordate in profile, 18–23 mm *l*, 12–16 mm *b*, fissure shallow, faces flush, elliptic-reniform, smooth, **Mar** distinct and sharply defined, irregularly dentate or sinuate, **Isl** few to many, small to large, irregular, usually very distinct, various shades of opaque creamy, yellowish or pinkish beige, **Win** usually ± occluded, **Chan** broad to narrow, irregular, various shades of translucent greyish, brownish or reddish green or greenish grey, **Rub** in the windows and channels, numerous bold lines, dots, hooks and stars, sometimes connected into a broken network, bright to dark blood-red or orange-red; **Fl** yellow, ∅ 20–42 mm; **Fr** 5 locules, profile boat-shaped, top ± flat, faces broadly elliptic to elliptic, with hinge-rim; **Ecol** fine-grained sheared quartz feldspar rock or feldspathic quartzite, reddish or yellowish brown, with some white; **Distr** Namaqualand, NC, S Africa.

Note: As shown by Cole (1970: 106–108) both descriptions are based on material from the same population, collected by Mrs. Eksteen.

L. eberlanzii ≡ **L. karasmontana** ssp. **eberlanzii**
L. edithae = **L. karasmontana** ssp. **eberlanzii**; = **L. pseudotruncatella** ssp. **pseudotruncatella**
L. eksteeniae ≡ **L. dorotheae**
L. elevata = **L. optica**
L. elisabethae = **L. pseudotruncatella** ssp. **pseudotruncatella**
L. elisae = **L. marmorata**
L. erniana = **L. karasmontana** ssp. **eberlanzii**
L. erniana var. *aiaisensis* = **L. karasmontana** ssp. **karasmontana**
L. erniana var. *witputzensis* = **L. karasmontana** ssp. **eberlanzii**
L. farinosa = **L. pseudotruncatella** ssp. **dendritica**
L. fossulifera ≡ **L. karasmontana** ssp. **karasmontana**
L. framesii = **L. marmorata**

L. francisci (Dinter & Schwantes) N.E.Br. GC 1926: 102 ≡ *M. francisci* Dinter & Schwantes ZSK 1925: 19 T Dinter 3743 (?).

3–6(–20) heads; **L** cordate-truncate, usually distinctly convex in profile, 20–30 mm *l*, 15–20 mm *b*, fissure deep, faces slightly elevated, elliptic-reniform, smooth to very slightly rugose, **Mar** usually not clearly distinguishable, very irregular, **Isl** very irregular, ragged and indistinct, faces opaque beige, grey-white, buff or cream, tinged with yellow, green or pink, **Win** ± occluded, occasionally absent, **Chan** slightly impressed, very irregular, narrow and indistinct, obscurely translucent dull greenish grey, **Rub** absent, **dusky dots** very numerous, dull grey-green; **Fl** yellow, ∅ 13–25 mm; **Fr** 5 locules, profile boat-shaped, top ± flat, faces broadly elliptic to elliptic; **Ecol** gneiss and schist, grey-white, light brown, some black; **Distr** Lüderitz-South, Namibia.

Note: This species can readily be distinguished by the clearly visible and numerous dots.

L. francisci var. *annae* = **L. gesineae**
L. francisci var. *gesinae* ≡ **L. gesineae**
L. friedrichiae = **Conophytum friedrichiae**
L. fulleri ≡ **L. julii** ssp. **fulleri**
L. fulleri var. *brunnea* = **L. julii** ssp. **fulleri**
L. fulleri var. *chrysocephala* = **L. julii** ssp. **julii**
L. fulleri var. *kennedyi* ≡ **L. villetii** ssp. **kennedyi**
L. fulleri var. *ochracea* = **L. hallii**
L. fulleri var. *rouxii* = **L. julii** ssp. **fulleri**
L. fulleri var. *tapscottii* = **L. julii** ssp. **fulleri**

L. fulviceps (N.E.Br.) N.E.Br. GC 1922: 55 ≡ *M. fulviceps* N.E.Br. 1914: 167 T Pearson s.n. (K) = *L. fulviceps* (N.E.Br.) N.E.Br. var. *lactinea* D.T.Cole Excelsa 1973: 56 T Cole 222 (PRE) = *L. lydiae* H.Jacobsen 1960: 1239 nomen invalid. = *M. ortendahlii* Dinter 1928: 70 nomen nudum.

2–3(–10) heads; **L** truncate in profile, 25–40 mm *l*, 20–27 mm *b*, fissure shallow, faces flush, elliptic-reniform, slightly rugose, **Mar, Isl, Win, Chan** usually absent or not clearly distinguishable, faces ± reticulated by slightly impressed grooves or furrows, generally opaque, various shades of fairly uniform rusty yellow-brown, orange-brown, red-brown, grey-brown, with greenish, pinkish or purplish tinges, **Rub** occasionally absent, usually few to many slightly impressed dots, dashes or slender and irregularly branched lines, deep yellow, orange-brown, red, **dusky dots**, the most distinctive characteristic of the species, usually very numerous and slightly raised, like miniature blisters over the whole face; **Fl** yellow, ∅ 20–35 mm; **Fr** 5 locules, profile boat-shaped, top flat to slightly convex, faces broadly elliptic to elliptic; **Ecol** quartzite, gneiss, schist, calcrete, light to dark brown or reddish brown with some white, grey-white; **Distr** Keetmanshoop, Warmbad, Namibia; Gordonia, NC, S Africa.

Note: Two varieties were distinguished by Cole (1988: 121–124): **L. fulviceps** var. *fulviceps*, var. *lactinea*.

L. fulviceps var. *lactinea* = **L. fulviceps**
L. geoffreyi = **L. verruculosa**

L. gesineae De Boer 1955: 2 (see **photo**) ≡ *L. francisci* (Dinter & Schwantes) N.E.Br. var. *gesinae* (De Boer) Fearn 1970: 92 **T** Erni A 10022 (AMD?) = *L. annae* De Boer 1956: 90 ≡ *L. gesinae* De Boer var. *annae* (De Boer) De Boer & Boom 1964: 51 nomen invalid. ≡ *L. gesineae* De Boer var. *annae* (De Boer) D.T.Cole Excelsa 1973: 45 ≡ *L. francisci* (Dinter & Schwantes) N.E.Br. var. *annae* (De Boer) Fearn 1970: 92 ≡ *L. herbertii* Schwantes 1955 in litt. nomen nudum **T** Geyer-Joubert s.n. (?) = *L. namtibensis* Tischer in litt. nomen nudum ≡ *L. namtipbergensis* Tischer in litt. nomen nudum.

1–2(–10) heads; **L** cordate-truncate in profile, 25–36 mm *l*, 15–27 mm *b*, fissure deep, faces flush to slightly elevated, elliptic to slightly reniform, smooth, **Mar** usually distinct but very irregularly dentate and sinuate, **Isl** usually distinct, small to large, irregular in shape, opaque beige tinged with pink, yellow or greyish pink, **Win** usually ± occluded, **Chan** narrow to broad, very irregular, translucent greyish green or brownish green, **Rub** absent; **Fl** yellow, ∅ 20–45 mm; **Fr** mostly 6 locules, profile boat-shaped, top flat to slightly convex, faces broadly elliptic to elliptic; **Ecol** reddish-brown weathered gneiss; **Distr** Bethanie, Lüderitz-North, Namibia.

Note: The names *L. namtibensis* and *L. namtipbergensis* were used in correspondence between Tischer, Erni, and others (Cole Excelsa 1973: 60–61). The name *L. herbertii* was proposed by Schwantes in letters, but material from the same locality was later collected by Geyer and subsequently named *L. annae.* Two varieties were distinguished by Cole (1988: 124–126): **L. gesineae** var. *annae* and var. *gesineae.*

L. gesineae var. *annae* = **L. gesineae**

L. geyeri Nel 1943: 11 ≡ *L. herrei* L.Bolus var. *geyeri* (Nel) De Boer & Boom 1961: 41 **T** Herre SUG (?) = *L. hillii* L.Bolus CSJGB 1958: 85 **T** Hill 26575 (BOL).

2–7(–12) heads; **L** cordate-truncate in profile, 18–29 mm *l*, 13–20 mm *b*, fissure fairly deep, faces flush, mostly elliptic, ± smooth, **Mar** usually distinct and fairly regular, **Isl** sometimes distinct, usually reduced to indistinct mottlings or flecks, various shades of opaque pale milky grey, buff or beige, **Win** usually ± occluded, sometimes ± open, **Chan** broad to narrow, irregular, various shades of translucent pale grey, pale bluish or greenish grey, greyish green or brownish grey-green, **Rub** absent; **Fl** yellow, often white-centred, ∅ 20–40 mm; **Fr** 5 locules, profile boat-shaped, top ± flat to slightly peaked or convex, faces broadly elliptic, capsule edges angular, valve rims very distinct; **Ecol** quartzite and schist or gneiss, calcrete with some quartzite, brown, yellowish grey, yellowish white, grey-white; **Distr** Namaqualand, NC, S Africa.

Note: The species occurs only in a small area in the Richtersveld.

L. glaudinae = **L. bromfieldii**

L. gracilidelineata Dinter in Graessner 1928: 18 ≡ *L. pseudotruncatella* (A.Berger) N.E.Br. var. *gracilidelineata* (Dinter) Fearn 1970: 92 **T** Rusch s.n. (?) = *L. streyi* Schwantes SKK 1951: 74 **T** Strey s.n. (?).

Seldom more than 1 head; **L** truncate in profile, usually slightly convex, 30–50 mm *l*, 20–38 mm *b*, fissure shallow, faces flush, round to mostly elliptic; **Fl** yellow, ∅ 20–45 mm; **Fr** 6(–7) locules, profile rotund to boat-shaped; **Ecol** in quartzite, feldspar, or in brown granitic gravel; **Distr** from Outjo to Karibib, Namibia.

Note: Two subspecies are recognized on the basis of facial colour and differences in rubrications.

Key to the Subspecies of L. gracilidelineata

1. Faces various shades of opaque greyish white or smudged with buff or pinkish brown .... **L. gracilidelineata** ssp. **gracilidelineata**
– Faces opaque beige to reddish brown .... **L. gracilidelineata** ssp. **brandbergensis**

L. gracilidelineata Dinter ssp. **brandbergensis** (De Boer) D.T.Cole Lithops 1988: 217 ≡ *L. pseudotruncatella* (A.Berger) N.E.Br. var. *brandbergensis* De Boer 1963: 53 **T** Rudner s.n. (?) **Neotype,** here designated: Cole 383 (PRE).

L face slightly rugose, **Mar** not clearly distinguishable, **Isl** only as fairly regular opaque areas within the channels, various shades of opaque beige or yellowish, pinkish or reddish brown, **Chan** narrow, mostly set in a fairly regular network of shallow valleys, sometimes faintly discernible as obscurely translucent dull greyish green, **Rub** usually a bold fairly regular network of lines, blood-red or brownish or purplish red; **Fr** rotund, top convex to peaked, faces broadly elliptic, with hinge-rim; **Ecol** weathered brown granitic gravel; **Distr** Omaruru, Namibia.

Note: Material of this subspecies was collected by Rudner, who sent plants to Schwantes, seedlings from which reached de Boer, who described the taxon.

L. gracilidelineata Dinter ssp. **gracilidelineata** = *L. gracilidelineata* Dinter var. *waldronae* De Boer 1963: 20 ≡ *L. pseudotruncatella* (A.Berger) N.E.Br. var. *gracilidelineata* Dinter forma *waldronae* (De Boer) Fearn 1970: 92 T Waldron (?).

L face usually distinctly rugose, occasionally smooth, **Mar** and **Win** rarely distinguishable, **Isl** manifest only as areas or humps enclosed by channels, faces usually ± uniform opaque pale greyish white, pearl grey, pale milky bluish, yellowish or pinkish grey, sometimes suffused with buff, beige or pinkish brown, **Chan** very slender lines, superficial or more usually impressed, thus dividing the face into a number of humps, obscurely translucent dark greyish green, greenish grey or brown, or reddish brown, **Rub** often absent, otherwise very obscure fine dots, dashes or short lines set in the channels; **Fr** top ± flat to slightly concave or convex, faces round to broadly elliptic, with hinge-rim; **Ecol** quartzite, fine feldspathic quartzite gravel, pegmatite gravel, yellowish white, white, pink, grey, grey-white, brown, sometimes black; **Distr** from Outjo to Karibib, Namibia.

Note: A widespread species.

L. gracilidelineata var. *waldronae* = **L. gracilidelineata** ssp. **gracilidelineata**
L. gulielmi = **L. schwantesii** ssp. **schwantesii**
L. halenbergensis Tischer MDK 1932: 281. The plants figured by Tischer with the original description resemble the white-flowering **L. karasmontana** ssp. **eberlanzii** from Haalenberg, but the flower of *L. halenbergensis* was reported to be yellow by F. Erni. Consequently, the identity of the name cannot be assessed.

L. hallii De Boer 1957: 85 (see **photo**) T Hall s.n. (?) = *L. hallii* De Boer var. *ochracea* (De Boer) D.T.Cole Excelsa 1973: 61 ≡ *L. fulleri* N.E.Br. var. *ochracea* De Boer 1962: 68 T Roux (?) = *L. salicola* L.Bolus var. *reticulata* De Boer 1960: 79 T Geyer (?).

2–3(–6) heads; **L** truncate in profile, 25–47 mm *l*, 17–30 mm *b*, fissure shallow, faces flush, mostly reniform, slightly rugose, **Mar** distinct, irregularly dentate, **Isl** few to many, mostly distinct, regular and fairly small, various shades of opaque grey or beige, faintly tinged with yellow, pink, blue, green, or yellowish, orange or reddish brown, **Win** usually ± occluded, occasionally ± open, **Chan** broad to narrow, sometimes reduced to a fine network of narrow grooves, various shades of opaque translucent grey or grey tinged with green, brown, blue, pink or yellow, **Rub** usually in the form of scattered dots, dashes and hooks in the windows and channels, bright to dull blood-red or shades of brown; **Fl** white, ∅ 20–45 mm; **Fr** 6 or 5 locules, profile shallow, boat-shaped, top flat to slightly convex, faces narrowly elliptic; **Ecol** calcrete with some jaspilite or fine-grained dark sandstone, yellow-brown sandy shale, grey-brown silty sandstone, grey-white, grey, brown, yellow-brown, grey-brown, with some red and black; **Distr** from Gordonia to Hopetown and Carnarvon, NC, S Africa.

Note: **L. hallii** belongs to the group of species with a rugose surface. Plants named *L. salicola* var. *reticulata* are almost certainly derived by hybridization of **L. hallii** and **L. salicola**.

L. hallii var. *ochracea* = **L. hallii**
L. helmae = **L. julii** ssp. **julii**

L. helmutii L.Bolus SAG 1933: 218 **T** Meyer 10106 (BOL).

2–6, also > 26 heads; **L** cordate to bicuneate in profile, 20–30 mm *l*, 13–19 mm *b*, fissure deep, lobes somewhat divergent, faces flush to elevated, ± elongate elliptic, smooth, **Mar** usually distinct slender lines, irregularly dentate, often extended down into fine points along the edges of the fissures, **Isl** few to many, small to large, irregular, usually ± distinct, sometimes reduced to flecks or mottling, various shades of cream tinged with yellow, pink or green, **Win** usually ± occluded, occasionally ± open, **Chan** broad to narrow, irregular, various shades of translucent pale grey, greyish green, pinkish or brownish green or grey, often with a rather glassy appearance, **Rub** absent; **Fl** yellow with white centre, ∅ 22–33 mm; **Fr** mostly 5 locules, profile boat-shaped, top flat to slightly convex or peaked, faces broadly elliptic to round, valves with numerous dark dots, valve rims very distinct; **Ecol** translucent white, red-stained and brownish quartzite; **Distr** Namaqualand, NC, S Africa.

Note: The species occurs in a very restricted area north of Steinkopf.

L. herbertii = **L. gesineae**

L. hermetica D.T.Cole 2001: 158 T Cole C 397 (WIND).

(1–)2 heads; **L** truncate in profile, 15–23 mm *l*, 11–16 mm *b*, fissure to 0.8 mm deep, faces slightly elevated, slightly rugose, **Mar** and **Isl** distinct, but very irregularly shaped, opaque pale grey, **Win** and **Chan** dark to greenish-grey, obscurely translucent, **Rub** absent, **dusky dots** numerous, greenish-grey; **Fl** yellow, ∅ 15–25 mm; **Fr** mostly 5 locules, profile boat-shaped, slightly convex; **Ecol** in calcrete beds between dark blue-grey dolomite rocks; **Distr** Lüderitz-South, Namibia.

Note: Further collections of this species are: Hammer 2003 and Graham & Francoise Williamson & Hammer 5444 (NBG).

L. herrei L.Bolus NM 1932: 346 ≡ *L. herrei* L.Bolus var. *plena* L.Bolus NM 1932: 346 T Herre 9176 (BOL) = *L. translucens* L.Bolus NM 1932: 347 T Schlechter 9844 (BOL).

2–7, also > 20 heads; **L** cordate-truncate in profile, 20–30 mm *l*, 15–21 mm *b*, fissure deep, faces flush to slightly elevated, elliptic-reniform, smooth to slightly rugose, **Mar** sometimes distinct and regular, usually irregular with numerous peninsulas, **Isl** few to many, small to large, usually very irregular, often reduced to indistinct mottling or flecks, various shades of opaque pale milky grey or pale grey tinged with green, blue, cream, yellow, pink or buff, **Win** usually ± occluded, sometimes ± completely open, **Chan** narrow to broad, irregular, sometimes reduced to a number of impressed irregular grooves and mini-windows, various shades of translucent pale to mid-grey, bluish, greenish, yellowish or pinkish grey or green, **Rub** absent; **Fl** yellow, sometimes bronze-yellow, with white centre, ∅ 12–23 mm; **Fr** mostly 5 locules, profile boat-shaped, top ± flat, usually with a dark line ± 1 mm wide just below the edge, faces broadly elliptic to elliptic, with hinge-rim; **Ecol** quartzite and schist, gneiss and schist, calcrete with quartzite and schist, amygdaloidal lava and calcrete, grey-white, black, white, grey, brown; **Distr** Lüderitz-South, Namibia; Namaqualand, NC, S Africa.

Note: The species occurs along the lower reaches of the Orange River.

L. herrei forma *albiflora* = **L. marmorata**
L. herrei var. *geyeri* ≡ **L. geyeri**
L. herrei var. *plena* ≡ **L. herrei**
L. hillii = **L. geyeri**

L. hookeri (A.Berger) Schwantes MDG 1928: 46 ≡ *M. hookeri* A.Berger 1908: 283 ≡ *M. turbiniforme* sensu N.E.Br. 1920: 137 GC 1921: 250 ≡ *L. turbiniformis* sensu N.E.Br. GC 1922: 55 **Lectotype**, here designated: Hook.f. CBM 1874: t. 6077 as *M. truncatellum* = *L. hookeri* (Haw.) N.E.Br. var. *elephina* (D.T.Cole) D.T.Cole 1988: 219 ≡ *L. turbiniformis* (Haw.) N.E.Br. var. *elephina* D.T.Cole NCSJ 1970: 9 T Cole 92 (PRE) = *L. hookeri* (A.Berger) Schwantes var. *lutea* (De Boer) D.T.Cole 1988: 225 ≡ *L. turbiniformis* (Haw.) N.E.Br. var. *lutea* De Boer 1964: 52 **NT** Cole 38 (PRE) = *L. hookeri* (A.Berger) Schwantes var. *subfenestrata* (De Boer) D.T.Cole 1988: 231 ≡ *L. turbiniformis* (Haw.) N.E.Br. var. *subfenestrata* De Boer 1964: 53 **NT** Cole 21 (PRE) = *L. hookeri* (A.Berger) Schwantes var. *susannae* (D.T.Cole) D.T.Cole 1988: 231 ≡ *L. turbiniformis* (Haw.) N.E.Br. var. *susannae* D.T.Cole 1970: 9 T Cole 91 (PRE) = *L. aurantiaca* L.Bolus NM 1932: 344 T Triebner 9180 (BOL) = *L. dabneri* L.Bolus JSAB 1965: 237 ≡ *L. turbiniformis* (Haw.) N.E.Br. var. *dabneri* (L.Bolus) D.T.Cole Excelsa 1973: 46 ≡ *L. hookeri* (A.Berger) Schwantes var. *dabneri* (L.Bolus) D.T.Cole 1988: 218 T Potgieter 79/65 = *L. marginata* Nel 1946: 112 ≡ *L. turbiniformis* (Haw.) N.E.Br. var. *marginata* (Nel) D.T.Cole Excelsa 1973: 59 ≡ *L. hookeri* (A.Berger) Schwantes var. *marginata* (Nel) D.T.Cole 1988: 142 **Lectotype**, here designated: Otzen in Nel 1946: t. 24.2 = *L. turbiniformis* (Haw.) N.E.Br. var. *brunneo-violacea* De Boer 1964: 52 **NT** Cole 19 (PRE).

2–4(–10) heads; **L** truncate in profile, 25–46 mm *l*, 20–35 mm *b*, fissure shallow, faces flush, elliptic-reniform, distinctly rugose, **Mar** usually not clearly distinguishable, usually incised by simple and forked lines tapering out from the grooves, **Isl** manifest only as very numerous areas, enclosed by grooves, various shades of opaque brown or buff, or greyish, reddish, pinkish or orange brown, **Win** absent, **Chan** an extensive irregular network of narrow impressed grooves, various shades of opaque or very obscurely translucent greenish brown or grey, greyish or brownish green, orange or reddish brown or red, **Rub** often absent, otherwise obscure and seldom visible without a lens; **Fl** yellow, ∅ 20–45 mm; **Fr** mostly 6 locules, profile boat-shaped, top flat to slightly convex or peaked, faces broadly elliptic to almost round; **Ecol** translucent quartzite, sandstone, amygdaloidal lava with or without some sandstone, sheared micaceous quartzite, grey-white, brown, pink, grey, some white, red; **Distr** from Gordonia to Kimberley and to Britstown, NC, S Africa.

Note: The names *M. turbiniforme* and *L. turbiniformis* were applied erroneously to plants of this species by Brown (Cole 1988: 38–41). Seven varieties were distinguished by Cole (1988: 138–146): **L. hookeri** var. *dabneri*, var. *elephina*, var. *hookeri*, var. *lutea*, var. *marginata*, var. *subfenestrata*, var. *susannae*.

L. hookeri var. *dabneri* = **L. hookeri**
L. hookeri var. *elephina* = **L. hookeri**
L. hookeri var. *hookeri* = **L. hookeri**
L. hookeri var. *lutea* = **L. hookeri**
L. hookeri var. *marginata* = **L. hookeri**
L. hookeri var. *subfenestrata* = **L. hookeri**
L. hookeri var. *susannae* = **L. hookeri**
L. inae = **L. verruculosa**
L. inornata = **L. schwantesii** ssp. **schwantesii**
L. insularis = **L. bromfieldii**
L. jacobseniana = **L. karasmontana** ssp. **karasmontana**

L. julii (Dinter & Schwantes) N.E.Br. GC 1926: 102 ≡ *M. julii* Dinter & Schwantes ZSK 1925: 26 T Dinter s.n. (?).

2–5(–8) heads; **L** truncate in profile, 20–38 mm *l*, 15–25 mm *b*, fissure fairly deep, lobes sometimes divergent, faces flush, mostly slightly reniform, smooth to slightly rugose; **Fl** white, ∅ 20–43 mm; **Fr** 5(–6) locules, profile boat-shaped; **Ecol** mostly in pegmatite, sometimes in granite or calcrete; **Distr** Warmbad, Namibia; Kenhardt, Namaqualand, NC, S Africa.

Note: Two subspecies are recognized on the basis of the presence or absence of a fuscous lip-smear.

Key to the Subspecies of L. julii

1. Typically with a fuscous lip-smear .... **L. julii** ssp. **julii**
– Usually with stitch-like marks in the outer margins, but without a lip-smear .... **L. julii** ssp. **fulleri**

L. julii (Dinter & Schwantes) N.E.Br. ssp. **fulleri** (N.E.Br.) Fearn NCSJ 1976: 59 ≡ *L. fulleri* N.E.Br. GC 1927: 70 **T** Fuller s.n. (K) = *L. julii* (Dinter & Schwantes) N.E.Br. ssp. *fulleri* (N.E.Br.) Fearn var. *brunnea* (De Boer) D.T.Cole 1988: 150 ≡ *L. fulleri* N.E.Br. var. *brunnea* De Boer 1962: 68 **T** without collector (?) = *L. julii* (Dinter & Schwantes) N.E.Br. ssp. *fulleri* (N.E.Br.) Fearn var. *rouxii* De Boer 1964: 139, also as *L. julii* (Dinter & Schwantes) N.E.Br. ssp. *fulleri* (N.E.Br.) Fearn var. *rouxii* De Boer in D.T.Cole 1988: 229[6] ≡ *L. fulleri* (N.E.Br.) Fearn var. *rouxii* (De Boer) D.T.Cole Excelsa 1973: 64 **T** Roux (?) = *L. fulleri* N.E.Br. var. *tapscottii* L.Bolus SAG 1929: 182 **T** Tapscott (BOL) = *L. maughanii* N.E.Br. KB 1929: 57 **T** Maughan Brown 708 (K).

Up to 15 heads or more; **Mar** distinct, narrow to broad, irregularly dentate or sinuate, inner and outer margins usually edged with a fuscous lining, but not a lip-smear, **Isl** usually fairly regular, few to many, mostly smallish, sometimes reduced to obscure flecks, various shades of opaque light grey, tinged with pink, yellow, blue, green, beige or buff, **Win** distinct, slightly impressed, usually ± occluded, sometimes completely open, **Chan** narrow to broad, slightly impressed, somewhat irregular, various shades of opaque milky blue, pale pinkish, bluish or yellowish grey, or mauve, pink or rust-brown, usually with a brownish smear thinly lining the inner and outer margins and sometimes also the islands, **Rub** may be absent, or appear as a broken network, but usually restricted to marginal indentations as dots, dashes, X, Y or V marks; **Fr** top flat to slightly convex, faces narrowly to broadly elliptic; **Ecol** fine- or coarse-grained pegmatite, granite, fine-grained pegmatitic granite, calcrete in granitic soil, white, grey-white, pink, brown, grey; **Distr** Kenhardt, Namaqualand, NC, S Africa.

Note: Three varieties were distinguished by Cole (1988: 146–152): **L. julii** var. *brunnea*, var. *fulleri*, and var. *rouxii*. This subspecies lacks the "lip-smear" which is characteristic of the other subspecies. The locality published for *L. fulleri* var. *tapscottii* is almost certainly incorrect.

L. julii ssp. *fulleri* var. *brunnea*
= **L. julii** ssp. **fulleri**
L. julii var. *fulleri* ≡ **L. julii** ssp. **fulleri**

L. julii (Dinter & Schwantes) N.E.Br. ssp. **julii** (see **photo**) = *L. julii* (Dinter & Schwantes) N.E.Br. var. *littlewoodii* De Boer 1963: 36 **T** Littlewood (?) = *L. chrysocephala* Nel 1943: 7 ≡ *L. fulleri* N.E.Br. var. *chrysocephala* (Nel) De Boer 1962: 69 **T** Dry s.n. (?) = *L. helmae* Triebner ex Schwantes 1957: 230 nomen nudum = *L. lactea* Schick & Tischer in Tischer 1933: 49 **T**?

Mar usually rather obscure, sometimes ± distinct but irregularly dentate or sinuate, along the inner margins, usually a fuscous lip-smear, sometimes prominent, sometimes fragmented, **Isl** usually large and distinct, sometimes small and obscure, various shades of opaque pale whitish grey, sometimes tinged with blue, mauve, pink or buff, lip-smear and marginal lining dark greenish or brownish grey, greenish brown or beige, **Win** usually ± occluded, sometimes ± open, **Chan** irregular, sometimes broad and distinct, often reduced to a coarse or fine network of narrow, slightly impressed grooves, usually opaque grey-white with various tints of blue, mauve, pink, beige or brown, sometimes obscurely translucent greenish grey, greenish or yellowish brown, or pinkish grey, **Rub** sometimes absent, usually a network of bright to dark red lines, often a series of short lines, hooks and/or dots; **Fr** top ± flat, faces elliptic to narrow elliptic; **Ecol** fine- or coarse-grained pegmatite, pegmatite gravel, white, grey-white, pink, brown, grey, red; **Distr** Warmbad, Namibia.

Note: The most distinctive feature of this subspecies is the "lip-smear".

L. julii var. *littlewoodii* = **L. julii** ssp. **julii**
L. julii var. *rouxii* = **L. julii** ssp. **fulleri**

L. karasmontana (Dinter & Schwantes) N.E.Br. GC 1926: 102 ≡ *M. karasmontanum* Dinter & Schwantes MK 1920: 36 **T** J. Dinter (?).

2–6, also > 60 heads; **L** truncate or cordate-truncate in profile, 15–38 mm *l*, 12–25 mm *b*, fissure fairly deep, faces flush to slightly elevated, elliptic-reniform, smooth to slightly rugose; **Fl** white, ∅ 20–45 mm; **Fr** 5(–6) locules, profile boat-shaped, top ± flat, faces elliptic; **Ecol** common in pegmatite, quartzite, and gneiss; **Distr** Bethanie, Keetmanshoop, Lüderitz-North and -South, Warmbad, Namibia.

Note: Three subspecies are recognized on the basis of differences in the colours of the faces and the structure of the channels.

[6] See Footnote 4 for authorship.

Key to the Subspecies of L. karasmontana

1. Relatively uniform in colour, channels distinct, pellucid green or grey-green .... **L. karasmontana** ssp. **bella**
- Extremely variable in colour **2**
2. Faces opaque, channels not distinguishable .... **L. karasmontana** ssp. **karasmontana**
- Channels narrow to broad, not pellucid green or grey-green .... **L. karasmontana** ssp. **eberlanzii**

L. karasmontana var. *aiaisensis*
= **L. karasmontana** ssp. **karasmontana**

L. karasmontana (Dinter & Schwantes) N.E.Br. ssp. **bella** (N.E.Br.) D.T.Cole 1988: 217 ≡ *L. bella* N.E.Br. GC 1922: 80 ≡ *M. bellum* (N.E.Br.) Dinter 1923: 122 **T** Phillips s.n. (K) = *M. delaetianum* Dinter 1923: 15 non Dinter 1923: 137, nomen nudum.

Mar distinct but usually irregularly dentate and sinuate, **Isl** distinct, usually fairly large, somewhat irregular, opaque creamy buff, beige, yellow-grey or greenish grey, **Win** usually ± occluded, seldom ± open, **Chan** slightly impressed, distinct, ± irregular, usually fairly broad, dark pellucid green or grey-green, sometimes with a brown or greyish red tint, **Rub** usually absent, occasionally a few obscure red dots and lines; **Ecol** weathered granite, feldspathic granitic gneiss, pegmatite, yellow-brown, brown, pink, grey-white, grey; **Distr** Bethanie, Lüderitz-South, Namibia.

Note: This subspecies occurs in a narrow crescent-shaped area including Helmeringhausen, Aus and Witputz; it can develop up to 60 heads in a single plant.

L. karasmontana (Dinter & Schwantes) N.E.Br. ssp. **eberlanzii** (Dinter & Schwantes) D.T.Cole 1988: 219 ≡ *M. eberlanzii* Dinter & Schwantes ZSK 1925: 25 ≡ *L. eberlanzii* (Dinter & Schwantes) N.E.Br. GC 1926: 80 ≡ *L. bella* N.E.Br. var. *eberlanzii* (Dinter & Schwantes) De Boer & Boom 1961: 38 **T** syntypes Dinter, Eberlanz (?) = *L. edithae* N.E.Br. CJ 1934: 70 non Schwantes 1957: 236 **T** without collector (K) = *L. erniana* Tischer ex H.Jacobsen 1933: 145 Loesch & Tischer 1936: 76 **T** Erni (?) = *L. erniana* H.Jacobsen var. *witputzensis* De Boer 1964: 5 **T** Triebner, Hall & Carp (?).

Mar usually not clearly distinguishable, with irregular indentations, occasionally fairly well-defined and regularly sinuate with bold peninsulas, **Isl** usually indistinct, occasionally bold and well-defined, various shades of opaque pale or dark grey-white, with tints of blue, mauve, pink, brown, buff, beige, yellow or green, **Win** usually ± occluded, sometimes completely opaque, rarely ± completely open, **Chan** broad to narrow, usually indistinct and irregularly branched, obscurely translucent or opaque pale grey tinted with blue, green, pink, brown, mauve, buff, beige or dark grey-green, **Rub** sometimes absent, usually a network of slender branched lines or dots, dashes and hooks; **Ecol** gneiss, biotitic gneiss and calcrete, pegmatite and calcrete, quartzite and calcrete, gneiss and calcrete, pegmatite; **Distr** Lüderitz-South, Namibia.

Note: This subspecies occurs mainly below the escarpment, but not at the coast.

L. karasmontana (Dinter & Schwantes) N.E.Br. ssp. **karasmontana** = *L. commoda* Dinter ex H.Jacobsen 1937: 163 nomen nudum = *M. damaranum* N.E.Br. 1920: 67 ≡ *L. damarana* (N.E.Br.) N.E.Br. GC 1922: 80 **T** collector unknown (K) = *L. fossulifera* Tischer ex H.Jacobsen 1955: 1484 nomen nudum ≡ *L. karasmontana* (Dinter & Schwantes) N.E.Br. var. *tischeri* D.T.Cole Excelsa 1973: 66 **T** Cole 182 (PRE) = *L. jacobseniana* Schwantes ex H.Jacobsen 1955: 1476 nomen invalid. ≡ *L. ursulae* Swüste 1936: 25 nomen nudum **T**? = *L. lateritia* Dinter in Graessner 1928: 16 **T** Rusch (?) = *M. lericheanum* Dinter & Schwantes ZSK 1925: 25 ≡ *L. lericheana* (Dinter & Schwantes) N.E.Br. GC 1926: 102 ≡ *L. bella* N.E.Br. var. *lericheana* (Dinter & Schwantes) De Boer & Boom 1961: 39 ≡ *L. karasmontana* (Dinter & Schwantes) N.E.Br. var. *lericheana* (Dinter & Schwantes) D.T.Cole Excelsa 1973: 56, also as *L. karasmontana* (Dinter & Schwantes) N.E.Br. ssp. *karasmontana* var. *lericheana* (Dinter & Schwantes) D.T.Cole in Cole 1988: 155 **T** Dinter (?) = *L. mickbergensis* Dinter in Graessner 1928: 16 **T** Rusch (?) = *L. karasmontana* (Dinter & Schwantes) N.E.Br. var. *aiaisensis* (De Boer) D.T.Cole 1987: 42, also as *L. karasmontana* (Dinter & Schwantes) N.E.Br. ssp. *karasmontana* var. *aiaisensis* De Boer in D.T.Cole 1988: 154[7] ≡ *L. erniana* H.Jacobsen var. *aiaisensis* De Boer 1964: 6 **T** Littlewood (?) = *M. opalinum* Dinter 1928: 23 nomen nudum ≡ *L. opalina* Dinter MDG 1927: 139 ≡ *L. karasmontana* (Dinter & Schwantes) N.E.Br. var. *opalina* (Dinter) De Boer & Boom 1961: 52 **T** Dinter (?) = *M. summitatum* Dinter 1928: 16 nomen nudum ≡ *L. summitatum* Dinter MDG 1927: 102 ≡ *L. karasmontana* (Dinter & Schwantes) N.E.Br. var. *summitatum* (Dinter) De Boer & Boom 1961: 52 **T** Dinter (?).

Mar usually obscure, sometimes well-defined but irregularly indented, **Isl** sometimes absent, usually surrounded by narrow channels or grooves, irregu-

[7] See Footnote 4 for authorship.

lar but fairly clearly defined, often manifest only as obscure humps within shallow valleys, faces various shades of opaque brick-red, beige, yellow, brown, pink, salmon, milky blue, grey-white, OR **Mar-Isl** opaque light grey tinged with blue, green, yellow, beige, brown, pink or red, **Win** usually absent, the faces then usually divided up by shallow valleys into a number of humps, otherwise ± smooth, **Chan** usually narrow, irregular, occasionally a bold network, with a number of dots, dashes or small triangular shapes set in shallow valleys, trenches or pits, **Win** and **Chan** (when apparent) very obscurely translucent grey, greyish, bluish or brownish green, greenish grey, greyish pink, brown or red-brown, **Rub** sometimes absent, usually a number of short lines, hooks, triangles and dots which may be linked into a network; **Ecol** pegmatite, quartzite, gneiss, quartzite and calcrete, pegmatite, quartzite and sandstone, grey-white, white, grey, pink, yellow-brown, brown, reddish brown, greenish white; **Distr** Warmbad, Namibia.

Note: Four varieties were recognized by Cole in **L. karasmontana** (Dinter & Schwantes) N.E.Br. ssp. **karasmontana:** var. *aiaisensis*, var. *karasmontana*, var. *lericheana*, and var. *tischeri*. The name *L. jacobseniana* has been used although it is still invalid; following Swüste (1936: 25), it was a substitute for the nomen nudum *L. ursulae*. Whether this is indeed the same as the *L. jacobseniana* Swüste referred to a little later (1936: 104) remains uncertain because it seems that different material was involved.

L. karasmontana var. *lericheana*
= **L. karasmontana** ssp. **karasmontana**
L. karasmontana var. *opalina*
= **L. karasmontana** ssp. **karasmontana**
L. karasmontana var. *summitatum*
= **L. karasmontana** ssp. **karasmontana**
L. karasmontana var. *tischeri*
= **L. karasmontana** ssp. **karasmontana**
L. koelemanii = **L. aucampiae** ssp. **aucampiae**
L. kuibisensis = **L. schwantesii** ssp. **schwantesii**
L. kunjasensis = **L. schwantesii** ssp. **schwantesii**
L. lactea = **L. julii** ssp. **julii**
L. lateritia
= **L. karasmontana** ssp. **karasmontana**
L. lericheana
= **L. karasmontana** ssp. **karasmontana**

L. lesliei (N.E.Br.) N.E.Br. GC 1922: 65 ≡ *M. lesliei* N.E.Br. in Burtt-Davy TRSSA 1912: 369 **Syntypes** Burtt-Davy 5613, Leslie 4937, Marloth 4870 (K).

2–5(–10) heads; **L** truncate in profile, 25–49 mm *l*, 20–33 mm *b*, fissure shallow, faces flush, smooth to very slightly rugose, **Rub** absent; **Fl** yellow, ∅ 25–55 mm; **Fr** 5(–6) locules, profile boat-shaped, top flat to slightly convex, faces broadly elliptic; **Ecol** very common in stony grassland; **Distr** SE parts of Botswana; FS, GAU, N, NW, S Africa.

Note: Two subspecies are recognized on the basis of differences in the colours and markings of the faces and channels on the leaves.

Key to the Subspecies of L. lesliei

1. Faces various shades of brown or grey, with translucent greenish or brownish windows and channels .... **L. lesliei ssp. lesliei**
– Faces various shades of whitish grey, with very dark slate-grey windows and channels, the latter with numerous fine clavate lines extending into the margins .... **L. lesliei ssp. burchellii**

L. lesliei var. *applanata* = **L. lesliei** ssp. **lesliei**

L. lesliei (N.E.Br.) N.E.Br. ssp. **burchellii** D.T.Cole 1988: 217 **T** Cole 302 (PRE).

Faces mostly ± elliptic, **Mar** often not clearly distinguishable, otherwise very irregular with fine indentations and laciniations, **Isl** usually small and somewhat irregular, various shades of opaque whitish grey, or pale grey tinged with cream or yellowish or pinkish beige, **Win** usually ± occluded, occasionally ± completely open, always with numerous fine clavate lines extending into the margins, **Chan** sometimes narrow to broad, usually very narrow slightly impressed grooves constituting a fine mesh, opaque very dark slate-grey, sometimes almost black; **Ecol** calcrete with some jaspilite, grey-white with some red, brown, black; **Distr** Herbert, NC, S Africa.

Note: This subspecies is restricted to a small area northeast of Douglas.

L. lesliei var. *hornii* = **L. lesliei** ssp. **lesliei**

L. lesliei (N.E.Br.) N.E.Br. ssp. **lesliei** = *M. ferrugineum* Schwantes 1922: 93 ≡ *M. lesliei minor* Schwantes 1922: 94 nomen illeg. **T** Chauvin (?) = *L. lesliei* (N.E.Br.) N.E.Br. var. *applanata* De Boer in litt. nomen nudum = *L. lesliei* (N.E.Br.) N.E.Br. var. *hornii* De Boer 1966: 37 also as *L. lesliei* (N.E.Br.) N.E.Br. ssp. *lesliei* var. *hornii* De Boer in D.T.Cole 1988: 163[8] **T** Horn (?) = *L. lesliei* (N.E.Br.) N.E.Br. var. *luteoviridis* De Boer 1962: 87 **NT** Cole 20 (PRE) = *L. lesliei* (N.E.Br.) N.E.Br. var. *maraisii* De Boer 1962: 84 **T** Marais (?)

[8] For authorship, see Footnote 4.

= *L. lesliei* (N.E.Br.) N.E.Br. var. *mariae* D.T.Cole 1970: 9 also as *L. lesliei* (N.E.Br.) N.E.Br. ssp. *lesliei* var. *mariae* D.T.Cole in D.T.Cole 1988: 165[9] T Cole 141 (PRE) = *L. lesliei* (N.E.Br.) N.E.Br. var. *minor* De Boer 1961: 53 nomen invalid. 1962: 83 ≡ *L. lesliei* (N.E.Br.) N.E.Br. forma *minor* (De Boer) Fearn 1981: 40 ≡ *L. lesliei* (N.E.Br.) N.E.Br. ssp. *lesliei* var. *minor* (De Boer) D.T.Cole 1988: 166 T Blom (?) = *L. lesliei* (N.E.Br.) N.E.Br. var. *rubrobrunnea* De Boer 1962: 87 also as *L. lesliei* (N.E.Br.) N.E.Br. ssp. *lesliei* var. *rubrobrunnea* De Boer in D.T.Cole 1988: 167[10] **NT** Cole 17 (PRE) = *L. orpenii* L.Bolus in litt. nomen nudum = *L. venteri* Nel 1940: 11 ≡ *L. lesliei* (N.E.Br.) N.E.Br. var. *venteri* (Nel) De Boer & Boom 1961: 53 also as *L. lesliei* (N.E.Br.) N.E.Br. ssp. *lesliei* var. *venteri* Nel in D.T.Cole 1988: 167[11] T Venter (?).

Faces elliptic-reniform, **Mar** usually distinct but very irregular, **Isl** usually distinct, very irregular, small to large, few to many, irregularly dotted with numerous mini-windows, producing a filigree effect, various shades of opaque greyish yellow, orange, rusty red or brown, also of greenish, pinkish or violaceous grey, **Win** ± occluded, rarely ± open, usually few to many mini-windows in the islands and peninsulas, **Chan** narrow to broad, very irregular, various shades of translucent greyish green, greenish grey, brownish or reddish green, greenish brown; **Ecol** coarse conglomeritic sandstone, brown shaly siltstone, quartzite, ferruginous quartzite, pegmatitic granite, fine-grained feldspathic sandstone, amygdaloidal lava, chert, shale, banded jasper, brown, yellow-brown, grey-brown, pink, grey, grey-white, greyish green; **Distr** Southern and Southeast Districts, Botswana; FS, GAU, eastern districts of NC, NP, eastern part of NW, S Africa.

Note: This species has the largest distribution area in the genus **Lithops** from Kimberley to Gaborone to Pietersburg and Bethlehem, extending as far south as Zastron District. Six varieties were distinguished by Cole (1988: 161–169): **L. lesliei** var. *hornii*, var. *lesliei*, var. *mariae*, var. *minor*, var. *rubrobrunnea*, and var. *venteri*.

L. lesliei var. *luteoviridis* = **L. lesliei ssp. lesliei**
L. lesliei var. *maraisii* = **L. lesliei ssp. lesliei**
L. lesliei var. *mariae* = **L. lesliei ssp. lesliei**
L. lesliei var. *minor* = **L. lesliei ssp. lesliei**
L. lesliei forma *minor* = **L. lesliei ssp. lesliei**
L. lesliei var. *rubrobrunnea* = **L. lesliei ssp. lesliei**
L. lesliei var. *venteri* = **L. lesliei ssp. lesliei**
L. lineata = **L. ruschiorum**

[9] For authorship see Footnote 4.
[10] For authorship see Footnote 4.
[11] For authorship see Footnote 4.

L. localis (N.E.Br.) Schwantes FR 1938: 230 (see **photo**) ≡ *M. locale* N.E.Br. 1920: 68 T Burke (K) = *L. peersii* L.Bolus NM 1929: 84 ≡ *L. localis* (N.E.Br.) Schwantes var. *peersii* (L.Bolus) De Boer & Boom 1961: 55 T Peers 1262/25 (BOL) = *L. terricolor* N.E.Br. GC 1922: 65 T Frith (K).

2–5, also > 20 heads; **L** cordate in profile, 15–30 mm *l*, 12–24 mm *b*, fissure fairly deep, faces flush, mostly ± elliptic, smooth, **Mar** often not distinguishable, sometimes fairly well-defined, **Isl** usually fairly small, few to many, very irregular, faces various shades of opaque pale pinkish, yellowish, greenish or bluish grey, or pinkish, yellowish or reddish brown, or greyish blue, green or lilac, **Win** sometimes ± completely open or partly occluded, but usually absent, **Chan** usually not distinguishable, otherwise obscure, narrow to broad, very irregular, various shades of obscurely translucent dark greenish grey or greyish green, **Rub** absent, **dusky dots** usually very numerous, superficial and clearly visible; **Fl** yellow, sometimes with white centre, ∅ 12–35 mm; **Fr** 5(–6) locules, profile boat-shaped, top ± flat to slightly convex or peaked, faces broadly elliptic to almost round; **Ecol** shale, weathered fine-grained feldspathic sandstone, quartzite, quartzite with fine-grained sandstone, siltstone, banded cherts and jasper, dark grey, grey-brown, grey-white, yellow-brown, brown, some red, pink, white, black; **Distr** from Laingsburg, NC, to Steytlerville, EC, S Africa.

Note: Hammer (Haseltonia 1995: 89–91) has recently demonstrated that the type of *M. locale* is a member of the genus **Lithops**: the stout rootstock and the smooth seeds agree with features found in *L. terricolor* from the same area. The nomenclature is adjusted here, amending Cole 1988.

L. localis var. *peersii* = **L. localis**
L. loganiae = **L. aucampiae ssp. aucampiae**
L. lydiae = **L. fulviceps**
L. marginata = **L. hookeri**
L. marlothii
= **Conophytum pellucidum ssp. pellucidum**

L. marmorata (N.E.Br.) N.E.Br. GC 1922: 80 (see **photo**) ≡ *M. marmoratum* N.E.Br. 1920: 68 T Pillans (?) = *L. diutina* L.Bolus NM 1932: 345 T Meyer 9794 (BOL) = *L. elisae* De Boer 1961: 107 ≡ *L. marmorata* (N.E.Br.) N.E.Br. var. *elisae* (De Boer) D.T.Cole Excelsa 1973: 48 T Geyer (?) = *L. framesii* L.Bolus SAG 1930: 182 T syntypes Frames 1605/29, Pillans 6222 (BOL) = *L. herrei* L.Bolus forma *albiflora* H.Jacobsen in Jacobsen & Rowley NCSJ 1955: 81 **Lectotype**, here designated: Jacobsen 1955: 1473 f 1204 = *L. umdausensis* L.Bolus NM 1932: 347 T Herre 8780 (BOL).

2–6, also > 26 heads; **L** truncate-cordate in profile, 20–36 mm *l*, 15–23 mm *b*, fissure deep, lobes usually somewhat divergent, faces flush to elevated, mostly elliptic, smooth to slightly rugose, **Mar** usually distinct, occasionally fairly regular, mostly irregularly dentate or sinuate, inner margins sometimes extended slightly downwards along the outer edges of the fissures and tapered to fine points, **Isl** few to many, small to large, mostly irregular, sometimes reduced to indistinct mottling or flecks, various shades of opaque pale grey, often tinged with pale blue, green, cream or pink, **Win** usually ± occluded, **Chan** very narrow to broad, irregular, various shades of ± translucent grey, often tinted with pink, green or blue, or greyish, bluish or brownish green, **Rub** absent; **Fl** white, ∅ 20–45 mm; **Fr** (5–)6 locules, profile boat-shaped, top ± flat, faces broadly elliptic to almost round; **Ecol** quartzite, gneiss, pegmatite and gneiss, pegmatite and calcrete, pegmatite and schist, grey-white, white, light brown, pink, brown, red-brown, some greyish-green; **Distr** Namaqualand, NC, S Africa.

Note: Two varieties were distinguished by Cole (1988: 170–171): **L. marmorata** var. *elisae* and var. *marmorata.*

L. marmorata var. *elisae* = **L. marmorata**
L. marthae = **L. schwantesii** ssp. **schwantesii**
L. maughanii = **L. julii** ssp. **fulleri**
L. mennellii = **L. bromfieldii**

L. meyeri L.Bolus SAG 1932: 102 T Meyer 9698 (BOL).

2–3(–5) heads; **L** bicuneate in profile, 25–35 mm *l*, 15–24 mm *b*, fissure deep, lobes distinctly divergent, faces usually somewhat elevated, elongated elliptic or slightly reniform, smooth, **Mar** not clearly distinguishable, **Isl** absent or barely perceptible, faces opaque pale whitish grey tinged with cream, yellow, milky green or pink, **Win** barely perceptible, cloudy and obscure, **Chan** absent or barely perceptible, various shades of dull greenish grey, **Rub** absent; **Fl** yellow with white centre, ∅ 20–40 mm; **Fr** 5(–6) locules, profile boat-shaped, top flat to slightly convex with a light brown line just below the edge, faces broadly elliptic to elliptic, with hinge-rim; **Ecol** quartzite, white, with some yellow-brown, brown and black; **Distr** Namaqualand, NC, S Africa.

Note: The species is restricted to a small area near Lekkersing in the Richtersveld.

L. mickbergensis
= **L. karasmontana** ssp. **karasmontana**
L. mundtii
= **L. pseudotruncatella** ssp. **pseudotruncatella**
L. namtibensis = **L. gesineae**
L. namtipbergensis = **L. gesineae**

L. naureeniae D.T.Cole 1980: 74 (see **photo**) T Cole 304 (PRE).

2–5, also > 28 heads; **L** bicuneate-cordate in profile, 20–35 mm *l*, 18–25 mm *b*, fissure deep, lobes somewhat divergent, faces flush to slightly elevated, mostly elongated elliptic, smooth, **Mar** distinct, narrow, sometimes dentate, usually very irregular with numerous peninsulas, inner margins usually extended deeply downwards and drawn to fine points, **Isl** usually well-defined but irregular, few to many, small to large, sometimes reduced to flecks, various shades of opaque pale grey or beige, sometimes tinged with pink, cream or green, **Win** ± occluded, infrequently ± completely open, **Chan** narrow to broad, irregular, various shades of translucent greyish green or greenish, brownish or reddish grey, **Rub** absent; **Fl** lustrous yellow with white centre, ∅ 20–35 mm; **Fr** 5(–6) locules, profile boat-shaped, top flat to slightly convex, faces broadly elliptic to almost round; **Ecol** reddish-brown and pink gneiss; **Distr** Namaqualand, NC, S Africa.

Note: Although the species occurs only in a small area near Springbok, it is very variable.

L. nelii = **L. ruschiorum**

L. olivacea L.Bolus NM 1929: 84 (see **photo**) T Fuller 1725/28 (BOL) = *L. olivacea* L.Bolus var. *nebrownii* D.T.Cole 1988: 227 T Cole 162 (PRE).

3–10(–30) heads; **L** truncate in profile, 15–24 mm *l*, 12–18 mm *b*, fissure fairly deep, lobes sometimes slightly divergent, faces flush, ± round, sometimes slightly elliptic, smooth, **Mar** very distinct, broad to narrow, regular to slightly dentate, **Isl** usually few, relatively small and quite distinct, various shades of opaque creamy grey, often tinged with yellow, green, pink, buff, pale salmon or mauve, **Win** mostly ± open, **Chan** usually absent, various shades of translucent green or bluish, greyish, or orange-green, or greenish or mauvish grey, **Rub** absent; **Fl** yellow with white centre, ∅ 20–45 mm; **Fr** 5(–6) locules, profile boat-shaped, top ± flat, faces broadly elliptic to elliptic, valves slightly concave, with very distinct raised rims; **Ecol** translucent white and red-stained quartzite, some grey, brown; **Distr** Namaqualand, NC, S Africa.

Note: Two varieties were distinguished by Cole (1988: 175–177): **L. olivacea** var. *nebrownii* and var. *olivacea* occurring in a small area west and southwest from Pofadder to Aggeneys.

L. olivacea var. *nebrownii* = **L. olivacea**
L. opalina = **L. karasmontana** ssp. **karasmontana**

L. optica (Marloth) N.E.Br. GC 1922: 80 ≡ *M. opticum* Marloth TRSSA 1910: 405 ≡ *M. marginatum* Marloth TRSSA 1910: 406 name used erroneously for *M. opticum* T Marloth 4675 (BOL) = *L. optica* (Marloth) N.E.Br. var. nova Triebner in Jacobsen 1955: 1488 nomen illeg. because "nova" is not permitted as an epithet = *L. elevata* L.Bolus NM 1932: 345 T Triebner 48 (BOL) = *L. rubra* (Tischer) N.E.Br. GC 1926: 116 ≡ *M. opticum* Marloth var. *rubrum* Tischer ZSK 1925: 65 ≡ *L. optica* (Marloth) N.E.Br. var. *rubra* (Tischer) Tischer MDG 1926: 330

≡ *L. optica* (Marloth) N.E.Br. forma *rubra* (Tischer) H.Jacobsen 1933: 148 T = *L. schlechteri* G.D.Rowley 1952: 61 nomen nudum T?

2–5, also > 20 heads; L cordate-truncate in profile, occasionally ± flat, 18–30 mm *l*, 13–20 mm *b*, fissure deep, lobes often somewhat divergent, faces slightly elevated, mostly ± reniform, smooth, **Mar** distinct, regular to slightly irregular, but in the maculate form often irregularly dentate or sinuate, **Isl** usually absent except in the maculate form, where they are often numerous and may be fairly large, opaque whitish grey, grey or beige, tinged with green, pink, blue or yellow, **Win** usually completely open, in the maculate form often reduced to narrow irregular channels, **Chan** mainly in the maculate form, usually broad but irregular, sometimes narrow and very irregular, various shades of translucent greenish grey, grey or bluish grey, **Rub** absent; **Fl** white, often with pink tips, ∅ 10–20 mm; **Fr** mostly 5 locules, profile boat-shaped, top ± flat, usually with a dark line just below the edge, faces broadly elliptic, with hinge-rim; **Ecol** biotitic gneiss, gneiss, quartzite, quartzite and gneiss, peg-matite and gneiss, sandstone, quartzite and calcrete, limestone and calcrete, calcrete and shale, grey-white, grey, pink, some yellow-brown, brown, red, greenish-grey, black; **Distr** Lüderitz-South, Namibia.

Note: The species occurs in a narrow strip ± 200 km long but only ± 30 km wide, extending southwards along the coast from ± 20 km north of Lüderitz. Cole (1988: 178) treated *L. optica* forma *rubra* as a cultivar: **L. optica** cv. **Rubra.**

L. optica var. nova = **L. optica**
L. optica var. *rubra* = **L. optica**
L. optica forma *rubra* = **L. optica**
L. orpenii = **L. lesliei** ssp. **lesliei**

L. otzeniana Nel 1937: 123 T Nel (?).

2–5, also > 23 heads; L cordate-truncate in profile, 20–30 mm *l*, 15–20 mm *b*, fissure deep, lobes slightly divergent, faces flush to slightly elevated, mostly ± elliptic, smooth to slightly rugose, **Mar** usually very distinct, typically scalloped with rounded peninsulas, **Isl** few to many, usually fairly large, regular, rounded, and mostly situated together with the peninsulas along the outer and inner margins, various shades of opaque light grey, usually tinged with pale pink, cream, green or blue, **Win** usually ± occluded, infrequently ± completely open, **Chan** broad to narrow, irregular, various shades of translucent bluish or greyish green, or greenish, bluish or pinkish grey, **Rub** absent; **Fl** yellow with white centre, ∅ 15–35 mm; **Fr** mostly 5 locules, profile boat-shaped, top flat, faces broadly elliptic to elliptic; **Ecol** gneiss, grey-white, grey, light brown, greenish grey-white; **Distr** Calvinia, NC, S Africa.

Note: The species is clearly distinguished by its scalloped edges. It occurs in a small area to the north and northwest of Loeriesfontein.

L. otzeniana var. *weberi* = **L. comptonii**
L. peersii = **L. localis**
L. pillansii = **L. ruschiorum**
L. pringlei = **L. dinteri** ssp. **multipunctata**

L. pseudotruncatella (A.Berger) N.E.Br. GC 1922: 65 ≡ *M. pseudotruncatellum* A.Berger 1908: 282, 289–90 ≡ *M. truncatellum* Dinter GC 1900: 211 nomen illeg. non Haw. 1803: 22, **Lectotype**, here designated:[12] Dinter in GC 1900: 211 f 68 as *M. truncatellum.*

(1–)2–4(–20) heads; L truncate in profile, fissure shallow, faces flush; **Fl** yellow, ∅ 20–52 mm; **Fr** 6(–8) locules, profile usually rotund, top usually convex or peaked, faces mostly round to elliptic, with hinge-rim; **Ecol** in hilly grassland; **Distr** from Okahandja to Maltahöhe, Namibia.

Note: Five subspecies are recognized on the basis of differences in the presence or relative absence of rubrications, the structure of the channels and the colours of the faces.

Key to the Subspecies of L. pseudotruncatella

1. Rubrications usually an extensive network 2
– Rubrications absent or obscure 3
2. Channels not clearly defined, obscurely translucent dull green or greyish-green .... **L. pseudotruncatella** ssp. **pseudotruncatella**
– Channels a fairly regular and distinct network, obscurely translucent shades of brown .... **L. pseudotruncatella** ssp. **dendritica**
3. Rubrications usually absent, faces opaque pale grey-white, channels absent or fine dots and/or dashes .... **L. pseudotruncatella** ssp. **groendrayensis**
– Rubrications usually present but indistinct or obscure 4
4. Faces opaque, milky grey-green .... **L. pseudotruncatella** ssp. **archerae**
– Faces opaque pale grey-white .... **L. pseudotruncatella** ssp. **volkii**

[12] Berger (1908: 289) cited the figure of Dinter 1900: 211, but did not designate it as the type.

L. pseudotruncatella forma *albiflora*
= **L. pseudotruncatella** ssp. **pseudotruncatella**
L. pseudotruncatella var. *albiflora*
= **L. pseudotruncatella** ssp. **pseudotruncatella**
L. pseudotruncatella var. *alpina*
= **L. pseudotruncatella** ssp. **pseudotruncatella**
L. pseudotruncatella var. *alta*
= **L. pseudotruncatella** ssp. **pseudotruncatella**

L. pseudotruncatella (A.Berger) N.E.Br. ssp. **archerae** (De Boer) D.T.Cole 1988: 217 ≡ *L. archerae* De Boer 1967: 122 ≡ *L. pseudotruncatella* (A.Berger) N.E.Br. var. *archerae* D.T.Cole Excelsa 1973: 45 **T** Archer (?).

1–2(–10) heads; **L** 30–40 mm *l*, 20–30 mm *b*, faces mostly round-elliptic, smooth to very slightly rugose, **Mar, Win, Chan, Isl** usually absent or very obscure, **Chan** occur only as a very slightly impressed broken network of narrow grooves, faces usually various shades of opaque milky grey-green, sometimes bluish or yellowish, **Rub** rarely absent, usually a number of broken lines, hooks and dots, but occasionally a coarse broken network; **Ecol** pinkish-grey and grey-white dolomite and limestone, some brown and grey; **Distr** Maltahöhe, Namibia.

Note: The inclusion of this taxon in synonymy of **L. schwantesii** by Friedrich (1970: 80) cannot be supported.

L. pseudotruncatella var. *archerae*
≡ **L. pseudotruncatella** ssp. **archerae**
L. pseudotruncatella var. *brandbergensis*
≡ **L. gracilidelineata** ssp. **brandbergensis**

L. pseudotruncatella (A.Berger) N.E.Br. ssp. **dendritica** (Nel) D.T.Cole 1988: 218 ≡ *L. dendritica* Nel 1946: 68 t. 6 ≡ *L. pseudotruncatella* (A.Berger) N.E.Br. var. *dendritica* (Nel) De Boer & Boom 1961: 67 **T** Otzen figured in Nel 1946: t. 6 f 2 = *L. pseudotruncatella farinosa* Dinter in Graessner 1928: 7 nomen nudum et illeg. ≡ *L. farinosa* Dinter ex H.Jacobsen 1933: 145 **T**? = *L. pseudotruncatella pulmonuncula* Dinter in Graessner 1928: 7 nomen nudum et illeg. ≡ *L. pseudotruncatella* (A.Berger) N.E.Br. forma *pulmonuncula* Dinter ex H.Jacobsen 1933: 150 ≡ *L. pseudotruncatella* (A.Berger) N.E.Br. var. *pulmonuncula* (Dinter ex H.Jacobsen) Dinter ex H.Jacobsen 1935: 203 ≡ *L. pseudotruncatella* (A.Berger) N.E.Br. ssp. *pulmonuncula* (Dinter ex H.Jacobsen) Dinter ex Schwantes 1957: 212 **T**?

Seldom more than 1 head, up to 4; **L** 30–45 mm *l*, 25–37 mm *b*, faces mostly round-elliptic, smooth to slightly rugose, **Mar** usually not clearly distinguishable, irregularly incised by simple or branched terminations of the channels, **Isl** manifest as numerous opaque areas enclosed by channels, faces (**Marg** and **Isl**) various shades of opaque pink, bluish pink, orange-brown or greenish, yellowish or pale grey, **Win** absent or very obscure, **Chan** usually a regular and distinct network of narrow and slightly impressed grooves, obscurely translucent light brown or greenish, yellowish or pinkish brown, **Rub** in the channels, sometimes absent, usually an extensive fairly regular network of connected lines, sometimes reduced to few or many dashes, hooks or dots; **Ecol** quartzite, white and red-stained white, brown, red-brown, some grey, black; **Distr** Rehoboth, Windhoek, Namibia.

Note: The very strongly marked and reddish-coloured forms are among the most attractive in the genus.

L. pseudotruncatella var. *dendritica*
≡ **L. pseudotruncatella** ssp. **dendritica**
L. pseudotruncatella var. *elisabethae*
= **L. pseudotruncatella** ssp. **pseudotruncatella**
L. pseudotruncatella farinosa
= **L. pseudotruncatella** ssp. **dendritica**
L. pseudotruncatella var. *gracilidelineata*
≡ **L. gracilidelineata** ssp. **gracilidelineata**
L. pseudotruncatella var. *gracilidelineata* forma *waldronae*
= **L. gracilidelineata** ssp. **gracilidelineata**

L. pseudotruncatella (A.Berger) N.E.Br. ssp. **groendrayensis** (H.Jacobsen) D.T.Cole 1988: 222 ≡ *L. pseudotruncatella* (A.Berger) N.E.Br. var. *groendrayensis* H.Jacobsen 1961: 169 ≡ *L. vallis-mariae* (Dinter & Schwantes) N.E.Br. var. *groendraaiensis* (H.Jacobsen) De Boer 1967: 179 **T** Triebner (?).

1 head, very rarely 2; **L** 30–53 mm *l*, 25–44 mm *b*, faces round-elliptic, smooth, minutely wrinkled or dotted with minute "pin-pricks", **Mar, Win, Isl** absent or very obscure, faces opaque pale grey-white, sometimes faintly tinted with blue, pink or green, **Chan** mostly absent or very obscurely visible only as fine dots, hooks and dashes, very obscurely translucent greenish grey, **Rub** usually absent, or visible only as very obscure subcutaneous dots or very short lines; **Ecol** quartzite, white, some red-stained white, pink, yellow-brown, brown, greyish green, red; **Distr** Rehoboth, Namibia.

Note: This subspecies has some features reminiscent of **L. vallis-mariae**, and it is possible that it may be an ancient, well established hybrid.

L. pseudotruncatella var. *groendrayensis*
≡ **L. pseudotruncatella** ssp. **groendrayensis**
L. pseudotruncatella ssp. *mundtii*
= **L. pseudotruncatella** ssp. **pseudotruncatella**
L. pseudotruncatella var. *mundtii*
= **L. pseudotruncatella** ssp. **pseudotruncatella**
L. pseudotruncatella forma *mundtii*
= **L. pseudotruncatella** ssp. **pseudotruncatella**

L. pseudotruncatella (A.Berger) N.E.Br. ssp. **pseudotruncatella** = *L. pseudotruncatella* (A.Berger) N.E.Br. forma *albiflora* H.Jacobsen in Jacobsen & Rowley NCSJ 1955: 181 ≡ *L. pseudotruncatella* (A.Berger) N.E.Br. var. *albiflora* (H.Jacobsen) H.Jacobsen 1955: 1489 nomen invalid. T? = *L. alpina* Dinter MDG 1927: 123 ≡ *L. pseudotruncatella* (A.Berger) N.E.Br. var. *alpina* Dinter MDG 1927: 123 T Rusch (?) = *L. pseudotruncatella* (A.Berger) N.E.Br. var. *alta* Tischer MDG 1926: 331 T Mundt (?) = *L. elisabethae* Dinter 1933: 50 ≡ *L. pseudotruncatella* (A.Berger) N.E.Br. var. *elisabethae* (Dinter) De Boer & Boom 1961: 66–67 also as *L. pseudotruncatella* (A.Berger) N.E.Br. ssp. *pseudotruncatella* var. *elisabethiae* (Dinter) De Boer & Boom in D.T.Cole 1988: 183[13] T Schneider s.n. (?) = *M. evexum* N.E.Br. nomen nudum on drawing at K = *L. mundtii* Tischer MDG 1926: 330 ≡ *L. pseudo-truncatella* (A.Berger) N.E.Br. forma *mundtii* (Tischer) H.Jacobsen 1933: 149 ≡ *L. pseudotruncatella* (A.Berger) N.E.Br. var. *mundtii* (Tischer) H.Jacobsen 1935: 203 ≡ *L. pseudotruncatella* (A.Berger) N.E.Br. ssp. *mundtii* (Tischer) Schwantes 1957: 211 nomen invalid. T Mundt (?) = *L. pseudotruncatella* (A.Berger) Schwantes var. *riehmerae* D.T.Cole 1987: 45 ≡ *L. pseudotruncatella* (A.Berger) Schwantes ssp. *pseudotruncatella* var. *riehmerae* D.T.Cole 1988: 183 ≡ *L. edithae* H.Jacobsen 1955: 1466 erroneously applied to this variety **T** Cole 97 (PRE).

2–4, also > 20 heads; **L** 20–50 mm *l*, 20–35 mm *b*, faces ± reniform, smooth to very slightly rugose, **inner Mar** usually fairly distinct, outer margins not clearly distinguishable, **Isl** irregular and not distinctly outlined, few to many, small to large, faces opaque pale grey or beige tinted with various shades of yellow-brown, orange-brown, brown, pink, greenish yellow, mauve or blue, **Win** usually not clearly distinguishable, **Chan** not clearly defined, very irregular, slightly impressed, obscurely translucent dull greenish or bluish grey or greyish green, **Rub** usually irregular and extensively branched, forming a broken network, but sometimes reduced to scattered short lines, dashes and dots; **Ecol** quartzite, mica schist, white, red-stained, yellow-brown, brown, some grey, dark brown, red-brown; **Distr** Gobabis, Okahandja, Windhoek, Namibia.

Note: Three varieties were distinguished by Cole (1988: 181–184): **L. pseudotruncatella** ssp. **pseudotruncatella** var. *elisabethiae*, var. *pseudotruncatella*, and var. *riehmerae*, mainly on the basis of surface colour of the leaves. The variety *albiflora* is considered a cultivar cv. Albiflora by Cole (1988: 216).

[13] For authorship, see Footnote 4.

L. pseudotruncatella pulmonuncula
= **L. pseudotruncatella** ssp. **dendritica**
L. pseudotruncatella ssp. *pulmonuncula*
= **L. pseudotruncatella** ssp. **dendritica**
L. pseudotruncatella var. *pulmonuncula*
= **L. pseudotruncatella** ssp. **dentritica**
L. pseudotruncatella forma *pulmonuncula*
= **L. pseudotruncatella** ssp. **dentritica**
L. pseudotruncatella var. *riehmerae*
= **L. pseudotruncatella** ssp. **pseudotruncatella**

L. pseudotruncatella (A.Berger) N.E.Br. ssp. **volkii** (De Boer & Boom) D.T.Cole 1988: 233 ≡ *L. volkii* Schwantes ex H.Jacobsen 1955: 1498 nomen invalidum ≡ *L. pseudotruncatella* (A.Berger) N.E.Br. var. *volkii* De Boer & Boom 1961: 68 T Herre (?).

2(–4) heads; **L** 25–38 mm *l*, 17–26 mm *b*, faces mostly slightly reniform, smooth to very slightly rugose, **Mar** and **Win** absent or not clearly distinguishable, **Isl** manifest only as opaque areas enclosed by channels, faces slightly polished and marble-like, opaque milky bluish white, sometimes faintly tinged with pink, **Chan** very obscurely visible as a coarse broken network or very slightly impressed slender lines, dashes and dots, obscurely translucent bluish grey, sometimes tinted pinkish or brownish, **Rub** sometimes absent, otherwise barely discernible as a faint broken network or short lines, dashes and dots; **Ecol** quartzite with some calcite, white, red-stained white, some yellow-brown, grey, dark brown; **Distr** Windhoek, Namibia.

Note: The polished marble-like surface of the leaves is a distinctive feature of this subspecies.

L. pseudotruncatella var. *volkii*
≡ **L. pseudotruncatella** ssp. **volkii**
L. rooi = **L. bromfieldii**
L. rouxii = **L. bromfieldii**
L. rubra = **L. optica**
L. rugosa = **L. schwantesii** ssp. **schwantesii**
L. ruschiana ≡ **L. ruschiorum** var. **ruschiorum**

L. ruschiorum (Dinter & Schwantes) N.E.Br. GC 1926: 194 ≡ *M. ruschiorum* Dinter & Schwantes ZSK 1925: 21 ≡ *M. ruschii* Dinter & Schwantes ZSK 1925: 22 orthographic variant ≡ *L. ruschiana* N.E.Br. GC 1926: 116 non *M. ruschianum* Dinter 1923: 25 T Rusch (?) = *L. ruschiorum* (Dinter & Schwantes) N.E.Br. ssp. *stiepelmannii* Schwantes 1957: 214 nomen nudum T Triebner (?) = *L. lineata* Nel 1946: 112 ≡ *L. ruschiorum* (Dinter & Schwantes) N.E.Br. var. *lineata* (Nel) De Boer & Boom 1964: 54 **Lectotype**, here designated: Nel 1946: t. 24 f 1 = *L. nelii* Schwantes BSKP 1939: 43 ≡ *L. ruschiorum* (Dinter & Schwantes) N.E.Br. ssp. *nelii* (Schwantes) Schwantes 1957: 214 nomen invalid. T Triebner (?) = *L. pillansii* L. Bolus NM 1929: 84 T Pillans 5952 (BOL).

2–5(–30) heads; L cordate in profile, usually very distinctly convex, 20–38 mm *l*, 15–27 mm *b*, fissure deep, faces somewhat elevated, elliptic-reniform, smooth to very slightly rugose, **Mar** and **Win** absent, **Isl** manifest only as areas ± enclosed by channels, faces opaque, mostly greyish white or cream, sometimes yellowish or pinkish white, **Chan** often absent, otherwise a broken network of slightly impressed grooves and pits, sometimes reduced to scattered short lines and dots, usually filled by the rubrications and seldom perceptible as greenish to reddish brown, **Rub** often absent, otherwise a number of lines, dashes or dots set in the grooves and pits, sometimes forming a coarse broken network; **Fl** yellow, ∅ 12–30 mm; **Fr** 5(–6) locules, profile boat-shaped, top ± flat, faces broadly elliptic; **Ecol** rose quartz and pegmatite, pegmatite, pegmatite and calcrete, pegmatite and black chert, veined quartz and calcite, veined quartz, calcite and calcrete, quartzite, calcite and some tourmaline, pink, grey-white, with some light brown, grey, black; **Distr** from Kaokoveld to Swakopmund, Namibia.

Note: Two varieties were distinguished by Cole (1988: 189–191): **L. ruschiorum** var. *lineata* and var. *ruschiorum*. The species occurs in a long narrow strip extending northwestwards along the coast from Swakopmund for ± 300 km, and inland up to ± 75 km.

L. ruschiorum var. *lineata* = **L. ruschiorum**
L. ruschiorum ssp. *nelii* = **L. ruschiorum**
L. ruschiorum ssp. *stiepelmannii* = **L. ruschiorum**

L. salicola L.Bolus NM 1936: 33 T Verdoorn 1622 (BOL).

2–5, also > 50 heads; L truncate in profile, 17–35 mm *l*, 11–26 mm *b*, fissure shallow, faces flush, elliptic-reniform, smooth, **Mar** very distinct, usually fairly regular, sometimes irregularly dentate or sinuate, **Isl** often absent, otherwise mostly fairly small, few to many, various shades of opaque light greenish, yellowish, pinkish or lilac grey, **Win** sometimes ± occluded, usually completely open, **Chan** usually absent, otherwise broad to narrow, various shades of obscurely translucent grey, greenish or bluish grey or greyish green or lilac, **Rub** usually absent, otherwise dots and short dashes mostly in the marginal indentations; **Fl** white, ∅ 20–50 mm; **Fr** 5(–6) locules, profile boat-shaped, top flat to slightly convex, faces elliptic to broadly elliptic; **Ecol** calcrete with calcareous shale and dolomite, also in the clay of very shallow pans, with little or no stone, grey-white, yellowish white, greenish grey, some grey, yellow, brown, black; **Distr** Fauresmith, Koffiefontein, FS, Hopetown, Philipstown, NC, S Africa.

Note: Extensive field studies (Cole 1988: 70) indicate that plants described under the name *L. salicola* L.Bolus var. *reticulata* De Boer 1960: 79 (T Geyer (?)) represent a hybrid population based on **L. salicola** and **L. hallii.**

L. salicola var. *reticulata* = **L. hallii**
L. schlechteri = **L. optica**

L. schwantesii Dinter in Graessner 1928: 14 T Rusch (?).

1–3, also > 15 heads; L truncate in profile, 20–40 mm *l*, 15–30 mm *b*, fissure shallow, faces flush, mostly elliptic-reniform, smooth to rugose; **Fl** yellow, often with longish pedicels, ∅ 20–36 mm; **Fr** 5(–6) locules, profile boat-shaped, top ± flat, faces elliptic to broadly elliptic; **Ecol** stony hills and low ridges; **Distr** Bethanie, Lüderitz-South, Maltahöhe, Namibia.

Note: Two subspecies are recognized on the basis of differences in structure and colours of windows and channels.

Key to the Subspecies of L. schwantesii

1. Windows and channels obscurely to pellucidly translucent grey-green, rubrications rarely absent, usually a prominent network of bold lines .... **L. schwantesii** ssp. **schwantesii**
– Windows absent, channels reduced to a network of grooves or furrows, largely taken up by rubrications, faces opaque to very obscurely translucent shades of brown or brownish grey .... **L. schwantesii** ssp. **gebseri**

L. schwantesii var. *christinae*
= **L. schwantesii** ssp. **schwantesii**

L. schwantesii Dinter ssp. **gebseri** (De Boer) D.T.Cole 1988: 221 ≡ *L. schwantesii* Dinter var. *gebseri* De Boer 1964: 110 T Gebser (?).

Mar usually not clearly distinguishable, **Isl** manifest as small opaque areas ± enclosed by grooves, faces various shades of opaque to very obscurely translucent purplish brown, grey-brown, reddish grey or yellowish grey, **Win** absent, **Chan** reduced to a fine network of slightly impressed grooves or furrows, obscure dark greenish or bluish grey, **Rub** occasionally indistinct, usually a broken network of fine branched lines, dashes, hooks and/or dots; **Ecol** dark brown ferruginous sandstone; **Distr** Bethanie, Maltahöhe, Namibia.

Note: Plants of this subspecies appear to be softer to the touch than those of the other subspecies.

L. schwantesii var. *gebseri*
≡ **L. schwantesii** ssp. **gebseri**
L. schwantesii var. *kunjasensis*
= **L. schwantesii** ssp. **schwantesii**
L. schwantesii var. *marthae*
= **L. schwantesii** ssp. **schwantesii**

L. schwantesii var. *nutupsdriftensis*
= **L. schwantesii** ssp. **schwantesii**
L. schwantesii var. *rugosa*
= **L. schwantesii** ssp. **schwantesii**

L. schwantesii Dinter ssp. **schwantesii** (see **photo**) = *L. schwantesii* Dinter var. *nutupsdriftensis* De Boer 1964: 110 **T** Gebser (?) = *L. christinae* De Boer 1958: 2 ≡ *L. schwantesii* Dinter var. *christinae* (De Boer) Fearn 1970: 92 **T** Geyer (?) = *L. gulielmi* L.Bolus NM 1937: 100–01 **T** Triebner 21976 (BOL) = *L. kuibisensis* Dinter ex H.Jacobsen 1933: 147 **T**? = *L. kunjasensis* Dinter in Graessner 1928: 14 ≡ *L. schwantesii* Dinter var. *kunjasensis* (Dinter) De Boer & Boom 1961: 75 **T** Rusch (?) = *L. marthae* Loesch & Tischer in Jacobsen 1933: 148 nomen nudum ≡ *L. inornata* Dinter ex H.Jacobsen 1935: 198 nomen invalid. ≡ *L. marthae* Loesch & Tischer 1936: 74 ≡ *L. schwantesii* Dinter var. *marthae* D.T.Cole Mad 1973: 34 ≡ *L. dinteri* Schwantes var. *marthae* (Loesch & Tischer) Fearn 1970: 92 **T** Erni (?) = *L. rugosa* Dinter in Graessner 1928: 15 ≡ *L. schwantesii* Dinter var. *rugosa* (Dinter) De Boer & Boom 1961: 75 **T** Rusch (?) = *L. sociorum* Triebner, Hall & Carp in hort. nomen nudum = *L. triebneri* L.Bolus NM 1933: 391 **T** Triebner 602 (BOL) = *L. urikosensis* Dinter in Graessner 1928: 15 ≡ *L. schwantesii* Dinter var. *urikosensis* (Dinter) De Boer & Boom 1961: 76 **T** Rusch (?).

Mar often prominently banded, but usually rather fuzzy and not clearly defined, **Isl** few to many, small to large, usually indistinct, sometimes clearly defined areas enclosed by grooves, shades of opaque yellowish, pinkish, orange, brownish or reddish grey, **Win** occasionally absent, usually ± occluded, sometimes ± completely open, **Chan** irregular, usually fairly broad, occasionally no more than slightly impressed grooves or furrows, obscurely to pellucidly translucent grey-green, often with red, orange-brown, purple-brown, greyish purple, grey-brown, blue or yellowish tints, **Rub** rarely absent, usually an entire or broken network of fairly regular bold lines, sometimes reduced to a number of isolated short lines, dashes and dots; **Ecol** sandstone, sandstone and quartzite, ferruginous quartzite, quartzite and gneiss, granitic gneiss, gneiss and pegmatite, calcrete and quartzite, calcrete and limestone, brown, red-brown, pinkish brown, red-stained white, greyish brown, greyish green, grey-white; **Distr** Bethanie, Lüderitz-South, Maltahöhe, Namibia.

Note: Four varieties were recognized by Cole (1988: 194–198): **L. schwantesii** var. *marthae*, var. *rugosa*, var. *schwantesii*, and var. *urikosensis*.

L. schwantesii var. *urikosensis*
= **L. schwantesii** ssp. **schwantesii**
L. sociorum = **L. schwantesii** ssp. **schwantesii**

L. steineckeana Tischer 1951: 37 8 **T** collector unknown, origin unknown (T?).

1–2(–4) heads; L ovate-cordate in profile, 15–23 mm *l*, 13–19 mm *b*, fissure very shallow, faces elevated, round-elliptic, smooth, **Mar** often absent, otherwise not clearly distinguishable, **Isl** absent or not clearly distinguishable, faces opaque pale grey, buff or cream, tinged with pink, pale yellow, or light orange-brown, **Win** often absent, otherwise reduced to small obscurely translucent areas, various shades of obscurely translucent pale grey or greenish grey, **Chan** absent or not clearly distinguishable, **Rub** usually absent, sometimes visible as a few dots, dashes and irregular lines; **Fl** yellow, ∅ 15–32 mm; **Fr** (6–)7 locules, profile boat-shaped to rotund, top strongly convex, faces almost round to broadly elliptic; **Ecol** not known; **Distr** not known.

Note: Almost certainly a horticultural hybrid, not a natural species (Cole 1988: 200–201).

L. streyi = **L. gracilidelineata** ssp. **gracilidelineata**
L. summitatum = **L. karasmontana** ssp. **karasmontana**
L. terricolor = **L. localis**
L. translucens = **L. herrei**
L. triebneri = **L. schwantesii** ssp. **schwantesii**

L. turbiniformis (Haw.) N.E.Br. 1920: 137 ≡ *M. turbiniforme* Haw. 1821: 84 non N.E.Br. 1920: 137 GC 1921: 250 **Lectotype**, here designated: drawing by Burchell (Africana Museum, Johannesburg) figured as frontispiece in Cole 1988.

Body obconical, truncate; **L** with obscure dots appearing more numerously around the margins of the face of the plant; **Ecol** on stony ground; **Distr** Prieska, NC, S Africa.

Note: There are no longer any plants to be found at the locality where Burchell reported having found his specimens (see Cole 1988: 38–41). The name was for many years used for plants of **L. hookeri**.

L. turbiniformis sensu N.E.Br. GC 1922: 55 = **L. hookeri**
L. turbiniformis var. *brunneo-violacea* = **L. hookeri**
L. turbiniformis var. *dabneri* = **L. hookeri**
L. turbiniformis var. *elephina* = **L. hookeri**
L. turbiniformis var. *lutea* = **L. hookeri**
L. turbiniformis var. *marginata* = **L. hookeri**
L. turbiniformis var. *subfenestrata* = **L. hookeri**
L. turbiniformis var. *susannae* = **L. hookeri**
L. umdausensis = **L. marmorata**
L. upingtonensis = **L. bromfieldii**
L. urikosensis = **L. schwantesii** ssp. **schwantesii**
L. ursulae = **L. karasmontana** ssp. **karasmontana**

L. vallis-mariae (Dinter and Schwantes) N.E.Br. GC 1926: 117 ≡ *M. vallis-mariae* Dinter & Schwantes ZSK 1925: 22 **T** Beetz (?) *L. vallis-mariae* (Dinter & Schwantes) N.E.Br. var. *margarethae* de Boer 1969: 9 **T** Geyer (?).

2–4(–10) heads; L truncate in profile, 25–40 mm *l*, 20–30 mm *b*, fissure shallow, faces flush, mostly ± reniform, slightly rugose, very minutely wrinkled or extensively dotted with pin-pricks, **Mar** and **Win** absent or very obscure, **Isl** manifest only as areas ± enclosed by channels, faces opaque pale grey-white, or grey-white tinged with creamy yellow, green, beige or pink, **Chan** often absent, otherwise a number of impressed dots, hooks and short lines usually less than 0,5 mm wide, sometimes quite deeply pitted, channels opaque to very obscurely translucent grey, bluish grey, pinkish grey or pinkish brown, **Rub** absent; **Fl** straw yellow, sometimes orange-yellow, may be tinged with bronze or pink, ∅ 15–30 mm; **Fr** mostly 5 locules, profile boat-shaped, flat, occasionally slightly convex, faces broadly elliptic; **Ecol** rounded quartzites and sandstone, calcrete with some quartzite, also in clay in a shallow pan with very few sandstone and quartzite pebbles, yellow-white, yellow-brown, grey-white, some brown, reddish brown, white, grey, black; **Distr** Gibeon, Keetmanshoop, Namibia.

Note: The minutely pitted top surface of the species is very distinctive. Occurring in the Hardap and Karas Regions, in a large quadrangular area bounded approximately by Berseba, a point 30 km northwest of Mariental, Stampriet and a point ± 40 km north of Aroab.

L. vallis-mariae var. *groendraaiensis*
≡ **L. pseudotruncatella** ssp. **groendrayensis**
L. vallis-mariae var. *margarethae* = **L. vallis-mariae**
L. vanzylii ≡ **Dinteranthus vanzylii**
L. venteri = **L. lesliei** ssp. **lesliei**

L. verruculosa Nel 1943: 9 **T** van der Westhuizen figured in Nel 1943: 9 f 8 (?) = *L. inae* Nel 1943: 3 **T** van der Westhuizen figured in Nel 1943: 4 f 2 = *L. geoffreyi* Pringle & McLachlan nomen nudum in litt. ≡ *L. verruculosa* Nel var. *glabra* De Boer 1966: 107 **T** syntypes McLachlan, Pringle & Roux (?).

2–4(–8) heads; **L** truncate in profile, 20–34 mm *l*, 15–26 mm *b*, fissure shallow, faces flush, usually distinctly reniform, almost smooth to somewhat rugose, **Mar** fairly well-defined in open-windowed specimens, otherwise not clearly distinguishable, **Isl** distinct or manifest as areas ± enclosed by grooves, few to many, small to large, mostly irregular, faces various shades of opaque grey or beige tinged with pink, yellow, light brown, lilac, green or blue, **Win** usually ± occluded, occasionally ± completely open, **Chan** broad to narrow, slightly impressed, irregular, often reduced to an irregular network of slightly impressed fine grooves, various shades of translucent greyish, brownish or reddish green, or greenish brown or bluish or greenish grey, **Rub** usually conspicuous and numerous, small to very small warts or pimples, up to 1.0 mm, usually slightly raised and round, often in the form of short dashes or commas, usually dark blood-red or purplish red, sometimes covered with a bluish film; **Fl** mostly straw-yellow, but also cream, lime yellow, golden yellow through dark yellow, light orange, bronze, carrot-coloured, shell-pink and rose-pink, often with an inner ring of orange, rose-pink, mauve or magenta, ∅ 15–35 mm; **Fr** 5(–6) locules, profile boat-shaped, top flat, faces elliptic to almost round; **Ecol** amygdaloidal lavas and banded ironstone with some calcrete, calcrete with fine-grained quartzite, sandstones, calcrete with cherts and jaspers, calcrete with cherts and fine-grained sandstone, calcrete with shaly sandstones, calcrete in granitic soil, grey-white, grey, brown, reddish brown, greyish green, some pink, red, black, blue; **Distr** southern central districts of NC, S Africa.

Note: Two varieties were distinguished by Cole (1988: 205–207): **L. verruculosa** var. *glabra* and var. *verruculosa*.

L. verruculosa var. *glabra* = **L. verruculosa**

L. villetii L.Bolus NM 1950: 234 **T** Villet 739/38 (BOL).

2–4(–7) heads; **L** 20–36 mm *l*, 15–28 mm *b*, faces flush, elliptic-reniform, smooth to slightly rugose, **Rub** absent; **Fl** white, ∅ 15–30 mm; **Fr** (4–)6(–7) locules, rarely 4, profile boat-shaped, top ± flat, faces broadly elliptic to elliptic; **Ecol** often in calcrete with shale or sandstone; **Distr** Calvinia, Namaqualand, NC, S Africa.

Note: Three subspecies are recognized.

Key to the Subspecies of L. villetii

1. Profile truncate, margins narrow and much indented, windows open to ± occluded, colours various shades of opaque grey or brown .... **L. villetii** ssp. **kennedyi**
– Profile truncate-cordate **2**
2. Margins distinct, windows mostly open, colours greenish grey or greyish green .... **L. villetii** ssp. **villetii**
– Margins irregular, windows more occluded, colours bluish or reddish grey .... **L. villetii** ssp. **deboeri**

L. villetii L.Bolus ssp. **deboeri** (Schwantes) D.T.Cole 1988: 218 ≡ *L. deboeri* Schwantes 1952: 42 ≡ *L. villetii* L.Bolus var. *deboeri* (Schwantes) D.T.Cole Excelsa 1973: 46 **T** Nel (?).

Mostly 2–3 heads; **L** truncate-cordate in profile, fissure shallow, **Mar** usually distinct but very irregularly dentate or sinuate, **Isl** usually numerous, mostly fairly small, usually distinct but irregular, various shades of opaque bluish, greenish, pinkish or reddish grey, **Win** usually ± occluded, **Chan** usually fairly narrow, irregular, occasionally reduced to narrow grooves, various shades of opaque or obscurely translucent bluish, greenish, pinkish, reddish or purplish-grey; **Fr** valves often wrinkled or slightly ribbed above; **Ecol** calcrete with some quartzite, calcrete with some sandstone and quartzite, shale and sandstone, pinkish grey-white with some brown, grey, pinkish brown with some red, black, grey; **Distr** Namaqualand, NC, S Africa.

Note: Jacobsen's suggestion that this species may be a form of **L. lesliei** (1955: 1462; 1960: 1237) cannot be supported.

L. villetii var. *deboeri* ≡ **L. villetii** ssp. **deboeri**

L. villetii L.Bolus ssp. **kennedyi** (De Boer) D.T.Cole 1988: 224 ≡ *L. villetii* L.Bolus var. *kennedyi* (De Boer) D.T.Cole Excelsa 1973: 55 ≡ *L. fulleri* N.E.Br. var. *kennedyi* De Boer 1967: 108 **T** Kennedy s.n., figured in De Boer 1967: 109 (?).

L truncate in profile, fissure shallow, **Mar** distinct, irregularly and extensively dentate, **Isl** few to many, mostly small, usually distinct, various shades of opaque grey, bluish or pinkish grey, or greyish pink, brown, orange-brown or red, **Win** usually ± occluded, infrequently ± completely open, **Chan** mostly fairly narrow, irregular, various shades of opaque greenish or brownish grey or greyish green, brown, orange-brown or purplish brown; **Ecol** fine-grained sandstone, silty sandstone, sandstone with some pegmatite, sandstone with some calcrete, calcrete with some sandstone, brown, reddish brown, greyish brown, with some grey, pink, black, white, pinkish grey-white with some grey, dark brown, black; **Distr** Calvinia, Namaqualand, NC, S Africa.

Note: Plants from the type locality differ considerably in colour from those in all other colonies of the subspecies.

L. villetii var. *kennedyi* ≡ **L. villetii** ssp. **kennedyi**

L. villetii L.Bolus ssp. **villetii.**

L truncate-cordate in profile, fissure fairly deep, **Mar** distinct, usually irregularly dentate or sinuate, **Isl** usually few and fairly small, mostly distinct, occasionally reduced to hazy flecks, various shades of opaque greenish, pinkish, brownish or creamy grey, or greyish green, **Win** occasionally ± occluded, mostly ± completely open, **Chan** usually fairly broad, irregular, various shades of opaque or obscurely translucent greenish grey or greyish, bluish, pinkish or brownish green; **Ecol** shale (some pyritic) and calcrete, sandstone shale and calcrete, grey, grey-white, brown, some red, black; **Distr** Calvinia, NC, S Africa.

Note: The subspecies occurs in a narrow strip some distance south of the other two subspecies.

L. viridis H.Lückhoff JSAB 1958: 159 ≡ *L. comptonii* L.Bolus var. *viridis* (H.Lückhoff) Fearn 1970: 92 **T** Hill & Lückhoff (?).

2–4(–8) heads; **L** cordate in profile, 15–25 mm *l*, 12–18 mm *b*, fissure deep, faces flush to slightly elevated, mostly ± elliptic, smooth, **Mar** distinct, regularly or very slightly dentate, various shades of opaque greyish pink, cream, beige or green, **Isl** rare, a few wispy flecks near the margins, indistinct pale grey or pinkish or greenish grey, **Win** nearly always completely open, various shades of translucent greyish or bluish green, **Chan** absent, **Rub** absent; **Fl** yellow with white centre, ∅ 20–35 mm; **Fr** 5 locules, profile boat-shaped, top ± flat, faces almost round to broadly elliptic; **Ecol** weathered chert and shale, grey, yellowish and brownish-grey, some reddish, pinkish and greenish-grey; **Distr** Calvinia, NC, S Africa.

Note: An amazingly uniform species from a very small area near Loeriesfontein.

L. volkii ≡ **L. pseudotruncatella** ssp. **volkii**
L. weberi = **L. comptonii**

L. werneri Schwantes & H.Jacobsen 1951: 69 **T** (?).

2–3(–10) heads; **L** cordate-truncate in profile, 15–24 mm *l*, 12–18 mm *b*, fissure relatively deep, faces flush, elliptic, somewhat rugose, **outer Mar** usually not clearly distinguishable, **inner Mar** often a ± continuous line, **Isl** occur only as irregular and slighly humped areas ± enclosed by channels, faces opaque light grey or light greenish grey, sometimes pinkish or yellowish grey, **Win** sometimes manifest as broad to narrow channels, often not distinguishable at all, **Chan** somewhat impressed, occasionally up to ± 4 mm wide, usually narrower and reduced to a network of slightly impressed grooves, obscurely translucent greenish grey or brown, orange brown or greyish green, **Rub** set in the channels or grooves, usually an entire or broken network of slender branched lines, occasionally reduced to a number of indistinct dashes and dots in the grooves, **dusky dots** numerous, scattered irregularly over the whole face; **Fl** yellow, ∅ 12–28 mm; **Fr** (5–)6 locules, profile rotund, top flat to peaked, faces ± round, with hinge-rim; **Ecol** weathered granitic gravel, light brownish white, light brown; **Distr** Karibib, Namibia.

Note: Plants of this species have very small bodies.

Litocarpus ≡ Aptenia

M. Gerbaulet

The name *Litocarpus* is illegitimate since Bolus (FPSA 1927: 261 f 11 & 12) based it on the same type as **Aptenia** N.E.Br., which has been published earlier (GC 1925: 412).

L. cordifolius ≡ **Aptenia cordifolia**

Ludolfia ≡ Tetragonia

H. E. K. Hartmann

Adanson (1763: 244) published the name *Ludolfia* Adanson as a substitute for **Tetragonia** L. (1753: 480) and *Tetragonocarpus* Commelijn (1701: t. 103), citing both names. Because **Tetragonia** is older and valid, *Ludolfia* is a younger illegitimate name. No species were provided.

Machairophyllum Ruschioideae

H. E. K. Hartmann

Machairophyllum Schwantes MDG 1927: 187 **Typus Ma. albidum** (L.) Schwantes; Bolus JSAB 1960: 155–158 **Etym** Gk *machaira*, dagger, *phyllon*, leaf, referring to the dagger-shaped leaves = *Perissolobus* N.E.Br. GC 1930: 278 **Typus** *Perissolobus bijliae* N.E.Br. ≡ **Ma. bijliae** (N.E.Br.) L.Bolus.

Compact, some species forming mats with age; **L** trigonous to triquetrous, dagger-shaped, old leaves persisting for several years on the plant, epidermis smooth or consisting of cobblestone-shaped cells, stomata thus either superficial or sunken, outer epidermal wall in all cases with a thick layer of calcium oxalate crystals of c. 2 µm ∅; **Fl** solitary or in cymes raised high above the plant, bracteolate, pedicels very long, petals yellow, filamentous staminodes none, filaments epapillate, nectary glands 7–15, often appearing as a ring divided into sections; **Fr** base funnel-shaped with concave sides, top with very high rims, later recurved, covering membranes complete, ± straight, translucent to whitish from a thick, spongy tissue, closing body rather small, expanding keels diverging from close contact at their bases, valve wings with a broad base, tapering into a fine awn distally, 7–14 locules; **Ecol** in crevices between rocks or among stones and gravel, in shrubby Fynbos vegetation with grassy undercover; **Distr** S districts of EC, E districts of WC, S Africa.

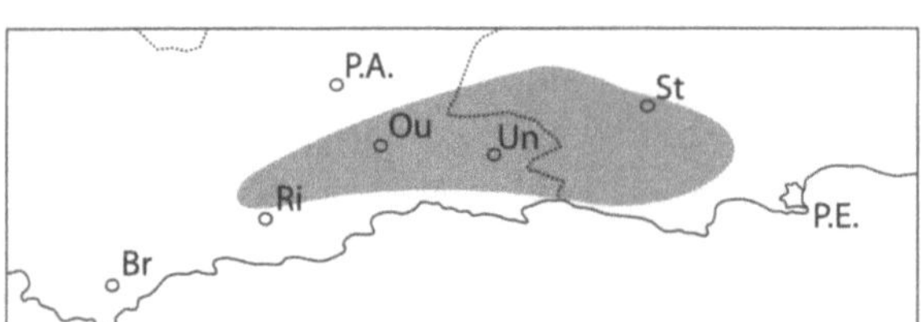

Distribution of **Machairophyllum** Schwantes

Note: This is one of the few genera of Ruschioideae settling almost exclusively in Fynbos. Although the plants are big, they are not easy to detect in the dense vegetation. In cultivation, most species seem to develop one to two flowers only of the dichasium, which can be richer in habitat. The cobblestone-shaped epidermal cells in some species resemble those on leaves of **Corpuscularia**. The genus is under study and four or five species may only be distinguishable: **Ma. albidum, M. bijliae, Ma. brevifolium, Ma. stayneri,** and possibly a new species (Kurzweil, Vlok, in litt.).

Ma. acuminatum L.Bolus NM 1935: 485 (see **photo**) **T** Fourcade 424/24 (BOL).

L tips truncate to acute, 40–55 mm *l*, 8–11 mm *b*, 6–10 mm *t*, sheaths 10 mm *l*; **Fl** solitary, pedicel basally bracteolate, 42–60 mm *l*, 27–38 mm ∅, **K** 6, petals golden, dorsally golden to reddish, filaments yellow, basally white, nectary glands 8; **Fr** top raised for c. 9 mm, ∅ 13 mm, expanded 24 mm; **S** brown, 1 mm *l*; **Ecol** in shrubby grassland on rocks; **Distr** Humansdorp, Joubertina, EC, S Africa.

Note: The species appears most similar to **Ma. stenopetalum.**

Ma. albidum (L.) Schwantes MDG 1927: 187 (see **photo**) ≡ *M. albidum* L. 1762: 699 ≡ *Bergeranthus albidus* (L.) Schwantes ZSK 1926: 180 ≡ *Carruanthus albidus* (L.) Schwantes ZSK 1926: 181 nomen invalid. **Lectotype**, here designated: Dillenius 1732: t. 189, f 232 = *Ma. cookii* (L.Bolus) Schwantes MDG 1927: 187 ≡ *M. cookii* L.Bolus ABH 1925: 2 ≡ *Bergeranthus cookii* (L.Bolus) Schwantes ZSK 1926: 180 ≡ *Carruanthus cookii* (L.Bolus) Schwantes ZSK 1926: 181 nomen invalid. **T** Compton & Cook 1995/23 (BOL).

Compact to caespitose; **L** with a ± rounded keel, epidermis cells elevated like cobblestones in a road, their anticlinal walls undulate, their upper edges with mostly 5 bulges, the centre often with a low bulge as well, wax cover continuous, with few amorphous particles, 35–60 mm *l*, 9 mm *t*; **Fl** in high reaching **Inf** with 3–7 **Fl**, the lowest bracteole at c. half the length of the stalk, **K** 5, c. 400 yellow petals, 500–520 stamens, nectary ring irregular; **Fr** covering membranes spongy, valve wings basally broader, tapering to the tips where they remain connate with the awns of the expanding keels, mostly 10–12 locules, values from 8 to 15 possible, 12–15 mm *l*, the top 4–6 mm *h*, 12–15 mm ∅; **Ecol** on rocky to gravelly slopes in rather open Fynbos with grass undercover; **Distr** in a zone from Swellendam to Oudtshoorn, WC, S Africa.

Note: Both species with five calyx lobes and at least double the number of locules are considered conspecific. **Ma. albidum** can further be recognized by the very high inflorescences, also visible when in fruit, on rather compact plants.

Ma. baxteri L.Bolus NM 1935: 486 **T** Baxter 3166/14 (BOL).

L 50–70 mm *l*, 8–14 mm *b*, 8–12 mm *t*, sheath 13 mm *l*; **Fl** solitary, nocturnal, pedicels basally bracteolate, 65–75 mm *l*, bracteoles 45 mm *l*, **K** 8, petals golden, reddish-copper outside, 17–26 mm *l*, to 1 mm *b*, filaments basally white, nectary glands 8; **Fr** 8 locules, 14 mm *l*, top elevated for 4–5 mm, ∅ 12 mm, expanded 22 mm; **Ecol** unknown; **Distr** George, WC, S Africa.

Note: Within the group of species with solitary flowers, **Ma. baxteri** shares the equal and high number of calyx lobes and locules with **Ma. bijliae** from further east and inland.

Ma. bijliae (N.E.Br.) L.Bolus NM 1938: 135 ≡ *Perissolobus bijliae* N.E.Br. GC 1931: 294 **T** van der Bijl 93 (K) = *Ma. vanbredai* L.Bolus JSAB 1964: 238 **T** van Breda 1749/62 (BOL).

Compact; **L** 20–60 mm *l*, 9–17 mm *b*, 9–11.3 mm *t*; **Fl** solitary, nocturnal, pedicels 13.5–17 mm *l*, on fruit to 60 mm *l*, bracteoles basal, 25–30 mm *l*, **K** 8, petals bright yellow, basally white, red outside, 8–18 mm *l*, filaments basally white, anthers and pollen yellow, nectary of connate glands?; **Fr** 8 locules; **Ecol** unknown; **Distr** Somerset East, EC, S Africa.

Note: The species can best be recognized by the agreement of eight calyx lobes and eight locules. The epithet was published as "bijlii", a mistake in gender because it was named after "Mrs. D. van der Bijl".

Ma. brevifolium L.Bolus NM 1938: 126 (see **photo**) **T** Lategan 2369/36 (BOL) = *Ma. latifolium* L.Bolus NM 1938: 126 **T** Lategan 2367/36 (BOL).

Compact, later forming small mats the centre of which dies off; **L** apiculate, the upper part rhombic, sharply keeled, epidermis smooth, stomata superficial, 25–30 mm *l*, 13–26 mm *b*, 12–15 mm *t*; **Fl** solitary, vespertine, pedicels 12–15 mm *l*, bracteoles basal 18–23 mm *l*, **K** 6, petals bright yellow, 13–20 mm *l*, nectary glands 6; **Fr** covering membranes translucent, valve wings with a lower half as broad as the expanding keels, ending in a narrow awn connate with the awn of the expanding keel, 6–11 locules, ∅ 9–12 mm, 10–12 mm *l*, top c. 4 mm *h*; **S** brown, 1 mm *l*; **Ecol** under shrubs in grass between rocks; **Distr** George, Oudtshoorn, WC, S Africa.

Note: Both species united here differ mainly in leaf width, which can vary rather widely in and between populations.

Ma. cookii = **Ma. albidum**
Ma. latifolium = **Ma. brevifolium**

Ma. stayneri L.Bolus JSAB 1960: 155 **T** Stayner 196/59 (BOL).

Mats to 120 cm ∅; **L** 75–150 mm *l*, 12–16 mm *b* and *t*; **Fl** in tall **Inf** raised high above the plant, ternate, nocturnal, 50–62 mm ∅, pedicels to 33 mm *l*, bracteoles basal, **K** 6–8, petals bright yellow to golden, basally pale yellow to almost whitish-green, externally pink, to 27 mm *l*, 0.5–1 mm *b*, small distant nectary glands present; **Fr** 7–8 locules; **Ecol** on the summit of a pass; **Distr** Uitenhage, EC, S Africa.

Note: The huge mats formed by this species make it easy to recognize.

Ma. stenopetalum L.Bolus NM 1931: 303 **T** Brauns 8030 (BOL).

L pale green, shining, to 40 mm *l*, 12 mm *b*, 8–10 mm *t*; **Fl** pedicels 30 mm *l*, bracteoles 34 mm *l*, **K** 6, 10–13 mm *l*, petals golden and deep copper-coloured, basally pale, to 16 mm *l*, to 0.75 mm *b*, filaments apically golden, anthers white, pollen pale yellow, nectary as a ring; **Fr** 7 locules; **Ecol** unknown; **Distr** Willowmore, EC, S Africa.

Note: In the number of calyx lobes and locules, **Ma. stenopetalum** is similar to **Ma. acuminatum**, from nearer the coast and further east.

Ma. vanbredai = **Ma. bijliae**

Macrocaulon = Carpanthea

H. E. K. Hartmann

When Brown (GC 1927: 12) described the genus as new, he did not contrast it with his older genus **Carpanthea**. Bolus (NM 1950: 231) saw "no justification for upholding the genus *Macrocaulon*, a very suitable name, however, for some very robust forms of **Carpanthea pomeridiana**". Ihlenfeldt & Gerbaulet (1990: 478) followed this decision.

M. candollii = **Carpanthea pomeridiana**

Malephora Ruschioideae

H. E. K. Hartmann

Malephora N.E.Br. GC 1927: 12 **Typus Ma. mollis** (Aiton) N.E.Br. **Etym** Gk *male*, armhole, *pherein* to bear, probably referring to the leaf sheaths through which the stem continues = *Crocanthus* L.Bolus FPSA 1927: t 255 **Typus** *Crocanthus thunbergii* (N.E.Br.) L.Bolus ≡ **Ma. thunbergii** (N.E.Br.) Schwantes = *Hymenocyclus* Schwantes MDG 1927: 27 **Typus** *Hymenocyclus herrei* Schwantes ≡ **Ma. herrei** (Schwantes) Schwantes.

Creeping, decumbent or erect shrubs, **I** ochre to grey, hard or spongy, growing numerous, branching stems per year, each potentially ending in a flower, appearing solitary or in cymes; **L** trigonous to terete, epidermis smooth, most cells small, but ± equidistant to the stomata, often a bigger cell not or inconspicuously raised above the other level epidermal cells: a reduced bladder cell, wax cover a thin layer adorned with a dense cover of threads or very thin; **Fl** solitary or in cymes, but often several solitary flowers developing on the growth of one season, **K** 4–5, petals yellow or purple and yellow, or saffron, changing during anthesis, filamentous staminodes absent, nectary composed of c. as many glands as there are locules, sometimes in a crenulate ring; **Fr** base elongate funnel-shaped, top low, but ± turret-shaped, rims low, covering membranes thin, in several species laterally and distally reduced, the placenta raised by a false septum for 1/4 to 2/3 of the height of the locules, placenta knobby, ending in a twin-row of knobs simulating closing bodies, expanding keels diverging a little, valve wings rectangular, a little broader than the expanding keels; **S** elongate pear-shaped, often almost tetrahedral, testa knobby from raised centres of the cells, covered to varying degrees by tiny rodlets standing together like sheaves put together in a field; **Ecol** in loamy, deep soil, requiring a rather high amount of moisture, hence often appearing in great quantities after extensive rainfall, and disappearing again after some time of drought; **Distr** Rehoboth, Namibia; western districts of EC, Calvinia, Namaqualand, Fraserburg, NC, FS, from Vanrhynsdorp to Murraysburg, WC, S Africa.

Note: **Malephora** is the only weedy genus in the subfamily Ruschioideae, settling quickly in deeper grounds after rain and along newly tarred roads, where a higher run-off occurs in comparison to the neighbouring areas. Several species are appreciated as ornamentals all over the world in suitable climates, and it cannot be excluded that some of the populations found in the wild have escaped from gardens, most probably in the form of seeds which remain viable for many years. With regard to its nomenclature, the genus is remarkable in so far as it has been described three times within the year 1927, **Malephora** having come out first on 1st of January, *Hymenocyclus* on 11th of January, and *Crocanthus* in October, reflecting that all three students of Aizoaceae active at that time came to identical conclusions.

Distribution of **Malephora** N.E.Br.

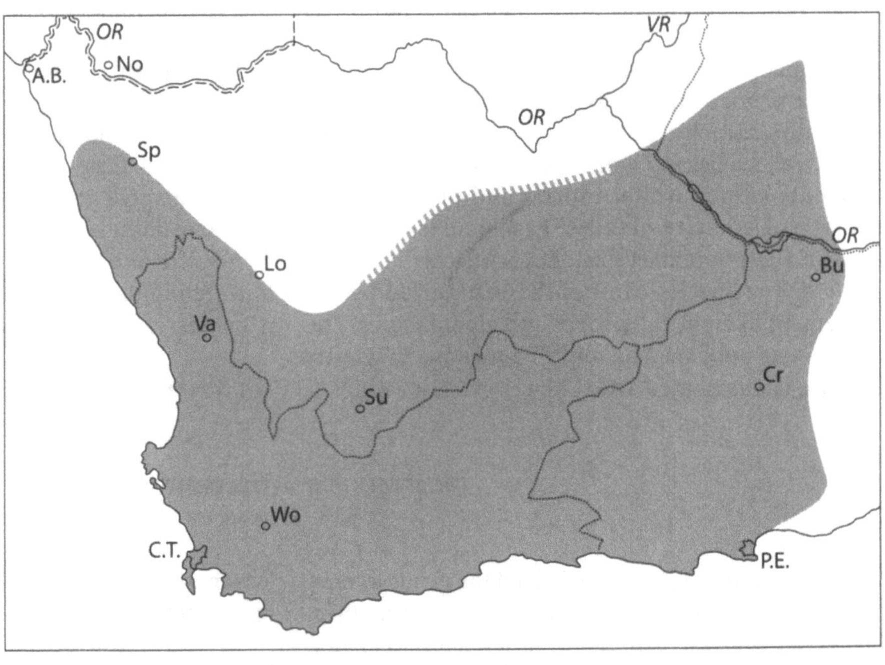

Ma. crassa (L.Bolus) H.Jacobsen & Schwantes in Jacobsen & Rowley NCSJ 1958: 77 (see **photo**) ≡ *Hymenocyclus crassus* L.Bolus NM 1932: 363 **T** Salter 2666 (BOL!).

Creeping, rooting at almost each node, forming densely leaved mats of up to 1 m ∅, **I** shining, spongy, ∅ > 4 mm; **L** much longer than **I**, trigonous to nearly club-shaped, bright green, epidermis smooth, wax cover a very thin layer crackling with age, to 40 mm *l*, 13 mm *b*, 14 mm *t*; **Fl** solitary on very short, erect shoots, pedicels 10–20 mm *l*, **K** 5, petals yellow, 13–29 mm *l*, nectary in a ring; **Fr** base funnel-shaped, top rather flat, covering membranes complete or nearly so, false septum > 1/2 to 2/3 of the height of the locule, ∅ 8–10 mm; **Ecol** in flats with very fine loam, probably of Ecca shale origin; **Distr** Ceres, WC, S Africa.

Note: **Ma. crassa** forms conspicuous green mats in the Tanqua Karroo; it shares its creeping growth and the bright green leaves with **Ma. uitenhagensis**, from which it differs by more extensive covering membranes and much higher false septa in the capsule.

Ma. crocea (Jacquin) Schwantes MDG 1928: 7 ≡ *M. croceum* Jacquin 1800: 17 ≡ *Hymenocyclus croceus* (Jacquin) Schwantes MDG 1927: 27 ≡ *Crocanthus croceus* (Haw.) L.Bolus FPSA 1927: t. 255 **Lectotype**, here designated: Jacquin 1800: t. 11, f 2.

Shrubs with terete, grey, glabrous, decumbent branches, young ones erect; **L** semi-cylindrical, glaucous, glabrous, 13.5–20 mm *l*, to 6 mm *t*; **Fl** solitary, **K** 5, petals twice the length of K, saffron-coloured on both sides, red below; **Fr** unknown; **Ecol** uncertain; **Distr** unknown, S Africa.

Note: In many features given by Jacquin (1800: 17, t. 11, f 2), this plant is similar to **Ma. veruculoides**, e.g. in the short leaves.

Very similar are also **Ma. engleriana**, **Ma. ochraceum**, and **Ma. purpureo-crocea**, differing by the purple colour in the petals. The complex seems to occcur only in high lying localities in the western parts of the distribution area.

Ma. crocea var. *purpureo-crocea*
≡ **Ma. purpureo-crocea**

Ma. engleriana (Dinter & A.Berger) Dinter & Schwantes MDG 1928: 7 ≡ *M. englerianum* Dinter & A.Berger 1914: 588 ≡ *Hymenocyclus englerianus* (Dinter & A.Berger) Dinter & Schwantes ZSK 1927: 16 ≡ *M. vernae* Dinter & A.Berger 1914: 590 **T** Dinter 2102 (B!).

Richly branched shrubs, c. 1/2 m^2 in extension, **I** brownish-grey, green when young, 15–40 mm *l*; **L** cylindrical to trigonous, glaucous green from a cover of fine wax, 25–35 mm *l*, 4–5 mm *b*; **Fl** solitary on numerous short side branches, rarely in cymes, pedicels 10–15 mm *l*, **K** 5, petals yellow; **Fr** base funnel-shaped, top only a little raised, false septum to c. 1/3 of the height of the locule, knobby, 10 locules; **Ecol** in loamy flats; **Distr** Rehoboth, Namibia.

Note: The name *M. vernae* was based on material preserved in liquid and sent separately to Berger, but of the same origin as the type (Dinter 1923: 189), hence only one species was involved from the beginning. The species has not been found again in the type locality, and it is possible that it was so abundant there at the time of its discovery because it had rained exceptionally well a few months earlier (Jürgens, pers. comm.). The species is very similar to **Ma. croceum**, **Ma. ochraceum**, and **Ma. veruculoides**.

Ma. flavo-crocea (Haw.) H.Jacobsen & Schwantes in Jacobsen & Rowley NCSJ 1958: 77 ≡ *M. flavo-croceum* Haw. 1826: 129 ≡ *M. purpureo-croceum* Haw. var. β *flavo-croceum* Haw. 1821: 155 ≡ *M. croceum* Jacquin var. *flavocroceum* (Haw.) DC. 1828: 438 **T?**

Shrubs with bent, decumbent branches, **I** woody; **L** somewhat crowded and distant, mealy-glaucous or glaucescent; **Fl** petals yellow on both sides, saffron-coloured in bud and decay, stamens and anthers yellow; **Fr** 7 locules; **Ecol** unknown; **Distr** unknown, S Africa.

Note: Brown (1920: 128–129) gave Burchell as the collector speculating also over the possible type, but no material or drawing has been found since.

Ma. framesii (L.Bolus) H.Jacobsen & Schwantes in Jacobsen & Rowley NCSJ 1958: 78 ≡ *Hymenocyclus framesii* L.Bolus NM 1929: 58 **T** Frames 18901 (BOL!).

Decumbent branches rooting occasionally, **I** very smooth, ∅ to 7 mm, spongy; **L** ± clavate, obtuse, epidermis covered by wax in small particles, to 40 mm *l*, 14 mm *b* and *t*; **Fl** solitary, ∅ 40 mm, pedicels 10 mm *l*, **K** 4, petals bright yellow inside, dull outside, to 19 mm *l*, to 2.5 mm *b*, filaments and pollen golden, nectary as a ring; **Fr** base funnel-shaped, top rather low, covering membranes complete, sometimes with a low distal recurving, placenta raised by a false septum for c. 1/3 of the height of the locule, ∅ 8 mm, 8 locules; **Ecol** near the sea; **Distr** Vredenburg, WC, S Africa.

Note: The fat leaves with a wax cover seem to distinguish this species well from all others.

Ma. herrei (Schwantes) Schwantes MDG 1928: 7 ≡ *Hymenocyclus herrei* Schwantes MDG 1927: 27 **Lectotype**, here designated: Herre s.n.: drawing of a fruit (HBG!).

Shrubs with decumbent and rooting branches, **I** 20–30 mm *l*; **L** trigonous, to 50 mm *l*, 5 mm *b* and *t*; **Fl** pedicels c. 30 mm *l*, flattish, **K** 4, petals to 18 mm *l*,

2.5 mm *b*, golden, filaments yellow, basally white, nectary ring crenulate; **Fr** covering membranes distally incomplete, valve wings very broad and their margins overlapping, 8 locules; **Ecol** unknown; **Distr** unknown, S Africa.

Note: The only material clearly referring to this species in Schwantes' collection at HGB is a drawing of a fruit, which is therefore chosen as the lectotype here. Rooting of stems occurs in several other species, from which **Ma. herrei** differs mainly in the slender leaves.

Ma. latipetala (L.Bolus) H.Jacobsen & Schwantes in Jacobsen & Rowley 1958: 78 ≡ *Hymenocyclus latipetalus* L.Bolus NM 1929: 118 T L.Bolus 18951 (BOL!).

Decumbent, long branches with many ± erect shorter ones, **I** firm and hard, brown, purple at first, 10–35 mm *l*; **L** terete, apically acute, bluish, glaucous, to 25 mm *l*, 3–5 mm *b*, sheaths c. 3 mm *l*; **Fl** pedicels 15 mm *l*, ∅ 34 mm, **K** 5, petals golden, 16 mm *l*, up to 3.5 mm *b*, nectary as a ring; **Fr** top almost flat, ∅ 6 mm, 9–10 locules; **Ecol** unknown; **Distr** Oudtshoorn, WC, S Africa.

Note: Differing from **Ma. lutea**, which has also big flowers, by the absence of rooting branches.

Ma. lutea (Haw.) Schwantes MDG 1928: 7 ≡ *M. luteum* Haw. 1826: 129 ≡ *Hymenocyclus luteus* (Haw.) Schwantes MDG 1927: 27 T?

Shrubs to 30 cm *h* or above, rooting at nodes touching the soil; **L** obtuse, longer than **I**, bluish-green; **Fl** solitary, **K** 4, petals bright yellow, richer in colour when withering, anthers yellow; **Fr** 8 locules; **Ecol** unknown; **Distr** unknown, S Africa.

Note: The plants are also reported to grow taller and branch more profusely than **Ma. flavo-crocea**, but in leaves and flowers, **Ma. lutea** is said to be three times smaller.

Ma. luteola (Haw.) Schwantes MDG 1928: 7 ≡ *M. luteolum* Haw. 1826: 129 ≡ *Crocanthus luteolus* (Haw.) L.Bolus FPSA 1927: t. 255 ≡ *Hymenocyclus luteolus* (Haw.) Schwantes MDG 1928: 7 **T?**

Shrubs with slender, densely placed leaves, nodes not rooting; **L** apiculate, recurved, glaucous blue or paler green; **Fl** small, petals yellow; **Fr** unknown; **Ecol** unknown; **Distr** unknown.

Note: For the distinction of the species it is important that Haworth (1826: 130) describes this species as being a quarter the size of *M. luteum*, now **Ma. lutea**.

Ma. mollis (Aiton) N.E.Br. GC 1927: 12 (see **photo**) ≡ *M. molle* Aiton 1789: 192 non Berger 1908: 130 nomen illeg. ≡ *Corpuscularia molle* (Aiton) Schwantes ZSK 1926: 186 **NT** Austin s.n. (K!).

Profusely branching shrubs to 50 cm *h*, ∅ 45–110 cm, **I** ochre, later red, hard, with fine grooves or ribs; **L** in groups at the base of the well developed side shoots, grey, recurved apically, to 20 mm *l*, c. 3 mm *t*, sheaths c. 6 mm *l*; **Fl** 1–2, sometimes in terminal cymes of 3, but flowers on each branch, thus giving the appearance of rich **Inf**, petals yellow; **Fr** pedicels 30–40 mm *l*, base funnel-shaped, top raised a little, roundish, rims low, covering membranes almost complete, false septum c. 1/2 the height of the locule, 7–10 locules, the maxima on the type sheet at 7 and 9; **S** elongate, flattish, often tetrahedral, testa cells raised prominently, on the sides and between the humps covered densely by epicuticular rodlets arranged in groups like sheaves on a field, dark brown, 1.2–1.25 mm *l*, 0.85 mm *b*; **Ecol** unknown; **Distr** Laingsburg, WC, S Africa.

Note: Aiton did not preserve a type, but Brown designated a neotype when he described **Malephora** (GC 1927: 12). This choice has to be followed because no original material has been found until now.

Ma. monticola ≡ **Ottosonderia monticola**

Ma. ochracea (A.Berger) H.E.K.Hartmann **comb. nov.** ≡ *M. ochraceum* A.Berger Mesembrianthemen und Portulacaceen 1908: 234 (as basionym) ≡ *Glottiphyllum ochraceum* (A.Berger) N.E.Br. GC 1922: 22 **Holotype** Wilmot s.n., collected by Strauβ from garden material (B!).

Shrubs with ascending branches, **I** compressed when young, smooth, spongy; **L** trigonous, very smooth and shining, 60–100 mm *l*, 6 mm *t*; **Fl** solitary, pedicels 25 mm *l*, **K** 5, petals saffron-coloured, base lighter, 16 mm *l*; **Fr** unknown; **Ecol** unknown; **Distr** unknown.

Note: The species was described after material from Willmot, England, which had been grown on in the garden at Berlin. Hence, a sheet collected there by Strauβ bearing the name is most probably part of the type plant, thus the sheet is not a neotype, as given by Hartmann & Gölling (1994: 47).

Ma. purpureo-crocea (Haw.) Schwantes GF 1928: 69 (see **photo**) ≡ *M. purpureo-croceum* Haw. 1795: 257 ≡ *Hymenocyclus pupureo-croceus* (Haw.) L.Bolus SAG 1927: 399 ≡ *Crocanthus purpureo-croceus* (Haw.) L.Bolus FPSA 1927: t. 255 ≡ *M. croceum* Jacquin var. *purpureo-croceum* DC. 1828: 438 ≡ *Ma. crocea* (Jacquin) Schwantes var. *purpureo-crocea* (Haw.) H.Jacobsen & Schwantes in Jacobsen & Rowley NCSJ 1958: 78 **T?** = *M. insititium* Willdenow 1809: 536 **T?**

Shrubs with many short, erect branches, which lie down later since they are too weak to carry their own weight, **I** ochre to later whitish, spongy from a great portion of cork tissue in the stem; **L** terete to subtrigonous, blunt keel and point, very glaucous from a dense layer of wax threads over a continuous thinner layer, 50–90 mm *l*, 8–10 mm *t*; **Fl** in erect, big cymes, **K** 5, petals bright purple outside, saffron-coloured inside; **Fr** stalks c. 100 mm *l*, base elongate, funnel-shaped, top rather low, covering membranes complete, false septum c. 1/4 of the height of the locules, 9–10 locules; **Ecol** in loamy flats, often along roadsides; **Distr** Namaqualand, NC, Vanrhynsdorp, WC, S Africa.

Note: All plants with purple flowers, often also inside, though sometimes only partially, have been placed in this species, which is also the most widespread one in cultivation, often escaping, e.g. in the Mediterranean area. Hence it is extremely difficult to determine the area of natural occurrence, but it appears that most collections far away from inhabited areas are centred in the districts given above.

Ma. smithii (L.Bolus) H.E.K.Hartmann **comb. nov.** (see **photo**) ≡ *Hymenocyclus smithii* L.Bolus South African Gardening & Country Life 17, 1927: 399 (as basionym) **Holotype** Smith 4101 (BOL).

Forming dense and compact mats, 3.5–4 cm *h*, creeping; **L** trigonous, obtuse, subclavate, 20–30 mm *l*, to 6 mm *t*, sheaths 4 mm *l*; **Fl** solitary, pedicels 15 mm *l*, ∅ to 42 mm, **K** 4, petals ochre, basally lighter in colour, exterior copper-coloured, filaments saffron-coloured, basally white, nectary in a ring; **Fr** unknown; **Ecol** unknown; **Distr** Fauresmith, FS, S Africa.

Note: The drawing published with the description shows clearly a plant of **Malephora**; the four sepals are similar to the also creeping **Ma. uniondalensis**, not yet reported from so far north.

Ma. thunbergii (Haw.) Schwantes GF 1928: 69 (see **photo**) ≡ *M. thunbergii* Haw. 1803: 86 ≡ *M. laeve* Thunberg 1791: 16 nomen illeg. non Aiton 1789: 187 ≡ *Corpuscularia thunbergii* (Haw.) Schwantes ZSK 1926: 186 ≡ *Crocanthus thunbergii* (Haw.) L.Bolus FPSA 1927: 255 ≡ *Hymenocyclus thunbergii* (Haw.) L.Bolus SAG 1927: 399 **T** Thunberg 11990 (UPS!).

Decumbent shrubs, **I** smooth outside, spongy, the cork comprising c. 2/3 of the diameter; **L** trigonous to terete, with an acute tip, epidermis covered with a whitish layer, probably wax, 20–33 mm *l*; **Fl** ebracteolate, solitary, but several on a branch with several **I**, **K** 4, petals yellow; **Fr** unknown; **Ecol** flowering in winter; **Distr** cf. Kirkwood, EC, S Africa.

Note: The species was collected by Thunberg at the Sundays River. The spongy stems are unusual for plants in this area, where thin, hard stems occur. The species differs from the widespread creeping **Ma. uitenhagensis** not only by the habit and the internodes, but also by the distinctly grey leaves.

Ma. uitenhagensis (L.Bolus) H.Jacobsen & Schwantes in Jacobsen & Rowley NCSJ 1958: 78 (see **photo**) ≡ *Hymenocyclus uitenhagensis* L.Bolus NM 1929: 118 **Lectotype**, here designated: Frames 2503/27 II (chosen from syntypes, BOL!).

Creeping, forming extensive mats to 2 m ∅ in favourable conditions, **I** woody, hard, purple, later brown, ∅ to 3 mm, 10–15 mm *l*; **L** trigonous, pointed, some a little incurved, epidermis smooth, wax cover thin, continuous, 11–25 mm *l*, 4–5 mm *b* and *t*; **Fl** solitary, stalks < **L**, ∅ 50 mm, **K** 4, petals deep golden or reddish-copper, to 24 mm *l*, 2 mm *b*, filaments pale, anthers and pollen golden; **Fr** base funnel-shaped, top rather low, covering membranes reduced laterally and distally, false septum 1/3 of the height of the locule, valve wings rectangular, broader than the expanding keels, 8–12 locules; **Ecol** in loamy flats, sometimes in flats with stony cover, always in open places; **Distr** Aberdeen, Kirkwood, Steytlerville, Uitenhage, Willowmore, EC, Beaufort West, Calitzdorp, Oudtshoorn, Prince Albert, WC, S Africa.

Note: The bright green mats resemble those of **Ma. crassa**, a species with thicker leaves, complete covering membranes and a much more highly raised placenta. **Ma. smithii** belongs to the same complex, but it is less well known.

Ma. veruculoides (Sonder) Schwantes GF 1928: 69 ≡ *M. veruculoides* Sonder in Harvey & Sonder 1862: 428 ≡ *M. verruculatum* var. β Thunberg 1823: 425 ≡ *M. verruculoides* (Sonder) Schwantes ex Jacobsen FR 1938: 148, name used erroneously **T** Thunberg 12043 β (UPS!) = *M. veruculoides* Sonder var. *minus* in Harvey & Sonder 1862: 428 **T** Ecklon & Zeyher 2110 (?).

Shrubs with many procumbent branches to 8 cm *h*, **I** light yellow, soft and spongy, cork c. 2/3 of the ∅; **L** subclavate, epidermis very smooth and apparently not waxy, c. 17 mm *l*, c. 5 mm *t*; **Fl** solitary, **K** 4, rarely 5, petals pale yellow; **Fr** 8–10 locules; **Ecol** flowering in winter; **Distr** Calvinia, NC, S Africa.

Note: The glabrous stems are remarkable in the genus.

Manettia ≡ Aridaria

M. Gerbaulet

Manettia Adanson (1763: 242) is an earlier homonym of **Manettia** Mutis ex L. 1771: 553 nomen conserv. Rubiaceae, and therefore an illegitimate name. No species names have ever been published for *Manettia* Adanson, but Adanson's citation of Dillenius 1732: t. 206 referred to the lectotype of **Aridaria**.

Marlothistella Ruschioideae

H. E. K. Hartmann

Marlothistella Schwantes GW 1928: 599 **Typus Ma. uniondalensis** Schwantes **Etym** *R. Marloth*, 1855–1931, professor of chemistry at Stellenbosch, botanist at Cape Town.

Compact plants with a thick tap-root dividing with age, the almost erect parts of the rootstock mainly under ground, leaves on them in heads of 6–8 to a branch, I enclosed; **L** subulate, the bases rounded, the tips finely acuminate, the sheath prominent, although only 6–7 mm *l*, the free parts 40–50 mm *l*, 2–5 mm *b* and ∅, epidermis very smooth, developing horizontal folds when drying, wax in horizontal plates, c. one per cell; **Fl** solitary, bracteoles at the bases of the pedicels, ∅ 20–25 mm, **K** 5, petals 28–40, pink in different shades, often vittate, but not regularly so, filamentous staminodes 30–45, white, 40–65 stamens with white filaments, nectary as a lophomorphic ring; **Fr** top low, but rims distinctly high, bases long to short, funnel-shaped, covering membranes almost straight, with a low closing ledge at their distal ends, closing body big and white, expanding keels basally widely separated by a spongy, whitish tissue and ending distally in awns, 5 locules; **S** yellow to light brown, testa cells a little elevated, 0.85–0.95 mm *l*, 0.65–0.8 mm *b*; **Ecol** on shaly soils, in low, shrubby Karoo vegetation or in grassy associations, > 250 mm rainfall p.a. at all seasons or in spring and autumn; **Distr** Joubertina, EC; Beaufort West, Oudtshoorn, Prince Albert, Uniondale, WC, S Africa.

Note: The genus was placed into synonymy with **Ruschia** soon after its description, based on the possession of closing bodies and the absence of valve wings. After the exclusion of **Acrodon** (Glen 1986) and **Antimima** (Dehn 1989) from **Ruschia** s.lat., **Marlothistella** is also re-established. It differs from **Ruschia** in the formation of a fat, branched rootstock and compact growth, and in the peculiar triangle of spongy tissue between the bases of the expanding keels of a valve. In these features, it resembles **Acrodon**, which again differs from **Marlothistella** by having dentate leaves, much stouter, darker, persisting fruits developed in cymes, and by the possession of very big, splitting closing bodies in the capsule. When Schwantes established the genus (GW 1928: 599), he also described a fruit which does not agree with the text above but is clearly a fruit of **Cephalophyllum subulatoides**. This confusion is not surprising because plants of both taxa occur at Uniondale, the reported place of origin, in very close vicinity, or even mixed, and it is very difficult to distinguish them in vegetative state. The part referring to the fruit has therefore to be deleted from the description of the genus.

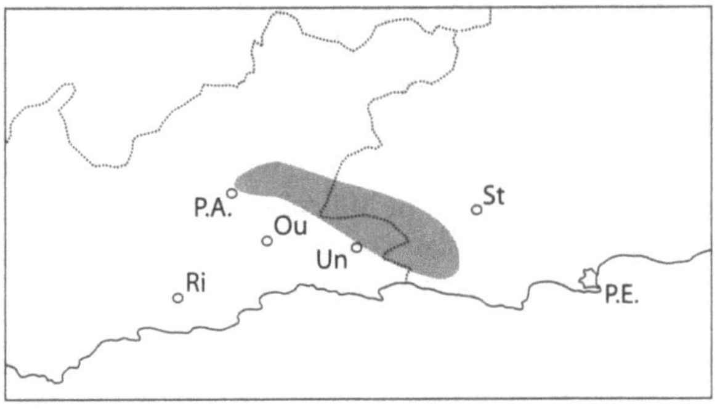

Distribution of **Marlothistella** Schwantes

Key to the Species of Marlothistella

1. Leaves only 2–3 mm broad and thick .... **Ma. stenophylla**
– Leaves c. 5 mm broad and 3 mm thick .... **Ma. uniondalensis**

Ma. stenophylla (L.Bolus) S.A.Hammer Bradleya 1995: 87 ≡ *M. stenophyllum* L.Bolus ABH 1925: 1 ≡ *Ruschia stenophylla* (L.Bolus) L.Bolus NM 1950: 219 T Fourcade 2992 (BOL!).

Subtuberous; **L** trigonous, acuminate, to 50 mm *l*, 2–3 mm *b* and ∅; **Fl** petals pinkish-purple, 16 mm *l*, otherwise as for genus; **Fr** as for genus; **Ecol** in grassland, rain at all seasons; **Distr** Joubertina, EC, S Africa.

Note: Grassland with Acacia-scrub determines the vegetation of the slopes where **Ma. stenophylla** grows, differing thus markedly from the localties of the other known populations.

Ma. uniondalensis Schwantes GW 1928: 600 (see **photo**) **Lectotype**, here designated: drawing of flower at HBG (!).

Distally branching tap-roots; **L** basally terete, sharply pointed above, c. 45 mm *l*, c. 5 mm *b*, c. 3 mm ∅; **Fl** often vittate; **Fr** as for genus; **Ecol** in open patches in grassland with some proteaceous and restionaceous vegetation almost to 2 m high, or in Karroo vegetation; **Distr** Uniondale, WC, also in Beaufort West, Oudtshoorn, Prince Albert, WC, S Africa.

Note: Kept under this name are at present also forms from more open, karroid vegetation types further north and west than Uniondale.

Maughania = Diplosoma

H.-D. Ihlenfeldt

When Brown (1931: 389–390) created the new genus *Maughania* with the type *Ma. insignis*, he had overlooked that the name had already been used by J.Saint-Hilaire (1813: 216) for a genus of the family Fabaceae. *Maughania* N.E.Br. is therefore illegitimate. Bolus (NM 1954: 260) transferred *Monilaria luckhoffii* L.Bolus to this genus making the new combination *Ma. luckhoffii* (L.Bolus) L.Bolus, which, however, is a nomen invalidum and illegitimum. Bolus (NM 1950: 46, footnote under a drawing) had already stated that there was "no reason for doubting" that *Ma. in-signis* and *Ma. luckhoffii* were conspecific. Bolus (1961: 264) replaced *Maughania* with *Maughaniella* L.Bolus.

Ma. insignis = **Diplosoma luckhoffii**
Ma. luckhoffii ≡ **Diplosoma luckhoffii**

Maughaniella = Diplosoma

H.-D. Ihlenfeldt

The name *Maughaniella* was explicitly established by Bolus (1961: 264) as a substitute for the illegitimate name *Maughania* N.E.Br. *Maughania insignis* is, therefore, automatically the type. A formal transfer of *Maughania insignis* to *Maughaniella* has never been made.

For a long time, the two species of the genus **Diplosoma** have been considered to represent two distinct genera, **Diplosoma** and *Maughaniella* (e.g. Bolus NM 1954: 260–261, Schwantes KuaS 1955: 135–138). Following a suggestion of Schwantes (1957: 127), Ihlenfeldt (1988: 397) united both genera.

Ma. luckhoffii ≡ **Diplosoma luckhoffii**

Mentocalyx = Gibbaeum

H. E. K. Hartmann

Brown (GC 1927: 251–252) established the genus for the species *M. muirii*, placing it near **Pleiospilos** at that time. Tischer (1937: 150–152) discussed the question of the position of *Mentocalyx* extensively and decided that it was rightly placed in **Gibbaeum**. Since both species possess hairs similar to those found in many species of **Gibbaeum**, and since flower features of both taxa agree as well, this view is followed here.

Me. muirii ≡ **Gibbaeum schwantesii**
Me. velutinum ≡ **Gibbaeum velutinum**

Mesembryanthemum Mesembryanthemoideae

M. Gerbaulet

Mesembryanthemum L. 1753: 480 nomen conserv.; Bittrich 1987: 5–116 ≡ *Gasoul* Adanson 1763: 243 nomen illeg. **Lectotypus M. nodiflorum** L., typus conserv. ≡ *Gasoul nodiflorum* (L.) Rothmaler **Etym** incorrectly from Gk *mesembria* (not mesembrya), midday, *anthemon*, flower, referring to the flowers which open during midday; or correctly from Gk *mes-*, middle, *embryo*, embryo or ovule, *anthemon*, referring to the free axile placentation of the flower = *Callistigma* Dinter & Schwantes in Schwantes GW 1928: 644 **Typus** *Callistigma inachabense* (Engler) Dinter & Schwantes ≡ **M. inachabense** Engler = *Cryophytum* N.E.Br. GC 1925: 412 ≡ *Pentacoilanthus* Rappa & Camarrone 1953: 32 nomen illeg. ≡ *Perapentacoilanthus* Rappa & Camarrone 1955: 6 nomen invalid. **Lectotypus** *Cryophytum crystallinum* (L.) N.E.Br. ≡ **M. crystallinum** L. = *Derenbergiella* Schwantes GW 1928: 599 **Typus** *Derenbergiella luisae* Schwantes = **M. guerichianum** Pax = *Eurystigma* L.Bolus NM 1930: 179 **Typus** *Eurystigma clavatum* (L.Bolus) L.Bolus ≡ **M. eurystigmatum** Gerbaulet = *Gynicidia* Necker 1790: 81 nomen invalid. **Typus** not given = *Halenbergia* Dinter KuaS 1937: 158 ≡ *Pteropentacoilanthus* Rappa & Camarrone 1960: 14 nomen illeg. **Typus** *Halenbergia hypertrophica* (Dinter) Dinter ≡ *Pteropentacoilanthus hypertrophicus* (Dinter) Rappa & Camarrone ≡ **M. hypertrophicum** Dinter = *Hydrodea* N.E.Br. GC 1925: 412 **Lectotypus** *Hydrodea cryptantha* (Hook.f.) N.E.Br. ≡ **M. cryptanthum** Hook.f. = *Opophytum* N.E.Br. GC 1925: 412 **Lectotypus** *Opophytum fastigiatum* (Thunberg) N.E.Br. ≡ **M. fastigiatum** Thunberg.

Prostrate, decumbent, or erect herbs, sometimes very large, mostly annuals, rarely longer living, with fibrous roots, **I** herbaceous, terete, angled or winged, cortex with additional vascular bundles, bladder cells mesomorphic, large to much flattened; **L** flat, cylindrical to slightly channelled, almost truncate, or almost globular, decussate and shortly connate at the base, becoming alternate in the **Inf** or alternate throughout, flat leaves without, cylindrical leaves with enlarged central water-storing cells, bladder cells mesomorphic, large to much flattened; **Fl** in cymes, ∅ 10–60 mm, **K** 5, petals pink, yellow or white, **K** and petals connate into a short tube or petals sometimes free to the base, petals filiform or narrowly lanceolate, rarely much reduced, filamentous staminodes present or absent, nectaries narrowly shell-shaped or tubular; **Fr** valve wings inflexed over the valves or reflexed and fused in pairs, 5 locules; **S** 0.5–1 mm *l*, D-shaped or triangular, black, brown, yellow or whitish, testa rough or more or less smooth, without a distinct crest; **Ecol** mostly in disturbed places and along roadsides, often on salty soils; **Distr** SW Angola; SW Australia; Azores; Canary Islands; Cape Verdes; Iraq; Iran; Mauritania; Mediterranean basin; Namibia; N Africa; Palestine; St. Helena; EC, FS, NC, WC, S Africa; Turkey; California, USA.

Note: As early as 1689 Jacob Breyne in his "Prodromus fasciculi rariorum plantarum" coined the name "Mesembrianthema" – "midday flowers". Some forty years later this name was cited by Dillenius (1732) and was eventually adopted by Linné (1753) when founding the genus **Mesembryanthemum**. For more than one and a half centuries this genus became a temporary repository for several hundred species, until the discovery of the fruit structure as a useful taxonomic tool by N.E.Brown in the early 1920's led to the rapid splitting of the old genus. Brown (GC 1925: 232) was also the first to designate a generic type, which became necessary in this period of nomenclatural upheaval. He chose **M. umbellatum** as the type of the genus, a species which is now placed in **Ruschia**. However, Schwantes (ZSK 1927: 24–28) pointed out that certain characters of **M. umbellatum** did not match those of **Mesembryanthemum** in Brown's generic key (GC 1925: 412, 413, 433). He chose *M. tenuifolium* as a new lectotype, a member of **Lampranthus** in current nomenclature. Bolus (NM 1939: 164–168), on the other hand, disclaimed the type selections of both, Brown and Schwantes. She proposed **M. nodiflorum** as nomenclatural type which she felt to be in closer agreement with the Linnean circumscription of **Mesembryanthemum.** This view attracted the most support. Eventually, Ihlenfeldt & Straka (1972: 537) proposed to have **Mesembryanthemum** L. emend. L.Bolus conserved. Their proposal was approved (McVaugh 1974: 820, 821) and **M. nodiflorum** was conserved as the type of the name **Mesembryanthemum** L.

The present generic circumscription and infrageneric subdivision follows Bittrich (1987). Only names of species and synonyms remaining in the revised genus **Mesembryanthemum** are listed here; names and synonyms of species now placed in other genera or of uncertain position follow in a separate list.

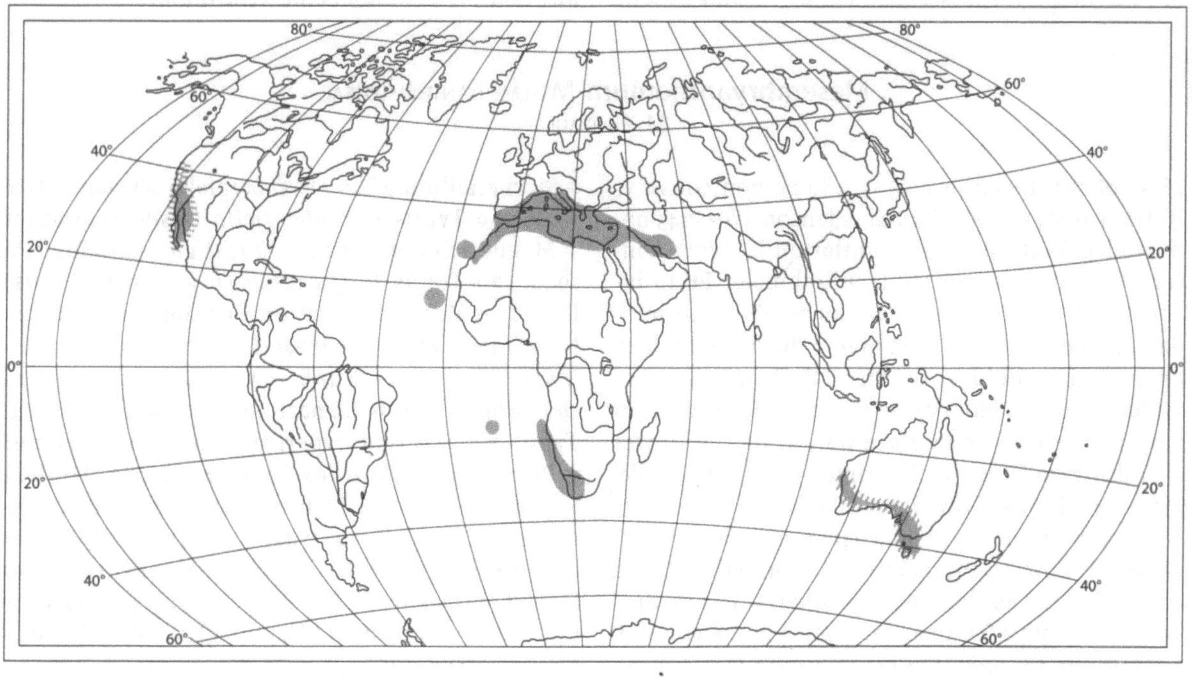

Distribution of **Mesembryanthemum** L.

Key to the Subgenera of Mesembryanthemum

1. Leaves flat, basal ones largest, usually at least 50 mm long, often much larger, becoming gradually smaller along the branches, margins usually undulating, flowers with filamentous staminodes, nectaries shell-shaped, sometimes deeply so but never tubular, valve wings of the fruits inflexed over the valves, not fused .... M. subg. **Cryophytum**
- Leaves flat or more or less cylindrical, basal ones not markedly larger, flat leaves never longer than 50 mm, usually much smaller, but margins may be undulating, flowers with or without filamentous staminodes, nectaries deep and usually tubular, valve wings of the fruits often reflexed and fused in pairs 2
2. Leaves cylindrical, very succulent, ∅ to 10 mm, bladder cells small or flattened, flowers with filamentous staminodes and ∅ to 60 mm, rarely much reduced, ∅ c. 10 mm .... M. subg. **Opophytum**
- Leaves flat or more or less cylindrical, cylindrical ones 2–5 mm in ∅, bladder cells distinct, flowers without or with some filamentous staminodes, rarely much reduced, ∅ 5–30 mm .... M. subg. **Mesembryanthemum**

M. subg. **Cryophytum** (N.E.Br.) Bittrich 1987: 72 ≡ *Cryophytum* N.E.Br. GC 1925: 412 ≡ *M.* subg. *Cryophytum* (N.E.Br.) Maire & Weiller in Maire 1962: 253–254 nomen invalid. ≡ *Pentacoilanthus* Rappa & Camarrone 1953: 32 nomen illeg. ≡ *Perapentacoilanthus* Rappa & Camarrone 1955: 6 nomen invalid. **Lectotypus M. crystallinum** L. = *Derenbergiella* Schwantes GW 1928: 599 **Typus** *Derenbergiella luisae* Schwantes = **M. guerichianum** Pax.

Annual or often biennial very large herbs, prostrate to decumbent or erect, **I** often angled or winged; **L** flat, sometimes very large, narrowed at the base, decussate and shortly connate at the base or sometimes alternate in the **Inf**, leaf margins usually undulating, without enlarged central water-storing cells, bladder cells conspicuous; **Fl** sometimes only on special secondary, annual long shoots and main stem without flowers, ∅ 15–60 mm, petals filiform or narrowly lanceolate, **K** and petals connate into a tube, filamentous staminodes present, nectaries narrow and flatly or more deeply shell-shaped, petals, filamentous staminodes and stamens mostly very numerous, placentas often with more than 2 rows of funicles; **Fr** valve wings inflexed over the valves; **S** dark or light brown, testa rough or rarely smooth, without a distinct crest; **Ecol** as for genus; **Distr** SW Angola; SW Australia; Azores; Canary Islands; Cape Verdes; Mauritania; Mediterranean basin; Namibia; N Africa; Palestine; St. Helena; EC, NC, WC, S Africa; California, USA.

Note: This subgenus comprises the conspicuously papillate and large-leaved so-called "ice-plants".

Key to the Species of Subgenus Cryophytum

1. Plants with a central rosette, not terminated by a flower 2
- Plants without a central rosette, main stem terminated by a flower 4
2. Plants usually annual, prostrate, to 8 cm high, internodes terete, c. 6 mm in ∅, basal leaves to 60 mm long and 35 mm broad, bladder cells very prominent, especially on the receptacle, flower ∅ 15–30 mm, petals, filamentous staminodes and stamens not particularly numerous .... **M. crystallinum**
- Plants annual or biennial in response to water supply, prostrate, decumbent or erect, plant height, internode ∅ and leaves larger, internodes usually 4–6-angled or winged, flower ∅ 30–60 mm, petals, filamentous staminodes and stamens very numerous 3
3. Plants decumbent to erect, to 100 cm high, internode ∅ to 40 mm, basal leaves to 400 mm long and 250 mm broad, bladder cells small .... **M. barklyi**
- Plants prostrate to decumbent, to 60 cm high but mostly much less, internode ∅ to 25 mm, basal leaves to 250 mm long and 100 mm broad, bladder cells conspicuous, sometimes large and sometimes somewhat hair-like .... **M. guerichianum**
4. Plants prostrate, annual, main stem short, ∅ c. 4 mm, leaves to 40 mm long and 20 mm broad, all bladder cells rounded .... **M. gariusanum**
- Plants decumbent to erect, annual to biennial in response to water supply, main stem rather long, ∅ c. 15 mm, leaves to 150 mm long and 80 mm broad, bladder cells rounded to much elongated and hair-like, all bladder cell morphs may occur on the same stem .... **M. pellitum**

M. subg. **Mesembryanthemum** = *Callistigma* Dinter & Schwantes in Schwantes GW 1928: 644 **Typus** *Callistigma inachabense* (Engler) Dinter & Schwantes ≡ **M. inachabense** Engler.

Annual herbs, prostrate to decumbent or erect, **I** terete; **L** flat, (almost) cylindrical to channelled, or almost truncate, flat ones narrowed at the base, decussate and shortly connate at the base, alternate in the **Inf** or all leaves alternate, leaf margins rarely undulating, usually not so, flat leaves without, cylindrical ones with enlarged central water-storing cells, bladder cells usually large, rarely flattened; **Fl** ∅ 10–30 mm, petals narrowly lanceolate or rarely much reduced, **K** and petals connate into a short tube or rarely petals free, filamentous staminodes absent, nectaries tubular; **Fr** valve wings usually reflexed and fused in pairs; **S** whitish or brown with dark warts, testa smooth or rough, without a distinct crest; **Ecol** as for genus; **Distr** SW Australia; Azores; Canary Islands; Cape Verdes; Iraq; Iran; Mauritania; Mediterranean basin; SW Namibia; N Africa; Palestine; St. Helena; EC, FS, NC, WC, S Africa; Turkey; California, USA.

Note: Bittrich (1987: 5–116) distinguished two groups of species within this subgenus, the **M. aitonis** group with flat-leaved species (**M. aitonis, M. inachabense**) and the **M. nodiflorum** group with almost cylindrically-leaved species (all other species).

Key to the Species of Subgenus Mesembryanthemum

1. Leaves flat, ovately to lanceolately spatulate or oblong **2**
- Leaves more or less cylindrical or somewhat channelled, linear to very narrowly obovate or oblong or almost truncate **3**
2. Petals pale to dark yellow, stigmas basally connate, only upper portion fertile, red, seeds whitish, testa smooth .... **M. inachabense**
- Petals white, sometimes suffused pink or yellow, stigmas basally not connate, pale yellow, seeds brown, testa rough .... **M. aitonis**
3. Bladder cells somewhat inconspicuous, leaves without marginal cilia **4**
- Bladder cells conspicuous, leaves usually with marginal cilia **5**
4. Stems basally slightly woody, plants annual or perhaps biennial in response to water supply, leaves linear to very narrowly oblong, petals white, often with pink tips and straw-coloured base .... **M. longistylum**
- Stems not woody, plants annual, leaves almost truncate, cylindrical, petals pink, basally white or straw-coloured .... **M. subtruncatum**
5. Leaves usually very narrowly obovate, petals yellow .... **M. excavatum**
- Leaves usually linear, petals white or pale pink or sometimes with pink tips only **6**
6. Flowers small, ∅ 5–10 mm .... **M. nodiflorum**
- Flowers large, ∅ to 30 mm .... **M. stenandrum**

M. subg. **Opophytum** (N.E.Br.) Bittrich 1987: 73 ≡ *Opophytum* N.E.Br. GC 1925: 412 ≡ *M.* subg. *Opophytum* (N.E.Br.) Maire & Weiller in Maire 1962: 253, 259 nomen invalid. **Lectotypus M. fastigiatum** Thunberg = *Eurystigma* L.Bolus NM 1930: 179 **Typus** *Eurystigma clavatum* (L.Bolus) L.Bolus ≡ **M. eurystigmatum** Gerbaulet = *Halenbergia* Dinter KuaS 1937: 158 ≡ *Pteropentacoilanthus* Rappa & Camarrone 1960: 14 nomen illeg. **Typus** *Halenbergia hypertrophica* (Dinter) Dinter ≡ *Pteropentacoilanthus hypertrophicus* (Dinter) Rappa & Camarrone ≡ **M. hypertrophicum** Dinter = *Hydrodea* N.E.Br. GC 1925: 412 **Lectotypus** *Hydrodea cryptantha* (Hook.f.) N.E.Br. ≡ **M. cryptanthum** Hook.f.

Annual herbs, prostrate, **I** terete; **L** cylindrical or globular, very succulent, decussate and shortly connate at the base or sometimes alternate in the **Inf**, becoming dry when plants in flower, with enlarged central water-storing cells, bladder cells small or flattened; **Fl** ∅ 20–60(–80) mm, petals filiform or rarely much reduced, **K** and petals connate into a short tube, petals, filamentous staminodes and stamens mostly very numerous, rarely not so, nectaries tubular or rarely shell-shaped, placentas often with more than 2 rows of funicles; **Fr** valve wings reflexed and fused in pairs or inflexed over the valves; **S** whitish to light brown, testa smooth or slightly rough in parts, without a distinct crest; **Ecol** as for genus; **Distr** SW Angola; SW Australia; Azores; Canary Islands; Cape Verdes; Mauritania; Mediterranean basin; Namibia; N Africa; Palestina; St. Helena; NC, WC, S Africa; California, USA.

Note: The four species belonging to this subgenus are very similar and are mainly distinguished by floral characters. All four species occur allopatrically.

Key to the Species of Subgenus Opophytum

1. Flowers small, ∅ to 10 mm, petals, staminodes and stamens few .... **M. cryptanthum**
- Flowers medium or large, petals, staminodes and stamens very numerous **2**
2. Petals straw-coloured to yellow, stigmas basally shortly connate with the ovary .... **M. eurystigmatum**
- Petals not straw-coloured to yellow, stigmas basally not connate with the ovary **3**

3. Leaves to 40 mm long, ∅ to 7 mm, flower ∅ c. 15 mm, petals white, testa at the back region rough .... **M. fastigiatum**
- Leaves to 70 mm long, ∅ to 10 mm, flower ∅ to 60 mm, petals basally greenish-yellow, upper parts pink, testa smooth throughout .... **M. hypertrophicum**

M. aitonis Jacquin 1776: t. 7 (see **photo**) subg. **Mesembryanthemum** ≡ *Cryophytum aitonis* (Jacquin) N.E.Br. GC 1928: 128 ≡ *Pentacoilanthus aitonis* (Jacquin) Rappa & Camarrone 1953: 30 ≡ *Perapentacoilanthus aitonis* (Jacquin) Rappa & Camarrone 1960: 23 nomen invalid. ≡ *Gasoul aitonis* (Jacquin) Eichler 1965: 133 **T** Jacquin 1776: t. 7 = *M. alsinifolium* G.Don in Loudon 1830: 480 nomen nudum **T** not given = *M. angulatum* Thunberg 1800: 91 ≡ *Cryophytum angulatum* (Thunberg) Schwantes GF 1928: 68 **Lectotype**, here designated: Thunberg 11942 (UPS, photograph!) = *Cryophytum bijliae* N.E.Br. GC 1929: 84 **T** van der Bijl 44 (K!) = *M. clandestinum* Haw. 1826: 129 ≡ *Cryophytum clandestinum* (Haw.) L.Bolus SAG 1927: 326 **Lectotype**, here designated: drawing (K, photograph BOL!) = *Cryophytum conjectum* N.E.Br. GC 1928: 153 **T** Page 17357 (BOL!) = *M. crystallophanes* Ecklon & Zeyher 1837: 322 **T** Ecklon & Zeyher 2099 (BOL!) = *M. lanceolatum* Haw. 1803: 45 **Lectotype**, here designated: drawing 966/688 of Duncanson (K, photograph BOL!) = *M. ovatum* Thunberg 1791: 8 ≡ *M. angulatum* Thunberg var. *ovatum* (Thunberg) Sonder in Harvey & Sonder 1862: 454 **Lectotype**, here designated: Thunberg 12001 (part 2) (UPS) = *Cryophytum paulum* N.E.Br. GC 1928: 211 ≡ *M. paulum* (N.E.Br.) L.Bolus NM 1939: 167 **T** Muir 4145 (K!) = *Cryophytum planum* L.Bolus NM 1928: 15 ≡ *M. louiseae* (L.Bolus) L.Bolus 1967: 308 ≡ *M. stratum* L.Bolus NM 1939: 167 nomen illeg. **T** Guthrie 18849 (BOL!) = *M. puberulum* Haw. 1831: 421 ≡ *Cleretum puberulum* (Haw.) N.E.Br. GC 1927: 228 ≡ *Micropterum puberulum* (Haw.) Schwantes FR 1938: 257 **Neotype**, here designated: Fourcade 3554 (BOL!) = *M. sessiliflorum* Aiton var. *album* Haw. 1812: 158 ≡ *M. sessiliflorum* Aiton var. β Haw. 1819: 93 ≡ *Micropterum sessiliflorum* Aiton var. *album* (Haw.) H.Jacobsen NCSJ 1949: 58 ≡ *Cryophytum burchellii* N.E.Br. GC 1928: 128 **Neotype**, here designated: Glen 1002 (BOL!) = *M. volckameri* Haw. 1795: 426 **Neotype**, here designated: Fourcade 4333 (BOL!).

Prostrate or ascending, without a basal vegetative rosette, annual or sometimes biennial in response to water supply, **I** terete or somewhat angular, ∅ to 3 mm; **L** flat, ovately or lanceolately spatulate, to 50 mm *l* and 15 mm *b* but often smaller, margins sometimes undulating, without enlarged central water-storing cells, bladder cells distinct, rounded; **Fl** ∅ 5–20 mm, petals white, yellowish in bud, sometimes suffused pink, connate into a short tube, filamentous staminodes more or less absent, petals and stamens not particularly numerous, nectaries tubular, stigmas pale yellow; **Fr** valve wings inflexed over the valves; **S** D-shaped, brown, testa rough; **Ecol** in disturbed places, winter and summer rainfall; **Distr** EC, Victoria West and Prieska, NC, S WC, S Africa; Australia.

Note: M. clandestinum is sometimes regarded as a separate species because of its minute flowers. However, the flowers of **M. aitonis** are particularly small in early anthesis and the petals grow longer while they continue to open and close. It is therefore inferred that the type specimen of *M. clandestinum* represents a plant early in flowering conspecific with **M. aitonis.** Moreover, populations from the western distribution area (described as *M. paulum*) show much smaller leaves than the eastern populations. Leaf size, however, proved to be a very variable character in response to radiation and moisture availability.

M. alatum = **M. guerichianum**
M. alkalifugum = **M. barklyi**
M. alsinifolium = **M. aitonis**
M. angulatum = **M. aitonis**
M. angulatum var. *ovatum* = **M. aitonis**
M. annuum = **M. stenandrum**
M. aquosum = **M. hypertrophicum**
M. arenarium = **M. inachabense**

M. barklyi N.E.Br. in Hook. 1889: t. 1820 (see **photo**) subg. **Cryophytum** ≡ *Cryophytum barklyi* (N.E.Br.) N.E.Br. ex L.Bolus SAG 1928: 83–84 ≡ *Platythyra barklyi* (N.E.Br.) Schwantes GF 1928: 68 **T** Barkly s.n. s.d. (?) = *M. barklyi* N.E.Br. var. *obtusifolium* L.Bolus ABH 1927: 79 **T** Pillans 4927 (BOL!) = *M. alkalifugum* Dinter FR 1923: 64 nomen nudum ≡ *Cryophytum alkalifugum* Dinter ex Range 1934: 13 nomen nudum **T** not given = *Cryophytum glaucum* Dinter 1935: 138 **T** Dinter s.n. s.d. (?) = *Cryophytum squamulosum* Dinter 1935: 138 non L.Bolus NM 1929: 71 nomen illeg. ≡ *M. squamulosum* (Dinter) Dinter ex Range FR 1938: 256 **T** Dinter 8249 (?) = *Cryophytum velutinum* L.Bolus NM 1930: 166 ≡ *M. macrophyllum* L.Bolus NM 1939: 168 **T** Parkins 18959 (BOL!).

Decumbent to erect, very large, forming first a basal vegetative rosette with flowering side branches, later the rosette internodes elongate until an additional rosette and new flowering side branches are formed, to 1.5 m *h*, usually biennial depending on water supply, **I** usually distinctly 4-angled and winged, rarely with less angles, ∅ to 40 mm; **L** flat, ovate, basal ones largest, to c. 400 mm *l* and c. 250 mm *b*, margins more or less undulating, without enlarged central water-storing cells, bladder cells small, rounded;

Fl open from noon till dusk, ∅ 40–60 mm, petals pink or pale pink, basally white or slightly greenish, connate into a tube, petals, filamentous staminodes and stamens very numerous, nectaries shell-shaped, stigmas pale yellow; **Fr** valve wings inflexed over the valves; **S** brown with darker spots, testa somewhat rough; **Ecol** in sandy plains, winter rainfall; **Distr** SW Namibia; NW Namaqualand, NC, S Africa.

Note: This species contains the largest plants of the genus, which are dubbed "elephant's toilet paper" on account of the huge leaves.

M. barklyi var. *obtusifolium* = **M. barklyi**
M. breve = **M. crystallinum**
M. caducum = **M. nodiflorum**
M. chrysum = **M. excavatum**
M. clandestinum = **M. aitonis**
M. copticum = **M. nodiflorum**
M. cryocalyx = **M. longistylum**

M. cryptanthum Hook.f. in Hook. 1867–1871: t. 1034 (see **photo**) subg. **Opophytum** ≡ *Hydrodea cryptantha* (Hook.f.) N.E.Br. JB 1928: 107 **T** Burchell 1807 (K) = *Hydrodea bossiana* Dinter KuaS 1937: 200 **T** Boss in Herb. Dinter s.n. s.d. (?) = *M. dactylinum* Welwitsch ex Oliver 1871: 582 ≡ *Opophytum dactylinum* (Welwitsch ex Oliver) N.E.Br. GC 1928: 253 **T** Welwitsch s.n. s.d. (?) = *M. forskahlii* Hochstetter ex Boissier 1872: 765 ≡ *Opophytum forskahlii* (Hochstetter ex Boissier) N.E.Br. GC 1928: 253 **Neotype**, here designated: Cramer s.n. April 1885 (G, Photograph BOL!) = *M. gaussenii* Leredde 1954: 246 ≡ *Opophytum gaussenii* (Leredde) H.Jacobsen ex Greuter & Burdet in Greuter 1980: 232 **T** Leredde 1895B (?) = *Hydrodea hampdenii* N.E.Br. JB 1928: 106 **T** Hobart-Hampden s.n. 1928 (BM) = *M. sarcocalycanthum* Dinter & A.Berger in Dinter FR 1923: 192 ≡ *Hydrodea sarcocalycantha* (Dinter & A.Berger) Dinter KuaS 1937: 199 **Lectotype**, here designated: Dinter s.n. s.d. (?) = *Aizoon theurkauffii* Maire 1936: 224 ≡ *M. theurkauffii* (Maire) Maire 1940: 343 ≡ *Opophytum theurkauffii* (Maire) Maire 1940: 343 **T** Theurkauff 43 (P?).

Annual, decumbent to erect, without a basal non-flowering rosette, to c. 25 cm *h*, prostrate when in flower, **I** terete, ∅ to 10 mm; **L** cylindrical, linear, to 50 mm *l*, ∅ to 15 mm, with enlarged central water-storing cells, bladder cells flattened; **Fl** ∅ to 10 mm, petals white or lemon yellow, narrow, connate into a tube, petals, filamentous staminodes and stamens very few, nectaries tubular, stigmas yellow; **Fr** valve wings reflexed and fused in pairs; **S** whitish to light brown, testa smooth; **Ecol** salty habitats, winter rainfall or fog zone of the northern Namib; **Distr** SW Angola; SW Australia; Azores; Canary Islands; Cape Verdes; Mauritania; Mediterranean basin; CW, NW Namibia; N Africa; Palestine; St. Helena; California, USA.

Note: **M. cryptanthum** differs from the other species of **M.** subgenus **Opophytum** by its reduced flowers. According to Zohary (1966: 77) the species is thought to be mentioned in the Bible. The author also states that reportedly bread is made from the seeds.

M. crystallinum L. 1753: 480 (see **photo**) subg. **Cryophytum** ≡ *Cryophytum crystallinum* (L.) N.E.Br. in Phillips 1926: 245 ≡ *Gasoul crystallinum* (L.) Rothmaler 1941: 413 ≡ *Pentacoilanthus crystallinus* (L.) Rappa & Camarrone 1953: 30 nomen invalid. *Perapentacoilanthus crystallinus* (L.) Rappa & Camarrone 1960: 23 nomen invalid. **Lectotype**, here designated: Dillenius 1732: t. 180 f 211 = *Cryophytum nanum* N.E.Br. GC 1929: 84 **T** Maughan Brown s.n. s.d.? (?) = *Cryophytum parvum* L.Bolus NM 1929: 98–99 ≡ *M. breve* L.Bolus NM 1939: 167 **T** Pillans 18941 (BOL!) = *M. glaciale* Haw. 1819: 92 **Neotype**, here designated: Page 17307 (BOL!).

Annual, prostrate, with a small basal non-flowering rosette, **I** terete, ∅ c. 6 mm; **L** flat, ovately spatulate, acute or acuminate, basal ones largest, c. 60 mm *l* and c. 35 mm *b*, margins often undulating, without enlarged central water-storing cells, bladder cells very large, rounded to elongated; **Fl** ∅ 15–30 mm, petals white, suffused pink, connate into a tube, petals, filamentous staminodes and stamens not particularly numerous, nectaries shell-shaped, stigmas white, bladder cells of the receptacle especially prominent; **Fr** valve wings inflexed over the valves; **S** brown, testa rough; **Ecol** along the coast, winter rainfall; **Distr** probably SW Angola; SW Australia; Azores; Canary Islands; Cape Verdes; Mauritania; Mediterranean basin; W Namibia; N Africa; Palestine; St. Helena; from Namaqualand, NC, to the Cape Peninsula, WC, and Port Elizabeth, EC, S Africa; California, USA.

Note: This species is very similar to **M. guerichianum**. However, **M. crystallinum** has smaller flowers than **M. guerichianum** and is always exceedingly papillose.

M. crystallinum var. *grandiflorum* = **M. guerichianum**
M. crystallophanes = **M. aitonis**
M. dactylinum = **M. cryptanthum**
M. dejagerae = **M. stenandrum**
M. erectum = **M. fastigiatum**

M. eurystigmatum Gerbaulet **nomen novum** (see **photo**) subg. **Opophytum** ≡ *Cryophytum clavatum* L.Bolus Notes on Mesembrianthemum 2, 1929: 151 (as replaced synonym) ≡ *Eurystigma clavatum* (L.Bolus) L.Bolus NM 1930: 179 **Holotype** 18999 (BOL!).

Annual, decumbent, without a basal non-flowering rosette, to 20 cm *h*, prostrate when in flower, **I** terete, ∅ to 4 mm; **L** cylindrical, linear, to 70 mm *l*,

∅ to 10 mm, with enlarged central water-storing cells, bladder cells flattened; **Fl** deeply funnel-shaped, ∅ to 60 mm, petals straw-coloured to yellow, very narrow, connate into a tube, petals, filamentous staminodes and stamens very numerous, nectaries tubular, stigmas basally shortly connate with the ovary, thick and broad, subulate, or rather filiform, yellow; **Fr** valve wings inflexed over the valves, small; **S** whitish to light brown, testa smooth; **Ecol** in disturbed places, winter rainfall; **Distr** Ceres, WC, S Africa.

Note: The combination of *Eurystigma clavatum* to **Mesembryanthemum** made a new name necessary, since the epithet has already been used in the genus by Haworth.

M. excavatum L.Bolus ABH 1923: 163 (see **photo**) subg. **Mesembryanthemum** ≡ *Cryophytum excavatum* (L.Bolus) L.Bolus NM 1928: 32 **T** Mathews 2392/17 (BOL!) = *Cryophytum aureum* L.Bolus NM 1928: 32 ≡ *M. chrysum* L.Bolus NM 1939: 167 **Lectotype**, here designated: L.Bolus 18850 (BOL!).

Annual, prostrate or decumbent, without a basal non-flowering rosette, **I** terete, ∅ to 4 mm; **L** channelled, usually very narrowly obovate, rarely linear, to 20 mm *l*, ∅ to 5 mm, with enlarged central water-storing cells, bladder cells distinct, rounded; **Fl** ∅ c. 15 mm, petals yellow, connate into a very short tube or almost free, filamentous staminodes absent, petals and stamens not particularly numerous, nectaries tubular, stigmas yellow; **Fr** valve wings reflexed and fused in pairs; **Ecol** very common along roadsides, winter to summer rainfall; **Distr** Beaufort West, Ceres, Laingsburg, WC, S Africa.

Note: This species is very similar to **M. nodiflorum**. However, **M. excavatum** has golden yellow – as opposed to white or slightly pinkish – flowers, and the leaves are usually very narrowly obovate as opposed to linear in **M. nodiflorum**.

M. fastigiatum Thunberg 1791: 7 (see **photo**) subg. **Opophytum** ≡ *Opophytum fastigiatum* (Thunberg) N.E.Br. GC 1928: 253 ≡ *Perapentacoilanthus fastigiatus* (Thunberg) Rappa & Camarrone 1960: 23 nomen invalid. ≡ *Pteropentacoilanthus fastigiatus* (Dinter) Rappa & Camarrone 1960: 23 nomen invalid. ≡ *M. erectum* Haw. 1803: 53 nomen illeg. ≡ *M. papuliferum* DC. 1828: 448 nomen illeg. **Lectotype**, here designated: Thunberg 11980 (UPS, photograph!) = *Cryophytum crassipes* L.Bolus NM 1929: 162–163 ≡ *M. pachypus* L.Bolus NM 1939: 167 **T** Pillans 6273 (BOL!).

Annual, decumbent to erect, without a basal vegetative rosette, to c. 10 cm *h*, prostrate when in flower, **I** terete, ∅ c. 7 mm; **L** cylindrical, linear, to 40 mm *l*, ∅ to 7 mm, with enlarged central water-storing cells, bladder cells usually quite conspicuous; **Fl** ∅ c. 15 mm, petals white, very narrow, connate into a tube, petals, filamentous staminodes and stamens not particularly numerous, nectaries tubular, stigmas yellow; **Fr** valve wings reflexed and fused in pairs; **S** light brown, testa smooth on the sides, rough at the back of the seed; **Ecol** in disturbed places, winter rainfall; **Distr** Vanrhynsdorp, WC, S Africa.

Note: This species is sometimes regarded as conspecific with **M. hypertrophicum**, both of which show a disjunct distribution. However, **M. fastigiatum** clearly differs in its smaller overall size, more conspicuous bladder cells, smaller flowers, and a rough strip along the back of the seed.

M. fenchelii = **M. guerichianum**
M. forskahlii = **M. cryptanthum**
M. fulleri = **M. inachabense**
M. galpinii = **M. stenandrum**

M. gariusanum Dinter FR 1928: 22 (see **photo**) subg. **Cryophytum** ≡ *Cryophytum gariusanum* Dinter ex Range 1934: 14 **Lectotype**, here designated: Dinter 4945 (BOL!) = *Cryophytum salteri* L.Bolus NM 1932: 316 ≡ *M. violense* L.Bolus NM 1939: 168 **T** Salter 1435 (BOL!).

Annual, prostrate, without a basal vegetative rosette, main stem short, **I** terete, ∅ c. 4 mm; **L** flat, obovate to rhomboid, basal ones largest, c. 50 mm *l* and 30 mm *b*, margins only slightly undulating, without enlarged central water-storing cells, bladder cells conspicuous, rounded; **Fl** ∅ 30–40 mm, petals white, connate into a tube, petals, filamentous staminodes and stamens very numerous, nectaries deeply shell-shaped, stigmas whitish; **Fr** valve wings inflexed over the valves; **Ecol** in sandy plains and along roadsides, border between winter and summer rainfall; **Distr** S Namibia; N Namaqualand, NC, S Africa.

Note: This species is sometimes regarded as conspecific with **M. guerichianum**, but **M. gariusanum** differs in its pure white flowers, in lacking a basal vegetative rosette and in having obovate to rhomboid leaves which are usually smaller than those of **M. guerichianum**.

M. gaussenii = **M. cryptanthum**
M. glaciale = **M. crystallinum**
M. grandifolium = **M. guerichianum**

M. guerichianum Pax 1895: 133 (see **photo**) subg. **Cryophytum** ≡ *Amoebophyllum guerichianum* (Pax) N.E.Br. JB 1928: 80 ≡ *Cryophytum guerichianum* (Pax) Schwantes GF 1928: 68 **T** Gürich 75 (B!) = *Cryophytum acuminatum* L.Bolus NM 1929: 54 ≡ *M. parvipapillatum* L.Bolus NM 1939: 167 **T** Frames 18887 (BOL!) = *Cryophytum alatum* L.Bolus NM 1929: 37 ≡ *M. alatum* (L.Bolus) L.Bolus NM 1939: 167 **T** Frames 18865 = *Cryophytum calycinum* L.Bolus NM 1929: 52

≡ *M. subtereticaule* L.Bolus NM 1939: 167 **T** Frames 18903 (BOL!) = *Cryophytum carinatum* L.Bolus NM 1929: 52 ≡ *M. rubroroseum* L.Bolus NM 1939: 167 **T** Frames 18902 (BOL!) = *Cryophytum crassifolium* L.Bolus SAG 1928: 83 ≡ *M. pingue* L.Bolus NM 1939: 167 **T** Pillans 6019 (BOL!) = *M. crystallinum* L. var. *grandiflorum* Ecklon & Zeyher 1837: 321 ≡ *Cryophytum grandiflorum* (Ecklon & Zeyher) N.E.Br. GC 1928: 153 nomen illeg. ≡ *M. magniflorum* L.Bolus NM 1939: 167 **T** Ecklon & Zeyher 2097b (?) = *M. fenchelii* Schinz 1897: 79 ≡ *Cryophytum fenchelii* (Schinz) N.E.Br. GC 1928: 153 **T** Fenchel 172 (?) = *Cryophytum framesii* L.Bolus NM 1928: 31 ≡ *M. intransparens* L.Bolus NM 1939: 167 **T** Frames 167/28 (BOL!) = *Cryophytum framesii* L.Bolus var. *laxum* L.Bolus NM 1929: 38 **T** Pillans 18870 (BOL!) = *M. grandifolium* Schinz 1897: 79 ≡ *Cryophytum grandifolium* (Schinz) Dinter & Schwantes GF 1928: 68 ≡ *M. grandiflorum* Schinz ex Range 1934: 13 orthographic error ≡ *Cryophytum grandiflorum* Dinter & Schwantes ex Range 1934: 13 orthographic error non (Ecklon & Zeyher) N.E.Br. GC 1928: 153 nomen illeg. **T** Fleck 341a (?) = *Cryophytum karrooicum* L.Bolus NM 1929: 55 ≡ *M. karrooense* L.Bolus NM 1939: 167 **T** Rogers 19806 (BOL!) = *Cryophytum latisepalum* L.Bolus NM 1930: 163 ≡ *M. latisepalum* (L.Bolus) L.Bolus NM 1939: 167 **T** Pillans 6275 (BOL!) = *Derenbergiella luisae* Schwantes GW 1928: 599 **Syntypes** Schlechter s.n. s.d. (?), Eberlanz s.n. s.d. (?) = *Cryophytum maxwellii* L.Bolus SAG 1928: 84 ≡ *M. macrostigma* L.Bolus NM 1939: 167 **T** Maxwell Bolus s.n. March 1927 (BOL!) = *Cryophytum neilsoniae* L.Bolus NM 1928: 32 ≡ *M. neilsoniae* (L.Bolus) L.Bolus NM 1939: 167 **T** Neilson 789/25 (BOL!) = *Cryophytum pentagonum* L.Bolus NM 1931: 228 ≡ *M. quinangulatum* L.Bolus NM 1939: 167 **T** Frames 19380 (BOL!) = *M. perlatum* Dinter FR 1928: 22 **Lectotype**, here designated: Dinter 4943 (BOL!) = *Cryophytum roseum* L.Bolus NM 1929: 38 ≡ *M. rhodanthum* L.Bolus NM 1939: 168 **T** Frames 18866 (BOL!) = *Cryophytum sessiliflorum* L.Bolus NM 1930: 165 nomen illeg. ≡ *M. sedentiflorum* L.Bolus NM 1939: 168 **Lectotype**, here designated: Parkins 18958 (BOL!) = *Cryophytum setosum* L.Bolus NM 1929: 53 ≡ *M. setosum* (L.Bolus) L.Bolus NM 1939: 168 nomen illeg. ≡ *M. horridum* Koutnik & Lavis in Koutnik & O'Connor-Fenton 1984: 369 **T** Frames 18886 (BOL!) = *Cryophytum squamulosum* L.Bolus NM 1929: 71 ≡ *M. squamulosum* (L.Bolus) L.Bolus NM 1939: 168 nomen illeg. non (Dinter) Dinter ex Range FR 1938: 256 **T** Frames 18896 (BOL!) = *Cryophytum subulatum* L.Bolus NM 1931: 229 ≡ *M. subrigidum* L.Bolus NM 1939: 168 **T** L.Bolus 19369 (BOL!) = *Cryophytum truncatum* L.Bolus NM 1930: 164 ≡ *M. purpureoroseum* L.Bolus NM 1939: 168 **T** Frames 19052 (BOL!).

Prostrate to decumbent, with a basal vegetative rosette, to c. 60 cm *h* but usually smaller, annual to biennial in response to water supply, **I** (3–)4–6-angled and sometimes winged, ∅ to 25 mm; **L** flat, narrowly to ovately spatulate, acute, basal ones largest, to 250 mm *l* and to 100 mm *b*, margins undulating, without enlarged central water-storing cells, bladder cells conspicuous, rounded or elongated; **Fl** open from noon till dusk, ∅ 25–55 mm, petals white to pale salmon or pink, pale greenish at the base, connate into a tube, petals, filamentous staminodes and stamens very numerous, nectaries deeply shell-shaped, stigmas whitish; **Fr** valve wings inflexed over the valves; **S** brown, testa rough; **Ecol** in sandy plains and along roadsides, winter and summer rainfall; **Distr** probably SW Angola; Namibia; in an area from Namaqualand, Kenhardt and Prieska to Malmesbury and Worcester, and to Cradock, EC, NC, WC, S Africa.

Note: This species is often referred to as **M. crystallinum**. However, the true **M. crystallinum** exhibits much smaller flowers and much larger bladder cells than **M. guerichianum**.

M. hesperanthum = **M. longistylum**;
≡ **Hereroa hesperantha**
M. horridum = **M. guerichianum**

M. hypertrophicum Dinter FR 1923: 63 (see **photo**) subg. **Opophytum** ≡ *Halenbergia hypertrophica* (Dinter) Dinter KuaS 1937: 200 = *Pteropentacoilanthus hypertrophicus* (Dinter) Rappa & Camarrone 1960: 23 nomen invalid. **T** Dinter 3875 (?) = *Opophytum ampliatum* L.Bolus 1964: 39 **T** Herre 14841 (BOL!) = *M. aquosum* L.Bolus ABH 1927: 72 ≡ *Opophytum aquosum* (L.Bolus) N.E.Br. GC 1928: 253 **T** Pillans 4928 (BOL!) = *Opophytum australe* L.Bolus 1961: 115 **T** Littlewood 961/60 (BOL!) = *Opophytum speciosum* N.E.Br. GC 1928: 253 **T** Pearson 5986 (BOL!).

Annual, decumbent, without a basal vegetative rosette, to 25 cm *h*, ∅ to 100 cm, prostrate when in flower, **I** terete, ∅ to 10 mm; **L** cylindrical, linear, to 70 mm *l*, ∅ to 10 mm, with enlarged central water-storing cells, bladder cells inconspicuous or flattened; **Fl** open day and night, scented at dusk, ∅ to 60 mm, petals whitish to pale yellow, upper parts sometimes pink, very narrow, connate into a tube, petals, filamentous staminodes and stamens very numerous, nectaries tubular, stigmas pale yellow; **Fr** valve wings reflexed and fused in pairs; **S** whitish to light brown, testa smooth; **Ecol** in disturbed places, winter rainfall; **Distr** SW Namibia; Namaqualand, NC, S Africa.

Note: This species is very similar to **M. fastigiatum**; however, **M. hypertrophicum** is bigger, often very much so, the bladder cells are inconspicuous, the flowers are much larger and the seeds are completely smooth. Both species show a disjunct distribution.

M. inachabense Engler 1909: 192 (see **photo**) subg. **Mesembryanthemum** ≡ *Callistigma inachabense* (Engler) Dinter & Schwantes GW 1928: 644 ≡ *Cryophytum inachabense* (Engler) N.E.Br. GC 1928: 211 **T** Dinter 952 (?) = *Cryophytum arenarium* N.E.Br. GC 1928: 192 ≡ *M. arenarium* (N.E.Br.) L.Bolus NM 1939: 167 **T** Pearson 8215 (BOL!) = *Cryophytum fulleri* L.Bolus NM 1931: 301 ≡ *M. fulleri* (L.Bolus) L.Bolus NM 1939: 167 **T** Fuller 130 (BOL!) = *M. longipapillosum* Dinter FR 1928: 19 **T** Dinter 4944 (BOL!).

Annual, decumbent, without a basal vegetative rosette, to c. 20 cm *h*, ∅ to 20 cm, **I** terete or 2-edged on the younger parts, ∅ c. 5 mm; **L** flat, oblong, 25–50 mm *l* and 10–15 mm *b*, margins not undulating, without enlarged central water-storing cells, bladder cells distinct, rounded; **Fl** open until late in the evening, ∅ 10–20 mm, petals pale to dark yellow, connate into a short tube, filamentous staminodes absent, petals and stamens not particularly numerous, nectaries deep, stigmas basally connate, only upper portion fertile, red; **Fr** valve wings erect or inflexed over the valves; **S** whitish, testa smooth; **Ecol** in disturbed places, along roadsides, summer rainfall; **Distr** S Namibia; Kenhardt, NE Namaqualand, NC, S Africa.

Note: **M. inachabense** is unique within the genus because of its bright red and basally connate stigmas. *M. longipapillosum*, here listed as a synonym, may on further investigation prove to be a separate species.

M. inornatum = **M. nodiflorum**
M. intransparens = **M. guerichianum**
M. karrooense = **M. guerichianum**
M. lanceolatum = **M. aitonis**
M. latisepalum = **M. guerichianum**
M. liebendalense = **M. stenandrum**
M. linearifolium = **M. longistylum**
M. longipapillatum = **M. stenandrum**
M. longipapillosum = **M. inachabense**

M. longistylum DC. 1804: t. 147 (see **photo**) subg. **Mesembryanthemum** ≡ *Aridaria longistyla* (DC.) Schwantes GF 1928: 69 ≡ *Nycteranthus longistylus* (DC.) Schwantes 1952: 50 nomen illeg. **T** DC. 1804: t. 147 = *Cryophytum intermedium* L.Bolus NM 1929: 53 ≡ *M. cryocalyx* L.Bolus NM 1939: 167 **Lectotype**, here designated: H.Bolus 13118 (BOL!) = *Cryophytum lineare* L.Bolus NM 1930: 165 ≡ *M. linearifolium* L.Bolus NM 1939: 167 **T** BOL 19053 (BOL!) = *Cryophytum neglectum* N.E.Br. GC 1928: 212 **Lectotype**, here designated: Cooper 1660 (K!) = *M. vigilans* L.Bolus in N.E.Br. 1920: 121 ≡ *Nycteranthus vigilans* (L.Bolus) Schwantes in Jacobsen 1960: 1326 ≡ *Sphalmanthus vigilans* (L.Bolus) L.Bolus in Jacobsen 1970: 515 nomen invalid. ≡ *M. hesperanthum* L.Bolus ABH 1915: 190 nomen illeg. non Dinter & A.Berger 1914: 589 ≡ *Aridaria hesperantha* (L.Bolus) N.E.Br. JB 1928: 142 **T** Rogers 769/13 (BOL!).

Decumbent to erect, basally slightly woody, without a basal vegetative rosette, 10–40 cm *h*, ∅ more than 40 cm, annual or biennial in response to water supply, **I** terete, ∅ 3–5 mm; **L** almost cylindrical, slightly channelled, or somewhat flattened, linear to very narrowly oblong, 20–35 mm *l* and 2–10 mm *b*, with enlarged central water-storing cells, bladder cells usually inconspicuous, rounded or somewhat flattened; **Fl** ∅ 10–20(–25) mm, petals white, often with pink tips and/or straw-coloured base, connate into a short tube, filamentous staminodes absent, petals and stamens not particularly numerous, nectaries tubular, stigmas yellow; **Fr** valve wings reflexed and fused in pairs; **S** brown, testa rough; **Ecol** in disturbed places and along roadsides, winter rainfall; **Distr** Williston, NC, from Vanrhynsdorp to Malmesbury, from Worcester to Riversdale and Laingsburg, WC, S Africa.

Note: **M. longistylum** was formerly included in **Aridaria** because of its slightly woody stems. However, the tubular nectaries in particular identify it as a member of **Mesembryanthemum**.

M. louiseae = **M. aitonis**
M. macrophyllum = **M. barklyi**
M. macrostigma = **M. guerichianum**,
≡ **Lampranthus macrostigma**
M. magniflorum = **M. guerichianum**
M. mollissimum = **M. pellitum**
M. neilsoniae = **M. guerichianum**

M. nodiflorum L. 1753: 480 (see **photo**) subg. **Mesembryanthemum** ≡ *Cryophytum nodiflorum* (L.) L.Bolus SAG 1927: 327 ≡ *Gasoul nodiflorum* (L.) Rothmaler 1941: 413 **T** Bauhin 1623: 289 = *M. caducum* Aiton 1789: 179 ≡ *Psilocaulon caducum* (Aiton) N.E.Br. GC 1928: 254 ≡ *Stigmatocarpum caducum* (Aiton) L.Bolus SAG 1927: 326 **T** drawing by Ann Lee (BM, photograph!) = *Cryophytum cleistum* L.Bolus NM 1929: 151 ≡ *M. inornatum* L.Bolus NM 1939: 167 **T** BOL 19053 (BOL!) = *M. copticum* L. 1762: 688 non Jacquin 1776: t. 6 nomen illeg. **T** Bauhin 1623: 286 = *Cryophytum gibbosum* N.E.Br. GC 1929: 84 **T** Maughan Brown s.n. Nov. 1928 (K!) = *Cryophytum rogersii* L.Bolus SAG 1928: 84 ≡ *M. paucandrum* L.Bolus NM 1939: 167 **T** Rogers 17177 (BOL!).

Annual, prostrate to decumbent, without a basal vegetative rosette, **I** terete, ∅ 2–6 mm; **L** almost cylindrical or slightly channelled, linear, 15–20 mm *l*, ∅ 2–5 mm, with enlarged central water-storing cells, bladder cells conspicuous, crystalline, rounded, usually ciliate along the basal parts of the leaf margins; **Fl** ∅ 5–10 mm, petals white or pale pink or white with pale pink tips, free, filamentous staminodes absent, petals and stamens very few, nectaries tubular, stigmas whitish to yellow; **Fr** valve wings reflexed and fused in pairs; **S** whitish to light brown, testa smooth; **Ecol** in disturbed places and along roadsides, mainly winter rainfall; **Distr** Arabia; SW Australia; Baluchistan; Canary Islands; S Europe; Iran; Kurdistan; Madeira; SW Namibia (only recorded from Lüderitzbucht); N Africa; Palestine; NW Namaqualand (recorded from Wallekraal), NC, from Piquetberg to the Cape Peninsula and from Worcester to Oudtshoorn, WC, S Africa; California, USA.

Note: In populations of **M. nodiflorum** from the Mediterranean basin and Asia tetracolpate pollen was recorded, whereas populations from S Africa show tricolpate pollen. Bittrich (1987: 5–116) established chromosome numbers $n = 18$ from N Africa, but $2n = 18$ from S Africa. These observations suggest that **M. nodiflorum** is naturally distributed in the northern hemisphere from where (Italy) it was originally described.

M. ovatum = **M. aitonis**
M. pachypus = **M. fastigiatum**
M. papuliferum = **M. fastigiatum**
M. parvipapillatum = **M. guerichianum**
M. paucandrum = **M. nodiflorum**
M. paulum = **M. aitonis**

M. pellitum Friedrich 1968: 211 (see **photo**) subg. **Cryophytum** ≡ *M. mollissimum* Dinter ex Friedrich 1959: 45 nomen invalid. T Dinter 8174 (M).

Decumbent to erect, without a basal vegetative rosette, main stem rather long, terminated by a flower, to c. 50 cm *h*, annual to biennial in response to water supply, **I** usually terete, ∅ to 15 mm; **L** flat, fiddle-shaped or oblong, basal ones largest, to 150 mm *l* and 80 mm *b*, margins undulating, without enlarged central water-storing cells, bladder cells conspicuous, hair-like on the leaves, but rounded, elongated, or hair-like on the stems; **Fl** ∅ to 50 mm, petals white, connate into a tube, petals, filamentous staminodes and stamens very numerous, nectaries shell-shaped, stigmas whitish; **Fr** valve wings inflexed over the valves; **S** brown, testa rough; **Ecol** in plains and along roadsides, winter rainfall; **Distr** SW Namibia; NW Namaqualand, NC, S Africa.

Note: In contrast to most other species of **Mesembryanthemum** subg. **Cryophytum**, **M. pellitum** shows an erect main stem which is terminated by a flower instead of a vegetative rosette. More flowers are produced on side branches and the plants may become fairly large. Another remarkable feature is found in the hair-like bladder cells of the leaves.

M. perlatum = **M. guerichianum**
M. pingue = **M. guerichianum**
M. puberulum = **M. aitonis**
M. purpureoroseum = **M. guerichianum**
M. quinangulatum = **M. guerichianum**
M. rhodanthum = **M. guerichianum**
M. rubroroseum = **M. guerichianum**
M. sarcocalycanthum = **M. cryptanthum**
M. sedentiflorum = **M. guerichianum**
M. sessiliflorum var. *album* = **M. aitonis**
M. sessiliflorum var. β = **M. aitonis**
M. setosum = **M. guerichianum**
M. squamulosum = **M. guerichianum**; = **M. barklyi**

M. stenandrum (L.Bolus) L.Bolus NM 1939: 168 (see **photo**) subg. **Mesembryanthemum** ≡ *Cryophytum stenandrum* L.Bolus NM 1929: 36 T Frames 18864 (BOL!) = *Cryophytum dejagerae* L.Bolus NM 1932: 315 ≡ *M. dejagerae* (L.Bolus) L.Bolus NM 1939: 167 T de Jager 19643 (BOL!) = *Cryophytum galpinii* L.Bolus NM 1931: 231 ≡ *M. galpinii* (L.Bolus) L.Bolus NM 1939: 167 T Galpin 19392 (BOL!) = *M. liebendalense* L.Bolus 1965: 309 ≡ *Cryophytum liebendalense* (L.Bolus) J.Ingram 1970: 46 T Hall 2882 (BOL!) = *Cryophytum longipapillatum* L.Bolus NM 1929: 38 ≡ *M. longipapillatum* (L.Bolus) L.Bolus NM 1939: 167 T Pillans 6139 (BOL!) = *Psilocaulon pomeridianum* L.Bolus NM 1928: 29 T Pillans 6143 (BOL!) = *Cryophytum wilmaniae* L.Bolus SAG 1928: 84 ≡ *M. annuum* L.Bolus NM 1939: 167 T Wilman s.n. Oct. 1921 (BOL!).

Annual, prostrate to decumbent, without a basal vegetative rosette, **I** terete, ∅ 2–6 mm; **L** almost cylindrical or slightly channelled, linear, 15–20 mm *l*, ∅ 2–5 mm, with enlarged central water-storing cells, bladder cells conspicuous, crystalline, rounded, usually ciliate along the basal parts of the margins; **Fl** ∅ 15–30 mm, petals white or pale pink or white with pale pink tips, connate into a short tube, filamentous staminodes absent, petals and stamens not particularly numerous, nectaries tubular, stigmas whitish to yellow; **Fr** valve wings reflexed and fused in pairs; **S** light brown, testa rough; **Ecol** disturbed places and along roadsides, winter and summer rainfall; **Distr** Fauresmith, FS, from Namaqualand and Calvinia to Gordonia and to Hopetown and De Aar, NC, Vanrhynsdorp, Clanwilliam, Beaufort West, WC, S Africa.

Note: **M. stenandrum** is very similar to **M. nodiflorum**. The characters of distiction are found in the flowers: the petals are much longer and more numerous in **M. stenandrum** than in **M. nodiflorum**. According to Bittrich (1987: 5–116), populations from Kenhardt (described as *M. longipapillatum*) have shorter nectaries and a punctate, microspinulose tectum versus a reticulate and spinulose tectum in **M. nodiflorum**.

M. stratum = **M. aitonis**
M. subrigidum = **M. guerichianum**
M. subtereticaule = **M. guerichianum**

M. subtruncatum L.Bolus NM 1939: 168 (see **photo**) subg. **Mesembryanthemum** ≡ *Cryophytum pusillum* L.Bolus NM 1929: 152 **Lectotype**, here designated: Compton 18990 (BOL!).

Annual, prostrate, without a basal vegetative rosette, to 2 cm *h*, ∅ 6–8 cm, **I** terete, ∅ 2 mm; **L** almost cylindrical, almost truncate, to 12 mm *l*, to 5 mm *b*, with enlarged central water-storing cells, bladder cells inconspicuous, rounded to flattened; **Fl** ∅ c. 20–25 mm, petals pink, basally white or straw-coloured, connate into a short tube, filamentous staminodes absent, petals and stamens not particularly numerous, nectaries tubular, stigmas yellow; **Ecol** in disturbed places in quartz, winter rainfall; **Distr** Ceres, Montagu, WC, S Africa.

Note: The species is most remarkable for its almost truncate leaves and calyx-lobes which give the plants an unmistakable appearance.

M. theurkauffii = **M. cryptanthum**
M. vigilans = **M. longistylum**
M. violense = **M. gariusanum**
M. volckameri = **M. aitonis**

Mesembryanthemum

H. E. K. Hartmann

Collected here are all those names published originally under **Mesembryanthemum** that cannot be included in the present circumscription of the genus. Most of the names have been transferred to other genera printed in bold letters behind the = (different types) or ≡ (the same type). Some species, however, are not only listed here but also described because they cannot be placed with certainty in any other genus. Names of these species appear in bold letters, too, when they are validly published. All names given in italics are either invalid or illegitimate.

M. abbreviatum = **Carpobrotus virescens**
M. aberdeenense ≡ **Delosperma aberdeenense**
M. abruptum ≡ **Octopoma abruptum**
M. abyssinicum ≡ **Delosperma abyssinicum;** ≡ **Delosperma schimperi**
M. acinaciforme ≡ **Carpobrotus acinaciformis;** = **Erepsia lacera**
M. acinaciforme var. β *flavum* ≡ **Carpobrotus edulis**
M. acinaciforme var. *purpureum* ≡ **Carpobrotus acinaciformis**
M. acrosepalum ≡ **Lampranthus acrosepalus**
M. aculeatum = **Ruschia spinosa**
M. acuminatum = **Phyllobolus splendens** ssp. **splendens**
M. acutangulum ≡ **Ruschia acutangula**
M. acutifolium ≡ **Lampranthus acutifolius**
M. acutipetalum ≡ **Khadia acutipetala**
M. acutisepalum = **Psilocaulon junceum**
M. acutum ≡ **Cephalophyllum subulatoides**
M. adscendens = **Glottiphyllum longum**
M. aduncum ≡ **Ruschia schollii;** ≡ **Ruschia willdenowii;** ≡ **Lampranthus aduncus**
M. aequilaterale ≡ **Carpobrotus aequilaterus**
M. aequilaterale var. β = **Carpobrotus aequilaterus**
M. aequilaterale var. *chiloënse* = **Carpobrotus chilensis**
M. aequilaterale var. *decagynum* = **Carpobrotus aequilaterus**
M. aequilaterum ≡ **Carpobrotus aequilaterus**
M. aereum ≡ **Delosperma aereum**
M. aereum var. *album* = **Delosperma aereum**
M. aestivum ≡ **Lampranthus aestivus**
M. aggregatum = **Conophytum piluliforme** ssp. **piluliforme**
M. agninum ≡ **Stomatium agninum**
M. agninum var. *integrifolium* = **Stomatium agninum**
M. albatum ≡ **Juttadinteria albata**
M. albertense = **Conophytum truncatum** ssp. **truncatum**
M. albertiniense = **Jordaaniella dubia**
M. albescens = **Conophytum bilobum** ssp. **bilobum**
M. albicaule = **Phyllobolus splendens** ssp. **splendens**
M. albicute ≡ **Ruschia bolusiae**
M. albidum ≡ **Machairophyllum albidum**
M. albiflorum ≡ **Drosanthemum albiflorum**
M. albinotum ≡ **Rabiea albinota**
M. albipunctum ≡ **Rabiea albipuncta**
M. albipunctum var. *majus* = **Nananthus vittatus**
M. alboroseum ≡ **Leipoldtia alborosea**
M. album ≡ **Oscularia alba**
M. aloides ≡ **Nananthus aloides**
M. alstonii ≡ **Cephalophyllum alstonii**
M. altigenum ≡ **Ruschia altigena**
M. altile = **Conophytum ficiforme**
M. amicorum ≡ **Ruschia amicorum**
M. amoenum ≡ **Lampranthus amoenus**
M. amplectens ≡ **Aspazoma amplectens**
M. amplexicaule = **Leipoldtia schultzei**
M. amplexum ≡ **Ruschia inclusa**
M. ampliatum ≡ **Ruschia ampliata**
M. anatomicum = **Sceletium emarcidum**

M. anatomicum var. *emarcidum* ≡ **Sceletium emarcidum**
M. anatomicum var. *fragile* = **Sceletium emarcidum**
M. anceps ≡ **Erepsia anceps**
M. anceps var. *pallidum* = **Erepsia anceps**
M. androsaceum ≡ **Antimima androsacea**
M. anemoniflorum = **Jordaaniella dubia**
M. angelicae ≡ **Conophytum angelicae** ssp. **angelicae**
M. angulipes ≡ **Gibbaeum angulipes**
M. angustipetalum = **Vanzijlia annulata**
M. angustum = **Glottiphyllum cruciatum**
M. angustum var. *heterophyllum* = **Glottiphyllum depressum; Glottiphyllum longum**
M. angustum var. *pallidum* = **Glottiphyllum cruciatum**
M. annulatum ≡ **Vanzijlia annulata**
M. antemeridianum ≡ **Lampranthus antemeridianus**
M. apetalum ≡ **Dorotheanthus apetalus;** = **Dorotheanthus clavatus**
M. apiatum = **Conophytum bilobum** ssp. **bilobum**
M. apiculatum ≡ **Braunsia apiculata**
M. apiculatum var. *muticum* = **Braunsia maximiliani**

M. apodanthum Schlechter & Diels in Schultze 1807: 692 T Schultze 222β (?).

Shrubs with erect branches, bark blackish, the side branches sometimes spinescent; **L** very dense, triquetrous, verrucose, glaucescent, 6–9 mm *l*, 4 mm *b*; **Fl** solitary, sessile, petals very narrow, 5–6 mm *l*, 0.8 mm *b*; **Fr** 5 locules; **Ecol** unknown; **Distr** Namaqualand, NC, S Africa.

Note: The species is said to be close to *M. schultzei*, now a member of the genus **Leipoldtia**, this latter species differing in having bigger leaves, longer side branches, and stalked flowers twice the size of those of **M. apodanthum**. The plant was collected in flower in August 1904.

M. approximatum ≡ **Ruschia approximata**
M. apricum = **Leipoldtia schultzei**
M. arboriforme ≡ **Mestoklema arboriforme**
M. arbuthnotiae ≡ **Lampranthus arbuthnotiae**
M. arenosum ≡ **Brownanthus arenosus;** ≡ **Lampranthus arenosus**
M. argenteum ≡ **Lampranthus argenteus**
M. aridum = **Sceletium tortuosum**
M. aristulatum ≡ **Ruschia aristulata**
M. armatum = **Arenifera stylosa**
M. arrectum = **Glottiphyllum surrectum**
M. articulatum ≡ **Psilocaulon articulatum**
M. asperulum ≡ **Delosperma asperulum**
M. asperum ≡ **Erepsia aspera**
M. asperum-caerulescens = **Erepsia bracteata**
M. assimile = **Conophytum ficiforme**
M. astridiae = **Titanopsis hugo-schlechteri**
M. atratum ≡ **Ruschia atrata**
M. atrocinctum ≡ **Antimima saturata**
M. attenuatum ≡ **Drosanthemum attenuatum**
M. attonsum ≡ **Trichodiadema attonsum**
M. augeiforme = **Conophytum chauviniae**
M. aurantiacum ≡ **Lampranthus glaucoides**
M. aurantium ≡ **Lampranthus glaucoides**
M. auratum = **Phyllobolus nitidus**
M. aureum ≡ **Lampranthus aureus;** = **Phyllobolus nitidus**
M. ausanum = **Prenia tetragona**
M. ausense ≡ **Juttadinteria ausensis**
M. australe ≡ **Disphyma australe**
M. austricola ≡ **Lampranthus austricola**
M. axillare (N.E.Br.) hort. Kew ex Hill 1931: 126 nomen nudum
M. axthelmianum ≡ **Ruschia axthelmiana**
M. barbatum ≡ **Trichodiadema barbatum**
M. barbatum var. β ≡ **Trichodiadema hirsutum**
M. barbatum var. γ ≡ **Trichodiadema densum**
M. barbatum var. *densum* ≡ **Trichodiadema densum**
M. barbatum var. *hirsutum* ≡ **Trichodiadema hirsutum**
M. barnardii ≡ **Ruschia barnardii**
M. beetzii ≡ **Stoeberia beetzii**
M. bellidiflorum ≡ **Acrodon bellidiflorus**
M. bellidiflorum var. *glaucescens* ≡ **Acrodon bellidiflorus**
M. bellidiflorum var. *simplex* = **Acrodon subulatus**
M. bellidiflorum var. *striatum* = **Acrodon bellidiflorus**
M. bellidiflorum var. *subulatum* ≡ **Acrodon subulatus**
M. bellidiflorum var. *viride* = **Acrodon bellidiflorus**
M. bellidiforme ≡ **Dorotheanthus bellidiformis**
M. bellum ≡ **Lithops karasmontana** ssp. **bella**
M. bergerianum = **Hereroa hesperantha**
M. berghiae ≡ **Lampranthus berghiae**
M. beswickii ≡ **Khadia beswickii**
M. bibracteatum = **Cheiridopsis rostrata**
M. bicolor ≡ **Lampranthus bicolor**
M. bicolor var. *inaequale* ≡ **Lampranthus inaequalis**
M. bicoloratum = **Stoeberia beetzii**
M. bicorne ≡ **Psilocaulon bicorne**
M. bidentatum = **Glottiphyllum difforme**
M. bifidum = **Cheiridopsis rostrata**
M. bifoliatum = **Ottosonderia monticola**
M. biforme ≡ **Antimima biformis**
M. bigibberatum = **Glottiphyllum difforme**
M. bijliae ≡ **Ruschia bijliae**
M. bilobum ≡ **Conophytum bilobum** ssp. **bilobum**
M. binum ≡ **Braunsia bina**
M. blandum ≡ **Lampranthus blandus**
M. blandum var. *curviflorum* ≡ **Lampranthus curviflorus**

M. boehmerianum = **Conophytum saxetanum**
M. bolusiae ≡ **Lampranthus recurvus**
M. bolusii ≡ **Pleiospilos bolusii**
M. bosscheanum ≡ **Faucaria bosscheana**
M. brachiatum = **Aptenia geniculiflora**
M. brachyandrum ≡ **Lampranthus brachyandrus**
M. brachypetalum = **Drosanthemum luederitzii**
M. brachyphyllum = **Drosanthemum luederitzii**; = **Lampranthus glaucus**
M. bracteatum ≡ **Erepsia bracteata**
M. brakdamense ≡ **Ruschia brakdamensis**
M. brevicaule = **Conicosia pugioniformis** ssp. **pugioniformis**
M. brevicolle ≡ **Antimima brevicollis**
M. brevifolium ≡ **Drosanthemum brevifolium**
M. brevipes = **Argyroderma fissum**
M. brevistaminum ≡ **Lampranthus brevistaminus**
M. britteniae = **Leipoldtia schultzei**
M. brownii ≡ **Lampranthus brownii**
M. brunnthaleri ≡ **Delosperma brunnthaleri**
M. buchubergense ≡ **Antimima buchubergensis**; = **Brownanthus pubescens**; = **Juttadinteria deserticola**
M. bulbosum ≡ **Trichodiadema bulbosum**
M. caespitosum ≡ **Lampranthus caespitosus**
M. caespitosum forma *luxurians* = **Lampranthus caespitosus**
M. calamiforme ≡ **Cylindrophyllum calamiforme**
M. calandrum ≡ **Leipoldtia calandra**
M. calcaratum ≡ **Lampranthus calcaratus**
M. calcareum ≡ **Titanopsis calcarea**; = **Titanopsis schwantesii**
M. calcicola ≡ **Ruschia calcicola**
M. calculus ≡ **Conophytum calculus** ssp. **calculus**
M. calculus element *minutiflorum* = **Conophytum pageae**
M. calendulaceum = **Carpanthea pomeridiana**
M. calycinum ≡ **Drosanthemum calycinum**
M. campestre ≡ **Ruschia campestris**
M. canaliculatum ≡ **Phyllobolus canaliculatus**
M. candens ≡ **Drosanthemum candens**
M. candidissimum = **Cheiridopsis denticulata**
M. candollii = **Carpanthea pomeridiana**
M. caninum ≡ **Carruanthus ringens**
M. caninum var. *paucidentatum* = **Carruanthus ringens**
M. caninum var. *vulpinum* = **Carruanthus ringens**
M. canonotatum ≡ **Ruschia canonotata**
M. canum ≡ **Pleiospilos compactus** ssp. **canus**
M. capillaceum ≡ **Lampranthus capillaceus**
M. capillare ≡ **Drosanthemum capillare**
M. capillare var. *acutifolium* ≡ **Drosanthemum acutifolium**
M. capitatum = **Conicosia pugioniformis** ssp. **pugioniformis**
M. capitatum var. *ramigerum* = **Conicosia pugioniformis** ssp. **pugioniformis**
M. capornii ≡ **Ruschia capornii**
M. capornii forma *ferum* = **Ruschia capornii**
M. capornii var. *longifolium* = **Ruschia capornii**
M. carinans ≡ **Hereroa carinans**
M. carinatum = **Erepsia lacera**
M. carneum = **Phyllobolus spinuliferus**
M. caroli ≡ **Ruschia caroli**
M. carolinense ≡ **Khadia carolinensis**
M. caroli-schmidtii ≡ **Cheiridopsis caroli-schmidtii**
M. catervum = **Conophytum truncatum** ssp. **viridicatum**
M. caudatum ≡ **Phyllobolus caudatus**
M. caulescens ≡ **Oscularia caulescens**
M. cauliculatum = **Cephalophyllum diversiphyllum**
M. cedarbergense ≡ **Oscularia cedarbergensis**
M. ceratophyllum = **Lampranthus curvifolius**
M. ceriseum ≡ **Lampranthus ceriseus**
M. chauviniae ≡ **Conophytum chauviniae**
M. chilense = **Carpobrotus chilensis**
M. chrysoleucum ≡ **Monilaria chrysoleuca**
M. cibdelum ≡ **Nananthus vittatus**
M. cigarettiferum = **Cheiridopsis namaquensis**
M. ciliatum ≡ **Brownanthus ciliatus**
M. ciliolatum = **Antimima dasyphylla**
M. cinctum ≡ **Ruschia cincta**
M. cinereum ≡ **Namibia cinerea**
M. citrinum ≡ **Lampranthus citrinus**
M. clausum ≡ **Eberlanzia clausa**
M. clavatum ≡ **Dorotheanthus clavatus**; ≡ **Aridaria noctiflora** ssp. **defoliata**
M. clavellatum ≡ **Disphyma clavellatum**
M. clavifolium ≡ **Jordaaniella clavifolia**
M. claviforme ≡ **Dorotheanthus clavatus**
M. clavulatum = **Psilocaulon subnodosum**
M. cleistum = **Eberlanzia sedoides**
M. clivorum ≡ **Mitrophyllum clivorum**
M. coccineum ≡ **Lampranthus coccineus**
M. cognatum = **Mitrophyllum clivorum**
M. collinum ≡ **Drosanthemum collinum**
M. coloratum ≡ **Ruschia versicolor**
M. commutatum = **Phyllobolus grossus**
M. compactum ≡ **Pleiospilos compactus**
M. compressum ≡ Ruschia virgata = **Erepsia bracteata**
M. comptonii ≡ **Oscularia comptonii**
M. comptonii var. *angustifolium* = **Lampranthus borealis**
M. comptonii var. *roseum* = **Oscularia comptonii**
M. comptum ≡ **Antimima concinna**
M. concavum = **Sceletium tortuosum**
M. concinnum = **Titanopsis schwantesii**
M. condensum ≡ **Antimima condensa**
M. confertum = **Lampranthus incurvus**
M. confusum ≡ **Cephalophyllum confusum**
M. congestum ≡ **Ruschia congesta**

M. connatum ≡ **Octopoma connatum**
M. conspicuum ≡ **Lampranthus conspicuus;**
≡ **Lampranthus dyckii**
M. constrictum = **Leipoldtia schultzei**
M. convexum ≡ **Lampranthus convexus**
M. cookii = **Machairophyllum albidum**
M. cooperi ≡ **Delosperma cooperi**
M. copiosum ≡ **Oscularia copiosa**
M. copticum = **Dorotheanthus apetalus**
M. coralliflorum ≡ **Lampranthus coralliflorus**
M. corallinum ≡ **Brownanthus corallinus;**
≡ **Lampranthus coralliflorus**

M. coralliforme Vest 1828: 224 T?

Erect shrubs, I two-edged; L triquetrous, compressed, smooth, acute or with a nutant apex, connate but not vaginate, 27 mm *l*; **Fl** solitary, sessile between the uppermost leaves, **K** with three horns, petals white or very pale pink, almost as long as the **K** lobes; **Fr** unknown; **Ecol** unknown; **Distr** unknown.

Note: Listed by Berger 1908: 292 under doubtful species, *M. coralliforme* has never been placed in any other genus. It was reported to be similar to *M. uncinatus* and the calyx similar to that of *M. insititium*, now understood as **Malephora purpureo-crocea.**

M. cordifolium ≡ **Aptenia cordifolia**
M. coriarium ≡ **Psilocaulon coriarium**
M. corniculatum ≡ **Cephalophyllum corniculatum**
M. coruscans
= **Conicosia pugioniformis** ssp. **pugioniformis**
M. cradockense ≡ **Ruschia cradockensis**
M. crassicaule ≡ **Sceletium crassicaule**
M. crassifolium ≡ **Disphyma crassifolium**
M. crassipes = **Aloinopsis spathulata**
M. crassulifolium = **Eberlanzia sedoides**
M. crassulinum ≡ **Delosperma crassuloides**
M. crassuloides ≡ **Delosperma crassuloides**
M. crassum ≡ **Ruschia crassa**
M. crateriforme ≡ **Argyroderma crateriforme**
M. criniflorum = **Dorotheanthus bellidiformis**

M. crispatum Haw. 1795: 170 ≡ *M. crispum* Haw. 1803: 44 nomen illeg. **Lectotype**, here designated: Petiver 1709: t. 88 f 5.

Probably annual, root a little thickish; **L** alternate, broad and flat, apparently smooth, the margins markedly undulate; **Fl** 2 on a stalk, not in a cyme, petals red; **Ecol** unknown; **Distr** unknown.

Note: Sonder (1862: 460) classified this species as a member of the Asteraceae, which may well be true, considering the figure of Petiver showing an almost globular bud, unlike any seen in **Aizoaceae.** The reported petals would then be marginal flowers. It seems that the name has not been considered in studies of South African Asteraceae.

M. crispum ≡ **M. crispatum**
M. croceum ≡ **Malephora crocea**
M. croceum var. *flavocroceum*
≡ **Malephora flavo-crocea**
M. croceum var. *purpureo-croceum*
≡ **Malephora purpureo-crocea**
M. cronemeyerianum
≡ **Delosperma cronemeyerianum**
M. cruciatum ≡ **Glottiphyllum cruciatum**
M. cryptopodium = **Gibbaeum nuciforme**
M. crystallino-papillosum = **Phyllobolus oculatus**
M. cultratum = **Glottiphyllum longum**
M. cuneifolium = **Dorotheanthus bellidiformis**
M. cupreum ≡ **Cephalophyllum caespitosum**
M. cupulatum ≡ **Ruschia cupulata**
M. curtophyllum ≡ **Cephalophyllum curtophyllum**
M. curtum ≡ **Ruschia curta**
M. curviflorum ≡ **Lampranthus curviflorus**
M. curvifolium ≡ **Lampranthus curvifolius**
M. curvifolium var. *minus* ≡ **Lampranthus flexifolius**
M. cyathiforme ≡ **Lampranthus cyathiformis**
M. cylindricum ≡ **Calamophyllum cylindricum**
M. cylindricum var. *teretifolium*
≡ **Calamophyllum teretifolium**
M. cymbifolium ≡ **Ruschia cymbifolia**
M. cymbiforme ≡ **Corpuscularia cymbiformis**
M. cymosum ≡ **Ruschia cymosa**
M. damaranum
= **Lithops karasmontana** ssp. **karasmontana**
M. dasyphyllum ≡ **Antimima dasyphylla**
M. dealbatum ≡ **Dracophilus dealbatus**
M. debile ≡ **Lampranthus debilis**
M. deceptum ≡ **Ruschia cymbifolia**
M. decipiens = **Cephalophyllum loreum**
M. decorum ≡ **Trichodiadema decorum**
M. decurrens ≡ **Ruschia decurrens**
M. decussatum = **Aptenia geniculiflora**
M. deflectum ≡ **Ruschia decumbens**
M. deflexum ≡ **Lampranthus deflexus**
M. defoliatum ≡ **Aridaria noctiflora** ssp. **defoliata**
M. dejagerae ≡ **Ruschia dejagerae**
M. dekenahi ≡ **Antimima dekenahi**
M. delaetianum ≡ **Dracophilus delaetianus;**
= **Lithops karasmontana** ssp. **bella**
M. delicatulum ≡ **Drosanthemum delicatulum**
M. deliciosum ≡ **Carpobrotus deliciosus**
M. deltoides ≡ **Oscularia deltoides**
M. deltoides var. γ ≡ **Oscularia caulescens**
M. deltoides var. *majus* ≡ **Oscularia major**
M. deltoides var. *muricatum* ≡ **Oscularia deltoides**
M. deltoides var. *pedunculatum*
≡ **Oscularia pedunculata**
M. deltoides var. *simplex* = **Oscularia caulescens**
M. delum ≡ **Phyllobolus delus**
M. demissum = **Disphyma australe**
M. densifolium ≡ **Lampranthus densifolius**

M. densipetalum ≡ **Lampranthus densipetalus**
M. densum ≡ **Trichodiadema densum**
M. dentatum = **Erepsia lacera**
M. denticulatum ≡ **Cheiridopsis denticulata**
M. denticulatum var. *canum* = **Cheiridopsis denticulata**
M. denticulatum var. *glaucum* = **Cheiridopsis denticulata**
M. denticulatum var. *candidissimum* = **Cheiridopsis denticulata**
M. dependens ≡ **Lampranthus dependens**
M. depressum ≡ **Glottiphyllum depressum**
M. depressum var. *lividum* = **Glottiphyllum depressum**
M. depressum var. *pallens* ≡ **Glottiphyllum depressum**
M. derenbergianum ≡ **Ebracteola derenbergiana**
M. derenbergianum var. *interiore* = **Ebracteola fulleri**
M. deserticola ≡ **Juttadinteria deserticola**
M. dichroum ≡ **Ruschia dichroa**
M. dielsianum ≡ **Ruschia dielsiana**
M. difforme ≡ **Glottiphyllum difforme**; ≡ **Cheiridopsis namaquensis**
M. diffusum ≡ **Lampranthus diffusus**
M. digitatum ≡ **Phyllobolus digitatus** ssp. **digitatus**
M. digitiforme = **Phyllobolus digitatus** ssp. **digitatus**
M. dilatatum ≡ **Ruschia haworthii**
M. dimidiatum ≡ **Carpobrotus dimidiatus**
M. diminutum = **Cephalophyllum subulatoides**
M. diminutum var. *cauliculatum* = **Cephalophyllum diversiphyllum**
M. diminutum var. *pallidum* = **Cephalophyllum subulatoides**
M. dimorphum = **Psilocaulon dimorphum**
M. dinterae = **Chasmatophyllum musculinum**
M. dinteri ≡ **Psilocaulon dinteri**
M. diplosum ≡ **Ruschia patens**
M. disgregum ≡ **Lampranthus disgregus**
M. dissimile = **Cephalophyllum corniculatum**
M. dissitum ≡ **Mitrophyllum dissitum**
M. distans ≡ **Antimima distans**
M. diutinum ≡ **Lampranthus diutinus**
M. divergens = **Antegibbaeum fissoides**
M. diversifolium ≡ **Cephalophyllum diversiphyllum**
M. diversipapillosum = **Brownanthus arenosus**
M. diversiphyllum ≡ **Cephalophyllum diversiphyllum**
M. diversiphyllum var. *congestum* = **Cephalophyllum diversiphyllum**
M. dolabriforme ≡ **Rhombophyllum dolabriforme**
M. dolomiticum ≡ **Antimima dolomitica**
M. dregeanum ≡ **Lampranthus dregeanus**
M. drepanophyllum ≡ **Esterhuysenia drepanophylla**
M. dressianum ≡ **Ruschia rigens**
M. duale ≡ **Antimima dualis**
M. dubitans ≡ **Lampranthus dubitans**
M. dubium ≡ **Jordaaniella dubia**; = **Aizoon canariense**
M. dulce ≡ **Lampranthus dulcis**
M. dumosum ≡ **Ruschia virens**
M. dunense ≡ **Erepsia dunensis**
M. duthiae ≡ **Ruschia duthiae**
M. dyckii ≡ **Lampranthus dyckii**
M. dyeri ≡ **Rhombophyllum albanense**
M. eberlanzii ≡ **Lithops karasmontana** ssp. **eberlanzii**
M. ebracteatum ≡ **Cephalophyllum ebracteatum**; = **Phyllobolus trichotomus**
M. echinatum ≡ **Delosperma echinatum**
M. ecklonis ≡ **Delosperma ecklonis**
M. edentulum ≡ **Ruschia edentula**
M. edule ≡ **Carpobrotus edulis**
M. edule var. β *flavum* ≡ **Carpobrotus edulis**
M. edule var. *equilaterum* ≡ **Carpobrotus aequilaterus**
M. edule var. *flavum* ≡ **Carpobrotus edulis**
M. edule var. *virescens* ≡ **Carpobrotus virescens**
M. edwardsiae ≡ **Lampranthus edwardsiae**
M. eendornense ≡ **Antimima eendornensis**
M. egregium ≡ **Lampranthus egregius**
M. elegans ≡ **Lampranthus elegans**
M. elineatum ≡ **Ruschia elineata**
M. elishae = **Conophytum bilobum** ssp. **bilobum**
M. elizae = **Juttadinteria deserticola**
M. elongatum ≡ **Conicosia elongata**; = **Aptenia haeckeliana**
M. elongatum var. *fusiforme* = **Conicosia elongata**
M. elongatum var. *grandiflorum* ≡ **Conicosia elongata**
M. elongatum var. *minus* = **Conicosia elongata**
M. emarcidum ≡ **Sceletium emarcidum**
M. emarginatoides ≡ **Lampranthus emarginatoides**
M. emarginatum ≡ **Lampranthus emarginatus**
M. emarginatum var. *puniceum* = **Lampranthus emarginatus**
M. englerianum ≡ **Malephora engleriana**
M. englishiae ≡ **Prenia englishiae**
M. enorme ≡ **Antimima tuberculosa**
M. equilaterum ≡ **Carpobrotus aequilaterus**
M. erectum ≡ **Ruschia erecta**
M. erigeriflorum ≡ **Drosanthemum erigeriflorum**
M. erminium ≡ **Stomatium erminium**
M. ernestii ≡ **Lampranthus ernestii**
M. evexum = **Lithops pseudotruncatella** ssp. **pseudotruncatella**
M. evolutum ≡ **Antimima evoluta**
M. exacutum ≡ **Ruschia acuminata**
M. excedens ≡ **Oscularia excedens**
M. exiguum ≡ **Cheiridopsis namaquensis**
M. exile ≡ **Ruschia gracilis**
M. expansum ≡ **Sceletium expansum**; = **Skiatophytum tripolium**
M. expersum ≡ **Drosanthemum expersum**
M. explanatum ≡ **Lampranthus explanatus**
M. exspersum ≡ **Drosanthemum exspersum**

M. falcatum ≡ **Lampranthus falcatus**; = **Erepsia lacera**
M. falcatum var. *galpinii* = **Lampranthus falcatus**
M. falciforme ≡ **Oscularia falciformis**
M. familiare = **Conophytum truncatum** ssp. **truncatum**
M. fastigiatum ≡ **M. fastigiatum**
= **Phyllobolus splendens** ssp. **splendens**
M. fastigiatum var. *pallens*
= **Prenia pallens** ssp. **pallens**
M. fastigiatum var. *reflexum*
= **Phyllobolus splendens** ssp. **splendens**
M. felinum ≡ **Faucaria felina**
M. fergusoniae ≡ **Lampranthus fergusoniae**;
≡ **Antimima fergusoniae**
M. ferrugineum = **Lithops lesliei** ssp. **lesliei**
M. festivum ≡ **Ruschia festiva**
M. fibuliforme ≡ **Conophytum fibuliforme**
M. ficiforme ≡ **Conophytum ficiforme**
M. filamentosum = **Erepsia forficata**
M. filamentosum var. *anceps* = **Erepsia forficata**
M. filicaule ≡ **Lampranthus filicaulis**
M. filiforme ≡ **Galenia filiformis**
M. filipetalum ≡ **Ruschia filipetala**
M. fimbriatum
= **Conophytum truncatum** ssp. **truncatum**
M. firmum ≡ **Ruschia brevipes**
M. fissoides ≡ **Antegibbaeum fissoides**
M. fissum ≡ **Argyroderma fissum**; = **Gibbaeum heathii**
M. flaccidum ≡ **Saphesia flaccida**;
≡ **Lampranthus marcidulus**
M. flanaganii = **Delosperma tradescantioides**
M. flavens ≡ **Rhinephyllum luteum**
M. flavo-croceum ≡ **Malephora flavo-crocea**
M. flavum ≡ **Drosanthemum flavum**
M. flexifolium = **Lampranthus flexifolius**
M. flexile ≡ **Lampranthus flexilis**
M. flexuosum = **Phyllobolus splendens** ssp. **splendens**
M. florentulum (N.E.Br.) hort. Kew ex Hill 1931: 127 nomen nudum
M. floribundum ≡ **Drosanthemum floribundum**
M. floriferum = **Eberlanzia schneideriana**
M. flos-solis = **Dorotheanthus bellidiformis**
M. foetens = **Dracophilus dealbatus**
M. foliosum ≡ **Ruschia foliosa**
M. forficatum ≡ **Erepsia forficata**
M. formosulum ≡ **Antimima peersii**
M. formosum ≡ **Lampranthus formosus**
M. fourcadei ≡ **Drosanthemum fourcadei**;
≡ **Ruschia fourcadei**
M. fragrans = **Glottiphyllum depressum**
M. framesii ≡ **Ruschia framesii**;
≡ **Lampranthus framesii**
M. francisci ≡ **Lithops francisci**
M. fraternum ≡ **Conophytum fraternum**
M. fredericii ≡ **Ruschia fredericii**
M. friedrichiae ≡ **Conophytum friedrichiae**
M. frithii ≡ **Peersia frithii**
M. frutescens ≡ **Stoeberia frutescens**
M. fulviceps ≡ **Lithops fulviceps**
M. fulvum = **Aridaria noctiflora** ssp. **straminea**
M. furvum ≡ **Lampranthus furvus**
M. fusiforme = **Conicosia elongata**
M. galpiniae ≡ **Lampranthus galpiniae**
M. geminatum ≡ **Braunsia geminata**;
≡ **Ruschia geminiflora**
M. geminiflorum ≡ **Ruschia geminiflora**
M. geminum ≡ **Cerochlamys gemina**
M. geniculiflorum ≡ **Aptenia geniculiflora**
M. gessertianum ≡ **Psilocaulon gessertianum**
M. gibbosum ≡ **Gibbaeum gibbosum**
M. giffenii ≡ **Drosanthemum giffenii**
M. giffenii var. *intertextum* = **Drosanthemum giffenii**
M. gigas ≡ **Ruschianthemum gigas**
M. glabrum ≡ **Hymenogyne glabra**
M. gladiatum = **Erepsia lacera**
M. glareosum = **Psilocaulon salicornioides**
M. glaucescens ≡ **Carpobrotus glaucescens**
M. glaucinum = **Erepsia forficata**
M. glaucoides ≡ **Lampranthus glaucoides**
M. glaucum ≡ **Lampranthus glaucus**
M. glaucum var. *tortuosum* = **Lampranthus glaucus**
M. glebula = **Conophytum minimum**
M. globosum ≡ **Conophytum globosum**;
≡ **Lampranthus globosus**
M. glomeratum ≡ **Lampranthus glomeratus**
M. godmaniae ≡ **Lampranthus godmaniae**
M. godmaniae var. *grandiflorum*
= **Lampranthus godmaniae**
M. goodiae ≡ **Ruschia goodiae**
M. gracile ≡ **Erepsia gracilis**
M. gracilipes ≡ **Lampranthus gracilipes**
M. gracilistylum
≡ **Conophytum bilobum** ssp. **gracilistylum**
M. gramineum = **Dorotheanthus apetalus**
M. grandiflorum ≡ **Glottiphyllum grandiflorum**
M. graniforme ≡ **Rhinephyllum graniforme**
M. graniticum ≡ **Antimima granitica**
M. granulatum ≡ **Hereroa granulata**
M. granulicaule ≡ **Psilocaulon granulicaule**
M. granulicaule var. *purpurascens*
= **Psilocaulon articulatum**
M. gratum = **Conophytum jucundum** ssp. **jucundum**
M. griquense ≡ **Ruschia griquensis**
M. griseum ≡ **Ruschia grisea**
M. grossum ≡ **Phyllobolus grossus**

M. guianense Klotzsch in Schomburgk 1848: 300 T?

Creeping, rooting, side branches very short; **L** opposite, terete, oblong, acute; **Fl K** lobes unequal, petals narrowly oblong, obtuse, deep purple, stamens enclosed, stigmas 5, spreading; **Fr** inferior, turbinate, 5 locules, 10-radiate; **Ecol** on sandy, disturbed open places; **Distr** Pirara, Guiana.

Note: Berger (1908: 293) and Jacobsen (e.g. 1981: 616) call this species a doubtful one, probably considering that many species named by Klotzsch in Schomburgk lack descriptions and are hence nomina nuda. In contrast to these, **M. guianense** is validly published and therefore needs attention.

It is obvious from the description above, which is mainly a translation of the Latin description given by Klotzsch, that this species is not a member of the genus **Mesembryanthemum** in the present circumscription. Into which genus the species ought to be transferred remains completely uncertain, however, because since its discovery no material seems to have been collected again, and no type material has been found until now. It is a challenge that after 150 years from its first report the species should be found again and identified.

M. guthriae ≡ **Oscularia guthriae**
M. gymnocladum = **Brownanthus arenosus**
M. haeckelianum ≡ **Aptenia haeckeliana**
M. halenbergense ≡ **Conophytum halenbergense**
M. hamatile ≡ **Antimima hamatilis**
M. hamatum ≡ **Ruschia hamata**
M. hantamense ≡ **Antimima hantamensis**
M. harazianum ≡ **Delosperma harazianum**
M. haworthii ≡ **Lampranthus haworthii**
M. heathii ≡ **Gibbaeum heathii**
M. hebes ≡ **Ruschia obtusa**
M. helianthoides ≡ **Apatesia helianthoides**
M. helianthoides var. *glabrum* = **Apatesia helianthoides**
M. helianthoides var. *glaucum* = **Apatesia helianthoides**
M. henricii ≡ **Lampranthus henricii**
M. herbeum ≡ **Delosperma herbeum**

M. hermanni Haw. 1795: 470 non Pax ex Engler 1909: 192, 193 **T?**

Annual; **Fl** blue.

Note: The name is based on the description of Hermann 1698: 169 cited as: "Ficoides Africana annua centauroides" by Haworth. The name appears to have been forgotten and cannot be assigned to any of the known taxa of **Aizoaceae**.

M. hermannii = **Drosanthemum luederitzii**
M. herrei ≡ **Antimima herrei**; ≡ **Ruschia sandbergensis**
M. hesperanthum ≡ **Hereroa hesperantha**
M. heteropetalum ≡ **Erepsia heteropetala**
M. heterophyllum = **Glottiphyllum difforme**
M. hexamerum ≡ **Ruschia hexamera**
M. hexaphyllum see **Aizoon sarmentosum**
M. hiemale ≡ **Lampranthus hiemalis**
M. hirsutum ≡ **Trichodiadema hirsutum**
M. hirtellum ≡ **Drosanthemum hirtellum**
M. hirtum ≡ **Delosperma hirtum**
M. hispidum ≡ **Drosanthemum hispidum**
M. hispidum var. β ≡ **Drosanthemum floribundum**
M. hispidum var. γ ≡ **Drosanthemum floribundum**
M. hispidum var. *pallidum* ≡ **Drosanthemum floribundum**
M. hispidum var. *platypetalum* = **Drosanthemum hispidum**
M. hispifolium ≡ **Drosanthemum hispifolium**
M. hoerleinianum ≡ **Lampranthus hoerleinianus**
M. hollandii ≡ **Lampranthus hollandii**
M. hookeri ≡ **Lithops hookeri**
M. horizontale = **Aridaria noctiflora** ssp. **defoliata**
M. hortenseae ≡ **Muiria hortenseae**
M. hospitale = **Ruschia spinosa**
M. hugo-schlechteri ≡ **Titanopsis hugo-schlechteri**
M. humifusum = **Dorotheanthus apetalus**

M. humile Haw. 1795: 471 non L.Bolus NM 1929: 73 nomen illeg. **Lectotype**, here designated: Petiver 1709: t. 88 f 8.

Probably annual, root a little thickened; **L** 2–3 per node, acuminate to subulate; **Fl** golden; **Fr** trigonous, each margin ending in a sharp point.

Note: Haworth based his description on Petiver's account and drawing which shows a plant most similar to **M. petiveri**. Both may belong to the family Asteraceae.

M. humile ≡ **Lampranthus perreptans**
M. hurlingii ≡ **Lampranthus hurlingii**
M. hutchinsonii = **Amphibolia laevis**

M. × hybridum Haw. 1812: 218 **T?**

L of the first year with some teeth, the ones formed later without teeth, whiter than those of *M. caninum* but less white than those of *M. albidum*.

Note: Haworth reports the plant to be a hybrid from *M. caninum* and *M. albidum*, come up from seed of *M. albidum* incidentally in the collection of Howard around 1808. The plant had not flowered until 1812, and it seems to have died without leaving offsprings behind.

M. ignavum Schwantes nomen nudum ≡ *Bergeranthus ignavus*
M. imbricans ≡ **Lampranthus imbricans**
M. imbricatum ≡ **Ruschia imbricata**
M. imbricatum var. *rubrum* = **Ruschia imbricata**
M. immelmaniae ≡ **Lampranthus immelmaniae**
M. impressum ≡ **Ruschia impressa**
M. inaequale ≡ **Lampranthus inaequalis**
M. inclaudens ≡ **Erepsia inclaudens**
M. inclusum ≡ **Octopoma inclusum**
M. incomptum ≡ **Delosperma incomptum**; ≡ **Delosperma invalidum**
M. incomptum var. *ecklonis* = **Delosperma invalidum**
M. inconspicuum ≡ **Lampranthus inconspicuus**

M. incurvatum ≡ **Ruschia incurvata**
M. incurvum ≡ **Lampranthus incurvus**
M. incurvum var. *multiradiatum* ≡ **Lampranthus multiradiatus**
M. incurvum var. *roseum* ≡ **Lampranthus roseus**
M. indecorum ≡ **Ruschia indecora**
M. induratum ≡ **Ruschia indurata**
M. inexpectatum ≡ **Dinteranthus inexpectatus**
M. inflexum = **Lampranthus glomeratus**
M. insidens ≡ **Ruschia insidens**
M. insigne ≡ **Erepsia insignis**
M. insititium = **Malephora purpureo-crocea**
M. insolitum = **Juttadinteria deserticola**
M. inspersum = **Cheiridopsis rostrata**
M. integrum = **Smicrostigma viride**
M. intermedium ≡ **Drosanthemum intermedium**
M. intervallare ≡ **Mossia intervallaris**
M. intonsum ≡ **Trichodiadema intonsum**
M. intonsum var. *album* = **Trichodiadema orientale**
M. intonsum var. *rubicundum* ≡ **Trichodiadema intonsum**
M. intricatum ≡ **Ruschia intricata**
M. introrsum ≡ **Trichodiadema introrsum**
M. intrusum ≡ **Ruschia intrusa**
M. invalidum ≡ **Delosperma invalidum**
M. iteratum ≡ **Ruschia britteniae**
M. ivori ≡ **Antimima ivori**
M. johannis-winkleri = **Conophytum pageae**
M. jucundum ≡ **Conophytum jucundum**
M. jugiferum = **Conophytum ficiforme**
M. julii ≡ **Lithops julii**
M. junceum ≡ **Psilocaulon junceum**
M. junceum var. *pauciflorum* = **Psilocaulon junceum**
M. juritzii = **Carpobrotus dimidiatus**
M. juttae ≡ **Synaptophyllum juttae**
M. karasbergense = **Hereroa hesperantha**
M. karasmontanum ≡ **Lithops karasmontana** ssp. **karasmontana**
M. karrooicum ≡ **Ruschia karrooica**
M. klaverense ≡ **Antimima klaverensis**
M. klinghardtianum ≡ **Delosperma klinghardtianum**
M. knysnanum ≡ **Ruschia knysnana**
M. kolbei ≡ **Jacobsenia kolbei**
M. kovisimontanum = **Juttadinteria simpsonii**
M. kuntzei ≡ **Brownanthus kuntzei**
M. labyrintheum = **Conophytum minimum**
M. lacerum ≡ **Erepsia lacera**
M. lacunatum = **Eberlanzia schneideriana**
M. laetum ≡ **Lampranthus laetus**
M. laeve ≡ **Malephora thunbergii**; ≡ **Amphibolia laevis**
M. laevigatum = **Carpobrotus acinaciformis**
M. lanceum ≡ **Prenia pallens** ssp. **lancea**
M. lapidiforme ≡ **Didymaotus lapidiformis**
M. latens ≡ **Ruschia inconspicua**
M. lateriflorum = **Drosanthemum erigeriflorum**
M. latum = **Glottiphyllum longum**; ≡ **Lampranthus plautus**
M. latum var. *breve* = **Glottiphyllum longum**

M. laurifolium Haw. 1795: 470 **Lectotype**, here designated: Petiver 1709: t. 88 f 4.

Probably perennial with a thick root; L greyish-blue, laurel-shaped; Fl blue, the centre "full of threads tip'd with purple".

Note: The bluish-green flat leaves and the sky-blue flowers suggest that the plant does not belong to the **Aizoaceae**. It seems that the name has never been used or considered.

M. lavisii ≡ **Lampranthus lavisii**
M. lawsonii ≡ **Antimima lawsonii**
M. laxipetalum ≡ **Ruschia laxipetala**
M. laxifolium ≡ **Lampranthus laxifolius**
M. laxum ≡ **Lampranthus sublaxus**; ≡ **Ruschia laxa**
M. lectum = **Cheiridopsis robusta**
M. lehmannii ≡ **Corpuscularia lehmannii**
M. leightoniae ≡ **Lampranthus leightoniae**
M. leipoldtii ≡ **Conophytum minusculum** ssp. **leipoldtii**; ≡ **Lampranthus leipoldtii**
M. lene ≡ **Ruschia mollis**
M. lepidum = **Lampranthus productus**
M. leptaleon ≡ **Lampranthus leptaleon**
M. leptarthron ≡ **Psilocaulon leptarthron**
M. leptosepalum ≡ **Lampranthus leptosepalus**
M. lericheanum = **Lithops karasmontana** ssp. **karasmontana**
M. lerouxiae ≡ **Ruschia lerouxiae**
M. lesliei ≡ **Lithops lesliei** ssp. **lesliei**
M. lesliei minus ≡ **Lithops lesliei** ssp. **lesliei**
M. leucantherum ≡ **Antimima leucanthera**
M. leviculum = **Conophytum minimum**
M. levynsiae = **Antimima pumila**
M. lewisiae ≡ **Lampranthus lewisiae**
M. liberale ≡ **Lampranthus liberalis**
M. limbatum ≡ **Antimima limbata**
M. limpidum = **Dorotheanthus bellidiformis**
M. lindequistii = **Aridaria noctiflora** ssp. **noctiflora**
M. lineare = **Dorotheanthus clavatus**
M. lineolatum ≡ **Ruschia lineolata**
M. linguaeforme = **Glottiphyllum longum**
M. linguaeforme var. *assurgens* = **Glottiphyllum longum**
M. linguaeforme var. *prostratum* = **Glottiphyllum longum**
M. linguaeforme var. *rufescens* = **Glottiphyllum depressum**
M. linguaeforme var. *subcruciatum* = **Glottiphyllum cruciatum**
M. linguiforme ≡ **Glottiphyllum linguiforme**; = **Glottiphyllum longum**

M. linguiforme var. *adscendens* = **Glottiphyllum longum**
M. linguiforme var. *angustum* = **Glottiphyllum depressum**
M. linguiforme var. *angustum* subvar. *heterophyllum* = **Glottiphyllum depressum**
M. linguiforme var. *cultratum* = **Glottiphyllum longum**
M. linguiforme var. *depressum* = **Glottiphyllum longum**
M. linguiforme var. *fragrans* = **Glottiphyllum depressum**
M. linguiforme var. *grandiflorum* ≡ **Glottiphyllum grandiflorum**
M. linguiforme var. *longum* ≡ **Glottiphyllum longum**
M. linguiforme var. *obliquum* = **Glottiphyllum longum**
M. linguiforme var. *pustulatum* = **Glottiphyllum longum**
M. linguiforme var. *scalpratum* = **Glottiphyllum linguiforme**
M. linguiforme var. *uncatum* = **Glottiphyllum longum**
M. lique ≡ **Drosanthemum lique**
M. litorale ≡ **Delosperma litorale**
M. locale ≡ **Lithops localis**
M. longifolium ≡ **Astridia longifolia**
M. longipes ≡ **Antimima longipes**
M. longisepalum ≡ **Hammeria meleagris**
M. longispinulum = **Phyllobolus grossus**
M. longistamineum ≡ **Lampranthus longistamineus**
M. longistylum ≡ **Lampranthus altistylus**
M. longistylum var. *purpurascens* = **Prenia pallens** ssp. **pallens**
M. longum ≡ **Glottiphyllum longum**
M. longum var. *angustius* = **Glottiphyllum depressum**
M. longum var. *attollens* = **Glottiphyllum longum**
M. longum var. *declive* = **Glottiphyllum depressum**
M. longum var. *flaccidum* = **Glottiphyllum depressum**; = **Glottiphyllum longum**
M. longum var. *purpurascens* = **Glottiphyllum depressum**
M. longum var. *uncatum* = **Glottiphyllum longum**
M. loratum = **Prenia pallens** ssp. **pallens**
M. loreum ≡ **Cephalophyllum loreum**
M. lucidum = **Glottiphyllum longum**
M. luederitzii ≡ **Drosanthemum luederitzii**
M. lunatum ≡ **Oscularia lunata**
M. lunulatum ≡ **Lampranthus lunulatus**
M. lupinum = **Faucaria felina**
M. luteolum ≡ **Malephora luteola**
M. luteoviride ≡ **Gibbaeum luteoviride**
M. luteum ≡ **Malephora lutea**
M. luxurians = **Jordaaniella cuprea**
M. macellum ≡ **Delosperma macellum**
M. macowanii ≡ **Ruschia macowanii**
M. macradenium ≡ **Peersia macradenia**
M. macrocalyx = **Erepsia dunensis**
M. macrocarpum ≡ **Lampranthus macrocarpus**
M. macrorhizum ≡ **Delosperma napiforme**; = **Mestoklema tuberosum**
M. macrosepalum ≡ **Lampranthus macrosepalus**
M. maculatum ≡ **Drosanthemum maculatum**
M. magnificum ≡ **Lampranthus magnificus**
M. magnipunctatum = **Pleiospilos compactus** ssp. **canus**
M. mahonii ≡ **Delosperma mahonii**
M. maleolens ≡ **Antimima maleolens**
M. malleoforme = **Conophytum truncatum** ssp. **truncatum**
M. mallesoniae = **Esterhuysenia drepanophylla**
M. margaretae ≡ **Lapidaria margaretae**
M. marginatum = **Lithops optica**
M. mariae ≡ **Lampranthus mariae**
M. marianae ≡ **Ruschia marianae**
M. marinum = **Psilocaulon dinteri**
M. maritimum = **Jordaaniella dubia**; = **Oscularia falciformis**
M. marlothii ≡ **Brownanthus marlothii**
M. marmoratum ≡ **Lithops marmorata**
M. martleyi ≡ **Lampranthus martleyi**
M. mathewsii = **Braunsia geminata**
M. matutinum ≡ **Lampranthus matutinus**
M. maximiliani ≡ **Braunsia maximiliani**
M. maximum ≡ **Ruschia maxima**
M. maxwellii ≡ **Antimima maxwellii**
M. medium = **Glottiphyllum longum**
M. megarhizum = **Mestoklema tuberosum**
M. melanospermum = **Aptenia geniculiflora**; = **Phyllobolus melanospermus**
M. meleagre ≡ **Hammeria meleagris**
M. mellei ≡ **Carpobrotus mellei**
M. mentiens = **Psilocaulon coriarium**
M. meyeri = **Antimima papillata**
M. micans ≡ **Drosanthemum micans**; ≡ **Drosanthemum exspersum**
M. micranthon = **Psilocaulon parviflorum**
M. micropetalum = **Stoeberia beetzii**
M. microphyllum ≡ **Antimima microphylla**
M. microspermum ≡ **Dinteranthus microspermus**
M. microstigmum ≡ **Lampranthus microstigma**
M. middlemostii ≡ **Lampranthus middlemostii**
M. milleri = **Erepsia lacera**
M. minimum ≡ **Conophytum minimum**
M. minusculum ≡ **Conophytum minusculum** ssp. **minusculum**; = **Conophytum truncatum** ssp. **truncatum**
M. minutiflorum = **Conophytum pageae**
M. minutum ≡ **Conophytum minutum**
M. mirabile ≡ **Trichodiadema mirabile**

M. miserum ≡ **Ruschia misera**
M. mitratum ≡ **Mitrophyllum mitratum**
M. modestum ≡ **Psammophora modesta**
M. molle ≡ **Malephora mollis**; ≡ **Ruschia mollis**
M. moniliforme ≡ **Monilaria moniliformis**
M. montaguense ≡ **Lampranthus montaguensis**
M. montanum = **Erepsia heteropetala**
M. monticola ≡ **Lampranthus monticola**; ≡ **Ottosonderia monticola**
M. montis-draconis = **Dracophilus dealbatus**
M. montis-moltkei ≡ **Ebracteola montis-moltkei**

M. muscosum Haw. 1795: 470 **T**?

Annual, moss-like, tiny.

Note: Hermann (1698: 170) describes this species as "Ficoides annua minima muscosa". The name has not been taken up again, and the identity of the plant remains obscure.

M. mucronatum ≡ **Antimima mucronata**
M. mucroniferum = **Ruschia spinosa**
M. mucronulatum = **Psilocaulon articulatum**
M. muelleri ≡ **Ruschia muelleri**
M. muirianum ≡ **Ruschia muiriana**
M. muirii ≡ **Carpobrotus muirii**
M. multiceps ≡ **Bergeranthus multiceps**
M. multiflorum ≡ **Ruschia multiflora**
M. multiflorum var. *rubrum* = **Ruschia imbricata**
M. multipunctatum = **Cheiridopsis rostrata**
M. multiradiatum ≡ **Lampranthus multiradiatus**
M. multiseriatum ≡ **Lampranthus multiseriatus**
M. munitum = **Ruschia intricata**
M. muricatum ≡ **Oscularia deltoides**
M. murinum ≡ **Stomatium murinum**
M. musculinum ≡ **Chasmatophyllum musculinum**
M. mustelinum ≡ **Stomatium mustelinum**
M. mutabile = **Erepsia forficata**
M. mutans ≡ **Lampranthus mutans**
M. muticum ≡ **Antimima mutica**
M. nakurense ≡ **Delosperma nakurense**
M. namaquanum = **Amphibolia rupis-arcuatae**
M. namaquense ≡ **Cheiridopsis namaquensis**
M. namibense ≡ **Brownanthus namibensis**
M. nanum = **Conophytum jucundum** ssp. **jucundum**; ≡ **Oophytum nanum**
M. napiforme ≡ **Delosperma napiforme**
M. nardouwense ≡ **Lampranthus nardouwensis**
M. nationae = **Khadia acutipetala**
M. necopinum = **Argyroderma × octophyllum**
M. neethlingiae ≡ **Delosperma neethlingiae**
M. neilii ≡ **Glottiphyllum neilii**
M. nevillei = **Conophytum obcordellum** ssp. **obcordellum**
M. nigrescens = **Carpobrotus aequilaterus**
M. nissenii ≡ **Psammophora nissenii**
M. nitidum ≡ **Phyllobolus nitidus**
M. niveum = **Ruschia rupicola**
M. nobile = **Pleiospilos compactus** ssp. **canus**
M. noctiflorum ≡ **Aridaria noctiflora**
M. noctiflorum var. *elatum* = **Aridaria noctiflora** ssp. **noctiflora**
M. noctiflorum var. *fulvum* = **Aridaria noctiflora** ssp. **straminea**
M. noctiflorum var. *phoeniceum* ≡ **Aridaria noctiflora** ssp. **noctiflora**
M. noctiflorum var. *stramineum* ≡ **Aridaria noctiflora** ssp. **straminea**
M. nodosum ≡ **Dicrocaulon nodosum**
M. nothum = **Phyllobolus splendens** ssp. **splendens**
M. nubigenum ≡ **Delosperma nubigenum**
M. nuciforme ≡ **Gibbaeum nuciforme**
M. nudicaule = **Erepsia aspera**
M. obconellum = **Conophytum obcordellum** ssp. **obcordellum**
M. obconicum ≡ **Lampranthus obconicus**
M. obcordellum ≡ **Conophytum obcordellum** ssp. **obcordellum**
M. obliquum ≡ **Drosanthemum lique**; = **Glottiphyllum longum**
M. obmetale = **Conophytum minimum**
M. obsubulatum ≡ **Cylindrophyllum obsubulatum**
M. obtusum ≡ **Antegibbaeum fissoides**
M. ochraceum ≡ **Malephora ochracea**
M. octojuge ≡ **Octopoma octojuge**
M. octonarium ≡ **Enarganthe octonaria**
M. octophyllum ≡ **Argyroderma × octophyllum**
M. oculatum ≡ **Phyllobolus oculatus**
M. odontocalyx ≡ **Ruschia odontocalyx**
M. odoratum = **Conophytum ficiforme**; ≡ **Hereroa odorata**
M. oehleri ≡ **Delosperma oehleri**
M. oligandrum = **Dorotheanthus apetalus**
M. olivaceum = **Phyllobolus tenuiflorus**
M. opalinum = **Lithops karasmontana** ssp. **karasmontana**
M. optatum = **Pleiospilos compactus** ssp. **compactus**
M. opticum ≡ **Lithops optica**
M. opticum var. *rubrum* = **Lithops optica**

M. opulentum L.Bolus ABH 1916: 28 **T** Smalbegu 4574/14 (BOL).

Shrubs with decumbent branches, the flowering branches erect, **I** 8–10 mm *l*; **L** conspicuously punctate from rough dots, 5–6 mm *l*, 2 mm *b*; **Fl** in rich **Inf**, ∅ 10 mm, petals pinkish-purple, inconspicuously veined, filamentous staminodes and stamens in a cone, filaments purple, anthers white, nectary a ring; **Fr** 5 locules; **Ecol** unknown; **Distr** Riversdale, WC, S Africa.

Note: Albeit the species has been known for many years, its placement is still uncertain. Jacobsen (1981: 620) suggested that it may be a synonym of **Ruschia virgata**, but gave no arguments to support this view.

M. ornatulum (N.E.Br.) hort. Kew ex Hill 1931: 124 nomen nudum
M. orpenii ≡ **Prepodesma orpenii**
M. ortendahlii = **Lithops fulviceps**
M. otzenianum ≡ **Lampranthus otzenianus**
M. oviforme ≡ **Oophytum oviforme**
M. oxysepalum ≡ **Erepsia oxysepala**
M. paardebergense ≡ **Oscularia paardebergensis**
M. pachycarpum = **Erepsia bracteata**
M. pachyphyllum ≡ **Cerochlamys pachyphylla**
M. pachypodium ≡ **Gibbaeum pachypodium**
M. pageae ≡ **Conophytum pageae**
M. pageanum ≡ **Delosperma pageanum**; ≡ **Oscularia ornata**
M. pakhuisense ≡ **Lampranthus pakhuisensis**
M. pallens ≡ **Prenia pallens**
M. pallescens = **Prenia pallens** ssp. **pallens**
M. pallidum = **Conophytum ficiforme**; ≡ **Lampranthus dilutus**
M. palustre ≡ **Lampranthus palustris**
M. pansifolium = **Amphibolia laevis**
M. papillatum ≡ **Antimima papillata**
M. papulosum ≡ **Cleretum papulosum**
M. paripetalum ≡ **Ruschia paripetala**
M. parvibracteatum ≡ **Eberlanzia parvibracteata**
M. parviflorum ≡ **Psilocaulon parviflorum**; ≡ **Ruschia parviflora**
M. parvifolium ≡ **Drosanthemum parvifolium**
M. parvipetalum = **Conophytum obcordellum** ssp. **obcordellum**
M. parvulum ≡ **Cephalophyllum parvulum**
M. parvum ≡ **Antimima pusilla**
M. patersoniae ≡ **Delosperma patersoniae**
M. patulum ≡ **Erepsia patula**
M. pauciflorum ≡ **Lampranthus pauciflorus**
M. paucifolium ≡ **Lampranthus paucifolius**
M. pauper ≡ **Drosanthemum pauper**
M. pauxillum = **Conophytum minimum**
M. paxianum = **Drosanthemum luederitzii**
M. paxii = **Drosanthemum luederitzii**
M. peacockiae ≡ **Lampranthus peacockiae**
M. pearsonii ≡ **Argyroderma pearsonii**
M. peersii ≡ **Deilanthe peersii**; ≡ **Lampranthus peersii**
M. pellanum ≡ **Ruschia muricata**
M. pellucidum ≡ **Conophytum pellucidum**
M. pentagonum ≡ **Erepsia pentagona**
M. perfoliatum ≡ **Ruschia perfoliata**
M. perfoliatum var. *integrifolium* = **Polymita albiflora**
M. perforatum ≡ **Antimima perforata**
M. perpusillum = **Conophytum minimum**
M. persistens ≡ **Lampranthus persistens**
M. perspicuum ≡ **Lampranthus schlechteri**
M. perviride ≡ **Gibbaeum perviride**
M. perviride var. β ≡ **Gibbaeum luteoviride**

M. petiveri Haw. 1795: 471 **Lectotype,** here designated: Petiver 1709: t. 88 f 9.

Probably annual; L 2–3 per node, acuminate to subulate; Fl purple; Fr triangular with 3 sharp points.

Note: The plant pictured is most similar to the one named **M. humile**, both are perhaps members of the family Asteraceae.

M. pfeilii = **Prenia tetragona**
M. phillipsii ≡ **Braunsia maximiliani**
M. phylicoides ≡ **Ruschia phylicoides**
M. pictum = **Conophytum minumum**
M. pillansii ≡ **Erepsia pillansii**
M. pilosulum ≡ **Gibbaeum pilosulum**
M. pilosum = **Carpanthea pomeridiana**
M. piluliforme ≡ **Conophytum piluliforme** ssp. **piluliforme**
M. pinnatifidum ≡ **Aethephyllum pinnatifidum**
M. pinnatum = **Aethephyllum pinnatifidum**
M. piquetbergense ≡ **Oscularia piquetbergensis**
M. piscodorum ≡ **Antimima piscodora**
M. pisiforme ≡ **Monilaria pisiformis**
M. pittenii ≡ **Lampranthus vanputtenii**
M. placitum = **Conophytum ficiforme**
M. planum ≡ **Hallianthus planus**
M. platycalyx = **Cephalophyllum curtophyllum**
M. plenifolium = **Phyllobolus splendens** ssp. **splendens**
M. plenum ≡ **Lampranthus plenus**
M. pocockiae ≡ **Lampranthus pocockiae**
M. pole-evansii ≡ **Dinteranthus pole-evansii**
M. politum ≡ **Erepsia polita**
M. polyanthon ≡ **Lampranthus polyanthon**
M. polypetalum ≡ **Erepsia polypetala**
M. polyphyllum = **Lampranthus emarginatus**
M. pomeridianum ≡ **Carpanthea pomeridiana**
M. pomeridianum var. *andrewsii* = **Carpanthea pomeridiana**
M. pomeridianum var. *curtisii* = **Carpanthea pomeridiana**
M. pomeridianum var. *glabrum* = **Carpanthea pomeridiana**
M. pomonae ≡ **Namibia pomonae**
M. ponderosum = **Namibia cinerea**
M. pottsii ≡ **Delosperma pottsii**
M. praecipitatum ≡ **Lampranthus praecipitatus**
M. praecox ≡ **Lampranthus maturus**; ≡ **Sarcozona praecox**
M. praecultum ≡ **Drosanthemum praecultum**
M. praepingue = **Glottiphyllum cruciatum**; = **Glottiphyllum difforme**
M. pressum = **Cheiridopsis robusta**
M. primivernum ≡ **Oscularia primiverna**
M. primulinum = **Cephalophyllum loreum**
M. prismaticum ≡ **Tanquana prismatica**
M. procumbens = **Jordaaniella dubia**
M. productum ≡ **Lampranthus productus**

M. productum var. *lepidum* = **Lampranthus productus**
M. productum var. *purpureum*
= **Lampranthus productus**
M. prominulum ≡ **Lampranthus prominulus**
M. promontorii ≡ **Lampranthus promontorii**
M. pronum ≡ **Ruschia radicans**
M. propinquum ≡ **Antimima propinqua**
M. prostratum ≡ **Antimima prostrata**
M. proximum = **Mitrophyllum dissitum**;
≡ **Mitrophyllum grande**
M. pruinosum = **Delosperma echinatum**
M. psammophilum = **Ruschia ruschiana**
M. pseudausanum = **Prenia tetragona**
M. pseudorupicola = **Ruschia barnardii**
M. pseudotruncatellum ≡ **Lithops pseudotruncatella** ssp. **pseudotruncatella**
M. puberulum = **Psilocaulon articulatum**
M. pubescens ≡ **Gibbaeum pubescens**
M. pugioniforme ≡ **Conicosia pugioniformis**
M. pugioniforme var. *bienne*
= **Conicosia pugioniformis** ssp. **pugioniformis**
M. pugioniforme var. *carneum*
= **Conicosia pugioniformis** ssp. **pugioniformis**
M. pugioniforme var. *purpureum*
= **Conicosia pugioniformis** ssp. **pugioniformis**
M. pulchellum ≡ **Ruschia pulchella**
M. pulchrum ≡ **Frithia pulchra**
M. pulverulentum ≡ **Drosanthemum pulverulentum**

M. pulvinatum O.Kuntze 1898: 109 T?

Plants with creeping branches forming cushions or mats, side branches 1–3 cm *h*; L thickish-triquetrous, apically obtuse or with three teeth, densely papillate, c. 10 mm *l*, 2 mm *b*; **Fl** solitary, pedicels short, no bracteoles present, **K** 5, petals dirty red, probably purple?; **Fr** 5 locules; **Ecol** unknown; **Distr** Beaufort West, WC, Burgersdorp, EC, S Africa.

Note: A type is not given, the identity can therefore not be determined at present. The combination of teeth and a papillate surface on the leaves appears unusual, as does the citation of Beaufort West and Burgersdorp for the localities, which lie several hundred miles apart. It cannot be excluded that mixed material was used to describe this species.

M. pumilum ≡ **Antimima pumila**
M. punctatum = **Cephalophyllum subulatoides**
M. punctulatum ≡ **Ruschia punctulata**
M. pungens ≡ **Ruschia pungens**
M. puniceum = **Lampranthus emarginatus**
M. purpurascens = **Cheiridopsis rostrata**
M. purpureo-album
≡ **Cephalophyllum purpureo-album**
M. purpureo-croceum ≡ **Malephora purpureo-crocea**
M. purpureo-croceum var. β *flavo-croceum*
≡ **Malephora flavo-crocea**
M. purpureostylum ≡ **Cerochlamys purpureostyla**
M. purpusii = **Conophytum truncatum** ssp. **truncatum**
M. pusillum = **Conophytum minimum**
M. pustulatum = **Glottiphyllum longum**
M. pustulatum var. *lividum* = **Glottiphyllum longum**
M. putterillii ≡ **Ruschia putterillii**
M. puttkamerianum ≡ **Hereroa puttkameriana**
M. pygmaeum ≡ **Antimima pygmaea**
M. pyrodorum = **Stomatium mustelinum**
M. pyrop(a)eum = **Dorotheanthus clavatus**
M. quadrifidum = **Cheiridopsis rostrata**
M. quaesitum
≡ **Conophytum quaesitum** ssp. **quaesitum**
M. quarziticum ≡ **Antimima quarzitica**
M. rabiesbergense ≡ **Lampranthus rabiesbergensis**
M. racemosum = **Erepsia bracteata**
M. radiatum = **Erepsia bracteata**
M. ramosissimum ≡ **Drosanthemum ramosissimum**
M. ramulosum = **Cheiridopsis rostrata**
M. rangei = **Cephalophyllum ebracteatum**
M. rapaceum ≡ **Caulipsolon rapaceum**
M. recumbens = **Chasmatophyllum musculinum**
M. recurvum ≡ **Lampranthus recurvus**;
≡ **Ruschia recurva**; ≡ **Ruschia schollii**
M. reductum ≡ **Octopoma subglobosum**;
≡ **Ruschia nana**
M. reflexum = **Phyllobolus splendens** ssp. **splendens**
M. rehmannii ≡ **Aizoanthemum rehmannii**
M. rehneltianum ≡ **Hereroa rehneltiana**
M. relaxatum = **Prenia pallens** ssp. **pallens**
M. renniei = **Ebracteola montis-moltkei**
M. reptans ≡ **Lampranthus reptans**
M. resurgens ≡ **Phyllobolus resurgens**
M. retroflexum ≡ **Lampranthus emarginatoides**
M. retroversum ≡ **Diplosoma retroversum**
M. rheolens = **Dracophilus dealbatus**
M. rhomboideum ≡ **Rhombophyllum rhomboideum**
M. rhopalophyllum ≡ **Fenestraria rhopalophylla**
M. rigidicaule ≡ **Ruschia rigidicaulis**
M. rigidum ≡ **Ruschia rigida**
M. ringens ≡ **Carruanthus ringens**
M. ringens var. β *felinum* ≡ **Faucaria felina**
M. robustum ≡ **Cheiridopsis robusta**
M. rogersii ≡ **Delosperma rogersii**
M. roseatum = **Drosanthemum pulverulentum**
M. roseolum ≡ **Antimima roseola**
M. roseum ≡ **Lampranthus roseus**
M. roseum var. *confertum* ≡ **Lampranthus incurvus**
M. rossii ≡ **Carpobrotus rossii**
M. rostellum ≡ **Ruschia rostella**
M. rostratoides = **Cheiridopsis rostrata**
M. rostratum ≡ **Cheiridopsis rostrata**
M. rostratum var. *bibracteatum* = **Cheiridopsis rostrata**
M. rostratum var. *brevibracteatum*
= **Cheiridopsis rostrata**
M. rosulatum ≡ **Aloinopsis rosulata**

M. rubricaule ≡ **Ruschia rubricaulis**
M. rubrocinctum = **Carpobrotus acinaciformis**
M. rubrolineatum ≡ **Aloinopsis rubrolineata**
M. rubroluteum ≡ **Lampranthus rubroluteus**
M. rubrum ≡ **Astridia rubra**
M. rufescens = **Glottiphyllum depressum**
M. rugulosum ≡ **Eberlanzia schneideriana**
M. rupestre ≡ **Lampranthus rupestris**
M. rupicola ≡ **Ruschia rupicola**
M. rupis-arcuatae ≡ **Amphibolia rupis-arcuatae**
M. rurale ≡ **Ruschia ruralis**
M. ruschianum ≡ **Ruschia ruschiana**
M. ruschii ≡ **Lithops ruschiorum**
M. ruschiorum ≡ **Lithops ruschiorum**
M. rustii ≡ **Lampranthus rustii**
M. sabulicola ≡ **Ruschia sabulicola**
M. sabulosum ≡ **Apatesia sabulosa**;
≡ **Lampranthus schlechteri**
M. salicola ≡ **Lampranthus salicola**
M. salicornioides ≡ **Psilocaulon salicornioides**
M. salmii ≡ **Glottiphyllum salmii**
M. salmii var. *angustifolium* = **Glottiphyllum salmii**
M. salmii var. *decussatum* = **Glottiphyllum salmii**
M. salmii var. *semicruciatum* = **Glottiphyllum salmii**
M. salmoneum = **Phyllobolus canaliculatus**
M. salteri ≡ **Lampranthus salteri**
M. sarmentosum ≡ **Ruschia sarmentosa**
M. sarmentosum var. *rigidius* = **Ruschia sarmentosa**
M. saturatum ≡ **Lampranthus saturatus**
M. sauerae ≡ **Lampranthus sauerae**
M. saxetanum ≡ **Conophytum saxetanum**
M. saxicola ≡ **Antimima saxicola**
M. scabrum ≡ **Lampranthus scaber**
M. scalpratum = **Glottiphyllum linguiforme**
M. scapiger ≡ **Bergeranthus scapiger**
M. schenckii ≡ **Brownanthus ciliatus** ssp. **schenckii**
M. schimperi ≡ **Delosperma schimperi**
M. schiazianum = **Cheiridopsis robusta**
M. schlechteri ≡ **Antimima schlechteri**;
≡ **Lampranthus schlechteri**
M. schlichtianum = **Brownanthus arenosus**
M. schneiderianum ≡ **Eberlanzia schneideriana**
M. schoenlandianum
≡ **Drosanthemum schoenlandianum**
M. schollii ≡ **Ruschia schollii**
M. schollii var. *caledonicum* = **Ruschia schollii**
M. schuhmannianum = **Drosanthemum longipes**
M. schultzei ≡ **Leipoldtia schultzei**
M. schwantesii ≡ **Titanopsis schwantesii**
M. scintillans = **Phyllobolus oculatus**
M. scitulum = **Conophytum minimum**
M. scutatum ≡ **Monilaria scutata**
M. secundum = **Psilocaulon articulatum**
M. sedoides ≡ **Eberlanzia sedoides**
M. semicylindricum = **Glottiphyllum difforme**
M. semidentatum ≡ **Ruschia semidentata**
M. senarium ≡ **Ruschia senaria**
M. separatum ≡ **Zeuktophyllum calycinum**
M. serpens ≡ **Lampranthus serpens**
M. serratulum ≡ **Ruschia serrulata**
M. serratum ≡ **Circandra serrata**
M. serrulatum ≡ **Ruschia serrulata**
M. sessile ≡ **Ruschia sessilis**
M. sessiliflorum
= **Cleretum papulosum** ssp. **papulosum**
M. sessiliflorum var. *luteum*
= **Cleretum papulosum ssp. papulosum**
M. setosum ≡ **Delosperma echinatum**
M. setuliferum ≡ **Trichodiadema setuliferum**
M. sexpartitum = **Corpuscularia lehmannii**
M. shandii ≡ **Gibbaeum shandii**
M. sibbettii ≡ **Neohenricia sibbettii**
M. signatum = **Conophytum minimum**
M. simile ≡ **Psilocaulon junceum**
M. simile var. *namaquense* = **Psilocaulon junceum**
M. simpsonii ≡ **Juttadinteria simpsonii**
M. simulans ≡ **Pleiospilos simulans**
M. sinuosum ≡ **Phyllobolus sinuosus**
M. sinus-redfordiani = **Psilocaulon salicornioides**
M. sladenianum ≡ **Prenia sladeniana**
M. sobrinum ≡ **Antimima sobrina**
M. sociorum ≡ **Lampranthus sociorum**
M. socium = **Argyroderma fissum**
M. solidum ≡ **Antimima solida**
M. solutifolium = **Brownanthus marlothii**
M. sororium ≡ **Pleiospilos compactus** ssp. **sororius**
M. spathulatum ≡ **Aloinopsis spathulata**;
= **Dorotheanthus bellidiformis**;
= **Lampranthus otzenianus**
M. spathulifolium = **Dorotheanthus bellidiformis**
M. speciosum ≡ **Drosanthemum speciosum**
M. spectabile ≡ **Lampranthus spectabilis**
M. spinescens = **Ruschia divaricata**
M. spiniforme ≡ **Lampranthus spiniformis**
M. spinosum ≡ **Ruschia spinosa**
M. spinosum var. *cradochense* ≡ **Ruschia cradochensis**
M. spinosum var. *micranthum* ≡ **Ruschia spinosa**
M. spinuliferum ≡ **Phyllobolus spinuliferus**
M. spinuliforme ≡ **Phyllobolus spinuliferus**
M. splendens ≡ **Phyllobolus splendens**
M. spongiosum ≡ **Jordaaniella spongiosa**
M. staminodiosum ≡ **Lampranthus staminodiosus**;
≡ **Ruschia staminodiosa**
M. stanleyi ≡ **Chasmatophyllum stanleyi**
M. stayneri ≡ **Lampranthus stayneri**
M. steenbergense ≡ **Oscularia steenbergensis**
M. steingröveri = **Ruschia rupicola**
M. stellans = **Antimima hantamensis**
M. stellatum = **Erepsia gracilis**;
= **Trichodiadema barbatum**
M. stelligerum ≡ ; = **Trichodiadema barbatum**
M. stenopetalum ≡ **Lampranthus stenopetalus**

M. stenophyllum ≡ **Marlothistella stenophylla**
M. stenum ≡ **Lampranthus stenus**
M. stephanii ≡ **Lampranthus stephanii**
M. stipulaceum ≡ **Lampranthus disgregus;**
≡ **Lampranthus stipulaceus**
M. stokoei ≡ **Esterhuysenia stokoei**
M. stramineum ≡ **Aridaria noctiflora** ssp. **straminea**
M. striatum ≡ **Drosanthemum striatum**
M. striatum var. *hispifolium*
≡ **Drosanthemum hispifolium**
M. striatum var. *pallens* ≡ **Drosanthemum pallens**
M. striatum var. *pallidum* ≡ **Drosanthemum pallens**

M. strictum Haw. 1803: 82 T?

Shrubs to 90 cm *h*, branches many, erect, stiff, I greyish; L trigonous, spreading widely, little connate, c. 14 mm *l*, epidermis with numerous smooth, green dots; **Fl** 1–3, probably yellow; **Ecol** unknown; **Distr** uncertain, S Africa.

Note: Haworth (1803: 82) grouped this species under the heading "Tardiflora", those that flower rarely in cultivation in England. The species has never been transferred to any other genus and must be considered as most insufficiently known.

M. strubeniae ≡ **Ruschia strubeniae**
M. strumosum ≡ **Trichodiadema strumosum**
M. styliferum ≡ **Arenifera stylosa**
M. stylosum = **Conophytum bilobum** ssp. **bilobum**
M. suaveolens ≡ **Phyllobolus lignescens;**
≡ **Stomatium suaveolens**
M. suavissimum = **Juttadinteria ausensis;**
≡ **Lampranthus suavissimus**
M. suavissimum forma *ferum*
= **Lampranthus suavissimus**
M. suavissimum var. *oculatum*
= **Lampranthus suavissimus**
M. subaequale ≡ **Lampranthus subaequalis**
M. subalatum = **Carpobrotus acinaciformis**
M. subalbum ≡ **Argyroderma subalbum**
M. subcompressum
≡ **Drosanthemum subcompressum**
M. subglobosum ≡ **Drosanthemum capillare;**
≡ **Lampranthus subrotundus**
M. subincanum ≡ **Delosperma subincanum**
M. sublunatum ≡ **Ruschia stricta**
M. sublunulatum ≡ **Ruschia stricta**
M. subnodosum ≡ **Psilocaulon subnodosum**
M. subrisum = **Conophytum pageae**
M. subspinosum ≡ **Drosanthemum subspinosum**
M. subulatoides ≡ **Cephalophyllum subulatoides**
M. subulatum ≡ **Acrodon subulatus**
M. succulentum ≡ **Amphibolia succulenta**
M. succumbens ≡ **Schwantesia succumbens**
M. sulcatum
= **Phyllobolus splendens** ssp. **splendens**
M. summitatum
= **Lithops karasmontana** ssp. **karasmontana**
M. superans ≡ **Oscularia superans**
M. suppositum ≡ **Zeuktophyllum suppositum**
M. surrectum ≡ **Glottiphyllum surrectum**
M. sutherlandii ≡ **Delosperma sutherlandii**
M. swartbergense ≡ **Lampranthus swartbergensis**
M. taurinum = **Glottiphyllum depressum**
M. taylori ≡ **Corpuscularia taylori**
M. taylorianum
≡ **Conophytum taylorianum** ssp. **taylorianum**
M. tegens ≡ **Lampranthus tegens**
M. tennellum ≡ **Ruschia tenella**
M. tenue ≡ **Lampranthus parcus;**
= **Psilocaulon parviflorum**
M. tenuicaule = **Erepsia gracilis**
M. tenuiflorum ≡ **Phyllobolus tenuiflorus**
M. tenuifolium ≡ **Lampranthus tenuifolius**
M. teretifolium ≡ **Calamophyllum teretifolium**
M. teretiusculum ≡ **Calamophyllum teretiusculum**
M. ternifolium = **Erepsia gracilis**
M. testaceum ≡ **Delosperma testaceum**
M. testiculare ≡ **Argyroderma testiculare**
M. testiculare var. β
≡ **Argyroderma × octophyllum;**
= **Argyroderma pearsonii;**
= **Argyroderma testiculare**
M. testiculare var. γ = **Argyroderma pearsonii**
M. testiculare var. ϑ = **Argyroderma testiculare**
M. testiculatum = **Argyroderma testiculare**
M. tetragonum ≡ **Prenia tetragona**

M. tetrasepalum Regel 1873: 434 T?

Plants prostrate to ascending, suffruticose and much branched, I terete; L decussate, petiolate, flat, spatulately lanceolate and acute, bladder cells small; **Fl** K usually 4-, rarely 5-lobed, petals pink; **Ecol** unknown; **Distr** unknown.

Note: **M. tetrasepalum** was described from cultivation, its origin being uncertain. The description, however, indicates a similarity with **Aptenia cordifolia. M. tetrasepalum** is probably one of the many hybrid forms of the latter species.

M. thecatum = **Conophytum minutum**
M. thermarum ≡ **Oscularia thermarum**
M. thunbergii ≡ **Malephora thunbergii**
M. tigrinum ≡ **Faucaria tigrina**
M. tinctum = **Meyerophytum meyeri**
M. torquatum ≡ **Drosanthemum floribundum**
M. tortuosum ≡ **Sceletium tortuosum**
M. tradescantioides ≡ **Delosperma tradescantioides**
M. translucens = **Stoeberia beetzii**
M. transvaalense = **Nananthus vittatus**
M. trichosanthum = **Phyllobolus viridiflorus**
M. trichotomum ≡ **Phyllobolus trichotomus**

M. tricolor = **Dorotheanthus clavatus;**
= **Erepsia forficata**
M. tricolor var. *album* = **Dorotheanthus clavatus**
M. tricolor var. *pyropaeum* = **Dorotheanthus clavatus**
M. tricolor var. *roseum* = **Dorotheanthus clavatus**
M. tricolorum ≡ **Cephalophyllum tricolorum**
M. triflorum ≡ **Ruschia triflora**
M. tripolium ≡ **Skiatophytum tripolium**
M. triquetrum ≡ **Antimima triquetra**
M. triticiforme
≡ **Ruschia cradockensis** ssp. **triticiformis**
M. trothai = **Psilocaulon salicornioides**
M. truncatellum
= **Conophytum truncatum** ssp. **truncatum;**
≡ **Lithops pseudotruncatella** ssp. **pseudotruncatella**
M. truncatum
≡ **Conophytum truncatum** ssp. **truncatum**
M. tuberculatum = **Cheiridopsis rostrata**
M. tuberculosum = **Faucaria tuberculosa**
M. tuberosum ≡ **Mestoklema tuberosum**
M. tugwelliae ≡ **Bijlia tugwelliae**
M. tuitum = **Ruschia intricata**
M. tulbaghense ≡ **Lampranthus argillosus;**
≡ **Lampranthus tulbaghensis**
M. tumidulum ≡ **Ruschia tumidula**
M. turbinatum ≡ **Lampranthus turbinatus**
M. turbiniforme ≡ **Lithops turbiniformis;**
≡ **Lithops hookeri**
M. turrigerum ≡ **Conophytum turrigerum**
M. uitenhagense ≡ **Ruschia uitenhagensis**
M. umbellatum ≡ **Ruschia umbellata**
M. umbelliflorum
= **Phyllobolus splendens** ssp. **splendens**
M. uncatum = **Glottiphyllum longum**
M. uncinatum ≡ **Ruschia uncinata**
M. uncinellum ≡ **Ruschia uncinata**
M. uncum ≡ **Lampranthus uncus**
M. uncum var. *gydouwense* = **Lampranthus uncus**
M. unidens ≡ **Ruschia unidens**
M. uniflorum ≡ **Lampranthus uniflorus**
M. urbanianum ≡ **Erepsia urbaniana**
M. utile ≡ **Stoeberia utilis**
M. uviforme ≡ **Conophytum uviforme**
M. vaginatum ≡ **Ruschia vaginata**
M. vaginatum var. *acutangulum*
≡ **Ruschia acutangula**
M. vaginatum var. *curtum* ≡ **Ruschia curta**
M. valens ≡ **Ruschia glauca**
M. validum = **Cephalophyllum corniculatum;**
≡ **Ruschia vetovalida**
M. vallis-gratiae ≡ **Lampranthus vallis-gratiae**
M. vallis-mariae ≡ **Lithops vallis-mariae**
M. vallis-pacis = **Ebracteola montis-moltkei**
M. vanputtenii = **Lampranthus vanputtenii**
M. vanzijliae ≡ **Lampranthus vanzijliae**
M. variabile ≡ **Lampranthus variabilis**
M. varians ≡ **Sceletium varians**
M. variifolium ≡ **Ruschia diversifolia**
M. velutinum ≡ **Astridia velutina;**
≡ **Gibbaeum velutinum**
M. ventricosum ≡ **Antimima ventricosa**
M. verecundum ≡ **Lampranthus verecundus**
M. vermeuleniae = **Ebracteola wilmaniae**
M. vernae ≡ **Malephora engleriana**
M. vernale ≡ **Lampranthus vernalis**
M. vernicolor ≡ **Oscularia vernicolor**
M. verruculatum ≡ **Scopelogena verruculata**
M. verruculatum var. β ≡ **Malephora veruculoides**
M. verruculosum ≡ **Antimima verruculosa**
M. veruculoides ≡ **Malephora veruculoides**
M. veruculoides var. *minus* = **Malephora veruculoides**
M. versicolor = **Lampranthus emarginatus**
M. vescum = **Cheiridopsis namaquensis**
M. vespertinum ≡ **Bergeranthus vespertinus**
M. viatorum ≡ **Lampranthus viatorum**
M. victoris ≡ **Ruschia victoris**
M. villiersii ≡ **Lampranthus villiersii**

M. villosum L. 1762: 692 T?
I with hairs hardly spreading; L pubescent, in particular along the margins and there more densely towards the base, lineal, semi-terete, canaliculate to flat; **Ecol** unknown; **Distr** uncertain, S Africa.
Note: Sonder (in Harvey & Sonder 1862: 460) suggested that this may be a species of **Aizoon.** A comparison with the descriptions of the species shows the greatest similarity of **M. villosum** with **Aizoon karooicum.**

M. vinaceum ≡ **Delosperma vinaceum**
M. violaceum ≡ **Lampranthus violaceus**
M. virens ≡ **Lampranthus erratus;**
≡ **Ruschia neovirens**
M. virescens ≡ **Carpobrotus virescens**
M. virgatum ≡ **Ruschia virgata**
M. viride ≡ **Smicrostigma viride**
M. viridicatum
≡ **Conophytum truncatum** ssp. **viridicatum**
M. viridiflorum ≡ **Phyllobolus viridiflorus**
M. vittatum ≡ **Nananthus vittatus**
M. vulpinum = **Carruanthus ringens**
M. vulvarium ≡ **Ruschia vulvaria**
M. watermeyeri ≡ **Lampranthus watermeyeri**
M. weigangianum ≡ **Leipoldtia weigangiana**
M. wettsteinii
≡ **Conophytum wettsteinii** ssp. **wettsteinii**
M. willianum = **Erepsia polypetala**
M. wilmaniae ≡ **Ebracteola wilmaniae**
M. wittebergense ≡ **Antimima wittebergensis**
M. woodburniae ≡ **Lampranthus woodburniae**
M. wordsworthiae ≡ **Lampranthus wordsworthiae**
M. zeyheri ≡ **Lampranthus zeyheri**
M. zygophylloides ≡ **Drosanthemum zypophylloides**

Mesembryanthus = Lampranthus

H. E. K. Hartmann

The genus *Mesembryanthus* was described by Necker (1790: 81) without any species or synonyms. Since the relevant publication has been taken up in the list of Opera oppressa (ICBN 2000: App. V), none of the names is validly published. Nevertheless, Rothmaler (1941: 413) validated the name of the genus, chose a type species, *Mesembryanthus tenuifolius*, and added two more species. Since *Mesembryanthus tenuifolius* is a species of the genus **Lampranthus**, as **L. tenuifolius**, *Mesembryanthus* Necker ex Rothmaler becomes a younger synonym of **Lampranthus**.

Me. bicolor ≡ **Lampranthus bicolor**
Me. glaucus ≡ **Lampranthus glaucus**
Me. tenuifolius ≡ **Lampranthus tenuifolius**

Mestoklema Ruschioideae

H. E. K. Hartmann

Mestoklema N.E.Br. ex Glen 1981: 454 ≡ *Mestoklema* N.E.Br. GC 1936: 164 **Typus Me. tuberosum** (L.) N.E.Br. ex Glen **Etym** Gk *mestos*, full, *klema*, a small branch, "in allusion to the abundantly branched habit of the bush" (Brown GC 1936: 164).

Shrubs to 120 cm *h*, ∅ to 60 cm, **l** light ochre when young, with a varying number of raised bladder cells not quite semi-globose in shape, turning into white, elongate bodies when dry, a narrow triangle of stem tissue between the bases of the leaf-pair turning into a somewhat sharp point, the old stems smoother and orange, red, or blackish with age, in some species a trunk developed basally resulting in a tree-like appearance, most species with huge underground organs, either as tap-roots when young, sometimes splitting with age, or in the shape of big potatoes; **L** trigonous, apically recurved, epidermis with many, dense, low papillae; **Fl** in rich to very rich pure cymes or with additional flowers forming a fan in some species, the flowering cymes raised well above last year's growth in most species, overtopping the remains of the fruits consisting of blunt spines, pedicels and base of flower with bladder cells like the stems, **K** 5, petals c. as long as the **K** lobes, from white to yellow in all shades, orange, or pink to purple with a reddish touch in all shades, these colours very often in a single population, filamentous staminodes 0–40, nectary glands 5; **Fr** base funnel-shaped with convex sides, top almost semi-globose from rounded valves, thus often ± globose and developing a tumble-fruit, the stalks becoming spines later, rims very low with a small top in the centre, covering membranes complete or reduced, closing body absent, expanding keels almost parallel or diverging and almost lunate, valve wings narrow to broad; **S** pear-shaped, brown; **Ecol** in deep soils, 200–400 mm rainfall p.a., mainly towards the summer; **Distr** SE Botswana; EC, FS, E part of NC, SE part of NW, S Africa.

Note: The genus has been described posthumously by Brown (GC 1936: 164–166) in English only, which was at that date not valid. Glen (1981: 454–455) supplied the necessary Latin descriptions without any alterations. Plants of **Mestoklema** can best be recognized by their huge inflorescences enlarging even after anthesis, by the breaking-off of the capsule from the stalk in most cases, the pedicels turning into blunt spines, and perhaps also by the small flowers of varying, bright colours; this last feature seems also to occur in some species of **Delosperma**. Typical are also the not quite semi-globose bladder cells on all inter-

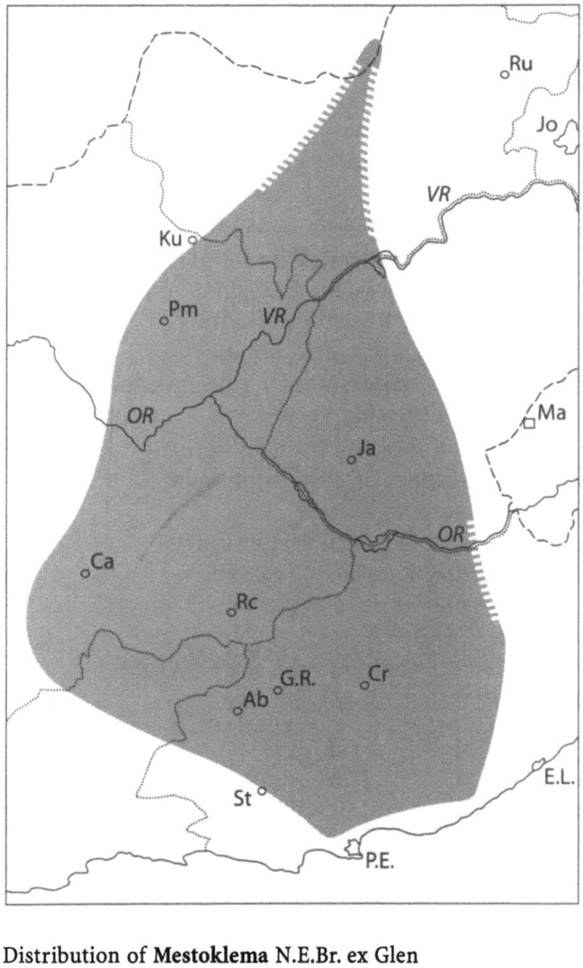

Distribution of **Mestoklema** N.E.Br. ex Glen

nodes, turning into narrow white bodies when dry. The roughish stems and sharpish triangles from tissue between the bases of a leaf-pair resemble conditions found in **Trichodiadema**, to which the fruit is also very similar in construction.

Me. albanicum N.E.Br. ex Glen 1981: 455 (see **photo**) ≡ *Me. albanicum* N.E.Br. GC 1936: 166 nomen invalid. **T** Dyer 1308 (K).

Shrubs 25–40 cm *h*; **L** 6.75–11.3 mm *l*, 1.2–2.8 mm ∅; **Fl** cymes 25–55 mm ∅, pedicels 9–16 mm *l*, ∅ 7 mm, petals 2.5–3 mm *l*, light magenta; **Fr** ∅ 4.5–5 mm; **Ecol** unknown; **Distr** Albany, EC, S Africa.

Note: The species possesses the broadest capsules and the longest pedicels in the genus.

Me. arboriforme (Burchell) N.E.Br. ex Glen 1981: 454 (see **photo**) ≡ *M. arboriforme* Burchell 1822: 343 ≡ *Me. arboriforme* (Burchell) N.E.Br. GC 1936: 165 nomen invalid. **T** Burchell 2004 (K).

Tree-shaped shrubs, 25–50 cm *h*, intricately branched, old stems purplish-brown or blackish-brown with age; **L** 5–15 mm *l*, 1–1.5 mm; **Fl** in cymes of 30–60 mm ∅, divided 3–6 ×, pedicels 1.5–4 mm *l*, ∅ of **Fl** 7–10 mm, petals 20–25, 3–4.5 mm *l*; **Fr** ∅ c. 3 mm, stalks c. 10 mm *l*; **S** 0.7 mm *l*, 0.5 mm *b*; **Ecol** in loamy or stony soil; **Distr** E districts of FS, NE districts of NC, S Africa.

Note: The species occurs in large populations in the area where Burchell collected the type. The plants appear to vary in the colour of the old branches, some turning more orange in colour than brown. Albeit the development of the big rootstock suggests distinctive habitats for this species, it has been found to settle immediately on bare, deep, loamy ground left devoid of plant cover after flooding. Two years later, a rich population dominated the aspect.

Me. copiosum N.E.Br. ex Glen 1981: 454 (see **photo**) ≡ *Me. copiosum* N.E.Br. GC 1936: 165 nomen invalid. **T** Anderson 720 (K, isotype BOL!).

Shrubs with a dense cover of very highly branched **Inf**, grey to brown with age; **L** 10–17 mm *l*, 1.5–2 mm ∅; **Fl** in extraordinarily rich cymes of up to 50 **Fl**, ∅ to 8 mm, pedicels 2–3 mm *l*, petals 3 mm *l*, 0.3–0.7 mm *b*; **Fr** tiny, whitish, base funnel-shaped, top semi-globose, covering membranes complete, expanding keels diverging widely, semi-lunate, valve wings c. half as broad as the expanding keels, ∅ 2–2.8 mm, stalks 6–6.5 mm *l*; **S** 0.63 mm *l*, 0.45 mm *b*; **Ecol** on stony slopes among boulders; **Distr** Hopetown, Kimberley, NC, S Africa.

Note: In appearance, the shrubs are similar to those of **Galenia**, dominated by the whitish capsules in the rich inflorescences.

Me. elatum N.E.Br. ex Glen 1981: 454 ≡ *Me. elatum* N.E.Br. GC 1936: 165 nomen invalid. **T** Dyer 890 (K, isotype BOL!).

Shrubs to 120 cm *h*, often lower and broad, **I** brown with age, the rootstock consisting of several thick, woody roots, each resembling a twisted carrot, 3–4 winding around each other, 30 cm *l*; **L** 8–15 mm *l*, 1.3–1.5 mm ∅; **Fl** in cymes of 25–40 mm ∅, 15–20 flowered, ∅ 6–7 mm, pedicels 2–6 mm *l*, petals purple; **Fr** breaking out of the base of the capsule when ripe, the parts consisting of units of half a locule or more are dispersed by the wind, base funnel-shaped with convex sides, top almost semi-lunate, covering membranes complete, expanding keels diverging widely, valve wings as broad as the expanding keels, 3.5–4 mm ∅, stalks 5–8.5 mm *l*; **S** c. 0.65 mm *l*, c. 4.5 mm *b*; **Ecol** in loamy flats in shrubby associations; **Distr** Albany, Uitenhage, EC, S Africa.

Note: The rather crowded inflorescences resemble those of **Me. copiosum**, but can be recognized well by the remains of bundles left over from a capsule that has lost all contents. Only later, the stalks develop into sharp spines.

Me. illepidum N.E.Br. ex Glen 1981: 454 ≡ *Me. illepidum* N.E.Br. GC 1936: 165 nomen invalid. **T** Dyer 2336 (K isotype BOL!).

Shrubs 23–25 cm *h*, **I** ochre to brown later; **L** 3–8 mm *l*, 1–1.5 mm ∅; **Fl** pedicels 2–4 mm *l*, ∅ c. 6 mm, petals 2 mm *l*, 0.4–0.5 mm *b*, submagenta; **Fr** base funnel-shaped with convex sides, top nearly semi-globose, breaking off the stalk when ripe, covering membranes complete, expanding keels diverging only a little, valve wings not seen, ∅ 2–3.5 mm, stalks 8.5 mm *l*; **S** 0.7 mm *l*, c. 0.5 mm *b*; **Ecol** in loamy soils in shrubby vegetation; **Distr** Bedford, EC, S Africa.

Note: Whether this species also forms compound rootstocks like **Me. elatum**, as discussed by Brown (GC 1936: 166), remains uncertain.

Me. macrorhizum ≡ **Delosperma napiforme**

Me. tuberosum (L.) N.E.Br. ex Glen 1981: 454 (see **photo**) ≡ *M. tuberosum* L. 1753: 484 ≡ *Me. tuberosum* N.E.Br. GC 1936: 165 ≡ *Delosperma tuberosum* (L.) Schwantes MDG 1927: 258 **LT** Dillenius 1732: t. 207, f 264 = *Me. tuberosum* (L.) N.E.Br. ex Glen var. *macrorhizum* (Haw.) N.E.Br. ex Glen 1981: 454 ≡ *M. macrorhizum* Haw. 1826: 331 non DC 1828: 425 nomen illeg. ≡ *M. megarhizum* Don 1834: 145 nomen illeg. non DC. 1828: 425 **LT** drawing of Duncanson (K).

Laxly branching shrubs, the main stem from a rootstock consisting of several very thick, potato-shaped roots, branches twisting and decumbent, **I** brown with age; **L** 9–20 mm *l*, 1.1–2.2 mm ∅; **Fl** in cymes, 4.5–6.7 mm ∅; **Fr** ∅ 3.5 mm; **Ecol** unknown; **Distr** EC, S Africa.

Note: Typical are the potato-shaped roots, shown on the type figure.

Me. tuberosum var. *macrorhizum* = **Me. tuberosum**

Meyerophytum Ruschioideae

H.-D. Ihlenfeldt

Meyerophytum Schwantes MDG 1927: 436; Bolus NM 1958: 345–346 **Typus Me. meyeri** (Schwantes) Schwantes **Etym** *G. Meyer*, 1867–1958, minister and missionary at Komaggas and Steinkopf, Namaqualand, botanical explorer of the area = *Depacarpus* N.E.Br. GC 1930: 71 **Typus** *Depacarpus tinctus* (L.Bolus) N.E.Br. = **Me. meyeri** (Schwantes) Schwantes.

Cushion-shaped or compact deciduous shrubs to 15 cm *h*, ∅ to 30 cm; I elongated or very short, 2–3 mm ∅ or 8–15 mm ∅, stems in young plants erect, in old plants procumbent; **L** epidermis mesomorphic with elliptic bladder-cells c. 0.25 mm *l*, 0.2 mm *b*, heterophyllous: 1st leaf-pair of the season ovate, slightly keeled on the lower side, connate to c. 2/3 of the length, forming a bilobed corpuscle, 2nd leaf-pair oblong, obtuse, connate to c. 1/2 of the length, towards the end of the season developing into a papery, only at the very base slightly sclerotic sheath enclosing the corpuscle of the first leaf-pair of the next season; **Fl** solitary, emerging towards the end of the season from the apical slit in the connate part of the 2nd leaf-pair, pedicellate, **K** 5, petals in c. 2–3 rows, some of the inner ones shorter, narrowly obovate, often emarginate, rose-purple throughout, or rose and white at the base, or white throughout, filamentous staminodes absent or very few, filaments papillate at the base, purple, yellow or white, holonectary, ∅ to 40 mm, VIII–X; **Fr** expanding keels merging into expanding sheets, valve wings broad, covering membranes over 9/10 of the locules, closing bodies present or absent, if absent, opening between the covering membranes and the rim of the capsule narrowed by a cluster of sterile funicles, locules (4–)5(–7), ∅ 4–13 mm; **S** testa smooth except for the micropylar region which may be slightly papillate, whitish to light brown; **Chr** $2n = 18$; **Ecol** in loamy soil, sometimes with a layer of quartz pebbles, < 100 mm rainfall p.a. mainly in winter; **Distr** NW and W Namaqualand, NC, S Africa.

Note: The presently recognized two species of **Meyerophytum** have originally been in included in **Mitrophyllum** (Schwantes MDG 1927: 150), *Conophyllum* (Bolus NM 1937: 52), or **Monilaria** (Bolus NM 1958: 341). However, in contrast to **Mitrophyllum** (and *Conophyllum*), in **Meyerophytum** the developing corpuscle of the 1st leaf-pair of the season does not fill up the whole connate part of the 2nd leaf-pair of the last season; moreover, after the start of the rainy season, the two leaves of the 1st leaf pair do not separate and do not spread. Those species formerly included in **Meyerophytum** in which the leaves of the 2nd leaf pair are only connate for less than 1/3 of their length and the capsules lack closing bodies have been transferred to the genus **Dicrocaulon**. As to the inclusion of *Monilaria globosa* in **Meyerophytum**, see under **Me. globosum**. Cultivation of **Meyerophytum** species is easy; they are winter growers and need a resting period without watering in summer.

Distribution of **Meyerophytum** Schwantes

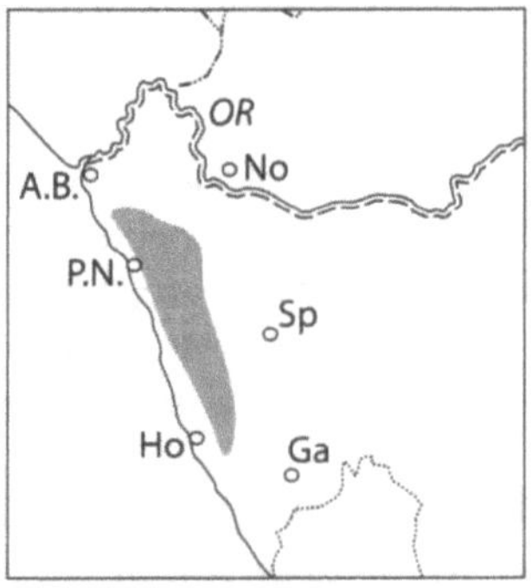

Key to the Species of Meyerophytum

1. 2nd leaf-pair 50–70 mm *l*, connate part to 15 mm ∅; fruit 9–13 mm ∅, without closing bodies .... **Me. globosum**
- 2nd leaf-pair to 35 mm *l*, connate part 3–4 mm ∅; fruit 3–10 mm ∅, with whitish closing bodies .... **Me. meyeri**

Me. globosum (L.Bolus) Ihlenf. **comb. nov.** (see **photo**) ≡ *Conophyllum globosum* L.Bolus Notes on Mesembryanthemum 3, 1937: 52 (as basionym) ≡ *Monilaria globosa* (L.Bolus) L.Bolus NM 1958: 341 **Holotypus** van Heerde 10428 (BOL!).

Compact, sparsely branched, to 10 cm *h* and to 20 cm ∅, **I** to 1.5 cm ∅, very short; **L** corpuscle of the 1st leaf-pair to 17 mm *l*, 14 mm *b*, 2nd leaf-pair 50–70 mm *l*, connate part to 25 mm ∅; **Fl** petals white, filaments yellow; **Fr** pedicels to 100 mm *l*, usually straight, locules 5(–6), rarely 7, without closing bodies, opening between the covering membranes and the rim of the capsule narrowed by a dense cluster of sterile funicles, 9–13 mm ∅; **S** > 20 per locule, c. 0.7 mm *l*, 0.5 mm *b*; **Ecol** in loamy soil with quartz pebbles; **Distr** W Namaqualand, NC, S Africa.

Note: **Me. globosum** is, in contrast to **Me. meyeri**, very uniform and occurs frequently in a restricted area forming large populations there. This species has to be excluded from **Monilaria** in the presently adopted circumscription for several reasons: deviating leaf-sequence, no "moniliform" internodes, the lack of a sclerotic sheath that completely covers the corpuscle of the 1st leaf-pair of the next season during the resting state, and non-succulent stems, although they are comparatively thick.

Me. meyeri (Schwantes) Schwantes MDG 1927: 437 (see **photo**) ≡ *Mitrophyllum meyeri* Schwantes MDG 1927: 150 **Neotype**, here designated: Jacobsen 1933: 161 f 162 = *Me. meyeri* (Schwantes) Schwantes var. *holgatense* L.Bolus NM 1958: 344 **T** Grobler 22938 (BOL!) = *M. tinctum* L.Bolus FPSA 1927: t. 279 ≡ *Me. tinctum* (L.Bolus) Schwantes ex L.Bolus NM 1958: 345 nomen invalidum ≡ *Depacarpus tinctus* (L.Bolus) N.E.Br. GC 1930: 71 **T** Pillans 5777 (BOL!).

Cushion-shaped, freely branched, to 30 cm ∅; **I** 2–3 mm ∅, those between the 1st and 2nd leaf-pair at the start of the season short, elongating later; **L** corpuscle of the 1st leaf-pair to 12 mm *l*, 10 mm *b*; 2nd leaf-pair to 35 mm *l*, connate part 3–4 mm ∅; **Fl** petals rose-purple throughout, or rose and basally white, or white turning pink with age, filaments purple, yellow or white; **Fr** pedicels to 45 mm *l*, usually curved, locules (4–)5(–7), with whitish closing bodies, 4–10 mm ∅; **S** 8–15 seeds per locule, c. 1 mm *l*, 0.6 mm *b*; **Ecol** in loamy soil, often with quartz pebbles; **Distr** NW Namaqualand, NC, S Africa.

Note: The neotype designated here probably represents material seen by Schwantes; it may, therefore, be a lectotype. **Me. meyeri** is a rather variable species, especially in the colour of the petals and filaments. In cultivation **Me. meyeri** may produce one or two additional leaf-pairs similar to the 2nd one.

Me. meyeri var. *holgatense* = **Me. meyeri**
Me. microstigma ≡ **Dicrocaulon microstigma**
Me. primosii = **Dicrocaulon microstigma**
Me. tinctum = **Me. meyeri**

Micropterum = Cleretum

H. E. K. Hartmann

As shown by Struck & Ihlenfeldt (1985) and Ihlenfeldt & Struck (1987), *Micropterum* agrees in its character states with **Cleretum** N.E.Br., the former name being, therefore, a younger synonym of the latter. The extensive and long use of the generic name is due to confusion in the typification.

Mi. cuneifolium = **Dorotheanthus bellidiformis**
Mi. herrei ≡ **Cleretum herrei**
Mi. limpidum = **Dorotheanthus bellidiformis**
Mi. longipes = **Cleretum papulosum** ssp. **schlechteri**
Mi. papulosum ≡ **Cleretum papulosum**
Mi. pinnatifidum ≡ **Aethephyllum pinnatifidum**
Mi. puberulum = **Mesembryanthemum aitonis**
Mi. schlechteri
≡ **Cleretum papulosum** ssp. **schlechteri**
Mi. sessiliflorum = **Cleretum papulosum**
Mi. sessiliflorum var. *album*
= **Mesembryanthemum aitonis**
Mi. sessiliflorum var. *luteum* = **Cleretum papulosum**

Millegrana = Cypselea

H. E. K. Hartmann

In his description of **Cypselea**, Turpin (1806: 88) mentioned the name *Millegrana* as occurring in the Herbarium of Surinam of de Jussieu, page 363, only. No species are given, nor a description cited. The name *Millegrana* is therefore a nomen nudum.

Mimetophytum = Mitrophyllum

H. E. K. Hartmann

L.Bolus based the genus *Mimetophytum* on the fact that the included species differed from those of **Mitrophyllum** and *Conophyllum* by the possession of closing bodies in the fruits. Poppendieck (1976) showed that presence and absence of closing bodies can occur within a single population; the feature can therefore not be used to distinguish a genus in this group.

Mi. crassifolium = **Mitrophyllum grande**
Mi. parvifolium = **Mitrophyllum dissitum**

Mitrophyllum Ruschioideae

H. E. K. Hartmann

Mitrophyllum Schwantes ZSK 1926: 181; Poppendieck 1976: 339–413 **Typus Mi. mitratum** (Marloth) Schwantes **Etym** Gk *mitra*, hat, in particular of a bishop or archbishop, *phyllos*, leaf, describing the pointed body formed by the leaves in the resting state = *Conophyllum* Schwantes ZSK 1928: 321 **Typus** *Conophyllum marlothianum* Schwantes = **Mi. clivorum** (N.E.Br.) Schwantes = *Mimetophytum* L.Bolus NM 1954: 252 **Typus** *Mimetophytum parvifolium* L.Bolus = **Mi. dissitum** (N.E.Br.) Schwantes.

Heterophyllous shrubs to 60 cm *h*; **L** mostly with one pair a little connate, free parts tongue-shaped to triangular, the other connate for 2/3 to 4/5 of its length, forming a corpuscle which, at the end of the rainy season, develops into a dry sheath enveloping the following leaf-pair during the resting state in habitat, only one species deviating from the scheme, epidermis mesomorphic, bladder cells inconspicuous, wax in a thin continuous layer with only few dispersed vertical platelets; **Fl** single, bracteoles differing from foliage leaves, petals yellow, white to pink or purple, holonectary; **Fr** with dominating expanding keels merging into expanding keels, covering membranes developed from endocarpal material available on the roof of the locule and hence of different shapes, closing bodies infrequently formed, mostly of spongy consistency, valve wings broad (Mitrophyllum-type) or fruits without covering membranes (Delosperma-type), locules shallow, mostly 5, roundish in outline as seen from above when closed; **S** triangular in side view to pear-shaped, papillate to smooth; **Chr** $2n = 18$; **Ecol** on quartzitic slopes, often in crevices of rocks; **Dist** N Namaqualand, NC, S Africa.

Note: In natural habitat mostly found on rather steep rocky slopes, often facing east, and forming conspicuous associations developing a spectacular sight when flowering in late summer. The genus is restricted to winter rainfall areas, indicating that summer rains cannot be borne readily. In cultivation, a strict summer rest is to be observed.

The taxonomic history of the genus was controversial for many years because neither the variability of the growth forms nor that of the fruits were recognized and classified until Poppendieck (1976) clarified the species in the complex. Further typifications offered here finalize the revision of the genus.

Distribution of **Mitrophyllum** Schwantes

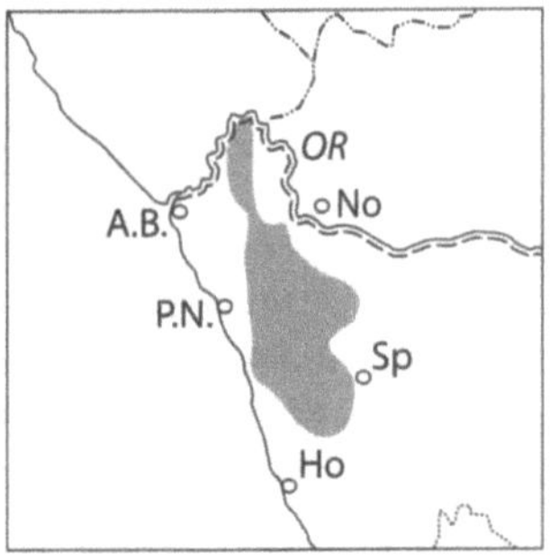

Key to the Subgenera of Mitrophyllum

1. Stalks of fruits bent, > 40 mm long, flowers purple, filaments basally papillate ... Mi. subg. **Kurkamium**
– Stalks of fruits erect and straight, flowers yellow or white, rarely pink suffused, filaments epapillate .... Mi. subg. **Mitrophyllum**

Mi. subg. **Kurkamium** Poppendieck 1976: 395 **Typus Mi. roseum** L.Bolus.

Low shrubs to 15 cm *h*, then hanging over, flowering and sterile branches similar; **L** of a pair basally connate for < 1/5, 3–6 pairs to a branch, mostly < 3 pairs formed per season; **Fl** formed after the 2nd or 3rd leaf-pair in a season, petals purple, filaments papillate near bases, flowering VIII–IX in habitat, II–III N of the equator; **Fr** stalks > 40 mm *l*, bent like a pipestem, covering membranes short, closing body absent, valve rims high, closed capsule distinctly pentangular in outline as seen from above; **Ecol** amongst rocks, < 100 mm rainfall p.a. in winter, fog; **Distr** N Namaqualand, NC, S Africa.

Note: Only a single species is known.

Mi. subg. **Mitrophyllum**, synonyms as given for the genus.

Shrubs to 60 cm *h*; **L** of the typical heteromorphic sequence, more leaf-pairs formed in wet cultivation; **Fl** yellow or white, rarely pink suffused, filaments smooth; **Fr** with erect, straight stalks, with rounded corners as seen from above when closed; **Ecol** in crevices on rocky slopes, predominantly in S and SE exposition; **Distr** N Namaqualand, NC, S Africa.

Note: Six distinct growth forms have been distinguished (Poppendieck 1976), but only **Mi. grande** expresses a single one; all other species vary greatly in habit, hence characters of flowers and fruits are used to delineate the species, most of which differ in their flowering season, at least in habitat.

Key to the Species of Mi. subg. Mitrophyllum

1. Corpuscles mostly > 45 mm long, > 15 mm thick, anthers golden yellow **2**
- Corpuscles < 45 mm long, < 15 mm thick, anthers whitish **3**
2. Fruit with narrow ledges as covering membranes or without them, flowering before the rainy season ..... **Mi. mitratum**
- Fruit with covering membranes extending c. 1/3 over the locules, flowering towards the end of the rainy season ..... **Mi. grande**
3. Fruit stalks basally corky and dark brown, distinct from the top ..... **Mi. dissitum**
- Fruit stalks basally not corky, changing gradually from the top **4**
4. Corpuscle c. as thick as the stem bearing it, flowering in the rainy season ..... **Mi. clivorum**
- Corpuscle thicker than the stem bearing it, flowering at the end of the rainy season .... **Mi. abbreviatum**

Mi. subg. *Monilaria* ≡ **Monilaria**

Mi. abbreviatum L.Bolus NM 1932: 368 (see **photo**) subg. **Mitrophyllum** ≡ *Conophyllum abbreviatum* (L.Bolus) L.Bolus NM 1958: 342 **T** Luckhoff 19656 (BOL) = *Conophyllum gracile* Schwantes 1939: 44 ≡ *Mi. gracile* (Schwantes) De Boer ex H.Jacobsen 1970: 469, nomen illeg. **Neotype**, here designated: Jacobsen 1970: t. 185, f 2[14].

Densely branched shrubs to 40 cm *h*, **I** 10–60 mm *l*, 2–4 mm ∅, basally to 8 mm ∅; **L** 1st leaf-pair 10–30 mm *l*, 5–8 mm *b*, 2nd leaf-pair connate for 4/5 of its length, round to oval in ∅, dry tips spreading; **Fl** following a leaf-pair of type 1, bracteoles 3–8 mm *l*, **I** below bracteoles short, 1–2 side branches, petals white, pink suffused, anthers pale yellow, bases smooth, flowering III–IV in habitat, IX–X N of the equator; **Fr** stalks erect, basally reddish, fruit 5–20 mm *l*, ∅ 6–8 mm, rims turned prominently outwards, centre lower than the outer edge of capsule; **S** smooth; **Ecol** in crevices of quartzitic rocks; **Distr** N Namaqualand, NC, S Africa.

Note: The species appears to occur in a restricted area only and may be threatened.

Mi. affine = **Mi. dissitum**
Mi. angustifolium = **Mi. clivorum**
Mi. articulatum = **Mi. dissitum**
Mi. brevisepalum = **Mi. grande**
Mi. carterianum = **Mi. grande**
Mi. chrysoleucum ≡ **Monilaria chrysoleuca**

Mi. clivorum (N.E.Br.) Schwantes ZSK 1926: 182 (see **photo**) subg. **Mitrophyllum** ≡ *M. clivorum* N.E.Br. 1920: 113 ≡ *Conophyllum clivorum* (N.E.Br.) Schwantes ZSK 1928: 323 ≡ *Schwantesia clivora* (N.E.Br.) L.Bolus SAG 1928: 279 nomen illeg. **T** Pearson 6200 (K) = *Conophyllum angustifolium* L.Bolus NM 1954: 244 ≡ *Mi. angustifolium* (L.Bolus) De Boer ex H.Jacobsen 1970: 468 nomen illeg. **T** Herre 8369 (BOL) = *M. cognatum* N.E.Br. 1920: 113 ≡ *Mi. cognatum* (N.E.Br.) Schwantes ZSK 1926: 182 ≡ *Conophyllum cognatum* (N.E.Br.) Schwantes ZSK 1928: 321 ≡ *Schwantesia cognata* (N.E.Br.) L.Bolus SAG 1928: 279 nomen illeg. **T** Pearson 6179 (K) = *Conophyllum compactum* L.Bolus NM 1954: 244 ≡ *Mi. compactum* (L.Bolus) De Boer ex H.Jacobsen 1970: 469 nomen illeg. **T** Hall 598/52 (BOL) = *Conophyllum compactum* L.Bolus var. *eenrietense* L.Bolus NM 1958: 344 ≡ *Mi. compactum* L.Bolus var. *eenrietense* (L.Bolus) De Boer ex H.Jacobsen 1970: 469 nomen illeg. **T** van Heerde 25629 (BOL) = *Mi. framesii* L.Bolus NM 1931: 289 ≡ *Conophyllum framesii* (L.Bolus) L.Bolus NM 1958: 343 **T** Frames 19641 (BOL) = *Conophyllum hallii* L.Bolus JSAB 1959: 251 ≡ *Mi. hallii* (L.Bolus) De Boer ex H.Jacobsen 1970: 469 nomen illeg. **T** Hall 696/53 (BOL) = *Conophyllum herrei* L.Bolus NM 1931: 274 ≡ *Mi. herrei* (L.Bolus) De Boer ex H.Jacobsen 1970: 469 nomen illeg. **T** 8752 (BOL) = *Mi. karrachabense* L.Bolus NM 1937: 98 **T** Herre 8783 (BOL) = *Mi. marlothianum* Schwantes ZSK 1928: 321 nomen nudum ≡ *Conophyllum marlothianum*[15] Schwantes MDK 1929: 14 ≡ *Mi. marlothianum* (Schwantes) De Boer ex H.Jacobsen 1970: 469 nomen illeg. **Lectotype**, here designated: Meyer s.n. (photograph of the type plant in collection Schwantes HBG!) = *Mi. niveum* L.Bolus NM 1934: 471 ≡ *Conophyllum niveum* (L.Bolus) Herre in Bolus NM 1954: 241 **T** Herre 8560 (BOL) = *Conophyllum ripense* L.Bolus NM 1954: 248 ≡ *Mi. ripense* (L.Bolus) De Boer ex H.Jacobsen 1970: 469 nomen illeg. **T** Hall 601/52 (BOL) = *Mi. tenuifolium* L.Bolus NM 1933: 409 **T** Herre 9941 (BOL) = *Mi. schickianum* Tischer in Brown et al. 1931: 268, **Lectotype**, here des-

[14] Here neotypified by Poppendieck.

[15] Published with the heading "*Mitrophyllum marlothianum*", but placed in the meanwhile described genus *Conophyllum* in a footnote by the author(Schwantes 1929). The earlier citation (Schwantes 1928) of "*C. marlothianum* = *Mitrophyllum marlothianum*" includes two nomina nuda because neither name had been published validly at that time. The later transfer to **Mitrophyllum** by de Boer is illegitimate because no clear reference to the original description is made.

ignated: f 126 = *Conophyllum vanheerdei* L.Bolus NM 1954: 246 ≡ *Mi. vanheerdei* (L.Bolus) De Boer ex H.Jacobsen 1970: 470 nomen illeg. **T** van Heerde 11449 (BOL).

Shrubs to 60 cm *h*, **I** of vegetative branches 10–50 mm *l*, 5–18 mm ∅, brown, later grey; **L** of 1st pair 10–50 mm *l*, 8–20 mm ∅, of 2nd pair connate for 1/2 to 2/3, corpuscles 10–40 mm *l*, 5–15 mm ∅ during resting state, dry tips hanging; **Fl** with 15–40 mm *l* bracteoles, **I** below **Fl** 10–40 mm *l*, petals yellow or white, pink suffused, anthers pale yellow, filaments smooth, flowering V–VI in habitat, X–XII N of the equator; **Fr** stalks ochre, later grey, erect, covering membranes almost complete, closing body rarely present, closed capsule round, as seen from above, 5–16 mm ∅, fruit 5–40 mm *l*; **Ecol** S and SE slopes with rocks, often in shady places; **Distr** N Namaqualand, NC, S Africa.

Note: This most widespread species of the genus shows also the widest variation in character states.

Mi. cognatum = **Mi. clivorum**
Mi. compactum = **Mi. clivorum**
Mi. compactum var. *eenrietense* = **Mi. clivorum**
Mi. conradii = **Mi. dissitum**
Mi. crassifolium = **Mi. grande**
Mi. cuspidatum = **Mi. grande**

Mi. dissitum (N.E.Br.) Schwantes ZSK 1926: 182 (see **photo**) subg. **Mitrophyllum** ≡ *M. dissitum* N.E.Br. 1920: 114 ≡ *Conophyllum dissitum* (N.E.Br.) Schwantes ZSK 1928: 322 ≡ *Schwantesia dissita* (N.E.Br.) L.Bolus SAG 1928: 279 nomen illeg. **T** Pearson 6116 (K) = *Mi. affine* L.Bolus NM 1937: 97 **T** Herre 9273 (BOL) = *Conophyllum articulatum* L.Bolus NM 1954: 248 ≡ *Mi. articulatum* (L.Bolus) De Boer ex H.Jacobsen 1970: 468 nomen illeg. **T** Pillans 5709 (BOL) = *Mi. conradii* L.Bolus NM 1937: 97 ≡ *Conophyllum conradii* (L.Bolus) 1954: 254 nomen illeg. **T** Herre 10188 (BOL) = *Mi. kubusanum* L.Bolus NM 1937: 99 ≡ *Conophyllum kubusanum* (L.Bolus) L.Bolus NM 1954: 247 **T** Herre 9220 (BOL) = *Conophyllum obtusipetalum* L.Bolus NM 1958: 343 ≡ *Mi. obtusipetalum* (L.Bolus) De Boer ex H.Jacobsen 1970: 469 nomen illeg. **T** Hall 428/54 (BOL) = *Mimetophytum parvifolium* L.Bolus NM 1954: 252 ≡ *Mi. parvifolium* (L.Bolus) G.D.Rowley 1978: 8 **T** Pillans 5708 (BOL) = *M. proximum* N.E.Br. 1920: 117 non L.Bolus SAG 1297: 239 nomen illeg. ≡ *Mi. proximum* (N.E.Br.) Schwantes ZSK 1926: 182 ≡ *Conophyllum proximum* (N.E.Br.) Schwantes ZSK 1928: 321 ≡ *Schwantesia proxima* (N.E.Br.) L.Bolus SAG 1928: 279 nomen illeg. **T**? non *M. proximum* sensu L.Bolus SAG 1927: 239.

Shrubs to 40 cm, rarely to 60 cm *h*; **I** of non-flowering branches reddish-brown, later grey, **L** to 60 mm long, ∅ 5 mm; 1st leaf-pair 10–50 mm *l*, basal ∅ 8–15 mm; 2nd leaf-pair connate for 4/5th of its length, corpuscle at rest 10–40 mm *l*, 5–15 mm *b*, oval in ∅, dry tips spreading; flowering branches with somewhat smaller leaves; **Fl** with bracteoles c. 12 mm *l*, **I** below 7–30 mm *l*, basally distinctly corky, dark brown, 1–2 side branches developing further, petals yellow, anthers whitish-yellow, filaments smooth, anthesis in V–VI mainly in habitat, X–XII N of the equator; **Fr** stalks straight, basally 1/3 corky, fruit 7–18 mm *l*, covering membranes well developed, closing bodies sometimes present, 6–13 mm ∅; **Ecol** on S and SE slopes, in crevices of quartzitic rocks or in coarser rubble on steeper slopes; **Distr** N Namaqualand, NC, S Africa.

Note: The species is variable in growth form: the dissitum-type develops into a rather laxly branched loose shrub, whereas the proximum-type often shows a high proportion of short internodes (Poppendieck 1976). Anthers are whitish-yellow, as in **Mi. clivorum**, from which the species differs by shorter, more pointed bracteoles and the distinct, dark brown, corky base of the stalk.

Mi. framesii = **Mi. clivorum**
Mi. gracile = **Mi. abbreviatum**

Mi. grande N.E.Br. GC 1929: 227 (see **photo**) subg. **Mitrophyllum** ≡ *Conophyllum grande* (N.E.Br.) L.Bolus NM 1954: 247 ≡ *M. proximum* L.Bolus SAG 1927: 239 nomen illeg. non N.E.Br. 1920: 117 **T** Pillans 4933 (K) = *Conophyllum brevisepalum* L.Bolus NM 1954: 247 ≡ *Mi. brevisepalum* (L.Bolus) De Boer ex H.Jacobsen 1970: 468 nomen illeg. **T** Schlechter 25206 (BOL) = *Conophyllum carterianum* L.Bolus NM 1954: 249 ≡ *Mi. carterianum* (L.Bolus) De Boer ex H.Jacobsen 1970: 468 nomen illeg. **T** Schlechter 524/34 B (BOL) = *Mimetophytum crassifolium* L.Bolus NM 1954: 252 ≡ *Mi. crassifolium* (L.Bolus) G.D.Rowley 1978: 8 **T** Schlechter 25205 (BOL) = *Conophyllum cuspidatum* L.Bolus NM 1954: 250 ≡ *Mi. cuspidatum* (L.Bolus) De Boer ex H.Jacobsen 1970: 468 nomen illeg. **T** Schlechter 524/34 (BOL) = *Conophyllum grande* (N.E.Br.) L.Bolus var. *compressum* L.Bolus NM 1954: 240 ≡ *Mi. grande* (N.E.Br.) L.Bolus var. *compressum* (L.Bolus) De Boer ex H.Jacobsen 1970: 469 nomen illeg. **T** Herre 8733 (BOL) = *Conophyllum latibracteatum* L.Bolus NM 1954: 245 ≡ *Mi. latibracteatum* (L.Bolus) De Boer ex H.Jacobsen 1970: 469 nomen illeg. **T** van Heerde 25340 (BOL) = *Conophyllum latibracteatum* L.Bolus forma *ferum* L.Bolus NM 1954: 246 ≡ *Mi. latibracteatum* (L.Bolus) de Boer ex H. Jacobsen forma *ferum* (L.Bolus) De Boer ex H.Jacobsen 1970: 469 nomen illeg. **T** van Heerde 25342 (BOL) = *Mi. pillansii* N.E.Br. GC 1929: 228 **Lectotype**, here designated: "*Mi. mitra-*

tum" L.Bolus non Schwantes FPSA 1927: t. 274 excluding the flowering material pictured[16].

Shrubs with compact centres with short shoots only, from which erect long shoots develop, 4–8 years old, bearing flowers; **I** in the centre markedly corky, ∅ 10–15 mm, with persisting remains of old leaves, 1st leaf-pair 6–12 cm *l*, 15–30 mm *b*, 2nd connate for 4/5th corpuscles at rest 25–100 mm *l*, 15–35 mm *b*, oval in ∅, yellowish-green in colour in habitat, dry tips spreading; flowering branches developing distinct ± dichasial perennial shoots, first **I** 80–140 mm *l*, following ones < 40 mm *l*, first bracteoles broad, 20–30 mm *l*, later ones 5–10 mm *l*, shoots terminated by a resting leaf-pair connate for 2/3, 10–35 mm *l*, 8–15 mm *t*, embracing the 1st leaf-pair connate for 2/3, 10–40 mm *l*, **I** below bracteoles 20–40 mm *l*; **Fl** petals white or yellow, anthers golden, filaments smooth, anthesis V–VII, X–XII N of the equator; **Fr** stalks straight, basal half corky, fruit length 15–45 mm, covering membranes complete, closing body sometimes present, ∅ 7–18 mm; **Ecol** on S to SE slopes with quartzitic stones; **Distr** N Namaqualand, NC, S Africa.

Note: Occurring in a restricted area together with **M. mitratum** (with reduced covering membranes) or **M. dissitum** (without a compact centre in growth).

Mi. grande var. *compressum* = **Mi. grande**
Mi. hallii = **Mi. clivorum**
Mi. herrei = **Mi. clivorum**
Mi. karrachabense = **Mi. clivorum**
Mi. kubusanum = **Mi. dissitum**
Mi. latibracteatum = **Mi. grande**
Mi. latibracteatum forma *ferum* = **Mi. grande**
Mi. marlothianum = **Mi. clivorum**
Mi. meyeri ≡ **Meyerophytum meyeri**

Mi. mitratum (Marloth) Schwantes ZSK 1926: 182 (see **photo**) subg. **Mitrophyllum** ≡ *M. mitratum* Marloth 1910: 35 **T** Alston 4690 (BOL) = *Mi. pillansii* N.E.Br. GC 1929: 228 p.p. **T** Pillans 5450 (BOL) = *Mi. mitratum* (Marloth) Schwantes var. *eburneum* L.Bolus NM 1931: 252 **T** Schlechter 8369 (BOL).

Shrubs variable in growth: from forms with laxly arranged branches to those with a compact centre, to 60 cm *h*, flowers on monochasial, elongated branches; **I** of non-flowering plants 15–50 mm *l*, 8–15 mm ∅, 1st leaf-pair 40–100 mm *l*, 15–22 mm *b*, 2nd leaf-pair connate for 2/3 to 4/5, corpuscles at rest 35–100 mm *l*, 15–35 mm *b*, cylindrical, dry tips spreading to hanging, flowering branches: 1st leaf-pair as above, bracteoles triangular, 50–80 mm *l*, **I** below enclosed; **Fl** appearing at the beginning of the growing season breaking laterally through the resting corpuscle, side branches 2 below the 1st flower, later only one continuing the stem, petals white, often purplish suffused, anthers golden, anthesis II–III, in the N hemisphere VIII–X; **Fr** stalks not corky at base, covering membranes absent or as narrow ledges, closing bodies absent, ∅ 7–13 mm, fruit length 7–10 mm; **S** flat, light ochre; **Ecol** on moderate S to SE slopes covered with quartzitic stones; **Distr** N Namaqualand, NC, S Africa.

Note: Known from a restricted area, growing sympatrically with **M. grande** (with complete covering membranes) and **M. clivorum** (with whitish anthers).

Mi. mitratum var. *eburneum* = **Mi. mitratum**
Mi. moniliforme ≡ **Monilaria moniliformis**
Mi. nanum = **Diplosoma retroversum**
Mi. niveum = **Mi. clivorum**
Mi. obtusipetalum = **Mi. dissitum**
Mi. parvifolium = **Mi. dissitum**
Mi. pillansii = **Mi. grande; Mi. mitratum**
Mi. pisiforme ≡ **Monilaria pisiforme**
Mi. proximum = **Mi. dissitum**
Mi. ripense = **Mi. clivorum**

Mi. roseum L.Bolus NM 1932: 349 (see **photo**) subg. **Kurkamium** ≡ *Conophyllum roseum* (L.Bolus) L.Bolus NM 1954: 242 **T** Herre 8914 (BOL).

Developing 3–5 leaf-pairs per season, non-flowering branches: 1st leaf-pair 15–20 mm *l*, 6–10 mm *b*, 2nd and subsequent leaf-pairs connate for 1/5, **I** long, reddish-brown, 15–45 mm *l*, ∅ c. 5 mm, lateral branches at first with 2 closely set leaf-pairs only, the 1st enclosed by the dry leaves of the main branch; **Fl** after the 2nd or 3rd leaf-pair, seemingly lateral to the main branch, but really at the end of the side branch, petals purple, anthers pale yellow, filaments purple, basally papillate, anthesis VIII–IX, II–III on the N hemisphere; **Fr** stalks 30–90 mm *l*, bent, base never corky, covering membranes short, closing body absent, 5–9 mm ∅; **S** c. 0.7 mm *l*; **Ecol** in rocks on W slopes receiving fog; **Distr** N Namaqualand, NC, S Africa.

Note: Known from two populations only, one of them rich in individuals, growing sympatrically with **M. dissitum**, difficult to flower in cultivation.

Mi. scutatum ≡ **Monilaria scutata**
Mi. tenuifolium = **Mi. clivorum**
Mi. schickianum = **Mi. clivorum**
Mi. vanheerdei = **Mi. clivorum**

---

[16] N. E. Brown based his new name on the cited drawing, of which only the non-flowering material is chosen here by Poppendieck as the lectotype of the species. The flowering material belongs to **Mi. grande**. The sheet Pillans 4690 cited also by N. E. Brown in the protologue cannot be found at BOL and must be considered lost. The citation of Pillans 5450 (BOL) as the type sheet by Poppendieck (1976) was an error corrected here.

Monilaria Ruschioideae

H.-D. Ihlenfeldt

Monilaria (Schwantes) Schwantes GW 1929: 69; Ihlenfeldt & Jörgensen 1973: 49–94 ≡ *Mitrophyllum* Schwantes subg. *Monilaria* Schwantes ZSK 1926: 181 ≡ *Conophyllum* Schwantes subg. *Monilaria* Schwantes ZSK 1928: 321 **Typus Mo. chrysoleuca** (Schlechter) Schwantes ≡ *Mitrophyllum chrysoleucum* (Schlechter) Schwantes ≡ *Conophyllum chrysoleucum* (Schlechter) Schwantes **Etym** L *monile*, necklace (of pearls), in allusion to the structure of the stems remotely resembling a thick string of pearls = *Schwantesia* L.Bolus p.p. SAG 1928: 279 nomen illeg. **Typus** *Schwantesia moniliformis* (Haw.) L.Bolus ≡ **Mo. moniliformis** (Thunberg) Ihlenf. & S.Jörgensen.

Sparsely branched, deciduous shrubs to 20 cm *h*, I thickened, with a conspicuous peripheral water storing tissue and with regular circular constrictions, I short, ∅ 10–20 mm, each covered with a persistent barrel-shaped or box-shaped sclerotic sheath; L 1st pair of the season ± hemispherical with a small tip, completely connate, forming a nearly spherical corpuscle to 10 mm ∅, hidden during the resting state inside the latest sclerotic sheath but ± emerging at the beginning of the season by elongation of the preceding internode, ephemeral, 2nd leaf-pair c. 100 mm *l* and c. 3–5 mm *b*, upper leaf surface flattened, connate only at the base for c. 1/10 of its length, basal part developing into a sclerotic jug-like persistent sheath at the end of the season, apical opening of the terminal sheath closed by a papery membrane originating from the upper parts of the 2nd leaf-pair, epidermis of both leaf types mesomorphic with bladder cells varying in shape and size and to 0.5 mm *l*, 0.3 mm *b*, 0.25 mm *h*, sometimes with a spike-like protrusion, rarely bladder cells inconspicuous; **Fl** solitary, terminal, emerging from the 2nd leaf-pair towards the end of the season, pedicellate, pedicels to 100 mm *l*, papillate, ∅ to 45 mm, **K** 5, spirally arranged, petals and filamentous staminodes white, yellow, salmon, rose-pink or purplish-red, filaments sometimes yellow, orange or purplish-red, inner filaments subbasally papillate, holonectary, VII–VIII; **Fr** expanding keels merging into expanding sheets, valve wings broad, covering membranes 1/2 to 3/4 of the locule, closing bodies absent, locules 5(–6), ∅ 5–13 mm; **Chr** $2n = 18$, 36; **S** testa usually smooth and dirty-white, occasionally papillate and brown; **Ecol** mostly in quartzitic flats, < 200 mm rainfall p.a. mainly in winter; **Distr** Namaqualand, NC, Vanrhynsdorp, Vredendal, WC, S Africa.

Note: All species exhibiting non-succulent (unthickened) stems, and/or deviating leaf sequences, formerly included in **Monilaria**, have been transferred to **Dicrocaulon, Meyerophytum**, or **Diplosoma**, respectively. In principle, all species of the genus **Monilaria** are allopatric excepting **Mo. pisiformis** which is sympatric with **Mo. chrysoleuca** and **Mo. moniliformis**, but usually occurring in separate populations.

Distribution of **Monilaria** (Schwantes) Schwantes

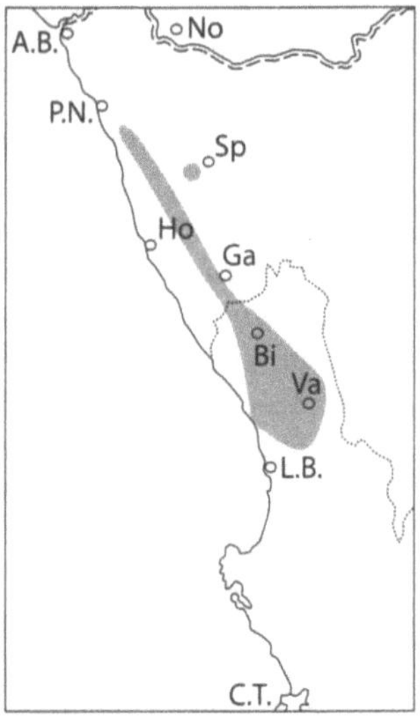

Key to the Species of Monilaria

1. Sclerotic sheaths campanulate (jug-shaped) or doliform (barrel-shaped), shape of terminal sheath during the resting state viewed from above ± circular or irregularly angled (but never rectangular!) 3
– Sclerotic sheaths flattened in one direction, therefore shape of terminal sheath during the resting state viewed from above rectangular 2
2. Sclerotic sheaths in side-view nearly rectangular or obovate, rims of old sclerotic sheaths tightly pressed to the stem .... **Mo. scutata**
– Sclerotic sheaths in side-view obconical, rims of old sclerotic sheaths spreading .... **Mo. obconica**
3. Terminal sclerotic sheath comparatively soft and short, covering the corpuscle of the 1st leaf-pair of the next season only incompletely .... **Mo. moniliformis**
– Terminal sclerotic sheath tough, covering the corpuscle of the 1st leaf-pair of the next season completely ... 4
4. Stems slender and crowded, ∅ c. 5–9 mm, sclerotic sheaths 4–9 mm *l*, 4–8 mm b .... **Mo. pisiformis**
– Stems strong and not crowded, ∅ to 15 mm, sclerotic sheaths 5–13 mm *l*, 6–10 mm b ... **Mo. chrysoleuca**

Mo. brevifolia ≡ **Dicrocaulon grandiflorum**

Mo. chrysoleuca (Schlechter) Schwantes GW 1929: 69 (see **photo**) ≡ *M. chrysoleucum* Schlechter 1899: 126 ≡ *Mitrophyllum chrysoleucum* (Schlechter) Schwantes ZSK 1926: 182 ≡ *Conophyllum chrysoleucum* (Schlechter) Schwantes ZSK 1928: 322 ≡ *Schwantesia chrysoleuca* (Schlechter) L.Bolus nomen illeg. SAG 1928: 279 **T** Schlechter 8292 (B!) = *Mo. chrysoleuca* var. *polita* (L.Bolus) Ihlenfeldt & S.Jörgensen 1973: 87 ≡ *Mo. polita* L.Bolus NM 1950: 224 **T** Nel 11902 (BOL!) = *Mo. chrysoleuca* (Schlechter) Schwantes var. *purpurea* L.Bolus NM 1950: 224 ≡ *Mo. chrysoleuca* (Schlechter) Schwantes forma *purpurea* (L.Bolus) G.D.Rowley in Jacobsen & Rowley 1973: 6 **T** Peers BOL 24225 (BOL!) = *Mo. salmonea* L.Bolus NM 1932: 355 **T** Pillans 6206 (BOL!).

Sclerotic sheaths campanulate or doliform, terminal sheath during the resting state viewed from above ± circular or irregularly angled, rim of expanded sheath often irregularly lacerate, sheaths 5–13 mm *l*, 6–10 mm *b*, **I** ∅ 10–20 mm; **Fl** petals white, yellow, orange, salmon, red, or purple, filaments sometimes in a different colour, even if petals are pure white; **Fr** ∅ 6–13 mm; **Chr** $2n = 18$; **Ecol** in quartzitic flats; **Distr** Vanrhynsdorp, WC, S Africa.

Note: In this species, the colour of the flowers varies even within a population. Some populations with very inconspicuous bladder cells on the 2nd leaf-pair and with a continuous, yet very restricted distribution area, are distinguished at the rank of variety, *Mo. chrysoleuca* var. *polita*.

Mo. chrysoleuca var. *polita* = **Mo. chrysoleuca**
Mo. chrysoleuca forma *purpurea* = **Mo. chrysoleuca**
Mo. chrysoleuca var. *purpurea* = **Mo. chrysoleuca**
Mo. globosa ≡ **Meyerophytum globosum**
Mo. luckhoffii ≡ **Diplosoma luckhoffii**
Mo. microstigma ≡ **Dicrocaulon microstigma**

Mo. moniliformis (Thunberg) Ihlenfeldt & S.Jörgensen 1973: 87 (see **photo**) ≡ *M. moniliforme* Thunberg 1791: 7 **T** Thunberg 11996 (UPS!) = *M. moniliforme* Haw. 1795: 132 nomen illeg. ≡ *Mitrophyllum moniliforme* (Haw.) Schwantes ZSK 1926: 182 nomen illeg. ≡ *Mo. moniliforme* (Haw.) Schwantes GW 1929: 69 nomen illeg. ≡ *Conophyllum moniliforme* (Haw.) Schwantes ZSK 1938: 322 nomen illeg. ≡ *Schwantesia moniliformis* (Haw.) L.Bolus nomen illeg. SAG 1928: 279 **T?** = *Mo. peersii* L.Bolus NM 1937: 58 **T** Peers 869/36 (BOL!).

In old plants **I** often ∅ > *l*, ∅ 10–20 mm, terminal sheath during the resting state viewed from above ± circular or irregularly angled, comparatively soft and short, covering the corpuscle of the first leaf-pair of the next season only incompletely, rim of expanded sheath often irregularly lacerate, sheaths 5–13 mm *l*, 6–10 mm *b*; **Fl** petals usually white, filaments white, orange or purple; **Fr** ∅ 6–13 mm; **Chr** $2n = 18$; **Ecol** mostly in quartzitic flats; **Distr** Vanrhynsdorp, Vredendal, WC, S Africa.

Note: **Mo. moniliformis** grows in the southernmost and thus least arid part of the distribution area of the genus. Haworth based his description on plants he had seen in two collections. No type was deposited, but the identity of Thunberg's and Haworth's taxa is doubtless.

Mo. nodosa ≡ **Dicrocaulon nodosum**

Mo. obconica Ihlenf. & S.Jörgensen 1973: 87 (see **photo**) **T** Ihlenfeldt & Hartmann 4538 (HBG!).

Terminal sheath during the resting state viewed from above rectangular, sheaths in side-view obconical, 5–11 mm *l*, 6–10 mm *b*, rim of old sheaths spreading, sometimes slightly lacerate, **I** ∅ 10–20 mm; **Fl** petals usually white, sometimes suffused with red; **Fr** ∅ 7–11 mm; **Chr** $2n = 36$; **Ecol** > 1 000 m altitude; **Distr** C Namaqualand, NC, S Africa.

Note: The distribution area of the tetraploid **Mo. obconica** is clearly separated from the rest and, being situated about 1 000 m above sea level, ecologically different from the areas of the remaining lowland species.

M. peersii = **Mo. moniliformis**

Mo. pisiformis (Haw.) Schwantes GW 1929: 69 (see **photo**) ≡ *M. pisiforme* Haw. 1803: 23 ≡ *Mitrophyllum pisiforme* (Haw.) Schwantes ZSK 1926: 182 ≡ *Conophyllum pisiforme* (Haw.) Schwantes ZSK 1928: 322 ≡ *Schwantesia pisiformis* (Haw.) L.Bolus nomen illeg. SAG 1928: 279 **NT** Littlewood 291/60 (BOL!).

In all parts smaller than the other species, to 12 cm *h*, stems crowded and only moderately thickened, sclerotic sheaths campanulate or doliform, terminal sheath during the resting state viewed from above ± circular or irregularly angled, rim of expanded sheath usually irregularly lacerate, sheaths 4–9 mm *l*, 4–8 mm *b*, **I** ∅ 8–12 mm; **L** tip of the 1st leaf-pair obscure or lacking; **Fl** petals white, yellow, orange, salmon, red or purple, filaments often different in colour, even if petals are pure white, ∅ 25–35 mm; **Fr** ∅ 5–9 mm; **Chr** $2n = 18$; **Ecol** in quartzitic flats; **Distr** Vanrhynsdorp, WC, S Africa.

Note: Though usually occurring in separate populations, hybrid-populations between **Mo. pisiformis** and the sympatrically occurring **Mo. chrysoleuca** have been recorded (Ihlenfeldt & Jörgensen 1973: 70).

Mo. polita ≡ **Mo. chrysoleuca**
Mo. primosii = **Dicrocaulon microstigma**
Mo. ramulosa ≡ **Dicrocaulon ramulosum**
Mo. salmonea = **Mo. chrysoleuca**

Mo. scutata (L.Bolus) Schwantes GW 1929: 69 ≡ *M. scutatum* L.Bolus ABH 1927: 73 ≡ *Mitrophyllum scutatum* (L.Bolus) Schwantes ZSK 1927: 105 ≡ *Conophyllum scutatum* (L.Bolus) Schwantes ZSK 1928: 322 ≡ *Schwantesia scutata* (L.Bolus) L.Bolus nomen illeg. SAG 1928: 322 T Pillans BOL 17794 (BOL!).

Terminal sheath during the resting state viewed from above rectangular, sheaths in side-view rectangular or obovate, 5–12 mm *l*, 6–12 mm *b*, rim of old sheaths pressed to the to the stem, **I** ∅ 10–20 mm; **Fl** petals usually white, filaments sometimes orange; **Fr** ∅ 7–11 mm; **Chr** $2n = 18$; **Ecol** in shallow soil with a surface layer of quartz pebbles; **Distr** CW, SW Namaqualand, NC, S Africa.

Note: In contrast to the other lowland species, **Mo. scutata** only occurs in scattered, often very small populations.

Key to the Subspecies of Mo. scutata

1. Sclerotic sheaths in side-view nearly rectangular, constrictions on the stem indistinct . **Mo. scutata** ssp. **scutata**
- Sclerotic sheaths in side-view obovate, constrictions on the stem distinct .... **Mo. scutata** ssp. **obovata**

Mo. scutata (L.Bolus) Schwantes ssp. **obovata** Ihlenf. & S.Jörgensen 1973: 88 (see **photo**) T Hall s.n. (BOL!).

Stems with regular, distinct constrictions, sheaths very hard, obovate in side view, 5–11 mm *l*, 6–12 mm *b*; **Distr** CW Namaqualand, NC, S Africa.

Note: **Mo. scutata** ssp. **obovata** possesses the hardest sheaths of the genus.

Mo. scutata (L.Bolus) Schwantes ssp. **scutata** (see **photo**) = *Mo. vestita* Schwantes ex H.Jacobsen 1933: 163 nomen nudum.

Stems with regular, rather indistinct constrictions, sheaths 5–12 mm *l*, 5–12 mm *b*; **Distr** SW Namaqualand, NC, S Africa.

Note: **Mo. scutata** ssp. **scutata** is remarkable for the often deep black stems, looking burnt.

Mo. vestita
= **Mo. scutata** ssp. **scutata**

Mo. watermeyeri
= **Dicrocaulon microstigma**

Mossia Ruschioideae

H. E. K. Hartmann

Mossia N.E.Br. GC 1930: 71 (25.1.1930); Brown 1930: 151 (22.2.1930); Smith & Hartzer 1997: 349; Brummit 1999: 368 **Typus Ms. intervallaris** (L.Bolus) N.E.Br. **Etym** *C.E.Moss*, 1870–1930, professor and South African plant collector.

Creeping perennials with numerous adventitious roots, short shoots with several leaf-pairs; **L** shortly trigonous, dark green to purple, epidermis cells a little elevated, the stomata mostly superficial, subsidiary cells raised, outer wall moderately thick with a layer of crystal sand in the middle, wax layer continuous with a dense cover of vertically arranged platelets; **Fl** solitary on short shoots, bracteolate, **K** 5, petals cream-coloured, spreading in the evening and remaining open at night, nectary in 5 lophomorphic groups placed at the top of a hypanthium where also the bases of the elements of the androecium are situated; **Fr** Delosperma-type, but with the expanding keels diverging subapically on the valve, the valve wings broadening from a narrow base, covering membranes broadest in the centre, narrowing towards the outer rim of the capsule, closing body absent or as a tiny elevation, most inconspicuous in some populations, 5 locules; **S** light brown, nearly round; **Ecol** in shallow pans on rocks, often reaching far over the rock surface, at altitudes around 1600 m, 300–500 mm rainfall p.a. in summer; **Distr** W Lesotho; E FS, GAU, MP, NE EC, S Africa.

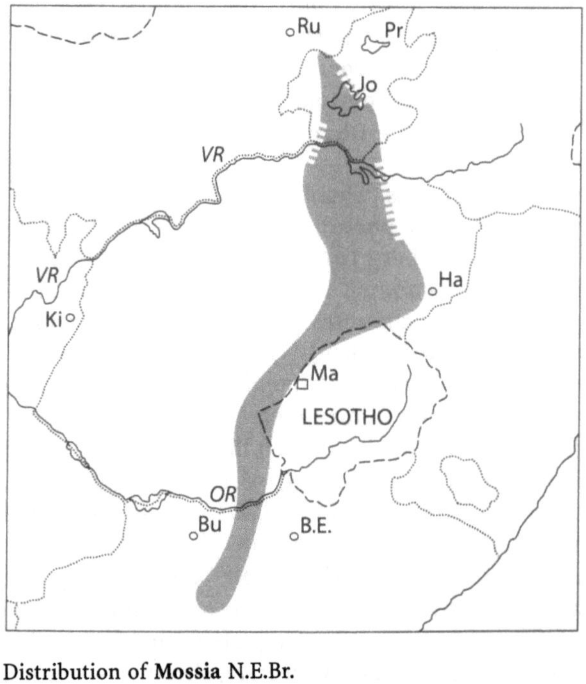

Distribution of **Mossia** N.E.Br.

Note: With his first description of the genus, Brown (25.1.1930) had cited *Ruschia intervallaris* erroneously as the basionym of the type species, a mistake he corrected less than a month later (22.2.1930), giving as the basionym of his type species *Mesembryanthemum intervallaris*. Smith & Hartzer (1997: 349) proposed to conserve the younger name against the older homonym. The decision of the Committee for Spermatophyta (Brummitt 1999: 368) in this unusual case was to suggest that the General Committee should give a statement that Brown's error of 25.1.1930 should be treated as a "correctable bibliographic error" and that "Mossia of 25 Jan. 1930 should then be treated as typified by *M. intervallare*". This view is followed here.

Although the genus is restricted to distinct ecological niches it occurs over a wide area in a broad corridor from c. 25°30' S to 31°45' S and between 26°15' and almost 30° E.

Ms. intervallaris (L.Bolus) N.E.Br. GC 1930: 71 (see **photo**); Smith et al. 1997: t. 310 ≡ *M. intervallare* L.Bolus ABH 1927: 81 ≡ *Ruschia intervallaris* N.E.Br. 1930: 71, nomen nudum, non **Ruschia intervallaris** L.Bolus **Syntypes** NBG 1430/25, Hatting 2191/24 (BOL).

Creeper extending over several square metres, **I** < **L** rarely **I** > **L**; **L** trigonous, sometimes almost crescent-shaped, 7–10 mm *l*, 6–7 mm *b* and *t*, turning rather soft and larger in wet conditions; **Fl** c. 14 mm ∅; **Ecol** submerged under water after rain, often for several days, completely dry for most of the time; **Distr** as for genus.

Note: The opening of the flower depends on low light intensity and high temperatures: flowers on branches lying on warm rocks open distinctly earlier than those placed on soil in the pan. Typical of a night flower are the hypanthium, the yellowish petals, and the odour of vanilla. Under less favourable conditions, the flowers are also self-fertile.

The citation of *Ruschia intervallaris* by Brown 1930: 71 was an error, as explained by Brown 1930: 151.

Muiria Ruschioideae

H. E. K. Hartmann

Muiria N.E.Br. GC 1927: 116; **Typus Mu. hortenseae** N.E.Br. **Etym** *Dr. J. Muir*, 1874–1947, South African plant collector.

Compact perennials with several short branches consisting of a single leaf-pair; **L** of a pair united for most of their length, leaving only a tiny apical slit open, epidermis even, but each cell elongated into a long, thin hair, outer wall not very thick, the body turning into a drying sheath towards the summer, enclosing the new developing body inside, the dry sheath splitting during March (in habitat) when the rains are starting; **Fl** solitary, at the top of the body, **K** 6, petals decreasing in width and length from the outer series, white or pale pink, stamens epapillate, nectary of 6–7 lophomorphic glands; **Fr** Delosperma-type, top raised from high valve rims, covering membranes very broad, rectangular wings, basally connate forming pockets, the septa completely without rims, mostly 6 locules; **S** brown, smooth, elevated cells around hilum and micropyle; **Ecol** in quartz flats, c. 200 mm rainfall p.a. mainly in March and November; **Distr** Riversdale, WC, S Africa.

Distribution of **Muiria** N.E.Br.

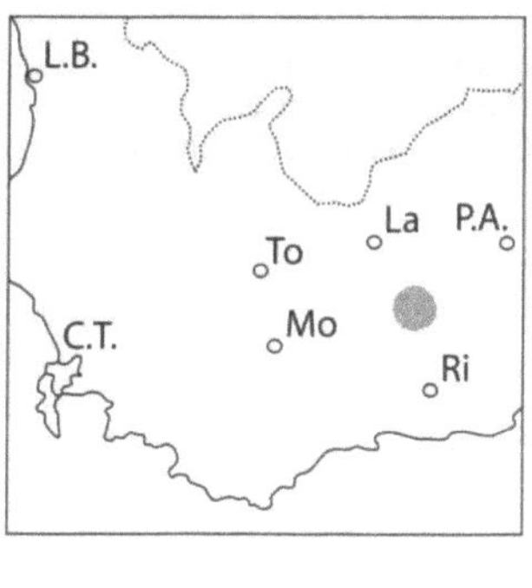

Note: Although the genus is restricted to a small area, it is well known in cultivation because it grows easily from seed. Like all plants with dry sheaths protecting the growth during the hot season, the adult plants tend to mould in wet situations not rare in cultivation.

Mu. hortenseae N.E.Br. GC 1927: 116 (see **photo**) ≡ *M. hortenseae* (N.E.Br.) hort. Kew ex Hill 1931: 128 **T** Muir 3885 (K).

Compact with 1–7 branches, each with a single body formed by the two connate leaves of the pair; **L** pair globose to ovate; **Fl** the gynoecium forming a roundish, balloon-shaped body above the leaf-pair; **Fr** also on top of the plant, the base of the flower having turned into a soft, fat fruit base ensuring the correct position of the capsule for opening; **Chr** $2n = 18$; **Ecol** as for genus; **Distr** as for genus.

Note: Brown (1927) saw similarities between **Mu. hortenseae** and **Oophytum**, which is a much smaller plant, bearing, however, the capsules in a similar way. Similar is also **Phyllobolus digitatus** in the resting state, when the fruits are borne terminally ready for expansion after rain. All three species represent examples of an adaptation in which protection is reached by a cover derived from the previous leaf-pair, a feature widespread in **Conophytum**. In very dry conditions, the plants can be observed growing sunken to almost half their height into the ground, sitting in the remaining bases of partly decayed leaf-pairs.

Namaquanthus Ruschioideae

H. E. K. Hartmann

Namaquanthus L.Bolus NM 1954: 257 **Typus N. vanheerdei** L.Bolus **Etym** *Namaqualand,* area in which the genus occurs; Gk *anthos,* flower.

Succulent shrubs to 30 cm *h*, **I** < **L**; **L** finger-shaped, dark green, later turning black (from a fungus?), those of a pair almost forming a U, epidermal cells a little elevated, stomata not sunken, but epidermis xeromorphic, wax cover continuous, breaking into horizontal plates, crystals in outer epidermal wall c. 2 µm ∅; **Fl** 1(–3), bracteolate, 4 or 6 calyx lobes, petals, filamentous staminodes and filaments deep purple, epapillate, anthers and pollen dark purple, nectary ring lophomorphic, light coloured; **Fr** stalks erect, modified Lampranthus-type: covering membranes with a radial spongy thickening, no closing devices, valve wings a little broader than the diverging expanding keels; **S** echinate, i.e. with long papillae along the back and short ones on the sides, brown; **Ecol** on rocky slopes facing S, < 100 mm rainfall p.a. in winter; **Distr** Namaqualand, NC, S Africa.

Note: After the removal of *N. farinosus,* only one species remains in the genus. The strikingly bright purple flowers resemble those of **Cheiridopsis speciosa** and **Ch. purpurea** in their very slow opening at anthesis: the inner petals unfold only after several days, hiding the interior in the beginning.

N. farinosus ≡ **Wooleya farinosa**

N. vanheerdei L.Bolus NM 1954: 257 (see **photo**) **T** van Heerde 25193 (BOL).

Somewhat loosely branched, to 30 cm *h*, ∅ to 60 cm; **L** c. 50 mm *l*, ∅ c. 12 mm, dark green turning black with age; **Fl** 120–135 purple petals, 0–15 filamentous staminodes, 130–150 stamens, nectary ring whitish, yellowish or pink; **Fr** top a little raised, base rounded to almost bell-shaped, (8–)12(–17) locules, 18–30 mm *l*, 12–15 mm ∅; **S** 1.4–1.7 mm *l*, 0.9–1.1 mm *b*; **Ecol** in crevices of quartzitic rocks; **Distr** as for genus.

Note: The black cover, reported already by van Heerde as being perhaps a fungus, was also visible when the plants were revisited 20 and 35 years later.

Distribution of **Namaquanthus** L.Bolus

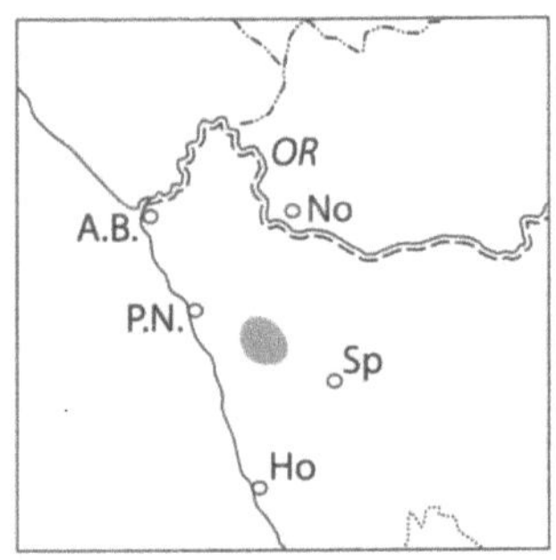

Namibia Ruschioideae

H. E. K. Hartmann

Namibia (Schwantes) Dinter & Schwantes ZSK 1927: 106 ≡ *Juttadinteria* Schwantes subg. *Namibia* Schwantes ZSK 1926: 184; Walgate NM 1939: 186–187; Friedrich 1970: 58–64 **Typus** *Juttadinteria cinerea* (Marloth) Schwantes ≡ **N. cinerea** (Marloth) Dinter & Schwantes **Etym** *Namib,* coastal desert in SW Namibia, where the genus occurs as an endemic.

Compact, forming a ± semi-globose body with up to 30 branches, each with 2–3 leaf-pairs; **L** fat, trigonous with bulging sides, epidermis with ± long papillae tapering to their tips, surface thus velvety to the touch to different degrees, 20–30 mm *l*, 15–20 mm *b* and *t*, wax cover in smooth grooves along the sides and on top of the papillae, but most of the surface covered densely by platelets, rodlets rare, outer epidermal wall with crystal sand in a thick layer; **Fl** solitary, ebracteolate, **K** (4–)5(–6), petals white to pink, stamens erect in a central column, nectary as a prominent, dark green lophomorphic ring; **Fr** of the Dracophilus-type, top raised in the centre, but base rounded to semi-globose, mostly with long papillae, covering membranes completely absent, expanding keels strictly radial and closely parallel to each other on the valve, 12–25 locules; **S** ovate to elongate pear-shaped; **Ecol** in crevices of dolomite or on slopes of various rocks, always very exposed, < 100 mm rainfall in winter, fog; **Distr** Lüderitz-South, Namibia.

Note: Like **Dracophilus**, **Namibia** has originally been united with **Juttadinteria**, from which it differs in growth, leaf shape, and epidermis, as well as in flower and fruit morphology.

Distribution of **Namibia** Dinter & Schwantes ex Schwantes

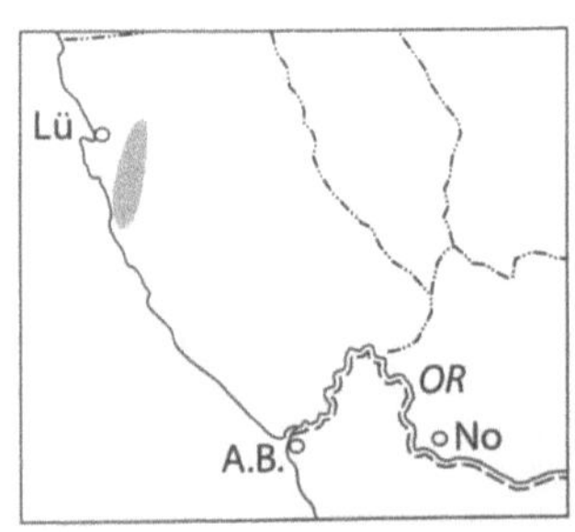

Key to the Species of Namibia

1. Locules 20–25 .... **N. pomonae**
- Locules 9–14, mostly 12 .... **N. cinerea**

N. cinerea (Marloth) Dinter & Schwantes ZSK 1927: 106 (see **photo**) ≡ *Juttadinteria cinerea* (Marloth) Schwantes ZSK 1926: 184 ≡ *M. cinereum* Marloth 1910: 407 **T** Marloth 4680 (BOL!) = *Juttadinteria longipetala* L.Bolus NM 1928: 12 **T** Pillans 5951 (BOL!) = *N. ponderosa* (Dinter) Dinter & Schwantes ex H.Jacobsen 1935: 222 ≡ *M. ponderosum* Dinter 1928: 103[17] **T** Dinter 3747.

L mostly markedly velvety, rarely smoother, keels and edges only prominent in desiccated state; **Fl** ∅ 45–65 mm, petals 50–114, mostly white, in few populations pink, sometimes with a white base, 260–460 stamens; **Fr** 8–14 locules, with a maximum at 12 in most populations, base always rounded to semi-globose; **Ecol** as for genus; **Distr** as for genus.

Note: Pink flowering forms have been treated as *N. ponderosa* or *Juttadinteria longipetala*. Walgate (1939: 175–176) stated already that the distinctions appear insufficient. Examination of the type and comparison with living material support the inclusion of *J. longipetala* in **N. cinerea.** Nevertheless, northern and southern populations differ a little: plants in the north prefer gneissic rocks, appear greener in colour in winter, and show keels more often; their flowers are open between 2 and 4.30 p.m. (Williamson, pers. comm.). In contrast, plants in the south grow in crevices of dolomite, often more compact than the northern ones, retain their grey colour in winter and appear unkeeled most of the time; their flowers remain open permanently (Williamson, pers. comm.).

N. pomonae (Dinter) Dinter & Schwantes ex Walgate[18] NM 1939: 187 ≡ *Juttadinteria pomonae* (Dinter) Schwantes ZSK 1926: 184 nomen invalid ≡ *M. pomonae* Dinter FR 1923: 125 **T** Dinter 3787 (?).

Cushions of up to 20 cm ∅, 10 cm *h*; **L** with more rounded papillae than **N. cinerea**, surface therefore less velvety to the touch; **Fl** ∅ 37 mm, petals white; **Fr** 25 locules; **Ecol** unknown **Distr** Lüderitz-South, Namibia.

Note: The species is insufficiently known since no type specimen can be found. Dinter (1923) noted that plants of **N. pomonae** occurred dispersed amongst those of **N. cinerea**, and in small number. It is undoubtedly difficult, at this state of knowledge, to verify whether a distinct species is involved or not. Certain remarks by Dinter (1923) in his original description cast some doubt even on the present placement of the species in **Namibia**. The species is unique in its high number of locules.

N. ponderosa = **N. cinerea**

Nananthus Ruschioideae

H. E. K. Hartmann

Nananthus N.E.Br. GC 1925: 433 **Typus N. vittatus** (N.E.Br.) Schwantes **Etym** Gk *nanos*, small, *anthos*, flower, "in allusion to the dwarf nature of these plants" (Brown GC 1931: 92).

Compact with a turnip-shaped, thick root of soft, succulent nature, edible, resembling fresh asparagus in taste, the root branching at the top, mostly 4 leaves to a branch; **L** erect or spreading to lying on the ground, upper leaf surface ± flat, linear to triangular, lower surface apically humped, epidermis with elevations above tannin idioblasts, thus rough to the touch from warts like leaves of **Aloinopsis** s.str., calcium oxalate crystals in a thin layer in the cells of the elevations giving the leaf a white-spotted appearance, crystals absent in the valleys, where the surface is dark green, wax in horizontal plates sometimes broken, only few amorphous particles on top; **Fl** solitary, bracteoles at base, almost as big as foliage leaves, **K** 5, petals yellow and red vittate to purplish-red and darker vittate, sometimes purely yellow, stamens always in a distinct, rather flat central cone, filaments white with white

[17] Walgate (1939) considered the publication as invalid. Yet, Dinter gave a name, a collection number and contrasted his find, *M. ponderosum*, with *M. cinerea* and *M. pomonae*, the two other species he places in his *Cinereum* group. *M. ponderosum* differs from *M. cinereum* by a smoother leaf surface and "red flowers" (rote Blüten), probably describing a darker pink and thus similar to *Juttadinteria longipetala*.

[18] The first mentioning of **Namibia** as a genus was in the description of the genus **Juttadinteria**, which was subdivided into three subgenera: **Juttadinteria, Dracophilus** and **Namibia.** Schwantes suggested that **Namibia** may perhaps be considered as a genus, with the species **N. cinerea** and **N. pomonae.** This is not a valid publication of the names because neither the genus nor the species were accepted here by the author. **N. cinerea** was validated some months later, as the type species of the therewith accepted genus **Namibia**, but **N. pomonae** was not mentioned. This was done for the first time by Walgate, as cited here.

Distribution of **Nananthus** N.E.Br.

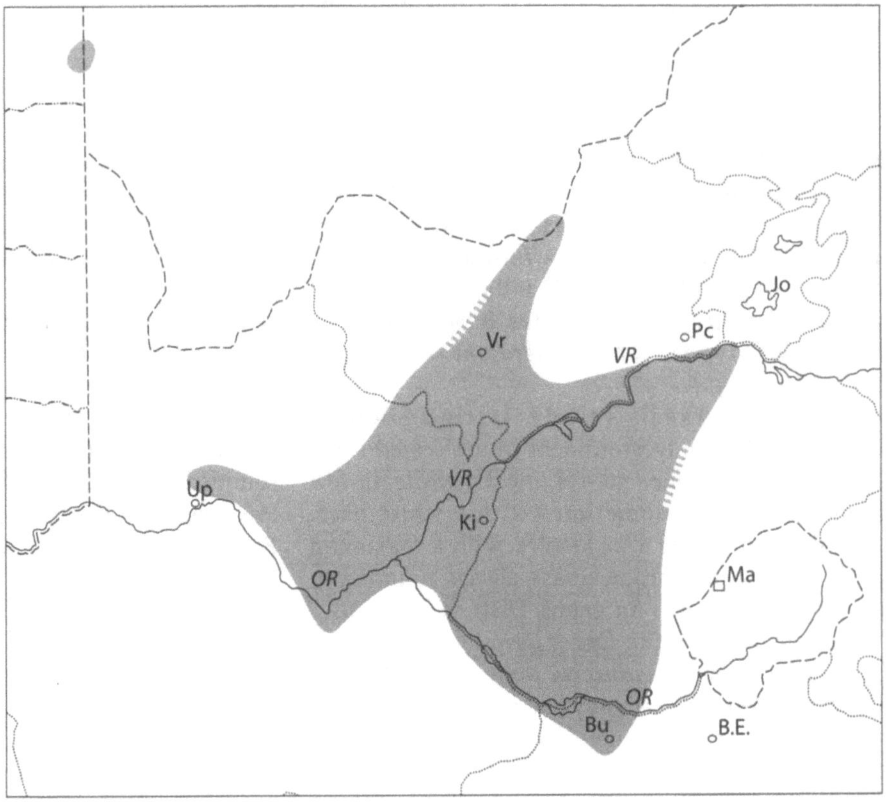

papillae near the base, nectary as a dark green ring, anthers mostly yellow; **Fr** breaking off the stalk and distributed as a tumble fruit, top flat to conically raised, rarely rounded, base semi-globular, interior of the Delosperma-type: covering membranes absent or as very narrow flaps, expanding keels diverging only a little at their tips, valve wings rectangular and mostly much broader than the expanding keels, columella very slender; **S** almost orbicular, the micropyle protruding, flat and big, light to dark brown, testa smooth, only the micropylar region with cushion-shaped cells adorned with amorphous epicuticular particles, 1.4–2 mm *l*, 1–1.6 mm *b*; **Chr** $2n = 18$; **Ecol** in plains with fine loam or on gentle slopes, often in soils with high lime contents, mostly > 300 mm rainfall p.a., mainly in summer; **Distr** SE Botswana; E Namibia; NE of NC, S of NW, FS, S Africa.

Note: Much confusion has accompanied the history of this genus, starting with a mistake in the original description, where covering membranes are said to be present whereas they are absent or only visible as very narrow ledges.

The distinction of species is particularly difficult in cultivated material which tends to develop longer and more slender leaves with less prominent warts than plants in habitat. Nevertheless, these latter can also vary considerably in size, depending on water regime. The six species recognized here differ mainly in leaf shapes and in some fruit characters.

Key to the Species of Nananthus

1. Leaves mostly erect, apparently almost as thick as broad 2
– Leaves spreading to different degrees, apparently flatter 3
2. Leaves without an apical tooth .... **N. margaretiferus**
– Leaves with an apical tooth at right angle to the length of the leaf .... **N. pole-evansii**
3. Leaves with the central longitudinal groove bordered by two lateral ledges .... **N. aloides**
– Leaves flat or concave above 4
4. Upper leaf surface ± flat, warts low .... **N. pallens**
– Upper leaf surface concave, warts in the upper parts prominent 5
5. Valve wings 2–3 times as broad as the expanding keels .... **N. vittatus**
– Valve wings 4 times as broad as the expanding keels .... **N. gerstneri**

N. acutus ≡ **Aloinopsis acuta**
N. albinotus ≡ **Rabiea albinota**
N. albipunctus ≡ **Rabiea albipuncta**

N. aloides (Haw.) Schwantes GF 1928: 68 (see **photo**) ≡ *M. aloides* Haw. 1819: 88 ≡ *Aloinopsis aloides* (Haw.) Schwantes ZSK 1926: 178 ≡ *N. aloides* (Haw.) N.E.Br. GC 1931: 133 **Lectotype**, here designated: unpublished drawing at K[19] (K) = *N. aloides* (Haw.) Schwantes var. *latus* L.Bolus NM 1958: 381 T Wilman (BOL) = *N. wilmaniae* (L.Bolus) L.Bolus NM 1938: 134 ≡ *Aloinopsis wilmaniae* L.Bolus NM 1933: 397 **T** BOL 23760 (BOL!).

L spreading widely so that only the upper leaf surfaces are visible, forming a visible triangle of almost equal sides of c. 10 mm width, the central longitudinal groove is bordered by two lateral ledges in the upper parts of the leaf, 5–7 mm *t*; **Fl** petals yellow with or without a central red stripe, even in one population; **Fr** ∅ 8.5 mm, capsules 6.7 mm *h*; **S** 1.5 mm *l*, 1.2 mm *b*; **Ecol** mostly at the edge of pans in finely decomposed limestone, the plants often sunken into the ground, 300–500 mm rainfall p.a., mainly in summer, plants then often flooded; **Distr** SE corner of Botswana; E part of NC, NW, S Africa.

Note: As already stated by Brown (e.g. 1920: 124), **N. aloides** has often been confused, mainly with the more slender-leaved **A. vittatus** in the area south and east of that of **A. aloides.** This confusion is due to the development of more narrow leaves in darker and wetter cultivation, but in habitat **N. aloides** can be recognized by its flat-lying, very broad leaves with ledges along the groove, a feature regularly absent in the other species.

N. aloides var. *latus* = **N. aloides**
N. aloides var. *striatus* = **N. vittatus**
N. broomii = **N. vittatus**
N. cibdelus = **N. vittatus**
N. comptonii ≡ **Rabiea comptonii**
N. cradockensis = **Aloinopsis rubrolineata**
N. crassipes = **Aloinopsis spathulata**
N. difformis ≡ **Rabiea difformis**
N. dyeri = **Aloinopsis rubrolineata**

N. gerstneri (L.Bolus) L.Bolus NM 1938: 134 (see **photo**) ≡ *Aloinopsis gerstneri* L.Bolus NM 1937: 46 **T** Gerstner 174/34 (BOL!).

L spreading, broadest at their middle, c. 35 mm *l*, 5–6 mm *b*, c. 4 mm ∅; **Fl** yellow with red longitudinal stripes; **Fr** valve wings very broad and overlapping in the open capsule; **Ecol** in fine soil in washes mainly, also near Dolerite slopes; **Distr** Aliwal North, Burgersdorp, EC, SE districts of FS, S Africa.

Note: In contrast to all other species in the genus, **N. gerstneri** seems not to prefer calcareous soils and rocks.

N. hilmarii ≡ **Deilanthe hilmarii**
N. jamesii ≡ **Rabiea jamesii;**
= **Aloinopsis rubrolineata**
N. lesliei ≡ **Rabiea lesliei**
N. lodewykii = **Aloinopsis luckhoffii**
N. loganii ≡ **Aloinopsis loganii**
N. luckhoffii ≡ **Aloinopsis luckhoffii**
N. malherbei ≡ **Aloinopsis malherbei**

N. margaretiferus L.Bolus JSAB 1960: 55 (see **photo**) **T** Giess 446/59 (BOL!).

L erect, trigonous, with a rounded acuminate tip, grey from white warts, to 30 mm *l*, to 8 mm *b*, to 7 mm *t*; **Fl** petals yellow with darker yellow stripes, pink on the outside; **Fr** mostly 8 locules; **Ecol** in crevices of hard limestone ledges; **Distr** W Botswana; Gobabeb, Namibia; Hay, NC, S Africa.

Note: Plants in habitat resemble those of **Prepodesma orpenii** very closely in the trigonous leaf shape and the erect position of the leaves, a feature also found in **N. pole-evansii**, which has thinner leaves with a distinct distal awn.

N. pallens (L.Bolus) L.Bolus NM 1938: 134 (see **photo**) ≡ *Aloinopsis pallens* L.Bolus NM 1935: 474 **T** Helm 579/33 (BOL!).

L slender, but margins not quite parallel, spreading widely, greyish-green from many low warts, mucro inconspicuous, 20–30 mm *l*, 6–8 mm *b*, 4–7 mm *t*; **Fl** petals yellow with conspicuous red stripes; **Fr** ∅ 9.5 mm, capsule 6.25 mm *h*; **S** 1.95 mm *l*, 1.6 mm *b*; **Ecol** in grassland with scrub, on loamy soils with a noticeable portion of pebbles, limestones among them, 200–300 mm rainfall p.a. mainly in summer; **Distr** Gordonia, Hay, Postmasburg, NC, S Africa.

Note: **N. pallens** appears to possess the biggest seeds in the genus.

N. peersii ≡ **Deilanthe peersii**

N. pole-evansii N.E.Br. GC 1931: 133 (see **photo**) **T** Pole-Evans 6440 (K!).

L erect, with a pronounced apical hump and an incurved leaf tip with an awn pointing inwards, the tip often laterally indented as if pressed together by fingers, greyish-green from prominent warts, 14–30 mm *l*, 5–8 mm *b*, 2.5–5 mm *t*; **Fl** petals yellow

[19] The two collections by Burchell (2197, 2249-8) on which the name had been based are neither at K nor at the Asa Gray Herbarium, as supposed by Brown (1920: 123–124). The drawing available at K is therefore chosen as the lectotype here.

with a deep red stripe, in some populations also purplish-red and darker vittate, rarely purely yellow; **Fr** 6.5–8 mm ∅, capsule 4.7–6.5 mm *h*; **Ecol** in open patches in grassland on exposed limestone rocks, also in limy rocks along pans or in brackish pans, 200–400 mm rainfall p.a. mainly in summer; **Distr** N parts of EC, FS, S Africa.

Note: **N. pole-evansii** possesses the thinnest leaves in the genus, it has been found in soils as brackish as pH 8.

N. orpenii ≡ **Prepodesma orpenii**
N. rosulatus ≡ **Aloinopsis rosulata**
N. rubrolineatus ≡ **Aloinopsis rubrolineata**
N. schoonesii ≡ **Aloinopsis schoonesii**
N. setiferus = **Aloinopsis luckhoffii**
N. soehlemannii = **Deilanthe peersii**
N. spathulatus ≡ **Aloinopsis spathulata**
N. tersus = **Prepodesma orpenii**
N. thudichumii ≡ **Deilanthe thudichumii**
N. transvaalensis = **Nananthus vittatus**
N. transvaalensis var. *griquensis* = **Nananthus vittatus**
N. transvaalensis var. *latus* = **Nananthus vittatus**
N. villetii = **Aloinopsis luckhoffii**

N. vittatus (N.E.Br.) Schwantes GF 1928: 68 (see **photo**) ≡ *M. vittatum* N.E.Br. GC 1876: 772 ≡ *N. vittatus* (N.E.Br.) N.E.Br. GC 1931: 92 nomen illeg. **T** Hutton s.n. (K) = *N. aloides* (Haw.) Schwantes var. *striatus* (L.Bolus) L.Bolus NM 1958: 385 ≡ *Aloinopsis aloides* (Haw.) Schwantes var. *striatus* L.Bolus SAG 1930: 48 **T** photograph in SAG 1920: 49 = *N. broomii* (L.Bolus) L.Bolus NM 1938: 133 ≡ *Aloinopsis broomii* L.Bolus NM 1935: 474 **T** Broom 2654/34 (BOL!) = *N. cibdelus* (N.E.Br.) Schwantes GF 1928: 68 ≡ *M. cibdelum* N.E.Br. 1920: 64 ≡ *Rabiea cibdela* (N.E.Br.) N.E.Br. GC 1931: 54 ≡ *Aloinopsis cibdela* (N.E.Br.) Schwantes ZSK 1926: 178 **LT** Salm-Dyck 1840: § 4 f 3[20] = *M. albipunctum* Haw. var. *majus* Haw. 1826: 127 ≡ *Rabiea albipuncta* (Haw.) N.E.Br. var. *major* (Haw.) L.Bolus NM 1958: 370 nomen nudum **Lectotype**, here designated: drawing at K! = *N. transvaalensis* (Rolfe) L.Bolus NM 1958: 382, 383[21] ≡ *M. transvaalense* Rolfe 1916: 8674 **A T** Burtt Davy s.n. (K) = *N. transvaalensis* Rolfe var. *griquensis* L.Bolus NM 1958: 381 nomen invalid. **T** Acocks s.n. (BOL) = *N. transvaalensis* Rolfe var. *latus* L.Bolus NM 1958: 381 nomen invalid. **T** Broom 2630/33 (BOL).

L ovate with an acuminate tip, spreading diagonally at most, often erect in habitat under dry conditions, often with a central channel, 25–30 mm *l*, 6–9 mm *b*, 3–4 mm *t*; **Fl** with a distinct red stripe; **Fr** ∅ 8–11 mm, capsules 6–8.5 mm *h*; **S** 1.4–1.95 mm *l*, 1.27–1.56 mm *b*; **Ecol** in grassveld often with some Acacia-bush, in often disturbed and overgrazed areas rich in lime stone, a high proportion of which can be found on the surface, also in loamy soils with a high lime contents, > 400 mm rainfall p.a. mainly in summer; **Distr** SE part of NC, SE part of NW, S Africa.

Note: A widespread species, often overlooked because of the sometimes dense vegetation cover in the localities. The placement of *M. transvaalensis* in synonymy of **N. vittatus** follows Brown (1931: 93). As described for **N. aloides**, the leaves can grow more slender in cultivation than in habitat, resembling those of **N. pallens**, whereas under dry conditions, in habitat, plants can resemble those of **N. pole-evansii**, all three from different areas and associations.

N. wilmaniae = **N. aloides**

Nelia Ruschioideae

H. E. K. Hartmann

Nelia Schwantes MDG 1928: 92 **Typus N. pillansii** (N.E.Br.) Schwantes **Etym** *G.C. Nel*, 1885–1950, professor of Botany at the University of Stellenbosch = *Sterropetalum* N.E.Br. GC 1928: 266 **Typus** *Sterropetalum pillansii* N.E.Br. ≡ **N. pillansii** (N.E.Br.) Schwantes.

Compact, up to 40 branches, mostly fewer, 2–3 leaf-pairs to a branch; **L** trigonous, yellowish to green, often black from a fungus, very smooth, wax as a layer with some platelets in places, stomata sunken to the bases of the subsidiary cells, epidermis with a thick outer epidermal wall filled with crystal sand, 15–50 mm *l*, 5–12 mm *b* and *t*; **Fl** 1–3 in cymes, with bracts and bracteoles, **K** 5, ∅ 14–22 mm, permanently open, the outer spreading petals at most as long as the calyx lobes, changing into erect filamentous staminodes inwards, inner ones bending over the gynoecium in the centre, all elements basally united

[20] The lectotype was designated by Stearn (1939: 73); this decision agrees well with the description, which is based on the drawing and on a herbarium specimen bearing the same name at K. The preserved plant, however, represents **N. pole-evansii** and can therefore not be taken into consideration for lectotypification.

[21] The publication of *N. transvaalensis* is invalid: Bolus used the combination without author (1958: 382) when reporting that the name had been sunken by Brown 1931: 89 (which was not the case, Brown only sunk *M. transvaalense*) in **N. vittatus**. Bolus added the name *M. transvaalense* and the literature citation only later in her discussion of the history of the species on p. 383 without giving an indication that a new combination was to be made there.

into a shallow hypanthium, stiff, white, later yellowish, nectary as a ring; **Fr** resembling fruits of **Dracophilus** with broad valve wings, no closing body, and reduced covering membranes, 5 locules, 4.5–8 mm ∅; **S** pear-shaped with a markedly protruding micropyle, light brown, smooth, only near the micropyle darker and rougher in surface sculpture, 1–1.4 mm *l*, 0.6–0.95 mm *b*; **Ecol** on quartzitic slopes or in flats, c. 125 mm winter rainfall p.a.; **Distr** N Namaqualand, NC, S Africa.

Note: The genus is remarkable for its permanently open flowers, arranged in cymes opening one after the other over several weeks. The plants grow and flower easily in cultivation. The genus was described without a species being named. This was valid at that date, especially so since only one collection was mentioned, implying that a single species was involved.

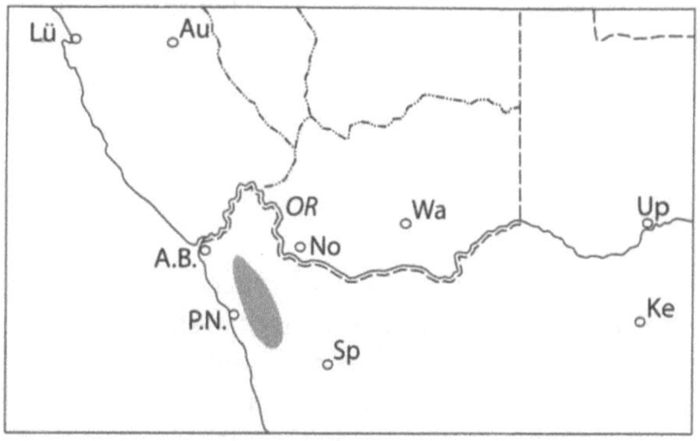

Distribution of **Nelia** Schwantes

Key to the Species of Nelia

1. Leaves narrowing gradually from the base to the tip, > 20 mm long, capsule ∅ 6–8 mm ...... **N. pillansii**
- Leaves widening subapically forming a triangle of equal sides at the tip, < 20 mm long, capsule ∅ c. 5 mm .... **N. schlechteri**

N. meyeri = **N. pillansii**
N. meyeri var. *longipetala* = **N. pillansii**

N. pillansii (N.E.Br.) Schwantes MDK 1929: 16 (see **photo**) ≡ *Sterropetalum pillansii* N.E.Br. GC 1928: 266 T Pillans 5705 (K!) ≡ *N. meyeri* Schwantes var. *longipetala* L.Bolus NM 1928: 151 T Pillans 5705 (BOL!) = *N. meyeri* Schwantes GF 1928: 129 **Lectotype**, here designated: Schwantes GF 1928: f on p 129 = *N. robusta* Schwantes 1929: 16 T Schlechter s.n. (HBG!).

L trigonous, ± acute at the tip, the upper side forming an elongate triangle, slender to fat, 20–50 mm *l*, 5–12 mm *b*; **Fl** in cymes of 1–3, developing one after the other, peduncles of varying length, bracteoles fat to slender; **Fl** with c. 280 petals merging into filamentous staminodes, c. 100 stamens; **Fr** variable in the lateral extension of the short, roundish covering membranes, those of a pair forming an oval, ∅ 6–8.5 mm; **S** 1–1.4 mm *l*, c. 0.65 mm *b*; **Ecol** as for genus; **Distr** as for genus.

Note: Species included as synonyms here have mainly been distinguished based on the number of flowers in the inflorescence and on sizes of leaves and bracteoles.

N. robusta = **N. pillansii**

N. schlechteri Schwantes 1929: 16 (see **photo**) T Schlechter s.n. (HBG!) = *Corpuscularia perdiantha* Tischer 1930: 20 **Lectotype**, here designated: Tischer 1930: f on p 20.

Semi-orbicular or lower; **L** trigonous, sides often oblique, the upper side basally with parallel sides, widened below the tip forming a triangle of equal sides above this point, keel and margins cartilaginous, undulate and minutely notched, < 20 mm *l*, c. 5 mm *b*; **Fl** always solitary, bracteoles immediately below base, 20–22 short, white petals merging into the inner filamentous staminodes, filaments green, anthers yellow; **Fr** without covering membranes or with narrow ledges only, ∅ < 5 mm; **S** c. 1 mm *l*, c. 0.8 mm *b*; **Ecol** on quartzite ridges; **Distr** W part of area of genus, N Namaqualand, NC, S Africa.

Note: Remarkable are the few, short, rather broad petals of this species.

Neogunnia = Gunniopsis

H. E. K. Hartmann

When recognizing that Mueller (1852) had used the name *Gunnia* F.Mueller as a younger homonym of **Gunnia** Lindley 1834, Pax & Hoffmann created the new name *Neogunnia* Pax & Hoffmann in Engler & Prantl 1934: 225, including two species. These have later been found to belong to one species only, being also a member of **Gunniopsis** (Chinnock 1983: 172).

N. drummondii = **Gunniopsis septifraga**
N. septifraga ≡ **Gunniopsis septifraga**

Neohenricia RUSCHIOIDEAE

H. E. K. HARTMANN

Neohenricia L.Bolus JSAB 1938: 51; Hammer CSJ 1994: 84–86 ≡ *Henricia* L.Bolus NM 1936: 39 nomen illeg. non **Henricia** Cassini 1817: 11 **Typus** *Henricia sibbettii* L.Bolus ≡ **N. sibbettii** (L.Bolus) L.Bolus **Etym** *Dr. M. Henrici*, 1892–1971, botanist at Fauresmith, FS, S Africa.

Caespitose to creeping plants, branches short, **I** < **L**, adventitious roots numerous; **L** club-shaped or triquetrous, epidermis with conspicuous warts, i.e. regions where cells rich in crystals are raised above groups of subhypodermal tannin idioblasts, the crystal layer thinner in the lower areas or valleys separating the warts; **Fl** solitary, bracteoles almost as big as the foliage leaves, **K** 5, ∅ 15–17 mm, opening at night, petals white to pale green, filamentous staminodes absent, filaments white to yellow, 5 nectaries at the upper rim of a low hypanthium, below the bases of the epapillate filaments; **Fr** base bowl-shaped, top almost flat, covering membranes absent, the septa split, but not spreading, expanding keels parallel except for the distal tips, valve wings absent, the placenta in basal and central position; **Ecol** in crevices or shallow pans on sandstone or dolerite > 200 mm rainfall p.a. at different seasons; **Distr** N EC, S FS, SW NC, S Africa.

Note: Only some populations are known from a rather wide area. Two species are recognized.

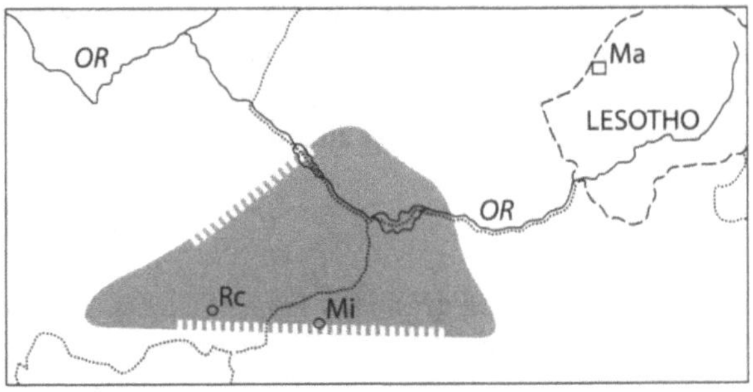

Distribution of **Neohenricia** L.Bolus

Key to the Species of Neohenricia

1. Leaves club-shaped with a flat or (in cultivation) roundish top .... **N. sibbettii**
– Leaves triquetrous with a pointed tip .... **N. spiculata**

N. sibbettii (L.Bolus) L.Bolus JSAB 1938: 51 (see **photo**) ≡ *M. sibbettii* L.Bolus NM 1931: 251 ≡ *Henricia sibbettii* (L.Bolus) L.Bolus NM 1936: 39 **T** Sibbett 2475/30 (BOL).

Plants with short creeping stems, the adventitious roots a little thickened; **L** nearly club-shaped, the top flat or rounded, the warts arranged in spirals or meanders on the top and on the upper 1/2 of the leaf length, or petering out into distant warts on the sides, < 10 mm *l*, ∅ c. 4 mm, wax as a continuous cover over the epidermis, very few tiny particles on the rugose surface; **Fl** with white petals, in some populations entirely or apically purple; **Fr** as for genus, ∅ c. 3 mm, falling off early; **Ecol** mainly on sandstone, rarely on dolerite, 200–400 mm rainfall p.a., mainly in March and November; **Distr** N parts of EC, S parts of FS, SE parts of NC, S Africa.

Note: As in other taxa with a patchy distribution, the populations can individually be recognized. The brain-like ornamentation on the top is particularly well expressed in the type population, but in other populations the aggregation of the warts can be less dense resembling **N. spiculata** in some parts.

N. spiculata S.A.Hammer CSJ 1994: 84 **T** Hammer & Marx 1403 (BOL).

Plants more caespitose, adventitious roots thin; **L** triquetrous, pointed, the warts dispersed at regular intervals over the upper half of the leaf, but not in a distinct pattern; **Fl** open at night from 9 p.m. to 5 a.m., **K** lobes thick, c. 95 petals pale lime-green, c. 65 pale yellow stamens; **Fr** ∅ c. 4 mm; **S** c. 0.8 mm *l*; **Ecol** in crevices of dolerite, c. 400 mm rainfall p.a. in summer; **Distr** Sterkstroom, EC, S Africa.

Note: Until now, only known from a restricted area at altitudes at about 2 000 m.

Neorhine ≡ Rhinephyllum

H. E. K. HARTMANN

Schwantes (MDK 1930: 22) named the genus *Neorhine* based on a figure published by Bolus (SAG 1928: 370) as **Rhinephyllum pillansii**. The plant shown is **Rhinephyllum muirii**, which is the type species of **Rhinephyllum**. As a consequence of this identification, *Neorhine* becomes an obligate synonym.

N. pillansii ≡ **Rhinephyllum muirii**

Nycteranthus = Aridaria

M. Gerbaulet

The genus *Nycteranthus* was published by Necker (1790: 82–83), who neither cited any species nor gave an unequivocal diagnosis. Notwithstanding, Rothmaler (1941: 413) effectively validated *Nycteranthus* by choosing *Nycteranthus noctiflorus* ≡ **Aridaria noctiflora** as nomenclatural type. Schwantes (KuaS 1952: 48–51) accepted this treatment and transferred all known species of **Aridaria** and *Sphalmanthus* to *Nycteranthus*. However, in 1960 the Nomenclatural Congress decided that *Nycteranthus* Necker was not validly published. As a corollary, *Nycteranthus* Necker ex Rothmaler becomes an illegitimate name, since it was based on the same type as **Aridaria.**

N. abbreviatus ≡ **Phyllobolus abbreviatus**
N. acuminatus
= **Phyllobolus splendens** ssp. **splendens**
N. albertensis = **Phyllobolus pumilus**
N. albicaulis
= **Phyllobolus splendens** ssp. **splendens**
N. anguineus = **Phyllobolus oculatus**
N. arcuatus = **Aridaria noctiflora** ssp. **noctiflora**
N. arenicola = **Phyllobolus oculatus**
N. aureus = **Phyllobolus nitidus**
N. ausanus = **Prenia tetragona**
N. bijliae = **Phyllobolus splendens** ssp. **splendens**
N. blandus = **Phyllobolus splendens** ssp. **splendens**
N. brevicarpus ≡ **Aridaria brevicarpa**
N. brevifolius
= **Phyllobolus splendens** ssp. **splendens**
N. calycinus = **Aridaria noctiflora** ssp. **noctiflora**
N. canaliculatus ≡ **Phyllobolus canaliculatus**
N. carneus = **Phyllobolus spinuliferus**
N. caudatus ≡ **Phyllobolus caudatus**
N. celans = **Phyllobolus splendens** ssp. **splendens**
N. commutatus = **Phyllobolus grossus**
N. compactus = **Aridaria serotina**
N. congestus ≡ **Phyllobolus congestus**
N. constrictus
= **Phyllobolus splendens** ssp. **splendens**
N. debilis = **Aridaria noctiflora** ssp. **noctiflora**
N. deciduus ≡ **Phyllobolus deciduus**
N. decurvatus ≡ **Phyllobolus decurvatus**
N. defoliatus ≡ **Aridaria noctiflora** ssp. **defoliata**
N. dejagerae = **Aridaria noctiflora** ssp. **straminea**
N. delus ≡ **Phyllobolus delus**
N. dinteri ≡ **Phyllobolus melanospermus**
N. dyeri = **Phyllobolus splendens**
N. ebracteatus = **Phyllobolus trichotomus**
N. ebracteatus var. *brevipetalus*
= **Phyllobolus trichotomus**
N. elongatus = **Phyllobolus prasinus**
N. englishiae ≡ **Prenia englishiae**
N. esterhuyseniae
= **Aridaria noctiflora** ssp. **straminea**
N. flexuosus
= **Phyllobolus splendens** ssp. **splendens**
N. floribundus = **Aridaria noctiflora** ssp. **noctiflora**
N. fourcadei
= **Phyllobolus splendens** ssp. **splendens**
N. fragilis = **Phyllobolus oculatus**
N. framesii = **Phyllobolus spinuliferus**
N. fulvus = **Aridaria noctiflora** ssp. **straminea**
N. geniculiflorus ≡ **Aptenia geniculiflora**
N. glanduliferus = **Phyllobolus sinuosus**
N. globosus = **Aridaria serotina**
N. godmaniae = **Phyllobolus sinuosus**
N. gracilis = **Aridaria serotina**
N. gratiae = **Phyllobolus grossus**
N. grossus ≡ **Phyllobolus grossus**
N. herbertii ≡ **Phyllobolus herbertii**
N. horizontalis = **Aridaria noctiflora** ssp. **defoliata**
N. inaequalis = **Phyllobolus nitidus**
N. intricatus = **Aridaria brevicarpa**
N. latipetalus ≡ **Phyllobolus latipetalus**
N. laxipetalus = **Phyllobolus grossus**
N. laxus = **Phyllobolus decurvatus**
N. leptopetalus
= **Phyllobolus splendens** ssp. **splendens**
N. ligneus = **Phyllobolus melanospermus**
N. longisepalus = **Aridaria noctiflora** ssp. **defoliata**
N. longispinulus = **Phyllobolus grossus**
N. longistylus ≡ **Mesembryanthemum longistylum**
N. longitubus = **Phyllobolus tenuiflorus**
N. luteoalbus = **Prenia tetragona**
N. macrosiphon = **Phyllobolus tenuiflorus**
N. meridianus = **Aridaria noctiflora** ssp. **defoliata**
N. meyeri = **Aridaria brevicarpa**
N. muirii = **Aridaria noctiflora** ssp. **defoliata**
N. multiseriatus = **Phyllobolus prasinus**
N. mutans = **Prenia tetragona**
N. noctiflorus ≡ **Aridaria noctiflora**
N. nothus = **Phyllobolus splendens** ssp. **splendens**
N. obtusus = **Phyllobolus decurvatus**
N. oculatus ≡ **Phyllobolus oculatus**
N. odoratus ≡ **Hereroa odorata**
N. oubergensis = **Phyllobolus pumilus**
N. ovalis = **Aridaria serotina**
N. parvisepalus = **Phyllobolus spinuliferus**
N. paucandrus = **Aridaria serotina**
N. paucandrus var. *gracillimus* = **Aridaria serotina**
N. peersii = **Phyllobolus tenuiflorus**
N. pentagonus
≡ **Phyllobolus splendens** ssp. **pentagonus**
N. pentagonus var. *occidentalis*
= **Phyllobolus splendens** ssp. **pentagonus**

N. platysepalus = **Phyllobolus grossus**
N. plenifolius
= **Phyllobolus splendens** ssp. **splendens**
N. pomonae = **Phyllobolus oculatus**
N. prasinus ≡ **Phyllobolus prasinus**
N. primulinus
= **Phyllobolus splendens** ssp. **splendens**
N. pumilus ≡ **Phyllobolus pumilus**
N. quartziticus ≡ **Phyllobolus quartziticus**
N. quaternus = **Phyllobolus spinuliferus**
N. rabiei ≡ **Phyllobolus rabiei**
N. rabiesbergensis
= **Phyllobolus splendens** ssp. **splendens**
N. radicans ≡ **Prenia radicans**
N. recurvus = **Phyllobolus sinuosus**
N. rhodandrus = **Phyllobolus nitidus**
N. roseus = **Phyllobolus splendens** ssp. **splendens**
N. salmoneus = **Phyllobolus canaliculatus**
N. saturatus ≡ **Phyllobolus saturatus**
N. scintillans = **Phyllobolus oculatus**
N. serotinus ≡ **Aridaria serotina**
N. sinuosus ≡ **Phyllobolus sinuosus**
N. spinuliferus ≡ **Phyllobolus spinuliferus**
N. splendens
≡ **Phyllobolus splendens** ssp. **splendens**
N. stramineus ≡ **Aridaria noctiflora** ssp. **straminea**
N. straminicolor = **Phyllobolus sinuosus**
N. striatus = **Phyllobolus splendens** ssp. **splendens**
N. strictus = **Phyllobolus spinuliferus**
N. subaequans
= **Phyllobolus splendens** ssp. **pentagonus**
N. subpatens
= **Phyllobolus splendens** ssp. **pentagonus**
N. subtruncatus
= **Aridaria noctiflora** ssp. **straminea**
N. suffusus = **Prenia tetragona**
N. sulcatus = **Phyllobolus splendens** ssp. **splendens**
N. tenuiflorus ≡ **Phyllobolus tenuiflorus**
N. tenuifolius = **Aridaria serotina**
N. tenuifolius var. *speciosus* = **Aridaria serotina**
N. tetragonus ≡ **Prenia tetragona**
N. tetramerus = **Phyllobolus trichotomus**
N. tetramerus var. *parviflorus*
= **Phyllobolus trichotomus**
N. trichotomus ≡ **Phyllobolus trichotomus**
N. umbelliflorus
= **Phyllobolus splendens** ssp. **splendens**
N. varians = **Phyllobolus oculatus**
N. vernalis = **Phyllobolus splendens** ssp. **splendens**
N. vespertinus ≡ **Aridaria vespertina**
N. vigilans = **Mesembryanthemum longistylum**
N. viridiflorus ≡ **Phyllobolus viridiflorus**
N. watermeyeri = **Phyllobolus spinuliferus**
N. willowmorensis = **Phyllobolus grossus**

Octopoma RUSCHIOIDEAE

H. E. K. HARTMANN

Octopoma N.E.Br. GC 1930: 72; GC 1930: 126; Hartmann Bradleya 1998: 70–74 **Typus O. octojuge** (L.Bolus) N.E.Br. **Etym** Gk *okto*, eight, *poma*, lid, cover "in allusion to the valves of the capsule, which are frequently eight in number".

Rather densely branched shrubs to 25 cm *h*, ∅ to 25 cm, **I** ochre to darker; **L** mostly connate, dark green; **Fl** solitary, bracteoles mostly embracing the base of the flower, **K** 4(–5), the inner a little smaller, ∅ 30–35 mm, the inner petals shorter than the outer, filamentous staminodes surrounding the stamens in a central cone, all elements epapillate or the inner with few, short papillae, nectary as a crenulate, dark green ring; **Fr** raised above the bracteoles, top convex with low rims, base funnel-shaped, covering membranes somewhat elevated in the centre, distally a little turned up, the closing ledge below very inconspicuous, closing body big, flat above, expanding keels with awns, valve wings narrow, broadest in their middle, 6–8 locules; **Ecol** in sand or granite, or in loam with quartzite pebbles, < 200 mm rainfall in winter or in March and November; **Distr** Namaqualand, NC; W districts of WC, S Africa.

Note: The genus was described in contrast to **Leipoldtia**, from which it was said to differ by the connate leaves, the "contiguous" expanding keels

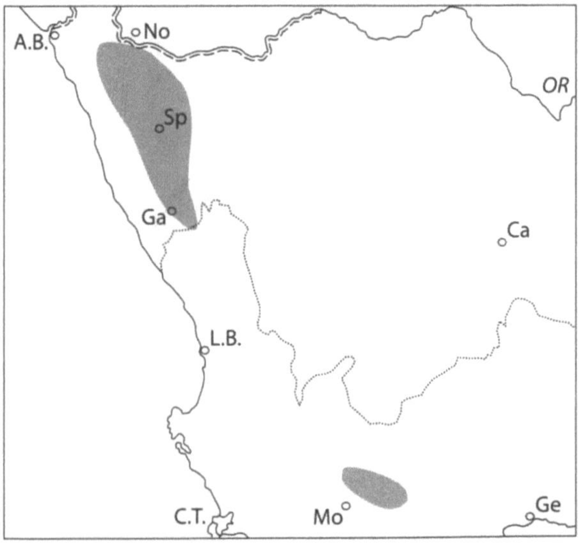

Distribution of **Octopoma** N.E.Br.

and the absence of valve wings (Brown 1930). Since **O. octojuge**, the type species, possesses narrow valve wings, this character expression is not true for the genus. An examination of the type material and many collections from habitat shows that most of the species placed in **Octopoma** have been put there only because they possess eight locules. They differ so widely in most other character states that at least two more genera can be distinguished. One species is transferred to **Zeuktophyllum**, the remaining ones, with the addition of two species added to here from **Ruschia**, can be arranged in two groups: one group of species around **O. octojuge** with capsules closer to the Leipoldtia-type, and the other group of species with fruits closer to the Ruschia-type. Both groups can be separated accordingly in a key.

Key to the Species of Octopoma

1. Capsules with narrow valve wings, covering membranes without or with small additional closing devices, closing body big 2
- Capsules without any valve wings, covering membranes with distinct, mostly protruding closing rodlets, closing body small 4
2. Leaves hardly connate, surface rough, with a strong tooth at the apex .... **O. subglobosum**
- Leaves basally connate, surface smooth, no tooth or mucro 3
3. Free parts of leaves almost globose, smooth, petals pale pink .... **O. quadrisepalum**
- Free parts of leaves spreading and somewhat trigonous, keels finely denticulate, petals white .... **O. octojuge**
4. Plants forming mats or cushions, the flowers and fruits raised well above the plants, fruit bases bell-shaped .... **O. connatum**
- Plants ± erect shrubs, flowers and fruits not conspicuously raised above the plants, fruit bases funnel-shaped, sometimes with convex sides 5
5. Leaves densely placed, in tetrastichous rows 6
- Leaves loosely arranged, not in distinct rows 7
6. Leaves c. 3 mm l .... **O. tetrasepalum**
- Leaves 6–9 mm l .... **O. inclusum**
7. Flowers solitary .... **O. rupigenum**
- Flowers in cymes .... **O. abruptum**

O. abruptum (A.Berger) N.E.Br. GC 1930: 126 ≡ *M. abruptum* A.Berger 1922: 638 ≡ *Ruschia abrupta* (A.Berger) G.D.Rowley NCSJ 1978: 8 **T** Schlechter 10828 (B, isotype BOL!).

Shrubs to 25 cm *h*, **I** greyish-brown, c. 20 mm *l*, rigid; **L** triquetrous, keel minutely denticulate; **Fl** in cymes, bracteolate, **K** 5, petals red (?); **Fr** base funnel-shaped with convex sides, top destroyed on isotype, covering membranes convex, closing rodlets prominent, closing body small, expanding keels ending in awns, without valve wings, 7–8 locules; **Ecol** on hills, < 125 mm rainfall p.a. in winter; **Distr** Clanwilliam, WC, S Africa.

Note: The species is insufficiently known.

O. calycinum ≡ **Zeuktophyllum calycinum**
O. conjunctum = **O. connatum**

O. connatum (L.Bolus) L.Bolus 1963: 49 (see **photo**) ≡ *Ruschia connata* L.Bolus NM 1928: 139 ≡ *M. connatum* N.E.Br. GC 1930: 32 **T** Pillans 5794 (BOL!) = *Ruschia conjuncta* L.Bolus NM 1931: 239 ≡ *O. conjunctum* (L.Bolus) L.Bolus 1966: 128 **T** Mathews 19381 (BOL!).

Plants forming mats above which the solitary flowers and fruits are borne; **L** trigonous to semi-terete, basal sheath with a line separating the two leaves of a pair, epidermis with central, rounded papillae, wax in a layer breaking with age into irregular thin horizontal plates, 13–25 mm *l*, to 5 mm *b* and ∅; **Fl** on long stalks, bracteoles more connate than the foliage leaves, placed in the upper half of the stalk, **K** 4, petals 40–50, purple, filamentous staminodes 0–32, purple, 120–205 stamens, purple, sometimes basally white; **Fr** base bell-shaped, sometimes semi-globose, top as a high dome, rims very low, covering membranes convex, distally with protruding closing rodlets, closing body oval, the long axis pointing into the locule, small, valve wings absent, 8 locules; **S** elongate pear-shaped, testa almost smooth, low epicuticular rodlets present; **Ecol** mostly in gravelly flats among bigger stones, often with quartzite rocks, < 125 mm rainfall p.a. in winter; **Distr** Namaqualand, NC, Vanrhynsdorp, WC, S Africa.

Note: O. conjuncta has been described from the Vanrhynsdorp District, but until now no differences supporting the separation of two species have been found. The flowers raised well above the plant are similar to those of **O. tetrasepalum**, forming more shrubby plants.

O. inclusum (L.Bolus) N.E.Br. GC 1930: 126 ≡ *M. inclusum* L.Bolus ABH 1926: 40 ≡ *Ruschia restituta* G.D.Rowley NCSJ 1978: 8 **LT** Pillans 17758 (BOL!).

Shrubs to 10 cm *h*, **I** invisible; **L** arranged in four distinct rows, cymbiform, epidermis with low papillae, wax in plates and platelets, free parts 6–9 mm *l*, sheaths c. 3 mm *l*, 4–5 mm *b* and ∅; **Fl** solitary, bracteoles almost enclosing the stalk, highly connate, **K** 4–5, petals pinkish-purple, 12 mm *l*, 1–1.5 mm *b*, filamentous staminodes present, filaments purple, basally white, anthers and pollen white; **Fr** base funnel-shaped, top dome-shaped, rims low, covering membranes convex, a distinct closing rodlet at the distal end, closing body oval, pointing with the long axis into the locule, valve wings absent, 7–8 locules; **Ecol** near the coast, < 100 mm winter rainfall p.a.; **Distr** Namaqualand, NC, S Africa.

Note: An apparently conspicuous plant.

O. octojuge (L.Bolus) N.E.Br. 1930: 72 (see **photo**) ≡ *M. octojuge* L.Bolus ABH 1927: 77 ≡ *Ruschia octojugis* (L.Bolus) L.Bolus NM 1930: 175 **T** Muir 3957 (BOL!).

Low shrubs to 8 cm *h*, ∅ to 15 cm; **L** free parts trigonous and a little spreading, keel with tiny teeth; **Fl** solitary, **K** 4, 70–85 petals, 30–40 filamentous staminodes and 180–240 filaments, all elements white and somewhat tousled; **Fr** base funnel-shaped, top raised with moderate rims, covering membranes undulate, no additional closing device, closing body big and reaching under the covering membrane, valve wings narrow, tapering a little towards tip and base, 8 locules; **Ecol** in more open patches of Broken Karroo, on loamy soils with some quartzite pebbles, 100–200 mm rainfall p.a. mainly in March and November; **Distr** Ladismith, Laingsburg, Swellendam, WC, S Africa.

Note: Plants of **O. octojuge** remain smaller than those of **O. quadrisepalum**. Both species occur in the same area, but have so long not been found sympatrically in one place.

O. quadrisepalum (L.Bolus) H.E.K.Hartmann Bradleya 1998: 73 (see **photo**) ≡ *Ruschia quadrisepala* L.Bolus NM 1930: 175 **LT** Ross Frames 19080 II (BOL!).

Shrubs to 10 cm *h*, ∅ to 25 cm; **L** free parts semiglobose, those of a pair appressed, keel smooth; **Fl** solitary, **K** 4, petals pale pink; **Fr** base funnel-shaped, covering membranes a little undulate, no additional closing devices, closing body big and reaching under the covering membrane, valve wings narrow, tapering towards tip and base, 8 locules; **Ecol** in shrubby associations, on loamy soils with some quartzite pebbles, 100–200 mm rainfall p.a. mainly in March and November; **Distr** Ladismith, Laingsburg, Swellendam, WC, S Africa.

Note: **O. quadrisepala** appears to settle more readily in deeper soils with fewer pebbles of quartzite among other shrubs.

O. rupigenum (L.Bolus) L.Bolus 1967: 306 (see **photo**) ≡ *Ruschia rupigena* L.Bolus NM 1935: 520 ≡ *Ruschia rupicola* L.Bolus NM 1933: 415 nomen illeg. non (Engler) Schwantes ZSK 1926: 187 **T** L.Bolus 1504/33 (BOL!).

Shrubs to 20 cm *h*, branches twisted, **I** 8–13 mm *l*, ∅ 4–9 mm; **L** triquetrous, epidermis rough from raised dots and finely papillate, 12 mm *l*, 3 mm *b*, 4 mm ∅; **Fl** solitary, pedicels 10–15 mm *l*, bracteoles touching the base of the flower, **K** 5, petals and filamentous staminodes bright pink, filaments pale or pink; **Fr** base funnel-shaped, top dome-shaped, covering membranes convex, with a small closing rodlet, closing body oval, the long axis pointing into the locule, no valve wings, 8 locules; **Ecol** among rocks; **Distr** Clanwilliam, WC, S Africa.

Note: A little known species.

O. subglobosum (L.Bolus) L.Bolus 1963: 49 (see **photo**) ≡ *Ruschia subglobosa* L.Bolus NM 1928: 140 ≡ *M. reductum* N.E.Br. GC 1930: 33 nomen illeg. non N.E.Br. GC 1930: 32 **T** Pillans 5844 (BOL!).

Shrubs to 20 cm *h*, to 25 cm ∅; **L** only a little connate, the base almost separated by an elongate ochre triangle of stem tissue, free parts spreading widely, fatly trigonous with a marked apical tooth and fine teeth on the keel; **Fl** solitary, **K** 5, petals purple, few purple filamentous staminodes surrounding the central cone of stamens, filaments with few lateral papillae; **Fr** base funnel-shaped, top semi-globose, covering membranes with distinct rodlets, closing body small, but reaching under the covering membrane, mostly 7–8 locules; **Ecol** in rough sand in pans in Namaqualand granite domes, < 100 mm rainfall p.a. in winter; **Distr** W Namaqualand, NC, S Africa.

Note: The species forms conspicuous stands on Namaqualand granites.

O. tetrasepalum (L.Bolus) H.E.K.Hartmann Bradleya 1998: 74 ≡ *Ruschia tetrasepala* L.Bolus NM 1932: 373 **T** Luckhoff 20203 (BOL!).

Shrubs to 13 cm *h*, branches in all directions, stems basally to 30 mm ∅, **I** dark brown; **L** cymbiform, c. 3 mm *l*, 4 mm *b*, to 5 mm ∅; **Fl** solitary, bracteoles below the middle of the pedicel, bigger than foliage leaves, **K** 4, petals pink, filamentous staminodes absent, filaments red, inner ones papillate, anthers white; **Fr** base funnel-shaped, covering membranes convex, distally with protruding closing rodlets, closing body a narrow hook, valve wings on type destroyed, 7 locules; **Ecol** unknown; **Distr** Vanrhynsdorp, WC, S Africa.

Note: In the elevated position of the flowering branches raised well above the plant it resembles **O. connatum**.

Odontophorus RUSCHIOIDEAE

H. E. K. HARTMANN

Odontophorus N.E.Br. GC 1927: 12; Hartmann 1976: 161–225; Hammer Bradleya 1994: 75–85 **Typus O. marlothii** N.E.Br. **Etym** Gk *odous, odontos*; tooth, *phora*, container, carrier of something, relating to the teeth on the leaves.

Compact, caespitose, or compact with long shoots, **I** long to very short and invisible, if visible, corky-spongy and smooth, or woody with a pleated to wrinkled surface; **L** trigonous to rhombic with a subapical widening and thickening, i.e. gibbose or humped, keel and margins dentate, epidermis elevated above the tannin idioblasts, stomata restricted to the valleys, cells papillate to hairy, surface therefore velvety, wax in fine erect platelets leaving smooth lateral grooves; **Fl** bracteolate, petals yellow in different shades, only the innermost stamina papillate near base, coilomorphic holonectary; **Fr** almost globose, top lower than basal part, the latter with very soft hairs, pedicels persisting, capsules of the Leipoldtia-type but with narrow valve wings and somewhat reduced closing ledges, 10-locular; **S** ovoid, light to darker brown, testa papillate to smooth; **Chr** $2n = 18$; **Ecol** on quartzitic or shaly slopes, winter rainfall < 100 mm p.a.; **Distr** Namaqualand, NC, S Africa.

Note: The velvety leaves and almost globose big fruits resemble some species of **Cheiridopsis** subg. **Odontophoroides**, where also teeth can occur along leaf keel and margins; yet, fruits of **Odontophorus** can be distinguished by their much narrower valve wings.

Distribution of **Odontophorus** N.E.Br.

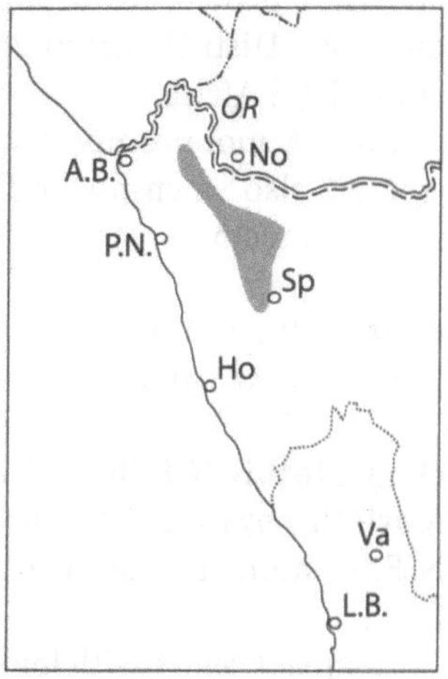

Key to the Species of Odontophorus

1. Plants consisting of a compact centre and long woody trailing, often climbing branches with long internodes .... **O. marlothii**
– Plants compact or caespitose, never a compact centre separated from climbing branches, internodes never woody **2**
2. Leaves slender: ∅ < 1/3 of the length **3**
– Leaves rather fat: ∅ c. 1/2 the length .... **O. nanus**
3. Leaves 30–50 mm *l*, ∅ up to 10 mm), teeth long and flexible .... **O. angustifolius**
– Leaves 22–25 mm *l*, ∅ 6–8 mm, teeth tiny .... **O. pusillus**

O. albus = **O. nanus**

O. angustifolius L.Bolus NM 1938: 128 T van Heerde s.n. 21845 (BOL!).

Caespitose, **I** spongy inside, smooth outside, light brown to orange or ochre, shorter than leaves, ∅ 3–5 mm; **L** slender, 30–50 mm *l*, 6–10 mm thick, irregularly dentate at keel and margins (0–5 teeth), teeth to 5 mm *l*, flexible, rising from a broad socle, elevations low, papillae short (to 150 µm), surface therefore soft rather than velvety; **Fr** with long pedicels and widely spreading bracteoles; **S** 0.7–1.2 mm *l*, pale brown, pear-shaped; **Ecol** on yellow or pinkish quartzite; **Distr** Namaqualand, NC, S Africa.

Note: The two subspecies share the spongy stems and the long, flexible teeth.

Key to the Subspecies of O. angustifolius

1. Plants prostrate, forming big mats, teeth at most reddish .... **O. angustifolius** ssp. **angustifolius**
– Plants rather compact, teeth black from the beginning .... **O. angustifolius** ssp. **protoparcoides**

O. angustifolius L.Bolus ssp. **angustifolius** (see **photo**).

Mat-forming, ∅ up to 3 m, though mostly smaller; **L** 40–50 mm *l*, 7–10 mm *t*, thus thicker than the other subspecies, teeth 0–4, green to reddish; **Fl** ∅ c. 40 mm; **Fr** as for species; **S** > 1 mm *l*, smooth with papillae on hilum only; **Ecol** on yellow quartzite; **Distr** S part of the area of the species, Namaqualand, NC, S Africa.

Note: In habitat, the plants can grow very big, but they can also die back considerably remaining visible as black mummies for several years.

O. angustifolius L.Bolus ssp. **protoparcoides** S.A.Hammer Bradleya 1994: 84 (see **photo**) T Jaarsveld 6138 (BOL!).

Compact to caespitose; **L** 30–40 mm *l*, 6–10 mm *t*, thus more slender than the other subspecies, teeth 3–5, black; **Fl** ∅ c. 50 mm; **Fr** as for genus; **S** < 1 mm *l*, chestnut brown, testa papillate; **Ecol** on pinkish quartzite; **Distr** N part of area of species, Namaqualand, NC, S Africa.

Note: A more compact version of the species, attractive also when not in flower through its black, recurved teeth.

O. areolatus = **O. nanus**
O. herrei = **O. nanus**

O. marlothii N.E.Br. GC 1927: 12 (see **photo**) **LT** Marloth 6974 (coll. G. Meyer) (BOL!) = *O. meyeri* N.E.Br. nomen nudum in schedis (Marloth 13299 BOL!).

Compact centre with long branches mostly climbing in other shrubs, **I** short in centre, longer than leaves on long shoots, dark reddish-brown, surface with longitudinal folds with age, thin: ∅ 1,5–2,5 mm; **L** 25–35 mm *l*, 7–10 mm thick, keel and margins with 1–6 brittle teeth, elevations moderate, papillae short (–150 µm), velvety to the touch; **Fl** petals bright yellow, often with a white base; **Fr** with dentate bracteoles, capsule slightly higher than those of the other species; **S** about 1 mm *l*, completely papillate; **Ecol** shaly slopes in shrubby succulent Karroo; **Distr** Namaqualand, NC, S Africa.

Note: Young plants grow with straight, erect stems, flowers occur mostly on long shoots, but in some populations flowering on the compact centre has been observed.

O. meyeri = **O. marlothii**

O. nanus L.Bolus NM 1929: 59 (see **photo**) **T** Pearson 4071 (BOL!) *O. albus* L.Bolus NM 1935: 490 **T** Herre 10111 (BOL!) = *O. areolatus* N.E.Br. nomen nudum in schedis (Marloth 13298 K!) = *O. herrei* L.Bolus NM 1935: 491 **T** Herre 10219 (BOL!) = *O. nudifolius* L.Bolus nomen nudum in schedis (Herre 10008 BOL!) = *O. primulinus* L.Bolus NM 1930: 223 **T** Marloth 13298 (BOL!).

Compact; **I** enclosed and invisible; **L** 25–50 mm *l*, 11–17 mm thick, very gibbous, margins with 1–6 teeth, keel with 0–1 teeth, teeth on elevations of leaves brittle, epidermis very velvety with high elevations and papillae about 400 µm, rarely smooth and shiny; **Fl** petals light yellow, often cream-coloured, sometimes white; **Fr** shorter, capsules lower than in the other species, base with long hairs, bracteoles parchment-like, thin; **S** < 0,9 mm *l*, testa smooth except for low papillae at hilum; **Ecol** on white quartzite rocks and slopes; **Distr** Namaqualand, NC, S Africa.

Note: The very high variability in leaf sizes and surfaces as well as in flower colour reflects the variation between populations (Hartmann 1976).

O. nudifolius = **O. nanus**
O. primulinus = **O. nanus**

O. pusillus S.A.Hammer 1996: 86 (see **photo**) **T** van Jaarsveld 11461 (BOL).

Shrubs with several almost equal, erect branches, to 8 cm *h*, ∅ to 8 cm, **I** corky and spongy, shining ochre; **L** trigonous, margins subapically with a tooth on each side rendering the tip of the upper leaf surface triangular, keel with several very tiny teeth if not serrulate, epidermis smooth and papillate; **Fl** petals yellow, basally white, acuminate; **Fr** base semi-globose, top almost flat, 10–11 locules; **S** dark brown, 0.75 mm *l*, 0.44 mm *b*; **Ecol** upper slopes on quartz gravel patch; **Distr** Namaqualand, NC, S Africa.

Note: Most remarkable is the erect growth of the short stems of the plants, the primary shoot being longer than the lateral ones, a rare conditions in the family. The spongy, bright ochre internodes resemble **O. angustifolius**.

Oophytum Ruschioideae

H.-D. Ihlenfeldt

Oophytum N.E.Br. GC 1925: 413; GC 1926: 48; Ihlenfeldt 1978: 303–328 ≡ *Conophytum* N.E.Br. subg. *Oophytum* (N.E.Br.) G.D.Rowley NCSJ 1978: 7 **Typus O. oviforme** (N.E.Br.) N.E.Br. **Etym** Gk *oon*, egg, *phyton*, plant, in allusion to the egg-shaped corpuscles formed by the leaf-pairs.

Compact, old plants caespitose, 1–2 cm *h*, 1–3 cm ∅, slightly heterophyllous, deciduous; **L** each stem with usually 2 leaf-pairs per season, completely connate, forming egg-shaped or nearly spherical pulpy corpuscles, 8–12 mm *h*, 4–10 mm *b*, 2nd leaf-pair of the season developing into a papery sheath covering the

1st pair of the next season during the resting state, epidermis mesomorphic, with small bladder cells and a thin continuous wax layer, bladder cells lacking on the upper surface of the leaf, dense along the free margins, connecting the margins of the developing leaves like a zip, corpuscles turning red under water stress in habitat; **Fl** solitary, stalk short, emerging towards the end of the season from the apical slit of the 2nd leaf-pair, the ovary remaining inside the corpuscle, **K** 6, the 2 outermost opposite, the remaining ones spirally arranged, petals in c. two rows, obtuse to emarginate, c. 2 rows of filamentous staminodes, petals and staminodes white, apically rose pink, or occasionally completely white, filaments subbasally papillate, holonectary, ∅ up to 25 mm, VII–IX; **Fr** fragile, expanding keels merging into expanding sheets, the latter rather indistinct, valve wings broad, covering membranes varying from narrow fringes to membranes covering 1/2 of the locule, closing bodies absent, locules 6, ∅ 3–6 mm; **Chr** $2n = 18$; **S** slightly papillate all over or smooth except for the micropylar region, testa black or light brown; **Ecol** in quartzitic flats or on slopes, < 200 mm rainfall p.a. mainly in winter; **Distr** Vanrhynsdorp, Vredendal, WC, S Africa.

Note: In its natural habitat, **Oophytum** may form very dense populations filling the gaps between the quartz pebbles. The two species are allopatric, **O. nanum** occurring in the drier eastern part of the dis-tribution area. In cultivation, plants may form a 3rd pair of leaves, similar to the 2nd one, and the corpuscles elongate considerably and will not turn red towards the end of the season. The inclusion of **O. oviforme** in **Conophytum** by Brown (GC 1922: 231, corrected in GC 1925: 48) and Rowley (all species as a new subg. *Oophytum*, 1978: 7) is erroneous (Ihlenfeldt 1971: 655–660).

Distribution of **Oophytum** N.E.Br.

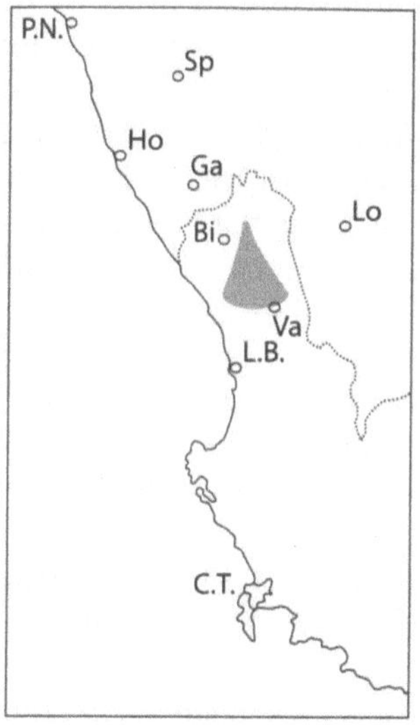

Key to the Species of Oophytum

1. Corpuscles nearly spherical, both leaf-pairs of the season visible, completely succulent and provided with a longitudinal keel on the lower surfaces; capsule viewed from above six-angled, broader than high, ∅ c. 4 mm; testa light brown, smooth .... **O. nanum**
- Corpuscles ovoid, 2nd leaf-pair included in the 1st, basal part of the 2nd leaf-pair membranous, keel lacking; capsule viewed from above nearly circular, higher than broad, ∅ c. 5 mm; testa black, papillate .. **O. oviforme**

O. nanum (Schlechter) L.Bolus NM 1929: 128 (see **photo**) ≡ *M. nanum* Schlechter 1899: 128 non L.Bolus 1913: 141 nomen illeg. ≡ *Conophytum rescriptum* G.D.Rowley 1978: 7 **T** Schlechter 8318 (B!).

Corpuscles nearly spherical; **L** both leaf-pairs of the season visible, both pairs completely succulent and with keels; **Fl** petals usually emarginate at the apex; **Fr** capsule viewed from above six-angled, broader than high, ∅ c. 4 mm, ridges on top of the capsule c. 0.6 mm *h*; **S** testa light brown, smooth except for the micropylar region; **Ecol** in the drier eastern part of the area of the genus; **Distr** Vanrhynsdorp, WC, S Africa.

Note: In cultivation, **O. nanum** flowers somewhat earlier in the season than **O. oviforme** (Hammer CSJ 1995: 238), but it is not sure whether this also applies to the natural habitat, where erratic rain showers apparently trigger flowering.

O. nordenstamii = **O. oviforme**

O. oviforme (N.E.Br.) N.E.Br. GC 1926: 48 (see **photo**) ≡ *M. oviforme* N.E.Br. 1920: 95 ≡ *Conophytum oviforme* (N.E.Br.) N.E.Br. GC 1922: 231 **LT** Brown GC 1921: 207 drawing f 84K = *O. nordenstamii* L.Bolus 1962: 291 ≡ *Conophytum mutatum* G.D.Rowley 1978: 7 **T** Nordenstam 811 (BOL!).

Corpuscles ovoid; **L** 2nd leaf-pair of the season included in the 1st, basal part of the 2nd leaf-pair membranous, **L** without keels; **Fl** petals white, apically rose pink, or pure white, obtuse; **Fr** capsule viewed from above nearly circular, higher than broad, ∅ c. 5 mm, ridges on top of the capsule c. 1.5 mm *h*; **S** testa black, papillate; **Ecol** in the moister W part of the area of the genus; **Distr** Vanrhynsdorp, Vredendal, WC, S Africa.

Note: White flowers, a feature described as distinctive for *O. nordenstamii*, have also been observed in **O. oviforme** (Ihlenfeldt 1978: 318).

Ophthalmophyllum = Conophytum

S. A. Hammer

The name *Ophthalmophyllum* Dinter & Schwantes appeared first in a key (ZSK 1927: 15) with the following distinctive character states: leaves connate except for the uttermost tips, soft, the apical area windowed, flower with long calyx- and petal-tube (translated from the German). The contrasting alternative in the key is "leaves not connate to form a body", but it must be remembered that the genus **Conophytum** does not appear in that key. In the protologue of the description proper of *Ophthalmophyllum*, it is clearly stated that the then only species, **C. friedrichiae**, agrees with **Conophytum** insofar as the flower "agrees completely with those of that genus". The main difference between **C. friedrichiae** and the other species of **Conophytum** lies in the windowed leaves and bracteoles, the partially subterranean growth form and the low tendency to branch; the species differs from **Lithops** in the hairy surface and the softness (after Dinter & Schwantes 1927: 22). Schwantes added only later (Kakteenkunde 1934: 58–60) that the flowers of *Ophthalmophyllum* possess a basally connate tube, the lower part consisting of the bases of the calyx-, petal- and stamen-elements and ending in a crenulate nectary (this part equivalent to the hypanthium of, e.g., **Argyroderma**), a middle part still connate but without the nectary, and an apical part consisting of a free tube formed by the calyx elements surrounding the inner tube of petal- and stamen-bases. Again, only *O. friedrichiae* was studied in detail and the new species added were placed there because of their leaves. Tischer (SUC 1928: 198) transferred *Ophthalmophyllum* as a section to **Conophytum**, based on floral similarities and the presence of windows on the leaf tips in other species of **Conophytum**. This view is followed by Hammer (Conophytum 1993: 230).

O. acutum ≡ **Conophytum acutum**
O. australe = **Conophytum caroli**
O. caroli ≡ **Conophytum caroli**
O. cornutum = **Conophytum subfenestratum**
O. dinteri = **Conophytum friedrichiae**
O. friedrichiae ≡ **Conophytum friedrichiae**
O. fulleri = **Conophytum longum**
O. haramoepense = **Conophytum lydiae**
O. herrei = **Conophytum longum**
O. jacobsenianum = **Conophytum praesectum**
O. latum ≡ **Conophytum maughanii** ssp. **latum**
O. latum var. *rubrum* = **Conophytum maughanii** ssp. **latum**
O. littlewoodii = **Conophytum devium**
O. longitubum = **Conophytum longum**
O. longum ≡ **Conophytum longum**
O. lydiae ≡ **Conophytum lydiae**
O. marlothii = **Conophytum pellucidum**
O. maughanii ≡ **Conophytum maughanii** ssp. **maughanii**
O. noctiflorum = **Conophytum maughanii** ssp. **latum**
O. praesectum ≡ **Conophytum praesectum**
O. pubescens ≡ **Conophytum pubescens**
O. rufescens = **Conophytum maughanii** ssp. **maughanii**
O. schlechteri = **Conophytum longum**
O. schuldtii = **Conophytum maughanii** ssp. **maughanii**
O. spathulatum = **Conophytum lydiae**
O. subfenestratum ≡ **Conophytum subfenestratum**
O. triebneri = **Conophytum friedrichiae**
O. vanheerdei = **Conophytum friedrichiae**
O. verrucosum ≡ **Conophytum verrucosum**
O. villetii = **Conophytum concordans**

Opophytum = Mesembryanthemum

M. Gerbaulet

Opophytum is regarded as a subgenus of **Mesembryanthemum: M.** subg. **Opophytum** (Bittrich 1987: 71). Together with *Eurystigma*, *Halenbergia* and *Hydrodea* it forms a monophyletic group which is probably nested in **Mesembryanthemum** subg. **Mesembryanthemum.**

O. ampliatum = **Mesembryanthemum hypertrophicum**
O. aquosum = **Mesembryanthemum hypertrophicum**
O. australe = **Mesembryanthemum hypertrophicum**
O. dactylinum = **Mesembryanthemum cryptanthum**
O. fastigiatum ≡ **Mesembryanthemum fastigiatum**
O. forskahlii = **Mesembryanthemum cryptanthum**
O. gaussenii = **Mesembryanthemum cryptanthum**
O. speciosum = **Mesembryanthemum hypertrophicum**
O. theurkauffii = **Mesembryanthemum cryptanthum**

Orthopterum Ruschioideae

H. E. K. Hartmann

Orthopterum L.Bolus SAG 1927: 281; NM 1928: 130 **Typus O. waltoniae** L.Bolus **Etym** Gk *orthos*, erect, straight, *pteron*, feather, wing, describing the fact that the septa stand erect in the fruit, "on account of the erect cell-wings".

Compact to caespitose, a little anisophyllous; **L** trigonous, epidermis smooth, perhaps without a crystal layer; **Fl** solitary, ebracteolate, **K** 5, petals yellow, filamentous staminodes none, 5 nectary glands; **Fr** pedicels short, laterally compressed, base broadly bell-shaped, top conical, as high as or higher than the base, septa splitting low down, hence their upper parts almost erect, but nevertheless covering up the interior of the locules, no horizontal parts of covering membranes present, closing body absent, expanding keels distant from one another, straight and ending in free awns, valve wings semi-lunate to very broadly rectangular, overlapping basally, 5 locules; **S** pear-shaped, papillate; **Ecol** in open patches in shrubby vegetation, rainfall mainly at all seasons; **Distr** Albany, Port Elizabeth, EC, S Africa.

Note: Although the first description of the genus is in English only, it was valid at that time. The genus differs from **Faucaria** mainly in the repeatedly opening and closing capsules.

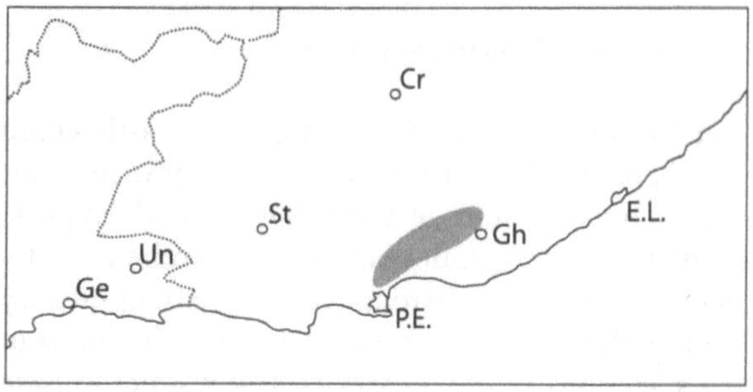

Distribution of **Orthopterum** L.Bolus

Key to the Species of Orthopterum

1. Leaves with 1–3 low teeth per margin or at least protrusions .... **O. coeganum**
– Leaves toothless, but finely rough .... **O. waltoniae**

O. coeganum L.Bolus NM 1937: 108 (see **photo**) **T** Duncan 1078/31 (BOL).

L spreading widely, margins with some undulate protrusions or few teeth, yellowish-green, 30–35 mm *l*, to 10 mm *b*, to 8 mm ∅; **Fl** petals golden, reddish outside, 17–22 mm *l*; **Fr** base very broadly bell-shaped, top higher than base, conical and pointed in the centre, septa ± erect, but hiding the interior of the locule, ∅ 10–12 mm; **Ecol** in shallow pockets on sandstone with lime; **Distr** Port Elizabeth, EC, S Africa.

Note: The leaves look extremely like those of **Faucaria subintegra.**

O. waltoniae L.Bolus SAG 1927: 281 (see **photo**); NM 1928: 130 **T** Walton 310/23 (BOL).

L slender, darkish-green, finely rough, 30–40 mm *l*, 6–8 mm ∅; **Fl** ∅ 25–45 mm, petals reddish-golden, filaments yellow, basally paler, pollen pale yellow; **Fr** broadly bell-shaped, top conical with its highest point in the centre, top almost as high as base, septa splitting at c. half their height, thus overarching the locules, but not forming any horizontal covering membrane, ∅ 8–9 mm; **Ecol** on gravelly ground in open patches at the edge of scrub; **Distr** Albany, EC, S Africa.

Note: The very slender leaves differ markedly from those of **O. coeganum** and any leaves in **Faucaria.** Plants in the type locality appear to be reduced by stock walking over them.

Oscularia Ruschioideae

H. E. K. Hartmann

Oscularia Schwantes MDG 1927: 187; Hartmann Bradleya 1998: 74–79 **Typus O. deltoides** (Haw.) Schwantes **Etym** La *osculum*, small mouth.

Shrubs with some erect, but mostly spreading to decumbent branches, **I** reddish-brown, mostly shining; **L** trigonous with convex sides, thicker than broad, markedly keeled, very light grey from a distinct layer of crystal sand in the outer epidermal wall and a thick wax cover of densely placed platelets petering out into narrow structures sometimes forming threads, margins and keel in several species dentate; **Fl** in terminal cymes, often with additional cymes below, thus forming impressive aggregate **Inf**, but in some spe-

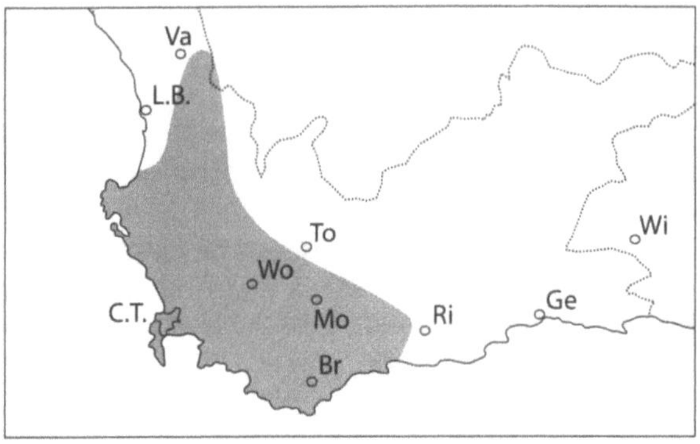

Distribution of **Oscularia** Schwantes

cies flowers solitary and arranged in a 3-flowered group, open day and night, with a fragrance of almonds or whitethorn, **K** 5, petals white to deep pink, filamentous staminodes variable in number, mostly in one circle, ovary surrounded by 5 distinct nectary glands, often so broad that they touch each other, but the 5 sections in most cases clearly visible; **Fr** base funnel-shaped, top a little raised, but the rims very low, grey to brown and softer in tissue than those of **Lampranthus**, covering membranes complete with an inconspicuous closing ledge at the distal end, the placenta ending in a tiny to small knob, valve wings very broad and rectangular, 5 locules; **Ecol** mostly in rocky places, often in crevices inhabiting shady niches; **Distr** western districts of WC, S Africa.

Note: When describing the genus, Schwantes (MDG 1927: 187) already noted a similarity with *M. lunatum*, a species placed later in **Lampranthus**. Since all species included at first in **Oscularia** possess teeth to a certain extend, this feature had been considered to represent an essential character expression of the genus. Based on this interpretation, other species agreeing in critical points with **Oscularia** were kept in **Lampranthus**, and it was even suggested to unite both genera (Glen 1980: 693–694), which required the conservation of the name **Lampranthus** against the older generic name **Oscularia** in order to prevent too many name changes. In view of the differences between **Lampranthus** and **Oscularia**, pointed out by Schwantes already, the latter genus is held up here. The thick wax layer and the existence of a rather thick layer of crystal sand in the outer epidermis wall, resulting in the whitish leaves, even in dry state, differ from most species of **Lampranthus**. The distinct 5 nectary glands are absent in **Lampranthus** as a rule, and instead an inconspicuous ring is formed in most species. Differences in the fruit are slight (only the low closing ledge against rodlets in **Lampranthus**); furthermore, fruits in **Lampranthus** are firm and persist for a long time whereas those of **Oscularia** are soft when pressed, disintegrating faster. The transfer follows for most species Glen (unpublished 1978) in his arrangement of the section *Lunata* (Haw.) Glen, nomen nudum, in **Lampranthus**. The lectotypifications follow Glen 1986: 55, where they are not cited as such, but as iconotypes and topotypes.

O. alba (L.Bolus) H.E.K.Hartmann Bradleya 1998: 75 ≡ *M. album* L.Bolus NM 1932: 365 ≡ *Lampranthus albus* (L.Bolus) L.Bolus NM 1939: 168 **T** Compton 1515/26 (BOL).

Shrubs to 20 cm *h*, **I** 10–20 mm *l*, ∅ 3 mm; **L** lunate, pale glaucous to almost grey, apiculate, to 20 mm *l*, to 3 mm *b*, to 6 mm ∅; **Fl** in several cymes aggregated apically in compound **Inf** spreading broadly, pedicels 10–20 mm *l*, bracteoles at their middle, petals white, 7 mm *l*, to 2 mm *b*, filamentous staminodes white, nectary ring conspicuously crenulate; **Fr** on stout stalks, base elongate funnel-shaped, top only moderately raised, c. 10 mm *l*, 7–7.5 mm ∅, 5 locules; **Ecol** in crevices of rocks; **Distr** Calvinia, NC, Vanrhynsdorp, WC, S Africa.

Note: The capsules are stout, broader and longer than any others in the genus.

O. caulescens (Miller) Schwantes MDG 1927: 187 ≡ *M. caulescens* Miller 1768: nr. 12 ≡ *M. deltoides* L. var. γ L. 1762: 690 **LT** Dillenius s.n. (OXF), figured in Dillenius 1732: t. 195 f 243 & 244 = *M. deltoides* L. var. *simplex* DC. 1804: t. 53 Lectotype, designated here: DC. 1804: t. 53.

Shrubs; **L** deltoid, keel smooth, margins finely dentate; **Fl** solitary, each with a pair of bracteoles, grouped as if in a cyme; **Fr** unknown; **Ecol** unknown; **Distr** unknown.

Note: The solitary flowers and the lack of teeth on the keel characterize this species.

O. cedarbergensis (L.Bolus) H.E.K.Hartmann Bradleya 1998: 75 ≡ *M. cedarbergense* L.Bolus NM 1935: 487 ≡ *Lampranthus cedarbergensis* (L.Bolus) L.Bolus NM 1939: 168 **Syntypes** Bolus 9007, L.Bolus 21021 (BOL).

Erect shrubs to 15 cm *h*; **L** trigonous, broadest near the tip, 15–20 mm *l*, 4 mm *b*, 4–6 mm ∅; **Fl** 3–7 per cyme, pedicels 5–10 mm *l*, bracteoles 7–8 mm *l*, petals pink, to 8 mm *l*, filamentous staminodes present, nectary glands touching each other, crenulate; **Fr** 5 locules; **Ecol** in mountains; **Distr** Clanwilliam, WC, S Africa.

Note: Placed here because of the expanded keel of the leaves and the big glands in the flowers.

O. compressa (L.Bolus) H.E.K.Hartmann Bradleya 1998: 76 ≡ *Lampranthus compressus* L.Bolus NM 1958: 329 **T** Leistner 12904 (BOL).

Diffusely branching shrubs to 18 cm *h*, ∅ to 60 cm, **I** 10–35 mm, ∅ to 5 mm; **L** sublunate, truncate, young ones apiculate, to 22 mm *l*, 3–4 mm *b*, ∅ 7–11 mm; **Fl** in rich cymes, pedicels 15–30 mm *l*, petals pale pink, the cone of filamentous staminodes constricted at the middle, filaments apically purplish-pink, anthers and pollen yellow, nectary of 5 connate glands, regularly crenulate with age; **Fr** 5 locules; **Ecol** unknown; **Distr** Clanwilliam, WC, S Africa.

Note: Transferred to here based on the sublunate leaves and the five glands in the flower.

O. comptonii (L.Bolus) H.E.K.Hartmann Bradleya 1998: 76 ≡ *M. comptonii* L.Bolus ABH 1923: 162 ≡ *Lampranthus comptonii* (L.Bolus) N.E.Br. GC 1930: 212 **T** Compton 295/22 (BOL) = *M. comptonii* L.Bolus var. *roseum* L.Bolus NM 1933: 409 ≡ *Lampranthus comptonii* (L.Bolus) N.E.Br. var. *roseus* (L.Bolus) L.Bolus ex H. Jacobsen FR 1938: 96 ≡ *Lampranthus comptonii* (L.Bolus) N.E.Br. forma *roseus* (L.Bolus) G.D.Rowley in Jacobsen & Rowley 1978: 77 **Syntypes** Herre 8566, Herre 8568, 8966 (BOL).

Diffusely branched shrubs, 15–23 cm *h*, **I** subcompressed; **L** turgidly trigonous, subtruncate, apically purplish-green, to 40 mm *l*, 6 mm *b*, 10 mm ∅; **Fl** in ternate cymes, ∅ 27 mm, pedicels to 30 mm *l*, bracteoles at their middle, petals white, basally inconspicuously pink vittate, filamentous staminodes pale, apically red-purple, filaments pale, apically red, nectary of 5 distinct semi-lunate glands; **Fr** unknown; **Distr** Vanrhynsdorp, WC, S Africa.

Note: Placed here because of the five distinct nectaries.

O. copiosa (L.Bolus) H.E.K.Hartmann Bradleya 1998: 76 ≡ *M. copiosum* L.Bolus NM 1932: 324 ≡ *Lampranthus copiosus* (L.Bolus) L.Bolus NM 1939: 168 **T** without collector 19923 (BOL).

Diffusely branched shrubs with decumbent branches, **I** 15–30 mm *l*, ∅ to 3 mm; **L** falcate, glaucous, 15–20 mm *l*, 3 mm *b*, 6 mm ∅; **Fl** in rich, aggregate cymes of 25–50 flowers open day and night, ∅ 26 mm, petals deep pink, to 11 mm *l*, to 1.75 mm *b*, filamentous staminodes few, pink, filaments urceolate, anthers and pollen golden, glands almost forming a crenulate ring; **Fr** unknown; **Distr** unknown.

Note: Thick, falcate leaves support the inclusion in **Oscularia.**

O. deltata ≡ **O. deltoides**
O. deltata var. *major* ≡ **O. major**

O. deltoides (L.) Schwantes MDG 1927: 187; Glen 1986: 55 ≡ *M. deltoides* L. 1753: 482 ≡ *O. deltata* (Miller) Schwantes MDG 1927: 187[22] nomen illeg. ≡ *Lampranthus deltoides* (L.) Glen 1986: 55 ≡ *O. muricata* (Haw.) Schwantes ex H.Jacobsen 1970: 478[23] nomen illeg. ≡ *M. muricatum* Haw. 1803: 75 ≡ *M. deltoides* L. var. *muricatum* A.Berger 1908: 190 ≡ *O. deltoides* (L.) Schwantes var. *muricata* (A.Berger) Schwantes in Jacobsen FR 1938: 158 **LT** Dillenius s.n. (OXF) figured in Dillenius 1732: t. 195 f 246.

Shrubs, **I** deep maroon, shining; **L** deltoid, keel and margins adorned with teeth, mostly 2–3 on the keel, 3–4 on the margins; **Fl** in rich **Inf** composed of a terminal and several additional cymes, petals "red" (probably deep pink), fragrant; **Fr** as for genus; **Ecol** in broad crevices among rocks, in shade for a long period of the day; **Distr** Paarl, Worcester, WC, S Africa.

Note: The species is most similar in the compound inflorescences and the dentate leaves to **O. major** with bigger leaves. Confusion in the names has occurred since Haworth (1803: 75) replaced the older name *M. deltoides* by *M. muricatum.*

O. deltoides var. *major* ≡ **O. major**
O. deltoides var. *muricata* ≡ **O. deltoides**
O. deltoides var. *pedunculata* ≡ **O. pedunculata**

O. excedens (L.Bolus) H.E.K.Hartmann Bradleya 1998: 77 ≡ *M. excedens* L.Bolus NM 1932: 324 ≡ *Lampranthus excedens* (L.Bolus) L.Bolus NM 1939: 168 **T** Mathews 2083/29 (BOL).

Erect shrubs to 16 cm *h*, **I** 15–30 mm *l*, 3 mm ∅, older ones to 6 mm ∅; **L** laterally compressed, falcate, to 20 mm *l*, to 4 mm *b*, to 6 mm ∅; **Fl** in 3–5 flowered cymes, pedicels to 10 mm *l*, ∅ to 23 mm, open day and night, petals bright pink, 8 mm *l*, almost 2 mm *b*, filamentous staminodes appressed to the stamens, filaments pale pink, anthers yellow, nectary a crenulate ring; **Fr** 5 locules; **Ecol** unknown; **Distr** Clanwilliam and/or Piquetberg, WC, S Africa.

Note: Although a nectary ring is reported, the species is placed here, based on the falcate, apically broadened leaves, a feature absent in **Lampranthus.**

O. falciformis (Haw.) H.E.K.Hartmann Bradleya 1998: 77 (see **photo**) ≡ *M. falciforme* Haw. 1819: 299 ≡ *Lampranthus falciformis* (Haw.) N.E.Br. GC 1930: 212 **LT** drawing 947/11 (K!) = *Lampranthus falciformis* (Haw.) N.E.Br. var. *maritimus* (L.Bolus) L.Bolus NM 1939: 170 ≡ *M. maritimum* L.Bolus NM 1937: 55 nomen illeg. non L.Bolus ABH 1925: 2 **T** Pillans 8352 (BOL).

Shrubs to 19 cm *h*, branches at first erect, later decumbent, **I** reddish; **L** acinaciform to falcate, with big dots, glaucous; **Fl** in terminal, fastigiate cymes, smelling of whitethorn, ∅ 20(–30) mm, pedicels terete, petals light pink, the few inner much smaller, filaments white, anthers yellow; **Fr** turbinate, 5 locular; **Ecol** unknown; **Distr** Piquetberg, WC, S Africa.

Note: The species is transferred to **Oscularia** because the flowers are reported to smell of hawthorn and the drawing kept at Kew shows typical grey, sickle-shaped leaves, and red stems.

[22] Schwantes in Jacobsen 1938: 158 explained that the name *O. deltata* was used in error, based on *M. deltoides* Miller, which is identical with *M. deltoides* L.

[23] The name *O. muricata* appears in the row of synonyms of **O. deltoides** only in that publication, it is therefore not validly published and illegitimate because the basionym is based on the same type as **O. deltoides.**

O. guthriae (L.Bolus) H.E.K.Hartmann Bradleya 1998: 77 ≡ *M. guthriae* L.Bolus ABH 1927: 101 ≡ *Lampranthus guthriae* (L.Bolus) N.E.Br. GC 1930: 212 **T** Guthrie 18531 (BOL).

Shrubs to 12 cm *h*, branches twisting and dense, old ones to 8 mm ∅; **L** subfalcate, but keel inconspicuous, glaucous green, to 22 mm *l*, 5–6 mm *b*, to 9 mm ∅; **Fl** ternate to solitary through abortion, ∅ to 20 mm, pedicels 40–80 mm *l*, bracteoles at their base, **K** 5–6, petals bright reddish-pink, to 9 mm *l*, 1.5 mm *b*, filaments pale, anthers and pollen golden, 5–6 conspicuous, connate glands; **Fr** 5–6 locules; **Ecol** unknown; **Distr** Ceres (?), WC, S Africa.

Note: The thick, grey leaves and the glands present the bases for the decision to transfer the species to here.

O. lunata (Willdenow) H.E.K.Hartmann Bradleya 1998: 77 (see **photo**) ≡ *M. lunatum* Willdenow 1809: 538 ≡ *Lampranthus lunatus* (Willdenow) N.E.Br. GC 1930: 212 **NT** Strauβ 27.3.1911 (B!).

Shrubs with decumbent branches, hence often broader than high; **L** acinaciform to lunate, impunctate, epidermis with lowly elevated, inconspicuous bladder cells covered densely by a pattern of threads and flakes of wax, stomata not sunken; **Fl** "red", probably pink to purple; **Fr** ∅ c. 5 mm, c. 4 mm *h*, breaking off their stalks with age, the pedicels turning into very blunt cane-shaped sticks; **S** broadly pear-shaped to triangular in side view, testa with short, erect rodlets, c. 0.8 mm *l*; **Ecol** in crevices shady for most of the time of the day; **Distr** Clanwilliam, Vanrhynsdorp, WC, S Africa.

Note: Two collections from the Berlin Botanical Garden, made later than the original description, are kept at B, both agreeing in the distinguishing features well with the original description. Salm-Dyck, who may also have had material from Willdenow closer to the time of the first description, figured a branch giving the leaf length as 11–13.5 mm only (Salm-Dyck 1836: § 29 f 3). This size is common but is regularly superceded by the leaves on the long shoots, where up to 25 mm have been measured. This arrangement is also visible in the two sheets at B, of which the older is chosen as the neotype here.

O. major (Weston) Schwantes in Jacobsen 1949: 58 ≡ *M. deltoides* L. var. *majus* Weston 1770: 169 ≡ *O. deltata* Schwantes var. *major* (Weston) Schwantes MDG 1927: 187 ≡ *O. deltoides* (L.) Schwantes var. *major* (Weston) Schwantes in Jacobsen FR 1938: 158 **LT** Dillenius s.n. (OXF) figured in Dillenius 1732: t. 196 f 247.

Shrubs; **L** deltoid, keel and margins dentate, bigger than in **O. deltoides**; **Fl** in rich, composed **Inf** consisting of a terminal and several lateral cymes; **Fr** as for genus; **Ecol** unknown; **Distr** uncertain.

Note: Following the description and the figure, the leaves of this species are bigger than those of **O. deltoides**, but otherwise no differences are visible. It should be examined whether the sizes may not represent different states in the variation in habitat.

O. muricata ≡ **O. deltoides**

O. ornata (L.Bolus) H.E.K.Hartmann Bradleya 1998: 78 (see **photo**) ≡ *Lampranthus ornatus* L.Bolus NM 1939: 169 ≡ *M. pageanum* L.Bolus NM 1929: 97 nomen illeg. non L.Bolus ABH 1927: 99 **T** collector unknown s.n. (drawing at BOL).

Robust shrubs with erect to diagonal branches to 25 cm *l*, **I** 15–30 mm *l*; **L** falcate, margins and keel near the tip reddish, smooth, pale glaucous green, 30–40 mm *l*, 7 mm *b*, to 12 mm ∅; **Fl** in terminal cymes with additional lateral ones below, the aggregate somewhat narrow, ∅ to 25 mm, petals pink, to 10 mm *l*, to 1.5 mm *b*, filamentous staminodes pink, apically deeper coloured, nectary ring crenulate; **Fr** base between funnel- and bell-shaped, the bundles prominent, top low, 7 mm *l*, c. 5 mm ∅; **Ecol** in crevices on rocky steps in shrubby associations; **Distr** Clanwilliam, WC, S Africa.

Note: The species is similar to **O. comptonii** which has also rather long falcate leaves. The illegitimate name *M. pageanum* is not used as a basionym here because a legitimate name at species rank is available.

O. paardebergensis (L.Bolus) H.E.K.Hartmann Bradleya 1998: 78 ≡ *M. paardebergense* L.Bolus NM 1930: 200 ≡ *Lampranthus paardebergensis* (L.Bolus) L.Bolus NM 1939: 169 **T** L.Bolus 19216 (BOL).

Shrubs to 20 cm *h*, rather compact, branches decumbent, **I** to 35 mm *l*; **L** lunate, conspicuously red at apex, 28 mm *l*, 5–7 mm *b*, 7–10 mm ∅; **Fl** in rich cymes of 3–14 flowers, bracteoles at the middle of the pedicels of 14 mm length, petals pink, not closing during the night, to 10 mm *l*, to 1.5 mm *b*, filamentous staminodes present, filaments pink, basally paler, pollen golden, glands nearing each other, conspicuously crenulate; **Fr** 5 locules; **Ecol** in fissures of rocks; **Distr** Malmesbury, WC, S Africa.

Note: The species is compared with **Lampranthus lunatus** and **Ruschia senaria** in the original text; leaves and glands agree well with other species in **Oscularia**.

O. pedunculata (N.E.Br.) Schwantes in Jacobsen 1949: 58 ≡ *M. deltoides* L. var. *pedunculatum* N.E.Br. 1920: 118 ≡ *O. deltoides* (L.) Schwantes var. *pedunculata* (N.E.Br.) Schwantes in Jacobsen FR 1938: 158 **LT** Dillenius 1732: t. 195 f 245.

Shrubs; **L** dentate on keel and margins; **Fl** solitary, but arranged in groups of threes very often, pedicels much longer than the leaves, all bracteolate at their middle; **Fr** unknown; **Ecol** unknown; **Distr** uncertain.

Note: The species shares the dentate leaves with **O. deltoides** and **O. major**, both with rich inflorescences. The solitary flowers are similar to those of **O. caulescens.**

O. piquetbergensis (L.Bolus) H.E.K.Hartmann Bradleya 1998: 78 ≡ *M. piquetbergense* L.Bolus NM 1932: 325 ≡ *Lampranthus piquetbergensis* (L.Bolus) L.Bolus NM 1939: 169 **T** Compton 1233/29 (BOL).

Diffusely branched shrubs to 10 cm *h*; **L** a little falcate, apiculate, margins rarely inconspicuously 1–2 dentate, smooth, glaucous, to 12 mm *l*, to 6 mm *b* and ∅; **Fl** in small cymes, ∅ 15–17 mm, petals pink, to 6 mm *l*, to 1.25 mm *b*, filamentous staminodes and filaments pink, anthers and pollen golden, glands subapproximate, crenulate; **Fr** 5 locules; **Ecol** on the mountain; **Distr** Piquetberg, WC, S Africa.

Note: The denticulate leaves remind of **O. caulescens.**

O. prasina (L.Bolus) H.E.K.Hartmann Bradleya 1998: 78 ≡ *Lampranthus prasinus* L.Bolus NM 1939: 157 **T** Pillans 8853 (BOL).

Richly and diffusely branched shrubs, **I** 10–20 mm *l*, rarely to 30 mm *l*, ∅ 1.5–3 mm; **L** falcate to almost semi-globose, apiculate, grass-green, very finely papillate, often to 8 mm *l*, 4 mm *b*, to 5 mm ∅; **Fl** in small cymes, pedicels 7–10 mm *l*, bracteoles near the middle, petals purplish-pink, 5–6 mm *l*, to 1.5 mm *b*, filamentous staminodes few, stigmas reddish-purple, glands subapproximate, conspicuous; **Fr** unknown; **Ecol** unknown; **Distr** Piquetberg, WC, S Africa.

Note: The green leaves present a problem in **Oscularia.**

O. primiverna (L.Bolus) H.E.K.Hartmann Bradleya 1998: 79 ≡ *M. primivernum* L.Bolus NM 1930: 199 ≡ *Lampranthus primivernus* (L.Bolus) L.Bolus NM 1939: 169 **T** Frames 145/28 (BOL).

Erect shrubs to 30 cm *h*, **I** 15–25 mm *l*, ∅ to 7 mm; **L** lunate, glaucous, reddish suffused, 24–28 mm *l*, to 3 mm *b*, to 9 mm ∅; **Fl** in rich cymes aggregated into compound **Inf**, ∅ 18 mm, pedicels 5 mm *l*, bracteolate, petals bright pink, later almost salmon-coloured, 7 mm *l*, to 1.5 mm *b*, filamentous staminodes bent towards the stamens, pale, apically rich pink, anthers and pollen yellow, glands separate, crenulate, conspicuous; **Fr** unknown; **Ecol** in crevices of rocks; **Distr** Piquetberg, WC, S Africa.

Note: Compared originally with **O. ornata**; placed here because of the lunate leaves and the distinct nectary glands.

O. steenbergensis (L.Bolus) H.E.K.Hartmann Bradleya 1998: 79 (see **photo**) ≡ *M. steenbergense* L.Bolus NM 1930: 199 ≡ *Lampranthus steenbergensis* (L.Bolus) L.Bolus NM 1939: 169 **T** Compton 1625/29 (BOL).

Decumbent branches to 10 cm *l*, **I** to 10 mm *l*, to 5 mm ∅; **L** falcate, to 23 mm *l*, 10 mm *b*, 13 mm ∅; **Fl** in cymes, ∅ 16 mm, pedicels to 9 mm *l*, bracteoles placed below the middle, petals pink, basally white, 6 mm *l*, to 2 mm *b*, filamentous staminodes covering the stamens, pink, basally white, anthers and pollen yellow, glands crenulate; **Fr** unknown; **Ecol** unknown; **Distr** Simonstown or Vredenburg, WC, S Africa.

Note: The originally given locality is corrected by hand; transfer based on the falcate leaves and the separate glands.

O. superans (L.Bolus) H.E.K.Hartmann Bradleya 1998: 79 ≡ *M. superans* L.Bolus NM 1932: 348 ≡ *Lampranthus superans* (L.Bolus) L.Bolus NM 1939: 170 **T** Godman 19917 (BOL).

Erect, diffusely branched shrubs to 25 cm *h*, **I** 15–30(–50) mm *l*, 2–3 mm ∅; **L** ± falcate, glaucous, 12–16 mm *l*, 4 mm *b*, to 6 mm ∅; **Fl** in cymes of up to 13 flowers, unpleasant smell, at first only open at noon, later day and night, ∅ to 18 mm, pedicels c. 10 mm *l*, bracteoles near their middle, petals bright pink, 7 mm *l*, to 1 mm *b*, filamentous staminodes white, apically pink, glands united into a crenulate ring; **Fr** 5 locules; **Ecol** unknown; **Distr** Clanwilliam, Vanrhynsdorp, WC, S Africa.

Note: Placed here because of the falcate leaves, although the nectary is reported to appear as a ring.

O. thermarum (L.Bolus) H.E.K.Hartmann Bradleya 1998: 79 ≡ *M. thermarum* L.Bolus NM 1932: 365 ≡ *Lampranthus thermarum* (L.Bolus) L.Bolus NM 1939: 170 **T** L.Bolus 20206 (BOL).

Erect shrubs, 10–20 cm *h*, **I** 10–30 mm *l*, to 3 mm ∅; **L** falcate, glaucous, to 16 mm *l*, 2 mm *b*, 5 mm ∅; **Fl** in compact cymes, pedicels to 7 mm *l*, bracteolate, ∅ 22 mm, petals pink, 8 mm *l*, to 1.5 mm *b*, filamentous staminodes apically recurved, filaments pale, pollen golden, crenulate glands connate; **Fr** 5 locules; **Ecol** unknown; **Distr** Clanwilliam, WC, S Africa.

Note: Reported to be very similar to **Lampranthus copiosa.**

O. vernicolor (L.Bolus) H.E.K.Hartmann Bradleya 1998: 79 ≡ *M. vernicolor* L.Bolus NM 1931: 304 ≡ *Lampranthus vernicolor* (L.Bolus) L.Bolus NM 1939: 170 **T** Frames 19872 (BOL).

Diffusely branched shrubs to 10 cm *h*, **I** grey, 2–10 mm *l*, ∅ to 3 mm; **L** falcate, apiculate, 18 mm *l*, 3–5 mm *b*, 5–7 mm ∅; **Fl** in cymes, developing tardily, ∅ to 18 mm, pedicels 6–10 mm *l*, bracteolate, petals bright pink, to 7 mm *l*, 2 mm *b*, filamentous staminodes recurved, filaments white, apically pink, anthers and pollen golden, glands conspicuously crenulate; **Fr** 5 locules; **Ecol** unknown; **Distr** Tulbagh, WC, S Africa.

Note: Falcate, thick leaves and separate glands agree well with features of **Oscularia.**

O. vredenburgensis (L.Bolus) H.E.K.Hartmann Bradleya 1998: 79 ≡ *Lampranthus vredenburgensis* L.Bolus 1965: 307 **T** Hall 2793 (BOL).

Diffusely branched shrubs, branches 15–24 cm *l*, **I** 4–30 mm *l*; **L** lunate, 25 mm *l*, 5–6 mm *b*, 5–8 mm ∅; **Fl** in small cymes, pedicels 14–30 mm *l*, bracteoles at their middle or below, petals bright pink to purplish-pink, to 8 mm *l*, 1.5–2 mm *b*, filamentous staminodes appressed to stamens, white, upper half pink, nectary of 5 semi-lunate, crenulate glands approaching each other; **Fr** 5 locules; **Ecol** "occasional near coast on outcrops"; **Distr** Vredenburg, WC, S Africa.

Note: The species is transferred based on lunate leaves and separate glands.

Ottosonderia Ruschioideae

H. E. K. Hartmann

Ottosonderia L.Bolus NM 1958: 292 **Typus O. monticola** (Sonder) L.Bolus **Etym** *Otto Sonder*, 1812–1881, botanist at the Kew Herbarium, London.

Very succulent, dense shrubs, to 35 cm *h*, ∅ to 60 cm, **I** < **L**; **L** trigonous to terete, with a distinct mucro, epidermis smooth, stomata sunken, the subsidiary cells overarching the stomata somewhat, outer epidermal walls moderately thick with a central layer of crystals of c. 2 µm ∅, wax cover continuous, breaking into horizontal plates with only few irregular small wax particles on top; **Fl** in well developed pure dichasia, enlarged every year by additional new cymes developing a perennial, persistent **Inf** forming the conspicuous top part of the plant, the foliage **L** often inconspicuous, although they are very long, bracteoles triquetrous and pointed, differing markedly from the foliage **L**, persisting as hard, dry remains for many years on the plant, **K** 4, petals ochre at tips, base white, or pink, filamentous staminodes white, apically reddish, more numerous than petals, surrounding the white, apically yellow stamens in a central cone, dark green lophomorphic holonectary; **Fr** base short, funnel-shaped, top nearly semi-globose, rims low, the interior of the Ruschia-type, closing body small but variable in size, covering membranes with distinct closing rodlets, (7–)8(–9) locules; **S** light brown, testa smooth, only the areas near hilum and micropyle with elevated cells; **Ecol** in crevices in Namaqualand granite, winter rainfall > 200 mm p.a.; **Distr** Namaqualand, NC, S Africa.

Distribution of **Ottosonderia** L.Bolus

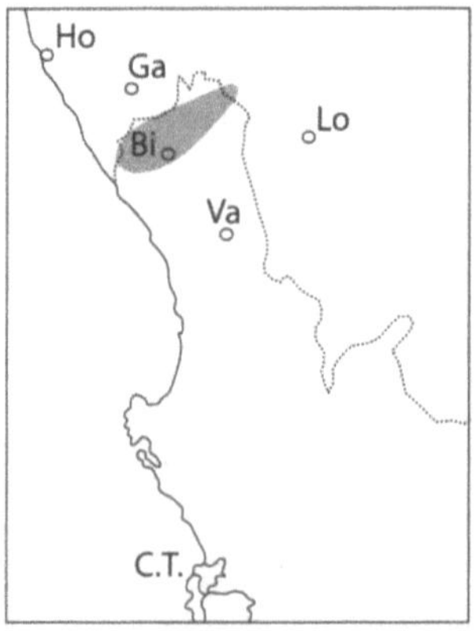

Note: The genus has been established based on the remarkable and unique feature of a big, persistent, annually enlarging inflorescence of pure cymose construction. The extraordinary drawing of it at BOL, placing the uppermost pairs of bracteoles just above soil level, has led to considerable confusion because the piece was thought to represent a plant. Yet, as described by Sonder (1862) and Bolus (1958), the foliage leaves are long and finger-shaped, thus completely different from the bracteoles.

O. monticola (Sonder) L.Bolus NM 1958: 292 (see **photo**) ≡ *M. monticola* Sonder in Harvey & Sonder 1862: 428 non L.Bolus 1930: 171 nomen illeg. ≡ *Hymenocyclus monticola* (Sonder) Schwantes 1949: 58 ≡ *Malephora monticola* (Sonder) H.Jacobsen & Schwantes 1958: 78 ≡ *Ruschia monticola* (Sonder) G.D.Rowley 1978: 8 **T** Zeyher (S, photograph!) = *M. bifoliatum* L.Bolus ABH 1922: 130 **T** Pillans 3686/15 (BOL!) = *O. obtusa* L.Bolus NM 1958: 293 ≡ *Ruschia ottosonderi* G.D.Rowley 1978: 8 **T** Hall 376/54 (BOL!).

L in the vegetative state c. 50 mm *l*, ∅ c. 8 mm, basally connate and a little decurrent, dark green; **Fl** bracteoles triangular as seen from above, 50–65 petals either pink to purple, or white and ochre, in four groups pointing into the directions of the 4 **K** lobes, 70–120 filamentous staminodes white, pink, or white and apically dark purple, 80–120 stamens, basally papillate; **Fr** without or with a recurving at the distal end of the covering membrane, ∅ 5–11 mm; **S** c. 1.3 mm *l*, c. 0.8 mm *b*; **Ecol** as for genus; **Distr** as for genus.

Note: The development of large, persisting inflorescences is related to a very low regeneration of the vegetative parts (e.g. from cuttings). This may have led to extinction of populations, probably less by grazing, since the plants are hard and firm, but probably by walking animals or men. Flowering is rare in cultivation. No reasons are given with the transfers of the species to **Malephora** and its synonym *Hymenocyclus*.

O. obtusa = **O. monticola**

Papularia = Trianthema

H. E. K. Hartmann

The genus *Papularia* was described by Forskål (1775: 69) for a single species which was soon transferred to **Trianthema**. Jeffrey (1960: 237) created the subgenus *Trianthema* L. subg. *Papularia* (Forskål) Jeffrey without giving a type and the basionym. The validation was done by Adamson, perhaps without intention; *Trianthema* L. subg. *Papularia* (Forskål) Jeffrey ex Adamson 1962: 247 is the correct name of the taxon at the rank of subgenus. The single species belongs into a complex of densely papillate plants centred in northern and eastern Africa.

P. crystallina ≡ **Trianthema crystallina**

Peersia Ruschioideae

H. E. K. Hartmann

Peersia L.Bolus FPSA 1927: t. 264 **Typus P. macradenia** (L.Bolus) L.Bolus **Etym** *V.S. Peers*, 1874–1940, plant collector in South Africa.

Compact to caespitose plants, sometimes forming huge cushions of 1–2 m ∅, the old leaves remaining on the stems for several years as black, dry fingers; **L** trigonous to terete, i.e. keel and margins often very inconspicuous, the sides convex rather than flat, apically acute, acuminate or obtuse, epidermis smooth, greyish-green, **L** 25–60 mm *l*, 6–9 mm *b* and ∅; **Fl** solitary, opening in the evening, pedicels > 20 mm *l*, compressed, bracteoles inconspicuous: c. 2/3 of the length and 1/2 of the width of the foliage leaves, of the same shape as the foliage leaves, therefore often overlooked, **K** 5, petals yellow or golden, filamentous staminodes absent, filaments yellow to paler, 5 big, high reaching nectary glands, truncate above; **Fr** stalks 20–40 mm *l*, base funnel-shaped with concave sides, top flat, the valves rising a little in the centre, in place of covering membranes only narrow edges from the septa spreading a little, closing bodies absent, but the end of the placenta occasionally raised into a tiny knob, expanding keels almost parallel to the tip of the valve, valve wings as broad as or broader than the expanding keels, 5 locules, ∅ 7–9.5 mm; **S** ovate to pear-shaped, testa cells with undulate anticlinal walls, with very fine granular sculptures; **Ecol** mostly on Ecca shales, rainfall mainly in March and November; **Distr** W part of EC, W part of NC, interior districts of WC, S Africa.

Note: Since 1938 (L.Bolus NM 1938: 135), **Peersia** has been treated as a synonym of **Rhinephyllum**, because "in floral characters the two genera are identical". Very similar features occur in the flowers of **Stomatium** and **Chasmatophyllum** as well. **Peersia** differs from **Rhinephyllum** most markedly in the very smooth leaves lacking any of the elevations typical of the latter genus and also of the genera **Stomatium** and **Chasmatophyllum**. Furthermore, leaves of **Peersia** are much bigger than those of **Rhinephyllum**, the fruits are distinctly longer stalked and the capsule itself much wider.

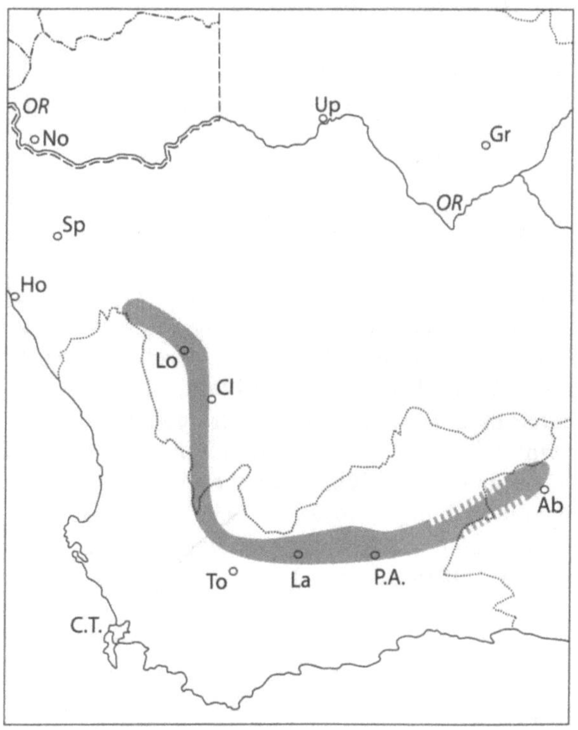

Distribution of **Peersia** L.Bolus

Key to the Species of Peersia

1. Leaves tapering from a broader base to an acute, pointed, often recurved tip .... **P. frithii**
– Leaves widening a little from the base, narrowing ± abruptly at the tip **2**
2. Leaf tips ± obtuse, calyx lobes almost equal in size .... **P. vanheerdei**
– Leaf tips ± acuminate, two calyx lobes distinctly bigger than the other three .... **P. macradenia**

P. frithii (L.Bolus) L.Bolus NM 1929: 94 (see **photo**) ≡ *M. frithii* L.Bolus ABH 1922: 125 ≡ *Rhinephyllum frithii* (L.Bolus) L.Bolus NM 1938: 135 **Lectotype**, here designated: Frith 431/18 (BOL!).

Compact or forming cushions; **L** triquetrous above, the keel rather sharp in the dry state, often the apical parts conspicuously spreading and recurved, 25–40 mm *l*, c. 8 mm *t*, thus stouter than the other species; **Fl** the outer 2 **K** lobes c. double the length of the 3 inner, petals golden, 6–12 mm *l*; **Fr** valve wings very broad, in the eastern populations almost semi-lunate, ∅ 7.5–9.5 mm, height 4–6 mm; **S** testa cells moderately raised over the entire seed; **Ecol** mostly on gentle slopes or in flats with rather fine Ecca shales, rarely on slopes with bigger stones, 100–200 mm rainfall p.a.; **Distr** Aberdeen, EC, Laingsburg, Prince Albert, WC, S Africa.

Note: The species can best be recognized by the combination of spreading, pointing leaves and the calyx lobes in two markedly different sizes.

P. macradenia (L.Bolus) L.Bolus FPSA 1927: t. 264 (see **photo**) ≡ *M. macradenium* L.Bolus ASAM 1913: 143 ≡ *Rhinephyllum macradenium* (L.Bolus) L.Bolus NM 1938: 135 **T** Pearson 3693 (BOL).

Caespitose plants developing into cushions; **L** acuminate, i.e. narrowing to the pointed tip only in the upper part of the leaf, semi-terete below, 25–60 mm *l*, 6–8 mm *t*; **Fl** with 2 taller **K** lobes, petals yellow; **Fr** valve wings broad, almost parallel to the sides of the valves, but tapering to the base, septa without or with very narrow ledges; **S** testa cells raised over the entire seed, the anticlinal walls markedly undulate; **Ecol** mostly in flats of rough, shaly Ecca, 100–200 mm rainfall p.a.; **Distr** Ceres, Laingsburg, Prince Albert, WC, S Africa.

Note: The combination of almost terete leaves with a rather abrupt acute tip and the differences in the sizes of the calyx lobes characterize this species.

P. vanheerdei (L.Bolus) H.E.K.Hartmann **comb. nov.** (see **photo**) ≡ *Rhinephyllum vanheerdei* L.Bolus Journal of South African Botany 29, 1963: 16 (as basionym) **Holotype** van Heerde 27272 (BOL!).

Plants forming mats of up to 2 m ∅; **L** obtuse, the lower leaf surface apically widened to almost forming a chin, **L** 30–65 mm *l*, 6–9 mm *b* and *t*; **Fl** 5 **K** lobes equal in size, petals yellow, 9–10 mm *l*; **Fr** valve wings broad to very broad, ledges on the septa very narrow; **S** testa cells flatter on the sides of the seed, these cells also with a smoother fine sculpture than the remaining, somewhat raised cells; **Ecol** in flats with Ecca shale, c. 100 mm rainfall p.a.; **Distr** Calvinia, Namaqualand, NC, Ceres, WC, S Africa.

Note: Rounded leaf tips and calyx lobes of equal size are typical of this species.

Pentacoilanthus = Mesembryanthemum

M. Gerbaulet

Pentacoilanthus (Rappa & Camarrone 1953: 32) was validly published without the designation of a type. Four species were included, one of which, namely **Mesembryanthemum crystallinum**, was chosen as lectotype by Hartmann & Bittrich (1990: 155). Consequently, in this circumscription *Pentacoilanthus* is illegitimate, since it is based on the same type as *Cryophytum* (N.E.Br. GC 1928: 128) which in turn is presently treated as **Mesembryanthemum** subg. **Cryophytum** (Bittrich 1987: 72).

Later, Rappa & Camarrone (1960: 13, 23, 24) changed their concept of *Pentacoilanthus* and transferred all four species to *Perapentacoilanthus* (Rappa & Camarrone 1955: 6, 1960: 23). In place of the former species of *Pentacoilanthus*, they combined three species of **Sceletium** to *Pentacoilanthus*. However, the name in the latter sense – beside being illegitimate as a younger homonym of *Pentacoilanthus* in its original circumscription – was not validly published.

P. aitonis ≡ **Mesembryanthemum aitonis**
P. crassicaulis ≡ **Sceletium crassicaule**
P. crystallinus ≡ **Mesembryanthemum crystallinum**
P. expansus ≡ **Sceletium expansum**
P. granulicaulis ≡ **Psilocaulon granulicaule**
P. splendens ≡ **Phyllobolus splendens**
P. tortuosus ≡ **Sceletium tortuosum**

Pentaschista = Stoeberia

H. E. K. Hartmann

The name *Pentaschista arborescens* appears on a sheet only (thus being a nomen nudum), collected in 1934, seven years after **Stoeberia** had been described. Friedrich (1960) took the epithet up for *Stoeberia beetzii* var. *arborescens*, a taxon included in **Stoeberia frutescens** here.

P. arborescens = **Stoeberia frutescens**

Perapentacoilanthus = Mesembryanthemum

M. Gerbaulet

Perapentacoilanthus (Rappa & Camarrone 1955: 6) was invalidly published without any species assigned. As shown by Hartmann & Bittrich (1990: 155, 156), the genus was created to receive all the species (Rappa & Camarrone 1960: 23) formerly included in *Pentacoilanthus* (Rappa & Camarrone 1953: 32), including the lectotype of the latter. Hence, *Perapentacoilanthus* is an illegitimate younger synonym of *Pentacoilanthus*, which in turn is an illegitimate younger synonym of **Mesembryanthemum** subg. **Cryophytum.**

P. acuminatus = **Phyllobolus splendens** ssp. **splendens**
P. aitonis ≡ **Mesembryanthemum aitonis**
P. brevifolius = **Phyllobolus splendens** ssp. **splendens**
P. crystallinus ≡ **Mesembryanthemum crystallinum**
P. delus ≡ **Phyllobolus delus**
P. fastigiatus ≡ **Mesembryanthemum fastigiatum**
P. granulicaulis ≡ **Psilocaulon granulicaule**
P. grossus ≡ **Phyllobolus grossus**
P. longispinulus = **Phyllobolus grossus**
P. otzenianus ≡ **Lampranthus otzenianus**
P. scintillans = **Phyllobolus oculatus**
P. splendens ≡ **Phyllobolus splendens**
P. sulcatus = **Phyllobolus splendens** ssp. **splendens**
P. vanrensburgii ≡ **Prenia vanrensburgii**
P. viridiflorus ≡ **Phyllobolus viridiflorus**

Peratetracoilanthus = Aptenia p.p.; = Aridaria p.p.; = Phyllobolus p.p.; = Prenia p.p.; = Psilocaulon p.p.

M. Gerbaulet

Peratetracoilanthus (Rappa & Camarrone 1960: 14, 15, 24) was invalidly published without the citation of a type. A total of eight species was transferred to this genus from various previously described genera. These combinations are invalid as well, since no basionyms were given.

P. defoliatus ≡ **Aridaria noctiflora** ssp. **defoliata**
P. geniculiflorus ≡ **Aptenia geniculiflora**
P. haeckelianus ≡ **Aptenia haeckeliana**
P. junceus ≡ **Psilocaulon junceum**
P. noctiflorus ≡ **Aridaria noctiflora**
P. parviflorus ≡ **Psilocaulon parviflorum**
P. spinuliferus ≡ **Phyllobolus spinuliferus**
P. tetragonus ≡ **Prenia tetragona**

Perissolobus = Machairophyllum

H. E. K. Hartmann

When Brown (GC 1930: 278) described this genus in a key, he had not noticed that the plants were most similar to those of **Machairophyllum**, because he was intrigued by the high number of eight calyx lobes in his only species, *P. bijliae*. A later detailed description (Brown GC 1931: 294) gave no reference to **Machairophyllum** either. Bolus (NM 1938: 135) was the first to recognize the identity and make the necessary transfer, although she did not alter the epithet to the correct gender.

P. bijliae ≡ **Machairophyllum bijliae**

Pherolobus = Dorotheanthus

H. E. K. Hartmann

Ihlenfeldt & Struck (1987) placed *Pherolobus* N.E.Br. as a subgenus in the genus **Dorotheanthus** Schwantes, based on similarities in the fruit: the expanding tissue consists mainly of a sheet, and the valve wings start with a broader base in the middle of the expanding tissue area, terminating distally in an awn.

P. maughanii ≡ **Dorotheanthus maughanii**

Phyllobolus Mesembryanthemoideae

M. Gerbaulet

Phyllobolus N.E.Br. GC 1925: 413; Bittrich 1987: 5–116; Gerbaulet 1994: 75–76; 1995: 385–399; 1997: 1–67; Gerbaulet & Struck 1996: 24–27; **Lectotypus P. resurgens** (Kensit) Schwantes **Etym** Gk *phyllon*, leaf, *bolos*, literally: cast of a die, intended is: shedding, referring to the deciduous leaves of the type species = *Amoebo-*

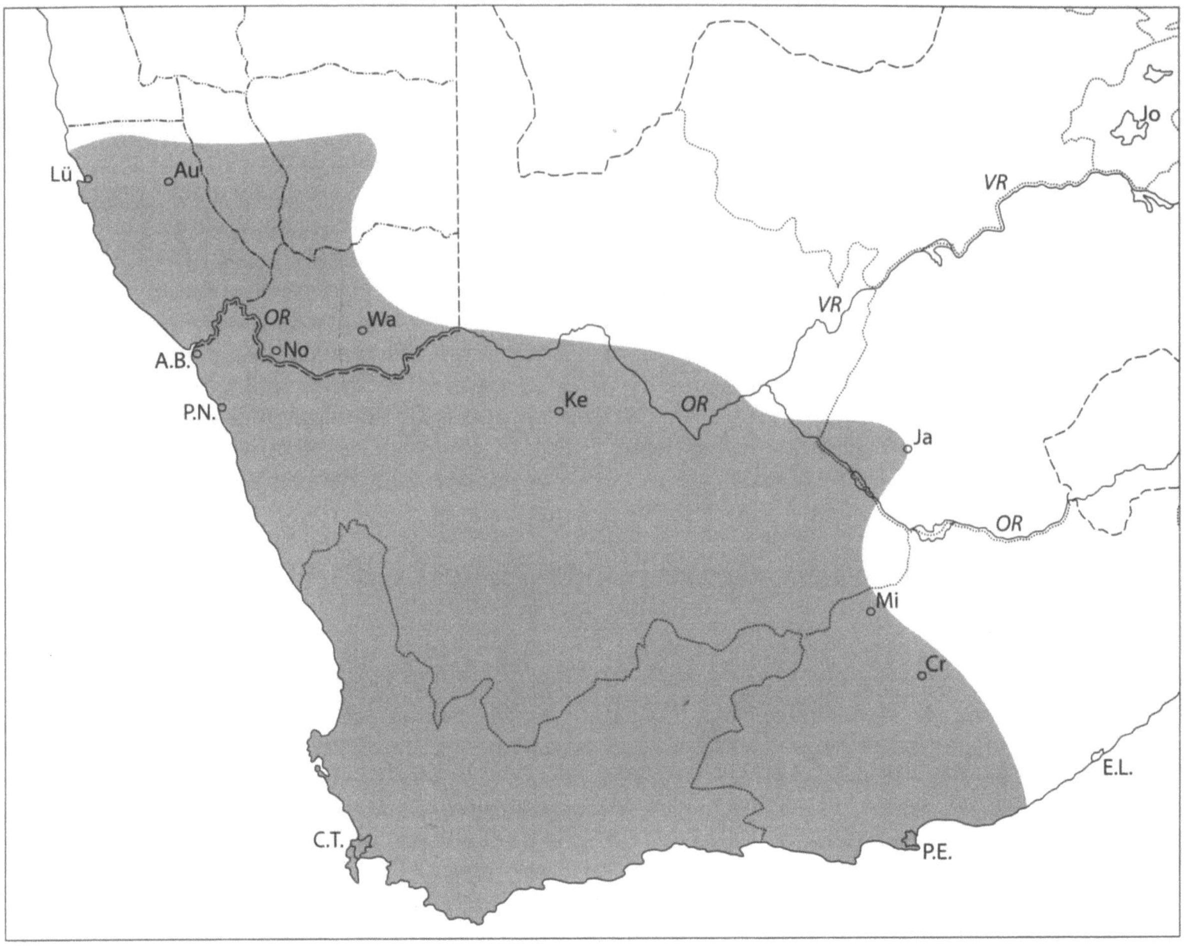

Distribution of **Phyllobolus** N.E.Br.

phyllum N.E.Br. GC 1925: 433 **Lectotypus** *Amoebophyllum angustum* N.E.Br. = **P. roseus** (L.Bolus) Gerbaulet = *Dactylopsis* N.E.Br. GC 1925: 413 **Typus** *Dactylopsis digitata* (Aiton) N.E.Br. ≡ **P. digitatus** (Aiton) Gerbaulet = *Sphalmanthus* N.E.Br. GC 1925: 433 ≡ *Phyllobolus* subg. *Sphalmanthus* (N.E.Br.) Bittrich 1987: 76 **Lectotypus** *Sphalmanthus canaliculatus* (Haw.) N.E.Br. ≡ **P. canaliculatus** (Haw.) Bittrich.

Decumbent or erect shrubs or prostrate to decumbent geophytes, roots often thickened, **I** herbaceous or corky or woody, long or much shortened, cortex usually with, rarely without additional vascular bundles, bladder cells usually mesomorphic, rarely somewhat xeromorphic, mostly distinct, large or small, sometimes hair-like, rarely indistinct and flat; **L** almost cylindrical or somewhat flattened, decussate and shortly connate at the base or becoming alternate in the **Inf** or alternate throughout the plants, deciduous, marcescent or persistent, central water-storing cells indistinct, bladder cells usually mesomorphic, rarely somewhat xeromorphic, mostly distinct, large or small, sometimes hair-like, rarely indistinct and flat; **Fl** in cymes or solitary, ∅ 10–40 mm, **K** (4–)5, petals in various shades of green, yellow, orange, red, or pink or almost white, **K** and petals connate into a short tube, filamentous staminodes present and reproductive parts not concealed, or filamentous staminodes absent and reproductive parts concealed, nectaries shell-shaped; **Fr** usually with valve wings inflexed over the valves or reflexed and fused in pairs, rarely without valve wings, locules (4–)5; **S** 1.1–1.8 mm *l*, D-shaped, rarely narrowly ovate to elliptic, black or brown, testa rough or rarely smooth, with or without a crest, sometimes with additional crests on the sides; **Ecol** karroid areas, winter and summer rainfall; **Distr** S Namibia; EC, FS, NC, WC, S Africa.

Note: Since its inception this genus was not only subject to some dramatic conceptual changes, but remained in nomenclatural limbo for much of its history. **Phyllobolus** originally comprised a couple of closely related species which are now classified under the name **P. resurgens** (Brown GC 1925: 412, in Phillips 1926: 245, GC 1928: 253). Almost four decades later, Bolus (1964: 42–43) included **Phyllobolus** in *Sphalmanthus*, disregarding the fact that the former name takes priority over the latter by 7 days. It was not after some 30 years of illegitimate application of the name *Sphalmanthus* that this error was rectified by Bittrich (1987: 5–116), who resurrected the name **Phyllobolus** for this genus. In addition, he included *Amoebophyllum*, **Aridaria**, **Prenia** and **Sceletium** in **Phyllobolus**. Questioning this approach, Gerbaulet (1995: 385–399) pointed out that **Phyllobolus** was not monophyletic in Bittrich's broad circumscription, but could be split into four coherent taxa. **Phyllobolus** was shown to form a monophyletic group after the inclusion of *Amoebophyllum*, *Dactylopsis* and *Sphalmanthus*. The revision of **Phyllobolus** (Gerbaulet 1997: 1–67) resulted in a most welcome reduction in the number of recognized species. It is as much unfortunate as inevitable that the rectification of the nomenclatural errors of the past required the resurrection of some unfamiliar specific names and the banishment into synonymy of several well established ones.

Key to the Species of Phyllobolus

1. Stems herbaceous and more or less completely deciduous, only basally becoming somewhat woody with age, or stems persistent with a conspicuous cork layer and rather succulent, sometimes very short **2**
- Stems always long and woody **23**
2. Filamentous staminodes absent, reproductive parts concealed **3**
- Filamentous staminodes present, reproductive parts not concealed **8**
3. Leaves alternate, stems short or not freely visible **4**
- Basal leaves decussate, becoming sometimes alternate in the inflorescences, stems short or elongated **6**
4. Leaves very thick and long, ∅ > 10 mm, more or less erect, heteromorphous like a finger and thumb, bladder cells of the leaves flat, seemingly absent, stems hidden by the leaf bases, viewed in cross section bladder cells of the stems forming a conspicuous layer, flowers small, ∅ 10–20 mm, petals tufted, white or cream .... **P. digitatus**
- Leaves much smaller, stems and leaves with distinct bladder cells, flower ∅ to 40 mm, not purely white or cream **5**
5. Bladder cells prominent, sometimes acuminate to hair-like, especially on the receptacle and pedicels; petals greenish yellow, flowers closing at night .... **P. resurgens**
- Bladder cells flattened; outer petals white to pale yellow, inner ones bright yellow, flowers remain open by day and night .... **P. herbertii**
6. Bladder cells large, sometimes flattened, sometimes elongated and tipped, especially on the receptacle, but not hair-like; petals greenish yellow .... **P. prasinus**
- Bladder cells hair-like **7**
7. Flowering branches prostrate to decumbent, outer petals greyish pink or (greenish) yellow, inner ones (greenish) yellow .... **P. tenuiflorus**
- Flowering branches erect, petals blue greenish white .... **P. viridiflorus**
8. Plants geophytic to hemicryptophytic with tuberous roots, stems with age often somewhat woody at the very base **9**
- Plants shrubby, roots not tuberous, stems often with a more or less conspicuous cork layer, leaves deciduous or persistent **16**
9. Plants creeping, rooting at the nodes .... **P. canaliculatus**
- Plants prostrate to decumbent or scrambling but never rooting at the nodes **10**
10. Leaves flattened, narrowly or broadly ovate **11**
- Leaves more or less cylindrical, linear to narrowly ovate **13**
11. Roots only slightly thickened into a tap-root, leaves basally decussate, otherwise mostly alternate, bladder cells large, often elongated on the receptacle and pedicels, flowers yellowish to sand-coloured or (salmon) pink with the centre somewhat darker .... **P. oculatus**
- Roots thickened into a distinct rounded tuber, leaves decussate throughout, bladder cells conspicuous, flowers yellow **12**
12. Leaves broadly ovate, bladder cells large .... **P. abbreviatus**
- Leaves narrowly ovate, bladder cells small .... **P. amabilis**
13. Leaves tufted, internodes usually not freely visible .... **P. pumilus**
- Leaves not tufted, internodes elongated and thus freely visible **14**

14. Leaves usually alternate almost throughout the inflorescences, bladder cells inconspicuous, conspicuous or hair-like, flowers in various shades of yellow, sand-coloured, copper, salmon or pink, seeds black without a crest .... **P. sinuosus**
– Leaves decussate throughout or only slightly alternate in the inflorescences, bladder cells conspicuous but never hair-like, flowers yellow or pale salmon or apricot **15**
15. Flower ∅ to 30 mm, yellowish to pale salmon or apricot, seeds brown with a crest .... **P. grossus**
– Flower ∅ to 40 mm, seeds black without a crest .... **P. caudatus**
16. Stems usually short and clustered, leaves deciduous **17**
– Stems usually elongated and not clustered, leaves more or less persistent **18**
17. Flowers yellow, dry leaves not becoming thorny .... **P. congestus**
– Flowers bright pink, dry leaves often becoming somewhat thorny .... **P. latipetalus**
18. Cork layer of the stems inconspicuous **19**
– Cork layer of the stems conspicuous **20**
19. Flowers bright pink, leaves flattened, triangular .... **P. gariepensis**
– Flowers in various shades of yellow to straw or sand-coloured, leaves more or less cylindrical, narrowly ovate .... **P. melanospermus**
20. Plants low, to 5 cm high, bladder cells conspicuous, locules 5, seeds brown with a crest at the back and additional crests on the sides .... **P. rabiei**
– Plants more than 10 cm, sometimes to 30 cm high, bladder cells conspicuous or inconspicuous, locules 4 or 5, seeds brown or black, with or without a crest at the back, but always without conspicuous crests on the sides **21**
21. Bladder cells conspicuous and usually very large, locules usually 5, sometimes 4, seeds brown with a crest .... **P. nitidus**
– Bladder cells inconspicuous **22**
22. Stems moderately thickened, locules as often 4 as 5, seeds black, without a crest .... **P. quartziticus**
– Stems very much thickened, locules usually 4, rarely 5, seeds brown, with a crest .... **P. spinuliferus**
23. Filamentous staminodes absent, reproductive parts concealed **24**
– Filamentous staminodes present, reproductive parts not concealed **26**
24. Herbaceous parts with large bladder cells, outer petals bright pink, inner ones bright yellow, only a few outer petals spreading, all other petals borne in a distinct cylinder .... **P. chrysophthalmus**
– Herbaceous parts with small, somewhat xeromorphic bladder cells, petals continuously spreading, inner ones not borne in a distinct cylinder **25**
25. Usually all branches erect, withered leaves becoming somewhat spiny, calyx lobes 4, outer petals pinkish red or cream, inner ones cream, rarely yellow .... **P. trichotomus**
– Younger branches somewhat decumbent, withered leaves not becoming spiny, calyx lobes 5, outer petals pinkish red, inner ones yellow .... **P. delus**
26. Leaves decussate, sometimes becoming alternate in the inflorescences, persistent, bladder cells small **27**
– Leaves alternate throughout, deciduous or marcescent (i.e. withering and not falling off), bladder cells large **30**
27. Leaves decussate throughout **28**
– Leaves becoming alternate in the inflorescences **29**
28. Branches not curving downward, withered leaves becoming somewhat spiny, flowers cream, yellow, or dull orange, locules 5 .... **P. saturatus**
– Branches curving downward, withered leaves not becoming spiny, flowers salmon pink, locules as often 4 as 5 .... **P. decurvatus**
29. Newly formed leaves tufted in old and dry leaf bases subtending short shoots, flowers cream to pink .... **P. suffruticosus**
– Leaves not tufted, flowers white, yellow, salmon, or pink .... **P. splendens**
30. Leaves marcescent and becoming somewhat spiny, recurring on the stems, flowers pale pink to pale salmon .... **P. roseus**
– Leaves deciduous, only the base becoming spiny or not spiny at all, not recurring on the stems, flowers pink with yellow centre or completely white **31**
31. Leaf bases not becoming spiny, flowers pink with yellow centre, open day and night, seeds greyish brown .... **P. deciduus**
– Leaf bases becoming spiny, flowers white, sweet-scented, opening in the evening, seeds black .. **P. lignescens**

P. subg. *Aridaria* ≡ **Aridaria**
P. subg. *Prenia* ≡ **Prenia**
P. subg. *Sceletium* ≡ **Sceletium**
P. subg. *Sphalmanthus* = **Phyllobolus**

P. abbreviatus (L.Bolus) Gerbaulet 1997: 26 (see **photo**) ≡ *Aridaria abbreviata* L.Bolus NM 1930: 177 ≡ *Nycteranthus abbreviatus* (L.Bolus) Schwantes KuaS 1952: 50 nomen illeg. ≡ *Sphalmanthus abbreviatus* (L.Bolus) L.Bolus in Jacobsen 1970: 510 nomen invalid. **T** Pillans 6213 (BOL!).

Prostrate geophytes with tuberous roots, **I** weakly lignified; **L** flattened, broadly ovate, decussate, with large bladder cells; **Fl** ∅ c. 20 mm, **K** 5, petals pale yellow, filamentous staminodes present, reproductive parts not concealed; **Fr** valve wings inflexed over the valves, 5 locules; **S** D-shaped, brown, testa rough, crest distinct; **Ecol** on shale covered with quartz pebbles, winter rainfall; **Distr** Vanrhynsdorp, WC, S Africa.

Note: This is a very delicate species with usually only a few leaves and flowers. However, after good rains, it may grow more vigorously.

P. amabilis Gerbaulet & Struck 1996: 25 (see **photo**) **T** Hammer 1405 (BOL!).

Small prostrate geophytes with tuberous roots, **I** weakly lignified; **L** flattened, narrowly ovate, decussate, becoming alternate in the **Inf**, with prominent bladder cells; **Fl** ∅ c. 20 mm, **K** 5, petals pale yellow, filamentous staminodes present, reproductive parts not concealed; **Fr** valve wings inflexed over the valves, 5 locules; **S** D-shaped, brown, testa rough, crest distinct; **Ecol** in rocky sites, border of the winter rainfall region; **Distr** Sutherland, NC, S Africa.

Note: This species is very similar to **P. grossus** from which it differs in smaller overall size and flattened leaves without a spinescent middle vein.

P. canaliculatus (Haw.) Bittrich 1987: 76 (see **photo**) ≡ *M. canaliculatum* Haw. 1795: 218 ≡ *Aridaria canaliculata* (Haw.) Friedrich 1959: 70 ≡ *Nycteranthus canaliculatus* (Haw.) Schwantes in Jacobsen 1960: 1314 nomen illeg. ≡ *Sphalmanthus canaliculatus* (Haw.) N.E.Br. GC 1928: 13 **LT** drawing (K!) = *Sphalmanthus calycinus* L.Bolus SAG 1927: 398 **LT** Ecklon & Zeyher 2924 (PRE!) = *M. salmoneum* Haw. 1821: 176 ≡ *Nycteranthus salmoneus* (Haw.) Schwantes in Jacobsen 1960: 1323 nomen illeg. ≡ *Sphalmanthus salmoneus* (Haw.) N.E.Br. GC 1928: 34 **NT** Salm-Dyck 1837: § 56 f 2.

Geophytes with tuberous roots and long, creeping branches, **I** weakly lignified, rooting at the nodes; **L** almost cylindrical or channelled, narrowly ovate to linear, decussate, becoming alternate in the **Inf**, with prominent bladder cells; **Fl** ∅ 20–30 mm, **K** 5, petals yellow or salmon, sometimes suffused red, filamentous staminodes present, reproductive parts not concealed; **Fr** valve wings inflexed over the valves, 5 locules; **S** small, D-shaped, black, testa rough, without a crest; **Ecol** in dunes along the S coast; **Distr** from the Cape Peninsula, WC, to Port Elizabeth, EC, S Africa.

Note: The stoloniferous habit of this species is unique within **Phyllobolus** and a special adaptation to sandy habitats.

P. caudatus (L.Bolus) Gerbaulet 1997: 27 ≡ *M. caudatum* L.Bolus ABH 1916: 29 ≡ *Aridaria caudata* (L.Bolus) L.Bolus SAG 1928: 127 ≡ *Nycteranthus caudatus* (L.Bolus) Schwantes in Jacobsen 1960: 1314 nomen illeg. ≡ *Sphalmanthus caudatus* (L.Bolus) N.E.Br. GC 1928: 35 **LT** Pillans 1673 (BOL!).

Prostrate to decumbent or scrambling geophytes with tuberous roots, **I** weakly lignified; **L** almost cylindrical, narrowly ovate to linear, decussate, becoming alternate in the **Inf**, with large bladder cells; **Fl** ∅ to 40 mm, **K** 5, petals yellow, filamentous staminodes present, reproductive parts not concealed; **Fr** valve wings inflexed over the valves, 5 locules; **S** D-shaped, dark brown or black, testa rough, without a crest; **Ecol** winter rainfall; **Distr** Worcester, WC, S Africa.

Note: **P. caudatus** was described from the Hex River Valley in the beginning of this century, but has not been found again ever since. The type specimen resembles **P. grossus** from which it differs in having much larger leaves and flowers as well as black seeds without a crest.

P. chrysophthalmus Gerbaulet & Struck 1996: 26 (see **photo**) **T** Gerbaulet & Struck 24733 (BOL!).

Erect, small shrubs to 15 cm *h*, roots not tuberous, **I** lignified; **L** almost cylindrical, narrowly ovate, decussate, persistent, with large bladder cells; **Fl** ∅ 10–20 mm, **K** 5, lobes remain ± erect during flowering, outer petals bright pink, inner ones golden yellow, only a few outmost petals spreading, other petals borne in a distinct cylinder, filamentous staminodes absent, reproductive parts concealed; **Fr** valve wings inflexed over the valves, 5 locules; **S** D-shaped, brown, testa rough, crest distinct; **Ecol** scrambling in other bushes, winter rainfall; **Distr** Vanrhynsdorp, WC, S Africa.

Note: This species is noteworthy for its beautiful pink and yellow flowers. The calyx lobes remain erect during flowering and most petals are borne in a distinct cylinder, whereas only a few outmost ones are spreading. The latter character state is also found in **P. herbertii**. On the other hand, erect calyx lobes occur in the very similar **P. delus** and – to a lesser extent – **P. trichotomus**. Otherwise within the Mesembryanthemoideae erect calyx lobes are only found in **Brownanthus**.

P. congestus (L.Bolus) Gerbaulet 1997: 28 (see **photo**) ≡ *Aridaria congesta* L.Bolus NM 1930: 178 ≡ *Nycteranthus congestus* (L.Bolus) Schwantes KuaS 1952: 50 nomen illeg. ≡ *Sphalmanthus congestus* (L.Bolus) L.Bolus in Jacobsen 1970: 511 nomen invalid. T Godman 19110 (BOL!).

Compact, to c. 5 cm *h*, roots not tuberous, I weakly lignified, becoming somewhat corky, usually short; L almost cylindrical, narrowly ovate, decussate, deciduous, with large bladder cells; **Fl** ∅ c. 20 mm, **K** 5, petals yellow, filamentous staminodes present, reproductive parts not concealed; **Fr** valve wings inflexed over the valves, 5 locules; **S** D-shaped, brown, testa rough, crest distinct; **Ecol** in rocky places, often on quartzite, winter rainfall; **Distr** Vanrhynsdorp, WC, S Africa.

Note: **P. congestus** is very similar to **P. latipetalus**, but differs by non-spinescent leaves and pale yellow flowers.

P. deciduus (L.Bolus) Gerbaulet 1997: 29 (see **photo**) ≡ *Aridaria decidua* L.Bolus SAG 1927: 256 ≡ *Nycteranthus deciduus* (L.Bolus) Schwantes KuaS 1952: 50 nomen illeg. ≡ *Sphalmanthus deciduus* (L.Bolus) L.Bolus in Jacobsen 1970: 511 nomen invalid. T Pillans 5289 (BOL!).

Large erect shrubs to c. 70 cm *h*, roots not tuberous, I lignified, stout; L cylindrical, narrowly ovate to linear, alternate except the first two leaf pairs, deciduous, with quite large bladder cells; **Fl** ∅ 20–30 mm, remaining open day and night, **K** 5, petals salmon pink, basally becoming yellow, filamentous staminodes present, reproductive parts not concealed; **Fr** valve wings inflexed over the valves, 5 locules; **S** small, D-shaped, greyish brown with darker spots, testa rough, without a crest; **Ecol** on rocky hills and in sandy plains, winter rainfall; **Distr** NW Namaqualand, WC, S Africa.

Note: **P. deciduus** is very similar to **P. lignescens** and **P. roseus**. All three species are large erect shrubs with periodically withering leaves which are basally spinescent in **P. lignescens** and usually marcescent and somewhat spinescent in **P. roseus**. Conversely, in **P. deciduus** the leaves are fully deciduous. Other character states of distinction of **P. deciduus** are the salmon-pink and yellow flowers which remain open day and night and the small greyish-brown seeds with dark spots.

P. decurvatus (L.Bolus) Gerbaulet 1997: 29 (see **photo**) ≡ *Aridaria decurvata* L.Bolus SAG 1927: 433 ≡ *Nycteranthus decurvatus* (L.Bolus) Schwantes KuaS 1952: 50 nomen illeg. ≡ *Sphalmanthus decurvatus* (L.Bolus) L.Bolus in Jacobsen 1970: 511 nomen invalid. T Pillans 5718 (BOL!) = *Aridaria laxa* L.Bolus SAG 1927: 433 ≡ *Nycteranthus laxus* (L.Bolus) Schwantes in Jacobsen 1960: 1319 nomen illeg. ≡ *Sphalmanthus laxus* (L.Bolus) N.E.Br. GC 1928: 35 T Pillans 5717 (BOL!) = *Aridaria obtusa* L.Bolus NM 1931: 227 ≡ *Nycteranthus obtusus* (L.Bolus) Schwantes KuaS 1952: 51 nomen illeg. ≡ *Sphalmanthus obtusus* (L.Bolus) L.Bolus in Jacobsen 1970: 513 nomen invalid. T SUG 9332 (BOL!).

Decumbent to erect shrubs to c. 40 cm *h*, branches curving downward, roots not tuberous, I lignified; L almost cylindrical, narrowly ovate, decussate, persistent, with small, somewhat flattened bladder cells; **Fl** ∅ 20–30 mm, **K** 4–5, petals salmon pink, basally cream, filamentous staminodes present, reproductive parts not concealed; **Fr** valve wings inflexed over the valves, 4–5 locules; **S** D-shaped, black, testa rough, without a crest; **Ecol** on lower rocky hillsides, winter rainfall; **Distr** NW Namaqualand, NC, S Africa.

Note: **P. decurvatus** is characterized by the downcurved branches.

P. delus (L.Bolus) Gerbaulet 1997: 30 (see **photo**) ≡ *M. delum* L.Bolus ABH 1925: 11 ≡ *Aridaria dela* (L.Bolus) L.Bolus FPSA 1927: t. 258 ≡ *Nycteranthus delus* (L.Bolus) Schwantes KuaS 1952: 50 nomen illeg. ≡ *Perapentacoilanthus delus* (L.Bolus) Rappa & Camarrone 1960: 23 nomen invalid. ≡ *Sphalmanthus delus* (L.Bolus) L.Bolus in Jacobsen 1970: 511 nomen invalid. **LT** Watermeyer 2040/23 (BOL!) = *Sphalmanthus hallii* L.Bolus 1962: 229 T Hall 2189 (BOL!) = *Sphalmanthus stayneri* L.Bolus 1962: 229 T Stayner 128/61 (BOL!).

Decumbent to erect small shrubs to 20 cm *h*, roots not tuberous, I lignified; L almost cylindrical, narrowly ovate, decussate, persistent, with small, somewhat xeromorphic bladder cells; **Fl** ∅ 10–20 mm, **K** 5, outer petals bright reddish pink, inner ones yellow, filamentous staminodes present, reproductive parts concealed; **Fr** valve wings inflexed over the valves, 5 locules; **S** D-shaped, brown, testa rough, crest distinct; **Ecol** in rocky sites, sometimes on quartzite, winter rainfall; **Distr** W Namaqualand, SW Calvinia, NC, Vanrhynsdorp, WC, S Africa.

Note: **P. delus** is strikingly similar to **P. trichotomus**, but differs most obviously by its less erect habit, somewhat softer leaves and 5 instead of 4 calyx lobes and locules.

P. digitatus (Aiton) Gerbaulet 1997: 30 ≡ *M. digitatum* Aiton 1789: 181 ≡ *Dactylopsis digitata* (Aiton) N.E.Br. in Phillips 1926: 244 **LT** drawing by Masson (BM).

Compact, to 20 cm *h*, roots somewhat thickened, I short and partly deciduous, weakly lignified but developing a huge cork layer, bladder cells slightly xeromorphic, about 1 mm *h* and narrowly packed; L cylindrical, narrowly oblong, alternate with sheath clasping the short stems, marcescent, the dry leaf providing a protective cover for the stem, heteromorphous, first leaf of a vegetation period long, second short, giving the appearance of "finger and thumb", bladder cells exceedingly flattened and thus virtually absent; **Fl** ∅ 10–20 mm, remaining open day and night, **K** 5, petals white to cream, filamentous staminodes absent, reproductive parts concealed; **Fr** valve wings reflexed and fused in pairs, 5 locules; **S** narrowly ovate to elliptical, light brown, testa smooth, without a crest; **Ecol** on shale covered with quartz pebbles, winter rainfall; **Distr** Vanrhynsdorp, Vredendal, WC, S Africa.

Note: Dactylopsis littlewoodii, previously treated as a species of its own, is included in **P. digitatus** as a subspecies. Although the morphological distinctness between the two subspecies is merely – yet consistently – a matter of dimensions, both taxa are geographically well separated. If properly tended, both subspecies will keep their shape and distinction even under cultivation.

Key to the Subspecies of P. digitatus

1. Leaves up to 100 mm long, flower ∅ to 20 mm .... **P. digitatus** ssp. **digitatus**
– Leaves up to 50 mm long, flower ∅ 10–15 mm .... **P. digitatus** ssp. **littlewoodii**

P. digitatus (Aiton) Gerbaulet ssp. **digitatus** (see **photo**) = *M. digitiforme* Thunberg 1791: 6 ≡ *Gibbaeum digitiforme* (Thunberg) N.E.Br. ex Range 1934: 15 **LT** Thunberg 11969 (UPS, photograph!).

Plants 10–20 cm *h*; **L** angle between "finger and thumb" < 90°, "finger" 50–100 mm *l* or even longer after ample rains; **Fl** ∅ to 20 mm; **Ecol** as for species; **Distr** S and C Vanrhynsdorp, Vredendal, WC, S Africa.

Note: In a good rainy season, the leaves of this subspecies may grow surprisingly long and chubby.

P. digitatus (Aiton) Gerbaulet ssp. **littlewoodii** (L.Bolus) Gerbaulet 1997: 32 (see **photo**) ≡ *Dactylopsis littlewoodii* L.Bolus 1963: 11 **T** Littlewood 308/60 (BOL!).

Plants < 10 cm *h*; **L** angle between "finger and thumb" > 90°, "finger" 30–50 mm *l*; **Fl** ∅ 10–15 mm; **Ecol** as for species; **Distr** N Vanrhynsdorp, WC, S Africa.

Note: This miniature subspecies retains its dwarf habit even after unusually good rains.

P. gariepensis Gerbaulet & Struck 1996: 26 (see **photo**) **T** Jürgens 93620 (BOL!).

Decumbent to erect shrubs c. 20–30 cm *h*, roots not tuberous, **I** weakly lignified; **L** somewhat flattened, triangular, decussate, becoming alternate in the **Inf**, persistent, with large bladder cells; **Fl** ∅ 20–30 mm, **K** 5, petals bright pink, filamentous staminodes present, reproductive parts not concealed; **Fr** valve wings inflexed over the valves, 5 locules; **S** D-shaped, black, testa rough, without a crest; **Ecol** on rocky mountain slopes, at border of the winter rainfall region; **Distr** NW Namaqualand, NC, S Africa.

Note: **P. gariepensis** is similar to **P. melanospermus**, but differs most obviously in its flattened, triangular leaves and bright pink flowers.

P. grossus (Aiton) Gerbaulet 1997: 32 (see **photo**) ≡ *M. grossum* Aiton 1789: 191 ≡ *Aridaria grossa* (Aiton) Friedrich 1959: 70 ≡ *Nycteranthus grossus* (Aiton) Schwantes in Jacobsen 1960: 1318 nomen illeg. ≡ *Perapentacoilanthus grossus* (Aiton) Rappa & Camarrone 1960: 23 nomen invalid. ≡ *Sphalmanthus grossus* (Aiton) N.E.Br. GC 1928: 35 **T** drawing by Taylor (BM) = *M. commutatum* A.Berger 1922: 631 ≡ *Nycteranthus commutatus* (A.Berger) Schwantes in Jacobsen 1960: 1314 nomen illeg. ≡ *Sphalmanthus commutatus* (A.Berger) N.E.Br. GC 1928: 35 **T** Salm-Dyck 1837: § 54 f 4 (as *M. longispinulum*) = *Aridaria gratiae* L.Bolus NM 1932: 312 ≡ *Nycteranthus gratiae* (L.Bolus) Schwantes KuaS 1952: 50 nomen illeg. ≡ *Sphalmanthus gratiae* (L.Bolus) L.Bolus in Jacobsen 1970: 512 nomen invalid. **T** Britten 11 (BOL!) = *Sphalmanthus gydouwensis* L.Bolus 1964: 75 **T** Leipoldt 4425 (BOL!) = *Aridaria laxipetala* L.Bolus NM 1930: 214 ≡ *Nycteranthus laxipetalus* (L.Bolus) Schwantes KuaS 1952: 50 nomen illeg. ≡ *Sphalmanthus laxipetalus* (L.Bolus) L.Bolus in Jacobsen 1970: 512 nomen invalid. **T** Leipoldt 19243 (BOL!) = *Sphalmanthus leipoldtii* L.Bolus 1964: 42 **T** Leipoldt 4483 (BOL!) = *M. longispinulum* Haw. 1824: 426 ≡ *Aridaria longispinula* (Haw.) L.Bolus SAG 1928: 127 ≡ *Nycteranthus longispinulus* (Haw.) Schwantes in Jacobsen 1960: 1319 nomen illeg. ≡ *Perapentacoilanthus longispinulus* (Haw.) Rappa & Camarrone 1960: 23 nomen invalid. ≡ *Sphalmanthus longispinulus* (Haw.) N.E.Br. GC 1928: 35 **T** drawing (K!) = *Aridaria platysepala* L.Bolus NM 1934: 417 ≡ *Nycteranthus platysepalus* (L.Bolus) Schwantes KuaS 1952: 51 nomen illeg. ≡ *Sphalmanthus platysepalus* (L.Bolus) L.Bolus in Jacobsen 1970: 513 nomen invalid. **T** Leighton & Lewis 21027 (BOL!) = *Aridaria subpetiolata* L.Bolus NM 1935: 476 ≡ *Sphalmanthus subpetiolatus* (L.Bolus) L.Bolus 1964: 42 **T** Leighton 21423 (BOL!) = *Aridaria willowmorensis* L.Bolus NM 1930: 216 ≡ *Nycteranthus willowmorensis* (L.Bolus) Schwantes KuaS 1952: 51 nomen illeg. ≡ *Sphalmanthus willowmorensis* (L.Bolus) L.Bolus in Jacobsen 1970: 516 nomen invalid. **T** SUG 8329 (BOL!).

Prostrate to decumbent or scrambling geophytes with tuberous roots, **I** slender, becoming somewhat lignified at the base of the plants; **L** almost cylindrical, narrowly ovate to linear, decussate, finally becoming alternate in the **Inf**, with prominent bladder cells, basal part of the middle vein somewhat lignified and persistent; **Fl** ∅ 20–30 mm, **K** 5, petals lemon yellow, straw-yellow, pale salmon or apricot, filamentous staminodes present, reproductive parts not concealed; **Fr** valve wings inflexed over the valves, 5 locules; **S** D-shaped, brown, testa rough, crest distinct; **Ecol** on slopes or in plains, often scrambling in other bushes, winter to summer rainfall; **Distr** from Ceres and Worcester, WC, to Cradock and Albany, EC, S Africa.

Note: All vegetative characters of this species are extremely variable depending on microclimatic and geological conditions.

P. herbertii (N.E.Br.) Gerbaulet 1997: 34 (see **photo**) ≡ *Sphalmanthus herbertii* N.E.Br. 1934: 70 (as *S. herberti*) ≡ *Aridaria herbertii* (N.E.Br.) Friedrich 1959: 55 ≡ *Nycteranthus herbertii* (N.E.Br.) Schwantes in Jacobsen 1960: 1318 nomen illeg. **T** Maughan Brown 1932 (K!) = *Sphalmanthus humilis* L.Bolus 1962: 291 **T** Hall 2497 (BOL!).

Compact, to 3 cm *h*, roots thickened, **I** short, flowering stems may be slightly elongated, weakly lignified but basally corky; **L** almost cylindrical, narrowly ovate, alternate, deciduous, with large but somewhat flattened bladder cells; **Fl** ∅ c. 40 mm, remaining open day and night, sweet-scented, **K** 5, outer petals white or light yellow, inner ones bright yellow, filamentous staminodes absent, reproductive parts concealed; **Fr** valve wings inflexed over the valves, 5 locules; **S** D-shaped, brown, testa rough, crest distinct; **Ecol** on quartzite, winter rainfall; **Distr** Vanrhynsdorp, WC, S Africa.

Note: **P. herbertii** is very similar to **P. resurgens**, but differs from the latter in the shape and size of the bladder cells, flower colour, time of flower presentation and in that the inner petals are borne in a distinct cylinder. In addition, **P. herbertii** produces only a few leaves during the growing season.

P. latipetalus (L.Bolus) Gerbaulet 1997: 34 (see **photo**) ≡ *Aridaria latipetala* L.Bolus NM 1929: 111 ≡ *Nycteranthus latipetalus* (L.Bolus) Schwantes KuaS 1952: 50 nomen illeg. ≡ *Sphalmanthus latipetalus* (L.Bolus) L.Bolus in Jacobsen 1970: 512 nomen invalid. **T** Pillans 6151 (BOL!).

Compact small plants to c. 5 cm *h*, roots not tuberous, **I** weakly lignified, becoming somewhat corky, usually short; **L** almost cylindrical, narrowly ovate to linear, decussate, becoming alternate in the **Inf**, finally deciduous but slightly spinescent at the base, with prominent bladder cells; **Fl** ∅ c. 20 mm, **K** 5, petals bright pink, whitish at the base, quite broad, filamentous staminodes present, reproductive parts not concealed; **Fr** valve wings inflexed over the valves, 5 locules; **S** D-shaped, black, testa rough, whithout a crest; **Ecol** on quartzite, winter rainfall; **Distr** NE Namaqualand, NC, S Africa.

Note: This species is very similar to **P. congestus**, but differs most prominently in persistent dry and somewhat spinescent leaves as well as in bright pink petals.

P. lesliei = **P. resurgens**

P. lignescens (L.Bolus) Gerbaulet 1997: 35 (see **photo**) ≡ *Sphalmanthus lignescens* L.Bolus 1963: 137 ≡ *M. suaveolens* L.Bolus NM 1958: 297 nomen illeg. non Schwantes 1926: 176, 254 ≡ *Cryophytum suaveolens* (L.Bolus) J.Ingram 1970: 46 **T** Bayer 26213 (BOL!).

Large erect shrubs to 50 cm *h*, roots not tuberous, **I** lignified, stout, without cortical bundles; **L** almost cylindrical, narrowly ovate to linear, alternate, finally deciduous but slightly spinescent at the base, with quite large bladder cells; **Fl** ∅ c. 30 mm, opening in the evening, sweet-scented, **K** 5, petals white, filamentous staminodes present, reproductive parts not concealed; **Fr** valve wings inflexed over the valves, 5 locules; **S** D-shaped, black, testa rough, without a crest; **Ecol** on rocky hills, summer rainfall; **Distr** SE Namibia, NE Namaqualand, Kenhardt, NC, S Africa.

Note: This species is similar to **P. deciduus** and **P. roseus**, but differs from both species in its pure white flowers, which open in the evening and in lacking cortical bundles. The latter feature is unique within the whole genus.

P. melanospermus (Dinter & Schwantes) Gerbaulet 1997: 35 (see **photo**) = *Sphalmanthus melanospermus* Dinter & Schwantes in Schwantes GW 1929: 69 ≡ *M. melanospermum* Dinter FR 1923: 133 nomen illeg. non A.Berger 1922: 633 ≡ *Aridaria dinteri* L.Bolus NM 1958: 298 nomen illeg. ≡ *Nycteranthus dinteri* (L.Bolus) Schwantes in Jacobsen 1960: 1316 nomen illeg. ≡ *Sphalmanthus dinteri* (L.Bolus) L.Bolus in Jacobsen 1970: 511 nomen invalid. **T** Dinter 3779 (BOL!) = *Aridaria lignea* L.Bolus NM 1929: 50 ≡ *Nycteranthus ligneus* (L.Bolus) Schwantes KuaS 1952: 50 nomen illeg. ≡ *Sphalmanthus ligneus* (L.Bolus) L.Bolus in Jacobsen 1970: 512 nomen invalid. **T** Pillans 6142 (BOL!).

Erect shrubs to 25 cm *h*, roots not tuberous, **I** weakly lignified; **L** almost cylindrical, narrowly ovate, decussate, becoming alternate in the **Inf**, persistent, with large bladder cells; **Fl** ∅ 20–30 mm, **K** 5, petals in different shades of yellow to straw or sand-coloured, filamentous staminodes present, reproductive parts not concealed; **Fr** valve wings inflexed over the valves, 5 locules; **S** D-shaped, black, testa rough, without a crest; **Ecol** in rocky sites, sometimes on quartzite, in the southern Namib and the dry Gariep (Orange) River Valley, winter rainfall; **Distr** SW Namibia; NW Namaqualand, NC, S Africa.

Note: **P. melanospermus** is very similar to **P. gariepensis**, but differs in the shape of the leaves and colour of the flowers. The epithet refers to the black seeds which, however, is a quite common character state in the genus.

P. nitidus (Haw.) Gerbaulet 1997: 36 (see **photo**) ≡ *M. nitidum* Haw. 1795: 412 ≡ *Aridaria nitida* (Haw.) N.E.Br. JB 1928: 143 ≡ *Drosanthemum nitidum* (Haw.) Schwantes FR 1938: 229 ≡ *Sphalmanthus nitidus* (L.Bolus) L.Bolus in Jacobsen 1970: 513 nomen invalid. **NT** DC. 1804: t. 129 (as *M. brachiatum*) = *M. auratum* Sonder in Harvey & Sonder 1862: 449 ≡ *Sphalmanthus auratus* (L.Bolus) L.Bolus in Jacobsen 1970: 510 nomen invalid. ≡ *M. aureum* Thunberg 1791: 16 nomen illeg. non L. 1759: 1060 ≡ *Aridaria aurea* (Thunberg) L.Bolus NM 1929: 68 ≡ *Nycteranthus aureus* (Thunberg) Schwantes KuaS 1952: 50 nomen illeg. **T** Thunberg 11948 (UPS, photograph!) = *Aridaria inaequalis* L.Bolus NM 1958: 295 ≡ *Nycteranthus inaequalis* (L.Bolus) Schwantes in Jacobsen 1960: 1318 nomen illeg. **T** Herre 12130 (BOL!) = *Sphalmanthus littlewoodii* L.Bolus JSAB 1960: 56 **T** Littlewood 449/59 (BOL!) = *Aridaria rhodandra* L.Bolus NM 1931: 225 ≡ *Nycteranthus rhodandrus* (L.Bolus) Schwantes KuaS 1952: 51 nomen illeg. ≡ *Sphalmanthus rhodandrus* (L.Bolus) L.Bolus in Jacobsen 1970: 514 nomen invalid. **T** Leipoldt 19372 (BOL!).

Decumbent to erect shrubs to 30 cm *h*, sometimes low and cushion-like or scrambling, roots not tuberous, **I** weakly lignified but with a conspicuous cork layer; **L** almost cylindrical, narrowly ovate, decussate, persistent, with small or large but always conspicuous bladder cells; **Fl** ∅ 20–30 mm, **K** (4–)5, petals (pale) yellow, sometimes reddish at the base, sometimes with pale pink or salmon-pink tips, filamentous staminodes present, reproductive parts not concealed; **Fr** valve wings inflexed over the valves, 4–5 locules; **S** D-shaped, brown, testa rough, crest distinct; **Ecol** in plains, winter rainfall; **Distr** Calvinia, NC, Vanrhynsdorp, Clanwilliam, Ceres, Worcester, Montagu, Laingsburg, Ladismith, Swellendam, Prince Albert, WC, S Africa.

Note: **P. nitidus** usually produces large bladder cells. Only in plants from the Ceres Karroo these are particularly small.

P. noctiflorus = **Aridaria noctiflora**

P. oculatus (N.E.Br.) Gerbaulet 1997: 37 (see **photo**) ≡ *M. oculatum* N.E.Br. KB 1911: 313 ≡ *Aridaria oculata* (N.E.Br.) L.Bolus NM 1930: 215 ≡ *Nycteranthus oculatus* (N.E.Br.) Schwantes KuaS 1952: 51 nomen illeg. ≡ *Sphalmanthus oculatus* (N.E.Br.) N.E.Br. GC 1928: 34 **T** Pearson 6172 (K!) = *Aridaria anguinea* L.Bolus SAG 1927: 256 ≡ *Nycteranthus anguineus* (L.Bolus) Schwantes KuaS 1952: 50 nomen illeg. ≡ *Sphalmanthus anguineus* (L.Bolus) L.Bolus in Jacobsen 1970: 510 nomen invalid. **T** Pillans 5694 (BOL!) = *Aridaria arenicola* L.Bolus NM 1932: 357 ≡ *Nycteranthus arenicola* (L.Bolus) Schwantes KuaS 1952: 50 nomen illeg. ≡ *Sphalmanthus arenicola* (L.Bolus) L.Bolus in Jacobsen 1970: 510 nomen invalid. **LT** Leighton 20157 (BOL!) = *M. crystallino-papillosum* L.Bolus ex Fedde & Schuster 1918: 92 **T** Pearson 6158 (SAM!) = *Sphalmanthus fragilis* N.E.Br. GC 1928: 34 ≡ *Aridaria fragilis* (N.E.Br.) Friedrich 1959: 70 ≡ *Nycteranthus fragilis* (N.E.Br.) Schwantes in Jacobsen 1960: 1317 nomen illeg. **T** Pillans 5734 (BOL!) = *Sphalmanthus longipapillatus* L.Bolus JSAB 1960: 163 **T** Herre 14238 (BOL!) = *Aridaria pomonae* L.Bolus NM 1934: 418 ≡ *Nycteranthus pomonae* (L.Bolus) Schwantes KuaS 1952: 51 nomen illeg. ≡ *Sphalmanthus pomonae* (L.Bolus) L.Bolus in Jacobsen 1970: 513 nomen invalid. **T** Frames 21029 (BOL!) = *Amoebophyllum rangei* N.E.Br. JB 1928: 79 ≡ *Aridaria rangei* (N.E.Br.) Friedrich 1959: 70 **T** Range 1056 (B!) = *M. scintillans* Dinter FR 1923: 128 ≡ *Aridaria scintillans* (Dinter) Friedrich 1959: 70 ≡ *Nycteranthus scintillans* (Dinter) Schwantes in Jacobsen 1960: 1323 nomen illeg. ≡ *Perapentacoilanthus scintillans* (Dinter) Rappa & Camarrone 1960: 23 nomen invalid. ≡ *Sphalmanthus scintillans* (Dinter) Dinter & Schwantes in Schwantes GW 1929: 69 **T** Dinter 3868 (B!) = *Aridaria varians* L.Bolus NM 1958: 296 ≡ *Nycteranthus varians* (L.Bolus) Schwantes in Jacobsen 1960: 1326 nomen illeg. **T** Eberlanz 12568 (BOL!).

Prostrate to decumbent or sometimes scrambling geophytes, roots thickened into a tap-root, **I** weakly lignified, basally becoming somewhat more lignified with age; **L** flattened, narrowly ovate to narrowly triangular, decussate at the base but soon becoming alternate, with large bladder cells, often much elongated, especially on the receptacle and pedicels, basal parts of the primary leaf veins somewhat lignified and persistent for a short while; **Fl** ∅ 20–30 mm, **K** 5, petals pale yellow to pale salmon pink or sand-coloured, filamentous staminodes present, reproductive parts not concealed; **Fr** valve wings inflexed over the valves, 5 locules; **S** D-shaped, black, testa rough, without a crest; **Ecol** in rocky sites or on sand, winter rainfall; **Distr** SW Namibia; NW Namaqualand, NC, S Africa.

Note: For the sake of nomenclatural legitimacy the commonly applied epithet *scintillans* had to concede priority to the preceding epithet **oculatus. P. oculatus** may form large mats of about 1 m in ∅.

P. pallens = **Prenia pallens**
P. pearsonii = **P. resurgens**

P. prasinus (L.Bolus) Gerbaulet 1997: 38 (see **photo**) ≡ *Aridaria prasina* L.Bolus NM 1928: 6 ≡ *Nycteranthus prasinus* (L.Bolus) Schwantes KuaS 1952: 51 nomen illeg. ≡ *Sphalmanthus prasinus* (L.Bolus) L.Bolus in Jacobsen 1970: 513 nomen invalid. T Pillans 17792 (BOL!) = *Sphalmanthus crassus* L.Bolus 1963: 18 T van Heerde 27271 (BOL!) = *Aridaria elongata* L.Bolus NM 1958: 297 ≡ *Nycteranthus elongatus* (L.Bolus) Schwantes in Jacobsen 1960: 1316 nomen illeg. T Herre 11337 (BOL!) = *Sphalmanthus herrei* L.Bolus 1964: 242 T Herre 14856 (BOL!) = *Aridaria multiseriata* L.Bolus NM 1931: 226 ≡ *Nycteranthus multiseriatus* (L.Bolus) Schwantes KuaS 1952: 50 nomen illeg. T L.Bolus 2504/30 (BOL!).

Decumbent small shrubs to 5 cm *h*, roots thickened, **I** mostly short, flowering stems longer, weakly lignified but basally with a thick cork layer; **L** almost cylindrical, narrowly ovate, decussate, becoming alternate in the **Inf**, deciduous, with large, sometimes flattened, sometimes elongated and tipped bladder cells, especially marked on the receptacles and pedicels; **Fl** ∅ c. 40 mm, **K** 5, petals greenish yellow, filamentous staminodes absent, reproductive parts concealed; **Fr** valve wings inflexed over the valves, 5 locules; **S** large, D-shaped, brown, testa rough, crest distinct; **Ecol** on quartzite plains or hills, winter rainfall; **Distr** NW Namaqualand, NC, S Africa.

Note: **P. prasinus** is very similar to **P. tenuiflorus** and **P. viridiflorus**. From both, it is distinct in the shape of the bladder cells, which is never hair-like. In addition, the species differs in a larger habit from **P. tenuiflorus** and in the flower colour from **P. viridiflorus**.

P. pubicalyx = **P. resurgens**

P. pumilus (L.Bolus) Gerbaulet 1997: 39 (see **photo**) ≡ *Aridaria pumila* L.Bolus NM 1930: 214 ≡ *Nycteranthus pumilus* (L.Bolus) Schwantes KuaS 1952: 51 nomen illeg. ≡ *Sphalmanthus pumilus* (L.Bolus) L.Bolus in Jacobsen 1970: 513 nomen invalid. T SUG 9175 (BOL!) = *Aridaria albertensis* L.Bolus NM 1954: 262 ≡ *Nycteranthus albertensis* (L.Bolus) Schwantes in Jacobsen 1960: 1313 nomen illeg. ≡ *Sphalmanthus albertensis* (L.Bolus) L.Bolus in Jacobsen 1970: 510 nomen invalid. T Acocks 17119 (BOL!) = *Aridaria oubergensis* L.Bolus NM 1934: 417 ≡ *Nycteranthus oubergensis* (L.Bolus) Schwantes KuaS 1952: 51 nomen illeg. ≡ *Sphalmanthus oubergensis* (L.Bolus) L.Bolus in Jacobsen 1970: 513 nomen invalid. T Leighton 21028 (BOL!).

Compact geophytes with tuberous roots, **I** short, longer on flowering stems, weakly lignified; **L** almost cylindrical, narrowly ovate, decussate, becoming alternate in the **Inf**, with large bladder cells; **Fl** ∅ c. 30 mm, **K** 5, petals yellowish to pale salmon, filamentous staminodes present, reproductive parts not concealed; **Fr** valve wings inflexed over the valves, 5 locules; **S** D-shaped, brown, testa rough, crest distinct; **Ecol** on rocky sites in shallow soil, winter to summer rainfall; **Distr** Sutherland, NC, Montagu, Laingsburg, Prince Albert, WC, S Africa.

Note: **P. pumilus** resembles **P. resurgens** in habit.

P. quartziticus (L.Bolus) Gerbaulet 1997: 40 (see **photo**) ≡ *Aridaria quartzitica* L.Bolus NM 1929: 69 ≡ *Nycteranthus quartziticus* (L.Bolus) Schwantes KuaS 1952: 51 nomen illeg. ≡ *Sphalmanthus quartziticus* (L.Bolus) L.Bolus in Jacobsen 1970: 513 nomen invalid. T Pillans 5034 (BOL!).

Erect shrubs to 30 cm *h*, roots not tuberous, **I** weakly lignified or more strongly lignified with age and furnished with a thick cork layer; **L** almost cylindrical, narrowly ovate, decussate, persistent, with small, flattened bladder cells; **Fl** ∅ 20–30 mm, **K** 4(–5), petals very pale pink or salmon pink, filamentous staminodes present, reproductive parts not concealed; **Fr** valve wings inflexed over the valves, 4(–5) locules; **S** D-shaped, black, testa rough, without a crest; **Ecol** in plains on sand or quartzite along the western coast, winter rainfall; **Distr** W Namaqualand, NC, Vredendal, WC, S Africa.

Note: **P. quarziticus** is similar to **P. spinuliferus**, but differs clearly in more slender stems and black seeds lacking a crest.

P. rabiei (L.Bolus) Gerbaulet 1997: 40 (see **photo**) ≡ *Aridaria rabiei* L.Bolus SAG 1927: 433 (as *A. rabei*) ≡ *Nycteranthus rabiei* (L.Bolus) Schwantes in Jacobsen 1960: 1322 nomen illeg. ≡ *Sphalmanthus rabiei* (L.Bolus) N.E.Br. GC 1928: 35 T Rabie & Smith 4099 (BOL!).

Decumbent small shrubs to c. 5 cm *h*, roots not tuberous, **I** short, weakly lignified but with a thick cork layer; **L** almost cylindrical, narrowly ovate, decussate, persistent, with small, somewhat flattened bladder cells; **Fl** ∅ 20–30 mm, **K** 5, petals pale salmon, filamentous staminodes present, reproductive parts not concealed; **Fr** valve wings absent, 5 locules; **S** large, D-shaped, brown, testa rough, crest distinct, additional crests on the sides; **Ecol** summer rainfall; **Distr** Fauresmith, FS, S Africa.

Note: **P. rabiei** is similar to **P. spinuliferus**, but differs in its much smaller and lower habit.

P. resurgens (Kensit) Schwantes GF 1928: 68 (see **photo**); Gerbaulet 1994: 76; 1997: 41 ≡ *M. resurgens* Kensit TRSSA 1909: 154 ≡ *Aridaria resurgens* (Kensit) L.Bolus SAG 1928: 127 ≡ *Sphalmanthus resurgens* (Kensit) L.Bolus 1964: 42 **LT** Pillans 969 (K!) = *P. lesliei* N.E.Br. GC 1928: 253 T Leslie s.n. 1926 (K!) = *Sphalmanthus micans* L.Bolus 1964: 43 T Middlemost 731/60 (BOL!) = *P. pearsonii* N.E.Br. in Phillips 1926: 246 nomen invalid. T Pearson 6441 (K!) = *P. pubicalyx* N.E.Br. GC 1928: 253 **LT** Schlechter 10913 (K!).

Compact, to 3 cm *h*, with a thick caudex, I not freely visible, weakly lignified but basally corky; **L** almost cylindrical, narrowly ovate to linear, alternate and tufted, deciduous, with large bladder cells, sometimes acuminate to hair-like, especially marked on the receptacle and pedicels; **Fl** often solitary, ∅ c. 40 mm, distinctly scented, **K** 5, petals greenish yellow to pale salmon, filamentous staminodes absent, reproductive parts concealed; **Fr** valve wings inflexed over the valves, 5 locules; **S** large, D-shaped, brown, testa rough, crest distinct; **Ecol** in shallow soil, winter rainfall; **Distr** C Namaqualand, SW Calvinia, NC, Ceres, Laingsburg, WC, S Africa.

Note: In a number of populations, a range of various floral perfumes have been discerned, even exuded in succession, starting with a lightly sweetish scent, which is followed by a distinct rose perfume and a final disagreeable smell. Other flowers exude a pepper-like scent. Furthermore, some variation has been observed in the overall size of the plants and in flower colour.

P. roseus (L.Bolus) Gerbaulet 1997: 42 (see **photo**) ≡ *Amoebophyllum roseum* L.Bolus SAG 1927: 365 **T** Pillans 5741 (BOL!) = *Amoebophyllum angustum* N.E.Br. JB 1928: 79 **T** Drège 2935 (K!) = *Sphalmanthus vanheerdei* L.Bolus 1964: 41 **T** van Heerde 27494 (BOL!).

Large erect shrubs to c. 1 m *h*, roots not tuberous, **I** lignified, stout; **L** almost cylindrical, narrowly ovate, alternate, marcescent and becoming somewhat spiny, recurring on the stems, with quite large bladder cells; **Fl** ∅ c. 30 mm, open until late at night, **K** 5, petals pink to salmon, filamentous staminodes present, reproductive parts not concealed; **Fr** valve wings inflexed over the valves, 5 locules; **S** large, D-shaped, black, testa rough, without a crest; **Ecol** on hills among rocks, winter rainfall; **Distr** C Namaqualand, NC, S Africa.

Note: **P. roseus** is widely known under the illegitimate synonym *Amoebophyllum angustum*. This species is very similar to **P. lignescens** from which it differs in that the withered leaves become usually more spiny. In addition, the pink to salmon flowers open during the day.

P. saturatus (L.Bolus) Gerbaulet 1997: 42 (see **photo**) ≡ *Aridaria saturata* L.Bolus NM 1930: 178 ≡ *Nycteranthus saturatus* (L.Bolus) Schwantes KuaS 1952: 51 nomen illeg. ≡ *Sphalmanthus saturatus* (L.Bolus) L.Bolus in Jacobsen 1970: 514 nomen invalid. **T** Frames 19109 (BOL!) = *Sphalmanthus acocksii* L.Bolus 1964: 75 **T** Acocks 17724 (BOL!) = *Sphalmanthus bayliss ii* L.Bolus 1963: 45 **T** Bayliss 920 (BOL!).

Decumbent to erect shrubs to 20 cm *h*, often scrambling, roots not tuberous, **I** lignified, slender; **L** almost cylindrical, narrowly ovate to linear, decussate, at least partly persistent, base somewhat spinescent, with small bladder cells, arranged in longitudinal rows; **Fl** ∅ c. 30 mm, **K** 5, petals yellow, coppery salmon, pale salmon, pale pink, or cream, filamentous staminodes present, reproductive parts not concealed; **Fr** valve wings inflexed over the valves, 5 locules; **S** D-shaped, brown, testa rough, crest distinct; **Ecol** often scrambling in other bushes, winter rainfall; **Distr** Namaqualand, Calvinia, NC, Vanrhynsdorp, Clanwilliam, Ceres, WC, S Africa.

Note: **P. saturatus** is similar to **P. splendens**, but is much smaller in overall size and the leaves are decussate throughout.

P. sinuosus (L.Bolus) Gerbaulet 1997: 43 (see **photo**) ≡ *M. sinuosum* L.Bolus ABH 1925: 14 ≡ *Nycteranthus sinuosus* (L.Bolus) Schwantes in Jacobsen 1960: 1323 nomen illeg. ≡ *Sphalmanthus sinuosus* (L.Bolus) L.Bolus in Jacobsen 1970: 514 nomen invalid. **T** Breach 518/21 (BOL!) = *Aridaria glandulifera* L.Bolus NM 1930: 215 ≡ *Nycteranthus glanduliferus* (L.Bolus) Schwantes KuaS 1952: 50 nomen illeg. ≡ *Sphalmanthus glanduliferus* (L.Bolus) L.Bolus in Jacobsen 1970: 511 nomen invalid. **T** SUG 9163 (BOL!) = *Aridaria godmaniae* L.Bolus NM 1929: 131 ≡ *Nycteranthus godmaniae* (L.Bolus) Schwantes KuaS 1952: 50 nomen illeg. ≡ *Sphalmanthus godmaniae* (L.Bolus) L.Bolus in Jacobsen 1970: 512 nomen invalid. **T** Godman 18981 (BOL!) = *Sphalmanthus praecox* L.Bolus 1963: 176 **T** Littlewood 359/59 (BOL!) = *Aridaria recurva* L.Bolus NM 1932: 359 ≡ *Nycteranthus recurvus* (L.Bolus) Schwantes KuaS 1952: 51 nomen illeg. ≡ *Sphalmanthus recurvus* (L.Bolus) L.Bolus in Jacobsen 1970: 514 nomen invalid. **T** Herre 9950 (BOL!) = *Aridaria straminicolor* L.Bolus NM 1938: 137 ≡ *Nycteranthus straminicolor* (L.Bolus) Schwantes KuaS 1952: 51 nomen illeg. ≡ *Sphalmanthus straminicolor* (L.Bolus) L.Bolus in Jacobsen 1970: 514 nomen invalid. ≡ *Aridaria straminea* L.Bolus NM 1932: 359 nomen illeg. **T** Herre 9256 (BOL!).

Prostrate to decumbent or scrambling geophytes with tuberous roots, **I** weakly lignified, most basal ones becoming somewhat more lignified; **L** almost cylindrical, narrowly ovate, decussate, soon becoming alternate, with small or larger and often hair-like bladder cells, most prominently so on the receptacle and pedicels, basal parts of the middle vein somewhat lignified and persistent; **Fl** ∅ 20–30 mm, **K** 5, petals pale yellow, straw-coloured, pale pink-straw, pale salmon, coppery salmon, or cream, basally often darker, filamentous staminodes present, reproductive parts not concealed; **Fr** valve wings inflexed over the valves, 5 locules; **S** D-shaped, black, testa rough, without a crest; **Ecol** on slopes and in plains, sometimes scrambling in other bushes, winter rainfall; **Distr** Namaqualand, NC, S Africa.

Note: **P. sinuosus** is very similar to **P. grossus** but is often larger in overall size and less brittle, has less conspicuously lignified middle veins of the leaves and black seeds without a crest. The bladder cells are remarkably variable in size and shape ranging from small and inconspicuous to large or even hair-like. Plants with larger bladder cells resemble small specimens of **P. oculatus** which differs by a much more slender tap-root. Similar to **P. resurgens**, the flowers of **P. sinuosus** exude various perfumes in succession changing from lightly sweetish to rose-like to slightly rotten.

P. spinuliferus (Haw.) Gerbaulet 1997: 44 (see **photo**) ≡ *M. spinuliferum* Haw. 1795: 206 ≡ *Aridaria spinulifera* (Haw.) N.E.Br. JB 1928: 144 ≡ *Nycteranthus spinuliferus* (Haw.) Schwantes KuaS 1952: 51 nomen illeg. ≡ *Peratetracoilanthus spinuliferus* (Haw.) Rappa & Camarrone 1960: 24 nomen invalid. ≡ *Sphalmanthus spinuliferus* (Haw.) L.Bolus in Jacobsen 1970: 514 nomen invalid. ≡ *M. spinuliforme* Harvey & Sonder 1862: 616 name used in error **LT** drawing (K!) = *Aridaria brevisepala* L.Bolus NM 1928: 26 ≡ *Sphalmanthus brevisepalus* (L.Bolus) L.Bolus in Jacobsen 1970: 510 nomen invalid. ≡ *Aridaria gibbosa* L.Bolus NM 1929: 70 nomen illeg. ≡ *Aridaria parvisepala* L.Bolus NM 1929: 129 name used in error ≡ *Nycteranthus parvisepalus* (L.Bolus) Schwantes KuaS 1952: 51 nomen illeg. **T** Pillans 17793 (BOL!) = *M. carneum* Haw. 1795: 206 ≡ *Nycteranthus carneus* (Haw.) Schwantes in Jacobsen 1960: 1314 nomen illeg. ≡ *Sphalmanthus carneus* (Haw.) N.E.Br. GC 1928: 34 **NT** Gerbaulet & Struck 50101 (BOL!) = *Aridaria framesii* L.Bolus NM 1928: 4 ≡ *Nycteranthus framesii* (L.Bolus) Schwantes KuaS 1952: 50 nomen illeg. ≡ *Sphalmanthus framesii* (L.Bolus) L.Bolus in Jacobsen 1970: 511 nomen invalid. **T** Frames 201/28 (BOL!) = *Aridaria quaterna* L.Bolus NM 1932: 337 ≡ *Nycteranthus quaternus* (L.Bolus) Schwantes KuaS 1952: 51 nomen illeg. ≡ *Sphalmanthus quaternus* (L.Bolus) L.Bolus in Jacobsen 1970: 513 nomen invalid. **T** Herre 8403 (BOL!) = *Sphalmanthus rejuvenalis* L.Bolus 1966: 125 **T** Hall 3038 (BOL!) = *Aridaria stricta* L.Bolus SAG 1927: 433 ≡ *Nycteranthus strictus* (L.Bolus) Schwantes KuaS 1952: 51 nomen illeg. ≡ *Sphalmanthus strictus* (L.Bolus) L.Bolus in Jacobsen 1970: 515 nomen invalid. **T** Pillans 5804 (BOL!) = *Aridaria watermeyeri* L.Bolus NM 1928: 5 ≡ *Nycteranthus watermeyeri* (L.Bolus) Schwantes KuaS 1952: 51 nomen illeg. ≡ *Sphalmanthus watermeyeri* (L.Bolus) L.Bolus in Jacobsen 1970: 516 nomen invalid. **T** Watermeyer 251/23 (BOL!).

Erect shrubs to 30 cm *h*, roots not tuberous, **I** very stout, weakly lignified but with a huge cork layer; **L** almost cylindrical, narrowly ovate, decussate, persistent, with small, flattened bladder cells; **Fl** ∅ 20–30 mm, **K** 4, petals pale yellow to straw or pale salmon to dull orange, filamentous staminodes present, reproductive parts not concealed; **Fr** valve wings inflexed over the valves, 4 locules; **S** D-shaped, brown, testa rough, crest distinct; **Ecol** in plains, winter rainfall; **Distr** W Namaqualand, NC, Vanrhynsdorp, Vredendal, WC, S Africa.

Note: This species is noteworthy for its exceedingly thick succulent stems.

P. splendens (L.) Gerbaulet 1997: 45 ≡ *M. splendens* L. 1753: 486 ≡ *Aridaria splendens* (L.) Schwantes GF 1928: 69 ≡ *Nycteranthus splendens* (L.) Schwantes KuaS 1952: 51 nomen illeg. ≡ *Pentacoilanthus splendens* (L.) Rappa & Camarrone 1953: 30 nomen invalid. ≡ *Perapentacoilanthus splendens* (L.) Rappa & Camarrone 1960: 23 nomen invalid. ≡ *Sphalmanthus splendens* (L.) L.Bolus in Jacobsen 1970: 514 nomen invalid. **T** Hort. Cliff. 220 (OXF, isotype K!) and all synonyms of the subspecies.

Large erect shrubs to 1 m *h*, roots not tuberous, **I** lignified; **L** almost cylindrical, narrowly ovate, decussate, becoming alternate in the **Inf**, persistent, with small bladder cells, rarely hair-like, especially marked on the receptacle and pedicels; **Fl** ∅ 30–40 mm, **K** 5, petals cream, pale pink, pale salmon or pale yellow, filamentous staminodes present, reproductive parts not concealed; **Fr** valve wings inflexed over the valves, 5 locules; **S** D-shaped, black or dark brown, testa rough, crest distinct or indistinct or absent; **Ecol** in plains or on lower slopes, often along roadsides, winter to summer rainfall; **Distr** EC, NC, WC, S Africa.

Note: In response to moisture availability and insolation **P. splendens** shows a considerable variation in the size and shape of its leaves, stems and bladder cells.

Key to the Subspecies of P. splendens

1. Seeds dark brown with a crest .... **P. splendens** ssp. **pentagonus**
– Seeds black, without or with an indistinct crest .... **P. splendens** ssp. **splendens**

P. splendens (L.) Gerbaulet ssp. **pentagonus** (L.Bolus) Gerbaulet 1997: 49 (see **photo**) ≡ *Aridaria pentagona* L.Bolus NM 1929: 35 ≡ *Nycteranthus pentagonus* (L.Bolus) Schwantes var. *pentagonus* KuaS 1952: 51 nomen illeg. ≡ *Sphalmanthus pentagonus* (L.Bolus) L.Bolus var. *pentagonus* in Jacobsen 1970: 513 nomen invalid. T Dyer 1833 (BOL!) = *Aridaria pentagona* L.Bolus var. *occidentalis* L.Bolus NM 1932: 357 ≡ *Nycteranthus pentagonus* (L.Bolus) Schwantes var. *occidentalis* (L.Bolus) Schwantes KuaS 1952: 51 nomen illeg. ≡ *Sphalmanthus pentagonus* (L.Bolus) L.Bolus var. *occidentalis* (L.Bolus) L.Bolus in Jacobsen 1970: 513 nomen invalid. T Bryant 720 (BOL!) = *Aridaria subaequans* L.Bolus NM 1936: 2 ≡ *Nycteranthus subaequans* (L.Bolus) Schwantes KuaS 1952: 51 nomen illeg. ≡ *Sphalmanthus subaequans* (L.Bolus) L.Bolus in Jacobsen 1970: 515 nomen invalid. T Leipoldt 9424 (BOL!) = *Aridaria subpatens* L.Bolus NM 1936: 3 ≡ *Nycteranthus subpatens* (L.Bolus) Schwantes KuaS 1952: 51 nomen illeg. ≡ *Sphalmanthus subpatens* (L.Bolus) L.Bolus in Jacobsen 1970: 515 nomen invalid. T Nel 10687 (BOL!).

Usually to 30 cm *h* and somewhat cushion-like; **S** dark brown with a large crest; **Ecol** as for species; **Distr** EC, FS, NC, WC, between Beaufort West, Steytlerville, Cradock, Fauresmith and Prieska, S Africa.

Note: Both subspecies occur sympatrically in a small area along the southern border of the distribution range of **P. splendens** ssp. **pentagonus**. Seeds of **P. splendens** ssp. **pentagonus** found in this contact zone exhibit a lower crest than those found in the remaining part of the distribution area. These low-crested seeds are intermediate between the otherwise large-crested seeds of **P. splendens** ssp. **pentagonus** and the indistinctly to non-crested seeds of **P. splendens** ssp. **splendens**. This may be the result of some, though limited, hybridization between both subspecies.

P. splendens (L.) Gerbaulet ssp. **splendens** (see **photo**) = *M. acuminatum* Haw. 1824: 426 ≡ *Aridaria acuminata* (Haw.) Schwantes GF 1928: 69 ≡ *Nycteranthus acuminatus* (Haw.) Schwantes KuaS 1952: 50 nomen illeg. ≡ *Perapentacoilanthus acuminatus* (Haw.) Rappa & Camarrone 1960: 23 nomen invalid. ≡ *Sphalmanthus acuminatus* (Haw.) L.Bolus in Jacobsen 1970: 510 nomen invalid. T drawing (K!) = *M. albicaule* Haw. 1826: 331 ≡ *Aridaria albicaulis* (Haw.) N.E.Br. JB 1928: 140 ≡ *Nycteranthus albicaulis* ((Haw.) Schwantes KuaS 1952: 50 nomen illeg. ≡ *Sphalmanthus albicaulis* (Haw.) L.Bolus in Jacobsen 1970: 510 nomen invalid. **NT** Salm-Dyck 1854: § 44 f 5 = *Aridaria bijliae* N.E.Br. JB 1928: 140 ≡ *Nycteranthus bijliae* (N.E.Br.) Schwantes in Jacobsen 1960: 1313 nomen illeg. ≡ *Sphalmanthus bijliae* (N.E.Br.) L.Bolus 1967: 306 T van der Bijl 20949 (BOL!) = *Aridaria blanda* L.Bolus NM 1929: 148 ≡ *Nycteranthus blandus* (L.Bolus) Schwantes KuaS 1952: 50 nomen illeg. ≡ *Sphalmanthus blandus* (L.Bolus) L.Bolus in Jacobsen 1970: 510 nomen invalid. T L.Bolus 18996 (BOL!) = *Aridaria brevifolia* L.Bolus NM 1928: 27 ≡ *Nycteranthus brevifolius* (L.Bolus) Schwantes KuaS 1952: 51 nomen illeg. ≡ *Perapentacoilanthus brevifolius* (L.Bolus) Rappa & Camarrone 1960: 23 nomen invalid. T Frames 2209/26 (BOL!) = *Aridaria celans* L.Bolus NM 1936: 1 ≡ *Nycteranthus celans* (L.Bolus) Schwantes KuaS 1952: 50 nomen illeg. ≡ *Sphalmanthus celans* (L.Bolus) L.Bolus in Jacobsen 1970: 511 nomen invalid. T Esterhuysen 21635 (BOL!) = *Aridaria constricta* L.Bolus NM 1928: 5 ≡ *Nycteranthus constrictus* (L.Bolus) Schwantes KuaS 1952: 50 nomen illeg. ≡ *Sphalmanthus constrictus* (L.Bolus) L.Bolus in Jacobsen 1970: 511 nomen invalid. T Pillans 17304 (BOL!) = *Aridaria dyeri* L.Bolus NM 1928: 28 ≡ *Nycteranthus dyeri* (L.Bolus) Schwantes KuaS 1952: 50 nomen illeg. ≡ *Sphalmanthus dyeri* (L.Bolus) L.Bolus in Jacobsen 1970: 511 nomen invalid. T Dyer 1292 (BOL!) = *M. flexuosum* Haw. 1803: 61 ≡ *Aridaria flexuosa* (Haw.) Schwantes GF 1928: 69 ≡ *Nycteranthus flexuosus* (Haw.) Schwantes KuaS 1952: 50 nomen illeg. ≡ *Sphalmanthus flexuosus* (Haw.) L.Bolus in Jacobsen 1970: 511 nomen invalid. **NT** Salm-Dyck 1840: § 44 f 7 = *Aridaria fourcadei* L.Bolus NM 1929: 69 ≡ *Nycteranthus fourcadei* (L.Bolus) Schwantes KuaS 1952: 50 nomen illeg. ≡ *Sphalmanthus fourcadei* (L.Bolus) L.Bolus in Jacobsen 1970: 511 nomen invalid. T Fourcade 4291 (BOL!) = *Aridaria leptopetala* L.Bolus NM 1931: 225 ≡ *Nycteranthus leptopetalus* (L.Bolus) Friedrich 1959: 70 nomen illeg. ≡ *Sphalmanthus leptopetalus* (L.Bolus) L.Bolus in Jacobsen 1970: 512 nomen invalid. T Frames 19377 (BOL!) = *M. nothum* N.E.Br. 1920: 109 ≡ *Aridaria notha* (N.E.Br.) N.E.Br. JB 1928: 143 ≡ *Nycteranthus nothus* (N.E.Br.) Schwantes KuaS 1952: 51 nomen illeg. ≡ *Sphalmanthus nothus* (N.E.Br.) L.Bolus in Jacobsen 1970: 513 nomen invalid. T Salm-Dyck 1854: § 44 f 4 (as *M. acuminatum*) = *M. plenifolium* N.E.Br. 1922: 155 ≡ *Aridaria plenifolia* (N.E.Br.) Stearn 1939: 76 ≡ *Nycteranthus plenifolius* (N.E.Br.) Schwantes KuaS 1952: 51 nomen illeg. ≡ *Sphalmanthus plenifolius* (N.E.Br.) L.Bolus in Jacobsen 1970: 513 nomen invalid. ≡ *M. fastigiatum* Haw. 1795: 210 nomen illeg. non Thunberg 1791: 7 ≡ *Aridaria fastigiata* (Haw.) Schwantes GF 1928: 69 nomen illeg. **NT** Salm-Dyck 1854: § 44 f 3 = *Aridaria primulina* L.Bolus NM 1928: 26 ≡ *Nycteranthus primulinus* (L.Bolus) Schwantes KuaS 1952: 51 nomen illeg. ≡ *Sphalmanthus primulinus* (L.Bolus) L.Bolus in Jacobsen 1970: 513 nomen invalid. **LT** Holland 686/25 (021/26) (BOL!) = *Aridaria rabiesbergensis* L.Bolus NM 1936: 1 ≡ *Nycteranthus rabiesbergensis* (L.Bolus) Schwantes KuaS 1952: 51 nomen illeg. ≡ *Sphalmanthus rabiesbergensis* (L.Bolus) L.Bolus in Jacobsen 1970: 514 nomen invalid. T Lewis 21636 (apparently lost) **NT** Esterhuysen 23882 (BOL!) = *M. reflexum* Haw. 1795: 213 ≡ *Aridaria reflexa* (Haw.) N.E.Br. JB 1928: 143

≡ *Sphalmanthus reflexus* (Haw.) L.Bolus in Jacobsen 1970: 514 nomen invalid. ≡ *M. fastigiatum* Haw. var. *reflexum* (Haw.) Haw. 1812: 256 **NT** Gerbaulet & Struck 50143 (BOL!) = *Aridaria rigida* N.E.Br. JB 1928: 144 **T** Muir 4231 (K!) = *Aridaria rosea* L.Bolus NM 1928: 27 ≡ *Nycteranthus roseus* (L.Bolus) Schwantes KuaS 1952: 51 nomen illeg. ≡ *Sphalmanthus roseus* (L.Bolus) L.Bolus in Jacobsen 1970: 514 nomen invalid. **T** L.Bolus 18859 (BOL!) = *Aridaria striata* L.Bolus NM 1936: 2 ≡ *Nycteranthus striatus* (L.Bolus) Schwantes KuaS 1952: 51 nomen illeg. ≡ *Sphalmanthus striatus* (L.Bolus) L.Bolus in Jacobsen 1970: 515 nomen invalid. **T** SUG 10702 (BOL!) = *M. sulcatum* Haw. 1821: 173 ≡ *Aridaria sulcata* (Haw.) Schwantes GF 1928: 69 ≡ *Nycteranthus sulcatus* (Haw.) Schwantes KuaS 1952: 51 nomen illeg. ≡ *Perapentacoilanthus sulcatus* (Haw.) Rappa & Camarrone 1960: 23 nomen invalid. ≡ *Sphalmanthus sulcatus* (Haw.) L.Bolus in Jacobsen 1970: 515 nomen invalid. **NT** Salm-Dyck 1840: § 44 f 1 = *M. umbelliflorum* Jacquin 1804: t. 478 ≡ *Aridaria umbelliflora* (Jacquin) Schwantes GF 1928: 69 ≡ *Nycteranthus umbelliflorus* (Jacquin) Schwantes KuaS 1952: 51 nomen illeg. ≡ *Sphalmanthus umbelliflorus* (Jacquin) L.Bolus in Jacobsen 1970: 515 nomen invalid. **T** Jacquin 1804: t. 478 = *Aridaria vernalis* L.Bolus NM 1932: 357 ≡ *Nycteranthus vernalis* (L.Bo-lus) Schwantes KuaS 1952: 51 nomen illeg. ≡ *Sphalmanthus vernalis* (L.Bolus) L.Bolus in Jacobsen 1970: 515 nomen invalid. **T** Compton 3726 (BOL!).

Most often to 1 m *h*, rarely cushion-like; **S** black without or with an indistinct crest; **Ecol** as for species; **Distr** from Albany, EC, to Worcester, WC, S Africa.

Note: Towards the eastern part of its distribution range, **P. splendens** ssp. **splendens** usually becomes less tall and attains a somewhat cushion-like habit similar to **P. splendens** ssp. **pentagonus**. For **P. splendens** ssp. **splendens** the bladder cells have been recorded to vary from rather inconspicuous to hair-like.

P. suffruticosus (L.Bolus) Gerbaulet 1997: 50 (see **photo**) ≡ *Cryophytum suffruticosum* L.Bolus NM 1931: 230 **T** Pillans 6286 (BOL!).

Large decumbent to erect shrubs to 70 cm *h*, roots not tuberous, **I** lignified; **L** almost cylindrical, narrowly ovate, decussate, becoming alternate in the **Inf**, persistent, with small bladder cells, new leaves tufted on short shoots; **Fl** ∅ 20–30 mm, opening in the afternoon about 4pm, closing in the evening about 6–8pm, **K** 5, petals cream to pink, filamentous staminodes present, reproductive parts not concealed; **Fr** valve wings inflexed over the valves, 5 locules; **S** D-shaped, black, testa rough, without a crest; **Ecol** on rocky slopes, winter rainfall; **Distr** Malmesbury, WC, S Africa.

Note: **P. suffruticosus** resembles **P. roseus** in growth and habit. However, both species differ in the leaves, which are mostly decussate and persistent in **P. suffruticosus**, but alternate and marcescent in **P. roseus.**

P. tenuiflorus (Jacquin) Gerbaulet 1997: 50 (see **photo**) ≡ *M. tenuiflorum* Jacquin 1800–1809: 44 ≡ *Nycteranthus tenuiflorus* (Jacquin) Schwantes in Jacobsen 1960: 1325 nomen illeg. ≡ *Sphalmanthus tenuiflorus* (Jacquin) N.E.Br. GC 1928: 34 **T** Jacquin 1800–1809: 44 = *Aridaria longituba* L.Bolus NM 1930: 193 ≡ *Nycteranthus longitubus* (L.Bolus) Schwantes KuaS 1952: 50 nomen illeg. ≡ *Sphalmanthus longitubus* (L.Bolus) L.Bolus in Jacobsen 1970: 512 nomen invalid. **T** Pillans 6211 (BOL!) = *Aridaria macrosiphon* L.Bolus NM 1932: 358 ≡ *Nycteranthus macrosiphon* (L.Bolus) Schwantes KuaS 1952: 50 nomen illeg. ≡ *Sphalmanthus macrosiphon* (L.Bolus) L.Bolus in Jacobsen 1970: 512–513 nomen invalid. **T** Herre 9488 (BOL!) = *Sphalmanthus nanus* L.Bolus 1962: 292 **T** Hall 2554 (BOL!) = *M. olivaceum* Schlechter 1899: 128 ≡ *Prenia olivacea* (Schlechter) H.Jacobsen 1960: 1283, 1348 nomen invalid. ≡ *Sphalmanthus olivaceus* (Schlechter) L.Bolus in Jacobsen 1970: 513 nomen invalid. **T** Schlechter 8208 (B!) = *Aridaria peersii* L.Bolus NM 1930: 193 ≡ *Nycteranthus peersii* (L.Bolus) Schwantes KuaS 1952: 51 nomen illeg. **T** Peers 1268/29 (BOL!).

Prostrate, to 3 cm *h* with a thick caudex, **I** mostly short, flowering stems longer, weakly lignified but basally corky; **L** almost cylindrical, narrowly ovate, decussate, becoming alternate in the **Inf**, deciduous, with large tipped or hair-like bladder cells, most prominently so on the receptacle and pedicels; **Fl** ∅ c. 40 mm, **K** 5, petals greenish yellow to yellow or outer petals greyish pink and inner ones greenish yellow, filamentous staminodes absent, reproductive parts concealed; **Fr** valve wings inflexed over the valves, 5 locules; **S** large, D-shaped, brown, testa rough, crest distinct; **Ecol** on quartzite, winter rainfall; **Distr** Vanrhynsdorp, WC, S Africa.

Note: Of particular interest are the flowers of **P. tenuiflorus** which may show greyish pink or greenish yellow outer petals within a single population.

P. tortuosus = **Sceletium tortuosum**

P. trichotomus (Thunberg) Gerbaulet 1997: 51 (see **photo**) ≡ *M. trichotomum* Thunberg 1791: 14 ≡ *Aridaria trichotoma* (Thunberg) L.Bolus FPSA 1927: t. 258 ≡ *Dicrocaulon trichotomum* (Thunberg) N.E.Br. JB 1928: 142 ≡ *Nycteranthus trichotomus* (Thunberg) Schwantes KuaS 1952: 51 nomen illeg. ≡ *Sphalmanthus trichotomus* (Thunberg) L.Bolus in Jacobsen 1970:

515 nomen invalid. T Thunberg 12033 (UPS, photograph!) = *M. ebracteatum* L.Bolus ABH 1915: 191 nomen illeg. non Schlechter & Diels in Schultze 1907: 693 ≡ *Aridaria ebracteata* N.E.Br. JB 1928: 141 ≡ *Nycteranthus ebracteatus* (N.E.Br.) Schwantes KuaS 1952: 50 T Pillans 32400 (BOL!) = *Aridaria ebracteata* N.E.Br. var. *brevipetala* L.Bolus NM 1931: 227 ≡ *Nycteranthus ebracteatus* (N.E.Br.) Schwantes var. *brevipetalus* (L.Bolus) Schwantes KuaS 1952: 50 T Leipoldt 2525/30 (BOL!) = *Aridaria tetramera* L.Bolus ABH 1927: 103 ≡ *Nycteranthus tetramerus* (L.Bolus) Schwantes KuaS 1952: 51 nomen illeg. ≡ *Sphalmanthus tetramerus* (L.Bolus) L.Bolus in Jacobsen 1970: 515 nomen invalid. T Pillans 17757 (BOL!) = *Aridaria tetramera* L.Bolus var. *parviflora* L.Bolus NM 1936: 3 ≡ *Nycteranthus tetramerus* (L.Bolus) Schwantes var. *parviflorus* (L.Bolus) Schwantes KuaS 1952: 51 nomen illeg. ≡ *Sphalmanthus tetramerus* (L.Bolus) L.Bolus in Jacobsen var. *parviflorus* (L.Bolus) L.Bolus in Jacobsen 1970: 515 nomen invalid. T Herre 10668 (BOL!).

Erect shrubs to 20 cm *h*, roots not tuberous, **I** lignified; **L** almost cylindrical, narrowly ovate, decussate, persistent, with small, xeromorphic bladder cells; **Fl** ∅ 10–20 mm, usually sessile, **K** 4, petals cream or pale salmon-cream, or outer petals bright reddish pink, inner ones cream or rarely yellow, filamentous staminodes absent, reproductive parts concealed; **Fr** valve wings inflexed over the valves, 4 locules; **S** D-shaped, brown, testa rough, crest distinct; **Ecol** in rocky sites, sometimes on quartzite, winter rainfall; **Distr** W Namaqualand, NC, Vanrhynsdorp, Clanwilliam, Piquetberg, WC, S Africa.

Note: **P. trichotomus** is very similar to **P. delus** from which it differs in the 4-lobed calyx, the 4-locular fruit and the more robust stems and leaves. Reddish flowers are most widespread in **P. trichotomus**, whereas cream flowers have been recorded from Namaqualand and Elandsbay.

P. viridiflorus (Aiton) Gerbaulet 1997: 52 (see **photo**) ≡ *M. viridiflorum* Aiton 1789: 196 ≡ *Aridaria viridiflora* (Aiton) L.Bolus NM 1929: 131 ≡ *Nycteranthus viridiflorus* (Aiton) Schwantes KuaS 1952: 51 nomen illeg. ≡ *Perapentacoilanthus viridiflorus* (Aiton) Rappa & Camarrone 1960: 23 nomen invalid. ≡ *Sphalmanthus viridiflorus* (Aiton) N.E.Br. GC 1928: 34 **NT** Jacquin 1800–1809: 43 = *M. trichosanthum* A.Berger 1922: 631 ≡ *Aridaria trichosantha* (A.Berger) N.E.Br. JB 1928: 145 **LT** Schlechter 8357 (BOL!).

Decumbent shrubs to 10 cm *h*, roots thickened, **I** weakly lignified but basal ones with a thick cork layer; **L** almost cylindrical, narrowly ovate, decussate, becoming alternate in the **Inf**, with large tipped or hair-like bladder cells, especially marked on the receptacle and pedicels; **Fl** ∅ c. 40 mm, **K** 5, petals blue greenish white, filamentous staminodes absent, reproductive parts concealed; **Fr** valve wings inflexed over the valves, 5 locules; **S** large, D-shaped, brown, testa rough, crest distinct; **Ecol** on hills or quartzite plains, winter rainfall; **Distr** Clanwilliam, WC, S Africa.

Note: **P. viridiflorus** is unique within the subfamily Mesembryanthemoideae for its large, most fascinating blue greenish white flowers. An endemic of the slopes above the Olifants River in the Clanwilliam district.

Piquetia = Erepsia

H. E. K. Hartmann

The name *Piquetia* N.E.Br. appeared in a key (GC 1925: 433), which would have been a valid publication had not Hallier published a genus of Theaceae earlier under the same name (Hallier 1921: 162). Fedde (1940: 11) created the new name *Kensitia* for *Piquetia* N.E.Br., the former becoming a synonym of **Erepsia** when the only species was transferred (Liede 1990: 433).

No species have been named in *Piquetia* N.E.Br.

Platythyra = Aptenia

M. Gerbaulet

Brown (GC 1925: 412) separated *Platythyra* from **Aptenia** on account of the connate versus free calyx lobes and the presence versus absence of valve wings. Otherwise, the two genera, both monotypic at that time, are very similar. Later, another species was placed in **Aptenia** with free calyx lobes on the one hand and fruits with valve wings on the other. On the basis of these and some other character states, the two genera are combined.

P. barklyi ≡ **Mesembryanthemum barklyi**
P. haeckeliana ≡ **Aptenia haeckeliana**
P. pallens ≡ **Prenia pallens**
P. relaxata = **Prenia pallens** ssp. **pallens**

Pleiospilos RUSCHIOIDEAE

H. E. K. HARTMANN

Pleiospilos N.E.Br. GC 1925: 433; Hartmann & Liede 1986: 433–485; 1987: 117–125 **Typus P. bolusii** (Hook.f.) N.E.Br. **Etym** Gk *pleio-*, many, more than usual, *spilos*, spot, smudge, describing the many dots visible on the leaves = *Punctillaria* N.E.Br. **Typus** *Punctillaria compacta* (Aiton) N.E.Br. ≡ **P. compactus** (Aiton) Schwantes.

Compact, single heads to caespitose; **L** trigonous to semi-globular, humped at the back and often laterally expanded as well, epidermis cells flat, without crystals, tannin idioblasts underneath therefore visible as small dots, stomata not sunken, wax cover visible to the naked eye, easily to be rubbed off, consisting of platelets or flat layers; **Fl** 1–3, petals yellow, rarely white as sports, opening in the late afternoon, stamens erect in centre, without papillae, nectary an inconspicuous ring; **Fr** robust, dark brown, top raised and rounded, valves with almost parallel expanding keels, valve wings basally broad, tapering to an awn at the tip, covering membranes firm, with a distal rim recurved adorned with a thick bulge at its base, closing body present, 9–15 locules; **S** dark brown, papillate with distant naps covered by long rodlets of c. 3 µm *l*, 1–1.5 mm *l*, 0.9–1.3 mm *b*; **Chr** $2n$=18; **Ecol** on N facing red-brown shaly slopes or in flats, 200–400 mm rainfall p.a. mainly in March and November; **Distr** western EC, southern NC and eastern WC, S Africa.

Note: A widespread genus with only 4 species in 2 subgenera. Species belonging to different subgenera can grow sympatrically, but species (and subspecies) of a subgenus occur allopatrically only.

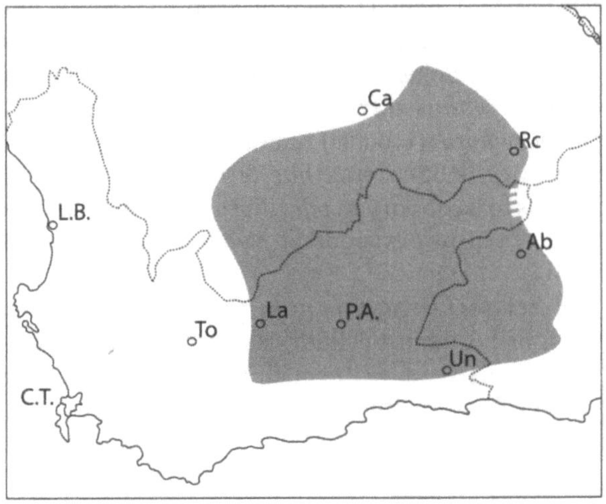

Distribution of **Pleiospilos** N.E.Br.

Key to the Subgenera of Pleiospilos

1. Branches > 4, compact to caespitose, capsules with big closing bodies, flowers solitary ..... P. subg. **Punctillaria**
– Branches 1–4, compact, capsules with tiny placental closing bodies, flowers ternate .... P. subg. **Pleiospilos**

P. subg. **Pleiospilos.**

1–4 branches per plant; **L** length < 2 × width; **Fl** mostly 2–3; **Fr** a tumble fruit, stalks disintegrated when ripe, with tiny placental closing bodies, covering membranes with an erect, short rim and a big basal bulge below; **Ecol** mostly on flats of Beaufort-shales in very open Karroo, 200–300 mm rainfall p.a.; **Distr** E part of the area of the genus, NC, WC, S Africa.

Note: Recent finds have enlarged the known distribution area, which may still not yet reflect the boundaries of the subgenus.

Key to the Species of P. subg. Pleiospilos

1. Leaves spreading widely (> leaf width), 2–4 branches .... **P. simulans**
– Leaves appressed or spreading less than the leaf width, mostly single 2
2. Leaves hood-shaped, spreading .... **P. bolusii**
– Leaves semi-globose, appressed .... **P. nelii**

P. subg. **Punctillaria** (N.E.Br.) H.E.K.Hartmann & Liede 1986: 475 = *Punctillaria* N.E.Br. GC 1925: 433 **Typus P. compactus** (Aiton) Schwantes.

Compact to caespitose; **L** with a dorsal hump and lateral subapical bulges (gibbous), length > 2 × width; **Fl** solitary; **Fr** mostly with a persistent stalk, closing bodies large, endocarpal, covering membranes with a markedly recurved distal rim and an inconspicuous bulge below; **Ecol** mostly on NE-slopes with reddish-brown shales, March and November rains; **Distr** as for genus, S Africa.

Note: To the north, populations get sparse and records are few.

Key to the Species of P. subg. Punctillaria

1. Leaves > 4 cm, width > 2 cm, hardly laterally widened (gibbous) .... **P. × purpusii**
- Leaves < 4 cm long, width > 2 cm, if longer, markedly laterally bulged (gibbous) .... **P. compactus**

P. archeri ≡ **Tanquana archeri**
P. barbarae = **P. bolusii**
P. beaufortensis = **P. bolusii**
P. borealis = **P. compactus** ssp. **canus**
P. brevisepalus = **P. compactus** ssp. **canus**
P. canus ≡ **P. compactus** ssp. **canus**
P. clavatus = **Tanquana archeri**

P. bolusii (Hook.f.) N.E.Br. GC 1926: 89 (see **photo**) subg. **Pleiospilos** ≡ *M. bolusii* Hook.f. 1882: t. 6664 T Peacock s.n. (K!) = *P. barbarae* Karrer 1932: 99 T? = *P. beaufortensis* L.Bolus NM 1939: 163 T Coutts 22118 (BOL!).

Unbranched, rarely up to 3 branches; **L** hood-shaped, spreading < leaf width, as thick as wide; **Fr** mostly 12 locules; **Ecol** in bands of shale of Beaufort-layers, 200–300 mm rainfall p.a. in March and November; **Distr** Aberdeen, Willowmore, EC, Beaufort West, WC, S Africa.

Note: A widespread species, but sparsely distributed in its localities, often in small populations over a wide area. In spite of their enormous size (c. 15 × 6 cm ∅ seen from above, to 7 cm high) the plants are difficult to find because they mimic the stones around them perfectly.

P. compactus (Aiton) Schwantes ZSK 1927: 23 subg. **Punctillaria** ≡ *M. compactum* Aiton 1789: 191 ≡ *Punctillaria compacta* (Aiton) N.E.Br. GC 1926: 228 **NT** SUG 10325 (BOL!) and all synonyms under the subspecies.

Compact to caespitose; **L** with a dorsal hump and lateral subapical bulges (gibbous), length > 2 × width; **Fl, Fr, Ecol** as for subgenus; **Distr** as for genus, S Africa.

Note: The only gradual differences between the included taxa led to the recognition as subspecies differentiated by geography and ecology beside leaf shapes and some more inconspicuous character states.

Key to the Subspecies of P. compactus

1. Fruits with mostly 12 locules (> 50% of a population) .... **P. compactus ssp. canus**
- Fruits with mostly 10 locules (> 50% of a population) **2**
2. Pedicels < 5 mm, withering, tumble fruit .... **P. compactus ssp. minor**
- Pedicels ≥ 10 mm, fruit persisting on plant **3**
3. Stalks of capsules > 30 mm, closing body white .... **P. compactus ssp. fergusoniae**
- Stalks of capsules < 20 mm, closing body brown **4**
4. Pedicels ≤ 4 mm .... **P. compactus ssp. compactus**
- Pedicels 10–20 mm long .... **P. compactus ssp. sororius**

P. compactus (Aiton) Schwantes ssp. **canus** (Haw.) H.E.K.Hartmann & Liede 1986: 476 (see **photo**) ≡ *M. canum* Haw. 1795: 158 ≡ *P. canus* (Haw.) L.Bolus NM 1929: 100 = *Punctillaria cana* (Haw.) L.Bolus SAG 1927: 326 **NT** Nel 9462 (BOL!) = *P. borealis* L.Bolus NM 1931: 288 T Archer 339 (BOL!) = *P. brevisepalus* L.Bolus NM 1931: 255 T Markoetter 8379 (BOL!) = *Punctillaria dekenahi* N.E.Br. GC 1928: 215 ≡ *P. dekenahi* (N.E.Br.) Schwantes MDG 1930: 22 T Muir 4102 (PRE!) = *P. framesii* L.Bolus NM 1929: 85 T Frames 18917 (BOL!) = *P. grandiflorus* L.Bolus NM 1931: 255 T H. Bolus 19488 (BOL!) = *P. kingiae* L.Bolus NM 1936: 33 T King 119 (BOL!) = *P. latifolius* L.Bolus NM 1939: 163 T Brauns 8250 (BOL!) = *P. latipetalus* L.Bolus NM 1931: 277 T Brauns 8032 (BOL!) = *P. leipoldtii* L.Bolus NM 1931: 278 T Leipoldt 586/31 (BOL!) = *P. longisepalus* L.Bolus NM 1939: 163 T Nel 9856 (BOL!) = *M. magnipunctatum* Haw. 1819: 87 ≡ *Punctillaria magnipunctata* (Haw.) N.E.Br. GC 1926: 212 ≡ *P. magnipunctatus* (Haw.) Schwantes ZSK 1927: 23 T Burchell 1402 = *P. magnipunctatus* (Haw.) Schwantes var. *inaequalis* L.Bolus NM 1931: 254 T Nel 8991 (BOL!) = *M. nobile* Haw. 1823: 381 = *P. nobilis* (Haw.) Schwantes ZSK 1927: 23 T? = *P. peersii* L.Bolus NM 1931: 278 T Peers 990/30 (BOL!) = *P. rouxii* L.Bolus NM 1939: 163 T Roux s.n. (BOL!) = *Punctillaria sesquiuncialis* N.E.Br. GC 1928: 492 ≡ *P. sesquiuncialis* (N.E.Br.) Schwantes 1930: 22 ≡ *P. magnipunctatus* (Haw.) Schwantes var. *sesquiuncialis* L.Bolus SAG 1927: 326 T F.Bolus 11511 (BOL!) = *P. willowmorensis* L.Bolus NM 1931: 256 T Brauns 8265 (BOL!).

L with a long keel, sides moderately bulging, 20–40 mm *l*; **Fl** pedicels < 10 mm *l*, **K** mostly 6, petals > 80; **Fr** stalks 8–20 mm, 12 locules, closing body brown; **Ecol** 200–500 mm rainfall p.a., March and November; **Distr** W EC, S NC, NE WC, S Africa.

Note: A very variable subspecies, especially in leaf shapes; however, the predominant number of 12 locules can always be relied on to identify the taxon.

P. compactus (Aiton) Schwantes ssp. **compactus** (see **photo**) = *M. optatum* N.E.Br. 1920: 92 ≡ *Punctillaria optata* (N.E.Br.) N.E.Br. GC 1926: 89 ≡ *P. optatus*

(N.E.Br.) Schwantes ZSK 1927: 23 T Simpson-Hayward 269–10 (K!) = *P. longibracteatus* L.Bolus NM 1931: 258 T SUG 9467 (BOL!).

L with a short keel, sides not laterally bulging, 25–40 mm *l*; **Fl** pedicels < 10 mm *l*, **K** 5, petals ≤ 80; **Fr** stalk 8–20 mm *l*, mostly 10 locules, closing body brown; **Ecol** c. 300 mm rainfall at all seasons: **Distr** Calitzdorp, Ladismith, Oudtshoorn, WC, S Africa.

Note: Typical of the subspecies are the absence of lateral subapical bulges on the leaves and the low number of petals.

P. compactus (Aiton) Schwantes ssp. **fergusoniae** (L.Bolus) H.E.K.Hartmann & Liede 1986: 478 (see **photo**) ≡ *P. fergusoniae* L.Bolus NM 1929: 99 T Ferguson 926/29 (BOL!).

L with a long keel, distinctly gibbous with lateral bulges, 28–40 mm *l*; **Fl** pedicels > 10 mm *l*; **Fr** stalks > 30 mm *l*, mostly 10 locules, closing body white; **Ecol** c. 200 mm rainfall p.a. in March and November mainly; **Distr** Ladismith, WC, S Africa.

Note: A long fruit stalk and white closing bodies separate this subspecies from all others.

P. compactus (Aiton) Schwantes ssp. **minor** (L.Bolus) H.E.K.Hartmann & Liede 1986: 479 (see **photo**) ≡ *P. minor* L.Bolus NM 1931: 258 T Schoonees SUG 8658 (BOL!).

L with a short keel and laterally gibbous, c. 25 mm *l*; **Fl** stalk < 5 mm *l*, withering, **K** mostly 5, > 80 petals; **Fr** tumble fruit when ripe, mostly 10 locules, closing body brown; **Ecol** 200–300 mm rainfall p.a. in March and November; **Distr** Steytlerville, EC, S Africa.

Note: Short, broad leaves (20–25 mm l) mm are typical of this subspecies with a tumble fruit like **P.** subg. **Pleiospilos.**

P. compactus (Aiton) Schwantes ssp. **sororius** (N.E.Br.) H.E.K.Hartmann & Liede 1986: 478 (see **photo**) ≡ *M. sororium* N.E.Br. 1920: 92 ≡ *Punctillaria sororia* (N.E.Br.) N.E.Br. GC 1926: 213 ≡ *P. sororius* (N.E.Br.) Schwantes ZSK 1927: 23 T hort. Kew 1.10.1918 (K!) = *P. dimidiatus* L.Bolus NM 1931: 257 T Nel 9166 (BOL!).

L with a long keel, distinctly laterally gibbous, 28–38 mm *l*; **Fl** pedicel > 10 mm *l*, **K** mostly 5, > 80 petals; **Fr** stalk 10–20 mm *l*, locules mostly 10, closing body brown; **Ecol** at the edge of the winter rainfall region, c. 200 mm rainfall p.a.; **Distr** Laingsburg, WC, S Africa.

Note: Geographically and in leaf shape near **P. compactus ssp. fergusoniae,** from which it can be distinguished by its shorter fruit stalks and brown closing body.

P. dekenahi = **P. compactus** ssp. **canus**
P. dimidiatus = **P. compactus** ssp. **sororius**
P. fergusoniae ≡ **P. compactus** ssp. **fergusoniae**
P. framesii = **P. compactus** ssp. **canus**
P. grandiflorus = **P. compactus** ssp. **canus**
P. hilmarii ≡ **Tanquana hilmarii**
P. kingiae = **P. compactus** ssp. **canus**
P. latifolius = **P. compactus** ssp. **canus**
P. latipetalus = **P. compactus** ssp. **canus**
P. leipoldtii = **P. compactus** ssp. **canus**
P. loganii = **Tanquana archeri**
P. longibracteatus = **P. compactus** ssp. **compactus**
P. longisepalus = **P. compactus** ssp. **canus**
P. magnipunctatus = **P. compactus** ssp. **canus**
P. magnipunctatus var. *inaequalis* = **P. compactus** ssp. **canus**
P. magnipunctatus var. *sesquiuncialis* = **P. compactus** ssp. **canus**
P. minor ≡ **P. compactus** ssp. **minor**

P. nelii Schwantes 1930: 146 (see **photo**) subg. **Pleiospilos LT** Schwantes 1920: f on p 147 = *P. pedunculatus* L.Bolus SAG 1930: 253 T Ryder 454/30 (BOL!) = *P. tricolor* N.E.Br. GC 1930: 296 T van der Bijl s.n.

Mostly unbranched; **L** semi-globose, appressed; **Fl** appearing after the leaves, stalked; **Fr** tumble fruit, mostly 11 locules; **Ecol** mostly on slopes with white quartz pebbles, < 200 mm rainfall p.a., mainly in March and November, **Distr** Graaff Reinet, Willowmore, EC, Beaufort West, Prince Albert, WC, S Africa.

Note: The late flower formation distinguishes this species from all others in the genus, as do the round bodies formed by the appressed leaf-pair.

P. nobile = **P. compactus** ssp. **canus**
P. optatus ≡ **P. compactus** ssp. **compactus**
P. pedunculatus = **P. nelii**
P. peersii = **P. compactus** ssp. **canus**
P. prismaticus ≡ **Tanquana prismatica**

P. × purpusii Schwantes MDG 1928: 46 subg. **Punctillaria** ≡ *Punctillaria purpusii* (Schwantes) N.E.Br. GC 1928: 493 **LT** f in Monatsschrift Kakteenkunde 17: 25 (1907) = *P. × rothii* H.Jacobsen 1933: 171 T? = **P. bolusii × P. compactus.**

L inconspicuously keeled, not nor inconspicuously gibbous, > 40 mm *l*, apex of upper leaf surface clearly triangular, but finally rounded; **Fl** 1–3, scented, > 50 mm ∅; **Fr** with a small closing body; **Ecol** and **Distr** known only from cultivation, but the plants can be expected to be found in the area of sympatric occurrence of **P. bolusii** and **P. compactus.**

Note: In the apically broadened upper leaf surface, the plants resemble seedlings.

P. roodiae = **Tanquana prismatica**
P. × rothii = **P. × purpusii**
P. rouxii = **P. compactus** ssp. **canus**
P. sesquiuncialis = **P. compactus** ssp. **canus**

P. simulans (Marloth) N.E.Br. GC 1926: 89 (see **photo**) subg. **Pleiospilos** ≡ *M. simulans* Marloth 1907: 43 **T** Marloth 4378 (PRE!).

Branches 1–3; **L** broader than thick, spreading widely, gap > leaf width; **Fr** tumble fruit, mostly 12 locules; **Ecol** in crevices of bands of Beaufort shales, c. 300 mm rainfall p.a. in March and November; **Distr** Jansenville, EC, S Africa.

Note: A threatened species in habitat, but widespread in cultivation, often hybridizing resulting in beautiful specimens.

P. sororius ≡ **P. compactus** ssp. **sororius**
P. tricolor = **P. nelii**
P. willowmorensis = **P. compactus** ssp. **canus**

Plinthus Aizooideae

H. E. K. Hartmann

Plinthus Fenzl 1839: 51; Verdoorn 1941: 177–179; Adamson 1961: 147–151 **Typus P. cryptocarpus** Fenzl **Etym** uncertain.

Low shrubs with ± erect to decumbent branches, young **I** with white or silvery bi-acuminate hairs; **L** imbricate, opposite or alternate, deciduous around anthesis, covered densely by bi-acuminate hairs; **Fl** in groups of 2–3, perigone lobes 4–5, stamens alternitepalous; **Fr** with reduced expanding tissue, 1–2 pendulous ovule per locule, 2–4 locules; **S** pear- to kidney-shaped; **Ecol** on sand or stones of Karroo rock origin; **Distr** from Namibia to the FS and N EC, S Africa.

Note: Similar to **Galenia** and **Aizoon**, the genus can best be recognized by the isomerous number of stamens. Although plants of the genus are important for farming, there has been very little study on the group. This treatment is based on the latest revision by Adamson (1961: 147–151), including the key.

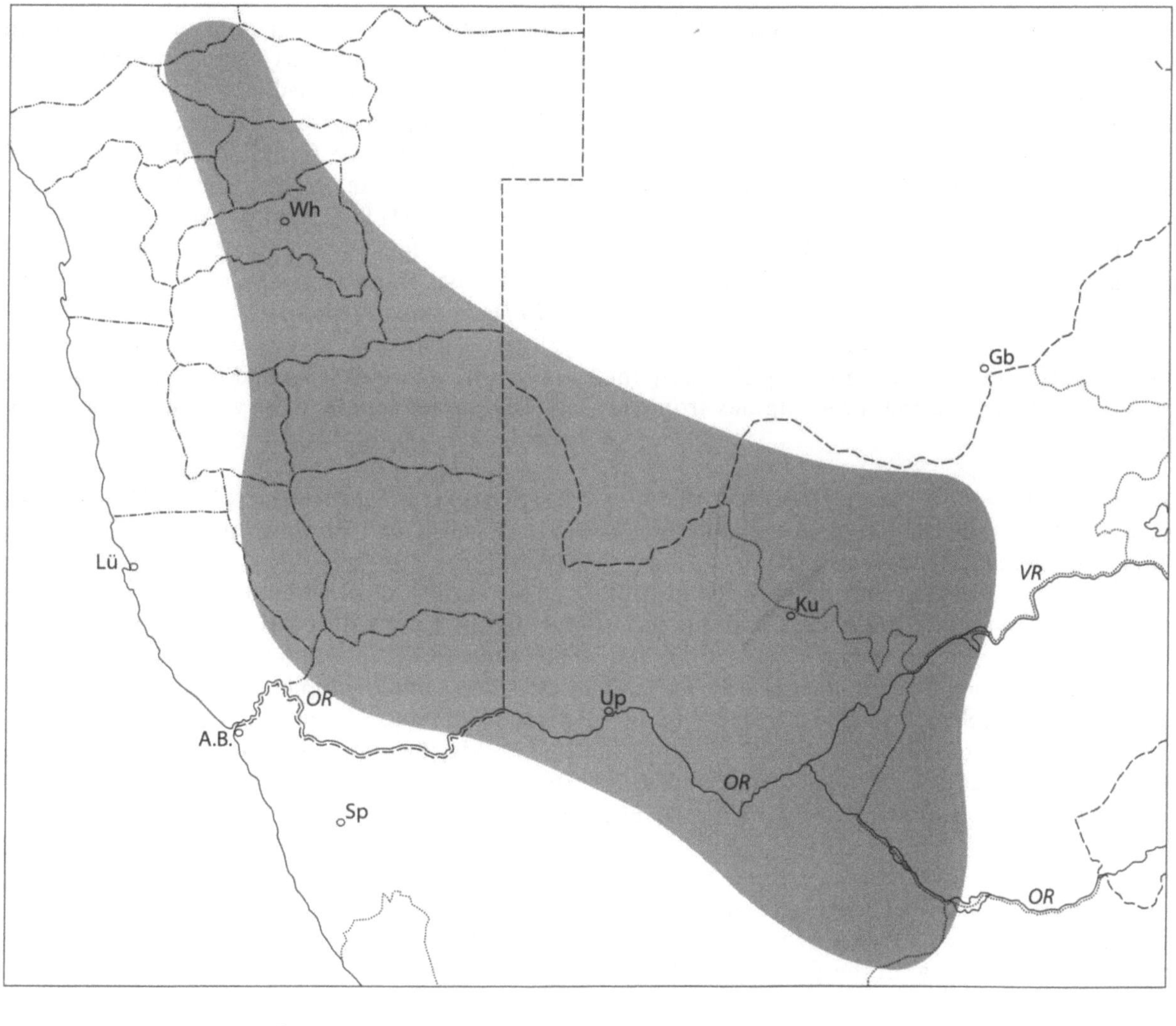

Distribution of **Plinthus** Fenzl

Key to the Species of Plinthus

1. Leaves opposite 2
- Leaves alternate 3
2. Leaves recurved, flowers with 5 perigone lobes .... **P. cryptocarpus**
- Leaves erect, flowers with 4 perigone lobes .... **P. karooicus**
3. Perigone with 5 lobes 4
- Perigone with 4 lobes 5
4. Leaves narrowing to both ends .... **P. sericeus**
- Leaves broadening to the tip and petiolate .... **P. rehmannii**
5. Leaves falcate or curved, widest near the tip .... **P. arenarius**
- Leaves widest at or below the middle .... **P. karooicus**

P. arenarius Adamson 1961: 149 T Leistner 1147 (K).

Shrubs with decumbent branches to 100 cm *l*, young **I** silvery, 4–8 mm *l*; **L** alternate, linear to almost spatulate, often falcate, concave above, 3–8 mm *l*; **Fl** in groups of 2–3, perigone lobes 4; **Ecol** sanddunes; **Distr** Warmbad, Namibia; Gordonia, NC, S Africa.

Note: Friedrich (1970: 92) placed this name as a synonym under **P. sericeus.**

P. cryptocarpus Fenzl 1839: 52 (see **photo**) T Drege 8024? (W).

Woody shrubs with ± erect branches, 20–45 cm *h*; **L** opposite, imbricate, broad-based, apically recurved, triangular-lanceolate, 4–10 mm *l*, 2–4 mm *b*; **Fl** mostly solitary, perigone with 5 lobes, 2–2.5 mm *l*; **Fr** 3 locules, 1–2 seeds developing; **Ecol** on Karroo rocks, among stones; **Distr** Keetmanshoop, Maltahöhe, Lüderitz-South, Namibia; Kenhardt, Victoria West, NC, S Africa.

Note: Very similar to **P. karooicus** in general appearance, this species can be distinguished by the spreading leaves and the three stigmas from the former.

P. karooicus I.Verdoorn 1941: 177 (see **photo**) T Smith 5286 (PRE) = *P. karooicus* I.Verdoorn var. *alternifolia* Adamson 1961: 149 T Acocks 5762 (K).

Woody base merging into the woody, splitting roots, young **I** with biforked hairs; **L** opposite or alternate, imbricate, erect, 3–6 mm *l*; **Fl** 2–3 in a group, the groups approaching each other closely, with 2–3 bracts and bracteoles, perigone lobes 4, 2–2.5 mm *l*, stamens 4; **Fr** 2 locules, only 1 seed developed; **Ecol** on Karroo rock among stones; **Distr** from Barkly West, NC, to Fauresmith, FS, and Middelburg, EC, S Africa.

Note: The species is browsed upon by sheep, representing even one of the preferred plants in feeding experiments (Preez 1970: 126–131); it is named "Karroo Ganna" in the Fauresmith District and "silwerganna" in the Middelburg District. The erect leaves and the 2 locules distinguish the species from **P. cryptocarpus.** Friedrich (1970: 92) considered the variety to be a synonym of **P. sericeus.**

P. karooicus var. *alternifolia* = **P. karooicus**
P. laxifolius = **P. sericeus**
P. psammophilus = **P. rehmannii**

P. rehmannii Schellenberg 1913: 499 **LT** Rehmann 5090 (Z) = *P. psammophilus* Dinter ex Adamson 1961: 151, nomen nudum in schedis **T?**

Sparsely branching shrubs, branches 15–60 cm *l*, young **I** grey or white from hairs, deeply furrowed with age; **L** alternate on the main shoots, elliptical, petiole 5 mm *l*, blade 10 mm *l*, 6 mm *b*; **Fl** solitary, sessile, perigone lobes 5; **Fr** 3 locules; **S** black; **Ecol** uncertain; **Distr** Transvaal, S Africa.

Note: Adamson (1961: 151) and Friedrich (1970: 92) gave Namibia as the area of distribution, which appears far and well separated from the type locality. The species appears to be very similar to **P. sericeus.**

P. sericeus Pax in Schellenberg 1913: 499 **T** Marloth 741 (B) = *P. laxifolius* I.Verdoorn 1941: 178 **T** Kotze 824 (PRE).

Low shrubs, old **I** without a cortex and deeply furrowed, young **I** silky-hairy; **L** alternate, lanceolate, pointed at tip and base, with silky hairs on both sides, 8 mm *l*, 2 mm *b*; **Fl** in groups of 3, sessile, perigone lobes 5, 2.5 mm *l*, silky hairy outside, yellow inside; **Fr** globose, 3 locules; **Ecol** in sand; **Distr** from Windhoek, Namibia, to Kimberley, NC, S Africa.

Note: The drawing by Pax in Engler (1910: 570, f 506) shows the furrowed basal woody stem well separated from the thinner fresh branches, suggesting that the latter may represent annual growths from a persisting woody branched base.

Polymita Ruschioideae

H. E. K. Hartmann

Polymita N.E.Br. GC 1930: 72; Hartmann Bradleya 1996: 29–56 **Typus P. albiflora** (L.Bolus) L.Bolus **Etym** Gk *poly*, many, *mitos*, thread, referring to the numerous linear petals.

Succulent shrubs to 45 cm *h*, mostly much broader, highly branched and dense; **L** > **I**, one pair to a branch in the resting state, basally connate for almost 50%, free parts spreading widely, trigonous to triquetrous, with a hard, white, cartilagenous keel and margins, epidermis with low papillae, raised around the sunken entire stomatal apparatus consisting of stomata and subsidiary cells, xeromorphic outer epidermis walls with a crystal layer up to 30 μm, wax cover continuous, breaking into irregular plates and horizontal platelets; **Fl** bracteolate, solitary, **K** 5, ± equal, petals very slender, white, in 5 groups, merging into the white, very slender filamentous staminodes, stamens white, apically often bright yellow, basally shortly papillate, nectary as a ring; **Fr** hard and robust, top raised a little to the centre, valves with valve wings broader than expanding keels, covering membranes complete, nearly straight, without a rim, but a thicker, spongy bulge below the distal end, closing bodies roundish, mostly 9–10 locules; **S** oval to pear-shaped, testa cells almost smooth; **Ecol** on quartz flats or slopes in open succulent Karroo or on gneiss, < 100 mm rainfall p.a. at the E edge of the winter rainfall area; **Distr** N Namaqualand, NC, S Africa.

Note: Originally described with the name of a single species only, the genus comprises two species now. The low cushions form monotypic stands on rather flat ground covered by quartzite or in flat pockets of granites. Cultivation is easy, but flowering occurs only rarely if enough light is provided in winter.

Distribution of **Polymita** N.E.Br.

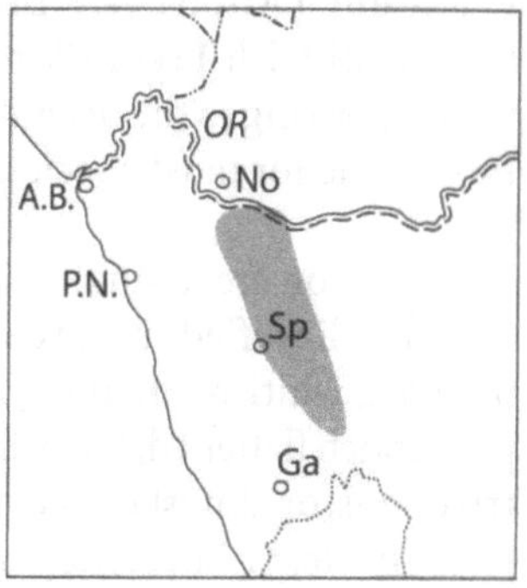

Key to the Species of Polymita

1. Leaves stiffly mucronate and therefore prickly to the touch .... **P. albiflora**
- Leaves without a mucro, not prickly, but firm to the touch .... **P. steenbokensis**

P. albiflora (L.Bolus) L.Bolus in Jacobsen 1960: 1348 (see **photo**) ≡ *Ruschia albiflora* L.Bolus JB 1928: 12 **T** Pillans 4930 (BOL) = *P. pearsonii* N.E.Br. GC 1930: 72 nomen nudum ≡ *M. perfoliatum* Haw. var. *integrifolium* L.Bolus 1913: 145 nomen nudum Pillans 3045 (BOL, K) = *P. diutina* (L.Bolus) L.Bolus 1963: 49 ≡ *Ruschia diutina* L.Bolus NM 1932: 356 **T** Pillans 6591 (BOL).

L with a big, hard, prickly mucro, keel and margins cartilagenous, white, 9–17 mm *l*, 4–6 mm *b*; **Fl** ∅ 25–30 mm, 50–100 white petals, 80–130 white filamentous staminodes, 200–250 white stamens; **Fr** valves always with broad valve wings narrowing only a little and gradually to their tips, closing bulge or rodlet below the distal end of the covering membrane, ∅ 5–10 mm, mostly 10 locules; **S** 0.7–1 mm *l*, 0.45–1 mm *b*; **Chr** 2*n* = 18; **Ecol** on flat field of quartzite or in flat pockets on granite, along the E boundary of the winter rainfall area, < 100 mm rainfall p.a.; **Distr** NE part of area of the genus, N Namaqualand, NC, S Africa.

Note: A characteristic plant of quartzitic or gneissic places in Namaqualand Broken Veld, often in conspicuous monotypic stands.

P. diutina = **P. albiflora**
P. pearsonii = **P. albiflora**

P. steenbokensis H.E.K.Hartmann Bradleya 1996: 42 (see **photo**) **T** Hartmann & Bayer 00.32 (HBG!).

L with a tiny, inconspicuous mucro, not prickly, firm, spreading, 10–12 mm *l*, 5–6 mm *t*; **Fl** with 70–140 white petals merging into 50–70 filamentous staminodes, 215–260 white stamens; **Fr** valves with basally very broad valve wings tapering distally almost into awns, covering membranes straight, without any closing device below the distal end, closing bodies small, round or hook-shaped, ∅ 6–7 mm, mostly 9 locules; **S** 1–1.1 mm *l*, 0.7–0.89 mm *b*; **Chr** 2*n* = 36; **Ecol** dominant on S slopes of rough quartzite, winter rainfall < 100 mm p.a.; **Distr** SW part of the area of the genus, N Namaqualand, NC, S Africa.

Note: The species is rarely seen in cultivation, where it flowers even more tardily than **P. albiflora**.

Portulacastrum ≡ **Trianthema**

H. E. K. Hartmann

Medikus (1789: 99–100) ascribed the genus *Portulacastrum* to De Jussieu, discussing in his text the description of "*T. monogyna* L." based on the description of Hermann, which is the base for Linnaeus' name, **Trianthema portulacastra** (≡ *T. monogyna*). The name *Portulacastrum* Jussieu ex Medikus is thus based on the same type as **Trianthema**. Medikus did not give a specific name at all, but *Portulacastrum monogynum* can be found cited.

Prenia Mesembryanthemoideae

M. Gerbaulet

Prenia N.E.Br. GC 1925: 412, 433; Bittrich 1987: 5–116; Gerbaulet BJS 1996: 25–40 ≡ *Phyllobolus* N.E.Br. subg. *Prenia* (N.E.Br.) Bittrich 1987: 75 **Lectotypus P. pallens** (Aiton) N.E.Br. ≡ *Phyllobolus pallens* (Aiton) Bittrich **Etym** Gk *prenes*, prone or bent forward, referring to the decumbent flowering branches; this is true for most species.

Prostrate or erect small shrubs with fibrous roots, I weakly lignified, cortex with additional vascular bundles, whitish when dry, bladder cells mesomorphic, much flattened; L flat, flatly triquetrous, obtusely trigonous, or almost cylindrical, circular, ovate to narrowly ovate, or linear, decussate and shortly connate at the base or becoming alternate in the **Inf**, persistent, with a thick wax layer which can easily be wiped off, without distinct central water-storing cells, bladder cells mesomorphic and much flattened; **Fl** in cymes, Ø 20–40 mm, **K** 4–5, petals white, yellow or pink, K and petals connate into a short tube, filamentous staminodes present, nectaries shell-shaped; **Fr** with valve wings, robust, with long lower part and deep locules, 4–5 locules; **S** to 1.3 mm *l*, D- or kidney-shaped, black, testa rough without or with an indistinct crest; **Ecol** in Karroo vegetation, winter and/or summer rainfall; **Distr** S Namibia; EC, NC, WC, S Africa.

Note: The circumscription of this genus has recently been amended (Gerbaulet BJS 1996: 25–40).

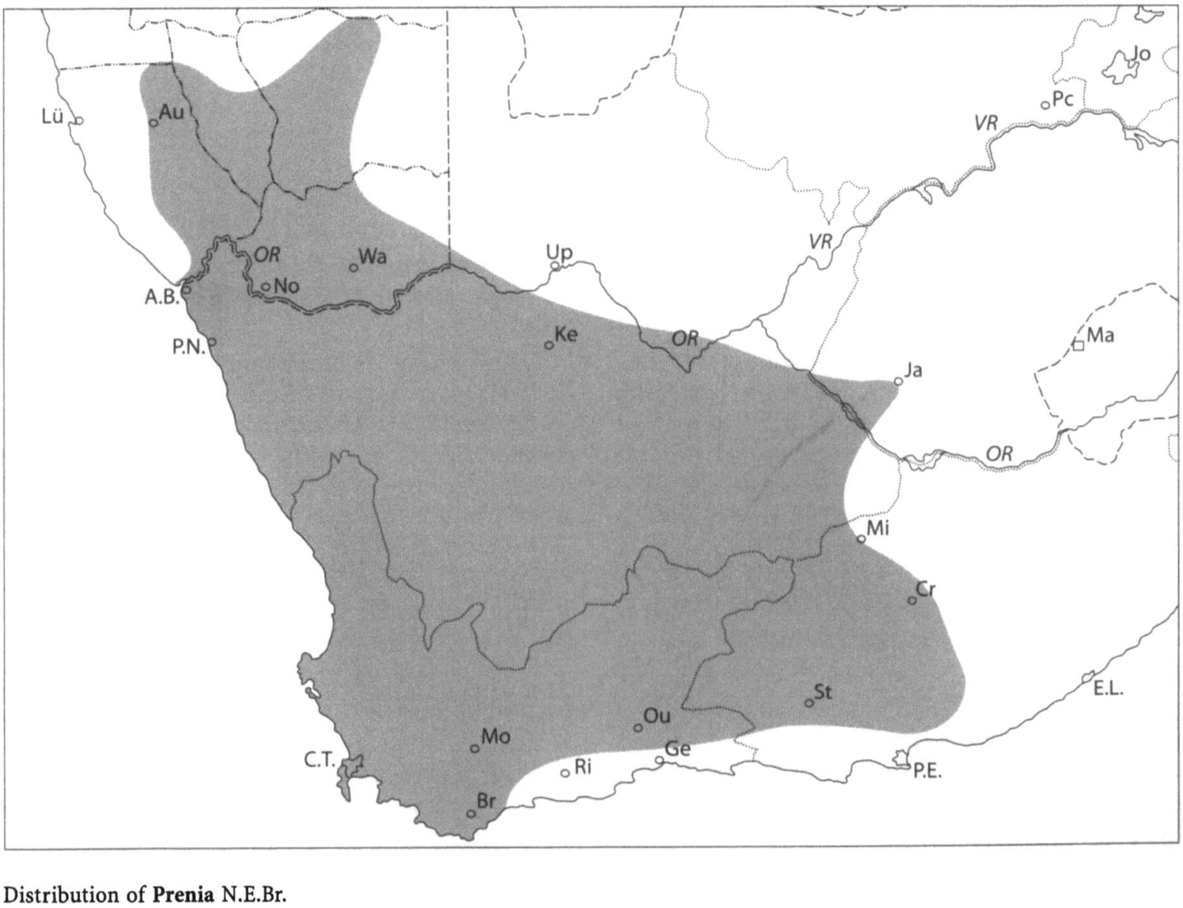

Distribution of **Prenia** N.E.Br.

Key to the Species of Prenia

1. Leaves flat or flatly triquetrous, circular or ovate or narrowly ovate, plants prostrate 2
– Leaves obtusely trigonous or almost cylindrical, linear, branches prostrate or erect 4
2. Leaves almost circular, margins of young leaf pairs sticking together and enclosing the leaf and flower buds .... **P. sladeniana**
– Leaves ovate to narrowly ovate, margins of young pairs not or only slightly sticking together 3
3. Plants very large, forming mats of up to several metres ∅; leaves ovate, sometimes slightly sticking together; internodes of the main stem as long as those of the side branches; flowers white or slightly yellowish; plants restricted to localities just above the shore line .... **P. vanrensburgii**
– Plants smaller, up to 1 or rarely 2 m ∅; leaves narrowly ovate to ovate, not sticking together; internodes of the main stem shorter than those of the side branches; flowers pink with a white or yellow centre or completely yellow .... **P. pallens**
4. Branches erect .... **P. tetragona**
– Branches prostrate 5
5. Branches with adventitious roots, leaves almost cylindrical .... **P. radicans**
– Branches without adventitious roots, leaves obtusely trigonous .... **P. englishiae**

P. englishiae (L.Bolus) Gerbaulet BJS 1996: 30 (see **photo**) ≡ *M englishiae* L.Bolus ABH 1922: 125 ≡ *Aridaria englishiae* (L.Bolus) N.E.Br. JB 1928: 142 ≡ *Nycteranthus englishiae* (L.Bolus) Schwantes in Jacobsen 1960: 1316 nomen illeg. ≡ *Sphalmanthus englishiae* (L.Bolus) L.Bolus in Jacobsen 1970: 511 nomen invalid. **T** English 2313/17 (BOL!).

Prostrate, **I** all of equal length; **L** obtusely trigonous, linear; **Fl** whitish to yellowish, **K** 4; **Fr** locules 4; **Ecol** often along roadsides or in disturbed places, winter rainfall; **Distr** Montagu, Robertson, Worcester, WC, S Africa.

Note: With the exception of the obtusely trigonous leaves, this species bears much resemblance to **P. pallens**. Both species grow in disturbed places and **P. englishiae** replaces **P. pallens** in distribution towards the east, presumably without much geographical contact.

P. olivacea = **Phyllobolus tenuiflorus**

P. pallens (Aiton) N.E.Br. GC 1928: 268 ≡ *M. pallens* Aiton 1789: 182 ≡ *Phyllobolus pallens* (Aiton) Bittrich 1987: 75 ≡ *Platythyra pallens* (Aiton) L.Bolus SAG 1927: 326 **LT** drawing by Ann Lee (BM).

Prostrate, **I** of the main stem shorter than those of the side branches; **L** flatly triquetrous, ovate or narrowly ovate; **Fl** pink with white centre or completely yellow, rarely pink with yellow centre, **K** 4–5; **Fr** locules 4–5; **Ecol** in disturbed places of the plains or along roadsides, winter rainfall; **Distr** NC, WC, S Africa.

Note: The basal leaves of **P. pallens** are always flatly triquetrous, but along the flowering branches they become somewhat more rounded and smaller. It is the shape of the leaves which renders **P. pallens** readily distinguishable from **P. englishiae**. Four subspecies are recognized.

Key to the Subspecies of P. pallens

1. Flowers completely yellow or with yellow centre and pink tips .... **P. pallens** ssp. **lutea**
– Flowers pink with white centre, sometimes almost pure white, never yellow 2
2. Leaves slender, narrowly ovate, borne in a distinct central rosette; flowers pink with white centre .... **P. pallens** ssp. **pallens**
– Leaves broader, narrowly ovate to ovate, borne in a loose central rosette; flowers light pink with white centre to almost pure white 3
3. Leaves very broad, more or less ovate; $2n = 18$ .... **P. pallens** ssp. **lancea**
– Leaves broad but usually still narrowly ovate; $2n = 36$ .... **P. pallens** ssp. **namaquensis**

P. pallens (Aiton) N.E.Br. ssp. **lancea** (Thunberg) Gerbaulet BJS 1996: 32 ≡ *M. lanceum* Thunberg 1800: 89 **LT** Thunberg 11991 (UPS, photograph K!).

L narrowly ovate to ovate, broad, basal ones more than 20 mm, often more than 60 mm *l*, borne in a loose rosette, which is often absent in older plants; **Fl** light pink with white centre or white; **Chr** $2n = 18$; **Ecol** as for species; **Distr** Calvinia, S Namaqualand, NC, Ceres, Clanwilliam, Vanrhynsdorp, Vredendal, Worcester, WC, S Africa.

Note: This broad-leaved subspecies bears a loose central rosette which vanishes with age.

P. pallens (Aiton) N.E.Br. ssp. **lutea** L.Bolus NM 1954: 259 (see **photo**) **LT** Hall 25281 (BOL!).

L narrowly ovate, small and slender, basal ones often only 20 mm *l*, borne in a distinct rosette; **Fl** yellow, sometimes with pink tips; **Chr** $2n = 18$; **Ecol** as for species; **Distr** Hopefield, Vredenburg, WC, S Africa.

Note: This small yellow-flowered subspecies has a very restricted distribution. To the east plants with yellow and pink flowers occur which indicates a certain degree of hybridization with the adjacent pink and white flowered subspecies **pallens.**

P. pallens (Aiton) N.E.Br. ssp. **namaquensis** Gerbaulet BJS 1996: 34 (see **photo**) **T** Gerbaulet & Struck 24916 (BOL!).

L narrowly ovate, broad and large, basal ones more than 60 mm *l*, borne in a loose rosette; **Fl** light pink with white centre; **Chr** $2n = 36$; **Ecol** as for species; **Distr** Namaqualand, NC, S Africa.

Note: **P. pallens** ssp. **namaquensis** is very similar to **P. pallens** ssp. **lancea.** Both subspecies differ in their chromosome numbers.

P. pallens (Aiton) N.E.Br. ssp. **pallens** = *M. fastigiatum* Haw. var. *pallens* Haw. 1812: 256 ≡ *M. pallescens* Haw. 1821: 174 **T** Jacquin 1798: t. 279 (as *M. pallens*) = *M. longistylum* DC. var. *purpurascens* DC. 1828: 446 **Neotype**, here designated: Gerbaulet & Struck 24749 (BOL!) = *M. loratum* Haw. 1821: 168 **NT** Page 17313 (BOL!) = *M. relaxatum* Willdenow 1814: 36 ≡ *Platythyra relaxata* (Willdenow) Schwantes GF 1928: 68 ≡ *Prenia relaxata* (Willdenow) N.E.Br. GC 1928: 268 **NT** drawing (K!).

L narrowly ovate, slender, basal ones more than 20 mm *l*, often 60 mm *l*, borne in a distinct rosette; **Fl** pink with white centre; **Chr** $2n = 18$; **Ecol** as for species; **Distr** from the Cape Peninsula to Piquetberg, WC, S Africa.

Note: In this subspecies the central rosette is most distinct.

P. radicans (L.Bolus) Gerbaulet BJS 1996: 34 (see **photo**) ≡ *Aridaria radicans* L.Bolus ABH 1927: 102 ≡ *Nycteranthus radicans* (L.Bolus) Schwantes KuaS 1952: 51 nomen illeg. ≡ *Sphalmanthus radicans* (L.Bolus) L.Bolus in Jacobsen 1970: 514 nomen invalid. **T** Coulter 1302/24 (BOL!).

Prostrate, rooting at the nodes, **I** all of equal length; **L** almost cylindrical, linear; **Fl** whitish to slightly yellowish, **K** 4; **Fr** locules 4; **Ecol** in plains or along roadsides, summer rainfall; **Distr** W EC, S Africa.

Note: This species is widely grown as a ground cover in public gardens or in bare and waste places. It differs from the remaining species of **Prenia** in its ability to form adventitious roots.

P. relaxata = **Prenia pallens** ssp. **pallens**

P. sladeniana (L.Bolus) L.Bolus SAG 1929: 49 (see **photo**) ≡ *M. sladenianum* L.Bolus ASAM 1913: 152 ≡ *Synaptophyllum sladenianum* (L.Bolus) N.E.Br. GC 1928: 254 **LT** Pearson 6019 (BOL!).

Prostrate, **I** all of equal length; **L** flat, almost circular, margins of young leaf pairs stick together and enclose the leaf and flower buds; **Fl** whitish to slightly pinkish, **K** 4; **Fr** locules 4; **Ecol** on lower hills or in sandy river beds, winter rainfall; **Distr** S Namibia; NW Namaqualand, NC, S Africa.

Note: The species is remarkable because of the young leaves adhering to each other along their margins to form an almost globular structure. Inside, the following leaf pair develops shielded against the harshness of the habitat. The leaves easily pop open when pressed. **P. sladeniana** bears a superficial resemblance to **Synaptophyllum juttae.**

P. tetragona (Thunberg) Gerbaulet BJS 1996: 36 (see **photo**) ≡ *M. tetragonum* Thunberg 1800: 91 ≡ *Aridaria tetragona* (Thunberg) L.Bolus FPSA 1927: t. 258 ≡ *Nycteranthus tetragonus* (Thunberg) Schwantes KuaS 1952: 51 nomen illeg. ≡ *Peratetracoilanthus tetragonus* (Thunberg) Rappa & Camarrone 1960: 24 nomen invalid. ≡ *Sphalmanthus tetragonus* (Thunberg) L.Bolus in Jacobsen 1970: 515 nomen invalid. **LT** Thunberg 12030 (UPS, photograph K!) = *M. ausanum* Dinter & A.Berger 1914: 586 ≡ *Aridaria ausana* (Dinter & A.Berger) Dinter & Schwantes GF 1928: 69 ≡ *Nycteranthus ausanus* (Dinter & A.Berger) Schwantes KuaS 1952: 50 nomen illeg. **LT** Dinter 1106 (BOL!) = *Aridaria luteoalba* L.Bolus NM 1928: 144 ≡ *Nycteranthus luteoalbus* (L.Bolus) Schwantes KuaS 1952: 50 nomen illeg. **T** Smith 482 (BOL!) = *Aridaria mutans* L.Bolus NM 1929: 149 ≡ *Nycteranthus mutans* (L.Bolus) Schwantes KuaS 1952: 50 nomen illeg. **T** Rogers 18947 (BOL!) = *M. pfeilii* Engler 1909: 196 ≡ *Psilocaulon pfeilii* (Engler) Schwantes GF 1928: 69 **T** Pfeil 136 (B!) = *M. pseudausanum* Dinter FR 1928: 29 **LT** Dinter 4755 (BOL!) = *Aridaria suffusa* L.Bolus ABH 1927: 102 ≡ *Nycteranthus suffusus* (L.Bolus) Schwantes KuaS 1952: 51 nomen illeg. ≡ *Sphalmanthus suffusus* (L.Bolus) L.Bolus in Jacobsen 1970: 515 nomen invalid. **T** Pillans 5237 (BOL!).

Erect, **I** all of equal length; **L** almost cylindrical, linear; **Fl** whitish to yellowish or pinkish, **K** 4; **Fr** locules 4; **Ecol** in plains or along roadsides, winter and summer rainfall; **Distr** in a triangle reaching from S Namibia to Fauresmith and Ladismith; S Namibia; FS, NC, WC, S Africa.

Note: This very widespread species shows some variation in the size of the leaves and floral colour. Specimens from the Free State are known to flower yellow, whereas from N Namaqualand only pink-flowering specimens have been recorded.

P. vanrensburgii L.Bolus NM 1954: 258 (see **photo**) ≡ *Perapentacoilanthus vanrensburgii* (L.Bolus) Rappa & Camarrone 1960: 23 nomen invalid. T van Rensburg 12945 (BOL!).

Prostrate, forming huge mats, **I** all of equal length; **L** flat, ovate; **Fl** whitish to slightly yellowish, **K** 5; **Fr** locules 5; **Ecol** next to the shoreline, winter rainfall; **Distr** Bredasdorp, Hermanus, WC, S Africa.

Note: This species is edible, locally known as "seepampoen". Its young leaves also stick together, but much less than in **P. sladeniana.**

Prepodesma Ruschioideae

H. E. K. Hartmann

Prepodesma N.E.Br. GC 1930: 278; N.E.Brown GC 1931: 389 **Typus P. orpenii** (N.E.Br.) N.E.Br. **Etym** Gk *prepo*, to be conspicuous; *desme*, a bundle, referring to the stamens forming an erect bundle in the centre of the flower.

Compact with an inconspicuous rootstock, 2–4 leaf-pairs formed per season; **L** trigonous, spreading a little, always distinctly mucronate, epidermis with elevations and papillate, but hardly velvety to the touch, wax in horizontal plates between the papillae, continuous on papillae, covered thickly by rodlets and platelets, thick outer epidermal wall without crystals, subhypodermal tannin idioblasts therefore visible as small dark dots; **Fl** solitary, bracteolate, **K** 5, subequal, petals yellow, whitish or pinkish outside, stamens in a central column, papillate near their bases, opening in the afternoon between 3 and 6 p.m., nectary ring lophomorphic; **Fr** long stalks persistent, Titanopsis-type with a thin columella only, covering membranes straight, endokarp raised a little lifting the placenta, but no separate closing body to be recognized, valve wings narrowing at base and tip, in the middle broader than the expanding keels, mostly 7 locules; **S** almost ovate in outline, light brown, mostly smooth, rarely with raised testa cells; **Ecol** on loamy shales or in crevices of quartzitic limestone rocks, in flat areas, 300–400 mm rainfall p.a.; **Distr** Hay, Herbert, Postmasburg, NC, S Africa.

Note: Following Bolus (NM 1938: 134–135), the genus has been sunken because she considered the difference between the erect bundle of stamens and the constriction seen in **Nananthus** not as distinctive enough to uphold a genus. This decision must be seen against the background that, at that time, she considered **Nananthus** and **Aloinopsis** as congeneric, a view she later abandoned (NM 1958: 372–385) placing **P. orpenii** in **Aloinopsis.** Based on differences in root formation, in leaf morphology and anatomy, as well as in fruit and seed morphology between **Prepodesma** and the two mentioned genera, **Prepodesma** is restored to genus rank. Only one species is recognized. The existence of bracteoles has been discussed controversially because the respective leaves are almost as big as foliage leaves. Since they wither with the ripening fruit and come off with the fruit stalk, they are classified as bracteoles here.

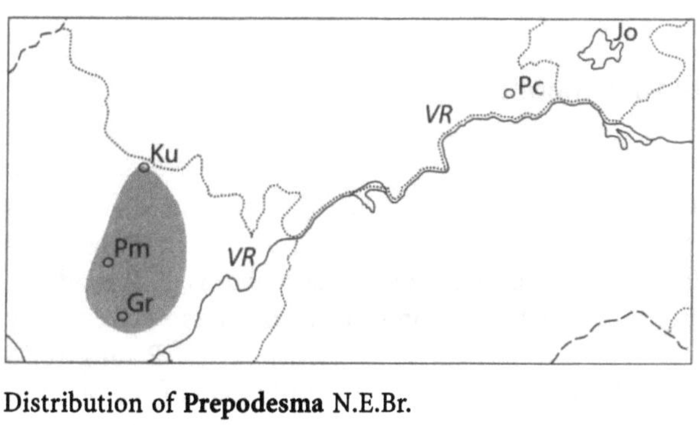

Distribution of **Prepodesma** N.E.Br.

P. orpenii (N.E.Br.) N.E.Br. GC 1931: 389 (see **photo**) ≡ *M. orpenii* N.E.Br. GC 1921: 303 ≡ *Aloinopsis orpenii* (N.E.Br.) L.Bolus SAG 1929: 288 ≡ *Nananthus orpenii* (N.E.Br.) NM 1938: 133 T Pole Evans 6910 (K!) = *Rabiea tersa* N.E.Br. GC 1931: 54 ≡ *Nananthus tersus* (N.E.Br.) G.D.Rowley NCSJ 1978: 6 T Pole Evans 7623 (K!).

I invisible; **L** with distinct but rounded margins and keel, a little humped at the back below the tip, papillae variable between populations, 20–30 mm *l*, 6.5–8 mm *t* and *b*; **Fl** bracteoles only a little shorter than the leaves, ∅ 40–50 mm, c. 100 bright yellow petals, often pink below, 250–270 stamens in a central column; **Fr** (5–)7(–9) locules; **S** 1–1.2 mm *l*, 0.6–0.8 mm *b*; **Ecol** as for genus; **Distr** as for genus.

Note: Although the type of *Rabiea tersa* consists of a flower and a broken leaf only, it shows on both organs the papillate epidermis typical of **Prepodesma** and unknown in **Rabiea.** The population agrees with other northern ones in the rather dense cover of papillae. In southern populations, older leaves are often less papillate, the papillae being worn off with age or not expressed as well from the beginning. Growing well from seed and flowering readily, the species presents an attractive addition to a collection. Its distribution area, far from winter rainfall and very often in loamy soils, is peculiar for a member of the Mesembryanthema.

P. uncipetala = **Hereroa wilmaniae**

Psammanthe = Sesuvium

H. E. K. Hartmann

Walpers (1852: 659–660) ascribed the new genus *Psammanthe* to Hance, citing an unpublished work of the latter, who had based his description on material from the island of Prata(s), off the southeastern coast of China. Although the author stated that the genus differs from **Trianthema**, **Sesuvium**, and *Pyxipoma*, he does not specify the distinction. *Psammanthe* Hance ex Walpers has subsequently been treated as a synonym of **Sesuvium**, with only one species having been described.

P. marina = **Sesuvium portulacastrum**

Psammophora Ruschioideae

H. E. K. Hartmann

Psammophora Dinter & Schwantes ZSK 1926: 188; Hartmann Bradleya 1996: 29–56 **Typus P. nissenii** (Dinter) Dinter & Schwantes **Etym** Gk *psammos*, sand, dune, *phora*, container, bearer, describing the feature that sand is caught by the leaf surfaces.

Compact to shrubby perennials; **L** trigonous to triquetrous, or crescent-shaped, hardly connate at their bases, at least some epidermis cells subdivided further by secondary, anticlinal walls, the outer wall covered by a thin cuticle when young, under which a sticky substance develops lifting the cuticle and bursting it finally exposing the glue to which sand particles adhere in great quantity forming a secondary protective layer over the leaf surface, the crystal sand layer below the sticky layer about as thick as the inner pure cellulose layer of the wall; **Fl** bracteolate, solitary, **K** 4–5, petals white to purple, nectary as a mostly dark green ring; **Fr** Dracophilus-type, with or without covering membranes, the valve-wings distally broader than proximally, the funnel-shaped base also sticky, (4–)5–6(–9) locules; **S** pear-shaped, testa cells with undulate anticlinal walls, in most species cushion-shaped, 0.6–1.1 mm *l*, 0.4–0.75 mm *b*; **Ecol** in sand and gravel rich in limestone, often in pockets of rocks, rarely on rocks, < 100 mm rainfall p.a. in winter, fog; **Distr** SW Namibia; NW Namaqualand, NC, S Africa.

Note: Sticky leaf surfaces occur occasionally in Aizoaceae, e.g. in **Galenia africana**, but in each case the stickiness is achieved in a different way. Of all genera with sticky surfaces, only **Psammophora** develops an extensive pattern of secondary anticlinal walls in the epidermis; even the hypodermis and lower layers can form secondary subdivisions, positioned in all directions. Four species are distinguished here forming two groups. **P. longifolia**, **P. modesta**, and **P. nissenii** share some features absent in **P. saxicola**: they possess special raised cells exceeding the others (between which the stomata are placed) by ten- to twelve-fold. These large cells are subdivided to different degrees in the three species, but the glue formed above them below the cuticle is united, developing an enormous layer before it is released through the burst cuticle. Sand sticks to this glue thickly, colouring the originally light green leaves reddish-brown and resulting in a perfect camouflage in habitat.

Distribution of **Psammophora** Dinter & Schwantes

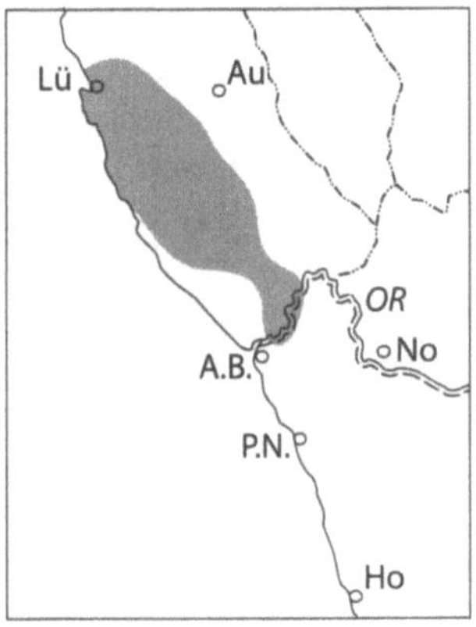

Key to the Species of Psammophora

1. Shrubs, calyx 4-lobed, filamentous staminodes 0–9 2
– Compact plants, calyx 5(–6) lobed, filamentous staminodes 17–20 3
2. Petals white, leaves crescent-shaped, epidermal cells with central papillae .... **P. saxicola**
– Petals purple or pink, leaves trigonous to triquetrous, some epidermal cells raised like cushions . **P. modesta**
3. Leaves slender, up to 10 mm broad and thick, straight, erect to diagonal, only the base covered by soil .... **P. longifolia**
– Leaves broadening to the tip, forming a recurved triangle there, which is often the only part of the plant to be seen in habitat .... **P. nissenii**

P. herrei = **P. longifolia**

P. longifolia L.Bolus NM 1930: 201 (see **photo**) T Erni 1093/30 (BOL) = *P. herrei* L.Bolus NM 1931: 305 T Herre 9201 (BOL).

Compact, with up to 12 branches, ∅ up to 12 cm; L erect to diagonal, slender to thickish, but the tip always tapering from parallel sides to a blunt end, to 40 mm *l*, to 10 (rarely 16) mm *b*, to 6 (rarely 13) mm *t*, elevated, subdivided epidermis cells rather densely packed like pavement stones, stomata in the cavities in between; **Fl** ∅ 20–23 mm, **K** 5, mostly < 40 white petals, 17–46 white filamentous staminodes, 75–233 basally papillate stamens; **Fr** with a central spongy tissue, sometimes forming a columella reaching the top, valve wings only a little broader than the expanding keels, septa with narrow ledges, no covering membranes proper, placenta in some populations raised by a low false septum, (4–)5(–6) locules; **S** testa cells cushion-shaped, epicuticular ornaments granular, micropylar fold with cuticular folds, 0.7–0.9 mm *l*, 0.55–0.75 mm *b*; **Ecol** in rough rubble or gravel, often in quartz, < 100 mm rainfall p.a. in winter and fog; **Distr** Lüderitz-South, Namibia; N Namaqualand, NC, S Africa.

Note: P. herrei differs only by slightly thicker leaves from **P. longifolia**. Most similar to **P. longifolia** is **P. nissenii**, the main difference being the more slender and erect leaves, the low false septum and the narrower valve-wings in the former. The subdivision of the epidermal cells appears less high in **P. nissenii** in comparison to **P. longifolia**. The species settle in adjacent areas in the Namib, **P. longifolia** inhabiting the southern habitats lower in lime contents.

P. modesta (Dinter & A.Berger) Dinter & Schwantes ZSK 1926: 189 (see **photo**) ≡ *M. modestum* Dinter & A.Berger 1914: 588 **T** Dinter 1012 (B).

Shrubs to 15 cm *h*, branches often decumbent and with leaves only in their terminal parts; **L** > **I**, trigonous to triquetrous, rarely roundish, 20–40 mm *l*, 7–10 mm *b*, 6–8 mm *t*, epidermis with prominent, raised, subdivided cells with fewer partitions than in the two compact species, humps well separated and single resulting in a rough appearance of the surface; **Fl** ∅ 15–25 mm, petals pink to purple, base sometimes white, mostly 0–4 white or pink filamentous staminodes, rarely up to 18, 70–190 stamens, dark green lophomorphic nectary ring; **Fr** with a spongy base, but no columella, sometimes a tiny false septum lifting the placenta above the bottom of the locule, valve wings very broad and rounded when the valves are expanded, septa without ledges, covering membranes therefore completely absent, 4–9 locules, maximum at 6 or 7; **S** testa cells cushion-shaped, no distinct epicuticular ornaments, micropylar fold with cuticular folds, 0.6–0.95 mm *l*, 0.45–0.7 mm *b*; **Ecol** on gravel or in sand-filled pockets, < 100 mm rainfall p.a. in winter and fog; **Distr** as for genus.

Note: A widespread species very variable in shape and size, in habit and leaf especially. Plants in cultivation seem to have a lifespan of up to 25 years, perhaps due the lack of regeneration from the base. In the very broad and rounded valve wings, **P. modesta** is similar to **P. nissenii**.

P. nissenii (Dinter) Dinter & Schwantes ZSK 1926: 188 (see **photo**) ≡ *M. nissenii* Dinter FR 1923: 123 **Lectotype**, here designated: Dinter 1019 (B).

Compact with up to 12 branches, ∅ to 10 cm, sunken in the ground for most of the leaf length, only the triangular leaf tips visible; **L** broadening towards the tip, forming a recurved triangle there, exhibiting the leaf surface in this part, 20–40 mm *l*, up to 15 mm *b*, the raised, subdivided prominent epidermis cells densely packed as if forming a pavement, the stomata in the sunken fissures in between; **Fl** ∅ 20–30 mm, **K** 5, mostly < 40 white petals, mostly 18–43 white filamentous staminodes, 74–138 white, basally papillate filaments; **Fr** spongy centre tiny or big, related to the depth of the capsule, septa with narrow ledges, no covering membranes proper, valve wings broader than the expanding keels, with a rounded outline, no basal false septum below the placenta, 5, rarely 6 locules; **S** testa cells cushion-shaped, surface rugose, but no granular ornaments, micropylar fold with cuticular folds, 0.7–0.9 mm *l*, 0.55–0.75 mm *b*; **Ecol** in sand and gravel rich in limestone, also in pockets in quartz or dolomite, < 100 mm rainfall p.a. in winter, fog; **Distr** Lüderitz-South, Namibia.

Note: The broad, recurved leaf tips and the wider and more rounded valve wings distinguish this species from **P. longifolia. P. nissenii** occurs in the northern part of the southern Namib and in soils with higher lime contents and receiving more fog than **P. longifolia.**

P. pillansii ≡ **Arenifera pillansii**

P. saxicola H.E.K.Hartmann Bradleya 1996: 44 (see **photo**) **T** Hartmann & Dehn 15052 (HBG!).

Shrubs to 60 cm *h*; **L** crowded near the apex of the branches, crescent-shaped, light to bright green, epidermis more uniform in appearance than the other species, cells with short, central, apically pointed papillae, subdivided only once or twice; **Fl** **K** 4, ∅ 35–40 mm, c. 40 white petals, white filamentous staminodes 0–1, 100–145 stamens in a narrow central cone; **Fr** Dracophilus-type, the covering membranes touching each other, but often reduced in width and length, 8 locules; **S** testa cells on sides flat, but anticlinal walls undulate, cushion-shaped cells over the micropylar and hilar end stretching over the dorsal region to some extent, granular epicuticular ornaments present, 0.95–1.15 mm *l*, 0.6–0.7 mm *b*; **Chr** 2*n* = 18; **Ecol** amongst stones on shaly slopes facing S, < 100 mm rainfall p.a., fog; **Distr** SE part of Lüderitz-South, Namibia.

Note: The species differs from the other three in its bright green leaves in habitat, probably due to the thinner sand cover, which in turn probably depends on the habitats: less sand is blown with the wind over the rocky places the plants grow in. In SEM-graphs, a reduced production of sticky substance is also obvious, probably related to the smaller epidermal cells. Although the white petals resemble those of **P. longifolia** and **P. nissenii**, growth form, leaf shape, and flower structure agree more with **P. modesta.**

Pseudobrownanthus = Brownanthus

M. Gerbaulet

The monotypic genus *Pseudobrownanthus* as established by Ihlenfeldt & Bittrich (1985: 319–320) was separated from **Brownanthus** because of its fruit being indehiscent and nut-like and not a hygrochastic capsule. However, since otherwise both genera have several derived character states in common, the exclusion of *Pseudobrownanthus* would leave **Brownanthus** paraphyletic. Consequently, *Pseudobrownanthus* is included here in **Brownanthus.**

P. nucifer ≡ **Brownanthus nucifer**

Psilocaulon Mesembryanthemoideae

C. Klak & M. Gerbaulet

Psilocaulon N.E.Br. GC 1925: 433; Ihlenfeldt & Bittrich 1985: 289–322; Bittrich 1987: 5–116; Klak & Linder 1998: 301–375; Klak 1999: 75–79 **Lectotypus P. articulatum** (Thunberg) N.E.Br. **Etym** Gk *psilos*, bare, bald, smooth, *kaulos*, stalk, referring to the leafless stems.

Prostrate, decumbent, or erect shrubs, some short-lived to annual, depending on water supply, with fibrous roots **I** at least to some extend with succulent green cortex, usually articulate, in most species nodes conspicuous with fine horizontal furrow, cortex with additional vascular bundles, bladder cells mesomorphic, hair-like, dome-shaped, or flattened; **L** almost cylindrical or slightly trigonous, decussate, shortly connate at the base or free, often with membranous margins at the leaf sheaths, annual and either deciduous or drying, without central water-storing cells, bladder cells mesomorphic, dome-shaped, hair-like, or flattened, stomata sometimes sunken; **Fl** in cymes, rarely solitary, ∅ 5–25 mm, **K** 4–5, connate into a short tube, petals pink, puce, yellow or white, few or rarely almost none, either very shortly connate or free, filamentous staminodes gathered into a cone, transparent except for a thickened middle vein and apically lacerate, very shortly connate or free, nectaries narrowly shell-shaped or absent; **Fr** valve wings inflexed over the valves, 4–5 locules, often deep (1.4–6 mm); **S** 0.5–1.2 mm *l*, D-shaped, ochre to brown, sometimes with dark warts, testa rough to smooth, without a distinct crest; **Ecol** often in disturbed places, winter to summer rainfall; **Distr** S Angola; Australia; Namibia; EC, FS, NC, WC, S Africa.

Note: Most species of **Psilocaulon** are fairly widespread pioneers. These are most often found in disturbed places on a wide variety of soils, and even in situations of high salinity. Some species, however, such as **P. foliosum**, **P. leptarthron**, or **P. parviflorum**, are more localized and specialized.

Key to the Species of Psilocaulon

1. Primary cortex soon withering and stems becoming woody, trunk to 50 mm in ∅, densely and regularly branched **2**
– Stems with a persistent green cortex, woody only at the base **3**
2. Leaves fine and narrow, locules 5 .... **P. foliosum**
– Leaves broad and thick, locules 4 .... **P. densum**
3. Nodes conspicuous, fine horizontal furrow at nodes present, leaves free to the base **4**
– Nodes inconspicuous, fine horizontal furrow at nodes usually absent or rarely present, leaves often shortly connate at the base **12**
4. Stems with hair-like papillae .... **P. articulatum**
– Stems glabrous or warty **5**

5. Plants prostrate, flowers white 6
- Plants decumbent to erect, flowers white, pink or pale yellow 7
6. Side branches very slender, < 2 mm in ∅, capsule 4-locular .... **P. parviflorum**
- Side branches > 3 mm in ∅, capsule 5-locular, stems forming a central clump and long annual shoots .... **P. bicorne**
7. Flowers usually solitary 8
- Flowers in few- to many-flowered cymes 9
8. Shrubs erect, 50–100 cm high, very regularly branched, stems straight, glabrous, flowers white, ∅ 18–25 mm, capsule ∅ 4.5–6 mm .... **P. leptarthron**
- Shrubs decumbent to erect, to 25 cm high, irregularly branched, with dome-shaped papillae, flowers pale yellow, ∅ 5–8 mm, capsule ∅ 3–3.5 mm .... **P. granulicaule**
9. Stems conspicuously constricted at the nodes, flowers puce, pink or rarely white, stems often warty ... **P. dinteri**
- Stems inconspicuously constricted at the nodes, flowers white or pink, stems smooth 10
10. Plants decumbent, forming small shrubs, to 30 cm high .... **P. dimorphum**
- Plants erect, forming medium to large shrubs, 30–150 cm high 11
11. Plants (50–)70–150 cm high, petals 0.5–1 mm broad .... **P. coriarium**
- Plants 30–50(–60) cm high, petals 1–2 mm broad .... **P. junceum**
12. Flowers yellow, petals more or less absent .... **P. gessertianum**
- Flowers white or pink, petals present 13
13. Small shrubs to 25(– 35) cm high with prostrate to decumbent stems, flower ∅ to 10 mm ... **P. salicornioides**
- Shrubs 50(–80) cm high with decumbent to erect stems, flower ∅ 10–15 mm .... **P. subnodosum**

Distribution of **Psilocaulon** N.E.Br.

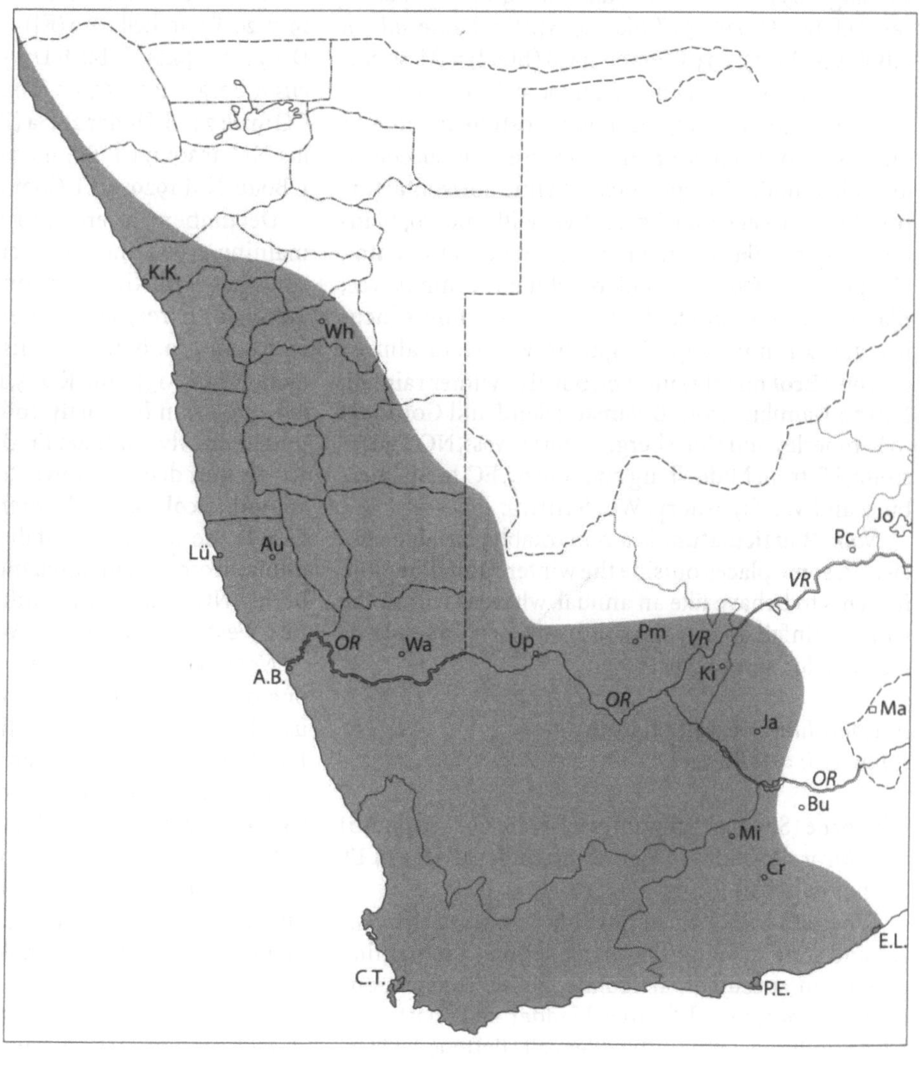

P. absimile = **P. coriarium**
P. acutisepalum = **P. junceum**
P. album = **P. leptarthron**
P. annuum = **P. articulatum**
P. arenosum ≡ **Brownanthus arenosus**

P. articulatum (Thunberg) N.E.Br. in Phillips 1926: 248 (see **photo**) ≡ *M. articulatum* Thunberg 1791: 10 **LT** Thunberg 11947a (UPS, photograph!) = *P. annuum* L.Bolus NM 1928: 29 **T** Gill 254 (BOL!) = *P. asperulum* N.E.Br. GC 1928: 254 **T** van der Bijl 31 (K!) = *P. bryantii* L.Bolus NM 1930: 173 **T** Bryant 708 (BOL!) = *P. dejagerae* L.Bolus NM 1931: 236 **T** de Jager 19382 (BOL!) = *P. duthiae* L.Bolus NM 1932: 331 **T** Duthie 9547 (BOL!) = *M. granulicaule* Haw. var. *purpurascens* A.Berger ex Range FR 1934: 18 nomen nudum **T** not given = *P. hirtellum* L.Bolus NM 1930: 191 **T** Pillans 19010 (BOL!) = *P. liebenbergii* L.Bolus 1964: 77 **T** Liebenberg 7390 (BOL!) = *M. mucronulatum* Dinter 1914: 42 ≡ *P. mucronulatum* (Dinter) N.E.Br. GC 1928: 254 **T** Jutta Dinter 3176 (SAM!) = *P. puberulum* Dinter FR 1931: 163 nomen nudum ≡ *M. puberulum* Dinter ex Range 1934: 17 nomen nudum non M. buberulum Dinter ex Range 1934: 17 nomen nudum **T** not given = *P. pubescens* N.E.Br. GC 1928: 254 **T** Muir 3978 (K!) = *P. roseoalbum* L.Bolus NM 1935: 493 **T** Power 21458 (BOL!) = *M. secundum* Thunberg 1827: 12 nomen nudum **T** not given.

Decumbent to erect, becoming prostrate in loose or sandy soil, 10–30 cm *h*, **I** remaining green and succulent, articulate, nodes conspicuous with fine horizontal furrow, bladder cells hair-like; **L** free, with mucro, bladder cells hair-like; **Fl** 1 to many per cyme, ∅ to 10 mm, **K** 5, petals dark or pale pink or white; 1–2 mm *b*, free, filamentous staminodes free; **Fr** ∅ 3.0–3.3 mm, 5 locules, 2.0–2.1 mm deep; **S** light brown, testa almost smooth; **Ecol** mainly summer, but also winter rainfall; **Distr** S Namibia; from E Namaqualand and Gordonia to Kimberley and Colesberg, Victoria West, NC, Fauresmith, FS, from Middelburg and Albany, EC, to Malmesbury and Vanrhynsdorp, WC, S Africa.

Note: **P. articulatum** is a remarkably variable species. In some places outside the winter rainfall region, it seems to behave like an annual, whereas within the winter rainfall area of the south-western Cape plants may live for several years.

P. asperulum = **P. articulatum**
P. baylissii = **P. dinteri**

P. bicorne (Sonder) Schwantes GF 1928: 69 (see **photo**) ≡ *M. bicorne* Sonder in Harvey & Sonder 1862: 434 **LT** Zeyher 2616 (S!).

Prostrate with a cluster of short stems in the centre and long, trailing flowering shoots, **I** remaining green and succulent, articulate, nodes conspicuous with fine horizontal furrow, bladder cells flattened; **L** free, with short mucro, bladder cells flattened; **Fl** few to many per cyme, rarely 1, ∅ to 15 mm, **K** 5, petals white, c. 1 mm *b*, free, filamentous staminodes free; **Fr** ∅ c. 3.5 mm, 5 locules, 2.8 mm deep; **S** brown, testa usually smooth, occasionally rough; **Ecol** winter and summer rainfall; **Distr** from Port Elizabeth to Willowmore, EC, Caledon, Uniondale and Worcester, WC, S Africa.

Note: **P. bicorne** forms a cluster of perennial short shoots in the centre of the plants. With the onset of the rainy season, long annual flowering shoots are formed, which may reach a length of up to 1 m.

P. bijliae = **P. junceum**
P. bryantii = **P. articulatum**
P. caducum = **Mesembryanthemum nodiflorum**
P. calvinianum = **P. junceum**
P. candidum = **P. junceum**
P. ciliatum ≡ **Brownanthus ciliatus**
P. clavulatum = **P. subnodosum**
P. corallinum ≡ **Brownanthus corallinus**

P. coriarium (Burchell ex N.E.Br.) N.E.Br. GC 1928: 254 (see **photo**) ≡ *M. coriarium* Burchell ex N.E.Br. 1920: 127 ≡ *Ruschia coriaria* (Burchell ex N.E.Br.) Schwantes ZSK 1927: 20 **T** Burchell 2679 (K!) = *P. absimile* N.E.Br. in Burtt Davy 1926: 49, 157 **T** Burtt Davy 10809 (PRE!) = *M. mentiens* A.Berger 1922: 630 ≡ *P. mentiens* (A.Berger) N.E.Br. GC 1928: 254 **T** Dinter 2148a (B!) = *P. stenopetalum* L.Bolus NM 1930: 171 **T** Bryant 712 (BOL!) = *P. uncinatum* L.Bolus NM 1929: 60 **T** Compton 1831/27 (BOL!).

Decumbent to erect, to 150 cm *h*, ∅ to 3 m, **I** remaining green and succulent, articulate, nodes conspicuous with fine horizontal furrow, bladder cells flattened; **L** free, short, often < 5 mm *l*, rarely with a short mucro, bladder cells flattened; **Fl** many per cyme, ∅ 10(–13) mm, **K** 4–5, petals white or pale pinkish, 0.5–1 mm *b*, shortly connate, filamentous staminodes shortly connate; **Fr** ∅ 3.1–4.0 mm, 4–5 locules, 3.1–3.5 mm deep; **S** brown, testa usually rough, rarely smooth; **Ecol** eastern border of the winter rainfall region to the summer rainfall region; **Distr** C and S Namibia; from E Namaqualand and Gordonia to Kimberley, NC, and to Fauresmith, FS, Cradock, EC, Beaufort West, Calvinia, Ceres, WC, S Africa.

Note: The extremely widespread **P. coriarium** is one of the largest species of **Psilocaulon**. Fruits are usually 5-locular in populations from Namibia (Rosh Pinah to Windhoek), whereas those from South Africa and the south-eastern corner of Namibia (Warmbad area) are usually 4-locular. *P. uncinatum* is said to be used by the 'Koi'koi to impart a reddish colour to leather. *P. absimile* has been reported to be poisonous to sheep unaccustomed to it. The ashes of these plants are used in soap-making.

P. dejagerae = **P. articulatum**
P. delosepalum = **P. junceum**

P. densum N.E.Br. GC 1928: 254 **T** Schlechter 1 (K!).

Decumbent to erect, **I** woody, indistinctly articulate, nodes inconspicuous, with or without fine horizontal furrow, bladder cells flattened; **L** shortly fused or free, without mucro, bladder cells flattened; **Fl** in cymes, ∅ c. 15 mm, **K** 4, petals free; **Fr** ∅ c. 5.0 mm, 4 locules, c. 2.6 mm deep; **S** ochre with a white perisperm region, testa almost smooth; **Ecol** winter rainfall; **Distr** Namaqualand, NC, S Africa.

Note: **P. densum** is only known from two specimens collected at the same locality near Steinkopf. The plants lack the succulent green stems typical for Psilocaulon. In addition, no information on the floral characters are available, so that the placement of this species in **Psilocaulon** remains uncertain.

P. dimorphum (Welwitsch ex Oliver) N.E.Br. GC 1928: 254 ≡ *M. dimorphum* Welwitsch ex Oliver 1871: 582 **T** Welwitsch 2377 (BM!).

Decumbent, to 30 cm *h*, **I** remaining green and succulent, articulate, nodes conspicuous with fine horizontal furrow, bladder cells usually flattened, occasionally papillate; **L** free, with a mucro, bladder cells usually flattened; **Fl** many per cyme, ∅ to 15 mm, **K** 4–5, petals white; **Fr** ∅ c. 3.0 mm, 4–5 locules, c. 3.0 mm deep; **S** ochre, testa smooth; **Ecol** in saline, sandy habitats, summer rainfall; **Distr** between Mossamedes and Porto Alexandre, S Angola.

Note: A little known species from the most northern parts of the distribution area of **Psilocaulon**. In the original description it is noted that "after 3 or 4 years the plants become rigid pubescent with much reduced leaves".

P. dinteri (Engler) Schwantes p.p. GF 1928: 69 (see **photo**) ≡ *M. dinteri* Engler p.p. 1909: 197 **LT** Dinter 954 (B!) = *P. baylissii* L.Bolus 1963: 47 **T** Bayliss 988 (BOL!) = *P. fasciculatum* N.E.Br. GC 1928: 254 **LT** Pillans 17767 (K!) = *P. godmaniae* L.Bolus var. *godmaniae* NM 1929: 143 **T** Godman 18979 (BOL!) = *P. godmaniae* L.Bolus var. *gracile* L.Bolus NM 1930: 174 **T** Godman 19049 (BOL!) = *P. herrei* L.Bolus 1963: 47 **T** Herre 14675 (BOL!) = *P. littlewoodii* L.Bolus 1961: 117 **T** Littlewood 591/59 (BOL!) = *P. littlewoodii* L.Bolus forma *laxum* L.Bolus 1961: 118 **T** Littlewood 629/59 (BOL!) = *M. marinum* A.Berger ex Engler 1915: 153 **NT** Klak 115 (BOL!) = *P. pageae* L.Bolus ABH 1927: 106 **T** Page 17167 (BOL!) = *P. pageae* L.Bolus var. *grandiflorum* L.Bolus NM 1928: 30 **T** Frames 159/28 (BOL!) = *P. variabile* L.Bolus NM 1930: 190 **T** Pillans 17769 (BOL!).

Prostrate to decumbent or erect, to 25 cm *h*, **I** remaining green and succulent, often very succulent on older stems, barrel-shaped in saline coastal areas, otherwise cylindrical, distinctly articulate, nodes conspicuous with fine horizontal furrow, bladder cells flattened or dome-shaped; **L** free, occasionally with a mucro, bladder cells flattened or dome-shaped; **Fl** few to many per cyme, ∅ 5–10 mm, **K** 5, petals puce, pink or occasionally white, 1–2 mm *b*, free, filamentous staminodes free; **Fr** ∅ 2.7–4.5 mm, 5 locules, 2.7–3.9 mm deep; **S** ochre, testa rough or rarely (almost) smooth; **Ecol** along the coast in saline habitats, further inland often found at edges of or in quartz fields, winter rainfall; **Distr** SW Namibia; Namaqualand, NC, Barrydale, Ceres, Clanwilliam, Piquetberg, Vanrhynsdorp, Vredendal, WC, S Africa.

Note: In saline habitats near the coast, the plants are always prostrate to decumbent with a bright red colour and often shiny epidermis; the internodes of the stems are up to 10 mm thick and barrel-shaped, the whole branch resembling a string of beads. In less saline conditions, the plants are decumbent to erect, green to faintly suffused with red, the epidermis is mostly covered with dome-shaped cells and, therefore, not shiny; the internodes are cylindrical.

P. distinctum = **Aptenia geniculiflora**
P. diversipapillosum = **Brownanthus arenosus**
P. duthiae = **P. articulatum**
P. fasciculatum = **P. dinteri**
P. filipetalum = **P. subnodosum**
P. fimbriatum = **P. salicornioides**

P. foliosum L.Bolus NM 1929: 100 (see **photo**); L.Bolus NM 1929: 142 **T** Pillans 17813 (BOL!).

Erect, to 100 cm *h*, ∅ 1–2 m, densely and regularly branched, **I** soon becoming woody, ∅ to 50 mm, young stems indistinctly articulate, nodes inconspicuous with or without fine horizontal furrow, bladder cells flattened; **L** shortly connate or free, remaining on the stems when dry, without mucro, bladder cells flattened; **Fl** several per cyme, ∅ c. 15 mm, **K** 5, petals white, c. 1 mm *b*, shortly connate, filamentous staminodes shortly connate; **Fr** ∅ c. 3 mm, 5 locules, 1.4–2.6 mm deep; **S** brown, testa rough; **Ecol** on reddish, often gneissic, loam, winter rainfall; **Distr** Namaqualand, NC, S Africa.

Note: A very large species of **Psilocaulon**, without persistent succulent green cortex. The plants are, nevertheless, permanently green, the leaves persisting even during the dry season.

P. framesii = **P. junceum**

P. gessertianum (Dinter & A.Berger) Dinter & Schwantes in Schwantes GF 1928: 69 (see **photo**) ≡ *M. gessertianum* Dinter & A.Berger in Dinter FR 1923: 189 **T** Dinter 1195 (B missing; **LT** BOL!) = *P. luteum* L.Bolus NM 1928: 141 **T** Coetzee 17171 (BOL!).

Decumbent to erect, to 25 cm *h*, **I** remaining green and succulent, indistinctly articulate, nodes inconspicuous without fine horizontal furrow, bladder cells finely hair-like; **L** shortly connate, with mucro, bladder cells flattened or with pointed tips when young; **Fl** few to many per cyme, ∅ c. 5 mm, **K** 5, petals much reduced, filamentous staminodes yellow, free; **Fr** ∅ 2.8–3.5 mm, 5 locules, 1.4–2.6 mm deep; **S** ochre, testa smooth; **Ecol** summer rainfall; **Distr** SE Namibia.

Note: **P. gessertianum** is noteworthy for its minute yellow flowers with reduced petals and bright yellow exerted anthers and stamens. When not in flower, the species may be confused with two other species growing in the same area. From **P. articulatum** it differs in having shortly connate leaves and inconspicuous nodes without fine horizontal furrow, and in being less hairy. From **P. salicornioides** it can only be distinguished when in flower.

P. glareosum = **P. salicornioides**
P. godmaniae = **P. dinteri**
P. godmaniae var. *gracile* = **P. dinteri**

P. granulicaule (Haw.) Schwantes GF 1928: 69 (see **photo**) ≡ *M. granulicaule* Haw. 1824: 427 ≡ *Pentacoilanthus granulicaulis* (Haw.) Rappa & Camarrone 1953: 30 nomen invalid. ≡ *Perapentacoilanthus granulicaulis* (Haw.) Rappa & Camarrone 1960: 23 nomen invalid. **T** drawing (K, photograph!) = *P. pauper* L.Bolus NM 1930: 191 **T** H.Bolus 11906 (BOL!).

Decumbent to erect, to 25 cm *h*, **I** remaining green and succulent, articulate, nodes conspicuous with fine horizontal furrow, bladder cells dome-shaped; **L** free, without mucro, bladder cells dome-shaped; **Fl** mostly solitary, rarely several per cyme, ∅ 5–8 mm, **K** 5, petals pale yellow, 1–2 mm *b*, free, filamentous staminodes free; **Fr** ∅ c. 3.5 mm, 5 locules, 2.5 mm deep; **S** brown, testa smooth; **Ecol** mainly summer, but also winter rainfall; **Distr** Cradock, Graaff Reinet, from Willowmore and Port Elizabeth to Albany and King Williamstown, EC, Bloemfontein, Fauresmith, FS, Colesberg, Kimberley, Victoria West, NC, from Uniondale to Worcester and Bredasdorp, WC, S Africa.

Note: This species is readily distinguishable by its pale yellow flowers, which arise on short side shoots along the longer axes. **P. granulicaule** has also been recorded from Australia (Victoria, Mildura, and Birchip) as well as from Namibia (Lüderitz and Okahandja), where it was presumably introduced.

P. gymnocladum = **Brownanthus arenosus**
P. herrei = **P. dinteri**
P. hirtellum = **P. articulatum**
P. imitans = **P. junceum**
P. implexum = **P. parviflorum**
P. inachabense = **P. salicornioides**
P. inconspicuum = **P. salicornioides**
P. inconstrictum = **P. subnodosum**

P. junceum (Haw.) Schwantes GF 1928: 69 (see **photos**) ≡ *M. junceum* Haw. 1803: 59 ≡ *Peratetracoilanthus junceus* (Haw.) Rappa & Camarrone 1960: 24 **T** drawing (K, photograph!) = *M. junceum* Haw. var. *pauciflorum* Sonder in Harvey & Sonder 1862: 434 ≡ *P. pauciflorum* (Sonder) Schwantes FR 1938: 229 **T** Zeyher 2957 (S!) = *M. acutisepalum* A.Berger 1922: 629 ≡ *P. acutisepalum* (A.Berger) N.E.Br. GC 1928: 253 **T** Bachmann 329 (B missing; part of **T** K!) = *P. bijliae* N.E.Br. GC 1928: 254 **T** van der Bijl 30 (K!) = *P. calvinianum* L.Bolus NM 1933: 410 **T** du Plessis 2267/32 (BOL!) = *P. candidum* L.Bolus 1961: 116 **T** Littlewood 631/59 (BOL!) = *P. delosepalum* L.Bolus NM 1930: 174 **T** L.Bolus 19050 (BOL!) = *P. framesii* L.Bolus NM 1930: 190 **T** Frames 19112 (BOL!) = *P. imitans* L.Bolus NM 1934: 429 **T** L.Bolus 21038 (BOL!) = *P. laxiflorum* L.Bolus NM 1929: 157 **T** Godman 19004 (BOL!) = *P. leightoniae* L.Bolus NM 1936: 14 **T** Leighton 21643 (BOL!) = *P. levynsiae* N.E.Br. GC 1928: 254 **T** Levyns 1722 (K!) = *P. lewisiae* L.Bolus NM 1935: 492 **T** Lewis 2011/33 (BOL!) = *P. oculatum* L.Bolus 1961: 118 **T** Littlewood 640/59 (BOL!) = *P. planisepalum* L.Bolus NM 1939: 160 **T** Esterhuysen 21855 (BOL!) = *P. rogersiae* L.Bolus NM 1928: 30 **T** Rogers 18856 (BOL!) = *P. semilunatum* L.Bolus NM 1930: 174 **T** L.Bolus 19051 (BOL!) = *M. simile* Sonder in Harvey & Sonder 1862: 435 ≡ *P. simile* (Sonder) Schwantes GF 1928: 69 **T** Zeyher 2618 (MEL!) = *M. simile* Sonder var. *namaquense* Sonder in Harvey & Sonder 1862: 435 ≡ *P. namaquense* (Sonder) Schwantes GF 1928: 69 **LT** Schlicht s.n s.d. (MEL!) = *P. simulans* L.Bolus p.p. 1961: 119 **T** Stayner 806/60a (BOL!) = *P. squamifolium* N.E.Br. GC 1928: 254 **T** Pearson 3273 (K!) = *P. stayneri* L.Bolus 1961: 119 **T** Stayner 893/59 (BOL!) = *P. subintegrum* L.Bolus NM 1934: 429 **LT** L.Bolus 21034 (BOL!) = *P. utile* L.Bolus NM 1929: 157 **T** Compton 19005 (BOL!).

Decumbent to erect, (10–)30–45(–60) cm *h*, **I** remaining green and succulent, articulate, nodes conspicuous with fine horizontal furrow, bladder cells flattened; **L** free, without mucro, bladder cells flattened; **Fl** few to many per cyme, ∅ 11–15 mm, **K** 4–5, petals white or (pale) pink, 1–2 mm *b*, free, filamentous staminodes free; **Fr** ∅ 2.1–4.7 mm, 4–5 locules, 2.1–3.6 mm deep; **S** dark brown, testa rough; **Ecol** mainly winter but also summer rainfall; **Distr** from Cradock and Albany, EC, to Malmesbury and Vanrhynsdorp, WC, Calvinia, Namaqualand, NC, S Africa.

Note: **P. junceum** is one of the most common, widespread, and variable species of **Psilocaulon**. In the coastal plains, the plants grow much larger than in the much higher altitudes of the South African plateau. In addition, plants usually become dwarfed under environmental stress. In the past, plants were burnt and the ashes used in soap-making.

P. kuntzei ≡ **Brownanthus kuntzei**
P. laxiflorum = **P. junceum**
P. leightoniae = **P. junceum**

P. leptarthron (A.Berger) N.E.Br. GC 1928: 254 (see **photo**) ≡ *M. leptarthron* A.Berger 1922: 628 **T** Schlechter 8151 (B!) = *P. album* L.Bolus SAG 1927: 365 **T** Compton 1475/26 (BOL!).

Erect, 50–100 cm *h*, with straight, regular branches, **I** remaining green and succulent, distinctly articulate, nodes conspicuous with fine horizontal furrow, bladder cells flattened; **L** free, without mucro, bladder cells flattened; **Fl** solitary, terminating long side branches, rarely few per cyme, ∅ 25 mm, **K** 5, petals white, 1–3 mm *b*, free, filamentous staminodes free; **Fr** ∅ 4.5–6.0 mm, 5 locules, 4.5–6.0 mm deep; **S** brown, testa rough; **Ecol** often found on red calcareous loams associated with the basin of the Sout River, Knersvlakte, winter rainfall; **Distr** Vanrhynsdorp, WC, S Africa.

Note: **P. leptarthron** is characterized by very straight side branches usually with a terminal solitary flower comparatively large for the genus.

P. levynsiae = **P. junceum**
P. lewisiae = **P. junceum**
P. liebenbergii = **P. articulatum**
P. lindequistii = **Aridaria noctiflora** ssp. **noctiflora**
P. littlewoodii = **P. dinteri**
P. littlewoodii forma *laxum* = **P. dinteri**
P. longipes = **Caulipsolon rapaceum**
P. luteum = **P. gessertianum**
P. marlothii ≡ **Brownanthus marlothii**
P. melanospermum = **Aptenia geniculiflora**
P. mentiens = **P. coriarium**
P. micranthon = **P. parviflorum**
P. mucronulatum = **P. articulatum**
P. namaquense = **P. junceum**
P. namibense ≡ **Brownanthus namibensis**
P. oculatum = **P. junceum**
P. otzenianum = **Lampranthus otzenianus**
P. pageae = **P. dinteri**
P. pageae var. *grandiflorum* = **P. dinteri**

P. parviflorum (Jacquin) Schwantes GF 1928: 69 (see **photo**) ≡ *M. parviflorum* Jacquin 1798: t. 278 non Haworth 1803: 95 nomen illeg. ≡ *Peratetracoilanthus parviflorus* (Jacquin) Rappa & Camarrone 1960: 24 ≡ *M. micranthon* Haw. 1812: 257 nomen illeg. ≡ *P. micranthon* L.Bolus SAG 1927: 365 nomen illeg. **T** Jacquin 1798: t. 278 = *P. implexum* N.E.Br. KB 1929: 60 **T** Muir 4234 (K!) = *M. tenue* Haw. 1821: 175 non L.Bolus NM 1928: 1 nomen illeg. ≡ *P. tenue* (Haw.) Schwantes GF 1928: 69 **T** drawing (K, photograph!).

Prostrate, mat-forming **I** remaining green and succulent, articulate, nodes conspicuous with fine horizontal furrow, young stems only 1–2 mm *t*, bladder cells flattened; **L** free, without mucro, bladder cells flattened; **Fl** few per cyme or solitary on shorter side branches, ∅ 5–10 mm, **K** 4, petals white, c. 1 mm *b*, free, filamentous staminodes free; **Fr** ∅ c. 2.5 mm, 4 locules, 2.7 mm deep; **S** brown, testa rough; **Ecol** in clayey-loamy and slightly saline soils, often covered with quartz gravel, winter rainfall; **Distr** from Clanwilliam to Malmesbury and Mosselbaai, WC, S Africa.

Note: Diagnostic features for this species include very slender side branches and short, narrow leaves as well as small, white, starry flowers. In the dryer areas about 150 km east of Cape Town, plants have usually shorter and narrower leaves than those from near Cape Town.

P. pauciflorum = **P. junceum**
P. pauper = **P. granulicaule**
P. peersii = **Brownanthus corallinus**
P. pfeilii = **Prenia tetragona**
P. pillansii = **Brownanthus pubescens**
P. planisepalum = **P. junceum**
P. planum = **P. salicornioides**
P. pomeridianum = **Mesembryanthemum stenandrum**
P. puberulum = **P. articulatum**
P. pubescens = **P. articulatum**
P. rapaceum ≡ **Caulipsolon rapaceum**
P. rogersiae = **P. junceum**
P. roseoalbum = **P. articulatum**

P. salicornioides (Pax) Schwantes GF 1928: 69 (see **photo**) ≡ *M. salicornioides* Pax 1895: 133 **T** Gürich 149 (B!) = *P. fimbriatum* L.Bolus ABH 1927: 105 **T** Pillans 5284 (BOL!) = *M. glareosum* A.Berger 1922: 630 ≡ *P. glareosum* (A.Berger) Dinter & Schwantes in Schwantes GF 1928: 69 **T** Dinter 1104 (B missing, part of **T** K!) = *P. inachabense* L.Bolus NM 1958: 336 **T** Gessert 13637 (BOL!) = *P. inconspicuum* L.Bolus NM 1929: 36 **T** Erni 18863 (BOL!) = *P. planum* L.Bolus NM 1930: 173 **T** Frames 19048 (BOL!) = *M. sinus-redfordiani* Dinter FR 1923: 44 nomen nudum ≡ *P. sinus-redfordiani* (Dinter) Dinter ex Range FR 1934: 17 nomen nudum **T** not given = *M. trothai* Engler 1909: 196 ≡ *P. trothai* (Engler) Schwantes GF 1928: 69 **T** Trotha 48a (K!) = *P. woodii* L.Bolus NM 1933: 393 **T** Harcourt-Wood 20449 (BOL!).

Decumbent to erect, occasionally prostrate, to 40 cm *h*, **I** remaining green and succulent, coastal form distinctly articulate, inland form indistinctly so, nodes inconspicuous without fine horizontal furrow, bladder cells flattened or with pointed tips; **L** shortly connate, with mucro, bladder cells flattened or occasionally with pointed tip; **Fl** few to many per cyme, rarely solitary, ∅ to 10 mm, **K** 4–5, petals white, pale pink or puce, 0.5–1 mm *b*, shortly connate, filamentous staminodes shortly connate; **Fr** ∅ 2.5–3.1 mm, 4–5 locules, 1.7–2.0 mm deep; **S** ochre or dark brown, testa rough or smooth; **Ecol** saline habitats, in sand or loam, or on granite outcrops, mainly summer rainfall; **Distr** C, NW, and S Namibia, Namaqualand, NC, S Africa.

Note: **P. salicornioides** is very similar to **P. subnodosum** and **P. gessertianum.** From the former it differs in a smaller overall size and smaller flowers, from the latter in character states of the flower only.

P. schlichtianum = **Brownanthus arenosus**
P. semilunatum = **P. junceum**
P. simile = **P. junceum**
P. simulans = **P. junceum**
P. sinus-redfordiani = **P. salicornioides**
P. squamifolium = **P. junceum**
P. stayneri = **P. junceum**
P. stenopetalum = **P. coriarium**
P. subintegrum = **P. junceum**

P. subnodosum (A.Berger) N.E.Br. GC 1928: 254 (see **photos**) ≡ *M. subnodosum* A.Berger 1922: 628 T Schlechter 11274 (B!) = *M. clavulatum* A.Berger 1922: 630 ≡ *P. clavulatum* (A.Berger) N.E.Br. GC 1928: 254 T Schlechter 11393 (B!) = *P. filipetalum* L.Bolus NM 1935: 491 T L.Bolus 21425 (BOL!) = *P. inconstrictum* L.Bolus NM 1939: 160 T Esterhuysen 21854 (BOL!).

Decumbent to erect, to 70 cm *h*, **I** remaining green and succulent, indistinctly articulate, nodes inconspicuous without fine horizontal furrow, bladder cells flattened or with pointed tips in the **Inf**, glaucous; **L** shortly connate, with short mucro, bladder cells flattened or sometimes with pointed tips; **Fl** few to many per cyme, ∅ 10–15 mm, **K** 5, petals white or pale pink, 0.5–1 mm *b*, shortly connate, filamentous staminodes shortly connate; **Fr** ∅ 2.8–3.2 mm, 5 locules, 2.1–2.5 mm deep; **S** ochre, testa smooth; **Ecol** mainly winter rainfall; **Distr** S Namibia; Gordonia, Kenhardt, Namaqualand, NC, S Africa.

Note: In the dryer part of the distribution area, plants produce fewer leaves which are more succulent and more connate. Superficially, these plants resemble certain species of **Ruschia**, e.g. **Ruschia grisea.**

P. tenue = **P. parviflorum**
P. trothai = **P. salicornioides**
P. uncinatum = **P. coriarium**
P. utile = **P. junceum**
P. variabile = **P. dinteri**
P. woodii = **P. salicornioides**

Pteropentacoilanthus = Mesembryanthemum

M. Gerbaulet

This name (Rappa & Camarrone 1960: 14) is illegitimate, since it was based on the same type as *Halenbergia* (Dinter KuaS 1937: 158). The genus comprises parts of **Mesembryanthemum** subg. **Opophytum.**

P. fastigiatus
≡ **Mesembryanthemum fastigiatum**
P. hypertrophicus
≡ **Mesembryanthemum hypertrophicum**

Pterotetracoilanthus

M. Gerbaulet

As discussed by Hartmann & Bittrich (1990: 155, 156), *Pterotetracoilanthus* (Rappa & Camarrone 1960: 14, 24) is one of the genera recognized by Rappa & Camarrone (1953, 1955, 1960) based on the number of nectaries and the absence or presence of valve wings. The mechanical nature of this approach is best shown by this genus, which combines character states the authors failed to find in any real plant. Hence, no species were assigned and the name is invalid.

Punctillaria = Pleiospilos

H. E. K. Hartmann

Brown (1925: 412, 413, 433) distinguished the genus *Punctillaria* in a key, describing it in more detail later (1926: 212). There he stressed the general similarity to **Pleiospilos** and pointed out that beside the more slender leaves *Punctillaria* possesses big closing bodies as a differentiating feature. The value of this distinction was discussed over several years in literature (e.g. Schwantes ZSK 1927: 23, MDG 1928: 238). Hartmann & Liede (1986) united both genera into one, **Pleiospilos**, distinguishing two subgenera agreeing in circumscription with the old genera *Punctillaria* and **Pleiospilos** sensu Brown.

P. cana ≡ **Pleiospilos compactus** ssp. **canus**
P. dekenahi = **Pleiospilos compactus** ssp. **canus**
P. magnipunctata
= **Pleiospilos compactus** ssp. **canus**
P. optata
= **Pleiospilos compactus** ssp. **compactus**
P. purpusii ≡ **Pleiospilos** × **purpusii**
P. roodiae = **Tanquana prismatica**
P. sororia ≡ **Pleiospilos compactus** ssp. **sororius**
P. sesquiuncialis
= **Pleiospilos compactus** ssp. **canus**

Pyxipoma = Sesuvium

H. E. K. Hartmann

When Fenzl (1839: 293) described the genus *Pyxipoma* Fenzl as new, he contrasted it against **Sesuvium** insofar as the stamens were reported to be inserted in a "hypogynous calyx" and against **Trianthema** insofar as the number of stamens is higher in *Pyxipoma* and the ovary does not remain adnate to the ovary. All these features occur in **Sesuvium**, and the genus *Pyxipoma* has been treated as a synonym of the former by later authors.

No species is cited in the genus *Pyxipoma*; the only species named, *Trianthema polyandra*, is not formally transferred with this citation following the description of the genus by Fenzl.

Rabiea Ruschioideae

H. E. K. Hartmann

Rabiea N.E.Br. GC 1930: 280; Brown GC 1931: 53–54; Bolus NM 1958: 369–372 **Typus R. albinota** (Haw.) N.E.Br. **Etym** *W.A. Rabie*, plant collector at Fauresmith in 1930.

Compact, with a tap-root when young, forming very tardily thick, short rhizomes or rootstocks; **L** of a pair often differing in size, semi-terete below, changing to trigonous or triquetrous above in several species, surface with conspicuously raised dots, often white, otherwise of a very dark green due to the lack of crystal sand in the outer epidermis wall and an only moderate layer of wax in horizontal, rugose plates; **Fl** solitary, opening in the afternoon, pedicels with basal bracteoles, short, petals and stamens together with the 5–10 glands of the nectary at the upper rim of the concave surface of the gynoecium; **Fr** breaking off the short and fragile stalk when ripe, base funnel- to bell-shaped, top conical or rounded, covering membranes at first straight, later the distal part falling obliquely into the empty locule, closing body absent, expanding keels diverging distally, valve wings broadest at their bases, tapering rather abruptly at about 2/3 into a distal awn, a narrow column in the centre, 8–11 locules, mostly 10 dominant in a population; **Ecol** among rocks on slopes, often in crevices between stones, rainfall mainly in summer; **Distr** Lesotho; EC, FS, S Africa.

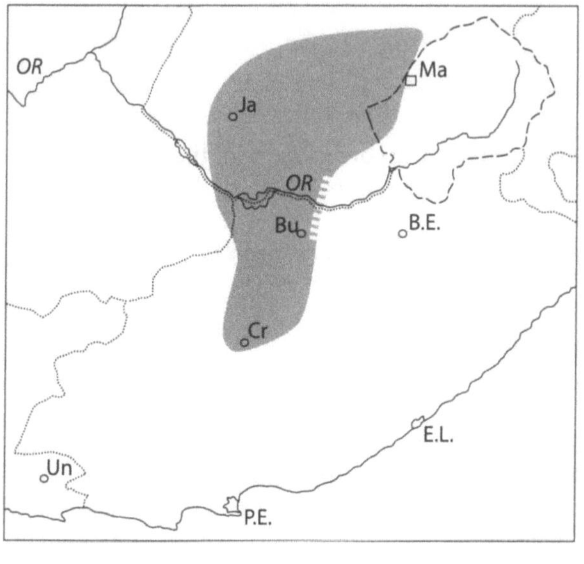

Distribution of **Rabiea** N.E.Br.

Note: Species of the genus have been confused with those of **Nananthus** or **Aloinopsis**, partly due to misinterpretations in earlier days. Both genera mentioned possess crenulate, complete nectary rings; **Rabiea** differs from both also by the concave surface of the ovary, a feature completely lost during fruit development. Species have mainly been distinguished based on leaf shapes, features difficult to assess in dry material.

R. albinota (Haw.) N.E.Br. GC 1931: 53 (see **photo**) ≡ *M. albinotum* Haw. 1826: 126 ≡ *Aloinopsis albinotus* (Haw.) Schwantes ZSK 1926: 178 ≡ *Nananthus albinotus* (Haw.) L.Bolus SAG 1927: 399 **T** drawing of a plant "raised in 1823 from Cape of Good Hope seed received from Mr. Bowie" (K) = *R. albinota* (Haw.) N.E.Br. var. *longipetala* L.Bolus NM 1958: 337 **T** Hurling & Neil 1936 (BOL!) = *R. albinota* (Haw.) N.E.Br. var. *microstigma* L.Bolus NM 1958: 337 **T** Frames 2541/30 (BOL!).

L upper surface broadening from the base to above the middle, then tapering to a pointed, mucronate recurved tip, the lower leaf surface acinaciform from a broader base, with a distinct keel, epidermis with sparse white dots, to 54 mm *l*, to 11.5 mm *b* at the broadest part; **Fl** ∅ 42 mm, stamens yellow; **Fr** unknown; **Ecol** unknown; **Distr** uncertain, S Africa.

Note: The subapically broadened upper leaf surface represents the diagnostic character state of the species.

R. albinota var. *longipetala* = **R. albinota**
R. albinota var. *microstigma* = **R. albinota**

R. albipuncta (Haw.) N.E.Br. GC 1931: 54 (see **photo**) ≡ *M. albipunctum* Haw. 1826: 126 ≡ *Aloinopsis albipuncta* (Haw.) Schwantes ZSK 1926: 178 ≡ *Nananthus albipunctus* (Haw.) N.E.Br. JB 1928: 78 **T** drawing at K = *R. albipuncta* (haw.) N.E.Br. var. *major* L.Bolus NM 1958: t.109 D nomen nudum T drawing NM 1958: t. 109 D (original at K).

Highly branched; **L** semi-terete, trigonous in the upper part, with white dots, c. 27 mm *l* or less, narrowing from the base to the tip; **Fl** as for genus; **Fr** unknown; **Ecol** unknown; **Distr** SE parts of FS, S Africa.

Note: Differing from **R. albinota** in the shorter, more erect leaves tapering from a broad base and the higher number of leaf-pairs per plant.

R. carolinensis ≡ **Khadia carolinensis**

R. comptonii (L.Bolus) L.Bolus NM 1938: 134 (see **photo**) ≡ *Nananthus comptonii* L.Bolus NM 1929: 120 **T** Compton 775/23 (BOL!).

L spreading widely, tapering to the tip from a broad base, the lower side a little widened below the apex, rough from dots, to 55 mm *l*, to 10 mm *b*; **Fl** ∅ 29–34 mm, petals golden, 70–100, filaments white, apically pale yellow, 210–270, 5 glands big and prominent, often tiny ones between them; **Fr** as for genus; **Ecol** in pans or crevices among rocks on slopes; **Distr** Bloemfontein, FS, S Africa.

Note: Brown (GC 1931: 53) placed the species as a synonym with **R. albinota**, a decision not upheld because the leaves spread widely and lack the subapical chin typical of **R. albinota.**

R. difformis (L.Bolus) L.Bolus NM 1938: 134 (see **photo**) ≡ *Nananthus difformis* L.Bolus NM 1933: 392 **T** without collector 9487 (BOL!).

L of a pair different: one leaf tapering gradually from a broad base towards the pointed tip, the other developing a moderate thickening below the apex, both rough from dots, to 40 mm *l*, to 9 mm *b* at the broadest part; **Fl** petals golden, 9–21 mm *l*, to 2 mm *b*, filaments and anthers golden, 5 glands; **Fr** unknown; **Ecol** unknown; **Distr** Cradock, EC, S Africa.

Note: The differences between the leaves of a pair appear not to be formed regularly; a similar condition is described for **R. lesliei.**

R. jamesii (L.Bolus) L.Bolus NM 1938: 134 ≡ *Nananthus jamesii* L.Bolus NM 1931: 304 non L.Bolus 1938: 133 nomen illeg. **T** James 1077/31 (BOL!).

L tapering from a broad base towards the pointed tips, almost terete, rough from dots, to 55 mm *l*, to 13 mm *b* in the middle, to 11 mm ∅; **Fl** petals golden, to 27 mm *l*, to 1.5 mm *b*, 5 glands; **Fr** unknown; **Ecol** unknown; **Distr** Cradock, EC, S Africa.

Note: In leaf size and shape, the species is very similar to **R. comptonii.**

R. lesliei N.E.Br. GC 1931: 54 ≡ *Nananthus lesliei* (N.E.Br.) G.D.Rowley NCSJ 1978: 6 **T** Leslie s.n. (K!).

Plants with very many shoots; **L** erect, those of a pair different: one leaf tapering to a pointed tip from a broad base, the other dilated at the keel near the apex, 20–22.5 mm *l*, 6.7–9 m *b*, 6.7–8 mm ∅; **Fl** ∅ 25–30 mm, petals c. 60, yellow withering orange, filaments c. 220, anthers bright yellow; **Fr** as for genus; **Ecol** on slopes with big stones, in grassland; **Distr** Clocolan and adjacent districts of FS, S Africa.

Note: **R. lesliei** differs from **R. difforme** only in the size of the leaves, the latter being described from dry material imported freshly from South Africa.

R. tersa = **Prepodesma orpenii**

Radiana ≡ Cypselea

H. E. K. Hartmann

Rafinesque (1814: 88) described the new genus *Radiana* based on material sent to him by Turpin, who had used the same material to establish the genus **Cypselea** Turpin (1806: 219) earlier. **Cypselea** is therefore the older and correct name of the taxon, *Radiana* an obligate synonym based on the same type.

R. petiolata ≡ **Cypselea humifusa**

Remè ≡ Trianthema

H. E. K. Hartmann

Adanson (1763: 245) replaced the name **Trianthema** L. by *Remè*, referring to Hermann (1698: 213), as Linnaeus had done. Thus, it is clear from the synonymy Adanson offers that the generic name *Remè* is based on the same material as **Trianthema,** which is the correct name for the genus because it is older. Adanson ascribed the name *Remè* to "Seneg".

Rhinephyllum Ruschioideae

H. E. K. Hartmann

Rhinephyllum N.E.Br. GC 1927: 92; Bolus NM 1936: 40–43 **Typus R. muirii** N.E.Br. **Etym** Gk *rhine*, a file, *phyllon*, a leaf, "in allusion to the rough surface of the leaves" ≡ *Neorhine* Schwantes MDK 1930: 22 **Typus** *Neorhine pillansii* (N.E.Br.) Schwantes = **R. muirii** N.E.Br.

Compact to mat-forming, some species with a thickish rootstock; **L** trigonous with convex sides to hood-shaped or club-shaped, often the leaves of a pair unequal in size and margins, the epidermis with numerous fine, raised dots, which are often white, formed by elevations above subhypodermal tannin idioblasts, the outer epidermis wall here with a very thick layer of fine calcium oxalate crystals, wax cover almost continuous on the elevations, breaking into horizontal plates in the intermediate areas; **Fl** solitary, opening in the late afternoon or evening, pedicels ebracteolate, petals yellow, filaments paler or yellow, nectary glands 5, separate, placed on the upper rim of the concave surface of the ovary; **Fr** on long, persisting stalks, base funnel-shaped, top flat, the valves rising a little in the centre when they have opened, covering membranes absent or as ledges of the septa, closing body absent, expanding keels parallel to the tip of the valve, valve wings broadest at their outer tips (in the open capsule), broader than the expanding keels, narrowing a little to their bases, 5 locules, ∅ 3–7 mm; **Ecol** mostly in flats, on Ecca shale, sandstone, or quartzite; **Distr** from Laingsburg, WC, to Victoria West, NC, Jansenville, EC, S FS, S Africa.

Note: In its original circumscription, the genus comprised late flowering plants with leaves with a finely rough surface and without teeth, and with capsules with broad valve wings and without covering membranes. Later additions included plants with teeth, and those with covering membranes in the fruit. After the transfer of smooth-leaved plants to **Peersia**, only rough-leaved forms remain in **Rhinephyllum**. Finely rough leaves, teeth on the margins, and absence of covering membranes occur also in species of **Chasmatophyllum** and **Stomatium**, albeit not in the type species of the latter genus, which lacks teeth. Eight species with compact growth and three with caespitose habit have been described.

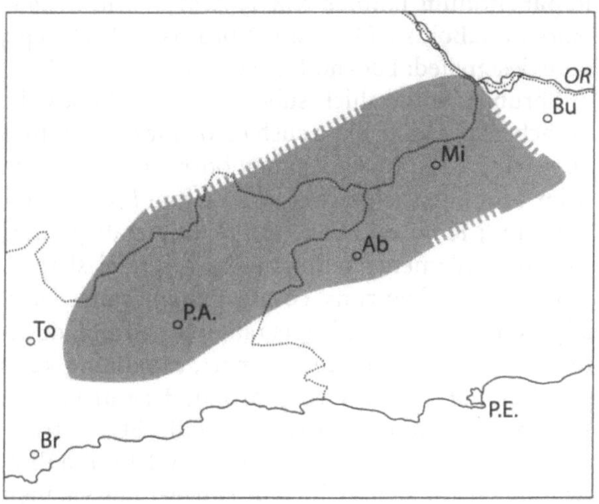

Distribution of **Rhinephyllum** N.E.Br.

R. broomii L.Bolus NM 1936: 14 (see **photo**) **T** Broom 34900 (BOL!).

Compact; **L** clavate, those of a pair only a little different, without teeth but with undulate margins, to 10 mm *l*, to 7 mm *b*; **Fl** opening between 5 and 8 p.m., petals bright yellow, to 7.5 mm, to 1.25 mm *b*; **Fr** unknown; **Ecol** unknown; **Distr** Fraserburg, NC, S Africa.

Note: The rather small leaves of the type material lack teeth.

R. comptonii L.Bolus SAG 1928: 370 **T** Compton 1788/28 (cited as 1718/28, BOL!) = *R. comptonii* L.Bolus var. *glandulis* nomen nudum in schedis, referring to a description without a name in SAG 1928: 371 **T** Pillans 10702 (BOL!).

Compact; **L** clavate, those of a pair a little different, some leaves dentate, to 21 mm *l*, to 10 mm *b*, to 9 mm *t*; **Fl** petals broad, filaments papillate, stigmas slender; **Fr** unknown; **Ecol** unknown; **Distr** Laingsburg, WC, S Africa.

Note: The species is very similar to **R. pillansii.**

R. comptonii var. *glandulis* = **R. comptonii**
R. frithii ≡ **Peersia frithii**

R. graniforme (Haw.) L.Bolus NM 1954: 275 (see **photo**) ≡ *M. graniforme* Haw. 1803: 82 **T** Bradley 1716: t. 20.

Rather flat cushions to 8 cm *h*, branches with old leaves of several years hidden inside the mat, all leaves appearing densely placed, but upon dissection the I visible inside the plants; **L** shaped like a grain, 6–7 mm *l*, rough from raised dotes; **Fl** petals yellow; **Fr** top markedly flat, valve wings very broad, covering membranes somewhat expanded but the pair over a locule not touching each other and short; **Ecol** in finely grained shaly soils in flats; **Distr** Beaufort West, Prince Albert, WC, S Africa.

Note: The species is very similar to **R. luteum**, as described by Bolus (1954: 276), differing from the latter not only in the smaller leaves but also in the more densely placed leaves which remain as dry relics inside the cushion for several years. Bolus (NM 1954: 275) described a "forma fera" without marking it with a taxonomic rank in print, yet with a Latin description, differing from **R. graniforme** in narrow ledges on their septa in the capsules, and leaves c. 10 mm long and 5–6 mm thick.

R. inaequale L.Bolus NM 1934: 430 (see **photo**) **T** SUG 9926 (BOL!) = *R. inaequale* L.Bolus var. *latipetalum* L.Bolus NM 1937: 59 **T** Villet 21823 (BOL!).

Compact mats, roots thickened and far-reaching under flat stones; **L** trigonous to clavate, rough, greyish-white, longest leaves to 22 mm *l*, 8–10 mm *b*, 4–7 mm *t*; **Fl** petals lemon-coloured, to 10 mm *l*, 0.5–1 mm *b*, filaments lemon-coloured, anthers and pollen yellow; **Fr** expanding keels parallel to each other, covering membranes broadest in the centre, tapering outwards into awns; **Ecol** in crevices of very light sandstone slabs; **Distr** Graaff Reinet, EC, S Africa.

Note: The species may be restricted to very gentle slopes with big slabs of light coloured stones; the leaves appear almost equal in habitat, but anisophylly can develop in cultivation.

R. inaequale var. *latipetalum* = **R. inaequale**

R. luteum (L.Bolus) L.Bolus NM 1950: 222 ≡ *Ruschia lutea* L.Bolus NM 1928: 138 ≡ *M. flavens* N.E.Br. GC 1930: 32 **T** van der Bijl 18777 (BOL!) = *R. luteum* L.Bolus var. *brevifolium* L.Bolus NM 1954: 276 **T** Acocks 15232 (BOL).

Caespitose, to 5 cm *h*, ∅ to 20 cm, the centre dying with age, the leaves placed distantly and persisting as dry relics inside the plant for a short time only; **L** semi-terete, obtuse, rough from distant elevations turning into prominent, white dots when dry, c. 10 mm *l*, 3 mm *b* and *t*; **Fl** solitary, borne above the plant forming a conspicuous show at anthesis, ∅ to 12 mm, pedicels 6–8 mm, in fruit to 12 mm *l*, petals yellow, c. 5 mm *l*, 1.5 mm *b*, filaments and anthers yellow; **Fr** base funnel-shaped, top with very low rims, rims from the septa so broad that they form narrow covering membranes touching each other in the centre, expanding keels parallel, valve wings as broad as the expanding keels, often with rounded ends, ∅ 4–5 mm; **Ecol** in shaly flats; **Distr** Laingsburg, Prince Albert, WC, S Africa.

Note: **R. luteum** differs from the superficially similar **R. graniforme** by a looser arrangement and shorter persistence of the leaves, by centrally broader covering membranes and narrower valve wings in the fruits. **R. luteum** settles further west and south than **R. graniforme**, but Bolus (NM 1954: 276) reported sympatrical occurrence, which can be expected along the contact line running from Prince Albert to Merweville.

R. luteum var. *brevifolium* = **R. luteum**
R. macradenium ≡ **Peersia macradenia**

R. muirii N.E.Br. GC 1927: 92 (see **photo**) **T** syntypes Muir 3790, 3884, 4023 (K) = *Neorhine pillansii* (N.E.Br.) Schwantes MDK 1930: 22 **T** f II B in Bolus SAG 1928: 370.

Compact; **L** trigonous to club-shaped, upper leaf surface only distally with raised dots, leaves of a pair only a little different, 11–20 mm *l*, 4.5–11 mm *b*, 3.5–6.5 mm *t*; **Fl** opening from 6 p.m. until 1 a.m., petals c. 35, pale lemon-yellow to white, 5.7–6.5 mm *l*, c. 1.1 mm *b*; **Fr** on a persistent stalk to 20 mm *l*, base funnel-shaped with concave sides, top flat, very narrow ledges from the septa, broad expanding keels, valve wings very broad, narrowing towards their base, ∅ 5–6 mm; **Ecol** in flats or on gentle slopes with bigger stones of Bokkeveld shales, mixed with quartzite, c. 200 mm rainfall p.a.; **Distr** Ladismith, Swellendam, WC, S Africa.

Note: This species settles at the south-western border of the area inhabited by the genus.

R. obliquum L.Bolus NM 1937: 59 **T** Broom 21824 (BOL!).

Compact; **L** clavate, those of a pair different: one fitting into a cavity in the other while both are young, the bigger leaf often with a pronounced chin and tip, hence somewhat hood-shaped, both leaves often oblique, 18–22 mm *l*, 7–8 mm *b*, to 5 mm *t*; **Fl** opening towards the night, fragrant, petals yellow, 6–10 mm *l*, c. 1 mm *b*; **Fr** unknown; **Ecol** unknown; **Distr** Fraserburg, NC, S Africa.

Note: Similar in the leaves ending in a rather pointed tip to **R. inaequale** from further east.

R. parvifolium L.Bolus NM 1933: 377 ≡ ?*R. willowmorense* L.Bolus NM 1933: 348 pro parte **Lectotype**, here designated: Luckhoff 20432 (BOL!).

Shrublets with a thick, subterranean rootstock, the branches spreading like a cushion on the ground, thus appearing caespitose; **L** subglobose, surface rough from prominent elevations, c. 4–9 mm *l*, 3–4 mm *b*, 2–3 mm *t*; **Fl** raised above the plant, petals golden, 5–7 mm *l*, filaments yellow; **Fr** base funnel-shaped, top with very low rims, covering membranes laterally reduced or: rims of septa rather broad and touching each other in the central part, expanding keels parallel, valve wings very broad, ± semi-lunate; **Ecol** in shaly ground; **Distr** Jansenville, EC, S Africa.

Note: The very thick rootstock in relation to thin, twining, fragile stems with only two leaf-pairs at their ends characterizes this species well within the genus.

R. pillansii N.E.Br. GC 1927: 92 T Pillans s.n. (K).

Compact to caespitose; **L** trigonous with a pronounced chin, those of a pair differing in size, light grey, 13.5–20 mm *l*, 13.5–18 mm *b*, 4.5–6.75 mm *t*; **Fl** unknown; **Fr** on persistent stalks to 15 mm *l*, base funnel-shaped, top flat, valves raised a little in the centre, narrow ledges extending from the septa, expanding keels parallel, valve wings very broad, apically rounded, basally narrowed, Ø 6–9 mm; **Ecol** in flats with grey shales; **Distr** Laingsburg, WC, S Africa.

Note: The light grey colour of the plants is very prominent in habitat.

R. rouxii (L.Bolus) L.Bolus NM 1936: 40 ≡ *Chasmatophyllum rouxii* L.Bolus NM 1936: 5 T Roux 155 (BOL!).

Compact; **L** club-shaped to spatulate, some margins dentate, those of a pair a little different, 16–20 mm *l*, 5.5–6 mm *b*; **Fl** petals bright yellow, 9 mm *l*, 1–1.5 mm *b*, filaments yellow; **Fr** on a persisting stalk of c. 15 mm length, base elongate, bell-shaped, top flat, septa with very narrow ledges, expanding keels broad and parallel, valve wings very broad, nearly rectangular; **Ecol** on a koppie; **Distr** Victoria West, NC, S Africa.

Note: A little known species.

R. schonlandii L.Bolus NM 1936: 15 T Broom 34901 (BOL!).

Compact with a thick rootstock; **L** of a pair unequal, one with a very broad, truncate top, the other more club-shaped and fitting into the hood of the bigger leaf when both are young, some leaf-pairs with broad teeth, longest leaves to 20 mm *l*, to 12 mm *b*; **Fl** petals yellow, apically red, 8 mm *l*, 0.75 mm *b*, filaments yellow, basally paler, nectary glands yellowish-green; **Fr** unknown; **Ecol** unknown; **Distr** Fraserburg, NC, S Africa.

Note: The species appears most similar to **Stomatium suaveolens**, the type species of **Stomatium.**

R. vanheerdei ≡ **Peersia vanheerdei**

R. willowmorense = **R. parvifolium** p.p.;
≡ **Chasmatophyllum willowmorense** p.p.

Rhombophyllum Ruschioideae

H. E. K. Hartmann

Rhombophyllum (Schwantes) Schwantes ZSK 1927: 16; Schwantes ZSK 1927: 23; Hartmann Bradleya 1998: 80–84 ≡ *Bergeranthus* Schwantes subg. *Rhombophyllum* Schwantes ZSK 1926: 180 **Typus** *Bergeranthus rhomboideum* (Salm-Dyck) Schwantes ≡ **R. rhomboideum** (Salm-Dyck) Schwantes **Etym** Gk *rhombos*, lozenge, *phyllon*, leaf.

Compact or shrubby plants; **L** rhomboid or hatchet-shaped, often bilobed to different degrees, but also with a straight keel, epidermis a little rough; **Fl** solitary or in cymes, peduncles and pedicels mostly long, **K** 5, petals yellow, filamentous staminodes absent, stamens in a central column, 5 distant nectary glands; **Fr** base bowl- to funnel-shaped, top with high rims, covering membranes complete with a distal rim with a closing ledge under it, touching the very broad but extremely flat closing body, which is formed by tissue descending from the expanding sheet and is of white, spongy consistence and divided into two sections visible from above in the open capsule, expanding keels mostly basally separate, valve wings very narrow, 5 locules; **Ecol** in shrubby associations on gravelly soils; **Distr** W parts of EC, S Africa.

Note: The genus was originally described to hold **R. rhomboideum** only, the critical feature being the very flat, divided closing bodies which appear almost invisible at first instance. Later, the bilobed leaves were added as typical character states (Schwantes GF 1928: 87) when plants with visible internodes were also included. Here, the genus is understood to comprise species with a divided closing body, in contrast to **Hereroa** with small closing bodies or none at all, and to **Bergeranthus** with a broad, undivided sill as a closing body. All three genera share a common flower morphology and similarities in fruit and growth form.

Key to the Species of Rhombophyllum

1.	Leaves bilobed in a vertical plane	3
–	Leaves not bilobed	2
2.	Plants compact with very tall inflorescences	**R. rhomboideum**
–	Plants mat-forming, leaves nearly finger-shaped	**R. albanense**
3.	Leaves c. 50 mm long, at least some of them deeply incised	**R. dyeri**
–	Leaves 15–30 mm long, most leaves almost bilobed at their ends	4
4.	Leaves 15–20 mm long	**R. nelii**
–	Leaves c. 30 mm long	**R. dolabriforme**

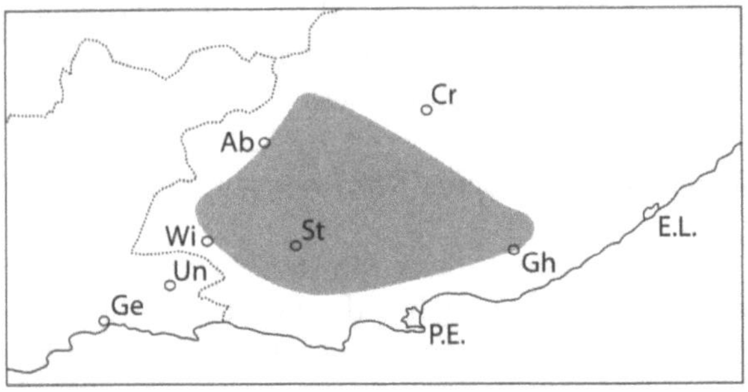

Distribution of **Rhombophyllum** (Schwantes) Schwantes

R. albanense (L.Bolus) H.E.K.Hartmann Bradleya 1998: 81 (see **photo**) ≡ *Ruschia dyeri* L.Bolus JB 1928: 230 ≡ *M. dyeri* (L.Bolus) N.E.Br. GC 1930: 32 ≡ *Hereroa albanensis* L.Bolus NM 1950: 222 **T** Dyer 1224 (BOL!).

Caespitose shrubs to 10 cm *h*, ∅ to 30 cm, **I** light grey with age, with scars where the old leaf bases have been, basally to 12 mm ∅; **L** trigonous, smooth, keel a little rounded, to 15 mm *l*, at most to 3 mm *b* and ∅; **Fl** solitary, 10–15 mm ∅, **K** 5, petals yellow, filaments white, anthers yellow, nectary glands blackish-green; **Fr** base funnel-shaped, top with high rims, otherwise flat, covering membranes complete, closing body consisting of two halves protruding well under the covering membranes, expanding keels widely diverging, valve wings broadening from a narrow distal part to a broader base, often adorned by a pronounced tip, 5 locules, ∅ c. 5 mm; **Ecol** in flats or on gentle slopes of Ecca shale, often in succulent vegetation with **Euphorbia quadrangularis**; **Distr** Albany, EC, S Africa.

Note: The fruit is clearly of the type belonging to **Rhombophyllum**, but the leaves are not indented like those of the other species.

R. dolabriforme (L.) Schwantes GF 1928: 87 (see **photo**) ≡ *M. dolabriforme* L. 1753: 487 ≡ *Hereroa dolabriformis* (L.) L.Bolus NM 1928: 143 **Lectotype**, here designated: Dillenius 1732: t. 191 f 237.

Low shrubs branching at their tips, the inner thick old branches becoming leafless with age; **L** hatchet-shaped, but always bilobed to different degrees, dark green, dotted, wax layer thin, few platelets; **Fl** mostly solitary, bracteoles acuminate, **K** 5, petals golden yellow; **Fr** with a broad, flat closing body; **Ecol** on withered sandstone with big, round stones; **Distr** Aberdeen, Graaff Reinet, Willowmore, EC, S Africa.

Note: The plants grow out into a leggy shrubs with age.

R. dyeri (L.Bolus) H.E.K.Hartmann Bradleya 1998: 81 ≡ *Hereroa dyeri* L.Bolus NM 1928: 143 **T** Dyer s.n. 1933 (BOL!).

Compact, anisophyllous; **L** of a pair differing: one deeply divided, the other with a pronounced hump on the keel, smooth or with somewhat raised dots, to 50 mm *l*, 3–10 mm ∅; **Fl** in cymes, raised well above the plant, bracteoles acuminate, petals golden, outer reddish outside; **Fr** closing bodies only as two adjacent distinct round areas visible, the covering membranes pressing against them with the bases of their distal rims which lack a ledge below, valve wings narrow and tapering towards tips and bases, but broader than those of other species, 5 locules; **Ecol** unknown; **Distr** Albany, EC, S Africa.

Note: The divided, hardly raised closing bodies and the deeply divided leaves support the placement of the species in **Rhombophyllum**.

R. nelii Schwantes GF 1928: 85 (see **photo**) **T** without collector 1433 (collection Schwantes, HBG!).

Shrubs branching distally, the inner stems blackish; **L** of a pair apically divided to various degrees, glaucous green, epidermis somewhat rough from raised dots which become prominent in the dry state, wax cover continuous, breaking into plates, one per cell, later, with amorphous sculptures or vertical platelets above, stomata only a little sunken; **Fl** 1–3, bracts and bracteoles with marked dorsal humps or divisions, bracteoles at c. half the length of the long pedicels, ∅ 20–30 mm, 50–80 yellow petals to 20 mm *l*, 1 mm *b*, purplish suffused outside, 110–160 stamens with white filaments and yellow anthers; **Fr** with rather low rims, covering membranes pressing against the flat, sill-shaped, mostly divided closing body, expanding keels widely separated basally, valve wings very narrow; **S** ovate, testa cells with distinctly undulate anticlinal walls, only a little elevated, c. 0.6 mm *l*, c. 0.4 mm *b*; **Ecol** in shrubby vegetation dominated by succulents on shaly soils; **Distr** Graaff Reinet, Jansenville, Steytlerville, EC, S Africa.

Note: The division of the leaf tips is not always as deep as shown by the photograph with the description, where the leaf is compared with leaf-pairs of **Conophytum**. The division of the rather small and inconspicuous closing body is also not regularly well expressed, varying even within a single fruit and often in fruits of one population.

R. rhomboideum (Salm-Dyck) Schwantes ZSK 1927: 23 (see **photo**) ≡ *M. rhomboideum* Salm-Dyck 1849: § 6 f 5 ≡ *Bergeranthus rhomboideus* (Salm-Dyck) Schwantes ZSK 1926: 180 **LT** Salm-Dyck 1849: § 6 f 5 = *R. rhomboideum* (Salm-Dyck) Schwantes var. *groppiorum* Heinrich ex H.Jacobsen 1970: 522 = *R. rhomboideum* (Salm-Dyck) Schwantes forma aut var. *groppiorum* Heinrich 1964: 54 nomen invalidum **LT** Heinrich 1964: f on p 54.

Compact, little branched; **L** hatchet-shaped above, broadly canaliculate in the lower half, margins cartilaginous, white, often twisted; **Fl** in high cymes, bracteoles acute, petals yellow, pink outside; **Fr** covering membranes very high, distal rim well developed, with a closing ledge below which presses against the closing body composed of two semi-orbicular halves, appearing as a sill in longitudinal section, expanding keels separated basally, valve wings very narrow, hardly varying in width along their extension; **Ecol** in open, gravelly patches in bushland; **Distr** Port Elizabeth, EC, S Africa.

Note: The type species does not show the common bilobed leaves, but shares the divided closing body with all other members of the genus.

Heinrich (1964: 54–57) published a forma or variety found in cultivation. Since he did not state the rank definitely, his description is invalid, being validated by Jacobsen (1970: 522), who chose the rank of variety for the taxon. The photograph published with the description shows very round, dense leaves resembling the situation in **Aloinopsis loganii,** for instance.

R. rhomboideum forma aut var. *groppiorum* = **R. rhomboideum**
R. rhomboideum var. *groppiorum* = **R. rhomboideum**

Rhopalocyclus = Leipoldtia

H. E. K. Hartmann

When Schwantes described the genus *Rhopalocyclus* (GW 1928: 599), naming it after the circle of club-shaped closing bodies in the fruit, he was unaware of the fact that Bolus (FPSA 1927: t. 256) had named this genus **Leipoldtia** already. The younger name is therefore superfluous and a synonym.

R. nelii = **L. schultzei**
R. weigangianus ≡ **L. weigangiana**
R. herrei = **L. schultzei**

Rimaria = Gibbaeum

H. E. K. Hartmann

The genus was first described in a key (Brown GC 1925: 413) without any species. In a more extensive presentation of the new genus, Brown (GC 1926: 134) included four species which today are considered to belong to three different genera. Since the type species, *R. heathii*, is a member of the genus **Gibbaeum**, the name *Rimaria* was superfluous from the beginning. Bolus (NM 1937: 69) used the name for a genus later to be known as **Vanheerdea** by adding species similar to *R. roodiae*. As a consequence, *Rimaria* L.Bolus non N.E.Br. is a younger homonym and illegitimate as well.

R. angusta = **Vanheerdea roodiae**
R. comptonii ≡ **Gibbaeum comptonii**
R. comptonii forma ≡ **Gibbaeum comptonii**
R. divergens = **Vanheerdea roodiae**
R. dubia = **Gibbaeum heathii**
R. heathii ≡ **Gibbaeum heathii**
R. heathii var. *major* = **Gibbaeum heathii**
R. heathii var. *elevata* = **Gibbaeum heathii**
R. luckhoffii ≡ **Gibbaeum luckhoffii**
R. microsperma ≡ **Dinteranthus microspermus**
R. pole-evansii ≡ **Dinteranthus pole-evansii**
R. primosii ≡ **Vanheerdea primosii**
R. roodiae ≡ **Vanheerdea roodiae**

Rocama = Zaleya

H. E. K. Hartmann

The genus *Rocama* Forskål was described (1775: 71) in another place than the species, *Rocama prostrata* Forskål (1775: CVIII, no. 200), albeit in the same book. The material had been collected by Forskål himself, who had also sent seed to Linnaeus, who grew plants from them (Christensen 1922: 18), naming the species *Trianthema pentandra* L. As a consequence, *Rocama prostrata* and *Trianthema pentandra*, now **Zaleya pentandra**, represent the same species (Melville 1952: 263), probably even based on the same collection. Gmelin (1791: 455) gave the name *Rocama africana* to the only species he cited under the generic name without a description but with the citation "Forsk. fl. aeg. arab. p. 71"; this alignment is not correct, and *R. africana* Gmelin must be considered a nomen nudum and, at the same time, a nomen illegitimum because it is based on the same type as *R. prostrata* Forskål. De Candolle (1828: 352) placed several species of **Trianthema** in his subgenus *Rocama*, attributing *Rocama digyna* to Forskål and citing also *Rocama arabica* Gmelin as a synonym for *Trianthema pentandra*. The subdivision of **Trianthema** in this sense has not been followed later.

R. arabica = **Zaleya pentandra**
R. digyna = **Zaleya pentandra**
R. prostrata = **Zaleya pentandra**

Roodia = Argyroderma

H. E. K. Hartmann

In the original description of *Roodia*, Brown (FPSA 1922: t. 78) gave three features as typical of the new genus: the formation of a calyx-tube (hypanthium), the stamens bent into the tube, and the possession of a sessile, undivided stigma. This last mentioned character state was also cited by Brown as typical of the genus **Argyroderma** (Brown GC 1922: 92), for which latter genus he gave silvery leaves as well as characteristic. Since both genera possess the described hypanthium and the stamens bent into the tube, too, they have been united, *Roodia* being the younger synonym, but kept at the rank of subgenus (Hartmann 1978: 207).

R. braunsii = **Argyroderma fissum**
R. brevipes = **Argyroderma fissum**
R. digitifolia = **Argyroderma fissum**

Ruschia Ruschioideae

H. E. K. Hartmann

Ruschia Schwantes 1926: 186; Friedrich 1961: 129–143; Dehn 1993: 91–198; Hartmann & Stüber 1993: 1–75, 1995: 353–373; Hartmann Bradleya 1998: 84–87; Hartmann 1999: 38–74 **Typus R. rupicola** (Engler) Schwantes **Etym** *E. Rusch*, 1867–1957, farmer in Namibia.

Erect, sometimes decumbent, even creeping or caespitose shrubs; L mostly longer than wide, rarely reduced, trigonous to triangular, often dentate, epidermis flat to papillate; **Fl** solitary or in ± rich cymes, small, petals often in five groups, purple to white, filamentous staminodes surrounding the stamens in a central cone, nectary a lophomorphic ring; **Fr** Ruschia-type, i.e. expanding keels diverging widely, covering membranes stout and persistent in shape with a distal, erect rim below which closing rodlets or closing ledges emerge, closing bodies as hooks, hoods, or rodlets, valve wings absent or narrow, 5(–6) locules; **S** smooth to papillate, brown, 0.6–1.6 mm *l*, 0.4–1.2 mm *b*; **Ecol** different for different species; **Distr** S Africa except the E and NE parts; S Namibia; Lesotho; SW part of Botswana.

Note: With more than 400 names available at species level, this is the one of the biggest genera in the family Aizoaceae. Since the original description of 18 species in the genus, no revision down to species level has been published. Friedrich (1961, 1970) investigated the species occurring in Namibia, and Hartmann (1999) presented an overview over all names at species level that have ever been placed in **Ruschia.**

Schwantes (MDK 1930: 68) gave a first division into subgenera and sections. These names lack descriptions entirely, but for most of them types are cited. Since a number of sections is taken from Haworth and Salm-Dyck, albeit without any citation of the references, which have been validly published at that date and must be taken into consideration when a revision of the genus is undertaken. Since the question of validity depends also on the species included, no attempt is made here to assess the subgeneric taxa published at that date. Schwantes himself seems to have considered all his subgenera and sections of 1930 invalid because in Jacobsen (1960: 1367–1368) he described 12 sections as "comb. nov."; no subgenera are mentioned in the text. Since these latter descriptions are in English only, they are all invalid at that date. When Dehn (1992) clarified the delimitation of the genus, he distinguished nine subgenera:

R. subg. **Cymbifoliae** Dehn 1993: 189 **Typus R. cymbifolia** (Haw.) L.Bolus
R. subg. **Maximae** Dehn 1993: 189 **Typus R. maxima** (Haw.) L.Bolus
R. subg. **Multiflorae** Dehn 1993: 189 **Typus R. multiflora** (Haw.) Schwantes
R. subg. **Ruschia**
R. subg. **Sarmentosae** (Haw.) Dehn 1993: 190 **Typus R. sarmentosa** (Haw.) Schwantes
R. subg. **Spinosae** (Salm-Dyck) Dehn 1993: 191 **Typus R. spinosa** (L.) Dehn
R. subg. **Strictae** Dehn 1993: 191 **Typus R. stricta** L.Bolus
R. subg. **Tumidulae** (Haw.) Dehn 1993: 192 **Typus R. tumidula** (Haw.) Schwantes
R. subg. **Vaginatae** (Haw.) Dehn 1993: 192 **Typus R. vaginata** (Haw.) Schwantes

Only one of the subgenera has been studied further, **R.** subg. **Spinosae** (Hartmann & Stüber 1993, 1995); only in this case, a description and a key to the species is given; the species are annotated accordingly. For all other subgenera, neither the character states given as typical nor the species composition, and consequently the validity, can be established with certainty at present. Descriptions and a key are therefore omitted here, nor are the species annotated as to the subgenus into which they might belong. Many of the species are important for agriculture, either as useful plants eaten by stock, or as settlers in ground

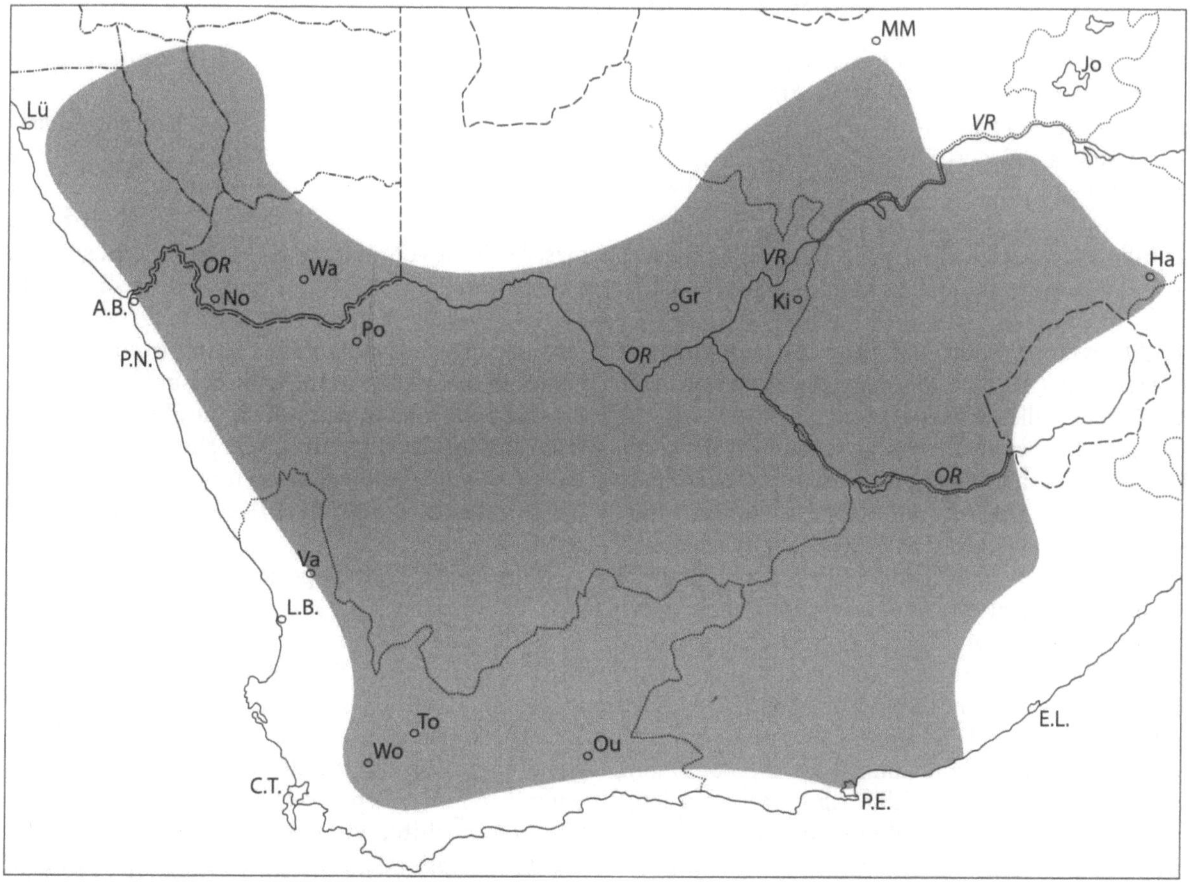

Distribution of **Ruschia** Schwantes

not bearing fodder plants, thus indicating possibly a deterioration of soil or other conditions.

Following Dehn (1989), a number of species has been separated from **Ruschia** and is placed in the genus **Antimima**. Furthermore, several species have been transferred to the genera **Acrodon, Arenifera, Ebracteola** (Glen 1986), **Stoeberia, Eberlanzia,** and **Amphibolia** (Hartmann 1996: 29–56).

R. § *Brevivaginatae* = **Eberlanzia**

Ruschia Schwantes subg. **Spinosae** (Salm-Dyck) Dehn 1992: 191 ≡ *M.* § *Spinosa* Salm-Dyck 1820: 25 Hartmann & Stüber 1993: 1–75 **Typus R. spinosa** (L.) Dehn.

Erect to flat shrubs, **I** whitish-grey or black or reddish; **L** hardly connate, mucronate, epidermis xeromorphic, homocellular; **Fl** in more or less rich **Inf** with regularly arranged, acute and firm spines derived from pedicels terminating in undeveloped tiny flower buds (vestigial calyces), bracteoles small, petals pink to purple, 25–32, bases of 0–30 filamentous staminodes and 30–110 stamens papillate; **Fr** Ruschia-type with a semi-globular top through high valve rims extending over the edge, the base elongate funnel-shaped, valves without or with narrow valve wings, closing bodies rodlet-shaped, opaque, covering membranes complete with a distal recurving and a closing rodlet below; **S** < 1 mm, smooth, pear-shaped; **Ecol** most species on flats in open associations with bimodal rainfall of 200–400 mm; **Distr** Lüderitz-South, Namibia; EC, FS, NC, NW, WC, S Africa.

Note: The species of the subgenus were treated as members of the genus **Eberlanzia** in the past, but they differ in too many characters to be kept there. The spiny inflorescences form particular attractions all year round, especially in young plants which lack flowers for several years. Older plants of some species present spectacular floral displays because of the high number of small flowers opening at the same time.

All species in **Ruschia** subg. **Spinosae** are appreciated as plants for browsing because their young, edible sprouts are available to animals when little else can be found, in particular in the area where summer and winter rainfalls meet and overlap at times. This assessment of farmers is supported by controlled feeding experiments (du Preez 1970: 126–131) showing that pellets of "*Ruschia vulnerans*" were well accepted in the absence of preferred plants, e.g. **Plinthus karroicus.**

Key to the Species of Ruschia subg. Spinosae

1. Cymes with a central fruit each and distal spines, often lacking fruits entirely .... **R. centrocapsula**
- Cymes with distal fruits and a central spine 2
2. Up to 5 spines and 4 fruits per cyme .... **R. cradockensis**
- > 7 spines and 4–256 fruits per cyme 3
3. Spines and younger stems whitish-grey to beige 4
- Spines and younger stems reddish to dark brown .... **R. intricata**
4. Inflorescences storeyed, i.e. with additional basal cymes, seeds 0.8–1 mm long .... **R. spinosa**
- Inflorescences in pure dichasia, not storeyed, seeds 0.6–0.8 mm long .... **R. divaricata**

R. abbreviata L.Bolus NM 1935: 493 (see **photo**) **T** Herre 10205 (BOL!) = *R. subaphylla* Friedrich 1968: 213 **T** Merxmüller & Giess 3373 (M).

Loosely growing shrubs to 30 cm *h*, ∅ to 50 cm, **I** enclosed by the sheath of the subsequent leaf-pair, succulent and green for some time, ∅ then c. 8 mm, 10–22 mm *l*, some **I** later drying into hard stem parts of c. 3 mm ∅, a feature that can occur in a branch that is still growing at its distal end, resembling **Psilocaulon** in this habit; **L** only prominent on young plants, where they are up to 10 mm *l*, on older stems the free parts 2–3 mm *l*, epidermis papillate, as that of the **I**, the terminal end smooth; **Fl** solitary, rarely 3, bracteoles as short, highly connate leaves, ∅ 8–20 mm, petals white; **Fr** top with high rims, base very long, bell-shaped, covering membranes convex with a distinct distal closing ledge, closing body a small hood, expanding keels short and roundish, well separated basally, 5 locules; **S** brown; **Ecol** on shaly, gravelly slopes, < 100 mm rainfall in winter; **Distr** Lüderitz-South, Namibia; N Namaqualand, NC, S Africa.

Note: In this species, the highest reduction of the lamina of the leaves can be observed for the genus **Ruschia**. Some similar species show also short free parts of leaves, but here the dominance of stem succulence is reached, a feature known to this extreme only from the genus **Psilocaulon.** The populations from north of the Orange River, described as *R. subaphylla* because Friedrich had not seen the type of **R. abbreviata**, following his remarks (1968: 215), appear to have smaller flowers.

R. abrupta ≡ **Octopoma abruptum**

R. acocksii L.Bolus NM 1954: 277 **T** Acocks 16958 (BOL!).

Shrubs to 30 cm *h*, **I** maroon to brown; **L** trigonous, acuminate, in the dry state, the tannin idioblasts clearly visible, but epidermis glabrous and without elevations; **Fl** 1–3, all bracteolate, the bracteoles in the middle of each stalk, **K** 5–6, petals purplish-pink, filamentous staminodes numerous, filaments pale, apically purplish-pink; **Fr** loosing the locules out of the persisting circle of bundles, top with very high rims reaching over the edge of the base, base long, funnel-shaped, covering membranes rising from a deeper centre in the middle of the fruit, closing ledge sharp, closing body a hook, into which the placenta is fixed from the inside, expanding keel short, rounded, no wings, 5–6 locules; **Ecol** c. 200 mm rainfall p.a. with a dry season in summer; **Distr** Sutherland, NC, S Africa.

Note: A little known species.

R. acuminata L.Bolus NM 1928: 20 ≡ *M. exacutum* N.E.Br. GC 1930: 32 **T** Bijl 18855 (BOL!).

Branches to 20 cm *l*, ascending to decumbent; **L** free parts trigonous, epidermis smooth, sheath long, basally connate with the stem, free parts 15–25 mm *l*, 7 mm *b*, 6 mm *t*; **Fl** pedicels 30 mm *l*, bracteoles basal, petals white, withering pink, 0.5 mm *b*, filamentous staminodes few; **Fr** base an elongate funnel, top with high rims, light grey, covering membranes convex, closing ledges low, closing body absent, the end of the placenta forming a rim bending inwards and probably mistaken for a closing body, expanding sheets distant at their bases, 5 locules; **Ecol** unknown; **Distr** Prince Albert?, WC, S Africa.

Note: The fruit agrees best with those of **R. burtoniae** and **R. muelleri**, but differs in the stem from both taxa, which do, most probably, not belong to **Ruschia**.

R. acutangula (Haw.) Schwantes FR 1938: 230 ≡ *M. acutangulum* Haw. 1824: 424 ≡ *M. vaginatum* Haw. var. *acutangulum* A.Berger 1908: 111 **LT** drawing by Duncanson 23.8.1826 (K!).

Erect to densely branched shrubs, **I** purple, later ochre; **L** triquetrous, margins and keel sharp and finely serrulate, sheath connate with the stem, probably < 10 mm *l*, somewhat bluish-green; **Fl** unknown; **Fr** unknown; **Ecol** unknown; **Distr** unknown.

Note: The connate leaf sheaths merging into the stem agree with many species of **Ruschia**.

R. addita ≡ **Antimima addita**

R. aggregata L.Bolus NM 1958: 339 **T** Esterhuysen 23636 (BOL!).

Erect shrubs, branches to 36 cm *l*, **I** 15–40 mm *l*, to 6 mm ∅; **L** terete to club-shaped, bases separated by a deep line, but sheath short, epidermis smooth, 10–25 mm *l*, 3–4 mm *b*, 5 mm *t*; **Fl** in aggregated, dense cymes, pedicels to 25 mm *l*, bracteoles basal, **K** 5, petals purplish-pink, 4–6 mm *l*, 0.5–1 mm *b*, filamentous staminodes absent, anthers and pollen straw-coloured; **Fr** base funnel-shaped, top raised, rims low, covering membranes with distal closing rodlets, closing body hook-shaped, expanding keels almost touching each other basally, 5 locules; **Ecol** unknown; **Distr** Namaqualand, NC, S Africa.

Note: The leaf-shape and the smooth, light epidermis are unusual for **Ruschia**, but the loose fruit with the type possesses the typical character states of the genus.

R. alata L.Bolus NM 1932: 350 **T** Luckhoff 20124 (BOL!).

Loose shrubs, branches to 18 cm *l*, **I** flat at first, reddish, later grey and finally brown; **L** triquetrous(?), margins and keel with long papillae, faces with shorter ones, c. 12 mm *l*, c. 3 mm *b* and *t*; **Fl** in terminal, ternate cymes, bracteoles near the middle of the pedicel, petals pink, filamentous staminodes red at tips; **Fr** 5 locules; **Ecol** < 100 mm rainfall p.a. in winter; **Distr** Vanrhynsdorp, WC, S Africa.

Note: The species is little known.

R. albertensis = **R. spinosa**
R. albiflora ≡ **Polymita albiflora**
R. alborubra ≡ **Antimima alborubra**

R. altigena (L.Bolus) L.Bolus NM 1950: 219 ≡ *M. altigenum* L.Bolus ABH 1927: 86 **T** Compton 3075 (BOL!).

Decumbent, branches to 13 cm *l*, **I** 4 mm ∅; **L** semi-terete, epidermis smooth, sheath with a short line, dark grey-green, to 27 mm *l*, 4 mm sheath, 4 mm *b* and *t*; **Fl** solitary, ∅ 23 mm, **K** 5, petals pinkish-purple, 11 mm *l*, to 1.5 mm *b*, filamentous staminodes pink, filaments pale pink, base white, anthers and pollen pale yellow; **Fr** firm, top almost semi-globose, base long, funnel-shaped, covering membranes with distinct closing ledges, closing body small, expanding keels with awns, no valve wings present, 5 locules; **Ecol** at 1700 m altitude; **Distr** Laingsburg, WC, S Africa.

Note: Reported to be similar to **R. campestris**, which is a bigger shrub.

R. amicorum (L.Bolus) Schwantes ZSK 1927: 20 ≡ *M. amicorum* L.Bolus ABH 1922: 127 **T** Kensit & Caporn 261/44 (BOL!).

Shrubs 30–40 cm *h*, **I** 6–15 mm *l*, reddish, later ochre; **L** semi-terete, glaucous green, 25–40 mm *l*, 5 mm *b*, 7 mm *t*; **Fl** 3 in a cyme, ∅ 15 mm, scented, pedicels 10–15 mm *l*, bracteoles below the middle, petals pinkish-purple, 5–6 mm *l*, 1.5 mm *b*, filamentous staminodes present, filaments pinkish-purple, anthers yellow; **Fr** base funnel-shaped, top raised, rims low, covering membranes with small closing rodlets pending from closing ledges, end of placenta knobby, but closing body absent, expanding keels broad, laciniate towards the ends, 5 locules; **Ecol** unknown; **Distr** Montagu, WC, S Africa.

Note: The absence of closing bodies and valve wings excludes the species from **Ruschia** and **Lampranthus**, but a genus to accept it is not available.

R. amoena ≡ **Antimima amoena**
R. amphibolia ≡ **Lampranthus amphibolia**

R. ampliata L.Bolus NM 1928: 145 ≡ *M. ampliatum* (L.Bolus) N.E.Br. GC 1930: 32 **T** Pillans 6070 (BOL!).

Shrubs to 25 cm *h* with erect branches; **L** "subglobose", to 15 mm *l*, 7 mm *b* and *t*; **Fl** 1–3, **K** 5, ∅ c. 18 mm, petals in 5 groups, 6 mm *l*, 0.5 mm *b*, filamentous staminodes white; **Fr** covering membranes with low closing ledges, otherwise as in Ruschia-type, 5 locules; **Ecol** unknown; **Distr** coastal Namaqualand, NC, S Africa.

Note: The report of six to eight locules cannot be verified in the type material, which shows five locules.

R. androsacea ≡ **Antimima androsacea**

R. approximata (L.Bolus) Schwantes ZSK 1926: 187 ≡ *M. approximatum* L.Bolus ABH 1926: 37 **T** Page 17170 (BOL).

Diffusely branched shrubs 9–14 cm *h*; **L** triquetrous, keel inconspicuously denticulate, 1–3 teeth, glaucous-green, 10–14 mm *l*, sheath of 5–7 mm, 4–5 mm *b*; **Fl** solitary, pedicels enclosed in bracteoles, **K** 5, enclosed by the bracteoles bearing 2–3 conspicuous teeth along the sharp keel, petals pink, 4–10 mm *l*, filamentous staminodes present, anthers and pollen yellow; **Fr** 5 locules; **Ecol** unknown; **Distr** Montagu, WC, S Africa.

Note: Reported to differ from **R. uncinata** in broader, thicker, and shorter leaves with less well pronounced teeth.

R. archeri L.Bolus ABH 1928: 127 **T** Archer 18748 (BOL!) = *R. archeri* L.Bolus var. *sexpartita* L.Bolus NM 1950: 219 **T** Hall 24062 (BOL!).

Branches to 33 cm *l*, basally enveloped in old leaf sheaths; **L** trigonous, epidermis almost smooth, free parts 6–12 mm *l*, 6 mm *b* and *t*, the sheath c. 20 mm *l*, connate basally with the stem; **Fl** solitary, pedicels 13 mm *l*, **K** 5, with 2–3 conspicuous teeth along the sharp keel, petals purplish-pink, filamentous staminodes not mentioned, anthers and pollen white, **Fr** base funnel-shaped, top with high rims, covering membranes convex, closing ledges distinct, closing body hook-shaped, expanding keels nearly touching each other basally, 5(–6) locules, ∅ 7 mm; **Ecol** unknown; **Distr** Ladismith, Swellendam, WC, S Africa.

Note: The fruit on the type sheet is unripe and can therefore not be assessed.

R. archeri var. *sexpartita* = **R. archeri**
R. arenosa ≡ **Lampranthus arenarius**
R. argentea ≡ **Antimima argentea**
R. aristata ≡ **Erepsia aristata**

R. aristulata (Sonder) Schwantes ZSK 1927: 19 ≡ *M. aristulatum* Sonder in Harvey & Sonder 1862: 424 **T** Ecklon & Zeyher 2023 (K?).

Prostrate branches somewhat rooting, **I** compressed; **L** triquetrous, glaucous, acute, mucronate to aristate, margins finely ciliate, 6.75–9 mm *l*, 2.25 mm *b*; **Fl** enclosed by the bracteoles, pedicel 4.5–6.75 mm *l*, petals pale red when dry, c. as long as the **K**; **Fr** 5 locules, otherwise unknown; **Ecol** in rocks; **Distr** Cape Town, WC, S Africa.

Note: The species is said to be similar to **R. rostella**.

R. armata = **Arenifera stylosa**

R. aspera L.Bolus NM 1939: 161 **T** Leipoldt 22111 (BOL!).

Shrubs with erect, to 22 cm high branches, **I** 5–13 mm *l*, ∅ 5–10 mm; **L** spreading, globose, but a little trigonous, epidermis with marked elevations above the tannin idioblasts, dark green, 3–4 mm *l*, 2–2.5 mm *t*; **Fl** solitary, enclosed by the bracteoles, **K** 5, petals purplish-pink, 4–5 mm *l*, 0.5–0.75 mm *b*, filamentous staminodes not mentioned, filaments pink; **Fr** 5 locules, otherwise not well known; **Ecol** unknown; **Distr** Namaqualand (only?), NC, S Africa.

Note: Several populations found in the area agreeing with the vegetative character states possess two different types of fruits.

R. atrata L.Bolus NM 1928: 144 ≡ *M. atratum* (L.Bolus) N.E.Br. GC 1930: 32 **T** Pillans 5743 (BOL!).

Erect shrubs to 28 cm *h*, **I** black, ∅ 0.5–3 mm; **L** trigonous, glaucous-green, epidermis cells a little raised, to 15 mm *l*, 4–5 mm *b*, 4–5 mm *t*; **Fl** 1–3, ∅ 10 mm, pedicels 3 mm, bracteoles clasping the base of the flower, **K** 5, petals pink, 4 mm *l*, 0.75 mm *b*, filamentous staminodes few, filaments white, apically purple, anthers white, stigmas purple to black; **Fr** 5 locules, otherwise unknown; **Ecol** unknown; **Distr** N Namaqualand, NC, S Africa.

Note: The leaves fall off easily, a feature rare in **Ruschia**. The lack of fruiting material prevents a definite placement of the species.

R. axthelmiana (Dinter) Schwantes ZSK 1926: 187 ≡ *M. axthelmianum* Dinter FR 19, 1923: 129 **T** Dinter 3881 (B).

Shrubs to 50 cm *h*, **I** grey to ochre, 20–30 mm *l*; **L** triquetrous, merging into the sheath connate with the stem, keel 2–4 denticulate, margins often with 1–2 teeth, mucronate, 20 mm *l*, 2–3 mm *b* and *t*; **Fl** solitary, ∅ 18 mm, pedicels 10 mm *l*, **K** 5, petals purplish-red, filamentous staminodes and filaments purplish-red, the latter basally white; **Fr** pentangular, 5 locules, 4–5 mm ∅; **S** rough, 1 mm *l*; **Ecol** in pebbly loam, rainfall in February; **Distr** Windhoek, Rehoboth, Keetmanshoop, Namibia.

Note: Friedrich (1970: 109–120) reported fleshy, thick yet spindle-shaped roots for plants of this species.

R. barnardii L.Bolus NM 1928: 19 (see **photo**) ≡ *M. barnardii* (L.Bolus) N.E.Br. GC 1930: 32 **LT** Barnard 660/25 (BOL!) = *R. kakamasensis* L.Bolus NM 1936: 15 **LT** Cilliers/Herre 10105 V (BOL!) = *M. pseudorupicola* Dinter FR, Beiheft 53, 1928: 12 nomen nudum.

At first open, later more densely branched low shrubs to 30 cm *h*; **L** trigonous with 1–2 keel teeth, epidermis with papillae of different sizes, 20–35 mm *l*, c. 5 mm *t*; **Fl** ∅ c. 35 mm, petals white to pale pink, filamentous staminodes white or apically pale pink, filaments white; **Fr** rims dipping in the centre, base long bell-shaped to funnel-shaped, closing bodies broadening to the interior of the locule, the funicles protruding from beneath into the locule, 5 very deep locules; **Ecol** in gravel between quartz rocks or in quartzitic fields, < 100 mm rainfall p.a. in March; **Distr** Keetmanshoop, Warmbad, Namibia; Gordonia, NC, S Africa.

Note: The habitats of **R. barnardii** and **R. odontocalyx** are similar in so far as gravelly areas between quartzitic blocks are preferred, but the much longer leaves and the larger flowers in **R. barnardii** permit a clear distinction.

R. beaufortensis L.Bolus NM 1934: 430 **T** Jager 807/28 (BOL!).

Branches to 12 cm *l*, **I** embraced by and connate with the leaf bases, ∅ 7 mm; **L** triquetrous, keel and margins with 2–6 stout teeth, epidermis cells papillate, free parts 10–20 mm *l*; **Fl** sessile, base laterally compressed, thus two-angled, **K** 5, petals pink, to 8 mm *l*, 0.5 mm *b*, filamentous staminodes pink, apically pale; **Fr** not available, 5 locules; **Ecol** in mountains; **Distr** Beaufort West, WC, S Africa.

Note: As noted by Bolus with the description, the plants look very similar to **R. perfoliata**.

R. bicolorata = **Stoeberia beetzii**
R. biformis ≡ **Antimima biformis**

R. bijliae L.Bolus NM 1928: 20 ≡ *M. bijliae* (L.Bolus) N.E.Br. GC 1930: 32 T Bijl 18857 (BOL!).

Mats 6–9 cm *h*, decumbent branches to 27 cm *l*; **L** free parts triquetrous, 1–4 teeth on the keel, sheath connate with the stem, with a line 2–3 mm *l*, free parts 8–14 mm *l*, 5 mm *b* and *t*; **Fl** solitary, sessile, bracteoles edentate, ∅ 18 mm, **K** 5, petals to 8 mm *l*, 0.5 mm *b*, filamentous staminodes deep pink, filaments apically pink, anthers yellow; **Fr** base an elongate funnel, top with high rims, covering membranes convex, closing rodlets protruding, closing body small, egg-shaped, expanding keels not touching each other basally, 5 locules; **Ecol** unknown; **Distr** Prince Albert, WC, S Africa.

Note: Very similar to **R. perfoliata** in leaf shape, but differing in its decumbent, mat-forming growth.

R. bina ≡ **Antimima bina**

R. bipapillata L.Bolus NM 1933: 412 T L.Bolus 1944/32 (BOL!).

Erect shrubs with rigid branches to 22 cm *l*, **I** 20–30 mm *l*, ∅ 3 mm; **L** semi-terete, sheath with a line, epidermis cells smooth, margins and keel cartilaginous, 15–30 mm *l*, 2–5 mm *b*, 3 mm *t*; **Fl** in rich pyramidal cymes, pedicels 5–30 mm *l*, bracts below their middle, **K** 5, petals pinkish-purple, filamentous staminodes few, purple, basally white, filaments pink; **Fr** 5 locules, otherwise unknown; **Ecol** on a hill near a river; **Distr** Vanrhynsdorp, WC, S Africa.

Note: Fruits are lacking.

R. bolusiae Schwantes FR 1938: 228 ≡ *R. rigida* L.Bolus NM 1929: 108 non (Haw.) Schwantes ZSK 1927: 19 ≡ *M. albicute* N.E.Br. GC 1930: 32 T Pillans 6171 (BOL!).

Robust shrubs to 12 cm *h*, **I** ochre, to 7 mm *l*; **L** glaucous-green, epidermis almost smooth, 20–30 mm *l*, 4 mm *b*, 5 mm *t*; **Fl** in cymes, ∅ 20 mm, pedicels 4–8 mm *l*, bracteoles near their middle, to 4 mm *l*, **K** 5, petals pink, vittate and deeper coloured along margins, filamentous staminodes with long papillae, basally subglobose, white, apically cylindrical, purple, anthers and pollen golden; **Fr** 5 locules, otherwise unknown; **Ecol** unknown; **Distr** Clanwilliam, WC, S Africa.

Note: The features given for the flower resemble those of **Amphibolia**, but fruits are lacking to support a transfer.

R. bracteata ≡ **Antimima bracteata**

R. brakdamensis (L.Bolus) L.Bolus NM 1950: 220 ≡ *M. brakdamense* L.Bolus ABH 1922: 127 T Pillans 5757 (BOL!).

Shrubs with decumbent branches to 23 cm *l*, **I** maroon, later grey, ∅ 6 mm; **L** triquetrous, apiculate, green, epidermis rough from elevations, cells raised, keel and margins finely serrulate, keel sometimes with an additional tooth, to 7 mm *l*, 4 mm *b*, 5 mm *t*; **Fl** 1, ∅ 13 mm, pedicel 3 mm, petals pinkish-purple, to 5 mm *l*, filamentous staminodes present, filaments white, anthers yellow; **Fr** base funnel-shaped, top raised, rims low, covering membranes with closing ledges, closing body hook-shaped, expanding keels distant a their bases; **Ecol** unknown; **Distr** Namaqualand, NC, S Africa.

Note: The rather small leaves merge into a sheath which is connate basally with the stem, a feature often found in **Ruschia**.

R. breekpoortensis L.Bolus NM 1931: 284 T Herre 8814 (BOL!).

Shrubs, **I** as long as **L**, grey; **L** triquetrous, epidermis with low papillae, 7–12 mm *l*, 3–4 mm *b* and *t*; **Fl** 1–3, bracteoles near the base of the stalk, ∅ 15–20 mm, **K** 5, petals pink, filamentous staminodes and filaments pale pink; **Fr** unknown; **Ecol** < 100 mm rainfall p.a. in winter; **Distr** N Namaqualand, NC, S Africa.

Note: An insufficiently known species, which may belong to **Antimima**, as suggested by Dehn.

R. brevibracteata L.Bolus NM 1929: 144 T L.Bolus 18982 (BOL!).

Decumbent to ascending shrubs straggling in other bushes, when the opportunity is there, branches to 10 cm *l*, **I** ∅ to 7 mm; **L** semi-terete, 30–35 mm *l*, sheath 3–5 mm with a line, 4 mm *b* and *t*; **Fl** in rich cymes and storeys below the terminal one, clearly separated from the vegetative part of the plant, bracteoles 2–4 mm *l*, ∅ 13 mm, **K** 5, petals purplish-pink, inconspicuously vittate, 6–7 mm *l*, 1 mm *b*, filamentous staminodes few, richly papillate, filaments apically red, anthers and pollen white; **Fr** base funnel-shaped, top with low rims, covering membranes convex, closing rodlet protruding, closing body a small hook, expanding keels touching each other basally, 5 locules; **Ecol** unknown; **Distr** Namaqualand, NC, S Africa.

Note: The species is a member of the group with thickish, swollen leaf sheaths.

R. brevicarpa ≡ **Antimima brevicarpa**
R. brevicollis ≡ **Antimima brevicollis**

R. brevicyma L.Bolus NM 1932: 370 T Sauer 8159 (BOL!).

Erect shrubs to 15 cm *h*; **L** triquetrous to trigonous, epidermis rough from elevations resulting in a finely serrate keel and similar margins, to 20 mm *l*, 3 mm *b* and *t*; **Fl** in cymes, often only 2, pedicels to 8 mm *l*, **K** 5–6, petals pink, deeply vittate, 4–6 mm *l*, 1.5–2 mm *b*, filamentous staminodes few, apically purple, anthers white; **Fr** falling off the long stalks when ripe, 5 locules, otherwise unknown; **Ecol** unknown; **Distr** unknown.

Note: As far as features of leaves and flowers show, the species may also be a member of the genus **Antimima**.

R. brevifolia L.Bolus 1964: 40 **T** Herre 14838 (BOL!).

Erect branches to 31 cm *l*, **I** deep brown, black with age, 10–15 mm *l*; **L** subovate, epidermis with low papillae, longer ones in an area above the sheath on the upper leaf surface, 6–10 mm *l*, to 6 mm *b*, 3–4 mm *t*; **Fl** 1–3, pedicels 3–5 mm *l*, bracteoles basal, connate > 50%, **K** 5, petals purplish-pink, base pale, to 8 mm *l*, filamentous staminodes not mentioned, filaments pale, anthers, pollen yellow; **Fr** 5 locules, otherwise unknown; **Ecol** unknown; **Distr** Namaqualand, NC, S Africa.

Note: Data from the fruit are lacking.

R. brevipes L.Bolus NM 1929: 74 ≡ *M. firmum* N.E.Br. GC 1930: 32 **T** H.Bolus 11502 (BOL!) = *R. brevipes* L.Bolus var. *gracilis* L.Bolus NM 1929: 74 **T** Krige 10469 (BOL!).

Erect shrubs 30–45 cm *h*, **I** 5–20 mm *l*; **L** cymbiform, obtuse, 8–11 mm *l*, 6 mm *t*; **Fl** solitary, pedicel 1–3 mm *l*, ∅ 18 mm, **K** 5, petals flesh-coloured, filamentous staminodes present, anthers and pollen brown; **Fr** base funnel-shaped, top with high rims, covering membranes convex, closing ledges present, closing body hook-shaped, expanding keels almost touching each other basally, 5 locules; **Ecol** in rocky hills; **Distr** Prince Albert, WC, S Africa.

Note: The cymbiform leaves resemble **R. cymbifolia.**

R. brevipes var. *gracilis* = **R. brevipes**

R. britteniae L.Bolus NM 1929: 45 ≡ *M. iteratum* N.E.Br. GC 1930: 32 **T** Britten 5542 (BOL!).

L triquetrous, epidermis with low papillae, longer ones along margins, 10 mm *l*, 4 mm *t*; **Fl K** 6, stigmas 6; **Fr** unknown; **Ecol** at edge of bush veld; **Distr** Albany, EC, S Africa.

Note: The species has only been described in contrast to and comparison with **R. parvifolia.**

R. buchubergensis ≡ **Antimima buchubergensis**

R. burtoniae L.Bolus NM 1931: 279 (see **photo**) **T** Burton 19358 (BOL!).

Erect shrubs to 60 cm *h*, **I** brown to blackish, shining and smooth, the inner third of the diameter, i.e. the medulla, spongy, yellow or red to maroon, 10–30 mm *l*, up to 10 mm ∅; **L** subterete, basal sheath a little swollen, greyish-green, sometimes purplish suffused, epidermis smooth, the subsidiary cells a little elevated, wax cover consisting of threads mainly, 35–50 mm *l*, 2.5 mm *b*, 3–4 mm *t*; **Fl** in cymes with thin stalks, bracteoles 4–6 mm *l*, **Inf** therefore distinct from the vegetative part of the plant, pedicels 5–15 mm *l*, ∅ 19 mm, **K** 5, petals pink, filamentous staminodes subpetaloid, pale, apically purple; **Fr** light grey, base markedly bell-shaped, top with low rims, covering membranes convex, closing ledge distinct, closing body absent, the end of the placenta erect and distally separated from the wall imitating a closing body, expanding keels diverging, no valve wings, 5 locules; **Ecol** in loamy flats and in rocky places; **Distr** Calvinia, Namaqualand, NC, Vanrhynsdorp, WC, S Africa.

Note: Material from loamy flats of the low lying Knersvlakte and from rocky places on the high plateau is placed in this species because most character states agree. **R. burtoniae** is most similar to **R. muelleri,** which also possesses thick stems with a spongy pith.

The fruits of both species do not agree with those of **Ruschia** because of the absence of a closing body.

R. calcarea L.Bolus NM 1935: 495 **T** du Toit 21426 (BOL!).

Compact to caespitose with decumbent branches to 5 cm *l*; **L** triquetrous, obliquely truncate, apiculate and with 1–2 recurved, white, hard teeth along the keel, grey, pinkish suffused, 10–13 mm *l*, c. 3 mm *b* and *t*; **Fl** almost sessile, **K** 5, petals purplish-pink, 6–7 mm *l*, 0.5 mm *b*, filamentous staminodes present, filaments pink to pale; **Fr** 5 locules, otherwise unknown; **Ecol** unknown; **Distr** Hopetown, NC, S Africa.

Note: The name suggests that the plants grow on limestone.

R. calcicola (L.Bolus) L.Bolus NM 1950: 220 ≡ *M. calcicola* L.Bolus ABH 1925: 5 **LT** Muir 3097 (BOL!).

Low shrub with decumbent branches, **I** 15–30 mm *l*, white with age; **L** triquetrous, apiculate, keel serrulate, epidermis cells raised, greyish-green, 25–30 mm *l*, 3 mm *b* and *t*; **Fl** 1, enclosed by bracteoles, ∅ 18 mm, pedicels 5–18 mm *l*, bracteoles at 1/2 of the length, petals pinkish-purple, base pale, 8 mm *l*, c. 2 mm *b*, filamentous staminodes basally white, filaments apically purple; **Fr** base elongate funnel-shaped, top low, rims low, covering membranes with protruding closing rodlets, closing body hook-shaped, expanding keels widely separated at their bases, 5 locules; **Ecol** in limestone on hills near the sea, rain at all seasons; **Distr** Riversdale, WC, S Africa.

Note: The species appears to be confined to limestone, where it forms small mats.

R. callifera L.Bolus NM 1932: 335 **T** Slabbert 8626 (BOL!).

Plants 13 cm *h*, branches long; **L** trigonous, keel with 2–3 teeth, margins with 1 small tooth and up to 3 hard small protrusions, merging into a long sheath connate with the stem, glaucous with deep green dots, epidermis cells papillate, 10–15 mm *l*, 5 mm *b* and *t*; **Fl** enclosed in pedicels, to 10 mm *l*, **K** 5, keels of lobes with up to 6 small teeth, petals pink, to 8 mm *l*, 0.75 mm *b*, filamentous staminodes present, filaments pink, anthers golden; **Fr** 5 locules, otherwise unknown; **Ecol** unknown; **Distr** Prince Albert, WC, S Africa.

Note: The dentate leaves show clearly that the species belongs to **Ruschia**.

R. calycina ≡ **Zeuktophyllum calycinum**

R. campestris (Burchell) Schwantes ZSK 1927: 20 ≡ *M. campestre* Burch. 1822: 259 **T** Burchell 1340 (K!).

Shrubs to 45 cm *h*, branches stiff, I grey, young ones enclosed by the **L** in dense placement like tiles on a roof; **L** triquetrous with a recurved apical hook, dry epidermis with elevations and somewhat raised epidermal cells, 6–10 mm *l*, sheaths c. 5 mm *l*, 2.5–3.5 mm *b*, 2–2.5 mm *t*; **Fl** solitary, pedicels < **L**, **K** 5, petals pink; **Fr** top raised with high rims, base funnel-shaped, covering membranes convex, with low distal closing rodlets, closing body round, small, expanding keels well separated from each other basally, 5 locules; **Ecol** 100–200 mm rainfall p.a. mainly in March and November; **Distr** Sutherland, NC, S Africa.

Note: The description follows that of Brown (1920: 126–127) of the type sheet at K. The species has been reported by Burchell (1822: 260) to occur "now in bloom, every where", yet, it seems to have been collected rarely only.

R. canonotata (L.Bolus) Schwantes ZSK 1926: 187 (see **photo**) ≡ *M. canonotatum* L.Bolus ABH 1926: 40 **T** Wilman 18501 (BOL!) = *R. orsmondiae* L.Bolus NM 1934: 433 **T** Orsmond 20992 (BOL!).

Low shrubs to 20 cm *h*, younger ones rather compact, older with longer branches, often eaten and the single fruit or flowers stalks remaining as protruding, blunt spines on the plant; **L** triquetrous, with very low papillae, leaves of a pair basally separated by a sharp white triangle (best visible on long vegetative shoots, not clear on short shoots), 12–22 mm *l*, 2–3 mm *b*, 2–3 mm *t*; **Fl** solitary, bracteoles in the middle of the stalk, reaching the base of the **K** tube, **K** 5, petals pink to bright pink, filamentous staminodes white and pink, often pale; **Fr** top with low rims, base funnel-shaped, grey, covering membranes convex, distal closing ledge clear and prominent, closing body a small hook, 5 locules; **Ecol** in open shrubby vegetation, most often on surface limestone along pans, > 300 mm rainfall p.a. mostly in summer; **Distr** W districts of FS, E districts of NC, S Africa.

Note: Young plants or those in crevices of solid rocks are small and covered in flowers when in season (e.g. type of *R. orsmondiae*), in deeper soils and in cultivation the growth expands considerably, mainly by elongation of internodes (type of **R. canonotata**). The whitish triangle cited as distinct for the species are not visible at all nodes.

R. capornii (L.Bolus) L.Bolus NM 1958: 333 ≡ *M. capornii* L.Bolus ABH 1916: 30 ≡ *Lampranthus capornii* (L.Bolus) L.Bolus NM 1939: 170 **T** Caporn 822/15 (BOL!) = *M. capornii* L.Bolus var. *longifolium* L.Bolus NM 1933: 405 **T** L.Bolus 1947/32 (BOL) = *M. capornii* L.Bolus forma *ferum* L.Bolus NM 1933: 405 **T** L.Bolus BOL 20746 (BOL).

Erect shrubs; **L** trigonous, falling off easily, epidermis glaucous, 4–10 mm *l*, 4 mm *b* and *t*; **Fl** in moderate lateral cymes, ∅ 9 mm, petals pinkish-purple, filamentous staminodes not mentioned; **Fr** unknown; **Ecol** unknown; **Distr** Namaqualand, NC, S Africa.

Note: Bolus (1958: 333) gave further data on the fruits taken from material collected later and assigned to this species.

R. capulata ≡ **R. cupulata**

R. caroli (L.Bolus) Schwantes ZSK 1927: 20 ≡ *M. caroli* L.Bolus ABH 1922: 129 **LT** Melle 3418/14 (BOL!).

Shrub with decumbent to ascending branches, I grey to maroon, 10–30 mm *l*, ∅ 5 mm; **L** trigonous, sheath tumid, glaucous-green with green dots, epidermis smooth, wax plates?, to 100 mm *l*, 4–5 mm *b*, 5–6 mm *t*; **Fl** in cymes, pedicels 15–25 mm *l*, bracteoles present, petals pink, vittate, 5–10 mm *l*, 1.5–2 mm *b*, filamentous staminodes pinkish-purple, anthers yellow; **Fr** unknown; **Ecol** unknown; **Distr** Clanwilliam, Montagu, Robertson, WC, S Africa.

Note: Although the description reports decumbent to ascending branches, the name has been used widely for shrubs with erect branches.

R. caudata L.Bolus NM 1938: 129 **T** Burton 21844 (BOL!).

Erect shrubs, 25–40 cm *h*, **I** maroon, 2–5 mm *l*, 5 mm ∅; **L** semi-terete, separated basally by a line along the basally part connate with the stem, epidermis cells somewhat raised; **Fl** in rich cymes, storeys below the terminal one, bracteoles at 1/2 the length of the pedicel, **K** 5, petals pinkish-purple, 5–9 mm *l*, 0.5–0.75 mm *b*, filamentous staminodes few, filaments white, anthers and pollen white,; **Fr** 5–8 locules, otherwise unknown; **Ecol** unknown; **Distr** Malmesbury, WC, S Africa.

Note: The species appears remarkable in the variation of 5–8 locules, a feature not known from other species in **Ruschia**.

R. cedarbergensis L.Bolus NM 1935: 496 **T** Leighton 21441 (BOL!).

Shrubs with erect branches to 25 cm *l*, **I** 10–30 mm *l*, ∅ to 5 mm; **L** of a pair almost in U-shape, separated at the sheath by a line of 4–6 mm, semi-terete, epidermis cells a little raised, 15–35 mm *l*, 4 mm *b* and *t*; **Fl** in rich cymes, all opening at ± the same time, pedicels 7–12 mm *l*, bracteoles near the middle, petals pinkish-purple, inconspicuously vittate, filamentous staminodes purple in the upper half, tips paler and recurved, filaments white, apically purple, anthers and pollen straw-coloured; **Fr** top low, rims lowest in the centre, base funnel-shaped, valve opening rather widely, covering membranes with distinct closing rodlets, closing body small, rodlet-shaped, expanding keels short, diverging widely, 5 locules; **Ecol** unknown; **Distr** Clanwilliam, WC, S Africa.

Note: Reported to resemble **R. incumbens.**

R. centrocapsula H.E.K.Hartmann & Stüber 1993: 62 (see **photo**) subg. **Spinosae T** Hartmann et al. 25672 (HBG!).

Flat shrubs to 30 cm high; **I** whitish-grey; **L** longish, 15–20 mm *l*, epidermis with convex to papillate cells and elevations above tannin idioblasts; **Fl** scattered in very spinescent **Inf** on lower orders of the dichasium, ∅ 20 mm, petals c. 30, 10–15 filamentous staminodes; **Fr** with narrow valve wings; **S** 0.8–0.9 mm *l*; **Chr** $2n = 36$; **Ecol** on rocky quartzitic slopes, rainfall < 100 mm p.a., mainly in winter; **Distr** N Namaqualand, NC, S Africa.

Note: Spines dominate the inflorescence, flowers are scarce, but the low cushion shape makes the species attractive for cultivation.

R. ceresiana L.Bolus NM 1938: 137 ≡ *R. congesta* NM 1937: 59 non (Salm-Dyck) L.Bolus SAG 1927: 239 nomen illeg. ≡ *R. ceresiana* Schwantes FR 1938: 229 nomen invalid. **T** Nel 8585 (BOL!).

Branches 14–25 cm *l*, **I** 20–25 mm *l*, ∅ to 5 mm; **L** nearly semi-terete, sheath a little swollen, 5 mm *l*, the leaves completely separated by a deep line along the sheath, 30–60 mm *l*, c. 5 mm *t*; **Fl** in rich terminal cymes, pedicels to 30 mm *l*, bracteoles basal, at most 10 mm *l*, thus distinctly different from the foliage leaves, **K** 5, petals in 5 groups, purplish-pink, inconspicuously vittate, to 8 mm *l*, c. 1 mm *b*, filamentous staminodes few to none, filaments pale pink, apically purplish-pink, anthers and pollen white to pale brown; **Fr** short funnel-shaped, top low, rims moderately high, covering membranes with protruding closing rodlets, closing body a moderately broad hood, expanding keels not touching each other, 5 locules; **Ecol** unknown; **Distr** Ceres, WC, S Africa.

Note: A little known species.

R. cincta (L.Bolus) L.Bolus NM 1950: 219 ≡ *M. cinctum* L.Bolus ABH 1922: 133 **LT** Pillans IV 2016 ≡ 493/16 (BOL!).

Decumbent branches to 20 cm *l*, **I** 10–20 mm *l*; **L** triquetrous, mucro recurved, keel and margins minutely serrate, glaucous-green, 20 mm *l*, 5 mm *b*, 6 mm *t*; **Fl** mostly solitary, ∅ 20–22 mm, pedicels 7–12 mm *l*, **K** 5, petals pink, margins purple, 9–10 mm *l*, 2 mm *b*, filamentous staminodes and filaments pinkish-purple, anthers pale; **Fr** base funnel-shaped, top raised, covering membranes with protruding closing rodlets, closing body a small hook, expanding keel not touching each other basally, 5 locules; **Ecol** unknown; **Distr** Riversdale, WC, S Africa.

Note: A sheet with a capsule on it has been chosen as the lectotype.

R. clavata L.Bolus NM 1929: 158 **T** Ross Frames 19006 (BOL!).

Diffusely, loosely branched shrubs up to 20 cm *h*, **I** ash-grey; **L** trigonous, epidermis with low papillae, up to 5 mm *l*, c. 3 mm *b* and ∅; **Fr** unknown; **Ecol** < 100 mm rainfall p.a. in winter; **Distr** Vanrhynsdorp, WC, S Africa.

Note: The species should perhaps be placed in **Antimima**, as suggested by Dehn.

R. clavellata ≡ **Disphyma clavellatum**
R. cleista = **Eberlanzia sedoides**
R. compacta ≡ **Antimima compacta**

R. complanata L.Bolus NM 1932: 376 **T** James 472 (BOL!).

Erect shrubs to 20 cm *h*, **I** 5–10 mm *l*; **L** trigonous to semi-terete, dirty green, epidermis somewhat rough from elevations, cells a little raised, mucro recurved, sheath basally connate with the stem, 4–6 mm *l*, 2 mm *b* and *t*; **Fl** 1, pedicels 10–20 mm *l*, **K** 5, petals white, c. 4 mm *l*, 0.25 mm *b*, filamentous staminodes glanduliferous (not visible in type), filaments white; **Fr** base elongate funnel-shaped, top almost semi-globose, rims low, covering membranes with protruding closing rodlets, closing body a small hook, expanding keels almost touching each other, 5 locules; **Ecol** unknown; **Distr** Cradock, EC, S Africa.

Note: The connate leaf sheaths merging into the stem represent a widespread feature in **Ruschia.**

R. compressa ≡ **Antimima compressa**
R. comptonii ≡ **Drosanthemum roseatum**
R. concava = **Antimima dasyphylla**
R. concinna ≡ **Antimima concinna**
R. condensa ≡ **Antimima condensa**

R. congesta (Salm-Dyck) L.Bolus SAG 1927: 239 non L.Bolus 1937: 59 ≡ *M. congestum* Salm-Dyck 1854: § 23 f 2 **T** Salm-Dyck 1854: § 23 f 2.

Shrubs with erect, compressed branches; **L** triquetrous, almost falcate, tips later recurved, glaucous, finely papillate or ciliate, 16–27 mm *l*, 4.5 mm *b*, 6.8–9 mm *t*; **Fl** solitary (in cultivation), pedicels c. 50 mm *l*, bracteoles in their middle, petals deep purple, filamentous staminodes pink, filaments white, anthers yellow; **Fr** 5 locules, obconical, otherwise unknown; **Ecol** unknown; **Distr** unknown.

Note: Described by Salm-Dyck from plants grown from seeds received from Hamburg, the origin is therefore unknown. The drawing resembles figures of **R. cymbifolia**, as already indicated by Salm-Dyck himself.

R. conjuncta = **Octopoma connatum**
R. connata ≡ **Octopoma connatum**
R. constricta = **Acrodon bellidiflorus**

R. copiosa L.Bolus NM 1939: 161 **T** L.Bolus 22112 (BOL!).

Erect shrubs 30–50 cm *h*, richly branched, **I** maroon 20–50 mm *l*, 5 mm ∅; **L** semi-terete, sheath basally a little inflated or swollen (tumid), epidermis smooth, 30–40 mm *l*, 3–4 mm *t*; **Fl** in rich terminal cymes with additional lower storeys, **K** 5, petals pinkish-purple, to 6 mm *l*, to 1 mm *b*, filamentous staminodes present; **Fr** 5 locules, otherwise unknown; **Ecol** unknown; **Distr** Clanwilliam, WC, S Africa.

Note: The tumid, swollen leaf sheaths suggest that the species belongs with the Tumidula-group of **Ruschia**.

R. coriaria ≡ **Psilocaulon coriarium**

R. costata L.Bolus NM 1937: 60 **T** Verwoerd 8053 (BOL!).

Plants with long, decumbent branches, **I** 25–35 mm *l*, ∅ 2–5 mm; **L** semi-terete, young ones with tumid sheaths, subrecurved, glaucous-green, epidermis smooth, 25–35(–60) mm *l*, to 4 mm *b*, to 5 mm *t*; **Fl** in cymes, pedicels 30–50 mm *l*, bracteoles 15–25 or 6–9 mm *l* (the ultimate ones), **K** 5, petals purplish-pink, 8–9 mm *l*, to 2.5 mm *b*, filamentous staminodes present, filaments apically purple, anthers, pollen, white; **Fr** top as a low turret, base elongate funnel-shaped, covering membranes with long, protruding closing rodlets, closing body a small hook, expanding keels almost touching each other; **Ecol** unknown; **Distr** Montagu, WC, S Africa.

Note: Described from type and description.

R. cradockensis (O.Kuntze) H.E.K.Hartmann & Stüber 1993: 63 subg. **Spinosae** ≡ *M. cradockense* O.Kuntze 1898: 109 ≡ *Eberlanzia cradockensis* (O.Kuntze) Schwantes ZSK 1926: 189 ≡ *M. spinosum* L. var. *cradockense* (O.Kuntze) A.Berger ex Range 1934: 19 nomen nudum **T** Kuntze s.n. (K!).

Erect shrubs to 30 cm *h*, long shoots dominating; **I** dark coppery, brown or grey; **L** –10 mm *l*, epidermis with convex cells, but neither papillate nor elevated, stomata superficial; **Fl** few per **Inf**, these mostly on short lateral shoots off the long shoots, ∅ 7–14 mm, petals < 30, 14–27 filamentous staminodes, 30–60 stamens and short stigmas forming a tulip-shaped centre; **Fr** without valve wings; **S** 0.6–0.8 mm *l*; **Ecol** in shrubby associations with rainfall below 400 mm p.a. **Distr** Namaqualand, Calvinia, NC, to Ceres, WC, and towards the east to Cradock, EC, S Africa.

Note: In habitat, plants of this species often have a marked blackish appearance, due to the dark internodes and dark leaves.

Key to the Subspecies of Ruschia cradockensis

1. Numerous spiny lateral cymes, leaf length > 2 × width, chromosomes $2n = 18$ .... **R. cradockensis** ssp. **cradockensis**
- Some to few spiny lateral cymes on long shoots, leaf length = width, chromosomes $2n = 36$ .... **R. cradockensis** ssp. **triticiformis**

R. cradockensis (O.Kuntze) H.E.K.Hartmann & Stüber ssp. **cradockensis** (see **photo**) = *R. horrescens* L.Bolus NM 1930: 204 ≡ *Eberlanzia horrescens* (L.Bolus) L.Bolus NM 1958: 387 **T** van der Merwe 74 (BOL!) = *R. horrescens* (L.Bolus) L.Bolus var. *densa* L.Bolus NM 1930: 204 ≡ *Eberlanzia horrescens* (L.Bolus) L.Bolus var. *densa* (L.Bolus) Jacobsen 1960: 1132 nomen invalid. **T?** = *R. horrida* L.Bolus NM 1930: 204 ≡ *Eberlanzia horrida* (L.Bolus) L.Bolus NM 1958: 387 **T** Dyer 2132 (BOL!).

L length c. 2 × width; spiny **Inf** numerous, but little branched, enveloping the long shoot conspicuously; **Fl** with an orange holonectary; **Chr** $2n = 18$; **Ecol** on loamy soils, often decomposed Ecca shales, 100–400 mm p.a. in March and November; **Distr** from Ceres, WC, to Cradock, EC, rarely outside this area, S Africa.

R. cradockensis (O.Kuntze) H.E.K.Hartmann & Stüber ssp. **triticiformis** (L.Bolus) H.E.K.Hartmann & Stüber 1993: 64 (see **photo**) ≡ *M. triticiforme* L.Bolus ABH 1925: 4 ≡ *Eberlanzia triticiformis* (L. Bolus) L.Bolus NM 1958: 387 **T** Juritz 119 (BOL!) = *Eberlanzia triticiformis* (L.Bolus) L.Bolus var. *subglobosa* L.Bolus JSAB 1959: 372 **T** du Plessis 26803 (BOL).

L length equals ± width; spiny **Inf** in moderate numbers, little branched; **Fl** with a dark green holonectary; **Chr** $2n = 36$; **Ecol** on rubble or pebbles of Dwyka diamictite along the E edge of the winter rainfall region, 100–200 mm rainfall p.a.; **Distr** Calvinia, Namaqualand, NC, S Africa.

R. crassa (L.Bolus) Schwantes ZSK 1926: 187 ≡ *M. crassum* L.Bolus ABH 1926: 38 **T** from the Municipal Garden, Cape Town, 17172 (BOL!).

Robust branches ascending, I to 9 mm Ø; L free parts 16–29 mm *l*, sheaths 10–20 mm *l*, basally connate with the stem, epidermis with long papillae, keel inconspicuously 1-dentate; **Fl** 1–2, Ø 22 mm, pedicel 5–10 mm *l*, petals white, 7 mm *l*, 1 mm *b*, filamentous staminodes present, filaments white, anthers pale, pollen yellow; **Fr** 5 locules, otherwise unknown; **Ecol** unknown; **Distr** unknown, perhaps from Laingsburg, WC, S Africa.

Note: The long papillae characterize this species well in the group with long, green leaf sheaths.

R. crassifolia ≡ **Antimima crassifolia**

R. crassisepala L.Bolus NM 1932: 334 T Mathews 2078/31 (BOL!) = *R. crassisepala* L.Bolus var. *major* L.Bolus NM 1954: 285 T Hall 25006 (BOL!).

Loosely branched shrubs, denser near the base, I reddish, later brown; L trigonous, 15–30 mm *l*, c. 4 mm *b* and *t*, epidermis smooth; **Fl** in ternate terminal cymes, rarely 5 flowers, but often additional flowers below the terminal inflorescence, those mostly as a pair, one to each side, Ø c. 17 mm, **K** 5, petals purplish-pink, basally vittate, filaments apically purplish-pink, bases white; **Fr** in elongated stands, bracteoles at or below the middle of the stalks, top raised with a truncate central part, thus turret-shaped, base elongate funnel-shaped, covering membranes rising convexly from a deeper centre, closing rodlet big, closing body a hollow hook, expanding keels short, lifting the valves into a vertical position only, Ø 4–9 mm; **Ecol** on slopes of quartzite, < 100 mm rainfall p.a. in winter; **Distr** Vanrhynsdorp, WC, S Africa.

Note: Rather big composed inflorescences characterize this species, which clearly belongs to **Ruschia** because the diagnostic features of the capsule-type are present on the type sheet. Remarkable is the reported absence of filamentous staminodes, also in the variety.

R. crassisepala var. *major* = **R. crassisepala**
R. crassuloides = **Eberlanzia sedoides**

R. cupulata (L.Bolus) Schwantes ZSK 1927: 106 ≡ *M. cupulatum* L.Bolus ABH 1927: 91 ≡ *R. capulata* (L.Bolus) Schwantes in Jacobsen 1949: 58 orthographic variant T Mathews 1720/24 (BOL!).

Shrubs to 20 cm *h*, I grey, 15–30 mm *l*, 6 mm Ø; L trigonous, keel finely papillate, yellow-green, sheath tumid, 3–5 mm *l*, line inconspicuous, 15–30 mm *l*, 3–4 mm *b*; **Fl** in cymes, bracteoles cupuliform, 69 mm *l*, **Inf** distinct from the basal vegetative part of the plant, **K** 5–6, petals 10–12 mm *l*, 1.5 mm *b*, filamentous staminodes few, brown, filaments white; **Fr** base funnel-shaped, top almost turret-shaped, rims very low, covering membranes with closing ledges, closing body a hook, expanding keels separated basally from each other, 5 locules; **Ecol** near the sea; **Distr** Vredenburg, WC, S Africa.

Note: Reported to be similar to **R. festiva**, which differs by its ivory-white flowers most prominently.

R. curta (Haw.) Schwantes FR 1938: 230 ≡ *M. curtum* Haw. 1819: 334 ≡ *M. vaginatum* Haw. var. *curtum* (Haw.) A.Berger 1908: 111 **T**?

Erect shrubs; L placed rather close together, spreading, sheaths short, often longer; **Fl** probably white; **Fr** unknown; **Ecol** unknown; **Distr** unknown.

Note: This species differs from **R. vaginata** mainly by its thicker leaves and the short leaf sheaths; the leaves are also not recurved so strongly as in **R. vaginata**. The drawing by Duncanson 1824 named *M. curtum* (K!) differs in several important points from the description: the leaf tips are recurved to reflexed, the sheaths are as long as the free parts and have a line running down them for their entire length. It can therefore not be considered to represent the type material, and the identity of the species remains doubtful.

R. cyathiformis ≡ **Eberlanzia cyathiformis**

R. cymbifolia (Haw.) L.Bolus FPSA 1927: t. 277 ≡ *M. cymbifolium* Haw. 1824: 424 ≡ *M. deceptum* N.E.Br. GC 1930: 32 nomen illeg. **LT** drawing 944/212 by Duncanson based on material from Bowie, 304 of 1822 (K!).

Suberect shrubs to c. 30 cm *h*, branches slender and not very firm; L almost boat-shaped, apically with a recurved white mucro, c. 13.5 mm *l*; **Fl** solitary or in composed cymes, petals purplish-pink, filamentous staminodes numerous; **Fr** typical of the genus; **Ecol** unknown; **Distr** Riversdale?, WC, S Africa.

Note: The description follows the texts of Haworth and Bolus.

R. cymosa (L.Bolus) Schwantes ZSK 1927: 20 ≡ *M. cymosum* L.Bolus ABH 1923: 170 **LT** Page 17168 (BOL!).

Shrubs 20–40 cm *h*, I 25–40 mm *l*, Ø to 4 mm; L semi-terete, acuminate, glaucous-green, epidermis smooth, 20–45 mm *l*, 4 mm *b* and *t*; **Fl** in rich cymes with additional storeys, well separated from the vegetative part of the plant by the possession of small bracteoles of up to 10 mm length, pedicels 10–20 mm *l*, petals purplish-pink, 8 mm *l*, 1–1.5 mm *b*, filamentous staminodes not mentioned, filaments pale pink, anthers pale, pollen yellow; **Fr** 5 locules, otherwise unknown; **Ecol** unknown; **Distr** Montagu, WC, S Africa.

Note: The species belong with the group with short, swollen sheaths at the bases of the leaf-pairs and a distinct inflorescence.

R. dasyphylla ≡ **Antimima dasyphylla**

R. decumbens L.Bolus ABH 1928: 126 ≡ *M. deflectum* N.E.Br. GC 1930: 32 T Archer 296 (BOL!).

Decumbent branches adorned with persisting, dry remains of old leaves for a long time; **L** trigonous to semi-terete, glaucous to purple, rough from elevations, sheaths 3–4 mm *l*, 10–18 mm *l*, 6 mm *b* and *t*; **Fl** solitary, ∅ 35 mm, pedicels 8 mm *l*, **K** 5, petals purplish-pink, vittate, 14 mm *l*, 2 mm *b*, filamentous staminodes and filaments pink; **Fr** hard, dark brown, base bell-shaped, top very high, but rims low, covering membranes with distal closing ledges, closing body a small hood, expanding keels distant from each other at their bases, 5 locules; **Ecol** unknown; **Distr** Ladismith, WC, S Africa.

Note: The hard, dark brown fruits look unusual for the genus.

R. decurrens L.Bolus NM 1929: 78 ≡ *M. decurrens* (L.Bolus) N.E.Br. GC 1930: 32 **T** Frames 196a/28 (BOL!).

Erect shrubs, branches to 13 cm *l*, **I** darkish purple to maroon, 10–20 mm *l*; **L** bases separated by an elongate triangle of stem tissue, persisting on the stem after the leaves have fallen off, trigonous, epidermis cells dome-shaped to papillate, to 18 mm *l*, 4 mm *b*, 8 mm *t*; **Fl** in ternate cymes, rarely solitary, ∅ 16 mm, **K** 5, petals 7 mm *l*, 1 mm *b*, filamentous staminodes white, apically purple, filaments white, apically purple; **Fr** 5 locules, otherwise not known; **Ecol** unknown; **Distr** Clanwilliam, Vanrhynsdorp, WC, S Africa.

Note: The type material has been taken from "rooted cuttings".

R. decurvans L.Bolus NM 1930: 211 T Godman 19244 (BOL!).

Erect shrubs to 45 cm *h*, **I** dark purple, older ones ochre, **I** 30–40 mm *l*; **L** almost semi-terete, sheath tumid, 3–7 mm *l*, free parts 4–65 mm *l*, to 5 mm *b* and *t*; **Fl** in rich pyramidal cymes raised above the plant, composed of a terminal pure dichasium and additional ones in up to three storeys, bracteoles small, **Inf** thus very distinctly separated from the vegetative part of the plant, ∅ 20–26 mm, petals pinkish-purple, inconspicuously vittate, 10 mm *l*, to 1.5 mm *b*, filamentous staminodes few, filaments pinkish-purple; **Fr** unknown; **Ecol** unknown; **Distr** Clanwilliam, Vanrhynsdorp, Vredendal?, WC, S Africa.

Note: The placement in **Ruschia** cannot be confirmed without fruiting material.

R. defecta ≡ **Antimima defecta**
R. deflecta ≡ **Antimima defecta**

R. dejagerae L.Bolus NM 1928: 139 ≡ *M. dejagerae* (L.Bolus) N.E.Br. GC 1930: 32 **T** de Jager 18568 (BOL!).

Compact, to 7 cm *h*, **I** reddish-brown, **I** ∅ 2–3 mm; **L** basal sheath merging into the stem, free parts glaucous-green, minutely papillate, 11 mm *l*, 3 mm *b* and *t*; **Fl** solitary, ∅ 17 mm, pedicels to 10 mm *l*, **K** 5, petals in 5 groups, pink, filamentous staminodes and filaments pink; **Fr** 5 locules, with valve wings, according to the description, but no fruit available on the type material, 5 locules; **Ecol** unknown; **Distr** Beaufort West, WC, S Africa.

Note: The reported presence of narrow valve wings suggests that the species may belong to **Antimima.**

R. dekenahi ≡ **Antimima dekenahi**

R. deminuta L.Bolus 1961: 260 **T** Hall 1847 (BOL!).

Dense cushions to 5 cm *h*; **L** trigonous with convex sides, keel, margins and upper parts of sides with high elevations: muricate, keel and margins somewhat cartilaginous, epidermis with low papillae, 3–5 mm *l*, 3–4 mm *b* and *t*; **Fl** bracteoles near the base, reaching well up to the base of the **K**, petals very pale pink, c. 6 mm *l*, 0.5 mm *b*, filamentous staminodes in a cone; **Fr** base broadly funnel-shaped, somewhat convex, top with rather low rims, covering membranes convex, with prominent broad closing rodlets at their distal end, closing bodies small, round, expanding keels well separated at their bases, no awns or valve wings, 5 locules; **Ecol** on the plateau of the escarpment, < 100 mm rainfall p.a. in March; **Distr** Lüderitz-South, Namibia.

Note: The possession of elevations resembling warts indicates a relationship to **R. muricata**, **R. primosii**, and **R. scabra**, from further south.

R. densiflora L.Bolus 1966: 202 **T** Hall 3052 (BOL!).

Shrubs to 45 cm *h*, ∅ to 120 cm, **I** 40–50 mm *l*, 4–5 mm *b*; **L** semi-terete, epidermis cells raised, 25–42 mm *l*, sheaths 4–5 mm *l*; **Fl** in dense cymes, pedicels to 10 mm *l*, bracteoles c. 4 mm *l*, at c. half the length of the pedicels, **Inf** thus clearly distinct from the vegetative part of the plant, petals brightly lilac-coloured, to 6 mm *l*, 0.75 mm *b*, filamentous staminodes absent; **Fr** base funnel-shaped, top wide and high, rims high, covering membranes with protruding closing rodlets, closing body a small hook, expanding keels not meeting each other at their bases, 5 locules; **Ecol** in coastal flats; **Distr** Clanwilliam, WC, S Africa.

Note: The type is reported to come from a population of 6–10 plants.

R. depressa L.Bolus NM 1932: 336 **T** James 28 (BOL!).

Decumbent branches to 13 cm *l*, **I** 3 mm ∅; **L** trigonous, keel with 1–2 teeth, epidermis cells raised, keel finely papillate, to 8 mm *l*, 3–4 mm *b* and *t*; **Fl** solitary, petals pale mauve (pink), to 7 mm *l*, 0.5 mm *b*, filamentous staminodes pale pink, outer filaments pink, inner white; **Fr** base funnel-shaped, top raised, rims moderately high, covering membranes convex with closing ledges at their distal ends, closing body a small hood, expanding keels distant from each other basally, 5 locules; **Ecol** unknown; **Distr** probably Uitenhage, EC, S Africa.

Note: Very similar to **R. nonimpressa.**

R. derenbergiana ≡ **Ebracteola derenbergiana**
R. dichotoma ≡ **Eberlanzia dichotoma**

R. dichroa (Rolfe) L.Bolus NM 1929: 103 ≡ *M. dichroum* Rolfe 1920: t. 8872 ≡ *Cylindrophyllum dichroum* (Rolfe) Schwantes ZSK 1927: 28 **T** Rolfe 1920: t. 8872 = *R. dichroa* (Rolfe) L.Bolus var. *alba* L.Bolus NM 1929: 103 **T** Compton 1508/26 (BOL!).

Suberect shrubs with mostly simple branches, 8–9 cm *l*; **L** triquetrous, glaucous, finely punctulate, 40–60 mm *l*, 6–9 mm *b*; **Fl** subsessile, ∅ 35–40 mm, petals white, apically bright pink, filamentous staminodes not mentioned, filaments whitish, anthers yellow; **Fr** 6 locules, otherwise unknown; **Ecol** unknown; **Distr** unknown.

Note: The species was described based on material sent to Kew by Mr. Simpson Hayward, thus no wild origin is known. At first placed in the group Calamiformia by Rolfe, a transfer to the genus **Cylindrophyllum** was undertaken by Schwantes later. In the absence of fruiting material and the lack of knowledge at this point, the placement of the species in **Ruschia** remains uncertain.

R. dichroa var. *alba* = **R. dichroa**

R. dielsiana (A.Berger) H.Jacobsen ex H.E.K.Hartmann Bradleya 1998: 84 ≡ *M. dielsianum* A.Berger 1922: 639 ≡ *R. dielsiana* (A.Berger) H.Jacobsen 1960: 1272, 1378 nomen invalid. **T** Diels 4365 (B!).

Branches lying on the ground, **I** grey, 30–50 mm *l*, ∅ 3 mm; **L** triquetrous, very acute towards the tips, epidermis smooth, but with distant, rounded elevations on keel and margins, 30–40 m *l*, c. 3 mm *b*; **Fl** in small cymes or solitary, but often grouped together in narrow aggregates, pedicels 20–35 mm *l*, upper bracteoles deltoid, **K** 5, petals longer than the **K** lobes, purple, stamens red; **Fr** base funnel-shaped, top round, covering membranes convex with protruding closing rodlets, closing body a narrow, small hook, expanding keels dark brown, valve wings absent, **Ecol** in sandy flats above granite, 25 m altitude; **Distr** Albany, Western Australia, Australia.

Note: The species has been placed in **Ruschia** by Jacobsen in 1960 already, but he did not give the full and direct reference, as required. Althought the type is reported to come from Australia, the species does not key out nor is it mentioned in the Flora of Australia (Prescott and Venning in George 1984: 19–61).

R. dilatata L.Bolus NM 1931: 240 **T** Herre 8490 (BOL!).

Shrubs with erect branches, probably ending in **Inf** and being replaced by new ones developing in the next season below the **Inf**, **I** > **L**, ochre; **L** triquetrous, keel and margins white, sharp, with very few papillae, epidermis otherwise smooth, up to 10 mm *l*, 2–3 mm *b* and *t*; **Fl** in terminal cymes, 3–5 flowers together, **K** 5, ∅ c. 20 mm, petals pink, filamentous staminodes white, filaments of the palest pink; **Fr** unknown; **Ecol** < 100 mm rainfall p.a. in winter; **Distr** Vanrhynsdorp, WC, S Africa.

Note: An insufficiently known species. The regeneration of erect branches from below the old remains of cymes suggests a placement in **Antimima**, as indicated by Dehn on the sheet.

R. disarticulata = **Antimima hantamensis**
R. dissimilis ≡ **Lampranthus dissimilis**
R. distans ≡ **Antimima distans**
R. diutina = **Polymita albiflora**

R. divaricata L.Bolus NM 1930: 202 (see **photo**) subg. **Spinosae** ≡ *Eberlanzia divaricata* (L.Bolus) L.Bolus NM 1958: 387 **T** Fuller 28 (BOL!) = *M. spinescens* Pax in Engler 1910: 540 nomen nudum = *Eberlanzia tatasbergensis* L.Bolus NM 1958: 326 **T** Herre SUG 9261 (BOL!) = *R. vulnerans* L.Bolus NM 1935: 504 ≡ *Eberlanzia vulnerans* (L.Bolus) L.Bolus NM 1958: 387 **T** du Toit 1723/34 (BOL!).

Shrubs to 35 cm *h*, often with fusiform thickened roots; **I** whitish-beige; **L** 10–15 mm *l*, epidermis cells convex to papillate, with elevations above the tannin idioblasts; **Fl** in rich terminal **Inf** with 7–64 spines, on terminal branches of high order in the cyme, ∅ 9–18 mm, 25–31 petals, 13–30 filamentous staminodes, 44–65 stamens and short (2–3 mm) aggregate stigmas; **Fr** with narrow valve wings; **S** 0.6–0.8 mm *l*; **Chr** $2n$=18; **Ecol** on sandy flats with limestone pebbles, sometimes in gneissic or granitic sand, with grass, rainfall 100–400 mm p.a., mostly in March and November, but also in winter; **Distr** Lüderitz-South, Warmbad, Namibia; FS, inland parts of NC, S Africa.

Note: Often found near **R. intricata** in the east and **R. spinosa** in the west.

R. diversifolia L.Bolus NM 1929: 103 ≡ *M. variifolium* N.E.Br. GC 1930: 32 **LT** Pillans 6079 I (BOL!).

Shrubs to 14 cm *h*, ∅ to 50 cm in cultivation, decumbent branches to 25 cm *l*, robust, **I** reddish when young, later grey, 30–40 mm *l*, ∅ 4 mm; **L** triquetrous, margins cartilaginous-serrulate, epidermis smooth, to 70 mm *l* on long shoots, shorter ones on short shoots, sheaths to 5 mm *l*, 12 mm *t*; **Fl** in cymes, ∅ 28 mm, pedicels to 20 mm *l*, bracteoles 6–9 mm, falcate, clearly distinct from foliage leaves, petals pinkish-purple, 13 mm *l*, 2 mm *b*, filamentous staminodes and filaments pale, apically deep pink, anthers white, pollen golden; **Fr** unknown; **Ecol** unknown; **Distr** Worcester, WC, S Africa.

Note: The diversity in leaf sizes on long and short shoots and during a season is a widespread feature not related to true heterophylly.

R. dolomitica ≡ **Antimima dolomitica**
R. drepanophylla ≡ **Esterhuysenia drepanophylla**
R. drepanophylla var. *sneeubergensis* ≡ **Esterhuysenia drepanophylla**
R. dualis = **Antimima dualis**
R. dubitans ≡ **Lampranthus dubitans**
R. duplessiae = **Acrodon bellidiflorus**

R. duthiae (L.Bolus) Schwantes in Jacobsen 1949: 58 ≡ *M. duthiae* L.Bolus ABH 1923: 162 **T** Duthie 171/15 (BOL!).

Creeping branches with erect short flowering branches 7–10 cm *h*, **I** grey to ochre, 20–35 mm *l*; **L** trigonous, epidermis smooth, green, 20–80 mm *l*, 7 mm *b*, 8 mm *t*; **Fl** solitary or in cymes, ∅ 24 mm, bracteoles 20–30 mm *l*, petals pale, apically pinkish-purple, vittate, 8 mm *l*, 2 mm *b*; filamentous staminodes and filaments white, apically purple, anthers white; **Fr** 5 locules, otherwise unknown; **Ecol** unknown; **Distr** Knysna, WC, S Africa.

Note: An insufficiently known species.

R. dyeri ≡ **Rhombophyllum albanense**
R. ebracteata ≡ **Eberlanzia ebracteata**

R. edentula (Haw.) L.Bolus NM 1950: 221 ≡ *M. edentulum* Haw. 1821: 125 ≡ *Echinus edentulus* (Haw.) N.E.Br. KB 1929: 57 ≡ *Braunsia edentula* (Haw.) N.E.Br. in Brown, Tischer & Karsten 1931: 126 **LT** Burman 1738: t. 26 f 3.

Shrubs to 30 cm *h*, old stems shrivelling and drying; **L** triquetrous, keel with a recurved hook and sometimes an additional tooth, thus often appearing truncate, sheath c. as long as the free parts of the leaves; **Fl** solitary, petals red; **Fr** base funnel-shaped with concave sides, top with very high rims, 5 locules; **Ecol** unknown; **Distr** unknown.

Note: Haworth established the species based on Burman 1738: t. 26 f 3, placing it in his § Uncinata. He said in a note that it does not answer the "preceding article", i.e. that on *M. perfoliatum*. Since the figure of Burman shows a member of the genus **Ruschia**, the transfer of Bolus (NM 1950: 221) must be followed. When Brown in his description, combining the name with those of **Braunsia acuminata** and **Braunsia bina**, gave the leaf surface as papillate, he undoubtedly only supported his own interpretation; the original texts of Haworth and Burman do not contain any information in that respect.

R. elevata ≡ **Antimima elevata**

R. elineata L.Bolus NM 1929: 105 ≡ *M. elineatum* (L.Bolus) N.E.Br. GC 1930: 32 **LT** Pillans 5775 III (BOL!).

Diffusely branched shrubs, climbing in other bushes, **I** 25–40 mm *l*, ∅ to 2 mm, blackish-grey; **L** trigonous, recurved, dirty glaucous-green, epidermis papillate, 15–20 mm *l*, 4 mm *b* and *t*; **Fl** in cymes, ∅ 12 mm, pedicels 10–25 mm *l*, petals pink, inconspicuously vittate, 5 mm *l*, 0.75 mm *b*, filamentous staminodes 3.5 mm *l*, white, anthers and pollen yellow; **Fr** unknown; **Ecol** in crevices in granite; **Distr** Namaqualand, NC, S Africa.

Note: The name refers to a line visible on the slightly tumidulous leaf sheath.

R. emarcescens ≡ **Antimima emarcescens**
R. emarcidens ≡ **Antimima emarcescens**

R. erecta (L.Bolus) Schwantes ZSK 1927: 105 ≡ *M. erectum* L.Bolus ABH 1927: 80 **T** Pillans 4961 (BOL!).

Shrubs to 60 cm *h*; **I** red, later grey, 20–35 mm *l*, ∅ 3 mm; **L** almost terete, subglaucous to green, epidermis smooth, keel with some minute teeth, 20–25 mm *l*, 3–4 mm ∅, sheaths tumid, 2 mm *l*; **Fl** in cymes, pedicels 6–7 mm *l*, bracteoles above their middle, **K** 5, petals purplish-pink, purplish vittate, filamentous staminodes very few, filaments pale pink, anthers pale, pollen yellow; **Fr** 5 locules, otherwise unknown; **Ecol** in rocky hills; **Distr** Namaqualand, NC, S Africa.

Note: In the absence of fruits, the placement of the species in **Ruschia** remains uncertain.

R. erosa ≡ **Antimima erosa**

R. esterhuyseniae L.Bolus NM 1950: 216 **LT** Esterhuysen 6356 II (BOL!).

Low shrubs to 20 cm *h* with erect, stiff branches, **I** reddish-brown; **L** triquetrous, the pointed tip bent back, epidermis ± smooth, 7–11 mm *l*, c. 2 mm *b* and *t*; **Fl** solitary, terminal, bracteoles at c. half the length of the pedicel, ∅ to 36 mm, **K** 5, petals pinkish-purple, outer ones 16 mm *l*, decreasing in length to the centre, where they merge into filamentous staminodes of 5–9 mm *l*; **Fr** with a nearly semi-globose top from high valve rims, base funnel-shaped, covering membranes convex, distal closing ledge sharp, closing body a hollow, short hook, expanding keels not touching each other basally, no awns or wings, 5 locules; **Ecol** unknown; **Distr** Uniondale, WC, S Africa.

Note: The decreasing length of the petals, the slender, recurved leaves, and the sharp closing ledge resemble **Erepsia**, but it is not known whether a hypanthium exists. Within **Ruschia**, the leaves are similar to those of **R. vaginata** and **R. hamata**, lacking the marked elevations of the latter two.

R. evoluta ≡ **Antimima evoluta**
R. excedens ° **Antimima excedens**

R. exigua L.Bolus NM 1939: 162 **T** Leighton 1016/35 (BOL!).

Apparently compact developing long shoots in cultivation; **L** almost terete or finger-shaped, epidermis glabrous, margins and keel smooth and whitish, c. 10 mm *l*, c. 3 mm *b*, c. 3.5 mm *t*; **Fl** solitary, bracteoles smaller than foliage leaves, placed in the middle of the stalk, **K** 5, petals pink, short, 0.5 mm, linear, filamentous staminodes white, apically pink, filaments white, apically dark purple; **Fr** unknown; **Ecol** c. 400 mm rainfall p.a. at all seasons; **Distr** Montagu, WC, S Africa.

Note: A very imperfectly known species.

R. exsurgens ≡ **Antimima exsurgens**

R. extensa L.Bolus NM 1930: 202 **LT** Pillans 5790 I (BOL!).

Diffusely branched shrubs, branches ascending to decumbent, to 85 cm *l*, **I** brown, 40–45 mm *l*, ∅ to 7 mm; **L** almost semi-terete, glaucous-green, epidermis rough when dry, 35 mm *l*, 3 mm *t*; **Fl** in contracted cymes, pedicels to 6 mm *l*, bracteoles 3–6 mm *l*, ∅ to 12 mm, **K** 5, petals purplish-pink, filamentous staminodes pale pink, apically brownish-golden, filaments pale, apically red; **Fr** 5 locules, otherwise unknown; **Ecol** unknown; **Distr** Namaqualand, NC, S Africa.

Note: The keel appears often as a line only in this species.

R. fenestrata ≡ **Antimima fenestrata**
R. fergusoniae ≡ **Antimima fergusoniae**
R. ferox = **R. intricata**

R. festiva (N.E.Br.) Schwantes ZSK 1927: 20 ≡ *M. festivum* N.E.Br. 1920: 122 **T** Zeyher 698 (K!).

Shrubs with erect branches to 30 cm *h*, **I** 30–80 mm *l*, 3–4 mm ∅; **L** semi-terete, glabrous, epidermis smooth, 30 mm *l*; **Fl** in cymes, pedicels 10–12 mm *l*, bracteoles 7–15 mm *l*, acuminate, forming a cupule-shaped sheath around the stem, **K** 5–6, petals 6–10 mm *l*, 0.3–0.5 mm *b*, inner filiform petals present (= filamentous staminodes?); **Fr** top turret-shaped, truncate above, base elongate funnel-shaped, covering membranes convex with distal closing rodlets, closing body hook-shaped, expanding keels touching each other basally, 5–7 locules; **Ecol** unknown; **Distr** Vanrhynsdorp, WC, S Africa.

Note: Remarkable is the variation in locule number, a rare feature in the genus.

R. filamentosa = **Erepsia forficata**

R. filipetala L.Bolus NM 1929: 94 ≡ *M. filipetalum* (L.Bolus) N.E.Br. GC 1930: 32 **T** Pillans 1675 (BOL!).

Erect shrubs with branches to 32 cm *l*, **I** round, 25–35 mm *l*; **L** trigonous, uncinate, epidermis smooth, 23 mm *l*, 3–4 mm *b* and *t*; **Fl** in terminal and lateral cymes forming a big **Inf**, bracteoles in the middle of the pedicels of 5–9 mm length, **K** 5, petals and filamentous staminodes filiform; **Fr** 5 locules, otherwise unknown; **Ecol** unknown; **Distr** Clanwilliam, WC, S Africa.

Note: Remarkable are the filiform petals in this species.

R. firma L.Bolus NM 1930: 213 **T** Frames 152/28 (BOL!).

Shrubs to 40 cm *h*, **I** grey, 30–45 mm *l*, ∅ 6 mm; **L** subfalcate, apically recurved, trigonous to semi-terete, yellowish-green, to 70 mm *l*, 4 mm *b*, 5 mm *t*; **Fl** in rich cymes, ∅ 18 mm, **K** 5, petals purlish-pink, 6 mm *l*, 1 mm *b*, filamentous staminodes present, filaments white, apically pink; **Fr** base funnel-shaped, top raised, rims moderately high, covering membranes with protruding closing rodlets, closing body prominent, a little laterally compressed and adorned apically by a narrow crest, all tissue white and endocarpal, expanding keels distant from each other basally, 5 locules; **Ecol** unknown; **Distr** Vanrhynsdorp, WC, S Africa.

Note: Elaborate closing bodies characterize this species.

R. floribunda L.Bolus NM 1935: 497 **T** L.Bolus 9617 (BOL!).

Erect shrubs to 30 cm *h*, **I** 30–50 mm *l*, ∅ to 7 mm; **L** glaucous-green, semi-terete to trigonous, to 75 mm *l*, sheath 6–7 mm *l*; **Fl** in cymes, bracteoles near the bases of the pedicels, **K** 5, petals pink, inconspicuously vittate, 9–10 mm *l*, 1.5 mm *b*, filamentous staminodes very few, anthers and pollen dirty brownish-yellow; **Fr** ∅ 6 mm, 5 locules; **Ecol** on hills; **Distr** Vanrhynsdorp, WC, S Africa.

Note: A fruit is described in the original publication, but the type lacks fruiting material.

R. foliolosa ≡ **R. namusmontana**

R. foliosa (Haw.) Schwantes ZSK 1927: 20 non Bolus 1962: 300 nomen illeg. ≡ *M. foliosum* Haw. 1803: 97 **LT** drawing of Duncanson 5.3.1824 (K!).

Shrubs to 60 cm *h*, branches erect, later decumbent from weight, **I** at first compressed and green, later brown to grey and round; **L** triquetrous, glaucous, c. 13.5 mm *l*, recurved or spreading, sheaths long and connate with the stem, but the bases of the leaves separated by a deep line; **Fl** 3–5, petals red, filamentous staminodes not mentioned, filaments red, anthers pale; **Fr** 5 locules, unknown otherwise; **Ecol** unknown; **Distr** unknown.

Note: Data on the flower have been added by Haworth later (1819: 286) and may not belong to the species. The drawing is very similar to that of **R. imbricata**.

R. foliosa ≡ **R. namusmontana**
R. forficata ≡ **Erepsia forficata**

R. fourcadei L.Bolus NM 1929: 46 ≡ *M. fourcadei* (L.Bolus) N.E.Br. GC 1930: 32 nomen illeg. non L.Bolus ABH 1927: 84 **LT** Fourcade 4303 III (BOL!).

Broad shrubs with diverging branches, **I** > **L**, dark brown to grey; **L** triquetrous with a recurved mucro, margins and keel very smooth, epidermis glabrous, 10–12 mm *l*, 4–5 mm *t*; **Fl** solitary or in ternate cymes, bracteoles as big as foliage leaves and reaching the base of the smooth **K** with their tips, ∅ to 26 mm, petals and filamentous staminodes pink; **Fr** with a high top from high valve rims, grey, base short, funnel-shaped, covering membranes forming a roof over each locule, the distal rim only a little elevated, and a low closing ledge below, closing body somwhat triangular as seen from above, small, expanding keels almost touching each other basally, no awns or wings; **Ecol** in loam in shrubby associations, c. 300 mm rainfall p.a. at various seasons; **Distr** Ladismith, Uniondale, WC, Uitenhage, EC, S Africa.

Note: The few known localities of the species appear far apart, but may reflect an adaptation to more shrubby associations along the edge of Broken Karroo. When transferring the name to *M. fourcadei*, Brown (GC 1930: 33) had overlooked that the latter name had already been used by Bolus earlier, for a species now placed in **Drosanthemum. R. fourcadei** "forma luxurians" (Bolus 1964: 81) is only a descriptive phrase for a well developed form, not a taxon with the rank of forma.

R. framesii L.Bolus JB 1928: 231 ≡ *M. framesii* (L.Bolus) N.E.Br. GC 1930: 32 nomen illeg. non L. Bolus NM 1928: 137 **T** Frames 2501/27 (BOL!).

Compact, to 7 cm *h*; **L** subglobose to trigonous, sheath merging into the stem, epidermis cells a little raised, 3–4 mm *l*, 3 mm *b*, 2 mm *t*; **Fl** 1, ∅ 18 mm, **K** 5, petals pink, 7–8 mm *l*, 0.5 mm *b*, filamentous staminodes pink; **Fr** unknown; **Ecol** unknown; **Distr** Jansenville, EC, S Africa.

Note: The leaves sit on the stem like pieces of meat on a spit, yet leaving some distance between them.

R. fredericii (L.Bolus) L.Bolus NM 1950: 219 ≡ *M. fredericii* L.Bolus ABH 1916: 31 **T** Rogers 16417 (BOL!).

Shrubs with climbing branches, **I** maroon, 30–35 mm *l*, glabrous; **L** subfalcate, subglobose, epidermis cells a little raised, to 20 mm *l*, 3 mm *t*; **Fl** 1–3, ∅ 13–17 mm, pedicels 10–20 mm *l*, petals to 0.7 mm *l*, filamentous staminodes not mentioned, filaments white, apically purple, anthers white; **Fr** unknown; **Ecol** unknown; **Distr** Worcester, WC, S Africa.

Note: The leaves have been described as subglaucous.

R. frutescens ≡ **Stoeberia frutescens**

R. fugitans L.Bolus NM 1937: 60 **T** G.Meyer 9160 (BOL!).

Elongate, decumbent branches, **I** 15–30 mm *l*, 3–4(–8) mm ∅; **L** subterete, epidermis smooth, glaucous-green, 40–65 mm *l*, sheath with a line, 3–5 mm *t*; **Fl** in cymes, **K** 5, petals finally reddish-purple, vittate, bracteoles tiny, to 5 mm *l*, filamentous staminodes few, filaments basally pale to brownish, apically purple; **Fr** top low, base funnel-shaped, covering membranes with long, protruding closing rodlets, closing body a tiny hook, expanding keels almost touching each other basally, 5 locules; **Ecol** unknown; **Distr** Namaqualand, NC, Vanrhynsdorp, WC, S Africa.

Note: Described after the type material and the original text.

R. fulleri ≡ **Ebracteola fulleri**
R. gemina ≡ **Cerochlamys gemina**

R. geminiflora (Haw.) Schwantes in Jacobsen 1949: 58 ≡ *M. geminiflorum* Haw. 1821: 114 ≡ *M. geminatum* Jacquin Fragmenta 1800–1809: t. 50 nomen illeg. non Haworth 1803: 92 ≡ *M. geminatum* Salm-Dyck 1820: 28 nec 31 **LT** Jacquin 1800–1809: t. 50.

Shrubs with weak, diffuse, long branches 60–90 cm *l*, **I** grey; **L** triquetrous, green, apically recurved, epidermis papillate, < 54 mm *l*, sheath with a distinct line separating the leaves of a pair; **Fl** in cymes or in pairs, **K** 5, basally dotted, petals purple, a little longer than the **K** lobes, vittate, filaments white, purple in the middle, anthers pale; **Fr** 5 locules; **Ecol** unknown; **Distr** unknown.

Note: Haworth (1821) described this species to replace the use of *M. geminatum* for other plants than his species, now **Braunsia geminata**. The present description follows mainly that of Jacquin and the accompanying figure, which shows a laxly growing plant with hairy leaves, not a creeper, as suggested by Haworth (1821: 114–115), who placed the species in his group Sarmentosa. Haworth received material of the species in flower from Salm-Dyck (Haworth 1821: 115) and may have had still another plant, different from the two named originally *M. geminatum*.

R. gibbosa = **Leipoldtia compacta**

R. glauca L.Bolus NM 1929: 121 ≡ *M. valens* N.E.Br. GC 1930: 32 **T** Pillans 5711 (BOL!).

Erect shrubs > 30 cm *h*, **I** grey to black, shining, spongy; **L** triquetrous, epidermis with papillae, c. 20 mm *l*, 5 mm *b*, 5 mm *t*; **Fl** in lateral and terminal cymes of 3 mostly, bracteoles basal, ∅ 18 mm, **K** 5, petals pink, filamentous staminodes apically purplish-pink; **Fr** unknown; **Ecol** < 1200 mm rainfall p.a. in winter; **Distr** N Namaqualand, NC, S Africa.

Note: Spongy, shiny internodes are rare in this genus.

R. globularis = **R. spinosa**

R. goodiae L.Bolus NM 1929: 78 ≡ *M. goodiae* (L.Bolus) N.E.Br. GC 1930: 32 **T** Good as Pillans 6129 (BOL!).

Dense shrubs with irregular, twisting branches, **I** ochre to dark blackish-brown, 2–40 mm *l*; **L** triquetrous, acute, glaucous-green, epidermis smooth or with low papillae, 20–25 mm *l*, 3 mm *t*; **Fl** ternate, pedicels to 13 mm *l*, **K** 5, petals beautifully pink, 12 mm *l*, 2–2.5 mm *b*, filamentous staminodes absent, filaments deep purple, anthers golden; **Fr** top a high turret, the rims of the valves forming protruding, high corners at the outer margin of the fruit when closed, a crater in the centre, base elongate funnel-shaped, covering membranes with protruding closing rodlets, closing body a small hook, expanding keels not meeting each other at their bases, 5 locules; **Ecol** in sand veld; **Distr** Namaqualand, NC, S Africa.

Note: Although not more material than the type is known, the fruit is very distinct with its very high top with a central crater and five marginal protrusions.

R. gracilipes L.Bolus NM 1958: 340 **T** Esterhuysen 23635 (BOL!).

Branches to 28 cm *l*, **I** 10–35 mm *l*, Ø 5 mm; **L** almost terete, apically rounded, epidermis cells smooth, **L** 16–21 mm *l*, 3–4 mm *b* and *t*; **Fl** in cymes, peduncle 30 mm *l*, pedicels 3–9 mm *l*, bracteoles at their middle, 2–3 mm *l*, **Inf** distinct from the basal vegetative part of the plant, **K** 5, petals pinkish-purple, 6–7 mm *l*, 1 mm *b*, filamentous staminodes absent, filaments purplish-pink; **Fr** base funnel-shaped, top raised, rims low, covering membranes with distal closing ledges, closing body a small hook, expanding keels distant from each other at their bases, 5 locules; **Ecol** unknown; **Distr** Namaqualand, NC, S Africa.

Note: A member of the group with distinct cymes.

R. gracilis L.Bolus NM 1929: 62 ≡ *M. exile* N.E.Br. GC 1930: 32 **LT** Muir 1030 (BOL!).

Shrubs to 15 cm *h*, branches to 60 cm *l*, decumbent, **I** 13 mm *l*, 3 mm Ø; **L** trigonous, tip somewhat recurved, green, epidermis cells a little raised, 15–40 mm *l*, 2–3 mm *t*; **Fl** in narrow cymes, pedicels to 20 mm *l*, bracteoles at their middle, **K** 5, petals pale pink to pink, inconspicuously vittate; **Fr** base funnel-shaped, top raised, rims low, covering membranes with protruding closing rodlets, closing body a small hook, expanding keels almost touching each other, 5 locules; **Ecol** at 120 m altitude near the sea; **Distr** Riversdale, WC, S Africa.

Note: The type material includes also a loose fruit of a species of **Antimima**.

R. gracillima ≡ **Antimima gracillima**
R. graminea = **Acrodon bellidiflorus**
R. granitica ≡ **Antimima granitica**
R. gravida ≡ **Eberlanzia gravida**

R. griquensis (L.Bolus) Schwantes ZSK 1926: 187 (see **photo**) ≡ *M. griquense* L.Bolus ABH 1926: 39 **T** Wilman 17888 (BOL!).

Mostly erect shrubs to 40 cm *h* in habitat, with regrowth from the base, probably as an adaptation to fire, decumbent with age in cultivation, **I** grey, later brown; **L** triquetrous, 1–2(–5) teeth along the keel, margins and keel whitish, smooth, with distant big papillae, 1–30 mm *l*, 2–3 mm *b* and *t*; **Fl** solitary, bracteoles near the middle of the stalk, the tips reaching the base of the **K**, **K** 5, petals pink to purple, c. 10 mm *l*, c. 0.5 mm *b*, filamentous staminodes pink to purple, surrounding the central cone of stamens; **Fr** top raised, but rims low, base funnel-shaped, grey, covering membranes convex, closing ledge distinct, closing body a small hollow hook, expanding keels not touching each other basally, no awns, no wings, Ø c. 7 mm, 5 locules; **Ecol** in Acacia-erioloba-grassland on rocky terrain, near exposed rock surfaces, 300–500 mm rainfall p.a. in summer; **Distr** Barkly West, Hay, Postmasburg, Kimberley, Kuruman, NC, Hoopstad, FS, S Africa.

Note: The type specimen was collected in Public Gardens, hence probably the denser and more decumbent growth described. In habitat, a distinct renewal of branches occurs from the base, probably in reaction to fire, but also to browsing or breaking. Because of the basal regeneration, the plants can be kept easily in cultivation since they resprout after cutting. Some populations possess beautifully bright purple flowers. In open places, germination is fast as can be seen from herds of seedlings in some places, but denser vegetation seems to lead to extinction.

R. grisea (L.Bolus) Schwantes ZSK 1926: 187 ≡ *M. griseum* L.Bolus ABH 1926: 39 **T** without collector 17786 (BOL!).

Shrubs to c. 30 cm *h*, branches spreading, **I** 10–45 mm *l*, grey; **L** free parts 2–4 mm *l*, sheaths to 5 mm *l*, basally connate with the stem, later turning light grey, 1 apical and 1 additional inconspicuous tooth on the keel, epidermis almost smooth; **Fl** solitary, Ø 14 mm, sessile, petals pale pink, 6 mm *l*, 0.5 mm *b*, filamentous staminodes pink, base pale, filaments apically pink, anthers and pollen yellow; **Fr** base elongate, funnel-shaped, top with very high rims, covering membranes convex, closing rodlets hidden underneath, closing body small, egg-shaped, expanding keels widely distant at their bases, 5 locules; **Ecol** unknown; **Distr** unknown.

Note: The shrubs are reported to look grey from the dry internodes.

R. gydouwensis ≡ **Lampranthus gydouwensis**
R. hallii ≡ **Antimima hallii**

R. hamata (L.Bolus) Schwantes ZSK 1927: 19 (see **photo**) ≡ *M. hamatum* L.Bolus ABH 1916: 29 **LT** Cruickshank 2520 (BOL!).

Shrubs with rigid, erect branches to 45 cm *h*, ∅ to 50 cm, but plants often slender, tap-root woody and thickened; **L** tapering from parallel sides to a prominent, markedly recurved white tip, the epidermis raised prominently above subhypodermal tannin idioblasts giving the leaves a rough appearance, the dots translucent in fresh leaves, often in rows along keel and rounded margins, resulting in finely serrate edges, wax cover continuous, breaking into horizontal plates with age, covered densely by numerous vertical platelets merging into long, fine threads, leaves near the base of the plant to 20 mm *l*, upper ones 6–10 mm *l*, c. 2 mm *t*; **Fl** solitary, but several on a season's growth, therefore often looking like a panicle, **K** 5, ∅ c. 16 mm, 30–40 pink to purple petals in a regular circle, filamentous staminodes purple with white tips, longer than the stamens in the interior of the central cone; **Fr** base bell-shaped, top semi-orbicular in side view from high valve rims, covering membranes of a locule forming a roof, the contact line being the ridge, closing rodlets big and protruding, closing body small, often touching the closing rodlets, expanding keels distant from each other, short; **S** roundish, testa with distinct, separated low papillae (naps), ornamented with epicuticular rodlets often standing together in groups; **Ecol** in grassveld, regularly with **Themeda triandra**, in deep or somewhat shallow soils, often with some lime, > 300 mm rainfall p.a. in summer; **Distr** W Lesotho; FS, S NW, NE NC, S Africa.

Note: A widespread species; the typical habitats are subject to two threats: floods and fire. Themeda grass veld is often burnt, and the older stems of **R. hamata** burn with it; regeneration occurs from the base, where numerous dormant buds are present. Some nutrients can be mobilized from the thickened root stock. **R. hamata** grows in flat areas, often on broad river terraces, in gravelly to fine loamy soils; floods can occur here after heavy rains, the water can stay for some weeks. No deaths of individuals have been observed some months after such an event. In cultivation, resprouting can be induced by cutting old, straggly stems, but growth from cuttings is extremely difficult and rarely successful.

R. hamatilis ≡ **Antimima hamatilis**

R. haworthii H.Jacobsen & G.D.Rowley 1973: 6 ≡ *M. dilatatum* Haw. 1803: 68 T?

Erect shrubs to 90 cm *h*, **I** reddish; **L** triquetrous, compressed, glaucous, gibbous, apically recurved, **Fl** ternate, pedicels c. 7 mm *l*, bracteoles above their middle, filiform, **K** 5, petals silvery-white, shiny, c. 7 mm *l*, filamentous staminodes not mentioned, filaments white; **Fr** 5 locules, otherwise unknown; **Ecol** unknown; **Distr** S Africa.

Note: The species is not known well.

R. herrei ≡ **Antimima herrei**

R. heteropetala L.Bolus NM 1954: 277 T Hall 514/53 (BOL!).

Shrubs to c. 20 cm *h*, **I** > **L**, dark brown; **L** densely placed on lateral, often paired, short shoots, triquetrous, keel and margins smooth, subhypodermal tannin idioblasts very conspicuous in the dry material, leaves of a pair separated basally by a triangle of stem tissue, c. 7 mm *l*, 2–3 mm *b*, 4 mm *t*; **Fl** solitary, bracteoles surrounding the **K**, petals bright pink, outer ones flat and spreading, inner apically recurved and ending in an awn, merging into the pink filamentous staminodes, filaments pink, with long papillae above the base; **Fr** unknown; **Ecol** 100–200 mm rainfall p.a. mainly in March and November; **Distr** Prince Albert, WC, S Africa.

Note: The name has been placed in synonymy with **R. fourcadei** (Bolus 1964), a decision that appears unfounded when the types are compared. In both species, the calyx is enclosed by the bracteoles, but **R. heteropetala** possesses an entirely different mode of branching, developing lateral, protruding short shoots annulate from the scars of old leaf-pairs which, in the latest growth, are set in 4–5 pairs like pieces of meat on a spit.

R. hexamera L.Bolus NM 1928: 144 ≡ *M. hexamerum* (L.Bolus) N.E.Br. GC 1930: 32 **T** Pillans 5703 (BOL) = *R. hexamera* L.Bolus var. *longipetala* L.Bolus NM 1931: 237 **T** M. Schlechter 8367 (BOL).

Erect, robust shrubs, vegetative branches short, flowering ones elongate, 8–9 cm *l*, **I** 25 mm *l*, ∅ 5 mm; **L** subcymbiform, glaucous, papillate, 25–40 mm *l*, 14–24 mm *b* and *t*; **Fl** solitary, pedicels 7 mm *l*, bracteoles basal, ∅ 16 mm, **K** 6, petals white or palest pink, 6 mm *l*, 0.5 mm *b*, filamentous staminodes and filaments white; **Fr** expanded 13 mm ∅, valve wings absent, closing body present, 6 locules; **Ecol** in hills; **Distr** Namaqualand, NC, S Africa.

Note: The number of six calyx lobes and six locules is rare in this genus.

R. hexamera var. *longipetala* = **R. hexamera**

R. holensis L.Bolus 1965: 84 T Hall 2864 (BOL!).

Loosely, but stiffly branched shrubs to c. 20 cm *h*, **I** reddish, flat at first, later brown, longer than leaves;

L connate basally, but separated by a line, trigonous, apically rounded, epidermis smooth, 10 mm *l*, 4–5 mm *b*, 5–6 mm *t*; **Fl** 1–3, bracteoles above the middle of the flower stalk, **K** 5, 28–33 purplish-pink petals inconspicuously vittate, filamentous staminodes absent, filaments pale, apically bright red; **Fr** stalk c. double the length of the flower stalk (30–50 mm *l*), bracteoles a little above the middle of the pedicel, top raised, but rims low, base deep, funnel-shaped, covering membranes convex from a deeper centre of the fruit, with prominent closing rodlets and a low rim at the distal end, closing body a small, hollow hook, expanding keels diverging tangentially, short, opening the valves only into a vertical position, 5 locules; **Ecol** < 100 mm rainfall p.a. in winter; **Distr** Vanrhynsdorp, WC, S Africa.

Note: "few plants seen, very widely scattered" (Hall in Bolus 1965: 84).

R. hollowayana = **Eberlanzia schneideriana**
R. horrescens = **R. cradockensis** ssp. **cradockensis**
R. horrescens var. *densa*
= **R. cradockensis** ssp. **cradockensis**
R. horrida = **R. cradockensis ssp. cradockensis**
R. hutchinsonii ≡ **Amphibolia laevis**

R. imbricata (Haw.) Schwantes ZSK 1927: 19 ≡ *M. imbricatum* Haw. 1795: 317 **LT** drawing of Duncanson 9501 (K!) = *M. imbricatum* Haw. var. *rubrum* Haw. 1819: 95 ≡ *M. multiflorum* Haw. var. *rubrum* (Haw.) Haw. 1821: 128 **T** Burchell (K).

Shrubs with erect branches, **I** light grey with age; **L** trigonous, glaucous, connate into a sheath which becomes connate with the stem further down, the upper part of the sheath with a deep line between the leaves; **Fl** in elongate cymes of which the pedicels and peduncles persist on the plant, adorned by small bracteoles, which, according to the original description, are imbricately placed, petals white or rich purple in the variety; **Fr** unknown; **Ecol** unknown; **Distr** unknown, perhaps Vredenburg, WC, S Africa.

Note: The description given here follows the original text and the drawing, which is chosen as lectotype. Haworth (1819: 95) stressed the difference of the variety from the true species, placing it later in *M. multiflorum*, which is also white flowered.

R. impressa L.Bolus NM 1929: 159 ≡ *M. impressum* (L.Bolus) Ingram 1970: 50 **T** Frames 18994 (BOL!).

Compact plants to 6 cm *h*, **I** ∅ 3 mm; **L** free parts broadly trigonous, 2–5 mm *l*, sheaths to 10 mm *l*, but lower part connate with the stem, epidermis cells a little elevated, keel inconspicuously 1-denticulate, pale glaucous-green; **Fl** solitary, ∅ 10–12 mm, **K** 5, bracteoles enclosing the base of the flower, petals pink, 5 mm *l*, 0.5 mm *b*, filamentous staminodes pink, filaments pinkish-purple; **Fr** base funnel-shaped, top with high rims, covering membranes convex, closing rodlets protruding from underneath them, closing body rodlet-shaped, expanding keels not touching each other basally, 5 locules; **Ecol** unknown; **Distr** Ladismith, WC, S Africa.

Note: The compact shape suggests that the species has a potential as garden or rockery plants.

R. inclaudens ≡ **Esterhuysenia inclaudens**

R. inclusa L.Bolus NM 1929: 44 ≡ *M. amplexum* N.E.Br. GC 1930: 32 **T** L.Bolus 18867 (BOL!).

Compact, 6–8 cm *h*, **I** 5–7 mm *l*, upper leaf-pair enclosing the base of the flower; **L** triquetrous, keel with 1 tooth beside the apical one, epidermal cells a little raised, sheath and free parts c. 4 mm *l*, to 5 mm *b*, 3 mm *t*; **Fl** ∅ 15 mm, petals purplish-pink, filamentous staminodes present, filaments rough, white, apically pink; **Fr** top raised, base funnel-shaped, covering membranes convex, distal rodlets protruding towards the small, hook-shaped closing body, expanding keels distant from each other basally, 5 locules; **Ecol** unknown; **Distr** Laingsburg, WC, S Africa.

Note: This is one of the small species, a rather rare growth form in **Ruschia**.

R. inconspicua L.Bolus JB 1928: 232 (see **photo**) ≡ *M. latens* N.E.Br. GC 1930: 32 **T** Pillans 5749 (BOL!).

Cushion-forming, to 15 cm *h*, ∅ to 25 cm, **I** grey with age; **L** basally connate, the sheath decurrent on the stem and connate with it, free parts trigonous, apically rounded, epidermis cells a little elevated, glaucous from wax plates on the cells adorned with vertical platelets, 7–10 mm *l*, 4–6 mm *b*, 5 mm *t*; **Fl** solitary, ∅ 12 mm, ca. 30 white petals to 5 mm *l*, equalling the **K** lobes in length at most, c. 0.75 mm *b*, c. 30 filamentous staminodes, c. 20 stamens; **Fr** top high from high rims, base funnel-shaped, covering membranes convex, distally with long closing ledges, the closing body reduced to a rim in the type material, but fresh fruits with small closing bodies in hook-shape, expanding keels short and widely diverging, 5 locules; **Ecol** on lower, gravelly slopes, < 100 mm rainfall p.a. in winter; **Distr** Lüderitz-South, Namibia; N Namaqualand, NC, S Africa.

Note: The species seems to have been overlooked since its description, also in the treatment of Friedrich (1970) of the Aizoaceae in Namibia. It differs from the superficially similar **R. odontocalyx** from further north in the absence of teeth on the keel and the low number of petals, filamentous staminodes and filaments.

R. incumbens L.Bolus NM 1930: 212 **T** Leipoldt 19246 (BOL!).

Shrubs to 45 cm *h*, **I** red, later ochre, to 35 mm *l*; **L** trigonous to semi-terete, epidermis drying into a rough surface, but no elevations, 25–40 mm *l*, 4 mm *b* and *t*; **Fl** in cymes up to 15 per **Inf**, ∅ 15 mm, **K** 5, petals pink, filamentous staminodes pale, pink at the middle; **Fr** 5 locules, otherwise unknown; **Ecol** unknown; **Distr** Clanwilliam (only?), WC, S Africa.

Note: The placement of the species in **Ruschia** remains uncertain.

R. incurvata L.Bolus NM 1929: 124 ≡ *M. incurvatum* (L.Bolus) N.E.Br. GC 1930: 32 **T** Peers 18950 (BOL!).

Shrubs to 50 cm *h*, **I** to 40 mm *l*, ∅ to 7 mm; **L** semi-terete, apiculate, somewhat swollen, epidermis smooth, 40 mm *l*, 5 mm *b* and *t*, sheath 6 mm *l*; **Fl** in rich cymes, pedicels 30–50 mm *l*, bracteoles to 20 mm *l*, thus the cyme distinct from the vegetative part of the plant, **K** 5, petals pale pink, filamentous staminodes white, apically pale pink, anthers and pollen yellow; **Fr** 5 locules, otherwise unknown; **Ecol** unknown; **Distr** Clanwilliam, WC, S Africa.

Note: A robust member of the group of species with a distinct inflorescence.

R. indecora (L.Bolus) Schwantes ZSK 1927: 106 ≡ *M. indecorum* L.Bolus ABH 1927: 91 **LT** L.Bolus 17418 II (BOL!).

Shrubs to 30 cm *h*, **I** reddish-brown, 7–25 mm *l*; **L** recurved, uncinate, sheaths tumid, 2–3 mm *l*, epidermis smooth, glaucous-grey, mostly 10–13 mm *l*, maximally 2 mm *b* and *t*; **Fl** in cymes of which old remains persist on the plant for several seasons, forming blunt spiny processes, ∅ 8–10 mm, pedicels 5–7 mm *l*, bracteoles at their middle, **K** 5, petals c. 20, white, linear, filamentous staminodes and filaments white, apically purple, pollen brown-purple; **Fr** unknown; **Ecol** unknown; **Distr** Wynberg (only?), WC, S Africa.

Note: Fruits are lacking on the type.

R. indurata (L.Bolus) Schwantes ZSK 1926: 187 ≡ *M. induratum* L.Bolus ABH 1922: 135 **LT** Pillans 383/16 (BOL!).

Branches prostrate, **I** enclosed, ∅ 5 mm; **L** triquetrous, keel and margins cartilaginous, ending in a pungent tooth at the apex, and with distant, big papillae, epidermis cells otherwise a little raised, 8–10 mm *l*, 3 mm *b*; **Fl** solitary, ∅ 20 mm, pedicels to 15 mm *l*, petals pink, 7–8 mm *l*, 1 mm *b*, filamentous staminodes present, anthers yellow; **Fr** falling out of their bases when ripe, leaving the firm stalk and five pungent bundles high above the plant, which, together with the pungent leaf tips make the plants a spiny cushion, 5 locules, otherwise unknown; **Ecol** unknown; **Distr** Steytlerville, EC, S Africa.

Note: Species with capsules losing their interior and becoming pungent with age are not rare in this genus.

R. insidens L.Bolus NM 1929: 77 ≡ *M. insidens* (L.Bolus) N.E.Br. GC 1930: 32 **T** Compton 1505/26 (BOL).

Shrubs 9–10 cm *h*, **I** grey, ∅ 2 mm; **L** trigonous, acute to subtruncate, apiculate, epidermis papillate, glaucous pinkish suffused, to 11 mm *l*, 3 mm *b* and *t*; **Fl** ∅ 18 mm, **K** 5, petals pink, 8 mm *l*, 1.25 mm *b*, filamentous staminodes few, filaments pale, anthers and pollen golden, stigmas purple; **Fr** unknown; **Ecol** unknown; **Distr** Calvinia, NC, S Africa.

Note: Bolus (1929: 77) described a difference in length of papillae on the lower leaf surface, in contrast to uniform papillae length on the upper leaf surface.

R. integra = **Smicrostigma viride**

R. intermedia L.Bolus NM 1950: 217 **T** Esterhuysen 16209 (BOL!).

Erect shrubs, 22–34 cm *h*, **I** 15–35 mm *l*, ∅ 2 mm; **L** trigonous to semi-terete and subfalcate, margins and keel finely serrulate from elevations, sheath with a deep line, epidermis smooth, 20–30 mm *l*, 3 mm *b*, 3–4 mm *t*; **Fl** in cymes, bracteoles just above the middle of the pedicel, **K** 5, petals pinkish-purple, vittate, 10–12 mm *l*, to 2 mm *b*, filamentous staminodes deep pinkish-purple, anthers and pollen pale yellow; **Fr** base elongate funnel-shaped, top highly raised, rims low, covering membranes convex, with protruding distal closing rodlets, closing body hook-shaped, expanding keels almost touching each other basally, 5 locules; **Ecol** on sandy, rocky plateau; **Distr** Piquetberg, WC, S Africa.

Note: The species is insufficiently known.

R. intervallaris ≡ **Mossia intervallaris**; ≡ **Antimima intervallaris**

R. intricata (N.E.Br.) H.E.K.Hartmann & Stüber 1993: 65 (see **photo**) subg. **Spinosae** ≡ *M. intricatum* N.E.Br. in Burtt Davy 1926: 160 ≡ *Eberlanzia intricata* (N.E.Br.) Schwantes ZSK 1927: 30 **T** Watt s.n. (K!) = *R. ferox* L.Bolus NM 1932: 371 ≡ *Eberlanzia ferox* (L.Bolus) L.Bolus NM 1958: 387 **T** James 20208 (BOL!) = *M. munitum* L.Bolus NM 1928: 4 ≡ *Eberlanzia munita* (L.Bolus) Schwantes in Jacobsen 1949: 58 **T** Peers 978/26 (BOL!) = *R. persistens* L.Bolus NM 1932: 334 non L.Bolus 1963: 16 ≡ *Eberlanzia persistens* (L.Bolus) L.Bolus NM 1958: 387 **T** Godman 19933 (BOL!) = *R. puniens* L.Bolus NM 1935: 502 ≡ *Eberlanzia puniens* (L.Bolus) L.Bolus NM 1958: 387 **T** Tapscott 3206 (BOL!) = *M. tuitum* N.E.Br. nomen nudum in schedis, Smith 464 (K!)[24].

[24] Other specimens named *M. tuitum* by N. E. Brown are **R. divaricata** (Smith 5470, K!) and **R. cradockensis** ssp. **cradockensis** (Verdoorn 1578, K!).

Shrubs to 70 cm *h*; I coppery to dark grey; L 12–30 mm *l*, with an even surface, but epidermis cells convex; **Fl** in very rich **Inf** with 7–63(–255) spines and 8–64 flowers due to the formation of lateral additional branching; **Fr** without valve wings; **S** 0.8–1.1 mm *l*; **Chr** $2n = 18, 36, 72$; **Ecol** in shallow, often saline soils in flats with grass, summer rainfall to 500 mm p.a.; **Distr** eastern districts of EC, FS, southern districts of NC, inland parts of WC, S Africa.

Note: Due to the high branching in the inflorescences, flowering of the plants is spectacular over some time.

R. intrusa (Kensit) L.Bolus NM 1950: 220 ≡ *M. intrusum* Kensit in H.Bolus & Kensit TRSSA 1909: 151 **LT** Kensit 6500 (BOL!).

Shrublet to 8 cm *h* with decumbent branches, I 5–7 mm *l*, enclosed by the remains of old leaves, greyish; **L** trigonous to falcate in side view, keel very rounded and very finely denticulate, epidermis cells a little raised, sheath c. 4 mm *l*, free parts c. 4 mm *l*, 5 mm *b*, 3 mm *t*; **Fl** ∅ 15 mm, base enclosed by the uppermost leaf-pair, petals purplish-pink, filamentous staminodes present, filaments scabrous, white, apically pink; **Fr** 5 locules; **Ecol** unknown; **Distr** Laingsburg, WC, S Africa.

Note: The epithet refers to the base of the flower being enclosed by the uppermost leaves, the bracteoles, which in this case do not differ from the foliage leaves. In growth and leaves the plants resemble those of **Antegibbaeum**.

R. ivori ≡ **Antimima ivori**
R. jacobseniana ≡ **Astridia longifolia**
R. kakamasensis = **R. barnardii**

R. karrachabensis L.Bolus NM 1931: 287 **T** Herre 8841 (BOL!).

Shrubs, I maroon; **L** trigonous with ± rounded tips, epidermis smooth, 10–15 mm *l*, 3.5 mm *b*, 5 mm *t*; **Fl** 1–3 in terminal position and on lateral shoots, bracteoles near base of stalk, **K** 5, ∅ 16 cm, petals pink, filamentous staminodes pale pink, filaments apically pink; **Fr** with a big tubercle; **Ecol** < 100 mm rainfall p.a. in winter; **Distr** N Namaqualand, NC, S Africa.

Note: A little known species.

R. karroidea ≡ **Antimima karroidea**

R. karrooica (L.Bolus) L.Bolus SAG 1927: 239 ≡ *M. karrooicum* ABH 1925: 9 **LT** H.Bolus 13454 (BOL!).

Shrubs to 30 cm *h*, branches almost erect, I ∅ c. 8 mm; **L** free parts trigonous with recurved tips, sheaths basally connate with the stem, epidermis smooth, 8–21 mm *l*, 3–5 mm *b* and *t*; **Fl** solitary, ∅ 50 mm, pedicel 20–45 mm *l*, bracteoles at c. half the length, petals pinkish-purple, purple vittate, 22 mm *l*, 2–3 mm *b*, filamentous staminodes present, filaments purple, anthers and pollen pale; **Fr** base a short funnel, top with medium high rims, covering membranes convex, closing rodlets hidden below them, closing body an open hood, triangular as seen from above in the open capsule, expanding keels widely separated basally, short, 5 locules; **Ecol** in dry plains; **Distr** Laingsburg, WC, S Africa.

Note: The lectotype chosen possesses a capsule of the Ruschia-type.

R. kenhardtensis L.Bolus NM 1935: 498 **T** van Niekerk 8649 (BOL!).

Erect shrubs to 35 cm *h*, I ochre with age, enclosed by a sheath merging into outer stem tissue downwards; **L** triquetrous, tips almost blunt, a tooth recurved to the back, sometimes another tooth on the keel, sheath seemingly 5–10 mm *l*, but its boundary to the stem tissue uncertain, free parts c. 5 mm *l*, 2.5 mm *b*, c. 3 mm *t*; **Fl** solitary, bracteoles also blunt, often a tooth at the back of the tip, in the middle of the stalk, **K** 5, petals pale pink, filamentous staminodes present; **Fr** falling out of the star-shaped ring of persistent bundles, pedicels long-lasting spines later, top with high rims, base a lone funnel, covering membranes convex, closing ledge conspicuous, closing body small, hook-shaped, expanding keels not touching each other at their bases, no awns or wings, 5 locules; **Ecol** in shrubby vegetation on limestone slopes, 200–400 mm rainfall p.a. mainly in March and November; **Distr** Hay, Kenhardt, NC, S Africa.

Note: The apical tooth sits at the back of the leaf tip, giving the leaf a truncate appearance in side view, which, together with the long leaf sheaths and the long funnel-shaped fruits, characterizes this species well.

R. klaverensis ≡ **Antimima klaverensis**

R. klipbergensis L.Bolus 1963: 175 **T** Hall 2227 (BOL!).

Erect shrubs to 33 cm *h*, branches ascending, I purple, 10–55 mm *l*, ∅ 1.5–4 mm; **L** straight or forming an elongate s, almost semi-terete, apically recurved, epidermis cells smooth, 25–35 mm *l*, 2–4 mm *b* and *t*; **Fl** in cymes, pedicels 13–25 mm *l*, bracteoles below their middle, c. 8 mm *l*, **K** 5, petals pink, with one central and two marginal pinkish-purple lines, 9–11 mm *l*, 0.75–1.25 mm *b*, filamentous staminodes pale pink, apically pinkish-purple, anthers and pollen white; **Fr** base funnel-shaped, top raised, rims low, covering membranes with protruding closing rodlets, closing body hook-shaped, expanding keels almost touching each other basally, ∅ 7 mm, 5 locules; **Ecol** unknown; **Distr** Malmesbury, WC, S Africa.

Note: A little known species from a well-known area.

R. knysnana (L.Bolus) L.Bolus NM 1928: 146 ≡ *M. knysnanum* L.Bolus ABH 1927: 97 T Duthie 671 (BOL!) = *R. knysnana* (L.Bolus) L.Bolus var. *angustifolia* L.Bolus NM 1928: 146 T Fourcade 2971 (BOL!).

Shrubs 30–40 mm *h*, **I** 30–60 mm *l*, ∅ to 5 mm; **L** triquetrous, apiculate, light and shiny green, margins and keel as if folded, translucent with dark dots when dry, **L** 33 mm *l*, 5 mm *b*, 6 mm *t*; **Fl** in cymes, pedicels to 25 mm *l*, ∅ 25 mm, **K** 4–6, mostly 6, petals pink to purplish-pink, 5–10 mm *l*, 1 mm *b*, filamentous staminodes and filaments white, anthers and pollen pale; **Fr** base funnel-shaped, top high from raised rims of the valves, covering membranes distally recurved and with distinct closing ledges below, expanding keels diverging and rather straight, ending in incurved, long awns, valve wings absent, closing body absent, but the placenta protruding a little, ∅ 8–13 mm when expanded, 6 locules; **Ecol** unknown; **Distr** Humansdorp, EC, Knysna, WC, S Africa.

Note: The reported presence of wings on the valves might confuse the understanding of the fruit morphology. The type material lacks valve wings, and the fruits resemble capsules of **Erepsia** or **Esterhuysenia.**

R. knysnana var. *angustifolia* = **R. knysnana**
R. koekenaapensis ≡ **Antimima koekenaapensis**
R. komkansica ≡ **Antimima komkansica**

R. kuboosana L.Bolus JSAB 1960: 53 T Herre 13250 (BOL!).

Branches 7–14 cm *l*, **I** 15–40 mm *l*, ∅ 3 mm; **L** ascending, linear to semi-terete, epidermis smooth, 30–45 mm *l*, 7 mm *b* and *t*; **Fl** solitary, ∅ 20–25 mm, **K** 5, bracteoles enclosing the flower basally, petals purplish-pink, 6–10 mm *l*, to 1.25 mm *b*, filamentous staminodes pale pink, filaments pale, apically pink, anthers and pollen yellow; **Fr** unknown, 5 locules; **Ecol** unknown; **Distr** Namaqualand, NC, S Africa.

Note: The lack of fruiting material makes it difficult to assess the position of this species.

R. langebaanensis L.Bolus NM 1932: 372 T L.Bolus 20204 (BOL!).

Decumbent, **I** 20–30 mm *l*, ∅ 5–7 mm; **L** ± erect, mucro recurved, green to greyish, 35 mm *l*, to 60 mm *l* in shade, 4–6 mm *b*, 5–6 mm *t*; **Fl** 3–5, pedicels 10–15 mm *l*, bracteoles at their bases, **K** 5, petals pinkish-purple, inconspicuously vittate, 5–10 mm *l*, 2 mm *b*, filamentous staminodes present, filaments white, apically purplish; **Fr** 5 locules, otherwise unknown; **Ecol** in fissure of granitic rocks near the sea, winter rainfall; **Distr** Vredenburg, WC, S Africa.

Note: A very little known species.

R. lapidicola L.Bolus NM 1954: 283 T Esterhuysen 21720 (BOL).

Erect shrubs to 30 cm *h*, **I** 4–25 mm *l*, ∅ 1–4 mm; **L** subfalcate, light green, 9–15 mm *l*, 3–6 mm *b*, 3–7 mm *t*; **Fl** in cymes, ∅ 15 mm, pedicels 6–15 mm *l*, bracteoles in their middle, **K** 5, petals purplish-pink, 4–6 mm *l*, 1 mm *b*, filamentous staminodes apically pinkish-purple, filaments apically pink, anthers and pollen pale brown or pale purplish-brown, stigmas blackish-purple; **Fr** top raised, valve wings obscure, closing body minimal, deeply placed, 5–6 mm *h*, ∅ 4–5 mm, 5 locules; **Ecol** stony soil and rock basins, 1200 m altitude; **Distr** Clanwilliam, WC, S Africa.

Note: The name refers to the rocky habitat.

R. lavisii L.Bolus NM 1930: 205 T Lavis 19211 (BOL!).

Shrubs to 40 cm *h*, **I** ochre, 35–50 mm *l*; **L** trigonous, acuminate, sheaths tumid, pale glaucous green, mostly 20–25 mm *l*, 5 mm *b*, 5 mm *t*; **Fl** 3–5 in cymes, ∅ 22 mm, pedicels 14 mm *l*, bracteoles c. 10 mm *l*, at the middle of the pedicels, petals to 9 mm *l*, 1.5 mm *b*, filamentous staminodes and stamens white, upper half purple; **Fr** with a conspicuous closing body, ∅ 5 mm, otherwise unknown; **Ecol** in crevices; **Distr** Strand, WC, S Africa.

Note: Although described, fruits are absent on the type sheet.

R. lawsonii ≡ **Antimima lawsonii**

R. laxa (Willdenow) Schwantes in Jacobsen 1949: 58 ≡ *M. laxum* Willdenow 1809: 536 non L.Bolus 1933: 406 nomen illeg. T?

Shrubs with weak, diffuse branches, young **I** compressed-terete; **L** triquetrous, tuberculate-punctate, keel and margins finely denticulate; **Fl** unknown; **Fr** unknown; **Ecol** unknown; **Distr** unknown, S Africa.

Note: No type of Willdenow exists at B.

R. laxiflora L.Bolus NM 1931: 237 T Pillans 19376 (BOL!).

Shrubs to 50 cm *h* and ∅, lateral branches as long shoots at right angles, loosely branched, **I** red to maroon, 20–50 mm *l*; **L** subterete, apically rounded, epidermis smooth, 20–30 mm *l*, 4 mm *t*; **Fl** in terminal and lateral cymes, forming a composed **Inf**, pedicels 15–20 mm *l*, bracteoles above their middle, 5–7 mm *l*, **K** 5, petals pink, 7 mm *l*, 1 mm *b*, filamentous staminodes and filaments pink, anthers yellow; **Fr** 5 locules, otherwise unknown; **Ecol** unknown; **Distr** Namaqualand, NC, S Africa.

Note: Up to 23 flowers per cyme are reported for this species.

R. laxipetala L.Bolus NM 1929: 44 ≡ *M. laxipetalum* (L.Bolus) N.E.Br. GC 1930: 32 T L.Bolus 18868 (BOL!).

Cushions 6–7 cm *h*, decumbent branches to 12 cm *l*, **I** 5–12 mm *l*; **L** sheath to 10 mm *l*, connate with the stem for most of its length, epidermis smooth, free parts 3–5 mm *l*, triquetrous, keel with 1–3 reflexed teeth, 3 mm *t*; **Fl** solitary, ∅ 18 mm, **K** 5, pedicels 5 mm *l*, petals pinkish-purple, 8 mm *l*, 1 mm *b*, filamentous staminodes present, filaments apically pink; **Fr** open 6 mm ∅, 5 locules; **Ecol** unknown; **Distr** Laingsburg, WC, S Africa.

Note: Reported to be similar to **R. framesii**, differing in a sharper keel with 2 teeth, laxer petals and longer stigmas.

R. leightonii ≡ **Lampranthus francesiae**
R. leipoldtii ≡ **Antimima leipoldtii**

R. leptocalyx L.Bolus NM 1931: 306 T L.Bolus 19874 (BOL!).

Shrubs with some long branches, **I** > **L**, greyish-yellow; **L** triquetrous with sharp margins and keel, the latter finely serrulate, epidermis glabrous, to 20 mm *l*, c. 5 mm *b*, c. 7 mm *t*; **Fl** solitary or in cymes, bracteoles near the base of the stalk, smaller than the foliage leaves, petals bright reddish-purple, filamentous staminodes and stamens pink; **Fr** grey, top with high valve rims, base funnel-shaped, covering membranes forming a roof over each locule, the distal rim only somewhat raised, but the closing ledge below it very sharp, closing body small, oval, the long end pointing into the locule, expanding keels not touching each other at their bases, no awns, no wings, 5 locules; **Ecol** near the coast, > 400 mm rainfall p.a. at all seasons; **Distr** Riversdale, WC, S Africa.

Note: An incompletely known species.

R. leptophylla = **Acrodon subulatus**

R. lerouxiae (L.Bolus) L.Bolus NM 1950: 219 ≡ *M. lerouxiae* L.Bolus ABH 1923: 170 ≡ *Lampranthus lerouxiae* (L.Bolus) N.E.Br. GC 1930: 212 T le Roux 15197 (BOL!).

Decumbent branches with erect or ascending branchlets, **I** 15–30 mm *l*, 3 mm ∅; **L** trigonous, apiculate, greyish-green, 60 mm *l*, 4–8 mm *b*, 6–8 mm *t*; **Fl** 3, ∅ 56 mm, petals purplish-pink, basally purplish vitatte, filamentous staminodes not mentioned, anthers and pollen yellow; **Fr** base elongate, funnel-shaped, top high from very high rims, covering membranes convex, big closing rodlets in vertical position under the distal end of the covering membranes, closing body tiny, expanding keels not touching each other basally, 5 locules; **Ecol** unknown; **Distr** Namaqualand, NC, S Africa.

Note: The species is a member of the sarmentose group with long, decumbent branches forming big mats as ground covers, but rooting tardily.

R. leucanthera ≡ **Antimima leucanthera**

R. leucosperma L.Bolus NM 1929: 160 T L.Bolus 19007 (BOL!).

Richly branched shrubs 10–20 cm *h*, **I** pubescent when young, to 5 mm ∅ with age basally, c. 10 mm *l*; **L** subglobose to elongate-obtuse, those of a pair hardly connate near their base, pale green, the upper leaf surface finely papillate, keel and margins somewhat translucent, 4–7 mm *l*, 4 mm *t*; **Fl** in 3-flowered cymes, bracteoles highly connate and almost enclosing the cymes, pedicels 3–4 mm *l*, later turning spiny, **K** 6, petals pink to purplish-pink, 5 mm *l*, 1 mm *b*, few fila-mentous staminodes, pinkish-purple, anthers white; **Fr** in cymes of 3 or 7, the capsule falling off the finely papillate stalk when ripe, leaving a spiny stalk behind, which can be overtopped by another cyme in the next season, base somewhat bell-shaped, top almost semi-globose and often with the bases of the stigmas erect in the centre, rims low, covering membranes with distinct closing rodlets, closing body a small hook, expanding keels diverging, short, 6 locules; **S** 2–3 per locule, light ochre, up to 1 mm *l*; **Ecol** unknown; **Distr** Namaqualand, NC, S Africa.

Note: Compared by Bolus (NM 1929: 160) with **R. stricta**. The persisting stigmas in the centre of the fruit resemble conditions found in **Arenifera**.

R. levynsiae = **Antimima pumila**
R. limbata ≡ **Antimima limbata**

R. lineolata (Haw.) Schwantes ZSK 1927: 20 ≡ *M. lineolatum* Haw. 1821: 130 **LT** drawing 397 by Duncanson based on material from Bowie s.n.1819 (K!).

Low shrubs to 15 cm *h*, somewhat mat-forming, younger branches compressed; **L** triquetrous, acuminate, subglaucous, with a line on the leaf sheath, margins pale, keel a little rough, < 13.5 mm *l*; **Fl** solitary, petals light purple, filamentous staminodes not distinguishable; **Fr** unknown; **Ecol** unknown; **Distr** unknown.

Note: The second drawing kept at K under the same name (Duncanson 969/215) shows considerably thicker, stouter leaves and no flowers, it was therefore not considered for the lectotypification. Haworth (1821: 130) named two more varieties, *M. l.* var. *minus* and *M. l.* var. *nitens*.

R. lisabeliae L.Bolus 1966: 125 **LT** Hall 3037 I (BOL!).

Shrubs with decumbent branches to 60 cm *l*, ∅ to 90 cm, **I** brown, black with age, 20–35 mm *l*, 2–4 mm ∅; **L** semi-terete, inconspicuously mucronate, green, shining, 20–32 mm *l*, 3–5 mm *b* and *t*; **Fl** in cymes, pedicels 20–35 mm *l*, bracteoles in their lower half, petals purplish-pink, to 11 mm *l*, 0.5–1.5 mm *b*, filamentous staminodes absent, filaments purplish-pink, anthers and pollen yellow; **Fr** with a closing body and closing rodlets; **Ecol** on loose stones; **Distr** Vredendal, WC, S Africa.

Note: "Sprawling over loose rocklets" (Hall in Bolus 1966: 125).

R. littlewoodii L.Bolus JSAB 1960: 51 T Littlewood 603/59 (BOL!).

Compact, with a thick tap-root and several branches at the top; **L** triquetrous, subulate, those of a pair with a smooth basal sheath, **L** 10–18 mm *l*, basally c. 4 mm *b* and *t*; **Fl** solitary, petals purplish-pink, filamentous staminodes absent, filaments basally white, apically pinkish-purple, ovary top with 5 high elevations; **Fr** unknown; **Ecol** "at the border of Protea Veld and Karoo", > 300 mm rainfall p.a. in winter; **Distr** Ceres, WC, S Africa.

Note: The very compact growth and the absence of filamentous staminodes cast doubt on the placement of the species in **Ruschia**.

R. lodewykii ≡ **Antimima lodewykii**
R. loganii ≡ **Antimima loganii**
R. lokenbergensis ≡ **Antimima lokenbergensis**
R. longifolia ≡ **Astridia longifolia**;
= **Acrodon bellidiflorus**
R. longipes ≡ **Antimima longipes**
R. luckhoffii ≡ **Antimima luckhoffii**
R. lutea ≡ **Rhinephyllum luteum**
R. macovanii ≡ **R. macowanii**

R. macowanii (L.Bolus) Schwantes in Jacobsen 1949: 58 ≡ *M. macowanii* L.Bolus ABH 1926: 37 ≡ *R. macovanii* (L.Bolus) Schwantes ex H.Jacobsen 1981: 533, orthographic variant **LT** Wolley-Dod 1470 (BOL!).

Diffusely branched shrubs, 15–20 cm *h*, the branches often hanging over cliffs, **I** 20–30 mm *l*, ∅ < 5 mm; **L** semi-terete, grey to glaucous, sheaths tumid, 5 mm *l*, **L** 20–35 mm *l*, 4 mm *b* and *t*; **Fl** in cymes, ∅ 22 mm, pedicels 10–20 mm *l*, bracteoles at their middle, petals pink, basally deeper in colour, 8–9 mm *l*, 1.5 mm *b*, filamentous staminodes present, filaments basally white, anthers and pollen pale yellow; **Fr** top low, base funnel-shaped, covering membranes with protruding closing rodlets, closing body a small hook, expanding keels almost touching each other at their bases, 5 locules; **Ecol** near the sea on rocks, in spray; **Distr** Cape, WC, S Africa.

Note: The lectotype possesses a fruit of the Ruschia-type.

R. macrophylla = **Acrodon bellidiflorus**
R. macroura = **R. spinosa**
R. maleolens ≡ **Antimima maleolens**
R. mallesoniae = **Esterhuysenia drepanophylla**
R. marginata ≡ **Lampranthus marginatus**

R. mariae L.Bolus NM 1937: 62 **T** M. Villet 21826 (BOL).

Caespitose, to 3 cm *h*, **I** enclosed, invisible; **L** trigonous, stout, apical tooth present, margins entire or 1–2 denticulate, light and shining glaucous, pinkish suffused, 6–9 mm *l*, sheaths 3 mm *l*, 2.5 mm *b* and *t*; **Fl K** 5, denticulate, petals purplish-pink, 4–7 mm *l*, 1 mm *b*, filamentous staminodes pale, filaments pink, anthers, pollen golden, nectary as if composed of 5 parts; **Fr** 5 locules, otherwise unknown; **Ecol** unknown; **Distr** Steytlerville, EC, S Africa.

Note: With only 3 cm height, the species is one of the few rather compact ones in **Ruschia**.

R. marianae (L.Bolus) Schwantes in Jacobsen 1949: 58 ≡ *M. marianae* L.Bolus ABH 1923: 169 **T** Marloth 17393 (BOL!).

Shrubs to 31 cm *h*; **L** triquetrous, sheaths tumid, glaucous-green to grey, 5–7 mm *l*, epidermis smooth, 50–100 mm *l*, 4–7 mm *b*; **Fl** in cymes, ∅ 40 mm, bracteoles truncate forming a pocket around the pedicel, petals scarlet, to 19 mm *l*, to 5 mm *b*, filamentous staminodes not mentioned, filaments red, ebarbate, anthers and pollen yellow; **Fr** 5 locules, otherwise unknown; **Ecol** unknown; **Distr** unknown.

Note: The species shares the formation of marsupiate bracteoles with **R. tardissima**, a species with a decumbent growth form.

R. maritima ≡ **Amphibolia hutchinsonii**
R. mathewsii ≡ **Antimima mathewsii**

R. maxima (Haw.) L.Bolus FPSA 1947: t. 1013 (see **photo**) ≡ *M. maximum* Haw. 1795: 402 ≡ *Astridia maxima* (Haw.) Schwantes ZSK 1927: 16 **T?**

Shrubs to 1.5 m *h*, young plants and branches arranged in a pyramid, as broad as 2 m with age; **L** lunate, epidermis almost smooth, a fine wax cover present, up to 70 mm *l*, c. 10 mm *b*, c. 20 mm *t*, leaving an elongate mark on the stem after falling off; **Fl** in rather rich cymes, the terminal one overtopping several lateral ones below it, bracteolate and bracteate, **K** 5, petals rose-pink with a darker central stripe, filamentous staminodes basally white, apically dark purple, surrounding the central stamens in a central cone; **Fr** top somewhat turret-shaped, rims low, light brown like the stalks, closing ledge present, covering membranes almost straight, a firm, white columella in the centre, c. 23–32 mm *l*, 6–7 mm ∅, opening widely when wet; **S** pear-shaped, testa cells elevated and with undulate walls, c. 1 mm *l*, c. 65 mm *b*; **Ecol** in crevices of rocks, c. 400 mm rainfall p.a. in winter; **Distr** Clanwilliam, WC, S Africa.

Note: This is probably the tallest species in **Ruschia**, growing well in cultivation and known there since the late 18th century. It flowers reluctantly and only after having reached a height of 60–70 cm.

R. maxwellii ≡ **Antimima maxwellii**
R. menniei ≡ **Antimima menniei**
R. mesklipensis ≡ **Antimima mesklipensis**
R. meyerae ≡ **Antimima meyerae**
R. meyeri = **Antimima papillata**
R. micropetala = **Stoeberia beetzii**
R. microphylla ≡ **Antimima microphylla**

R. middlemostii L.Bolus 1962: 297 **LT** Middlemost 424/61 I (BOL!).

Distally branching shrubs to 40 cm *h*, **I** greyish-brown, dark; **L** trigonous with convex sides, epidermis papillate, yellowish-green, 10–15 mm *l*, 7–10 mm *b*, c. 10 mm *t*, longer and more slender in cultivation; **Fl** solitary, but often grouped near the end of the branch, bracteoles in the middle of the stalk, **K** 6, petals pink, filamentous staminodes present, filaments white, apically purple, surface of gynoecium with densely placed, long papillae; **Fr** top with low rims, base elongate, funnel- to bell-shaped, with long papillae, withering slowly so that the harder bundles persist as spiny stars over several years, covering membranes convex, touching each other at their bases, a pronounced closing rodlet at the distal opening almost touching the hook-shaped, small closing body, expanding keels ending in short awns, no signs of valve wings, 6 locules; **Ecol** on slopes below rocky parts, < 100 mm rainfall p.a. in winter; **Distr** N Namaqualand, NC, S Africa.

Note: Superficially similar to other species with yellow-green leaves, **R. middlemostii** can be recognized by the persistent bundles of the fruits and the papillate gynoecium surface.

R. milleflora ≡ **Eberlanzia dichotoma**
R. minutifolia ≡ **Antimima minutifolia**

R. misera (L.Bolus) L.Bolus in Jacobsen 1960: 1391 ≡ *M. miserum* L.Bolus ABH 1923: 164 (BOL!) **T** without collector BOL 24882 (BOL!).

Shrubs to 37 cm *h*, branches erect, stiff, **I** maroon, 15–30 mm *l*, ∅ 2 mm, **L** subfalcate, trigonous, basally separated in the lower half of the sheath by a triangle of stem tissue, 17 mm *l*, 2.5 mm *b*; **Fl** in rich cymes, the terminal one enlarged by additional, lower storeys, ∅ 7 mm, petals absent, filamentous staminodes present, filaments white, apically yellowish; **Fr** 5 locules, otherwise unknown; **Ecol** unknown; **Distr** uncertain.

Note: **R. misera** differs from all other species by the absence of petals, a feature known from **Eberlanzia** so far only. Both taxa differ markedly in growth and leaves.

R. modesta ≡ **Antimima modesta**
R. modesta f. *glabrescens* = **Antimima modesta**

R. mollis (A.Berger) Schwantes GW 1929: 69 ≡ *M. molle* A.Berger 1908: 130 nomen illeg. non Aiton 1789: 192 ≡ *M. lene* N.E.Br. GC 1930: 32 **LT** Berger 1908: 130 f 23.

Low shrubs, branches decumbent to spreading, **I** two-edged, brownish, later grey; **L** trigonous, spreading, with rounded sides, mucronate, margins marked by lines, greyish-green, finely dotted, 12–15 mm *l*, 4–6 mm *b*; **Fl** solitary, bracteoles at c. half the pedicel, ∅ c. 15 mm, **K** 5, petals red, white towards the base, filamentous staminodes present, filaments dark red, basally white; **Fr** 5 locules; **Ecol** unknown; **Distr** unknown.

Note: Berger (1908: 130–131) described the plant under the incorrect name of *M. molle*, which is a species of **Malephora** flowering yellow. The illegitimate name was used as a basionym by Schwantes (1929: 69), who recognized the error.

R. montaguensis L.Bolus NM 1934: 432 **T** Leighton 20986 (BOL!).

Shrubs with erect branches to 30 cm *l*, **I** 20–40 mm *l*, 3–4 mm ∅; **L** erect, semi-terete, connate base with a separating line of 5–7 mm, epidermis cells a little raised, margin and keel as faint edges only, 13–20 mm *l*, 4 mm *b* and *t*; **Fl** in terminal cymes of 3–5 and lower flowers, pedicels to 10 mm *l*, **K** 5, petals purplish-pink, inconspicuously vittate, 7 mm *l*, 1 mm *b*, filamentous staminodes present, filaments white, apically purple; **Fr** top raised and broader than the top of the base, base elongate, funnel-shaped, covering membranes with distinct, big closing rodlets, closing body a small hook, expanding keels touching each other basally, short and divergent, no valve wings or awns, 5 locules, ∅ 6 mm; **Ecol** unknown; **Distr** Montagu, WC, S Africa.

Note: The connate leaf bases with a distinct line resemble those of **R. tumidula.**

R. monticola ≡ **Ottosonderia monticola**
R. mucronata ≡ **Antimima mucronata**

R. muelleri (L.Bolus) Schwantes ZSK 1927: 20 (see **photo**) ≡ *M. muelleri* L.Bolus ABH 1922: 130 **T** Müller 57/14 (BOL!).

Shrubs to 120 cm *h* with erect branches, **I** shiny, orange-brown, later grey and finally blackish, the pith yellow and spongy, 10–13 mm ∅, 25–30 mm *l*, to 60 mm *l* in young ones; **L** almost terete, epidermis almost smooth, 35–50 mm *l*, 5 mm *b*, to 6 mm *t*; **Fl** in rich cymes, bracteoles tiny, ∅ 10 mm, **K** 5, petals deep pink, basally reddish-vittate on paler ground, 4 mm *l*, filamentous staminodes apically dark red, basally golden or white, anthers pale, pollen golden; **Fr** light grey, base markedly bell-shaped, top round and reaching over the outer edge, rims low, covering membranes almost straight, distally with rather sharp closing ledges or broad closing rodlets, expanding keels diverging, closing body absent, valve wings absent;

Ecol in high shrubby associations in loamy shallow pockets of granite or sandstone; **Distr** Lüderitz-South, Namibia (?), Namaqualand, NC, S Africa.

Note: Two incidents have made it difficult to identify fresh material of **R. muelleri**: a probable error in the place of the origin, and a subsequent error in the synonymy. On the type sheet, the origin of the material is given as Pillans' garden at Rosebanks, Cape Town, where it had been grown from a cutting from material at Kirstenbosch collected by Müller and reported to have come from Lüderitzbucht. Extensive searches in Lüderitz-South have not yielded any plants of this species; yet, plants agreeing in all critical features with the type have been found in Namaqualand, as far north as Kosies. Friedrich (1961: 133, 1970: 116) cited **R. psammophila** and **R. ruschiana** as synonyms of **R. muelleri**, probably without having seen the type. Material of all three species shows only one character state common to all: the absence of closing bodies in the fruit. As a consequence of both incidents, **R. muelleri** has most probably been reported wrongly for Namibia. The species is most similar to **R. burtoniae** from further south, differing from the latter mainly by a thicker pith of spongy tissue which allows the stems to be pressed and squashed easily, even in herbarium material. Neither **R. burtoniae** nor **R. muelleri** possess closing bodies.

R. muiriana (L.Bolus) Schwantes ZSK 1927: 106 ≡ *M. muirianum* L.Bolus ABH 1927: 94 **T** Muir 3938 (BOL!).

Shrubs to 25 cm *h*, **I** reddish to grey; **L** subulate with a recurved tip, epidermis rough from raised dots, i.e. elevations above tannin idioblasts, c. 25 mm *l*, c. 3 mm *b* and *t*; **Fl** solitary, bracteoles near the middle of the pedicel, **K** 5, ∅ c. 19 mm, 15–20 petals purplish-pink, purplish vittate, 8 mm *l*, to 2 mm *b*, few filamentous staminodes surrounding the stamens in a central cone; **Fr** top with low rims, base elongate, funnel-shaped, covering membranes convex, with long closing rodlets at their distal ends, closing body a small hook, expanding keels well separated basally, valve wings absent, 5 locules; **Ecol** c. 300 mm rainfall p.a. mainly in March and November; **Distr** Ladismith, WC, S Africa.

Note: The species is very similar to **R. vaginata** in leaf shapes and sizes and in the epidermis rough from elevations, which show dominantly along the keel often. This leaves the purplish flowers of **R. muiriana** as the main difference between the two species.

R. multiflora (Haw.) Schwantes ZSK 1927: 20 ≡ *M. multiflorum* Haw. 1795: 318 **T**?

Shrubs; **L** triquetrous, perfoliate; flowers in rich **Inf**, subsessile, petals white; **Fr** unknown; **Ecol** unknown; **Distr** unknown.

Note: The species is likened to **R. imbricata** in the size and shape of the flowers, differing from the latter by the richer cymes.

R. muricata L.Bolus NM 1928: 21 (see **photo**) ≡ *M. pellanum* N.E.Br. GC 1930: 32 **LT** Hutchinson 1071 II (BOL!).

Cushion-shaped plants, decumbent but rarely creeping, **I** enclosed by remains of old leaves for several years; **L** trigonous but markedly rounded on top, keel and margins with roundish teeth, roundish elevations on some leaves on the sides, free parts 4–7 mm *l*, sheaths to 5 mm *l*, c. 4 mm *b*, c. 4 mm *t*; **Fl** solitary, bracteoles in the lower half of the stalk, free parts shorter than those of foliage leaves, 12 mm ∅, petals pale pink, filamentous staminodes present; **Fr** top rather high from high rims, base elongate, funnel-shaped, covering membranes declining from a high centre, closing ledges thickening towards the expanding keels, closing body small, hood-shaped, expanding keels basally parallel, but not touching each other, apically ending in awns merging into awn-shaped ends of narrow valve wings, 5 locules; **S** ± smooth, pale brown, 0.65–0.85 mm *l*, 0.5–0.6 mm *b*; **Ecol** in loamy to finely grained gravelly soils of flats with a grass cover after good rains, < 100 mm rainfall p.a. in March or March and November; **Distr** Calvinia, Kenhardt, Namaqualand, NC, S Africa.

Note: Friedrich (1970) reported the occurrence of the species in SE Namibia. It is conspicuous in habitat because the greyish cushions grow in swarms together. The possession of narrow valve wings is rare in **Ruschia**.

R. mutata ≡ **Lampranthus mutatus**
R. mutica ≡ **Antimima mutica**
R. namaquana = **Amphibolia rupis-arcuatae**
R. namaquana var. *quinqueflora*
≡ **Eberlanzia ebracteata**

R. namusmontana Friedrich 1968: 218 (see **photo**) ≡ *R. foliosa* L.Bolus 1962: 300 nomen illeg. non (Haw.) Schwantes ZSK 1927: 20 ≡ *R. foliolosa* L.Bolus ex H.Jacobsen 1981: 529 orthographic variant **T** Hall 1927 (BOL!).

Shrubs to 35 cm *h*, **I** brown; **L** crowded at the ends of branches, crescent-shaped, yellowish-green, epidermis smooth, but covered by wax, 15–35 mm *l*, 5–10 mm *b*, c. 10 mm *t*; **Fl** solitary between the last leaf-pair, bracteoles at base of stalk, **K** 6, petals pink, filamentous staminodes few; **Fr** top a low cone with low rims, base short, funnel-shaped, covering membranes convex, closing ledge low and long, closing body absent, expanding keels ending in long awns, valve wings narrow and tapering to base and tip, 6 locules; **Ecol** on rocky slopes on the escarpment, < 100 mm rainfall p.a. in winter; **Distr** Lüderitz-South, Namibia.

Note: Similar in appearance to **R. middlemostii.** The extremely narrow valve wings exclude this species from **Ruschia.**

R. nana L.Bolus NM 1929: 75 ≡ *M. reductum* N.E.Br. GC 1930: 32 non 33 nomen illeg. **T** H. Bolus 13417 (BOL!).

Caespitose, to 4 cm *h*, **I** ∅ to 2 mm; **L** subovate to trigonous, mucro occasionally recurved, margins and keel partly cartilaginous, 3–5 mm *l*, 3 mm *t*; **Fl** solitary, ∅ 26 mm, pedicels 4–7 mm, bracteoles embracing the base of the flower, **K** 6, petals white, 12 mm *l*, 1 mm *b*, filamentous staminodes 6 mm *l*; **Fr** 6 locules, otherwise unknown; **Ecol** unknown; **Distr** Laingsburg, WC, S Africa.

Note: The compact growth and the number of 6 calyx lobes and locules are an unusual combination in this genus.

R. neilii ≡ **Stayneria neilii**

R. nelii Schwantes MDK 1930: 66 **T**?

Dwarf shrubs with very short **I**; **L** ovate, upper surface flat, lower leaf surface almost semi-globose, keel inconspicuous, bluish-green, dotted, to 5 mm *l*, 5 mm *b*, 3 mm *t*; **Fl** unknown; **Fr** unknown; **Ecol** unknown; **Distr** unknown.

Note: No material referring to this name has been found in the collection of Schwantes at HGB.

R. neovirens Schwantes in Jacobsen & Volk 1950: 124 nomen invalid ≡ *M. virens* Haw. 1821: 121 non Salm-Dyck 1840: § 24 f 1 **T**?

Erect shrubs c. 15 cm *h*, branches later decumbent; **L** compressed-triquetrous, subacinaciform, green with dots, similar to those of **R. maxima**; **Ecol** unknown; **Distr** Port Elizabeth?, S Africa.

Note: A new name was required because **R. virens** exists already. No type can be found, however, and the name remains invalid until material has been identified.

R. nieuwerustensis L.Bolus NM 1930: 210 **T** Pillans 6226 (BOL!).

Decumbent branches 10–15 cm *l*, **I** ochre, 15–20 mm *l*, ∅ to 4 mm; **L** subterete, obtuse, light green, sheath with a line, 2.5 mm *l*, **L** 25–30 mm *l*, 6 mm *t*; **Fl** in cymes, ∅ c. 30 mm, pedicels 10–20 mm *l*, bracteoles at their middle, **K** 5, petals pinkish-purple, inconspicuously vittate, to 14 mm *l*, 2.5 mm *b*, filamentous staminodes not mentioned, filaments pinkish-purple, anthers and pollen white; **Fr** 5 locules, otherwise unknown; **Ecol** unknown; **Distr** Vanrhynsdorp, WC, S Africa.

Note: Decumbent growth in combination with well-developed cymes are typical of a group of species in **Ruschia.**

R. nivea = **R. rupicola**
R. nobilis ≡ **Antimima nobilis**

R. nonimpressa L.Bolus NM 1932: 373 **T** James 188 (BOL!).

Mat-forming, **I** 15 mm *l*; **L** triquetrous, keel with 1–2 teeth, sheath merging into the stem, epidermis cells raised, **L** 10–12 mm *l*, 3–4 mm *b* and ∅: **Fl** 1, bracteoles placed above the middle of the pedicels, **K** 5, petals purplish-pink, 5–7 mm *l*, 0.75 mm *b*, filamentous staminodes pink, pollen golden; **Fr** base funnel-shaped, top raised, rims low, covering membranes convex with closing ledges at their distal ends, closing body a small hood, expanding keels distant from each other basally, 5 locules; **Ecol** unknown; **Distr** Cradock, EC, S Africa.

Note: Similar to **R. depressa.**

R. nordenstamii ≡ **Antimima nordenstamii**

R. obtusa L.Bolus NM 1929: 79 ≡ *M. hebes* N.E.Br. GC 1930: 32 **T** Pillans 5750 (BOL!).

Erect shrubs to 40 cm *h*, **I** later blackish-brown, 35–40 mm *l*, ∅ to 5 mm; **L** light greyish-green, obtusely carinate, epidermis smooth, 20–45 mm *l*, 5–8 mm *b* and *t*; **Fl** 1–3, ∅ 18 mm, bracteoles hooded, embracing the base of the flower, **K** 6, petals purplish-pink, 8 mm *l*, to 1.5 mm *b*, filamentous staminodes absent, anthers white; **Fr** base short, funnel-shaped, top raised, but rims low, closing body hook-shaped, 6 locules; **S** 1.1 mm *l*, 0.8 mm *b*; **Ecol** on rocky slopes in shallow places, < 100 mm rainfall p.a. in winter, fog; **Distr** NW Namaqualand, NC, S Africa.

Note: An unripe fruit on the type sheet agrees in critical features with those of **Ruschia.**

R. obtusifolia = **Antimima watermeyeri**
R. octojugis ≡ **Octopoma octojuge**
R. octonaria ≡ **Enarganthe octonaria**

R. odontocalyx (Schlechter & Diels) Schwantes ZSK 1926: 187 (see **photo**) ≡ *M. odontocalyx* Schlechter & Diels in Schultze 1907: 693 **T** Schultze 440 (B!).

Low, broad shrubs; **L** trigonous, regularly with 1–2 keel teeth, epidermis with low papillae, few long ones on margins, free parts c. 15 mm *l*, c. 4 mm *t*; **Fl** c. 50 white petals, c. 45 white filamentous staminodes surrounding the 150–170 stamens in a central cone; **Fr** rims dipping in the centre, base long bell- to funnel-shaped, closing bodies broadening to the interior of the locule, the funicles protruding from beneath into all possible directions, 5 very deep locules; **Ecol** in gravel among quartzitic rocks, < 100 mm rainfall p.a. in March; **Distr** Bethanien, Namibia.

Note: **R. odontocalyx** is understood here in its narrow sense, excluding **R. rupicola.** Both species are very similar in their fruit structure, but they differ clearly: leaves of **R. odontocalyx** are more slender than those of **R. rupicola**, and they possess keel teeth regularly.

R. orientalis L.Bolus FPSA 1927: t. 277 (see **photo**) **T** Britten 5528 (BOL!).

Rather dense, erect to later spreading shrubs, to 35 cm *h*, ∅ to 50 cm,; **L** bright green, triquetrous, epidermis almost smooth, to 20 mm *l*, to 3 mm *b*, to 4 mm *t*; **Fl** solitary, bracteolate, **K** 5, ∅ 20–30 mm, petals rose-purple, pink filamentous staminodes surrounding the central white filaments; **Fr** base funnel-shaped, top almost semi-orbicular from high valve rims, covering membranes convex, closing ledge short or closing rodlet long, closing body broadening a little to the centre, expanding keels touching each other at their bases, ∅ 7–8 mm; **S** roundish, dark brown, testa cells elongate-elevated; **Ecol** in gravel in scrubby associations and in grass, 400–500 mm rainfall p.a. in all seasons; **Distr** Albany, Alexandria, Kirkwood, Port Elizabeth, Uitenhage, EC, S Africa.

Note: The plants look conspicuous in habitat, due to their bright green leaves and their rather big flowers, which suggest also that they could well be used in gardening in appropriate climates.

R. orsmondiae = **R. canonotata**
R. ottosonderi ≡ **Ottosonderia monticola**
R. oviformis ≡ **Antimima oviformis**
R. pakhuisensis ≡ **Lampranthus pakpassensis**

R. pallens L.Bolus NM 1929: 146 **T** Pillans 5739 (BOL!).

Erect shrubs to 60 cm *h*, **I** grey, 25–35 mm *l*, c. 5 mm ∅; **L** pale glaucous-green, trigonous to semi-terete, apically recurved, sheath tumid, epidermis cells raised, 35–50 mm *l*, 4–5 mm *b* and *t*; **Fl** in cymes, pedicels 25–30 mm *l*, bracteoles at their middle, 11 mm *l*, the **Inf** distinct from the vegetative part of the plant, **K** 5, petals palest pink, 8 mm *l*, 1 mm *b*, filamentous staminodes very few, filaments white, apically pink; **Fr** 5 locules, otherwise unknown; **Ecol** near the sea; **Distr** Namaqualand, NC, S Africa.

Note: Rich cymes of up to 11 flowers are reported for this species.

R. papillata ≡ **Antimima papillata**

R. paripetala (L.Bolus) L.Bolus NM 1950: 221 ≡ *M. paripetalum* L.Bolus ABH 1927: 87 **T** Pillans 17759 (BOL!).

Shrubs 30–60 cm *h*, **I** grey, 10–25 mm *l*, ∅ to 8 mm; **L** spreading, epidermis papillate, sheath 2–7 mm *l*, semi-cymbiform, 20–32 mm *l*, to 12 mm *b* and *t*; **Fl** solitary, ∅ 27 mm, pedicel 5–14 mm *l*, **K** 6, petals 35, linear, 12 mm *l*, 1–1.5 mm *b*, filamentous staminodes white, filaments white, apically golden to red; **Fr** top raised moderately, base funnel-shaped, covering membranes with distinct, long closing ledges, closing body small, expanding keels ending in elongate awns, 6 locules; **Ecol** near the coast in winter rainfall area; **Distr** Namaqualand, NC, S Africa.

Note: Characterized by the possession of hexamerous flowers, the species is remarkable for its big, stout fruits.

R. paripetala var. *occultans* = **Antimima perforata**
R. parvibracteata ≡ **Eberlanzia parvibracteata**

R. parviflora (Haw.) Schwantes ZSK 1927: 19 ≡ *M. parviflorum* Haw. 1803: 95 nomen illeg. non Jacquin 1798: 15 **LT** drawing of Duncanson 989/566 (K!).

Shrubs with erect branches to 90 cm *h*; **L** triquetrous, margins and keel finely serrate, tip recurved, sheath connate with the stem, a deep line separating the two leaves of a pair, 9–13.5, or in younger ones, 27 mm *l*; **Fl** solitary, beautiful and small, white to pearl-coloured; **Fr** unknown; **Ecol** unknown; **Distr** unknown.

Note: The drawing named *M. parviflorum* at K is not dated, but it represents most probably a branch from the type plant.

R. parvifolia L.Bolus JB 1928: 12 **T** L.Bolus 18588 (BOL!).

Cushions to 8 cm *h*, **I** ∅ 2–3 mm; **L** free parts triquetrous, keel with 2 teeth, one of them apical, epidermis with short papillae, elevations on the sides, greyish-glaucous, sheath to 5 mm *l*, basally connate with the stem, free parts 6–8 mm *l*, 3 mm *b* and *t*; **Fl** solitary, ∅ 20 mm, pedicel enclosed by bracteoles, **K** 5, petals in 5 groups, pink, 8 mm *l*, 0.5 mm *b*, filamentous staminodes and stamens pink; **Fr** 5 locules, otherwise unknown; **Ecol** on stony hills; **Distr** Beaufort West, WC, S Africa.

Note: Very similar to **R. unidens**, which forms higher shrubs up to 16 cm tall, at least in cultivation.

R. patens L.Bolus NM 1928: 145 ≡ *M. diplosum* N.E.Br. GC 1930: 32 **T** Immelman 2205/27 (BOL!).

Shrubs with erect branches to 25 cm *l*, **I** 25–35 mm *l*; **L** spreading-reflexed, keel and margins serrulate; **Fl** 1–3, ∅ 24 mm, **K** 5, petals purplish-pink, deeply vittate, 10 mm *l*, 1.5 mm *b*, filamentous staminodes pink, apically deeper in colour, anthers and pollen pale yellow; **Fr** top low, base funnel-shaped, covering membranes with distinct closing rodlets, closing body small, valve wings absent, 5 locules; **Ecol** unknown; **Distr** Piquetberg?, WC, S Africa.

Note: The recurved leaves resemble **R. vaginata**.

R. patulifolia L.Bolus NM 1930: 206 **T** Frames 19210 (BOL!).

Shrubs to 30 cm *h* with long shoots dominating and many lateral short shoots, **I** ochre, later grey, 10 mm *l*, ∅ 3 mm; **L** trigonous, obtuse, green, to

7 mm *l*, 2–3 mm *b* and *t*; **Fl** solitary, ∅ 12 mm, pedicels 8 mm *l*, bracteolate in their middle, bracteoles 4 mm *l*, petals in 5 groups, pinkish-purple, filamentous staminodes pink, filaments white, apically red; **Fr** unknown; **Ecol** unknown; **Distr** Vanrhynsdorp, WC, S Africa.

Note: The dominating long shoots characterize the plants in habitat.

R. pauciflora L.Bolus NM 1932: 352 **T** L.Bolus 20126 (BOL!).

Creeping branches to 40 cm *l*, **I** winged, 15–50 mm *l*, pale brown; **L** triquetrous, keel serrulate, epidermis almost smooth, to 25 mm *l*, 4 mm *b*, to 6 mm *t*; **Fl** solitary, lateral ones aborted (always?), bracteoles in the middle of the pedicel reaching to the base of the flower, **K** 5, petals purplish-pink, filamentous staminodes apically red, stigmas deep red; **Fr** 5 locules; **Ecol** unknown; **Distr** Piquetberg, WC, S Africa.

Note: Remarkable are the leaves said to be thicker than broad.

R. paucifolia ≡ **Antimima paucifolia**

R. paucipetala L.Bolus JSAB 1960: 51 **T** Littlewood 659/59 (BOL!).

Shrubs to 24 cm *h*, ∅ to 28 cm, densely branched, **I** 5–10 mm *l*, ∅ 3 mm; **L** ovate to subglobose, epidermis almost smooth, 5–10 mm *l*, 4 mm *b*, 5 mm *t*; **Fl** 1–3, ∅ 12 mm, pedicels 5–6 mm *l*, **K** 5, petals purplish-pink, filamentous staminodes absent, filaments pale pink; **Fr** 5–6 locules, otherwise unknown; **Ecol** unknown; **Distr** Namaqualand, NC, S Africa.

Note: The species may not belong to **Ruschia**.

R. pauper ≡ **Antimima pauper**
R. peersii ≡ **Antimima peersii**

R. perfoliata (Miller) Schwantes ZSK 1926: 187 ≡ *M. perfoliatum* Miller 1768: nr 19 **LT** Dillenius 1732: f 240.

Sublignose shrubs with some basal branches developing further lateral branches, **I** white with age; **L** merging into the stems, free parts with 2–4 stout teeth along keel and margins, triquetrous, grey; **Fl** solitary, terminal, but several developed on subsequent sections of stems in a season, **K** 5, petals pale purple, filamentous staminodes present; **Fr** 5 locules, otherwise unknown; **Ecol** unknown; **Distr** uncertain.

Note: The perfoliate Mesembryanthema have been known since the 17th century in Europe, but they are still rare in cultivation, probably due to their rather spreading, loose growth requiring some space.

R. perforata ≡ **Antimima perforata**
R. persistens ≡ **Antimima persistens**; = **R. intricata**

R. phylicoides L.Bolus NM 1929: 75 ≡ *M. phylicoides* (L.Bolus) N.E.Br. GC 1930: 32 **T** Pillans 17925 (BOL!).

Shrubs to 15 cm *h*, **I** 5–15 mm *l*, 5 mm ∅; **L** triquetrous, keel and margins marked by dark idioblasts under the surface, thus finely crenulate or rough, epidermis smooth on sides, **L** 5–10 mm *l*, 1.5 mm *t*; **Fl** 3–5 per cyme, pedicels 2–4 mm *l*, to 7 mm in fruit, **K** 5, petals dull white, to 1.5 mm *l*, 0.75 mm *b*, filamentous staminodes basally with brown flecks (from age?), filaments white; **Fr** brown, hard, base funnel-shaped with convex sides, top semi-globose, covering membranes convex, closing ledge distinct, closing body absent, expanding keels touching each other basally, 5 locules; **Ecol** in sandy plains; **Distr** Namaqualand, NC, S Africa.

Note: The lack of a closing body in the locule excludes this species from **Ruschia**.

R. pillansii = **Eberlanzia schneideriana**
R. pilosula ≡ **Antimima pilosula**

R. pinguis L.Bolus NM 1929: 146 **T** L.Bolus 18978 (BOL!).

Shrubs to 10 cm *h*, branches ascending to erect, **I** 10 mm *l*, 3–4 mm ∅,; **L** terete to inconspicuously club-shaped to subfalcate, glaucous-green, epidermis with short papillae, 25–30 mm *l*, 9 mm *b*, 13 mm *t*; **Fl** in cymes, ∅ 22 mm, peduncle 30 mm *l*, **K** 5, petals in 5 groups, white, 7 mm *l*, to 1 mm *b*, filamentous staminodes numerous, white, pollen yellow, stigmata purplish-brown; **Fr** 5 locules, otherwise unknown; **Ecol** unknown; **Distr** Namaqualand, NC, S Africa.

Note: The remains of a capsule on the type sheet do not permit a more detailed description, all inner parts having been destroyed.

R. piscodora ≡ **Antimima piscodora**

R. polita L.Bolus NM 1932: 332 **T** Nel 9095 (BOL).

Cushion-shaped, to 7 cm *h*; **L** trigonous, keel and margins cartilaginous, apically crenulate, shining, pale glaucous, to 11 mm *l*, 5 mm *b* and *t*, sheaths to 6 mm *l*; **Fl** bracteoles clasping the pedicel, **K** 5, petals pink, to 10 mm *l*, 0.75 mm *b*, anthers yellow; **Fr** 5 locules, otherwise unknown; **Ecol** unknown; **Distr** Worcester, WC, S Africa.

Note: Bolus (1954: 260) supported her preliminary placement (indicated by a question mark in the original description) of this species in **Ruschia** later (Bolus 1954: 260), when she had seen a fruit. Klak (2000: 41) placed the taxon recently as a synonym with **Braunsia geminata**, based on the cartilaginous margins and keel. Studies in leaf surfaces and capsules are needed to confirm this view.

R. pollardii Friedrich 1961: 136 **T** Merxmüller & Giess 2664 (M!).

Shrubs to 30 cm *h*, short shoots mainly opposite, **I** brown, the upper ones flattened and bi-angular; **L** club-shaped, surface "farinaceous", probably from a marked cover of wax, 10–30 mm *l*, 4–7 mm *b*, 4–10 mm *t*; **Fl** solitary, ± falcate bracteoles placed near the base of the stalk, **K** 6, petals in 6 groups, white, filamentous staminodes also white; **Fr** top with low rims, base funnel-shaped, covering membranes convex, closing rodlet pronounced, closing body slender, oval to roundish, very variable in type material, expanding keels ending in short awns, 6 locules; **Ecol** < 100 mm rainfall p.a. in winter; **Distr** Lüderitz-South, Namibia.

Note: The species is similar in appearance to **R. middlemostii** and **R. namusmontana**, differing from the first by the absence of papillae and from the second by the possession of a closing body.

R. primosii L.Bolus NM 1936: 44 (see **photo**) **T** Primos 21463 (BOL!).

Compact, c. 3 cm *h*; **L** trigonous, apically rounded, bright green, epidermis cells smooth, but elevations above subhypodermal tannin idioblasts conspicuous in dry leaves, whitish, up to 11 mm *l*, perhaps heterophyllous, because sheaths differ in length between 1/3 and 2/3 of the entire leaf length, 5 mm *b*, 3 mm *t*; **Fl** solitary, bracteoles near the base, **K** 5, petals pink, filamentous staminodes pale, filaments pale pink; **Fr** base bowl-shaped, covering membranes very high in the centre, rims very high, too, closing ledges distinct, closing body medium sized, white, crenulate, expanding keels not touching each other at bases, ending in long awns, 5 locules; **Ecol** < 100 mm rainfall p.a. in March and November mainly; **Distr** Namaqualand, NC, S Africa.

Note: **R. primosii** belongs with the group of species possessing whitish elevations on the leaves; it differs from most others of this group by its bright green leaf colour. In habit, it is most similar to **R. deminuta** from Namibia.

R. profunda ≡ **Lampranthus profundus**
R. prolongata ≡ **Antimima prolongata**
R. promontorii = **Amphibolia laevis**
R. propinqua ≡ **Antimima propinqua**
R. prostrata ≡ **Antimima prostrata**
R. psammophila = **R. ruschiana**

R. pulchella (Haw.) Schwantes ZSK 1927: 20 ≡ *M. pulchellum* Haw. 1803: 72 **T**? = *R. pulchella* (Haw.) Schwantes var. *caespitosa* L.Bolus NM 1929: 123 **T** Compton 18956 (BOL!).

Caespitose with creeping branches; **L** triquetrous, connate, placed densely together, ending in fine points, c. 5 mm *l*, margins ciliate-pubescent; **Fl** solitary, pedicels with bracteoles more hairy than the foliage leaves at their middle, **K** 5, petals pale inside, apically darker purple, purple outside, filaments basally white, anthers yellowish; **Fr** 5 locules; **Ecol** unknown; **Distr** uncertain, S Africa.

Note: The combination of decumbent branches and solitary flowers is rare in **Ruschia** and may be due to cultivation.

R. pulvinaris L.Bolus NM 1931: 241 **T** du Plessis 2004/29 (BOL!).

Cushion-forming, to 7 cm *h*, **I** enclosed when young, to 8 mm ∅ at the base of the plant; **L** trigonous to semi-terete, acuminate and pungently apiculate, keel sometimes toothed, epidermis almost smooth, but keel and margins with distinct, low papillae, glaucous-green, to 13 mm *l*, sheaths to 2 mm *l*; **Fl** solitary, ∅ 15–20 mm, pedicels to 10 mm *l*, bracteoles in their lower half, **K** 5, petals pinkish-purple, filamentous staminodes pinkish-purple, filaments pink, anthers and pollen golden; **Fr** unknown; **Ecol** unknown; **Distr** Steynsburg, EC, S Africa.

Note: Reported to be similar to **R. indurata**.

R. pumila ≡ **Antimima pumila**

R. punctulata (L.Bolus) L.Bolus ex H.E.K.Hartmann Bradleya 1998: 85 ≡ *M. punctulatum* L.Bolus NM 1928: 17 ≡ *R. unca* (L.Bolus) L.Bolus var. *punctulata* (L.Bolus) L.Bolus NM 1950: 220 ≡ *Lampranthus punctulatus* (L.Bolus) L.Bolus NM 1939: 169 **T** Levyns 1301/26 (BOL!).

Low shrubs to 10 cm *h*, ∅ to 30 cm, **I** ochre to light grey; **L** subulate with a recurved, whitish tip, old leaves persisting on the plant, surface rough from elevations above tannin idioblasts, to 14 mm *l*, c. 3 mm *b* and *t*; **Fl** solitary, bracteoles near the base of the pedicel, c. 14 mm ∅, petals purplish-pink, filamentous staminodes pink, filaments pink, pollen golden; **Fr** top with low rims, base funnel-shaped, covering membranes convex with marked closing rodlets at their outer rim, closing body a small hook, expanding keels not touching each other basally, valve wings absent, 5 locules; **Ecol** c. 200 mm rainfall p.a. mainly in March and November; **Distr** Calvinia, NC, S Africa.

Note: The species has been united with **Lampranthus uncus** in the past. Fruits on the type material and the rough leaf surfaces clearly place the species in **Ruschia**, close to **R. vaginata**, from which it differs mainly by its smaller leaves and the purplish flowers. Following the description, **R. punctulata** is similar to **R. campestre**, stouter in growth and from an area further south east. The name has been attributed to L. Bolus, but no valid publication has been found untis this date.

R. pungens (A.Berger) H.Jacobsen NCSJ 1958: 78 ≡ *M. pungens* A.Berger 1922: 633 **T** Rogers 102 (B, isotype BOL!).

Erect shrubs, **I** ochre to grey, with scars from the sheath after it has fallen off, 20–25 mm *l*; **L** trigonous, recurved, obtuse, epidermis cells well raised, 20–30 mm *l*, sheath tumid; **Fl** in elongate, compound cymes of several dichasia at right angles, pedicels turning into blunt spines after the fruits have fallen off, petals longer than **K**, filamentous staminodes not mentioned; **Fr** 5 locules, otherwise unknown; **Ecol** unknown; **Distr** Albany (?), EC, S Africa.

Note: Most remarkable is the report of a leaf sheath persisting for some time after the leaves have fallen off.

R. puniens = **R. intricata**
R. purpureostyla ≡ **Cerochlamys purpureostyla**
R. pusilla ≡ **Antimima pusilla**

R. putterillii (L.Bolus) L.Bolus NM 1950: 220 (see **photo**) ≡ *M. putterillii* L.Bolus ABH 1925: 4 **LT** Putterill 1443 (BOL!).

Low growing shrubs with a rather dense centre and long shoots from it, branches erect and later decumbent, or erect in other bushes, rootstock woody and thick; **L** almost lunate in side view, triquetrous, the white tip bent back, epidermis with prominently raised dots above the tannin idioblasts, 8–10 mm *l*, 3–4 mm *b* and *t*; **Fl** solitary, distant, **K** 5, petals rose-purple, filamentous staminodes and stamens purple; **Fr** base short funnel-shaped, top semi-globose in side view from high valve rims, covering membranes over a locule forming a roof, big closing rodlets placed at the opening, closing body a small hook, expanding keels distant at their bases, short; **S** brown, testa with lowly elevated cells, > 1 mm *l*, c. 0.6 mm *b*; **Ecol** in crevices on exposed rock plates, at altitudes > 1 600 m, > 500 mm rainfall in summer; **Distr** W Lesotho; E EC, E FS, S Africa.

Note: The species grows in deep crevices where water can be retained in cisterns after rain, the long shoots reaching far over the sheer rock surface. More rarely, the plants grow under and in bushes, in gravelly soil. Regrowth from the centre is rich, the long shoots showing often internodes much longer than the leaves.

R. pygmaea ≡ **Antimima pygmaea**
R. quadrisepala ≡ **Octopoma quadrisepalum**
R. quarzitica ≡ **Antimima quarzitica**

R. radicans L.Bolus NM 1928: 146 ≡ *M. pronum* N.E.Br. GC 1930: 32 **LT** Leipoldt 1411/25 III (BOL!).

Creeping, branches to 33 cm *l*, **I** to 35 mm *l*; **L** trigonous, finely papillate, 10–15 mm *l*, 5 mm *t*; **Fl** solitary, ∅ 18 mm, pedicels enclosed by bracteoles, petals 8 mm *l*, 1 mm *b*, filamentous staminodes pink, inner white, apically deep purple, anthers and pollen white; **Fr** 5 locules, otherwise unknown; **Ecol** unknown; **Distr** Clanwilliam, WC, S Africa.

Note: Differing from the similar **Antimima watermeyeri** by the creeping habit.

R. rariflora L.Bolus NM 1933: 414 **T** Pillans 6071 (BOL!).

Branches spreading to decumbent, **I** 20–50 mm *l*, 5 mm ∅; **L** subterete, apically recurved, glaucous green, epidermis cells smooth to somewhat raised, **L** sheath 8 mm *l*, with a line, 40–80 mm *l*, (3–)5 mm *t*; **Fl** in terminal and lateral cymes forming a compound **Inf**, pedicels 20–25 mm *l*, ∅ 24 mm, **K** 5, petals pink, inconspicuously vittate, 7–10 mm *l*, 2 mm *b*, filamentous staminodes present, filaments white, apically pink, anthers and pollen dirty golden; **Fr** 5 locules, otherwise unknown; **Ecol** unknown; **Distr** Clanwilliam, WC, S Africa.

Note: The very smooth leaves are a remarkable feature of this species.

R. renniei = **Ebracteola montis-moltkei**

R. recurva (Moench) H.E.K.Hartmann Bradleya 1998: 85 (see **photo**) ≡ *M. recurvum* Moench 1802: 190 non Haw. 1819: 90 nec L.Bolus 1932:367 **NT** Hartmann & Dehn 19507 (HBG!).

Shrubs to 35 cm *h*, but the branches decumbent later; **L** subtriquetrous, papillate, a little rough from elevations, the tip markedly recurved; **Fl** solitary, **K** 5, petals white, filamentous staminodes white; **Fr** 5 locules; **Ecol** on calcareous, gravelly soil in shrubby grassland, > 400 mm rainfall p.a. at all seasons; **Distr** Uitenhage, EC, S Africa.

Note: As already described and discussed by Brown (1920: 123) this name has been overlooked ever since its first publication. Solitary, white, seemingly ebracteate flowers occur in other species of **Ruschia** close to **R. uitenhagensis** or **R. hamata**. White flowers of **Lampranthus** possess bracteoles always.

R. restituta ≡ **Octopoma inclusum**

R. rigens L.Bolus NM 1931: 259 (see **photo**) ≡ *M. dressianum* Ingram 1970: 50 nomen illeg. erroneously based on *M. ringens* L.Bolus **LT** Smith 4103 II (BOL!).

Decumbent shrubs to 15 cm *h*, branches often prostrate, rarely creeping; **L** old ones persisting for several years, rather densely placed and embracing the internodes for most of their length, triquetrous, the marked keel unidentate, the tooth often bent back and placed away from the apical tooth, very prickly to the touch, greyish-green, smooth, epidermis cells with low central papillae, bigger ones dispersed along keel and margins, wax cover continuous with fine threads or irregularly shaped platelets, L 25–30 mm *l*, c. 4 mm *b*, c. 5 mm *t*; **Fl** pink, filamentous staminodes present, filaments apically and basally paler; **Fr** top high from high rims, base funnel-shaped, the inner fruit falling out leaving a ring of hard, dry bundles raised by a rather long stalk, covering membranes convex, the distal closing rodlet broad, closing body tiny, round, expanding keels almost touching each other at their bases, 5 locules; **S** roundish; **Ecol** in open space on finely grained soils, either in grassland or in pans above rocks, 400–600 mm rainfall p.a. mainly in summer; **Distr** Bloemfontein, Fauresmith, Ventersburg, Virginia, FS, S Africa.

Note: Plants of **R. rigens** form conspicuous dense grey cushions, very prickly to the touch from leaves and fruit-bundles. They seem to grow more readily in open space in finely grained soils, the depth of which is of minor importance. When transferring the species to *Mesembryanthemum* in the sense of Brown, which is equivalent to that of **Ruschia** Schwantes, Ingram wrongly cited "*R. ringens* L.Bolus NM 1931: 259" as the basionym; this error led to his conclusion that under **Mesembryanthemum** a new name would be required, since *M. ringens* L. was already available. Therefore, *M. dressianum* is an illegitimate name, based on a spelling mistake.

R. rigida (Haw.) Schwantes ZSK 1927: 19 non L.Bolus NM 1929: 108 nomen illeg. ≡ *M. rigidum* Haw. 1803: 95 T?

Bushy shrubs to 30 cm *h*, branches stiff, later decumbent; **L** triquetrous, acuminate, but not with a recurved tip, glabrous except for the upper rough end of the keel; **Fl** solitary, but several on a branch in one season suggesting a panicle, petals shining white; **Fr** unknown; **Ecol** unknown; **Distr** unknown, S Africa.

Note: Described in the group "Paniculata" by Haworth (1803: 94–97) in comparison to **R. tenella** and together with **R. parviflora**, **R. vaginata** and others like **R. multiflora**.

R. rigida ≡ **R. bolusiae**

R. rigidicaulis (Haw.) Schwantes in Jacobsen 1949: 58 ≡ *M. rigidicaule* Haw. 1821: 116 T?

Procumbent with robust branches; **L** triquetrous, margins and keel rough; **Fr** unknown; **Ecol** unknown; **Distr** S Africa.

Note: Very little is known about this species.

R. robusta L.Bolus NM 1929: 147 non Schwantes MDK 1930: 66 T L.Bolus 18986 (BOL!).

Shrubs with erect branches to 50 cm *h*, lateral flowering shoots later persisting as spiny branched systems, stems dark brown; L trigonous, dark glaucous-green, appearing blackish in habitat, epidermis with raised dots and low papillae, 6–9 mm *l*, 3.5 mm *b* and *t*; **Fl** in small lateral cymes or solitary, pedicels 2 mm *l*, ∅ 18 mm, petals pinkish-purple, filamentous staminodes and filaments pale pink, anthers and pollen straw-coloured; **Fr** falling off and dispersed as tumble fruits, base short, funnel-shaped, top semi-globose with high rims, covering membranes convex, closing rodlets protruding, closing body hook-shaped, expanding keels almost touching each other at their bases, 5 locules; **Ecol** in gravelly flats, < 100 mm rainfall p.a. mainly in winter; **Distr** Namaqualand, NC, S Africa.

Note: The dark, rather high plants are prominent in habitat, resembling those of **R. cradockensis**, with spines from aborted buds instead of those developed here after the fruits have developed and have fallen off.

R. robusta ≡ **R. valida**
R. roseola ≡ **Antimima roseola**

R. rostella (Haw.) Schwantes ZSK 1927: 20 ≡ *M. rostellum* Haw. 1821: 123 **LT** drawing 1002 at K!

Thin, decumbent branches; **L** of a pair forming a beak, developing into a dry basal sheath persisting for some time around the fresh foliage leaves' sheath, green, semi-terete with a pointed, acuminate tip; **Fl** solitary, petals whitish to pinkish; **Fr** 5 locules, otherwise unknown; **Ecol** unknown; **Distr** unknown.

Note: The drawing at K resembles **Cephalophyllum rostellum** closely, which differs in the number of locules at first sight. The formation of a basal persisting sheath resembles heterophyllous plants in **Antimima**.

R. rubra ≡ **Astridia rubra**

R. rubricaulis (Haw.) L.Bolus NM 1928: 132 ≡ *M. rubricaule* Haw. 1803: 77 **LT** drawing 1003 by Duncanson (K!).

Shrubs with young red, glabrous **I**; **L** triquetrous, somewhat laterally compressed, dotted, margins and keel serrulate; **Fl** solitary, pedicels 13 mm *l*, ∅ 13.5 mm, **K** 5, petals purple, inconspicuously vittate, filamentous staminodes white, reddish above, ending in white, recurved tips, anthers yellow; **Fr** 5 locules; **Ecol** unknown; **Distr** unknown.

Note: The leaves appear almost boat-shaped in the drawing, in which a number of undeveloped flowers on short side branches is also visible.

R. rupicola (Engler) Schwantes ZSK 1926: 187 (see **photo**) non L.Bolus NM 1933: 415 nomen illeg. ≡ *M. rupicola* Engler BJS 1909: 191 T Range 131 (B, lost) **LT** Pax in Engler 1910: 545, f 480, based on Range 131 = *M. steingröveri* Pax in Engler 1910: 545 nomen nudum ≡ *R. stein-*

gröveri (Pax) Schwantes ZSK 1926: 187 nomen illeg. **T** Steingröver 25 (B) = *R. nivea* L.Bolus NM 1929: 45 ≡ *M. niveum* (L.Bolus) N.E.Br. GC 1930: 32 **LT** Pillans 6104 I (BOL!).

Roundish shrubs to 1 m ∅; **L** triquetrous, keel hardly dentate, but often a tiny recurved tooth at the tip, sometimes denticulate, free parts c. as long as sheaths, epidermis with papillae of different sizes, the smaller around the stomata, wax cover continuous with vertical platelets, c. 10 mm *l*, 5 mm *b*, 4 mm *t*; **Fl** petals pale pink; **Fr** rims dipping in the centre, base long, bell- to funnel-shaped, closing bodies broadening to the interior of the locule, the funicles protruding from beneath into all possible directions, 5 very deep locules; **Ecol** in granitic gneiss fields, < 100 mm rainfall p.a. in March; **Distr** Lüderitz-South, Namibia.

Note: In contrast to the treatment by Friedrich (1961, 1970), **R. rupicola** is kept as a separate species here. The inclusion of *R. nivea* into the synonymy is mainly based on the fact that the more slender and darker leaves of the type sheet are understood as reactions to the wetter cultivation the plants experienced in Pillans' garden near Cape Town.

R. rupicola ≡ **Octopoma rupigenum**
R. rupigena ≡ **Octopoma rupigenum**
R. rupis-arcuatae ≡ **Amphibolia rupis-arcuatae**

R. ruralis (N.E.Br.) Schwantes FR 1938: 230 (see **photo**) ≡ *M. rurale* N.E.Br. in Burt Davy 1926: 160 **T**?

Low shrubs to 15 cm *h*, branches erect, dense, I ochre to grey; **L** triquetrous, keel with 0–2(–5) teeth, connate and forming a basal sheath of 3–8 mm, merging unnoticeably into the stem below, 10–25 mm *l*, 2–3 mm *b* and *t*; **Fl** 1–2, bracteoles in the middle of the stalk, often touching the base of the **K**, **K** 5, petals purple, filamentous staminodes purple; **Fr** top with low rims, base short, funnel-shaped, covering membranes convex, closing ledge distinct, closing body a medium-sized hook, expanding keels without awns, no valve wings, 5 locules; **Ecol** in grassland with few trees, on gravelly soils, 400–600 mm rainfall p.a. mainly in summer; **Distr** Bophutatswana, Christiania, Schweizer Reneke, Wolmaranstad, Vryburg, NW, S Africa.

Note: In habit very similar to **R. canonotata**, which grows in places with limestone on the surface and further south; **R. ruralis** can be recognized by the absence of a triangle between the leaf bases, having a short green sheath instead. The distribution is incompletely known.

R. ruschiana (Dinter) Dinter & Schwantes ZSK 1927: 20 ≡ *M. ruschianum* Dinter FR 19, 1923: 179 **T** Dinter 3740 (B!) = *M. psammophilum* Dinter FR Beiheft 23, 1923: 52 ≡ *R. psammophila* (Dinter) Dinter & Schwantes ZSK 1927: 20 **LT** Dinter 3864 II (BOL!).

Shrubs to 60 cm *h*, with a main stem, but branching from the base as well, I 20–25 mm *l*; **L** club-shaped, keel and margins as translucent lines, epidermis dark bluish-green, light grey when dry, epidermis cells a little raised, 40 mm *l*, 9 mm *b* and *t*; **Fl** in ± rich cymes, bracteoles cymbiform, 10 mm *l*, **K** 5, petals 40, purple, 7 mm *l*, filamentous staminodes not mentioned; **Fr** base funnel-shaped, top raised, low rims, covering membranes convex, distal closing rodlets protruding, closing body a small hook, expanding keels touching each other basally, 5 locules; **Ecol** unknown; **Distr** Lüderitz-South, Namibia.

Note: The type material differs too markedly from the type of **R. muelleri** to be included in one species, as suggested by Friedrich (1961: 133, 1970: 116). *M. psammophilum* has traditionally been treated as a nomen nudum, but in the original text 60 cm height and red flowers are mentioned, albeit not in a formal description. Since material distributed by Dinter under the name is available as well, the name is here considered to have been validly published.

R. sabulicola Dinter FR 1932: 190 ≡ *M. sabulicola* Dinter FR Beiheft 53, 1928: 12, 98 nomen nudum **T** Dinter 4763 (B!).

Densely branched shrubs to 30 cm *h*, rooting basally in the covering sand; **L** bluish-green, epidermis cells a little elevated, a recurved tooth below the apex on the keel, sometimes 2–3 teeth, sheaths to 25 mm *l*, free parts 10–15 mm *l*; **Fl** solitary, pedicels to 25 mm *l*, bracteoles at their middle with short free parts, petals white, 6–8 mm *l*, 0.75–1.5 mm *b*; **Ecol** in coarse sand in dry river beds with **Aristida namaquensis**; **Distr** Keetmanshoop, Warmbad, Namibia.

Note: As already noted by Friedrich (1970: 119), the species differs from **R. vulvaria** only by the white flowers.

R. saginata ≡ **Amphibolia saginata**
R. salteri ≡ **Hammeria meleagris**

R. sandbergensis L.Bolus NM 1958: 334 ≡ *M. herrei* L.Bolus NM 1931: 276 nomen illeg. non (Schwantes) N.E.Br. GC 1930: 32 ≡ *Lampranthus herrei* (L.Bolus) L.Bolus NM 1939: 169 **T** Herre 8862 (BOL!).

Shrubs to 16 cm *h*, inner branches subaggregate; **L** trigonous, broadest near their tips, between club-shaped and falcate, epidermis smooth, to 30 mm *l*, 6 mm *b*, 15 mm *t*; **Fl** solitary, ∅ 50–60 mm, pedicels compressed, **K** 5, petals pale salmon, to 25 mm *l*, to 2.5 mm *b*, filamentous staminodes absent, filaments bright yellow; **Fr** 5 locules, otherwise unknown; **Distr** Namaqualand, NC, S Africa.

Note: Data from the description and the type form the base of this description, material added on later has not been considered.

R. sarmentosa (Haw.) Schwantes in Jacobsen 1949: 58 ≡ *M. sarmentosum* Haw. 1812: 238 T? = *R. sarmentosa* (Haw.) Schwantes var. *rigida* (Salm-Dyck) Schwantes in Jacobsen 1949: 58 ≡ *M. sarmentosum* Haw. var. *rigidius* Salm-Dyck 1840: § 17 f 3 T Salm-Dyck 1840: § 17 f 3.

Creeping branches to 45 cm *l*; **L** triquetrous, epidermis finely rugulose, margins subcartilaginous, to 54 mm *l*; **Fl** 1–3, bracteoles in the middle of the pedicels which become clavate and tuberculate in their upper parts, **K** 5, petals reddish, margins deeper coloured, filaments whitish, apically pink, anthers yellow; **Fr** 5 locules, otherwise unknown; **Ecol** unknown; **Distr** "Nova Hollandia", Australia.

Note: The rugulose leaves, the thickened pedicel and the origin from "Nova Hollandia" suggest that a species of **Sarcozona** might be involved.

R. sarmentosa var. *rigida* = **R. sarmentosa**
R. saturata ≡ **Antimima saturata**
R. saxicola ≡ **Antimima saxicola**

R. scabra H.E.K.Hartmann 1996: 46 (see **photo**) **T** Hartmann, Bruckmann & Hansen 31718 (HBG!) = *R. muricata* L.Bolus forma 1963: 48 **T** Stayner 741/62 (BOL!).

Low, cushion-forming shrubs, ∅ to 12 cm, but mostly only up to 5 cm, the main root woody, thick, branching and descending straight into the soil, proximally forming several fibrous roots in horizontal arrangement; **L** trigonous, keel, margins, and convex sides denticulate, c. 5 mm *l*, c. 4 mm *b* and *t*; **Fl** 1, bracteoles touching the base of the **K**, petals purplish-pink; **Fr** top raised moderately, rims low, base broadly bell-shaped to funnel-shaped, covering membranes very high in the centre, closing ledges protruding prominently, broad, closing body hood-shaped, expanding keels short, without awns and valve wings, 5 locules; **S** brown, 0.55–0.6 mm *l*, 0.4–0.5 mm *b*; **Ecol** in flats rich in limestone gravel, < 100 mm rainfall p.a. in March or March and November mainly; **Distr** Calvinia, Carnarvon, Williston, NC, S Africa.

Note: The possession of whitish elevations on the leaves unites this species with several others of **Ruschia** in a distinct group including also with **R. muricata, R. deminuta**, and **R. primosii**.

R. schlechteri ≡ **Antimima schlechteri**
R. schneideriana ≡ **Eberlanzia schneideriana**

R. schollii (Salm-Dyck) Schwantes in Jacobsen 1949: 58 ≡ *M. schollii* Salm-Dyck 1820: 10 ≡ *M. aduncum* Jacquin 1804: t. 51 f 2 non Haw. 1803: 87 nec Willdenow 1809: 534 nomen illeg. ≡ *M. recurvum* Haw. 1819: 90 nomen illeg. non Moench 1802: 190 nec L.Bolus NM 1932: 367 **LT** Jacquin 1804: t. 51, f 2 = *R. schollii* (Salm-Dyck) Schwantes var. *caledonica* (L.Bolus) Schwantes in Jacobsen 1949: 58 ≡ *M. schollii* Salm-Dyck var. *caledonicum* L.Bolus ABH 1923: 161 **LT** Glover 1901/17 (BOL!).

Decumbent plants with some erect branches, the long ones to 90 cm *l*, **I** grey to ochre; **L** trigonous, recurved, epidermis cells convex, elevations above the tannin idioblasts, these dense and confluent on keel and margins, which are often whitish, to 100 mm *l*, 4.5–7 mm *b*; **Fl** 1–3 in cymes, bracteoles connate and with papery white margins, ∅ c. 25 mm, petals rosy pink, filamentous staminodes white with reddish-purple tips or purple; **Fr** base a broad funnel, top with high rims, covering membranes convex, high rims distally, below these distinct closing ledges, closing body a small hook, expanding keels distant at their bases, short; **Ecol** 300–400 mm rainfall p.a.; **Distr** Caledon, WC, S Africa.

Note: Described and named first by Jacquin by a name already used before for a different species, this species received a second name which had also been employed before for yet a different species. The third name stays with the plant, which can be recognized as a member of the Sarmentosa-group by the formation of long, decumbent branches. As is usual in the group, flowering branches grow erect from the lying ones, the cyme developing additional flowers every years over 3–5 years, starting with a single flower often, as drawn by Jacquin (1804).

R. schollii var. *caledonica* = **R. schollii**
R. scopelogena ≡ **Scopelogena gracilis**
R. sedoides = **Eberlanzia sedoides**

R. semidentata (Haw.) Schwantes ZSK 1926: 187 ≡ *M. semidentatum* Haw. 1819: 95 ≡ *M. semidentatum* Salm-Dyck 1816: 22 nomen nudum **LT** drawing of Duncanson 1008/281, 22.3.1823 (K!).

Shrubs with firm, simple branches; **L** triquetrous, keel with 1–3 teeth and rough, leaf sheaths with a distinct line separating the leaves of the pair, connate with the stem further down, epidermis punctate; **Fl** solitary, **K** 5, petals bright pink (bright red); **Fr** unknown; **Ecol** unknown; **Distr** unknown.

Note: The species was first named by Salm-Dyck (1816) without a description, the name is therefore not validly published at that date. Haworth (1819) gave the first valid description, albeit based on "few dying branches from the ... Prince of Salm-Dyck". More extensive descriptions are published by Salm-Dyck (1820: 9 and 1836: § 33.6), the latter with a drawing. The lectotype was chosen from two drawings at K of material sent to there by Salm-Dyck, thus representing original material.

R. semiglobosa L.Bolus NM 1950: 217 **T** L.Bolus 24333 (BOL!).

Shrubs to 20 cm *h*, **I** 2–5 mm *l*, ∅ 2–4 mm; **L** connate for 1/4, semi-terete, epidermis cells a little elevated, 18–32 mm *l*, 8 mm *b*, 7 mm *t*; **Fl** 1–3, pedicels 10–15 mm *l*, bracteoles connate for 1/3 of their length, **K** 5, petals purplish-pink 8–10 mm *l*, to 1.25 mm *b*, filamentous staminodes absent; **Fr** base bell-shaped, top high, but valve rims very low, almost turret-shaped, covering membranes convex, with big distal closing rodlets, closing body small, expanding keels diverging, no valve wings, 5 locules; **Ecol** 125–250 mm rainfall p.a. in winter; **Distr** Vanrhynsdorp, WC, S Africa.

Note: Bell-shaped fruit bases and turret-shaped tops separate **R. semiglobosa** from the other species of **Ruschia**, resembling most closely **Octopoma conjunctum** from the same area, but with eight locules and with filamentous staminodes.

R. senaria L.Bolus NM 1929: 108 ≡ *M. senarium* (L.Bolus) N.E.Br. GC 1930: 32 **T** Pillans 5826 (BOL!).

Branches to 30 cm *l*, **I** blackish-grey, 25–30 mm *l*, ∅ to 7 mm; **L** terete to club-shaped, "glaucous-green", apically rounded, epidermis with long papillae, 20–35 mm *l*, 5–8 mm *b* and *t*; **Fl** solitary, ∅ 26–30 mm, **K** 6, petals purplish-pink, vittate, to 12 mm *l*, to 1 mm *b*, filamentous staminodes and filaments arranged in pitcher-shape, apically purple, basally pink, anthers and pollen straw-coloured; **Fr** 6 locules, otherwise unknown; **Ecol** unknown; **Distr** Namaqualand, NC, S Africa.

Note: The possession of 6 calyx lobes and locules in combination with finely velvety leaves are rare features in **Ruschia**.

R. serrulata (Haw.) Schwantes FR 1938: 230 ≡ *M. serrulatum* Haw. 1803: 77 ≡ *M. serratulum* (Haw.) Schwantes in Jacobsen 1970: 502 orthographic variant **LT** drawing of Duncanson s.n. (K!).

Main stem erect, side branches decumbent to prostrate, **I** angulate, purplish; **L** compressed-triquetrous, margins finely serrulate or rough, 27 or even 54 mm *l*; **Fl** unknown; **Ecol** unknown; **Distr** S Africa.

Note: The species is very little known.

R. sessilis (Thunberg) H.E.K.Hartmann Bradleya 1998: 85 ≡ *M. sessile* Thunberg 1791: 14 ≡ *Drosanthemum sessile* (Thunberg) Schwantes ZSK 1927: 106 **T** Thunberg 12018 (UPS!).

Shrubs with some erect, long branches bearing short, lateral shoots in great numbers, each ending in a flower or a cyme, **I** dark red, smooth; **L** trigonous to globose, epidermis rough from elevations, papillae low; **Fl** solitary, almost sessile, **K** 5, petals red; **Fr** base funnel-shaped, top with very high valve rims, 5 locules; **Ecol** unknown; **Distr** Vanrhynsdorp, WC, S Africa.

Note: The fruits on the type are clearly those of a species of **Ruschia**, making the new combination necessary.

R. simulans ≡ **Antimima simulans**

R. singula L.Bolus NM 1954: 282 **T** SUG 11553 (BOL!).

Compact; **L** subterete to acuminate, keel and margins cartilaginous, epidermis smooth, grey, glaucous, to 50 mm *l*, c. 7 mm *b* and *t*; **Fl** solitary, bracteolate at the bases of the up to 25 mm *l* stalks, **K** 6, petals pale pink, 15–17 mm *l*, filamentous staminodes pale, filaments epapillate, nectary of 6 separate glands; **Fr** top high from very high valve rims, base funnel-shaped, covering membranes complete with an inconspicuous distal closing ledge, the placenta ending in a hood-shaped, pronounced body, round as seem from above, thus simulating an endocarpal closing body, expanding keels of one valve well separated basally, apically ending in an awn, 6 locules; **Ecol** unknown; **Distr** "Griqualand West", but actually unknown, S Africa.

Note: The possession of a purely placental closing body resembles most closely conditions found in fruits of **Ebracteola** and **Khadia**, both occurring in the wider area given tentatively as the origin of the type material.

R. sobrina ≡ **Antimima sobrina**
R. socia = **Argyroderma fissum**
R. solida ≡ **Antimima solida**
R. solida var. *stigmatosa* = **Antimima solida**

R. solitaria L.Bolus NM 1958: 337 **T** Hall 26223 (BOL!).

Shrubs to 25 cm *h*, **I** 20–40 mm *l*, to 7 mm ∅; **L** trigonous to semi-terete, apically rounded, mucronate, darkish-green with age, 15–32 mm *l*, 3–5 mm *b* and *t*; **Fl** 1–3, pedicels 6–10 mm *l*, bracteoles at half their length, **K** 5(–6), petals purplish-pink, inconspicuously vittate, 5–6 mm *l*, 0.5–0.75 mm *b*, filamentous staminodes present, filaments white, apically reddish-purple, anthers and pollen yellow; **Fr** base funnel-shaped, top moderately high, covering membranes convex, closing rodlets big and protruding, closing body a small rodlet, expanding keels distant from each other at their bases, 5(–6) locules; **Ecol** unknown; **Distr** Vredendal, WC, S Africa.

Note: 5–6 stigmas are reported for this species.

R. spathulata = **Eberlanzia schneideriana**
R. spinescens ≡ **Arenifera spinescens**

R. spinosa (L.) Dehn 1993: 191 (see **photo**) subg. **Spinosae** ≡ *M. spinosum* L. 1753: 483 ≡ *Eberlanzia spinosa* (L.) Schwantes 1926: 189 **LT** Dillenius 1732: t. 208, f 263 = *M. aculeatum* N.E.Br. 1922: 163 ≡ *Eberlanzia aculeata* (N.E.Br.) Schwantes 1927: 30 **LT** Thunberg 12019 (UPS!) = *R. albertensis* L.Bolus NM 1954: 282 ≡ *Eberlanzia albertensis* (L.Bolus) L.Bolus NM 1958: 387 **T** Acocks 17112 (BOL!) = *R. globularis* L.Bolus NM 1933: 413 ≡ *Eberlanzia globularis* (L.Bolus) L.Bolus NM 1958: 387 **T** Luckhoff 20822 (BOL!) = *M. hospitale* Dinter FR 19, 1923: 130 ≡ *Eberlanzia* *hospitalis* (Dinter) Schwantes 1926: 189 **LT** Dinter 3636 (B!) = *R. macroura* L. Bolus NM 1954: 281 ≡ *Eberlanzia macroura* (L. Bolus) L.Bolus NM 1958: 387 **T** Acocks 17115 (BOL!) = *M. mucroniferum* Haw. 1823: 381 ≡ *Eberlanzia mucronifera* (Haw.) Schwantes 1926: 189 **T**? = *M. spinosum* L. var. *micranthum* Pax 1889: 12 ≡ *Eberlanzia micrantha* (Pax) Schwantes 1926: 189 **T** Marloth 1507 lost?

Shrubs to 35 cm *h*, in the N populations to 70 cm *h*; **I** whitish-grey turning coppery or black with age; **L** 10–20 mm *l*, epidermis with elevations above tannin idioblasts, cells convex to papillate; **Fl** in storeyed **Inf** with up to 31 spines and 32 flowers, enriched by additional lateral branches, ∅ 22–28 mm, c. 30 purple petals, 0–8 filamentous staminodes, c. 100 stamens, stigmas long; **Fr** in distal position, without or with narrow valve wings; **S** 0.8–1 mm *l*; **Chr** $2n = 36$; **Ecol** in open Karroo vegetation in flats, along the E border of the winter rainfall region, to 300 mm rainfall p.a.; **Distr** Lüderitz-South, Warmbad, Namibia; W part of EC, W part of NC, interior districts of WC, S Africa.

Note: The species is best recognized by the additional flowers in the rich cymes.

R. staminodiosa L.Bolus NM 1928: 140 ≡ *M. staminodiosum* (L.Bolus) N.E.Br. GC 1930: 32 non L.Bolus NM 1937: 76 nomen illeg. **T** Muir 1821/23 (BOL!).

Richly branched shrubs with erect branches, to 15 cm *h*; **L** sharply triquetrous, keel and margins very smooth, epidermis glabrous, greyish, c. 15 mm *l*, c. 5 mm *b* and *t*; **Fl** solitary, the bracteoles as big as the foliage leaves, reaching the base of the **K**, ∅ c. 25 mm, petals purplish-pink, linear, decreasing in length from the outer c. 10 mm *l* ones to the inner which merge into pale filamentous staminodes, filaments epapillate; **Fr** hard, reddish, top rather high, but rims low, base elongate, funnel-shaped, covering membranes roof-shaped, the distal recurving low, but the closing ledge below it distinct, no closing body, but the rounded end of the placenta protruding being lifted away from the wall, expanding keels not touching each other basally, no awns or wings, 5 locules; **Ecol** c. 400 mm rainfall p.a. at all seasons; **Distr** Riversdale, WC, S Africa.

Note: The species agrees in leaf character states with **R. fourcadei**, but differs from the Ruschia-type of fruit so markedly that it cannot be kept in the genus. As noted in the original description by Bolus (1928) already, the flowers remind of **Erepsia**: the petals decrease centripetally in length, they are linear in shape, the filaments are epapillate; in addition, the absence of a closing body supports the possible relationship to **Erepsia**. Yet, the conically arranged stamens in the centre contradict a placement in **Erepsia** as presently circumscribed.

R. stayneri ≡ **Antimima stayneri**
R. steingröveri = **R. rupicola**
R. stellata = **Antimima hantamensis**
R. stenopetala = **Antimima watermeyeri**
R. stenophylla ≡ **Marlothistella stenophylla**
R. stokoei ≡ **Antimima stokoei**

R. stricta L.Bolus NM 1928: 146 ≡ *M. sublunatum* N.E.Br. GC 1930: 32 (erroneously as *M. sublunulatum* in Jacobsen 1970: 503, 584) **T** Pillans 17789 (BOL!).

Shrubs to 1 m *h*, **I** reddish when young, later brown to blackish; **L** sublunate in side view, keel and margins cartilaginous, pale glaucous-green, an elongate triangle separating the bases of each leaf-pair, 15–25 mm *l*, 7–12 mm *b*, 9 mm *t*; **Fl** petals pale pink; **Fr** without closing bodies, very low closing ledges, 6 locules; **Ecol** in crevices on northern slopes; **Distr** Namaqualand, NC, S Africa.

Note: The "big closing bodies" cannot be verified on the type material nor on fresh plants.

R. stricta var. *turgida* = **Amphibolia saginata**

R. strubeniae (L.Bolus) Schwantes in Jacobsen 1949: 58 ≡ *M. strubeniae* L.Bolus ABH 1920: 5 **T** (BOL!).

Erect shrubs to 75 cm *h*, **I** brownish-grey; **L** subacinaciform to falcate, bright green, keel serrulate, epidermis smooth, 40–60 mm *l*, 5–6 mm *b*, to 12 mm *t*; **Fl** in cymes, ∅ 32 mm, bracteoles connate and truncate in some flowers, petals pink, veins purple, filamentous staminodes present, filaments white, apically purple, anthers white, stigmas purple; **Fr** unknown; **Ecol** unknown; **Distr** unknown.

Note: The bracteoles in their pocket-shape resemble those of **R. tardissima** and **R. marianae** closely, the erect growth differs from both.

R. stylosa ≡ **Arenifera stylosa**

R. suaveolens L.Bolus NM 1932: 352 **T** Mathews 2079/31 (BOL!).

Erect shrubs to 50 cm *h*, **I** 15–45 mm *l*, 5 mm ∅; **L** trigonous, epidermis in dry state with elevations and somewhat raised cells, 30–60 mm *l*, 3–4 mm *b*; **Fl** in terminal and lateral cymes forming a compound **Inf**, ∅ 15 mm, pedicels 25 mm *l*, bracteoles 13 mm, compressed, **K** 5, petals mauve, 6 mm *l*, 1 mm *b*, filamentous staminodes absent, anthers straw-coloured; **Fr** 5 locules, otherwise unknown; **Ecol** in east facing crevices; **Distr** Vanrhynsdorp, WC, S Africa.

Note: The type lacks a capsule, making an assessment difficult.

R. subaphylla = **R. abbreviata**
R. subglobosa ≡ **Octopoma subglobosum**

R. subpaniculata L.Bolus NM 1935: 503 **T** Leighton 21428 (BOL!).

Erect shrubs, branches to 29 cm *l*, **I** 20–60 mm *l*, ∅ c. 4 mm; **L** semi-terete, separated basally by a line of 3–6 mm, epidermis smooth, **L** to 20 mm *l*, 3 mm *t*; **Fl** in rich cymes, pedicels to 10 mm *l*, **K** 5, petals pink, inconspicuously vittate, to 6 mm *l*, to 1.5 mm *b*, filamentous staminodes acuminate, filaments pale, apically reddish, anthers and pollen straw-coloured; **Fr** 5 locules; **Ecol** near the sea, winter rainfall; **Distr** Clanwilliam, WC, S Africa.

Note: A little known species.

R. subsphaerica L.Bolus NM 1930: 208 **LT** L.Bolus 19214 I (BOL!).

Shrubs to 25 cm *h*, long shoots dominating, many lateral short shoots present, **I** grey, later brown, 5–8 mm *l*, 3 mm ∅; **L** subglobose, falling off easily, greyish-green, rough from minute dots, 4–5 mm *l*, 3–4 mm *t*; **Fl** solitary on branched, lateral short shoots, the ends of the branches developing into spines after the fruits have fallen off, pedicels 2 mm *l*, ∅ 13 mm, **K** 5, petals purplish-pink, to 5 mm *l*, 1 mm *b*, filamentous staminodes few to none, filaments white, apically pink, anthers and pollen white; **Fr** covering membranes with distal closing rodlets, closing body small, ∅ 10 mm when expanded; **Ecol** unknown; **Distr** Namaqualand, NC, S Africa.

Note: The nearly round leaves resemble those of **R. sessilis.**

R. subteres L.Bolus NM 1934: 434 **T** Lavis 21033 (BOL!).

Shrubs with erect branches to 21 cm *l*, **I** to 40 mm *l*; **L** erect to spreading, subterete, epidermis almost smooth, 25–30 mm *l*, to 5 mm *t*; **Fl** 5–15 per **Inf**, **K** 5, petals purplish-pink, to 6 mm *l*, to 1 mm *b*, filamentous staminodes few, filaments pink, anthers and pollen dark brown-golden; **Fr** top high from high rims, broader than the top of the base, base funnel-shaped, covering membranes with distinct closing rodlets, closing body narrowly oval, the long axis pointing into the locules, expanding keels broadening distally (in the open capsule) developing a loose end like a wing, but of expanding keel tissue, 5 locules; **Ecol** unknown; **Distr** Worcester, WC, S Africa.

Note: Almost round leaves with almost round tips are rare in **Ruschia.**

R. subtruncata ≡ **Antimima subtruncata**
R. subtruncata var. *minor* = **Antimima subtruncata**
R. succulenta ≡ **Amphibolia succulenta**

R. tardissima L.Bolus NM 1932: 353 **T** Leipoldt 20125 (BOL!).

Plants with a compact centre with decumbent branches to 40 cm *l*, **I** laterally compressed, 50–60 mm *l*, 5 mm ∅; **L** triquetrous, base of sheath with a small triangle of stem tissue; **Fl** 3, bracteoles forming a broad, truncate sheath resembling a pocket around the pedicel, **K** 5, petals pinkish-purple, vittate, 14–17 mm *l*, 2 mm *b*, filamentous staminodes basally white, anthers and pollen yellow; **Fr** 5 locules, otherwise unknown; **Ecol** unknown; **Distr** Calvinia, NC, Ceres, WC, S Africa.

Note: Compared in the original description with **R. rigidicaulis**, from which it is said to differ in the bracteoles and the dense petals. The formation of broad, truncate pockets below the terminal dichasium occurs also in **R. strubeniae** and **R. marianae.**

R. tatasbergensis = **R. divaricata**

R. tecta L.Bolus NM 1932: 332 **T** L.Bolus 19642 (BOL!).

Shrubs to 57 cm *h*, **I** 25–80 mm *l*; **L** in the shape of an elongate s, semi-terete, base connate and somewhat swollen, 95 mm *l*, sheath 15 mm *l*, 3 mm *b* and *t*; **Fl** in rich cymes, **K** 6, petals very many, purplish-pink, basally white, to 17 mm *l*, 1 mm *b*, filamentous staminodes longer than stamens, both bending inwards and hiding the inner stamens, white, apically pinkish, anthers and pollen white, surface of gynoecium concave, almost forming a hypanthium, stigmas short and spreading; **Fr** base elongate, funnel-shaped, top turret-shaped, low rims, covering membranes convex, distal closing device a broad, thick ledge at the distal reflexion, closing body rather big, white, expanding keels not separated from the expanding sheets, only raised to a sharp edge, 6 locules; **Ecol** in sandy places; **Distr** Malmesbury, WC, S Africa.

Note: With very narrow petals and hidden inner stamens the flowers mimic those of **Erepsia**, even to the extent of forming a recess flower with short, spreading stigmas. The rather big closing body excludes the plants from **Erepsia**, but they remain unusual in **Ruschia** as well.

R. tenella (Haw.) Schwantes ZSK 1927: 19 ≡ *M. tenellum* Haw. 1795: 315 **T?**

Shrubs to 30 cm *h*, branches wiry and slender; **L** perfoliate, basally connate, but separated by a line along the connate sheath, free parts 6–12.5 mm *l*, sharply triquetrous, slightly subulate, recurved, keel almost serrulate; **Fl** 1–2, small and white; **Fr** unknown; **Ecol** unknown; **Distr** unknown.

Note: Distinguished from **R. viridis** by less high growth, slightly glaucous and roughly punctate leaves (Haworth 1795: 316).

R. testacea L.Bolus NM 1934: 435 **T** Salter 3836 (BOL!).

Branches to 24 mm *l*, **I** light red to maroon, 40–45 mm *l*, 5 mm ∅; **L** subfalcate to almost terete, light glaucous, epidermis smooth, 35–55 mm *l*, 4–5 mm *b*, 6 mm *t*; **Fl** in terminal and lateral cymes forming a compound **Inf** of up to 50 flowers, bracteoles 5–25 mm *l*, petals filiform, brick red, to 5 mm *l*, 1 mm *b*, filamentous staminodes not well visible, anthers pale; **Fr** 5 locules, otherwise unknown; **Ecol** unknown; **Distr** Vanrhynsdorp, WC, S Africa.

Note: Brick red flowers are very rare in **Ruschia**.

R. tetrasepala ≡ **Octopoma tetrasepalum**
R. thomae ≡ **Esterhuysenia stokoei**
R. thomae var. *microstigma* = **Esterhuysenia stokoei**
R. translucens = **Stoeberia beetzii**

R. tribracteata L.Bolus NM 1931: 238 **T** Pillans 5840 (BOL!).

Shrubs to 30 cm *h*, **I** 20–50 mm *l*, 3–4 mm ∅; **L** trigonous to semi-terete, light green, densely punctulate, epidermis smooth, 5 mm *b* and *t*; **Fl** in terminal and lateral cymes, forming a compound **Inf**, pedicels 20–40 mm *l*, bracteoles in lower half 6–14 mm *l*, ∅ 14 mm, **K** 5, petals in 5 groups, filamentous staminodes pink, filaments apically pinkish-purple, anthers yellow; **Fr** unknown; **Ecol** unknown; **Distr** Namaqualand, NC, S Africa.

Note: The name refers to the observation that each pedicel possesses a pair of bracteoles.

R. triflora L.Bolus NM 1929: 109 ≡ *M. triflorum* (L.Bolus) N.E.Br. GC 1930: 32 **T** Leipoldt 18942 (BOL!).

Compact, to 10 cm *h*, **I** grey, 20–50 mm *l*, often enclosed; **L** subterete, glaucous green, epidermis smooth, 30 mm *l*, to 3 mm *t*; **Fl** ternate, pedicels 13–18 mm *l*, bracteoles below the middle, **K** 5, petals pink, basally vittate, to 8 mm *l*, to 1.5 mm *b*; filamentous staminodes present, apically reddish-purple, anthers almost brown; **Fr** 5 locules, otherwise unknown; **Ecol** unknown; **Distr** Clanwilliam, WC, S Africa.

Note: A rather compact species.

R. triquetra ≡ **Antimima triquetra**

R. truteri L.Bolus 1962: 299 **T** van Breda 1711/61 (BOL!).

Erect shrubs with long shoots in shrubby associations, more caespitose in open places on shallow ground; **L** trigonous, the bases of a pair hardly connate and almost separated by a triangle of stem tissue, the leaf bases leaving triangular scares on the older stem, epidermis smooth; **Fl** solitary, but often in groups of 3–5, petals purplish-pink, filamentous staminodes few, filaments basally pale, apically pink; **Fr** unknown; **Ecol** > 400 mm rainfall p.a. mainly in summer; **Distr** Somerset East, EC, S Africa.

Note: At least three different species have been collected in the area, answering the vegetative characters of the type material, as far as that goes.

R. tuberculosa ≡ **Antimima tuberculosa**

R. tumidula (Haw.) Schwantes ZSK 1927: 20 ≡ *M. tumidulum* Haw. 1812: 286 **NT** drawing by Duncanson 1030/569 (K!).

Shrubs with ascending branches; **L** semi-terete, apically recurved, tumid sheath with a distinct line between the leaves of a pair, not connate with the stem below, c. 20 mm *l*, greenish to greyish; **Fl** petals red; **Fr** unknown; **Ecol** unknown; **Distr** unknown.

Note: The drawing chosen as the neotype has been assigned to belong to the species in 1824, grown from seed collected by Bowie 1818 but does not represent material on which the description was based, which had been introduced in 1802 and sent to Haworth by Howard. In the absence of older material or a type, this appears the best choice since it has been named by Haworth himself.

R. turneriana ≡ **Antimima turneriana**

R. uitenhagensis (L.Bolus) Schwantes ZSK 1927: 105 ≡ *M. uitenhagense* L.Bolus ABH 1927: 82 **T** Muir 3949 (BOL!).

Dense, round, scrubby shrubs, branches elongating in cultivation; **L** ending in a markedly recurved, white tooth from parallel margins, epidermis with elevations above the subhypodermal tannin idioblasts, surface thus rough, c. 9 mm *l*, c. 2 mm *t*; **Fl** solitary on short shoots, several of these formed per season, thus developing a paniculoid structure, **K** 5, ∅ c. 14 mm, petals pink, 28–30, filamentous staminodes and filaments pale pink; **Fr** 5 locules, otherwise unknown; **Ecol** in shrubby associations on gravelly soil, > 400 mm rainfall p.a. at all seasons mostly; **Distr** Uitenhage and adjacent districts, EC, S Africa.

Note: As described by Bolus (1927), **R. uitenhagensis** differs from **R. hamata** only in the round shape of the shrubs.

R. umbellata (L.) Schwantes ZSK 1927: 20 ≡ *M. umbellatum* L. 1753: 481 **LT** Dillenius 1732: t. 208 f 266.

Shrubs with erect branches 30–60 cm *l*, **I** dark brown to maroon; **L** trigonous, a little recurved, the bases of a pair separated by a short line and a basal triangle of stem tissue, 17 mm *l*, c. 5 mm *t*; **Fl** in rich cymes, terminal and additional lower ones, distinctly dinsiguished from the vegetative basal parts of the shrubs, the subulate bracteoles much smaller than the leaves, petals short and white; **Fr** 5 locules, otherwise unknown; **Ecol** in coastal sand; **Distr** Cape Town, Goodwood, Simonstown, WC, S Africa.

Note: This species was most probably figured at first by Plukenet (1691: t.117, f. 1) showing an inflorescence with small bracteoles and, below it, a pair of short shoots with leaves more than double the length and distinctly pointed; the description appeared later (Plukenet 1696: 149, seen in the 2nd reprint of 1796).

Hermann (1698: t.166) figured an entire plant with clearly distinguishable rich inflorescences; he cited Plukenet's figure and gave data on the ecology of the species, reporting also that the plants come up spontaneously in their natural habitat. Bradley (1727: 12) mentioned very white flowers with light yellow tips and a smell of bitter almonds, depicting a branch with long lateral shoots below the rich apical inflorescence, the leaves being more than three time the length of the bracteoles and those of a pair separated apparently at their bases by a distinct triangle, a feature not visible in the figures published earlier.

R. unca ≡ **Lampranthus uncus**
R. unca var. *punctulata* ≡ **R. punctulata**

R. uncinata (L.) Schwantes ZSK 1926: 187 ≡ *M. uncinatum* L. 1753: 483 ≡ *R. uncinella* (Haw.) Schwantes ZSK 1926: 187 nomen illeg. ≡ *M. uncinellum* Haw. 1821: 125 nomen illeg. **LT** Dillenius 1732: t. 193 f 239.

Densely branched shrubs, apparently articulate; **L** triquetrous, recurved markedly, keel with 2 teeth mostly, sheath merging into the stem; **Fl** solitary, but several in apparent panicles, petals red (purple); **Fr** unknown; **Ecol** unknown; **Distr** unknown.

Note: The name is occasionally attributed to Miller (1868: nr 18), a later publication than Linnaeus 1753. *R. uncinella*, based on *M. uncinellum*, is a homotypic synonym because Haworth (1821: 125) cited the drawing representing the type of **R. uncinata** as the only traceable reference. The drawing at K (Duncanson 1033/482!), made in 1822, shows somewhat stouter leaves than the drawing of Dillenius (1732: f 239).

R. uncinella ≡ **R. uncinata**

R. unidens (Haw.) Schwantes ZSK 1926: 187 ≡ *M. unidens* Haw. 1826: 330 **LT** drawing 238/169 by Duncanson (K!).

Shrubs to 16 cm *h*; **L** subacinaciform, apically cuspidate and with a tooth on the keel a little below the tip, the connate sheath with a short, inconspicuous line; **Fl** unknown; **Fr** unknown; **Ecol** unknown; **Distr** unknown.

Note: Distinguished from **R. semidentata** by the less prominent line on the sheath.

R. utilis ≡ **Stoeberia utilis**
R. utilis var. *gifbergensis* = **Stoeberia frutescens**

R. vaginata (Haw.) Schwantes ZSK 1927: 19 ≡ *M. vaginatum* Haw. 1821: 126 **LT** drawing by Duncanson 3.3. 1824, 1024/565 (K!).

Shrubs to 60 cm *h*, often lower and broad; **L** ending in a white, recurved tip from parallel margins, base forming a long sheath connate with the stem, a line separating the leaves of a pair, cartilaginous-tuberculate keel and margins, not minutely serrulate, to 25 mm *l*; **Fl** solitary on short shoots, several per season on a long shoot, therefore giving the impression of a panicle; **Fr** 5 locules; **Ecol** unknown; **Distr** unknown.

Note: Although known in England since the early 19th century, the species seems to be rare in cultivation, perhaps due to its reluctance to flower and its untidy, straggly growth into a big, loose bush.

R. valida Schwantes FR 1938: 228 ≡ *R. robusta* Schwantes MDK 1930: 66 nomen illeg. non L.Bolus 1929: 147 **LT** photograph of M. Schlechter s.n. (collection Schwantes, HBG!).

Branches to 20 cm *l*, **I** invisible; **L** cymbiform, epidermis papillate, dark green, sheaths 5 mm *l*, free parts to 25 mm *l*, 5 mm *b*, 10 mm *t*; **Fl** solitary, ∅ 18 mm; **Fr** bracteoles at the middle of the stalks of 15 mm *l*, valves only opening into an erect position, covering membranes with distal closing rodlets, closing body hood-shaped, ∅ 10 mm, open 13 mm, 6 locules; **Ecol** unknown; **Distr** Namaqualand, NC, S Africa.

Note: In growth, plants of this species resemble some species of **Antimima**, also velvety to the touch.

R. vanbredai L.Bolus 1962: 296 **T** van Breda 1688 (BOL).

Shrubs to 12 cm *h*, richly and diffusely branched, **I** reddish-brown, 25–35 mm *l*, ∅ 2 mm; **L** falcate to subfalcate, 10–17 mm *l*, 3–4 mm *b*, 4–6 mm *t*; **Fl** in cymes, bracteoles near bases of pedicels, subglobose, **K** 5, petals purplish-pink, vittate, 3–8 mm *l*, 0.5–1.25 mm *b*, filamentous staminodes few, paler basally, filaments paler, apically purplish-pink, anthers and pollen white; **Fr** top raised, valves expanding into an erect position only, closing body present, touching the covering membranes, 15–20 mm *h*, 5 locules; **Ecol** unknown; **Distr** Uniondale, EC, S Africa.

Note: Rather unusual is the reported big closing body, touching the covering membranes.

R. vanderbergiae L.Bolus NM 1935: 503 **T** van der Berg 3256 (BOL!).

Erect shrubs to 20 cm *h*, **I** grey, shining pale brown with age, 10–15 mm *l*, ∅ 3 mm; **L** trigonous, glaucous-green, apiculate, margins with 1–2 fine teeth, 5–9 mm *l*, sheath 4 mm, 2–3 mm *b* and *t*; **Fl** pedicels 8 mm *l*, petals purplish-pink, to 7 mm *l*, 0.75 mm *b*, inner petals merging into pale pink filamentous staminodes, pollen golden; **Fr** base funnel-shaped, top low, covering membranes with closing ledges at their distal ends, closing body a small hook, expanding keels just not touching each other basally, 5 locules; **Ecol** unknown; **Distr** Graaff Reinet, EC, S Africa.

Note: The species is named after Miss Vanderberg.

R. vanheerdei L.Bolus 1963: 15 **T** van Heerde 27273 (BOL!).

Decumbent branches, **I** ochre-reddish, grey to white later, 30–50 mm *l*, 3 mm ∅; **L** subterete, obtuse, epidermis smooth, 30–40 mm *l*, 6–8 mm *b* and *t*; **Fl** in cymes, pedicels to 30 mm *l*, bracteoles at 1/2 their length, upper bracteoles forming a pocket of connate, truncate leaves, **K** 5, petals white, 9–20 mm *l*, 0.5–2 mm *b*, filamentous staminodes absent, filaments white, anthers and pollen greyish-pink or reddish-golden, later pale; **Fr** top turret-shaped, base elongate, funnel-shaped, covering membranes with distal, protruding closing rodlets, closing body a hook, expanding keels well separated from each other basally, 5 locules, stalks to 50 mm *l*; **Ecol** unknown; **Distr** Namaqualand, NC, S Africa.

Note: Bolus (1963: 16) remarked that "the presence of wing-like appendages near the margin of the lower surface of the covering wings of the locules has not been recorded in any other species described in these Notes", which appears surprising in view of the numerous species of **Ruschia** with similar protuberances she had described before that year.

R. vanniekerkiae L.Bolus NM 1932: 335 **T** van Niekerk 9074 (BOL!).

Low shrubs, **I** 1–2 mm ∅; **L** trigonous, keel 1-denticulate, tip bent back, glaucous-green, pink suffused, sheath connate with stem, 4–7 mm *l*, 2–3 mm *b* and *t*; **Fl** solitary, subsessile, **K** 5, petals pink, 8 mm *l*, 0.5 mm *b*, filamentous staminodes few, pink, filaments apically red, anthers yellow; **Fr** 5 locules, otherwise unknown; **Ecol** unknown; **Distr** Ladismith, WC, S Africa.

Note: Reported to be similar to **R. impressa**.

R. vanzylii ≡ **Antimima vanzylii**
R. varians ≡ **Antimima varians**
R. velutina = **Eberlanzia schneideriana**
R. ventricosa ≡ **Antimima ventricosa**
R. verruculata ≡ **Scopelogena verruculata**
R. verruculosa ≡ **Antimima verruculosa**

R. versicolor L.Bolus NM 1928: 7 ≡ *M. coloratum* N.E.Br. GC 1930: 32 **T** Pillans 17754 (BOL!).

Ascending branches to 30 cm *l*, **I** 20–25 mm *l*, to 7 mm ∅; **L** semi-terete, acutely pointed, greyish-green, connate base swollen, 95 mm *l*, 8 mm *t*; **Fl** in rich cymes and lower storeys, petals 26–30, pink, withering reddish, to 7 mm *l*, 1.5 mm *b*, filamentous staminodes white, apically pink, filaments finally reddish-golden; **Fr** base elongate, funnel-shaped, top broader than upper rim of base, raised, but with low rims, covering membranes convex, closing rodlets protruding, closing body hook-shaped, expanding keels almost touching each other basally; **Ecol** in flats; **Distr** Namaqualand, NC, S Africa.

Note: Named after the different colours in the flower, partly probably due to wilting.

R. vetovalida H.E.K.Hartmann Bradleya 1998: 85 ≡ *M. validum* Haw. 1826: 329 non Salm-Dyck 1836: §15 f 8 nomen illeg. ≡ *Cephalophyllum validum* (Haw.) Schwantes GF 1928: 69 **T** Bowie (K).

Short, ascending branches; **L** erect, pale green, the margins rough, **L** < **I**; **Fl** 1–3 in cymes, pedicels rather short, petals pink with red margins; **Fr** 5 locules; **Distr** unknown.

Note: The transfer of *M. validum* to **Cephalophyllum** was probably based on the drawing of the species by Salm-Dyck (1836), which does indeed represent a member of that genus. It deviates, however, from the original description by not having 5 locules nor rough margins of leaves. Since Haworth in his protologue compared the plant closely with *M. rigidicaule* ≡ **R. rigidicaulis** and placed the species in his group "Sarmentosa", the transfer of *M. validum* to **Ruschia** has been undertaken. A new name had to be given because **R. valida** has been used before, **R. vetovalida** referring to the "old valida".

R. victoris (L.Bolus) L.Bolus NM 1929: 94 ≡ *M. victoris* L.Bolus ABH 1925: 7 **T** Peers 1440/21 (BOL!).

Densely branched shrubs, **I** to 6 mm ∅; **L** ascending and incurving or spreading, finely papillate, 20–65 mm *l*, 6 mm *b* and *t*; **Fl** 3 in a terminal cyme, additional ones below, thus often 5, petals pink, purplish vittate, filamentous staminodes not mentioned, filaments purple, basally white; **Fr** very hard, opening tardily, covering membranes with prominent, protruding closing rodlets, closing body as a small hook, expanding keels not touching each other basally, 5 locules; **Ecol** in flats; **Distr** Clanwilliam, WC, S Africa.

Note: A little known species.

R. villetii = **Antimima dualis**

R. virens L.Bolus NM 1928: 21 ≡ *M. dumosum* N.E.Br. GC 1930: 32 **LT** Muir 4247 (BOL!).

Shrubs with erect branches 25–28 cm *l*, **I** ochre to brown, 15–20 mm *l*, ∅ to 12 mm basally, **L** falcate, truncate apically, epidermis smooth, 20–28 mm *l*, 6 mm *b*, 10–12 mm *t*; **Fl** 1–3, ∅ 28 mm, bracteoles the same shape and size as the leaves, **K** 4(–5), petals pink, 15 mm *l*, to 1 mm *b*, filamentous staminodes pale, filaments white, anthers and pollen dirty yellow; **Fr** top low, base short, funnel-shaped, covering membranes with closing ledges, closing body a tiny protuberance, expanding keels almost touching each other basally, 4–6 locules; **Ecol** at 360 m altitude; **Distr** Mosselbay, WC, S Africa.

Note: The lectotype was chosen for its possession of fruiting material.

R. virgata (Haw.) L.Bolus SAG 1927: 239 and NM 1927: 72, both on 1st of July ≡ *M. virgatum* Haw. 1803: 88

≡ *M. compressum* Haw. 1795: 416 nomen illeg. non 326 ≡ *Antimima virgata* (Haw.) Dehn 1989: 213 nomen illeg. **LT** Bortenschlag s.n. (K!, designated by Hartmann Bradleya 1998: 71).

Erect, thin branches from a denser centre, often in other bushes; **I** > **L**; **L** triquetrous, hardly connate, the pair at its base leaving a triangle of stem visible between them, epidermis smooth, glabrous; **Fl** solitary, bracteolate, pedicel flattened, **K** 5, petals bright purple, filamentous staminodes purple with lighter bases, filaments dark purple; **Fr** base elongate, funnel- to bell-shaped, top almost semi-orbicular from high valve rims, covering membranes forming a roof, the ridge being the contact line, closing ledges sharp and long, closing body nearly hook-shaped, hollow, expanding keels distant at their bases; **S** light brown, pear-shaped, testa almost smooth; **Ecol** in gravel in disturbed Rhenosterbosveld, > 300 mm rainfall p.a. all seasons; **Distr** Uniondale, WC, S Africa.

Note: Material matching the collection of Bortenschlag, which has been seen by Haworth, possesses fruits of the Ruschia-type, with deep locules and small, hook-shaped closing bodies. The species is therefore placed back in **Ruschia**.

R. viridifolia L.Bolus NM 1929: 147 **T** L.Bolus 18983 (BOL!).

Shrubs with erect branches; **L** subterete, apiculate, smooth, apparently with a wax layer, otherwise green, 20–40 mm *l*, 4–7 mm *t*; **Fl** ternate plus two lower set flowers often, thus 5 flowers forming a composed **Inf**, **K** 5, petals pinkish-purple, 14 mm *l*, 2 mm *b*, filamentous staminodes absent, filaments purple, anthers and pollen yellow; **Fr** top with 5 angles from almost erect rims, base funnel- to bell-shaped, covering membranes with big closing rodlets at their distal ends, closing body oval, narrow, the long axis pointing into the locule, valve wings absent, 5 locules; **Ecol** in rocky hills, < 100 mm rainfall p.a. in winter; **Distr** Namaqualand, NC, S Africa.

Note: Reported to be seen frequently between Springbok and Bitterfontein.

R. viridis ≡ **Smicrostigma viridis**
R. vulnerans = **R. divaricata**

R. vulvaria (Dinter) Schwantes ZSK 1926: 187 ≡ *M. vulvaria* Dinter 1914: 44 **T** Dinter 2104 (B!).

Densely branched shrubs of semi-orbicular shape, 30–40 cm *h*, smelling of fish, more specifically, of pickled herring, **I** grey; **L** bluish-green, trigonous, keel with a prominently recurved tooth, epidermis cells raised, longer ones along keel and margins, free parts 10–20 mm *l*, sheaths connate with the stem; **Fl** solitary, pedicels 7 mm *l*, ∅ 20 mm, **K** 5, petals purplish-pink, filamentous staminodes and filaments purplish-pink, apically paler to whitish; **Fr** opening completely, covering membranes with distal closing ledges, closing body a small hook, ∅ 5 mm, 5 locules; **Ecol** in hard loamy soil; **Distr** Gibeon, Keetmanshoop, Rehoboth, Namibia.

Note: The characteristic smell remains with the plants even after they have been in the herbarium for several years. On the type, the only fruit sits on a long pedicel of amost 80 mm, indicating an elongation after anthesis.

R. watermeyeri ≡ **Antimima watermeyeri**

R. willdenowii Schwantes FR 1938: 230 ≡ *M. aduncum* Willdenow 1809: 534 nomen illeg. non Haw. 1803: 87 nec Jacquin 1804: t. 51 f 2 nomen illeg. **T** Willdenow 9754 (B-W, xerocopy!).

Shrubs with diffuse stems; **L** terete, curved like a bow, and recurved apically, placed densely together in the lower part of the stem, separated by **I** of 10–13 mm length in the upper part, 25 mm *l*; **Fl** solitary, petals equalling the **K** lobes in length, outside yellow, purplish suffused, interior pale purple, stigmas blackish-purple; **Ecol** unknown; **Distr** unknown.

Note: M. aduncum Haw. has been placed as a synonym of *M. curvifolium* (now **Lampranthus curvifolius**) by Salm-Dyck already (1836: § 42.2), albeit in the variety β minus. On which grounds Schwantes (1938: 230) based his deviating decision including the raise of rank to that of species cannot be traced. The copy of the type shows a branch with two solitary flowers and no fruit.

R. wilmaniae ≡ **Ebracteola wilmaniae**
R. wilmaniae var. *angustifolia* = **Ebracteola wilmaniae**
R. wilmaniae var. *vermeuleniae* = **Ebracteola wilmaniae**
R. wittebergensis ≡ **Antimima wittebergensis**

Ruschianthemum Ruschioideae

H. E. K. Hartmann

Ruschianthemum Friedrich 1960: 563 **Typus R. gigas** (Dinter) Friedrich **Etym** *Ernst Rusch jun.*, farmer and botanical explorer in Namibia.

Branches erect to ascending, **I** whitish; **L** crescent-shaped, appearing inflated often, hardly connate at base, spreading, falling off easily, apex rounded, yellowish-green, epidermis almost smooth, stomata sunken a little; **Fl** in very rich terminal dichasia, with additional cymes below, petals white, c. as long as the 5 **K** lobes, filamentous staminodes and stamens in a distinct, low cone; **Fr** base elongate, bowl- to almost inversely bottle-shaped, yellowish-grey, top flat, valves raised only once and incompletely, the reddish-brown expanding keels separating from them remaining connate with the mericarp enclosing 1–2 seeds, the walls of the mericarp consisting of the interior tissue of the capsule base and the locules sclerified during the ripening of the fruit, the outer tissue of the mericarp appearing spongy and torn, the five mericarps fall out of the cage formed by the woody, persistent 5 dorsal bundles of the locules and the connecting apical ring surrounding the top of the capsule, a very big, spongy closing body overarching the locule disintegrates with age leaving a cavity when the capsule is opened, 1 seed often placed above the closing body, the only one to be released in the hygroscopic opening process, 5 locules; **S** rarely free, mostly enclosed in mericarps as dispersal units; **Ecol** in sandy pockets amongst stones and rocks on slopes, < 100 mm rainfall p.a. in winter, fog; **Distr** Lüderitz-South, Namibia; Namaqualand, NC, S Africa.

Note: Plants of **Ruschianthemum** are very similar to those of **Stoeberia beetzii**, both occurring in the same region and the latter differing only in the possession of free seeds in the permanently open locules. The extreme similarity makes it difficult to place flowering or vegetative material in the right genus. It remains therefore still uncertain whether *Ruschia micropetala*, known only as a flowering plant, is correctly placed in synonymy under **Stoeberia beetzii**, following Friedrich (1970), or rather with **R. gigas**, following Friedrich (1960). The very short expanding keels of **Ruschianthemum**, not reaching into the locules at all, could be seen as reduced forms of the short expanding keels of **Stoeberia**. Plants of **Eberlanzia** s.str. show also similarities to both, **Stoeberia** and **Ruschianthemum**.

Distribution of **Ruschianthemum** Friedrich

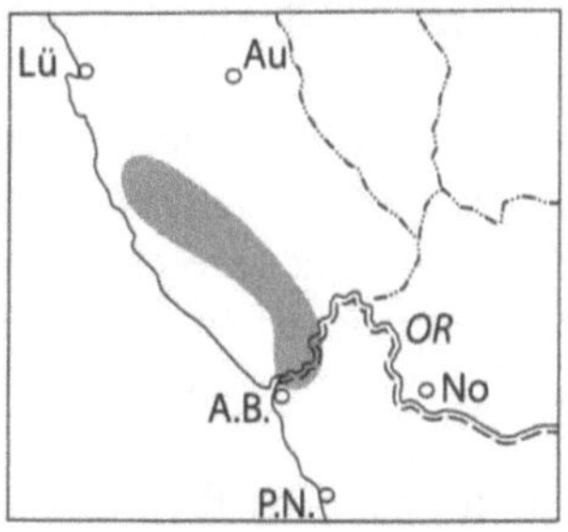

R. gigas (Dinter) Friedrich 1960: 563 (see **photos**) ≡ *M. gigas* Dinter FR 1923: 153 ≡ *Stoeberia gigas* (Dinter) Dinter & Schwantes ZSK 1927: 17 **T** Dinter 3791 (B).

To 80 cm *h*; **L** 30–50 mm *l*, 7–10 mm *b*, 12–16 mm *t*; **Infl** to 10 cm in ∅, enlarging moderately during the ripening of the fruits, **Fl** ∅ c. 5 mm, petals narrow, in 5 groups, filamentous staminodes and filaments often apically purple to red; **Fr** 4–5 mm *h*, ∅ c. 4 mm, persistent bundles of locules and apical rim remaining on the plant for at least two seasons; **Ecol** as for genus; **Distr** as for genus.

Note: Like in **Stoeberia**, the highly inconspicuous flowers appearing only rarely in cultivation make the plants less attractive. In habitat, the cymes are very rich and prominent, flowering over a long period and developing a beautiful filigree of woody bundles when the fruits have lost their contents. In habitat, **Ruschianthemum gigas** can be found on more rocky places, in comparison to the sympatrically growing **Stoeberia beetzii** preferring gravelly, flatter expositions. Yet, in each single case, the capsules have to be examined to determine the genus.

R. merenskyanum ≡ **Stoeberia carpii**

Ruschianthus Ruschioideae

H. E. K. Hartmann

Ruschianthus L.Bolus JSAB 1960: 62 **Typus R. falcatus** L.Bolus **Etym** *E. Rusch jun.*, farmer and botanical explorer in Namibia.

Compact; **L** erect to spreading, falcate, greyish-green, epidermis cells a little elevated, stomata a little sunken, outer epidermal wall very thick with a thick central layer of crystal sand, wax cover thick, one horizontal plate per cell, breaking into smaller floes on older leaves, with only some small particles rarely forming crystals; **Fl** solitary, rarely 2, bracteolate, **K** 5, petals spreading only a little, as permitted by the enclosing leaves, green to whitish, filamentous staminodes apically papillate, placed together with petals and stamens on a cushion surrounding the lower lying surface of the gynoeceum, thus forming a short hypanthium and a recess flower; **Fr** on a long pedicel, of the Delosperma-type, but with valve wings narrowing towards their base, placenta mainly basal, 5 locules; **Ecol** < 100 mm rainfall p.a. in winter, fog; **Distr** Lüderitz-South, Namibia.

Note: One species is known.

Distribution of **Ruschianthus** L.Bolus

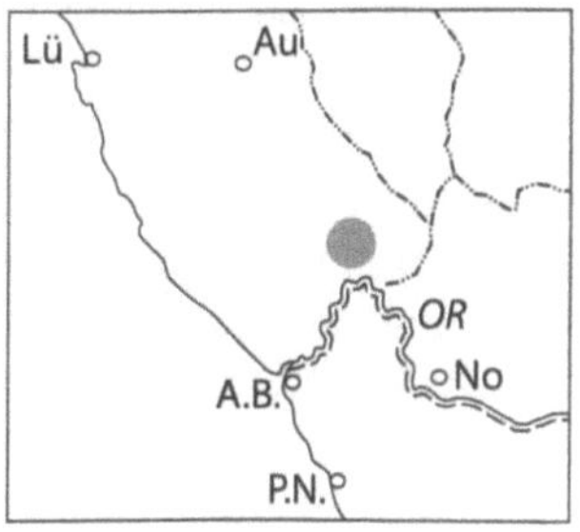

R. falcatus L.Bolus JSAB 1960: 63 (see **photo**) T Rusch 259/60 (BOL).

Up to 10 branches; **L** hardly connate at base and spreading diagonally, 30–45 mm *l*, 5–6 mm *b*, to 18 mm *t*; **Fl** not opening widely in natural conditions because the leaves surround it too closely, 80–90 conspicuously green petals, 110–120 greenish-white filamentous staminodes in at least two rows, entirely or only apically papillate, 150–160 stamens, the inner almost horizontally covering the centre of the recess flower, all androecial elements and the five conspicuously dark green nectaries placed on a prominent cushion forming a ring at the bottom of the flower, stigmas c. 1 mm *l*; **Fr** greyish, ∅ c. 6 mm; **Ecol** as for genus; **Distr** as for genus.

Note: Green is a rare flower colour in Mesembryanthema, even rarer than recess flowers, both features probably representing an ecological adaptation which may be related to the fact that the leaves enclose the flowers rather narrowly so that they can never open widely, as is usual within the family.

Saphesia Ruschioideae

H. E. K. Hartmann

Saphesia N.E.Br. GC 1932: 205; Ihlenfeldt & Gerbaulet 1990: 457–498 **Typus S. flaccida** (Jacquin) N.E.Br. **Etym** Gk *saphes*, distinct, "because the plant is quite distinct from every other known member of the family" (Brown 1932: 205).

Perennial or biennial, thick, branched rootstock, deciduous branches; **L** opposite to alternate, flat, slender, light green, mesomorphic, with dispersed bladder cells; **Fl** solitary (?) **K** 5, petals white, nectary as a slightly lophomorphic ring; **Fr** xerochastic, valves not separating from the septa, in contrast to all other members of the Ruschioideae, forming a distal ring along their outer margin into which septa and placentas protrude forming sections into which single seeds are growing, dispersal as castor type, i.e. by shaking the capsule; **S** dark brown, testa with low papillae; **Ecol** near the sea, rain at all seasons, > 400 mm p.a.; **Distr** SW districts in WC, S Africa.

Note: The genus is probably closely related to **Skiatophytum** and **Caryotophora**, in which the compartments containing seeds are developed into closets and finally into a woody container (Ihlenfeldt & Gerbaulet 1990).

Distribution of **Saphesia** N.E.Br.

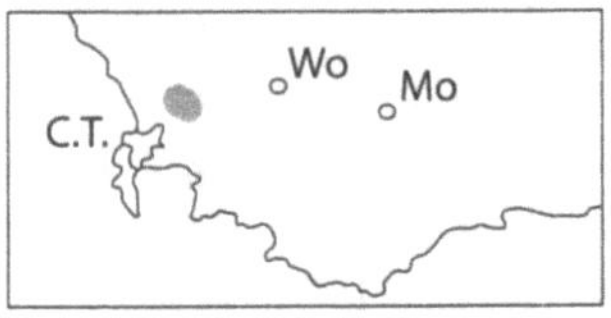

S. flaccida (Jacquin) N.E.Br. GC 1932: 205 ≡ *M. flaccidum* Jacquin 1804: 38, t. 475 non L.Bolus ABH 1923: 164 nomen illeg. **T** Jacquin 1804: t. 475.

Stems to 1 m *l*, branching, **L** opposite or alternate, free at their bases; **Fl** ∅ 35–45 mm, stalks 45–210 mm *l*; **Fr** 5 locules; **Ecol** as for genus; **Distr** as for genus.

Note: The unique dispersal mechanism separates the species from all others.

Sarcozona Ruschioideae

H. E. K. Hartmann

Sarcozona J.Black TPRSSA 1934: 176; Blake 1969: 32–39; Venning in Prescott & Venning 1984: 28–30 **Typus** *S. pulleinei* (J.Black) J.Black = **S. praecox** (F. Mueller) S.T.Blake ex H.Eichler **Etym** Gk *sarcos*, flesh, *zonae*, zone, ring, involucrum, referring to the complete sheath of the bracteoles surrounding closely the base of the flower.

Shrubs with erect to decumbent branches, rooting very tardily near the stem, I apparently spongy; **L** triquetrous to trigonous, often one side convex, the other concave, basal part of sheath persisting as a collar, epidermis with dots or small warts, keel and margins finely denticulate in the upper parts; **Fl** 1–3, bracteoles sheathing the base of the flower closely, appearing almost connate, the pair therefore being called "involucrum", **K** 4–5, petals pale pink to white, always white towards the base, filaments white, anthers yel-

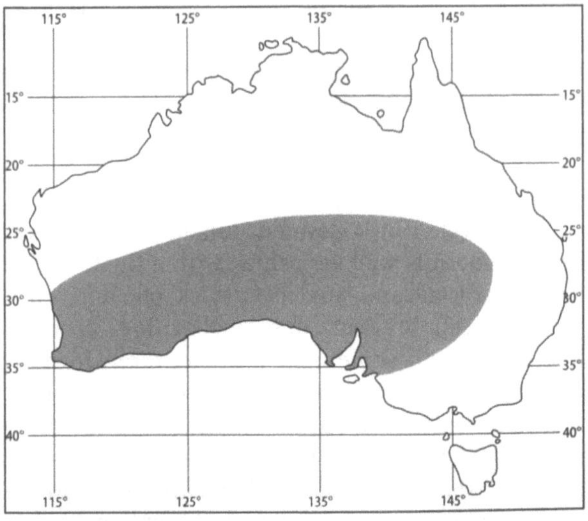

Distribution of **Sarcozona** J.Black

low, placenta parietal and on the central axis, nectary as an inconspicuous ring; **Fr** remaining succulent like those of **Carpobrotus,** 4–5 locules; **Ecol** mostly in sand; **Distr** New South Wales, South Australia, West Australia, Victoria, Australia.

Note: In following the treatment in Flora of Australia (ed. George 1984), **Sarcozona** is kept here as a separate genus, although it has been suggested to unite it with **Carpobrotus** (Rowley 1978: 9) because of the succulent fruit and the very similar, big triquetrous leaves. The more shrubby, erect growth, the possession of an involucral sheath of bracteoles around the base of the flower, and the low number of locules in **Sarcozona** appear distinct enough to support the position taken in the Flora of Australia.

Key to the Species of Sarcozona

1. Base of flower terete, enclosed for half its length in the sheath formed by the bracteoles ...... **S. praecox**
– Base of flower 2-keeled, enclosed for less than half the length by the sheath formed by the bracteoles .... **S. bicarinata**

S. bicarinata S.T.Blake 1969: 37 (see **photo**) T Blake 21514 (BRI).

To 10 cm *h*, branches erect to decumbent later, **I** at first 2-edged, later polygonal, brown to pale, 3–7 mm ∅; **L** obliquely triquetrous, dull green, often suffused with purple, 25–50 mm *l*, 5–9 mm *b*; **Fl** solitary, rarely ternate, sheath of bracteoles enclosing the base of the flower only basally, **K** (4–)5, base 2-edged, ribs to 2.5 mm *h*, 20–55 petals, 5–14 mm *l*, 0.4–1.2 mm *b*, 20–50 stamens; **Fr** turning purplish, globose to ovoid with persisting two ridges, 7–15 mm *l*, 7–15 mm ∅, (4–)5 locules; **Ecol** in sand; **Distr** coastal South Australia.

Note: The bicarinate base of the flower resembles calyces found in some species of **Carpobrotus**.

S. praecox (F.Mueller) S.T.Blake ex H.Eichler 1965: 134 (see **photo**) ≡ *M. praecox* F.Mueller 1852: 384 non L.Bolus NM 1928: 3 nomen illeg. ≡ *Carpobrotus praecox* (F.Mueller) G.D.Rowley 1978: 9 **T** F.Mueller s.n. (MEL) = *Carpobrotus pulleinei* J.Black 1932: 40 ≡ *S. pulleinei* (J.Black) J.Black 1934: 176 **LT** Morris s.n. (AD).

To 30 cm, branches erect, angled, often pink; **L** trigonous to triquetrous, surface a little rough from raised dots, keel smooth or rough, **L** 30–100 mm *l*, 4.5–7 mm *b*, 5–8 mm ∅; **Fl** almost enclosed by the sheath of the bracteoles, **K** 4(–5), base oblong, terete; **Fr** enclosed, 10–20 mm *l*, 8–16 mm ∅, 4(–5), locules; **Ecol** in sandy and/or saline soils; **Distr** New South Wales, South Australia, West Australia, Victoria, Australia.

Note: Mueller (1852) already noted that plants of this species do not creep like those of **Carpobrotus** but remain rather shrubby.

S. pulleinei = **S. praecox**

Sceletium Mesembryanthemoideae

M. Gerbaulet

Sceletium N.E.Br. GC 1925: 412; Bittrich 1987: 5–116; Gerbaulet BJS 1996: 9–24 ≡ *Phyllobolus* subg. *Sceletium* (N.E.Br.) Bittrich 1987: 75 **Lectotypus S. tortuosum** (L.) N.E.Br. ≡ *Phyllobolus tortuosus* (L.) Bittrich **Etym** La *sceletus*, skeleton; referring to the "skeletonized" persistent dry leaves which retain their shape through the lignified vascular bundles.

Prostrate, decumbent, or rarely erect small shrubs with fibrous roots, often scrambling in other bushes, **I** weakly lignified, with additional vascular bundles in the cortex, branches slender, basal ones often robust with age, bladder cells mesomorphic, conspicuous or inconspicuous; **L** flat, ovate, decussate and shortly connate at the base or becoming alternate in the **Inf**, dry leaves persistent and "skeletonized" through lignified veins and enclosing the young buds, tips recurved or incurved, without distinct central water-storing cells, bladder cells mesomorphic, conspicuous or inconspicuous; **Fl** in few-flowered cymes, sometimes solitary, stalked or sessile, ∅ 20–40 mm, **K** 4–6, persistent, petals white, pale to bright yellow, pale pink, or pale salmon, 1 or 2 mm *b*, pointed, emarginate or irregularly denticulate, **K** and petals basally connate into a short tube, filamentous staminodes present, sometimes concealing the reproductive parts, nectaries shell-shaped; **Fr** with or without valve wings, very light, breaking off readily to be dispersed by wind as tumble fruits, locules 4, 5 or 6; **S** to 2 mm *l*, D-shaped, black or brown, testa rough with a distinct or an indistinct crest; **Ecol** karroid areas, all species often underneath bushes and somewhat scrambling, winter and summer rainfall; **Distr** EC, NC, WC, S Africa.

Note: The most striking feature of **Sceletium** is the skeleton-like appearance of the dry leaves enclosing and protecting the young buds during the dry season. The venation patterns differ between the species and are of taxonomic value.

Distribution of **Sceletium** N.E.Br.

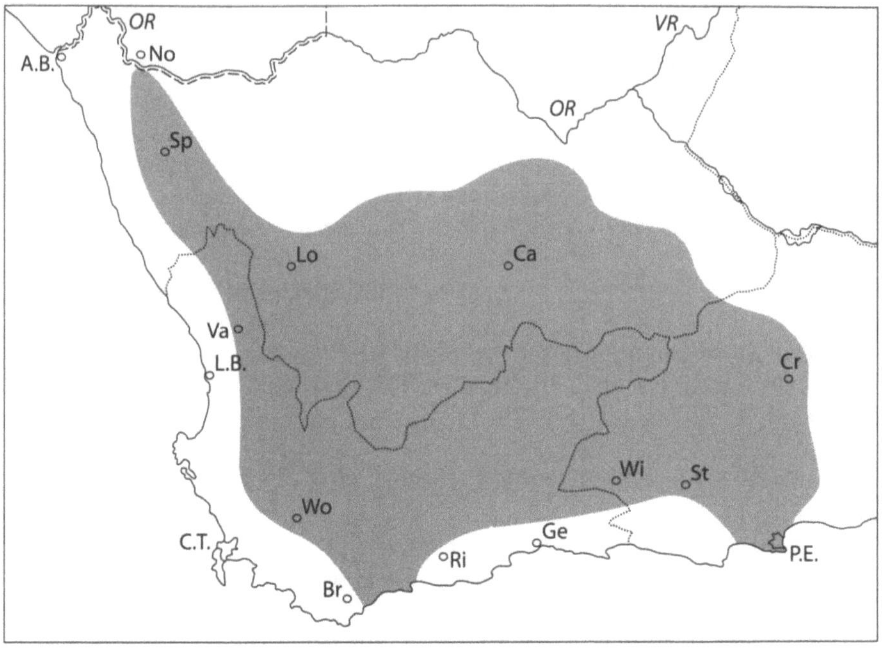

Key to the Species of Sceletium

1. Leaves with one distinctly straight main vein and several equally conspicuous lateral veins which curve back to the main vein 2
- Leaves with 3, rarely 5, straight veins, lateral veins less conspicuous 4
2. Fruits with valve wings ..... **S. emarcidum**
- Fruits without valve wings 3
3. Plants becoming robust, branches often erect, leaves to 25 mm long ..... **S. rigidum**
- Plants delicate, branches decumbent, leaves to 35 mm long ..... **S. exalatum**
4. Flowers white or pale yellow with bright yellow centre, ∅ c. 40 mm, petals c. 2 mm broad, emarginate or irregularly denticulate, reproductive parts concealed by filamentous staminodes 5
- Flowers pale yellow, pale pink, or pale salmon, centre somewhat darker than the periphery but not bright yellow, ∅ 20–30 mm, petals c. 1 mm broad, pointed, reproductive parts not concealed 6
5. Leaves narrowly ovate, to 20 mm long ..... **S. varians**
- Leaves ovate, to 65 mm long ..... **S. expansum**
6. Bladder cells prominent, leaf tips incurved, leaves imbricate ..... **S. tortuosum**
- Bladder cells flattened, leaf tips recurved, leaves not imbricate 7
7. Leaves narrowly ovate, calyx lobes and locules 4 or 5 ..... **S. strictum**
- Leaves ovate, calyx lobes and locules 5, rarely 6 ..... **S. crassicaule**

S. albanense = **S. crassicaule**
S. anatomicum = **S. emarcidum**
S. archeri = **S. rigidum**
S. boreale = **S. tortuosum**
S. compactum = **S. tortuosum**
S. concavum = **S. tortuosum**

S. crassicaule (Haw.) L.Bolus SAG 1927: 399 (see **photo**) ≡ *M. crassicaule* Haw. 1824: 425 ≡ *Pentacoilanthus crassicaulis* (Haw.) Rappa & Camarrone 1960: 23 nomen invalid. T drawing (K!) = *S. albanense* L.Bolus NM 1933: 396 **T** Dyer 2137 (BOL!).

Prostrate to decumbent, rarely scrambling, **I** becoming robust; **L** not imbricate, tips recurved, to 40 mm *l*, with 5(-3) straight veins, lateral veins less conspicuous, bladder cells completely flat; **Fl** stalked, ∅ 20–30 mm, K 5–6, petals white to pale yellow or pale pink, to 1 mm *b* with pointed apex, stigmas obtuse, less than 2 mm *l*; **Fr** ∅ 10–15 mm, with valve wings, valves do not open completely, locules 5–6; **S** brown with a distinct crest; **Ecol** coastal and inland karroid areas, summer and winter rainfall; **Distr** Willowmore to Port Elizabeth and Albany, EC, Bredasdorp, Swellendam, WC, S Africa.

Note: **S. crassicaule** is similar to **S. tortuosum**, but is distinguished by the recurved leaf tips and the completely flat bladder cells.

S. dejagerae = **S. emarcidum**

S. emarcidum (Thunberg) L.Bolus ex H.Jacobsen 1960: 1411 (see **photo**) ≡ *M. emarcidum* Thunberg 1791: 9 ≡ *M. anatomicum* Haw. var. *emarcidum* (Thunberg) DC. 1828: 450 **LT** Thunberg 11974 A (UPS, photograph BOL!) = *M. anatomicum* Haw. 1803: 50 ≡ *S. anatomicum* (Haw.) L.Bolus SAG 1927: 399 ≡ *Tetracoilanthus anatomicus* (Haw.) Rappa & Camarrone 1960: 24 nomen invalid. **LT** drawing (K!) = *M. anatomicum* Haw. var. *fragile* Haw. 1821: 169 **NT** Dyer 1024 (BOL!) = *S. dejagerae* L.Bolus NM 1929: 126 **T** de Jager 18953 (BOL!).

Prostrate to decumbent, often scrambling, **I** usually weak; **L** not imbricate, tips recurved, to 35 mm *l*, with 1 distinctly straight and several equally conspicuous, curved lateral veins, bladder cells small, but conspicuously vaulted; **Fl** stalked, ∅ c. 20 mm, **K** 4–5, petals whitish, to 1 mm *b* with pointed apex, stigmas subulate, 2–5 mm *l*; **Fr** ∅ 5–10 mm, with valve wings, locules 4–5; **S** black with an indistinct crest; **Ecol** summer and winter rainfall; **Distr** between Cradock, EC, Prieska, NC, Beaufort West and Vanrhynsdorp, WC, S Africa.

Note: This species is very similar to **S. exalatum** and **S. rigidum**, but differs from both in the presence of valve wings in the fruit.

S. exalatum Gerbaulet BJS 1996: 17 (see **photo**) **T** Struck 186 (STE!).

Prostrate to decumbent, often scrambling, **I** weak; **L** not imbricate, tips recurved, to 35 mm *l*, with 1 distinctly straight and several equally conspicuous, curved lateral veins, bladder cells small, but conspicuously vaulted; **Fl** stalked, ∅ c. 20 mm, **K** 4, petals whitish, to 1 mm *b* with pointed apex, stigmas subulate, 2–5 mm *l*; **Fr** ∅ 5–10 mm, without valve wings, locules 4; **S** black with an indistinct crest; **Ecol** winter rainfall; **Distr** Namaqualand, NC, S Africa.

Note: **S. exalatum** is very similar to **S. emarcidum** and **S. rigidum**. From the former it differs in lacking valve wings, from the latter in being delicate.

S. expansum (L.) L.Bolus SAG 1927: 399 (see **photos**) ≡ *M. expansum* L. 1759: 1059 non Thunberg 1800: 88 nomen illeg. ≡ *Pentacoilanthus expansus* (L.) Rappa & Camarrone 1960: 23 nomen invalid. **LT** Dillenius 483 (ex Herb. Oxford, K!, photograph ex Herb. Sherard, BOL!) = *S. regium* L.Bolus JSAB 1959: 260 **T** Leipoldt 19914 (BOL!).

Prostrate to decumbent or scrambling, **I** becoming robust; **L** not imbricate, tips recurved, to 65 mm *l*, with 3 ± straight veins, lateral veins somewhat less conspicuous, bladder cells small; **Fl** stalked, ∅ c. 40 mm, **K** 5, petals pale yellow, to 2 mm *b*, emarginate, filamentous staminodes bright yellow, concealing the reproductive parts, stigmas obtuse, to 2 mm *l*; **Fr** ∅ c. 10 mm, with valve wings, locules 5; **S** brown with a distinct crest; **Ecol** winter rainfall; **Distr** Clanwilliam, Malmesbury, Piquetberg, WC, S Africa.

Note: **S. expansum** and **S. varians** share large flowers which are characterized by broad petals concealing the reproductive parts. **S. expansum**, however, has much larger and broader leaves.

S. framesii = **S. tortuosum**
S. gracile = **S. tortuosum**
S. joubertii = **S. tortuosum**
S. namaquense = **S. tortuosum**
S. namaquense var. *subglobosum* = **S. tortuosum**
S. ovatum = **S. tortuosum**
S. regium = **S. expansum**

S. rigidum L.Bolus SAG 1928: 279 (see **photo**) **T** Hofmeyer de Villiers 1135/28 (BOL!) = *S. archeri* L.Bolus NM 1928: 16 **T** Archer 18848 (BOL!).

Decumbent to erect, often scrambling, **I** often robust; **L** not imbricate, tips recurved, to 25 mm *l*, with 1 distinctly straight and several equally conspicuous, curved lateral veins, bladder cells small but very conspicuously vaulted, glittering; **Fl** sessile, ∅ c. 20 mm, **K** 4, petals whitish, to 1 mm *b* with pointed apex, stigmas subulate, 2–5 mm *l*; **Fr** ∅ 5–10 mm, without valve wings, locules 4; **S** black with an indistinct crest; **Ecol** summer and winter rainfall; **Distr** Laingsburg, Prince Albert, WC, S Africa.

Note: **S. rigidum** is closely allied to **S. emarcidum** and **S. exalatum**. It differs from both species by a more robust habit and from the former in lacking valve wings.

S. strictum L.Bolus SAG 1927: 399 (see **photo**) **LT** Bolus 145 (BOL!).

Prostrate to decumbent, rarely scrambling, **I** becoming robust; **L** not imbricate, tips recurved, to 40 mm *l*, with 3 straight veins, lateral veins less conspicuous, bladder cells flattened; **Fl** stalked, ∅ 20–30 mm, **K** 4–5, petals white to pale yellow, to 1 mm *b* with pointed apex, stigmas obtuse, less than 2 mm *l*; **Fr** ∅ 10–15 mm, with valve wings, locules 4–5; **S** brown with a distinct crest; **Ecol** on rocks, winter rainfall; **Distr** Ladismith, WC, S Africa.

Note: **S. strictum** is very similar to **S. crassicaule** from which it differs by narrower leaves and fewer calyx lobes and locules.

S. subvelutinum = **S. varians**
S. subvelutinum forma *luxurians* = **S. varians**

S. tortuosum (L.) N.E.Br. 1926: 245 (see **photo**) ≡ *M. tortuosum* L. 1753: 487 ≡ *Pentacoilanthus tortuosus* (L.) Rappa & Camarrone 1960: 24 nomen invalid. ≡ *Phyllobolus tortuosus* (L.) Bittrich 1987: 75 ≡ *M. aridum* Moench 1802: 192 nomen illeg. **LT**

Dillenius 1732: t. 181 f 222 = *S. boreale* L.Bolus NM 1932: 375 **T** Collins 896/31 (BOL!) = *S. compactum* L.Bolus NM 1933: 393 **T** Lewis 20424 (BOL!) = *M. concavum* Haw. 1821: 168 ≡ *S. concavum* (Haw.) Schwantes GF 1928: 68 ≡ *Tetracoilanthus concavus* (Haw.) Rappa & Camarrone 1960: 24 nomen invalid. **T** Burman 1738: t. 26 f 1 = *S. framesii* L.Bolus NM 1930: 176 **T** Frames 1115/26 (BOL!) = *S. gracile* L.Bolus NM 1932: 375 **T** SUG 9134 (BOL!) = *S. joubertii* L.Bolus NM 1934: 435 **T** Joubert 45 (BOL!) = *S. namaquense* L.Bolus NM 1932: 374 **T** Herre 8818 (BOL!) = *S. namaquense* L.Bolus var. *subglobosum* L.Bolus NM 1932: 375 **T** Herre 9382 (BOL!) = *S. ovatum* L.Bolus NM 1954: 288 **T** Human 13015 (BOL!) = *S. tugwelliae* L.Bolus NM 1933: 394 **T** Tugwell 20425 (BOL!).

Prostrate to decumbent, rarely scrambling, **I** becoming robust; **L** imbricate, tips incurved, to 40 mm *l*, with 3(–5) straight veins, lateral veins less conspicuous, bladder cells large, but rather plane; **Fl** more or less sessile, ∅ 20–30 mm, **K** 4–5, petals white to pale yellow, pale salmon, or pale pink, to 1 mm *b* with pointed apex, stigmas obtuse, less than 2 mm *l*; **Fr** ∅ 10–15 mm, with valve wings, locules 4–5; **S** brown with a distinct crest; **Ecol** often on quartzite, winter to summer rainfall; **Distr** EC, NC, WC, S Africa; from Namaqualand to Montagu and Aberdeen.

Note: **S. tortuosum** is the most widespread among the species of **Sceletium**. It is readily distinguishable on account of its imbricate leaves with the tips incurved. Known as "kougoed", **S. tortuosum** is widely used as a drug against stomach ache, to sooth small children, or just against depression.

S. tugwelliae = **S. tortuosum**

S. varians (Haw.) Gerbaulet BJS 1996: 20 (see **photo**) ≡ *M. varians* Haw. 1803: 51 **T** Petiver 1709: t. 78 f 10 = *S. subvelutinum* L.Bolus NM 1930: 223 **T** Herre 9345 (BOL!) = *S. subvelutinum* L.Bolus forma *luxurians* L.Bolus NM 1930: 224 **T** Leipoldt 19245 (BOL!).

Decumbent or becoming erect, scrambling, **I** becoming robust; **L** not imbricate, tips recurved, to 20 mm *l*, with 3 straight veins, lateral veins less conspicuous, bladder cells small, sometimes pointed; **Fl** stalked, ∅ c. 40 mm, **K** (4–)5, petals white to pale yellow, to 2 mm *b*, emarginate or irregularly denticulate, filamentous staminodes bright yellow, concealing the reproductive parts, stigmas subulate, to 2 mm *l*; **Fr** ∅ c. 10 mm, with valve wings, locules (4–)5; **S** brown with a distinct crest; **Ecol** winter rainfall; **Distr** Robertson, Worcester, WC, S Africa.

Note: Because of its large flowers with concealed reproductive parts and broad petals, **S. varians** is very similar to **S. expansum**. It differs from the latter in a more erect habit and smaller and narrower leaves with sometimes pointed bladder cells. When describing the species, Haworth cited also catalogue number 478, in which Petiver (1709) gives "7.10" as the reference to the figure. This latter figure is a butterfly, and Petiver 78.10 has still to be found.

Schlechteranthus Ruschioideae

H. E. K. Hartmann

Schlechteranthus Schwantes MDK 1929: 16 **Typus S. maximiliani** Schwantes **Etym** *Maximilian Schlechter*, 1874–1960, trader and plant collector in Southern Africa, brother of the botanist R. Schlechter.

Succulent shrubs to 40 cm *h*, highly branched and dense; **L** longer than **I**, one pair to a branch in the resting state, basally connate for almost 50%, free parts erect to spreading a little, triangular to triquetrous when drying, apically mucronate, sometimes serrulate along the keel, epidermis with low papillae, raised around the sunken stomata and subsidiary cells, xeromorphic outer epidermis walls with a crystal layer up to 30 µm ∅, wax cover continuous, breaking into irregular plates and horizontal platelets; **Fl** solitary, bracteolate, **K** 6, ± equal, petals white to pink, filamentous staminodes white to pink, stamens white to pink, basally shortly papillate, nectary ring; **Fr** hard and robust, top elevated, dome-shaped, valves with or without valve wings, covering membranes complete, often spongy for a long time, the distal rim bent back to various degrees, sometimes covering the covering membrane below as a second outer layer, a sharp closing ledge developed below the rim, closing bodies oval, the long axis pointing into the locule, 8–12 locules; **S** almost oval, testa cells elongate, anticlinal walls undulate; **Ecol** on quartz slopes in open succulent Karroo, < 100 mm rainfall p.a. in winter; **Distr** Namaqualand, NC, S Africa.

Note: Both species are most often found in monotypic stands on exposed slopes of rough quartzitic stones. **Schlechteranthus** is very similar to **Polymita** in appearance, but has less prickly leaves, smaller flowers, and more narrow valve wings than the latter.

Distribution of **Schlechteranthus** Schwantes

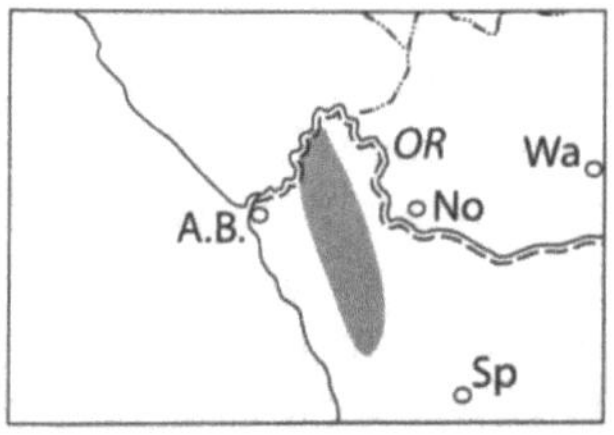

Key to the Species of Schlechteranthus

1. Leaves 20–30 mm long, most fruits of a population with 11–12 locules .... **S. hallii**
- Leaves 10–18 mm long, most fruits of a population with 10 locules .... **S. maximiliani**

S. hallii L.Bolus 1964: 241 (see **photo**) **T** Hall 1358 (BOL).

L spreading only a little, 20–30 mm *l*, 8–10 mm *t*, darkish greyish-green, often purplish suffused; **Fl** ∅ 20–25 mm, 40–80 white to pinkish suffused petals, 30–80 white filamentous staminodes, 100–150 white stamens; **Fr** valves always with narrow valve wings, rims of covering membranes moderately recurved, closing ledge below very distinct and sharp, ∅ 9–10 mm, mostly 11–12 locules; **S** 1.2–1.7 mm *l*, 0.75–1 mm *b*; **Ecol** on N slopes of quartzitic stones; **Distr** N part of the area of the genus, Namaqualand, NC, S Africa.

Note: A characteristic plant of moderately sloping, north facing sections of mountains in the Richtersveld, often blackened by a fungus. Plants in a population appear to be of one age group, dying within a short time standing as black mummies for several years.

S. maximiliani Schwantes MDK 1929: 16 (see **photo**) **T** M.Schlechter 51 (HBG!).

L spreading, 10–18 mm *l*, 5–6 mm *t*, light yellowish-green, often orange to purplish suffused; **Fl** ∅ 25–28 mm, 50–60 pink petals, 30–130 filamentous staminodes often with white bases and purple tips, 90–170 white stamens; **Fr** valves without, very rarely with valve wings, rims of covering membranes mostly widely recurving thus forming a second upper layer above the covering membranes proper, interstice often filled with a spongy tissue, ∅ 5–9 mm, mostly 9–11 locules, maximum at 10; **S** 1.2–1.5 mm *l*, 0.65–0.9 mm *b*; **Chr** $2n = 18$; **Ecol** dominant on S slopes of rough quartzite; **Distr** S part of the area of the genus, Namaqualand, NC, S Africa.

Note: **S. maximiliani** prefers southern slopes, growing a little lower but mostly more highly branched than those of **S. hallii**.

Schonlandia ≡ Corpuscularia

H. E. K. Hartmann

Bolus (FPSA 1927: t. 259) described the new genus *Schonlandia* with the type *S. lehmanni*, which is the type of the earlier described genus **Corpuscularia**, to which *Schonlandia* must therefore be placed as a synonym. It can also be argued that the name is too similar to *Schoenlandia* Cornu (1896: 24, not 21 as in Index Kewensis) = **Cyanastrum** Oliver in the family Cyanastraceae to be kept up, but neither confusion has occurred nor is either name legitimately in use at present.

S. lehmannii ≡ **Corpuscularia lehmannii**

Schwantesia Ruschioideae

H. E. K. Hartmann & N. Zimmermann

Schwantesia Dinter MDG 1927: 234 non L.Bolus SAG 1928: 279 nomen illeg.; Zimmermann CSJ 1996: 257–266 **Typus S. ruedebuschii** Dinter **Etym** *G. Schwantes*, 1881–1960, professor of prehistory and botanist at Kiel, Germany.

Compact plants; **L** erect to spreading, trigonous, symmetrical when young, becoming asymmetrical when starting to flower, often very skewed due to an oblique keel, often broader than thick, with or without teeth, epidermis of elevated cells shaped like cobblestones or papillate to different degrees, cell sides smooth or with 6–10 bulges; **Fl** solitary, seemingly ebracteolate, **K** 5, the keels of the 2 outer, bigger lobes running down the pedicel as ± broad wings, petals bright yellow, often basally white, filamentous staminodes absent, stamens in a central column, nectary as a lophomorphic ring; **Fr** winged pedicels persistent, keeping the fruit erect between the leaves, top a little conical, rims rather high, rising to the centre, leaving a central deepening like a crater free in some species, base bowl-shaped, inner morphology of the Delosperma-type, the septa with a very narrow rim spreading a little over the locules, placenta basal and parietal, also axial in some species, expanding keels reaching down to the level of the upper end of the placenta, 5 locules; **S** ± pear-shaped, testa cells smooth or a little elevated, 0.65–0.9 mm *l*, 0.45–0.6 mm *b*; **Ecol** in crevices of quartzite rocks, rarely in shaly soils, < 100 mm rainfall p.a. mainly in March, rarely also in November and in summer; **Distr** Lüderitz-South, Warmbad, Namibia; Gordonia, Kenhardt, Namaqualand, NC, S Africa.

Note: Together with the genera **Dinteranthus**, **Lapidaria**, and **Ihlenfeldtia**, **Schwantesia** belongs to the genera typical of the Pofadder centre. The ebracteate flowers and the fruits are most similar to those of **Lithops**, which has been understood as a relative in derived form because of its sunken growth. The name *Schwantesia* has been used later again, being an illegitimate name for species of **Dicrocaulon**, **Mitrophyllum**, and **Monilaria**.

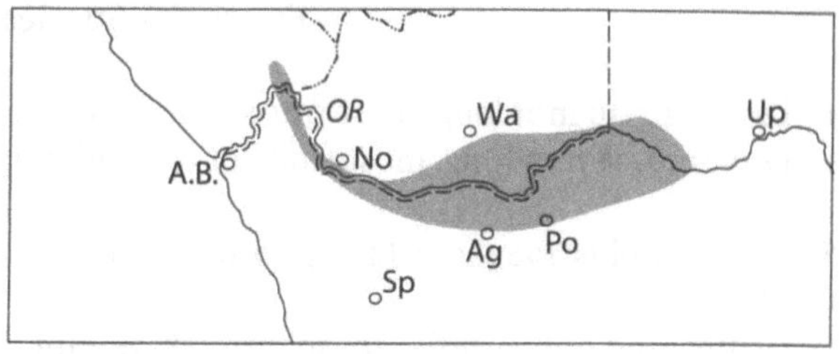

Distribution of **Schwantesia** Dinter

Key to the Species of Schwantesia

1. Leaves with thick, flexible teeth on keel and margins 2
– Leaves edentate 3
2. Leaves with long papillae .... **S. marlothii**
– Leaves with cobblestone-shaped cells with very short central papillae .... **S. ruedebuschii**
3. Leaves uniformly greyish-green 4
– Leaves with white irregular dots as if marbled .... **S. borcherdsii**
4. Leaves slender, margins parallel for most of their length, ± erect 5
– Leaves fattish, margins not visibly parallel for any length, ± spreading 7
5. Margins and keel sharp, marked by very finely crenulate corky edges turning dark brown soon .... **S. constanceae**
– Margins and keel rounded, no conspicuous dark edges 6
6. Leaves with long papillae .... **S. triebneri**
– Leaves with low papillae .... **S. pillansii**
7. Leaves shorter than 35 mm 10
– Leaves longer than 40 mm 8
8. Capsules with very high valve rims (up to 2.5 mm) leaving a free space like a crater in the centre .... **S. acutipetala**
– Capsules with moderately high valve rims (up to 1.5 mm), mostly meeting in the centre 9
9. Leaf margins with a brown line, testa with elevated cells .... **S. speciosa**
– Leaf margins light in colour, testa completely smooth .... **S. succumbens**
10. Leaf length to maximum leaf width > 1.5, stamens > 4.5 mm long .... **S. herrei**
– Leaf length to maximum leaf width < 1.5, stamens < 4.5 mm long .... **S. loeschiana**

S. acutipetala L.Bolus NM 1937: 62 (see **photo**) **T** Heerde 10767 (BOL!).

L broadening only a little from the base, broadest at c. 4/5th of the length, apically triangular, keel shifted a little to one side, epidermis cells cobblestone-shaped with 6–9 bulges around the upper edge, centre with a thickened knob of wax, but no papilla, **L** 25–50 mm *l*, 13–18 mm *b*; **Fl** ∅ 38–45, 40–50 petals, 165–190 stamens, rims of later valves well visible as high ridges between which the stigmas sit in the centre; **Fr** top with very high rims rising towards the centre where they leave a free space from which the stigmas have disappeared, resembling a crater, placenta very variable in extension; **S** testa cells smooth, only the dorsal end towards and at the micropyle with elongate, elevated cells, 0.6–0.75 mm *l*, 0.48–0.55 mm *b*; **Chr** $2n = 18$; **Ecol** in crevices of quartzitic rocks; **Distr** N Namaqualand, NC, S Africa; Warmbad, Namibia.

Note: The remarkable rims on the edges of the valves of the fruit distinguish the species from all others clearly.

S. australis = **S. ruedebuschii**

S. borcherdsii L.Bolus NM 1929: 86 (see **photo**); Zimmermann & Hartmann 1995: 162–165 **T** Borcherds 557/29 (BOL!).

L flattish with undulate margins, only a certain number of epidermal cells with a central layer of crystals in the outer wall, surface therefore with a marbled pattern, epidermis cells with a smooth, round base each, leaving no gap between them, and low central papillae, wax cover a ± continuous layer with some amorphous particles and platelets at the base of the papilla, 30–45 mm *l*, 13–15 mm *b*; **Fl** ∅ 32–40 mm, 34–46 petals, 113–170 stamens; **Fr** rims c. 1.2 mm *h*, meeting in the centre, placenta mostly basal; **S** testa almost smooth, c. 0.8 mm *l*, c. 0.6 mm *b*; **Chr** $2n = 18$; **Ecol** in rocky places on ridges; **Distr** Gordonia, Kenhardt, NC, S Africa.

Note: The species has only recently been found again, but meanwhile several localities are known. It is best recognized by the marbled leaf surfaces.

S. constanceae N.Zimmermann KuaS 1996: 84 (see **photo**) T Zimmermann 26/94 (JE!).

L with parallel margins, acuminate, keel and margins sharp, minutely crenulate and maroon from tannin idioblasts in dense packing below, epidermis cells like cobblestones with almost vertical, plain sides and edges, near the margins some rows of cells with bulges on their sides and low central elevations, the cells above the margin elongate, placed in 1–2 rows arranged horizontally, like loaves in the oven, 17–27 mm *l*, 9–14 mm *b*; **Fl** ∅ 30–40 mm, 40–45 petals, 130–150 stamens; **Fr** rims low, meeting in the centre, placenta basal and parietal reaching higher up on the outer wall than in any other species; **S** elongate and pear-shaped, testa cells almost smooth, elevated ones on the back and in the micropylar region, 0.8–0.85 mm *l*, 0.5–0.55 mm *b*; **Chr** $2n = 18$; **Ecol** on gneissic shales; **Distr** Warmbad, Namibia.

Note: This is the only species known until now that grows sympatrically with another species, **S. succumbens**.

S. herrei L.Bolus NM 1931: 287 (see **photo**) T Herre 9184 (BOL!).

L narrowing from a broad base, the upper leaf surface forming an almost equal-sided triangle, epidermis cells cobblestone-shaped with 6–8 lateral bulges around the upper edge, a low papilla in the centre, **L** 19–30 mm *l*, 12–16 mm *b*; **Fl** ∅ c. 35 mm, c. 50 petals, c. 165 stamens, rims on ovary well visible; **Fr** rims to 2 mm *h*, the central crater less deep than in **S. acutipetala** because the rims are connate basally in the crater, placenta basal-parietal; **S** 0.8–0.9 mm *l*, 0.5–0.55 mm *b*; **Ecol** in crevices of quartzitic ridges; **Distr** N Namaqualand, NC, S Africa.

Note: In fruit morphology and the number of petals, **S. herrei** resembles **S. acutipetala** most closely, but in the shorter leaf it comes close to **S. loeschiana**.

S. herrei var. *minor* = **S. loeschiana**
S. herrei forma *major* = **S. loeschiana**

S. loeschiana Tischer Jahrbuch der Deutschen Kakteengesellschaft 1936: 71 ≡ *S. herrei* L.Bolus forma *major* G.D.Rowley 1973: 6 T Erni s.n., lost **LT** Tischer 1936: 71 f = *S. herrei* L.Bolus var. *minor* L.Bolus NM 1958: 341 T Herre 13611 (BOL!).

L narrowing from a broad base, the upper leaf surface forming an almost equal-sided triangle, epidermis cells cobblestone-shaped with 6–8 prominent lateral bulges along the sides, a very low papilla in the centre, papillae on cells along the margins a little longer, **L** 14–24 mm *l*, 10–17 mm *b*; **Fl** ∅ 30–40 mm, 43–48 petals, 130–140 stamens; **Fr** with low rims to 1.2 mm *h*, meeting centrally, placenta basal to parietal; **S** c. 0.8 mm *l*, 0.45 mm *b*; **Ecol** on steep slopes among quartzitic rocks; **Distr** Lüderitz-South, Namibia.

Note: The species has shorter leaves than **S. herrei**, appearing therefore broader and stouter, the shape is very similar. The fruit resembles those of the species from further east more closely by the lack of a central crater between the rims.

S. marlothii L.Bolus NM 1932: 354 (see **photo**) T Marloth 13283 (BOL!).

L widening gradually from the base, the apical part with broad, flexible teeth, flattish, keel oblique, epidermis cells with a long, central papilla each, on a smooth, roundish base of the elevated cell, **L** 35–42 mm *l*, 12–16 mm *b*, 7–10 mm ∅; **Fl** ∅ 30–45 mm, 32–45 petals, 115–160 stamens, rims on ovary to 1.4 mm *h*; **Fr** rims meeting in the centre, to 1.5 mm *h* here, placenta basal mainly; **S** roundish, testa almost smooth, elongate elevated cells in the dorsal region to the micropyle, 0.63–0.7 mm *l*, 0.45–0.475 mm *b*; **Ecol** among big quartzitic stones and in crevices; **Distr** N Namaqualand, NC, S Africa.

Note: The species differs from the also dentate **S. ruedebuschii** by its more gradually narrowing leaves and the smooth, long papillae on the epidermis cells.

S. pillansii L.Bolus NM 1929: 86 (see **photo**) T Pillans 6159 (BOL!).

L erect, keel and margins roundish, tapering into a pointed apex, epidermis cells densely placed, 6–8-angular as seen from above, no bulges, a very low central papilla per cell, wax cover ± continuous, but often scaly and with some platelets, **L** 25–48 mm *l*, 9–13 mm *b*; **Fl** ∅ 45–60 mm, 38–60 petals, 130–160 stamens; **Fr** rims on valves low, but on slightly conically rising valves, so that a high centre can be seen, lateral wings on stalk narrow; **S** ovate, all testa cells elevated, most with undulate anticlinal walls, 0.7–0.75 mm *l*, 0.45–0.5 mm *b*; **Ecol** in crevices of rocks; **Distr** Kenhardt, N Namaqualand, NC, S Africa.

Note: Very similar to **S. triebneri** with long papillae on the epidermis cells, but differing from the latter also in features of the flower.

S. ruedebuschii Dinter MDG 1927: 234 (see **photo**) **Lectotype**, here designated: Dinter 5190 (B!) = *S. australis* L.Bolus NM 1937: 63 T Heerde 10772 (BOL!).

L broadening laterally below the apex, becoming almost spoon-shaped, the margins there with broad, flexible teeth, epidermis cells like cobblestones, with 6–8 bulges around the upper edges and a central, low papilla per cell, **L** 33–37 mm *l*, 12–13 mm *b*, 7–8 mm ∅; **Fl** ∅ 40–45 mm, 30–42 petals, 130–199 stamens; **Fr** rims meeting in the centre, up to 1 mm *h*, placenta basal-parietal; **S** testa cells almost smooth, their anticlinal walls undulate, 0.65–0.7 mm *l*, 0.45–0.5 mm *b*; **Ecol** among big quartzitic stones on slopes; **Distr** Warmbad, Namibia; Namaqualand, NC, S Africa.

Note: The dentate margins of this species agree with those of **S. marlothii**, but the low papillae of the epidermis of the leaves differ from the latter.

S. speciosa L.Bolus NM 1958: 340 (see **photo**) **T** Heerde 10768 (BOL!).

L broadening subapically, the keel very often shifted to lie below one of the margins, leaves therefore apparently flat due to the asymmetry, silver grey, keel and margins sharp, with a dark line, epidermis cells like cobblestones, with a low central elevation and 5–6 lateral bulges around the upper edge, **L** 26–31 mm *l*, 13–17 mm *b*; **Fl** ∅ 24–30 mm, 47–53 petals, 150–160 stamens; **Fr** rims on flat valves to 2 mm *h*, meeting in the centre, placenta basal to parietal; **S** ovate, testa cells elevated and undulate along the anticlinal walls, 0.65–0.7 mm *l*, c. 0.45 mm *b*; **Ecol** in crevices of yellow quartz; **Distr** N Namaqualand, NC, S Africa.

Note: The species has been distinguished mainly on orange petal colour.

S. succumbens (Dinter) Dinter in Jacobsen 1933: 178 ≡ *M. succumbens* Dinter MDG 1927: 198 **T** Dinter 5212 (not 5215, as cited by Dinter, B!).

L keel and margins sharp, broadest at their middle, narrowing to the tip, ending in a sharp, brown awn, greenish-grey, epidermis cells cobblestone-shaped, up to 12 lateral bulges on the sides, a very low elevation in the centre indicative of a papilla, **L** 22–32 mm *l*, 14–16 mm *b*; **Fl** ∅ 38 mm, c. 36 petals, 160–170 stamens; **Fr** rims low, < 1 mm *h*, meeting in the centre and often laterally recurved or spreading, placenta basal to parietal; **S** ovate, completely smooth, c. 0.7 mm *l*, c. 5 mm *b*; **Ecol** on slopes with gneissic shale; **Distr** Warmbad, Namibia.

Note: The species has been described from almost dead plants, and recollections have been very rare.

S. triebneri L.Bolus NM 1937: 100 (see **photo**) **T** Triebner 10858 (BOL!).

L ± erect, narrowing from the base into a sharp, dark awn, epidermis cells each with a long, central papilla, base of cells rounded and smooth, not bulging, 25–50 mm *l*, 8–12 mm *b*; **Fl** ∅ 24–40 mm, 27–38 petals, 90–140 stamens; **Fr** rims high from flat valves, meeting in the centre; **S** testa cells elevated, with undulate anticlinal walls, 0.65–0.7 mm *l*, c. 0.5 mm *b*; **Ecol** among big stones of quartzite; **Distr** Namaqualand, NC, S Africa.

Note: Similar in leaf shape to **S. pillansii**, but differing from the latter in the long papillae and the number of elements of the flower mainly. The smooth, long papillae on the leaves are very similar to those of **S. marlothii**.

Schwantesia = Monilaria

H.-D. Ihlenfeldt

The generic name *Schwantesia* L.Bolus (SAG 1928: 279) is illegitimate since it is a later homonym of **Schwantesia** Dinter (MDG 1927: 234). Bolus delimited her new genus *Schwantesia* against **Mitrophyllum** on the base of fruit characters, which, however, she did not specify. Moreover, she did not designate unambiguously a type for the new genus. From the context one can deduce that either **Monilaria moniliformis** or **Monilaria pisiformis** had been envisaged by Bolus to represent the type species of the genus, though **Monilaria scutata** (figured and described by Bolus as *M. scutatum* SAG 1929: 126–127) cannot be excluded either. Since today all these species are included in **Monilaria**, one is allowed to conclude that *Schwantesia* L.Bolus is a synonym of **Monilaria**, although the list of species given by Bolus comprises also species which today are included in **Mitrophyllum** and **Dicrocaulon**, as can be seen from the list of synonyms.

S. chrysoleuca ≡ **Monilaria chrysoleuca**
S. clivorum ≡ **Mitrophyllum clivorum**
S. cognata = **Mitrophyllum clivorum**
S. dissita ≡ **Mitrophyllum dissitum**
S. moniliformis ≡ **Monilaria moniliformis**
S. pisiformis ≡ **Monilaria pisiformis**
S. proxima = **Mitrophyllum dissitum**
S. ramulosa ≡ **Dicrocaulon ramulosum**
S. scutata ≡ **Monilaria scutata**
S. watermeyeri = **Dicrocaulon microstigma**

Scopelogena Ruschioideae

H. E. K. Hartmann

Scopelogena L.Bolus 1962: 9; Klak 2000: 39 **Typus S. verruculata** (L.) L.Bolus **Etym** Gk *scopelos*, rock, *genos*, offspring, child, in allusion to rocks "being the wanted habitats of this genus".

Shrubs to 60 cm *h*, ∅ to 2 m with age, the branches decumbent with age, often hanging from crevices in steep rock faces; **L** terete, acuminate, grey with a reddish tip, those of a pair forming a u, epidermis al-

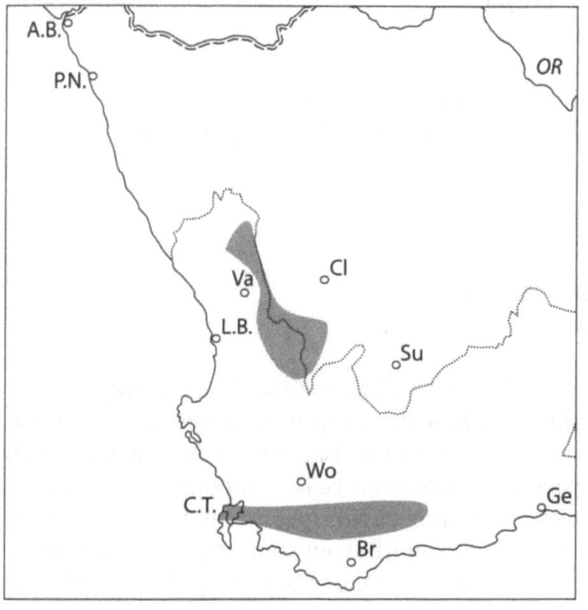

Distribution of **Scopelogena** L.Bolus

most smooth, covered densely by thread-shaped wax protrusions which can be united to develop erect platelets, stomata not sunken, outer epidermis walls without any crystals; **Fl** in very rich terminal dichasia enlarged by additional cymes below, petals dull yellow, "honey coloured", rarely white, filamentous staminodes in a central cone; **Fr** persistent on the plants over more than one season, top with low rims, base elongate, funnel-shaped, covering membranes nearly straight but short or longer, expanding keels almost touching each other basally, very high, valve wings as narrow ledge in fruits opening for the first time, closing body absent, but the placenta raised a little at its end, 5 locules, the 5 valves opening but once and remaining in an erect position, rarely closing again to some degree; **S** dark brown, tuberculate; **Ecol** in crevices of rocks or on small plateaus with gravel below steep falls, rainfall in winter; **Distr** Khamieskroon, NC, to Ceres and in districts around Cape Town, WC, S Africa.

Note: Although plants of the genus have been known in Europe since more than 250 years, they are rare in cultivation. This is probably due to their loose, rather open growth and their reluctance to flower, a fact that appears surprising because they are reportedly growing in less sunny situations than many species of Mesembryanthema. Very recent work has revealed that young capsules possess very narrow valve wings. This has resulted in the description of a new species, based mainly on the yellow petals, in **Scopelogena**, **S. bruynsii**. This species deviates from a critical feature of the genus in the repeated closing of the valves, and it remains open at present which character states should be considered essential to circumscribe **Scopelogena**. In contrast to Klak (2000: 39–40), both old species of the genus are kept separate here, leaving a final decision to a revision of the complex.

Key to the Species of Scopelogena

1. Capsules remaining open after the first opening, valves erect **2**
– Capsules closing again after the first opening, valve therefore closed .... **S. bruynsii**
2. Leaves about as thick as a swan's quill .... **S. verruculata**
– Leaves at most the thickness of a chicken's quill .... **S. gracilis**

S. bruynsii Klak 2000: 39 **T** Klak & Bruyns 462 (BOL).

Flat shrubs to 30 cm *h*, ∅ to 1 m, **I** grey to brown; **L** erect, incurved, mucronate, whitish grey to reddish, c. 45 mm *l*, 5–8 mm *b*; **Fl** in rich cymes, petals yellow, salmon or pink; **Fr** soft, not woody, opening repeatedly, ∅ 3–4.5 mm; **S** dark brown, 1–1.1 mm *l*; **Ecol** on low sandstone cliffs, 100–200 mm rainfal p.a.; **Distr** from Khamieskroon, NC, to Ceres, WC, S Africa.

Note: Remarkable is not only the soft, repeatedly opening capsules of this species, but also the variability of petal colour, a feature known well from **Mestoklema**.

S. gracilis L.Bolus 1962: 10 ≡ *Ruschia scopelogena* G.D.Rowley NCSJ 1978: 9 **Lectotype**, here designated: Hall 1506 I (chosen from two sheets, BOL!).

Loose to pendant shrubs; **L** c. 30 mm *l*, c. 4 mm *t*; **Ecol** in almost permanent shade, > 500 mm rainfall; **Distr** Swellendam, WC, S Africa.

Note: The species has most often been included in **S. verruculata**.

S. vereculata ≡ **S. verruculata**

S. verruculata (L.) L.Bolus 1962: 10 (see **photo**) ≡ *M. verruculatum* L. 1753: 486 ≡ *Lampranthus verruculatus* (L.) L.Bolus in Adamson & Salter 1950: 385 ≡ *Ruschia verruculata* (L.) G.D.Rowley 1957: 7 ≡ *S. vereculata* (L.) L.Bolus ex H.Jacobsen 1970: 509 orthographic variant **LT** Dillenius 1732: t. 203 f 259.

Mostly low, broad shrubs; **L** 20–30 mm *l*, 6–8 mm *t*; **Ecol** mostly on small terraces of gravel below steep rock faces, > 500 mm rainfall; **Distr** on and probably near the Cape Peninsula, W WC, S Africa.

Note: The occurrence of the species outside the Cape Peninsula is still under discussion.

Semnanthe = Erepsia

H. E. K. Hartmann

Liede (1990a) found that *Semnanthe lacera* was not separable from the genus **Erepsia** containing several fat-leaved species with five or more locules. The genus *Semnanthe* is, therefore, included in **Erepsia** subg. **Crassifoliae.**

S. lacera ≡ **Erepsia lacera**
S. lacera var. *densipetala* = **Erepsia lacera**

Sesuvium Sesuvioideae

H. E. K. Hartmann

Sesuvium L. 1759: 1058; Adamson 1962: 244–246; Friedrich 1970: 124–126 ≡ *Halimum*[25] Loefling 1758: 191 nomen illeg. non **Halimus** P.Browne 1756 nec *Halimus* Wallroth 1822: 117 nomen illeg. ≡ *Halimus* Rumph ex O.Kuntze 1891: 263[26] nomen illeg. **Typus S. portulacastrum** (L.) L. **Etym** *sesuvium*, a word of uncertain origin = *Diplochonium* Fenzl 1839: 57 **Typus** *Diplochonium sesuvioides* Fenzl ≡ **S. sesuvioides** (Fenzl) Verdcourt = *Psammanthe* Hance ex Walpers 1852: 659 **Typus** *Psammanthe marina* Hance ex Walpers = **S. portulacastrum** (L.) L. = *Pyxipoma* Fenzl 1839: 293 **Typus** no species named in the genus, cited only *Trianthema polyandra* Blume = **S. portulacastrum** (L.) L. = *Squibbia* Rafinesque 1838: 16 nomen nudum **Typus** *Squibbia maritima* (Walter) Rafinesque ≡ **S. maritimum** (Walter) Britton et al.

Annual to perennial, prostrate to ascending herbs or shrubs with a short tap-root, sometimes rooting at the nodes, I often reddish, young ones sometimes with conspicuous bladder cells; **L** opposite, ± equal, petioles broadening basally and forming a papery sheath clasping the stem, but not stipulate, blades almost cylindrical to elliptical or obovate, epidermis with low or higher bladder cells; **Fl** solitary, pedicels absent to long, 2 bracteoles at the base of each flower, perigone with 5 lobes ending in long, prominent apical-dorsal appendages, pink to purple inside, more rarely white, the 5 or numerous stamens free or united basally into a low ring, connate with the perigone, hence the Fl perigynous; **Fr** enclosed in the persisting perigone, circumscissile near the middle, 2–5 locules; **S** numerous in each locule, shining, black, ± round, enveloped

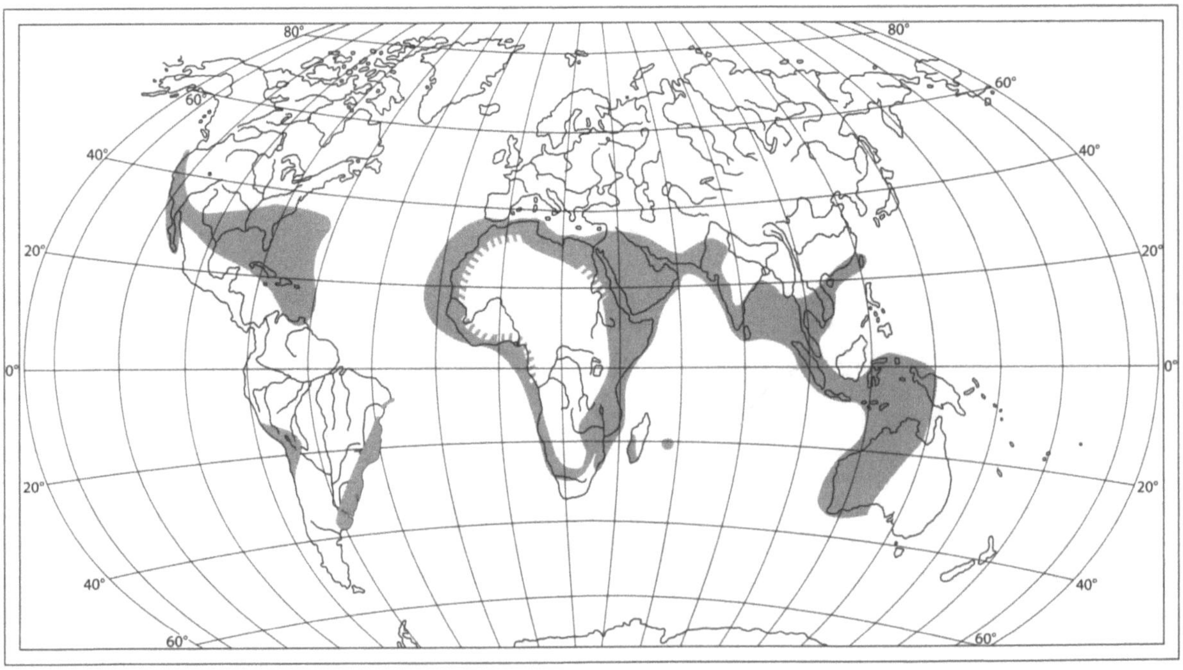

Distribution of **Sesuvium** L.

[25] The names *Halimum* and *Halimus* have always been considered the same, see Hiern 1898: 411–414.

[26] Kuntze (1891: 263) cited "Rumpf" 1750: 165 and t. 72, f 1 as the author for *Halimus*, but Rumph described the plant, which is also listed by Linnaeus (1759: 1058) for his **S. portulacastrum**, as **Crithmus indicus.**

in a membranous aril; **Ecol** in warmer, mostly tropical climates, most often in saline soils, coastal or inland; **Distr** worldwide in the tropics.

Note: Linnaeus (1759: 1058) clearly established the name **Sesuvium** as a substitute for *Halimum* Loefling by citing "(Loefl.)" behind his new genus name. Whether this indicates that Linnaeus was aware of **Halimus** P.Browne (1756), cannot be proved, but Linnaeus had used the name also, for a genus placed in synonym with **Atriplex**. The present treatment is based on available flora treatments and the original descriptions. The possession of 2–5 stigmas and many seeds clearly separates the genus from **Trianthema**, with which some authors have united it.

S. acutifolium Miquel 1843: 75 **T**?

Branches to 15 cm *l*, **I** terete; **L** narrowly lanceolate to spatulate, basally narrowing into a short petiole, almost sessile, acute, up to 50 mm *l*, 5–7 mm *b*; **Fl** pedicels 15 mm *l*, perigone 10 mm *l*; **Fr** shorter than the perigone, fragile, 5 locules; **Ecol** near the sea; **Distr** Surinam.

Note: The species differs from the very similar **S. portulacastrum** mainly by the acute leaf tips.

S. ayresii Marais 1978: 483 **T** Ayres s.n. (K).

Perennial herbs with a strong tap-root, branches prostrate, 15–45 cm *l*, not usually rooting at the nodes, **I** glabrous, deep green tinged purple, yellow or red, shiny; **L** petiolate, blade ovate, roundish, 8–25 mm *l*; **Fl** subsessile, perigone lobes 5, pink or purplish inside, stamens 12–20, anthers purplish-white, pollen white; **Fr** 3 locules, c. 25 seeds; **S** reniform to circular, black, hardly up to 1 mm *l*; **Ecol** near the sea on coral rocks, rarely on sand, near salt water or in spray; **Distr** Madagascar.

Note: The species differs from **S. portulacastrum** by the non-rooting branches, the subsessile flowers and the smaller seeds.

S. brevifolium Schumacher & Thonning 1829: 7 **T** Thonning s.n. (C).

Creeping herbs, **I** glabrous, coloured; **L** petioles short, blade linear-cuneate; **Fl** subsessile, otherwise as in **S. portulacastrum**; **Fr** not mentioned; **Ecol** along salty rivers; **Distr** Guinea.

Note: The species is said to differ from **S. portulacastrum** in the shorter leaf.

S. congense Welwitsch ex Oliver 1871: 586 ≡ *Halimus congensis* (Welwitsch ex Oliver) O.Kuntze 1891: 263 **T** Welwitsch 2382 (BM).

Shrubs with prostrate, diffuse, twisted branches, **I** terete, minutely scabrous-papillate; **L** linear, revolute, papillate, 7–25 mm *l*; **Fl** subsessile, perigone lobes ovate-lanceolate, stamens numerous; **Fr** 3 locules; **Ecol** near the sea; **Distr** Benguela, Luanda, Moçamedes, Angola; Ambriz, Congo.

Note: The species differs from others in the area by its shrubby growth and three locules; it also seems to occur on more tropical shores than other species.

S. crithmoides Welwitsch 1858: 586 ≡ *Halimus portulacastrum* (L.) O.Kuntze var. *crithmoides* (Welwitsch) Hiern 1898: 412 **T** Welwitsch 2386 (BM).

Perennial herbs; **I** deep red, branches prostrate; **L** semi-cylindrical, glaucous; **Fl** pink, rather big for the genus; **Fr** unknown; **Ecol** in sand at the sea; **Distr** Luanda, Angola; Georgia, USA.

Note: Gonçalves (1970: 323) placed **S. mesembryanthemoides** in synonymy with **S. crithmoides**, keeping *S. crystallinum* as a separate species. The report of **S. crithmoides** from Georgia in 1902 (Bogle 1970: 451) is a single find.

S. crystallinum = **S. mesembryanthemoides**
S. digynum = **S. sesuvioides**
S. digynum var. *angustifolium* = **S. sesuvioides**

S. distylum Ridley 1891: 38 **T** without collector, s.n. (?).

Prostrate, bright green herbs, branches 80–160 mm *l*; **L** cuneate to spatulate, obtuse, 13.5 mm *l*, 3.5 mm *b*, basally scariously vaginate; **Fl** pedicels short, 3.5 mm ∅, perigone white inside, stamens < perigone, ovary shortly conical, obtuse, apically depressed, 2 short stigmas; **F** reniform, black; **Ecol** in spray from the sea; **Distr** Rat Island, Fernando Noronho, off the shore of Brazil.

Note: This species differs from the vegetatively similar **S. portulacastrum** by the white colour of the perigone and the only two stigmas, 3–5 occurring in the latter species.

S. eastwoodianum Howell 1966: 352 **T** Crocker 207606 (?).

Mat-forming herbs with creeping branches, not woody, **I** with globose, crystalline papillae; **L** narrowly oblanceolate, covered densely by globose papillae, 15–30 mm *l*, 2–4 mm *b*; **Fl** pedicels 5–10 mm *l*, erect when in fruit, perigone 7–8 mm *l*, stamens numerous; **Fr** 4–5 mm *l*, 3 mm ∅, 3 locules; **S** subrotund, matt, conspicuously rugose, crests interrupted and adorned with papillae, c. 1 mm *l*; **Ecol** on low sand islets; **Distr** Galapagos Islands, Equador.

Note: The species occurs on the Galapagos Islands together with the also endemic **S. edmonstonei** and the widespread **S. portulacastrum**, differing in its markedly ornamented seeds from both.

S. edmonstonei Hook.f. 1847: 221 ≡ *Halimus edmonstonii* (Hook.f.) O.Kuntze 1891: 263 **T** Goodridge no.? (K?).

Shrubs with woody basal parts, branches ascending, **I** covered by scales when dry, most probably collapsed papillae or bladder cells; **L** > **I**, linear, obtuse,

13.5 mm *l*, 2.25–3.4 mm *b*; **Fl** petioles c. 7 mm *l*, stamens many; **Fr** oblong; **Ecol** on rocks; **Distr** Galapagos Islands, Equador.

Note: An endemic species of the Galapagos Islands, on rocks.

S. edule = **S. portulacastrum**

S. erectum Correll 1966: 421 **T** Correll & Johnston 17868 (LL).

Perennial herbs with erect to spreading branches to 50 cm *l*, **I** ± densely covered with globose papillae, brownish-black or smoky grey with age; **L** oblanceolate to spatulate, obtuse, covered with globose papillae, up to 40 mm *l*, to 8 mm *b*; **Fl** pedicels up to 5 mm *l*, perigone lobes 4–8 mm *l*, stamens numerous; **Fr** ovoid, truncate at apex, c. 5 mm *h*, 3 mm ∅, 3–5 locules; **S** black, plump, c. 1 mm *l*; **Ecol** in depression or at the edge of a pond; **Distr** inland Texas, USA.

Note: The species is very similar to **S. sessile** with a lower, more decumbent growth occurring near the coast.

S. hoepfnerianum = **S. sesuvioides**
S. hoepfnerianum var. *brevifolium* = **S. sesuvioides**
S. hydaspicum = **S. sesuvioides**

S. longifolium Humboldt & Bonpland ex Willdenow 1809: 521 **T** Humboldt & Bonpland (?).

L linear-spatulate, as long as the **I**; **Fl** pedicellate; **Ecol** unknown; **Distr** uncertain, S America.

Note: The species is characterized by the leaves being as long as the internodes.

S. maritimum (Walter) Britton, Sterns & Poggenburg 1888: 20 ≡ *Pharnaceum maritimum* Walter 1788: 117 ≡ *S. pentandrum* Elliott 1821: 556 ≡ *Mollugo maritima* (Walter) Seringe ex DC. 1824: 393 ≡ *Squibbia maritima* Rafinesque 1838: 16 nomen nudum ≡ *Halimus maritimus* (Walter) O.Kuntze 1891: 263 **T** Squibb?

Annual, branches procumbent to erect, **I** glabrous; **L** obovate or spatulate, 6–20 mm *l*, 2–8 mm *b*; **Fl** sessile or nearly so, perigone lobes 5, elliptical, 2–2.2 mm *l*, stamens 5; **Fr** ovoid, 4–4.5 mm *h*, 2–2.5 mm ∅; **S** smooth, lustrous, 0.7–0.8 mm *l*; **Ecol** on the seashore; **Distr** Bahamas; Cuba; Porto Rico; from New York to Florida and to Texas, USA.

Note: Although this is a widespread species, it seems not to occur in big masses (Elliott 1821: 556). It is the only species in the genus in northern America with only five stamens.

S. mesembryanthemoides Wawra & Peyritsch 1860: 564 **T** Wawra 210 (W) ≡ *Halimum mesembrianthemoides* (Wawra & Peyritsch) Hiern 1898: 413 = *S. mesembryanthemoides* Welwitsch 1858: 557 nomen nudum **T**? = *S. crystallinum* Welwitsch ex Oliver 1871: 586 ≡ *Halimus crystallinus* (Welwitsch ex Oliver) O.Kuntze 1891: 263 **T** Welwitsch 2389 (BM).

Prostrate subshrubs forming large clumps, **I** with densely placed, watery-glossy papillae like **Mesembryanthemum crystallinum**; **L** in dense arrangement, terete, concave above, basally broader and enclosing the stem, papillate like the stem, 6–15 mm *l*, 2–4 mm ∅; **Fl** sessile, perigone papillate outside, pink to purple inside, stamens in groups, many; **Fr** globose, glabrous, 3–4 locules; **S** black, minutely punctate; **Ecol** on sand and gravel on the seashore; **Distr** Benguela, Mossamedes, Angola; Lüderitz-South, Namibia.

Note: The dense cover of well developed papillae characterizes this species, which seems to occur in Angola and Namibia only.

S. microphyllum Humboldt & Bonpland ex Willdenow 1809: 521 **T** Humboldt & Bopland no.? (B?) = *S. spathulatum* Humboldt, Bonpland & Kunth 1823: 87 **T** Humboldt & Bonpland? (P?).

Branches procumbent, < 30 cm *l*, **I** glabrescent, reddish; **L** petioles short, blade obovoid, elliptic to terete in ∅, 0.4–16 mm *l*, 2–7 mm *b*; **Fl** pedicellate, perigone lobes 5, 3–4 mm *l*, 2–2.5 mm ∅, whitish to purplish inside, stamens numerous; **Fr** ovoid, 4.5 mm *l*, 3–3.5 mm ∅, 3–4 locules; **S** lustrous, black, c. 1 mm *l*; **Ecol** at the sea; **Distr** Cuba.

Note: Both description are based on material collected by Humboldt and Bonpland and may refer to a single collection. Riha (1976: 122–124) reported that plants identified as **S. microphyllum** grow in low lying crevices or holes of coral reef always in sea water with an osmotic potential of up to 60 atmospheres, whereas plants identified as **S. portulacastrum** occur in sand and deeper soil of somewhat high-lying places. In transplanting experiments, both in situ and in a garden, the plants adapted in leaf shape to the new habitat within a year. Riha concluded that both species represent extreme ecoforms of one species, reporting also intermediate plants from intermediate habitats.

S. nyasicum (J.G.Baker) M.L.Gonçalves 1965: 381 ≡ *Trianthema nyasica* J.G.Baker 1897: 268 **T** Whyte s.n. (K?).

Perennial herbs, branches decumbent, thin, **I** glabrous; **L** sessile, oblanceolate, obtuse, glabrous, 9 mm *l*; **Fl** several (per cyme?), sessile, perigone tube 4.5 mm *l*, 5 lobes 4.5 mm *l*; **Fr** conical, membranous, 2 locules; **S** rugose, black, reniform; **Ecol** uncertain; **Distr** Nyassa, Zambia; also Angola, Malawi, Tanzania, Zimbabwe?

Note: Gonçalves (1970: 326) identified material from Angola etc. with **S. nyasicum**, but he did not cite the type as being seen. Remarkable are the glabrous stems and leaves of the species.

S. ortegae = **S. revolutifolium**

S. parviflorum DC. 1828: 453 **T** Commerson s.n. (P?) = *Trianthema americana* Gillies ex Arnott 1831: 354 **T** Gillies s.n. (?) = *Aizoon montevidense* Sprengel ex Rohrbach in Martius 1872: 310 nomen nudum **T**? = *S. sessiliflorum* Dombey ex Rohrbach in Martius 1872: 310 nomen nudum **T**? = *S. sessile* Larrañaga 1922: 23 nomen illeg. **T**?

Herbs; **L** linear-spatulate, **I** terete; **Fl** sessile, perigone red inside, c. 5 mm, stamens numerous; **Fr** unknown; **Ecol** near the sea; **Distr** Montevideo, Uruguay.

Note: Collected under this name are all synonyms from the area around Montevideo, which are reported to have sessile reddish flowers half the size of those of **S. portulacastrum.**

S. pedunculatum = **S. portulacastrum**
S. pentandrum ≡ **S. maritimum**
S. portulaca ≡ **S. portulacastrum**

S. portulacastrum (L.) L. 1759: 1058 (see **photo**) ≡ *Portulaca portulacastrum* L. 1753: 446 ≡ *S. portulaca* Crantz 1766: 136 **T** Hermann s.n. (BM) = *S. portulacastrum* (L.) L. var. *griseum* O.Degener and Fosberg 1952: 47 ≡ *Trianthema grisea* (O.Degener and Fosberg) O.Degener and I.Degener 1960: family 116 **Lectotype,** here designated: Bryan 24 (BISH) = *S. edule* Wight ex Wallich 1832: 229 nomen nudum[27] = *Psammanthe marina* Hance ex Walpers 1852: 660 **T** Hance? = *S. pedunculatum* Persoon 1807: 39 **Lectotype,** here designated: Lamarck 1794: t.434 = *Trianthema polyandra* Blume 1826–1827: 1137, this name cited also by Fenzl 1839: 293 with the description of *Pyxipoma* **T**? = *Trianthema portulacastrum* L. var. *hillebrandii* O.Degener and I.Degener 1960: family 116 **T** Degener and Degener Oahu 1958 (BISH) = *Crithmus indicus* Rumph 1750: 165 ≡ *S. repens* Willdenow 1809: 521 **Lectotype,** here designated: Rumph 1750: t. 72 f 1.

Shrubs with long, prostrate branches rooting irregularly at intervals, **I** often purplish, glabrous; **L** ± elongate-elliptical, concave above, margins sometimes evolute, 30–80 mm *l*, 2–7 mm *b*; **Fl** pedicels 5–20 mm *l*, perigone pink inside, 5–15 mm *l*, appendages 4–6 mm *l*, stamens numerous; **Fr** obovate, 3–5 locules; **S** black, shiny; **Ecol** on seashores; **Distr** world-wide in the tropical regions.

Note: Considering the very wide distribution of the species, it is not surprising that the identity has been discussed and disputed since its first description (e.g. Haworth 1819: 86–87, under **S. revolutum**). The figure by Hermann (1698: t. 212, as Portulaca corassavica angusto longo lucidoque folio), showing a plant with very long pedicels and flattish elongate-elliptic leaves, is cited by Linnaeus and Crantz, indicating these features as being typical of the taxon. The same characteristics are visible in Plukenet 1692: t. 216, f.1, cited together with the figure of Hermann by Haworth (1819: 86) as examples of **S. pedunculatum.** When describing that latter species, Persoon (1807: 39) cited three figures: "Dec. pl. grass. (ined) S. portulacastrum, L.", which has been published as DC 1799: t. 9, showing a flower on the erect branch with hardly a pedicel visible; "Lam. ill. t. 434, f.1" is a branch with two long-stalked flowers, and "Aizoon canariense Andrew. t. 201" depicts a somewhat twining branch with several flowers on medium long pedicels and the broadest leaves of all three (Andrews 1802: t. 201). The figure of Lamarck (1794: t. 434) is chosen as the lectotype here, exhibiting the longest stalks of the flowers. If it can be shown that long and short pedicels in this complex are specific features, a separation of *S. pedunculatum* from **S. portulacastrum** may be justified.

With his description of *S. repens*, Willdenow (1809: 521) cited a figure of Rumph (1750: t. 72, f.1) showing an erect branch with a long-stalked flower; in the accompanying text by Rumph, the habit is mentioned as "prorepens", from which feature Willdenow drew the epithet. The picture agrees well with those, e.g., of Plukenet, Hermann, or Andrews. The name *S. repens*, however, has later been used for plants with a more expressed creeping, fragile habit, much smaller and apically rounded leaves, and pedicles shorter than the flowers themselves (Hooker 1836: t. 203, figuring a drawing kept at K!). No further information is available with the named figures.

S. portulacastrum is listed as a weed in Holm et al. (1979: 332) for Fiji and Vietnam, but no specification as to the degree of weediness is given.

S. portulacastrum var. *griseum*
= **S. portulacastrum**
S. portulacastrum var. *crithmoides* = **S. crithmoides**
S. quadrifidum ≡ **Gunniopsis quadrifida**
S. repens = **S. portulacastrum**

S. revolutifolium Ortega 1797/1800: 19 **T** in hort. Madrid from seed sent by D. Sess. (?) = *S. ortegae* Sprengel 1801: 36 **T** Grensel? (?) = *S. revolutifolium* Vahl ex Willdenow 1809: 521 **T**? = *S. revolutifolium* Colla ex Cesati 1873: 10 nomen illeg. et nudum **T** Colla 100 (?) = *S. revolutum* Persoon 1807: 39 **Lectotype,**

[27] The notation in Wallich 1832: 229 is repeated here in order to show that *S. edule*, cited in Index Kewensis as given above, has not been intended to be a binary name at all:
6836 Sesuvium repens Röttl. Hb Madras (Hb = herbarium)
B – – – Hb. Wight prope Irangvebar
C – portulacastrum haud & Date
D – – edule H. Wight, Trincomale.
It appers from this that the addition "edule" was only made to mark plants of **S. portulacastrum** that were edible.

here designated: De Candolle 1837?: t. (179)[28] = *S. portulacastrum* (L.) L. var. *revolutum* Sims CBM 1815: t. 1701 T?

Annual to perennial herbs, stem tetragonal; **L** subsessile, margins revolute, linear to spatulate, finely papillate on both sides; **Fl** upper ones sessile, lower ones pedicellate, perigone red, stamens numerous, purple, inner gradually shorter; **Fr** 3–6 locules; **S** reniform, black with a white aril; **Ecol** in saline places; **Distr** Cuba; S America.

Note: In the original description of Ortega, Cuba is given as the place of origin; the other species placed in synonymy here come from South America. In the present treatment, all forms with revolute leaves are seen as members of one species. Whether, however, the revolute leaves and the upper sessile flowers distinguish the species clearly from **S. portulacastrum** remains open.

S. revolutifolium = **S. revolutifolium**
S. revolutum = **S. revolutifolium**
S. sessile = **S. parviflorum**

S. sessile Persoon 1807: 39 T?

L linear-oblong, flat; **Fl** sessile; **Distr** not reported.

Note: It appears very difficult to identify the species described as given here. The name is used for plants with sessile flowers and a dense cover of papillae occurring along the coast of Texas (Correll & Johnston 1970: 605).

S. sessiliflorum = **S. parviflorum**

S. sesuvioides (Fenzl) Verdcourt 1957: 349 (see **photo**) ≡ *Diplochonium sesuvioides* Fenzl 1839: 58 ≡ *Halimus sesuvioides* (Fenzl) O.Kuntze 1891: 263 T Drege 2938 (W) = *S. sesuvioides* (Fenzl) Verdcourt var. *angustifolium* (Schinz) M.L.Gonçalves 1965: 381 ≡ *S. digynum* Welwitsch ex Oliver var. *angustifolium* Schinz 1897: 74 ≡ *Halimus sesuvioides* (Fenzl) O.Kuntze var. *angustifolius* (Schinz) Hiern 1898: 414 T Hoepfner 111a (Z?) = *Halimus sesuvioides* (Fenzl) O.Kuntze var. *reduplicatum* Welwitsch ex Hiern 1898: 414 T Welwitsch 2388 (BM?) = *Halimus sesuvioides* (Fenzl) O.Kuntze var. *welwitschii* Hiern 1898: 414 T Welwitsch 2394 (BM?) = *S. digynum* Welwitsch ex Oliver 1871: 586 T Welwitsch 2391 (BM) = *Trianthema dinteri* Engler 1915: 144 nomen nudum T? = *Diplochonium hochstetteri* Stocks ex Aitchinson 1869: 66 nomen nudum T Stocks s.n. (K)[29] = *S. hoepfnerianum* Schinz 1897: 75 T Hoepfner 25 (Z?) = *S. hoepfnerianum* Schinz var. *brevifolia* Schinz 1897: 75 T Steingröver 38 (Z?) = *S. hydaspicum* (Edgeworth) M.L.Gonçalves 1965: 381 ≡ *Trianthema hydaspica* Edgeworth 1862: 203 T? = *Trianthema polysperma* Hochstetter ex Oliver 1871: 588 T Kotschy 68 (BM?) = *Trianthema salarium* Bremekamp 1933: 239 T Bremekamp 21.1.31 (PRE?).

Herbs with only slightly woody stems at their base, **I** with prominent papillae when young, rough from white remains of papillae when older, pale, branches to 50 cm *l*; **L** petioles 2–5 mm *l*, blades elliptical, margins revolute, 7–20 mm *l*, 3–10 mm *b*; **Fl** mostly sessile, perigone rosy inside, appendages 2–4 mm *l*, stamens numerous; **Fr** ellipsoid, 2 locules; **S** black, shiny, smooth or finely rough; **Ecol** in sand, inland or on the seashore, in dry subtropical climates; **Distr** worldwide in dry areas of the tropics and subtropics.

Note: All species described with two stigmas and two locules have been included in this species in most floras. The species is probably best identified by the numerous seeds as a member of **Sesuvium**, and by the rough young internodes in combination with two locules within the genus.

S. sesuvioides var. *angustifolium* = **S. sesuvioides**
S. spathulatum = **S. microphyllum**

S. trianthemoides Correll 1966: 422 T Tharp 47431 (TEX).

Annual herbs, branches to 35 cm *l*, **I** covered densely with globose papillae; **L** oblanceolate or spatulate, with scattered big papillae, 30 mm *l*, 10 mm *b*; **Fl** sessile, perigone 3.5 mm *l*, stamens 5; **Fr** ovoid-ellipsoid, pointed apically, 4–5 mm *l*, c. 10 seeds per capsule, 2 locules; **S** conspicuously rugose with brownish irregular ridges, additional light patches extending from the hilum, c. 1.5 mm *l*; **Ecol** on dunes; **Distr** Texas, USA.

Note: The species differs in being annual from most others in the area, it also has distinctive seeds similar to those of **Trianthema portulacastra** (Correll 1966: 422).

S. uvifolium Sessé & Moçiño 1894: 126 **T?**

Creeping perennial herbs, **I** terete, glabrous, deep red; **L** terete-obovate, like the berries of grape, glabrous, often red; **Fl** pedicellate, margins of the perigone lobes purple outside, stamens purple; **Fr** unknown; **Ecol** among rocks near the sea; **Distr** Mexico.

Note: The species appears distinct because of its leaves shaped like the berries of grape.

[28] In some copies of De Candolle's Plantes grasses exists a sheet named *Sesuvium revolutum*, agreeing in the revolute leaves and the sessile flowers with the two critical features of the species, thus answering the reference of Persoon. It seems appropriate to choose this figure as the lectotype, which is unnumbered and has been seen as no. 179 in one copy, probably from 1837, but this cannot be determined with certainty.

[29] Index Kewensis gives **Oryga decumbens** Forskål as the correct name of this species, but Verdcourt (1957: 349) stated that the specimens of Stocks agree with **S. sesuvioides**.

S. verrucosum Rafinesque 1838: 16 **T** collector not given (?).

Branches prostrate to erect, 30–90 cm *l*, **I** covered with globose papillae turning into warts when dry; **L** petiolate, blades obovate, also covered densely by globose papillae, blades 10–30 mm *l*, 2–10 mm *b*; **Fl** 1–3 in cymes, ± sessile, perigone lobes 4.5 mm *l*, white inside, stamens numerous; **Fr** conical, c. 5 mm *l*, 3 mm ∅; **S** smooth, lustrous, black, 1 mm *l*; **Ecol** uncertain; **Distr** N Mexico; from California to Arkansas, USA.

Note: The dense cover of globose bladder cells as papillae on leaves and stem gives the plant a distinctive greyish to yellowish appearance. Correll (1966: 421) pointed out that the plants dry to a light brown or a tan colour. The species is also reported as introduced in Bahrain (Verdcourt 1985: 208).

Sialodes = Galenia

H. E. K. Hartmann

The genus *Sialodes* Ecklon & Zeyher has been described for a single species (1837: 329), which soon was found to agree in critical features with **Galenia** (Walpers 1843: 232).

S. glauca = **Galenia pallens**

Sineoperculum = Dorotheanthus

H. E. K. Hartmann

Ihlenfeldt & Struck (1987) consider the absence of covering membranes, used to establish the genus *Sineoperculum*, as a feature of specific value only in the genus **Dorotheanthus**, with which the former genus agrees in the expanding sheets and valve wings ending in awns from a broader base.

S. rourkei ≡ **Dorotheanthus rourkei**

Skiatophytum Ruschioideae

H. E. K. Hartmann

Skiatophytum L. Bolus SAG 1927: 435; Leistner 1958b: 89–102; Bittrich 1987: 50–53; Ihlenfeldt & Gerbaulet 1990: 489–492 **Typus S. tripolium** (L.) L.Bolus **Etym** Gk *skias*, shade, *phyton*, plant, i.e. plants preferring shady habitats ≡ *Gymnopoma* N.E.Br. GC 1928: 193 **Typus** *Gymnopoma tripolium* ≡ **S. tripolium** (L.) L.Bolus N.E.Br.

Biennial or perennial with fleshy, stout, branching roots, persistent vegetative shoots and lateral annual **Inf**; **L** alternate or opposite, flat, dark green, without bladder cells, mesomorphic; **Fl** in cymes of 2–5, **K** 5, petals white, nectary as a flat ring; **Fr** xerochastic, valves opening incompletely in so far as only the central parts are lifted leaving a firm ring around the outer rim of the fruit under which closets with a seed each are developed, but most seeds are placed in the big central cavity, (4–)5(–7) locules; **S** black, kidney-shaped, flattened, testa with low papillae; **Ecol** in humus-sand in shade on lower slopes of mountains, < 250 m, 500–1 000 mm rainfall p.a. in winter; **Distr** SW districts in WC, S Africa.

Distribution of **Skiatophytum** L.Bolus

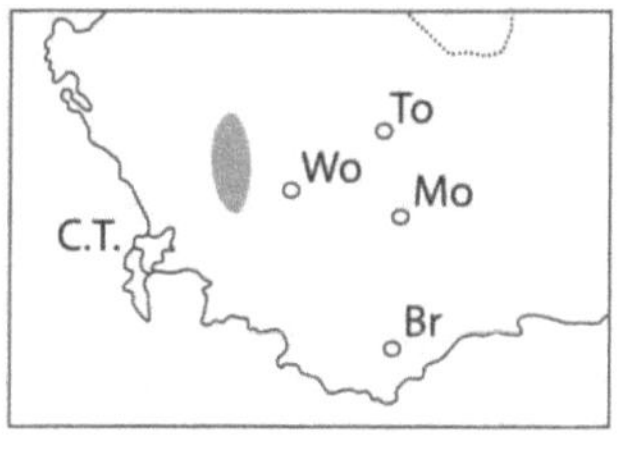

Note: Very similar to **Caryotophora** in habit and leaves, differing in some fruit characters and the fact that the seeds are dispersed by a censer mechanism.

S. tripolium (L.) L.Bolus SAG 1927: 435 (see **photo**) ≡ *M. tripolium* L. 1753: 481 ≡ *Gymnopoma tripolium* (L.) N.E.Br. GC 1928: 194 **LT** Dillenius 1732: 230 t. 179, f 220 = *M. expansum* Thunberg 1800: 88 nomen illeg. non L. 1759: 1059 **T** Thunberg 11976 (UPS!).

L with undulate margins when young; **Fl K** with dentate ribs on base; **Fr** ∅ 16–24 mm; **S** 1.8–2.2 mm *l*; **Ecol** as for genus; **Distr** as for genus.

Note: Dispersal occurs in three phases, like in **Conicosia** subg. **Herrea**: 1. the seeds from the locules are distributed by shaking the fruit on the long stalk like a censer, 2. the capsules break off being rolled away by the wind as tumble fruits losing seeds, 3. the seeds from the closets are set free at last by disintegration of the fruit. Reports of a very first phase when the valves are said to open hygrochastically cannot be confirmed, no expanding tissue swelling in water has been found (Gerbaulet pers. comm.).

Smicrostigma Ruschioideae

H. E. K. Hartmann

Smicrostigma N.E.Br. GC 1930: 72; Liede 1990a: 391–479 ≡ *Erepsia* subg. *Pleiocarpa* L.Bolus FPSA 1927: t. 270 **Typus Smicrostigma viride** (Haw.) N.E.Br. ≡ *Erepsia viridis* (Haw.) L.Bolus **Etym** Gk *smicros, micros*, small, *stigma*, stigma, referring to the supposedly short stigmas.

Shrubs with erect branches; **L** triquetrous, basally connate forming a long sheath with the stem, free parts recurving, acuminate and mucronate, with a small pustule at their bases, epidermis smooth, stomata sunken, wax in horizontal plates breaking into smaller, rugose units later, outer epidermal wall c. 10 µm in ∅, calcium oxalate crystals of 3 µm ∅, layer to 4 µm *t*; **Fl** solitary, permanently open, pedicellate, bracteoles not different from foliage leaves, **K** 5, ± equal, petals mostly purple, filamentous staminodes and the ± erect stamens basally papillate, white, apically purple in most cases, their bases terete, holonectary dark green, pollen longiaxial, tectate; **Fr** top low, bowl-shaped, base funnel-shaped, expanding keels basally almost parallel, apically spreading, tapering into short awns in place of valve wings, covering membranes complete, closing ledge inconspicuous; **S** pear-shaped, brown, testa cells raised to cushion-shape, covered by minute rodlets forming groups by joining their tips; **Ecol** in shrubby associations of Rhenosterbosveld mainly, 200–400 mm rainfall p.a. mainly in March and November; **Distr** SE parts of WC, S Africa.

Note: Liede (1990) was able to show that an inclusion in the superficially similar genus **Erepsia** is not correct because **Smicrostigma** lacks the hypanthium, the almost free leaves and the presence of narrow valve wings in the fruit. In leaf anatomy and flower morphology, **Smicrostigma** is similar to **Octopoma.**

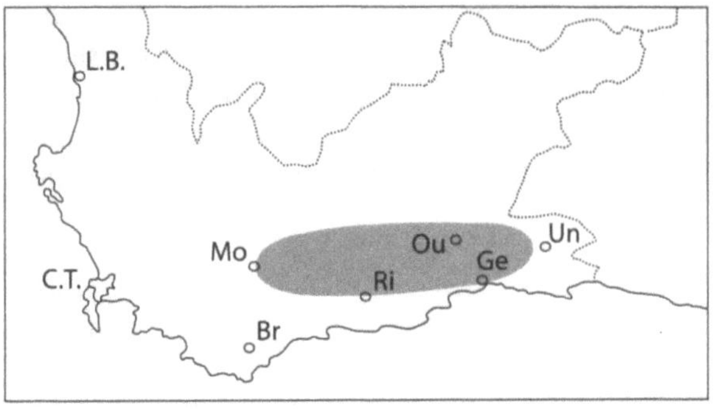

Distribution of **Smicrostigma** N.E.Br.

S. viride (Haw.) N.E.Br. GC 1930: 72 (see **photo**) ≡ *M. viride* Haw. 1795: 314 ≡ *Erepsia viridis* (Haw.) L.Bolus FPSA 1927: t. 270 ≡ *Ruschia viridis* (Haw.) G.D.Rowley NCSJ 1978: 9 **Lectotype,** here designated: Salm-Dyck 1837: § 33 f 2 = *M. integrum* L.Bolus ABH 1922: 134 ≡ *Ruschia integra* (L.Bolus) Schwantes ZSK 1926: 187 **T** Rogers 17175 (BOL!).

To 60 cm *h* with few to many erect branches; **L** ∅ < 5 mm, dark green due to many tannin idioblasts in the layer of assimilating tissue around the central water tissue; **Fl** ∅ c. 30 mm, mostly deep purple, rarely pink, very rarely pure white 25–50 petals, ∅ c. 30 mm, 200–250 filamentous staminodes, 150–170 stamens; **Fr** as for genus, ∅ 5–8 mm; **S** as for genus, 0.5–0.8 mm *l*, 0.4–0.6 mm *b*; **Ecol** as for genus; **Distr** as for genus.

Note: The species represents an element of Rhenosterbosveld, growing often also along roads and in disturbed places. The brightly coloured, continuously open flowers contrast conspicuously with the dark green leaves.

Sphalmanthus = Phyllobolus

M. Gerbaulet

In its original circumscription *Sphalmanthus* comprised some prostrate to decumbent geophytic species with tuberous roots (Brown GC 1925: 433, Brown GC 1928: 13, 34–35). This circumscription was decisively broadened when Bolus (1962: 228–229) transferred several shrubby species to *Sphalmanthus*, which had formerly been placed in **Aridaria** (Brown JB 1928: 140–145), and added some new taxa. Incidentally, all combinations from **Aridaria** to *Sphalmanthus* were not validly published, albeit they were continuously used ever since. Finally, Bolus amalgamated **Phyllobolus** and *Sphalmanthus* (1964: 42–43), but disregarded the fact that the first name takes priority over the latter by 7 days. This error was rectified by Bittrich (1987: 5–116), who resurrected the name **Phyllobolus** for this genus. The most recent revision of **Phyllobolus** by Gerbaulet (1995: 385–399) follows Bolus' taxonomic concept with a slight amendment of the generic circumscription.

S. abbreviatus ≡ **Phyllobolus abbreviatus**
S. acocksii = **Phyllobolus saturatus**
S. acuminatus
= **Phyllobolus splendens** ssp. **splendens**
S. albertensis = **Phyllobolus pumilus**
S. albicaulis
= **Phyllobolus splendens** ssp. **splendens**
S. anguineus = **Phyllobolus oculatus**

S. arenicola = **Phyllobolus oculatus**
S. auratus = **Phyllobolus nitidus**
S. baylissii = **Phyllobolus saturatus**
S. bijliae = **Phyllobolus splendens** ssp. **splendens**
S. blandus = **Phyllobolus splendens** ssp. **splendens**
S. brevisepalus = **Phyllobolus spinuliferus**
S. calycinus = **Phyllobolus canaliculatus**
S. canaliculatus ≡ **Phyllobolus canaliculatus**
S. carneus = **Phyllobolus spinuliferus**
S. caudatus ≡ **Phyllobolus caudatus**
S. celans = **Phyllobolus splendens** ssp. **splendens**
S. commutatus = **Phyllobolus grossus**
S. congestus ≡ **Phyllobolus congestus**
S. constrictus = **Phyllobolus splendens** ssp. **splendens**
S. crassus = **Phyllobolus prasinus**
S. deciduus ≡ **Phyllobolus deciduus**
S. decurvatus ≡ **Phyllobolus decurvatus**
S. decussatus = **Aptenia geniculiflora**
S. delus ≡ **Phyllobolus delus**
S. dinteri ≡ **Phyllobolus melanospermus**
S. dyeri = **Phyllobolus splendens** ssp. **splendens**
S. englishiae ≡ **Prenia englishiae**
S. flexuosus = **Phyllobolus splendens** ssp. **splendens**
S. fourcadei = **Phyllobolus splendens** ssp. **splendens**
S. fragilis = **Phyllobolus oculatus**
S. framesii = **Phyllobolus spinuliferus**
S. geniculiflorus ≡ **Aptenia geniculiflora**
S. glanduliferus = **Phyllobolus sinuosus**
S. godmaniae = **Phyllobolus sinuosus**
S. gratiae = **Phyllobolus grossus**
S. grossus ≡ **Phyllobolus grossus**
S. gydouwensis = **Phyllobolus grossus**
S. hallii = **Phyllobolus delus**
S. herbertii ≡ **Phyllobolus herbertii**
S. herrei = **Phyllobolus prasinus**
S. humilis = **Phyllobolus herbertii**
S. latipetalus ≡ **Phyllobolus latipetalus**
S. laxipetalus = **Phyllobolus grossus**
S. laxus = **Phyllobolus decurvatus**
S. leipoldtii = **Phyllobolus grossus**
S. leptopetalus = **Phyllobolus splendens** ssp. **splendens**
S. lignescens ≡ **Phyllobolus lignescens**
S. ligneus = **Phyllobolus melanospermus**
S. littlewoodii = **Phyllobolus nitidus**
S. longipapillatus = **Phyllobolus oculatus**
S. longispinulus = **Phyllobolus grossus**
S. longitubus = **Phyllobolus tenuiflorus**
S. macrosiphon = **Phyllobolus tenuiflorus**
S. melanospermus ≡ **Phyllobolus melanospermus**
S. micans = **Phyllobolus resurgens**
S. nanus = **Phyllobolus tenuiflorus**
S. nitidus ≡ **Phyllobolus nitidus**
S. nothus = **Phyllobolus splendens** ssp. **splendens**
S. obtusus = **Phyllobolus decurvatus**
S. oculatus ≡ **Phyllobolus oculatus**
S. olivaceus = **Phyllobolus tenuiflorus**
S. oubergensis = **Phyllobolus pumilus**
S. pentagonus ≡ **Phyllobolus splendens** ssp. **pentagonus**
S. pentagonus var. *occidentalis* = **Phyllobolus splendens** ssp. **pentagonus**
S. platysepalus = **Phyllobolus grossus**
S. plenifolius = **Phyllobolus splendens** ssp. **splendens**
S. pomonae = **Phyllobolus oculatus**
S. praecox = **Phyllobolus sinuosus**
S. prasinus ≡ **Phyllobolus prasinus**
S. primulinus = **Phyllobolus splendens** ssp. **splendens**
S. pumilus ≡ **Phyllobolus pumilus**
S. quartziticus ≡ **Phyllobolus quartziticus**
S. quaternus = **Phyllobolus spinuliferus**
S. rabiei ≡ **Phyllobolus rabiei**
S. rabiesbergensis = **Phyllobolus splendens** ssp. **splendens**
S. radicans ≡ **Prenia radicans**
S. recurvus = **Phyllobolus sinuosus**
S. reflexus = **Phyllobolus splendens** ssp. **splendens**
S. rejuvenalis = **Phyllobolus spinuliferus**
S. resurgens ≡ **Phyllobolus resurgens**
S. rhodandrus = **Phyllobolus nitidus**
S. roseus = **Phyllobolus splendens** ssp. **splendens**
S. salmoneus = **Phyllobolus canaliculatus**
S. saturatus ≡ **Phyllobolus saturatus**
S. scintillans = **Phyllobolus oculatus**
S. sinuosus ≡ **Phyllobolus sinuosus**
S. spinuliferus ≡ **Phyllobolus spinuliferus**
S. splendens ≡ **Phyllobolus splendens**
S. stayneri = **Phyllobolus delus**
S. straminicolor = **Phyllobolus sinuosus**
S. striatus = **Phyllobolus splendens** ssp. **splendens**
S. strictus = **Phyllobolus spinuliferus**
S. subaequans = **Phyllobolus splendens** ssp. **pentagonus**
S. subpatens = **Phyllobolus splendens** ssp. **pentagonus**
S. subpetiolatus = **Phyllobolus grossus**
S. suffusus = **Prenia tetragona**
S. sulcatus = **Phyllobolus splendens** ssp. **splendens**
S. tenuiflorus ≡ **Phyllobolus tenuiflorus**
S. tetragonus ≡ **Prenia tetragona**
S. tetramerus = **Phyllobolus trichotomus**
S. tetramerus var. *parviflorus* = **Phyllobolus trichotomus**
S. trichotomus ≡ **Phyllobolus trichotomus**
S. umbelliflorus = **Phyllobolus splendens** ssp. **splendens**
S. vanheerdei = **Phyllobolus roseus**
S. vernalis = **Phyllobolus splendens** ssp. **splendens**
S. vigilans = **Mesembryanthemum longistylum**
S. viridiflorus ≡ **Phyllobolus viridiflorus**
S. watermeyeri = **Phyllobolus spinuliferus**
S. willowmorensis = **Phyllobolus grossus**

Squibbia = Sesuvium

H. E. K. Hartmann

The generic name *Squibbia* was only used once by Rafinesque (1838: 16) when he mentioned "the *S. pentandrum* of Elliot which is my G. *Squibbia maritima*" in comparison with **S. verruculosum**. As a consequence, *Squibbia* is a nomen nudum, which has not been used again.

Sq. maritima ≡ **Sesuvium maritimum**

Stayneria Ruschioideae

H. E. K. Hartmann

Stayneria L.Bolus JSAB1960: 47 **Typus** *S. littlewoodii* = **S. neilii** (L.Bolus) L.Bolus **Etym** *W.F. Stayner*, 1907–1981, curator of the Karoo Botanic Garden at Worcester.

Shrubs to 1.5 m *h*; **L** triquetrous, basally connate and adnate to the reddish-brown stem, triquetrous, epidermis smooth, stomata superficial, wax in horizontal plates covered by vertical platelets, particularly around the slightly sunken stomata; **Fl** in rich dichasia, often with additional cymes at their bases, **K** 4, petals white, filamentous staminodes papillate in their lower halves, surrounding the central stamens, nectary in a lophomorphic ring; **Fr** opening completely, but not closing completely again, due to the expanding keels reaching only for a short distance into the locule, where they are lifted from the abutment losing contact with this underlying tissue so that no connection exists when in the drying process the expanded keel should be fetched back into the locule, expanding keels diverging, tapering into a short awn, covering membranes ± straight, with a radial thickening ending in a pronounced closing rodlet, but rim not well developed, closing body small, hook-shaped, 5–8 locules; **S** pear-shaped, brown, testa cells raised; **Ecol** among rocks in high, shrubby Rhenosterbosveld, 300–400 mm rainfall p.a. mainly in March and November; **Distr** Caledon, Robertson, Worcester, WC, S Africa.

Distribution of **Stayneria** L.Bolus

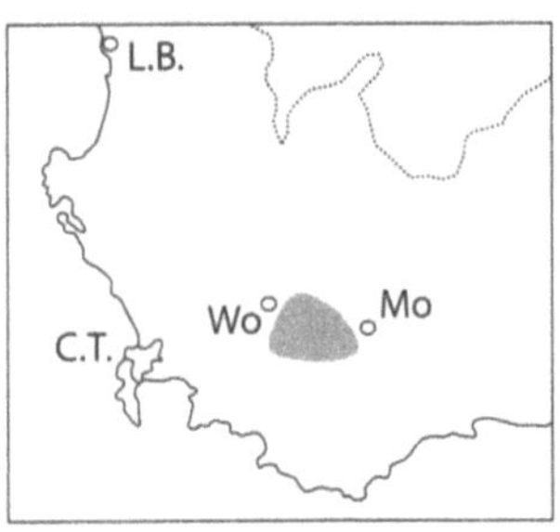

Note: The genus comprises a single species only. The incomplete closing of the once opened fruit resembles conditions found in **Stoeberia** and **Scopelogena**, where entirely different mechanisms are at work. The remarkable thickening of the covering membranes is also found in **Schlechteranthus** and **Namaquanthus**. Characteristic of the genus are its high growth and the big, stout, woody fruits.

S. littlewoodii = **S. neilii**

S. neilii (L.Bolus) L.Bolus 1967: 306 (see **photo**) ≡ *Ruschia neilii* L.Bolus NM 1934: 433 **T** Hurling & Neil 20983 (BOL!) = *S. littlewoodii* L.Bolus JSAB 1960: 48 **T** Stayner & Littlewood 438/60 (BOL).

Stiff, nearly erect branches forming a funnel-shaped bush keeping its form well; **L** dark green, to 80 mm *l*, ∅ 8–10 mm; **Fl** ∅ c. 40 mm, 30–40 white petals, ca. 60 filamentous staminodes white, apically cream coloured, the same colouring in the 160–170 stamens; **Fr** base long, funnel shaped, c. 10 mm *l*, ∅ 10–15 mm; **S** c. 0.8 mm *l*, c. 0.7 mm *b*; **Chr** $2n = 18$; **Ecol** as for genus **Distr** as for genus.

Note: The species occurs in vegetation of about the same height or higher, thus the plants are very inconspicuous in spite of their remarkable height for the whole subfamily. The fact that the valves open repeatedly indicates that dispersal takes place by raindrops mainly. Cultivation from seed is possible, but the plants have to reach a height of c. 70 cm before they start flowering. Cuttings also take well.

Sterropetalum = Nelia

H. E. K. Hartmann

Sterropetalum was published on 14th of April, 1928, by Brown (GC 83: 266), a month later than **Nelia** Schwantes (MDG 43: 92, 11th of March, 1928), both clearly being the same genus; *Sterropetalum* becomes, therefore, a synonym of **Nelia**.

S. pillansii ≡ **Nelia pillansii**

Stigmatocarpum = Dorotheanthus

H. E. K. Hartmann

Published on 1st of September 1927, *Stigmatocarpum* L.Bolus was described as having persistent stigmas, a feature also typical of the genus **Dorotheanthus** Schwantes, published 11th of August, 1927. *Stigmatocarpum* is therefore superfluous and a younger synonym of **Dorotheanthus.**

S. caducum
= **Mesembryanthemum nodiflorum**
S. criniflorum
= **Dorotheanthus bellidiformis ssp. bellidiformis**
S. copticum = **Dorotheanthus apetalus**

Stoeberia Ruschioideae

H. E. K. Hartmann

Stoeberia Dinter & Schwantes ZSK 1927: 14; Friedrich 1960: 554–567; Jaarsveld 1994: 68–76; Hartmann Bradleya 1996: 29–56 **Typus S. beetzii** (Dinter) Dinter & Schwantes **Etym** *E. Stöber*, teacher and botanical explorer at Lüderitzbucht = *Pentaschista* Dinter nomen nudum **T** *Pentaschista arborescens* Dinter nomen nudum = **S. frutescens** (L.Bolus) Van Jaarsveld.

Flat, low to erect, high shrubs, up to 3 m *h*, I to 50 mm *l*, Ø to 20 mm; **L** club-shaped to almost trigonous, hardly connate at base, falling off easily, apically round, greyish- to yellowish-green, 20–50 mm *l*, to 12 mm *t*; **Fl** in dichasia, sometimes enlarged by basal or lateral additional cymes, with bracts and bracteoles, **K** 5, petals white to pink, inconspicuous and mostly not longer than **K** lobes, filamentous staminodes few to many, nectary as a lophomorphic ring; **Fr** pedicels and peduncles elongating after anthesis, capsules light greyish-yellow to dark brown, top raised a little through low valve rims, base funnel- to bowl-shaped, valves opening but once, expanding keels short, not fixed to the locule wall and breaking off their bases on the septum in the opening process, valve wings very broad and hard, preventing the closing of the valves by their radial position, covering membranes short, covering the locules only in the centre of the fruit partially, firm and stiff, closing body absent or placed well inside the locule, often almost half way down it, 5(–6) locules; **S** on short, almost straight funicles which break off their bases in some species when the seeds are ripe; **Ecol** on coastal sands, in gravel inland or on rocks in mountains, < 100 to 300 mm rainfall p.a. in winter, fog; **Distr** SW Namibia and W coastal parts of the NC and WC, S Africa.

Note: The genus comprises the tallest plants in the whole family, developing tree-like shapes and reaching heights of 3 m. It is most similar to **Ruschianthemum**, sharing with it the nearly club-shaped leaves, the rich dichasia, and the inconspicuous flowers, a feature also typical of **Eberlanzia.** Dispersal appears to occur mainly anemochorically in **Stoeberia**: the open, dry capsules are shaken by the wind, the seeds are spilt out like salt from a castor. The evolution of flat seeds that can be carried better by wind can be seen as a further development in the direction of improved anemochory. Connected with anemochory is also the feature that the short, stiff funicles break off their bases and are also blown away together with the seeds. Alternatively, some species develop deciduous capsules which are dispersed as tumble fruits. In these, the erect, short funicles remain inside the fruit, the seeds fall out when the capsule is tumbling around driven by wind.

Dinter wanted to name the new genus *Pentaschista*, a nomen nudum used on a herbarium sheet only. Friedrich (1960) took the epithet up, placing the taxon in **S. beetzii**, but it belongs with **S. frutescens.**

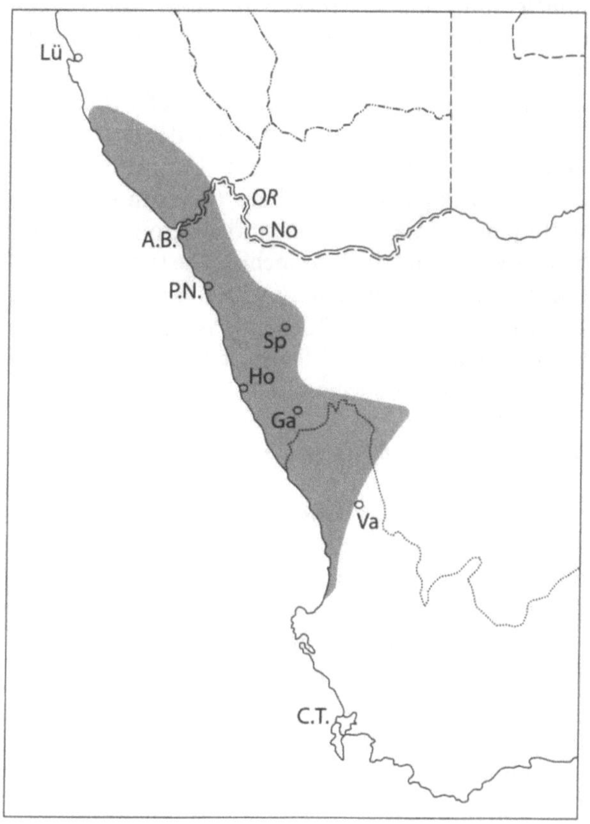

Distribution of **Stoeberia** Dinter & Schwantes

Key to the Species of Stoeberia

1. Capsules with elongate, funnel-shaped bases 8–10 mm *l*, Ø 9–10 mm ..... **S. carpii**
- Capsules with rounded or shortly funnel-shaped bases 2–7 mm *l*, Ø 3–8 mm 2
2. Small cymes on lateral short shoots, seeds 0.6–0.75 mm *l*, 0.4–0.5 mm *b*, almost smooth, yellow ..... **S. utilis**
- Rich cymes terminal, often with additional ones below the terminal one, seeds ≥ 0.75 mm *l*, ≥ 0.5 mm *b*, at least partially papillate, brown 3
3. Cymes enlarged by additional flowers, pedicels elongating only moderately after anthesis, the fruiting arrangement therefore resembling a besom or fan ..... **S. porphyrea**
- Pure cymes, sometimes more below the terminal one, but not enriched by flowers within the dichasium, pedicels and peduncles elongating up to 5 × their original length, the final fruiting arrangement therefore loose and fragile 4
4. Seeds papillate only in the region above the embryo, the area above the reduced endosperm smooth, always flat ..... **S. frutescens**
- Seeds papillate all over, mostly thick and pear-shaped, rarely flattish 5
5. Seeds 0.75 mm *l*, big shrubs or trees ..... **S. × arborea**
- Seeds > 0.75 mm *l*, low shrubs, mostly broader than high ..... **S. beetzii**

S. apetala = **S. beetzii**

S. × arborea Van Jaarsveld ≡ *S. arborea* Van Jaarsveld 1994: 69 **T** van Jaarsveld, Sardien, van der Walt & Bezuidenhout 13453 (NBG).

Shrubs with a stout stem of c. 20 cm in Ø; **L** club-shaped, 20–55 mm *l*, basally a little more connate than **S. frutescens**, **Fl** in rich cymes, the terminal one often pyramidal in shape, Ø 5–7 mm; **Fr** Ø 5 mm, closing body absent; **S** pear-shaped, c. 0.75 mm *l*; **Ecol** in rocks, c. 100 mm rainfall in winter; **Distr** N Namaqualand, NC, S Africa.

Note: Described as a species, the populations included comprise forms with the growth- and leaf-forms of **S. frutescens** and character expressions of fruits and seeds of **S. porphyrea**. It can be suggested that these populations are of hybrid origin, hence the change in naming (Hartmann Bradleya 1996: 47–55), as also indicated by Jaarsveld (1994). Some of the hybrids show a combination of character states different from the one given for **S. × arborea**, e.g. the fan-shaped inflorescence and reddish-brown stems of **S. porphyrea** but features and sizes of fruits and seeds of **S. frutescens.**

S. beetzii (Dinter) Dinter & Schwantes ZSK 1927: 14 (see **photo**) ≡ *M. beetzii* Dinter FR 1923: 180 **LT** Dinter 3786 (B) = *S. apetala* L.Bolus 1961: 120 **T** Hall 2144 (BOL) = *Ruschia bicolorata* L.Bolus NM 1929: 101 ≡ *M. bicoloratum* (L.Bolus) N.E.Br. GC 1930: 32 **Lectotype**, here designated: Pillans 17775 II (chosen from 3 sheets of that number, BOL!) = *Ruschia micro-petala* L.Bolus NM 1929: 76 ≡ *M. micropetalum* (L.Bolus) N.E.Br. GC 1930: 32 ≡ *S. micropetala* (L.Bolus) L.Bolus ex Friedrich 1960: 564, nomen nudum **T** Pillans 5737 (BOL) = *Ruschia translucens* L.Bolus NM 1929: 92 ≡ *M. translucens* (L.Bolus) N.E.Br. GC 1930: 33 **T** Erni 18919 (BOL).

Flat to erect shrubs, 20–50 cm *h*, **I** grey, 20–30 mm *l*; **L** club-shaped, mostly 15–20 mm *l*, greyish-green; **Fl** in terminal cymes, petals in 5 groups, thin, translucent to white, numerous filamentous staminodes, Ø 0.7–1.4 mm; **Fr** in big, spreading cymes, base shortly funnel- to bowl-shaped, closing body at c. 3/4 of the depth of the locule, compact to spiral-shaped to different degrees, seeds below and above the closing body; **S** pear-shaped to somewhat flattened, papillate all over, dark brown, 0.75–1.2 mm *l*, 0.5–0.8 mm *b*, funicles loose and dispersed with seeds; **Ecol** in sandy to gravelly flats near the coast, < 100 to 300 mm rainfall p.a. in winter, fog; **Distr** SW Namib, Namibia; adjacent areas of NW Namaqualand, NC, S Africa.

Note: The mostly low plants with decumbent branches ending in rich cymes are typical of the flats of the southern Namib, where they can dominate some places. Inland forms tend to grow into taller bushes with more erect branches. The synonymy follows Friedrich (1970) mainly, with some doubts about *Ruschia micropetala*. **S. beetzii** is very similar to **S. frutescens**, seeds and fruits have to be examined to distinguish the species.

S. beetzii var. *arborescens* = **S. frutescens**

S. carpii Friedrich 1960: 563 **T** Dinter 8148 (M) ≡ *Ruschianthemum merenskyanum* Dinter nomen nudum in schedis = *Jensenobotrya vanheerdei* L.Bolus 1966: 201 **T** Heerde 27741 (BOL).

To 1.5 m *h*, also straggling in other bushes, **I** greyish-white, 10–90 mm *l*; **L** broadly club-shaped, 20–50 mm *l*, to 10 mm *b* and *t*, whitish-grey; **Fl** in terminal cymes branching once or twice plus one or two more storeys of cymes below the terminal one, petals white, longer than the 5(–6) **K** lobes, Ø up to 24 mm, filamentous staminodes surrounding the central cone of stamens; **Fr** occasional additional flowers in the distal dichasia forming a flat besom, pedicels and peduncles elongating only moderately after anthesis, base elongate, funnel-shaped, grey, 8–10 mm *l*, ∅ at top 8–10 mm, closing bodies absent, 5(–6) locules; S shortly pear-shaped, dark yellow with dark brown naps, 0.8–0.9 mm *l*, 0.6–0.7 mm *b*, funicles persisting in c. two rows in empty fruit; **Chr** $2n = 18$; **Ecol** in rocky places on mountains, < 100 mm rainfall p.a. in winter, fog; **Distr** SE corner of Namib, Namibia; N parts of NW Namaqualand, NC, S Africa.

Note: In leaf and fruit dimensions, this species is the biggest of all in the genus. In habit, it resembles **S. frutescens** because the dichasia form an upper layer over the plant until the next season. Similar are also the additional storeys of dichasia below the terminal one. Unique in the genus are the rather round seeds with prominent brown naps on the dark yellow surface. The absence of closing bodies in the capsules and the persistence of the erect, stiff funicles in the ripe fruit agree with features seen in **S. utilis** and **S. porphyrea.**

S. frutescens (L.Bolus) Van Jaarsveld 1994: 74 (see **photo**) ≡ *M. frutescens* L.Bolus ASAM 1913: 151 ≡ *Ruschia frutescens* (L.Bolus) L.Bolus NM 1954: 260 T Pearson 6433 I (BOL!) = *S. beetzii* (Dinter) Dinter & Schwantes var. *arborescens* Friedrich 1960: 560 ≡ *Pentaschista arborescens* Dinter nomen nudum T Dinter 8215 (B) = *Ruschia utilis* L.Bolus var. *gifbergensis* L.Bolus NM 1954: 285 T Esterhuysen 22051 (BOL).

To 2.5 m *h*, **I** greyish; **L** club-shaped, sometimes up to 50 mm *l*, greyish-green; **Fl** in rich cymes branching up to 5 ×, often further lateral cymes below the terminal one, next years **Inf** overtopping the persisting old ones, forming a new top layer over the plant, petals whitish to pale pink, a little longer than the **K** lobes; **Fr** pedicels and peduncles elongating markedly after anthesis thus exposing the capsules more widely and well above the plant at that time, base shortly funnel-shaped, closing bodies at c. 4/5 th of the depth of the locule, protruding straightly or reduced to a small hump, variable even in one capsule, seeds placed only below the closing bodies, 5(–6) locules; **S** brown, flat, the endosperm region smooth, the embryo region papillate, the embryo mostly straight or its micropyle recurved, almost hook-shaped, funicles loose and dispersed with the seeds; **Ecol** in gravelly flats (in the S) or in rocky places > 300 m altitude (in the N), < 100 to c. 300 mm rainfall p.a. in winter, fog in S; **Distr** SE Namib, Namibia; W Namaqualand, NC, Vanrhynsdorp, Vredendal, W WC, S Africa.

Note: The northern populations grow exclusively in rocky places at higher altitudes than the southern ones, from where only *S. utilis* var. *gifbergensis* is reported as growing amongst rocks in higher altitudes. Seeds and fruits of the northern populations are bigger and have more prominent papillae than those of the southern plants, but no clear limits can be drawn at present. The flat, almost rectangular seeds with papillae only on the embryo region are typical of this species.

S. gigas ≡ **Ruschianthemum gigas**
S. hallii ≡ **Lampranthus amphibolius**
S. littlewoodii ≡ **Lampranthus mutatus**
S. micropetala = **S. beetzii**

S. porphyrea H.E.K.Hartmann Bradleya 1996: 51 (see **photo**) T Hartmann & Dehn 19265 (HBG!).

Erect, tree-like shrubs to 3 m *h*, **I** reddish dark brown; **L** thickly club-shaped, greyish-green, to 25 mm *l*, ∅ to 7 mm; **Infl** in dense cymes enriched distally by additional, lateral flowers in one plain, thus forming a loose fan, further dichasia of the same type below the terminal one, **Fl** ∅ to 8 mm; **Fr** in flabelliform besoms, pedicels only moderately longer than in anthesis, bases bowl-shaped to shortly funnel-shaped, 2–3.5 mm *h*, apical ∅ 3–5 mm, closing bodies absent, 5 locules; **S** pear-shaped, finely papillate all over, light brown, 0.75–0.85 mm *l*, 0.5–0.6 mm *b*, the short, stiff funicles persisting when the fruit is ripe; **Ecol** amongst rocks, > 300 m, c. 125 mm rainfall p.a. in winter; **Distr** SE Namib, Namibia; N Namaqualand, NC, S Africa.

Note: The species differs in its dark reddish-brown internodes and the dense besom-shaped bundles of fruits in terminal and subterminal, lateral position distinctly from all other species of the genus. It also forms the highest plants seen up to now in the genus, developing prominent stems resembling a tree overtopping the surrounding dense vegetation. In open, more rocky places, it can grow into broad, high bushes.

Brown internodes occur also in younger parts of **S. frutescens**, but the cymes of that species spread widely after anthesis and the bases of the capsules are grey, distinctly higher, and mostly funnel-shaped. **S. porphyrea** is most similar to **S. utilis**, from which it differs in the arrangement of the besom-shaped, enriched cymes and the reddish internodes mainly.

S. rupis-arcuatae ≡ **Amphibolia rupis-arcuatae**

S. utilis (L.Bolus) Van Jaarsveld 1994: 75 (see **photo**) ≡ *M. utile* L.Bolus NM 1929: 99 ≡ *Lampranthus utilis* (L.Bolus) Schwantes in Jacobsen NCSJ 1949: 58 ≡ *Ruschia utilis* (L.Bolus) L.Bolus NM 1950: 219 T Pillans 17928 (BOL).

To 2 m *h* with erect branches developing short shoots only, new long shoots from base, **I** greyish; **L** greyish-green, club-shaped, 10–15 mm *l*, 3–5 mm *t*; **Fl** in cymes on lateral short shoots, terminal ones inconspicuous, Ø c. 15 mm, petals as long as **K** lobes, white, few filamentous staminodes; **Fr** pedicels and peduncles not elongating markedly, ripe fruits deciduous, base bowl-shaped to shortly funnel-shaped, 3–5 mm *h*, Ø at top 4.5–7.5 mm, closing bodies absent; **S** nearly pear-shaped, almost smooth, yellow, 0.6–0.75 mm *l*, 0.4–0.5 mm *b*, funicles short, persisting in the empty capsule after maturity of the seeds; **Ecol** in sandy to gravelly coastal flats, 100–350 mm rainfall p.a. in winter, fog; **Distr** SW Namib, Namibia; W coast of Namaqualand, NC, Vanrhynsdorp, Vredendal, Clanwilliam, Piquetberg, WC, S Africa.

Note: High, long shoots dominate plants of this species making them conspicuous in the coastal flats of Namaqualand. The formation of lateral cymes and the small, almost smooth, yellow seeds characterize the species reliably in the genus. In capsule morphology and the formation of tumbling fruits, **S. utilis** agrees well with **S. porphyrea**, which differs in the structure of the cymes and colour of internodes from the former.

Stomatium Ruschioideae

H. E. K. Hartmann

Stomatium Schwantes ZSK 1926: 175 **Lectotypus S. suaveolens** Schwantes **Etym** Gk *stoma*, mouth, describing the spreading, dentate leaves resembling an open mouth = *Agnirictus* Schwantes MDK 1930: 21 nomen invalid. ≡ *Stomatium* § *Agnirictus* O.H.Volk in Jacobsen & Volk 1950: 83 nomen invalid. **Typus** not chosen.

Compact to caespitose with age, often anisophyllous, especially when young; **L** trigonous to flattish, in particular when young, older leaves mostly isophyllous, keel ± distinct, epidermis with conspicuous elevations, thus rough, or smoother, margins dentate to various degrees, keel smooth to dentate, the teeth either small and composed of confluent or pointed elevations, or big and conical, composed of several elevations with a number of tannin idioblasts below forming the base of an emergence, always crowned with a sharply pointed elevation, the outer epidermis wall with marked, thick crystal layers, these often only developed above the elevations which in this way turn into prominent grains on the leaves, the areas in between often green, the underlying tannin idioblasts visible as dark dots here; **Fl** on rather short stalks, often enclosed by the leaf sheaths, bases of **K** lobes, petals and stamens growing from a thickish cushion-shaped basal ring raised well above the top of the ovary and called a tube by Schwantes, the 5 glands at the base of the tube, reaching up to half the height of the tube, **K** 5, petals yellow, white or pink, inner often shorter, mostly very narrow, filaments erect to bending over the short stigmas, covering them for most of the anthesis, ovary raised in the centre, short stigmas on top, but not united, opening at night; **Fr** base globose to bell-shaped, top flat to somewhat raised in

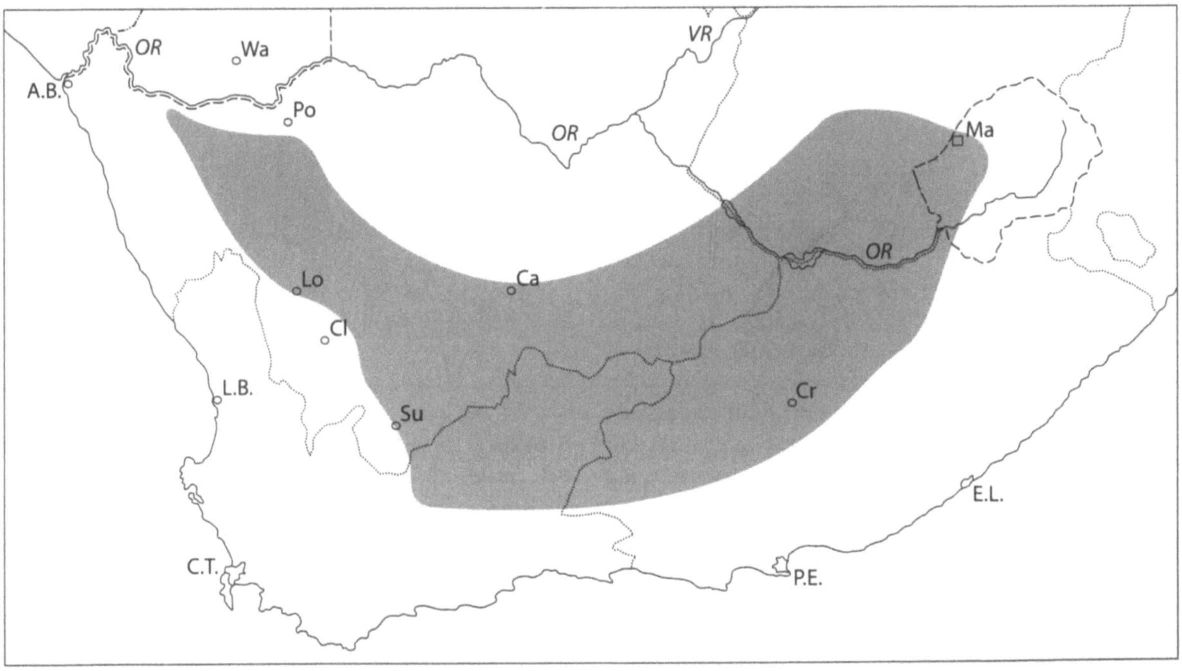

Distribution of **Stomatium** Schwantes

the centre from the remains of the bases of the stigma, dispersed as tumble fruits with their stalks on, septa with narrow to very narrow rims or none at all, the placenta either flat on the base of the locule or raised by a false septum to various degrees, expanding keels always parallel, valve wings broad and somewhat diverging in shape between the species, 5, very rarely 6, locules; **S** 0.6–0.95 mm *l*, 0.45–0.75 mm *b*; **Ecol** mostly in flat or on very gentle slopes in finely grained soils of different origin or in crevices of shales, mostly > 100 mm rainfall p.a. in March and November; **Distr** Lesotho; W districts of EC, SW districts of NC, FS, N and NE districts of WC, S Africa.

Note: The type species, though only described later, had been named by the author with the original description of the genus, it was therefore known to him, the choice can be accepted (Schwantes ZSK 1927: 105). Although the presence of teeth has been given as typical of the genus, some species lack them, at least at times, and often in habitat. This is particularly obvious in the type species, **S. suaveolens**, which differs also from many other species by developing a tumble fruit breaking off the stalk. Remarkable is also the fact that the species possesses a tube of 5 mm length, a value not reached by any of the other species described. The distinction against **Rhinephyllum**, also without covering membranes proper, remains difficult. It appears, e.g., that **R. schonlandii** may be **S. suaveolens**. In cultivation, all species tend to develop much more spreading, longer leaves, mostly by elongating the base of the leaf.

S. acutifolium L.Bolus NM 1938: 130 **T** Meglaughlin 21983 (BOL!).

L ± erect, keel with 0–3 small teeth, margins with big teeth, 2 or 3 on each side, epidermis moderately rough with rather low elevations, **L** 20–30 mm *l*, 5–6 mm *b*, 4–6 mm *t*; **Fl** tube 1.25 mm *l*, petals 12 mm *l*, 0.5 mm *b*; **Fr** unknown; **Ecol** unknown; **Distr** Beaufort West, WC, S Africa.

Note: Reported to be very similar to **S. angustifolium**.

S. agninum (Haw.) Schwantes 1926: 175 (see **photo**) non 1926: 176 ≡ *M. agninum* Haw. 1826: 127 ≡ *Agnirictus agninus* (Haw.) Schwantes MDK 1930: 21 **Lectotype**, here designated: drawing 126 by Duncanson 28.4.1826 of material grown from seed collected by Bowie in 1823 (K!) = *M. agninum* Haw. var. *integrifolium* Salm-Dyck 1840: § 5 f 8 ≡ *S. agninum* (Haw.) Schwantes var. *integrifolium* (Salm-Dyck) O.H.Volk in Jacobsen & Rowley 1955: 80 **Lectotype**, here designated: Salm-Dyck 1840: § 5 f 8.

Compact to caespitose with age; **L** triangular as seen from above, spreading widely, margins undulate and with teeth, **L** to 54 mm *l*, 13.5 mm *b*; **Fl** sessile; **Fr** unknown; **Ecol** unknown; **Distr** Cradock, EC, S Africa.

Note: The drawing chosen as the lectotype here was made before Haworth's description was published on 28.8.1826.

S. agninum = **S. murinum**
S. agninum var. *integrifolium* = **S. agninum**

S. alboroseum L.Bolus NM 1931: 242 **T** Herre 8919 (BOL!) = *S. niveum* L.Bolus NM 1931: 263 **T** Luckhoff 19495 (BOL).

Compact and semi-globose or flattish and forming mats with age, depending on the depth of the soil; **L** spatulate-trigonous, fat in the upper part, rounded above, keel inconspicuous and almost smooth, margins undulate, elevations confluent and forming 3–6 small, white teeth like warts, elevations rather pointed and sharp, leaves therefore very rough, wax cover in horizontal thin plates with numerous ascending platelets emerging from the plates, **L** 20–25 mm *l*, c. 10 mm *b*, c. 7 mm *t*; **Fl** tube 1.5–2 mm *l*, petals white to pink, 8.5 mm *l*, 1 mm *b*, anthers white; **Fr** base ± semi-globose, top almost flat, rims raised a little in the centre, septa with very narrow rims, placenta basally flat or raised only a little, valve wings c. as broad as the expanding keels, narrowing towards their base, ⌀ 4.5–5.5 mm; **S** 0.6–0.7 mm *l*, 0.45–0.57 mm *b*; **Ecol** in flats with sand or fine limestone pebbles, apparently in soils with high contents of lime; **Distr** Namaqualand, NC, Vanrhynsdorp, WC, S Africa.

Note: The species is common on the calcareous flats of south-western Bushmanland, sometimes also growing in sand overlying the limestone. It is most similar to **S. meyeri** and **S. fulleri**.

S. angustifolium L.Bolus NM 1935: 505 **T** Rennie 639/34 (cited as 639/24, BOL!).

L erect, keel rounded, the leaf therefore appearing somewhat incurved, keel with some confluent elevations, but edentate, margins with stout, big teeth, epidermis margaretiferous, i.e. finely grainy, base of leaf thinner than apical part, 18–26 mm *l*, 5–7 mm *t*; **Fl** tube 2 mm *l*, petals yellow, 6–12 mm *l*, to 0.4 mm *b*; **Fr** unknown; **Ecol** in flats with shales; **Distr** Somerset East, EC, Murraysburg, WC, S Africa.

Note: The rounded, incurved, erect, finely pearly, slender leaves permit an easy identification of the species in habitat.

S. beaufortense L.Bolus NM 1937: 63 **T** Dyer 549/32 (BOL!).

L trigonous, very fat, teeth on margins tiny, consisting of prominent, pointed elevations only, keel smooth, epidermis with elevations a little lower than those on the margins, **L** 32–34 mm *l*, 14 mm *b*, 5–9 mm *t*; **Fl** tube 2 mm *l*, petals bright yellow, 13–15 mm *l*, 0.5 mm *b*, anthers and pollen golden, nectary glands reaching half way up the tube; **Fr** unknown; **Ecol** unknown; **Distr** Beaufort West, WC, S Africa.

Note: The tiny teeth are said to be among the smallest in the genus.

S. bolusiae Schwantes MDK 1930: 22 (see **photo**) **Lectotype**, here designated: Moss 1206/27 (BOL!).

L margins with teeth with broad bases and an elevation at the end, keel smooth or with 1–2 smaller teeth, the lower leaf surface protruding forming a chin; **Fl** tube very short; **Fr** unknown; **Ecol** at the edges of huge rock plates in shallow ground; **Distr** Jagersfontein, FS, S Africa.

Note: The name was used to describe a figure of the type material which had been identified as **S. erminium** by Bolus (SAG 1928: 370).

S. braunsii L.Bolus NM 1932: 309 **T** Brauns 9389 (BOL!).

L trigonous with convex sides, somewhat incurved, teeth on margins on very broad bases ending in pointed elevations, keel rounded, rough from acute elevations, i.e. tiny teeth, epidermis also with white elevations, **L** 22–28 mm *l*, 11 mm *b*, 4–7 mm *t*; **Fl** tube 1 mm *l*, petals bright yellow, to 12 mm *l*, 0.5 mm *b*, filaments pale, apically bright yellow like the anthers; **Fr** base funnel-shaped, top flat, septa with very narrow rims, placenta raised to almost half the height of the locule by a false septum, valve wings a little broader than the expanding keels; **Ecol** unknown; **Distr** given as Clanwilliam to Vanrhynsdorp, WC, S Africa.

Note: Data taken from description and the type. The geographical information seems to be wrong because until now, no populations of **Stomatium** have been found in the area.

S. bryantii L.Bolus NM 1935: 507 **T** Bryant 724 (BOL!).

L trigonous, acuminate to the tip, epidermis with white elevations, 2–6 teeth on margins ending in pointed elevations from a broad base each, keel mostly with pointed elevations only, these rarely with a broader base, **L** 18–30 mm *l*, 10–11 mm *b*, 6–8.5 mm ∅; **Fr** unknown; **Ecol** unknown; **Distr** Prieska, NC, S Africa.

Note: No collection of this species has been made recently in the Prieska district.

S. deficiens L.Bolus NM 1935: 508 **T** James 2213/31 (BOL!).

L trigonous with convex sides, elevations white, margins and keel very rarely with few more pointed elevations than the sides, **L** 12–22 mm *l*, 5–8 mm *b*, to 5.5 mm *t*; **Fl** tube short, petals very few, anthers yellow; **Fr** unknown; **Ecol** unknown; **Distr** Cradock, EC, S Africa.

Note: Bolus noted that the plant may be a hybrid because of the reduced number of petals and the lack of fruits.

S. difforme L.Bolus NM 1931: 263 **T** Goeldner 9338 (BOL!).

L spatulate to trigonous, markedly broadened subapically, margins with 3–6(–18) teeth consisting of a very broad base and a pointed elevation on top, to 1.5 mm *l*, keel smooth on the type, reported to have 1–3 teeth, **L** to 26 mm *l*, to 16 mm *b*, to 9 mm *t*; **Fl** tube 2.5 mm *l*, petals bright yellow, apically red, to 11 mm *l*, 0.5 mm *b*; **Fr** unknown; **Ecol** unknown; **Distr** Laingsburg, WC, S Africa.

Note: The number of teeth given in the description and found on the type material differs. From the same area is **S. loganii** known, which is reported to have fewer teeth and more slender leaves.

S. duthiae L.Bolus NM 1932: 309 **T** Duthie 9664 (BOL!).

Caespitose; **L** trigonous, keel without any teeth, margins with 3–4 pointed elevations on broad bases; **Fl** tube 3 mm *l*, petals bright yellow, filaments yellow, anthers golden; **Fr** base deep, funnel-shaped, top flat, septa with very narrow rims, the outer wall of the locule protruding a little under the placenta, which runs down to the bottom of the locule, valve wings apically angular, basally rounded, distinctly broader than the expanding keels; **Ecol** unknown; **Distr** Middelburg, EC, S Africa.

Note: This species is reported to differ from **S. middelburgense** only in flower morphology and colour.

S. erminium (Haw.) Schwantes ZSK 1926: 176 ≡ *M. erminium* Haw. 1826: 127 **T**?

Densely caespitose; **L** margins with 4–6 stout, acute, spreading teeth, keel smooth; **Fl** petals c. 20 mm *l*, narrow, yellow, anthers and pollen almost golden; **Fr** unknown; **Ecol** unknown; **Distr** unknown.

Note: Data taken from Haworth's original description. This refers clearly to a plant with marginally dentate leaves with a smooth keel, in contrast to **S. murinum**. Whether the plant figured by Salm-Dyck (1836: § 5 f 6) represents the original material, cannot be concluded with certainty, the figure is therefore not chosen as a lectotype here. How widely the description can be interpreted is obvious from the figure published by Bolus (SAG 1930: 370) under that name, transferred by Schwantes to *Neorhine* and representing **Rhinephyllum muirii**.

S. fulleri L.Bolus NM 1929: 90 **T** Fuller 10 (BOL!).

Compact, forming mats later; **L** spatulate-trigonous, rounded above, keel inconspicuous, margins with up to 7 broad teeth, not very stout, the tips breaking off with age, to 20 mm *l*, 7 mm *b*, 6 mm *t*; **Fl** incompletely known; **Fr** base semi-globose, top with valves raised in the centre, rims of septa very narrow, placenta flat on the base of the locule, valve wings a little broader than the expanding keels, rectangular, ∅ 5–6 mm; **S** c. 0.7 mm *l*, 0.55 mm *b*; **Ecol** in flats with sand and some quartzite stones; **Distr** Kenhardt, NC, S Africa.

Note: The species differs from the similar **S. alboroseum** mainly by the more pronounced bases of the teeth and the occurrence in sandy flats, from **S. meyeri** by the smoother leaf surfaces.

S. geoffreyi L.Bolus NM 1934: 436 **T** James 126 (BOL!).

L triangular upper leaf surface, trigonous and pointed, broadest in the middle, margins with 3–6 teeth with a broad base and a pointed top each, keel smooth, to 20 mm *l*, 9–12 mm *b*, 4–5 mm *t*; **Fl** tube 1 mm, **K** 6, petals dirty reddish or pale straw-coloured and apically purple, to 9 mm *l*, 0.75 mm *b*, filaments golden, anthers yellow; **Fr** base long, funnel-shaped, top flat, septa with narrow rims, placenta reaching deeply into the locules, raised a little apically from a bulging of the outer wall, but no closing body formed, valve wings c. double the width of the expanding keels, basally narrowing, 6 locules; **Ecol** on flat rocks; **Distr** Cradock, EC, S Africa.

Note: Reported to grow on top of a mountain at c. 2700 m.

S. gerstneri L.Bolus NM 1935: 505 (see **photo**) **T** Gerstner 2638/33 (BOL!).

L distinctly different in a pair, trigonous, acuminate, the upper surface triangular as seen from above, margins with 1–3 teeth with a broad base and a pointed elevation on top, keel smooth, 20–24 mm *l*, 8–9 mm *b*, 6 mm *t*; **Fl** tube 1.5–2 mm *l*, **K** 5, petals yellow, 7–11 mm *l*, 0.25–0.5 mm *b*; **Fr** base ± bell-shaped, top flat, the valves raised a little in the centre after the first opening, locules deep, the placenta up to 2/3 of the depth, basally on a false septum raised only for < 1/4 of the depth of the locule, valve wings c. as broad as the expanding keels which reach the tip of the valve in the open capsule, ∅ c. 6 mm, base c. 7 mm *h*; **Ecol** on shale covered with sand; **Distr** Lady Grey, Zastron, EC, S Africa.

Note: The leaves of this species resemble those of **S. geoffreyi** from Cradock very much, but **S. gerstneri** has only 5 calyx lobes and locules, and much more narrow valve wings.

S. grandidens L.Bolus NM 1935: 506 **T** collector unknown 21429 (BOL!).

L trigonous, upper leaf surface either spoon-shaped or almost ovate, margins with 2–4 teeth with a very broad base and ending in white, firm elevations, keel smooth or with few elevations as tiny teeth only, **L** to 14 mm *l*, 5–12 mm *t*; **Fl** tube 3 mm *l*, petals yellow or lemon-coloured, 13 mm *l*, 0.4 mm *b*; **Fr** unknown; **Ecol** unknown; **Distr** unknown, S Africa.

Note: The species was described from material grown in a garden, hence probably the enormous plumpness and the wide teeth.

S. integrum L.Bolus NM 1950: 225 **T** Armstrong 22198 (BOL!).

Caespitose with age, to 17 cm ∅; **L** with very low elevations only, keel and margins with few somewhat higher elevations, but no teeth proper, 15–25 mm *l*, 6–11 mm *b*, 4–6 mm *t*; **Fl** tube 1.5 mm *l*, **K** 6, petals bright yellow, 10 mm *l*, 0.75 mm *b*; **Fr** ∅ 6 mm, expanded 10 mm; **Ecol** unknown; **Distr** Adelaide, Tarka, EC, S Africa.

Note: In the absence of teeth, the species resembles **S. deficiens** most closely differing in the number of 6 lobes of the calyx and leaf measurements.

S. jamesii L.Bolus NM 1931: 308 **T** James 885/31 (BOL!).

Caespitose; **L** trigonous, sides convex, epidermis with acute elevations, thus very rough, margins with c. 3 teeth with broad bases and an apical elevation each, keel with confluent elevations, but no teeth formed, **L** to 19 mm *l*, to 9 mm *b*, to 6 mm *t*; **Fl** tube 2 mm *l*, petals bright yellow, 6–10 mm *l*, to 0.5 mm *b*; **Fr** base a shallow funnel-shape, septa with very narrow rims, valve wings almost double the width of the expanding keels, nearly rectangular in shape, ∅ 5–8 mm; **Ecol** unknown; **Distr** Cradock, EC, S Africa.

Note: The shallow capsule base may be typical of the species.

S. latifolium L.Bolus NM 1932: 310 **T** Engelbrecht 9649 (BOL).

L erect, apically widened, margins with 2–4 teeth, keel inconspicuously dentate with 1–2 teeth, glaucous, rough from raised dots and with 2–3 tubercles, to 18 mm *l*, to 15 mm *b*, to 8 mm *t*; **Fl** tube 1 mm *l*, petals yellow, stamens yellow; **Fr** unripe to 10 mm ∅, 6 mm *l*; **Ecol** unknown; **Distr** Williston, NC, S Africa.

Note: Reported to be similar to **S. trifarium**.

S. lesliei (Schwantes) O.H.Volk in Jacobsen & Rowley 1955: 80 ≡ *Agnirictus lesliei* Schwantes MDK 1930: 21 **T** Leslie s.n. (?).

L triangular, spreading widely, margins with 2–3 distant, small teeth, to 40 mm *l*, c. 15 mm *b*; **Fl** sessile, ∅ c. 25 mm; **Fr** unknown; **Ecol** unknown; **Distr** uncertain, S Africa.

Note: The species was described after a photograph of a flowering plant.

S. leve L.Bolus NM 1936: 17 **T** Logan 21646 (BOL).

L widening in the upper part, margins here with 2–4 teeth, surfaces smooth, **L** 25–30 mm *l*, to 12 mm *b* apically, 7 mm *t*; **Fl** tube 2 mm, petals golden yellow, outer apically red, 6–11 mm *l*, 0.4 mm *b*; **Fr** unknown; **Ecol** unknown; **Distr** Cradock, EC, S Africa.

Note: The smooth leaf surfaces are most unusual in the genus.

S. loganii L.Bolus NM 1936: 39 **T** Logan 21 (BOL).

L apically widening and forming a chin, epidermis with prominently raised dots, keel almost smooth, margins with 2–3 teeth, 12–20 mm *l*, to 11 mm *b*, to 6.5 mm *t*; **Fl** tube 2–3 mm *l*, petals yellow, to 8 mm *l*; **Fr** unknown; **Ecol** unknown; **Distr** Laingsburg, WC, S Africa.

Note: The species is similar to **S. difforme**, which has more teeth.

S. meyeri L.Bolus NM 1931: 262 (see **photo**) **T** Meyer 8797 (BOL).

Forming mats with age; **L** spatulate, apically forming a lunate, round tip as seen from above, surface rough from raised dots, margins and keels with 3–4 broadish, low, white teeth, single warts of this type also on the sides occasionally, **L** 20–26 mm *l*, 7–12 mm *b* apically, 7–9 mm *t*; **Fl** tube 2 mm *l*, petals white, pink outside, to 11 mm *l*, 0.5–1 mm *b*; **Fr** base almost funnel-shaped, top flat, septa with very narrow rims, placenta raised a little by a false septum, valve wings a little broader than the expanding keels, apically angled, basally semi-lunate, 5.3 mm ∅; **S** 0.7 mm *l*, 0.6 mm *b*; **Ecol** on slopes densely covered with quartzite stones, < 100 mm rainfall p.a.; **Distr** Namaqualand, NC, S Africa.

Note: Differing from **S. alborosea** and **S. fulleri** in the low false septum.

S. middelburgense L.Bolus NM 1932: 309 **T** Duthie 1738/31 (BOL).

L as described for **S. duthiae**; **Fl** tube 1.5 mm *l*, petals pale lemon-coloured or almost straw-coloured, filaments very pale, anthers yellow; **Fr** unknown; **Ecol** unknown; **Distr** Middelburg, EC, S Africa.

Note: Reported to be very similar to **S. duthiae**.

S. murinum (Haw.) Schwantes ex H.Jacobsen 1933: 180 ≡ *M. murinum* Haw. 1795: 165 ≡ *S. agninum* Schwantes ZSK 1926: 176 nomen illeg. non Schwantes 1926: 175 **Lectotype**, here designated: Salm-Dyck 1849: § 5 f 4.

Compact, developing longer stems with age; **L** trigonous, erect, the lower leaf surface forming a keeled chin, margins and keel with teeth, epidermis rough from raised dots; **Fl** petals yellow; **Fr** unknown; **Ecol** in shallow soil with some quartz; **Distr** probably Fraserburg, NC, S Africa.

Note: The possession of three rows of teeth characterizes this species.

S. musculinum ≡ **Chasmatophyllum musculinum**

S. mustelinum (Haw.) Schwantes ZSK 1926: 176 (see **photo**) ≡ *M. mustelinum* Haw. 1819: 87 ≡ *M. mustelinum* Salm-Dyck 1816: 18 nomen nudum **Lectotype**, here designated: Salm-Dyck 1836: § 5 f 5 = *M. pyrodorum* Diels 1909: 119 ≡ *S. pyrodorum* (Diels) L.Bolus FPSA 1927: t. 265 **T** Diels 741 (B).

Compact, developing short thickish stems with age; **L** spreading widely so that only the upper, spatulate leaf surface is visible, base narrow, margins with up to 8 big, stout teeth, keel not marked, epidermis a little rough from very fine elevations, shaly grey, 15–20 mm *l*, 8–22 mm *b*, 4.5–7 mm *t*; **Fl** pear-scented, ∅ 25–30 mm, 10–150 very narrow petals yellow, 125–200 yellow stamens, glands yellowish-green; **Fr** base semi-globose, top flat with distinctly elevated bases of the stigmas in the centre, rims of septa very narrow, placenta raised for c. 1/4 of the height of the locule by a false septum, the upper end of the placenta pushed forward by a bulging in the outer wall, reminding of a closing sill, but much smaller, ∅ 6–8.5 mm, 10.15 mm *l*; **S** 0.8–0.95 mm *l*, 0.64–0.7 mm *b*; **Ecol** on very gentle slopes with fine Ecca shale; **Distr** Calvinia, NC, S Africa.

Note: **S. mustelinum** is also wrongly spelt with a double l, an invention of Haworth (1821: 89–90). The name was originally used by Salm-Dyck in his Plantae Succulentae Horti Dyckensis 1816: 18, without a description. The latter was supplied by Haworth (1819: 87), he is therefore the author of the name. Nevertheless, the lectotype is chosen from Salm-Dyck (1836: § 5 f 5) because the plant figured there is undoubtedly the one already in cultivation with Salm-Dyck in 1816. The plants keep well in cultivation flowering readily. In habitat, the spreading leaves mimic the small slaty pieces of shale around it so well that they are perfectly camouflaged and very difficult to be found.

S. niveum = **S. alboroseum**

S. patulum L.Bolus ex H.Jacobsen 1935: 248 **Lectotype**, here designated: Jacobsen 1935: f 234, 2.

L trigonous, widening from a narrow base, margins with 2–9 prominent, fat teeth, keel with lower, less conspicuous teeth, **L** 17–19 mm *l*, to 8 mm *b*; **Fl** unknown; **Fr** unknown; **Ecol** unknown; **Distr** uncertain, S Africa.

Note: Data taken from description and figure chosen as the lectotype.

S. paucidens L.Bolus NM 1935: 507 **T** of unknown origin 2016/34 (BOL).

L narrow, margins near the apex with 1–2 teeth, keel with 1–3 obscure teeth, widening a little to the tip, 22–32 mm *l*, to 11 mm *b*, to 8 mm *t*; **Fl** tube 2 mm, petals bright yellow, to 14 mm *l*, to 0.5 mm *b*; **Fr** unknown; **Ecol** unknown; **Distr** unknown.

Note: The species was described from plants flowering in Bolus' garden.

S. peersii L.Bolus NM 1936: 17 **T** Peers 21644 (BOL).

L subspatulate, margins with 2–3 teeth of 1.5–2 mm length, keel near the apex 2-denticulate, 15–17 mm *l*, 5–8 mm *b*, 5–6 mm *t*; **Fl** tube 2 mm, petals bright yellow, to 11 mm *l*, to 0.25 mm *b*; **Fr** unknown; **Ecol** unknown; **Distr** Victoria West, NC, S Africa.

Note: The possession of few teeth on the keel near the apex of the leaf may be typical of this species.

S. pluridens L.Bolus NM 1937: 64 **T** van Heerde s.n., cited as 21837, marked 22127 (BOL!).

Compact; **L** erect with a rounded lower leaf surface ending in a chin, margins parallel for most of their length, with 3–6 big, broad-based, long teeth, keel smooth or only a little undulate as if forming bases of teeth, epidermis with distant, raised, rather sharp elevations, thus white margaretiferous, **L** 32–42 mm *l*, 7–9 mm *b* apically, 6–9 mm *t*; **Fl** tube 3 mm *l*, 6–10 mm *l*, 0.25–0.75 mm *b*; **Fr** ∅ 5.5–6.5 mm; **S** 0.65–0.68 mm *l*, 0.53–0.59 mm *b*; **Ecol** on slopes with big, round stones; **Distr** Calvinia, Namaqualand, NC, S Africa.

Note: This species differs from the nearby **S. alboroseum** in the erect leaves with prominent, big teeth on the margins and from species with similar leaf shapes occurring further in the east by the smaller seeds.

S. pyrodorum = **S. mustelinum**

S. resedolens L.Bolus NM 1937: 64 **T** Logan 28 (BOL!).

L flattish, short and broad, margins with 3–4 stout teeth composed of a well developed base and a pointed tip, keel smooth or 1–3 tiny teeth, **L** 20–25 mm *l*, to 2 mm *b*, 6 mm *t*; **Fl** tube to 2.5 mm *l*, petals yellow, 12 mm *l*, 0.25–0.5 mm *b*, nectary glands 1/3 up the tube; **Fr** unknown; **Ecol** unknown; **Distr** Sutherland, NC, S Africa.

Note: The species appears most similar to **S. difforme** from a little further south.

S. ronaldii L.Bolus NM 1932: 310 **T** James 6 (BOL!).

L upper leaf surface acute to the tip from parallel margins, these with pointed elevations only, not raised on a broader base, keel smooth, **L** 7–15 mm *l*, 4–6 mm *b*, 4–6 mm *t*; **Fl** tube 2 mm *l*, petals filamentous, 10 mm *l*; **Fr** unknown; **Ecol** on flat rocks on summit of mountain; **Distr** Cradock, EC, S Africa.

Note: In the tiny teeth along the margins, the species is very similar to **S. deficiens**, **S. integrum**, and **S. ryderae**.

S. rouxii L.Bolus NM 1936: 17 **T** Roux 156 (BOL!).

L trigonous, epidermis somewhat rough, margins with 3–4 teeth consisting of a broad base and a pointed elevation at the tip, keel with few pointed elevations, **L** 20–26 mm *l*, 9–12 mm *b*, 5–7 mm *t*; **Fl** tube 1–1.5 mm *l*, petals yellow, 9 mm *l*, 0.5 mm *b*; **Fr** unknown; **Ecol** unknown; **Distr** Carnarvon, Victoria West, NC, S Africa.

Note: The leaf surfaces are reported to become smoother in cultivation.

S. ryderae L.Bolus NM 1936: 18 **T** Ryder 21645 (BOL!).

L with parallel margins, acute to the tip, margins with 2–3 teeth with flat bases and a pointed elevation on top, often only the elevation present, keel smooth, **L** to 24 mm *l*, 7–11 mm *b*, 5–8 mm *t*; **Fl** tube 2.5 mm *l*, petals 8–13 mm *l*, 0.25–0.3 mm *b*; **Fr** base funnel-shaped, top raised in the centre, septa with very narrow rims, placenta elevated to 1/2 the height of the locule by a false septum, valve wings almost semilunate, c. double the width of the expanding keels; **Ecol** unknown; **Distr** Cradock, EC, S Africa.

Note: The tiny teeth on the leaf margins resemble those of **S. deficiens**, **S. integrum**, and **S. ronaldii**, all from the same wider area.

S. suaveolens (Schwantes) Schwantes ZSK 1926: 254 (see **photo**) ≡ *M. suaveolens* Schwantes ZSK 1926: 176 non L.Bolus 1958: 297 nomen illeg. ≡ *S. suaveolens* Schwantes ZSK 1926: 176 nomen nudum non L.Bolus 1958: 297 **T** Marloth s.n. (HBG!).

Compact, flat to semi-globose in habitat; **L** spatulate with a thick, prominent chin at the apex, the leaf tips almost semi-lunate as seen from above, the keel hardly visible in habitat, more prominent in cultivation, teeth hardly visible, sometimes as confluent elevations on margins and keel, more so in cultivation, elevations numerous and dense, giving the leaf a rough appearance, epidermis with a thick layer of crystal sand, stomata only very little sunken, **L** 10–20 mm *l*, 9–15 mm *b*, 7–9 mm *t*; **Fl** tube 5 mm *l*, petals yellow, anthers yellowish; **Fr** base bowl-shaped, breaking off the stalk when ripe, ochre with red dots, top rising a little to the centre, septa with extremely narrow rims, rising to the centre, valve wings as broad as the expanding keels, narrowing to their bases, 5–5.3 mm ∅; **S** 0.75–0.85 mm *l*, 0.6–0.72 mm *b*; **Ecol** in shallow pans and along their edges; **Distr** Fraserburg, Sutherland, Williston, NC, S Africa.

Note: When Schwantes described the genus, he included this species but gave no description. This was only published later in the year, all older names are therefore invalid.

S. suricatinum L.Bolus NM 1931: 242 **T** Luckhoff 1886/30 (BOL!).

L almost erect, trigonous, sides convex, epidermis moderately rough, margins with 4–6 small teeth consisting of pointed elevations only, keel smooth, **L** 25–30 mm *l*, 7–8 mm *b*, 6 mm *t*; **Fl** tube 3 mm *l*, petals lemon-coloured, to 10 mm *l*, 0.5 mm *b*; **Fr** unknown; **Ecol** on a plateau between foot hills; **Distr** Laingsburg, WC, S Africa.

Note: The leaves of this species keep their almost erect position for most of their lifetime.

S. trifarium L.Bolus NM 1929: 90; Bolus NM 1934: 437 (see **photo**) **T** Hutchinson s.n. (BOL!).

L trigonous, margins and keel with teeth, those on the margins stout with a very broad base and ending in pointed elevations, those on the keel with a broad but lower base, also ending in pointed elevations, but lower on the leaf the teeth only consisting of pointed elevations, **L** 15–25 mm *l*, 6–9 mm *b*, 8 mm *t*; **Fl** tube 2–3 mm *l*, petals 6–12 mm *l*, to 0.75 mm *b*, stamens golden; **Fr** unknown; **Ecol** unknown; **Distr** Fraserburg, NC, S Africa.

Note: Very similar to **S. murinum**, differing mainly in the shorter leaves whereas width and thickness are ± the same.

S. villetii L.Bolus NM 1938: 129 **T** Villet 21984 (BOL!).

L trigonous, sides convex, margins on young leaves with teeth with broad bases ending in pointed elevations, on older leaves pointed elevations only, keel smooth or with few somewhat pointed elevations, 21–25 mm *l*, 5–7 mm *b*, 5–6 mm *t*; **Fl** tube 2 mm *l*, petals bright yellow, apically red, 9 mm *l*, 0.25 mm *b*, anthers and pollen golden, nectary glands reaching c. 1/3 up the tube; **Fr** unknown; **Ecol** unknown; **Distr** Beaufort West, WC, S Africa.

Note: Data taken from type and description; from these, the species appears very similar to **S. viride** from the same area, which has longer and thicker leaves.

S. viride L.Bolus NM 1936: 39 **T** Dyer 550/32 (BOL!).

L trigonous, sides convex, margins with 2–4 teeth composed of rather broad, low bases and pointed elevations on top, keel smooth on type material in contrast to the 1–3 teeth recorded, 25–35 mm *l*, 11–14 mm *b*; **Fl** tube 3 mm *l*, petals bright yellow, to 13 mm *l*, 0.5–0.75 mm *b*; **Fr** unknown; **Ecol** unknown; **Distr** Beaufort West, WC, S Africa.

Note: The species differs from **S. villetii** only in its longer and fatter leaves. In leaf size, the species is similar to **S. beaufortense**, which has a shorter tube in the flower.

Synaptophyllum MESEMBRYANTHEMOIDEAE

M. GERBAULET

Synaptophyllum N.E.Br. GC 1925: 412; Bittrich 1987: 5–116 **Lectotypus S. juttae** (Dinter & A.Berger) N.E.Br. **Etym** Gk *synaptos*, gnarled, *phyllon*, leaf.

Prostrate to decumbent annual herbs branching from the base, with fibrous roots, **I** herbaceous, with additional vascular bundles in the cortex, epidermis smooth; **L** flat, roundish or oval, decussate, those of a pair connate into a disc or a cone, upper ones much smaller and free, without enlarged central water-storing cells, epidermis smooth or with small mesomorphic bladder cells at the margins, cuticular folds in a net-like structure; **Fl** in cymes, Ø to 15 mm, **K** 4, connate into a short tube, petals white, basally free, filamentous staminodes absent, nectaries narrow; **Fr** with valve wings, 4 locules; **S** whitish, testa smooth, without a distinct crest; **Ecol** in the fog zone of the southern Namib; **Distr** Lüderitz-South, Namibia.

Note: Only one species is known. On account of absent filamentous staminodes, free petals and small whitish seeds with a smooth testa, this annual is most similar to **Mesembryanthemum**.

Distribution of **Synaptophyllum** N.E.Br.

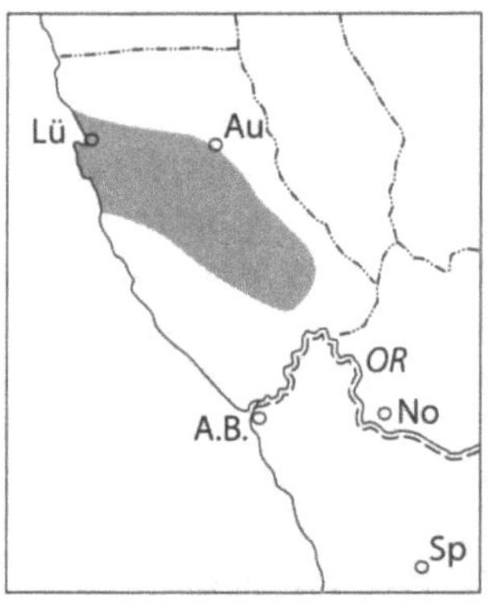

S. juttae (Dinter & A.Berger) N.E.Br. GC 1928: 254 (see **photo**) ≡ *M. juttae* Dinter & A.Berger 1914: 589 **Lectotype**, here designated: Dinter 1016 (BOL!).

Plants often fresh-purple or salmon-red; **Fl** Ø 7–15 mm; **S** to 0.6 mm *l*; **Ecol** as for genus; **Distr** as for genus.

Note: The tube-shaped connate leaves probably aid the collection of dew droplets, which, according to Bittrich (1987), can effectively be absorbed by the leaves.

S. sladenianum ≡ **Prenia sladeniana**

Tanquana Ruschioideae

H. E. K. Hartmann

Tanquana H.E.K.Hartmann & Liede 1986: 479; 1987: 117–125 **Typus T. archeri** (L.Bolus) H.E.K.Hartmann & Liede **Etym** *Tanqua Karroo*, vegetation type in the Ceres Karroo where the plants occur mainly.

Distribution of **Tanquana** H.E.K.Hartmann & Liede

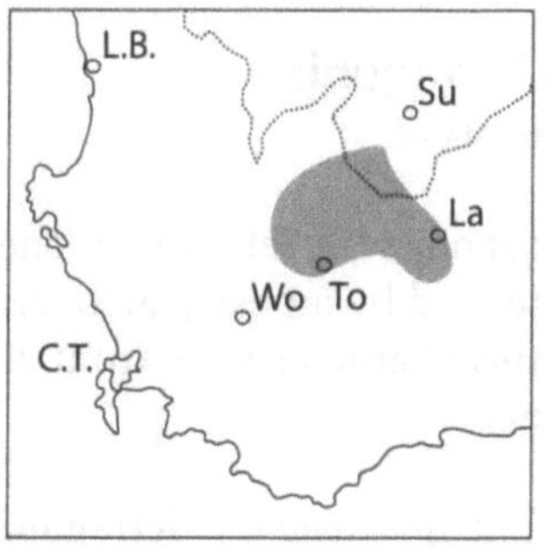

Compact, branched to unbranched, above or in the ground, 1 leaf-pair persisting during the resting period, anisophyllous; **L** thick with rounded tips, punctate through idioblasts visible underneath the epidermis lacking crystals; **Fl** 1, bracteolate, **K** 4–5, 30–70 yellow petals, stamens basally papillate, flowering midday; **Fr** light brown, base funnel-shaped, ∅ < 10 mm, of the Titanopsis-type with a small endocarpal closing body, 9–10 locules; **S** testa cells somewhat raised, light brown, < 1 mm *l*; **Ecol** on Dwyka and Ecca shales, near the SE border of the winter rainfall region, ≤ 200 mm rainfall p.a.; **Distr** Ceres, Ladismith, Laingsburg, WC, Africa.

Note: Formally included in the genus **Pleiospilos** because of the dotted leaves, separated on the base of differences found in all other characters.

Key to the Species of Tanquana

1. Unbranched .... **T. hilmarii**
- Branched **2**
2. Leaves almost isophyllous, ≤ 25 mm long, sunken in the ground .... **T. archeri**
- Leaves anisophyllous, > 25 mm long, cushions above ground .... **T. prismatica**

T. archeri (L.Bolus) H.E.K.Hartmann & Liede 1986: 480 (see **photo**) ≡ *Pleiospilos archeri* L.Bolus NM 1931: 276 T Archer 342 (BOL!) = *Pleiospilos clavatus* L.Bolus NM 1934: 454 T Dyer 537/32 (BOL!) = *Pleiospilos loganii* L.Bolus NM 1934: 453 T Logan 1871/33 (BOL!).

Sunken in the ground, c. 10 branches, isophyllous or only a little anisophyllous; **L** 20–25 mm *l*; **Ecol** on NE slopes of little decomposed Ecca shales, rainfall mainly in March and November; **Distr** Laingsburg, WC, S Africa.

Note: Restricted to shaly slopes.

T. hilmarii (L.Bolus) H.E.K.Hartmann & Liede 1986: 481 (see **photo**) ≡ *Pleiospilos hilmarii* L.Bolus NM 1931: 253 T Luckhoff 19494 (BOL!).

Sunken in the ground, unbranched, isophyllous; **L** only the tips visible, < 20 mm *l*; **Ecol** in crevices of shales of the Witteberg formation; **Distr** Ladismith, WC, S Africa.

Note: In shale bands.

T. prismatica (Schwantes) H.E.K.Hartmann & Liede 1986: 481 (see **photo**) ≡ *M. prismaticum* Marloth ex Schwantes 1919: 45[30] ≡ *Pleiospilos prismaticus* (Schwantes) Schwantes MDG 1928: 46 **LT** Schwantes 1919: f on p 43 = *Pleiospilos roodiae* (N.E.Br.) Schwantes ZSK 1927: 23 ≡ *Punctillaria roodiae* N.E.Br. GC 1926: 229 T Rood s.n.

Above ground, to 30 branches, the new leaf-pairs developing peripherally, fruits borne in the centre of the plant, distinctly anisophyllous; **L** 25–40 mm *l*; **Ecol** in flats or on slopes of Dwyka and Ecca shales, salt tolerant, winter rainfall c. 200 mm p.a.; **Distr** Ceres, WC, S Africa.

Note: Turning bright orange, red or purple in sunlight.

Tephras = Galenia

H. E. K. Hartmann

The name *Tephras* has been used by E. Meyer on several sheets he identified, without ever having been published. It was taken up by Sonder in Harvey & Sonder 1862: 475 as *T. papulosa* as a synonym only. Hence, the name of the genus is a nomen nudum. Later, Adamson (1956: 104) cited a sheet named *T. pubescens* by E. Meyer, also a nomen nudum.

T. papulosa = **Galenia pruinosa**
T. pubescens = **Galenia affinis**

[30] Schwantes, assuming that Marloth had published a description for the species he was distributing, did not give a formal description himself. His remarks, in particular since they are accompanied by Marloth's photograph, can be taken as a description.

Tetracoilanthus ≡ Aptenia

M. Gerbaulet

The name *Tetracoilanthus* is illegitimate since Rappa & Camarrone (1953: 34) based it on the same type as **Aptenia**, published by N. E. Brown (GC 1925: 412) nearly 30 years earlier.

T. anatomicus = **Sceletium emarcidum**
T. concavus = **Sceletium tortuosum**
T. cordifolius ≡ **Aptenia cordifolia**

Tetragonella = Tetragonia

H. E. K. Hartmann

The genus *Tetragonella* Miquel (in Lehmann 1845: 245) was described in contrast to **Tetragonia** and **Galenia**, differences being the stigmas and free ovary in the first case, and the position of the stamens in the second. The distinction against **Tetragonia** has soon been considered to be insufficient (e.g. Hooker 1856: 148), and *Tetragonella* has been included as a synonym in **Tetragonia**. The single species is characterized by a nearly globose, fleshy fruit, differing in this character state from all other species of **Tetragonia.**

T. amplexicoma ≡ **Tetragonia implexicoma**
T. implexicoma ≡ **Tetragonia implexicoma**

Tetragonia Tetragonioideae

H. E. K. Hartmann

Tetragonia L. 1753: 480; Adamson 1955: 109–154; Prescott 1984: 37–42; Taylor 1994: 575–589 ≡ *Ludolfia* Adanson 1763: 244 nomen illeg. ≡ *Tetragonocarpus* Commelijn 1701: 203 nomen invalid. ≡ *Tetragonocarpus* Miller 1754: no page ≡ *T.* subg. *Tetragonocarpus* Reichenbach 1828: 166 **Typus T. fruticosa** L. ≡ *Tetragonocarpus fruticosus* Commelijn 1701: t. 102 (for *Tetragonocarpus*) **Etym** Gk *tetra*, four, *gone*, reproductive organs, to describe the four-angled or four-winged fruits present in many species = *Anisostigma* Schinz 1897: 78 ≡ *T.* § *Anisostigma* (Schinz) Engler 1915: 148 ≡ *T.* subg. *Anisostigma* (Schinz) Adamson 1955: 146 **Typus** *Anisostigma schenckii* (Schinz) Schinz ≡ **T. schenckii** Schinz = *Demidovia* Pallas 1781: 150 **Typus** *Demidovia tetragonioides* Pallas ≡ **T. tetragonioides** (Pallas) O.Kuntze = *Tetragonella* Miquel in Lehmann 1845: 245 **Typus** *Tetragonella amplexicoma* Miquel ≡ **T. implexicoma** (Miquel) Hook. f.

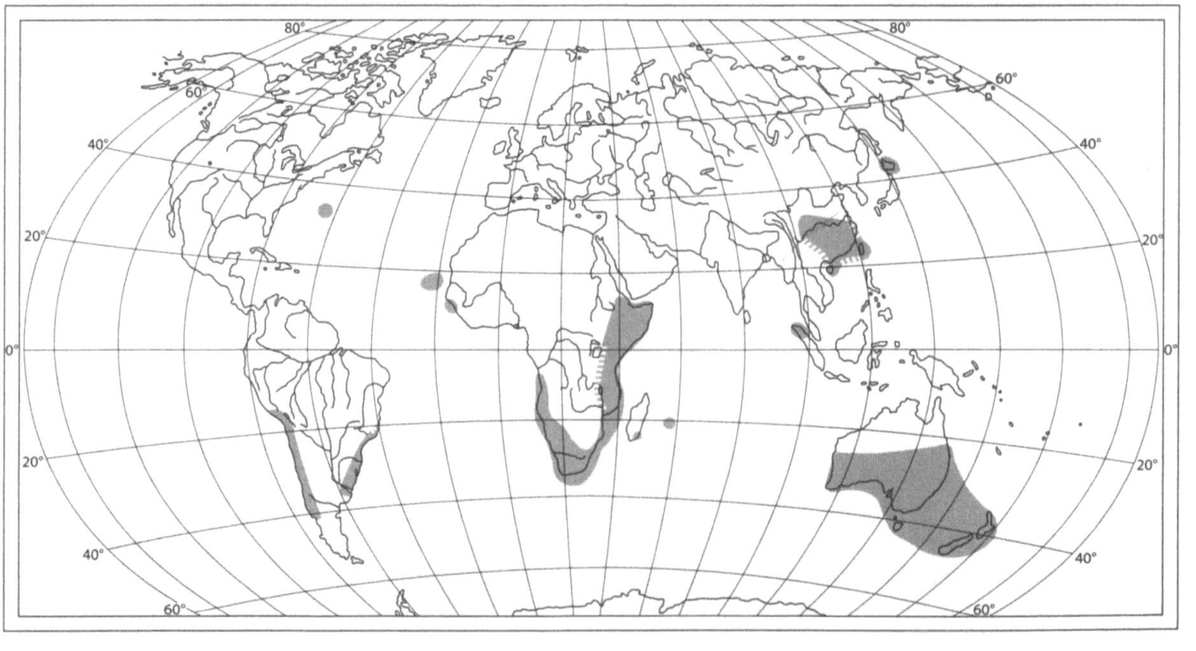

Distribution of **Tetragonia** L.

Annuals to perennials with herbaceous or woody branches, I mostly covered by papillae to different degrees when young, often with a decurrent ridge or line from the base of the leaf above; **L** basally often opposite, but in most of the plants alternate, petioles short to long, blades ovate to almost linear, tips often obtuse, epidermis with papillae of various shapes, often two different types, one of them elongate and hairy; **Fl** solitary, in small to bigger cymes, or in well developed, elongated cymes in which each flower bears a bract, these latter **Inf** often appearing as spikes or raceme-like in shape, cymes often with a peduncle from which 1 to several pedicels arise as fascicles, if a peduncle and a pedicel occur in a solitary flower, the latter also named "stipe" for South American species when it adheres to the fruit after ripening, perigone mostly 4-lobed, ovary inferior, stamens mostly twice the number of the perigone lobes; **Fr** a woody nut of several locules in which the seeds lie above one another, winged, ridged, or tuberculate, the ornaments basically in 4 rows, additional, often different, protrusions between them, most often papillate when young, glabrous with age, brown to black, the pedicel often dispersed with the fruit; **S** pear-shaped, enclosed in the fruit and not dispersed separately; **Ecol** many species in sand and near the sea, but others in rocky terrain, in dry climate; **Distr** in tropical climates worldwide, in the southern hemisphere also in drier climates.

Note: Various authors have subdivided the genus into subgenera, sections, and series, one of the earlier being Haworth (1803: 113–123). He distinguished two sections: *T.* § *Sessiles* Haw. 1803: 119, with sessile flowers, and *T.* § *Pedunculatae* Haw. 1803: 119, with always "pedunculate" flowers. Some time later, Reichenbach (1828: 166) established two ranks below genus, which can be understood as subgenera: under **Tetragonia** "a. *Tetragonioides* DC, b. *Tetragonocarpus* Comm." were given without further descriptions. The latter name cannot, as is often done, be taken as the description of a new genus, but only as a subgenus (Adamson 1955: 111), based on *Tetragonocarpus* Commelijn (1701: 203–206), which was published before the starting date for the nomenclature of the group. As indicated by Druce (1914: 435), *Tetragonocarpus* Miller (1754: no page) is the oldest validation of that name at genus rank. In the latest treatment for southern Africa, Adamson (1955: 109–154) subdivided the genus into four subgenera, one of which is treated here as the separate genus **Tribulocarpus**.

- T. subg. **Tetragonia** = *T.* subg. *Tetragonocarpus* Reichenbach 1828: 166 **Typus T. fruticosa** L.
- series **Fruticulosae** Fenzl 1862: 461 (as *Fruticosae* in Adamson 1955: 112) ≡ *T.* § *Pedunculatae* Haw. 1803: 119 **Type T. fruticosa** L.
  Species included by Adamson:
  T. arbuscula, T. arbusculoides, T. calycina, T. fruticosa, T. glauca, T. macrocarpa, T. rangeana, T. reduplicata, T. spicata, T. virgata.
- series **Robustae** Adamson 1955: 123 **Type T. robusta** Fenzl.
  Species included by Adamson:
  T. distorta, T. robusta, T. sarcophylla.
- series **Salignae** Adamson 1955: 127 ≡ § *Haplogonia* Fenzl 1862: 461 **Type T. saligna** Fenzl.
  Species included by Adamson:
  T. erecta, T. haworthii, T. lasiantha, T. saligna.
- series **Decumbentes** Adamson 1955: 130 **Type T. decumbens** Miller.
  Species included by Adamson:
  T. decumbens, T. namaquensis, T. verrucosa.
- series **Macranthae** Fenzl 1862: 461 **Type T. hirsuta** L.f.
  Species included by Adamson:
  T. hirsuta, T. rosea.
- series **Pedunculares** Fenzl 1862: 460 **Typus T. herbacea** L.
  Species included by Adamson:
  T. halimoides, T. herbacea, T. nigrescens, T. portulacoides, T. sphaerocarpa.
- series **Chenopodinae** Fenzl 1862: 460 **Typus T. chenopodioides** Ecklon & Zeyher.
  Species included by Adamson:
  T. chenopodioides, T. galenioides, T. pillansii.

- T. subg. **Tetragonioides** DC. 1828: 452 ≡ *Demidovia* Pallas 1781: 150 ≡ *T.* subg. *Tetragoniodes* Reichenbach 1828: 166 ≡ *T.* § *Sessiles* Haw. 1803: 119 **Typus T. tetragonioides** (Pallas) O.Kuntze.
  Species included by Adamson:
  T. acanthocarpa, T. caesia, T. echinata, T. microptera, T. tetragonioides.

- T. subg. **Anisostigma** (Schinz) Adamson 1955: 148 ≡ *Anisostigma* Schinz 1897: 78 ≡ *T.* § *Anisostigma* (Schinz) Engler 1915: 148 **Typus T. schenckii** Schinz.
  Species included by Adamson:
  T. schenckii.

T. subg. *Tetragonioides* ≡ T. subg. **Tetragonia**
T. subg. *Tetragonocarpus* = T. subg. **Tetragonia**
T. subg. *Tribulocarpus* ≡ **Tribulocarpus**
T. § *Pedunculatae*
≡ T. subg. **Tetragonia** series **Pedunculatae**
T. § *Sessiles* ≡ T. subg. **Tetragonioides**

T. acanthocarpa Adamson 1955: 146 T Gill 206 (BOL).

Herbs with prostrate branches 10–25 cm *l*; **L** petioles 4–7 mm *l*, blades elliptical or rhombic, papillate, 5–10 mm *l*, 4–7 mm *b*; **Fl** solitary, pedicels 1–3 mm *l*, perigone 4 lobes, stamens 4, alternate with the perigone lobes, stigmas 3, rarely 2 or 4; **Fr** pedicel 6 mm *l*, covered with rigid spines, not ridged, truncate at base, papillate all over, 4–5 mm *l*, 3 mm ∅; **Ecol** on open ground; **Distr** Middelburg, Somerset East, EC, Fauresmith, FS, Hanover, NC, S Africa.

Note: The species is also recorded from Nakuru, Kenya, as having been introduced to there.

T. ameghinoi Spegazzini 1899: 52
≡ **Halophytum ameghinoi** (Spegazzini) Spegazzini 1902: 153 **T** Ameghino s.n. (?)

T. angustifolia Barnéoud in Gay 1847: 470 (see **photo**) **T** Gay s.n. (P).

Shrubs to 1 m *h*, branches erect or as erect side branches from prostrate ones, **I** glabrous, sometimes papillate when young; **L** alternate, sessile, petioles with narrow wings, blades elliptic or oblanceolate, truncate; **Fl** dioecious, male flowers 1–3 per axil in fasciculate cymes, stalks 1.5–9 mm *l*, base of perigone 4-angled or 4-winged, wings 0.1–0.3 mm *b*, perigone lobes 4, 2–4 mm *l*, yellow-green, stamens 8–15, vestigial stigmas 2–5, in females flowers 2–5 stigmas; **Fr** subsessile, sharply 4-angled or 4-winged, smooth, sides sometimes somewhat concave, glabrous, perigone persisting, 5 mm *l*, 4 mm ∅; **Ecol** in sand, gravel or cracks in rocks, 50–1 200 m altitude; **Distr** N Chile.

Note: This is one of the dioecious shrubby species of the genus in Chile.

T. arbuscula Fenzl in Harvey & Sonder 1862: 467 **LT** Drege s.n. (S) = *T. arbuscula* Fenzl var. *latifolia* Fenzl 1862: 467 **T** Drege 7048 (W) = *T. arbuscula* Fenzl var. *linearis* Fenzl 1862: 467 **T**? = *T. karasmontana* Dinter nomen nudum **T** Dinter 4922 (?).

Woody shrubs to 40 cm *h*, branches erect with ascending side branches, **I** pale, with raised decurrent lines from the leaf bases; **L** petioles 2–3 mm *l*, blades linear to lanceolate, obtuse tip, 5–12 mm *l*, 1–3 mm *b*, some broader and longer; **Fl** 1–3, pedicels as long as or longer than the bracts, perigone lobes bright yellow inside, stamens many, anthers bright yellow, stigmas 1–3, rarely 4–5; **Fr** pendulous, wings thin, smooth, fruit rounded at the base, cordate at the top, 6–15 mm *l* and ∅; **Ecol** inland; **Distr** Namibia; from Gordonia, NC, to FS and Tarka, EC, S Africa.

Note: From **T. spicata**, with which it has been confused, it differs mainly in the pendulous fruits.

T. arbuscula var. *latifolia* = **T. arbuscula**
T. arbuscula var. *linearis* = **T. arbuscula**

T. arbusculoides Engler 1913: 503 (see **photo**) **T** Dinter 200 (B).

Shrubs with woody stems, 35–50 cm *h*, **I** thickly covered with vesicular papillae, later yellowish-brown; **L** petioles 3–6 mm *l*, blades obovate, acute, papillate on both sides, 13–18 mm *l*, 4–7 mm *b*; **Fl** in small cymes, petioles 1–1.5 mm *l*, perigone with vesicular papillae outside, the 3 perigone lobes yellow inside, stamens c. 12, 3 stigmas; **Fr** with 3–4 wings 10 mm *l* and 6 mm *b*, ridges between the wings; **Ecol** in dry, sandy places; **Distr** S Angola; Namibia; Namaqualand to Prieska, NC, S Africa.

Note: The species differs from **T. arbuscula** by its obovate leaves, and from **T. macroptera** in the broader, more prominently papillate leaves.

T. borealis Battandier & Trabut 1917: 226 **T** Ducellier (P?).

Shrubs with herbaceous branches, basally woody, **I** greyish-green from hyaline, minute papillae; **L** alternate, blades narrowing gradually into the petiole, ovate-hastate, apically attenuate, margins recurved, 50–60 mm *l*, up to 25 mm *b*; **Fl** solitary or in small cymes, pedicels short, perigone lobes 5, obtuse, stamens c. 25 in 5 groups alternating with the perigone lobes, stigmas 3–4; **Fr** hard, thin and brittle, without wings, 3–4 locules, each with one seed; **S** subreniform; **Ecol** unknown; **Distr** Casablanca, Morocco.

Note: Maire (1962: 288) placed the name as a synonym with **T. tetragonioides**, but the described greyish-green colour of the leaves due to papillae casts some doubt on this alignment.

T. caesia Adamson 1955: 147 **T** Adamson 4569 (BOL).

Herbs with prostrate branches to 40 cm *l*, **I** covered with glistening papillae and some transparent scales; **L** petioles as long as the blades, these latter dark green when fresh, grey-green when dry, papillate, deltoid, rhombic or ovate, 5–30 mm *l*, 5–20 mm *b*; **Fl** mostly solitary, pedicels 1–6 mm *l*, perigone lobes 4, densely papillate outside, with a tuft of scales at the tip, stamens 4–6, mostly 5, the ovary projecting as a cone c. half as high as the perigone lobes are long, stigmas 2; **Fr** barrel-shaped, 3–4 obtuse ridges, 3 mm *l*, 2.5 mm ∅; **Ecol** open sand moist in spring; **Distr** Mitchells Plain, Wynberg, WC, S Africa.

Note: Few stamens are a rare feature in the genus.

T. calycina Fenzl in Harvey & Sonder 1862: 466 **T** Drege 7059 (W).

Shrubs to 50 cm *h*, **I** papillate and with decurrent ridges when young, later terete, pale, almost smooth; **L** petioles distinct but short, blades lanceolate, folded or revolute, lower leaf surface papillate, upper less so, 20–25 mm *l*, 2–6 mm *b*; **Fl** solitary, pedicels 10–20 mm *l*, perigone lobes 4, papillate outside, stamens c. 15; **Fr** 4-winged, apically notched, wings firm and with distinct intermediate ridges, 16–25 mm *l* and ∅; **Ecol** among rocks; **Distr** Namibia; Namaqualand, NC, S Africa.

Note: Friedrich (1967: 4) stressed the similarity of this species with **T. arbuscula** und **T. spicata**.

T. chenopodiodes Ecklon & Zeyher 1837: 324 **LT** Ecklon & Zeyher 2118 (B?) = *T. macrostylis* Schlechter 1899: 124 ≡ *T. haplostylis* Schlechter ex Schumann 1901: 475 orthographic variant T Schlechter 8529 (B).

Herbaceous branches flaccid, i.e. not really holding themselves up, sprawling, lateral short shoots erect, **I** papillate; **L** petiolate, blades elliptical to oblong, narrowing to both ends, 20–55 mm *l*, 7–17 mm *b*; **Fl** 2–5 per cyme, sessile or on very short pedicels, perigone papillate, stamens numerous, yellow or red, stigmas 3–5, red; **Fr** with 8 distinct, projecting ridges, but not winged, truncate above, usually papillate, 6 mm *l* and ∅; **Ecol** open sand on coastal plains; **Distr** Clanwilliam, Hopefield, Malmesbury, WC, S Africa.

Note: Ecklon & Zeyher (1837: 324) had two different plants from two localities included in their number 2118. Fenzl (1862: 463) separated one as **T. portulacoides**, choosing this type for the original name of the species.

T. chisimajensis = **Trianthema portulacastrum**

T. copiapina R.A.Philippi 1893: 169 T F.Philippi 48431 (SGO).

Annual herbs to 10 cm *h*, **I** glabrous to densely papillate with crinkled hairs to 1.5 mm *l*; **L** alternate, sessile, linear to oblanceolate, 3–50 mm *l*, 0.5–7 mm *b*; **Fl** hermaphroditic, solitary, pedicels 1–5 mm *l*, ebracteolate, perigone base terete, lobes 4, 3–4.2 mm *l*, yellow-green outside, yellow inside, stamens 15–20; **Fr** ovoid to oblong, pedicels 8–20 mm *l*, the 4 ridges with some rather low, bulging tubercles, glabrous to densely papillate, tepals persisting; **Ecol** in sand or gravel, at 0–900 m altitude; **Distr** N Chile.

Note: The fruit of this species resembles that of **T. macrocarpa** in the bulbous ornamentation of the four lateral ridges, differing in the slender shape and the persisting perigone lobes.

T. cornuta = **T. tetragonioides**

T. coronata Rye & Trudgen 1996: 459 T Trudgen 8000 (PERTH).

Annual herbs with sometimes an erect central stem to 20 cm *h* and decumbent branches, vegetative parts with simple, spreading papillae, stems changing from rather densely hairy at first to sparsely hairy later; **L** with 0–25 mm *l* petioles, blades ovate to broadly ovate, acute or obtuse, 30–100 mm *l*, 15–50 mm *b*, papillae on main veins mainly; **Fl** solitary, pedicels 2–5 mm *l*, (4–)5 tepals yellow inside, persisting, densely hairy outside; **Fr** conspicuously hairy, ovoid and horned, 14–23 mm *l*, 10–15 mm ∅; **S** 4 mm *l*, 1.5 mm *b*, pale brown; **Ecol** in reddish sandy soils or loam, sometimes near calcrete; **Distr** Western Australia, Australia.

Note: The fruit is reported to be larger than that of **T. tetragonioides** and have mostly five persisting tepals; **T. coronata** is similar to **T. cristata** (Rye & Trudgen 1996: 460–462) with wings in place of horns on top of the fruit.

T. cristata Prescott 1983: 181 (see **photo**) T Gardner 244467 (PERTH).

Annual herbs with prostrate branches to 60 cm *l*, plants crystalline on all parts, **I** glabrous or with few vesicular hairs when young; **L** alternate, petioles 5–20 mm *l*, blades ovate to rhomboid, acute, densely papillate, 5–50 mm *l*, 3–25 mm *b*; **Fl** solitary, sessile, perigone lobes 4, papillate outside, yellow and glabrous inside, c. 3 mm *l*, stamens 12–30, ovary top-shaped, densely papillate, stigmas 4–8; **Fr** top-shaped with 4 wings extending above the fruit, narrowing from the middle, with convoluted ridges between the wings, densely papillate, dry bony or woody, 8–15 mm *l*, 8–12 mm *b*, purplish when fresh; **S** several, separated by wooden parts of the wall, pear-shaped, pale brown, smooth, c. 3 mm *l*; **Ecol** red sand to granite outcrops; **Distr** Western Australia, Australia.

Note: Vegetatively, the species is similar to **T. tetragonioides** and **T. eremea**; the ornamented fruit is typical.

T. crystallina L'Heritier 1784: 81 **LT** L'Heritier 1784: t. 39.

Annual herbs to 25 cm *h*, **I** with crystalline, papillate hairs, later glabrous; **L** alternate, lower ones opposite, sessile, ovate, tapering to the base, more distinctly so in the upper part of the plant, epidermis crystalline-papillate, 10–90 mm *l*, 3–60 mm *b*; **Fl** solitary, subsessile, ebracteolate, perigone lobes 4, 2.5–3 mm *l*, green and papillate outside, yellow inside, stamens 8–12; **Fr** peduncles 3–8 mm *l*, ovate, perigone lobes persisting, connivent and pointed, 6–9 mm *l*, 4–6 mm ∅; **Ecol** near the coast, 100–2 000 m altitude; **Distr** S Peru.

Note: The combination of large leaves and small fruits characterizes this species.

T. cynocrambe = **T. microptera**

T. decumbens Miller 1768: no. 2 (see **photo**) **LT** Miller 1760: t. 263 = *T. obovata* Haw. 1821: 73 T? = *Aizoon perfoliatum* L.f. 1781: 261 ≡ *T. perfoliata* (L.f.) Druce 1917: 650 T Thunberg 12055 (UPS!) = *T. tetrapteris* Haw. 1803: 121 T? = *T. zeyheri* Fenzl in Harvey & Sonder 1862: 465 T Ecklon & Zeyher 2125 (?).

Shrubs with prostrate branches, **I** covered with glistening papillae and some hairs, woody when old, with narrow decurrent lines, 3–5 mm ∅; **L** blades narrowing to the base, but not petiolate, blades ovate to nearly orbicular, covered with glistening papillae and some hairs, 10–60 mm *l*, 5–30 mm *b*; **Fl** 3–5 per cyme, pedicels 3–10 mm *l*, densely papillate, perigone papillate like the leaves, 4 lobes, stamens many; **Fr** 4-winged with lower ridges between them, flat on top, 8–15 mm ∅; **Ecol** most common on the sea shore; **Distr** from Namaqualand, NC, to Cape Town, WC, and to Transkei, EC, S Africa.

Note: The description and the synonymy follow Adamson (1955: 130–132) with some reservation. The type of *Aizoon perfoliatum* seen at UPS possesses opposite leaves, a feature not agreeable with **T. decumbens**. The original description of Miller gave only ovate leaves and a woody, trailing stem. Heyligers (1999) reported the occurrence of this species as an adventitious one in Western Australia and New South Wales, where it had been misidentified as **T. nigrescens**. The capsules are said to be buoyant in seawater.

T. diptera F.Mueller 1878: 8 (see **photo**) **T** F.Mueller s.n. (isotype MEL).

Herbs with prostrate to semi-erect branches, **I** papillate when young, later glabrous; **L** petioles to 10 mm *l*, blades lanceolate to rhomboid; **Fl** solitary, rarely paired, sessile, perigone lobes 4, c. 1 mm *l*, lanceolate, stamens 4, alternating with the perigone lobes; **Fr** narrowing to the base, heart-shaped in side view, 2 lateral wings extending over the fruit and to the base, 5–8 mm *l*, 5–10 mm ∅, 4 seeds per fruit; **S** pear-shaped, honey coloured, c. 1.5 mm *l*; **Ecol** near the sea; **Distr** Western Australia, Australia.

Note: Differing from the similar **T. tetragonioides** in the number of only four stamens.

T. distorta Fenzl in Harvey & Sonder 1862: 467 **T** Drege 7044 (W).

Rather compact shrubs, branches often divaricate, **I** pale, strongly papillate when young, almost 4-angled from decurrent ridges; **L** subsessile, oblanceolate, revolute, papillate, 10–60 mm *l*, 2–4 mm *b*; **Fl** solitary, pedicels 5–8 mm *l*, perigone densely papillate, lobes cucullate, stigmas 1–3; **Fr** 4-winged, with a small tubercle or ridge between the wings, 10–15 mm *l* and ∅; **Ecol** unknown; **Distr** Namaqualand, NC, Vanrhynsdorp, WC, S Africa.

Note: According to Adamson (1955: 126) the species is very similar to **T. sarcophylla.**

T. echinata Aiton 1789: 177 (see **photo**) **T** Masson s.n. (BM).

Annuals with ± prostrate branches to 40 cm *l*, **I** papillate, with raised decurrent lines from the leaf base; **L** petioles 3–8 mm *l*, blades ovate, rhombic or deltoid, occasionally almost orbicular, 10–25 mm *l*, 8–17.5 mm *b*; **Fl** 2–4 per cyme, pedicels very short, perigone 2 mm *l*, densely papillate, 3–5 lobes, stamens 3–5, stigmas usually 3; **Fr** papillate, 3–5 projecting ridges with spiny outgrowths, rounded or truncate, 3–5 mm *l* and ∅; **Ecol** in open places as settlers on disturbed ground; **Distr** from Caledon, WC, to Graaff Reinet, EC, S Africa.

Note: The echinate fruits are typical of this species.

T. erecta Adamson 1955: 129 **T** 5155 (BOL).

Shrubs, 10–15 cm *h*, erect stems from the base, **I** pale, distinctly ridged; **L** petiole 1–2 mm *l*, blades oblong, 3–10 mm *l*, 3–4 mm *b*; **Fl** solitary, sessile, stamens 4–10, stigma 1, on a cone c. 1/2 as long as the perigone; **Fr** ovoid, obscurely 4-ridged when dry, 4–5 mm *l*, 2–3 mm ∅; **Ecol** unknown; **Distr** Riversdale, Swellendam, WC, S Africa.

Note: Distinct from other species by the erect stems.

T. eremaea Ostenfeld 1921: 59 (see **photo**) **T** Ostenfeld 640 (C?).

Annuals with prostrate branches to 7 cm *l*, the whole plant looking pruinose, **I** with papillae and vesicular hairs when young; **L** petioles 2–15 mm *l*, lanceolate to rhomboid, 10–50 mm *l*, 4–15 mm *b*; **Fl** solitary, rarely 2, sessile or almost so, perigone lobes 4, ovate, c. 1 mm *l*, stamens 4, stigmas 4–6; **Fr** with 4 wings, widest lateral extension in the middle of the height, wings not overtopping the fruit, 2 wings wider than the others, sometimes with ridges between the wings, c. 4 mm *l*, 4–8 mm ∅, 3–8 seeds; **S** gourd- or pear-shaped, honey coloured, c. 2 mm *l*; **Ecol** in dry areas; **Distr** South Australia, Western Australia, Australia.

Note: Similar to **T. diptera,** this species differs in the broader fruit with at least 4 lateral wings not extending over the top of the fruit.

T. espinosae Muñoz 1939: 248 **T** Muñoz B-212 (GH).

Annual herbs to 15 cm *h*, **I** glabrous; **L** alternate or sometimes basal ones opposite, sessile, elliptic to oblong, acute, base rounded to attenuate, 10–45 mm *l*, 4–18 mm *b*; **Fl** hermaphroditic, 1–3, stipes 0.1–3 mm *l*, sometimes 2 deciduous bracteoles c. 1 mm *l*, perigone tube 1.5–2 mm *l*, terete to 4-angular, perigone lobes 4, 1.5–2 mm *l*, yellowish-green outside, yellow inside, stamens 4; **Fr** on stipes 3–12 mm *l*, rhomboidal in side view, markedly 4-angular, the sides between the angles also with distinct tubercles, sometimes confluent into a ridge, glabrous, perigone lobes persisting, 4–10 mm *l*; **Ecol** sand near the coast, 0–550 m altitude; **Distr** N Chile.

Note: Distinctive of the species are the four stamens and the small fruit with prominent corners almost forming a skewed cube. The species is named after Mr. Espinosa.

T. expansa ≡ **T. tetragonioides**
T. expansa var. *strongylocarpa* = **T. implexicoma**

T. fruticosa L. 1753: 480 T LINN 648: 1 (LINN) = *T. linearis* Haw. T?

Shrubs with woody stems and prostrate branches to 200 cm *l*, branchlets erect, **I** with decurrent ridges when young, later glabrescent; **L** linear, mostly revolute or rolled, obtuse at the apex, narrowed to the base but not petiolate, papillate on the lower surface, 10–45 mm *l*, 2–10 mm *b*; **Fl** 2–5 per cymes, crowded, the **Inf** apparently spicate, bisexual, male and transitional, pedicels spreading, papillate, 4–15 mm *l*, longer on the fruit, perigone yellowish and tinged with red, densely papillate outside, stamens numerous; **Fr** 4-winged, rounded at the base, flat or emarginate at the top, a distinct tubercle or small ridge between the wings in the upper half, finally black, 8–25 mm *l*, 10–30 mm *b*; **Ecol** in sand or on rocks; **Distr** from Namaqualand, NC, to Albany, EC, S Africa.

Note: A widespread species with much variation in leaves and stems. **T. fruticosa** is considered a valuable crop, which can be found browsed down to a stump, e.g. in areas near water places; most easily to be found on the road side of a fence, out of reach of animals.

T. galenioides Fenzl in Harvey & Sonder 1862: 463 T Drege 7034 (W).

Herbs with yellow-green prostrate branches 15–35 cm *l*, the short lateral shoots ± erect, **I** covered by coarse, hair-like scales; **L** petioles of the lower leaves as long as the blades, upper ones shorter to sessile, blades ovate to oblong, also scaly, 20–50 mm *l*, 8–18 mm *b*; **Fl** almost sessile in cymes, sometimes forming a spike-shaped mass, perigone densely covered with hair-like scales, stamens 5–8, stigma 1; **Fr** 4-winged, with acute, projecting tubercles between the wings, wings papillate when young, **Fr** rounded at the base, indented at the top, 7–12 mm *l*, 4–8 mm ∅; **Ecol** usually on rocks; **Distr** Clanwilliam, WC, S Africa.

Note: An apparently rare species.

T. glauca Fenzl in Harvey & Sonder 1862: 468 **LT** Drege 7030 (S).

Shrubs to 50 cm *h*, branches spreading, **I** pale, papillate to hirsute when young, older ones glabrous and ridged; **L** petioles 1–3 mm *l*, blades lanceolate to elliptical, acute, glaucous, papillate, 20–25 mm *l*, 4–6 mm *b*; **Fl** 2–4 per cyme, the uppermost flowers on the branch without bracts and male only, pedicels 4–10 mm *l*, perigone thinly papillate, stamens many, stigmas 2–3; **Fr** usually pendulous, 4-winged, with small intermediate wings in between, 5–15 mm *l*, 4–20 mm ∅; **Ecol** hills; **Distr** George, WC, S Africa.

Note: Adamson (1955: 121) cited several collections from the western districts in South Africa for this species as well.

T. halimifolia = **T. tetragonioides**

T. halimioides Fenzl in Harvey & Sonder 1862: 463 T Ecklon & Zeyher 87 (S) = *T. halimioides* Fenzl var. *procumbens* Adamson 1955: 139 T Adamson 4550 (BOL).

Herbs with stout, decumbent branches, **I** covered with coarse, spreading hairs; **L** petioles 5–12 mm *l*, upper **L** sessile, blades green, oblong to lanceolate, acute, papillate, scaly when dry, 20–50 mm *l*, 15–30 mm *b*; **Fl** solitary, in the upper half of the branch, pedicels hairy, 10–30 mm *l*, perigone green or brown outside, bright yellow inside, 4–5 lobes, stamens very numerous; **Fr** 4-winged, with distinct bulges between the wings, glabrous, widest near the base, notched at the top, 20–25 mm *l*, 20–30 mm ∅; **Ecol** in sand or clay; **Distr** Calvinia, Vanrhynsdorp, S Africa.

Note: The herbaceous branches resemble those of **T. nigrescens** and **T. portulacoides**, but plants of **T. halimioides** possess bigger solitary fruits.

T. halimioides var. *procumbens* = **T. halimioides**
T. haplostylis = **T. chenopodioides**

T. haworthii Fenzl in Harvey & Sonder 1862: 468 T?

Shrubs, branches ascending or erect, **I** angular, obsoletely papillate; **L** petiolate, blades lanceolate to elliptical, subpapillate, with evolute margins, 20–30 mm *l*, 3–9 mm *b*; **Fl** 2–3 per cyme, pedicels short, perigone lobes 4, stamens very numerous; **Fr** subglobose, with 7–8 blunt angles, 3–4 locules; **Ecol** unknown; **Distr** uncertain, S Africa.

Note: Fenzl (1862: 468) suggested that the species may be only a form of **T. saligna.**

T. herbacea L. 1753: 480 **LT** Commelijn 1701: t. 102.

Herbs with prostrate branches to 50 cm *l* and a tuberous root, **I** hairy, scaly or glabrous; **L** mostly opposite, upper ones alternate, petioles 10–20 mm *l*, blades dark green, smooth above, papillate below, rotund to broadly ovate, 10–16 mm *l* and *b*; **Fl** 2–7 in terminal umbel-shaped cymes and 1–2 in apparently lateral cymes, pedicels hairy or scaly, 20–60 mm *l*, perigone green or red, covered with scales outside, bright yellow inside, stamens numerous, anthers orange-yellow; **Fr** obtusely 4-angled, the angles protruding as sharp riges when dry, conical or pear-shaped, 10–15 mm *l*, 6–10 mm ∅; **Ecol** mostly on dry clay soils; **Distr** from Vanrhynsdorp to Bredasdorp, WC, S Africa.

Note: With its tuberous root, **T. herbacea** is similar to **T. nigrescens** and **T. portulacoides.**

T. heterophylla = **T. nigrescens**

T. hirsuta L.f. 1781: 258 T Thunberg 11938 (UPS!) = *T. villosa* Poiret 1806: 604 T of unknown origin ex hort. Paris (P?).

Shrubs to 50 cm *h* with ascending, herbaceous, annual branches woody at their bases, **I** with stiff, long hairs, 3–5 mm ∅; **L** length of petioles decreasing to the top of the branch, blades ovate, coarsely and densely hairy, 20–60 mm *l*, 10–20 mm *b*; **Fl** 2–4 per cyme, pedicels densely hairy, perigone brownish and hispid outside, bright yellow inside, stamens numerous, stigmas 3; **Fr** orbicular in outline, with stiff, later distant hairs, 1.2–2.5 mm *l*, 9–2.2 mm ∅; **Ecol** in sandy soils; **Distr** Calvinia, Clanwilliam, Piquetberg, Vanrhynsdorp, Worcester, WC, S Africa.

Note: The stiff, erect hairs appear typical of the species. The plants are available for grazing until late in the season, providing an important component for farming.

T. horrida Britton 1893: 122 **T** Morong 917 (NY)
= **Sclerophylax spinescens** Miers 1848: 19.
Note: Gereau (1993: 411–412) confirmed the identity of both species and the exclusion of the type of *T. horrida* from **Tetragonia**.

T. implexicoma (Miquel) Hook. f. 1856: 148 (see **photo**) ≡ *Tetragonella implexicoma* Miquel 1845: 246, as *Tetragonella amplexicoma*, corrected by Miquel 1846/7: 235 **T** Preiss 2393 (?) = *T. expansa* Murray var. *strongylocarpa* Endlicher 1833: 73 ≡ *T. strongylocarpa* (Endlicher) W.R.B.Oliver 1917: 137 **T** Bauer in Endlicher 1833: t. 98 (not seen) = *Trianthema maidenii* S.Moore 1920: 207 **LT** Maiden s.n. (BM).

Branches prostrate to climbing in other bushes, **I** rough from papillae, later glabrous; **L** alternate, sessile, obovate to rhomboid, obtuse, epidermis rough, later glabrescent, 10–50 mm *l*, 2–40 mm *b*; **Fl** solitary or in pairs, pedicels 5–30 mm *l*, bracteoles linear, perigone lobes 4, longer than the campanulate connate base, papillate outside, yellow inside, stamens 10–20, ovary glabrous; **Fr** globular, red, succulent, 5–8 mm *l*; **S** 1, truncate, c. 2 mm *l*; **Ecol** on limestone and sand dunes; **Distr** S coast of Australia.

Note: Prescott (1984: 37) placed *Trianthema maidenii* as a synonym with **T. implexicoma**, but a comparison of the descriptions shows some discrepancies like a papillate ovary and a leathery fruit in *Trianthema maidenii*.

T. inermis = **T. tetragonioides**
T. ivaefolia L.f. 1781: 257 = **Haloragis alata** Jacquin, a member of the Haloragaceae
T. japonica = **T. tetragonioides**
T. karasmontana = **T. arbuscula**
T. kuntzei = **T. microcarpa**

T. lanceolata Burman 1768: 14 **T**?

Branches herbaceous; **L** petioles half as long as the blades, the latter ovate-lanceolate; **Distr** S Africa.

Note: The species is not treated by Adamson (1955: 109–154). Wijnands (in litteris 1991) said that Linnaeus in a letter of 1768 wondered whether this could be **T. herbacea**, but a specimen has not been found at G to settle the problem.

T. lasiantha Adamson 1955: 129 **T** Adamson 5115 (BOL).

Low shrubs with branches to 60 cm *l*, **I** densely papillate-hirsute, older glabrescent; **L** petioles 2–4 mm *l*, blades lanceolate, oblong, thinly papillate-hirsute below, 15–30 mm *l*, 2–7 mm *b*; **Fl** 2–5 per cyme, bracts in the upper part of the flowering stem very small, subsessile, densely covered with hair-like papillae, perigone lobes cucullate, stigma 1; **Fr** with 4 ridges when dry, 3–5 mm *l*, 3–4 mm ∅; **Ecol** sheltered bushy places; **Distr** Heidelberg, Montagu, Riversdale, WC, S Africa.

Note: Differing from **T. saligna** by the denser cover of longer hairs.

T. linearis = **T. fruticosa**

T. macrocarpa R.A.Philippi 1860: 19 (see **photo**) **LT** R.A.Philippi 48406 (SGO).

Annual herbs to 15 cm *h*, **I** sparsely villose or hairy when young, later glabrous; **L** alternate, basally opposite, sessile, elliptic to oblong, obtuse, attenuate at base, 6–25 mm *l*, 2–8 mm *b*; **Fl** 1–2, sessile, ebracteolate, base of perigone terete to 4-angular, 2–3 mm *l*, perigone lobes 4, 2–3 mm *l*, 2–4 mm *b*, yellow-green outside, yellow inside, stamens c. 15; **Fr** stipes 3–5 mm *l*, turret-shaped with a broader base, the 4 ridges tuberculate, additional tubercles on the sides sometimes in rows, 5–7 mm *l*, 4–6 mm ∅; **Ecol** in sand and gravel near the coast, 0–550 m altitude; **Distr** N Chile.

Note: The tuberculate fruits are similar to those of **T. copiapina**, but markedly thicker at the base.

T. macrocarpa var. *microcarpa* ≡ **T. microcarpa**

T. macroptera Pax 1889: 11 **T** Marloth 1427 (B).

Shrubs to 60 cm *h*, branches ascending, **I** papillate when young, later glabrescent, pale, ridged when dry; **L** petioles short, blades ovate, often folded or revolute, glabrous, 15–30 mm *l*, 4–6 mm *b*; **Fl** solitary, pedicels 5–15 mm *l*, perigone papillate outside, stamens many, stigmas 2–3; **Fr** smooth, flat or slightly emarginate at the top, 4-winged, with distinct ridges between the wings, 15–20 mm *l*; **Ecol** in rocky places; **Distr** S Namibia; NW S Africa.

Note: The original material came from Namibia, but Adamson (1955: 123) cited also collections from Hay, NC, and Clanwilliam, WC, S Africa.

T. macrostylis = **T. chenopodioides**

T. maritima Barnéoud in Gay 1847: 469 (see **photo**) **LT** Gay 1330 (PO).

Dioecious or rarely monoecious shrubs to 120 cm *h*, branches erect or erect from prostrate branches, **I** papillate when young, later glabrous; **L** alternate, occasionally opposite, sessile, oblanceolate or obovate, truncate or rounded apically, base straight to cuneate or rounded, 5–45 mm *l*, 1.5–25 mm *b*; **Fl** male ones 1–10, on fasciculate peduncles or stipes 0.5–18 mm *l*, or with peduncles 2–4 mm *l*, and pedicels 0.5–8 mm *l*, perigone tube 0.5–2.5 mm *l*, perigone lobes 4–5, 1–4 mm *l*, yellow-green outside, bright yellow-green to yellow inside, stamens 8–15, female flowers solitary, stipes 0.1–2.5 mm *l*, perigone tube 2.5–5 mm *l*, 4-winged, perigone lobes 4, 2–3 mm *l*, staminodes 8–12, stigmas 2–5; **Fr** stipes 4 mm *l*, 4-angled, base and the persisting, connivent perigone lobes tapering to their ends, 11–13 mm *l*, 4.5–7 mm ∅; **Ecol** in sand, gravel or cracks in rocks, 0–1 250 m altitude; **Distr** N Chile.

Note: This is one of the species in Chile with a shrubby growth form, differing from the also shrubby **T. angustifolia**, sometimes occurring sympatrically, in vegetative features mainly.

T. microcarpa R.A.Philippi 1872: 717 (see **photo**) ≡ *T. macrocarpa* R.A.Philippi var. *microcarpa* (R.A.Philippi) Reiche 1898a: 366; Reiche 1898b: 370[31] **T** R.A. Philippi 48426 (SGO) = *T. kuntzei* J.Buchwald in Kuntze 1898: 110 **T** Buchwald s.n. (B) = *T. tenella* Johnston 1929: 34 **T** Johnston 5677 (B) = *T. trigona* R.A.Philippi 1891: 24 **LT** F.Philippi 1908 a (SGO).

Annual herbs to 30 cm *h*, **I** glabrous; **L** alternate, sessile, narrowly elliptic or oblong, obtuse tip rounded, base cuneate; **Fl** hermaphroditic, 1–3, on fasciculate stipes or peduncles 0.1–1.5 mm *l*, ebracteate, perigone tube 1–2 mm *l*, lobes 3–4, 1–2 mm *l*, yellow-green outside, yellow inside, stamens 4–20; **Fr** elongate-ovate, peduncle or stipes 2–6 mm *l*, 3–4 angled, sometimes laterally compressed and 2–3 angled, smooth, angles rounded to acute, glabrous, perigone lobes persisting; **Ecol** in sand, 50–3 100 m altitude; **Distr** N Chile; S Peru.

Note: The persisting, connivent perigone lobes on top give the fruit an almost fusiform shape in side view.

T. microptera Fenzl in Harvey & Sonder 1862: 461 ≡ *T. microptera* Fenzl var. *trisperma* Fenzl in Harvey & Sonder 1862: 461 **T** Drege 2932 (W) = *T. microptera* Fenzl var. *monosperma* Fenzl in Harvey & Sonder 1862: 461 **T**? = *T. microptera* Fenzl var. *platycarpa* Adamson 1955: 146 **T** Herre s.n. (BOL) = *T. cyno-crambe* Dinter ex Adamson 1955: 145 nomen nudum **T** Dinter 3905 (B).

Annuals with branches up to 50 cm *l*, **I** usually papillate; **L** petioles 4–12 mm *l*, blades glistening from papillae, broadly ovate or orbicular, 5–20 mm *l*, 5–18 mm *b*; **Fl** sessile, 2–5 per cyme, perigone very papillate, 3–4 lobes, stamens as many as perigone lobes, alternating with the latter, 2–3 stigmas; **Fr** papillate, obpyramidal, base rounded, top truncate, 3–4 projecting ridges forming flat triangular extensions at the top, 2–4 mm *l*; **Ecol** a pioneer in open places in arid regions; **Distr** Bethanien, Lüderitz-South, Namibia; Namaqualand, Kenhardt, NC, south to Knysna, WC, S Africa.

Note: This description follows that of Adamson (1955: 144).

T. microptera var. *trisperma* ≡ **T. microptra**
T. microptera var. *monosperma* = **T. microptra**
T. microptera var. *platycarpa* = **T. microptera**

T. namaquensis Schlechter 1899: 124 **T** Schlechter 8184 (B).

Shrubs with erect branches woody at their bases, c. 20 cm *h*, **I** papillate-hirsute when young, not glistening; **L** sessile, lanceolate, tapering to base and tip, sometimes the margins revolute, papillate on both sides, up to 60 mm *l*, 4–6 mm *b*; **Fl** in cymes arranged in a spike-like **Inf**, pedicels papillate, 3–6 mm *l*, perigone densely papillate-hirsute, stamens many, stigmas 3, sometimes 4–5; **Fr** 4-winged, densely papillate, < 10 mm *l*; **Ecol** shaly rocks; **Distr** Namaqualand, NC, to Piquetberg, WC, S Africa.

Note: The species is similar to **T. verrucosa**, differing in the narrower leaves, pedicellate flowers and smaller fruits.

T. nigrescens Ecklon & Zeyher 1837: 323 **T** Ecklon & Zeyher 2117 (B) ≡ *T. nigrescens* Fenzl var. *hirsuta* Fenzl in Harvey & Sonder 1862: 463 **T** Ecklon & Zeyher 2117 (B) = *T. nigrescens* Fenzl var. *pruinosa* Fenzl in Harvey & Sonder 1862: 463 **T** Drege 2932 (?) = *T. nigrescens* Fenzl var. *maritima* Sonder in Harvey & Sonder 1862: 463 **T**? = *T. heterophylla* Ecklon & Zeyher 1837: 324 = *T. nigrescens* Fenzl var. *hirta* Fenzl in Harvey & Sonder 1862: 463 **T** Ecklon & Zeyher 2119 (B?).

Perennial herbs with decumbent branches and a fusiform tap-root, **I** papillate to glabrous or with hair-like scales; **L** petioles short, blades broadly ovate or almost orbicular, papillate, 10–80 mm *l*, 6–35 mm *b*; **Fl** 3–7 in umbellate cymes in terminal position on long shoots or on short, lateral shoots widely distributed along the stems, pedicels papillate or hairy, 5–25 mm *l*, perigone green, papillate or hairy outside, 4 or 5 lobes, stamens numerous, anthers yellow, stigmas 3–4; **Fr** black, 4-winged, indented at the top, wings thin, smooth, with an inconspicuous tubercle between

[31] The choice of reference for the oldest publication follows Skottsberg (1937: 30).

the wings, 5–15 mm *l* and ∅; **Ecol** in sandy, low lying places; **Distr** from Namaqualand, NC, to Bredasdorp, WC, S Africa.

Note: The whole plant is reported to turn black when dry. Reports of this species as adventitious in New South Wales have been shown to have been misidentifications of plants of **T. decumbens** by Heyligers (1999).

T. nigrescens var. *hirsuta* ≡ **T. nigrescens**
T. nigrescens var. *hirta* = **T. nigrescens**
T. nigrescens var. *pruinosa* = **T. nigrescens**
T. nigrescens var. *maritima* = **T. nigrescens**
T. obovata = **T. decumbens**

T. ovata R.A.Philippi 1893: 168 (see **photo**) **T** Borchers 48397 (SGO) = *T. robusta* R.A.Philippi 1893: 169 nomen illeg. non Fenzl ex Sonder in Harvey & Sonder 1862: 467 **T** King 48403 (SGO).

Annual herbs to 20 cm *h*, **I** glabrous; **L** alternate, basally opposite, sessile, narrowly to broadly elliptical or oblong, apex rounded, base cuneate, 15–65 mm *l*, 3–40 mm *b*; **Fl** 1–3, stipes or peduncles 0.1–2 mm *l*, 1–3 bracteoles to 1 mm *l*, perigone base terete to 4-angled, lobes 4, 1.8–2.2 mm *l*, 1–2 mm *b*, yellow green outside, yellow inside, stamens 4–8; **Fr** peduncles 5–20 mm *l*, ovoid, basally a little thicker, 4-angular, sides smooth, angles rounded to sharp, glabrous, perigone lobes persisting and connivent or deciduous; **Ecol** in sand, silt, or mud, 0–650 m altitude; **Distr** N Chile; Peru?

Note: Plants of this species differ in their more expanded basal parts and smaller apical parts from most others in the area. The identification of material from Peru does not verify earlier reports from there (Taylor 1994: 585).

T. pedunculata R.A.Philippi 1872: 717 **T** King 48402 (SGO).

Annual herbs to 12 cm *h*, **I** densely covered with hairs to 1 mm *l* when young; **L** alternate, sometimes basally opposite, sessile, elliptical to oblong, obtuse or rounded at apex, attenuate at base, 6–30 mm *l*, 3–8 mm *b*; **Fl** 1–2, stipes or peduncles 0.1–5 mm *l*, base of perigone cylindrical, 1.5–2 mm *l*, terete to 4-angular, perigone lobes 4, 2–3 mm *l*, 2–2.5 mm *b*, stamens c. 12; **Fr** stipes 5–10 mm *l*, semi-prismatic, broadest at base, the 4 angles and the sides ornamented with tubercles and covered densely with long, crinkling hairs, 10–12 mm *l*, 8–12 mm ∅, perigone lobes persisting; **Ecol** in sand or gravel near the coast, 0–550 m altitude; **Distr** N Chile.

Note: The fruit of this species is most similar to that of **T. macrocarpa**, which lacks the hairy cover and which is smaller in size than **T. pedunculata**.

T. pentandra Balfour f. 1884: 404 **T** Balfour? (?).

Annuals with procumbent branches, **I** glabrous; **L** petioles up to 18 mm *l*, blades ovate-triangular, basally ± truncate, c. 25 mm *l*, c. 25 mm *b*; **Fl** 1–2, subsessile, perigone lobes 5, stamens 5, ovary 3-locular; **Fr** pentagonal, smooth, without horns, c. 2 mm *l*; **Ecol** in limestone gravel; **Distr** Sokotra, N Somalia.

Note: The species was only found twice, originally on Sokotra, and again in northern Somalia (Thulin 1993: 165).

T. perfoliata = **T. decumbens**
T. peruviana Haw. 1827: 124 nomen nudum
≡ *Phacosperma peruviana* Haw. 1827: 124
≡ **Calandrinia phacosperma** DC. 1828: 359 **T** material from hort. Chelsea.

Note: When describing the species as *Phacosperma peruviana*, Haworth mentioned the name *T. peruviana* only as the one under which the plant had been kept at the Chelsea garden, where he got his material from.

T. pillansii Adamson 1955: 142 **T** Adamson 5020 (BOL).

Herbs, **I** papillate; **L** petioles 20 mm *l*, blades oblong, lanceolate, narrowing basaslly, 20–30 mm *l*, 5–8 mm *b*; **Fl** in rather compact cymes, sessile, densely papillate, the upper ones male only, perigone lobes triangular, stamens c. 20, stigmas 2–3, rarely 4; **Fr** round to cylindrical, truncate, with 8 longitudinal obtuse ridges, 4–9 mm *l*, c. 4 mm ∅; **Ecol** open sand in river beds; **Distr** Namaqualand, NC, S Africa.

Note: The species appears to be restricted to a narrow area near the coast.

T. portulacoides Fenzl in Harvey & Sonder 1862: 462 **T** Ecklon & Zeyher 2118 Rietvalley (S).

Perennials with prostrate, herbaceous branches with persisting woody bases, **I** papillate-pruinose; **L** petiolate, obovate or oblanceolate, obtuse, 10–25 mm *l*, 4–7 mm *b*, upper ones smaller; **Fl** 2–7 in terminal, umbel-shaped cymes, 2–3 in lateral cymes on shorter upper shoots, pedicels 10–18 mm *l*, perigone light brown, sometimes reddish outside, 4–5 lobes, densely papillate outside, the tips incurved, stamens numerous; **Fr** 4-winged, smooth between the wings, 6–17 mm *l*, 4–14 mm ∅; **Ecol** open sand at low altitudes; **Distr** from Calvinia, WC, to Humansdorp, EC, S Africa.

Note: Ecklon & Zeyher (1837: 324) included material of two species from different places under their number 2118. Fenzl (1862: 462) established a new name on the material of 2118 from Rietvalley. The plants are similar to **T. nigrescens** in their growth form but do not blacken when drying.

T. psiloptera = **T. robusta**

T. purpurea Otto ex Sweet 1830: 234 **T?**

Annual herbs, purple; **Fl** anthesis in June and July in cultivation in England; **Distr** unknown.

Note: In a list of species of **Tetragonia**, Sweet had this name filed with the above given information, which may be insufficient to present a proper description.

T. pusilla R.A.Philippi 1872: 716 **T** Philippi 38912 (SGO).

Low annual herbs with erect branches, to 43 mm *h*; **L** oblong, lower ones to 26 mm *l*; **Fl** incompletely known; **Fr** base conical, with 2 prominent tubercles; **Ecol** on sand near the sea; **Distr** Huasco, Chile.

Note: Taylor (1994: 588) excluded the species from her revision of the genus in South America because the type possesses young fruits only which may develop either into those of **T. espinosae** or those of **T. ovata**. In both cases, **T. pusilla** would be an older name for the taxon.

T. quadricornis = **T. tetragonioides**

T. rangeana Engler in Schellenberg 1913: 504 **T** Range 2 (B).

Shrubs to 40 cm *h* with erect branches, **I** yellowish-white, covered by vesicular papillae; **L** petioles 5–8 mm *l*, blades obovate, obtuse, 20–28 mm *l*, 10–15 mm *b*; **Fl** few per cyme, pedicels c. 1 mm *l*, perigone lobes 3, vesicular papillae outside; **Fr** pedicels c. 5 mm *l*, with 6–8 wings consisting of 3–4 bigger ones of 9 mm length and 3–4 smaller ones 5 mm *l*, all 1–2 mm *b*; **Ecol** on rocks near the sea; **Distr** Lüderitz-South, Namibia.

Note: Friedrich (1967: 5) placed the name as a synonym with **T. decumbens.**

T. reduplicata Welwitsch ex Oliver 1871: 583 **T** Welwitsch s.n. (K).

Shrubs to 80 cm *h*, **I** glabrous; **L** ovate, markedly reduplicate along the midrib, both lower halves pressed together, papillate, 10–20 mm *l*, 10–15 mm *b*; **Fl** 2–4 in cymes, pedicels c. 5 mm *l*, perigone lobes 4, obtuse, stamens 12–20; **Fr** winged, smooth between the wings, 10–15 mm *l*; **Ecol** in dry river beds; **Distr** S Angola; Lüderitz-Süd, Swakopmund, Namibia; Kenhardt, S Africa.

Note: Friedrich (1967: 5) placed **T. arbusculoides** as a synonym with this species.

T. retusa Thulin 1993: 165 **T** Hemming 3423 (FT).

Shrubs to 50 cm *h*, **I** minutely papillate when young, terete; **L** blades cuneate, tapering into the petiole without a distinct limit of either, apically marginate or truncate, densely and minutely papillate on both sides, 17 mm *l*, 4 mm *b*; **Fl** solitary, pedicels 2.5–5.5 mm *l* and papillate, basally with 2 papillate bracteoles, perigone basally connate for 4 mm, the 5 deciduous lobes 8 mm *l*, subapically horned with a unifacial protrusion from the lower leaf surface, stamens numerous, style c. 2.5 mm *l*, spreading apically into 2 stigmas; **Fr** ± orbicular, 5-winged, emarginate at base and top, papillate, 20–25 mm *l* and ∅; **S** 2, shaped like a falling drop of water, pale brown; **Ecol** on limestone gravel; **Distr** C Somalia.

Note: Thulin (1993: 167) discussed the relation of the species to **Tribulocarpus** and **Tetragonia.**

T. robusta Fenzl in Harvey & Sonder 1862: 467 **T** Drege 7041 (W) = *T. psiloptera* Fenzl in Harvey & Sonder 1862: 466 ≡ *T. robusta* Fenzl var. *psiloptera* (Fenzl) Adamson 1955: 124 **T** Drege 7052 (W).

Shrubs from 10 cm to 100 cm *h*, branches ascending, **I** 2–3 mm ∅, papillate, pale when young, older ones powdery, terete, with decurrent lines; **L** sessile, basally with a clasping sheath, blades ovate to oblanceolate, papillate, revolute or rolled, 5–10 mm *l*, 2–5 mm *b*; **Fl** 2–5 per cyme, rarely solitary, pedicels very short, perigone lobes 4, densely papillate outside, stamens c. 15; **Fr** pedicels 2–5 mm *l*, wings thin, olive-green to brown, papillate when young, powdery with age, finally glabrous, deeply notched at top; **Ecol** rocky places; **Distr** from Namaqualand, NC, to Albany, EC, S Africa.

Note: Long shoots are often formed after flowering, and in dry conditions the lateral branches may turn spiny.

T. robusta = **T. ovata**
T. robusta var. *psiloptera* = **T. robusta**

T. rosea Schlechter 1899: 124 **T** Schlechter 7973 (B).

Shrubs with decumbent, trailing or climbing branches to 60 cm *l*, **I** with white to pale pink, stiff, flexuose, dense hairs, later glabrescent; **L** petioles ciliate near their bases, blades dark green to purplish, elliptical or oblong, obtuse, with dense, long hairs, 30–60 mm *l*, 8–20 mm *b*; **Fl** 1–3 per cyme, upper bracts very small, perigone villose outside, pink inside, lobes 5–9 mm *l*, stamens numerous, stigmas 2–3; **Fr** 4-winged, ridges between them, the wings with coars hairs on their margins, 15–20 mm *l*, 15–18 mm ∅; **Ecol** in rocky places; **Distr** Calvinia, Clanwilliam, Piquetberg, Vanrhynsdorp, WC, S Africa.

Note: Pink flowers are rare in the genus.

T. saligna Fenzl in Harvey & Sonder 1862: 469 **T** Drege 7037 (W) = *T. saligna* Fenzl var. *extrusa* Adamson 1955: 128 **T** Adamson 5072 (BOL) = *T. saligna* Fenzl var. *latifolia* Adamson 1955: 128 **T** Adamson 4998 (BOL).

Shrubs with lax, sprawling branches to 60 cm *l*, I pale, with faint decurrent lines; L petioles 1–4 mm *l*, shorter in the upper parts of the branches, the blades linear- or oblong-lanceolate, 10–40 mm *l*, 3–10 mm *b*; **Fl** in small cymes, pedicels up to 1.5 mm *l*, perigone covered with very short papillae, the lobes basally connate forming a short tube, stamens 15–20; **Fr** ovoid, dark green, black when ripe, truncate at top, rounded at base, with 4 prominent and 4 small ridges, 3–4 mm *l*, 2–3 mm ∅; **Ecol** among rocks; **Distr** Vanrhynsdorp to George, WC, S Africa.

Note: The upper cymes are closely set resembling a racemose inflorescence.

T. saligna var. *extrusa* = **T. saligna**
T. saligna var. *latifolia* = **T. saligna**

T. sarcophylla Fenzl in Harvey & Sonder 1862: 468 T Drege 8018 (W) = *T. sarcophylla* Fenzl var. *glabrata* Sonder in Harvey & Sonder 1862: 468 T Zeyher 714 (W) = *T. sarcophylla* Fenzl var. *saxatilis* (Phillips) Adamson 1955: 126 ≡ *T. saxatilis* Phillips 1913: 106 T Phillips 7500 (SAM).

Densely branched shrubs with ascending or sometimes decumbent branches to 40 cm *l*, I pale, smooth or slightly papillate, with small decurrent ridges; L petioles short, blades oblanceolate to linear, finely papillate, flat, revolute or rolled, 10–25 mm *l*, 2–6 mm *b*; **Fl** in terminal **Inf** 40–60 mm *l*, bracts smaller than leaves, pedicels 2–5 mm *l*, perigone densely papillate, stamens 15–20, stigmas 2–3; **Fr** 3–4-winged, wings smooth, a small oblong bulge between them but no ridge, rounded in outline, papillate when young, later glabrous, black when ripe, 12–20 mm *l*, 14–20 mm ∅; **Ecol** stony places; **Distr** from Namaqualand, NC, to Albany, EC, S Africa.

Note: Adamson (1955: 127) contrasted the species with **T. robusta**, a taller shrub with broader leaves and lighter coloured fruits.

T. sarcophylla var. *glabrata* = **T. sarcophylla**
T. sarcophylla var. *saxatilis* = **T. sarcophylla**
T. saxatilis = **T. sarcophylla**

T. schenckii Schinz 1894: 204 (see **photo**) ≡ *Anisostigma schenckii* (Schinz) Schinz 1897: 79 T syntypes Schenck 1d, Schinz 2a (Z).

Woody shrubs with erect branches bearing numerous short shoots, I greyish-brown; L sessile to petiolate, blades obovate to spatulate, leathery, with low, round papillae, 10–40 mm *l*, 3–15 mm *b*; **Fl** in rich, cymose **Inf** placed closely and thus raceme-like in a compound aggregate, these up to 12 cm *l*, pedicels 1–2 mm *l*, bracts 1–1.5 mm *l*, perigone lobes 5–6, 6–8 stamens, stigmas 4; **Fr** brown, 3-winged, narrowed to the base, 5–10 mm *l*, 3–6 mm ∅; **Ecol** in sand of dry river beds; **Distr** Namibia.

Note: When describing the species a second time in the genus *Anisostigma*, Schinz (1897: 79) did not mention the older description of **T. schenckii**, but he cited at least one sheet in both texts, namely "Schinz 2a". The originally second collection "Schenck 1d" is replaced by "Schenck 371" in Schinz 1897: 79, which appears to be a mistake because another collection is cited at the end: Fleck 371. It seems that the number of the latter specimen has been incidentally transferred to the collection Schenck 1d. There is no doubt, though, that both descriptions refer to the same species.

T. somalensis = **Tribulocarpus dimorphanta**

T. sphaerocarpa Adamson 1955: 141 T Bolus 13558 (BOL).

Herbs with sprawling branches to 45 cm *l*, I hispid to papillate when young; L petioles 2–4 mm *l*, blades ovate, acute, 10–40 mm *l*, lower ones 15–25 mm *l*, upper ones 5–10 mm *l*; **Fl** solitary, the upper arranged in raceme-like **Inf**, pedicels 5–15 mm *l*, densely papillate-hirsute, perigone densely hispid-papillate, the upper hairs divaricate, 4 lobes, bright yellow inside, stamens numerous; **Fr** wingless, spherical, densely covered with short, spreading semi-rigid scales; **Ecol** in sand among bushes; **Distr** Piquetberg, WC, S Africa.

Note: Similar in habit and leaves to **T. herbacea**, but with wingless solitary fruits.

T. spicata L.f. 1781: 258 ≡ *T. spicata* L.f. var. *angustifolia* Fenzl in Harvey & Sonder 1862: 465 **Lectotype**, here designated: Thunberg 11940 (UPS!) = *T. spicata* L.f. var. *latifolia* Fenzl in Harvey & Sonder 1862: 465 T Ecklon & Zeyher 2121 (?) = *T. spicata* L.f. var. *laxa* Adamson 1955: 117 T Adamson 4630 (BOL).

Shrubs with at least basally woody stems, branches erect or sprawling, up to 100 cm *l*, I papillate-hairy when young, glabrescent later; L petioles 2–10 mm *l*, linear to lanceolate, also ovate, obtuse at the apex, with densely placed, round papillae, 10–40 mm *l*, 4–20 mm *b*; **Fl** 2–5 in cymes, these arranged in 7–8 storeys at the end of branches suggesting a spicate **Inf**, pedicels hairy-papillate, 3–8 mm *l*, male and bisexual flowers together, perigone papillate-hirsute outside, lobes 2–2.5 mm *l*, anthers 12–15, anthers bright yellow, stigmas 1–3; **Fr** 4-winged, flat or indented on the top, basally narrowing, with a distinct tubercle or ridge between the wings; **Ecol** at low altitudes; **Distr** Namibia; from Namaqualand, NC, to Albany, EC, S Africa.

Note: The apparently spicate inflorescences characterize this species, which is otherwise variable in growth form and size.

T. spicata var. *angustifolia* ≡ **T. spicata**
T. spicata var. *latifolia* = **T. spicata**
T. spicata var. *laxa* = **T. spicata**
T. striata = **T. tetragonioides**
T. strongylocarpa = **T. implexicoma**
T. tenella = **T. microcarpa**

T. tetragonioides (Pallas) O.Kuntze 1891: 264 (see **photo**) ≡ *Demidovia tetragonioides* Pallas 1781: 150 ≡ *T. expansa* Murray 1784: 208 T Pallas 1781: t.1 = *T. cornuta* Gaertner 1788: 483 T Gaertner 1791: t.179, f 3 = *T. halimifolia* Forster 1786: 39 **Lectotype**, here designated: UPS 11931 (UPS!) = *T. inermis* F.Mueller 1852: 384 T? = *T. japonica* Thunberg 1784: 208 **Lectotype**, here designated: UPS 11932 (UPS!) = *T. quadricornis* Stokes 1812: 127 T material gathered by Murray in hort. Göttingen (?) = *T. striata* Larrañaga 1923: 173 T? = *T. trigyna* Banks & Solander ex Hook.f. 1864: 84 T Banks & Solander? (?).

Annual herbs, branches prostrate to ascending, up to 50 cm *h*, **I** with densely placed globose to flattish papillae when young; **L** alternate, petioles up to 30 mm *l*, decurrent as a line of papillae, epidermis with somewhat distant, big, globose papillae, fewer along the margins, blades 5–107 mm *l*, 25–80 mm *b*; **Fl** solitary, rarely 3, sessile or shortly pedicellate, perigone lobes up to 2 mm *l*, papillate and green outside, yellow and only minutely papillate inside, stamens at least twice as many as perigone lobes; **Fr** top-shaped, apically with ± conspicuous horns at the upper margin, hard, seeds several in two rows, **Fr** c. 12 mm *l*, c. 10 mm ∅; **S** pear-shaped, smooth, amber to light brown; **Ecol** on sandy or rocky seacoasts, also in fields; **Distr** worldwide in tropical to mild climates.

Note: The species was at first collected in New Zealand and became known as a vegetable "New Zealand spinach". It spread all over the world, constituting a weed in many places now (Holm et al. 1979: 360). Cultivation requires warm temperatures for germination, restricting its use in temperate climates to some extent. Taylor (1994: 589) excluded *T. striata* from the genus **Tetragonia** because the type has been lost. The description may apply to a member of **Tetragonia**, but it may as well belong to a different genus.

T. tetrapteris = **T. decumbens**
T. trigona = **T. microcarpa**
T. trigyna = **T. tetragonioides**

T. verrucosa Fenzl in Harvey & Sonder 1862: 465 ≡ *T. verrucosa* Fenzl var. *latifolia* Fenzl in Harvey & Sonder 1862: 465 T Drege 7031 (W) = *T. verrucosa* Fenzl var. *angustifolia* Fenzl in Harvey & Sonder 1862: 465 nomen nudum T?

Shrubs with erect to ascending branches woody at their bases and often divaricate, **I** pale, densely papillate and with few scales, 2–4 mm ∅; **L** petioles winged, long persisting, ovate-oblong, upper ones lanceolate, pale green, papillate, up to 35 mm *l*, up to 8 mm *b*; **Fl** sessile or nearly so, 2–3 per cyme, the uppermost male, often crowded, perigone densely papillate, stamens numerous, stigmas 2; **Fr** 4-winged, the wings with reticulated veins when dry, papillate, green, round in outline but indented at top, 15–20 mm *l* and ∅; **Ecol** shaly rocks; **Distr** from Namaqualand, NC, to Prince Albert, WC, S Africa.

Note: Friedrich (1967: 5) pointed out that the citation of the species for Namibia by Adamson (1955: 132) is erroneous.

T. verrucosa var. *latifolia* ≡ **T. verrucosa**
T. verrucosa var. *angustifolia* = **T. verrucosa**

T. vestita Johnston 1928: 88 T Johnston 3553 (GH).

Annual herbs to 20 cm *h*, **I** moderately to densely hairy when young; **L** alternate, sessile, elliptical to rhombic, 10–60 mm *l*, 5–30 mm *b*; **Fl** solitary, stipes or peduncles 1–4 mm *l*, ebracteolate, base of perigone 3–5 mm *l*, 4-angular, perigone lobes 4, 2–5 mm *l*, 2–4 mm *b*, yellow-green outside, yellow inside, stamens 15–20; **Fr** stipes or peduncles 2–5 mm *l*, obovoid, the persisting perigone lobes clearly distinct on top, erect, covered in erect, long hairs, 6–8 mm *l*, 4–8 mm ∅; **Ecol** near the coast, 100–900 m altitude; **Distr** S Peru.

Note: Within the group with rather round fruits, the species can be singled out by its dense cover of long hairs appressed slightly and pointing to the apex.

T. villosa = **T. hirsuta**

T. virgata Schlechter 1899: 125 T Schlechter 8255 (B).

Shrubs to 50 cm *h*, **I** smooth, the younger pale, ridged with age; **L** petioles very short, blades elliptical or ovate-lanceolate, margins revolute when dry, 20–40 mm *l*, 3–7 mm *b*; **Fl** 2–4 per cyme, pedicels glabrous, pendulous, 5–10 mm *l*, perianth lobes 4, stamens many, stigmas mostly 3; **Fr** 4-winged, the wings with a thickened edge, minutely papillate, pendulous, 9–30 mm *l* and ∅; **Ecol** in dry, sandy places at c. 500 m altitude; **Distr** Vanrhynsdorp, WC, S Africa.

Note: Adamson (155: 122) gave also collections from Clanwilliam in the south and up to Gordonia in the north.

T. zeyheri = **T. decumbens**

Tetragonocarpus = Tetragonia

H. E. K. Hartmann

When establishing the genus name **Tetragonia** L. (1753: 480), Linnaeus cited *Tetragonocarpus* Commelijn 1701: 203, t. 102 and 205, t.103 beside other references. **Tetragonia** was, therefore, clearly a new name for that genus and is the correct one because *Tetragonocarpus* was published before 1753, the earliest date for valid publication for Spermatophyta (ICBN 2000: Art.13.1). The first use of the name *Tetragonocarpus* (as *Tetragonocarpos*) at generic level after 1753 seems to have been by Miller (1754). Hence, the correct citation is *Tetragonocarpus* Miller 1754 (without page or number); this book is sometimes cited as being printed in 1743, adding to the confusion. When the name "*Tetragonocarpus* Comm." was used by Reichenbach later (1828: 166), it was done so under the genus **Tetragonia** L., under which "a. Tetragonioides DC and *b. Tetragonocarpus* Comm." were given, indicating probably the rank of subgenus, but certainly not the rank of genus, as is often cited. **Tetragonia** and *Tetragonocarpus* are both based on the species **Tetragonia fruticosa** L., but on different types for the species. *Tetragonocarpus fruticosus* can only be lectotypified by Commelijn 1701: t. 103, which is cited by Linnaeus (753: 480) in his original description of **Tetragonia fruticosa**, but the lectotype of the latter species is given as LINN 648.1 (LINN) by Adamson (1955: 113).

T. fruticosa = **Tetragonia fruticosa**

Thyrasperma = Hymenogyne

H. E. K. Hartmann

Although *Thyrasperma* was described as being extremely similar to **Hymenogyne**, the citation of the invalidly published combination *Thyrasperma helianthoides* N.E.Br. (1926: 246), suggesting that it could have been based on *M. helianthoides* Aiton, confused the application of the name. Following Gerbaulet & Ihlenfeldt (1990), *Thyrasperma helianthoides* is a synonym of **H. conica**, based on their choice of the lectotype.

T. helianthoides ≡ **Hymenogyne conica**

Tischleria = Carruanthus

H. E. K. Hartmann

When Schwantes described *Tischleria* (1951: 142) he contrasted the genus mainly by the lack of covering membranes in the fruit against the very similar genus **Carruanthus**. This difference cannot be supported by material in Schwantes' collection (HBG!), in which capsules from several collections share the fruit morphology of **Carruanthus**. Schwantes' opinion may at that date still have been influenced by his original inclusion of species of **Machairophyllum** under *Bergeranthus* subg. *Carruanthus*, which possess well developed covering membranes indeed.

T. peersii = **Carruanthus ringens**

Titanopsis Ruschioideae

H. E. K. Hartmann

Titanopsis Schwantes 1926: 178 **Typus T. calcarea** (Marloth) Schwantes **Etym** Gk *titanos*, chalk, *-opsis*, appearance, referring to the similarity with limestone gravel = *Verrucifera* N.E.Br. GC 1930: 278 **Typus** *Verrucifera schwantesii* (Schwantes) N.E.Br. ≡ **T. schwantesii** (Schwantes) Schwantes.

Compact, sunken in the soil, rootstock branching 6–10 times; **I** enclosed by leaves; **L** 4–8 to a branch forming a rosette, 20–25 mm *l*, triangular to clavate, or apically flattened, surface apically warty, bases smooth, in habitat often the tips only visible, warts consisting of a group of several tannin idioblasts covered by several layers of small cells, crystal layers in the outer walls of the epidermal cells decreasing basipetally in the wart, wax layer rugose, thick, breaking into thin plates, with amorphous particles or platelets; **Fl** 1–3, ebracteolate, 6 almost equal calyx-lobes, petals yellow, often apically orange to purple, stamens collected into a broad central cone, filaments basally bearded, lophomorphic, big holonectary; **Fr** stalked, but capsule breaking away and distributed as a tumble fruit, top flat to conical, covering membranes almost complete, translucent, valve wings broad at base, rectangular or tapering towards

Distribution of **Titanopsis** Schwantes

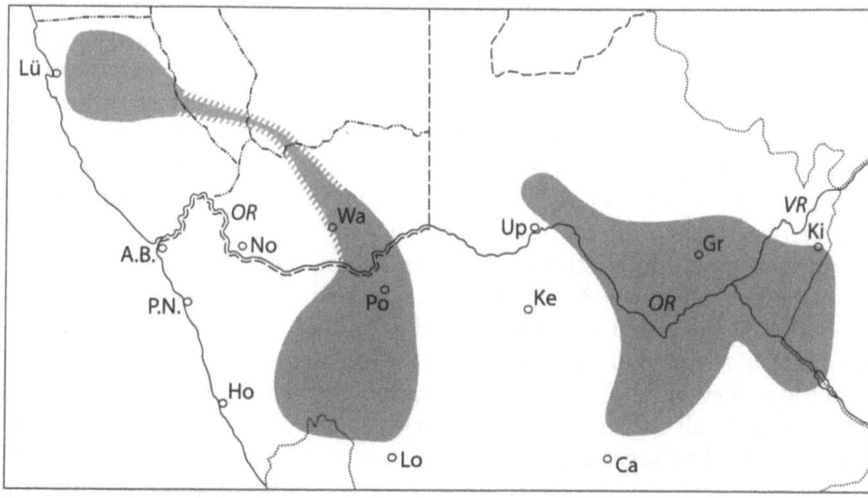

their tips, small closing bodies present or absent, 6-locular, Ø 6,5–8 mm; **S** white to light brown, 0.5–0.9 mm *l*, 0.4–0.7 mm *b*; **Ecol** on limestone flats or soft slopes, rarely in sand; **Distr** S Namibia; NC, FS, S NW, S Africa.

Note: A widespread genus, but often difficult to find because the leaf tips resemble the calcareous pieces of stones so closely. Three species are distinguished.

Key to the Species of Titanopsis

1. Leaves flattened, c. half as thick as wide, base of capsule bell-shaped .... **T. calcarea**
- Leaves thicker to triangular, more than half as thick as wide, base of capsule funnel-shaped, short **2**
2. Warts well elevated with a rounded top, whitish, pearly, hard, densely packed, of similar sizes, valve wings almost rectangular .... **T. schwantesii**
- Warts flat on top, translucent, not pearly, in darker ground tissue, distantly arranged, valve wings narrowing a little towards their tips .... **T. hugo-schlechteri**

T. calcarea (Marloth) Schwantes 1926: 178 (see **photo**) ≡ *M. calcareum,* Marloth 1907: 45 T Marloth 4384 = *T. fulleri* Tischer in Jacobsen 1935: 249 nomen invalid.

Mostly sunken into the ground; **L** spatulate, flat, tip with big, separated warts covered by short papillae of equal length, crystals in upper part of wart; **Fr** stalks tougher than in the other species, base bell-shaped, deep, valve wings tapering towards their tips, closing bodies very rarely present, capsule > 5 mm high; **S** testa cells somewhat elevated, light brown, 0.7–0.9 mm *l*, 0.55–0.7 mm *b*; **Chr** $2n = 18$; **Ecol** in limestone flats, amongst pebbles, mostly > 200 mm rainfall p.a. in March to summer; **Distr** E FS, E and N NC, S NW, S Africa.

Note: The thickness of the crystal layer in the outer epidermal wall decreases rapidly from the tip into the valleys between the warts, which therefore stand out more prominently than in the other species. Material of *T. fulleri* matches **T. calcarea** in fruit shapes and seed characters completely, both features separating this complex from the other species. Nevertheless, forms with more erect leaves exhibiting warts on a margin rather than on a terminal triangle are often treated as *T. fulleri.* Plants with these deviating leaves can form populations of their own or occur mixed with plants typical of **T. calcarea.**

T. crassipes = **Aloinopsis spathulata**
T. fulleri = **T. calcarea**

T. hugo-schlechteri (Tischer) Dinter & Schwantes 1927: 30 (see **photo**) ≡ *M. hugo-schlechteri* Tischer 1926: 203 ≡ *Verrucifera hugo-schlechteri* (Tischer) N.E.Br. GC 88 1930: 513 **Neotype**, here designated: Dinter & Schwantes 1927: f on p 156, lower plant (not upper, as in caption) = *M. astridiae* Dinter 1928: 93 nomen nudum = *T. hugo-schlechteri* (Tischer) Dinter & Schwantes var. *alboviridis* Dinter 1927: 158 T Dinter s.n. 13.2.1924 (HBG!).

Compact, sunken into the ground; **L** spatulate, flattish to triangular, warts flatter than in the other species, translucent-opaque, not pearly, on a dark green or purplish surface, warts of irregular sizes, often purplish or rosy, rarely dark green, low papillae with crystal sand in the outer wall only present in a ring above the base of the wart, dark dots between the warts; **Fr** stalks breaking early from capsule, base shortly funnel-shaped, top mostly flat, valve wings more narrow towards the tip, but not tapering, closing bodies always absent, capsule < 5 mm high; **S** smooth, white, only the micropyle light brown, 0.5–0.6 mm *l*, 0.4–0.5 mm *b*; **Ecol** in red sand overlying limestone, or in limestone flats, < 100 mm rainfall p.a. in March and November; **Distr** Warmbad, Namibia; Calvinia, Kenhardt, Namaqualand, NC, S Africa.

Note: A more whitish form named *T. hugo-schlechteri* var. *alboviridis* can very rarely been found. The species grows more strictly sunken in the soil than the other two, in which elevated forms can also occur. The darker colour of the warts is due to the absence of crystal sand in the upper part of the warts, permitting a view of the often orange-green coloured chloroplasts. Of all species, the thinnest crystal layers and the smallest seeds are found here.

T. hugo-schlechteri var. *alboviridis*
= **T. hugo-schlechteri**
T. luckhoffii ≡ **Aloinopsis luckhoffii**
T. lüderitzii = **T. schwantesii**
T. primosii = **T. schwantesii**

T. schwantesii (Schwantes) Schwantes 1926: 179 (see **photo**) ≡ *M. schwantesii* Dinter ex Schwantes 1916: 82 ≡ *Verrucifera schwantesii* (Schwantes) N.E.Br. GC 1930: 513 **Lectotype**, here designated: Schwantes 1916: f on p 83, flowering plant = *M. calcareum* Dinter 1914: 38 nomen illeg. non Marloth 1907: 45 = *M. concinnum* N.E.Br. GC 1916: 204 **Lectotype**, here designated: Brown 1921: 126, f 49 = *Verrucifera lüderitzii* N.E.Br. GC 88 1930: 513 ≡ *T. lüderitzii* Tischer 1927: 156 **Lectotype**, here designated: Tischer 1927: f on p 157 = *T. primosii* L.Bolus nomen nudum in schedis (Primos NBG 1971/30 BOL).

Compact, often well above the ground, sometimes almost orbicular in shape; **L** triangular, with a regular pattern of fine warts on the tips, the warts resembling densely packed pearls, papillae very short on top of the warts, longer on the lower flanks, crystal sand present on warts; **Fr** stalks breaking early from the capsules which are dispersed as tumble-fruits, base shortly funnel-shaped, top a little elevated, valve wings narrowing only a little to their tips, closing bodies present in northern populations, capsule < 5 mm high; **S** smooth, white in N populations, light brown with slightly elevated testa cells in S populations, 0.6–0.75 mm *l*, 0.5–0.65 mm *b*; **Chr** $2n = 18$; **Ecol** in limestone flats, ≤ 100 mm rainfall p.a. in winter, partly fog; **Distr** Lüderitz-South, Namibia; W Namaqualand, NC, S Africa.

Note: The species has a disjunct distribution, the northern forms (with white seeds) tend to grow into semi-globose humps, the southern ones (with light brown seeds) grow more often sunken in the ground. This species possesses the thickest layer of crystals in the outer epidermal walls in the genus.

T. setifera = **Aloinopsis luckhoffii**
T. spathulata ≡ **Aloinopsis spathulata**

Trianthema SESUVIOIDEAE

H. E. K. HARTMANN

Trianthema L. 1753: 223; Jeffrey 1960: 235–237 ≡ *T.* § *Rocama* (Forskål) DC. 1828: 352 **Typus T. portulacastra** L. **Etym** Gk *tri-*, prefix for three, *anthemon*, flower, referring to the flowers in crowded cymes, sometimes three in a group ≡ *Portulacastrum* Jussieu ex Medikus 1789: 99 **Typus** *Portulacastrum monogynum* Medikus = **T. portulacastra** L. ≡ *Remè* Adanson 1763: 245 **Typus** not given[32] = *Ancistrostigma* Fenzl 1839: 293 **Typus** *Ancistrostigma cypseloides* Fenzl ≡ **T. cypseloides** (Fenzl) Bentham = *Papularia* Forskål 1775: 69 ≡ *T.* subg. *Papularia* (Forskål) Jeffrey 1960: 247 nomen nudum ≡ *T.* subg. *Papularia* (Forskål) Jeffrey ex Adamson 1962: 247 **Typus** *Papularia crystallina* Forskål ≡ **T. crystallina** (Forskål) Vahl.

Annual to perennial herbs, branches basally sometimes woody, prostrate and diffuse; **L** those of a pair unequal in size, petioles widening at the base forming papery sheaths, terete to flat, epidermis ± glabrous, papillate or hairy; **Fl** solitary or in ± compact cymes, each **Fl** with 2 bracteoles, perigone connate basally, perigone lobes 5, stamens 5 to many, adnate to the lower part of the perigone, ovary with a single stigma; **Fr** circumscissile, the operculum falling off containing 1–2 seeds, the base persisting on the plant with 1 to several seeds; **S** triangular, reniform or curved, smooth or ornamented, often papillate, black; **Ecol** in dry areas, often in saline places in tropical and subtropical climates; **Distr** in a broad belt around the equator from 40° N to 40° S.

Note: Delimitation of the genus follows Jeffrey (1960: 235–238), who stated that, in spite of the mentioning of two "styles" by Linnaeus in the original description, a monogynous plant was meant when **T. portulacastra**, and the genus, were described (Linnaeus 1753: 223). The genera **Sesuvium** and **Zaleya** possess at least two stigmas per ovary, **Sesuvium** differing, furthermore, from **Trianthema** and **Zaleya** by the possession of numerous seeds in an axillary po-

[32] Adanson (1763: 245) cited the synonyms **Portulaca** based on Hermann 1698: 213, t. 213, and **Trianthema** L. Therefore, *Remè* is a substitute for both aforementioned names and thus superfluous and illegitimate.

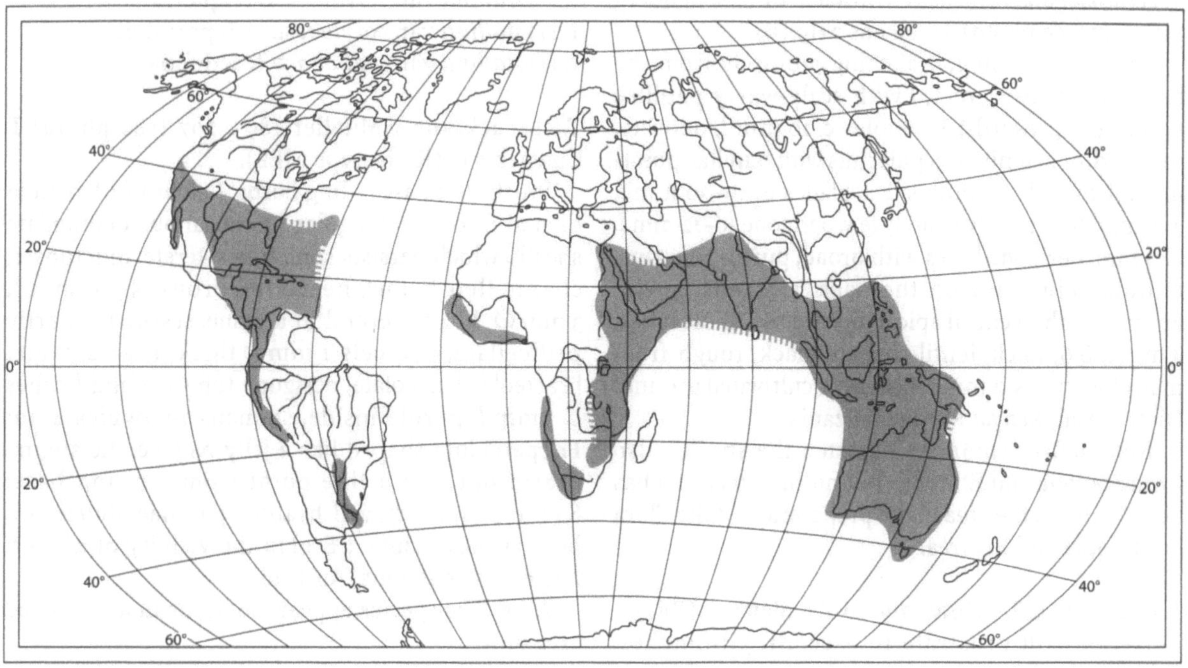

Distribution of **Trianthema** L.

sition. **Zaleya** can be distinguished from the other two genera by the possession of an operculum which breaks into two units representing two clausae, each containing one seed.

De Candolle (1828: 352) divided the genus into two sections:

- *T.* § *Zaleya* (Burman) DC. 1828: 352 with 10 stamens and *T.* § *Rocama* (Forskål) DC. 1828: 352 with 5 stamens. Since **T. portulacastra**, as *T. monogyna*, is placed in *T.* § *Rocama*, that section is synonymous with **Trianthema.**

Jeffrey (1960: 237) divided the genus into two subgenera:

- Ovules 4 to many, flowers usually solitary *T.* subg. *Trianthema*
- Ovules 2, one above the other, flowers usually in cymes *T.* subg. *Papularia*

These names are not validly published because they lack references to the original descriptions and the types (Bittrich 1990: 120). It appears that the validation of the subgenus excluding the type of the genus was undertaken by Adamson (1962: 247): *T.* subg. *Papularia* (Forskål) Jeffrey ex Adamson 1962: 247.

A problem not solved finally for the genus name is its gender. Following a detailed discussion, Tjaden (1995: 213–216) arrived at the conclusion that the name of the genus should rather be *Trianthemum*, a word in singular neuter. In a note in the article, Nicolson argued that **Trianthema** can also be understood as a word in singular feminine, based on similar samples cited. The present treatment follows Nicolson's arguments, taking also into consideration that in most floras and other treatments **Trianthema** is used as a word in feminine singular.

T. americana = **Sesuvium parviflorum**
T. anceps ≡ **Acrosanthes anceps**

T. argentina Hunziker & Cocucci 1959: 17 **T** Hunziker, Cocucchi & di Fulvo 13865 (CORD).

Annual herbs with decumbent to erect branches, anisophyllous, **I** papillate; **L** petioles 3–4.5 mm *l*, basally with rectangular pergamentaceous wings, obovate to spatulate, obtuse, densely papillate, 11–13 mm *l*, 6–8.5 mm *b*; **Fl** in rather compact cymes, perigone lobes c. as long as the tube, papillate outside, 1 stamen, rarely 2–3, 2 locules; **Fr** obconical, operculum truncate with a wing-like margin surrounding the periphery, only a short ring of tissue below this ornamentation, both seeds superposed; **S** pea-shaped, the lower one lenticular, with some low ridges, black, c. 1.4 mm *l*; **Ecol** apparently along the road?; **Distr** Cordoba, La Riocha, Argentine.

Note: The species is remarkable in two aspects: it possesses mostly a single stamen only in the flower, and it occurs in a restricted area in South America, where few members of the genus have been found.

T. australis ≡ **Zaleya galericulata** ssp. **australis**
T. camillii = **Zaleya camillii**

T. ceratosepala Volkens & Irmscher in Schellenberg 1913: 497 (see **photo**) T Volkens 2219 (B).

Shrubs to 50 cm *h*, branching profusely from the base, branches ± erect, **I** with scaly-warty papillae pointing backwards; **L** petioles c. 10 mm *l*, obovate, rounded, or acuminate, epidermis with papillae pointing downwards on both sides, blades up to 30 mm *l*, 4–15 mm *b*; **Fl** mostly 1, rarely few, perigone 10–12 mm *l*, ∅ 4 mm, perigone lobes with broad, purple margins, middle vein hairy on the back, stamens 20–30; **Fr** ovate with a central apical dip and two short lobes, 8 mm *l*; **S** 6, thick, lentil-shaped, black, rough from many low crests, 2 mm *l*; **Ecol** near cultivated ground; **Distr** Masai, Kenya; Moshi, Tanzania.

Note: Jeffrey (1961: 25) reported the species also from Somalia and Ethiopia, but no such material has been found in the area in the preparation of the Flora of Ethiopia and Erytrea.

T. compacta C.T. White 1919: 10 T Bailey s.n. (BRI).

Herbs with prostrate to ascending branches, anisophyllous, **I** glabrous; **L** the bigger one of a pair with a well developed denticulate basal sheath on the petiole, the smaller consisting of blade and sheath only, blades oblong to nearly orbicular, 4–7 mm *l*, 4–7 mm *b*; **Fl** solitary, pedicels 2–4 mm *l*, the single bract ovate, perigone tube 1.5 mm *l*, lobes 2 mm *l*, pink inside, stamens 5–7, irregularly arranged, many ovules; **Fr** operculum subglobular, c. 2 mm *l*; **S** several, elongate ovate, shiny, black, smooth, 0.5 mm *l*; **Ecol** on shores; **Distr** Northern Territory, Queensland, Australia.

Note: The combination of smooth, small seeds with a nearly globose fruit permit an instant identification of this coastal species with almost round leaves.

T. corymbosa = **T. parvifolia**

T. crystallina (Forskål) Vahl 1790: 32 ≡ *Papularia crystallina* Forskål 1775: 69 T Forskål C III. 14 (C) = *T. crystallina* (Forskål) Vahl forma *suberosa* Terracciano 1894: 108 T Terrachiano 187 (FT) = *T. crystallina* (Forskål) Vahl forma *obscura* Armari 1904: 146 **Lectotype**, here designated: Terrachiano 51 bis (FT).

Branches procumbent, **I** papillate, terete, the whole plant looking crystalline; **L** petiole with basal, adnate, lanceolate membranes, blades ovate, covered densely by papillae, 7–8 mm *l*; **Fl** perigone calyx-shaped, stamens 5, anthers red, pollen pale yellow; **Fr** cylindrical, apically concave; **Ecol** unknown; **Distr** Arabia; Ethiopia; Somalia; Sudan.

Note: The species is similar to **T. sedifolia** from the same region.

T. crystallina forma *obscura* = **T. crystallina**
T. crystallina forma *suberosa* = **T. crystallina**
T. crystallina var. *clavata* = **T. triquetra**
T. crystallina var. *corymbosa* = **T. parvifolia**
T. crystallina var. *sedifolia* ≡ **T. sedifolia**

T. cussackiana F.Mueller 1895: 207 (see **photo**) T Cussack s.n. (K?, isotype MEL).

Prostrate to ascending herbs, **I** glabrous; **L** petiole dilated into a rectangular membranous denticulate sheath, which persists sometimes, terete, mucronate, channelled below, nearly glabrous, 5–40 mm *l*, 3 mm ∅; **Fl** 2 to several in a cyme, sessile to shortly pedicellate, pedicels 1 mm *l*, thick, bracts ovate, bracteoles lanceolate, perigone tube 2–3 mm *l*, lobes c. 3 mm *l*, purple inside, stamens 10, ovules 4–10; **Fr** operculum shaped like a jelly-bag, i.e. the stigma drawn out into a fine point from a round base; **S** reniform, black, aril hyaline, papillae shorter on both sides, c. 2 mm *l*; **Ecol** in the vicinity of a river; **Distr** West Australia, Australia.

Note: The species appears to occur in a restricted area only.

T. cypseloides (Fenzl) Bentham 1866: 331 ≡ *Ancistrostigma cypseloides* Fenzl 1839: 293 T Bauer s.n. (W?).

Perennial herbs with prostrate branches rooting finally, **I** filiform, glabrous, densely covered by the **L**; **L** petiolate, opposite and alternate, blades ovate to subrotund, up to 5 mm *l*, up to 2.5 mm *b*, ovary with 7–10 ovules; **Fl** solitary, perigone c. 2.5 mm *l*, green, stamens 7–9; **Fr** globose to ellipsoid, 4–8 seeds, stigma sigmoid, i.e. shaped like a σ; **S** reniform, black, smooth; **Ecol** unknown; **Distr** New South Wales, Australia.

Note: The description in Prescott (1984: 54) follows Bentham (1866: 331) differing in some points from the original description of Fenzl (1839: 293) used here as the base of the text.

T. debilis Sprengel ex Turczaninow 1843: 60 nomen nudum = **Adenocline ovalifolia** Turczaninow.

Note: The name *T. debilis* was only mentioned as a synonym in the original description of **Adenocline ovalifolia.**

T. decandra ≡ **Zaleya decandra**
T. decandra var. *flava* = **Zaleya govindia**
T. decandra var. *rubra* = **Zaleya govindia**

T. diffusa Miller 1768: no.2 T?

Diffuse branches, to 50 cm *h*; **L** petiolate, oval, less succulent than **T. portulacastra**; **Fl** in clusters, sessile, perigone white inside; **Fr** with several seeds; **Ecol** unknown; **Distr** "East India".

Note: Little attention has been payed to this species, as far as floras are concerned.

T. digyna = **Zaleya pentandra**
T. dinteri = **Sesuvium sesuvioides**

T. dubia Sprengel ex Turczaninow 1852: 179 nomen nudum = **Adenocline pauciflora** Turczaninow 1843: 60. The name *T. dubia* is listed in Kew Index as given above, but there is no publication of Turczaninow in the cited Journal in 1852. Instead, the new name **Adenocline pauciflora** is published in Turszaninow 1843: 60 without any indication of a synonym in *Trianthema*. But the species **Adenocline ovalifolia** is described in that same paper (1843) with *T. debilis* as a synonym.
T. flexuosa = **T. portulacastra**
T. fruticosa Vahl 1790: 32 ≡ *Gymnocarpon fruticosum* Persoon 1805: 262 nomen illeg. ≡ **Gymnocarpos decandrum** Forskål 1775: 65. A species of a genus in the family **Illecebraceae** R.Br. (e.g. Brummitt 1992: 192), sometimes also kept in the subfamily **Paronychioideae** (A. L. de Jussieu) Meisner of the family **Caryophyllaceae** A. L. de Jussieu, and even as a synonym under the genus **Paronychnia** Miller (Bittrich in Kubitzki 1993: 223).
T. galericulata ≡ **Trianthema galericulata**

T. glandulosa Peter 1932: 30 T Peter 42741 b (B).
Herbs with prostrate branches to 15 cm *l*, I covered densely by small papillae; **L** triangular-linear, obtuse, basally broadly vaginate; **Fl** in rich, ± compact cymes, perigone densely papillate outside, 2.2 mm *l*; **Fr** unknown; **Ecol** unknown; **Distr** Masai, Tanzania.
Note: Jeffrey (1961: 26) placed the species as a synonym with **T. triquetra**.

T. glaucifolia F.Mueller 1859: 172 T?
Perennial herbs turning black when dry, branches prostrate, up to 30 cm *l*, **I** glabrous, on branchlets sparsely and minutely papillate; **L** petioles short, basally on both sides with dentate membranes, blades obtuse, linear to terete, glaucous; **Fl** solitary, with short pedicels, perigone with 5, rarely 6 lobes, green with white margins; **Fr** round, depressed, with 2 seeds; **S** lentil-shaped, black, almost smooth; **Ecol** in fertile basaltic soils; **Distr** Australia.
Note: The species has been placed as a synonym with **T. triquetra** by Prescott (1984: 58), but Bittrich (1990: 122) pointed out that the variability of **T. triquetra** is still insufficiently known.

T. glinoides = **Zaleya decandra**

T. glossistigma F.Mueller in Wing 1884: 282 T Crossland (MEL).
Prostrate branches, **I** glabrous; **L** petioles 5–6 mm *l*, with small scarious wings, blades glabrous, 3–20 mm *l*; **Fl** 1–3, pedicels as long as the bracteoles, perigone tube very short, lobes 6 mm *l*, stamens 15–20, c. 10 ovules; **Fr** operculum shaped like an elongate helmet, c. 4 mm *l*, the stigma a little shifted to one side, recurved; **S** 4–8 per ripe fruit, black, pear-shaped, ornamented with ridges which appear as parallel walls in the dorsal part, c. 1 mm *l*; **Ecol** uncertain; **Distr** Western Australia, Northern Territory, Australia.
Note: The pear-shaped seeds ornamented with prominent ridges characterize this species well.

T. govindia ≡ **Zaleya govindia**
T. grisea = **Sesuvium portulacastrum**

T. hecatandra Wingfield & M.F.Newman 1994: 115 T Wingfield 14706 (CORO).
Perennial herbs with decumbent branches rooting finally near the edge of the mat, centre dying off with age, the rooting offsprings growing on forming a circle, anisophyllous, **I** usually pink, 10–35 mm *l*; **L** opposite, the broadened base of the petiole of the small leaf overlapping that of the bigger leaf, petioles 4–5 mm *l*, base and proximal half red, blades orbicular to ovate or rotund, glabrous, but rough to the touch when dry, 15–25 mm *l*, 10–25 mm *b*; **Fl** solitary, perigone 12.5–14.5 mm *l*, lobes 9.5–11 mm *l*, pale pink or mauvish inside, white in older flowers, stamens 81–100, filaments pink, anthers pale yellow in young **Fl**, c. 5 ovules per locule; **Fr** operculum 4-lobed, 2–4 seeds; **S** shiny black with shallow pits all over, aril pale brown, turning sticky when wetted; **Ecol** on bare, firm soil in coastal desert, 300 mm rainfall p.a.; **Distr** Miranda, Democracia, Venezuela.
Note: Similar in some aspects of the growth form and leaves to **T. portulacastra**, this species differs in the peculiar formation of rings around a dead plant and the numerous stamens in the big flowers mainly. It is reported that is not eaten by goats.

T. hereroensis Schinz 1897: 76 T Stapf s.n. (Z?).
Low shrubs, developing a "dune-cushion" with age when blown over with sand, **I** inside the cushion woody, tough; **L** petioles with broad wings, blades elliptical to ovate, acute, 3–5 mm *l*, 2–3 mm *b*; **Fl** 2–5 in ± compact cymes, perigone lobes 2 × as long as broad, c. 2 mm *l*, stamens 5; **Fr** operculum rather high; **Ecol** in sandy desert near the sea; **Distr** Swakopmund, Namibia.
Note: Friedrich (1970: 132) pointed out that most plants develop into an almost semi-globose cushion shape, the stems being very robust and continuing to grow when they are overblown by sand.

T. humifusa ≡ **Acrosanthes humifusa**
T. humillima
= **Gigaspermum repens** (Hook.) Lindb., a moss
T. hydaspida = **Sesuvium sesuvioides**

T. kimberleyi Bittrich & Jenssen in Bittrich 1990: 118 (see **photo**) T Bittrich & Jenssen 18618 (HBG).

Annual herbs with prostrate branches, ∅ of plants c. 20 cm, 5 cm *h*, branching in the vegetative part monopodial, in the flowering region sympodial anisophyllous and with side branches differing in thickness, the thicker in the axil of the bigger leaf, **I** glabrous; **L** opposite in the vegetative region, petioles 1–2 mm *l*, broadening basally into a membranous sheath with 2 acuminate lobes, blades elliptical to obovate, epidermis weakly papillate, water storing tissue in epidermis and hypodermis; **Fl** solitary, smelling intensely of honey during anthesis, pedicels up to 2 mm, elongating to 5 mm when fruiting, bracteoles lanceolate and denticulate, perigone c. 5 mm ∅, tube 2 mm *l*, lobes whitish-pink inside, stamens 10, 5 alternate, 5 opposite, nectary disc whitish, 10–12 ovules; **Fr** operculum subovoid; **S** 6–8, broadly ovoid to pyriform, brownish to black, smooth, 1.1 mm *l*, 0.9 mm *b*; **Chr** $2n = 48$; **Ecol** on flat hills in schistose rocks; **Distr** Western Australia, Australia.

Note: The species is classified as rare by the authors.

T. littoralis J.Cordemoy 1895: 290 T Cordemoy s.n.? (P?).

Herbs with terete **I** with a line of hairs along them; **L** petioles sparsely hairy, basally with a broader membrane, blades ovate, obtuse, glabrous, 20–30 mm *l*; **Fl** solitary, perigone whitish-pink, 20 stamens, alternating short and long; **Fr** with few seeds, truncate; **S** black, striped, helmet-shaped; **Ecol** on the shore; **Distr** Réunion.

Note: In several features the description reads very similar to that of **T. portulacastra**, but the line of hairs along the internodes seems to be a distinguishing character state not reported for other species.

T. maidenii = **Tetragonia implexicoma**

T. megasperma Prescott 1983: 183 (see **photo**) T Lazarides 7990 (CANB).

Perennial herbs with prostrate branches to 100 cm *l*, **I** densely hirsute, hairs 2–3 mm *l*, branchlets regularly alternate anisophyllous; **L** opposite, petioles c. 6 mm *l*, on the bigger leaf of the pair basally broadened into a sheath, blades obovate to spatulate, acute or obtuse, pale green, covered with long hairs like the **I**, 3–30 mm *l*, 3–15 mm *b*; **Fl** solitary, pedicels 2–10 mm *l*, perigone tube very short, lobes spreading, hirsute outside, glabrous, white to pink with mauve tips inside, c. 6 mm *l*, stamens 5–15, 2 ovules; **Fr** a dry ovoid capsule, operculum semi-globose, with an undulating ridge near the top, tightly enclosing the single big seed, the stigma rolled up; **S** somewhat compressed pear-shaped, with small, distant papillae, black, c. 3 mm *l*; **Ecol** in sandy soil; **Distr** Northern Territory, Australia.

Note: The single big seed and the hairy pubescens of the entire plant permit an easy identification.

T. monogyna ≡ **T. portulacastra**

T. multiflora Peter 1932: 29 **LT** Peter 42741 a (B).

Annuals with erect to spreading branches 12–30 cm *l*, to 5 cm *h*, **I** with small, densely placed papillae; **L** linear-lanceolate, basally broader; **Fl** in rich cymes, sessile; **Fr** unknown; **Ecol** unknown; **Distr** Masai, Tanzania.

Note: The species was placed by Jeffrey (1961: 26) as a synonym with **T. salsoloides**.

T. nigricans Peter 1932: 30 T Peter 42741 c (B).

Erect annual herbs, to 25 cm *h*, **I** covered densely by papillae; **L** fragile, narrowly linear, covered by papillae, basally developing a scaly sheath; **Fl** in compact, dense cymes, blackening when drying, perigone 2.8 mm *l*, the margins of the lobes finely serrate; **Fr** unknown; **Ecol** unknown; **Distr** Masai, Tanzania.

Note: Jeffrey (1961: 26) placed the species as a synonym with *T. triquetra* ssp. *triquetra* var. *sanguinea*.

T. nyasica = **Sesuvium nyasica**
T. obcordata = **T. portulacastra**

T. oxycalyptra F.Mueller 1859: 173 (see **photo**) ≡ *T. oxycalyptra* var. *pedunculata* F.Mueller 1859: 174 nomen illeg. **LT** MEL 99967 (MEL) = *T. oxycalyptra* var. *sessiliflora* F.Mueller 1859: 174 **NT** MEL 99968 (MEL).

Herbs with prostrate to ascending branches, **I** glabrous, sparsely hirsute when young; **L** petioles 3–8 mm *l*, blades obovate to spatulate, 3–25 mm *l*, 1–8 mm *b*; **Fl** mostly 3, rarely 1–2, pedicels 0–10 mm *l*, bracts linear-lanceolate, apiculate, perigone tube 2–3 mm *l*, lobes 5–6 mm *l*, stamens 10–20, ovules 10–12; **Fr** c. 6 mm *l*, operculum a short, low cone with a long, thin, terminal stigma, 6–9 seeds; **S** ovoid with the micropyle protruding, black, papillae prominent all over or only in the marginal parts, c. 2.mm *l*; **Ecol** in the desert; **Distr** Northern Western Australia, Australia.

Note: The taxonomy follows Prescott (1984: 57, 319), who made the lecto- and neotypifications.

T. oxycalyptra var. *pedunculata* ≡ **T. oxycalyptra**
T. oxycalyptra var. *sessiliflora* = **T. oxycalyptra**

T. patellitecta Prescott 1983: 185 (see **photo**) T Perry 3067 (CANB).

Herbs with branches to 100 cm *l*, branchlets either one per node or one longer and the other shorter, **I** hirsute with spreading hairs; **L** petioles c. 6 mm *l*, with a broad, membranous sheath, blades ovate to spatulate, covered with stiff, spreading hairs, 4–30 mm *l*, 3–15 mm *b*; **Fl** 1–2, sessile, perigone purple inside, c. 2.5 mm *l*, stamens 15–25, 2 locules; **Fr** globose, operculum dish-shaped with a fine hyaline cover over the apex, stigma persisting, ∅ c. 4 mm, 1 seed in the operculum, 1 in the basal part; **S** pear-shaped, faintly reticulate, shiny black, c. 1.5 mm *l*; **Ecol** on sandy river embankments; **Distr** Northern Territory, Western Australia, Australia.

Note: It appears that the operculum has the shape of a dish placed upside down on the basal part of the fruit.

T. parvifolia E.Meyer ex Sonder in Harvey & Sonder 1862: 598 ≡ *T. parvifolia* E.Meyer in Drege 1843: 63, 226 nomen nudum ≡ *T. triquetra* Willdenow ex Sprengel ssp. *parvifolia* (Sonder) Jeffrey 1960: 237 T Drege (S) = *T. parvifolia* Sonder var. *annulata* Adamson 1962: 251 T Pearson 9231 (K) = *T. crystallina* (Forskål) Vahl var. *corymbosa* Sonder in Harvey & Sonder 1862: 598 ≡ *T. corymbosa* E.Meyer in Drege 1843: 93 nomen nudum ≡ *T. parvifolia* Sonder var. *corymbosa* (Sonder) Adamson 1962: 251 T Drege s.n. (S) = *T. rubens* E.Meyer ex Sonder in Harvey & Sonder 1862: 598 ≡ *T. rubens* E.Meyer in Drege 1843: 92 nomen nudum ≡ *T. parvifolia* Sonder var. *rubens* (Sonder) Adamson 1962: 250 T Drege (S).

Perennial herbs with prostrate branches, somewhat woody at the base, anisophyllous, **I** papillate or glabrous, shining; **L** petiolate, blades oblong to sub-rotund, obtuse; **Fl** in ternate cymes, perigone 1.5–2 mm *l*, stamens 5; **Fr** operculum conical, apically truncate with a thickened ring around, 1–2 seeds; **Ecol** on stony hills; **Distr** Warmbad, Namibia; Namaqualand, NC, Beaufort West, WC, S Africa.

Note: The circumscription of the species follows Adamson (1962: 249–251), including all varieties described there. Excluded is **T. hereroensis**, for which Friedrich (1970: 132) gave the stout, woody stems growing on when covered by sand as the most obvious distinct feature.

T. parvifolia var. *annulata* = **T. parvifolia**
T. parvifolia var. *corymbosa* = **T. parvifolia**
T. parvifolia var. *rubens* = **T. parvifolia**
T. pentandra ≡ **Zaleya pentandra**
T. pentandra var. *flava* ≡ **Zaleya govindia**
T. pentandra var. *hirtula* = **Zaleya pentandra**
T. pentandra var. *rubra* ≡ **Zaleya govindia**

T. pilosa F.Mueller 1859: 174 (see **photo**) T Mueller s.n. (isotype MEL).

Herbs with procumbent, long main branches bearing short lateral branchlets, **I** densely hispid, especially when young; **L** petioles 1–15 mm *l*, basally broadening into an oval sheath, blades ovate to obovate, obtuse or acute, 3–30 mm *l*, 2–10 mm *b*; **Fl** 3–4 in a dense cluster, perigone tube c. 3 mm *l*, lobes 2–3 mm *l*, white, pinkish, or red inside, stamens 15–20, 3–4 ovules; **Fr** operculum cylindrical, truncate, centre depressed, sides often reticulate to striate, 3–5 mm *l*, 1 seed in the operculum, 1 in the basal part; **S** disc-shaped, black, finely papillate all over, c. 1.7 mm *l*; **Ecol** in desert; **Distr** Northern Territory, Western Australia, Queensland, South Australia, Australia.

Note: The species occurs in a wide area, mainly in sand in dry habitats.

T. polyandra = **Sesuvium sesuvioides**
T. polysperma = **Sesuvium sesuvioides**

T. portulacastra L. 1753: 223 (see **photo**) ≡ *T. monogyna* L. 1767: 69 ≡ *T. procumbens* Miller 1768: no 1 **LT** Hermann 1698 t. 213 = *Tetragonia chimajensis* Chiovenda 1932: 226 **T** Gorini 437 (FT?) = *T. flexuosa* Schumacher & Thonning 1827: 221 **T** Thonning s.n. (C?) = *T. obcordata* Roxburgh 1832: 445 **T** Roxburgh s.n. (after Wallich 1832: no 6837 A, K).

Herbs with procumbent branches to 60 cm *l*, **I** glabrous or sparsely hairy, especially when young; **L** petioles 5–25 mm *l*, leaves of a pair differing, the basal broader part connate with that of the opposite leaf, forming a cup-shaped sheath around the stem, blades elliptic to obovate, 4–50 mm *l*, 4–45 mm *b*; **Fl** solitary, sessile, almost hidden in the cup formed by the bracts, perigone tube fused to leaf sheath, lobes 2.5–5 mm *l*, a few hairs outside, pink or white inside, stamens 10–20, ovules 10–15; **Fr** operculum turret-shaped, truncate with a marked distal denticulate rim, 2–3 mm *l*, of the 3–12 seeds 1–2 in the operculum; **S** reniform, black, with low crests, 1.5–2 mm *l*; **Ecol** in tropical climates as weeds in fields; **Distr** worldwide in the tropics.

Note: Since in his original description Linnaeus gave two stigmas for this species, some discussion has surrounded the decision of Jeffrey (1960: 235–236) to choose Hermann's text and figure (1698: 213 and *t.* 213) as the lectotype. Yet, this choice is in agreement with Linnaeus' later action (1767: 69), when he placed **T. portulacastra** as a synonym with the new name *T. monogyna*. Note should be taken that in several cases the names **Sesuvium portulacastrum** and **Trianthema portulacastra** have been used synonymously. When consulting floras it should therefore be checked whether either name is applied correctly.

Osbornová-Kosinová (1984: 279–285) stressed the occurrence of the species as a weed in Egypt comparing its frequency with that reported for other countries based on data from Holm (1979) mainly.

T. portulacastra var. *hillebrandii*
= **S. portulacastrum**
T. procumbens ≡ **T. portulacastra**
T. redimita = **Zaleya redimita**

T. rhynchocalyptra F.Mueller 1859: 174 (see **photo**) T F.Mueller s.n. (isotype MEL).

Perennial herbs with prostrate branches from a woody rootstock, **I** densely hairy; **L** clustered, petioles

c. 4 mm *l*, expanding basally into a denticulate sheath, blades lanceolate to obovate, acute, densely hairy or glabrous with ciliate margins, shiny above, pale below, 4–20 mm *l*, 2–8 mm *b*; **Fl** sessile, solitary, perigone tube c. 4 mm *l*, lobes c. 3 mm *l*, white, pink, or green inside, stamens 10, 4–6 ovules per ovary; **Fr** operculum somewhat elongate semi-ovoid, the recurved stigma placed a little to the side of the apex; **S** nearly globose to pear-shaped, black, finely papillate, longer papillae along dorsal and basal sides, c. 2 mm *l*; **Ecol** on hills and in plains with rocky and sandy soil; **Distr** Northern Territory, Queensland, Australia.

Note: The long, persisting, beak-shaped stigma has given the species its name.

T. rubens = **T. parvifolia**
T. salarium = **Sesuvium sesuvioides**

T. salsoloides Fenzl ex Oliver 1871: 588 T Kotschy 137 (K) = *T. transvaalensis* Schinz 1915: 396 ≡ *T. salsoloides* Oliver var. *transvaalensis* (Schinz) Adamson 162: 248 T Schlechter 4876 (Z) = *T. salsoloides* Oliver var. *stenophylla* Adamson 1962: 249 T van der Schijff & Marais (K).

Shrubs with prostrate to ascending branches woody at base, **I** hispid or papillate when young, glabrous with age; **L** petioles short, with lateral dentate wings, blades linear to lanceolate, papillate, 10–25 mm *l*, 2–3 mm *b*; **Fl** in ± compact cymes on long shoots or lateral short shoots, perigone < 5 mm *l*, stamens 5 or less?; **Fr** operculum with a prominent, smooth thick ring, with 1 seed; **Ecol** in sand; **Distr** Sudan; NP, S Africa.

Note: The three varieties distinguished by Adamson (1962: 248–249) inhabit either the northern part of the distribution range or the southern, a connection seems to be absent.

T. salsoloides var. *transvaalensis* = **T. salsoloides**
T. salsoloides var. *stenophylla* = **T. salsoloides**

T. sanguinea Volkens & Irmscher in Schellenberg 1912: 498 ≡ *T. triquetra* Willdenow ex Sprengel ssp. *triquetra* var. *sanguinea* (Volkens & Irmscher) Jeffrey 1960: 237 T Volkens 458 (B).

Branches procumbent, **I** with globose papillae when young, later the whole plant turning red; **L** petioles with lanceolate wings over its length, blades linear, margins rolled backwards and papillate, blades 10–15 mm *l*, 2 mm *b*; **Fl** in compact cymes, perigone 2–3 mm *l*, densely covered with globose papillae outside, stamens 5; **Fr** truncate with a thick marginal ring, base cylindrical, 2 seeds in the ovary; **S** black, rough, irregularly cristate; **Ecol** in grassland with trees, steppe; **Distr** Kenya, Tanzania.

Note: The description suggests a similarity with **T. crystallina** from further north, but Jeffrey (1960: 237, 1961: 26) placed this species as a variety with **T. triquetra**.

T. sedifolia Visiani 1836: 19 ≡ *T. crystallina* (Forskål) Vahl var. *sedifolia* (Visiani) Hiern 1898: 415 **Lectotype**, here designated: Visiani 1836: t. 3, f 1 = *T. sedifolia* Visiani var. *microphylla* Courbon 1862: 156 T Courbon s.n. (P).

Annual herbs with prostrate branches 5–8 cm *l*, **I** conspicuously papillate; **L** petioles with broad basal wings clasping the stem, blades linear-oblong, with prominent, densely placed papillae turning into scales with age; **Fl** 3–4 in cymes, perigone with 5 lobes which bend over the ripe fruit, stamens 3; **Fr** operculum shaped like a cupula; **S** orbicular to reniform, black; **Ecol** uncertain; **Distr** Egypt.

Note: The species has repeatedly been placed in synonymy with **T. crystallina**, also originally described from Egypt, from which it differs in the three stamens and possibly the number of four seeds. This item is doubtful because Visiani cited Brocchi as mentioning four seeds but stated that the material he (Visiani) has examined had two seeds only.

T. sedifolia var. *microphylla* = **T. sedifolia**
T. senniei ≡ **Zaleya senniei**
T. transvaalensis = **T. salsoloides**

T. triquetra Willdenow ex Sprengel 1825: 381 (see **photo**) ≡ *T. triquetra* Willdenow in Rottler 1803: 181 nomen nudum **T** Rottler s.n. (B) = *T. crystallina* var. *clavata* J.M.Black 1923: 369 ≡ *T. triquetra* Willdenow ex Sprengel var. *clavata* (J.M.Black) H.Eichler 1965: 136 **LT** Cockburn s.n. (AD) = *T. triquetra* Willdenow ex Sprengel var. *oblongifolia* Gamble 1919: 199 T Bournes s.n. (K).

Herbs with prostrate branches 20–60 cm *l*, with short lateral branchlets, **I** finely papillate or glabrous; **L** petioles c. 2 mm *l*, broadening basally into a denticulate sheath, blades terete to clavate, the water storing tissue either below the epidermis or centrally, epidermis with ± low papillae, 3–30 mm *l*, 2–4 mm *b*; **Fl** 3–6 in compact cymes, sessile, bracts lanceolate, bracteoles ovate, perigone tube 1–2 mm *l*, lobes c. 2 mm *l*, persisting on the fruit and moving hygrochastically, stamens 5, alternating with the perigone lobes, 2 ovules; **Fr** operculum flat, without remains of the style, containing 1 of the 2 seeds; **S** black, disc-shaped, rows of papillae around the flat edge, otherwise smooth, 1–1.5 mm *l*; **Ecol** in dry areas in tropical climates; **Distr** Australia; India; Namibia; S Africa.

Note: The authorship follows Daniel & Umamaheswari (1999: 409–410), who demonstrated that Sprengel attributed the name to Willdenow in Rottler, who had only given the name a such and placed it in synonymy with **T. crystallina.** Bittrich (1990: 120–122) described the variation in leaf shapes observed, i.e. mainly differences in water storing tissue, pointing also out that only a thorough study on a worldwide basis can solve the observed problems in identification and delimitation of taxa. Noticeable is that Sprengel placed the species in a group with 10 stamens.

T. triquetra var. *oblongifolia* = **T. triquetra**
T. triquetra ssp. *parvifolia* ≡ **T. parvifolia**
T. triquetra ssp. *triquetra* var. *clavata* ≡ **T. triquetra**
T. triquetra ssp. *triquetra* var. *sanguinea* ≡ **T. sanguinea**

T. turgidifolia F.Mueller 1876: 83 (see **photo**) T Crouch s.n. (isotype MEL).

Perennial herbs or shrubs, branches semi-erect or sprawling, **I** glabrous, nodes often hairy; **L** petioles with small scarious wings at their bases, blades semiterete to almost clavate, obtuse, glabrous, 3–20 mm *l*, 3 mm *b*; **Fl** 1–3, pedicels hairy, 2–10 mm *l*, bracts dentate, bracteoles fringed, perigone tube 2–3 mm *l*, lobes 3–5 mm *l*, white or purple inside, stamens 10 in a ring, ovules 7 to many; **Fr** operculum semi-ovoid, papery, c. 2.5 mm *l*, 2–3 seeds; **S** flattish ovoid, dark brown, with parallel rows of papillae, c. 2 mm *l*; **Ecol** mainly in sand; **Distr** Western Australia, Australia.

Note: A widespread species in very dry areas of Australia.

Tribulocarpus TETRAGONIOIDEAE

H. E. K. HARTMANN

Tribulocarpus S.Moore 1921: 228 ≡ *Tetragonia* L. subg. *Tribulocarpus* (S.Moore) Adamson 1955: 148 **Typus T. dimorphanta** (Pax) S.Moore ≡ *Tetragonia dimorphanta* Pax **Etym** *Tribulus*, a plant genus with spiny fruits, Gk *carpos*, fruit, describing the spinose fruit of the only species of the genus.

Shrubs, branches basally woody, **I** papillate, canescent, glabrous; **L** petiolate, obovate, obtuse, narrowing into the petiole; **Fl** in compact groups, the terminal one bisexual, with a long perigone tube in which the stamens are inserted at c. 2/3 of the length, lobes 4–5, short, 3–4 lateral male flowers protruding from the base of the terminal one; **Fr** appearing composed, with numerous spines, globose; **Ecol** in dry grassland; **Distr** Namibia; Somalia.

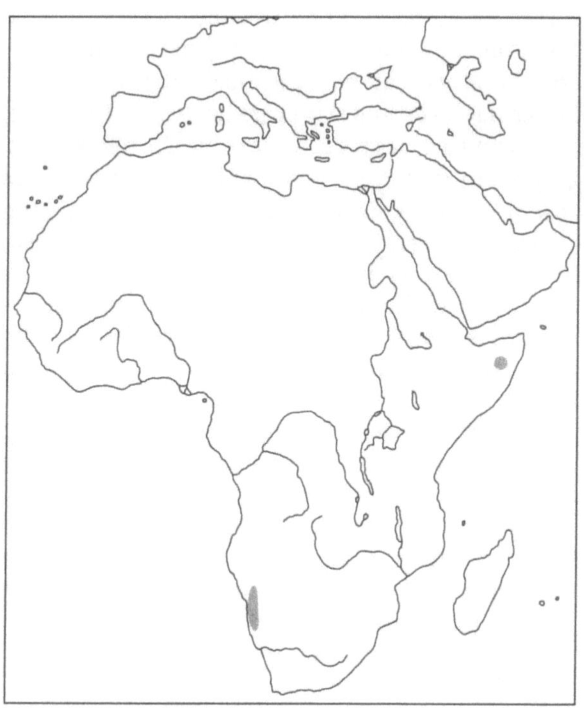

Distribution of **Tribulocarpus** S.Moore

Note: When Pax (1889: 12) described the type species as one of **Tetragonia,** he already noted that it differed markedly from the other species of the genus by the heteromorphy of the flowers. As a consequence, the genus **Tribulocarpus** has been erected, which has later also been reduced to the rank of subgenus (Adamson 1955: 148). The recent detection of **Tetragonia retusa** with a tube in the flower (Thulin 1993: 165) is discussed by the author in relation to **Tribulocarpus** as well. It is necessary that the problem is studied again.

The genus contains only one species, albeit it occurs in widely separated parts of Africa.

T. dimorphanta (Pax) S.Moore 1921: 288 ≡ *Tetragonia dimorphanta* Pax 1889: 12 T Marloth 1249 (B) = *Tetragonia somalensis* Engler in Schellenberg 1913: 502 **Syntypes** Brichetti 31 and 507 (RO).

Shrubs to 50 cm *h* (in Namibia) and 100–150 cm *h* (in Somalia), apically herbaceous; **L** blades tapering basally into a petiole, apically truncate to obtuse, covered by globose papillae and appearing whitish, 10–20 mm *l*, 10–15 mm *b*; **Fl** the terminal group with a very long tube that can be mistaken as a pedicel when not sectioned, 3–4 male, bracteate flowers emerging from the lower part of the outer wall of the inferior ovary, all perigone lobes finely papillate outside, stamens numerous; **Fr** spines 6–8 mm *l* in Namibian material, 2–3 mm *l* in Somalian plants; **Ecol** as for genus; **Distr** Namibia; Somalia.

Note: The species is remarkable in its distribution in two widely separated areas.

Trichocyclus = Brownanthus

M. Gerbaulet & S. M. Pierce

Trichocyclus N.E.Br. (1922: 151) is a later homonym of *Trichocyclus* Dulac (1867: 31), a genus of the fern family Aspleniaceae. The name *Trichocyclus* Dulac itself is illegitimate because it was based on the same type as **Woodsia** R.Br. However, since the name *Trichocyclus* Dulac was validly published, its further use is prohibited.

T. buchubergensis = **Brownanthus pubescens**
T. ciliatus ≡ **Brownanthus ciliatus**
T. marlothii ≡ **Brownanthus marlothii**
T. namibensis ≡ **Brownanthus namibensis**
T. pillansii = **Brownanthus pubescens**
T. pubescens ≡ **Brownanthus pubescens**
T. schenckii
≡ **Brownanthus ciliatus** ssp. **schenckii**
T. simplex = **Brownanthus ciliatus** ssp. **schenckii**

Trichodiadema Ruschioideae

I. Niesler

Trichodiadema Schwantes ZSK 1927: 106; Oberstein 1912: 298–302; Ihlenfeldt 1980: 145–163; Niesler 1997: 13–27 **Typus** *T. stelligerum* (Haw.) Schwantes = **T. barbatum** (L.) Schwantes **Etym** Gk *trix*, hair, *diadema*, crown, allusive to the leaf tip consisting of bristles rising from cup-celled bases.

Erect to decumbent perennial shrubs, climbing to compact, 1–30 cm *h*, thick rootstocks common in compact plants, spindle-shaped storing-roots, single or as a bundle and very long thin roots developing storing tubers almost 30 cm under ground, **I** terete or angular, younger ones often papillate showing modified shapes of the leaf bladder cells, longer **I** often with white papillae, visible as white dots, preferably below the **L**, 5–43 mm *l*, 1–5 mm ∅; **L** cylindrical, ovate or rounded-triangular in section, green to greyish, with a characteristic bladder cell epidermis: densely arranged huge bladder cells in rows, often with thickened apical walls, raised above the level of normal smaller epidermal cells and stomata, shapes including roundish globular bladder cells with or without a small papillate appendix in central or apical position, more ovate or rhomboid bladder cells or elongated forms, then often ending basally and apically in a more or less obtuse papilla giving the leaves a rough touch, frequently the basal marginal bladder cells developed into elongated papillae, resulting in a ciliate lower margin, the leaf tip is differentiated into a diadem in most species, a unit consisting of one to several rows of basal cup-cells, above which apical bristles form an erect-inclining or radiating bundle, this type of diadem can be removed as a whole from the leaf; in other species, leaf tips can possess some elongated papillae only, which do not differ in colour from those of the laminae, or coloured bristles much more elongated than those of the leaf; **L** 4.5–44 mm *l*, 1–6 mm *b*; **Fl** solitary, overtopped by subsequent branches during the season forming a monochasium, or a cyme of 3, rarely 2–5 flowers per cyme, in one case bracteolate, pedicels hairy or papillate like **I** or papillae smaller, subsessile or up to 45 mm *l*, **Fl** 14–45 mm ∅, **K** (4–)5–6(–7), often bladder cells more elongated into long papillae or hairs, sometimes sclerotized, especially at the base of **K**, and coloured like the diadem, lobes unequal and with pergamentaceous margins, petals 1 to 4 series, emarginate or unequally tipped, lanceolate, white, cream-coloured, apricot-coloured with a red middle vein (vittate), light pink to dark pinkish-red, rarely yellow, filamentous staminodes present, rarely absent, stamens in 2–3(5) series, filaments occasionally ciliate, often inner circle at half their length or basally tufted with string-of-pearl-like papillae, (4–)5–6(–7) lophomorphic dark green nectaries, unequal in shape, stigmas (4–)5–6(–7), flowering almost all year, in most cases fully expanded at noon, sometimes in the afternoon, visitors may be Lycaenidae (Lepidoptera) or Masaridae (Vespoidea); **Fr** pedicels short or long, erect or angled, base of capsule conical to bell-shaped, greyish to light yellow-brownish, closing bodies absent in the typical forms, the pair of covering membranes derived from one septum heart-shaped, rhombic, or very narrowly ovate, almost reduced, often with white dots from raphide bundles, expanding keels of one valve as semi-circles or more erect, light yellow to sometimes basally dark brown, closed 4–9 mm ∅, open 6.5–11.8 mm ∅, 3–7 mm *h*; **S** flattened roundish to ovate, light to dark brown, in most cases the dorsal part darker in colour, the perisperm region may even be transparent, c. 0.8 mm *l*, 0.5 mm *b*; **Chr** $2n = 18, 36$; **Ecol** in fissures of rocks (decumbent forms), in white quartz fields (compact forms), in stony to sandy ground (erect to decumbent forms), grass veld, seldom in shady localities, on pastures often browsed; **Distr** SE part of Namibia; NC, NW, WC, FS, EC, S Africa.

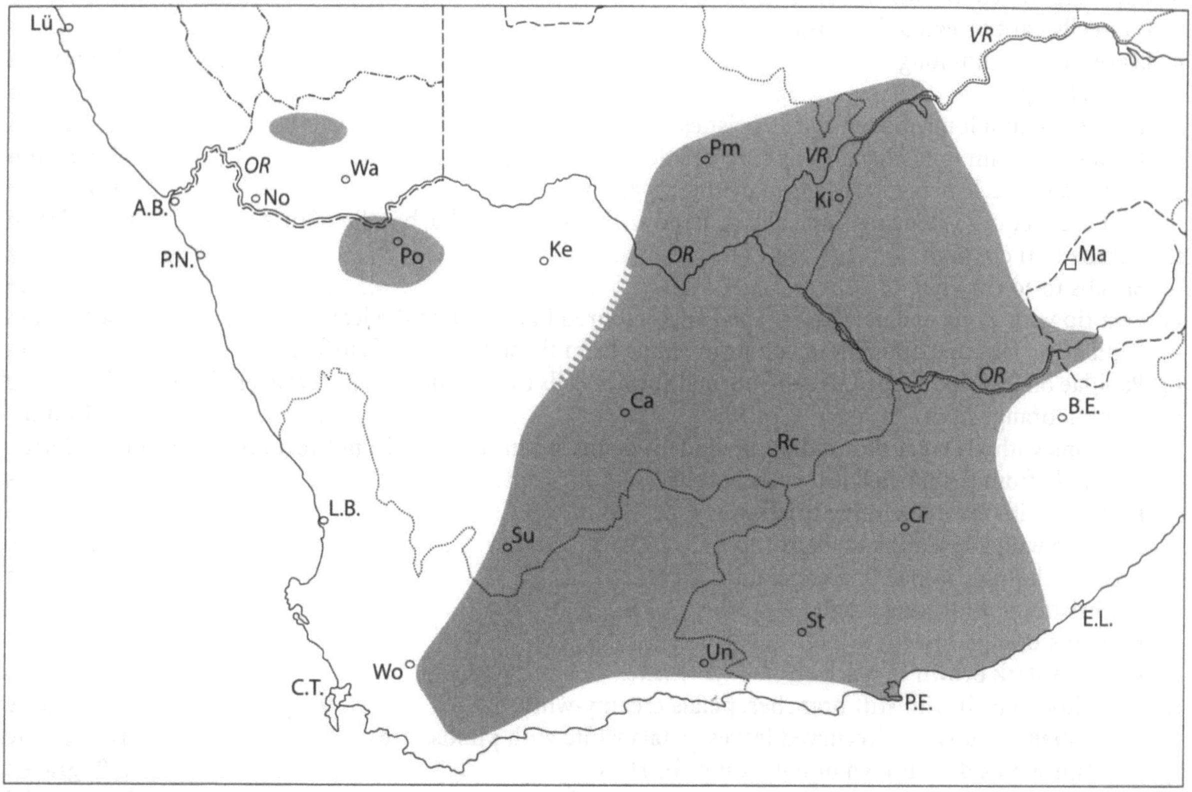

Distribution of **Trichodiadema** Schwantes

Note: The characteristic of the genus is obviously the diadem, which is in its simplest form difficult to distinguish from diadem-like structures in other genera, e.g. **Delosperma, Drosanthemum**, and **Mestoklema** (Niesler 1997: 13–27). Species of **Trichodiadema** play an important role as a pasture element in low, shrubby associations because in particular the species developing long shoots grow very quickly even after only moderate rainfalls. These are soon available after droughts and can be browsed until the more slowly growing, bigger shrubs or the grasses provide any fodder. Thus, regeneration from the base is an important feature of the growth form of several species, giving the farmer an indication on the situation the plant cover is in when having to decide about cycles of usage. These members of the genus are called "vye veld", representing one the valuable members of the family for farming.

Key to the Species of Trichodiadema

1. Very compact 2
– Erect, semi-erect to decumbent, internodes visible 12
2. Leaf-tip with elongated papillae or hairs of the same colour as those of the leaf 3
– Leaf-tip with a diadem proper, bristles rising from one to several rows of yellow to dark brown or blackish cup-cells 7
3. Papillae on the leaf-bladder cells long and hairy 4
– Papillae on the leaf-bladder cells not hairy 5
4. Roots thin tubers .... **T. fergusoniae**
– Roots thick tubers .... **T. strumosum**
5. Papillae of the bladder cells of the leaves elongated, leaves to 6 mm long .... **T. pygmaeum**
– Bladder cells roundish to ovate, leaves > 6 mm long 6
6. Leaves 10–20 mm long, bladder cells roundish with a very small central papilla .... **T. calvatum**
– Leaves to 7.5 mm long, their bladder cells ovate with a small apical papilla .... **T. occidentale**
7. Cup-cells of the diadems light yellow .... **T. emarginatum**
– Cup-cells of the diadems orange-brown or dark blackish brown 8

8. Cup-cells orange-brown 9
- Cup-cells dark brown to blackish 11
9. Leaves up to 5 mm long ..... **T. rupicola**
- Leaves longer 10
10. Leaves c. 6 mm long, diadems of 12 bristles ..... **T. burgeri**
- Leaves to 10 mm long, diadems of c. 20 bristles ..... **T. densum**
11. Leaf bladder cells irregularly edged, without papillae ..... **T. stayneri**
- Leaf bladder cells elongated, apically spurred with a small papilla, basally obtuse ..... **T. marlothii**
12. Shrubs 6–11 cm high 13
- Shrubs to 30 cm high 26
13. Leaf tip with elongated papillae, shaped and coloured like those of the leaf ..... **T. attonsum**
- Papillae on leaf tips different in colour or shape from those on the leaf surface 14
14. Papillae on leaf tips shaped like the normal surface cells but smaller, drying early on the assimilating leaf turning orange ..... **T. imitans**
- Leaf tips with an erect or a radiating diadem or the bristles at the tip not forming a unit but differing distinctly from the normal leaf surface cells 15
15. Leaf tips with erect, inclining bristles 16
- Leaf tips with a radiating diadem 20
16. Erect, inclining papillae forming a unit 17
- 5–8 bristles not forming a unit 19
17. Diadems orange-brown ..... **T. hallii**
- Diadems dark brown to blackish 18
18. Caespitose shrub with stiff branches, petals creamy-white ..... **T. mirabile**
- Decumbent shrubs with recurved leaves, petals white with pinkish tips ..... **T. orientale**
19. Leaf tips with 5 dark brown or 6–8 orange bristles ..... **T. gracile**
- Leaf tips with c. 12 brown bristles, sometimes spreading ..... **T. fourcadei**
20. Diadems with one row of cup cells 21
- Diadems with several rows of cup cells 28
21. Diadems with 1–6 short yellow bristles < 1 mm long 22
- Diadems with longer bristles 23
22. Branches to 14 cm long ..... **T. obliquum**
- Branches to 24 cm long ..... **T. ryderae**
23. Petals red outside, yellow inside ..... **T. introrsum**
- Petals in shades of pink 24
24. Internodes < 10 mm long 25
- Internodes > 10 mm long 26
25. Flowers light pink, leaves 15 mm long, 1.2 mm broad ..... **T. pomeridianum**
- Flowers pinkish-purple, leaves 16 mm long, 5 mm broad ..... **T. rogersiae**
26. Flowers pale pink to whitish ..... **T. littlewoodii**
- Flowers bright pink, mauve, or purple 27
27. Leaves green, pedicels 25–38 mm long ..... **T. setuliferum**
- Leaves bluish-green, pedicels 6–32 mm long ..... **T. decorum**
28. Flowers white or yellow 29
- Flowers in shades of purple 31
29. Flowers golden yellow ..... **T. aureum**
- Flowers white 30
30. Calyx base bristly from dark brown papillae ..... **T. olivaceum**
- Calyx base hispid from elongated, soft, white hairs ..... **T. peersii**
31. Diadem of 4–6 white bristles ..... **T. concinnum**
- Diadem of 6–10 whitish bristles 32
32. Flowers pale pink 33
- Flowers bright pink 34
33. Locules 5 ..... **T. intonsum**
- Locules 6(–8) ..... **T. hirsutum**
34. Diadem with 5–7 bristles, cup cells black ..... **T. barbatum**
- Diadem with c.16 bristles, cup cells brown ..... **T. bulbosum**

T. attonsum (L.Bolus) Schwantes ZSK 1927: 106 ≡ *M. attonsum* L.Bolus ABH 1927: 78 **T** Muir 3969 (BOL!).

Erect shrubs 10–15 cm *h*, **I** with very short and small distant papillae, **I** 7–10 mm *l*, 2 mm ∅; **L** bladder cells with one apical papilla, papillae at leaf tip elongated, not forming a diadem, not different in colour from the leaf, **L** 10–13 mm *l*, 3–4 mm *b*; **Fl** solitary, pedicels with longer papillae than the leaf, base of those papillae enlarged in contour, pedicels 3–12 mm *l*, **K** 5, 16 white petals in 3 series, filamentous staminodes present, 16–20 mm ∅, XI in cultivation; **Fr** covering membranes well developed, expanding keels narrowing, (4–)5 locules, open 6.5–7 mm ∅, closed 4–4.5 mm ∅, 3 mm *h*; **Ecol** on white quartzite; **Distr** locality uncertain, Riversdale, WC, S Africa.

Note: The leaves are lacking the typical diadem.

T. aureum L.Bolus 1962: 11 **T** Stayner 628/61 (BOL!).

Erect to semi-erect shrubs 12–13 cm *h*, roots 13 cm *l*, tuberous after 3–4 cm, tuber 9 cm *l*, 3.5 cm ∅, base of stem 8 mm ∅, **I** brownish-yellow, distant papillae along the edges, 5–10 mm *l*, 1–2.5 mm ∅; **L** dorsally rounded, bladder cells as elongated papillae, apical and basal parts obtusely elongated, the apical a little longer, diadem radiating, cup-cells brown, bristles 5–12, lighter than the cup-cells, **L** 5–6 mm *l*, c. 3 mm *b*; **Fl** pedicels 2–8 mm *l*, papillae more elongate than those on **I**, **K** 5, base with rough hairs from even more elongated papillae than on the pedicels, 31 mm ∅, petals 3-seriate, golden-yellow, base pale, outside pale pink, filamentous staminodes pale, stamens 4-seriate, filaments pale, inner circle papillate at half their length, anthers yellow, IX; **Fr** presumably 5 locules; **Ecol** unknown; **Distr** Willowmore, EC, S Africa.

Note: In the genus only two species possess yellow flowers, **T. aureum** and **T. introrsum**.

T. barbatum (L.) Schwantes ZSK 1926: 188 ≡ *M. barbatum* L. 1753: 482 ≡ *M. stelligerum* Haw. 1812: 278 nomen illeg. **Lectotype**, here designated: Dillenius 1732: t. 190 f 234 = *M. stellatum* Miller 1768: nr. 14 non Haworth 1795: 416 nomen illeg. ≡ *T. stellatum* (Miller) Schwantes ZSK 1926: 188 ≡ *M. stelligerum* Haw. 1824: 61 nomen illeg. non Haw. 1812: 278 ≡ *T. stelligerum* (Haw.) Schwantes ZSK 1926: 188 **Lectotype**, here designated: Bradley 1716: p 6, t. 5.

Erect shrubs, branches slender, sublignose, partially procumbent, **I** rough from papillae, fragile, green when young and greyish when older; **L** green to greyish, trigonous, bladder cells rather big, globular with edged sides as the result of the arrangement in narrow rows, diadem radiating, 5–7 bristles, black cup-cells; **Fl** solitary, pedicels c. 2 × as long as the **I**, starting to open at 7–8 a.m., fully expanded at noon and closing soon afterwards, **K** 5, petals 1-seriate, bright purple, notched at tip, appearing almost metallic in full sun, filamentous staminodes shorter and of paler colour, anthers yellow, stigmas 5; **Fr** 5 locules; **Ecol** unknown; **Distr** uncertain, "Cape", S Africa.

Note: Both paintings cited by Linnaeus (1753) resemble each other very much, differing in the contour of the calyx base only: Dillenius' plant shows a more cone-shaped calyx base, Bradley's (1716: 6 t. 5) a more bell-shaped, rounded form. The drawing of Dillenius is chosen as lectotype because it appears to represent the more widespread form. The lectotypification follows the arguments of Stearn (1939: 72, 74).

T. bulbosum (Haw.) Schwantes ZSK 1926: 188 ≡ *M. bulbosum* Haworth 1824: 428 **T?**

Erect shrubs with several basal branches, 1–3 oblong "bulbous" tubers rising 1–2 cm above the ground and continuing c. 5 cm downwards, **I** smooth, light brownish-yellowish, 2–4.5 mm *l*; **L** cylindrical, covered with ovate bladder cells of which both, apical and basal parts, terminate in a papilla, the apical one c. twice the length of the basal one, diadem radiating, 4 mm ∅, c. 16 white bristles from orange-brown cup-cells, dark brown on older leaves, the long bristles of all diadems, contrasting with the short ones on the rest of the leaf, produce a cover like that of a cobweb over the entire plant, **L** 4 mm *l*, 1.8 mm *b*; **Fl** terminal, solitary, pedicels 0–2 mm *l*, **K** 5, base with a ring of enlarged papillae, petals bright dark pink, uniseriate; **Fr** 5 locules; **Ecol** unknown; **Distr** uncertain, "Cape", S Africa.

Note: The described tuberous root, rising above the ground, combined with erect branches and the cobweb-like cover are most distinctive of this species.

T. burgeri L.Bolus 1963: 17 **T** Burger 27274 (BOL!).

Erect shrubs 2–6 cm *h*, branch bases 2 mm ∅, rootstock tuberous, 33 mm ∅, > 54 mm *l*; **I** overlapped, c. 2 mm *l*; **L** imbricate, the elongate bladder cells ending in a basal obtuse part, the apical part elongated as papilla, radiate diadem, 14 white bristles from yellow-orange cup-cells, 5–6 mm ∅, **L** 5–12 mm *l*, 4 mm ∅; **Fl** pedicels densely covered by thick hairs, 3–6 mm *l* while flowering, 8 mm *l* while fruiting, **K** base as hairy as the pedicels, c. 38 mm ∅, petals in 2 series, obtuse, reddish, filamentous staminodes present, filaments white, 5 series, 5 mm *l*, outer circle obscurely ciliate-papillate, inner circle with papillate hairs at the bases, stigmas 5, 2.8 mm *l*; **Fr** pale, locules 5, covering membranes well developed, the area near the septum whitish from stiff white cells, expanding keels sickle-shaped at the inner side, outer side almost right-angled, half length of the outer proximal side dark brown, closed 7 mm ∅, open 12 mm ∅, 5.2 mm *h*; **Ecol** unknown; **Distr** Oudtshoorn, WC, S Africa.

Note: There is a strong similarity to **T. densum**, which differs in the number of six locules and the more compact growth form.

T. calvatum L.Bolus NM 1936: 19 **T** Esterhuysen 2747/35 (BOL!).

Erect shrubs 9–13 cm *h*, roots very thick just below the branching and soon bifurcating downwards, 3.8 cm ∅, **I** papillate, 10–18 mm *l*; **L** bladder cells globular to ovate with an apical papilla, tip-papillae elongated, no diadem, **L** 10–20 mm *l*, 3.8 mm *b*; **Fl** pedicels 10–15 mm *l*, hairy by elongation of the apical papilla of bladder cells, 30 mm ∅, petals purplish-pink, 1–2 series, 10–20 mm *l*, 2.5 mm *b*, filamentous staminodes none, stamens in 2 series, bases of the inner circle of filaments adaxially with a tuft of transparent string-of-pearl-like hairs, stigmas 5, 2.5 mm *l*, IX in cultivation; **Fr** 5 locules, covering membranes overlapping above the locules, closed 6.5 mm ∅; **Ecol** unknown; **Distr** Robertson, WC, S Africa.

Note: The leaves are lacking the typical diadem and the flower the filamentous staminodes.

T. concinnum L.Bolus 1964: 79 **T** Hall 1559/50 (BOL!).

Erect shrubs 3.5–13 cm high, some branches with long **I** decumbent, other plants more compact, roots consisting of a bundle of long and thin parts and several tubers, each c. 6.5 cm *l*, 4–13 mm ∅, **I** yellow-grey, almost smooth, only sometimes several very short papillae at the edges, **I** 10 mm *l*; **L** leaf tips recurved, bladder cells ovate, with an obtuse apical papilla, radiate diadem, c. 4–6 white bristles from brown-blackish cup-cells, 4.5 mm *l*, 1.5 mm *b*; **Fl** pedicels densely covered with obtuse hairy papillae, 15 mm *l*, 21 mm ∅, **K** base with a ring of obtuse hairs, petals purplish-pink, 28–32, 1–2 series, 11 mm *l*, filamentous staminodes white, filaments 3-seriate, white, those of the inner circle papillate at half their length, anthers bright yellow, stigmas 5, V; **Fr** 5 locules, closed 6 mm ∅; **Ecol** unknown; **Distr** Alexandria, EC, S Africa.

Note: Bolus (1964: 80–81) suggested that this species was very closely related to **T. bulbosum**, which develops more erect branches.

T. decorum (N.E.Br.) Stearn ex H.Jacobsen 1939: 9 ≡ *M. decorum* N.E.Br. GC 1912: 403 **T** Pearson 4221 (K!).

Shrubs with erect branches c. 18 cm *h*, **I** dotted from white papillae, up to 20 mm *l*; **L** bluish-green, bladder cells obtusely rounded to 5–6-edged, with a central papilla, tip with a simple radiating diadem, 7–9 white bristles from orange brown cup-cells, **L** up to 23 mm *l*, up to 2.5 mm ∅; **Fl** pedicels with few papillae, 6–32.5 mm *l*, terminating in a single flower or a cyme of three flowers, **K** 5, ∅ c. 15 mm, petals bright rosy-mauve, filamentous staminodes > stamens, V in cultivation; **Fr** 5 locules; **Ecol** unknown; **Distr** Albany, WC, S Africa.

Note: Following Brown (GC 1912: 403), the type plant had been figured in Curtis' Botanical Magazine 1873: t. 6057. Brown agreed with the drawing as representing **T. decorum** (as *M. decorum*) except for the flower colour and the colour of the leaf tips. Since the colour of the flower and the colouring of the diadem appear to be constant features for species in **Trichodiadema**, Brown's view as discussed above is not followed here but the plant is understood to represent a different species: **T. introrsum.**

T. densum (Haw.) Schwantes ZSK 1926: 188 ≡ *M. densum* Haw. 1795: 302 ≡ *M. barbatum* L. var. γ L. 1753: 482 ≡ *M. barbatum* L. var. *densum* Willdenow 1799: 1046 **Lectotype**, here designated: Dillenius 1732: t. 248 f 236.

Compact, **I** not visible; **L** bladder cells apically and basally elongated as papillae, radiating diadem 10 mm ∅, c. 20 white bristles, cup-cells orange, 8–10 mm ∅, **L** c. 11 mm *l*, 4 mm *b*; **Fl** pedicels hairy, 7 mm *l* while fruiting, **K** 6, petals purplish-red, nectaries (5–)6(–8), stigmas 6; **Fr** covering membranes from one septum heart-shaped, expanding keels brown, shaped like an irregular triangle, 6 locules, closed 7.5 mm ∅, open 11 mm ∅, 6.5 mm *h*; **Ecol** in white quartz fields; **Distr** Willowmore, EC, S Africa.

Note: **T. densum** possesses the most compact growth form in the genus and the diadems with the largest diameter; its habit has been compared to species of the genus **Mammilaria** in Cactaceae.

T. echinatum ≡ **Delosperma echinatum**

T. emarginatum L.Bolus NM 1928: 151 **T** Pillans 467/16 (BOL!).

Compact, to 3.5 cm *h*, stem base 14 mm ∅, **I** invisible; **L** imbricate, bladder cells ovate, apical and basal parts ending in a papilla each, apical papilla a little longer, radiating diadem of orange-brown cup-cells and 5–10 whitish bristles, 3.2 mm ∅, **L** 15 mm *l*, 6 mm *b*; **Fl** pedicels densely covered with hairs on voluminous bases, 7 mm *l*, **K** 5, base with longer hairs than pedicels, 30 mm ∅, petals pink, emarginate, c. 15 mm *l*, uniseriate, filamentous staminodes as long as stamens, base papillate, inner filaments at half their length with white papillae, anthers pale yellow, stigmas 5, VII; **Fr** unknown; **Ecol** unknown; **Distr** unknown.

Note: The leaves resemble those of **T. mirabile**, but differ in the diadem-type, whereas the similarity to **T. marlothii**, differing in the number of locules in the fruit, is evident in growth form.

T. fergusoniae L.Bolus NM 1928:9 **T** Ferguson 18822 (BOL!).

Compact, 3 cm *h*, root thin, side-roots rare, 12 cm *l*, up to 2.5 cm ∅, **I** invisible; **L** densely arranged, bladder cells with a central or more apical hair 1 mm *l*, hairs at the leaf tip not longer, **L** 9–16 mm *l*, 2.5–5 mm *b*; **Fl** pedicels hairy, hairs slightly longer towards the calyx base, pedicels 5(–8.5) mm *l*, **K** 5, bases hairy, petals apricot-coloured with a red middle vein, filamentous staminodes as long as stamens, VII; **Fr** 5 locules, reddish brown, closing bodies 2-lobed, covering membranes triangular overlapping, those derived from one septum forming a rhomb, closed 6 mm ∅, open 9 mm ∅, 3 mm *h*; **Ecol** unknown; **Distr** Riversdale, WC, S Africa.

Note: Bolus supposed this to be the same species as **T. strumosum** (Bolus SAG 1928: 127). The typical diadem is lacking, the reddish-brown colour of the fruits is not typical of the genus and closing bodies do not occur in forms with typical diadems.

T. fourcadei L.Bolus NM 1928: 138 T Fourcade 3327 (BOL!).

Erect to decumbent shrubs, 8–11 cm *h*; roots with tubers 30 mm under ground, tubers 22 mm *l*, 11 mm ∅, **I** with hairy papillae, 5–19 mm *l*; **L** bladder cells with an apical papilla, at the apex c. 12 completely brown bristles which cannot be removed as a unit, **L** 15–25 mm *l*, 2–3 mm ∅; **Fl** pedicels hairy, 5–26 mm *l*, solitary, **K** 5, base hairy, bristly brown papillae in 1–2 rings prominent, 20 mm ∅, petals white, in 3 series, filamentous staminodes present, stamens 4 mm *l*, filaments white, anthers golden-yellow, inner circle papillate, II in cultivation; **Fr** covering membranes with small white dots of raphide-bundles, not touching each other, expanding keels light yellow, 5 locules, closed 8 mm ∅, open 11 mm ∅, 7 mm *h*; **Ecol** hillsides, at 100 m altitude; **Distr** Humansdorp, EC, S Africa.

Note: This diadem shows longer bristles than that of other members of the group with erect, inclining diadems, the bristles are irregular in shape, not erect but bent or undulate; cup-cells are absent; the growth form resembles a stouter version of **T. gracile**.

T. gracile L.Bolus NM 1929: 126 T Neethling 1998/24 (BOL!) = *T. gracile* L.Bolus var. *piliferum* L.Bolus 1964: 78 T Ferguson 21440 (BOL!) = *T. gracile* L.Bolus var. *setiferum* L.Bolus 1964: 79 T Ridley 402/14 (BOL!).

Erect shrubs with some decumbent branches, 20 cm *h*, roots tuberous, 100 mm *l*, 30 mm ∅, **I** with some short papillae along the edges, 7–11 mm *l*, 1.2 mm *b*; **L** bladder cells with an apical and a basal papilla, lateral basal papillae elongated, laminal papillae reduced, about 6 papillate bladder cells elongated at the leaf-tip, the whole tip orange, **L** 10–40 mm *l*, 3 mm ∅; **Fl** solitary, pedicels c. 15 mm *l*, **K** 5, 18 mm ∅, petals pink, 1-seriate, 7 mm *l*, 1.2 mm *b*, purple filamentous staminodes as long as the stamens, anthers with whitish pollen grains, VII in cultivation; **Fr** unknown; **Ecol** dry hillside, very stony; **Distr** Caledon, Mosselbay, WC, S Africa.

Note: The papillae of the leaf tip are not at one level and although they are coloured they are not forming typical diadems with a base of cup-cells. **T. gracile** and especially the apricot-coloured flowering *T. gracile* var. *setiferum* belongs to the more doubtful taxa in the genus.

T. gracile var. *piliferum* = **T. gracile**
T. gracile var. *setiferum* = **T. gracile**

T. hallii L.Bolus NM 1950: 214 T Hall 1125/47 (BOL!).

Erect shrubs 6 cm *h*, in cultivation up to 20 cm, roots tuberous, **I** covered by the leaves, 4 mm ∅; **L** strong and stiff, with elongate bladder cells, apical part forming a thick papilla, erect inclining orange-brown diadem, c. 10 elongated bristles, **L** 11–18 mm *l*, 4 mm ∅; **Fl** sessile or on very short pedicels with bladder cells with an obtuse apical papilla, 46 mm ∅, petals white to light pink, 2-seriate, inner white, 20–30 mm *l*, 2.5 mm *b*, outer filaments ciliate, inner basally papillate, nectaries 5–6, XII; **Fr** (4–)5–6 locules; **Ecol** unknown; **Distr** Ladismith, WC, S Africa.

Note: The leaves show the only diadem of the erect-inclining type in orange-brown colour, the same shape in **T. mirabile** and **T. orientale** is dark brown.

T. hirsutum (Haw.) Stearn 1939: 80 ≡ *M. hirsutum* Haw. 1795: 297 ≡ *M. barbatum* L. β 1753: 481 ≡ *M. barbatum* L. var. *hirsutum* Willdenow 1799: 1045 **Lectotype**, here designated: Dillenius 1732: t. 190 f 235.

Low, erect shrubs, some branches decumbent, **I** with horizontal hair-like papillae; **L** oblong, semicylindric, bladder cells roundish-oblong with an apical papilla, radiating diadem of 20(–40) white bristles from pale brown cup-cells; **Fl** **K** 6(–7), lobes unequal, petals pale purple, filaments with apical crystalline papillae, anthers pale yellow; **Fr** 6 locules; **Ecol** unknown; **Distr** uncertain, "Cape", S Africa.

Note: The plant described resembles **T. marlothii** very much, even in the 6 locules of the fruit. The high number of diadem bristles is unique in the genus.

T. imitans L.Bolus NM 1950: 214 T without collector 11928 (BOL!).

Shrubs with erect strong branches to 33 cm *h*, **I** dotted by small vertical hairs, 28 mm *l*, 2.5–3 mm *b*; **L** recurved, bladder cells elongate with an apical papilla, early drying, tip of the recurved leaf abaxial, without a diadem, terminal papillae of the same shape as those of the leaves, but the tip-zone orange, old leaves may loose the complete tip, basal margins with elongated papillae, **L** 15–25 mm *l*, 2–3 mm *b*; **Fl** pedicels with short hairs as on **I**, but in a dense arrangement, 3–4.5 mm *l*, **K** 5, 34 mm ∅, petals white, 3 series, 10–15 mm *l*, filamentous staminodes and filaments white, 5 mm *l*, inner circle with long papillae, anthers golden yellow, I–VII in cultivation; **Fr** covering membranes laterally narrowed, expanding keels yellow,

5 locules, closed 7.5 mm Ø, open 8.8 mm Ø, 6.7 mm *h*; **Ecol** unknown; **Distr** unknown.

Note: The yellow leaf-tips without a diadem, even without elongated papillae, represent a deviating structure in the genus.

T. inornatum ≡ **Drosanthemum inornatum**

T. intonsum (Haw.) Schwantes ZSK 1928: 188 ≡ *M. intonsum* Haw. 1824: 62 **Lectotype**, here designated: drawing of Duncanson 963/699, 16.8.1824 (K!) = *M. intonsum* (Haw.) Schwantes var. *rubicundum* Haw. 1824: 62 **Lectotype**, here designated: drawing of Duncanson 962/968 (K!).

Shrubs erect to decumbent; **L** semi-erect, base with lateral elongated papillae, thus appearing ciliate, tip with a radiating diadem of < 10 bristles rising from black cup-cells; **Fl** solitary, pedicels short, hispid, **K** 5, base hispid from enlarged black papillae, diadem of the calyx-lobes with 5 white bristles rising from black cup-cells, petals pale pink; **Fr** 5 locules; **Ecol** unknown; **Distr** WC, S Africa.

Note: A little known species.

T. introrsum (Haw. ex Hook.f.) Niesler **comb. nov.** (see **photo**) ≡ *M. introrsum* Haw. ex Hook. f. Curtis Botanical Magazine 99 (29 in series 3), 1873: t. 6057 (as basionym) ≡ *T. introrsum* (Haw. ex Hook.f.) Niesler 1997: 16 nomen invalid. **Holotype** Mac Owan 6057 (K!).

Erect to decumbent shrubs, 25 cm *h*, branched from the base, some branches > 30 cm *l*, 1–2 mm *b*, **I** rough from white papillae below the leaves, 6–14.5 mm *l*; **L** somewhat recurved, semi-terete with very rounded sides, bladder cells rounded without papillae, simple orange-brown diadem of up to 10 bristles from one row of cup-cells, **L** up to 25 mm *l*, c. 5 mm *b*; **Fl** pedicels slender, more or less papillate like the **I**, 10–18 mm *l*, **K** 5, Ø c. 17 mm, c. 60 petals in 2–3 series, golden-yellow with red tips and light pink bases inside, reddish outside, c. 40 filamentous staminodes in 2 series, light pink with yellow tip, stamens c. 50 in 2–3 series, inner circle of filaments tufted at half their length with string-of-pearl-like papillae, stigmas 5; **Fr** pedicels up to 18 mm *l*, 5 locules, covering membranes inserting in one septum obtuse-ovate, white dots of raphide-bundles, expanding keels very light yellow, not diverging; **Ecol** unknown; **Distr** Alexandria, Kirkwood, EC, S Africa.

Note: Although the name *M. introrsum* originated in a printer's error (Brown GC 1912: 403), the plant described is neither **T. intonsum** nor **T. decorum**, as named by Brown. Both latter species flower pink with **T. intonsum** possessing a hairy calyx base. In order not to hide this wonderful yellow-flowering species, it is kept separate here and the new combination is made.

T. littlewoodii L.Bolus 1962: 302 **T** Littlewood 267/61 (BOL!) = *T. littlewoodii* L.Bolus forma *alba* L.Bolus 1962: 303 **T** without collector, Karroo Garden 267/61 (BOL).

Shrubs 19–25 cm *h* with rather stiff branches, roots in a bundle of some primary tubers, each c. 60 mm *l*, 2.5 cm Ø, secondary tubers c. 16, 90 mm *l*, 22 mm Ø, **I** strong, edged, grey brown with small bladder cells with a very small papilla, all visible as rough white dots on younger parts, 10–30 mm *l*, 1–3 mm Ø; **L** only on apical parts of branches, bladder cells rounded to 5- or 6-edged, no papillae, simple radiating diadem, cup-cells yellow, 6–9 lighter bristles, **L** 10–27 mm *l*, 3–5 mm Ø; **Fl** bases of pedicels with very small bladder cells but with longer papillae than on **I**, the bladder cells bigger towards the **K**, the apical papilla more and more obtuse, 10–20 mm *l*, 22 mm when fruiting, 2–5 flowers in terminal cymes, 15–20 mm Ø, pale pink to whitish, petals 4 series, 8 mm *l*, filamentous staminodes present, filaments 4 series, 4 mm *l*, inner circle at about half length papillate, anthers pale yellow, stigmas 5, III; **Fr** 5 locules, covering membranes of one septum forming a narrow oval, expanding keels light yellow, closed 6.5 mm Ø, open 10.5 mm Ø, 5 mm *h*; **Ecol** unknown; **Distr** Warmbad, Namibia.

Note: This rather high shrub with strong branches is very rich in flowers, more so than any other species in the genus.

T. littlewoodii forma *alba* = **T. littlewoodii**

T. marlothii L.Bolus NM 1928: 150 **T** Marloth 17371 (BOL!).

Shrubs 4 cm *h*, longer branches 12 cm *l*, almost decumbent, stem base 12 mm Ø, **I** invisible; **L** bladder cells spurred with an apical papilla, base obtusely elongated, radiating diadem parted to two sides like hair, 5–10 white bristles from brown cup-cells, 2.5–5 mm Ø, **L** 17 mm *l*, 6 mm *b*; **Fl** pedicels with brownish hairs, 10 mm *l*, 30–35 mm Ø, **K** whole base with long brownish hairs, petals dark pink, many of them slightly emarginate, 1–2 series, 13 mm *l*, 1.5 mm *b*, filamentous staminodes with a white base and a pink tip, 7 mm *l*, filaments laxly papillate-pubescent, anthers pale yellow, nectaries 6, stigmas 6, VI; **Fr** 6 locules, covering membranes well developed; **Ecol** unknown; **Distr** Swellendam, WC, S Africa.

Note: **T. marlothii** resembles **T. emarginatum** very closely, which shows in contrast 5 locules in the fruit, a lower growth form, and more emarginate petals.

T. mirabile (N.E.Br.) Schwantes ZSK 1926: 188 ≡ *M. mirabile* N.E.Br. GC 1903: 131 **T** Pillans 10696 (BOL!) = *T. mirabile* (N.E.Br.) Schwantes var. *leptum* L.Bolus NM 1950: 215 **T** Esterhuysen 7779 (BOL!).

Erect shrubs 2.5–8 cm *h*, branches densely leaved, **I** bladder cells only with a more enlarged apical papilla in contrast to the leaf, I to 20 mm *l*; L erect, stiff, bladder cells rhomboid with a basal and an apical papilla, the latter more enlarged, lateral basal bladder cells with a long apical papilla, erect inclining blackish-brown diadem of c. 12 stiff bristles, their tips paler, 8–12 mm *l*, 1–2 mm *b*; **Fl** subsessile, bracteolate, pedicels when fruiting 4 mm *l*, **K** 6, white to cream-coloured petals c. 18 mm *l*, 2–3-seriate, filamentous staminodes present, filaments white with a tuft of hairs at their base, anthers yellow, VI in cultivation; **Fr** the last leaf-pair persisting, 6 locules, covering membranes each distally with a V-shaped cut, closed 5.8 mm ∅, open 9 mm ∅, 6 mm *h*; **Ecol** on hillside, southern exposition; **Distr** Laingsburg, WC, S Africa.

Note: The erect, inclining dark diadem, the white to cream-coloured petals, and the covering membranes are typical of **T. mirabile**. *T. mirabile* var. *leptum* possesses more slender branches, but grows taller than the typical form of the species.

T. mirabile var. *leptum* = **T. mirabile**

T. obliquum L.Bolus 1961: 121 T Stayner 773/61 (BOL!).

Erect shrubs 11 cm *h*, decumbent branches longer, I with white ovate to oblong bladder cells, 10–20 mm *l*, 1.5 mm ∅; L bladder cells rounded to 5–6-edged, lateral basal bladder cells with a central papilla, pointing upwards, small, simple diadem of c. 5 short yellow bristles from at most 1 row of yellow cup-cells, L 15–20 mm *l*, 3 mm ∅; **Fl** pedicels very thin, few papillae, 10–20 mm *l*, 15 mm ∅, petals dark pink, 4–5 series, 5–6 mm *l*, filamentous staminodes 3 mm *l*, filaments white, outer basally ciliate, inner papillate at medium height, anthers yellow, stigmas 5, II; **Fr** covering membranes from one septum heart-shaped, expanding keels yellow to greyish-brown, 5 locules, closed 7 mm ∅, open 9 mm ∅, 6 mm *h*; **Ecol** unknown; **Distr** Beaufort West, WC, S Africa.

Note: **T. obliquum** is very similar to other species with long branches and long, often climbing, internodes.

T. occidentale L.Bolus NM 1928: 150 T Smith 5140 (BOL!).

Very dense, c. 2.2 cm *h*, roots below first branch very tuberous, 3.5 cm *l*, 1–1.5 cm *b*, I covered by leaves, invisible; L bladder cells very narrow, irregular in shape, bulged, with an apical papilla, bladder cells at tip with an elongated papilla, 0.5 mm *l*, L 11 mm *l*, 3 mm *b*; **Fl** pedicels hairy, 8 mm *l*, solitary, **K** 5, base hairy, 15 mm ∅, petals apricot-coloured, 8 mm *l*, 2 mm *b*, filamentous staminodes as long as the stamens, stamen 2.5 mm *l*, filaments purplish, stigmas 5, 1.5 mm *l*, IV in cultivation; **Fr** covering membranes triangular, overlapping, those derived from one septum forming a rhomb, 2 globular closing bodies, 5 locules, closed 6 mm ∅, open 9.5 mm ∅, 5.2 mm *h*; **Ecol** 110 m altitude; **Distr** Bredasdorp, WC, S Africa.

Note: Closing bodies are present in the fruits of this species, and the leaf tips shows no difference in colour, resembling erect inclining-diadem only in shape.

T. olivaceum L.Bolus NM 1931: 261 (see **photo**) T Grobbelaar 9365 (BOL!).

Erect shrubs up to 11 cm *h*, branches either densely leaved and I invisible, or I up to 10 mm *l* and branches then decumbent, bladder cells terminating apically in a rough papilla, roots 7 cm *l*, 3–8 mm ∅; L bladder cells somewhat elongate, apical and basal parts developing a papilla each, the basal one more obtuse, radiating diadem 7 mm ∅, cup-cells brown, the 5–8 bristles lighter, L 5–12 mm *l*, 2 mm *b*; **Fl** bladder cells on pedicels with very long apical hairs, thus appearing pubescent, 7 mm *l*, **K** base hairy from dark brown hairs, 20–40 mm ∅, petals white, 2 series, 11 mm *l*, 2 mm *b*, filamentous staminodes white, 6 mm *l*, inner filaments at medium height with papillae, straw-coloured, stigmas 5, III/IV; **Fr** closed 7 mm ∅, open 11.8 mm ∅, 3 mm *h*, covering membranes very narrow, expanding keels narrow, distal brown, proximal yellow; **Ecol** unknown; **Distr** Somerset East, EC, S Africa.

Note: The local name is "Kerrimoer".

T. orientale L.Bolus NM 1928: 150 (see **photo**) T Paterson 212 (BOL!) = *M. intonsum* Haw. var. *album* Haw. 1824: 62 **Lectotype**, here designated: drawing of Duncanson 962/35 (K!).

Erect shrubs 6.5–10 cm *h*, some branches decumbent, roots thickened like a spindle, 9 mm ∅, thin, branched, to 19 cm *l*, I younger parts with bladder cells, resembling those of the leaf, but smaller and more flattened, 2–12 mm *l*; L bladder cells enlarged, only an apical papilla formed, base rounded, basal-lateral papillae enlarged, erect inclining diadem dark brown, c. 16 stiff, erect bristles, L 10–14 mm *l*, 5 mm ∅; **Fl** pedicels papillate like I, c. 6 mm *l*, **K** base with a ring of dark brown bristles, 22 mm ∅, petals light pink to whitish with pink tips, densely arranged in 4 series, filamentous staminodes > stamens, inner filaments with hairs in apical region, stigmas 5, XI; **Fr** covering membranes from one septum forming a narrow oval, expanding keels light yellow, 5 locules, closed 5 mm, open 7–12 mm ∅, 3 mm *h*; **Ecol** among rocks, 360 m altitude; **Distr** Albany, Uitenhage, Ciskei, EC, S Africa.

Note: In contrast to the superficially similar **T. mirabile**, the bladder cells of the leaves of **T. orientale** are not so enlarged, the plants grow less high, and the flowers are light pink or white with pink tips.

T. peersii L.Bolus NM 1929: 42 T Peers 1265/25 (BOL!).

Suberect shrubs 7–9 cm *h*, **I** greyish-beige with some transparent papillae, 2.5 mm *l*, up to 2.7 mm *b*, flowering branches densely leaved, hence **I** invisible; **L** bladder cells oblong, almost of rhombic shape, basally and apically papillate, apical papilla a little more obtuse and curved, radiating diadem, 4–9 lighter bristles rising from dark brown cup-cells, **L** 5–8 mm *l*, 4 mm ∅; **Fl** pedicels with bladder cells with an apical enlarged papilla, brownish, transparent, up to 10 mm *l* or sessile, **K** 5, base with bladder cells with smaller papillae than on pedicels, diadems of the lobes with deeper brown bristles than those of the leaves, c. 38 mm ∅, petals white, 1 row, 18 mm *l*, filamentous staminodes white, 6 mm *l*, stamens at half their length papillate, 4 mm *l*, anthers yellow, stigmas 5, summer; **Fr** 5–6 locules, covering membranes from one septum heart-shaped, expanding keels orange-brown to yellow, sharp-angled, closed 4–5 mm ∅, open 7 mm ∅, 3 mm *h*; **Ecol** unknown; **Distr** Willowmore, EC, S Africa.

Note: The coloured diadem bristles on the calyx lobes are unusual.

T. pomeridianum L.Bolus NM 1928: 138 T Smith 3912 (BOL!).

Erect shrubs c. 30 cm *h*, higher in dense grass, root long and thin, roots after 9–15 cm with 1 or 2 tubers each, 3.3 cm *l*, 15 mm *b*, root continuing, **I** with white rounded papillae enlarging towards the node, **I** c. 9 mm *l*; **L** bladder cells dense, without papillae, 5–6 edged, leaf base lateral with elongated apical papillae on the bladder cells, tip with a simple radiating diadem, 3–8 yellowish bristles rising from not very distinct cup-cells of the same colour, some bladder cells lateral of the diadem more enlarged than the bladder cells of the leaves, but yellowish in colour like diadem bristles, **L** 15 mm *l*, 2 mm ∅; **Fl** pedicels with fewer and smaller papillae than on **I**, 10–13 mm *l*, c. 15 mm ∅, petals purplish-pink, 2 series, obtuse to denticulate, white filamentous staminodes longer than the 4 mm *l* stamens, filaments white, at half their length papillate, anthers pale yellow, stigmas 5, III; **Fr** covering membranes reduced, very narrow, expanding keels yellow-orange, ovate, closed 6 mm ∅, open 9 mm ∅, c. 3 mm *h*; **Ecol** on site of grazing plots, 1250 m altitude; **Distr** Fauresmith, FS, S Africa.

Note: The local name is "Perde Vygie".

T. pygmaeum L.Bolus 1966: 335 (see **photo**) T Thomas 27823 (BOL!).

Small shrubs 1–3 cm *h*, c. 5 cm ∅, root 25 cm *l*, 1–3 mm *b*, **I** invisible, c. 1 mm ∅; **L** ovate bladder cells with apical hairy papillae, tip with hairs 0.4 mm *l*, **L** 6 mm *l*, 3 mm ∅; **Fl** pedicels hairy, c. 3 mm *l* or almost sessile, **K** 5, base with longer papillae, ∅ c. 20 mm, petals pink, c. 28 in 2 series, inside purple, outside lighter, 5–8 mm *l*, no filamentous staminodes, filaments 3 series, inner lighter pink, outer purplish, c. 2 mm *l*, inner circle with a tuft of papillae at half their length, anthers yellow, stigmas 5, VII in cultivation; **Fr** covering membranes rhomboid, closing bodies present; **Ecol** unknown; **Distr** Swellendam, WC, S Africa.

Note: The elongated papillae of the leaf tip resemble an erect inclining diadem only in shape.

T. rogersiae L.Bolus NM 1930: 177 T Rogers 2686/29 (BOL!).

Erect shrubs 11 cm *h*, **I** below the leaves dotted with white bladder cells of irregular shape, up to 9 mm *l*, 2 mm ∅; **L** bladder cells rounded up to 5–6 edged, without papillae, small diadem of 5–7 yellow orange bristles, different in shape, cup-cells not very distinct, about 1 row, **L** 16 mm *l*, 5 mm ∅; **Fl** pedicels more densely papillate than **I**, 2–8.5 mm *l*, 32 mm ∅, petals pinkish-purple, densely 2–3 series, 14 mm *l*, 1.5 mm *b*, filamentous staminodes pale, stamens 6–6.5 mm *l*, filaments 5 mm *l*, inner circle at half their length with white papillae, anthers yellow, XI–XII; **Fr** covering membranes from one septum heart-shaped, expanding keels light yellow, each an irregular triangle, 5 locules, closed 7 mm ∅, open 8 mm ∅, 3.5 mm *h*, XII in cultivation; **Ecol** unknown; **Distr** Cradock, EC, S Africa.

Note: The low growth with strong branches, bladder cells of the leaves without papillae, and the yellow-orange diadem are typical of this species.

T. rupicola L.Bolus NM 1928: 150 T Paterson 1896 (BOL!).

Erect to decumbent shrubs to 28 cm *h*, roots thin, spindle-shaped, > 40 mm *l*, 2.5 mm *b*, **I** invisible; **L** imbricate, bladder cells as elongated papillae, apical papilla a little bigger, elongated papillae at lateral leaf-base, radiate diadem of 3 mm ∅, (6–)8(–10) whitish bristles from orange-brown cup-cells, **L** 4 mm *l*, 1.3 mm *b*; **Fl** pedicels and whole **K** covered with elongate, hairy papillae, becoming shorter towards the calyx lobes, pedicels 2–4.5 mm *l* or **Fl** almost sessile, 13 mm ∅, **K** 5, petals uniseriate, white to very pale pink, 6 mm *l*, filamentous staminodes > stamens, filaments of the inner circle at half their length papillate, stigmas 5, 1.5 mm *l*, X; **Fr** unknown; **Ecol** growing on rocks; **Distr** Uitenhage, EC, S Africa.

Note: The remarkably plants of this species possess small leaves.

T. ryderae L.Bolus NM 1931: 244 T Ryder 265/30 (BOL!).

Shrubs with erect to decumbent branches 28 cm *l*, **I** papillae white, small and flattened, densely placed, **I** 3–43 mm *l*; **L** bladder cells with double the volume of the internodial papillae, ovate, without papillae, tip with 1–6 light yellow to orange bristles, short, 3 perhaps longer, cup-cells of same colour, **L** 13–38 mm *l*, 3.5 mm *b*; **Fl** pedicels 20–45 mm *l*, thin, fewer papillae than on I, 14 mm ∅, petals purplish-pink, 2–3 series, 6 mm *l*, 0.75 mm *b*, inner circle pale, filamentous staminodes shorter, pale, filaments 2.5 mm *l*, inner circle papillate at medium height, anthers whitish, stigmas 2 mm *l*, XI–XII; **Fr** unknown; **Ecol** in rocks, growing in shade; **Distr** Bedford, EC, S Africa.

Note: The long leaves, equipped with rather simple and small diadems, help to distinguish this species from other erect, climbing, or decumbent shrubs.

T. schimperi ≡ **Delosperma schimperi**

T. setuliferum (N.E.Br.) Schwantes ZSK 1926: 188 ≡ *M. setuliferum* N.E.Br. GC 1876: 8 **T** Bolus before 1875 s.n. (K!) = *T. setuliferum* (N.E.Br.) Schwantes var. *niveum* L.Bolus NM 1931: 243 **T** Marloth 14093 (BOL!).

Erect shrubs 23 cm *h*, some branches decumbent, roots of branching woody fibres, **I** with rough papillae, obliterated on old wood, 11–24 mm *l*; **L** basal ones longer, bladder cells globular, with a central papilla each curved towards the leaf tip, diadem of 7–16 white to light yellow bristles, more radiating on older leaves, from 1–2 rows of orange cup-cells, **L** c. 22 mm *l*, 3–5 mm *b*; **Fl** pedicels up to 25–38 mm *l*, papillate, solitary or in a monochasium, **K** 5, petals bright magenta, many series, decreasing in size inwards, 14 mm ∅, XII–I, in cultivation VI; **Fr** 5 locules; **Ecol** 450 m altitude; **Distr** uncertain, Somerset East?, EC, S Africa.

Note: Brown (GC 1876: 8) placed this plant nearest to **T. barbatum** and **T. intonsum**. *T. setuliferum* var. *niveum* differs in the colour of the flower, the shape of the bladder cells, and the lower number of bristles from the typical form of the species.

T. setuliferum var. *niveum* = **T. setuliferum**

T. stayneri L.Bolus NM 1961: 121 **T** Stayner 903/60 (BOL!).

Compact to semi-erect shrubs 4–6 cm *h*, branches densely leaved, decumbent, roots tuberous, 18 cm *l*, 7 mm ∅, **I** only papillate when very young, 1–6 mm *l*, some 10 mm *l*, 1.5–2 mm ∅; **L** bladder cells irregular in shape, rounded to cornered like cobblestones in a pavement, sometimes very tiny papillae apically and basally on the bladder cells, radiating diadem, 5 light bristles from brown to dark brown cup-cells, diadem 2–3 mm ∅, **L** 7–11 mm *l*, 2.5 mm ∅; **Fl** pedicels 3–6 mm *l*, over the whole length with small bladder cells with an apical papilla, preferably along the edges, 17 mm ∅, petals purplish-pink, 2 series, emarginate, denticulate, 6–9 mm *l*, filamentous staminodes white, 7 mm *l*, filaments 4–5 series, white, inner circle at half length papillate, anthers pale yellow, stigmas 5, XI; **Fr** pedicels 10 mm *l*, pale brown, 4–5 locules, 9 mm ∅; **Ecol** unknown; **Distr** Uitenhage, EC, S Africa.

Note: The material of the herbarium sheet is completely covered with fine red sand and dust.

T. stellatum = **T. barbatum**
T. stelligerum = **T. barbatum**

T. strumosum (Haw.) L.Bolus SAG 1928: 127 ≡ *M. strumosum* Haw. 1821: 190 **T**?

Compact, roots tuberous; **L** depressed cylindrical, hairy, tip without a diadem; **Fl** pedicels with short hairs, solitary, **K** 5, petals pale straw-coloured or whitish, VIII; **Fr** unknown; **Ecol** unknown; **Distr** unknown, S Africa.

Note: Haworth (1821: 190) noted the term "hedgehog" to describe the growth form. This habit is seen in an undated drawing of Duncanson (1011/222, K!), which differs from the description by having a well-developed white diadem of erect hairs. A second drawing (Duncanson 1002/999, K!), showing leaves without a diadem, was based on material collected by Bowie in 1823 and produced in 1825 only, both dates younger than the description and therefore perhaps not the original plant. Bolus considered **T. fergusoniae** to be synonymous with **T. strumosum** (L.Bolus SAG 1928: 127), but the latter lacks a diadem, the former species possesses a diadem-like structure.

T. tenue ≡ **Delosperma neethlingiae**

Vanheerdea Ruschioideae

H. E. K. Hartmann

Vanheerdea L.Bolus ex H.E.K.Hartmann Bradleya 1992: 15 ≡ *Rimaria* L.Bolus NM 1937: 69 nomen illeg. non N.E.Br. GC 1925: 413 ≡ *Vanheerdea* L. Bolus NM 1938: 136 nomen illeg. **Typus V. roodiae** (N.E.Br.) L.Bolus ex H.E.K.Hartmann **Etym** *P. van Heerde*, 1893–1979, school principal in Springbok, a collector of Mesembryanthema.

Compact, 20–40 branches, anisophyllous, I short and enclosed, new leaves developing over the entire

Distribution of **Vanheerdea** L.Bolus ex H.E.K.Hartmann

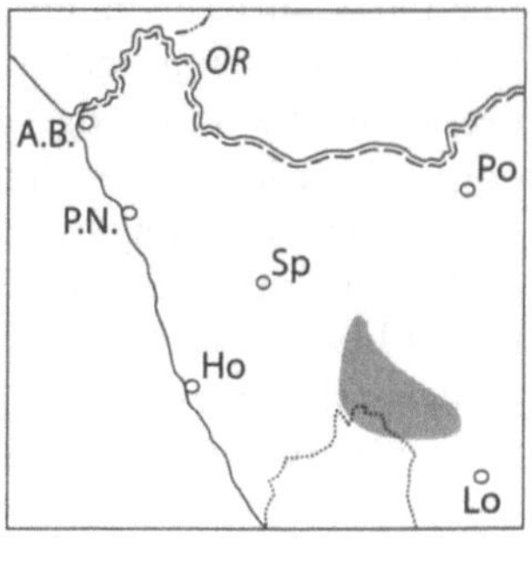

plant; **L** yellowish-green, often yellow to orange, gibbous with convex sides, margins and keel dentate, papillate; **Fl** 1–3, deep golden petals; **Fr** Titanopsis-type with distally narrowing valve wings, mostly 10-locular; **S** small, almost globose; **Ecol** on decomposed calcareous soils, edge of area of winter rains, c. 125 mm rainfall p.a. mainly in March and November; **Distr** Namaqualand, NC, S Africa.

Note: The genus is probably related to **Tanquana** and **Ihlenfeldtia.** The species settle in adjacent areas and have not yet been found sympatrically. They flower mainly in autumn, being apparently independent from watering or rains.

Key to the Species of Vanheerdea

1. Leaves with apical, flat windows .... **V. primosii**
- Leaves without windows .... **V. roodiae**

V. angusta = **V. roodiae**
V. divergens = **V. roodiae**

V. primosii (L.Bolus) L.Bolus ex H.E.K.Hartmann Bradleya 1992: 16 (see **photo**) ≡ *Rimaria primosii* L.Bolus NM 1933: 411 ≡ *V. primosii* L.Bolus NM 1938: 136 nomen invalid. **T** Primos NBG 1970/30 (BOL!).

Sunken growth; **L** with a window on the flattened top, but also with teeth along the keel line over the window, papillae shorter than diameter of cells, absent on window; **Fl** mostly smaller than **V. roodiae,** Ø < 25 mm; **Fr** also smaller, Ø < 8 mm, < 25 mm *l*; **S** almost globular, micropyle attached; **Chr** $2n = 18$; **Ecol** mostly in flats of highly decomposed limestone; **Distr** NE part of the area of the genus, NC, S Africa.

Note: In cultivation, the basal leaf sheath can elongate when watered well, lifting the plants high above the ground; the species flowers less readily than **V. roodiae** in Europe.

V. roodiae (N.E.Br.) L.Bolus ex H.E.K.Hartmann Bradleya 1992: 15 (see **photo**) ≡ *Rimaria roodiae* N.E.Br. GC 1926: 155 ≡ *V. roodiae* (N.E.Br.) L. Bolus NM 1938: 136 nomen illeg. **T** Rood s.n. (K!) = *Rimaria angusta* L.Bolus NM 1937: 67 ≡ *V. angusta* (L.Bolus) L.Bolus NM 1938: 136 nomen illeg. **T** van Heerde BOL 21825 (BOL!) = *Rimaria divergens* L.Bolus NM 1937: 67 ≡ *V. divergens* (L.Bolus) L.Bolus NM 1938: 136 nomen illeg. **T** van Heerde SUG 10853 (BOL!).

Mostly well above ground level; **L** markedly hunched, mostly keeled, papillae as high as broad; **Fl** Ø to 45 mm; **Fr** Ø > 8 mm, > 25 mm *l*; **S** with a protruding micropyle resembling a handle (lens!); **Chr** $2n = 18, 36$; Ecol on slopes or in flats with rather rough mixed gravel with a high proportion of calcareous pebbles; **Distr** SE part of the area of the genus, NC, S Africa.

Note: Variation in keel, colouring, and size is remarkable, depending on the origin of the material.

Vanzijlia Ruschioideae

H. E. K. Hartmann

Vanzijlia L.Bolus FPSA 1927: t 262; Hartmann 1983: 499–538 **Typus V. annulata** (A.Berger) L.Bolus **Etym** *Dorothy van Zijl,* a plant collector and grower from the Clanwilliam area.

Developing 5–10 long, climbing to straggling, rarely erect branches from a compact centre, heterophyllous, I Ø 1,5–2 mm; **L** first leaf-pair on a branch connate for > 50%, later developed leaves connate for < 50% and long free parts, surface smooth, wax cover in horizontal plates breaking into smaller units, thus rough, stomata overarched by cutin ledges from the subsidiary cells, crystal layer in outer epidermal wall thin; **Fl** solitary, erect, terminal on long shoots, the uppermost leaf-pair of the long leaf type, petals white, filamentous staminodes lilac to purple, erect and elongating during anthesis; **Fr** erect, of the Leipoldtia-type, valves broad and rectangular, covering membranes with a distal rim and a ledge or bulge below, closing bodies big and

Distribution of **Vanzijlia** L.Bolus

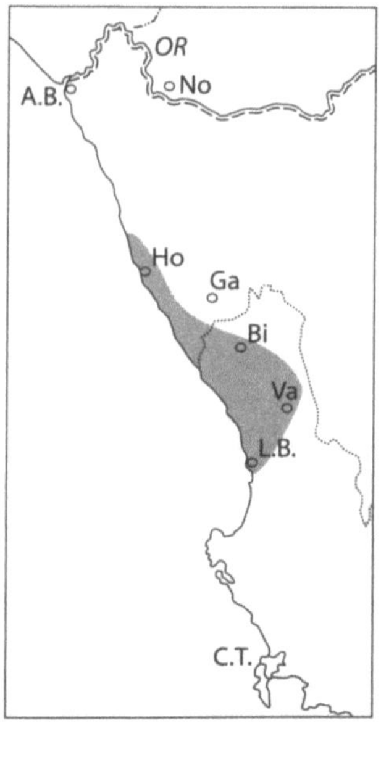

white; **S** light brown, darker at hilum and micropyle, testa cells flat to a little elevated; **Chr** $2n = 36$; **Ecol** on costal sands or, inland, on loam over calcareous soils, in Succulent Karroo or Strandveld, 100–200 mm rainfall p.a. in winter; **Distr** S Namaqualand, NC, Clanwilliam, Vanrhynsdorp, Vredendal, WC, S Africa.

Note: Only one species, inconspicuous in habitat due to its regular growth inside other shrubs. Cultivation in hanging baskets is possible, but seems to reduce readiness to flower.

V. angustipetala = **V. annulata**

V. annulata (A.Berger) L.Bolus FPSA 1927: 262 (see **photo**) ≡ *M. annulatum* A.Berger 1922: 627 **LT** Schlechter 8280 (BOL!) non *V. annulata* (A.Berger) N.E.Br. KB 1929: 62 = *V. angustipetala* (L.Bolus) N.E.Br. KB 1929: 62 ≡ *M. angustipetalum* L.Bolus ABH 1923: 166 **T** van Zijl BOL 17391, cited as 15199 (BOL!).

Short leaves at the base of the branch 8–14 mm *l*, long ones higher up 20–40 mm *l*, without tannin idioblasts; **Fl** ∅ 25–40 mm, 50–90 petals, 80–140 filamentous staminodes, 140–220 stamens, most similar in flower structure to **Hallianthus**, flowering after the beginning of the rainy season; **Fr** grey, 25–50 mm long, top raised, base mostly funnel-shaped, 9–10 locules; **Ecol** as for genus; **Distr** as for genus.

Note: Discrepancies in naming arose from the fact that the original type sheet of Berger (B!) contained two different plants: the piece Schlechter 8280 sent to BOL agrees with **Vanzijlia**, from this the correct naming is derived. The piece remaining at B, and studied by N.E.Brown later, is a species of **Dicrocaulon**. It is therefore not surprising that Bolus (1927) and Brown (1929) could not agree about the set of characters states of the genus and its species. The plants set flower only after a rather entangled mass of long shoots has been formed, not lending itself to neat cultivation. Yet, the showy flowers with their soft lilac filamentous staminodes make up for the vegetative mess.

V. rostella ≡ **Cephalophyllum rostellum**

Verrucifera = Titanopsis

H. E. K. Hartmann

In a key, Brown gave for *Verrucifera* (GC 1930: 278) the more erect and spreading stamens, the half-superior ovary, and different expanding keels as diagnostic features of the genus. These differences have subsequently not been found to exceed the variation seen in **Titanopsis**, with which *Verrucifera* is therefore placed as a synonym. In a later treatment (Brown GC 1930: 513), *V. schwantesii* was chosen as the type species.

V. hugo-schlechteri ≡ **Titanopsis hugo-schlechteri**
V. lüderitzii = **Titanopsis schwantesii**
V. schwantesii ≡ **Titanopsis schwantesii**

Veslingia ≡ Aizoon

H. E. K. Hartmann

The name *Veslingia* Heister 1748: 11 has been created by the author in a list of plants with dry fruits, as a replacement "pro Ficoidea T. & Aizoon Linnaei". Since this publication appeared before the starting point for nomenclatural regulations, Linnaeus' Species Plantarum (1753), *Veslingia* Heister is not available as a name at genus level. Instead, the later publication of Fabricius contains the correct name: *Veslingia* Fabricius 1759: 201; the name should, therefore, be cited as *Veslingia* Heister ex Fabricius. The citations of Fabricius 363 by Willdenow (1799) and Moench (1802) refer to the second edition of the work: Fabricius 1763: 363. Fabricius did not cite any species under *Veslingia* Heister, but added immediately after the genus name: "Aizoon Linn.", stressing the fact that both generic names are synonyms.

The name *Veslingia* has been used again by Willdenow (1799: 1053), who listed "Veslingia Heisteri. Fabric. helmst. p. 363" as a synonym under **Aizoon canariense**, creating in this way perhaps unintentionally a new superfluous synonym.

Moench (1802: 299–300) gave a description of the genus *Veslingia* and named one species: *V. caulifloris* Moench. Behind the description of this species, Moench cited the figures of Nissol as "1711. icon.t. 19" (correctly: 1730: t. 13) and Volckamer "p. 236 icon. 236" (correctly: Volckamer 1718: t. 236), each showing a plant of **Aizoon canariense**. Furthermore, Moench (1802: 300) referred to Linnaeus' description of **Aizoon canariense** (Linnaeus 1762: 1053), pointing out that *Veslingia* differed by free stamens and the number of stamens widely from the genus **Aizoon**. Of the cited figures, Nissol 1730: t. 13 is chosen as the lectotype here, based on the fact that additional drawings of details permit a well-founded identification of the species.

V. caulifloris = **Aizoon canariense**
V. heisteri = **Aizoon canariense**

Vlokia Ruschioideae

H. E. K. Hartmann

Vlokia S.A.Hammer CSJ 1994: 256 **Typus V. ater** S.A.Hammer **Etym** *J.H.J. Vlok*, * 1957, environmental advisor with a special interest in plants.

Compact to caespitose perennials with short shoots with mostly one leaf-pair; **L** shortly trigonous, dark green to black, persisting on the plant for several years, epidermis with marked cuticular folds covered thickly by wax breaking into horizontal plates; **Fl** solitary on short shoots, bracteolate, **K** 6, petals pink, filamentous staminodes pink, nectary a dark green ring; **Fr** dark, hard and persistent over several seasons, bracteoles just above the middle of the stalk, top almost flat, base broad, forming a deep bowl, covering membranes complete, but thin, closing body absent, but placenta sometimes lifted from the wall, expanding keels not touching each other basally, 6 locules; **S** very dark brown, nearly round; **Ecol** in decomposing Witteberg quartzite at the edge of exposed rocks in shallow pans, in open spaces in Fynbos vegetation, c. 1300 m altitude, c. 300 mm rainfall mostly in March and November; **Distr** Montagu, WC, S Africa.

Note: As one of the younger genera in Aizoaceae, **Vlokia** will certainly be the object of much critical observation to establish its position within the family. Some similarities in fruit morphology call to mind genera like **Erepsia**, **Esterhuysenia**, **Antegibbaeum**, or **Cerochlamys**. To the last named genus, more similarities exist than to any of the others, the differences being that **Vlokia** lacks the very high valve wings of **Cerochlamys** and that it possesses a distinct nectary ring.

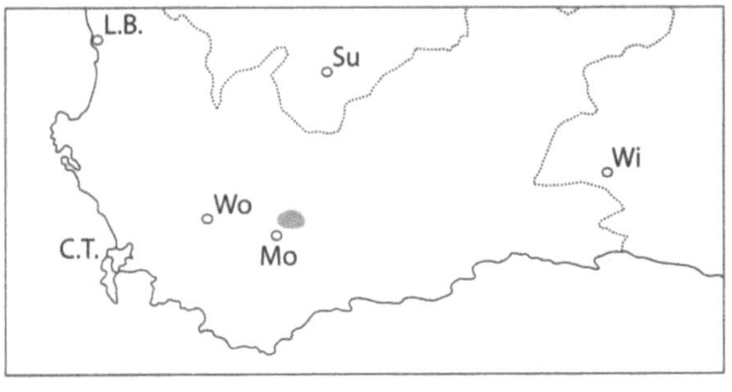

Distribution of **Vlokia** S.A.Hammer

V. ater S.A.Hammer CSJ 1994: 256 (see **photo**) (description for genus and species together) **T** Hammer & Vlok 1181 (BOL).

Growth compact when young, more spreading with age; **L** punctate, c. 10 mm *l*, 8 mm *b* and *t*, blackening with age; **Fl** not opening widely, 22–25 pink petals, 1.5 mm *b*, c. 80 pink to white filamentous staminodes, c. 16 basally papillate stamens; **Fr** often black from fungus with age, (5–)6(–7) locules; **S** 0.46 mm *l*, 0.42 mm *b*, 0.35 mm *t*; **Ecol** in high lying Fynbos in open patches along exposed rocks; **Distr** Montagu, WC, S Africa.

Note: Mesembryanthema occur rarely in Fynbos, and this species came certainly as a surprise when it was found. The black colour is most probably a fungus, as is the black cover of the fruits.

Vossia

H. E. K. Hartmann

Adanson (1763: 242–243) created four generic names for members of the genus **Mesembryanthemum** L., among them *Vossia* Adanson (1763: 243). A description is lacking, and no type is given, but character states are listed in the form of entries in a table. These indicate that compact or creeping plants with multilocular fruits are included. No species are cited, but figures of Dillenius (1732: 183–186, 199, 210–212) are listed as references in a block under the heading *Vossia*. These cited figures represent plants belonging to the genera **Carpanthea, Carpobrotus, Cephalophyllum, Cheiridopsis, Cylindrophyllum, Glottiphyllum**, and **Lampranthus**. A comparison of the character states of the species figured by Dillenius with those in the list shows that in each single case a contradiction can be found, e.g. higher numbers of stamens occur in **Cephalophyllum** and **Glottiphyllum**, the fruits of **Carpobrotus** possess no valves but are berries. The one species with only a minor deviation from the data given by Adanson is **Cheiridopsis robusta** with 138–426 stamens, thus exceeding the noted limits of 30–150 stamens. As a consequence, it is impossible to determine a type species from the cited names. The comparisons show also that Rothmaler (1941: 414), when equating *Vossia* with **Glottiphyllum**, had not considered the relevant features of both genera. In addition to these uncertainties, the name **Vossia** has been used again by Wallroth & Griffith for a genus of Poaceae (1836: 572 t. 23), with a single species, **V. procera** Wallroth & Griffith nomen illeg. At that time, the name **Vossia** was illegitimate, because of the older, validly published name by Adanson, but **Vossia** Wallroth & Griffith has been accepted as a nomen conservandum (ICBN 2000: App. IIIA).

Wooleya Ruschioideae

H. E. K. Hartmann

Wooleya L.Bolus JSAB 1960: 48 **Typus W. farinosa** (L.Bolus) L.Bolus **Etym** *C.H.F. Wooley*, Major and plant collector.

Densely leaved succulent shrubs to 50 cm *h*, **I** < **L**; **L** thick, clavate, grey from a thick wax cover of broad bands surrounding the stomata almost like chimneys, apparently unique in the family, **L** terete and apically rounded, later black, epidermis flat, stomata not sunken, outer epidermal wall thicker than lumen, filled with crystals of calcium oxalate of c. 2 µm ∅; **Fl** single, bracteolate, **K** 4, pale pink petals and filamentous staminodes, the latter in a broad cone surrounding the white filaments adorned with papillae in their lower parts, nectary ring prominently lophomorphic; **Fr** short-stalked, modified Lampranthus-type: valve wings c. as broad as the nearly parallel expanding keels, covering membranes straight, raised in the centre, no closing devices, but the placenta raised by a spongy tissue forming pockets at the base of the locule containing some seeds (paraspermy), rims moderately raised and reflexed, (9–)12(–16) locules; **S** opaque to light brown, the perisperm region globular, the embryo attached in a distinct curved roll with a testa of brown, raised cells, the micropyle spreading; **Ecol** on sandy soil in shrubby associations of Strandveld, near the sea in regular fog, < 100 mm rain p.a. in winter; **Distr** NW Namaqualand, NC, S Africa.

Note: Described at first as a species of the genus **Namaquanthus,** the taxon was separated and raised to the rank of genus later, based on differences in fruit, flower, and leaf characters, namely the particular wax cover. Only one species.

Distribution of **Wooleya** L.Bolus

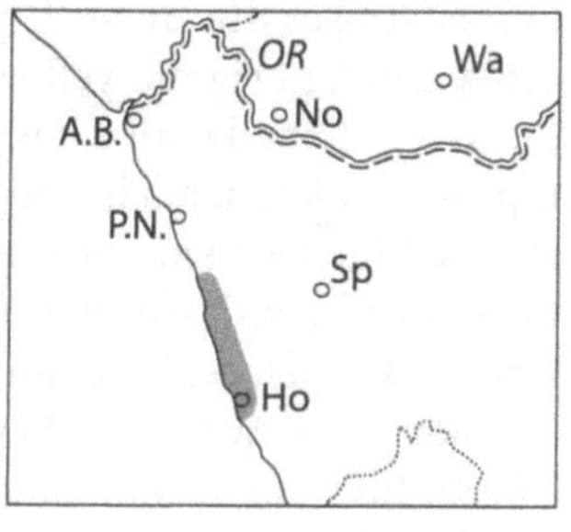

W. farinosa (L.Bolus) L.Bolus JSAB 1960: 48 (see **photo**) ≡ *Namaquanthus farinosus* L.Bolus NM 1958: 334 **T** Wooley 583/43 (BOL).

To 70 cm ∅; **L** 2 grey pairs per season, older ones persisting, but turning black with age, c. 30 mm *l*, c. 10 mm *t*; **Fl** stalks erect, pale pink to white; **Fr** 10–20 mm *l*, ∅ 9–11 mm; **S** 0.6–0.7 mm *l*, 0.47–0.53 mm *b*; **Ecol** as for genus; **Distr** as for genus.

Note: Restricted to a coastal strip in N Namaqualand, where it occurs sympatrically with species of **Jordaaniella**, e.g. **Jordaaniella spongiosa**. Easy to grow from cuttings and flowering moderately well after a certain size has been reached.

Zaleya Sesuvioideae

H. E. K. Hartmann

Zaleya Burman f. 1768: 110; Melville 1952: 266–278; Jeffrey 1960: 235–238 **Typus Z. decandra** (L.) Burman f. **Etym** *zaleya*, a name of uncertain origin = *Rocama* Forskål 1775: 71 **Typus** *Rocama prostrata* Forskål = **Z. pentandra** L.

Annual to perennial herbs, **I** mostly longer than **L**, often rough from papillae; **L** opposite, those of a pair unequal in size, petiolate, the petiole with a basal papery broadening reminding of stipules, but never separate from the petiole, blades obovate to lanceolate, epidermis finely papillate, some species with rough hairs; **Fl** in ± dense cymes, often congested, very rarely solitary, sessile or shortly pedicellate, bracts and bracteoles mostly subulate and very narrow, perigone connate for 1/2 the length, apical unifacial appendages short, stamens 5–15; **Fr** turret-shaped, the operculum often with ornamentations and elongations around the apical flattish top, breaking into two halves with one seed in each half, 2 more seeds remaining in the bottom part of the ripe fruit, 2 locules; **S** one above the other in the locule, rugose, black; **Ecol** in tropical to subtropical climates; **Distr** in a belt between c. 30° N and S around the equator.

Note: Jeffrey (1960: 237) re-established the genus **Zaleya,** based on the possession of two locules with four ovules and two stigmas. In contrast to **Trianthema**, it can be added that the opercula of **Zaleya** split when ripe, releasing two halves, each with a single seed inside; these units represent clausae formed one beside the other, each derived from the upper half of one of the two locules. The two seeds remaining in the basal part of the fruit are released as free dispersal units without a wall, unlike the clausae in which the seeds are kept inside during dispersal. It seems that the species occur to a great extent allopatrically in different parts of the drier subtropical and tropical zones of the world, but floras do not always distinguish between **Trianthema** and **Zaleya**. The

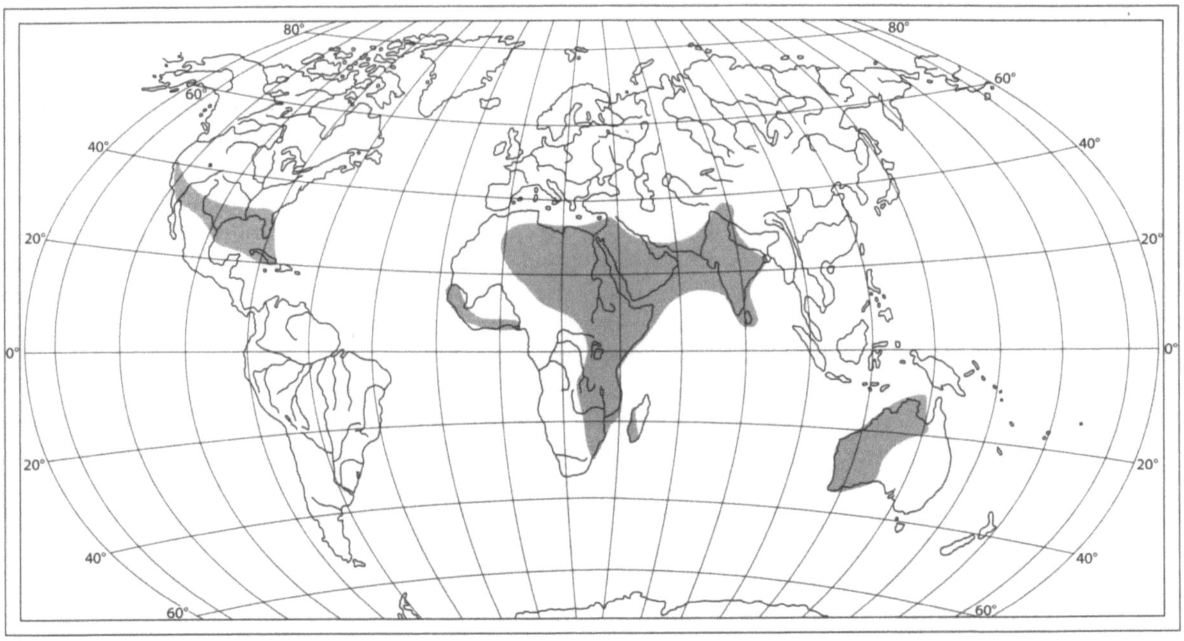

Distribution of **Zaleya** Burman f.

geographical distribution presented is drawn from data in c. 110 floras, but it appears highly incomplete. A number of species is cited as being of medical importance (e.g. Samuelsson et al. 1991: 41), and some species are reported to be edible, but it is impossible to be certain which taxa are meant at this state.

Z. camillii (J.Cordemoy) H.E.K.Hartmann **comb. nov.** ≡ *Trianthema camillii* J.Cordemoy Flore de l'ile de Réunion: 291 (as basionym) **Holotype** C. J. Cordemoy s.n. (MARS!).

Herbs with prostrate branches, **I** with fine, rounded papillae; **L** ovate to obovate, finely papillate; **Fl** in cymes of 3–6, tepals pinkish inside, papillate and green outside, stamens 5, alternitepalous, filaments and gynoecium pink; **Fr** with a bilobed, rounded top, flat or with warty-papillate rims at two side, the sides of the top part straight; **Ecol** in open places; **Distr** Madagascar, Réunion.

Note: The species has been included as a synonym under **Z. pentandra**; typical of **Z. camillii** are the straight sides of the top part of the fruit and the rounded, warty-papillate top surface.

Z. decandra (L.) Burman f. 1768: 110 ≡ *Trianthema decandra* L. 1767: 70 **T** LINN 527.4 (LINN); Melville 1952: 261–263 = *Trianthema glinoides* Persoon 1805: 487 **Lectotype**, here designated: Burman 1768: t.31, f 3.

Annual herbs with prostrate branches to 30 cm *l*, **I** with hyaline papillae twice to three times as long as broad; **L** narrowly to broadly elliptical, petiolate, papery broadening at base extending for 1/2 to 3/4 of length, upper leaf surface covered by white papillae, lower subglabrous, 8–20 mm *l*, 3–18 mm *b*; **Fl** 3–5 per cyme, subsessile or with pedicels and peduncles 0.5–3 mm *l*, perigone yellow inside?, stamens 10–15; **Fr** turret-shaped with a low margin around the ± flat top, incomplete septa leaving a small triangular space at basal interior side of the clausa; **S** ± pear-shaped to roundish, with elevated undulate crests, 1.7 mm *l*; **Ecol** in dry places; **Distr** India; Sri Lanka.

Note: The treatment follows Melville (1952: 261–263). The addition of *Trianthema glinoides* is based on the citation of Burman 1768: 110, t. 31 f 3, which is the text and drawing on which the genus **Zaleya** was based.

Z. decandra var. *flava* = **Z. govindia**
Z. decandra var. *rubra* = **Z. govindia**

Z. galericulata (Melville) H.Eichler 1965: 136 ≡ *Trianthema galericulata* Melville 1952: 267 **T** Morris 2758 (K).

Herbs with prostrate branches, **I** with hyaline papillae twice as long as broad; **L** apically broadly rounded or acute, epidermis with papillae on the upper surface, subglabrous below, 15–45 mm *l*, 4–20 mm *b*; **Fl** cymes dense or lax, perigone lobes broadly ovate, stamens 5–12; **Fr** turret-shaped with ± high teeth around the ± flat top, operculum breaking easily into two halves; **S** markedly crested to crenulate, c. 2 mm *l*; **Ecol** in dry areas; **Distr** Australia.

Note: Jacobs (1988: 139) distinguished two subspecies.

Z. galericulata (Melville) H.Eichler ssp. **australis** (Melville) Jacobs 1988: 139 ≡ *Trianthema australis* Melville 1952: 266 **T** White 13559 (SYD).

Fl stamens 5, extremely rarely 10; **Fr** with septa in the operculum meeting at the very tip, leaving free an area between the locules that has the shape of a narrow-based triangle; **Distr** New South Wales, Queensland, Australia.

Note: The subspecies settles mainly in the south-eastern part of the area of the genus.

Z. galericulata (Melville) H.Eichler ssp. **galericulata**.

Fl stamens 5 or 10, c. 50% of a population with each number; **Fr** with septa in the operculum meeting c. 0.5 mm below the tip, leaving free an area shaped like a key-hole, the opening basally not reaching the level of the circumscissile line; **Distr** central and eastern parts of Australia.

Note: The subspecies occurs in a very wide area, also near places where the other subspecies has been found.

Z. govindia (Buchanan-Hamilton ex G.Don) N.C.Nair 1966: 86 ≡ *Trianthema govindia* Buchanan-Hamilton ex Wallich 1832: no. 6838 nomen nudum ≡ *Trianthema govindia* Buchanan-Hamilton ex G.Don 1834: 72 **T** Buch.-Ham. in Wallich 6838 (K) = *Z. decandra* L. var. *flava* (Blatter & Hallberg) N.C.Nair 1966: 86 ≡ *Trianthema pentandra* L. var. *flava* Blatter & Hallberg 1919: 531 ≡ *Trianthema decandra* L. var. *flava* (Blatter & Hallberg) Santapau 1959: 280 **T** syntypes Blatter & Hallberg 6788, 6789 (?) = *Z. decandra* L. var. *rubra* (Blatter & Hallberg) N.C.Nair 1966: 86 ≡ *Trianthema pentandra* L. var. *rubra* Blatter & Hallberg 1919: 531 ≡ *Trianthema decandra* L. var. *rubra* (Blatter & Hallberg) Santapau 1959: 280 **LT** Blatter & Hallberg 6772 (?).

Branches prostrate, **I** covered densely with hyaline papillae c. 1.5 times as long as broad when young; **L** petioles 1/3 to 2/3 of the length of the blades, blades elliptical to orbicular, epidermis with fine papillae on both leaf surfaces, 4–40 mm *l*, 4–30 mm *b*; **Fl** 7–15 per cyme, subsessile, bracts triangular-acuminate, denticulate, perigone lobes marginally scarious, stamens 5, shorter than the perigone lobes; **Fr** top with a pronounced, high rim, each valve ending in 1–2 rather sharp teeth; **S** undulate-costate, c. 1.5 mm *l*; **Ecol** in dry areas; **Distr** Arabia; India; Iran.

Note: When separating **Z. govindia** from **Z. pentandra** (then a species of **Trianthema**), Melville (1952: 261–262) said that the valves were too different to belong to one species. His drawings show that **Z. govindia** possesses higher and tooth-shaped protrusions at the flat end of the top, whereas **Z. pentandra** has a more rounded top without any tooth-shaped appendages. Furthermore, **Z. govindia** possesses smaller septa which leave a very broadly based triangle free between the locules.

Z. pentandra (L.) Jeffrey 1960: 238 (see **photo**) ≡ *Trianthema pentandra* L. 1767: 70 **LT** LINN 572.2 (LINN) = *Trianthema pentandra* L. var. *hirtula* Battandier & Trabut 1913: 245 **T?** figure in Battandier & Trabut 1913: t. 6, based on material sent by officers of the forces = *Limeum kenyense* Suessenguth 1950: 46 **T** Quartin (cited as Martin) 17 (K) = *Rocama prostrata* Forskål 1775: CVIII, no. 200 ≡ *Rocama arabica* Gmelin 1791: 455 nomen nudum et nomen illeg. ≡ *Rocama digyna* Forskål ex DC. 1828: 352 ≡ *Trianthema digyna* DC. 1828: 352, nomen nudum **T** Forskål C III. 22[33] (C).

Perennial shrubs, branches erect to decumbent, **I** covered sparsely to densely by papillae longer than broad, later glabrescent; **L** elliptical to ovate, papillae rather dense and long, often pointed, 6–35 mm *l*, 1.5–18 mm *b*; **Fl** 6–20 in compact cymes, sessile, bracts entire, perigone with appendages 0.3–1 mm *l*, stamens 5; **Fr** turret-shaped with a rim of slightly varying height around the flat top, septa well developed, leaving a small triangle free which is mostly higher than broad; **S** crested, crests often divided into tubercles, 1.5–2 mm *l*; **Ecol** in dry areas; **Distr** from Senegal to Socotra; Réunion; along the eastern coast and in the south to Moçambique; NP, SA.

Note: Deviating from Jeffrey (1960: 238), **Z. camillii** and **Z. redimita** are considered to represent different species, based on extensive material from Africa and the islands. Samuelsson et al. (1991: 31) give detailed information on the medicinal properties of the species and its application in Somalia, including preparation and dosage of the remedy gained from roots, leaves, or the entire plants.

Z. redimita (Melville) H.E.K.Hartmann **comb. nov.** ≡ *Trianthema redimita* Melville 1952: 268 (as basionym) **Holotype** Dale 700 (K).

Perennial herbs with basally woody branches, decumbent; **L** moderately papillate; **Fl** tepals pink inside; **Fr** turret-shaped, the two lateral wings on top long and dentate to varying degrees; **Ecol** in disturbed areas; **Distr** NE and SW Kenya.

Note: In contrast to **Z. pentandra**, the asymmetrical wings on top of the fruit are much longer than the stigmas.

Z. sennii (Chiovenda) Jeffrey 1960: 238 ≡ *Trianthema sennii* Chiovenda 1932: 225 **T** Senni 666 (FI).

Branches low on the ground, **I** with papillae when young, blackish with age; **L** petioles broadly membranous, basally stipulate?, blades obovate to lanceolate, finely papillate on both sides, longer hairs on the vein protruding markedly on the lower leaf surface, 8–15 mm *l*, 2–6 mm *b*; **Fl** 2–7 per cyme, rarely solitary, pedicels 5–15 mm *l*, perigone with long hairs outside, dorsal appendages 0.2–0.4 mm *l*, stamens 10; **Fr** not described; **Ecol** unknown; **Distr** Somalia.

Note: Melville did not include the species in his study of the other species.

---

[33] The systematic numbering of Forskål's collections was only undertaken by Christensen (1922: part 3, 1–54), albeit material had earlier been identified by various botanists. De Candolle (1828: 352) used the first word of Forskål's description as an epithet as well as the name *R. prostrata*, given in a different place.

Zeuktophyllum RUSCHIOIDEAE

H. E. K. HARTMANN

Zeuktophyllum N.E.Br. GC 1927: 12; Hartmann Bradleya 1998: 87 **Typus Z. suppositum** (L.Bolus) N.E.Br. **Etym** Gk *zeuktos*, closely joined, connected; *phyllon*, leaf, referring to the closely set leaf-pairs appearing like spouts pushed together.

Compact succulents shrubs to 12 cm *h*, 2–4 leaf-pairs formed per season, the flowering branches erect and raised well above the vegetative part; **L** triquetrous, almost thicker than broad, hard to the touch, keel apically ± serrate, the leaf margins smooth and somewhat rounded, epidermis cells only a little raised, stomata sunken only a little, wax cover breaking into horizontal plates smaller than the underlying cells, surface of small plates rugose, with only few vertical platelets on them; **Fl** solitary at the end of a season's growth, bracteoles enclosing it, withering with the ripening of the fruit, but not differing in size or shape from the foliage leaves, calyx lobes 5, 2 of them distinctly longer (to 12 mm), petals white, only 1 mm *b*, filamentous staminodes present, filaments basally papillate, nectary in a lophomorphic ring; **Fr** of the Titanopsis-type, but stouter in tissue, covering membranes complete or reduced to rims, depending on the extension of the expanding keels, these running radially for their lower portion, spreading towards the tips which can extend as a triangle over the edge of the valve, valve wings very narrow or absent, in particular with age, 8–9(–10) locules; **S** light brown, testa cells a little convex; **Ecol** on quartzitic slopes between gravel, together with low shrubs, 150–200 mm rainfall p.a. in March and November; **Distr** Ladismith, Laingsburg, NC, S Africa.

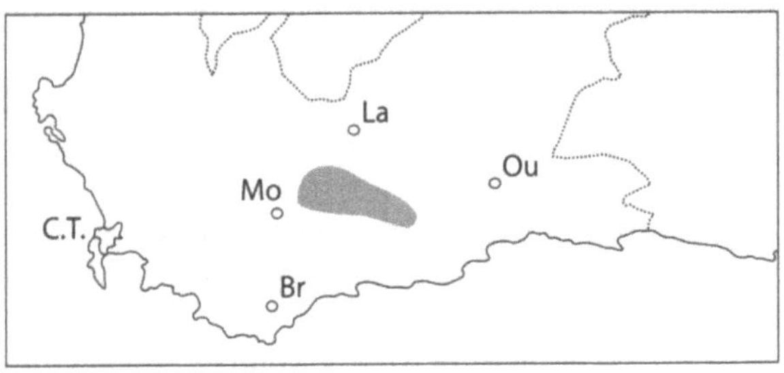

Distribution of **Zeuktophyllum** N.E.Br.

Note: The genus has been lost since its discovery, inspite of extensive searches. Based on comparisons of fruits, the position in the Titanopsis-group appears well founded now, in contrast to Hartmann (1991). Two species can be distinguished, one of them transferred from the inhomogeneous genus **Octopoma.** Both species have been confused since the first one had been described, even by the original collector, Muir, as is obvious from his letters and the material kept at K.

Key to the Species of Zeuktophyllum

1. Expanding keels on valves reaching well over the edge of the valves, forming a prominent triangle; between these and the bases of the valves extremely narrow valve wings can occur .... **Z. suppositum**
– Expanding keels narrowing to their tips, reaching only marginally over the edge of the valves, valve wings proper as narrow bands along the valve bases and the expanding keels .... **Z. calycinum**

Z. calycinum (L.Bolus) H.E.K.Hartmann Bradleya 1998: 88 (see **photo**) ≡ *Ruschia calycina* L.Bolus NM 1929: 47 ≡ *Octopoma calycinum* (L.Bolus) L.Bolus JSAB 1963: 49 ≡ *M. separatum* N.E.Br. GC 1930: 32 **T** Muir 4024 (BOL!).

Compact with regular branching, to 9 cm *h*, to 15 cm ∅; **L** 10–20 mm *l*, to 6 mm *b*, to 9 mm *t*; **Fl** white or straw-coloured petals; **Fr** distally spreading expanding keels, valve wings as very narrow bands, broadening a little to the base, covering membranes always nearly complete; **S** 1–1.15 mm *l*, 0.75–0.9 mm *b*; **Ecol** as for genus; **Distr** in the W part of the area of the genus.

Note: The species has been placed in **Octopoma** because it possesses eight locules, a feature found in several genera. In leaf shape and surface, in growth form as well as in flower and fruit morphology, the species is very similar to **Z. suppositum.**

Z. suppositum (L.Bolus) N.E.Br. GC 1927: 12 ≡ *M. suppositum* L.Bolus ABH 1925: 1 **T** Muir 3069 (BOL!).

Compact with some erect branches bearing flowers raised well above the basal part of the plant; **L** in cultivation to 40 mm *l*, trigonous, the sides convex, mucronate; **Fl** with white petals and filamentous staminodes; **Fr** with expanding keels developing triangular distal wings, valve wings proper only at the bases of the valves, extremely narrow, covering membranes complete or reduced to very narrow rims along the septa, different states in one population possible; **S** 1.15–1.25 mm *l*, 0.7–0.85 mm *b*; **Ecol** as for genus; **Distr** E part of the area inhabited by the genus.

Note: Plants of this species have probably been often mistaken for **Z. calycinum** in the field, based on similarities of leaf shapes, growth form, and gross fruit morphology.

References

Frequently used references are abbreviated:

ABH	The Annals of the Bolus Herbarium 1–4. Cambridge University Press. London. 1914–1928
ASAM	Annals of the South African Museum
BB	Bibliographia Botanica. E. Schweizerbart'sche Verlagsbuchhandlung. Stuttgart
BBP	Beiträge zur Biologie der Pflanzen
BCSJ	British Cactus and Succulent Journal
BJS	Botanische Jahrbücher für Systematik
BSKP	Beiträge zur Sukkulentenkunde und -pflege
CBH	Contributions from the Bolus Herbarium
CBM	Curtis's Botanical Magazine
CJ	Cactus Journal (Croyden)
CSJ	Cactus and Succulent Journal (U.S.)
CSJGB	Cactus and Succulent Journal of Great Britain
FPSA	Flowering Plants of South Africa
FR	Feddes Repertorium
GC	Gardeners' Chronicle ser. 3
GF	Gartenflora
GW	Gartenwelt
ICBN	International Code of Botanical Nomenclature (see Greuter et al.)
JB	Journal of Botany
JSAB	Journal of South African Botany
KB	Kew Bulletin of Miscellaneous Information
KuaS	Kakteen und andere Sukkulenten
MBMUZ	Mitteilungen des Botanischen Museums der Universität Zürich
MBSM	Mitteilungen der Botanischen Staatssammlung München
MDG	Möllers Deutsche Gärtnerzeitung
MDK	Monatsschrift der Deutschen Kakteengesellschaft
MIBH	Mitteilungen aus dem Institut für Allgemeine Botanik Hamburg
MKK	Monatsschrift für Kakteenkunde
NCSJ	The National Cactus and Succulent Journal
NM	Notes on Mesembrianthemum and Allied Genera 1–2. 1928–1935. University of Cape Town, Cape Town
NM	Notes on Mesembryanthemum and Allied Genera 3. 1936–1958. University of Cape Town, Cape Town
SAG	South African Gardening and Country Life
SAJB	South African Journal of Botany
SKK	Sukkulentenkunde. Jahrbuch der Schweizer Kakteengesellschaft
SUC	Succulenta. Nederlands-Belgische Vereniging van Liefhebbers van Cactussen en ander Vetplanten
TRSSA	Transactions of the Royal Society of South Africa
VNVH	Verhandlungen des naturwissenschaftlichen Vereins Hamburg (NF)
ZSK	Zeitschrift für Sukkulentenkunde

Adamson R.S. (1956) The South African species of Aizoaceae. JSAB 22: III. Galenia: 87–127
Adamson R.S. (1959) The South African species of Aizoaceae. JSAB 25: VI. Acrosanthes: 23–28
Adamson R.S. (1959) The South African species of Aizoaceae VII. Aizoon. JSAB 25: 29–51
Adamson R.S. (1961) The South African species of Aizoaceae XI. Plinthus. JSAB 27: 147–151
Adamson R.S. (1962) The South African species of Aizoaceae XII. Sesuvium, Trianthema and Zaleya. JSAB 28: 243–253
Adamson R.S., Salter T. (1950) Flora of the Cape Peninsula:385. Juta, Cape Town
Adanson M. (1763) Familles des Plantes 2: 242–245. Vincent, Paris
Aellen P. (1938) Halimione Aellen, eine rehabilitierte Chenopodiaceen-Gattung. Verhandlungen der Naturforschenden Gesellschaft in Basel 49: 118–137
Aitchinson J.E.T. (1869) A catalogue of the plants of the Punjab and Sindh: 66. Taylor & Francis, London
Aiton W. (1789) Hortus Kewensis, ed. 1, 2: 177–196. Nicol, London
Aiton W. (1811) Hortus Kewensis, ed. 2, 3: 212–253. Longman, Hurst, Rees, Orme, & Brown, London
Andrews H.C. (1799) The botanist's repository 1: 57. Bensley, London
Andrews H.C. (1802) The botanist's repository 3: t. 201. Bensley, London
Armari B. (1904) Aizoaceae, Umbelliferae, Rubiaceae, Borraginaceae. In: Pirotta R. (ed) Flora della Colonia Eritrea. Annuario del. R. Istituto Botanico di Roma 8, 2: 144–162
Arnott G.A. (1931) On some new species of Portulacaceae. Journal of Natural and Geographical Science 3: 354
Arvidsson L., Lidén M. (1977) Galenia secunda new to Europe. Botaniska Notiser 130: 130
Augier A. (1801) Essai d'une nouvelle classification des végétaux. Bruyset Ainé, Lyon
Baker J.G. (1897) Diagnoses africanae X. KB 1897: 268
Balfour B. (1884) Diagnoses plantarum novarum Phanerogamarum Socotrensium ... 4 (Supplementum). Proceedings of the Royal Society of Edinburgh 12: 402–411
Barnéaud F.M. (1847) Portuláceas. In: Gay C.: Historia Física y política de Chile ... Botanica (Flora chileana) 2: 469–470. The editor, Paris
Bartling F.T. (1832) Ficoideae auctore Fr. Th. Bartling. Linnea 7: 541
Battandier J.A., Trabut L.C. (1917) Plantes du Tassili des Azdjer. Bulletin de la Société d'histoire naturelle de l'Afrique du Nord 8: 226–227
Bauer F. (1833) Illustrations to Endlicher, S. 1833: Prodromus Florae Norfolkiae ...: t. 98, 99. Beck, Vienna (not seen)
Bauhin C. (1623) Pinax theatri botanici. Sumptibus & typis Ludovici Regis. no place given
Bentham G. (1866) Flora Australiensis 3:327–331. Reeve, London
Berger A. (1908) Mesembrianthemen und Portulacaceen. Ulmer, Stuttgart
Berger A. (1910) Einige neue afrikanische Sukkulenten. BJS 45: 223–233. (13.12.1910, often cited as 1911)
Berger A. (1922) Mehrere neue Mesembrianthemum und eine Aloe. BJS 57: 627–640
Berger A., Dinter C. (1914) Succulenta Dinteriana. BJS 50, Suppl.: 586–592
Berger A., Schlechter R. (1922) In: Berger A. BJS 1922
Bittrich V. (1987) Untersuchungen zu Merkmalsbestand, Gliederung und Abgrenzung der Unterfamilie Mesembryanthemoideae (Mesembryanthemaceae FENZL). MIBH 21: 5–116
Bittrich V. (1990) Systematic studies in Aizoaceae. MIBH 23b: 491–507
Bittrich V. (1990) Aizoaceae. A new species of Galenia from the Great Karoo. Bothalia 20: 217–219
Bittrich V. (1990) A new species of Trianthema (Aizoaceae) from the Kimberley region and a note on T. triquetra. Nuytsia 7: 117–122
Bittrich V. (1993) Caryophyallaceae. In: Kubitzki K. (ed) The families and genera of vascular plants:206–236. Springer, Berlin Heidelberg New York
Bittrich V., Hartmann H.E.K. (1988) The Aizoaceae – a new approach. Botanical Journal of the Linnean Society 97: 239–254
Bittrich V., Jenssen K. (1990) In: Bittrich V. 1990: 118
Black J.M. (1923) Additions to the flora of South Australia 21. Transactions and Proceedings of the Royal Society of South Australia 47: 367–370
Black J.M. (1932) Additions to the flora of South Australia 30. Transactions and Proceedings of the Royal Society of South Australia 56: 40–42
Black J.M. (1934) Additions to the flora of South Australia no. 32. Transactions and Proceedings of the Royal Society of South Australia 58: 176–177, t. 11
Black J.M. (1952) Flora of South Australia 2nd ed. 2: 531. Hawes, Adelaide
Blake S.T. (1969) A revision of Carpobrotus and Sarcozona in Australia, genera allied to Mesembryanthemum (Aizoaceae). Contributions from the Queensland Herbarium 7: 1–65
Blatter E., Hallberg F. (1919) The flora of the Indian desert (Jodhpur and Jaisalmer). Journal of the Bombay Natural History Society 26: 525–531
Blume C.L. (1826–1827) Bijdragen tot die Flora van Niederlandsch-Indie 17: 1137
Boer H.W. de (1955) Lithops gesinae de Boer sp. nov. SUC 1955: 2–3
Boer H.W. de (1956) Lithops annae de Boer sp. nov. SUC 1956: 90–92
Boer H.W. de (1957) Lithops hallii de Boer sp. nov. SUC 1957: 85–86
Boer H.W. de (1958) Lithops christinae de Boer sp. nov. SUC 1958: 2–3
Boer H.W. de (1960) Lithops koelemanii de Boer spec. nov. SUC 1960: 28–29
Boer H.W. de (1960) Lithops salicola var. reticulata de Boer var. nov. SUC 1960: 79–81
Boer H.W. de (1960) Lithops glaudinae de Boer spec. nov. SUC 1960: 129–132
Boer H.W. de (1961) Lithops elisae de Boer sp. nov. SUC 1961: 107–108
Boer H.W. de (1962) Enige variëteiten van Lithops fulleri N.E.Br. SUC 1962: 67–69
Boer H.W. de (1962) Enige variëteiten van Lithops lesliei (N.E.Br.) N.E.Br. SUC 1962: 81, 83–87
Boer H.W. de (1963) Lithops julii var. littlewoodii var. nov. de Boer. SUC 1963: 33, 35–37
Boer H.W. de (1963) Lithops pseudotruncatella var. brandbergensis var. nov. de Boer. SUC 1963: 51–54
Boer H.W. de (1964) Enige nieuwe variëteiten van Lithops erniana Loesch et Tischer. SUC 1964: 1, 4–7
Boer H.W. de (1964) Lithops turbiniformis (Haw.) N.E.Br. en enige nieuwe variëteiten. SUC 1964: 50–53
Boer H.W. de (1964) Twee nieuwe variëteiten van Lithops schwantesii Dtr. SUC 1964: 105, 107–110
Boer H.W. de (1964) Lithops julii var. rouxii de Boer var. nov. SUC 1964: 137, 139–140
Boer H.W. de (1965) Conophytum wittebergense de Boer sp. nov. SUC 44: 21–22
Boer H.W. de (1966) Lithops dinteri var. multipunctata de Boer var. nov. SUC 1966: 21–22
Boer H.W. de (1966) Lithops lesliei var. hornii de Boer var. nov. SUC 1966: 37–38
Boer H.W. de (1966) Lithops aucampiae var. euniceae de Boer var. nov. SUC 1966: 49–54
Boer H.W. de (1966) Lithops verruculosa var. glabra de Boer var. nov. SUC 1966: 107–108
Boer H.W. de (1967) Lithops fulleri N.E.Br. var. kennedyi de Boer var. nov. SUC 1967: 108–109
Boer H.W. de (1967) Lithops archerae De Boer spec. nov. SUC 1967: 113, 122–123
Boer H.W. de (1967) Lithops vallis-mariae var. groendraaiensis (Jacobsen) De Boer comb. nova SUC 1967: 179–181
Boer H.W. de (1968) Conophytum smorenskaduense de Boer spec. nov. SUC 47: 166–168

Boer H.W. de (1969) Lithops vallis-mariae var. margarethae De Boer var. nov. SUC 1969: 9–10

Boer H.W. de (1969) Conophytum taylorianum (Dtr. & Schwant.) var. ernianum en Conophytum var. ernerianum (Loesh et Tisch.) de Boer comb. nov. SUC 48: 51–55

Boer H.W. de (1970) Conophytum rolfii de Boer Spec. nov. SUC 49: 9–10

Boer H.W. de (1970) Conophytum anjametae de Boer spec. nov. SUC 49: 43–44

Boer H.W. de (1970) In: Jacobsen H. 1970: 469

Boer H.W. de, Boom B.K. (1961) Notities over Lithops. I. SUC 1961: 37–44

Boer H.W. de, Boom B.K. (1961) Notities over Lithops. II. SUC 1961: 52–55

Boer H.W. de, Boom B.K. (1961) Notities over Lithops. III. SUC 1961: 65–69

Boer H.W. de, Boom B.K. (1961) Notities over Lithops. IV. SUC 1961: 73–77

Boer H.W. de, Boom B.K. (1964) An analytical key for the genus Lithops. NCSJ 19: 34–37, 51–55. Reprints also distributed as mededeling no. 239, 1965 by: Instituut voor de Veredeling van Tuinbouwgewassen, Wageningen

Bogle A.L. (1970) The genera of Molluginaceae and Aizoaceae in the Southeastern United States. Journal of the Arnold Arboretum 51: 431–462

Boissier P.E. (1872) Flora Orientalis 2: 765. Georg. Basel, Geneva, Leiden

Bolus H., Kensit L. (1909) Contributions to the African flora. TRSSA 1: 147–163, t. 21

Bolus L. (1913) Ficoideae. In: Pearson H.H.W.: List of the plants collected in the Percy Sladen Memorial Expedition 1908–9, 1910–11, September 1911, continued. ASAM 9: 140–158

Bolus L. (1914–28) Novitates Africanae. ABH 1: 17 (1914), 129, 190–191 (1915); 2: 28–31 (1916); 3: 4–6 (1920); 71–72 (1921); 123–139 (1922), 159–172 (1923); 4: 1–35 (1925); 37–42 (1926); 72–109 (1927); 126–130 (1928)

Bolus L. (1927) Hereroa tugwelliae. Nature Notes 2: 5; 70: 6

Bolus L. (1927) Carpobrotus fourcadei FPSA 7: t. 247, Mesembryanthemum stipulaceum t. 253, Erepsia aspera, E. restiophila, E. gracilis t. 254, Crocanthus thunbergii: t. 255, Leipoldtia amplexicaule: t. 256, Aridaria dela, Aridaria tetragona, Aridaria trichotoma: t. 258, Schonlandia lehmannii: t. 259, Leipoldtia plana: t. 260, Delosperma tradescantioides t. 261 with f 1–9, Litocarpus cordifolius: t. 261 with f 10,11, Vanzijlia annulata: t. 262, Peersia macradenia: t. 264, Stomatium pyrodorum: t. 265, Braunsia apiculata: t. 266, Argyroderma crateriforme: t. 268, Erepsia viridis: t. 270, Juttadinteria proxima: t. 273, Disphyma crassifolium: t. 276, Ruschia cymbifolia: t. 277, Delosperma pergamentaceum: t. 278, Psammophora pillansii: t. 280

Bolus L. (1927) South African succulents ...: SAG 17: 134, 238–239, 256, 280–281, 326–327, 365, 398–400, 433–435

Bolus L. (1927) Sphaeroids, Dumplins, Buttons. Conophytum, N.E.Brown. SAG 17: 294–295

Bolus L. (1927–1928) Notes on Mesembrianthemum and allied genera. (NM) 1: 1–156 (1.7.1927–1.7.1928). Latin descriptions of species: 128–152 (29.5.1928). The speciality press of S.A. Cape Town

Bolus L. (1928) Mesembrianthemum and allied genera. SAG 18: 15–16, 83–84, 127, 140, 156, 249–250, 278–279, 370–371

Bolus L. (1928) Novitates Africanae. JB 66: 9–16, 126, 195–200, 229–233

Bolus L. (1928–1935) Notes on Mesembrianthemum and allied genera. (NM) 2: 1–508. 1–16 (9.11.1928), 17–32 (21.12.1928), 33–48 (24.1.1929), 49–64 (12.4.1929), 65–80 (3.5.1929), 81–94 (6.6.1929), 95–110 (4.7.1929), 111–129 (16.8.1929), 130–146 (4.11.1929), 147–160 (22.11.1929: the last two texts end in the middle of a description), 161–176 (20.2.1930), 177–192 (9.5.1930), 193–208 (15.8.1930), 209–224 (12.11.1930), 225–244 (12.2.1931), 245–268 (1.5.1931), 269–292 (3.7.1931), 293–308 (24.9.1931), 309–336 (29.1.1932), 337–356 (24.6.1932), 357–376 (5.12.1932), 377–396 (19.5.1933), 397–416 (16.10.1933), 417–436 (26.1.1934), 437–452 (23.5.1934), 453–472 (17.8.1934), 473–508 (11.2.1935). 517–520 Corrigenda. University of Cape Town. Bolus Herbarium. Cape Town

Bolus L. (1929) Mesembrianthemum and allied genera. SAG 19: 49, 181, 244–245, 288

Bolus L. (1929) Lithops fulleri var. Tapscottii. SAG 19: 182. (figure and legend only)

Bolus L. (1930) Mesembrianthemum and allied genera. SAG 20: 18–19, 121–122, 182, 212–214

Bolus L. (1932) A new species of Lithops from Namaqualand. SAG 22: 102

Bolus L. (1932) A new Lithops from Griqualand West. SAG 22: 306

Bolus L. (1932) Plants – new or noteworthy. SAG 22: 330–331

Bolus L. (1933) Plants – new or noteworthy. SAG 23: 22

Bolus L. (1933) A new Lithops from Namaqualand. SAG 23: 218

Bolus L. (1934) Plants – new or noteworthy. SAG 24: 197–198, 221

Bolus L. (1936–1958) Notes on Mesembryanthemum and allied genera. (NM) 3: 1–19 (5.2.1936), 21–44 (31.7.1936), 45–78 (14.1.1937), 79–114 (12.8.1937), 115–138 (24.3.1938), 139–188 (20.7.1939), 189–236 (25.8.1950), 237–288 (30.8.1954), 289–417 (30.4.1958). University of Cape Town. Cape Town

Bolus L. (1938) Neohenricia. JSAB 4: 51

Bolus L. (1947) Ruschia maxima. FPSA 26: t. 1013, Aloinopsis malherbei. FPSA 26: t. 1035

Bolus L. (1949) Lampranthus haworthii. FPSA 27: t. 1064.

Bolus L. (1953) Drosanthemum strictifolium. FPSA 29: t. 1152

Bolus L. (1958) Carpanthea pilosa. FPSA 32: t. 1264

Bolus L. (1958) Notes on Mesembryanthemum and allied genera. CSJGB 20: 82–88

Bolus L. (1959) Notes on Mesembryanthemum and allied genera. JSAB 25: 150, 251–262, 371–374

Bolus L. (1959) Semnanthe lacera var. densipetala. FPSA 33: t. 1286

Bolus L. (1960) In: Jacobsen H 1960: A handbook of succulent plants. 1174, 1197, 1391

Bolus L. (1960) Notes on Mesembryanthemum and allied genera. JSAB 26: 155–164

Bolus L. (1960) Notes on Mesembryanthemum and allied genera. JSAB 27: 47–63 (published 30.12.1960, often cited as 1961)

Bolus L. (1961) Notes on Mesembryanthemum and allied genera. JSAB 27: 111–123, 169–180, 255–265

Bolus L. (1962) Notes on Mesembryanthemum and allied genera. JSAB 28: 9–19, 121–122, 138–139, 219–230, 291–303

Bolus L. (1963) Notes on Mesembryanthemum and allied genera. JSAB 29: 11–19, 45–49, 131–139, 145, 169, 171–179

Bolus L. (1964) Notes on Mesembryanthemum and allied genera. JSAB 30: 33–44, 75–79, 115–121, 231–243

Bolus L. (1965) Notes on Mesembryanthemum and allied genera. JSAB 31: 83–88, 169–174, 237–240, 307–311

Bolus L. (1966) Notes on Mesembryanthemum and allied genera. JSAB 32: 125–128, 201–203, 229–234, 335–339

Bolus L. (1967) Notes on Mesembryanthemum and allied genera. JSAB 33: 65–69, 305–309

Bolus L. (1969) Notes on Mesembryanthemum and allied genera. JSAB 35: 143–4

Boom B.K. (1959) Nomenklatuurvraagstukken bij Succulenten. SUC 11: 141–145

Boom B.K. (1971) Een nieuwe tabel voor het geslacht Lithops. Instituut voor de Veredeling van Tuinbouwgewassen, Wageningen

Boom B.K. (1973) Het geslacht Conophytum 5. SUC 52: 102–105

Boom B.K. (1973) Het geslacht Conophytum 7. SUC 52: 142–145

Boom B.K. (1973) Het geslacht Conophytum. Instituut voor veredeling van tuinbouwgewassen. Mededeling 375: 1–47. Instituut voor veredeling van tuinbouwgewassen, Wageningen

Bradley R. (1716–1727) Historia plantarum succulentarum ...: Decades I–V. W. Mears, London

Bremekamp C.E.B. (1933) New or otherwise noteworthy plants from the Northern Transvaal. Annals of the Transvaal Museum 40: 239

Breyne J.P. (1678) Exoticarum aliariumque minus cognitarum plantarum centuria prima: 160–162, t. 79,80,81. The author? Danzig

Breyne J.P. (1689) Prodromus fasciculi rariorum plantarum. J. P. Breyne, Danzig
Britton N.L. (1893) An enumeration of the plants collected by Dr. Thomas Morong in Paraguay, 1888–1890. The Annals of the New York Academy of Sciences 7: 122
Britton N.L., Sterns E.E., Poggenburg J.F. (1888) Preliminary catalogue of Anthophyta and Pteridophyta ...: 20. Torrey Botanical Club, New York
Brown N.E. (1876) Mesembryanthemum ficiforme, Haw. Mesembryanthemum vittatum, N.E.Br. GC ser. 2, 2: 772
Brown N.E. (1876) New garden plants. Mesembryanthemum setuliferum. GC ser. 2, 6: 8
Brown N.E. (1878) New garden plants. Mesembryanthemum hirtum, N.E.Br. GC ser. 2, 10: 138
Brown N.E. (1902) New or noteworthy plants. GC ser. 2, 32: 190, 350
Brown N.E. (1903) Mesembryanthemum mirabile GC 34: 131
Brown N.E. (1908) Diagnoses africanae XXIV. KB 1908: 290–291; XXV: 407
Brown N.E. (1911) Diagnoses africanae XXXIX, XLV. KB 1911: 82–83, 313, 356–359
Brown N.E. (1912) Diagnoses africanae XLIX. KB 1912: 277
Brown N.E. (1912) In: Burtt-Davy J. (1912) A new species of Mesembryanthemum from the Transvaal. TRSSA 2: 369–371
Brown N.E. (1912) New or noteworthy plants. GC 51: 403
Brown N.E. (1913) Diagnoses africanae LIII. KB 1913: 118–123
Brown N.E. (1914) Diagnoses africanae LIX. KB 1914: 167–168
Brown N.E. (1915) Mesembryanthemum stylosum. CBM 141: t. 8595A
Brown N.E. (1915) Mesembryanthemum taylori, N.E.Brown (new species) GC 58: 197
Brown N.E. (1916) New and noteworthy plants. GC 60: 204
Brown N.E. (1920) New and old species of Mesembryanthemum. Journal of the Linnean Society, Botany 45: 53–140, t. 5–10
Brown N.E. (1921) Mesembryanthemum and some new genera separated from it. GC 70: 126, 138, 151–152, 172–173, 207, 223, 250, 273, 289, 290, 303, 311, 327, 336
Brown N.E. (1922) The genera Aloe and Mesembryanthemum as represented in Thunberg's herbarium. Bothalia 1: 139–169
Brown N.E. (1922) Mesembryanthemum and some new genera separated from it. GC 71: 9, 22, 44, 55, 65, 80, 92–93, 105, 129, 151, 214, 231, 261–262, 307; GC 72: 8, 24, 54, 83–84, 124
Brown N.E. (1922) Roodia gen. nov. FPSA 2: t. 78
Brown N.E. (1925) Mesembryanthemum ... GC 78: 232, 412–413, 433, 451, 468, 484, 500
Brown N.E. (1926) Ficoidaceae. In: Burtt Davy J.: A manual of the flowering plants of the Transvaal ... 1: 49, 156–161. Longmans, Green & Co. Ltd, London
Brown N.E. (1926) Mesembranthemum s.lat. In: Phillips E.P.: The genera of South African plants: 241–249
Brown N.E. (1926) Mesembryanthemum ... GC 79: 12, 30, 80, 102, 116–117, 135, 155–156, 194, 215–216, 225, 234, 268–269, 406–407, 425–426. GC 80: 9, 35, 72–73
Brown N.E. (1926) Plants new or noteworthy. GC 80: 150
Brown N.E. (1927) Mesembryanthemum ... GC 81: 12, 31–33, 52–53, 70, 84–85, 113–114, 116, 251–253, 430; GC 82: 92–93, 113–114, 228, 251, 263, 290–291, 348–349, 369, 390, 409, 429–430, 449–450, 470–471, 489–490, 509–511
Brown N.E. (1927) Two new Mesembryanthemeae. JB 65: 348–349
Brown N.E. (1928) Mesembryanthemum ... GC 83: 13, 49, 123–124, 193–194, 339, 391, 419; GC 84: 13, 34–35, 128, 153, 192, 211–212, 253–254, 266–268, 302, 313, 472, 492–493
Brown N.E. (1928) Mesembryanthemum and allied genera. JB 66: 75–80, 106–107, 138–145, 171–172, 267–268, 322–327
Brown N.E. (1928) Ergänzungen zu dem Mesembriaceen-Aufsatz. MDG 43: 400–401
Brown N.E. (1929) Mesembryanthemum ... GC 85: 34, 84; GC 86: 32–34, 208, 244, 248–249, 264
Brown N.E. (1929) Mesembryanthemum and allied genera. JB 67: 17–20
Brown N.E. (1929) New species of and critical notes on Mesembryanthemum and allied genera. KB 1929: 56–62
Brown N.E. (1930) Mesembryanthemum ... GC 87: 13–14, 32–33, 71–72, 126, 151, 186, 211–212, 267, 515–516; GC 88: 8, 278–280, 474, 513, 516
Brown N.E. (1931) Mesembryanthemum ... GC 89: 13–15, 53–54, 92–93, 133–135, 259, 279, 294, 389; GC 90: 13–14, 136–137, 335
Brown N.E. (1931) Conophytum lucipunctum N.E.Br.; Conophytum miserum N.E.Br. CJ 1: 58
Brown N.E. (1931) A strange new plant. CSJ 2: 389–390
Brown N.E. (1931) Some new species of Conophytum. CSJ 2: 427–429
Brown N.E. (1931) In: Brown N.E., Tischer A., Karsten M. 1931: 210
Brown N.E. (1932) Mesembryanthemum. GC 91: 144, 205, 262–263
Brown N.E. (1933) Conophytum germanum N.E.Br.; C. rarum N.E.Br.; C. subtilis N.E.Br. CJ 2: 7
Brown N.E. (1933) Conophytum doornense N.E.Br.; Conophytum namibense N.E.Br.; Conophytum decoratum N.E.Br. CJ 2: 39
Brown N.E. (1934) Conophytum novicium N.E.Br.; Conophytum parviflorum N.E.Br.; Conophytum pumilum N.E.Br. CJ 2: 53
Brown N.E. (1934) Sphalmanthus herberti, Lithops edithae. CJ 2: 70
Brown N.E. (1934) Conophytum compressum N.E.Br.; Conophytum laxipetalum N.E.Br. CJ 3: 37
Brown N.E. (1936) Mestoklema. GC 100: 164–166
Brown N.E., Tischer A., Karsten M. (1931) Mesembryanthema. ed. EJ Labarre. L Reeve & Co. Ashford, Kent
Browne P. (1756) The civil and natural history of Jamaica, ed. 1: 206. Osborne, London
Brummitt R.K. (1992) Vascular plant families and genera. Royal Botanic Gardens, Kew
Brummitt R.K. (1999) Report of the Committee for Spermatophyta: 48. Taxon 48: 359–371
Brummitt R.K., Powell C.E. (1992) Authors of plant names. Royal Botanic Gardens, Kew London
Brusse F.A. (1996) A new species of Dorotheanthus (Aizoaceae, Ruschioideae) from limestone areas of the south-western Cape, South Africa. Bradleya 14: 89–93
Bruyns P.V. (1997) A new combination in the genus Ruschia and the correct name for "Cynanchum capense". SAJB 63: 241
Buchwald J. (1898) In: Kuntze O. 1898: 110
Bullock A.A. (1937) Notes on Mesembryanthemaceae I. The genus Brownanthus Schwantes. KB 1937: 492–497
Bullock A.A. (1939) Notes on Mesembryanthemaceae: II. The lectotypes of Nananthus and Aloinopsis. KB 52: 153–161
Burchell J. (1822) Travels in the interior of Southern Africa 1: 259, 343. Longman, Hurst, Rees, Orme & Brown, London
Burgoyne P.M. (1998) Finding a place in the sun: where does Ruschia purpureostyla belong? Aloe 35: 60–61
Burgoyne P.M., Smith G.F., Chesselet P. (1998) Hammeria, a new genus of Aizoaceae from South Africa. CSJ 70: 203–208
Burman J. (1738) Rariorum africanorum plantarum..., decas prima: 64, t. 26, f 1–3. Boussiere, Amsterdam
Burman N. (1768) Flora indica: 110, t. 31 and: Prodromus florae capensis (as appendix): 14–15. Haak. Leiden & Schouten, Amsterdam
Candolle A.P. de (1799–1832) Plantarum historia succulentarum. t. 72 (1801), t. 135 (1803). Garnery, Paris
Candolle A.P. de (1800–1804) Plantes grasses. t. 26, 53 (1800); 60 (1800 or 1801); 83, 84 (1801); 89 (1801 or 1802); 95 (1802); 129, 130 (1803); 147 (1804). Garnery, Paris (all dates after Rowley, GD 1956–1957)
Candolle A.P. de (1824) Prodromus systematis naturalis regni vegetabilis, sive ... 1: 393. Treuttel & Würtz. Paris
Candolle A.P. de (1828) Prodromus systematis naturalis regni vegetabilis, sive ... 3: 352, 415–456. Treuttel & Würtz, Paris
Candolle A.P. de (1829) (text), 1832 (plate): Plantes grasses. t. 164. Garnery, Paris
Candolle A.P. de (1837) Plantes grasses. t. (179). Probably not published
Candolle A.P. de (1849) Prodromus systematis naturalis regni vegetabilis, sive ...: Phytolacceae. 13,2: 16–19. Masson, Paris

Cassini A.H.G. (1817) Aperçu des genres ou sous-genres nouveaux formes par M. Henri Cassini dans la famille des Synantherées. 2 nd part. Bulletin par la Société Philomatique de Paris 2: 10–13

Cesati V. (1873) Illustrazione de alcune piante raccolte dal Sig. Prof. Strobel sul versante orientale delle Ande Chilene dal passo del Planchon sino a Mendoza, attraverso del Pampa del Sud. Atti dell'Academia delle Scienze Fisiche et Mathematiche 5: 10

Chesselet P., Hartmann H.E.K. (1995) Khadia alticola Chess. & H.E.K. Hartm. spec. nov. (Mesembryanthema, Aizoaceae). Aloe 32: 46–49

Chesselet P., Hartmann H.E.K., Hahn N., Smith G.F. (1998) Taxonomic notes on the genus Khadia (Mesembryanthemaceae/ Aizoaceae). Bothalia 28: 25–33

Chesselet P., Smith G.F., Burgoyne P.M., Klak C., Hammer S.A., Hartmann H.E.K., Kurzweil H., Jaarsveld E.J. van, Wyk B.-E. van, Leistner O.A. (2000) Mesembryanthemaceae. In: Leistner O.A. (ed) Seed plants of Southern Africa: families and genera. Strelitzia 10: 360–410

Chinnock R.J. (1971) Studies in Disphyma - A genus related to Mesembryanthemum 1. New Zealand Journal of Botany 9: 331–344

Chinnock R.J. (1976) Studies in Disphyma - A genus related to Mesembryanthemum 2. New Zealand Journal of Botany 14: 77–78

Chinnock R.J. (1983) The Australian genus Gunniopsis Pax (Aizoaceae). Journal of the Adelaide Botanical Garden 6: 133–179

Chinnock R.J. (1986) In: Jessop PJ: Aizoaceae. In: Jessop P.J., Toelken H.R. (ed) Flora of South Australia 1: 194

Chinnock R.J. (1996) To the limits of Disphyma (Aizoaceae: Ruschioideae) and beyond. Aloe 33: 59–61

Chiovenda E. (1932) Flora Somala 2: 225–228. The Author, Modena

Christensen C. (1922) Index to Pehr Forsskål: Flora aegyptiaco-arabica 1775 with a revision of Herbarium Forskålii ... Dansk Botanisk Arkiv 4, part 3: 1–54

Clark J.Y. (1996) A key to Lithops N.E.Br. (Aizoaceae). Bradleya 14: 1–9

Cole D.T. (1970) Four new varieties of Lithops. NCSJ 25: 7–10

Cole D.T. (1970) Lithops dorotheae Nel - A tragedy of errors. NCSJ 25: 106–108

Cole D.T. (1973) Lithops dinteri Schwantes: A review & a new variety. Madoqua ser. 1 (7): 33–38

Cole D.T. (1973) Lithops gulielmi L.Bolus & Lithops fossulifera nomen nudum: 35 years of confusion resolved. Madoqua ser. 1 (7): 39–43

Cole D.T. (1973) Lithops: A checklist & index. Excelsa 3: 37–72. Aloe, Cactus & Succulent Society of Rhodesia [Zimbabwe]

Cole D.T. (1980) A new species of Lithops. NCSJ 35: 74–75

Cole D.T. (1987) Lithops of SWA/Namibia. Madoqua Memoir 1

Cole D.T. (1988) Lithops - Flowering stones. Acorn Books & Russel Friedman, Randburg

Cole D.T. (1997) Lithops - Sassi Fioriti. Translation: Sezzi & Swierczewski. Montanari, Casalgrande Alto

Cole D.T. (2001) : Lithops hermetica, Cole - A new species from Namibia. Cactus & Co. 4,4: 156–161 (published as volume IV 2000 in February 2001)

Commeli(j)n C. (1701) Horti medici amsteldamensis.. 2: 203–206, t. 102–103. P.&J.Blaeu; van Someren, Amsterdam

Compton H. (1931) The flora of the Whitehill district. TRSSA 19: 283

Cordemoy E.J. (1895) Flore de l'ile de la Réunion: 290–291. Klincksieck, Paris (reprint 1972: Cramer, Lehre)

Cornu M. (1896) Note sur un genre noueau de Pontédériacées d'Afrique: Schoenlandia (Sch. gabonensis Cornu). Bulletin de la Société botanique de France 43: 21–24

Correll D.S. (1966) Some additions and corrections to the flora of Texas 3. Rhodora 68: 421–422

Correll D.S., Johnston M.C. (1966) Manual of the vascular plants of Texas: Aizoaceae: 602–605. Texas Research Foundation, Renner, Texas

Crantz H.I.N. (1766) Institutiones rei herbariae iuxta..1: 135–136. Kraus, Vienna

Curtis W. (1794) Mesembryanthemum aureum CBM 8: t. 262

Curtis W. (1798) Mesembryanthemum spectabile. CBM 11: t. 396

Daniel P., Umamaheswari P. (1999) Did Willdenow validly publish the name Trianthema triquetra (Aizoaceae)? Indian Journal of Forestry 22: 409–410

D'Antonio C.M., Mahall B.E. (1991) Root profiles and competition between the invasive, exotic perennial, Carpobrotus edulis, and two native shrub species in California coastal scrub. American Journal of Botany 78: 885–894

Deflers A. (1889) Voyage au Yemen. Journal d'une excursion botanique: Ficoideae: 140–141. Klincksieck, Paris

Degener O., Degener I. (1960) Aizoaceae. Flora Hawaiiensis: family 116

Degener O., Fosberg F.R. (1952) A Central Pacific Sesuvium. Occasional Papers of Bernice P. Bishop Museum, Honolulu, Hawaii: 21: 45–47

Dehn M. (1989) Untersuchungen zum Merkmalsbestand und zur Stellung der Gattung Antimima N.E.Br. emend. Dehn (Mesembryanthemaceae Fenzl). MIBH 22: 189–215

Dehn M. (1993) Untersuchungen zum Verwandtschaftskreis der Ruschiinae (Mesembryanthemaceae) MIBH 24: 91–198

Derenberg J. (1926) Mesembrianthemum microspermum Dtr. et Derenbg. ZSK 2: 264–268

Diels L. (1904) In: Diels L., Pritzel E. 1904: Fragmenta phytographae australiae occidentalis. BJS 35: 196–197

Diels L. (1909) Formationen und Florenelemente im nordwestlichen Kapland BJS 44: 91–124

Diels L., Schlechter R. (1907) Aizoaceae. In: Schultze L: Aus Namaland und Kalahari: 692–694

Dietrich D. (1842) (not 1843 or 1847, as cited elsewhere): Synopsis plantarum ...: Aizoon: 128–130. Voigt, Weimar

Dillenius J.J. (1732) Hortus elthamensis ... The author, London

Dinter K. (1900) "Karroo Plants". GC 27: 210–211

Dinter K. (1914) Neue und wenig bekannte Pflanzen Deutsch-Südwest-Afrikas. Selbstverlag, Okahandja

Dinter K. (1922) Mesembrianthemum montis-moltkei. MKK 32: 113 (1922)

Dinter K. (1923) Beiträge zur Flora von Südwestafrika I. FR 19: 122–160

Dinter K. (1923) Beiträge zur Flora von Südwestafrika II. FR 19: 177–186

Dinter K. (1923) Index der aus Deutsch-Südwestafrika bis zum Jahre 1917 bekannt gewordenen Pflanzenarten XI. FR 19: 187–192 (30.7.1923)

Dinter K. (1923) Index der aus Deutsch-Südwestafrika bis zum Jahr 1917 bekannt gewordenen Pflanzenarten XII. FR 19: 235–236 (30.9.1923)

Dinter K. (1923) Succulentenforschung in Südwestafrika. I. FR Beiheft 23: 1–80

Dinter K. (1927) Neue Sukkulenten-Arten meiner Südwestafrika-Reise I. Lithops summitatum Dtr. sp. n.: MDG 42: 101–102. III. Lithops alpina Dtr. sp. n.: 123. IV. Dinteranthus margaretae Schw.: 150–151. V. Titanopsis hugo-schlechteri Dtr. & Schw.: 156–158. VI. Mesembryanthemum succumbens Dtr. sp. n.: 198–199

Dinter K. (1927) Schwantesia rüdebuschii Dtr. gen. nov. Dtr. MDG 42: 234

Dinter K. (1927) Sukkulenten-Studien. MDG 42: 301–302, 315–316

Dinter K. (1928) Südwestafrikanische Lithopsarten. R. Graessner, Perleberg

Dinter K. (1928) Sukkulentenforschung in Südwestafrika II: 12, 93, 98, 102–103, 105–106. Gustav Winter, Herrnhuth i.Sa. (this print is also bound as FR Beiheft 53: 1–145)

Dinter K. (1930) Die Sukkulenten der Buchuberge. MDK: 215–220

Dinter K. (1931) Kurzer Bericht über meine Reise 1929 in der Küstenwüste SW-Afrikas, spez. die Buchuberge. FR 29: 163–170

Dinter K. (1932) Diagnosen neuer südwestafrikanischer Pflanzen FR 30: 180–205

Dinter K. (1933) Lithops Elisabethae Dtr. n. sp. Kakteenkunde 1933: 50

Dinter K. (1935) Aizoanthemum membrumconnectens Dtr. n. gen. Kakteenkunde 1935: 27–28

Dinter K. (1935) Brief aus den Auasbergen. Kakteenkunde 1935: 112–116, 138
Dinter K. (1937) In: Jacobsen H 1937: 156–159, 199–200
Dinter K. & Berger A. (1914) In: Berger A., Dinter C. 1914: 586–592
Dinter K. & Derenberg J. (1926) In: Derenberg J. 1926: 264–268
Dinter K. & Schwantes G. (1920) In: Schwantes G. 1920: 36
Dinter K. & Schwantes G. (1925) In: Schwantes G. ZSK 1925: 12–20, 21–28
Dinter K. & Schwantes G. (1926) In: Schwantes G. ZSK 1926: 188–189
Dinter K. & Schwantes G. (1927) In: Schwantes G. ZSK 1927: 14–30, 105–106
Dinter K., Schwantes G. (1927) XI. Dracophilus, Dinter & Schwantes gen. nov. MDG 42: 187
Dinter K. & Schwantes G. (1928) In: Schwantes G. GF 77: 68–69
Dinter K. & Schwantes G. (1929) In: Schwantes G. GW 1929: 69
Dinter K. & Schwantes G. (1933) In: Jacobsen H. 1933: 138
Dinter K. & Schwantes G. (1935) In: Jacobsen H. 1935: 222
Dinter K. & Schwantes G. (1937) In: Jacobsen H. 1937: 159–163
Don G. (1834) A general system of Gardening and Botany ...: 72. Rivington et al., London
Donn J. (1800) Hortus Cantabrigiensis or a catalogue of plants ... ed. 2: 66. Burges, Cambridge
Drege J.F. (1843) Zwei Pflanzengeographische Documente. Besondere Beigabe zur Flora 1843, vol. 2: 63, 92, 93, 226, 227
Druce G.C. (1914) Supplement to botanical exchange club report for 1913. Part II. The abridgement of Miller's gardener's dictionary of 1754. The Botanical Society and Exchange Club of the British Isles: 426–435. Arbroath
Druce G.C. (1917) Nomenclatural notes: Chiefly African and Australian. Report for 1916. Second supplement to Botanical Society and Exchange Club: 602–655. Bunde & Co, Arbroath
Druce G.C. (1926) Plant notes for 1925. The Botanical Society and exchange Club of the British Isles 7: 771
Dulac J. (1867) Flore du Département des Hautes Pyrénées. F. Savy, Paris
Ecklon C.F., Zeyher C. (1837) Enumeratio plantarum africae australis extratropicae ... 3: Ficoideae: 308–329. The authors, Hamburg
Edgeworth M.P. (1862) Florula Mallica. Journal of the Linnean Society. Botany 6: 203–204
Edwards S.T. (1818) The botanical register 3: t. 260. Ridgway, London
Edwards S.T. (1820) The botanical register 6: t. 493, 494. Ridgway, London
Edwards S.T. (1825) The botanical register 10: t. 863. Ridgway, London
Eichler H.J. (1965) Supplement to J. M. Black's flora of South Australia (second edition 1943–1957): 134–136. Hawes, Government Printer, Adelaide
Elliott S. (1821) A sketch of the botany of South-Carolina and Georgia 1: 556. Schenck, Charleston (reprint in: Classica botanica Americana 6. Hafner, New York)
Endlicher S. (1833) Prodromus Florae Norfolkiae ...: 72–73. Beck, Vienna
Engler A. (1892) Über die Hochgebirgsflora des tropischen Afrika. Abhandlungen der preußischen Akademie der Wissenschaften 11: 208–209
Engler A. (1909) Aizoaceae africanae. BJS 43: 189–198
Engler A. (1910) Vegetation der Erde 1.2. Die Pflanzenwelt Afrikas I. 41. Das westliche Namaqua: und Hereroland: 517, 540. Engelmann, Leipzig
Engler A. (1913) In: Schellenberg G. 1913: 502–504
Engler A. (1915) Die Pflanzenwelt Afrikas 3,1: 144, 148, 153. Engel-mann, Leipzig
Esler K., Rundel P., Connors P. (1995) All that glisters ... Veld & Flora 81: 12–13
Ewart E.J. (1908) Contributions to the flora of Australia 7. Proceedings of the Royal Society of Victoria n.s. 20,2: 128–129
Ewart E.J., White J. (1908) Contributions to the flora of Australia No. 10. Proceedings of the Royal Society of Victoria 21 (N.S.) part 2: 540–549
Fabricius P.C. (1759) Enumeratio methodica plantarum horti medici Helmstadiensis ...: 201. Drimborn, Helmstedt
Fabricius P.C. (1763) Enumeratio methodica plantarum horti medici Helmstadiensis ... ed. 2: 363. Drimborn, Helmstedt
Fearn B. (1970) New combinations and an analytical key for the genus Lithops. CSJ 42: 89–93
Fearn B. (1976) The Lithops julii – L. fulleri – L. salicola complex. NCSJ 31: 58–61
Fearn B. (1981) Lithops. National Cactus & Succulent Society, Oxford
Fedde F. (1940) Nomenklatorische Notizen I. FR 48: 11
Fedde F., Schuster K. (1918) XVI. Novorum generum, specierum, varietatum, formarum, nomimum Siphonogamorum Index. Aizoaceae. Just's Botanischer Jahresbericht 41: 92
Fenzl E. (1836) Monographie der Mollugineen und Steudelieen, zweier Unterabteilungen der Familie Portulaceen. Annalen des Wiener Museums für Naturgeschichte 1: 337–386
Fenzl E. (1839) Monographie der Mollugineen. Annalen des Wiener Museums der Naturgeschichte 2: 243–310
Fenzl E. (1839) Novarum Stirpium decades. Decas 6: 51–52; Decas 7: 57–58. Sollinger, Wien
Fenzl E. (1862) Tetragonia. In: Harvey W.H., Sonder W. 1862: 460–469
Forskål P. (1775) Flora Aegytiaco-arabica sive descriptiones plantarum ...: 25, 69–71, 95, CVIII, no. 200, t. 10. Möller, Kopenhagen
Forster J.G.A. (1786) Florulae insularum australium prodromus: 39. Dieterich, Göttingen
Friedrich H.-C. (1955) Beiträge zur Kenntnis einiger Familien der Centrospermae. MBSM 12: 56–60
Friedrich H.-C. (1957) Aizoanthemum Dinter ex Friedr., eine wenig beachtete Gattung der Ficoideae aus Südwestafrika. MBSM 2: 339–347
Friedrich H.-C. (1959) In: Herre H., Friedrich H.-C.: Die geographische Verbreitung der Ficoidaceae – Mesembryanthemoideae in Südafrika. 1. Teil: Aptenieae. MBSM 3: 44–70
Friedrich H.-C. (1960) Mesembrianthemenstudien I. Beitrag zur Kenntnis der Gattungen Stoeberia Dtr. et Schw. und Ruschianthemum Friedr. gen. nov. MBSM 3: 554–567
Friedrich H.-C. (1960) Ficoidaceae. MBSM 3: 613–614
Friedrich H.-C. (1961) Mesembrianthemenstudien II. Die Gattung Ruschia Schw. in Südwestafrika. MBSM 4: 129–143
Friedrich H.-C. (1967) Tetragoniaceae. In: Merxmüller H. (ed) Prodromus einer Flora von Südwestafrika 28: 1–7
Friedrich H.-C. (1968) Mesembrianthemenstudien III. Neue Sippen und Kombinationen. MBSM 7: 211–218
Friedrich H.-C. (1970) Aizoaceae. In: Merxmüller H. (ed) Prodromus einer Flora von Südwestafrika 27: 1–135
Gabali S.A. (1998) Studies in the flora of Yemen. 4. The endemic species. Candollea 53: 73–92
Gaertner J. (1791) De fructibus et seminibus plantarum 2: 483, t. 179. Schramm, Tübingen (unchanged reprint 1974. Asher & Co, Amsterdam)
Gamble J.S. (1919) Flora of the Presidency of Madras 3. London (not seen)
Gardner C.A. (1930) Enumeratio plantarum Australiae occidentalis: 42. Government Printer, Perth
Gay C. (n.d.) Historia Física y política de Chile ... Botanica (Flora chileana) 2: 469–470. The editor, Paris
Gerbaulet M. (1994) Lectotypification of Phyllobolus N.E.Br. (Aizoaceae s.str.). Taxon 43: 75–76
Gerbaulet M. (1995) Phyllobolus N.E.Br. emend. Bittrich (Aizoaceae): A reassessment of generic boundaries. BJS 117: 385–399
Gerbaulet M. (1996) Revision of the genus Sceletium N.E.Br. (Aizoaceae). BJS 118: 9–24
Gerbaulet M. (1996) Revision of the genus Prenia N.E.Br. (Aizoaceae). BJS 118: 25–40
Gerbaulet M. (1996) Revision of the genus Aridaria N.E.Br. (Aizoaceae). BJS 118: 41–58
Gerbaulet M. (1997) Revision of the genus Phyllobolus N.E.Br. (Aizoaceae). BJS 119: 145–211
Gerbaulet M., Ihlenfeldt H.-D. (1990) Lectotypification of the genus Thyrasperma. Taxon 39: 511–512
Gerbaulet M., Struck M. (1996) Three new species of Phyllobolus (Mesembryanthemoideae, Aizoaceae) from the arid parts of South Africa. Haseltonia 4: 24–27

Gereau R.E. (1993) The identity of Tetragonia horrida Britton. Taxon 42: 411–412
Gilbert M.D. (1993) Aizoaceae. In: Thulin M.: Flora of Somalia 1: 111–117. Royal Botanic Gardens Kew, London
Gilbert M.D., Hartmann H.E.K., Edwards S. (2000) Aizoaceae. In: Edwards S.B., Tadesse M., Demissew S., Hedberg I. (eds) Flora of Ethiopia and Eritrea 2,1: 240–248. Biology Department, University of Addis Ababa & Department of Systematic Botany, University of Uppsala, Addis Ababa, Uppsala
Glen H.F. (1974) A revision of the Gibbaeinae (Mesembryanthemaceae). Unpublished MSc thesis, University of Cape Town
Glen H.F. (1978) A taxonomic monograph of Lampranthus and allied genera. Unpublished PhD thesis, University of Cape Town
Glen H.F. (1980) Proposal (537) to conserve the generic name 2405 Lampranthus N.E.Br. (1930) against Oscularia Schwantes (1927) (Mesembryanthemaceae). Taxon 29: 693–694
Glen H.F. (1981) Mesembryanthemaceae: Nomenclature in the genus Mestoklema. Bothalia 13: 454–455
Glen H.F. (1986) Mesembryanthemaceae: a new combination in Lampranthus. Bothalia 16: 55
Glen H.F. (1986) Numerical taxonomic studies in the subtribe Ruschiinae (Mesembryanthemaceae) - Astridia, Acrodon and Ebracteola. Bothalia 16: 203–226
Gmelin J.F. (1791) Caroli a Linné Systema Naturae ... 2: 455, 845. Beer, Leipzig
Gonçalves M.L. (1965) Sobsidios para o conhecimento da flora de Angola - I. Garcia de Orta 13: 381–382
Gonçalves M.L. (1970) Aizoaceae. In: Conspectus Florae Angolensis 4: 302–333. Junta de investigaçoes do ultramar e ... Lisbon
Gonçalves M.L. (1978) Aizoaceae sensu lato. Flora Zambesiaca 4: 508–555
Graessner R. (1937) Mesembranthemen. Samenliste 1937: 14–20. F. Grunick Nachf., Perleberg
Graessner R. (1937) Mesembranthemen - Mesembryanthema. Hauptverzeichnis: 22–29. Publisher? Perleberg
Gray A. (1848) Genera florae Americae boreali-orientalis illustrata 1: t. 100. Munroe, Boston
Greuter W. (1980) Med-Checklist Notulae 2. Willdenowia 10: 227–232
Greuter W., McNeill J., Barrie F.R., Burdet H.M., Demoulin V., Figueiras T.S., Nicolson D.H., Silva J.E., Skog P.C., Trehane P., Turland N.J., Hawksworth D.L. (2000) International code of botanical nomenclature (Saint Louis Code). Koeltz, Königstein
Groen L.E. (1994) Faucaria nemorosa L.Bolus ex Groen. SUC 73: 39–42
Haage F.A. (1930) Kakteenpreisverzeichnis: 108. The author, Erfurt
Haage F.A. (1937) Kakteen-Samen 1937: 16. The author, Erfurt
Hallier H. (1921) Beiträge zur Kenntnis der Linaceae (DC. 1819) Dumort. Beihefte zum Botanischen Centralblatt 39.2: 1–178
Hammer S. (1988) A new Conophytum from Steinkopf. Aloe 25: 10–11
Hammer S. (1988) A new Conophytum from Bushmanland. CSJ 60: 217–218
Hammer S. (1988) A new Conophytum from the Richtersveld. CSJ 60: 262–263
Hammer S. (1988) Conophytum: an annotated checklist (A–C). Bradleya 6: 101–120
Hammer S. (1989) A new Conophytum from Namaqualand. CSJ 61: 285–286
Hammer S. (1989) Conophytum: an annotated checklist (D–K). Bradleya 7: 41–62
Hammer S. (1990) A new blood-red Conophytum CSJ 62: 25–27
Hammer S. (1990) Conophytum: an annotated checklist (L–R). Bradleya 8: 53–84
Hammer S. (1991) Brown, Bolus, and Bijlia. Aloe 28: 84–88
Hammer S. (1991) Conophytum: an annotated checklist (S–W). Bradleya 9: 105–128
Hammer S. (1991) Tantalus et Tantillum. CSJ 63: 139–141
Hammer S. (1992) Esterhuysenia L.Bolus, or a Gordian knot. BCSJ 10: 59–60
Hammer S. (1993) Red and speckled plums. CSJ 65: 114
Hammer S. (1993) A new dwarf Delosperma from the Cape Province. CSJ 65: 123–125
Hammer S. (1993) Dwarf dwarfs: the section Minuscula. Haseltonia 1: 19
Hammer S. (1993) The genus Conophytum. A conograph. Succulent Plant Publications, Pretoria
Hammer S. (1994) Macro-observations on Cheiridopsis N.E.Br. and Odontophorus N.E.Br. (Aizoaceae). Bradleya 11: 75–85
Hammer S. (1994) A novel Neohenricia. CSJ 66: 84–86
Hammer S. (1994) Vlokia, a new genus in Aizoaceae. CSJ 66: 255–258
Hammer S. (1995) New nothogenera, and a new combination, in Mesembryanthema. CSJ 67: 172–173
Hammer S. (1995) Mastering the art of growing mesembs. CSJ 67: 195–247
Hammer S. (1995) Hartmanthus, a new genus in Aizoaceae. Haseltonia 3: 77–82
Hammer S. (1995) Notes on the generic placement of Mesembryanthemum locale N.E.Br. (Aizoaceae). Haseltonia 3: 89–91
Hammer S. (1995) Marlothistella Schwantes: A rising Star. Bradleya 13: 82–87
Hammer S. (1996) Piante Grasse Speciale 1995: 1–97. Supplementum to Piante Grasse 15, 4
Hammer S. (1997) New chrysogenic species of Conophytum N.E.Br. from the Richtersveld. CSJ 69: 127–131
Hammer S. (1998) New taxa from Namaqualand: Aizoaceae and Asphodelaceae. Cactus & Co. 2: 5–6
Hammer S., Barnhill C. (1997) Odes part two: Descriptions of new Mesembryanthema. Bradleya 15: 42–47
Hammer S., Lavis M. (1989) A new Conophytum from Namaqualand. CSJ 61: 3–4
Hammer S., Opel M. (2000) Correction. CSJ 72: 302 (validation of Conophytum § Batrachia Opel & Hammer)
Hammer S., Uijs R. (1994) Lithops coleorum S.A. Hammer & R. Uijs sp. nov., a new species of Lithops N.E.Br. from the Northern Transvaal. Aloe 31: 36–38
Hammer S., Opel M., Smale T., Rodgerson C., Juergens N., Witt T., Schmiedel U. (2001) Conophytum - New views of the genus. Dumpling and his wife. East Anglia Engravings, Oxford
Hartmann H.E.K. (1973) New combinations and a key for the genus Argyroderma N.E.Br. NCSJ 28: 48–50
Hartmann H.E.K. (1976) Monographie der Gattung Odontophorus N.E.Br. (Mesembryanthemaceae Fenzl). Monographien der Subtribus Leipoldtiinae Schw. I. BJS 97: 161–225
Hartmann H.E.K. (1978) Monographie der Gattung Argyroderma N.E.Br. (Mesembryanthemaceae Fenzl). MIBH 15: 121–235
Hartmann H.E.K. (1978) Zur Kenntnis der Gattung Cephalophyllum. BJS 99: 264–302
Hartmann H.E.K. (1979) De naam Argyroderma ovale en het probleem van een "cline" binnen de soort A. pearsonii (N.E.Br.) Schwantes. SUC 58: 277–281
Hartmann H.E.K. (1980) Die Gattung Argyroderma I. Die natürlichen Lebensbedingungen. KuaS 31: 292–296. II. Bestimmungsschlüssel. KuaS 32: 9–12
Hartmann H.E.K. (1981) Die Gattung Argyroderma III. Argyroderma testiculare (Aiton) N.E.Brown. KuaS 32: 62–65. IV. Die Argyroderma-delaetii-Gruppe, 1. and 2. KuaS 32: 228–232, 296–301
Hartmann H.E.K. (1981) Notes on the genus Cephalophyllum 3. The status of Cheiridopsis cuprea (L.Bol.) N.E.Br. CSJGB 43: 96–100
Hartmann H.E.K. (1982) Die Gattung Argyroderma V. Die Argyroderma-pearsonii-Gruppe. 1. KuaS 33: 172–175
Hartmann H.E.K. (1982) Monographien der Subtribus Leipoldtiinae. II. Monographie der Gattung Fenestraria (Mesembryanthemaceae). BJS 103: 145–183
Hartmann H.E.K. (1982) Zur Frage von Gattungsabgrenzungen in der Familie Mesembryanthemaceae Fenzl am Beispiel der Gattungen Cephalophyllum N.E.Br. und Vanzijlia L.Bol. MIBH 18: 85–101

Hartmann H.E.K. (1983) Monographien der Subtribus Leipoldtiinae. IV. Monographie der Gattung Vanzijlia (Mesembryanthemaceae). BJS 103: 499–538
Hartmann H.E.K. (1983) Monographien der Subtribus Leipoldtiinae. V. Monographie der Gattung Hallianthus (Mesembryanthemaceae). BJS 104: 143–169
Hartmann H.E.K. (1983) Interaction of ecology, taxonomy, and distribution in some Mesembryanthemaceae. Bothalia 14: 653–659
Hartmann H.E.K. (1983) Untersuchungen zum Merkmalsbestand und zur Taxonomie der Subtribus Leipoldtiinae (Mesembryanthemaceae). BB 136: 1–67
Hartmann H.E.K. (1983) Die Gattung Argyroderma V. Die Argyroderma-pearsonii-Gruppe. 2. KuaS 35: 2–5
Hartmann H.E.K. (1984) Die Gattung Argyroderma N.E.BR. VI. Die Untergattung Roodia. KuaS 35: 26–30
Hartmann H.E.K. (1984) Zur Biologie und Taxonomie des Cephalophyllum-curtophyllum-Komplexes. MIBH 19: 141–163
Hartmann H.E.K. (1984) Monographien der Subtribus Leipoldtiinae. VI. Monographie der Gattung Jordaaniella (Mesembryanthemaceae). BJS 104: 321–360
Hartmann H.E.K. (1989) Monographien der Subtribus Leipoldtiinae. VIII. Monographie der Gattung Cephalophyllum (Mesembryanthemaceae). MIBH 22: 93–187
Hartmann H.E.K. (1991) Mesembryanthema. CBH 13: 75–157
Hartmann H.E.K. (1992) Ihlenfeldtia, a new genus in Mesembryanthema (Aizoaceae). BJS 114: 29–50
Hartmann H.E.K. (1992) A new name for Bijlia cana (Mesembryanthema). CSJ 64: 173–180
Hartmann H.E.K. (1992) Revision of the genus Vanheerdea L.Bolus ex H.E.K. Hartmann. Bradleya 10: 5–16
Hartmann H.E.K. (1993) Aizoaceae. In: Kubitzki K.: The families and genera of vascular plants: 37–69. Springer, Berlin Heidelberg New York
Hartmann H.E.K. (1994) On the phytogeography and evolution of Mesembryanthema (Aizoaceae). In: Seyani JH, Chikuni AC (eds) Proceedings of the XIIIth Plenary meeting of AETFAT, Malawi 1991, 2: 1165–1180
Hartmann H.E.K. (1996) Miscellaneous taxonomic notes on Aizoaceae. Bradleya 14: 29–56
Hartmann H.E.K. (1996) Antimima aurasensis H.E.K.Hartmann (Aizoaceae), eine neue Art aus Namibia. KuaS 47: 229–234
Hartmann H.E.K. (1998) Clés d'identification des genres de Mésembryanthmacées. Succulentes (France) 21, n° spécial: 14–42
Hartmann H.E.K. (1998) New combinations in Antimima N.E.Br. (Ruschioideae, Aizoaceae). Bothalia 28: 67–82
Hartmann H.E.K. (1998) New combinations in Ruschioideae, based on studies in Ruschia (Aizoaceae). Bradleya 16: 44–91
Hartmann H.E.K. (1999) An annotated list of the species placed in Ruschia (Aizoaceae). Bradleya 17: 38–74
Hartmann H.E.K. (2000) The genus Delosperma in Gauteng. I. A new species in the group with white flowers: D. gautengense H.E.K.Hartmann. ALOE 38: 4–8
Hartmann H.E.K. (2000) The genus Delosperma in Gauteng. II. A new species in the group with purple flowers: D. purpureum H.E.K.Hartmann. ALOE 38: 9–12
Hartmann H.E.K. (2001a) The genus Delosperma in Gauteng. I. A new species in the group with white flowers: D. gautengense H.E.K.Hartmann. ALOE 38: 4–8
Hartmann H.E.K. (2001b) The genus Delosperma in Gauteng. II. A new species in the group with purple flowers: D. purpureum H.E.K.Hartmann. ALOE 38: 9–12
Hartmann H.E.K., Bittrich V. (1990) Nomenclature in Mesembryanthema (Aizoaceae): The generic names by Rappa & Camarrone. Bothalia 20: 153–157
Hartmann H.E.K., Bittrich V. (1991) Typification of suprageneric names – some nomenclatural changes in Aizoaceae. SAJB 57: 73–74
Hartmann H.E.K., Bruckmann C. (2000) The capsules of Drosanthemum Schwantes (Ruschioideae, Aizoaceae). Bradleya 18: 75–112
Hartmann H.E.K., Dehn M. (1987) Monographien der Leipoldtiinae. VII. Monographie der Gattung Cheiridopsis (Mesembryanthemaceae). BJS 108: 567–663
Hartmann H.E.K., Dehn M. (1989) A re-examination of the genus Amphibolia. Bothalia 19: 179–182
Hartmann H.E.K., Gölling H. (1994) A monograph of the genus Glottiphyllum. Bradleya 11: 1–49
Hartmann H.E.K., Newton L.E. (1991) Delosperma oehleri – an interesting Mesemb from East Africa. CSJ 63: 191–193
Hartmann H.E.K., Niesler I. (1991) On the identity of Trichodiadema schimperi (Engler) Herre (Mesembryanthema). CSJ 63: 143–149
Hartmann H.E.K., Rust S. (1995) Monographien der Leipoldtiinae IX. Monographie der Gattung Leipoldtia L.Bolus s.lat. (Aizoaceae). VNVH 34: 275–351
Hartmann H.E.K., Stüber D. (1993) On spiny Mesembryanthema and the genus Eberlanzia (Aizoaceae). CBH 15: 1–75
Hartmann H.E.K., Stüber D. (1995) Erweiterung der Gattung Leipoldtia um zwei dornentragende Arten (Mesembryanthema, Aizoaceae). VNVH 34: 353–372
Harvey W.H., SonderW. (1862) Flora Capensis 2: 386–478, 596–598, 616. Hodges, Smith & Co, Dublin
Haworth A.H. (1795) Observations on the genus Mesembryanthemum ..., part 2. The author, London
Haworth A.H. (1797) Mesembryanthemum inclaudens. Andrews Botanical Repertorium 4: t. 388
Haworth A.H. (1803) Miscellanea naturalia, ... J. Taylor, London
Haworth A.H. (1812) Synopsis plantarum succulentarum. R. Taylor, London
Haworth A.H. (1819) Supplementum plantarum succulentarum. R. & A. Taylor. London
Haworth A.H. (1821) Revisiones plantarum succulentarum. R. & A. Taylor, London
Haworth A.H. (1823) Plantae Rarae Succulentae. Mesembryanthemum. Linn. &c. The Philosophical Magazine and Journal 62: 381
Haworth A.H. (1824) Observations on the Mesembryanthema barbata. The Philosophical Magazine and Journal 64: 61–62
Haworth A.H. (1824) An account of the Mesembryanthema Ringentia. The Philosophical Magazine and Journal 64: 109–111
Haworth A.H. (1824) Decas tertia novarum Plantarum Succulentarum. The Philosophical Magazine and Journal 64: 423–428
Haworth A.H. (1826) Decas sexta novarum Plantarum Succulentarum. The Philosophical Magazine and Journal 68: 125–132. Decas septima ...: 328–331
Haworth A.H. (1827) Decas octava Plantarum Novarum Succulentarum. The Philosophical Magazine and Journal, New series 1: 124–125
Haworth A.H. (1831) Mr. Haworth's thirteenth decade of new succulent plants. The Philosophical Magazine and Journal, New series 10: 414–424
Heath P.V. (1993) The type of Gibbaeum Haworth. Calyx 4: 27–28
Heine R. (1986) Lithops – lebende Steine. Neumann-Neudamm, Melsungen
Heinrich W. (1964) Rhombophyllum rhomboideum (S.-D.) Schwantes. Cactus (France) 82: 54–56
Heinrich W. (1970) In: Jacobsen H. 1970: 522
Heister L. (1748) Systema plantarum ...: 11. Drimborn, Helmstedt
Hermann P. (1698) Paradisus Batavus: 166–168, 212–214, t. 155, 212. Elzevier, Leiden
Herre H. (1948) Arenifera Herre gen. nov. SKK 2: 35–38; Beiträge zur Kenntnis der Mesembryanthemaceen. SKK 2: 38–39
Herre H. (1951) Jensenobotrya lossowiana Herre gen. et spec. nov. SKK 4: 79–81
Herre H. (1954) In: Bolus L. NM 1954: 241–242
Herre H. (1971) The Genera of Mesembryanthemaceae. Tafelberg, Cape Town
Herre H., Friedrich H.-C. (1959) Die geographische Verbreitung der Ficoidaceae – Mesembryanthemoideae in Südafrika. 1. Teil: Aptenieae. MBSM 3: 44–70
Heyligers P.C. (1999) The occurrence and history of Tetragonia decumbens (Aizoaceae) in new South Wales. Telopea 8: 371–373
Hiern W.P. (1898) Catalogue of the African plants collected by F. Welwitsch 1,2: 411–415. Trustees of the British Museum, London

Hill A.W. (1931) Mesembryanthemum. Handlist of tender dicotyledons and Gnetaceae, ed. 2: 126–131. Royal Botanic Gardens Kew, London

Hill J. (1756) Aizoon. British Herbal: 53–54, f s.n. Osborne & Shipton, London

Hill J. (1768) Hortus Kewensis: 154–155. Baldwin & Ridley, London

Hofmann U. (1973) Morphologische Untersuchungen zur Umgrenzung und Gliederung der Aizoaceae. BJS 93: 247–324

Holm L.G., Pancho J.V., Herberger J.P., Plucknett D.L. (1979) A geographical atlas of world weeds. Wiley, New York Chichester Brisbane Toronto

Hooker J.D. (1847) Dr. J.D. Hooker's enumeration of the plants of the Galapagos Archipelago. Transactions of the Linnean Society, Botany 20: 221

Hooker J.D. (1864) Handbook of the New Zealand flora: 84. Reeve and Co., London

Hooker J.D. (1873) Mesembryanthemum introrsum. CBM 99 (29 of the new series): t. 6057

Hooker J.D. (1874) Mesembryanthemum truncatellum. CBM 30: t. 6077

Hooker J.D. (1877) Mesembryanthemum Sutherlandii CBM 33: t. 6299. Mesembryanthemum Cooperi. CBM 33: t. 6312

Hooker J.D. (1882) Mesembryanthemum Bolusii. CBM 40: t. 6664

Hooker J.D. (1888) Mesembryanthemum brownii. CBM 44: t. 6985

Hooker W.J. (1867–1871) Icones Plantarum 11. Longman, Orme, Brown, Green & Longmans, London

Hooker W.J. (1889) Icones Plantarum 19. Longman, Orme, Brown, Green & Longmans, London

Horaninow P.F. (1834) Primae Lineae Systematis Naturae, ...: 83. Krajan, St. Petersburg

Horaninow P.F. (1843) Tetractys naturae seu systema quadrimembre omnium naturalium, quod primis lineis systematis naturae, a se editis: 29. Wienhöber, St. Petersburg

Horaninow P.F. (1847) Characteres essentiales familiarum ac tribuum regni vegetabilis et amphorganici ad leges tetracydis naturae conscripti ...: Portulacaceae: 138–140. Wienhöber, St. Petersburg

Howell J.T. (1966) Plantae galapaganae excerptae. Leaflets of western Botany 10: 351–352

Humboldt F.A.H., Bonpland A.J.A., Kunth C.S. (1823) Voyage de Humboldt et Bonpland 6: Nova genera et species plantarum: 87. Librairie de Gide Fils, Paris

Hunziker AT, Cocucci AE (1959) El genero Trianthema (Aizoaceae) en la Republica Argentina. Boletin de la Academia National de Ciencias Cordoba 41: 17–28

Ihlenfeldt H.-D. (1971) Zur Morphologie und Taxonomie der Mitrophyllinae Schwantes (Mesembryanthemaceae). Berichte der Deutschen Botanischen Gesellschaft 84: 655–660

Ihlenfeldt H.-D. (1973) Blattmorphologie und Systematik der Mitrophyllinae Schwantes (Mesembryanthemaceae). IOS Bulletin 2: 28–33

Ihlenfeldt H.-D. (1975) Some trends in the evolution of the Mesembryanthemaceae. Boissiera 24: 249–254

Ihlenfeldt H.-D. (1978) Morphologie und Taxonomie der Gattung Oophytum N.E.Br. (Mesembryanthemaceae). BJS 99: 303–328

Ihlenfeldt H.-D. (1980) Der Haarapparat ("Diadem") der Gattung Trichodiadema Schwant. (Mesembryanthemaceae) MIBH 17: 145–163

Ihlenfeldt H.-D. (1988) Morphologie und Taxonomie der Gattungen Diplosoma Schwantes und Maughaniella L.Bolus (Mesembryanthemaceae). BBP 63: 375–401

Ihlenfeldt H.-D. (1994) Diversification in an arid world: The Mesembryanthemaceae. Annual Review of Ecology and Systematics 25: 521–546

Ihlenfeldt H.-D. (1997) The systematic position of the genus Drosanthemopsis Rauschert and a revision of the genus Jacobsenia L.Bolus & Schwantes (Mesembryanthemaceae). MIBH 27: 109–126

Ihlenfeldt H.-D., Bittrich V. (1985) Morphologie, Gliederung und Abgrenzung der Gattung Psilocaulon N.E.Br. s.l. (Mesembryanthemaceae). BJS 105: 289–322

Ihlenfeldt H.-D., Gerbaulet M. (1990) Untersuchungen zum Merkmalsbestand und zur Taxonomie der Gattungen Apatesia N.E.Br., Carpanthea N.E.Br., Conicosia N.E.Br., Herrea Schwantes und Hymenogyne Haw. (Mesembryanthemaceae Fenzl) BJS 11: 457–498

Ihlenfeldt H.-D., Jörgensen S. (1973) Morphologie und Taxonomie der Gattung Monilaria (Schwantes) Schwantes s.str. (Mesembryanthemaceae). MIBH 14: 49–94

Ihlenfeldt H.-D., Straka H. (1972) Proposal 335. Mesembryanthemum. Taxon 21: 537

Ihlenfeldt H.-D., Struck M. (1987) Morphologie und Taxonomie der Dorotheanthinae Schwantes (Mesembryanthemaceae). BBP 61: 411–453

Ihlenfeldt H.-D., Hartmann H., Poppendieck H.-H. (1978) Chromosomenzahlen der Mitrophyllinae Schwantes (Mesembryanthemaceae) MIBH 16: 171–182

Ihlenfeldt H.-D., Schwantes G., Straka H. (1962) Die höheren Taxa der Mesembryanthemaceae. Taxon 11: 52–56

Ingram J. (1968) Nomenclatural observations on miscellaneous Aizoaceae. Baileya 16: 141–142

Ingram J. (1970) Nomenclatural observations on miscellaneous Aizoaceae. II. Baileya 17: 46–50

Jaarsveld E. van (1982) Sineoperculum van Jaarsveld: 'n nuwe monotipiese genus in die familie Mesembryanthemaceae. JSAB 48: 5–10

Jaarsveld E. van (1994) A synopsis of the genus Stoeberia. Aloe 31: 68–76

Jaarsveld E. van (1997) Argyroderma theartii van Jaarsv. spec. nov. (Mesembranthemaceae), a new species from the Knersvlakte. Aloe 34: 12–13

Jaarsveld E. van, Hammer S.A. (1996) Gibbaeum johnstonii, a new species from the Little Karoo. Bradleya 1996: 14–16

Jackson G. (1908) In: Andrews H.C.: The botanists's repository 8: t. 540. Bensley, London

Jacobs S.W.L. (1988) Notes on Aizoaceae and Chenopodiaceae. Teleopea 3: 139–142

Jacobsen H. (1933) Die Sukkulenten. Parey, Berlin

Jacobsen H. (1935) Succulent plants: Mesembryanthemum L. Williams & Norgate, London

Jacobsen H. (1937) Halophyten – Pseudosukkulenten. KuaS 10: 156–159; 199–200

Jacobsen H. (1937) Unstimmigkeiten in der Bezeichnung der Mesembrianthema. KuaS 10: 159–163

Jacobsen H. (1938) Verzeichnis der Arten der Gattung Mesembryanthemum L. ... FR Beiheft 106: 1–198

Jacobsen H. (1938) Unstimmigkeiten in der Bezeichnung der Mesembryanthema. FR 43: 222–230

Jacobsen H. (1938) Cleretum N.E.Br. oder Micropterum Schwant.? FR 43: 231–232

Jacobsen H. (1938) Strauchige Eiskrautgewächse (Aizoaceen) für den Steingarten. BSKP 1938: 60–64

Jacobsen H. (1949) More nomenclatural alterations in the Mesembryanthemaceae. NCSJ 4: 58

Jacobsen H. (1955) Handbuch der sukkulenten Pflanzen 3. Fischer, Jena

Jacobsen H. (1955) In: Jacobsen H., Rowley G.D. 1955: 80–85

Jacobsen H. (1958) In: Jacobsen H., Rowley G.D. 1958: 77–78

Jacobsen H. (1960) A handbook of succulent plants 3. Mesembryanthemums. Blandford, London

Jacobsen H. (1961) Betrachtungen über Lithops. KuaS 12: 165–171

Jacobsen H. (1970) Das Sukkulentenlexikon. Fischer, Jena

Jacobsen H. (1974) Lexicon of succulent plants. Blandford, London

Jacobsen H. (1981) Das Sukkulentenlexikon, ed.2. Fischer, Jena

Jacobsen H., Rowley G.D. (1955) Some name changes in succulent plants – part II. NCSJ 10: 80–85

Jacobsen H., Rowley G.D. (1958) Some name changes in succulent plants IV. NCSJ 13: 75–78

Jacobsen H., Rowley G.D. (1973) Some name changes in succulent plants V. NCSJ 28: 1–6

Jacobsen H., Schwantes G. (1958) In: Jacobsen H., Rowley G.D. 1958: 77–78

Jacobsen H., Volk O.H. (1950) Mesembryanthemaceae. Ulmer, Stuttgart (also often cited as: Jacobsen H., Volk O.H., Herre H. (1950))

Jacquin N.J. (1776) Hortus botanicus vindobonensis 3: t. 6, t. 7. Kaliwoda, Wien

Jacquin N.J. (1789) Collectanea 2: 319. Wappler, Wien

Jacquin N.J. (1797–1804) Plantarum rariorum horti caesarei schoenbrunnensis ... 1: t. 108 (1797), 3: t. 278, 279 (1798), 4: t. 436–442, 475–478 (1804). Wappler, Wien

Jacquin N.J. (1800–1809) Fragmenta botanica, ...: 17, 44, 50, 51, 53. Schmidt, Wien

Jeffrey C. (1960) Notes on tropical African Aizoaceae. KB 14: 235–238

Jeffrey C. (1961) Aizoaceae. In: Hubbard C.E., Milne-Redhead E. (eds) Flora of tropical East Africa: 1–37

Johnston I.M. (1928) Some undescribed American spermatophytes. Contributions from the Gray Herbarium of Harvard University 81: 88

Johnston I.M. (1929) Aizoaceae. In: Papers on the flora of Northern Chile. I. The coastal flora of the departments of Chañaral and Taltal. Contributions from the Gray Herbarium of Harvard University 85: 34

Joubert A.J. (1982) Die Gibbaeums. Wonderplante van die Klein Karoo. Landbouweekblad No. 250: 82–89

Jump J.A. (1981) The seed as a criterion in Lithops classification. CSJ 53: 197–200

Jürgens N. (1990) A life form concept including anatomical characters, adapted for the description of succulent plants. MIBH 23a: 321–341

Jussieu A.L. de (1789) Genera Plantarum 15: 312–317. Herissont & Barrois, Paris

Karrer S. (1932) Pleiospilos Barbarae, eine Mimikry treibende Art aus Südafrika. MDG 47: 99

Kensit L. (1908) Mesembryanthemum pillansii. In: Wildeman 1908: 3, t. 57

Kensit L. (1909) In: Bolus H., Kensit L. 1909

Ker J.B., Gawler J. (1821) Mesembryanthemum blandum. In: Edwards S.T. 1821: The Botanical Register 7: t. 582

Kerner J.S. (1795)1830: Hortus sempervirens, exhibens ... 23: t. 26. Academiae Carolinae, Stuttgart

Klak C. (1999) Psilocaulon N.E.Br.: a common but little known genus from southern Africa. Aloe 36: 75–79

Klak C. (2000) Taxonomic studies in the Aizoaceae from South Africa: Three new species and some new combinations. Bothalia 30: 35–42

Klak C., Linder H.P. (1998) Systematics of Psilocaulon N.E.Br. and Caulipsolon Klak gen. nov. (Mesembryanthemoideae, Aizoaceae). BJS 120: 301–375

Klotzsch J. (1848) Mesembryanthemum guianense. In: Schomburgk R.: Reisen in Britisch-Guiana ...: 300. J.J. Weber, Leipzig

Koutnik D.L., O'Connor-Fenton M. (1984) Notes on the genus Mesembryanthemum L. JSAB 50: 365–371

Koutnik D.L., O'Connor-Fenton M. (1985) Lectotypification of the genus Delosperma (Mesembryanthemaceae). SAJB 51: 197–198

Kühn U. (1993) Chenopodiaceae. In: Kubitzki K.: The families and genera of vascular plants: 253–281. Springer, Berlin Heidelberg New York

Kuntze O. (1886) Plantae Rehnelianae Hereroensis. Jahrbuch des königlichen botanischen Gartens und des botanischen Museums zu Berlin 9: 264

Kuntze O. (1891) Revisio generum plantarum ... 1: 263–264. A. Felix, Leipzig; Dulau & Co. London etc.

Kuntze O. (1898) Revisio generum plantarum ... 3(2): 109–110. A. Felix, Leipzig; Dulau & Co. London etc.

Kunze G. (1847) Hortus universitatis literarum Lipsiensis seminum anno 1845 perceptorum offert delectum. Linnaea 19: 405

Lamarck J.-B.P.A. (1792) Ficoide. Encyclopédie méthodique. Botanique ... 2,2: 474–488. Panckoucke, Paris

Lamarck J.-B.P.A. (1792) Aizoon. Encyclopédie méthodique. Botanique ... 3,2: 418. Panckoucke, Paris

Larrañaga D.A. (1922) Escritos de Don Damasco Antonio Larrañga.1. Los publicaciones del Instituto Histórico y geografico del Uruguay 1: 23–24. Imprenta National, Montevideo

Larrañaga D.A. (1923) Escritos de Don Damasco Antonio Larrañga. 2. Los publicaciones del Instituto Histórico y geografico del Uruguay 1: 23–24. Imprenta National, Montevideo

Lavis M. (1929) Mesembrianthemum and allied genera. SAG 19: 144

Lavis M. (1930) Mesembrianthemum and allied genera. SAG 19: 153, 212, 214

Lavis M. (1930) Novitatea Africanae. JB 68: 77–78

Lavis M. (1932) A new Conophytum from Namaqualand. SAG 22: 257

Lavis M. (1933) Conophytum colorans. FPSA 13: t. 509

Lavis M. (1966) Notes on the genus Delosperma (Mesembrieae). JSAB 32: 209–210, 341–345

Lavis M. (1967) Notes on the genus Delosperma (Mesembrieae). JSAB 33: 71–74, 311–314

Lavis M. (1969) Notes on the genus Delosperma (Mesembrieae). JSAB 35: 145–147

Lehmann C. (1845) Plantae Preissianae sive enumeratio plantarum ...: 245–246. Meissner, Hamburg

Leistner O.A. (1958) A new monotypic genus of the Mesembryanthemaceae. NM 3: 289–291

Leistner O.A. (1958) Skiatophytum L.Bol.: 'n morfologies-taksomomiese studie. JSAB 24: 89–102

Leredde C. (1954) Première note sur quelques plants du Tassili des Ajjer. Bulletin de la Société d'Histoire Naturelle de l'Afrique du Nord 44: 243–247

L'Héritier C.L. (1784) Stirpes novae aut minus cognitae: 81–82, t. 39. Pierre, Paris

Liede S. (1990) Untersuchungen zum Merkmalsbestand und zur Taxonomie der "Erepsiinae" (Mesembryanthemaceae). BBP 64: 391–479

Liede S. (1990) Untersuchungen zur Ökologie der Gattung Erepsia N.E.Br. (Mesembryanthemaceae). BBP 65: 1–98

Liede S. (1992) Nomenclatural types of the sections of the genus Erepsia N.E.Br. (Aizoaceae – Ruschioideae). BBP 66: 155–156

Liede S. (1994) Again: Nomenclatural types of the sections of the genus Erepsia N.E.Br. (Aizoaceae – Ruschioideae). Bradleya 11: 50

Liede S., Meve U. (1990) What is Lampranthus scaber? Bradleya 8: 39–44

Lindley J. (1824) Mesembryanthemum obliquum. In: Edwards ST, 1824: 863

Lindley J. (1834) Oncidium ampliatum. Edwards' Botanical Register 20 (NS 7): t. 1699

Lindley J. (1853) The vegetable kingdom 3rd ed.: 527. Bradbury & Evans, London

Link H.F. (1829–1831) Handbuch zur Erkennung der nutzbarsten und am häufigsten vorkommenden Gewächse 2: 12–18. Haude & Spenersche Buchhandlung (S. J. Josephy), Berlin (following Stafleu & Cowan, 1981, Taxomomic Literature III: 69, this volume came out in Jan.–Aug. 1829; following Brizicky G. (1969) In Taxon 18: 651, the volume appeared in 1831)

Link H.F., Otto F. (1823) Abbildungen auserlesener Gewächse des königlichen botanischen Garten zu Berlin, ...: t. 43, S. 93–94. Decker, Berlin

Linnaeus C. von (1753) Species plantarum. Facs. 1957. Ray Society, London

Linnaeus C. von (1759) Systema naturae, ed. 10: 1058–1060. Laurentius Salvius, Stockholm

Linnaeus C. von (1762) Species plantarum, ed. 2: 515, 687–700. Salvius, Stockholm

Linnaeus C. von (1767) Mantissa plantarum prima: 69–70. Salvius, Stockholm

Linnaeus C. von (1771) Mantissa plantarum altera: 553. Salvius, Stockholm

Linnaeus C. von (1774) In: Murray JA 1774: 392

Linnaeus C. von fil. (1763) Decas secunda plantarum rariorum horti upsalensis ...: 25, f 27. Salvius, Stockholm

Linnaeus C. von fil. (1781) Supplementum plantarum. Systematis vegetabilium - ed. 13 generum plantarum: 227, 258–261, 392. At the expenses of the orphanage, Braunschweig
Littlewood R. (1963) A new species of Conophytum and notes on the subgenus Fenestrata and the series Minuscula and Minuta of Euconophytum. JSAB 29: 139–142
Littlewood R. (1966) Notes on Conophytum ficiforme (Haw.) N.E.Br. and Conophytum placitum (N.E.Br.) N.E.Br. JSAB 32: 205–207
Littlewood R. (1966) A new species of Conophytum. JSAB 32: 235
Loddiges C. (1821) Mesembryanthemum blandum. The botanical cabinet 6: t. 599
Loddiges C. (1828) Mesembryanthemum incomptum. The botanical cabinet 14: t. 1311
Loefling P. (1758) Iter hispanicum, eller resa ...: 106, 191. Salvius, Stockholm
Loesch A., Tischer A. (1933) In: Tischer A. 1933: 47–48
Loesch A., Tischer A. (1936) In: Tischer A. 1936: 12–14
Loudon J.C. (1830) Hortus britannicus. Longman, Rees, Orme, Brown & Green, London
Loureiro J. (1790) Flora cochinchinensis 2: 633. Typis academicis, Lisboa
Lückhoff H.A. (1958) A new Lithops from the Calvinia Karoo. JSAB 24: 159–161
Maass C.A. (1928) Argyroderma delaetii spec. nov. ZSK 3: 180–183, 294–295
Maass C.A. (1928) Mesembryanthemum. Genus Trichocyclus N.E.Br. ZSK 3: 233–234, 322
Maire R. (1936) Contribution à l'étude de la Flore de l'Afrique du Nord 24. Bulletin de la Société d'Histoire naturelle de l'Afrique du Nord 27: 203–237
Maire R. (1940) Mesembryanthemum theurkauffii. Opophyum theurkauffii. Bulletin de la Société d'Histoire naturelle de l'Afrique du Nord 30: 343
Maire R. (1962) Flore de l'Afrique du Nord. Aizoaceae: 246–289
Marais W. (1978) A new mascarene Sesuvium (Aizoaceae). KB 32: 483
Marais W. (1979) A new species of Delosperma (Mesembryanthemaceae) from Madagascar. KB 33: 521–522
Marloth R. (1907) New South African succulents. Transactions of the South African Philosophical Society 18: 43–46, 281–282, f s.n.
Marloth R. (1910) Some new South African succulents: Aizoaceae. TRSSA 1: 405–407, t. 27; 2,1: 34–36, t. I
Marloth R. (1910) Some new South African succulents 3. TRSSA 2: 33–40
Marloth R. (1914) A new mimikry plant. TRSSA 4: 137–138
Marloth R. (1948) Mesembryanthemum crassipes. FPSA t. 73
Martius C.F.P. (1872) Flora brasiliensis 14(2): 310. Fleischer, Leipzig München
Martynov I.I. (1820) Tekhno-Botanicheskii Slovar, ia Latinskom i Rossilskom Yazuikakh: 15. Rossilska Akademia, St. Peterburgh
McVaugh R. (1974) Proposal 335. Mesembryanthemum. Taxon 23: 820–821
Medikus F.K. (1789) Philosophische Botanik mit kritischen Bemerkungen: 99. Neue Hof- und Akademische Buchhandlung, Mannheim
Melville R. (1952) Trianthema pentandra L. and some related species. KB 7: 266–278
Miers J. (1848) Contributions to the botany of South America: Sclerophylax. London Journal of Botany 7: 17–26
Miller P. (1754) The gardeners dictionary ... abridged from the last folio edition, ... 2: Ficoides, 3: Tetragonocarpos (no pages given). Rivington, London
Miller P. (1760) Figures of the most beautiful, useful, and uncommon plants described in the gardeners' dictionary 2: t. 263. Rivington, Millar & Dodsley, London
Miller P. (1768) Gardeners' dictionary ed.8: Mesembryanthemum: no. 1–32. Trianthema: no. 1–2. The author, London
Miller P. (1772) Abbildungen der nützlichsten, schönsten und seltensten Pflanzen welche in seinem Gärtner-Lexikon vorkommen 2: t. 259 (i.e. t. 263 of Miller, P. 1760)
Miquel F.A.W. (1843) Animadversiones in Herbarium Surinamense, quod in Colonia Surinamensi legit H.C. Focke. Tijdschrift voor Natuurlijke Geschiedenis en Physiologie 10: 75–93
Miquel F.A.W. (1845) Tetragonella. In: Lehmann J.G.C. 1845: 245–246
Mitchell A.R. (1984) A newcomer from Namaqualand. BCSJ 2: 45–46
Mitchell A.R., Hammer S. (1988) Conophytum buysianum. Aloe 25: 34–35
Moench C. (1794) Methodus Plantarum horti botanici et agri Marburgensis: 698. Cotta, Marburg
Moench C. (1802) Supplementum ad methodum plantas ...: 190–193, 299–300. Cotta, Marburg (reprint 1966: Koeltz, Königstein)
Molina G.I. (1910) Saggio sulla Storia Naturale de Chile, ed. 2: 133. Masi Brothers, Bologna
Moore S. (1921) Alabastra diversa 34. 1. Plantae Rogersianae - VI. JB 59: 226–232
Morison R. (1699) Plantarum historiae universales Oxoniensis 3, sect. 12: t. 6 f 13. Sheldon, Oxford
Moss C.E. (1914) Mesembryanthemum edule. Cambridge British Flora 2: 151, t. 153
Mueller C. (1847) Systema muscorum ordinis "Cleistocarpi" Brid. Botanische Zeitung 5: 97–102
Mueller F. von (1852) Portulaceae. Linnea 25: 384–385
Mueller F. von (1859) Fragmenta Phytographiae Australiae 1: 172–175. Government Printer, Melbourne
Mueller F. von (1859) Report on the plants collected during Mr Babbage's expedition into the North West interior of South Australia in 1858: 9–10. Government Printer, Melbourne
Mueller F. von (1861) Fragmenta Phytographiae Australiae 2: 148–149. Government Printer, Melbourne
Mueller F. von (1866) Fragmenta Phytographiae Australiae 5: 157–158. Government Printer, Melbourne
Mueller F. von (1871) Fragmenta Phytographiae Australiae 7: 129. Government Printer, Melbourne
Mueller F. von (1876) Fragmenta Phytographiae Australiae 10: 72–83. Government Printer, Melbourne
Mueller F. von (1878) Fragmenta Phytographiae Australiae 11: 8. Ferris, Melbourne
Mueller F. von (1884) Definitions of some new Australian plants. Southern Science Record 3: 282
Mueller F. von (1895) Description of hitherto unknown plants from north-western Australia. The Chemist and Druggist of Australia 10: 207
Mueller K. (1908) Beiträge zur Systematik der Aizoaceen. BJS 42: Beiblatt 97: 68, 72
Muñoz C. (1939) Dos especies nuevas para el Norte de Chile. Journal of the Arnold Arboretum 20: 248
Murray J.A. (1774) Caroli a Linné equitis Systema vegetabilium secundum classes ... ed. 13: 388–392. Dieterich, Göttingen
Nair N.C. (1966) Three new combinations in Aizoaceae. Bulletin of the Botanical Survey of India 8: 86
Nakai T. (1942) Notulae ad plantas Asiae Orientalis (18). Journal of Japanese Botany 18: 91–103
Necker N.J. de (1790) Elementa Botanica 2: 81–83. Societas Typographica, Paris (1791: Neuwied)
Nel G.C. (1937) Lithops otzeniana Nel nov. spec. Kakteenkunde 1937: 123–127
Nel G.C. (1939) Lithops dorotheae Nel sp. nov. SUC 1939: 77–79
Nel G.C. (1940) Lithops venteri Nel sp. nov. SUC 1940: 11–12
Nel G.C. (1940) Lithops weberi Nel sp. nov. SUC 1940: 16–18
Nel G.C. (1943) Species novae vel minus cognitae: Lithops and Caralluma. Annals of the University of Stellenbosch 21 Sect. A, No. I: 1–17
Nel G.C. (1953) Gibbaeum handbook: 81. Batsford, London
Nel G.C. (1946) Lithops. Author, Stellenbosch
Niesler I. (1997) Is Trichodiadema Schwantes one genus only? - Problems in delimitation or "What is a diadem?" Bradleya 15: 13–27
Nissole G. (1730) E'Tablissement De Quelques Nouveaux Genres De Plantes. Histoire de l'Académie Royale des Sciences of the year 1711: 316–320, 324, t. 13, Paris

Oberstein O. (1912) Über den Bau der Blattspitzen der Mesembryanthema-Barbata. Beihefte zum Botanischen Centralblatt 29.1: 298–302
Oesterreich G. (1937) Die Lithopsarten und ihre Fundorte. KuaS 1937: 97–100
Oliver D. (1871) Flora of tropical Africa 2: 582–588. Reeve & Co., London
Oliver W.R.B. (1917) Vegetation and flora of the Lord Howe Island. Transactions and Proceedings of the New Zealand Institute 49: 137
Opel M.R., Hammer S.A. (1999) Batrachia: a new section of the genus Conophytum N.E.Brown (Aizoaceae). CSJ 71: 153–158
Ortega C. Gómez de (1797–1800) Novarum, aut rariorum plantarum horti reg. botan. matritio, decriptionum decades, decas secunda: 19. Iberriaria (1797); Mariniana (1800), on two subsequent title pages, Madrid
Osbornová-Kosinová J. (1984) Notes on the synanthropic plants of Egypt. 1. Trianthema portulacastrum L. (Aizoaceae). Folia geobotanica et phytotaxonomica 19: 279–285
Ostenfeld C.H. (1921) Contributions to West Australian botany. Det kongelike Danske Videnskabernes Selskaps Biologiske Meddelser 3,2: 59
Pallas P.S. (1781) Appendix continens Bigam plantarum descriptam. Enumeratio planatrum quae in horto viri illustris atque Demidof: 150–163, t. I. Acad. Imper. Scientiarium. St. Petersburg
Pavelka P. (1998) Expedice Jihoafrická republika 1997 aneb Gibbaeum johnsonii a jiné zázitky. Astridia 97, 3: 35–43
Pavelka P. (1999) Rod Conophytum N.E.Brown – vybrané druhy. Kaktusy 35, Speciál 1999/1: 1–32
Pax F. (1889) Aizoaceae (Ficoidaceae). In: Engler A: Plantae Marlothianae. Ein Beitrag zur Kenntnis Südafrikas. Englers Botanische Jahrbücher 10: 10–13
Pax F. (1889) Aizoaceae. In: Engler A., Prantl K.: Natürliche Pflanzenfamilien ed 1: 3,1b: 33–51. Engelmann, Leipzig
Pax F. (1892) Aizoaceae. In: Engler A.: Über die Hochgebirgsflora des tropischen Afrika. Abhandlungen der preußischen Akademie der Wissenschaften 11. 1891–1892 (reprint 1975: 208–209. Otto Koeltz, Koenigstein)
Pax F. (1895) Aizoaceae. In: Engler A.: Plantae Gürichianae. BJS 19: 133–134
Pax F. (1910) Das Sommerregengebiet Westafrikas. 41. Namaqua- und Hereroland. In: Engler A.: Vegetation der Erde: Die Pflanzenwelt Afrikas 1.2: 517, 532–564. Engelmann, Leipzig
Pax F. (1913) In: Schellenberg G 1913: 499
Pax F. (1915) In: Engler A., Drude O.: Die Vegetation der Erde 9: Die Pflanzenwelt Afrikas 3.1: 147. Engelmann, Leipzig
Pax F., Hoffmann K. (1934) Aizoaceae. In: Engler A., Prantl K.: 2. ed 16c: 179–233. Engelmann, Leipzig
Perrottet G.S. (1825) Catalogue raisonné des plantes introduites dans les colonies francaises de Madagascareigne et de Cayenne, et de celles rapportées vivantes des mers d'Asie et de la Guyane, au Jardin des plantes de Paris. Mémoires de la Société Linnéenne de Paris ... 3: 89–151
Persoon C.H. (1805) Synopsis plantarum, seu enchiridium botanicum ... 1: 262, 487. Cramer, Paris; Cotta, Tübingen
Persoon C.H. (1807) Synopsis plantarum, seu enchiridium botanicum ... 2: 39. Treuttel & Würtz, Paris; Cotta, Tübingen
Peter A. (1932) Descriptiones plantarum novarum Africae orientalis. FR Beiheft 40, 2: 29–30
Petit-Thouart A. du (1811) Mélanges de Botanique et de Voyages: 6. 37. Bertrand, Paris
Petiver J. (1703)4: Gazophylacii naturae et artis decades X ... 1: Decas IV/V: t. 39. Christophorus Bateman, London (copy seen from Göttingen)
Petiver J. (1709) Gazophylacii naturae et artis decades X ... 2: Decas IX: t. 77; 78 not seen. Christophorus Bateman, London (copy seen from Berlin)
Petiver J. (1764) Opera historiam naturalem spectantia or Gazophylacium. 2, vol. 1: t. 88. Millan, London (copy seen from Dublin)
Philippi R.A. (1860) Florula Atacamensis ...: 19. E. Anton, Halle
Philippi R.A. (1872) Memorias científicas i literaturias. Anales de la Universidad de Chile 41: 716–717
Philippi R.A. (1891) Verzeichnis der von Friedrich Philippi auf der Hochebene von Antofagasta und Tarapacá gesammelten Pflanzen: 24. Brockhaus. Leipzig. Facsimile in: Anales del' Museo Nacional de Chile, Botanica. ser. 2: 24. 1891 (it appears impossible to determine which publication precedes the other one)
Philippi R.A. (1893) Plantas nuevas Chilenas. Anales de la Univsidad de Chile 85: 168–170
Phillips E.P. (1913) Descriptions of new plants from the Giftberg collected by the Percy Sladen Memorial Expedition. Annals of the South African Museum 9: 106
Pierce S., Gerbaulet M. (1997) Brownanthus Schwantes (Mesembryanthemoideae, Aizoaceae): two new species and a new combination from the Richtersveld and southwestern Namibia. Aloe 34: 42–44
Plukenet L. (1691) Phytographia sive Stirpium Illustrorium ... 1: t. 117. The author, London
Plukenet L. (1692) Phytographia sive Stirpium Illustrorium ... 3: t. 216. The author, London
Plukenet L. (1696) Almagestum botanicum sive Phytographia Plukenetianae Onomasticum ..., Ficoides Africanum: 148–150. Seen: 2nd reprint: Davies T et al. (eds). The author, London
Poellnitz K. von (1935) In: Jacobsen H. 1935
Poiret J.L.M. (1806) Encyclopédie méthodique, Botanique 7: 604. Panckoucke, Paris
Poppendieck H.-H. (1976) Untersuchungen zur Morphologie und Taxonomie der Gattung Mitrophyllum Schwantes s.lat. BJS 97: 339–413
Poppendieck H.-H., Ihlenfeldt H.-D. (1978) Delosperma harazianum (Deflers) Poppend. & Ihlenf., eine wenig bekannte Mesembryanthemaceae aus dem Südjemen. MIBH 16: 183–187
Prain D. (1912) Mesembryanthemum pearsonii. CBM 138: t. 8463
Preez C.M.R. du (1970) The relative preference of Merino sheep for a number of pelleted Karoo bush species. Hand. Weidingsverenigingen van Suid Africa 5: 126–131
Preez G. du, Esler KJ (1997) How to conserve Carpobrotus – the edible way. Veld & Flora 83: 45
Prescott A.A. (1983) Three new species of Tetragonia and Trianthema (Aizoaceae). Journal of the Adelaide Botanical Garden 6: 181–188
Prescott A.A. (1984) Trianthema. In: Prescott A.A., Venning J. 1984: 52–60. In: George A.S. (ed) Flora of Australia 4. And: 319
Prescott A.A., Venning J. (1984) Aizoaceae. In: George A.S. (ed) Flora of Australia 4: 19–61
Presl K.B. (1831) Symbolae botanicae sive descriptiones .. 1,2: 24, t. 14. Calve, Prague
Preston C.D., Sell R.D. (1988) The Aizoaceae naturalized in the British Isles. Watsonia 17: 217–245
Rafinesque C.S. (1814) Botanica. Specchio delle Scienze o Giornale encecopledico di Sicilia 1: 86–88
Rafinesque C.S. (1838) New flora and botany of North America 4: 16. The Author, Philadelphia (not 1836, as often cited for vol. 4)
Range P. (1934) Die Flora des Namalandes 4. Aizoaceae. FR 36: 10–19
Range P. (1938) Die Flora des Namalandes 26. 11. Aizoaceae. FR 43: 255–256
Rappa F. (1912) Per una classificazione naturale dei Mesembrianthemi. Bolletin Reale Orto Botanico Giardino Coloniale Palermo 11: 21–36
Rappa F., Camarrone V. (1953) La nova sistematica di Mesembrianthemum. Lavori dell'Istituto Botanico e del Giardino Coloniale di Palermo 14: 1–39
Rappa F., Camarrone V. (1955) Divinazioni mendelejeffiane in sistematica botanica. Lavori dell'Istituto Botanico e del Giardino Coloniale di Palermo 15: 1–16
Rappa F., Camarrone V. (1960) La Classificazione naturale delle Mesembriantemacee. Lavori dell'Istituto Botanico e del Giardino Coloniale di Palermo 18: 1–24
Rauschert S. (1982) Nomina nova generica et combinationes novae Spermatophorum et Pteridophytorum. Taxon 31: 554–563

Rawe R. (1969) Conophyten – nomenklatorische Stiefkinder unter den Mesems. KuaS 20: 58–59
Rawe R. (1970) Two new species (and a variety) of Conophytum. CSJ 42: 258–260
Rawe R. (1970) Untersuchungen über die Gattung Conophytum. KuaS 21: 198–200, 207–209
Rawe R. (1971) Beschreibungen und Aufschlüsselung der Sektion Minuscula Schwant. emend. Tisch. der Gattung Conophytum. KuaS 22: 104–107, 123–127
Rawe R. (1973) Conophytum laetum L.Bolus. KuaS 24: 197–199
Rawe R. (1974) Conophytum ficiforme (Haworth) N.E.Br. KuaS 25: 246–247
Rawe R. (1975) Conophytum ceresianum L.Bolus. KuaS 26: 198–200
Rawe R. (1975) The genus Conophytum: Section Costata Schwantes emended Tischer, subsection Verrucosa. CSJ 47: 126–132
Rawe R. (1975) The genus Conophytum section: Cataphracta Schwantes. CSJ 47: 180–186
Rawe R. (1975) The genus Conophytum: Section 3: Wettsteiniana Schwantes, subsection 2: Minuta (Littlewood) Tischer. CSJ 47: 209–212
Rawe R. (1980) Conophytum uvaeforme. Aloe 18: 13
Rawe R. (1980) New taxa of Conophytum. CSJ 52: 230–231
Rawe R. (1980) Conophytum piluliforme N.E.Br. and its forms. CSJ 52: 283–284
Rawe R. (1981) Conophytum, section Saxetana. CSJ 53: 183–196
Rawe R. (1981) Conophytum hirtum var. baradii Rawe var. nova. CSJ 53: 273
Rawe R. (1982) Conophytum: The section Truncatella. CSJ 54: 165–169
Rawe R. (1982) Conophytum: The section Ovigera. CSJ 54: 218–222
Rawe R. (1983) Conophytum: The section: Tuberculata. CSJ 55: 3–9
Regel E. (1873) Descriptiones plantarum novarum in regionibus Turkestanicis a … cum adnotationibus ad plantas vivas in horto Imperiali botanico Petropolitano cultis. Acta Horti Petropolitani 2: 401–437. W. W. Pratz. St. Peterburg
Regel E. (1873) Mesembrianthemum abyssinicum Rgl. GF 1873: 299–300, t. 774
Reiche C. (1898) Aizoáceas. In: Flora de Chile 2: 362–367. Imprenta Cervantes, Santiago de Chile
Reiche C. (1898) Estudios críticos sobre la Flora de Chile. Anales de la Universidad de Chile 100: 369–370
Reichenbach H.G.L. (1827) Mössler's Handbuch der Gewächskunde … ed. 2, 1: 51–52. Hammerich, Altona
Reichenbach H.G.L. (1828) Conspectus regni vegetabilis: 166. Cnobloch, Leipzig
Reveal J.I. (1998) Comments on Art. 14, footnote 2: first place of publication for seed plant family names. Taxon 47: 851–856
Ridley H.N. (1891) On the botany of Fernando Noronha. Ficoideae. Journal of the Linnean Society, Botany 27: 38
Riha J. (1976) Sesuvium portulacastrum L. (Aizoaceae). Ashingtonia 2: 122–124
Rohrbach P. (1872) Ficoidaceae. In: Martius CFP: Flora brasiliensis 14,2: 310
Rolfe R.A. (1916) Mesembryanthemum tuberculosum. CBM 142: t. 8674B
Rolfe R.A. (1920) Mesembryanthemum dichroum. CBM (146) 16: t. 8872
Rothmaler W. (1941) Zur Nomenklatur der Mesembranthemen. Notizblatt des Botanischen Gartens und Museums zu Berlin-Dahlem 15: 410–414
Rottler J.B. (1803) Botanische Bemerkungen auf der Hin- und Rückreise von Trankenbar nach Madras von Herrn Missionar Rottler mit Anmerkungen von Herrn Prof. C. L. Willdenow. Neue Schriften der Gesellschaft Naturforschender Freunde zu Berlin 4: 181
Rowley G.D. (1951) Studies in Ficoidaceae. I. The family name. NCSJ 6: 27–28
Rowley G.D. (1952) Succulents in the Bates collection. NCSJ 7: 61–62
Rowley G.D. (1956, 1957) Pierre-Joseph Redouté – "Raphael of the succulents". CSJGB 18: 91–93, 101 (1956); 6–8, 30–32, 36, 54–57, 77, 89–93 (1957) (Preprint 1956. Strange the Printer, London & Eastborne)
Rowley G.D. (1958) In: Jacobsen H., Rowley G.D. 1958: Some name changes in succulent plants IV. NCSJ 13: 75–78
Rowley G.D. (1973) In: Jacobsen H., Rowley G.D. 1973: Some name changes in succulent plants V. NCSJ 28: 1–6
Rowley G.D. (1978) Reunion of some genera of Mesembryanthemaceae. NCSJ 33: 6–9; – A postcript: 62
Rowley G.D. (1979) The botanical name of the Livingstone Daisy (Aizoaceae). Baileya 20: 137–140
Roxburgh W. (1832) Flora indica 2: 445. Thacker & Co. Calcutta; Parbury, Allen & Co., London
Rumph G.E. (1750) Herbarium amboinense 6: 165, t. 72. Changuion, Catuffe, Uytwerf, Amsterdam
Rye B.L., Trudgen M.E. (1996) Tetragonia coronata, a new species of Aizoaceae from Western Australia. Nuytsia 10: 459–462
Saint-Hilaire J. (1813) Errata du N° 63. Nouveau Bulletin des Science par la Société Philomanique de Paris 2: 216
Salisbury R.A. (1796) Prodromus stirpium in horto ad chapel Allerton vigentium: 283. The author? London
Salm-Reifferscheidt-Dyck J Duke of (1816) Plantae Succulentae horti Dyckensis: 22. The author?
Salm-Reifferscheidt-Dyck J. Duke of (1820) Observationes botanicae in Horto Dyckensi notatae: 7–35. Thiriart, Köln
Salm-Reifferscheidt-Dyck J. Duke of (1822) Observationes botanicae in Horto Dyckensi notatae: 10, 34. Thiriart, Köln
Salm-Reifferscheidt-Dyck J. Duke of (1836–63) Monographia generum Aloe et Mesembryanthemi. I–V: Arnz, Düsseldorf; V, VI: Henry & Cohen, Bonn. VII: M. Cohen & fil., Bonn
Samuelsson G., Farah M.H., Claeson P., Hagos M., Thulin M., Hedberg O., Warfa A.M., Hassan A.O., Elmi A.H., Abdurahman A.D., Elmi A.S., Abdi Y.A., Alin M.H. (1991) Inventory of plants used in traditional medicine in Somalia. I. Plants of the families Acanthaceae-Chenopodiaceae. Journal of Ethnopharmacology 35: 25–63
Santapau H. (1959) Lectotypes of the species and varieties described by Blatter and Hallberg in their 'flora of the Indian desert'. Journal of the Bombay Natural History Society 56: 276–281
Sauer N. (1975) Dinteranthus. Aloe 13: 9–24
Sauer N. (1978) Dinteranthus pole-evansii, D. microspermus, D. wilmotianus. FPSA 1778 A & B, 1779 A & B, 1780 A & B
Schellenberg G. (1913) Aizoaceae africanae 2. BJS 48: 499–504
Schellenberg G. & Schlechter R. (1913) In: Schellenberg G. 1913
Schick C. (1927) Conophytum tischeri Schick. ZSK 3: 158–159
Schick C. & Tischer A. (1927) In: Tischer A.: SUC 1927: 168–169
Schick C. & Tischer A. (1927) In: Tischer A.: ZSK 1927: 151–155
Schick C. & Tischer A. (1928) Conophytum violaciflorum Schick et Tisch. spec. nov. MDG 43: 238–239
Schick C. & Tischer A. (1928) In: Tischer A. 1928: ZSK 3: 202–203
Schick C. & Tischer A. (1928) In: Tischer A. 1928: Aus N. E. Browns Mesembryanthemum-Studien. ZSK 3: 347–351
Schick C. & Tischer A. (1931) In: Tischer A. 1931: 52–54, 222–226
Schick C. & Tischer A. (1933) In: Tischer A. 1933
Schinz H. (1894) Ficoideae. Bulletin de l'Herbier Boissier 2: 204–205
Schinz H. (1897) Die Pflanzenwelt Deutsch-Südwest-Afrikas (mit Einschluß der westlichen Calahari). Bulletin de l'Herbier Boissier 5, Appendix 3: 5–101
Schinz H. (1898) Beiträge zur Kenntnis der afrikanischen Flora (Neue Folge) VII. Mitteilungen aus dem botanischen Museum der Universität Zürich V. Bulletin de l'Herbier Bossier, Genève 6: 523
Schinz H. (1915) Mitteilungen aus dem botanischen Museum der Universität Zürich LXXI: Aizoaceae. Vierteljahresschrift der Naturforschenden Gesellschaft in Zürich 60: 396–397
Schlechter R. (1896) Mesembrianthemum L. BJS 24: 445
Schlechter R. (1897) Decades Plantarum Novarum Austro-Africanarum. Decas iv. JB 35: 280
Schlechter R. (1898) Decades Plantarum Novarum Austro-Africanarum. Decas viii. JB 36: 314–318
Schlechter R. (1899) Beiträge zur Flora von Afrika XVIII. Plantae Schlechterianae novae vel minus cognitae describuntur II. BJS 27: 124–131
Schlechter R. & Berger A. (1922) In: Berger A. 1922: 627–637

Schlechter R. & Diels L. (1907) In: Diels L., Schlechter R.: Aizoaceae. In: Schultze L 1907: Aus Namaland und Kalahari: 692–694
Schoenland S. & Berger A. (1910) In: Berger A. 1910
Schultze L.S. (1907) Aus Namaland und Kalahari: Bericht an die königliche Akademie der Wissenschaften zu Berlin über eine Forschungsreise im westlichen und zentralen Südafrika, ausgeführt in den Jahren 1903–1905. G. Fischer, Jena
Schumacher F.C., Thonning E. (1827) Beskrivelse af Guineiske planter som ...: 221. Preprint from: Det Kongelige Danske Videnskabernes Selskabs naturvedenskabelige og Mathematiske afhandlinger 3: 241–242 (print of 1828 seen)
Schumacher F.C., Thonning E. (1827) Beskrivelse af Guineiske planter som ...: 233. Preprint from: Det Kongelige Danske Videnskabernes Selskabs naturvedenskabelige og Mathematiske afhandlinger 4: 7 (print of 1829 seen)
Schumann K. (1901) Neue Arten der Siphonogamen 1899. Justs's Botanischer Jahresbericht 27: 475
Schwantes G. (1916) Mesembrianthemum Schwantesii Dinter spec. nov. MKK 26: 82–83
Schwantes G. (1919) Mesembrianthemum margaretae. MKK 29: 55–57
Schwantes G. (1920) Mesembrianthema sphaeroidea. MKK 30: 35–40, 77–78, 81–82
Schwantes G. (1920) Mesembrianthema sphaeroidea. MKK 30: 117–121, 129–132
Schwantes G. (1921) Zur Stammesgeschichte der Sphäroidea. MKK 31: 23–29, 34–38, 165–173
Schwantes G. (1922) Mesembrianthemum ferrugineum Schwantes spec. nov. MKK 32: 93–94
Schwantes G. (1925) Neue Mesembrianthemen aus Südwestafrika. ZSK 2: 17–20. (25.1.1925)
Schwantes G. (1925) Neue Mesembrianthemen aus Südwestafrika. ZSK 2: 21–28. (5.3.1925)
Schwantes G. (1925) Die Mesembrianthema cordiformia. ZSK 2: 137–138
Schwantes G. (1926) Zur Systematik der Mesembrianthemen. ZSK 2: 173–189, 242, 277
Schwantes G. (1926) Stomatium suaveolens. ZSK 2: 254
Schwantes G. (1927) Conophytum concinnum S. sp. nov., Hymenocyclus Herrei spec. nov. MDG 42: 27, Ophthalmophyllum Dinter et Schwantes: 64, Conophytum (Derenbergia) frutescens Schwantes spec. nov.: 123–124
Schwantes G. (1927) Einige neue Mesembriaceen. MDG 42: 139, 150, 154, 158–159, 187, 234, 283
Schwantes G. (1927) Delosperma N.E.Br. emend. MDG 42: 234–235, 245, 258–259
Schwantes G. (1927) Dinteranthus. MDG 42: 223–224
Schwantes G. (1927) Conophytum pellicidum Schwant. nom. nov. MDG. 42: 316
Schwantes G. (1927) Herrea Schwantes gen. nov. MDG 42: 436
Schwantes G. (1927) Meyerophytum Schwantes gen. nov. MDG 42: 436–437
Schwantes G. (1927) Zur Systematik der Mesembrianthemen. ZSK 3: 14–30, 105–106
Schwantes G. (1927) Lithops dinteri ZSK 3: 97–99
Schwantes G. (1927) Zur Kultur von Conophytum. GF 76: 422–426
Schwantes G. (1928) Mesembriaceen unserer Kulturen in neuer Benennung. GF 77: 68–69; Nelia meyeri spec. nov.: 129–130
Schwantes G. (1928) Eine neue interessante Mesembriazee. Rhombophyllum nelii. GF 77: 85–87
Schwantes G. (1928) Malephora N.E.Br. MDG 43: 7; Micropterum Schwantes gen. nov.: 17; Conophytum Poellnitzianum Schwantes spec. nov.: 31–32; Pleiospilos Purpusii Schwantes spec. nov.: 46; Nelia Schwantes gen. nov.: 92; Glottiphyllum nelii Schwantes spec. nov.: 92
Schwantes G. (1928) Zur Mesembriaceen-Benennung. MDG 43: 238–239
Schwantes G. (1928) Mesembriaceen-Studien. GW 32: 37–38
Schwantes G. (1928) Mesembriaceen-Studien I–III. GW 32: 514–515, 599–600, 644
Schwantes G. (1928) Neue Mesembriaceen I. ZSK 3: 178–180, 275–279; III: 299–302, 321–323
Schwantes G. (1929) Mesembriaceen-Studien IV. GW 33: 25
Schwantes G. (1929) Mesembriaceen-Studien V. GW 33: 68–69
Schwantes G. (1929) Neue Mesembriaceen IV. MDK 1: 14–17; Argyroderma: 27–32; Corpuscularia: 35–36
Schwantes G. (1930) Neue Mesembriaceen V, VI. MDK 2: 21–22, 64–69; Pleiospilos nelii: 146–147
Schwantes G. (1933) In: Jacobsen H. 1933
Schwantes G. (1934) Die Gattung Ophthalmophyllum. Kakteenkunde 1934: 58–60
Schwantes G. (1935) In: Jacobsen H. 1935
Schwantes G. (1937) In: Jacobsen H. 1937
Schwantes G. (1938) Beiträge zur Kenntnis der Mesembryanthemen. BSKP 1938: 14–16, 88–91
Schwantes G. (1938) In: Jacobsen H. 1938: FR Beiheft 106: 1–198
Schwantes G. (1938) In: Jacobsen H. 1938: FR 43: 222–232, 257
Schwantes G. (1939) Beiträge zur Kenntnis der Mesembryanthemen. BSKP 1939, 2. Teil: 43–45
Schwantes G. (1947) System der Mesembryanthemaceen. Sukkulentenkunde 1: 34–40
Schwantes G. (1949) In: Jacobsen H. 1949
Schwantes G. (1950) Micropterum Schwantes. KuaS 1(2): 4–5, (3) 6–8
Schwantes G. (1950) Zur Systematik von Conophytum N.E.Br. KuaS 1(4): 7–11
Schwantes G. (1951) Zur Systematik von Conophytum N.E.Br. KuaS 2(1) 11–12, 28
Schwantes G. (1951) The Lithopinae. NCSJ 6: 1–3
Schwantes G. (1951) Bemerkungen zu einigen Mesembryanthemaceen. SKK 4: 73–79
Schwantes G. (1952) Die Früchte der Mesembryanthemaceen. MBMUZ 193: Serie Botanik Nr. 39: 1–38
Schwantes G. (1952) Lithops de Boerii Schwantes spec. nov. NCSJ 7: 42
Schwantes G. (1952) Vergessene Gattungen: Nycteranthus Necker. KuaS 3: 48–51
Schwantes G. (1954) Jacobsenia L.Bol. et Schwant. gen. nov. KuaS 5: 69–71
Schwantes G. (1954) Einige neue Glottiphyllum-Arten. KuaS 5: 71–79
Schwantes G. (1955) Maughania N.E.Br. KuaS 6: 135–138
Schwantes G. (1957) Ficoidaceae (Juss.) em. Hutchinson. KuaS 8: 157–158, 167–169
Schwantes G. (1957) Flowering stones and mid-day flowers. Benn, London
Schwantes G. (1958) In: Jacobsen H., Rowley G.D. 1958: 75–78
Schwantes G. & Jacobsen H. (1951) Lithops werneri Schwant. et Jacobs. spec. nov. CSJGB 13: 69
Sessé M., Moçiño J.M. (1894) Flora mexicana, ed. 2: 126. Escalante, Mexico
Sims J. (1811) Mesembryanthemum minutum. CBM 34: t. 1376
Sims J. (1813) Mesembryanthemum testiculare (?). CBM 38: t. 1573
Sims J. (1814) Mesembryanthemum obcordellum. CBM 40: t. 1647
Sims J. (1815) Sesuvium portulacastrum (?) var. revolutum. CBM 41: t. 1701
Sims J. (1816) Mesembryanthemum depressum. CBM 44: t. 1866
Sims J. (1820) Mesembryanthemum pyropaeum. CBM 47: t. 2144
Skottsberg C. (1937) Die Flora der Desventuras-Inseln. Göteborgs kungl. Vetenskaps- och Vitterhets-samhälles handlinger 5. Följden. Ser. B. Band 5. No. 6: 28–30
Smale T. (2000) Conophytum roodiae and its relatives. BCSJ 18: 101–107
Smith G.F., Chesselet P., Jaarsveld E.J. van, Hartmann H.E.K., Hammer S., Wyk B.-E. van, Burgoyne P. Klak C., Kurzweil H. (1998) Mesembs of the world. Briza, Pretoria
Smith G.F., Hartzer P. (1997) (1293–1294) Proposals to conserve the names Mossia and M. intervallaris (Aizoaceae) against their earlier homonyms. Taxon 46: 349–350
Smith G.F., Hartzer P., Wyk A.E. van (1997) Mossia intervallaris. CBM ser. nov. 14: 16–22; t. 310
Sonder O.W. (1862) Aizoon. Galenia. Mesembryaceae. Trianthema. In: Harvey W.H., Sonder W. (eds) Flora Capensis
Spegazzini C. (1899) Nova addenda as Floram Patagonicam, pars 1. Anales de la Sociedad científica Argentina 48: 52–53
Spegazzini C. (1902) Nova addenda as Floram Patagonicam, pars 3. Anales del Museo Nacional de Buenos Aires 7: 153–154

Sprague T.A. (1922) The nomenclature of plant families. JB 60: 69–73
Sprengel C. (1801) Erster Nachtrag zu der Beschreibung des Botanischen Gartens der Universität zu Halle: 36–37. Kümmel, Halle
Sprengel C. (1818) Novi proventus hortorum academicorum halensis et berolinensis: 18–19. Gebauer & filii, Halle
Sprengel C. (1825) Linnaei systema vegetabilium ed.16, 2: 381–382. Dieterich, Göttingen
Sprengel C. (1827) Linnaei systema vegetabilium ed.16, curae posteriores 4, pars 2: 195–196. Dieterich, Göttingen
Stafleu F.A., Cowan R.S. (1981) Taxonomic literature 2nd ed. 3: 77, 670. Bohn, Scheltema & Holkema, Utrecht
Stearn W.T. (1938–9) An annotated index to Salm-Dyck's "Monographia generum Aloes et Mesembryanthemi". CJ 7: 34–44 (1938), 66–85 (1939)
Steudel E.T. (1841) Nomenclator botanicus ed. 2. 1: 46. Cotta, Stuttgart
Stevens P.F. (1983) Augustin Augier's "arbre botanique" (1801), a remarkable early botanical representation of the natural system. Taxon 32: 203–211
Stokes J. (1812) A botanical materia medica: 127. Johnson & CO., London
Straka H. (1955) Anatomische und entwicklungsgeschichtliche Untersuchungen an Früchten paraspermer Mesembryanthemen. Nova Acta Leopoldina Neue Folge 118: 127–190
Strohschneider I. (1935) Eine neue Aizoaceae aus Südafrika. Österreichische Botanische Zeitschrift 84: 292–294
Struck M. (1998) Petiver's Gazophyllacii and the identity of Dorotheanthus gramineus (Haw.) Schwantes (Aizoaceae). FR 109: 5–6
Struck M., Ihlenfeldt H.-D. (1985) The correct names of genera in the subtribe Dorotheanthinae (Mesembryanthemaceae). Taxon 34: 520–522
Suessenguth K. (1950) Diagnoses novae plantarum in Africa australi a Dr. S. Rehm et aliis collectarum. MBSM 2: 46–56
Sweet R. (1830) Sweet's Hortus britannicus ed.2: 234. Ridgway. London
Swüste F. (1936) Lithops ursulae. SUC 1936: 25
Swüste F. (1936) Lithops jacobseniana. SUC 1936: 104
Taylor C.M. (1994) Revision of Tetragonia (Aizoaceae) in South America. Systematic Botany 19: 575–589
Taylor N., Eggli U. (1986) The lectotype of Delosperma N.E.Brown (Aizoaceae). Taxon 35: 709–711
Terraciano A. (1894) Contribuzioni alla conoscenza della flora dell' Africa orientale II. Annuario del R. Istituto Botanico di Roma 5: 89–121
Thulin M. (1993) Notes on Tetragonia (Aizoaceae-Tetragonioideae) in Somalia. Nordic Journal of Botany 13: 165–167
Thunberg C.P. (1827) Mesembryanthemum secundum. Museum Naturaleum Academiae Upsalensis 2: 12
Thunberg P. (1784) Flora japonica: 208–209. J. G. Müller, Leipzig
Thunberg P. (1791) Descriptiones Mesembryanthemorum quorundam in Capitis Bonae Spei Africes ... Novorum Actorum medico-physicorum academiae Caesareae Leopoldino-Carolinae naturae curiosorum in Germania 8: App.: 1–18
Thunberg P. (1794) Prof. Thunberg's observations on the flora Japonica. Transactions of the Linnean Society, Botany 2: 335
Thunberg P. (1800) Prodromus plantarum capensium, quas in Promontorio Bonae Spei Africae annis 1772–1775 collegit: 77–91. Edman, Uppsala
Thunberg P. (1823) Flora Capensis 1: 384, 2: 384, 411–427. Cotta, Stuttgart
Tischer A. (1925) Mesembrianthemum opticum var. rubrum. ZSK 2: 65–66
Tischer A. (1926) Conophytum brownii. MDG 41: 257–258
Tischer A. (1926) Interessante Mesembrianthemen VIII. MDG 41: 330–331
Tischer A. (1926) Mesembrianthemum hugo-schlechteri Tisch. ZSK 2: 203
Tischer A. (1927) Twee nieuwe Mesembrianthema. SUC 9: 144–145
Tischer A. (1927) Titanopsis lüderitzii Tisch spec. nov. SUC 11: 156–158
Tischer A. (1927) Conophytum elongatum Schick & Tisch. sp. nov. SUC 9: 168–169
Tischer A. (1927) Interessante Mesembryanthemen. XVI. Conophytum longifissum Tisch. (Neue Art). MDG 42: 343
Tischer A. (1927) Neue Mesembryanthemen aus Klein=Namaqualand. ZSK 3: 120–122
Tischer A. (1927) Vier neue Mesembrianthemen. ZSK 3: 151–155
Tischer A. (1928) Conophitum chloratum Tisch. spec. nov. MDG 43: 150; Glottiphyllum haagei Tisch.: 394–395
Tischer A. (1928) Conophytum globuliforme Schick & Tischer spec. nov. ZSK 3: 202–203
Tischer A. (1928) Conophytum multicolor Tisch. spec. nov. SUC 10: 127–128
Tischer A. (1928) De bilobe vomern van Conophytum. SUC 10: 198–200
Tischer A. (1929) Faucaria haagei Tisch. spec. nov. MDK 1: 20–21
Tischer A. (1929) Aus N. E. Browns Mesembrianthemum-Studien IX. Glottiphyllum, Haw. MDK 1: 106
Tischer A. (1930) Corpuscularia perdiantha. MDK 2: 20–21
Tischer A. (1930) Conophytum graessneri Tisch. spec. nov. MDK 2: 63
Tischer A. (1931) Conophytum, N.E.Br. Nieuwe soorten, nieuwe vindplaatsen, Bloemen. SUC 13: 51–54
Tischer A. (1931) Conophytum, N.E.Br. Nieuwe soorten, nieuwe vindplaatsen, bloemen. SUC 13: 221–226
Tischer A. (1931) In: Brown N.E. et al. 1931: 268
Tischer A. (1932) Conophytum, N.E.Br. SUC 14: 113–114
Tischer A. (1932) Einige neue Mesembryanthema. MDK 4: 280–283
Tischer A. (1933) Einige neue Mesembrianthema. Kakteenkunde 1933: 47–49
Tischer A. (1934) Conophytum, N.E.Br. SUC 16: 54–55
Tischer A. (1934) Tagesfragen für Mesemfreunde im Mai. Kakteenkunde 1934, 5: 99–100
Tischer A. (1935) Conophytum. CJ 3: 63
Tischer A. (1935) Conophytum teguliflorum. CJ 4: 38–39
Tischer A. (1935) Conophytum quartziticum Tisch sp. nov. CSJ 7: 9
Tischer A. (1935) In: Jacobsen H. 1935: 249
Tischer A. (1936) Conophytum calitzdorpense Tisch sp. nov. CJ 5: 13
Tischer A. (1936) Conophytum N.E.Br. SUC 18: 12–15
Tischer A. (1936) Conophytum N.E.Br. SUC 18: 33–36
Tischer A. (1936) Lithops N.E.Br. SUC 1936: 74–77
Tischer A. (1936) Schwantesia Loeschiana Tisch. spec.nov. Jahrbuch der Deutschen Kakteengesellschaft 1: 71–72
Tischer A. (1937) Conophytum N.E.Br. CJ 6: 37–38
Tischer A. (1937) Conophytum nutaboiense. Kakteenkunde 1937: 16
Tischer A. (1937) Gibbaeum N.E.Br. KuaS 1937, 10: 150–152
Tischer A. (1938) Einige Neue Mesembryanthemen. Kakteen und Sukkulenten (Berlin) 2: 33–34
Tischer A. (1951) Lithops steineckeana Tisch. spec. nov. SUC 1951: 37
Tischer A. (1951) New Conophytums. NCSJ 6: 32–31 (end on p 31)
Tischer A. (1952) New Conophytums Part II. NCSJ 7: 28–29
Tischer A. (1953) Alte und neue Conophyten. KuaS 4: 30–31
Tischer A. (1953) Ophthalmophyllum. CSJGB 15: 33
Tischer A. (1954) Conophytum speciosum Tisch sp. nov. CSJGB 16: 10
Tischer A. (1954) Conophytum praegratum Tisch sp. nov. CSJGB 16: 43
Tischer A. (1955) New Conophyta. CSJGB 17: 18
Tischer A. (1955) Conophytum orbicum Tisch sp. nov. CSJGB 17: 78
Tischer A. (1955) Conophytum cupreiflorum Tisch sp. nov. CSJGB 17: 79
Tischer A. (1956) Conophytum craterulum Tisch sp. nov. CSJGB 18: 31
Tischer A. (1956) Conophytum fragile Tisch sp. nov. CSJGB 18: 32
Tischer A. (1956) Conophytum fragile Tisch sp. nov. CSJGB 18: 55
Tischer A. (1957) Beiträge zur Kenntnis der Arten von Conophytum N.E.Br. KuaS 8: 75–77

Tischer A. (1957) Beiträge zur Kenntnis der Arten von Conophytum N.E.Br. KuaS 8: 169–170
Tischer A. (1957) Conophytum impressum Tisch sp. nov. CSJGB 19: 10
Tischer A. (1957) Conophytum rostratum Tisch sp. nov. CSJGB 19: 33
Tischer A. (1957) Conophytum extractum Tisch sp. nov. CSJGB 19: 78
Tischer A. (1958) Conophytum ellipticum Tisch sp. nov. CSJGB 20: 9
Tischer A. (1958) Ophthalmophyllum latum Tisch sp. nov. CSJGB 20: 33
Tischer A. (1958) Conophytum viride Tisch sp. nov. CSJGB 20: 54
Tischer A. (1958) Conophytum nudum Tisch. spec. nov. KuaS 9: 31–32
Tischer A. (1958) Conophytum ricardianum Loesch et Tisch. subsp. rubriflorum Tisch. subsp. nov. KuaS 9: 85–86
Tischer A. (1958) Conophytum paucipunctum Tisch. spec. nov. KuaS 9: 157
Tischer A. (1959) Beiträge zur Kenntnis der Arten von Conophytum N.E.Br. KuaS 10: 40
Tischer A. (1959) Beiträge zur Kenntnis der Arten von Conophytum N.E.Br. KuaS 10: 102
Tischer A. (1959) Beiträge zur Kenntnis der Arten von Conophytum N.E.Br. KuaS 11: 35–36
Tischer A. (1959) Beiträge zur Kenntnis der Arten von Conophytum N.E.Br. KuaS 11: 68–69
Tischer A. (1959) Conophytum pulchellum Tisch sp. nov. CSJGB 21: 30
Tischer A. (1959) Conophytum steytlervillense Tisch sp. nov. CSJGB 21: 31
Tischer A. (1959) Conophytum semilunulum Tisch sp. nov. CSJGB 21: 78
Tischer A. (1960) Conophytum koupense Tisch sp. nov. CSJGB 22: 27
Tischer A. (1960) Conophytum subincanum Tisch sp. nov. CSJGB 22: 28
Tischer A. (1960) Conophytum dilatatum Tisch sp. nov. CSJGB 22: 75
Tischer A. (1960) Conophytum giftbergense Tisch sp. nov. CSJGB 22: 76
Tischer A. (1962) Conophytum pardicolor Tisch sp. nov. CSJGB 24: 80
Tischer A. (1962) Conophytum dolomiticum Tisch sp. nov. CSJGB 24: 81
Tischer A. (1966) Beiträge zur Kenntnis der Arten von Conophytum N.E.Br. KuaS 17: 71–75
Tischer A. (1966) Beiträge zur Kenntnis der Arten von Conophytum N.E.Br. KuaS 17: 202–205
Tischer A. (1967) Beiträge zur Kenntnis der Arten von Conophytum N.E.Br. Conophytum rauhii. KuaS 18: 130–131
Tischer A. (1968) Eine Pflanze und ihre Varietäten. Conophytum truncatum (Thbg,) N.E.Br. KuaS 19: 115–118
Tischer A. (1969) Some contributions to the study of Conophytum N.E.Br. Conophytum obcordellum (Haw.) N.E.Br. and its variants. NCSJ 24: 2–4
Tischer A. (1969) Some contributions to the study of Conophytum N.E.Br. Conophytum parviflorum var. impressum. NCSJ 24: 35
Tischer A. (1970) Conophytum wettsteinii (Brgr.) N.E.Br. and its variants. NCSJ 25: 5–76
Tischer A. (1971) Some contributions to the study of Conophytum (Haw.) N.E.Br. Conophytum uvaeforme. NCSJ 26: 50–51
Tischer A., Jacobsen H. (1956) Genre Conophytum N.E.Br. Conophytum marnierianum. Cactus 51: 92–93
Tischer A., Loesch A. (1931) Conophytum ricardianum. MDK 3: 74–76
Tischer A., Loesch A. (1932) Conophytum udabibense. MKK 4: 16
Tjaden W.L. (1995) Regarding gender and number of generic names supposedly ending in -anthema. Taxon 44: 213–216
Toelken H.R. (1996) Surprises in Australian Carpobrotus. Aloe 33: 28–29
Toelken H.R., Jessop J.P. (1976) Mesembryanthemaceae. Nomenclature of the genus Amphibolia. Bothalia 12: 64
Tournefort J.P. de (1730) Etablissement de quelques nouveaux genres de plantes. Histoire de l'Académie Royale des Science 1705: 236–241, t. 4
Turczaninow N.S. (1843) Decas generum plantarum huiusque non descriptorum. Bulletin de la Société des naturalistes à Moscou 16: 51–62
Turpin P.J.F. (1806) Cypselea. Annales du Muséum d'histoire naturelle 7: 219–221, t. 121 f 5
Urban I. (1929) Plantae Haitienses et Domingensies novae vel rariores VII. a cl. E. L. Ekman 1924–1928 lectae. Arkiv för Botanik 22 A: 15
Vahl M. (1790) Symbolae botanicae 1: 32. Möller, Copenhagen
Venning J. (1984) In: Prescott A., Venning J. 1984
Ventenat E.P. (1805) Jardin de la Malmaison ...: t. 109. Crapelet, Paris
Verdcourt B. (1957) Notes from the east African Herbarium V. KB 12: 348–350
Verdcourt B. (1985) An introduced Sesuvium (Aizoaceae) in Arabia. KB 40: 208
Verdoorn I.C. (1941) The genus Plinthus. Bothalia 4: 177–179
Vest L. (1828) Plantarum novarum vel minus cognitarum descriptiones. Sylloge plantarum novarum itemque minus cognitum 2: 224. Vidua Brenck, Regensburg
Visiani R. de (1836) Plantae quaedam Aegypti ac Nubiae ...: 19, t. 3, f 1. Minerva, Padova
Volckamer J.G. (1718) Flora Noribergensis sive Catalogus Plantarum in agro norenbergensi ...: 236, t. 236. Monath, Nürnberg
Volk O. (1950) In: Jacobsen H., Volk O.H. 1950: 83
Volk O. (1955) In: Jacobsen H., Rowley G.D. 1955
Volkens G., Irmscher G. (1913) In: Schellenberg G. 1913: 497–499
Wagner R. (1904) Ein neues Aizoon aus Südaustralien
Walgate M. (1939) A revision of Juttadinteria Schwant., Dracophilus Dint. et Schwant., and Namibia Dint. et Schwant. NM 3: 171–188
Wallich N. (1832) A numerical list of dried specimens of plants in the East Indian Company Museum, collected ...: 229, nrs 6837, 6838. The author, London
Wallroth C.F.W., Griffiths W. (1836) Description of some grasses which form part of the vegetation in the Jheels of the district of Sylhet. The Journal of the Asiatic Society of Bengal. Calcutta. 5: 570–575, t. 23
Walpers G.G. (1843) Repertorium Botanices Systematicae 2: 232. Hofmeister, Leipzig
Walpers G.G. (1852) Annales Botanices Systematicae 2: 659–660; 3: 195. Hofmeister, Leipzig
Walter T.A. (1788) Flora caroliniana, secundum ...: 117. Fraser, London
Wawra H., Peyritsch J.J. (1860) Sertum benguelense. Aufzählungen und Beschreibungen ... Sitzungsberichte der math.-naturwiss. Classe der kaiserlichen Akademie der Wissenschaften 38: 564. K.&K. Hof- und Staatsdruckerei, Vienna
Weber C. (1968) Notes on the nomenclature of some Aizoaceae. Baileya 16: 9–12
Welwitsch F.M.J. (1859) Annaes do Conselho Ultramarino. Apotamentos phytogeographicos ...: 557, 586. Imprensa nacional, Lisbon
Welwitsch F.M.J. (1873) On an undescribed species of Mesembryanthemum from the south of Portugal. JB 11: 289–290, t. 136
Weston R. (1770) The universal botanist and nurseryman: ... Mesembranthemum: 167–173. Bell et al., London
White C.T. (1919) Contributions to the Queensland flora. Queensland's Department of Agriculture Botanical Bulletin 21: 10–12
Wijnands D.O. (1983) The botany of the Commelins. Balkema, Rotterdam
Wildeman E. de (1908) Icones Selectae Horti Thenensis: 109–111, t. 230. Veuve Monnom, Bruxelles
Wildeman E. de (1908) Plantae novae vel minus cognitae ex herbario Horti Thenensis 2: 3–5. Veuve Monnom, Bruxelles
Willdenow C.L. (1799) Species plantarum, ed. 2: 1025–1027, 1051–1053. G. C. Nauk, Berlin

Willdenow CL (1803) In: Rottler J.B., Willdenow C.L. 1803: 181
Willdenow C.L. (1804) Hortus Berolinensis sive ...: t. 22. Schüppel, Berlin
Willdenow C.L. (1809) Enumeratio plantarum ... part 1: Mesembryanthemum, Sesuvium: 521–539. Taberna Libraria Scholae Realis, Berlin
Willdenow C.L. (1814) Enumeratio plantarum ... Supplementum: 35–37. Taberna Libraria Scholae Realis, Berlin
Williamson G., Kennedy H.C. (1997) A new Conophytum from the north-east Richtersveld. CSJ 69: 205–208
Wing A. (1884) Southern science record 3: 282
Wingfield R., Newman M. (1994) A new species of Trianthema (Aizoaceae) from Venezuela. KB 47: 115–117
Winter P.J.D., Hahn N. (1999) A new species of Khadia N.E.Br. (Mesembryanthemaceae) from the Northern Province of South Africa. Aloe 36: 16–19
Wisura W., Glen H.F. (1993) The South African species of Carpobrotus (Mesembryanthema – Aizoaceae). CBH 15: 76–107
Wolley-Dod A.H. (1900) New Cape plants. JB 38: 170–171
Wood J.M. (1907) Ficoideae. Handbook of the flora of Natal: 55. Bennett & Davis, Durban
Wulff H.D. (1944) Untersuchungen zur Zytologie und Systematik der Aizoaceen-Subtribus Gibbaeinae Schwant. Botanisches Archiv 45: 149–189
Wulff H.D. (1949) Gibbaeum Tischleri (Schwant. ex Jacobsen) Wulff nom. nov. SKK 3: 55–56
Zahlbruckner A. (1900) Plantae pentherianae. Annalen des k.u.k. Naturhistorischen Hofmuseums 15: 35
Zanoni G. (1675) Istoria Botanica: 35, t. 13. Longhi, Bologna
Zimmermann N. (1996) Schwantesia constanceae N. Zimmermann spec. nov. – eine neue Art aus dem südlichen Namibia. KuaS 47: 81–86
Zimmermann N. (1996) Frithia pulchra N.E.Brown – eine Reise zu zwei Populationen im Transvaal mit Besprechung der sukkulenten Begleitvegetation. KuaS 47: 146–152
Zimmermann N. (1996) The genus Schwantesia Dinter. CSJ 68: 257–266
Zimmermann N., Hartmann H.E.K. (1995) Studies in Schwantesia. The rediscovery of S. borcherdsii. CSJ 67: 162–165
Zohary M. (1966) Aizoaceae. Aptenioideae. In: Flora Palaestina 1: Aptenioideae: 76–77. The Israel Academy of Sciences and Humanities, Jerusalem

Faucaria felina (Weston) Schwantes

H.E.K. Hartmann

Faucaria gratiae L.Bolus

H.E.K. Hartmann

Faucaria nemorosa L.Bolus ex L.E.Groen

B. Groen

Faucaria subintegra L.Bolus

B. Groen

Faucaria tigrina (Haw.) Schwantes

H.E.K. Hartmann

Fenestraria rhopalophylla ssp. **aurantiaca** (N.E.Br.) H.E.K.Hartmann

H.E.K. Hartmann

Frithia humilis Burgoyne

H.E.K. Hartmann

Frithia pulchra N.E.Br.

H.E.K. Hartmann

Galenia africana L.

Galenia papulosa (Ecklon & Zeyher) Sonder

Gibbaeum album N.E.Br.

Gibbaeum angulipes (L.Bolus) N.E.Br.

Gibbaeum dispar N.E.Br.

Gibbaeum geminum N.E.Br.

Gibbaeum gibbosum (Haw.) N.E.Br.

Gibbaeum heathii (N.E.Br.) L.Bolus

Gibbaeum nebrownii Tischer

Gibbaeum nuciforme (Haw.) Glen & H.E.K.Hartmann

Gibbaeum pachypodium (Kensit) L.Bolus

Gibbaeum petrense (N.E.Br.) Tischer

Gibbaeum pubescens (Haw.) N.E.Br.

Gibbaeum shandii N.E.Br.

Gibbaeum velutinum (L.Bolus) Schwantes

Glottiphyllum carnosum N.E.Br.

Glottiphyllum cruciatum (Haw.) N.E.Br.

Glottiphyllum depressum (Haw.) N.E.Br.

Glottiphyllum difforme (L.) N.E.Br.

Glottiphyllum fergusoniae L.Bolus

Glottiphyllum grandiflorum (Haw.) N.E.Br.

Glottiphyllum linguiforme (L.) N.E.Br.

Glottiphyllum longum (Haw.) N.E.Br.

Glottiphyllum neilii N.E.Br.

H.E.K. Hartmann
Glottiphyllum nelii Schwantes

H.E.K. Hartmann
Glottiphyllum oligocarpum L.Bolus

H.E.K. Hartmann
Glottiphyllum peersii L.Bolus

H.E.K. Hartmann
Glottiphyllum regium N.E.Br.

H.E.K. Hartmann
Glottiphyllum salmii (Haw.) N.E.Br.

H.E.K. Hartmann
Glottiphyllum suave N.E.Br.

H.E.K. Hartmann
Glottiphyllum surrectum (Haw.) L.Bolus

V. Bittrich
Gunniopsis calcarea Chinnock

R. Chinnock

Gunniopsis calva Chinnock

R. Chinnock

Gunniopsis glabra (Luehmann ex Ewart) Gardner

R. Chinnock

Gunniopsis intermedia Diels

V. Bittrich

Gunniopsis kochii (R.Wagner) Chinnock

R. Chinnock

Gunniopsis papillata Chinnock

R. Chinnock

Gunniopsis propinqua Chinnock

R. Chinnock

Gunniopsis quadrifida (F.Mueller) Pax

R. Chinnock

Gunniopsis rodwayi (Ewart) Gardner

Gunniopsis rubra Chinnock

R. Chinnock

Gunniopsis septifraga (F.Mueller) Chinnock

R. Chinnock

Gunniopsis tenuifolia Chinnock

R. Chinnock

Gunniopsis zygophylloides (F.Mueller) Diels

V. Bittrich

Hallianthus planus (L.Bolus) H.E.K.Hartmann

H.E.K. Hartmann

Hammeria meleagris (L.Bolus) Klak

H.E.K. Hartmann

Hartmanthus hallii (L.Bolus) S.A.Hammer

H.E.K. Hartmann

Hartmanthus pergamentaceus (L.Bolus) S.A.Hammer

H.E.K. Hartmann

H.E.K. Hartmann

Hereroa aspera L.Bolus

H.E.K. Hartmann

Hereroa glenensis (N.E.Br.) L.Bolus

H.E.K. Hartmann

Hereroa hesperantha (Dinter & A.Berger) Dinter & Schwantes

H.E.K. Hartmann

Hereroa joubertii L.Bolus

H.E.K. Hartmann

Hereroa pallens L.Bolus

H.E.K. Hartmann

Hereroa puttkameriana (Dinter & A.Berger) Dinter & Schwantes

H.E.K. Hartmann

Hereroa stanfordiae L.Bolus

H.E.K. Hartmann

Hereroa wilmaniae L.Bolus

Ihlenfeldtia excavata (L.Bolus) H.E.K.Hartmann

Ihlenfeldtia vanzylii (L.Bolus) H.E.K.Hartmann

Jacobsenia hallii L.Bolus

Jacobsenia kolbei (L.Bolus) L.Bolus & Schwantes

Jacobsenia vaginata (L.Bolus) Ihlenf.

Jensenobotrya lossowiana Herre

Jordaaniella clavifolia (L.Bolus) H.E.K.Hartmann

Jordaaniella cuprea (L.Bolus) H.E.K.Hartmann

Jordaaniella dubia (Haw.) H.E.K.Hartmann

H.E.K. Hartmann

Jordaaniella spongiosa (L.Bolus) H.E.K.Hartmann

H.E.K. Hartmann

Jordaaniella uniflora (L.Bolus) H.E.K.Hartmann

H.E.K. Hartmann

Juttadinteria albata L.Bolus

H.E.K. Hartmann

Juttadinteria attenuata Walgate

H.E.K. Hartmann

Juttadinteria ausensis (L.Bolus) Schwantes

H.E.K. Hartmann

Juttadinteria deserticola (Marloth) Schwantes

H.E.K. Hartmann

Juttadinteria simpsonii (Dinter) Schwantes

H.E.K. Hartmann

H.E.K. Hartmann

Khadia acutipetala (N.E.Br.) N.E.Br.

H.E.K. Hartmann

Khadia alticola Chesselet & H.E.K.Hartmann

H.E.K. Hartmann

Khadia beswickii (L.Bolus) N.E.Br.

H.E.K. Hartmann

Khadia borealis L.Bolus

H.E.K. Hartmann

Khadia carolinensis (L.Bolus) L.Bolus

H.E.K. Hartmann

Lampranthus flexifolius (Haw.) N.E.Br.

H.E.K. Hartmann

Lampranthus fugitans L.Bolus

H.E.K. Hartmann

Lampranthus otzenianus (Dinter) Friedrich

Lampranthus productus (Haw.) N.E.Br.

Lampranthus scaber (L.) N.E.Br.

Lampranthus uniflorus (L.Bolus) L.Bolus

Lapidaria margaretae (Schwantes) Dinter & Schwantes

Leipoldtia alborosea (L.Bolus) H.E.K.Hartmann & Stüber

Leipoldtia frutescens (L.Bolus) H.E.K.Hartmann

Leipoldtia klaverensis L.Bolus

Leipoldtia rosea L.Bolus

Leipoldtia schultzei (Schlechter & Diels) Friedrich

Leipoldtia uniflora L.Bolus

Leipoldtia weigangiana grandifolia (L.Bolus) H.E.K.Hartmann & Rust

Leipoldtia weigangiana ssp. **littlewoodii** (L.Bolus) H.E.K.Hartmann & Rust

Leipoldtia weigangiana (Dinter) Dinter & Schwantes ssp. **weigangiana**

Lithops aucampiae L.Bolus ssp. **aucampiae**

Lithops coleorum S.A.Hammer & Uijs

Lithops gesineae De Boer

D. Cole

Lithops hallii De Boer

D. Cole

Lithops julii (Dinter & Schwantes) N.E.Br. **ssp. julii**

D. Cole

Lithops localis (N.E.Br.) Schwantes

D. Cole

Lithops marmorata (N.E.Br.) N.E.Br.

D. Cole

Lithops naureeniae D.T.Cole

D. Cole

Lithops olivacea L.Bolus

D. Cole

Lithops schwantesii Dinter **ssp. schwantesii**

H.E.K. Hartmann

Machairophyllum acuminatum L.Bolus

Machairophyllum albidum (L.) Schwantes

Machairophyllum brevifolium L.Bolus

Malephora crassa (L.Bolus) H.Jacobsen & Schwantes

Malephora mollis (Aiton) N.E.Br.

Malephora purpureo-crocea (Haw.) Schwantes

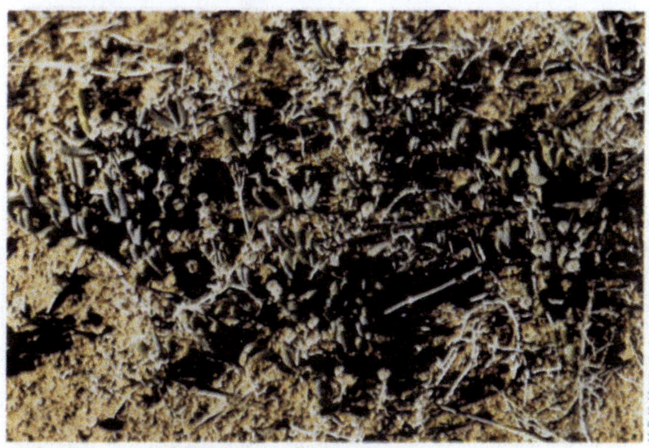

Malephora smithii (L.Bolus) H.E.K.Hartmann

Malephora thunbergii (Haw.) Schwantes

Malephora uitenhagensis (L.Bolus) H.Jacobsen & Schwantes

Marlothistella uniondalensis Schwantes

H.E.K. Hartmann

Mesembryanthemum aitonis Jacquin

M. Struck

Mesembryanthemum barklyi N.E.Br.

M. Struck

Mesembryanthemum cryptanthum Hook.f.

M. Struck

Mesembryanthemum crystallinum L.

M. Struck

Mesembryanthemum eurystigmatum Gerbaulet

M. Struck

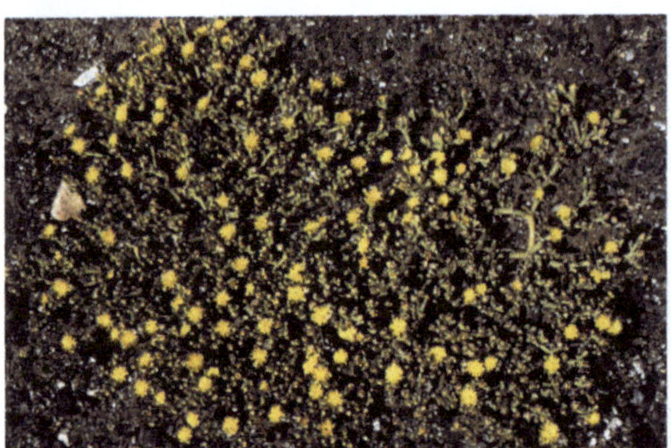

Mesembryanthemum excavatum L.Bolus

M. Struck

Mesembryanthemum fastigiatum Thunberg

M. Struck

M. Struck
Mesembryanthemum gariusanum Dinter

M. Struck
Mesembryanthemum guerichianum Pax

M. Struck
Mesembryanthemum hypertrophicum Dinter

V. Bittrich
Mesembryanthemum inachabense Engler

M. Struck
Mesembryanthemum longistylum D.C.

M. Struck
Mesembryanthemum nodiflorum L.

M. Struck
Mesembryanthemum pellitum Friedrich

M. Struck
Mesembryanthemum stenandrum (L.Bolus) L.Bolus

U. Schmiedel

Mesembryanthemum subtruncatum L.Bolus

H.E.K. Hartmann

Mestoklema albanicum N.E.Br. ex Glen

H.E.K. Hartmann

Mestoklema arboriforme (Burchell) N.E.Br. ex Glen

H.E.K. Hartmann

Mestoklema copiosum N.E.Br. ex Glen

H.E.K. Hartmann

Mestoklema tuberosum (L.) N.E.Br. ex Glen

H.-D. Ihlenfeldt

Meyerophytum globosum (L.Bolus) Ihlenf.

H.-D. Ihlenfeldt

Meyerophytum meyeri (Schwantes) Schwantes

H.-D. Ihlenfeldt

Mitrophyllum abbreviatum L.Bolus

Mitrophyllum clivorum (N.E.Br.) Schwantes

H.-D. Ihlenfeldt

Mitrophyllum dissitum (N.E.Br.) Schwantes

H.-D. Ihlenfeldt

Mitrophyllum grande N.E.Br.

H.-D. Ihlenfeldt

Mitrophyllum mitratum (Marloth) Schwantes

H.-D. Ihlenfeldt

Mitrophyllum roseum L.Bolus

H.-D. Ihlenfeldt

Monilaria chrysoleuca (Schlechter) Schwantes

H.-D. Ihlenfeldt

Monilaria moniliformis (Thunberg) Ihlenf. & S.Jörgensen

H.-D. Ihlenfeldt

Monilaria obconica Ihlenf. & S.Jörgensen

H.-D. Ihlenfeldt

H.-D. Ihlenfeldt

Monilaria pisiformis (Haw.) Schwantes

H.-D. Ihlenfeldt

Monilaria scutata (L.Bolus) Schwantes **ssp. obovata** Ihlenf. & S.Jörgensen

H.-D. Ihlenfeldt

Monilaria scutata (L.Bolus) Schwantes ssp. **scutata**

H.E.K. Hartmann

Mossia intervallaris (L.Bolus) N.E.Br.

H.E.K. Hartmann

Muiria hortenseae N.E.Br.

M.B. Bayer

Namaquanthus vanheerdei L.Bolus

H.E.K. Hartmann

Namibia cinerea (Marloth) Dinter & Schwantes

H.E.K. Hartmann

Nananthus aloides (Haw.) Schwantes

Nananthus gerstneri (L.Bolus) L.Bolus

H.E.K. Hartmann

Nananthus margaretiferus L.Bolus

H.E.K. Hartmann

Nananthus pallens (L.Bolus) L.Bolus

H.E.K. Hartmann

Nananthus pole-evansii N.E.Br.

H.E.K. Hartmann

Nananthus vittatus (N.E.Br.) Schwantes

H.E.K. Hartmann

Nelia pillansii (N.E.Br.) Schwantes

H.E.K. Hartmann

Nelia schlechteri Schwantes

H.E.K. Hartmann

Neohenricia sibbettii (L.Bolus) L.Bolus

H.E.K. Hartmann

Octopoma connatum (L.Bolus) L.Bolus

Octopoma octojuge (L.Bolus) N.E.Br.

Octopoma quadrisepalum (L.Bolus) H.E.K.Hartmann

Octopoma rupigenum (L.Bolus) L.Bolus

Octopoma subglobosum (L.Bolus) L.Bolus

Odontophorus angustifolius L.Bolus ssp. **angustifolius**

Odontophorus angustifolius L.Bolus ssp. **protoparcoides** S.A.Hammer

Odontophorus marlothii N.E.Br.

Odontophorus nanus L.Bolus

H.E.K. Hartmann

Odontophorus pusillus S.A.Hammer

H.E.K. Hartmann

Oophytum nanum (Schlechter) L.Bolus

H.-D. Ihlenfeldt

Oophytum oviforme (N.E.Br.) N.E.Br.

H.-D. Ihlenfeldt

Orthopterum coeganum L.Bolus

H.E.K. Hartmann

Orthopterum waltoniae L.Bolus

H.E.K. Hartmann

Oscularia falciformis (Haw.) H.E.K.Hartmann

H.E.K. Hartmann

Oscularia lunata (Willdenow) H.E.K.Hartmann

H.E.K. Hartmann

Oscularia ornata (L.Bolus) H.E.K.Hartmann

H. E. K. Hartmann

Oscularia steenbergensis (L.Bolus) H.E.K.Hartmann

H. E. K. Hartmann

Ottosonderia monticola (Sonder) L.Bolus

H. E. K. Hartmann

Peersia frithii (L.Bolus) L.Bolus

H. E. K. Hartmann

Peersia macradenia (L.Bolus) L.Bolus

H. E. K. Hartmann

Peersia vanheerdei (L.Bolus) H.E.K.Hartmann

H. E. K. Hartmann

Phyllobolus abbreviatus (L.Bolus) Gerbaulet

M. Struck

Phyllobolus amabilis Gerbaulet & Struck

M. Struck

M. Struck

Phyllobolus canaliculatus (Haw.) Bittrich

M. Struck

Phyllobolus chrysophthalmus Gerbaulet & Struck

M. Struck

Phyllobolus congestus (L.Bolus) Gerbaulet

M. Struck

Phyllobolus deciduus (L.Bolus) Gerbaulet

M. Struck

Phyllobolus decurvatus (L.Bolus) Gerbaulet

M. Struck

Phyllobolus delus (L.Bolus) Gerbaulet

M. Struck

Phyllobolus digitatus (Aiton) Gerbaulet ssp. **digitatus**

M. Struck

Phyllobolus digitatus ssp. **littlewoodii** (L.Bolus) Gerbaulet

Phyllobolus gariepensis Gerbaulet & Struck

Phyllobolus grossus (Aiton) Gerbaulet

Phyllobolus herbertii (N.E.Br.) Gerbaulet

Phyllobolus latipetalus (L.Bolus) Gerbaulet

Phyllobolus lignescens (L.Bolus) Gerbaulet

Phyllobolus melanospermus (Dinter & Schwantes) Gerbaulet

Phyllobolus nitidus (Haw.) Gerbaulet

Phyllobolus oculatus (N.E.Br.) Gerbaulet

Phyllobolus prasinus (L.Bolus) Gerbaulet

Phyllobolus pumilus (L.Bolus) Gerbaulet

Phyllobolus quartziticus (L.Bolus) Gerbaulet

Phyllobolus rabiei (L.Bolus) Gerbaulet

Phyllobolus resurgens (Kensit) Schwantes

Phyllobolus roseus (L.Bolus) Gerbaulet

Phyllobolus saturatus (L.Bolus) Gerbaulet

Phyllobolus sinuosus (L.Bolus) Gerbaulet

Phyllobolus spinuliferus (Haw.) Gerbaulet

Phyllobolus splendens ssp. **pentagonus** (L.Bolus) Gerbaulet

Phyllobolus splendens (L.) Gerbaulet ssp. **splendens**

Phyllobolus suffruticosus (L.Bolus) Gerbaulet

Phyllobolus tenuiflorus (Jacquin) Gerbaulet

Phyllobolus trichotomus (Thunberg) Gerbaulet

Phyllobolus viridiflorus (Aiton) Gerbaulet

Pleiospilos bolusii (Hook.f.) N.E.Br.

Pleiospilos compactus ssp. **canus** (Haw.) H.E.K.Hartmann & Liede

Pleiospilos compactus (Aiton) Schwantes ssp. **compactus**

Pleiospilos compactus ssp. **fergusoniae** (L.Bolus) H.E.K.Hartmann & Liede

Pleiospilos compactus ssp. **minor** (L.Bolus) H.E.K.Hartmann & Liede

Pleiospilos compactus ssp. **sororius** (N.E.Br.) H.E.K.Hartmann & Liede

Pleiospilos nelii Schwantes

Pleiospilos simulans (Marloth) N.E.Br.

Plinthus cryptocarpus Fenzl

Plinthus karooicus I.Verdoorn

Polymita albiflora (L.Bolus) L.Bolus

Polymita steenbokensis H.E.K.Hartmann

Prenia englishiae (L.Bolus) Gerbaulet

Prenia pallens (Aiton) N.E.Br. **ssp. lutea** L.Bolus

Prenia pallens (Aiton) N.E.Br. **ssp. namaquensis** Gerbaulet

Prenia radicans (L.Bolus) Gerbaulet

Prenia sladeniana (L.Bolus) L.Bolus

Prenia tetragona (Thunberg) Gerbaulet

M. Struck

Prenia vanrensburgii L.Bolus

M. Struck

Prepodesma orpenii (N.E.Br.) N.E.Br.

H.E.K. Hartmann

Psammophora longifolia L.Bolus

H.E.K. Hartmann

Psammophora modesta (Dinter & A.Berger) Dinter & Schwantes

H.E.K. Hartmann

Psammophora nissenii (Dinter) Dinter & Schwantes

H.E.K. Hartmann

Psammophora saxicola H.E.K.Hartmann

H.E.K. Hartmann

Psilocaulon articulatum (Thunberg) N.E.Br.

M. Struck

M. Struck

Psilocaulon bicorne (Sonder) Schwantes

M. Struck

Psilocaulon coriarium (Burchell ex N.E.Br.) N.E.Br.

M. Struck

Psilocaulon dinteri (Engler) Schwantes

M. Struck

Psilocaulon foliosum L.Bolus

M. Struck

Psilocaulon gessertianum (Dinter & A.Berger) Schwantes

M. Struck

Psilocaulon granulicaule (Haw.) Schwantes

M. Struck

Psilocaulon junceum (Haw.) Schwantes

M. Struck

Psilocaulon junceum (Haw.) Schwantes

Psilocaulon leptarthron (A.Berger) N.E.Br.

M. Struck

Psilocaulon parviflorum (Jacquin) Schwantes

M. Struck

Psilocaulon salicornioides (Pax) Schwantes

M. Struck

Psilocaulon subnodosum (A.Berger) N.E.Br.

M. Struck

Psilocaulon subnodosum (A.Berger) N.E.Br.

M. Struck

Rabiea albinota (Haw.) N.E.Br.

H.E.K. Hartmann

Rabiea albipuncta (Haw.) N.E.Br.

V. Bittrich

Rabiea comptonii (L.Bolus) L.Bolus

H.E.K. Hartmann

Rabiea difformis (L.Bolus) L.Bolus

Rhinephyllum broomii L.Bolus

Rhinephyllum graniforme (Haw.) L.Bolus

Rhinephyllum inaequale L.Bolus

Rhinephyllum muirii N.E.Br.

Rhombophyllum albanense (L.Bolus) H.E.K.Hartmann

Rhombophyllum dolabriforme (L.) Schwantes

Rhombophyllum nelii Schwantes

Rhombophyllum rhomboideum (Salm Dyck) Schwantes

Ruschia abbreviata L.Bolus

Ruschia barnardii L.Bolus

Ruschia burtoniae L.Bolus

Ruschia canonotata (L.Bolus) Schwantes

Ruschia centrocapsula H.E.K.Hartmann & Stüber

Ruschia cradockensis ssp. **cradockensis**

Ruschia cradockensis ssp. **triticiformis** (L.Bolus) H.E.K.Hartmann & Stüber

Ruschia divaricata L.Bolus

Ruschia griquensis (L.Bolus) Schwantes

Ruschia hamata (L.Bolus) Schwantes

Ruschia inconspicua L.Bolus

Ruschia intricata (N.E.Br.) H.E.K.Hartmann & Stüber

Ruschia maxima (Haw.) L.Bolus

Ruschia muelleri (L.Bolus) Schwantes

Ruschia muricata L.Bolus

H.E.K. Hartmann

Ruschia namusmontana Friedrich

H.E.K. Hartmann

Ruschia odontocalyx (Schlechter & Diels) Schwantes

H.E.K. Hartmann

Ruschia orientalis L.Bolus

H.E.K. Hartmann

Ruschia primosii L.Bolus

H.E.K. Hartmann

Ruschia putterillii (L.Bolus) L.Bolus

H.E.K. Hartmann

Ruschia recurva (Moench) H.E.K.Hartmann

H.E.K. Hartmann

Ruschia rigens L.Bolus

H.E.K. Hartmann

Ruschia rupicola (Engler) Schwantes

Ruschia ruralis (N.E.Br.) Schwantes

Ruschia scabra H.E.K.Hartmann

Ruschia spinosa (L.) Dehn

Ruschianthemum gigas (Dinter) Friedrich

Ruschianthemum gigas (Dinter) Friedrich

Ruschianthus falcatus L.Bolus

Sarcozona bicarinata S.T.Blake

Sarcozona praecox (F.Mueller) S.T.Blake ex H.Eichler

Sceletium crassicaule (Haw.) L.Bolus

Sceletium emarcidum (Thunberg) L.Bolus ex H.Jacobsen

Sceletium exalatum Gerbaulet

Sceletium expansum (L.) L.Bolus

Sceletium expansum (L.) L.Bolus

Sceletium rigidum L.Bolus

Sceletium strictum L.Bolus

Sceletium tortuosum (L.) N.E.Br.

Sceletium varians (Haw.) Gerbaulet

M. Struck

Schlechteranthus hallii L.Bolus

H.E.K. Hartmann

Schlechteranthus maximiliani Schwantes

H.E.K. Hartmann

Schwantesia acutipetala L.Bolus

N. Zimmermann

Schwantesia borcherdsii L.Bolus

H.E.K. Hartmann

Schwantesia constanceae N.Zimmermann

N. Zimmermann

Schwantesia herrei L.Bolus

H.E.K. Hartmann

Schwantesia marlothii L.Bolus

H.E.K. Hartmann

Schwantesia pillansii L.Bolus

H.E.K. Hartmann

Schwantesia ruedebuschii Dinter

N. Zimmermann

Schwantesia speciosa L.Bolus

H.E.K. Hartmann

Schwantesia triebneri L.Bolus

H.E.K. Hartmann

Scopelogena verruculata (L.) L.Bolus

R. Kabutz

Sesuvium portulacastrum (L.) L.

V. Bittrich

Sesuvium sesuvioides (Fenzl) Verdcourt

H.E.K. Hartmann

Skiatophytum tripolium (L.) L.Bolus

U. Schmiedel

Smicrostigma viride (Haw.) N.E.Br.

Stayneria neilii (L.Bolus) L.Bolus

Stoeberia beetzii (Dinter) Dinter & Schwantes

Stoeberia frutescens (L.Bolus) Van Jaarsveld

Stoeberia porphyrea H.E.K.Hartmann

Stoeberia utilis (L.Bolus) Van Jaarsveld

Stomatium agninum (Haw.) Schwantes

Stomatium bolusiae Schwantes

Stomatium gerstneri L.Bolus

Stomatium meyeri L.Bolus

Stomatium mustelinum (Haw.) Schwantes

Stomatium suaveolens (Schwantes) Schwantes

Stomatium trifarium L.Bolus

Synaptophyllum juttae (Dinter & A.Berger) N.E.Br.

Tanquana archeri (L.Bolus) H.E.K.Hartmann & Liede

Tanquana hilmarii (L.Bolus) H.E.K.Hartmann & Liede

H.E.K. Hartmann

Tanquana prismatica (Schwantes) H.E.K.Hartmann & Liede

C. Taylor

Tetragonia angustifolia Barnéoud

H.E.K. Hartmann

Tetragonia arbusculoides Engler

V. Bittrich

Tetragonia cristata Prescott

V. Bittrich

Tetragonia decumbens Miller

V. Bittrich

Tetragonia diptera F.Mueller

V. Bittrich

Tetragonia echinata Aiton

V. Bittrich

Tetragonia eremaea Ostenfeld

Tetragonia implexicoma (Miquel) Hook.f.
V. Bittrich

Tetragonia macrocarpa R.A.Philippi
C. Taylor

Tetragonia maritima Barnéoud
C. Taylor

Tetragonia microcarpa R.A.Philippi
C. Taylor

Tetragonia ovata R.A.Philippi
C. Taylor

Tetragonia schenckii Schinz
C. Taylor

Tetragonia tetragonioides (Pallas) O.Kuntze
V. Bittrich

Titanopsis calcarea (Marloth) Schwantes
H. E. K. Hartmann

H.E.K. Hartmann

Titanopsis hugo-schlechteri (Tischer) Dinter & Schwantes

H.E.K. Hartmann

Titanopsis schwantesii (Schwantes) Schwantes

H.E.K. Hartmann

Trianthema ceratosepala Volkens & Irmscher

V. Bittrich

Trianthema cussackiana F.Mueller

V. Bittrich

Trianthema kimberleyi Bittrich & Jenssen

V. Bittrich

Trianthema megasperma Prescott

V. Bittrich

Trianthema oxycalyptra F.Mueller

V. Bittrich

Trianthema patellitecta Prescott

Trianthema pilosa F.Mueller

V. Bittrich

Trianthema portulacastra L.

V. Bittrich

Trianthema rhynchocalyptra F.Mueller

V. Bittrich

Trianthema triquetra Willdenow

V. Bittrich

Trianthema turgidifolia F.Mueller

V. Bittrich

Trichodiadema introrsum (Hook.f.) Niesler

H. E. K. Hartmann

Trichodiadema olivaceum L.Bolus

H. E. K. Hartmann

Trichodiadema orientale L.Bolus

I. Niesler

Trichodiadema pygmaeum L.Bolus

Vanheerdea primosii (L.Bolus) H.E.K.Hartmann

Vanheerdea roodiae (N.E.Br.) H.E.K.Hartmann

Vanzijlia annulata (A.Berger) L.Bolus

Vlokia ater S.A.Hammer

Wooleya farinosa (L.Bolus) L.Bolus

Zaleya pentandra (L.) Jeffrey

Zeuktophyllum calycinum (L.Bolus) H.E.K.Hartmann

MIX
Papier aus verantwortungsvollen Quellen
Paper from responsible sources
FSC® C105338

If you have any concerns about our products,
you can contact us on
ProductSafety@springernature.com

In case Publisher is established outside the EU,
the EU authorized representative is:
Springer Nature Customer Service Center GmbH
Europaplatz 3, 69115 Heidelberg, Germany

Printed by Libri Plureos GmbH
in Hamburg, Germany